HOW PRIMATES EAT

HOW PRIMATES EAT

A Synthesis of Nutritional Ecology across a Mammal Order

Joanna E. Lambert,
Margaret A. H. Bryer, and
Jessica M. Rothman

*With a foreword by
T. H. Clutton-Brock and
an afterword by Alison Richard*

The University of Chicago Press
Chicago and London

The University of Chicago Press, Chicago 60637
The University of Chicago Press, Ltd., London
© 2024 by The University of Chicago
All rights reserved. No part of this book may be used or reproduced in any manner whatsoever without written permission, except in the case of brief quotations in critical articles and reviews. For more information, contact the University of Chicago Press, 1427 E. 60th St., Chicago, IL 60637.
Published 2024
Printed in the United States of America

33 32 31 30 29 28 27 26 25 24   1 2 3 4 5

ISBN-13: 978-0-226-82973-9 (cloth)
ISBN-13: 978-0-226-82975-3 (paper)
ISBN-13: 978-0-226-82974-6 (e-book)

DOI: https://doi.org/10.7208/chicago/9780226829746.001.0001

Library of Congress Cataloging-in-Publication Data

Names: Lambert, Joanna E., editor. | Bryer, Margaret A. H., editor. | Rothman, Jessica M., editor. | Clutton-Brock, T. H., writer of foreword. | Richard, Alison F., writer of afterword.
Title: How primates eat : a synthesis of nutritional ecology across a mammal order / Joanna E. Lambert, Margaret A. H. Bryer, and Jessica M. Rothman ; with a foreword by T. H. Clutton-Brock and an afterword by Alison Richard.
Description: Chicago : The University of Chicago Press, 2024. | Includes bibliographical references and index.
Identifiers: LCCN 2023053668 | ISBN 9780226829739 (cloth) | ISBN 9780226829753 (paperback) | ISBN 9780226829746 (ebook)
Subjects: LCSH: Primates—Nutrition. | Primates—Feeding and feeds. | Primates—Physiology. | BISAC: SCIENCE / Life Sciences / Zoology / Primatology | SCIENCE / Life Sciences / Zoology / Ethology (Animal Behavior)
Classification: LCC QL737.P9 H785 2024 | DDC 599.815—dc23/eng/20240125
LC record available at https://lccn.loc.gov/2023053668

♾ This paper meets the requirements of ANSI/NISO Z39.48-1992 (Permanence of Paper).

For the animals

The whole of nature, as has been said, is a conjugation of the verb *to eat*, in the active and in the passive.

WILLIAM RALPH INGE (1922), *Outspoken Essays*, 2nd ser.
(Longmans, Green)

Contents

Foreword xi
T. H. Clutton-Brock

Preface xv
Joanna E. Lambert, Margaret A. H. Bryer, and Jessica M. Rothman

Introduction: From Diets to Disturbance: The Evolution of Primate Feeding Studies 1
David J. Chivers and Kim R. McConkey

PART I   FINDING, BUILDING, AND USING A DIET 19

1. **The Role of Macro- and Micronutrients in Primate Food Choice** 21
Annika Felton and Joanna E. Lambert

2. **What Extant Primates Eat: A Global Survey** 35
Joseph E. Hawes, Carlos A. Peres, and Andrew C. Smith

3. **The First Diet: Mother's Milk** 52
Katie Hinde, Lauren A. Milligan, and Gregory E. Blomquist

4. **Diet and the Energetics of Reproduction** 68
Melissa Emery Thompson

5. **Primate Energy Requirements: Brains, Babies, or Behavior?** 82
Alex R. DeCasien, Mary H. Brown, Stephen R. Ross, and Herman Pontzer

6. **Primate Senses: Finding and Evaluating Food** 95
Amanda D. Melin and Carrie C. Veilleux

7. **Seasonality in Food Availability and Energy Intake** 114
Cheryl D. Knott and Andrea L. DiGiorgio

PART II   **NUTRIENTS, NUTRITION, AND FOOD PROCESSING** 135

8 **Enzymes and Microbes of the Mammalian Gut: Toward an Integrated Understanding of Digestion** 137
Joanna E. Lambert, Richard Mutegeki, and Katherine R. Amato

9 **Secondary Compounds in Primate Foods: Time for New Approaches** 150
Eleanor M. Stalenberg, Jörg U. Ganzhorn, and William J. Foley

10 **Hormonally Active Phytochemicals in Primate Diets: Prevalence across the Order** 176
Michael D. Wasserman, Marie-Lyne Després-Einspenner, Richard Mutegeki, and Tessa Steiniche

11 **Nutrition and Immune Function in Primates** 198
Erin R. Vogel, Astri Zulfa, Sri Suci Utami Atmoko, and Lyle L. Moldawer

12 **Nutrition and Primate Life History** 222
Carola Borries and Andreas Koenig

PART III   **FOOD ACQUISITION AND NUTRITION IN SOCIAL ENVIRONMENTS** 241

13 **Social Food Competition, Then and Now** 243
Charles H. Janson

14 **Applying a Framework of Social Nutrition to Primate Behavioral Ecology** 261
Margaret A. H. Bryer and Moreen Uwimbabazi

15 **Primate Cognitive Ecology: Challenges and Solutions to Locating and Acquiring Resources in Social Foragers** 272
Paul A. Garber

16 **Feeding-Related Tool Use in Primates: A Systematic Overview** 299
Jill D. Pruetz, Landing Badji, Stephanie L. Bogart, Stacy M. Lindshield, Papa Ibnou Ndiaye, and Kristina R. Walkup

17 **Hunting by Primates** 327
David Watts

18 **Movement Ecology and Feeding Neighborhoods** 355
Margaret C. Crofoot and Shauhin E. Alavi

19 **Foraging in a Landscape of Fear** 364
Russell Hill

20 **Behavioral Flexibility and Diet** 381
AJ Hardie and Karen B. Strier

PART IV   METHODS, PRACTICE, AND APPLICATION   397

21 **Measuring Food in the Field**   399
Eckhard W. Heymann

22 **Wild Plant Food Chemistry**   417
Nancy Lou Conklin-Brittain

23 **Evaluating Primate Diets with Stable Isotopes**   431
Matt Sponheimer and Brooke Crowley

24 **Mechanical Properties of Primate Foods**   446
Adam van Casteren and Peter Lucas

25 **Modeling Primate Nutrition**   463
David Raubenheimer

26 **Reconstructing Fossil Primate Diets: Dental-Dietary Adaptations and Foodprints for Thought**   498
Peter S. Ungar

27 **Food and Primate Carrying Capacity**   515
Andrew J. Marshall

28 **Climate Change and Primate Nutritional Ecology**   532
Jessica M. Rothman, John B. Makombo, and Mitchell T. Irwin

29 **Primate Foraging Strategies Modulate Responses to Anthropogenic Change and Thus Primate Conservation**   544
Colin A. Chapman, Kim Valenta, Fabiola Espinosa-Gómez, Amélie Corriveau, and Sarah Bortolamiol

Afterword   555
Alison Richard

Acknowledgments   557
Literature Cited   559
List of Contributors   715
Index   723

Foreword

T. H. Clutton-Brock

How Primates Eat is, at the same time, a masterly synthesis of current knowledge of primate nutritional ecology, a celebration of all that has been achieved over the last 50 years, and a road map for future research. As the introduction recounts, descriptions of primate feeding behavior extend back to the earliest primate field studies by Clarence Ray Carpenter and others—but were mostly nonquantitative and seldom focused on particular questions. Stimulated by the work of John Crook and his colleagues, a regular discussion group of primate fieldworkers met irregularly in Cambridge (then one of the principal centers for primate research in the UK) to discuss particular aspects of primate ecology and social organization and different ways of quantifying variation in feeding behavior. This eventually led to a multiauthor book, *Primate Ecology* (1977), which contained chapters on the feeding behavior of four prosimians, three New World monkeys (howling monkeys, yellow-handed titis, and Colombian spider monkeys), six African and Asian monkeys (mangabeys, rhesus macaques, gelada baboons, guerezas, and gray and purple-faced langurs), and four apes (siamangs, orangutans, mountain gorillas, and chimpanzees). At the end, I tried to synthesize our current understanding of inter- and intraspecific differences in primate feeding behavior and activity budgets.

It is difficult to provide a realistic impression of how far studies of primate nutritional ecology have advanced since then. In the early 1970s, we spent an inordinate time discussing different techniques of quantifying feeding behavior—so much so that in the viva of my own PhD exam, my internal examiner (a distinguished neurobiologist) started off the exam with "You know, Tim, most of this is all rather boring—you need to stick your neck out more." The generalizations that emerged were simple and unsurprising: species differences in foraging time were positively related to body weight and negatively to reliance on foliage, and home range size and day range length increased with body weight and group size.

The advances and achievements of research on primate nutritional ecology since then are remarkable. *How Primates Eat* summarizes them under four headings: "Finding, Building, and Using a Diet" provides a systematic overview of primate diets, energy requirements, and feeding behavior and their association with life history parameters; "Nutrients, Nutrition, and Food Processing" covers the components of primate foods, digestive processes, the effects of secondary metabolites and endocrine-disrupting chemicals, and the consequences of variation in diet for immune function; "Food Acquisition and Nutrition in Social Environments" synthesizes the results of research on the cognitive challenges and solutions involved in meeting energetic and nutritional requirements and the interaction between social organization, social structure, and foraging behavior; and "Methods, Practice, and Application" explores methodological advances, including recent techniques of measurement of food availability and food intake and the application of modeling techniques. The book ends with four chapters that examine the implications of research on primate nutritional ecology for reconstructing the diets of fossil primates, explaining contrasts in population density, and predicting the causes and consequences of climate change and other anthropogenic factors for primate ecology and primate populations.

As well as enormously expanding the range of species studied and topics explored, *How Primates Eat* shows how research in primate feeding ecology has been successfully integrated both with other lines of research on primates (including studies of primate phylogeny and morphology) and with research on social evolution and cognitive processes. In addition, primate research is now far more extensively integrated with similar work on other animals than it was in the 1970s—an important development, for answers to many ecological questions about primates require an understanding of their differences from other orders. However, despite its successful integration with research on other animals, research on primate ecology retains a distinctive feel related to its underlying focus on the evolution of species differences within a single order. While this may in the past have been one of its weaknesses, it now represents one of its strengths, because the taxonomic coverage that studies of primates now provide makes it possible to compare associations between diet, physiology, morphology, and behavior across multiple radiations of related species. Similar coverage of species differences in ecology does not exist for any other order of mammals or birds, and its presence in primates makes it possible to answer questions about evolutionary sequences and processes that cannot be explored in orders where taxonomic coverage is much less extensive. Over the last 50 years, comparative studies of primate social behavior have commonly acted as a seedbed for research on similar topics in other animals while relevant data are more accessible—and I suspect that the studies of species differences in primate feeding behavior will stimulate related research on a wide range of other animals in the coming decades.

Despite the enormous advances that have been made, studies of primate feeding ecology still face substantial logistic obstacles. Primate diets are quantitatively complex, and changes in food quality and quantity are frequently difficult to quantity. Variation in growth, breeding success, and survival is often difficult to measure so that it is often hard to assess the fitness consequences of variation in diet. There are relatively few species where it is easy to sample blood, skin, or feces, making it difficult to construct the multigenerational pedigrees that have allowed field studies of a number of other mammals to measure the heritability of different traits and the intensity of selection on genetic variation. And as changes in many important ecological and evolutionary processes occur over decades rather than over individual years, studies are needed that can monitor changes in diet, behavior, life history parameters, and demography in the same populations over periods of multiple decades. This is particularly true for studies that aim to measure the consequences of climate change or other anthropogenic factors likely to have protracted consequences for primate ecology. As a result, the maintenance of field teams on site throughout multiple years and secure access to particular sites and populations are of primary importance. Finally, as the editors note in the preface, we need to involve more local scientists in our field studies and find ways to increase their number and to support their careers. The Organization for Tropical Studies (in the US) and the Tropical Biology Association (in Europe) both play an important role in providing field training to local ecologists and in supporting their careers. Both organizations need to extend the scope of their activities and deserve all the support that we can give them.

Species differ widely in the extent of the logistical obstacles they present for field research. As the specificity and complexity of questions increase, field studies will need to be increasingly strategic in matching the species they plan to study to the questions they wish to answer. Relatively visible, diurnal species that can be easily caught, are comparatively short-lived, and do not range too widely offer substantial advantages.

The evolution of research on primate nutritional ecology over the last 50 years suggests that future studies will continue to develop methodologies that avoid or minimize these problems and constraints. Remote sampling techniques and biologging devices will continue to develop and are likely to play an increasingly important role. So, too, will genetic and genomic analysis of differences between individuals, groups, populations, and species. However, the interpretation of remote data frequently presents difficulties unless it is combined with the intuitive understanding provided by experienced, on-site observers. As a result, remote recording devices will never replace the need for curious, experienced field observers who are strongly motivated to identify and answer ecological questions of fundamental importance.

Reading through the chapters of *How Primates Eat*, I was gratified to see the extent, complexity, and success of current research on primate feeding ecology that has, in part, developed from our early studies in the 1970s. I have every confidence that the studies that will develop over the next 50 years from the research documented in this book will generate even greater advances in our understanding of the nutritional ecology of primates and other mammals.

Tim Clutton-Brock
October 2022

Preface

Joanna E. Lambert, Margaret A. H. Bryer, and Jessica M. Rothman

In his 1942 synthesis of evolutionary theory, Julian Huxley suggested that all biological facts about an organism can be queried by way of three aspects: mechanistic-physiological, adaptive-functional, and historical-evolutionary (Huxley 1942). In 1961, Ernst Mayr further clarified how we address biological questions by distinguishing between proximate causation (How?), the domain of the functional biologist, and ultimate causation (Why?), the domain of the evolutionary biologist (Mayr 1961). Together, Huxley's and Mayr's organizing principles provided a scaffolding for asking questions about animals so robust that its application to animal behavior and ecology at mechanistic, developmental, functional, and evolutionary scales was ultimately awarded the Nobel Prize in Physiology and Medicine (awarded to Lorenz, Tinbergen, and von Frisch in 1973; Burkhardt 2014; Grodwohl 2019). The synthesis of theory that occurred in the last century represents nothing less than a road map for all research related to animal biology and ecology today—including investigation into feeding, diet, and nutrition (S. Gould 2002; Grodwohl 2019; see fig. P.1).

Our goal in this volume is to synthesize our understanding of what primates eat (biological fact), how primates eat (proximate), and why primates eat what they eat (ultimate). Using new tools and the knowledge provided by decades of boots-on-the-ground (and in the lab) research on primate species across the order, the 29 chapters of this volume integrate all aspects of Huxley's organizing tenets, variously emphasizing mechanism, development, function, or evolution as they relate to proximate or ultimate causation as posed by Mayr.

As editors of this book, our training spans three decades (Lambert—PhD, 1997; Rothman—PhD, 2006; Bryer—PhD, 2020). Our academic foundation not only is positioned within the broader context of a 20th-century synthesis of evolutionary theory but also stands on the shoulders of those who have been working toward knowledge of mechanism, development, function, and evolution of primates from the outset. Key volumes such as those produced by Clutton-Brock (1977), Chivers et al. (1984), Richard (1985), Rodman and Cant (1984), and Smuts et al. (1987) are landmark achievements in primate ecology that provide the backdrop against which all future work in primate feeding biology and diet unfolded. Scholars such as Katherine Milton, Ken Glander, Richard Wrangham, Tom Struhsaker, Deborah Baranga, and Rich Kay, among many others, asked questions at proximate and ultimate levels on extant and extinct species alike.

Importantly, what distinguishes this volume from previous work is that we chose the sobriquet "nutritional ecology" rather than "feeding biology." Though the two concepts are inextricable in practical terms of how an animal makes a living, only recently have we had the field and laboratory methods along with a substantive body of wet laboratory chemistry data to claim "nutrition" rather than just "diet" or "feeding." Indeed, the past two decades of primate research has moved beyond asking questions about the intra- and interspecific variability in primate diets to understanding how

Figure P.1. Redtail monkeys (*Cercopithecus ascanius*) foraging for insects in the bark of a *Prunus africana* tree, Kibale National Park, Uganda. This picture captures many aspects of feeding biology discussed in this volume that can be queried from proximate (e.g., how do redtail monkeys digest the chitin of insect carapaces?) to ultimate scales (e.g., why do redtail monkeys forage in cohesive social groups?).

Photo: Jessica M. Rothman.

primates meet nutritional needs when faced with complex mixtures of micro- and macronutrients, plant toxins, and digestion inhibitors in seasonally and spatially variable nutritional landscapes and in complex social contexts.

We embarked on this book project to bring together scientists studying diet and nutrition on different continents with different (albeit complementary) goals and at various scales spanning proximate and ultimate causation. We aimed to produce a synthetic volume that graduate students and colleagues alike can use as a resource when embarking on studies of feeding ecology and nutrition—regardless of taxon—that describe the state of the field to the present.

Why primates? Arguably, primates have always been investigated through a different lens than other taxa—

with an aim toward interpretation of our own ancestry and biology. There are both idiosyncratic reasons for this history and theoretically driven rationale (DeVore 1965; Haroway 1990; Kinzey 1987). From the perspective of diet, feeding, and nutrition, several facts about primates are relevant. First, most primates are by and large primary consumers (Chivers and Hladik 1980). Along with their trophic counterparts in other largely herbivorous mammalian orders (e.g., Perissodactyla, Artiodactlya, and Rodentia), this means that they confront—and cope with—the various challenges posed by plants, including wide and unpredictable amplitude in nutritional content and a host of chemical and physical defenses (Glander 1982). Primates are also a tropical radiation, with most species found at lower latitudes (Le Gros Clark 1959). Among other things, this can mean a high likelihood of spatiotemporal variance in the availability of any given plant species. Perhaps because of this variance, many, though certainly not all, primate species have diets comprising hundreds of plant species and parts and the guts (literally) for coping with such breadth along with the brains and cognition to find them (Milton 1981b). A hallmark distinction of primates is having to cope with these challenges as both highly encephalized species (with commensurately high energetic requirements) and intensely social species, navigating social politics at every scale from individual to species.

In this volume, we address each of these biological facts about primates and do so in five sections. The volume is introduced with discussion of the historical context by which to understand the study of primate feeding, diet, and nutrition (Chivers and McConkey, introduction). Part 1, "Finding, Building, and Using a Diet," is initiated with an evaluation of the micronutrients (vitamins and minerals) and macronutrients (carbohydrates, proteins, and lipids) required by primates and their roles in shaping primate food choice (Felton and Lambert, chap. 1) and then moves on to a systematic overview of primate diets, with an analysis of the breadth of foods and food types consumed by primates around the globe (Hawes, Peres, and Smith, chap. 2). Hinde, Milligan, and Blomquist (chap. 3) evaluate the first diet, integrating life history and behavioral ecology perspectives to discuss the state of knowledge of primate mother's milk. Chapters follow on the energetics of reproduction (Emery Thompson, chap. 4), requirements for energy (DeCasien, Brown, Ross, and Pontzer, chap. 5), an evaluation of the sensory modalities involved in procuring an adequate diet (Melin and Veilleux, chap. 6), and the role of seasonality in primate energy intake (Knott and DiGiorgio, chap. 7).

Part 2, "Nutrients, Nutrition, and Food Processing," opens with a description of the enzymes and microbes of the primate gut (Lambert, Mutegeki, and Amato, chap. 8). In chapter 9, Stalenberg, Ganzhorn, and Foley consider the role of secondary metabolites in primate diets. In chapter 10, Wasserman, Després-Einspenner, Mutegeki, and Steiniche review the costs and benefits of dietary endocrine disrupting chemicals. Vogel, Zulfa, Utami Atmoko, and Moldawer (chap. 11) examine the multidimensional and bidirectional relationship between nutrition and host immunity. The section concludes with a review of how growth, maturation, and reproductive performance vary predictably with nutrient availability (Borries and Koenig, chap. 12).

The chapters comprising part 3, "Food Acquisition and Nutrition in Social Environments," untangle the social and cognitive challenges and solutions to meeting energetic and nutritional requirements. The section opens with a comprehensive review of the history and status of the study of primate social feeding competition (Janson, chap. 13) and then moves on to a discussion of what the "social nutrition" framework used for other taxa could bring to the study of primate social feeding (Bryer and Uwimbabazi, chap. 14). Garber (chap. 15) then examines social foraging strategies in primates from a cognitive perspective, presenting evidence for species differences in the types of social and ecological information used by individuals in making foraging decisions. Pruetz, Badji, Bogart, Lindshield, Ndiaye, and Walkup (chap. 16) provide a synthesis of tool use and diet, and Watts (chap. 17) describes hunting by primates and emphasizes that though primates are largely primary consumers, there are important exceptions. The section moves on to evaluate how primates' movement ecology both shapes and is shaped by their dietary choices, specifically how motion and navigation capacity and environmental and social factors interact to drive primate foraging patterns (Crofoot and Alavi, chap. 18). Hill (chap. 19) explores how spatial variation in predation risk generates a primate "landscape of fear," with prey animals expected to modify their distribution and feeding behavior in response to this variable predation risk. An overview of the dynamic interactions among behavioral adjustments in feeding, ranging, and grouping patterns in wild primates (Hardie and Strier, chap. 20) concludes this section.

Part 4, "Methods, Practice, and Application," first explores methodological advances in primate feeding biology, including recent techniques for measuring food availability and abundance in the wild, with consideration of "foodscapes" (Heymann, chap. 21), nutritional composition (Conklin-Brittain, chap. 22), the use of stable isotope analyses to gain insight into primate diets (Sponheimer and Crowley, chap. 23), and mechanical properties in the context of primate foods (van Casteren and Lucas, chap. 24). Raubenheimer (chap. 25) addresses the need for formal models in nutritional ecology and reviews the modeling approaches that have been applied to feeding in primates and other taxa. We close the volume with four chapters that discuss the applied significance of primate feeding research, specifically how knowledge of extant diet and dental morphology of living animals can be used to infer feeding biology in fossil species (Ungar, chap. 26), the link between food availability and its implications for carrying capacity (Marshall, chap. 27), how human-induced shifts in climate have affected primate nutrition (Rothman, Makombo, and Irwin, chap. 28), and how information on primate foraging strategies and nutritional needs are requisite for conservation planning (Chapman, Valenta, Espinosa-Gómez, Corriveau, and Bortolamiol, chap. 29).

We aimed to produce as exhaustive a volume as possible, but we acknowledge that there are gaps—indeed, in primate nutritional ecology writ large. For example, the application of primate nutritional biology to questions regarding biodiversity and food web theory are largely lacking. And though powerful new molecular methods (e.g., genome-wide analyses) have yielded important insights into wildlife nutrition, this work is not addressed here. Also missing is discussion regarding diet-induced impacts on gut-brain interactions and analyses of individual variance as a function of epigenetics, phenotypic plasticity, and environmental input (e.g., foods consumed).

Moreover, another deficit of this book relates to questions of access and equity in science. Though we set out to assemble this volume (well over five years ago) in as fair and inclusive a fashion as possible and aimed for balanced authorship, we succeeded only in the realm of gender, with a majority of the authors female. The low representation of scholars of color and from primate-habitat countries in Africa, Asia, and Central and South America highlights structural discrepancy at all scales in the science of primates and their wild diets. We acknowledge our role in this and will do better.

In closing, though we encourage the reader to consider the content of this volume through the theoretical and empirical lenses provided by our academic science, we would be remiss if we did not also acknowledge two issues of gravest consequence existing beyond the ivory tower of academia: a global pandemic and the extinction crisis. Simply put, our science will not stand the test of time unless we can extend it beyond academic conversations. Using the data collected over literally millions of hours of field and laboratory work to create actionable conservation solutions as they apply to extant primates and other species is a call to us all, not least our student readers at the outset of their careers in nutritional ecology of primates and other wildlife.

Introduction
From Diets to Disturbance: The Evolution of Primate Feeding Studies

David J. Chivers and Kim R. McConkey

From the seeds germinated in field studies of birds in the middle of the last century arose early primate field research, which blossomed into more diverse recent studies. The seeds of these studies have dispersed widely to include the interactions of primates with plant and animal communities, encouraging conservation biology, molecular research, and other new disciplines. As habitat loss escalates (first it was forest clearance for tin and rubber, and now it is oil palm in many regions) and hunting for sport and trade increases, so does the risk of primate extinction. Primate researchers are now obligated to focus more on the ecological needs of primates for survival, including considerations of carrying capacities and habitat management.

In this introduction, we summarize the progression of feeding studies on primates from the early qualitative studies of diet to the broad, multifaceted approaches that are used to understand primate feeding patterns today. Our goal is to show how approaches to understanding primate feeding have evolved from projects narrow in scope to projects that include more variables and greater detail, as researchers became increasingly aware of the complicated, interconnected communities and habitat mosaics in which primates live. We highlight the earliest primate feeding studies, from which the latter topics and studies have been developed. An additional goal is to show how the focus in primate feeding studies has shifted to incorporate approaches in which primates play an integral part but are not the main focus, as researchers investigate the ecological role of primates in a wider context.

EARLY DIETARY ANALYSES

Clarence Ray Carpenter should be regarded as the pioneer of primate field studies, with his studies of howler monkeys (*Alouatta palliata*) in the Panama Canal Zone and of lar gibbons (*Hylobates lar*) in Thailand in the 1930s (C. Carpenter 1934, 1940; figs. I.1, I.2). He was inspired, as were Harold Bingham (1932) and Henry Nissen (1956) in their search for African apes, by the psychologist Robert Yerkes, a pioneer in studies of human and nonhuman primate intelligence, animal behavior, and comparative psychology. This early work was driven by research on animal and human behavior and cognitive processes, with studies of captive chimpanzees in the US in relation to human behavior, hence the focus on apes. Thereafter, little progress was made until the late 1950s, when there was a burgeoning of primate field studies throughout the tropics, especially in Africa. The first studies were led by anthropologists who focused on models of human evolution, relationships within social groups, and the maintenance of social structure; these studies were collected in Irven DeVore's (1965) *Primate Behavior*, followed by volumes edited by Stuart Altmann (1967) and Phyllis Jay (1968). Descriptions of diet and feeding behavior demonstrated clear differences between species, but it was initially a surprise that there could be significant variation within species, such as gray langurs (*Presbytis entellus*, now *Semnopithecus entellus*) and olive baboons (*Papio anubis*). There was a noticeable shift in the 1960s toward

Figure I.1. One of the first primate studies was by Clarence Ray Carpenter on lar gibbons in Thailand in the 1930s. Photos of lar gibbons eating leaves (*left*) and fruit (*right*) in Khao Yai National Park, by Kulpat Saralamba.

zoologists seeking a fuller understanding of primate socioecology and of integrating primates into their environments, which consist of variables including climate, habitat structure, and food availability.

ECOLOGICAL GRADES

Crook and Gartlan (1966) and Eisenberg et al. (1972) organized primate species into ecological grades based on all available socioecological data known at the time. This was the first attempt to integrate social and ecological parameters, based on frameworks of previous studies of birds and other mammals. Under both classifications, primate species were grouped according to their dietary categories, gross differences in the habitats occupied, and social systems (solitary, monogamous pairs, one-male groups, and multimale groups).

John Crook and Steve Gartlan's (1966) classification system grouped primate societies into five evolutionary grades that were correlated with species ecology. Grade I were mostly strepsirrhines—nocturnal, solitary or paired, forest-dwelling insectivores. Grade II contained diurnal, vegetarian primates in small family groups (one male). Grades III and IV included arboreal and terrestrial species with more sexual dimorphism and larger multimale groups in a range of habitats. Grade V was a curious variant of grade II, described as open-country one-male units, with the largest males, often aggregating into large herds for feeding and sleeping. Crook and Gartlan (1966) argued that the evolution of these diverse grades was driven by ecology as well as phylogenetic inertia. Problems with this system include the lack of true dependent variables and weak correlations between variables. For example, as more data accumulated, ceboids were found to be scattered

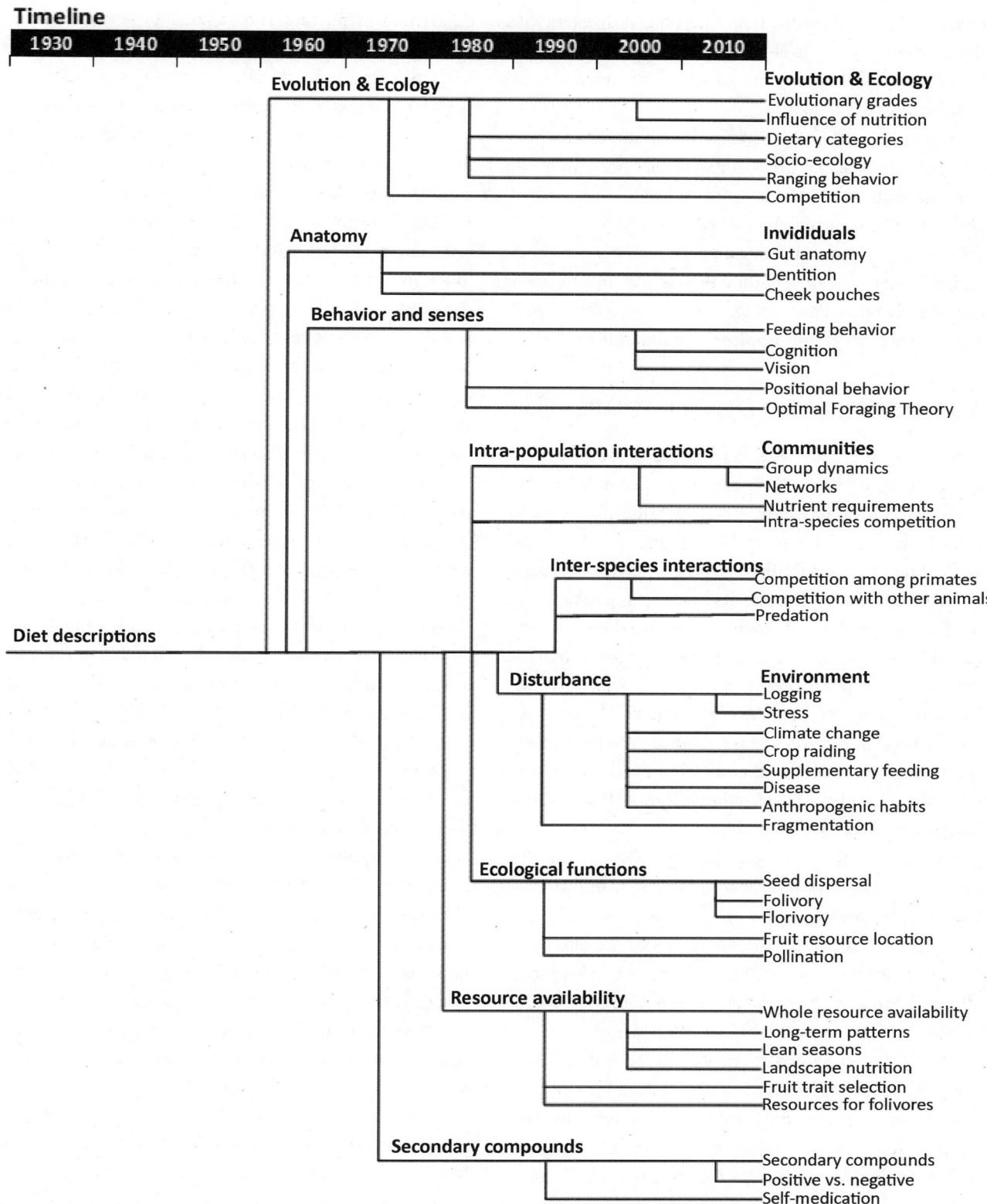

Figure I.2. Timeline showing the evolution of primate feeding studies. The first primate studies describing primate diets, conducted in the 1930s, have expanded into multiple subfields within the last 80 years.

though grades I–III rather than solely forest dwellers with similar diets. Hence, useful though the grades are, at this level of analysis they are not mutually exclusive, and there is overlap between them.

Eisenberg et al. (1972) focused their analysis on male involvement in social life and its correlation with group size, the nature of the dominance system, and territoriality. They introduced the category of age-graded male groups as a significant intermediate between one-male and multimale groups. Although this classification was not directly related to feeding, they followed the ecological features detailed by Crook and Gartlan (1966).

SOCIOECOLOGICAL SYNTHESIS

A focus on diets and feeding was developed during a conference in July 1975 in the magnificent Edwardian castle on the Isle of Rhum, Inner Hebrides. In the landmark volume resulting from this meeting (Clutton-Brock 1977), the authors described feeding behavior in relation to quantitative analyses of diets across primate taxa. The book contains three chapters on strepsirrhine primates (*Lemur catta* and *L. fulvus* [Robert Sussman]), *Indri indri* [John Pollock], and *Propithecus verreauxi* [Alison Richard]), three on New World monkeys (*Alouatta palliata* [Chris Smith], *Callicebus torquatus* [Warren Kinzey], and *Ateles belzebuth* [Lewis and Deborah Klein]), five on Old World monkeys (*Lophocebus albigena* [Peter Waser], *Macaca mulatta* [Don Lindburg], *Theropithecus gelada* [Robin Dunbar], *Colobus guereza* [John Oates], and *Presbytris senex* and *P. entellus* [Marcel Hladik]), and six chapters on apes (*Symphalangus syndactylus* [David Chivers], *Pongo pygmaeus* [Peter Rodman], two on *Gorilla gorilla beringei* [Dian Fossey and Sandy Harcourt; Alan Goodall], and *Pan troglodytes* in Gabon and Tanzania [Marcel Hladik] and in Tanzania [Richard Wrangham]).

There are two synthetic chapters at the end of the volume. The first, by Tim Clutton-Brock, examines intraspecific variation in feeding and ranging behavior in primates, looking at (1) food choice and ecological segregation within species, in relation to differences in food choice across species; (2) age and sex differences in feeding levels and sites, activity budgets, and food choice; (3) diurnal variation in feeding levels, activity patterns, and food choice; and (4) seasonal variation. The second concluding chapter (Clutton-Brock and Harvey) discusses species differences in feeding and ranging behavior, including time spent feeding, dietetic diversity, activity budgets, day range length, home range size, and population density and biomass. Finally, there are three appendices: (1) methods and measurement, (2) measurement of dietetic diversity, and (3) field methods of processing field samples. Thus, this volume illustrates the nature, scope, and breadth of primate field studies by the mid-1970s, with excellent quantitative and comparative analyses (fig. I.1). The significance of this volume is that it contains the first thorough and quantitative analysis of feeding behavior and the first objective comparisons across primate taxa.

The importance of a socioecological approach to understanding foraging behavior was extended further by Wrangham (1979) after he determined that grouping and ranging patterns of primates appeared to be linked directly to foraging strategies and that these patterns often differed between sexes. His key message was that the social system is a compromise between females maximizing feeding efficiency (for rearing young) and males seeking access to as many females as possible, to maximize breeding efficiency (easier at some times than others). Only by remaining in social contact can a breeding unit split into foraging units, which requires efficient communication and a long memory. Wrangham (1979) argued that the ability of monkeys to cope better with toxic substances may be a major distinction between monkeys and apes, with implications for their respective social organizations. Though a diet focused on small, scattered, ripe, pulpy fruit would make solitary behavior optimal for gibbons (*Hylobates*), gibbon females face the costs of rearing offspring and need to be associated with males for defense. Such compromises lead to the monogamous territorial groups of gibbons, which contrast with the loose one-male groups of orangutans (*Pongo*), the cohesive one-male groups of gorillas (*Gorilla*), and the multimale, multifemale groups of chimpanzees (*Pan*); ape social diversity stems from spatial and temporal availability of food and mates.

In the 1980s, Carel van Schaik addressed the idea that competition for food resources was an important factor in the evolution of social organization of primate groups (Janson and van Schaik 1988; van Schaik 1983; van Schaik and van Hooff 1983). Ganzhorn (1988, 1989a) took this idea further by showing how competition for foods can structure habitat use by sympatric primate species. In his study of seven lemur species, Ganzhorn showed that species in the same habitat choose foods with different chemical properties, whereas species eating the same foods exploit different microhabitats.

IMPORTANCE OF RANGING BEHAVIOR

Familiarity with an area of habitat, termed "philopatry" (which literally means "love of the fatherland"), is crucial to the survival of primates living in groups. Because primates depend on knowledge of their habitat, especially the location and timing of availability of the various foods on which they depend for their survival, ranging behavior is integral to feeding (Chivers 1969, 1974; Clutton-Brock 1977). Thus, social groups of all species have a home range—what differs is the pattern of use. In many species, the ranging patterns of a particular area are variable with a central core area and no clear boundaries evident, while in some species the pattern of use is more even and there are distinct boundaries, which are defended as territories (e.g., Chivers 1969). What is important for the integrity of the social group is that, at any time, the group has exclusive use of at least part of their home range. Intimate knowledge of an area of habitat, of the location of food trees and timing of the availability of food, is essential for survival (Clutton-Brock 1977). Hence, it was recognized early in the 1970s that a thorough documentation of ranging behavior was needed to understand feeding ecology in primates. Initially, ranging behavior was considered solely in the context of primate movements in relation to access to available foods and other essential resources (as well as in relation to predation risk; e.g., Clutton-Brock 1977). More recently, limited research has been conducted on how primates locate their foods, focusing on cognitive and sensory processes (Asensio, Brockelman, et al. 2011; Dominy, Lucas, et al. 2001; Janson 2016; Janson and Byrne 2007; Trapanese et al. 2019; Melin and Veilleux, this volume; Garber, this volume).

DIETARY CATEGORIES

In 1980, David Chivers and Marcel Hladik modernized the antiquated dietary categories that had until that time been in common usage: carnivore/insectivore (with their taxonomic connotations), herbivore (too broad and confusing), and omnivore (essentially wrong) were shifted to faunivore (animal matter, vertebrate or invertebrate), frugivore (reproductive parts of plants), and folivore (vegetative parts, including gums and bark). Dietary analyses had revealed that no mammal could eat significant quantities (10%–20%) of each of these main categories; hence, "omnivores" (foli-fauni-frugivores) can rarely exist, for anatomical and physiological reasons. Fruit is often the main dietary component of animals—such as smaller bears, pigs, and peccaries—that were commonly called omnivores (Fredricksson and Wich 2006; Sridhara et al. 2016). Humans, evolved from the frugivorous apes, owe their omnivory to cooking and food processing! More recently, researchers have moved from a description of diets in terms of food types and physical properties to studies of nutrient balancing and the nutritional contents of foods ingested (e.g., Ganzhorn 1992; J. Lambert and Rothman 2015; I. Wallis et al. 2012; Wasserman and Chapman 2003).

Dietary indices were derived from quantitative feeding data displayed on a three-dimensional graph (Chivers and Hladik 1980), with all species following a crescentic path from 100% faunivory through 55%–80% frugivory to 100% folivory (fig. I.3a). Gums are long-chain sugars comparable to leaves for digestive purposes, and bark is similar; the latter and fungi are such rare dietary components that they never confound the trichotomy described above (and are included in folivory). Gums fit into folivory as nectar and flowers fit into frugivory. Since 100% frugivory is not possible because certain essential amino acids are absent from fruit (Waterman 1984), smaller frugivores (fauni-frugivores) must supplement fruit with animal matter, and larger frugivores (foli-frugivores) supplement with leaves. Kay's threshold defines the relationship between primate body weight and main dietary category, with species below 500 g mainly fauni-frugivores and those above 500 g mainly foli-frugivores (Fleagle 1988). Access to foliage is allowed by the elaboration of chambers for the bacterial fermentation of fiber—the stomach in monkeys from the family Colobinae and the caecum and colon in, for example, indriids (Indriidae), *Lepilemur*, *Callithrix*, and *Gorilla*.

Primates are unique among mammalian orders in focusing on the reproductive parts of plants, thereby avoiding the specializations that characterize other mammals, which have either small and simple guts to process animal matter or large, complex guts to process the vegetative parts of plants (Chivers and Hladik 1980). Exceptions are

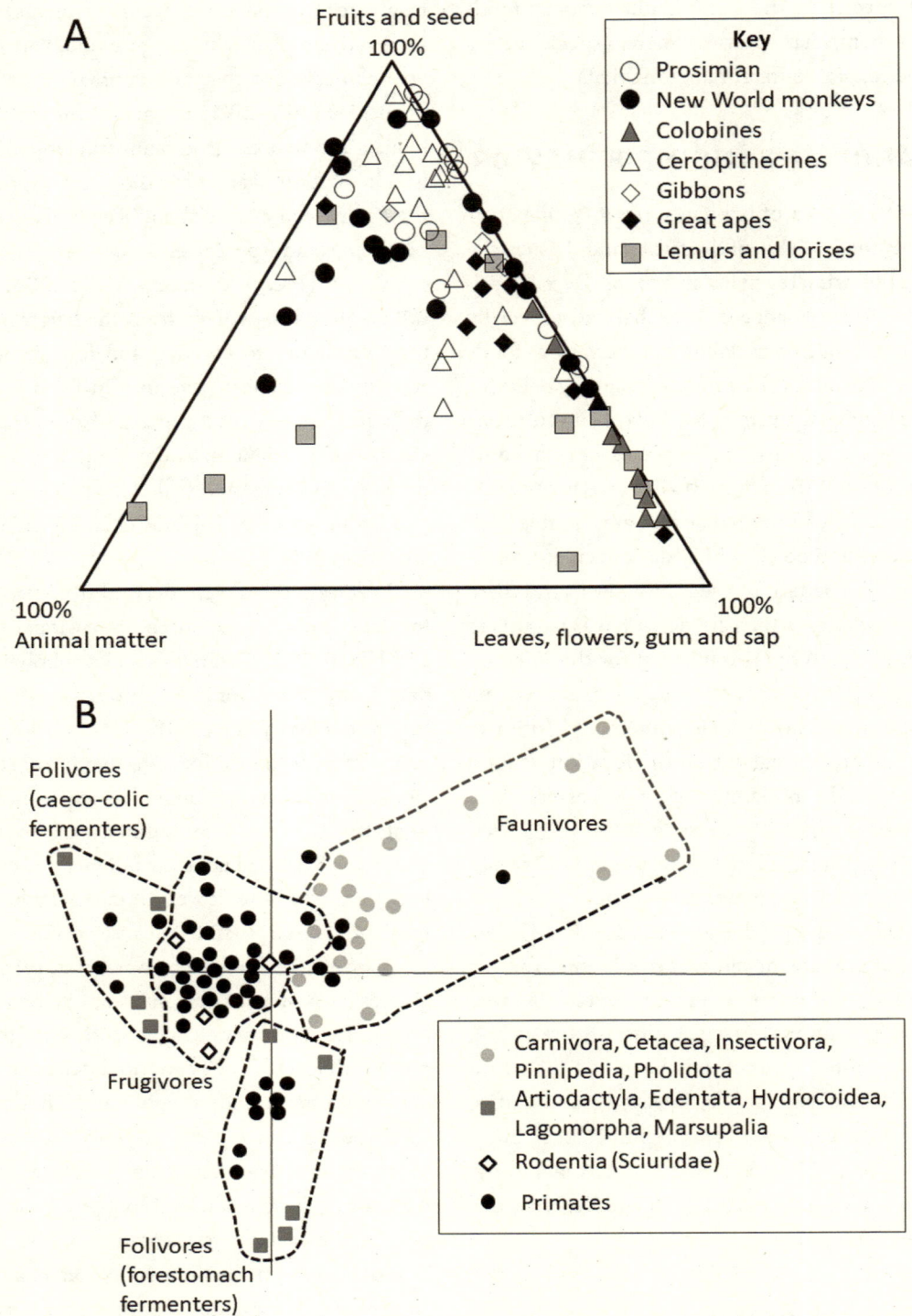

Figure I.3. Diets of primates in terms of the extent of frugivory, folivory, and faunivory. *A*, Average diets of 80 primate species are expressed as the percent consumption of fruit (and seeds), leaves, flowers and exudates, and animal matter. Diets are presented in the form of a triangular diagram, with 100% faunivory, 100% folivory, and 100% frugivory at the three corners; fauni-frugivores are to the left and foli-frugivores to the right. *B*, Multidimensional scaling of gut compartments (plot of indices for surface areas of stomach, small intestine, and caecum + colon) for 80 primates and other mammals, showing the radiation of morphology from a more generalized frugivorous gut or from the specialized faunivorous gut (dominated by small intestine) to folivorous guts dominated by caecum and colon or by a sacculated stomach. Figures are adapted from MacLarnon et al. (1986) and Chivers (1992, 1998) (Cambridge University Press).

colobines and indriids, which have complex guts to overcome the chemical problems of digesting leaves and seeds, allowing them to exploit food items in forest canopies that are underused by other primate species (J. Campbell et al. 2000; Chivers 1994; Matsuda et al. 2019).

The focus on the reproductive parts of plants, which are intermediate in abundance and digestibility, requires unique dietary flexibility. This feeding flexibility, coupled with reproductive flexibility and aseasonal breeding, allowed haplorrhine primates (most obviously in Old World taxa) to become especially successful among forest-dwelling mammals, extending their distribution as species evolved to occupy more open habitats (Chivers 1991; Clutton-Brock 1988). In forest habitats, breeding usually occurs throughout the year, with peaks relating to rainy periods followed by plant growth (leaf flush, flowering, and fruiting; e.g., Chivers 1980). When females resume cycling after lactation in any habitat, they will become pregnant again in their first or second menstrual cycle, given the efficiency of their social system. It is in more open habitats, with marked seasonality, that there are distinct breeding seasons, again relating to the rains and plant growth; pregnancy lasts six months in monkeys and lactation for a further six months, with infants weaned at the next annual burst of food availability. Because of this pronounced seasonality, students of baboons (and chimpanzees) and other savanna-dwelling primates erroneously wrote about estrus cycles and estrous females, missing a key feature of primate biology.

DIET AND GUT MORPHOLOGY

The primate radiation provides unique opportunities to compare gut dimensions across specialist and generalist mammals, and progress on this was reviewed by Chivers and Hladik (1980). From a frugivorous base that still describes most primates, colobines converge on the ruminants; great apes, large atelids, and some strepsirrhines (e.g., *Indri* and *Lepilemur*) converge on the equids; and smaller cebids (including *Callithrix* and *Saguinus*) converge on carnivores and insectivores. These patterns are revealed by a multidimensional analysis of three gut quotients for the sizes of stomach, small intestine, and caecum and colon larger or smaller than expected from Kleiber's law (R. Martin et al. 1985). Foregut fermenters have stomachs that are much larger than expected, caeco-colic fermenters have large intestines larger than expected, and faunivores, with guts dominated by small intestine, have gut compartments smaller than expected (fig. I.3b). The foundations for this analysis were laid by Chivers and Hladik (1980) in their thorough analysis of gut dimensions in many primates and nonprimate mammals, in relation to their diets; 50 of the 78 mammalian species and 117 of the 180 individuals included were primates. Following a description of the gastrointestinal tracts in the different mammalian taxa based on W. Hill (1958), indices were calculated for diets, small intestine areas (in relation to absorptive capacity), and stomach and large intestine volumes (in relation to fermentation capacity, actual or potential). While there are more recent data on diet and some on gut dimensions, it is very unlikely that the basic relationships described in Chivers and Hladik (1980) will be altered; it would be useful, however, to determine how more detailed knowledge on species' diets and guts, with the addition of new species, might alter specific relationships and patterns.

Interspecific comparisons were developed for the whole data set, allowing for allometric corrections. The volumes of stomach and large intestine were related to actual body size in faunivores (in which microbial fermentation is minimal), whereas these chambers were more voluminous in larger frugivores (e.g., *Ateles*, *Mandrillus*, *Pongo*, and *Pan*) and caeco-colic-fermenting folivores (e.g., *Indri*, *Lepilemur*, *Alouatta*, and *Gorilla*); foregut fermenters showed a marked decrease in fermentation capacity with increasing body size, given the problems of absorption where surface area does not increase at the same rate as volume. Gastrointestinal surface areas for absorption were directly related to metabolic body size in frugivores, while area for absorption was relatively less in larger faunivores and more in larger folivores.

Milton (1981c) discussed how the digestion of food is a compromise between two conflicting forces—efficiency of nutrient extraction and food passage rate. There is a neat contrast between the folivorous howler monkey (*Alouatta palliata*) and the frugivorous spider monkey (*Ateles geoffroyi*). Food passage rate is 20.4 hours in howler monkeys and 4.4 hours in spider monkeys; in the former, with a more capacious colon, energy return from leaves is maximized by more efficient fermentation, whereas in the latter, the fruit (too low in protein to support howlers) are processed faster. She stresses that once a particular digestive strategy evolved, with attendant morphological, physiological, and behavioral adaptations, switching diet, at least in the short term, is not possible. Ganzhorn, Arrigo-Nelson, Boinski, et al. (2009) argue that the fruit in the Neotropics contain enough protein to satisfy the protein needs of primates, so that there has been less selection pressure to digest the less digestible leaves in the Neotropics, compared with Madagascar, where the low concentration of nitrogen in fruit seems to have contributed to the virtual absence of frugivores on the island.

DIET AND ORAL ANATOMY

While the gastrointestinal tract is primarily concerned with the chemical processing of food, the mouth is concerned with mechanical processes to fragment food, producing maximal surface areas for digestive enzymes. Nevertheless, peristalsis aids the mixing of foods along the gut, and saliva starts chemical processes in the mouth, especially in those species with cheek pouches. Studies on dental anatomy (Kay 1975b) and the masticatory cycles (opening, closing, and power strokes) of jaws (puncture-crushing and chewing; Hiiemae 1978) that were conducted in the late 1970s further advanced our knowledge of primate feeding anatomy (fig. I.2), and these studies continue to be expanded. Ross, Iriarte-Diaz, and Nunn (2012) have developed innovative studies of mandibular morphology in relation to diet, concluding that there is some support for the hypothesis that larger primates, mostly the more folivorous ones, chew more than smaller ones. Godfrey, Samonds, et al. (2001) show that teeth develop more rapidly in folivores than in frugivores to promote food processing and foraging independence.

It emerged during the 1970s that chewing is more elaborate in primates than most other mammals apart from folivores, with smaller amplitude and greater transverse movement. The selective influences on tooth shape are body size (with smaller animals having to prepare proportionately more food than large ones), the energy content of food (which affects the amount of food and mechanical efficiency and wear resistance in tooth design), and food consistency (physical properties affecting tooth design) (Kay 1975b). While gut anatomy differs between faunivores and folivores, with frugivores showing intermediate gut anatomy, the teeth of faunivores and folivores are similar, and both contrast with those of frugivores; faunivore and folivore teeth perform similar cutting roles for chitin and cellulose, while frugivore teeth perform a grinding or crushing role. Cutting, puncturing, and shearing ("multiple-chopping") involve a small contact area and high occlusal pressure, whereas crushing and grinding ("pestle-and-mortar") involve a large contact area and low occlusal pressure. Faunivores and folivores have short, pointed incisors, and their premolars and molars have shearing blades and thin enamel, whereas frugivores have larger incisors and low, rounded premolars and molars.

Lucas and Luke (1984) and Lucas (1994) have demonstrated that teeth "recognize" different physical features and do not necessarily distinguish between animal matter, fruit, and leaves. Hence, from measures of deformability, strength, toughness, and strain at fracture, they categorize foods as "hard/brittle" (seeds, nuts, and bone), "juicy" (fruit and insect parts), and "tough/soft" (animal soft tissues, leaves, and grasses). The first requires thick enamel and crushing basins (frugivores); the last requires many sharp blades, high cusps, and thin enamel (faunivores and folivores); and the "juicy" category is intermediate.

Molar morphology very clearly reflects the diets of primates (Kay 1975b). The folivory of the siamang (*Symphalangus syndactylus*), the largest gibbon, is reflected by more elaborate molar crests and cusps than those of the Kloss gibbon (*Hylobates klossii*), which cannot eat leaves because of leaf chemical defenses on the Mentawai Islands, leading to more frugivorous dentition. Similarly, the dentitions of the more folivorous langurs (*Trachypithecus*) contrast with those of the more frugivorous (seed-eating) langurs (*Presbytis*; A. Davies, Bennett, and Waterman 1988). In the African cercopithecines, J. Lambert et al. (2004) found that differences in the thickness of tooth enamel of gray-cheeked mangabeys (*Lophocebus albigena*) and redtail monkeys (*Cercopithecus ascanius*) could not be explained by the hardness of their fruit diets, generally; however, the fallback foods of these two species differed in hardness, with *L. albigena* consuming the hardest foods. They concluded that the mechanical properties of fallback foods may have served as the selective pressure for thick enamel in *Lophocebus*. Among the frugivorous New World monkeys, dental anatomy is clearly reflected in those consuming more leaves or invertebrates (e.g., *Callicebus torquatus* cf. *C. moloch*; Kay 1975b).

Cheek pouches are a unique feature of cercopithecine monkeys, and research into the implications of cheek-pouch use for foraging has been conducted sporadically since the 1970s. The use of cheek pouches allows individuals to harvest food rapidly in risky situations (J. Lambert 2005; P. Murray 1975), and these pouches are larger and used more frequently in semiterrestrial species (Albert et al. 2013). There is no evidence of fermentation in the cheek pouches, but some salivary predigestion of unripe fruits and seeds is likely (J. Lambert 2005; Rahaman et al. 2015). Cheek pouches would seem to provide clear advantages in the energetically costly retrieve-and-retreat foraging strategy, which is an important aspect of the competitive, dominance-based social systems of many Old World monkeys, in contrast to the cooperation and food sharing that is sometimes observed in apes (Dunbar 1988).

THE BEGINNINGS OF A MULTIDISCIPLINARY PERSPECTIVE

During the 1970s, studies of digestive physiology had not developed to the same extent as those of dentition (fig. I.2). Diet and feeding behavior cannot be explained fully without studies integrating digestion with patterns of food acquisition and food processing. Knowledge of forest productivity—cycles of leaf flushing, flowering, and fruiting—and the nutritional content of each food is similarly essential for interpreting food choice, and real progress has been made in such studies. As far back as the 1960s, Marcel and Annette Hladik (botanist; e.g., C. Hladik 1975, 1978, 1979) were combining research on diet and feeding behavior of primates in Gabon, Panama, and Sri Lanka with studies of gut anatomy and food content and availability—real interdisciplinary pioneers. It became apparent that a multidisciplinary approach was the only way to elucidate patterns of primate grouping and movement (fig. I.4). Primate societies are structured mostly according to the availability and composition of foods in time and space and the ability of individuals and groups to defend food, and the most productive research over the last 40 years has been directed toward such understanding.

SECONDARY COMPOUNDS

It is not just the nutrient and energetic content of food that determines food choice but also the presence of secondary compounds (e.g., Milton 1979). A major breakthrough in our understanding of diet selection among primates came in the 1970s with the appreciation that plant secondary compounds—alkaloids (toxins) and phenolics (digestion inhibitors)—affected food choice, including the specific part of the plant that is eaten (e.g., Janzen 1979b; reviewed in Glander 1982; Stalenberg et al., this volume). In the coevolution of plants and animals, plants presented attractive food packages for seed dispersal, for which they depended on animals rather than wind or other mechanisms. Primates benefited from the succulent fruit containing the seeds, while the seeds benefited by being moved away from the parent plant by primates (Razafindratsima et al. 2018; Russo and Chapman 2011). Some primates developed an adaptive evolutionary response, consuming unripe fruit including the seeds. The plants responded by loading the unripe fruit with secondary compounds to deter such consumption as well as evolving mechanical defenses (Cipollini and Levey 1997; Norconk, Grafton, and McGraw 2013; Oates, Swain, and Zantovska 1977). Leaves, especially mature leaves, are similarly loaded with secondary compounds to deter consumption by folivores (Glander 1982), mainly the colobine monkeys in Africa and Asia, as discussed below.

Examples from folivorous and frugi-folivorous primates advance our understanding of the varying contribution of leaves to primate diets. In the 1970s, it was noted that even colobine monkeys could not consume the mature (and even sometimes young) leaves of many trees in forests growing on poor soils. It was initially believed that these trees were defended much more effectively by secondary compounds, so that the main dietary component of black colobus monkeys (*Colobus satanas*) in Cameroon was seeds (McKey 1978; Waterman, Ross, and McKey 1984); to a lesser extent, this was also observed in Borneo, where seeds accounted for a much larger component of the diet of maroon langurs (*Presbytis rubicunda*; 30%) on poorer soils than the usual diet of mostly leaves (A. Davies, Bennett, and Waterman 1988). The genus *Presbytis* (banded langurs) coexists throughout Southeast Asia with *Trachypithecus* (dusky langurs) by eating more seeds than young leaves (Kool 1993). Nevertheless, it is now accepted that seed predation is a valid feeding strategy in its own right, following van Roosmalen and colleagues' (1988) description of seed eating in platyrrhines, thereby confirming it to occur on all three continents. The diet of

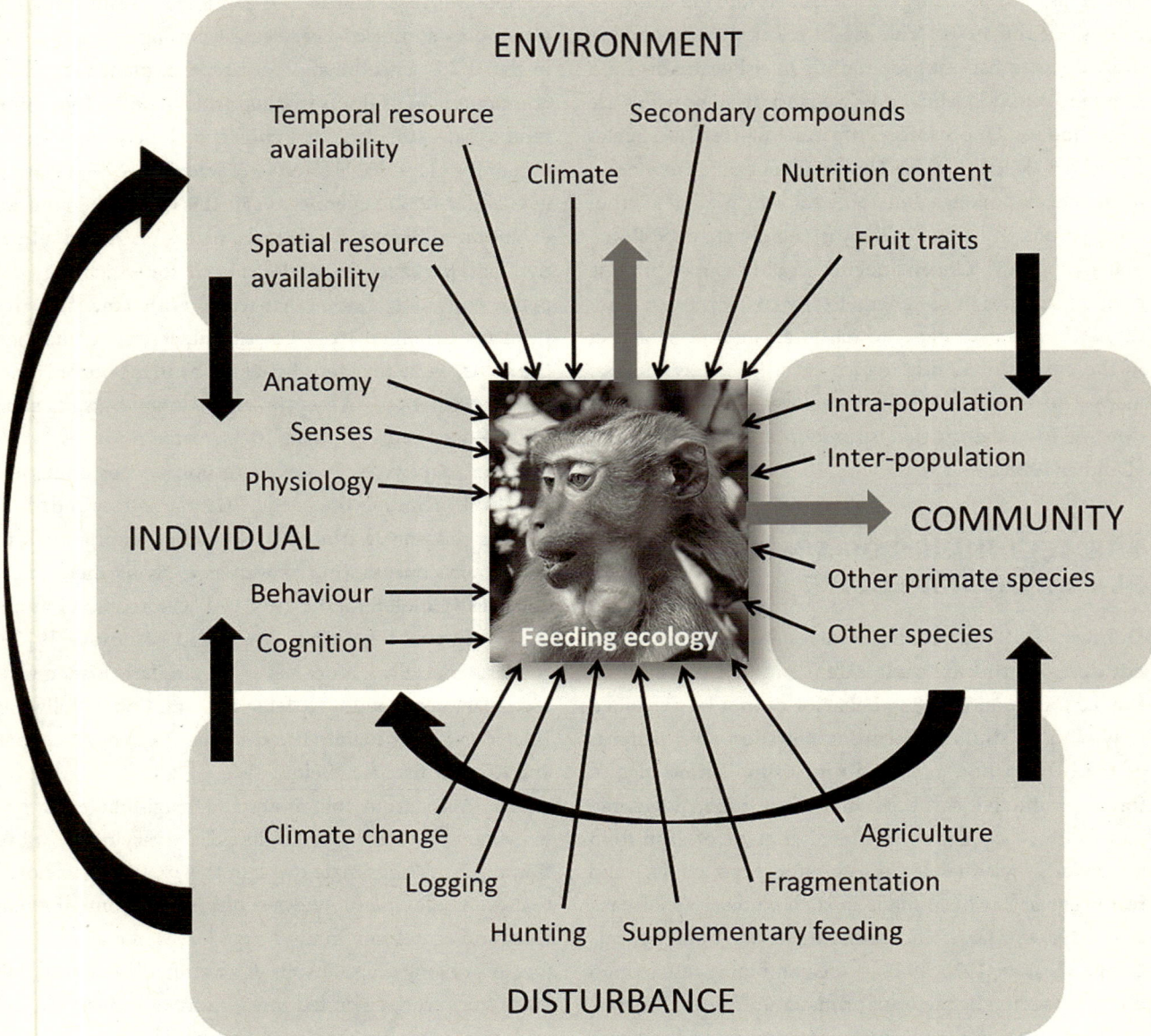

Figure I.4. Factors that determine the feeding ecology of a primate. The feeding choices a primate makes are influenced by multiple factors of the individual animal, the community, the environment, and disturbance effects. These factors also influence each other, and the foraging patterns can feed back to the environment and community. Primate feeding studies have evolved from being focused on individual components of these factors to investigating the interactions between multiple factors. Photo is of a southern pig-tailed macaque (*Macaca leonina*) using its cheek pouches, by Kulpat Saralamba.

the bearded saki (*Chiropotes satanas*) in Surinam is primarily made up of immature seeds (66%) and ripe fruit (30%) of the Brazil nut family (Lecythidaceae). Collared titi monkeys (*Callicebus torquatus*) also eat mainly nuts, seeds, and hard fruit (Kinzey et al. 1977).

In the Mentawai Islands (off Sumatra), Kloss gibbons (*Hylobates klossii*), lacking a sacculated stomach for fermentation, are unable to eat the young leaves of the well-protected trees (A. Whitten 1982). They instead obtain necessary protein from invertebrates. Elsewhere in Southeast Asia, except on Borneo, young leaves are a significant component of the gibbon diet, albeit secondary to ripe fruit pulp (Chivers and Hladik 1984). In Madagascar, the ratio of protein to fiber in leaves correlated positively with the biomass of folivorous lemurs, who selected leaves with high protein and low fiber (Ganzhorn 1992).

Food selection by male and female red-tailed sportive lemurs (*Lepilemur ruficaudatus*) did not differ in relation to the chemical composition of leaves, but females consumed fruit with lower fiber content (Ganzhorn, Pietsch, et al. 2004).

In the 1990s, Michael Huffman began investigating the potential role of secondary compounds in plants for self-medication by primates. Hence, interest in how primates were able to tolerate some secondary compounds in plants shifted to the suggestion that some plants may be specifically selected for as treatment for parasite infection or gastrointestinal upset (Huffman 1997). Various food items are considered to have medicinal properties, including fruits, leaves, macronutrient-poor bark and wood, clay soils, and invertebrates (e.g., Morrogh-Bernard et al. 2017; Peckre et al. 2018). Certain feeding behaviors have also been linked with self-medication, such as the folding and swallowing of rough hispid leaves by chimpanzees (*Pan troglodytes*) to physically expel intestinal parasites (Huffman and Hirata 2004). Most evidence for secondary compounds having medicinal benefits comes from work on African apes (Huffman 1997), with some work on Neotropical primates (Glander 1994) and, more recently, lemurs (Carrai et al. 2003; Peckre et al. 2018) and orangutans (Morrogh-Bernard et al. 2017).

A more recent focus in studies on secondary compounds in consumed foods has been to understand the balance between the positive and negative aspects of consumption. For example, many wild plants consumed by primates contain phytoestrogens (Wasserman, Taylor-Futt, et al. 2012; Wasserman et al., this volume). Phytoestrogens can protect against estrogen-dependent cancers, but they may have reproductive costs by altering hormone levels and reducing fertility (Wasserman, Milton, and Chapman 2013). This cost-benefit system remains to be fully investigated in primates.

SENSORY CUES TO FEEDING

Investigations into how primate anatomy reflects diet choices were followed, in the 1990s, by an increase in studies on the sensory perceptions of primates. The discovery of trichromatic vision in all Old World monkeys (Cercopithecidae), apes, and howler monkeys (*Alouatta*) (Dominy and Lucas 2001; Osorio and Vorobyev 1996) enhanced our understanding of food selection. Red-green discrimination is most important for distinguishing palatable young leaves from nonpalatable, whereas fruit discrimination can also be made along the yellow-blue channel (Dominy and Lucas 2001). Because primates are usually regarded as visual animals, research into other senses has been comparatively lacking (Laska, Seibt, and Weber 1999). Recently, Nevo and Heymann (2015) reviewed the evidence for use of the olfactory sense in food acquisition by primates and concluded that olfaction is probably primarily used for assessing feeding items. Some primate species might have heightened olfactory sensitivity, but this may reflect their social behavior as well as food-finding behavior (Barton 2006; Laska, Seibt, and Weber 1999). Taste perception in primates is poorly understood but reflects an adaptation to perceiving the nutrient and toxin compounds present in food plants (C. Hladik and Simmen 1998).

POSITIONAL BEHAVIOR AND FEEDING

The ability of an arboreal primate to feed successfully is determined by the structure of its habitat and its ability to move and posture efficiently therein, which depends on the size, stability, and orientation of the branch and whether the primate positions itself above or below the substrate. Positional behavior when feeding has been investigated in a diverse range of primates since the 1980s, revealing important relationships between positional anatomy and the different locations of food within the tree. In this context, feeding on the ground is more straightforward. The respective requirements for suspensory or supportive behavior are clearly major sources of anatomical adaptation among arboreal primates, which in turn limits their foraging efficiency on the ground (Fleagle 1984).

Most fruit and flowers—the most nutritious plant foods—are at the ends of the flexible terminal branches of forest trees, as are new leaves (Grand 1972). Thus, there is a great mechanical advantage in suspending to feed at the ends of branches (with the food bending toward the primates) rather than trying to balance on top (when the food bends away). This places the small apes (Hylobatidae) and larger New World monkeys (spider [*Ateles*], woolly [*Lagothrix*], and howler [*Alouatta*], with their prehensile tails) at a great advantage: their long and powerful arms (and tail) allow them to brachiate through the forest canopy and hang from the ends of branches, using their feet for support (Bergeson 1998; Fleagle 1980; Youlatos 2002). While such foraging postures are energetically efficient, brachiation is costly, which partially leads to the relatively short day range and relatively small territory size of gibbons (*Hylobates*) compared to other primates of similar size (Bertram 2004; Fleagle 1980; Grand 1972; Parsons and Taylor 1977). Furthermore, the greater efficiency of brachiation leads to a reduction in travel distance (Parsons and Taylor 1977).

Posture is also an important factor in the diverse foraging behaviors of the small-bodied New World primates (Boinski 1989). Callitrichine primates exhibit behavior and morphology that enable them to exploit resources that would otherwise be unavailable (Garber 1980, 1992). These small primates have clawlike nails that allow them to use large vertical supports, such as tree trunks, for foraging on gums and insects (Garber 1992). Similarly, quadrumanual positional behavior enables orangutans (*Pongo pygmaeus*)—by far the largest mammal to use the forest canopy—to solve problems posed by the tapering of branches and gaps between trees (Cant 1987).

OPTIMAL FORAGING THEORY

Garber (1987) presented a most effective assessment of our state of knowledge of foraging behavior in the late 1980s. He examined how diet, group size, food quality, resource availability, patch size and distribution, social organization, resource defense, sensory capabilities, within-group competition, and digestive morphology affect foraging patterns in primates. While there is an unusually broad range of foraging strategies and dietary preferences among primate species, within-species patterns are generally conservative and constrained by particular aspects of body size, morphology, physiology, behavior, and ecology.

Studies of primate social organization, subgroup formation, and within-group feeding competition, at about the time Garber (1987) was writing, stimulated a reassessment of traditional ideas concerning the evolution of social-group living and primate sociobiology (Janson and van Schaik 1988; van Schaik 1983; van Schaik and van Hooff 1983). At this time, there was a need to expand such studies to develop a comprehensive framework from which to examine detailed relationships among the size, composition, and organization of particular social groups, feeding ecology, foraging strategies, and reproductive success. Inter- and intrasexual and age-related differences in foraging strategies also needed to be a major focus in future studies.

In his review of foraging strategies among living primates, Garber (1987) shares our view that optimal foraging theory is of limited value for species with such diverse and variable diets. Fundamental differences in nutritional content, renewal rate, digestibility, abundance, distribution, and seasonal availability of leaves, corms, grasses, insects, fruits, flowers, leaves, gums, and vertebrate prey have a major impact on primate feeding patterns, so that general assumptions regarding the relationship between feeding rate and energy intake, when applied to folivores in particular, are poor predictors of foraging behavior. The concentrations of protein, fiber, and secondary metabolites in leaves vary with maturity and often the successional stage of the plant. Furthermore, the rates at which primates encounter different food types are rarely constant or independent. In forests with high species diversity and low species density, with clumping of tree species and predictability in fruiting patterns, nonrandom and goal-directed foraging patterns are likely to be promoted (Asensio, Brockelman, et al. 2011; Chapman, Rothman, and Lambert 2012). Thus, among primates, foraging efficiency may be more dependent on their ability to locate foods; the well-known ability of primates to remember the location and availability of seasonal foods is an important factor in foraging success (Garber 1987). The application of optimal forging models to primate feeding behavior is impaired by the lack of data on partial feeding preferences and resource sampling, but these are beginning to be addressed in current studies (J. Lambert and Rothman 2015; Rothman 2015). The complexities of seasonal strategies and dietary preferences of primate diets and foraging patterns mean that such key aspects of feeding ecology have not been addressed in most foraging models.

Despite the multitude of factors that influence foraging behavior, Garber (1987) identified seven main patterns: (1) body size, dental morphology and volume (and area), and anatomy of digestive tracts are strong predictors of primate foraging behavior; (2) primates consuming high levels of structural carbohydrates are selective feeders, preferring foods low in condensed tannins and with a high protein:fiber ratio; (3) the ability to store and process information on spatial distribution, productivity patterns, and productivity of feeding sites significantly reduces foraging costs and increases foraging efficiency; (4) the size, distribution, and renewal rate of food patches influence the size and cohesiveness of foraging groups; (5) with food scarcity, subordinate group members may have higher foraging costs and increased levels of interference competition; (6) to accommodate the nutritional costs of lactation (and pregnancy), females increase feeding time, exploit higher-quality foods, or both; and (7) individuals in larger social groups may increase foraging efficiency by cooperatively defending productive and clumped feeding sites and may share them with members of other species (conferring safety in numbers).

FEEDING RESPONSES TO RESOURCE AVAILABILITY

As it became apparent to researchers that there was substantial intraspecific and intrapopulation variation in primate diets, there was a burst in studies (in the 1980s) in which changes in resource availability—both temporally and then spatially—were examined in relation to how they influenced diets (Peres 1994b; reviewed in Chapman, Rothman, and Lambert 2012). Today, measuring resource availability is an integral part of most primate field studies, particularly for frugivorous primate species that preferentially consume ripe fruits that are patchily available. Frugivorous primates have been shown to alter their diet choices, dietary diversity, and behavior in relation to fruit availability (A. Davies, Oates, and Dasilva 1999; Kaplin et al. 1998; Lawes 1991; McConkey, Ario, et al. 2003; J. G. Robinson 1986; A. Stone 2007; Terborgh 1983; Yeager 1989). Such studies were later expanded to examine more finely why primates might choose particular fruit resources (Julliot 1996a; M. Leighton 1993; McConkey, Aldy, et al. 2002; Suwanvecho et al. 2018). Long-term studies demonstrate just how patchy fruit supplies can be, with plant species often fruiting on supra-annual timescales and forcing primates to be generalist frugivores (Chapman, Chapman, Zanne, et al. 2005; P. Wright 2006; Brockelman 2011). Studies on resource abundance and resource renewal on diet in folivores indicate that, contrary to initial expectations, these primates also suffer resource limitation with associated impacts on competition within groups (Dasilva 1994; T. Harris et al. 2010; Hemingway 1998; D. Watts 1998). Research into the impacts of resource availability on the foraging of nonherbivores is limited, but similar patterns—for example, in spectral tarsiers (*Tarsius spectrum*)—have been shown with resource availability influencing diet as well as behavior (Gursky 2000). The impact of climate change on primate resource availability is an important developing research area (Chapman, Chapman, Zanne, et al. 2005; Rothman, Chapman, Struhsaker, et al. 2015).

INTERACTIONS WITHIN AND BETWEEN SPECIES

Food choices made by primates are clearly influenced by a multitude of factors (fig. I.4), including the behavior and foraging patterns of interacting individuals and species. In the late 1980s, researchers began to study the impact that intrapopulation competition and group dynamics may have on feeding behavior (or conversely, how food distribution influences group dynamics; fig. I.2; Barton and Whiten 1993; Isbell 1991; Janson 1988b). In these studies, complex relationships between group size, interactions among individuals, and foraging behavior were highlighted.

It was also recognized that interactions among group members may enhance food acquisition through opportunities for information sharing, although the relationship is modified by food-resource distribution (Chapman and Lefebvre 1990; Garber, Bicca-Marques, and Azevedo-Lopes 2009). Studies of conspecific interactions were expanded in the first decade of the 2000s to examine interspecific competition for resources. Evidence was found for direct competition over fruit resources among sympatric primate species during the periods when fruit was most abundant, with niche divergence during the lean seasons (Guillotin et al. 2009; B. Schreier et al. 2009; P. Stevenson et al. 2000; Terborgh 1983; Wrangham, Conklin-Brittain, and Hunt 1998). Interspecific competition has more rarely been investigated between primate and nonprimate species, despite the obvious overlap of resource use (A. Marshall, Cannon, and Leighton 2009; Russak 2014; Kamilar et al. 2015). That primate feeding behavior is influenced by nonprimate communities as well as other primate species has been confirmed by these studies.

Predation risk as perceived by primates adds further complexity to their foraging behavior (Cowlishaw 1997; Gogarten, Jacob, et al. 2014; C. Richter et al. 2015; Teichroeb et al. 2015; Hill, this volume). For example, the type of food that is being consumed influences polyspecific associations and the spatial distribution of individuals in redtail monkeys (*Cercopithecus ascanius*), and these patterns may be driven by predation risk (Bryer, Chapman, and Rothman 2013). Network analysis has been used to unravel the foraging strategies and interactions among individual group members within a group of howler monkeys (*Alouatta palliata*; Dattilo et al. 2014).

IMPORTANCE OF LEAN SEASONS

Once the importance of resource abundance for primate foraging patterns was established, some researchers considered the myriad ways in which the period of lowest resource abundance might affect primate populations. In the simplest form, foods were identified that sustain primates through the lean season and might be of critical importance to maintaining populations generally (A. Marshall, Boyko, et al. 2009). John Terborgh was the first to propose this idea in 1983, and complementary frameworks have since been developed to identify these fallback foods. J. Lambert (2007) suggested that fallback foods may be relatively abundant, low-quality foods that require specific anatomical adaptations to eat them, or scarce, high-quality foods that require behavioral adaptations to exploit them efficiently. A different form of classification was developed by A. Marshall and Wrangham (2007), who categorized these foods based on their importance in the diet, terming them "staple" or "filler" foods. The most severe periods of resource scarcity (referred to as "ecological crunches" and which may affect primates more rarely) lead to heightened resource competition that may impact masticatory and digestive anatomy, grouping and ranging behavior, speciation, extinction, and adaptation (reviewed in A. Marshall, Boyko, et al. 2009). More recently, the quantity and quality of fallback foods have been considered to influence population abundance directly, although this relationship remains unclear due to a paucity of studies in which this was specifically tested (Hanya and Chapman 2013; J. Lambert and Rothman 2015).

COGNITIVE PROCESSES IN LOCATING FOOD RESOURCES

Wild primates must navigate their home ranges to find sufficient food, and this can present a substantial cognitive challenge, particularly when searching for patchily distributed resources such as fruit (Asensio, Brockelman, et al. 2011; Janson 2016; Janson and Byrne 2007; Garber, this volume). There has been a recent move (since the first decade of the 2000s) to understand how food distribution influences primate movements and the cognitive processes that determine primate foraging (Chapman, Rothman, and Lambert 2012; Trapanese et al. 2018). While most researchers accept that primates retain spatial knowledge of the resources in their home range (e.g., Asensio, Brockelman, et al. 2011; Normand and Boesch 2009; Noser and Byrne 2010; Valero and Byrne 2007), there is continued debate over which type of cognitive map primates might use. A "mental map" can be topological or route-based, in which the primates use landscape features to orient themselves, or it can be a Euclidean map, in which a geometric representation of food resource locations is retained without necessarily following landscape features. Orientation using a Euclidean map has been supported for two ape species (gibbons [*Hylobates lar*], Asensio, Brockelman, et al. 2011; chimpanzees [*Pan troglodytes*], Normand and Boesch 2009), while, so far, there is only evidence for ranging related to topography for monkey species (spider monkeys [*Ateles belzebuth*] and woolly monkeys [*Lagothrix poeppigii*], Di Fiore and Suarez 2007; baboons [*Papio ursinus*], Noser and Byrne 2010).

PRIMATE NUTRITION AT LARGE SCALES

One of the most important recent advances (mainly since 2010) in our understanding of primate feeding ecology has come from the research groups of Jorg Ganzhorn, Jessica Rothman, Colin Chapman, and Joanna Lambert. Improved efficiency in measuring the amounts of macro- and micronutrients in foods has allowed investigations into primate nutrition at detailed and over large scales. This research includes detailed studies of how the nutritional content of minor or major diet items may influence feeding decisions and potentially primate evolution (Bryer,

Chapman, et al. 2015; Felton, Felton, Wood, Foley, et al. 2009; Masette et al. 2014; Rothman, Raubenheimer, et al. 2014), specific energetic or nutrient requirements of different sexes and age groups (Rothman Dierenfeld, Hintz, et al. 2008; reviewed in Chapman, Rothman, and Lambert 2012; Emery Thompson, this volume), and a broader understanding of food selection to balance nutrient intake across seasonally and spatially heterogeneous landscapes ("nutritional geometry"; Felton, Felton, Lindenmayer, et al. 2009; Rothman 2015; Rothman, Raubenheimer, and Chapman 2011; Raubenheimer, this volume). At the largest temporal scale, the nutritional quality of leaves eaten by colobus monkeys (*Procolobus*) was shown to have declined over 15–30 years in Kibale, Uganda, perhaps due to climate change (Rothman, Chapman, Struhsaker, et al. 2015). Over large spatial scales, researchers have found that higher protein contents of Neotropical fruit, compared to fruit from Madagascar, could explain differences in the proportion of primates that are frugivorous across biogeographic regions (Ganzhorn, Arrigo-Nelson, Boinski, et al. 2009). These two large-scale studies have yet to be expanded.

Stable isotope biogeochemistry has also advanced in recent years and is now being used to investigate aspects of primate foraging ecology that were not previously detectable (Crowley, Reitsema, et al. 2016; Sponheimer and Crowley, this volume). For example, stable isotope analysis has distinguished differences among chimpanzee (*Pan troglodytes*) populations occupying different habitats in their carbon and nitrogen isotope space—reflecting different diets (Loudon, Sandberg, et al. 2016). Comparisons with fossil hominins (*Sivapithecus*, *Gigantopithecus blacki*, *Ardipithecus ramidus*, and *Australopithecus anamensis*) have also enhanced our understanding of the ecology of past primate communities (Loudon, Sandberg, et al. 2016).

ECOLOGICAL ROLES

In the 1990s, researchers began to consider how the feeding ecology of primates might shape the structure of plant communities through frugivory and the subsequent dispersal of seeds. The focus of this research has been primarily on specific seed-dispersal roles played by seed-swallowing primates (Dew and Wright 1998; Garber 1986; Julliot 1996b; McConkey 2000; Wehncke et al. 2003) but also includes the functions of seed-spitting primates (Albert, Hambuckers, et al. 2013; Corlett and Lucas 1990; J. Lambert 1999) or, more rarely, seed-eating primates (Gautier-Hion, Gautier, and Maisels 1993; Peres 1991; van Roosmalen et al. 1988). It is now recognized that primates play very important roles in seed dispersal in tropical rainforests (McConkey and Brockelman 2011; Poulsen, Clark, et al. 2002; Russo and Chapman 2011) and other habitats (Tsujino and Yumoto 2008) because they are efficient seed dispersers, are often unspecialized frugivores, and can form significant proportions of the frugivore biomass (Chapman 1995). Network analyses in Neotropical and Paleotropical rainforests have confirmed that frugivorous primates are critical components of the frugivore community (Donatti et al. 2011; Ong et al. 2022). Once the importance of primates in seed dispersal was realized, interest in how they choose and locate specific fruit resources developed, since it directly relates to their role in the frugivore community and might provide information on frugivore-fruit coevolution (Dominy 2004; Julliot 1996a; J. Lambert and Garber 1998; M. Leighton 1993; McConkey, Aldy, et al. 2002; Tonos et al. 2022). In these studies, it is indicated that primates choose fruits and locate specific fruit resources by knowledge of their home range and cues such as bird activity.

The importance of folivory by primates, in terms of its impact on ecosystems, has been considered only more recently (after 2010). The available evidence suggests that leaf-eating primates may also play important ecological functions when they kill or slow the growth rate of trees by eating the bark or leaves (Chapman, Bonnell, Gogarten, et al. 2013). While folivory by Japanese macaques (*Macaca fuscata*) had discernible impacts on individual tree species, these effects were negligible at the community level when compared to those of insects (Hanya, Fuse, et al. 2014). The functions of florivory (Chapman, Bonnell, Sengupta, et al. 2013) and pollination (Gautier-Hion and Maisels 1994) are also performed by primates, and it has been proposed that the type of primate-flower interaction (pollination or florivory) that ensues might depend on body mass and dietary strategy (Heymann 2011).

DIET AND PHYLOGENY

In "Dietary Categories," we started with the traditional view that the ancestral primates were small faunivores, evolving grasping abilities and binocular vision for predation on invertebrates (Cartmill 1974). Sussman (1991) and Sussman et al. (2013) argue that, while insects were important components of the diets of the earliest euprimates, visual predation was not the major impetus for the radiation. Instead, a major evolutionary event in the Eocene, involving the angiosperms (flowering plants) and their dispersers, with long-term coevolutionary interactions between trees and primates, bats, and birds, initiated pollination and seed dispersal by tropical fauna. The major adaptations of modern primates resulted from the abundance of resources available in the terminal branches of the forest canopy (e.g., Fuzessy et al. 2022). Eriksson (2016) spelled out the start of diversification of fruit and seeds 80 million years ago (mya), with the peak of diversity of seed and fruit sizes and fruit types peaking 50 mya with rapid, diverse coevolution. The Eocene/Miocene boundary 34 mya saw a dramatic climatic shift, creating more open habitats with patchy food sources. Indeed, we have always considered the Miocene to represent the peak of angiosperm radiation, resulting in the dramatic radiation of apes and then monkeys.

Matrices of dietary overlap and phylogenetic divergence times in five communities across the tropics—two in South America (Manu and Surinam), two in Africa (Uganda and Ivory Coast), and one in Asia (Malaysia)—showed a weak relationship between phylogeny and gross dietary categories (L. Porter, Gilbert, and Fleagle 2014; see also Lim et al. 2021). This is an area that requires further investigation. The problem is that, while one would expect strong geographical relationships between phylogeny and diet, diet is a topic where one would expect divergence to allow coexistence, as with langurs and gibbons in Southeast Asia (Chivers 1980).

FEEDING ECOLOGY AND ECOSYSTEM DISTURBANCE

Most primate species face multiple forms of ecosystem disturbance and must adjust their foraging patterns if they are to adapt to these changes and maintain the health of the population. Changes in feeding and movement patterns in relation to logging, fragmentation, and intensification of agriculture have been investigated by researchers mostly in the last decade, although the ground-breaking study of A. D. Johns (1986) in selectively logged forests was conducted much earlier than this. Primates use multiple strategies to cope with ecosystem degradation, including consuming crops and other food resources provided by humans (Bessa et al. 2015; Campbell-Smith et al. 2011; C. Hill 2017; Naughton-Treves et al. 1998; Pozo-Montuy et al. 2013; Sengupta et al. 2015) and supplementing the diet with wild fallback foods or other foods that were previously underused, such as the case of sifakas (*Propithecus diadema*) consuming mistletoe (*Bakerella clavata*) year-round rather than just during the lean season (Irwin 2008; Serckx et al. 2015). While dietary changes may occasionally be nutritionally beneficial to primates (such as supplementary feeding and crop raiding; Naughton-Treves et al. 1998), overall food availability or nutritional quality of foods is usually reduced in disturbed areas (Chaves, Stoner, and Arroyo-Rodrígues 2012; Rode, Chapman, McDowell, et al. 2006; Simmen, Bayart, et al. 2007). In some situations, this can increase stress levels in affected primates, with consequences for population health (e.g., howler monkeys [*Alouatta pigra*], Behie and Pavelka 2013; red colobus [*Procolobus rufomitratus*], Chapman, Schoof, et al. 2015). However, such health consequences will depend on the type of disturbance and the feeding adaptations of the species concerned, as many primates find enough food of the appropriate quality in forests with anthropogenic and nonanthropogenic disturbance. The effects of forest fragmentation and, specifically, crop raiding have been linked directly to bacterial transmission among primates, livestock, and humans (T. Goldberg et al. 2008), promoting new directions in research.

Feeding studies have also been conducted in regions that are completely human-modified, such as agricultural matrices and urban areas (Jaman and Huffman 2013; Kwiatt 2017; Munoz et al. 2006; Thatcher et al. 2020). Our understanding of the feeding choices primates make in these disturbed habitats (Rothman, Chapman, Struhsaker, et al. 2015; Rothman, Raubenheimer, and Chapman 2011) will be improved by further nutritional studies of macro- and micronutrient contents of consumed foods. Indeed, climate change (changes in temperature and rainfall and elevated CO_2) is predicted to impact the nutritional content of some primate foods, opening up new study areas for primate feeding ecology (Rothman, Chapman, Struhsaker, et al. 2015).

FUTURE OF PRIMATE FEEDING STUDIES

Over the past 85 years, our knowledge on primate feeding has burgeoned from the first descriptive studies of diet to detailed, quantitative analyses of the multitudes of subcomponents that determine primate feeding ecology. These subcomponents were initially developed independently, with a focus on the anatomy and behavior of primates, but as the interaction between internal (within-primate) and external factors quickly became apparent, researchers adopted a more integrative approach. In the 1980s, the importance of resource availability and plant secondary compounds was firmly established, and the basic fields of behavior and anatomy diverged as researchers focused on different aspects of each. As habitat destruction became an alarming concern for the future of primate populations in the 1990s, many researchers focused on the impacts of disturbance on feeding ecology. Since the turn of the century, these main topic areas have diverged into many subcomponents reflecting our increasing understanding of feeding processes as well as advances in technology that have allowed certain areas to expand that previously could not (such as detailed nutritional analyses).

The future of primate studies may lie in increasingly integrative approaches as researchers attempt to understand the multiple factors influencing feeding in a systematic way (fig. I.4). Exciting developments in understanding food nutrition may open many new areas of multidisciplinary research, as researchers apply this technology more widely—such as to study minor food items or the specific effects of disturbance on nutrient intake and foraging decisions made by primates. We still have a relatively poor understanding of how climate change might impact primate physiology and feeding, and this issue must be addressed as its impacts progress. Furthermore, while there is a wealth of information documenting the importance of primates in seed dispersal, we still lack information on other functions performed by primates (aside from a handful of studies) and on how primates interact with nonprimates in the ecosystems they inhabit. An understanding of primate feeding ecology is essential for determining their nutritional requirements in their increasingly disturbed habitats. Perhaps the most important future of these studies is to bring the threads of information together to determine the long-term carrying capacity of habitats for primates (Chapman, Snaith, and Gogarten 2014).

SUMMARY AND CONCLUSIONS

- Studies of primate feeding ecology began with descriptions of diets in the 1930s and then in 1960s and 1970s, and they have expanded over the past 85 years to cover many subdisciplines.
- From early on, the interactions between different aspects of primate morphology, primate behavior, and attributes of the environment were realized, and there have been increasing efforts to integrate multiple components.
- As concerns over habitat loss and declines in primate populations intensified, more researchers studied the impacts of disturbance on primate feeding ecology and began to investigate how declines in primates may impact functions that depend on their feeding behavior, such as seed dispersal.
- New technologies have opened areas of research that could not previously be achieved, such as developments in determining the nutritional values of food.
- By integrating the available information on primate feeding ecology, we can begin to understand the carrying capacity of disturbed habitats for primates.

PART I

Finding, Building, and Using a Diet

Animal diets come packaged in various forms. Yet, animals choose to consume foods not because of their packaging (e.g., fruit, leaves, or insects) but because of their nutritional composition: proteins, carbohydrates, fats (the macronutrients), vitamins, and minerals (the micronutrients). This is the case regardless of whether a consumer's diet is constrained to a single trophic level (e.g., herbivores or carnivores) or spans trophic levels (i.e., true omnivores) such as with most primates. This section illustrates the breadth of foods consumed by primates at both scales: the macro- and micronutritional composition of foods (Felton and Lambert) and the packages that yield a diet (Hawes, Peres, and Smith). A primate's very first diet consists of mother's milk (Hinde, Milligan, and Blomquist); subsequent chapters evaluate the allocation of nutrients to reproduction (Emery Thompson) and to brains and babies (DeCasien, Brown, Ross, and Pontzer). This is, of course, assuming that the foods are found in the first place (Melin and Veilleux) and that the seasonality of their availability is successfully tracked and not overly limiting (Knott and DiGiorgio).

1 The Role of Macro- and Micronutrients in Primate Food Choice

Annika Felton and Joanna E. Lambert

Whether a primate feeds predominantly on leaves, fruit, nectar, or animal matter, its diet must include some combination of protein, carbohydrates, lipids, vitamins, and minerals (Oftedal 1991). The range of foods a primate includes in its diet is influenced by numerous variables, including body size, metabolic requirements, and digestive strategy (Milton 1993; Parra 1978). These variables, in turn, shape species' feeding adaptations (J. Lambert 2014) and nutritional strategies (Felton, Felton, Lindenmayer, et al. 2009). Determining nutritional strategies is a challenging task, as most primate species are omnivorous, flexible feeders and live in diverse tropical habitats with a plethora of potential food items from which they can select.

The field of science that grapples with this complexity is nutritional ecology. Research into wild mammal nutritional ecology began in the 1970s, and during the intervening decades several hypotheses have been forwarded regarding which nutritional factors best explain diet choice. An influential hypothesis is that herbivores select foods to maximize energy intake (Belovsky 1978; Moen, Pastor, and Cohen 1997). Alongside the related protein maximization hypothesis (Mattson 1980), the energy maximization hypothesis originates from optimization and optimal foraging theory (Emlen 1966; Pyke et al. 1977; Schoener 1971). Other proposed models include alternative currencies for explaining diet choice, such as micronutrients (Nagy and Milton 1979a), the ratio between plant protein and fiber (Milton 1979), or a parameter assumed to have a negative effect on food selection (a toxin or a digestion inhibitor; Oates 1978). Furthermore, the nutrient balancing hypothesis suggests that animal diet selection is governed by the primary goal of getting an optimal balance among multiple different nutritional parameters (S. Simpson and Raubenheimer 2012; Westoby 1974).

In this chapter, we use these various hypotheses as a theoretical platform by which to tease apart the nutritional factors shaping primate food choice. We summarize knowledge regarding the roles of macro- and micronutrients in the diet choice of wild primates and evaluate evidence for the hypotheses that have shaped investigation into primate nutritional ecology.

MACRONUTRIENTS

The nutrients that mammals ingest in relatively large amounts every day are known as macronutrients and include proteins, carbohydrates, and lipids (fats). These compounds are used in different ways by the mammalian body, but the feature they all share is providing energy to the consumer. Concentrations of macronutrients vary widely among the foods available to wild primates. For example, food items consumed by spider monkeys (*Ateles chamek*) during a year in Bolivia contained anywhere from 1% to 28% protein, 2% to 72% carbohydrate, and 0% to 75% fat (percentage of dry matter; Felton, Felton, Raubenheimer, Simpson, Foley, et al. 2009). Furthermore, fruit species in the same family or genus do not necessarily show the same macronutrient pattern (Milton 2008), and there can be significant differences in nutrient composition among fruits growing on different individuals of the same tree species in the same forest (Chapman, Chapman, Rode, et al. 2003). Nonetheless, there are patterns to the distribution of the macronutrients, and below we summarize how macronutrients influence primate diet choice.

MACRONUTRIENTS: PROTEIN

The Nature of Proteins and Where Primates May Find Them

Proteins are large molecules composed of amino acids, of which nitrogen (N) is a key element. The nitrogen concentration in a food can be used as a rough measure of the amount of protein in that food. So-called crude protein, for example, is often estimated as total N multiplied by a conversion factor of 6.25 (Rothman, Chapman, and Van Soest 2012). Proteins are involved in virtually every cellular and metabolic process in the body. While each protein form (e.g., enzymes, hormones, or antibodies) has specialized functions, in general they are necessary for metabolism, cell structure, growth and development, endocrine interaction, and muscle repair (Mattson 1980). While primates can synthesize de novo many of the 20 required amino acids, 9 of them must be obtained from their diet and are hence called "essential."

Primates obtain protein from leaves, buds, fruit, flowers, insects, and other prey animals. Indeed, almost everything primates consume (besides water) contains at least some protein. However, the amount of protein that is available for the consumers varies widely among food categories. Vertebrate and invertebrate prey are particularly rich in digestible protein. For example, the protein concentration in caterpillars, locusts, and flies is around 50%–70% of dry matter, while beetles and termites generally have less (Raubenheimer and Rothman 2013). Insects can be a major source of protein for some primate species (Raubenheimer and Rothman 2013), such as capuchin monkeys (*Cebus capucinus*; McCabe and Fedigan 2007). Ripe fleshy fruits, on the other hand, generally have concentrations of crude protein between 2% and 6%, while leaves and buds often have >10%. Furthermore, a significant portion of the protein in plant tissue is bound within cell walls and is thus unavailable to most herbivores (Rothman, Chapman, and Pell 2008). In addition, plant secondary metabolites (e.g., tannins) create insoluble complexes with proteins, rendering them difficult to digest (W. Foley and McArthur 1994). To further complicate matters, the measure of total nitrogen may also include some soluble nonprotein nitrogen and lignified nitrogen that occur in most plants and that are not digestible by most consumers (Conklin-Brittain, Dierenfeld, et al. 1999; Van Soest 1994). It is therefore very important to quantify how much of the crude protein in a food is actually available protein when studying a species' nutritional ecology (DeGabriel, Moore, Felton, et al. 2014; Felton, Felton, Lindenmayer, et al. 2009; Rothman, Chapman, and Van Soest 2012), as illustrated with examples below.

Do Primates Maximize Protein Intake?

Because nitrogen is the fundamental building block of amino acids and can be a relatively scarce component in the diet of many organisms, nitrogen (and therefore protein) has been proposed as a limiting factor for growth, health, reproduction, and survival (Elser et al. 2000; Mattson 1980; T. C. White 1993). There are different ways to reach such a conclusion when studying particular species and populations. For example, the intake of protein by a particular primate social group or population may be estimated as low relative to protein consumption in other groups or populations and thus determined as limiting for the population in question (Elser et al. 2000; Milton 1979). Similar conclusions may be reached when fitness indices of populations living in different regions are correlated positively with protein levels in available foods (McArt, Spalinger, Collins, et al. 2009), when fitness is observed to improve after adding nitrogen to the diet (Boonstra and Krebs 2006), or when protein concentrations correlate positively with the consumer's observed preferences for different food items (McKey, Gartlan, et al. 1981). In the latter case, it is assumed that the species has evolved a foraging behavior geared toward prioritizing protein intake, due to a consistent protein limitation in diets at an evolutionary timescale.

Closely related to this line of argument is the protein maximization hypothesis: the idea that the driving factor behind herbivore food choice is to maximize daily protein intake, due to its limited availability in habitats

(Mattson 1980). The idea of maximizing a single nutritional currency—and the presumed significance of that nutrient to fitness—has its roots in optimal foraging theory (Emlen 1966; Pyke et al. 1977; Schoener 1971). This theory suggests that an animal's foraging behavior can be predicted by reducing the costs and benefits of foraging into a single best currency that links to the fitness of the animal, such as energy or protein. A prediction of the protein maximization hypothesis is that, when individuals are given free choice, they should select the most protein-rich food items available to them, provided they also meet other constraints imposed on them by their physiology and environment (e.g., hydration, temperature regulation, and predator avoidance).

Dietary protein may be a limiting factor for some primate populations, although not many studies have tested this directly. A study of urinary metabolites of wild Bornean orangutans (*Pongo pygmaeus*) revealed that they at times live on "the brink of protein bankruptcy" but have evolved protein recycling and assimilation mechanisms to survive such periods (E. Vogel, Knott, et al. 2012). In contrast, the diets of several other primate species (e.g., cercopithecoids and *Gorilla beringei*) have been found to have protein levels higher than what is probably required (Conklin-Brittain, Wrangham, and Hunt 1998; Oftedal 1991; Rothman, Raubenheimer, and Chapman 2011). But rather than comparing protein intakes to metabolic requirements, most primate studies assess the role of protein in the context of food selection ranking, or food item selection versus rejection. Mixed results are reported with regard to protein being the main driver of food choice. Possible maximization of protein intake by careful selection of foods has been reported for folivorous colobine monkeys (A. Davies, Bennett, and Waterman 1988; McKey, Gartlan, et al. 1981; Waterman, Ross, et al. 1988) and macaques (Hanya, Kiyono, Takafumi, et al. 2007) and for omnivorous baboons (Barton and Whiten 1994). For example, the avoidance by black colobus monkeys (*Colobus satanas*) of some seeds was explained by their low nitrogen contents (McKey, Gartlan, et al. 1981). That some folivorous species (e.g., *Alouatta* spp. and Colobinae) often select protein-rich plant parts or add insects to their otherwise leaf-dominated diet may be viewed as another indicator of the significance of protein in diet selection (Milton 1979; Oftedal 1991). Gorillas also have been found to select foods with relatively high protein concentrations compared to those foods uneaten (Calvert 1985; Ganas et al. 2008; Rogers, Maisels, et al. 1990; Rothman, Dierenfeld, Hintz, et al. 2008).

However, it is difficult to determine a conclusive pattern of protein maximization among primate species or even among populations of the same species. Plants eaten by even highly folivorous primates are not consistently characterized by high protein concentrations (Gaulin and Gaulin 1982; Kool 1992; Oates, Waterman, and Choo 1980). Neither silver leaf monkeys (*Trachypithecus auratus sondaicus*) in Indonesia (Kool 1992) nor red howler monkeys (*Alouatta seniculus*) in Colombia (Gaulin and Gaulin 1982) select leaves on the basis of their crude protein content. In the case of mantled howler monkeys (*Alouatta palliata*; Milton 1979) and black colobus monkeys (*Colobus satanas*; McKey, Gartlan, et al. 1981), the protein content of leaves is a factor, but not the only factor, that influences the monkeys' leaf selection.

Furthermore, the few attempts that have been made to directly test the protein maximization hypothesis through detailed observations of individual food intake show no support for protein maximization (Behie and Pavelka 2012a; Felton, Felton, Raubenheimer, Simpson, Foley, et al. 2009; Rothman, Raubenheimer, and Chapman 2011) but instead highlight the very different roles that protein can play in food selection among primate species and populations. For example, adult spider monkey individuals appear to regulate their intake of protein around a particular daily target amount by careful selection of fruit and leaves, ingesting neither too much nor too little protein, regardless of seasonal fluctuation in dietary options (Felton, Felton, Raubenheimer, Simpson, Foley, et al. 2009). Similar to spider monkeys, chimpanzees (*Pan troglodytes*) are ripe-fruit specialists and maintain a stable protein intake relative to the large variation in carbohydrate and fat intake across seasons (Conklin-Brittain, Wrangham, and Hunt 1998; Uwimbabazi et al. 2021). For these two species, the protein content of the diet is prioritized, in the sense that it is maintained within tight limits compared with nonprotein energy. An alternative nutritional strategy that likewise challenges the perception that primates maximize protein intake comes from a detailed study of mountain gorillas (Rothman, Raubenheimer, and Chapman 2011). This study showed that mountain gorillas neither maintain a stable protein intake nor maximize their protein intake but instead prioritize the stable daily intake of carbohydrates and fats (Rothman, Raubenheimer, and Chapman 2011). By doing so, they allow daily protein intake to fluctuate widely when the diet composition changes with the seasons, often ingesting much more nitrogen than is estimated to be required by gorillas (Rothman, Raubenheimer, and Chapman 2011). Hence, for mountain gorillas, who appear capable of eating and excreting excess nitrogen, protein is neither limiting nor a primary factor regulating food intake.

In summary, although dietary protein is essential, support for the protein maximization hypothesis and for the argument that protein is limiting for wild primates is not uniform. Inconsistencies in results even for the same species or closely related taxa are probably due to three main reasons. First, we should expect nutritional limitations to differ between habitats as a function of differences in resource availability, soil chemistry, and other abiotic factors. Second, different measures are used among investigations, and crude protein (or total N), for example, may show a very different intake pattern than "available protein" (I. Wallis et al. 2012). Third, and perhaps most important, the limiting and therefore sought-after nutrient can change from one hour to the next as the nutritional state of the individual changes with each consecutive meal and the animal's physiological state (S. Simpson and Raubenheimer 2012). Because a high-protein item may be a perfect choice at one point in time but less suitable at another, it is difficult to extrapolate from observations conducted during a small part of the day.

MACRONUTRIENTS: CARBOHYDRATES

The Nature of Carbohydrates and Where Primates May Find Them

Carbohydrates—molecules consisting of carbon, hydrogen, and oxygen—come in many forms. **Monosaccharides** (simple sugars such as glucose and fructose) are the smallest carbohydrate molecules (monomers) and are readily absorbed by the cells (J. Lambert 2010). The monosaccharide ribose is a fundamental building block of RNA, and other small carbohydrate monomers play key roles in the immune system and other vital biological processes. **Disaccharides** are composed of two monosaccharide molecules. Sucrose, for example, comprises glucose and fructose and is a common sugar in fruits favored by primates. Monosaccharides and disaccharides are often referred to as "soluble sugars." **Polysaccharides** are the most complex carbohydrates, made up of long, complex polymers of monosaccharides, and include carbohydrates such as starch, glycogen, and—as described below—fiber.

Carbohydrates that can be metabolized via endogenous enzymes are often collectively called "nonstructural carbohydrates." This is in contrast to those polysaccharides that provide structure to plant cell walls—so-called structural carbohydrates, which include pectin, cellulose, hemicellulose, and lignin. "Dietary fiber" is a collective name for structural carbohydrates, and no mammal is known to produce the enzymes for directly digesting these complex molecules. Instead, primates—like all mammals—must rely on anaerobic, symbiotic microbes housed in the gut (in either a sacculated stomach or an expanded region of the large intestine) to ferment plant cell wall components, yielding short-chain fatty acids and other fermentation by-products of direct use by the host for energy (Cork and Foley 1991; J. Lambert and Fellner 2012).

Primate foods that commonly have the highest concentrations of readily digestible, nonstructural carbohydrates are sugar-rich plant exudates (such as saps, but not gums) and ripe fruit (Bearder and Martin 1980; J. Lambert 2010; L. Nash 1986). Immature fruit, buds, flowers, and leaves are also valuable sources of starches, sugars, and potentially digestible fibers, available at varying concentrations. The large variation in readily available, nonstructural carbohydrate sources of energy in primate foods has resulted in a diversity of feeding and digestive strategies in primates. For example, highly folivorous primates (e.g., colobines, howlers, and lemurs) have adapted to the relatively high levels of fiber and low levels of nonstructural carbohydrates in their leafy diets via behavioral, morphological, and physiological adaptations (e.g., *Hapalemur meridionalis*; Eppley, Verjans, and Donati 2011). Relative to species that rely more heavily on sugar-rich fruit, folivorous species tend to have higher shearing crests on their molar cusps for masticating plant cell walls, relatively long digestive retention times for fermenting fiber, adaptations in the gut for housing symbiotic fermenting microbes, shorter day ranges, and long bouts of rest during the day (Milton 1978).

The Role of Carbohydrates in Primate Nutritional Ecology

Macronutrients are often lumped together to obtain an estimate of total digestible energy. There are thus relatively few primate studies that have focused on the role of carbohydrates per se in food intake and selection; these studies are valuable for the insights they provide for understanding diet choice.

Conklin-Brittain, Wrangham, and Hunt (1998) found that chimpanzees ingested dramatically more simple sugars during peak fruit season than at other times of the year, while sympatric blue monkeys (*Cercopithecus mitis*), redtail monkeys (*Cercopithecus ascanius*), and mangabeys (*Lophocebus albigena*) ingested similar amounts of simple sugars throughout the year. Wild mountain gorillas select leaves and stems that are relatively rich in sugars, and it appears that the balance between carbohydrates and protein in the food is an important cue for food selection (Ganas et al. 2008; Rothman, Raubenheimer, and Chapman 2011). When food options fluctuate with the seasons and gorillas are not able to compose a diet with the preferred balance, they try to maintain a stable intake of carbohydrates and lipids but allow protein intake to fluctuate (Rothman, Raubenheimer, and Chapman 2011). These large apes have a leaf-dominated diet, live in a relatively protein-rich environment, and have the digestive strategy (capacious large intestine and long gut retention times) expected for a large-bodied folivore (Rothman, Raubenheimer, and Chapman 2011). Howler monkeys share some of these characteristics, and the balance between simple sugars and protein likewise explains food selection by black howler monkeys (*Alouatta pigra*; Behie and Pavelka 2012a). Although it might be assumed that sugars and starches would influence diet choice in ripe-fruit specialists, spider monkeys do not in fact appear to use carbohydrates as the foremost cue in food selection (Felton, Felton, Raubenheimer, Simpson, Foley, et al. 2009).

THE PROTEIN-TO-FIBER RATIO AND THE INFLUENCE OF TANNINS ON PROTEIN DIGESTIBILITY

Any discussion of the importance of protein and carbohydrates in primate food selection must consider the protein-to-fiber ratio of leaves. First proposed by Milton (1979), the rationale behind using such a ratio is that folivorous primates should deal with their digestive limitations by selecting leaves that are rich in protein but relatively low in dietary fiber (A. Davies, Bennett, and Waterman 1988; Milton 1979; Waterman, Ross, et al. 1988). Thus, in theory, primates with access to a diet with higher protein-to-fiber ratios should have better fitness than primates without. Although this index has not been explicitly tied to fitness per se, an assessment of the protein and fiber in leaves of dominant feeding tree species within a species' habitat has successfully predicted the biomass of small-bodied folivorous primates at both local (Chapman and Chapman 2002; Ganzhorn 2002; Simmen, Tamaud, and Hladik 2012) and regional scales (Chapman, Chapman, Naughton-Treves, et al. 2004; Fashing et al. 2007; Oates, Whitesides, et al. 1990; Wasserman and Chapman 2003; Waterman, Ross, et al. 1988). For example, the protein-to-fiber ratios of mature leaves available to groups of black-and-white colobus monkeys (*Colobus guereza*) accounted for 87% of the variance in group biomass (Chapman, Chapman, Naughton-Treves, et al. 2004).

A consumer's decision to select one leaf over another is driven by processes acting on a smaller temporal and spatial scale than at a population level in different habitats. The value of the protein-to-fiber ratio as a measure of food quality when explaining food item selection has also been tested, and several leaf-eating primate species (*Alouatta palliata, Presbytis rubicunda, Colobus satanas, Nasalis larvatus, Gorilla gorilla gorilla, Macaca fuscata,* and lemur spp.) appear to select or reject leaves according to the principle of ensuring a relatively high protein-to-fiber ratio (A. Davies, Bennett, and Waterman 1988; Ganzhorn 1992; Hanya, Kiyono, Takafumi, et al. 2007; McKey, Gartlan, et al. 1981; Milton 1979, 1998; Rogers, Maisels, et al. 1990; Yeager et al. 1997). In some cases, this ratio has also been useful in explaining different degrees of preference among leaf species in a consumer's diet as demonstrated in *Piliocolobus tephrosceles, Colobus guereza,* and *Papio* spp. (Chapman and Chapman 2002; Chapman, Chapman, Naughton-Treves, et al. 2004; Fashing et al. 2007; Whiten, Byrne, et al. 1991).

However, support for the protein-to-fiber ratio as a predictor of biomass and diet selection is not consistent, and exceptions occur. Changes in the density of red colobus (*Colobus rufomitratus*) over 36 years in Kibale National Park could not be predicted by changes in the protein-to-fiber ratios of mature leaves in their habitat over this time (Chapman, Chapman, Jacob, et al. 2010; Chapman, Struhsaker, et al. 2010). This ratio also does not predict differences among red colobus groups in the number of infants per female (Gogarten, Guzman, et al. 2012). Furthermore, under some circumstances colobines prefer leaves with a low protein-to-fiber ratio (McKey, Gartlan, et al. 1981; Mowry et al. 1996; Oates, Whitesides, et al. 1990), and the protein-to-fiber ratio of food patches of *Colobus guereza* cannot explain their patch residence time as would be predicted by the hypothesis (J. Johnson et al. 2015). Similarly, howler monkeys (*Alouatta pigra*) living in heavily hurricane-disturbed habitat actually preferred mature leaves over young leaves, despite lower protein-to-fiber ratios (Behie and Pavelka 2012a). Inconsistencies also occur between different study groups in the same area (Chapman and Chapman 2002; Chapman, Chapman, Naughton-Treves, et al. 2004) and between closely related taxa (Dasilva 1994; Ganzhorn 1992; Kool 1992). It is possible that these inconsistencies result from differences in competitive pressures that can affect the monkeys' feeding behavior (Chapman and Chapman 2002; Ganzhorn 1992; Mowry et al. 1996). The apparent mismatch between dietary selection and the effect of the protein-to-fiber ratio on primate biomass may also be because some studied primate populations live in habitats with protein-to-fiber ratios above a critical threshold, which results in a relative lack of feeding selectivity along this gradient (Simmen, Tarnaud, Marez, et al. 2014). The inconsistencies we see in the data, which have led to questions about the generality of this model (Felton, Felton, Lindenmayer, et al. 2009; Gogarten, Guzman, et al. 2012), can also be due to fundamental problems with the computation of the ratio itself (I. Wallis et al. 2012), which can surface when this approach is applied at the finer scale of food item selection.

For example, researchers have often used total N as a currency when calculating the ratio, when in fact it is available N in the food that is nutritionally important for the animal (I. Wallis et al. 2012). Commonly, around 30%–40% of the nitrogen in plants is bound up in cell walls and tannin complexes and as such is largely unavailable to the consumer (Milton and Dintzis 1981; Van Soest 1994). Further complicating the use of a protein-to-fiber ratio is that the fiber fraction normally used for calculating the ratio (neutral detergent fiber or acid detergent fiber) contains a complex mixture of fiber, protein, and tannins (Makkar et al. 1995). Tannins—which bind closely with protein, thereby rendering the proteins unavailable to the consumer—can therefore make the resultant ratio misleading (I. Wallis et al. 2012). It may be more fruitful to estimate the available fraction of N when trying to understand primate food choice (Felton, Felton, Lindenmayer, et al. 2009; Rothman, Chapman, and Van Soest 2012)—for example, by estimating the fiber-bound portion of N (Licitra et al. 1996; Rothman, Chapman, and Pell 2008) or by quantifying the influence of tannins on N digestibility (DeGabriel, Wallis, et al. 2008). Thus, although the protein-to-fiber ratio may be a reasonable predictor of the biomass of some leaf-eating primates and may sometimes correlate with types of foods that are selected or preferred, the value of the approach for fine-scale food selection research requires closer consideration.

The influence of tannins on primate macronutrient intake and food selection is an interesting and complicated topic in itself (Stalenberg et al., this volume). Tannins are a complex collection of compounds with differing biological properties (W. Zucker 1983) and effects on the consumer (M. Ayres et al. 1997). Some tannins reduce the digestibility of protein, minerals, and dry matter (C. Robbins, Hanley, et al. 1987; Santos-Buelga and Scalbert 2000) and can also cause acute toxicity, resulting in necrosis and kidney and liver failure (Shimada 2006). There are thus reasons to expect primates to actively reject items that are rich in such tannins. Indeed, the presence of condensed and hydrolyzable tannins in foods has been found to negatively influence primate diet choice (Glander 1982; Marks et al. 1988; Masette, Isabirye-Basuta, Baranga, and Chemurot 2015; Takemoto 2003; Wrangham and Waterman 1981). However, it is important that tannin concentrations be considered not alone but in the context of the balance between macronutrient and tannin concentrations when trying to understand food choice. This is because animals can trade off these components and are often more tolerant of tannins and similar antifeedants when macronutrient levels are high (S. Simpson and Raubenheimer 2001; Villalba and Provenza 2005). Interesting examples of such trade-offs have been reported for mangabeys (Masette, Isabirye-Basuta, Baranga, and Chemurot 2015), gorillas (Takemoto 2003), and spider monkeys (Felton, Felton, Raubenheimer, Simpson, Foley, et al. 2009). Moreover, some primate species have evolved physiological adaptations to tannin-rich diets, such as tannin-binding salivary proteins (e.g., hamadryas baboons; Mau, de Almeida, et al. 2011) and tannin-degrading microbes in the gastrointestinal tract (e.g., gorillas; Frey et al. 2006). To make matters even more complex, some tannins appear to have direct positive effects on the consumer via antioxidant and antipathogen properties (Min et al. 2003; Mueller-Harvey 2006; Santos-Buelga and Scalbert 2000). See Stalenberg and colleagues' chapter in this volume for further discussion about plant secondary metabolites.

MACRONUTRIENTS: LIPIDS

The Nature of Lipids and Where Primates May Find Them

Lipids (fats) are the most energy-dense of the macronutrients. One gram of fat provides the consumer with approximately twice as many calories as one gram of carbohydrate or protein (J. Lambert 2010). Fats not only provide mammals with high amounts of energy but also influence neurotransmitter levels that regulate reproductive hormones (C. Robbins 1993). While fat concentrations in leaves and buds are normally low, primates can find richer sources of fat in insects, other kinds of animal flesh, seeds, and the arils of some ripe fruit (e.g., *Virola* spp., *Sapium* spp., and some palm fruit; Castellanos 1995; Felton, Felton, Wood, Foley, et al. 2009; Milton 2008; Norconk and Conklin-Brittain 2004).

The Role of Fats in Primate Nutritional Ecology

The analysis of fatty acid composition in food items is both time consuming and expensive (Rothman, Chapman, and Van Soest 2012). These constraints likely explain why relatively few primate studies have done such analyses (J. Chamberlain et al. 1993; Reiner et al. 2014). What is more common is that studies instead report on crude fat contents in primate diets and their role in food selection. Some ateline species, whose diets are dominated by ripe fruit, have been observed to preferentially select and ingest large volumes of lipid-rich fruit (Castellanos 1995; Dew 2005). For example, *Ateles belzebuth* in Venezuela spent a disproportionate amount of time during drought seasons in energy-rich feeding patches where lipid-dense fruit provided a large proportion of the energy (Castellanos 1995). Interestingly, while a group of *Ateles belzebuth* in Ecuador showed similar preferences for lipid-rich fruit species, sympatric woolly monkeys (*Lagothrix lagotricha*) avoided these particular fruit species and instead obtained a larger proportion of their daily fat from invertebrates (Dew 2005). This study demonstrates that sympatric ripe-fruit specialists with overlapping diets can occupy distinct nutritional niches. Like the woolly monkeys, western lowland gorillas (*Gorilla g. gorilla*) eat a diverse mix of leaves, fruit, and stems and appear to also avoid fatty fruit (Rogers, Maisels, et al. 1990). However, as we discuss below, an apparent preference for or against fatty fruits does not necessarily mean that fat or energy content is shaping consumption.

Nutritional niche partitioning has also been demonstrated by Conklin-Brittain, Wrangham, and Hunt (1998), with outcomes of direct relevance to understanding dietary fats. Over the course of a full year, sympatric chimpanzees (*Pan troglodytes*), blue monkeys (*Cercopithecus mitis*), redtail monkeys (*C. ascanius*), and gray-cheeked mangabeys (*Lophocebus albigena*) consumed a frugivorous diet significantly higher in fat during times of ripe-fruit abundance in Kibale National Park, Uganda. However, while chimpanzees mainly obtained fat from ripe fruit, the cercopithecine monkey species also derived fat from unripe fruit pulp and seeds (Conklin-Brittain, Wrangham, and Hunt 1998). We suspect that crude fat and certainly the concentration of specific fatty acids (Reiner et al. 2014) play a larger role in food selection by primates than is thus far understood. Because fat's contribution to the diet is most often incorporated into the compound measure of energy, its functional role in diet selection is obscured. In the next section, we discuss energy as a factor in primate diet selection and what is potentially lost in the use of this measure.

ENERGY—WHAT IS THE CURRENCY AND WHAT DOES IT TELL US?

Primates, like all mammals, need to obtain energy in order to survive, and a direct link between energy intake and fitness has been shown in some primate species (S. Altmann

1998). It is important to note that energy is not a nutrient per se but a property of macronutrients. Many factors influence species and individual energy requirements, including body mass, age, sex, activity level, reproductive status, and climatic conditions (C. Robbins 1993).

There are different ways to quantify the energy value of a food item. **Gross energy**, the energy contained in all components of a food, can be estimated through bomb calorimetry, in which food material is combusted and the heat produced is measured as calories per gram of dry matter (Rothman, Chapman, and Van Soest 2012). This measure, however, does not represent the available energy accessible to a mammal consumer as it includes nondietary calories (as an example, a piece of coal contains a lot of energy, none of which can be used by a primate). Nevertheless, it is not uncommon for gross energy to be used as a measure of the energy consumed by a primate (Ganas et al. 2008; L. Gould, Power, et al. 2011; Wasserman and Chapman 2003), and these measures should be interpreted with extreme caution. **Total energy**, the energy contained in the macronutrient components of a food, can be estimated by first identifying how much crude protein, crude fats, and total nonstructural carbohydrates there are in the item and then multiplying each of these fractions by their respective energy contribution per unit weight. This measure thus represents the food's dietary energy but may include significant fractions of macronutrients that are not available for the consumer to digest due to them being bound by cell wall structures or tannins, or because they are present in otherwise indigestible forms. **Digestible energy**, in contrast, is an estimate of the number of calories that are actually available for the animal to digest and is therefore a recommended measure (Rothman, Chapman, and Van Soest 2012). Furthermore, if one also subtracts metabolic losses (energy lost via urine and methane) from digestible energy, an estimate of **metabolizable energy** is yielded, which is a common basis for evaluating feed quality in monogastric animal nutrition (Van Soest 1994). Despite its relevance, few studies in primate ecology actually quantify metabolizable energy because it would require more invasive studies. Instead, most studies estimate the metabolizable energy calorie contributions of crude fat, carbohydrates, and crude protein (Conklin-Brittain, Knott, and Wrangham 2006; Rothman, Chapman, and Van Soest 2012). It is important to note, though, that regardless of which variant of energy one uses, this currency has the potential to hide underlying patterns attributable to the different sources of calories in the animal's diet.

Studies of optimal foraging theory (OFT) typically operate with the assumption that total digestible energy is the most useful currency to predict foraging decisions, particularly of herbivores (Belovsky 1978; Belovsky and Jordan 1978; Moen, Cohen, and Pastor 1998; Moen, Pastor, and Cohen 1997; Shipley, Illius, et al. 1999). More specifically, OFT posits that individuals should maximize energy yield per unit time feeding, within stipulated constraints (such as gut capacity and mineral requirements), while minimizing costs such as time spent foraging or processing foods (Belovsky 1978; D. Stephens and Krebs 1986). While energy maximization models may acknowledge that animals can select a food to obtain rare nutrients, energy is typically a prioritized measure of overall food quality, suggesting that a lot of digestible energy equates with high quality. This particular conception of food quality has also influenced the field of primate nutritional ecology (Cowlishaw and Dunbar 2000; J. Lambert 2010; M. Leighton 1993; Strier 2007).

Digestible energy has indeed been shown to be an important factor in herbivore foraging decisions, and OFT is used widely as a predictive tool in analyses of ungulate nutritional ecology (Moen, Cohen, and Pastor 1998; Moen, Pastor, and Cohen et al. 1997; vanWieren 1996). For example, the mean diameter of twigs selected by free-ranging ungulates was predicted with high accuracy by a model using energy maximization as an assumption (Shipley, Blomquist, and Danell 1998). Similarly, empirical support for energy maximization has been provided for several species of grassland herbivores (Belovsky 1986). However, because these studies describe a diet's quality only in terms of energy (by pooling all macronutrients), it is difficult to discern the role that energy plays relative to nutritional factors (Pyke 1984; Raubenheimer, Simpson, and Mayntz 2009). In actuality, when all macronutrients were separated a priori in a recent study, even the moose (*Alces alces*)—the primary model species in OFT development and a classic "energy maximizer" (Shipley 2010)—was not found to be following an energy-maximizing strategy when selecting food (Felton, Felton, Raubenheimer, Simpson, Krizsan, et al. 2016).

So what about primates? Attempts have been made only relatively recently to empirically test the energy maximization hypothesis in wild primates, in part due to the complexity of primate diets and foraging circumstances (Milton 1979; Post 1984). Instead, researchers have suggested that some primate species are energy maximizers based on comparisons of anatomical, behavioral, and ecological observations. For example, because of their short digestive retention times (Milton 1981b), preferences for fatty and sugary fruit (Castellanos 1995; Dew 2005; Laska, Salazar, and Luna 2000), large territories, and fluid social structure, spider monkeys (*Ateles* spp.) and some other atelines (e.g., *Brachyteles* spp.) have been suggested to follow an energy-maximizing strategy (Di Fiore and Rodman 2001; Rosenberger and Strier 1989; Strier 1992). However, more recent research involving detailed and continuous observations of focal individuals' daily nutrient intake indicates that the underlying nutritional goal of spider monkeys (*Ateles chamek*) is not to maximize daily energy intake but instead to ingest a set amount of protein each day (Felton, Felton, Raubenheimer, Simpson, Foley, et al. 2009).

In summary, even though energy is an important factor for herbivore fitness and a rational choice of currency in many investigations, using energy alone as a measure of food quality obscures the relative importance of the macronutrients, and errors can therefore be introduced in predictive models. Whether some primate species have an energy-maximizing strategy or not when deciding which food items to eat remains an open question, and we encourage more primatologists to investigate this area further.

MICRONUTRIENTS AND THEIR ROLE IN PRIMATE NUTRITIONAL ECOLOGY

The micronutrients—minerals and vitamins—are essential to primate health, growth, and reproduction. While they do not provide the body with energy per se, they do facilitate the use of energy-yielding macronutrients and perform many other functions as well. Very little is known about micronutrient consumption by wild primates, although of the two forms, minerals have received greater attention. Minerals are inorganic elements that are critical for maintaining physiological function and are therefore essential to life (C. Robbins 1993). The broad category of minerals can be subdivided into two mineral types that are relevant for animal health and nutritional ecology: **macrominerals**, which are needed in relatively large amounts and are expressed as a percentage of the diet (calcium, phosphorus, magnesium, potassium, sodium, chlorine, and sulfur), and **trace elements**, whose concentrations are normally expressed as parts per million (iron, copper, manganese, zinc, iodine, selenium, chromium, and cobalt; J. Lambert 2010). These elements play diverse roles in the mammalian body and can serve as structural components in biological molecules (e.g., iron in hemoglobin and iodine in thyroxine), activate hormonal and enzymatic processes, regulate cell activity, and act as constituents of body fluids (NRC 2003). Requirements for minerals will vary over the lifetime of a mammal according to a variety of physiological factors, such as age, growth, body composition, genetics, pregnancy, lactation, wound healing, infections, and diseases (Freeland-Graves et al. 2015). For example, lactating marmoset females (*Callithrix jacchus*) require more calcium than nonlactating females (M. L. Power, Tardif, et al. 1999).

The sources of minerals for wild primates are diverse. Young leaves and petioles normally have higher concentrations of minerals than fruit and may for this reason be important complementary food items for primates who have a fruit-dominated diet (J. Lambert 2010; Rode, Chapman, Chapman, et al. 2003). However, some fruit species offer uniquely high levels of minerals. Figs, for example (the syconium of *Ficus* trees), have more than three times as much calcium as most other fruits (O'Brien, Kinnaird, and Dierenfeld 1998), which is likely one of several reasons why they are such important foods for primates (Felton, Felton, Wood, Foley, et al. 2009). Primates may also obtain minerals from nondietary sources, such as soil and termitaria, which we describe further below.

Minerals may be a limiting factor for wild primate fitness, at least in some habitats. Minerals have been suggested to be critical in the nutrition of omnivorous primates such as some cercopithecines in Uganda (e.g., redtail monkeys, *Cercopithecus ascanius*), as populations lacking sufficient sources of Cu, Zn, Mn, Mg, and Ca in heavily logged areas have much lower population densities than populations in unlogged forests with plenty of mineral-rich food items available (Rode, Chapman, McDowell, et al. 2006). It also appears that the consumption of particularly sodium-rich stems and calcium-rich leaves may have been a crucial mechanism for folivorous howler monkeys (*Alouatta pigra*) to avoid deficiencies in the years after a hurricane greatly reduced their normal sources of a nutritionally balanced diet (Behie and Pavelka 2012b). On a similar note, the relatively high densities of *Colobus guereza* in Kibale National Park in Uganda, compared to other study areas, may be due to the presence of swamps containing sodium-rich plants that are consumed by these monkeys (Oates 1978). A more recent study on the same *C. guereza* also found that movement patterns were dictated by the presence of *Eucalyptus* trees (T. R. Harris and Chapman 2007), the plant parts of which have high sodium concentrations (Rode, Chapman, Chapman, et al. 2003). Sodium and copper may also be limiting for *Alouatta palliata*, as their natural diet is quite low in these minerals in relation to their estimated requirements (Nagy and Milton 1979a).

It is reasonable to assume that long-term limitation of a certain mineral would be an important selective pressure for a preference for items that are relatively rich in this mineral. Mountain gorillas appear to provide an example of this (*Gorilla gorilla beringei*). In Bwindi Impenetrable National Park, Uganda, mountain gorillas regularly consume decaying wood—a substrate with little macronutritional value: it is low in protein and sugar and high in lignin that cannot be digested or fermented (Rothman, Van Soest, and Pell 2006). However, the wood is high in sodium content, and although the wood comprised only 3.9% of the total wet weight food intake of the study subject, it contributed over 95% of dietary sodium (Rothman, Van Soest, and Pell 2006; see also Magliocca and Gautier-Hion 2002). Similarly, diademed sifakas (*Propithecus diadema*) in Madagascar appear to select young leaves of *Impatiens* sp. due to their unusually high mineral content (Ca, Zn, P, and Fe), despite the leaves being relatively poor in macronutrients and requiring the monkeys to risk predation by descending to the ground to eat them (Irwin, Raharison, Chapman, et al. 2017). Moreover, there is some evidence that mineral concentrations (Cu, Zn, Mn, Mg, and Ca) may at least partly guide redtail monkeys' selection of some rare food items, such as petioles, bark, and caterpillars (Rode, Chapman, Chapman, et al. 2003); similar indications exist for the black-and-white colobus monkey (*Colobus guareza*; Oates 1978) and the Japanese macaque (*Macaca fuscata*; Hanya, Kiyono, Takafumi, et al. 2007). However, contrary to expectations, researchers could not find evidence for sodium selection by redtail monkeys (Rode, Chapman, Chapman, et al. 2003). It is important to note, though, that it can be difficult to identify such targeted selection statistically, as the mineral in question may vary only slightly in concentration among the plants eaten. Furthermore, the mineral levels may also be correlated with concentrations of secondary metabolites that the monkeys simultaneously try to avoid (Rode, Chapman, Chapman, et al. 2003).

It is a little easier to tease out such issues when dealing with other types of food items than leaves and fruit. For example, primates that are predominantly insectivorous but also eat fruit and tree exudates, such as marmosets (*Callithrix* spp.) and tamarins (*Saguinus* spp.), provide interesting study subjects when assessing the roles of calcium and phosphorous in diet selection (M. L. Power, Tardif, et al. 1999). Insects and many other invertebrates are poor sources of digestible calcium but are rich in phosphorous (Garber 1984). In contrast, tree gums are rich in calcium but poor in phosphorus. This suggests that a balanced intake of both insects and gums provides insectivorous monkeys with suitable proportions of these minerals and that this is why they select both types of items on a daily basis (Garber 1984). Such a mix between food item categories may be equally important for the aye-aye (*Daubentonia madagascariensis*), which eats seeds, nectar, fungus, and insect larvae and for whom the larvae provide a significant dietary source of sodium (Sterling et al. 1994).

The ingestion of soil (**geophagy**) by primates, at salt licks, mud puddles or termite mounds (termitaria), can sometimes indicate a need for mineral supplementation (Krishnamani and Mahaney 2000). Other plausible (and not mutually exclusive) reasons for geophagy by primates may be the absorption of toxins (Oates 1978), treatment of diarrhea (Mahaney et al. 1995), and stomach pH adjustment (A. Davies and Baillie 1988). However, the literature is inconclusive regarding evidence for mineral supplementation as a primary selective pressure for geophagy. While it appears that soil consumption by Yunnan snub-nosed monkeys (*Rhinopithecus bieti*) is a response to mineral deficiency (D. Li et al. 2014), the elemental composition of soil eaten by howler monkeys and spider monkeys cannot clearly explain this behavior (Izawa 1993). Similarly, even though the termitaria soil eaten by red leaf monkeys (*Presbytis rubicunda*) has higher concentrations of Ca and K than surrounding ground soil, it was not certain that the monkeys ingested this matter due to any deficiencies in their diet (A. Davies and Baillie 1988). It is important to note here, though, that animals do not have to be deficient in order to use a micro- or macronutrient as a cue to food selection if the underlying driving factor is to obtain an overall balanced nutrient intake (Felton, Felton, Raubenheimer, Simpson, Krizsan, et al. 2016).

While our understanding of minerals and their role in wild primate nutritional ecology and diet choice requires much further investigation, our knowledge of vitamins in wild primate diets is almost nonexistent, in part because of the challenges of collecting samples for vitamin analysis and the actual analysis itself. Wild plants intended for vitamin analysis must be stored in liquid nitrogen for transport to a laboratory with the capacity to undertake high-performance liquid chromatography (HPLC)—a form of column chromatography that is expensive and time-consuming. To date, there have been only a handful of assessments of the vitamin content of wild primate diets. Prominent examples include studies of vitamin intake by mantled howler monkeys (*Alouatta palliata*) on Barro Colorado Island, Panama (Milton 2003b), and of aye-ayes (*Daubentonia madagascariensis*) in Nosy Mangabe, Madagascar (Sterling et al. 1994).

Vitamins are typically categorized into those that are water soluble and hence must be consumed on a near-daily basis (e.g., vitamins B and C) and those that are fat-soluble (e.g., A, D, E, and K) and stored by the body. With the exception of B_{12}, all vitamin requirements can be readily met by a plant diet, and it is generally understood that these requirements are not limiting. As with minerals, most of what is known derives from a few model primate species held in experimental, captive circumstances in which requirements are determined by manipulating vitamin loads in nonwild diets and then documenting the physiological response, typically by determining deficiency thresholds rather than optimal intake levels. The little work that has been done on wild plants demonstrated that they have significantly higher mineral and vitamin levels than their domesticated counterparts (Milton 2003b). Of wild primate foods in particular, research has emphasized vitamin C (ascorbic acid), probably because anthropoid primates lack the enzyme critical to the last step of endogenously synthesizing ascorbate from glucose: L-gulonolactone oxidase (Milton 2003b). Thus, unlike most mammals, anthropoids—including humans—must derive this critical antioxidating micronutrient from their plant diet.

This suggests that primates have had a long evolutionary history of consuming high levels of vitamin-related antioxidants (A. D. Johns 1996), and this certainly has implications for human and captive primate health. Indeed, in work conducted by Sterling et al. (1994), the diets of wild aye-ayes comprised an order of magnitude more vitamin A (a powerful antioxidant) than captive diets (720 versus 14.8 IU/kg).

NUTRIENTS COMBINED: THE IMPORTANCE OF GETTING THE BALANCE RIGHT

In the 1970s, it was suggested that herbivores select their food to obtain a nutritionally balanced diet (Westoby 1974), and some primatologists hypothesized that this may be the case for primates (e.g., Gaulin and Gaulin 1982; Milton 1979). The nutrient balancing hypothesis suggests that, when sufficient food is available, animal diet selection is governed by the primary goal of getting an optimal balance among nutrients (S. Simpson and Raubenheimer 2012; Westoby 1974) rather than concentrating on one particularly important currency, such as energy or some macro- or micronutrient. Underlying this line of thinking is the awareness that nutrition is best understood in a multidimensional context, as foraging is a dynamic process that involves balancing the intake of many different nutrients to satisfy complex nutritional needs that change over short timescales (S. Simpson and Raubenheimer 2012). Analytical tools are now available to empirically deal with several nutritional currencies at the same time (see Raubenheimer, this volume). Nutritional geometry is a modeling approach designed specifically for this purpose (Raubenheimer and Simpson 1999; S. Simpson and Raubenheimer 1995; Raubenheimer, this volume). With this analytical framework, one can unify several nutritional measures using simple geometric models that contrast observed and predicted patterns of nutrient intake and allow the researcher to assess interactions among nutrients. Owing to work using this approach, organisms belonging to a large variety of taxonomic groups—from slime molds to brown bears—are now thought to mix foods to achieve a particular target balance (e.g., Dussutour et al. 2010; Raubenheimer, Zemke-White, et al. 2005; C. Robbins, Fortin, et al. 2007; S. Simpson, Sibly, et al. 2004). Scholars interested in the diets of primates in particular are increasingly using these analytical tools. Examples include studies of spider monkeys (*Ateles chamek*; Felton, Felton, Raubenheimer, Simpson, Foley, et al. 2009), chimpanzees (Uwimbabazi et al. 2021), gorillas (*Gorilla beringei*; Ganas et al. 2008; Raubenheimer and Rothman 2013; Rothman, Raubenheimer, and Chapman 2011), sifakas (*Propithecus diadema*; Irwin, Raharison, Raubenheimer, et al. 2015), chacma baboons (*Papio hamadryas ursinus*; C. A. Johnson, Raubenheimer, Rothman, et al. 2013), rhesus macaques (*Macaca mulatta tcheliensis*; Cui, Shao, et al. 2019; Cui, Wang, Shao, et al. 2018), blue monkeys (*Cercopithecus mitis*), and guereza monkeys (*Colobus guereza*; J. Johnson et al. 2015). These studies all provide evidence that the primates in question balance their macronutrient intake in one way or another. Such findings may reveal, for example, that ripe-fruit specialists are not necessarily energy maximizers as previously thought (Felton, Felton, Raubenheimer, Simpson, Foley, et al. 2009; Uwimbabazi et al. 2021), that nitrogen does not need to be limiting for terrestrial herbivores (Rothman, Raubenheimer, and Chapman 2011), or that the time monkeys spend in a particular food patch can be explained by the macronutritional balance of the food in the patch (J. Johnson et al. 2015).

Animals not only need to balance their intake of macronutrients and energy but also must balance their intake of minerals, vitamins, and plant secondary metabolites (e.g., Raubenheimer and Simpson 2006). Although this complexity is challenging to assess, by evaluating food choice through the lens of nutrient balancing, it becomes clear why animals do not consistently maximize their intake of a rare mineral, for example, or completely avoid intake of a digestion inhibitor, but instead mix complementary items to obtain a suitable balance. The nutrient balancing framework also makes it a little easier to understand the influence on food choice of the complex interactions between the many constituents in the food (Bjorndal 1991; Eppley, Tan, et al. 2017). For example, the balance between fiber and minerals in the diet is very important, as a high fiber intake may have a considerable inhibitory effect on mineral absorption (Freeland-Graves et al. 2015). It is hypothesized that population density of redtail monkeys (*Cercopithecus ascanius*) is positively related to their ability to obtain a diet with a beneficial balance between fiber and minerals and that differences in their capacity to achieve this balance can at least partially explain reproductive success in undisturbed versus disturbed habitat (Rode, Chapman, McDowell, et al. 2006). Another interesting example of complex interactions of this sort is how mangabeys do not have to actively avoid tannins when their diet is rich in macronutrients (Masette, Isabirye-Basuta, Baranga, and Chemurot 2015). Only by approaching diet selection from a multidimensional perspective can such patterns be identified.

The framework of nutrient balancing can sometimes serve to explain peculiar and unexpected observations. For example, the frequent intake of mature leaves by howler monkeys in Belize (despite poor protein-to-fiber ratios and the abundance of young leaves) is likely helping balance their energy and protein intake (Behie and Pavelka 2012a). Nutrient balancing also provides a good explanation as to why spider monkeys consume a diverse array of ripe fleshy fruits to overcome periods of fig scarcity rather than consuming figs only to overcome periods of ripe fruit scarcity, as was expected (Felton, Felton, Wood, Foley, et al. 2009). We suggest that many of the inconsistencies we see between studies in identifying the significance of a particular currency or nutrient (e.g., protein and energy) can be explained by nutrient balancing. When an animal is regulating toward a balanced nutritional intake, the limiting and therefore desired component can change quite quickly. Furthermore, the theory of nutrient balancing goes hand-in-hand with the idea of diet mixing, which helps us understand inconsistencies in results relating to the protein-to-fiber ratio of leaves. This ratio is often used to explain preferences for some leaves over others without including an estimation of the intake of other food items on the same day—even though folivorous monkeys often mix their leaf-based diet with fruit and seeds (Dasilva 1994). Because the ingestion of one diet item can affect the digestion of another item (Bjorndal 1991; Villalba and Provenza 2005), an intake of seeds in the morning can influence the choice of leaf in the afternoon.

We see a promising trend now in the field of primate nutritional ecology to view diet selection within a multidimensional framework, with an increasing use of the analytical tools and chemical assays available (e.g., Raubenheimer, Machovsky-Capuska, Chapman, et al. 2015). By identifying an animal's target nutrient balance and demonstrating the means by which that balance is achieved, we are better able to predict their foraging behavior and thereby improve the management of both wild and captive populations. Such approaches can provide unexpected insights into human evolution and physiology (Felton, Felton, Raubenheimer, Simpson, Foley, et al. 2009; Raubenheimer and Rothman 2013; Rothman, Raubenheimer, and Chapman 2011; S. Simpson and Raubenheimer 2005), game management (Felton, Felton, Raubenheimer, Simpson, Krizsan, et al. 2016; Felton, Wam, et al. 2021), habitat restoration and conservation (C. A. Johnson, Raubenheimer, Rothman, et al. 2013; Raubenheimer and Simpson 2006), and even cannibalism (S. Simpson, Sword, et al. 2006).

HOW DO WE DEFINE FOOD QUALITY? AND SHOULD WE?

It is common practice in herbivore nutritional ecology to classify food items as being either "high quality" or "low quality" for the consumer in question. As a general rule, it is assumed that low-quality foods are characterized by low concentrations of available energy or protein, or high levels of dietary fiber and therefore low digestibility. This is exemplified by the many studies that use the protein-to-fiber ratio, as we discussed previously. We wish to emphasize, though, that when animals' daily and frequent selections between available food items are considered in terms of nutrient balancing, such categorical classifications (high versus low quality) make less sense. When an animal is regulating toward a balanced nutritional intake, as mediated by physiological feedback loops between the senses and the gut (Provenza 1995), the limiting and therefore desired component can change rapidly. Because the animals appear to correct for previous intakes in order to reach a particular target balance, a so called high-quality fruit may be rejected in favor of a rather mediocre leaf, if the leaf will take the individual a little bit closer to the target.

Only foods that closely match the nutritional goal of the animal can be argued to have a relatively stable quality value, meaning that the consumer can spend most of its feeding time consuming this food type and remain in a balanced trajectory in nutritional space, within the limits of how much of the item's plant secondary metabolites can be tolerated (S. Simpson and Raubenheimer 2012). Such foods are therefore likely to become staple components in their diet, as exemplified by the nutritionally well-balanced figs for spider monkeys (Felton, Felton, Raubenheimer, Simpson, Foley, et al. 2009; Felton, Felton, Wood, Foley, et al. 2009). Food items that are particularly well balanced do not necessarily have to have high absolute concentrations of nutrients or energy but can conventionally be seen as low quality. A diet low in fat may, for example, be better for the health of apes than a high-fat diet (Reiner et al. 2014). We do not argue against the value of contrasting seasonal diets with each other in the categorical manner (i.e., the period of fruit abundance offers a higher-quality diet for a primate species than fruit-poor periods; Conklin-Brittain, Wrangham, and Hunt 1998) or classifying some habitats as offering a generally higher-quality diet than others (Rode, Chapman, McDowell, et al. 2006). However, when dealing with food item selection on a finer scale, we suggest that instead of adhering to the categorical way of classifying foods as high or low quality, we should embrace the idea of complementarity (K. Parker et al. 2009; Provenza 1995). This approach to nutritional ecology is likely to provide valued insights for both theory and practice.

SUMMARY AND CONCLUSIONS

Macro- and micronutrients are central elements in diet selection by wild primates. Studies on optimal foraging using single currencies (such as energy or protein) have taught us a large amount about primate foraging behavior and nutrient requirements. However, when dietary complexity is reduced to one nutritional currency, a large part of the dietary variation remains unexplained. Some of those gaps can be filled by looking at primate food choice through the lens of nutrient balancing—where we assume that multiple nutritional variables present in the diet interact both synergistically and antagonistically—which provides a better understanding of the animals' priorities and behavior.

- Despite previous predictions that protein is limiting for wild primates and that they therefore should preferentially select high-protein food items, the evidence is inconsistent. Some primates aim to maintain their daily protein intake as stably as possible, while others allow for a large fluctuation in daily protein intake, letting other nutritional factors govern their food choice.
- Primates obtain most of their energy from carbohydrates and lipids in fruits, leaves, and insects. Even

though energy may be a rational choice of currency in many investigations of primate ecology, using energy alone in studies of food selection obscures the relative importance of the individual macronutrients. By separating the macronutrients a priori, we can see that some primates select their foods in a way to obtain a stable intake of carbohydrates and fats every day, while others do not use the concentrations of these nutrients as cues.

- The protein-to-fiber ratio in mature leaves sometimes correlates with types of foods that are selected or preferred by leaf-eating primates. However, due to inherent problems with the way this ratio is calculated, the value of the approach for fine-scale food selection research requires closer consideration.

- A current trend in the field of primate nutritional ecology involves viewing diet selection within a multidimensional framework. New models reveal interesting aspects of primate feeding behavior and food quality from the perspective of nutrient balancing, providing insights into an animal's feeding tactics and into species-level adaptations. These results yield important insights for primate evolution, physiology, habitat restoration, and conservation.

2 What Extant Primates Eat
A Global Survey

Joseph E. Hawes, Carlos A. Peres, and Andrew C. Smith

As primates are arguably the best-studied mammalian order, there is an accumulated wealth of information on their feeding ecology. The basic dietary profile is now relatively well understood for most primates, following extensive long-term observational field studies (Garber, Estrada, et al. 2009; Kappeler and Watts 2012). Combining the results of these efforts reveals an impressive diversity of trophic strategies, with primate diets ranging from resource specialists to generalists or opportunists, and from feeders on plant-based resources such as fruit pulp, seeds, gums, leaves, and nectar to consumers of insects and meat (fig. 2.1). Many meta-analytic reviews have attempted to classify diets across this spectrum (e.g., Fleagle 1998; A. Hladik and Hladik 1969; Rosenberger 1992), but assigning generalist or omnivorous diets to strict or broad categories inevitably involves substantial loss of information.

Simplifying this complexity is useful to aid comparisons across species but does not reliably inform us of a diet's nutritional quality (J. Lambert 2007; Rothman, Plumptre, et al. 2007) and conceals details such as the type of invertebrate prey, the growth stages of leaves or fruit, and the dental or gut treatment of seeds ingested. Furthermore, diets may vary within a species both across populations throughout their geographic range and between individuals, and most primates exhibit pronounced seasonal variation in dietary profiles, primarily in response to resource phenology (Chapman and Chapman 1990). These variations are again obscured when simplifying diets to an average annual summary for comparative purposes.

Despite these issues, our level of knowledge is now reaching the point where large-scale compilations can provide considerable insights into broad-scale patterns in primate diets and the relationships among diet and other anatomical, behavioral, and ecological traits (Hawes and Peres 2014; J. Smith and Smith 2013).

In this chapter, we review some of these broad dietary categories to provide an overview of the full range of resources exploited by primates worldwide, with examples from the Americas, Asia, continental Africa, and Madagascar, before going on to propose a more flexible approach that considers variation in diets along a continuous spectrum. Considering the large number of potential examples to draw on, this review is admittedly superficial, but we aimed to provide an illustrative synthesis and an introductory background for more focused assessments of individual dietary classes. We cover the major dietary classes of frugivory sensu lato (fruits and seeds), faunivory (invertebrate and vertebrate prey), folivory (leaves from trees, herbs, and grasses), florivory (floral parts), and exudativory (gums and saps) and follow this with a brief summary of less commonly exploited food items, before discussing the term "omnivory" in relation to the considerable degree of overlap existing across these simplified traditional categories (Chivers 1998).

Improving the resolution of primate dietary data and examining both inter- and intraspecific variation, including spatial and temporal fluctuations, remains important to advance our understanding of the drivers of food selection within and between primate species. We therefore

Figure 2.1. Major dietary classes for nonhuman primates worldwide. *Clockwise from top left*: chimpanzee (*Pan troglodytes*) feeding on figs (*Ficus sur*) in Kibale National Park, Uganda; Toppin's titi (*Callicebus toppini*) feeding on *Inga* sp. seeds in Tambopata National Reserve, Peru; Gursky's spectral tarsier (*Tarsius spectrumgurskyae*) feeding on an orthopteran (Orthoptera) in Tangkoko National Park, Indonesia; chacma baboon (*Papio ursinus*) feeding on guinea fowl (*Numida melegaris*) in Mashatu Game Reserve, Botswana; ring-tailed lemur (*Lemur catta*) feeding on tamarind (*Tamarindus indica*) leaves in Tsimanampetsotsa, Madagascar; golden bamboo lemur (*Hapalemur aureus*) feeding on giant bamboo (*Cathariostachys madagascariensis*) shoots in Ranomafana National Park, Madagascar; Tarai gray langur (*Semnopithecus hector*) feeding on Indian elm (*Holoptelea integrifolia*) flowers in Shiwalik Forest Division, India; and moustached tamarin (*Saguinus mystax*) feeding on unidentified tree exudate at Estación Biológica Quebrada Blanco, Peru.

Photo by Alain Houle, BMC Ecology image competition 2014: the winning images [http://www.biomedcentral.com/1472-6785/14/24], CC BY [https://creativecommons.org/licenses/by/4.0/]; Davidwfx [https://www.flickr.com/photos/davidwfx/], Flickr [https://www.flickr.com/photos/davidwfx/4439703820/], CC BY-SA [https://creativecommons.org/licenses/by-sa/2.0/]; Viglianti, Wikimedia Commons [https://commons.wikimedia.org/wiki/File:Tarsius_eating_locust.JPG], CC BY-SA [https://creativecommons.org/licenses/by-sa/3.0/]; William R Ranch, Predation on Helmeted Guineafowl by Grey-footed Chacma Baboon in Mashatu Game Reserve, north east Botswana [http://bo.adu.org.za/content.php?id=345], CC BY [https://creativecommons.org/licenses/by/4.0/]; Frank Vassen [https://www.flickr.com/people/42244964@N03], Flickr [https://www.flickr.com/photos/42244964@N03/4309671260/], CC BY [https://creativecommons.org/licenses/by/2.0/]; Charles J. Sharp [https://www.wikidata.org/wiki/Q54800218], Sharp Photography [https://www.sharpphotography.co.uk/], CC BY-SA [https://creativecommons.org/licenses/by-sa/4.0/]; Rohitjahnavi [https://commons.wikimedia.org/wiki/User:Rohitjahnavi], Wikimedia Commons [https://commons.wikimedia.org/wiki/File:Gray_Langur_feeding_on_flowerbeds_of_Holoptelea_integrifolia_DSCN3819_06.jpg], CC BY-SA [https://creativecommons.org/licenses/by-sa/4.0/]; Eckhard W. Heymann, Personal permission.

also consider some of these potential drivers of dietary decisions, which are examined in more detail in other chapters of this volume, including a focus on the role of body size. We then describe the association between body size and diet, and particularly the degree of frugivory, in a case study focusing on New World monkeys (Platyrrhini). Such comparative analyses of existing dietary data are particularly useful to illustrate broad patterns across taxa and geographic regions and to highlight knowledge gaps that require further study. While gaps still exist that require filling, it is already possible to begin collating studies across the entire primate order, and this chapter therefore also provides a coarse provisional global summary of primate diets as a starting point for future refinements.

Body size is clearly not the only variable responsible for determining dietary profiles across primates. In addition to metabolic and energetic requirements (DeCasien et al., this volume), feeding decisions reflect the challenges of acquiring the appropriate balance of protein and other nutrients (Felton and Lambert, this volume). Rather than focusing on the drivers of food acquisition, perhaps it is more informative to think of the interactive and reciprocal relationships between diet and cognition (Garber, this volume), reproduction (Hinde et al., this volume; Emery Thompson, this volume), digestion (Lambert, Mutegeki, and Amato, this volume), and movement (Crofoot and Alavi, this volume). As highlighted in this book, all of these areas of primate biology are intimately connected, both shaped by and, in turn, shaping dietary strategies.

MAJOR DIETARY CLASSES

Frugivory (including Granivory)

Fruits are present in the diets of many primates yet represent a resource that is patchily distributed in space and time (Fleming et al. 1987; Herrera 1998; Levey 1988), and fruit pulp typically provides low protein value compared to both animal prey (e.g., arthropods) and foliage (Oftedal et al. 1991). Primates are therefore most frequently broad herbivores (folivore/frugivore) or herbivore/faunivores rather than entirely frugivorous, with fruit consumed in varying ratios in combination with alternative food sources. Key tree families consumed on a pantropical scale include Moraceae, Arecaceae, Sapotaceae, Clusiaceae, and Burseraceae. Primates may eat either fruit pulp (strict frugivory) or seeds (granivory) or both, although these provide distinct nutritional resources and present unique challenges and are therefore usually considered separately.

Fruit (Frugivory Sensu Stricto)

Ripe-fruit specialists in Neotropical forests are best represented by spider monkeys (*Ateles* spp., 82%; Russo et al. 2005) and woolly monkeys (*Lagothrix* spp., 67%; Peres 1994a). Fleshy fruits dominate the diet of these canopy dwellers, whose semibrachiating locomotion enables access to a wide diversity of canopy to subcanopy fruit species from across large forest areas (Di Fiore and Suarez 2007). Similar morphological and behavioral adaptations are also apparent in the hylobatids (gibbons and siamangs—*Hoolock* spp., *Hylobates* spp., *Nomascus* spp., and *Symphalangus syndactylus*) of Southeast Asia, which share this preference for ripe fruit pulp (53%–89%; P. Fan, Ai, et al. 2013; Islam and Feeroz 1992; S. Kim et al. 2011; Ni et al. 2014; Palombit 1997), particularly figs (*Ficus* spp.) where and when they are available (P. Fan, Ai, et al. 2013; M. E. Harrison and Marshall 2011; Kinnaird and O'Brien 2005; Ni et al. 2014). At approximately twice the size of other gibbons, the siamang (*S. syndactylus*) consumes fewer nonfig fruits and correspondingly more leaves (Chivers 1974; Palombit 1997).

Orangutans (*Pongo* spp.) are highly frugivorous, with a smaller proportion of their diet comprising leaf material and bark cambium (54%–68% fruit; E. Fox, van Schaik, et al. 2004; A. Taylor 2006; Wich, Utami-Atmoko, Mitra Setia, Djoyosudharmo, et al. 2006). Sustaining their large body size primarily through frugivory requires the daily coverage of a large area and the associated development of complex cognitive maps, particularly to successfully track seasonal fruiting patterns in the dipterocarp forests of Southeast Asia (Tomoko et al. 2010). This degree of intelligence and planning are similarly found in chimpanzees and bonobos (*Pan* spp.), both of which are recognized as ripe-fruit specialists, though with broad diets including a wide range of other food sources (49%–71%; Badrian et al. 1981; Georgiev, Thompson, et al. 2011; Goodall 1986; Tweheyo et al. 2004; Wrangham, Conklin-Brittain, and Hunt 1998). Chimpanzees and bonobos differ from other frugivorous species mentioned above as their foraging behavior includes both arboreal and terrestrial activity (Doran 1996), and their diets are correspondingly more diverse across trophic levels. This pattern of increased omnivory in conjunction with increased terrestrial activity is also demonstrated in African forests by drills and mandrills (*Mandrillus* spp.), although fruits remain their dominant food source (Astaras et al. 2008; Hoshino 1985). Terrestriality is associated with a more generalized diet in other regions too (Eppley et al. 2022), and a similar situation can be observed in Madagascar with ruffed (*Varecia* spp.) and ring-tailed lemurs (*Lemur catta*). Both combine arboreal and terrestrial locomotion, but, while ruffed lemurs are the most frugivorous of all extant lemurs (88%; Vasey 2000), the more terrestrial ring-tailed lemurs (70% fruit; C. Hladik 1979) are more commonly described as opportunistic omnivores (A. Jolly 2003).

Seeds (Granivory)

While consuming fruits, primates often also ingest seeds either intentionally or unintentionally, particularly where seeds have a thin layer of well-adhered pulp (e.g., *Abuta*) or where seeds are very small (e.g., *Ficus*). With many of the primates mentioned in the previous section, ripe fruit pulp appears to be the main attraction, with smaller seeds often ingested inadvertently and passed whole through the digestive system (Fuzessy et al. 2016), where they may function as dietary roughage (Heymann 2013). In contrast, other primates specifically target either unripe fruits or mature seeds in isolation, even avoiding the fruit pulp, and are well adapted to cope with their associated chemical and physical defenses. For example, the Pitheciidae are among the most frugivorous (*sensu lato*) of all Neotropical primates (Hawes and Peres 2014), but their specialized mandibles and dentition (Kinzey 1992) identify them more accurately as granivores or seed predators (Boyle et al. 2016; Norconk and Conklin-Brittain 2016; Palminteri, Powell, and Peres 2012). This is the case for uakaris (*Cacajao* spp.), bearded saki monkeys (*Chiropotes* spp.), and saki monkeys (*Pithecia* spp.), and to a lesser degree titi monkeys (*Callicebus* spp.) (Barnett, Boyle, Pinto, et al. 2012; Norconk 2011). Pithecid dental adaptations include robust canines, nearly featureless molars, and procumbent incisors that allow pithecids to overcome the mechanical protection of hard and tough pericarps to access mature and young seeds (Kinzey 1992; Norconk and Veres 2011; Norconk, Wright, and Conklin-Brittain 2009). It remains unclear if this is also accompanied by morphological or physiological specialization to aid digestion, particularly to cope with the high levels of toxins found in seeds (Janzen 1971), although this requirement may be reduced by preference for immature seeds or mechanically protected mature seeds (Shaffer 2013a).

Comparable craniodental adaptations for seed eating, such as increased leverage, have also been revealed in the Colobinae, although this subfamily was previously considered predominantly folivorous (Koyabu and Endo 2009, 2010). Indeed, the specialized digestive anatomy of colobines may confer advantages for consuming seeds as well as leaves, with substantial granivory observed both in Asian species such as the red (*Presbytis rubicunda*, 30.1%; G. Davies 1991) and Phayre's leaf monkeys (*Trachypithecus phayrei*, 23.2%; Gupta and Kumar 1994) and in African species such as the black (*Colobus satanas*, 60.1%; Tutin, Ham, et al. 1997), Angola (*C. angolensis*, 50%; Maisels et al. 1994), and king colobus monkeys (*C. polykomos*, 33%; Dasilva 1994). Seed eaters are also represented within the more frugivorous Cercopithecinae—for example, by the sooty (*Cercocebus atys*, 56.9%; McGraw, Vick, and Daegling 2011), Tana river (*C. galeritus*, 37.6%; Wieczkowski 2009), and gray-cheeked mangabeys (*Lophocebus albigena*, 29%–42%; Cash 2013; Poulsen, Clark, and Smith 2001; Tutin, Ham, et al. 1997). In Madagascar, the aye-aye (*Daubentonia madagascariensis*) relies heavily on *Canarium* seeds (47.4%; Sterling et al. 1994), in addition to its better-known manipulative foraging for insect larvae, and Milne-Edwards's sifaka (*Propithecus edwardsi*) spends 35% of feeding time on seeds, in addition to whole fruit (Hemingway 1996).

Faunivory (including Insectivory and Carnivory)

Animal prey is one of the principal food sources complementing fruit. This can take the form of vertebrate prey, typically arboreal amphibians, reptiles, nestling birds, and even other primates, although faunivorous diets more typically comprise invertebrates, with insects in particular consumed by most extant primate species (Hamilton and Busse 1978). Commonly consumed invertebrate taxa include Araneae (spiders and scorpions), Blattodea (cockroaches), Coleoptera (beetles), Diptera (flies), Isoptera (termites), Hemiptera (cicadas and scale insects), Hymenoptera (ants), larval Lepidoptera (caterpillars), and Orthoptera (crickets, grasshoppers, and katydids); their nutritional importance was systematically reviewed by Rothman, Raubenheimer, et al. (2014). This dietary class is frequently termed "insectivory" despite including non-insect arthropod prey. It is often considered to be closest to the ancestral primate diet, being especially common among Tarsiidae and the strepsirrhines (McGrew 2014; Rothman, Raubenheimer, et al. 2014).

Invertebrates (Insectivory)

The highest levels of insectivory are found among the Galagidae, the Lorisidae, and particularly the Tarsiidae. Indeed, tarsiers (*Tarsius* spp., *Carlito syrichta*, and *Cephalopachus bancanus*) are the only primates without any plant parts confirmed in their diet. Adaptations that may relate closely to this diet profile include nocturnal activity, acute hearing, finger pads, and a leaping locomotion. Spectral tarsiers (*T. tarsier*) are the most insectivorous primate species (Gursky 2000), feeding mainly on Lepidoptera, Orthoptera, Hymenoptera, Isoptera, and Coleoptera. Among the galagos or bushbabies (*Galago* spp., *Galagoides* spp., *Euoticus* spp., and *Otolemur* spp.), the highest levels of insectivory are seen in dwarf galagos (*Galagoides* spp., 52%–70%; Charles-Dominique 1977; Nekaris and Bearder 2007), which are the smallest-bodied bushbabies and display some similar characteristics to tarsiers. Similarly, within the Lorisidae, it is the smaller-bodied slender lorises (*Loris* spp., 95%–100%) and angwantibos (*Arctocebus* spp., 87%) that represent the most specialized insectivorous primates in Asia and Africa, respectively (Nekaris and Bearder 2007). In contrast to galagos, which use their long tails for leaping to catch prey and escape predators, lorisids have vestigial tails and are adapted instead for stretching between branches and using camouflage to avoid detection (Nekaris and Bearder 2007). The loris diet is also remarkable among primates for including a high proportion of toxic prey, which may contribute to chemical defenses against predators, parasites, and competing conspecifics (Nekaris, Moore, et al. 2013).

Predation of invertebrates appears to be prohibitively inefficient for primates of much larger size, unless they can be accessed in large quantities with minimal effort. For example, beyond opportunistic feeding on scorpions, baboons preferentially consume locusts (*Schistocerca* spp.) or scale insects (Hemiptera: Coccidea) during massive population outbreaks but not at other times (Hamilton et al. 1978). The only invertebrates that provide sufficient returns to justify regular exploitation by large primates are colonial hymenopterans and isopterans, with chimpanzees well documented to use tools to conduct ant-dipping or termite-fishing (Boesch and Boesch 1990). The relationship between body size and insectivory is less clear in Madagascar, as even the smallest-bodied Cheirogaleidae, including mouse lemurs (*Microcebus* spp.) and hairy-eared dwarf lemur (*Allocebus trichotis*), consume a variety of additional food sources (Biebouw 2009; Dammhahn and Kappeler 2008). In addition to the unique percussive foraging for insect larvae by aye-ayes (Erickson 1991), another interesting dietary strategy found in lemurs is the exploitation of homopteran secretions by mouse lemurs and giant mouse lemurs (*Mirza* spp.), although this feeding behavior could be more accurately aligned with exudativory than insectivory (Dammhahn and Kappeler 2008; Petter et al. 1971).

Insectivory in the Paleotropics is typically more prevalent in nocturnal primates, as evidenced by the examples above, but Neotropical insectivores do not continue this trend and are commonly active during the day, as in the Callitrichidae and Cebidae (Mittermeier and Van Roosmalen 1981; Terborgh 1983). Orthopterans, in particular katydids (Tettigonidae), represent the main prey of many callitrichids such as tamarins (*Saguinus* spp.; Peres 1992b; A. Smith 2000), while capuchins (*Cebus* spp. and *Sapajus* spp.) and squirrel monkeys (*Saimiri* spp.) include a variety of invertebrates in their wide-ranging diets (Janson and Boinski 1992), in addition to vertebrate prey. There is also a potential crossover with frugivory, as invertebrate larvae may be ingested either deliberately or inadvertently during the consumption of fruit. Larvae are more likely to be encountered in unripe fruit since they typically emerge before fruit ripen, and there is some evidence that unripe-fruit specialists such as uakaris positively select infested fruits (Barnett, Ronchi-Teles, Silva, et al. 2017). If confirmed, this may demonstrate a specialized strategy to overcome the high time and energy costs more commonly associated with searching and hunting for invertebrate prey (Redford et al. 1984). Consumption of insect-infested fruit has now also been documented for spider monkeys, potentially showing the importance of larvae for ripe-fruit specialists too (dos Santos-Barnett et al. 2022).

Vertebrate Meat (Carnivory)

Hunting of vertebrate prey is far less common than insectivory but is also widespread among primates, even in taxa that are primarily frugivorous or folivorous (Butynski 1982). Small-bodied animals such as bats, frogs, lizards (including chameleons and agamids), and small birds (and eggs) are prey items for primates as taxonomically and morphologically diverse as mouse lemurs (Dammhahn and Kappeler 2008), galagos and pottos (*Perodicticus* spp.; Charles-Dominique 1977), lorises (Nekaris, Moore, et al. 2013; Nekaris and Rasmussen 2003), tamarins (*Saguinus* spp.; A. Smith 2000), macaques (*Macaca* spp.; D. Hill 1997), and gibbons (P. Fan, Ai, et al. 2013; P. Fan, Fei, et al. 2011). Even the tiny Horsfield's tarsier (*Cephalopachus bancanus*) has been observed eating small vertebrates such as bats, birds, and snakes in addition to insect prey (Nimeitz 1979).

The slow loris (*Nycticebus* spp.) is the only primate to produce venom, but, although this could be used to help kill small prey, no supporting evidence has yet been found for this hypothesis (Nekaris, Moore, et al. 2013). Relevant adaptations instead include powerful jaws and sharp teeth, which are also found in faunivorous primates in the Americas such as the capuchins (*Cebus* spp. and *Sapajus* spp.) and squirrel monkeys (*Saimiri* spp.), although perhaps the most important platyrrhine adaptations for the capture of small vertebrates are behavioral rather than morphological (Janson and Boinski 1992). Indeed, groups of capuchins conduct coordinated chases of squirrels, although their common style of predation, described as destructive foraging (Fedigan 1990), has more in common with the close-range detection typical of small-bodied primates than the stalking pursuit and active chases typical of larger-bodied taxa such as baboons (*Papio* spp.), chimpanzees (*Pan troglodytes*), and bonobos (*P. paniscus*). Olive baboons (*Papio anubis*), for example, conduct coordinated hunts of Thomson's gazelles (*Eudorcas thomsonii*) and other small antelopes (Strum 1981), and, while the most common prey of chimpanzees is the red colobus (*Piliocolobus tephrosceles*), they have also been recorded to hunt duikers (Cephalophinae), bushbucks (*Tragelaphus* spp.), bushpigs (*Potamochoerus larvatus*), and baboons (Stanford 1998). Carnivory was previously presumed to be much rarer or absent in bonobos, but this likely resulted from comparably undetailed data, as confirmed bonobo prey now also include duikers, flying squirrels (Anomaluridae), and monkeys (e.g., *Lophocebus aterrimus*; Hohmann and Fruth 2007; Ihobe 1992; Surbeck et al. 2009).

Folivory (including Graminivory)

In addition to insect prey, leaves are the other major dietary complement to fruit. Given the almost ubiquitous importance of fruit, the early frugivore/folivore/insectivore models have since been refined to classify primates as either frugivore-insectivores or frugivore-folivores (Rosenberger 1992). However, leaf material often represents the principal dietary component in some primates, and this is typically more prevalent in larger-bodied species. A distinction is normally made between folivores that consume either new or mature leaves from woody plants and graminivores (not to be confused with granivores), which feed on grasses. With the latter, plant parts such as flowers or seeds are consumed together with leaf blades and are therefore not considered separately.

Foliage

The most folivorous primates are probably the Indriidae, comprising the sifakas (*Propithecus* spp.), woolly lemurs (*Avahi* spp.), and indri (*Indri indri*). In apparent contradiction of the Jarman-Bell principle (Gaulin 1979), the highest levels of folivory in this case are found in the smallest of these, the woolly lemurs, which overlook fruits when available and feed almost exclusively on leaves (95%), supplemented only by flowers (Norscia, Ramanamanjato, and Ganzhorn 2012). The indri, the largest-bodied extant species in the family, is also a young-leaf specialist (72%) with associated dental and digestive adaptations (Powzyk and Mowry 2006), while sifakas (35%–65% leaves; Sato, Santini, et al. 2016) show a greater preference for fruits and seeds, as reflected in their shorter large intestine and longer small intestine compared to the indri (Powzyk and Mowry 2003). Like the woolly lemurs, the sportive lemurs (*Lepilemur* spp.) are nocturnal and also highly folivorous (90%; Dröscher and Kappeler 2014), again despite the restrictions imposed on digestion efficiency by their small body size (Dröscher et al. 2016). Locomotion style and ranging behavior are likely mechanisms for reducing potential competition with indris (R. Warren and Crompton 1997).

Beyond Madagascar, other well-known folivores are the Colobinae, represented in Asia by the proboscis monkey (*Nasalis larvatus*), gray langurs (*Semnopithecus* spp.), lutungs (*Trachypithecus* spp.), douc langurs (*Pygathrix* spp.), pig-tailed langurs (*Simias* spp.), and surilis (*Presbytis* spp.; Erb et al. 2012; Vandercone et al. 2012; Workman 2010), and in Africa by the colobus monkeys (*Colobus* spp., *Piliocolobus* spp., and *Procolobus verus*; A. Davies, Oates, and Dasilva 1999). Morphological differences in dental and facial features related to chewing ability appear to separate colobine seed-eaters from leaf specialists such as *T. obscurus*, *T. vetulus*, *C. guereza*, *Piliocolobus badius*, and *Procolobus verus* (Koyabu and Endo 2009, 2010). Because of the challenges in digesting leaf material posed by numerous plant structural and chemical defenses, folivorous primates also exhibit gastrointestinal and behavioral adaptations; one example of a behavioral adaptation is often focusing on new leaf growth, which tends to be less challenging to digest (Chivers and Hladik 1980; C. Hladik 1978).

It is partly because of the understanding that complex guts, capable of digesting leaves with high levels of fiber, tannins, and toxins, are possible only at larger body sizes (Chivers 1994) that the levels of folivory exhibited by the relatively small-bodied lemurs is so remarkable (C. Hladik 1979). The more usual scenario is demonstrated in the platyrrhines, where the howler monkeys (*Alouatta* spp.) and muriquis (*Brachyteles* spp.) are the most folivorous, although similarly sized atelids (*Ateles* spp. and *Lagothrix* spp.) are much more frugivorous. Digestive differences between howler and spider monkeys are even more pronounced than between indris and sifakas (Milton 1981b; Powzyk and Mowry 2006). However, howler monkey populations vary along the folivory-frugivory gradient in relation to fruit availability (Dias and Rangel-Negrín 2015), and digestive passage rate and metabolism suggest that folivory in muriquis is a secondary adaptation following range restriction (Ford and Davis 1992; Milton 1984; Talebi et al. 2005). No extant New World primate is much larger than 10 kg, although it is unclear whether the extinction of larger-bodied platyrrhines, such as *Protopithecus brasiliensis*, can help explain the apparently vacant niche for obligate folivores in the Neotropics (Rosenberger, Cooke, et al. 2015).

Large-bodied folivores are best exemplified in extant primates by the gorillas (*Gorilla* spp.), which have a highly selective folivorous diet including stems, pith bamboo, and herbaceous shoots, in addition to other foliage. They consume a diverse range of fruits too, and, as in howler monkeys, there is substantial variation along the folivore-frugivore spectrum. Eastern gorillas (*G. beringei*) are much more folivorous than the more frugivorous western gorillas (*G. gorilla*), and there is variation between different conspecific populations, particularly among eastern gorillas (Rogers, Abernethy, et al. 2004; Rothman, Plumptre, et al. 2007).

Grasses

Grass can be included as a food item by baboons (Hamilton et al. 1978), macaques (D. Hill 1997), and vervet monkeys (*Chlorocebus pygerythrus*; P. Lee and Hauser 1998), although usually only as part of a more generalist omnivorous diet. Olive baboons (*Papio anubis*) have been documented to spend over half their feeding time grazing on grasses (DeVore and Washburn 1963; R. S. O. Harding 1976), but in most studies this value is lower, with fruit eaten preferentially whenever available (e.g., Okecha and Newton-Fisher 2006). The only primates to consistently feed primarily on grasses are geladas (*Theropithecus gelada*), which use a shuffling sitting position to feed and represent true grazers comparable to ungulates (Dunbar and Bose 1991; Iwamoto and Dunbar 1983).

In contrast to the gelada grazing lifestyle on the open plains, the only other primates that may count as graminivorous are bamboo lemurs (*Hapalemur* spp.) and greater bamboo lemurs (*Prolemur simus*), which specialize, as their name suggests, on bamboo (Poaceae). Indeed, the greater bamboo lemur feeds almost entirely on young and mature leaves from a single bamboo (*Cathariostachys madagascariensis*), while the smaller bamboo lemurs (*H. aureus* and *H. griseus*) include up to four bamboo species, preferentially selecting young leaf bases and new sprouts, supplemented by nonbamboo foliage and fruits (C. Tan 1999). The Lac Alaotra bamboo lemur (*H. aloatrensis*) is unique among the bamboo lemurs, and indeed primates worldwide, in having a distribution restricted to reed and papyrus marshland, where it subsists almost exclusively on just four species of nonbamboo grasses and sedges (Mutschler 2002). The closely related southern gentle lemur (*H. meridionalis*) has now been confirmed to replace bamboo with other terrestrial grasses (83%) from the swamp and marsh around littoral forest (Eppley, Verjans, and Donati 2011).

Florivory (including Nectarivory and Palynivory)

Consumption of flowers is less common, in comparison to other dietary components such as fruit, invertebrates, and leaves, but has been documented in 165 primate species, and the importance of florivory is still likely to have been underestimated, at least seasonally (Heymann 2011). Flowers are low in secondary metabolites and high in protein content, with levels sometimes higher than fruit and comparable to leaves (J. Lambert and Rothman 2015; Oftedal et al. 1991). They provide various valuable resources, including nectar, pollen, and petals or other flower parts, but there is only limited evidence for adaptations in tongue and snout morphology for specialization in nectarivory and similarly mainly indirect evidence for the role of primates in pollination (Heymann 2011). In most primates, flowers tend to contribute less than other food categories, although Heymann's thorough review revealed flowers or nectar to rank second in the diet of some primates generally classified as frugivore-folivores or frugivore-faunivores, and there are a few cases where flowers may even represent the most important food resource, at least seasonally, if not overall (Powzyk and Mowry 2003; Terborgh 1983; F. Wiens et al. 2006).

Prominent consumers of flower resources in Asia include the slow lorises (e.g., *Nycticebus coucang*, 31.7%; F. Wiens et al. 2006), pig-tailed langur (*Simias concolor*, 25.5%; Erb et al. 2012), gray langurs (e.g., *Semnopithecus entellus*, 18.7%; Vandercone et al. 2012), and douc langurs (e.g., *Pygathrix nigripes*, 3.9–14.6%; Duc et al. 2009; Phiapalath et al. 2011). In Africa, flowers can represent a substantial proportion of the overall diet of red colobuses (e.g., *Piliocolobus tephrosceles* 30.1%; Wachter et al. 1997), while galagos and pottos have been recorded as potential pollinators for several plants (Heymann 2011). Larger-bodied primates are more likely to consume whole flowers; nectar-feeding is typically the preserve of smaller-bodied species (Heymann 2011). This tendency is well demonstrated in Madagascar, with the sifakas preferring whole flowers (Powzyk and Mowry 2003) and the dwarf lemurs nectar (Lahann 2007). Similarly, in the Americas, smaller-bodied lion tamarins (*Leontopithecus* spp.; Dietz, Peres, and Pinder 1997) and night monkeys (*Aotus* spp.; Marín-Gómez 2008) are more likely to feed on nectar rather than whole flowers, although even larger-bodied species such as spider monkeys and muriquis have also been recorded feeding on nectar (Heymann 2011).

Florivory has been described as the intersection between pollination and herbivory, and it is important to distinguish between the destructive feeding behavior that damages floral buds or mature flowers and nondestructive flower visits that may provide pollination services (McCall and Irwin 2006). However, as for the comparable situation of seed dispersal versus seed predation (J. Lambert and Garber 1998), this is not straightforward to achieve and will be specific to each primate-plant interaction. For example, white-faced capuchins (*Cebus imitator*) have been proposed as a likely pollinator of *Luhea speciosa*, despite detrimental effects of their flower feeding in other plant species (Hogan, Melin, et al. 2016), and the few studies to date suggest that florivory in primates is more likely to inhibit plant reproduction (Chapman, Bonnell, Gogarten, et al. 2013).

Exudativory (including Gumnivory/Gummivory)

Gums and saps represent another group of resources that have been previously overlooked when considering the major dietary categories of frugivory, folivory, and faunivory (L. Nash 1986). A comprehensive review of exudate feeding in primates (A. Smith 2010a) shows consumption by 69 species, while acknowledging this as underrepresented due to observational challenges associated particularly with cathemeral and nocturnal species. As for florivory, the scant study effort accumulated to date probably explains a large degree of the variation in exudate feeding records across species, but the identification of those species considered to be specialist gummivores is likely to be accurate. Adaptations for the exploitation of exudates include sharp claws or nails to scale vertical tree trunks, as shown in the callitrichids (e.g., *Cebuella pygmaea*, *Callithrix* spp., *Mico* spp., and *Saguinus* spp.), the galagos (e.g., *Galago* spp., *Euoticus* spp., and *Otolemur* spp.), and some lemurs (e.g., *Phaner* spp.), and modified dental morphology to allow either gouging of trunks or scraping of gum (Burrows et al. 2015; L. Nash 1986; J. Smith and Smith 2013).

There is an apparent distinction between species that opportunistically exploit gum deposits resulting from insect or mechanical damage and those that gouge holes in trees to stimulate the flow directly (L. Nash 1986). The pygmy marmoset (*Callithrix pygmaea*) is a specialized tree-gouging gummivore (76.7%; Yépez et al. 2005), with *Callithrix*, *Mico*, *Euoticus*, and *Phaner* also capable of gouging to stimulate exudate flow. Seven species of slender (*Loris*) and slow loris (*Nycticebus*) are now also recognized to procure exudates via active gouging (Starr and Nekaris 2013). In the slender loris, exudates represent a supplement (2.8%; Nekaris and Jayewardene 2003; Nekaris and Rasmussen 2003), but they can be the dominant food source for the slow loris (43%–87%; Starr and Nekaris 2013). Exudates are considered poor-quality food sources, low in protein and lipids but high in carbohydrates of variable digestibility (Hausfater and

Bearce 1976; C. Hladik et al. 1980; L. Nash 1986), and their digestion/fermentation is aided by low food passage and metabolic rates (Cabana et al. 2017c) and gastrointestinal adaptations such as an enlarged caecum (Cabana et al. 2017a). Larger-bodied species can therefore specialize to a greater degree in exudativory (e.g., *N. bengalensis*, 1.1–2.4 kg, 67.3%–94.3%; Swapna et al. 2010), while increased insectivory may compensate for the lower contribution of exudates to the diet of smaller-bodied species (e.g., *N. pygmaeus*, 0.2–0.7 kg, 50%; Starr and Nekaris 2013).

It is important to note that exudates can also account for a significant proportion of the diet in less specialized species that lack the ability to gouge. These include relatively large-bodied exudativores such as the patas monkey (*Erythrocebus patas*, max. 13 kg, 39.4%; Isbell 1998), vervet monkey (*C. pygerythrus*, 8 kg; P. Lee and Hauser 1998), and greater galagos (*Otolemur* spp., 1.8 kg) and some of the smallest, such as the mouse lemurs that, in addition to plant exudates, include homopteran insect secretions in their diet (Dammhahn and Kappeler 2008). Tamarins (*Saguinus* spp.) are also unable to stimulate gum production (A. Smith 2010b) and therefore utilize holes gouged by pygmy marmosets (Ramírez 1985) or rely on other sources of exudates, including those produced by leguminous pods such as *Parkia* spp. (Fabaceae), which are also heavily exploited by other callitrichids and a wide range of other Neotropical primates (A. Smith 2010a). *Parkia* seedpod exudates represent a unique and important resource and, as opposed to gums produced as a result of damage to the plant, are likely to serve as an attractant to encourage seed dispersal (H. Hopkins 1983; Peres 2000b).

MINOR DIETARY CLASSES

Although the sections above cover the main food sources for the majority of primates, they do not represent an exhaustive list of all documented dietary items. An obvious omission is the dependence of all infant primates on maternal milk (Hinde and Milligan 2011; Hinde et al., this volume). However, many additional foods are also occasionally included in adult primate diets, often as supplementary items, and can be consumed opportunistically or actively sought out. These include plant material such as buds, bark, pith, roots, stems, stalks, and wood, as well as fungi, lichens, honey, birds' eggs, dung, and soil. The consumption of such resources typically occurs as rare events, but several long-term studies have now revealed their importance to primate diets, at least on a seasonal basis.

Feeding on fungi (mycophagy) has been shown to be relatively important within the callitrichids, comprising up to 65% of total feeding records in buffy-headed marmosets (*Callithrix flaviceps*; Hilário and Ferrari 2010), and up to 63% of dry-season records in Goeldi's monkeys (*Callimico goeldi*; L. Porter 2001). Lichens have been found to be similarly important to snub-nosed monkeys, comprising the primary food for *Rhinopithecus bieti* (Kirkpatrick et al. 1998) and *R. roxellana* (29.9%–50%; Guo, Li, and Watanabe 2007), particularly in the winter. In contrast, honey is an infrequent but highly valued resource for chimpanzees, as evidenced by the range of tool-using strategies developed for its exploitation (Sanz and Morgan 2009). Consumption of decaying wood, observed in spider monkeys, chimpanzees, and gorillas, can provide a source of sodium and other minerals (Chaves, Stoner, Ángeles-Campos, Arroyo-Rodríguez, et al. 2011; Reynolds, Lloyd, et al. 2009; Rothman, Van Soest, and Pell 2006). Reingestion of feces (caecotrophy) by sportive lemurs (*Lepilemur* spp.) is considered to be an adaptive strategy to compensate for the constraints imposed by their small body size on their folivorous lifestyle (Dröscher et al. 2016). Finally, soil ingestion (geophagy) is a widespread behavior with various suggested health and nutrition benefits (Krishnamani and Mahaney 2000). Switching between different food groups to include these less common dietary items may partly reflect a deliberate strategy to achieve an optimal diet with a properly balanced range of nutrients (Felton, Felton, Lindenmayer, et al. 2009) or to avoid the suite of chemical compounds employed by plants in their defense (Koricheva and Barton 2012). Alternatively, such feeding behavior may reflect necessity, in response to seasonal dips in the availability of preferred food sources, and some of these minor dietary classes may therefore qualify as fallback foods (Grueter, Li, et al., "Fallback Foods," 2009; A. Marshall, Boyko, et al. 2009). However, differentiating between these forces is not easy, and classifying foods as either optimal or fallback may not even be advisable considering the dynamic value of food items relative to other sources (J. Lambert and Rothman 2015). Regardless of the mechanism or combination of mechanisms, the ability of primates to shift their diet according to a range of variables including spatial and temporal availability (van Schaik, Terborgh, and Wright 1993) and pressures from predation (Isbell 1994a) and competition (Janson 1988b) demonstrates a remarkable flexibility that is one of the distinguishing characteristics of this order.

OMNIVORY

There is a spectrum of specialization across each of the dietary modes described here, including the major categories of frugivory, faunivory, and folivory and some of the more minor categories. At one extreme lie those species that are highly specialized within one particular category, with the caveat that primates rarely fit neatly into any one category and tend to exploit resources with diverse characteristics from multiple categories. At the other end of the scale are those species with the broadest diets, with the term "omnivory" commonly used to describe diets spanning more than one trophic level.

Among the Platyrrhini, the best example of a generalist omnivore is the brown capuchin (*Sapajus apella*), which consumes both invertebrate and vertebrate prey in addition to leaves, flowers, fruits, and seeds. Macaques (*Macaca* spp.) are among the most generalist consumers in Asia (Yeager 1996), while in Africa, the kipunji (*Rungwecebus kipunji*) is recognized to have a diverse diet more in common with baboons (*Papio* spp.) than with the more closely related mangabeys (*Cercocebus* spp. and *Lophocebus* spp.; Davenport et al. 2010). Baboons are generally recognized as the least specialized nonhuman primates, representing the most extreme case of omnivory, with a variable diet including bulbs and tubers in addition to fruit, invertebrates, vertebrates, leaves, and flowers (Norton et al. 1987; Whiten, Byrne, et al. 1991).

However, if "omnivore" is defined to include any animal that exploits more than one trophic level as a food source, all primates can be classified as omnivores (R. A. Harding 1981; Milton 1987; Sussman 1987). The considerable variation described across and within primate diets indicates that the omnivore label is not particularly informative. All primates, including great apes, have been observed to consume at least insects in addition to plant material (Hamilton and Busse 1978), and this is not restricted to inadvertent consumption along with vegetation or fruits (Harcourt and Harcourt 1984). Ants have been recorded in the diet of mountain gorillas (*G. b. beringei*; Ganas and Robbins 2004), to accompany the more commonly recognized exploitation of social insects by other gorillas and chimpanzees (Cipolletta et al. 2007; Deblauwe and Janssens 2008; Tutin and Fernandez 1992), and orangutans have been observed to eat meat on rare occasions (Hardus et al. 2012) in addition to insects (Galdikas 1988).

Although such examples of rarer feeding events may still be nutritionally important, a more restrictive definition of omnivory would limit the use of this term to species that can digest substantial amounts of material across both animal and plant matter (Chivers 1994). Under this definition, rather than all primates qualifying as true omnivores, perhaps only humans can be judged to have achieved this state, and then only via the process of cooking and processing food prior to ingestion (Chivers 1998). Apparently contradicting positions on the prevalence of omnivory in primates may therefore stem entirely from different definitions of the term.

CONTINUUM OF DIETARY STRATEGIES

As made apparent in the above descriptions of major dietary categories and the difficulties in defining omnivory, trophic classification of species in such broad strokes is perhaps reaching the limit of its practical usefulness. Most species usually make use of a variety of food resources that vary across space and time in relation to environmental heterogeneity (e.g., Rothman, Plumptre, et al. 2007; Russo et al. 2005) and seasonal fluctuations (e.g., Cabana et al. 2017b; P. Fan, Ai, et al. 2013; Paim et al. 2017; Poulsen, Clark, and Smith 2001; Talebi et al. 2005). Even those displaying the highest degree of specialization, with adaptations to focus primarily on a single resource, are not without dietary flexibility.

For example, although strictly insectivorous, spectral tarsiers modify their diet according to seasonal resource availability by consuming more Coleoptera and Hymenoptera and fewer Orthoptera and Lepidoptera during the dry season when resource abundance is low (Gursky 2000). Similarly, although highly specialized graminivores, bamboo lemurs consume more liana bamboo during the dry, cool season, perhaps in response to seasonal new growth and reproductive cycles (Overdorff et al. 1997). Specialist frugivores such as orangutans also face phenological fluctuations in fruit abundance, particularly during periods of low availability that follow mast fruiting events, and will exploit bark at sites if fruiting figs are also absent (Knott 1998; Wich, Utami-Atmoko, Mitra Setia, Djoyosudharmo, et al. 2006). Most species are therefore generalists to varying degrees, and the boundaries between the traditional dietary classes are also inherently fuzzy. This calls for a new interpretation and understanding of primate diets as points along a continuous spectrum, particularly as available data from multiple species, populations, study sites, and study periods can increasingly be efficiently compiled and analyzed.

A comprehensive quantitative summary of primate diets has yet to be achieved on a global scale, partly as a result of varying field methods employed by primatologists and severe inequalities and systematic biases in the distribution of sampling effort. However, an example of what is increasingly possible using a growing array of primate field studies conducted worldwide is provided by a compilation of data from studies across Central and South America, taking into account biases in sampling effort (Hawes et al. 2013), to produce the first systematic review of platyrrhine diets (Hawes and Peres 2014). The following section expands on one component of this approach to present an updated assessment of the proportional contributions of different food sources to primate genera worldwide, which can be used to explore the role of potential drivers of dietary profiles, such as body size (Chivers 1998).

DRIVERS OF DIETARY PROFILES

Geographical influences such as climate and habitat type can impose limits on the dietary lifestyles available to an animal, with specialized frugivory, for example, restricted to tropical and subtropical rainforests (Pineda-Munoz et al. 2016). Eventual niche partitioning within each habitat is also a response to any potentially competing species. However, the anatomical constraint of body size is often of overriding importance in determining an animal's trophic ecology, including dietary choices (Chivers 1998; Ford and Davis 1992; Milton and May 1976; Pineda-Munoz et al. 2016; J. G. Robinson and Redford 1986). This is particularly interesting because of the many other traits that are related to body size, such as metabolic rate, fecundity, population density, home range size, locomotion, vertical stratification of forest use, and social behavior (Clutton-Brock and Harvey 1977; Milton and May 1976; J. G. Robinson and Redford 1986; Rosenberger 1992), although traits may also vary along a fast-slow continuum even when controlling for body size (Bielby et al. 2007).

Insects provide a high-quality source of nutrients and calories (with variation by insect species and developmental stage), ideal for the high metabolic requirements of small primates (Kleiber 1947). In contrast, large primates require a greater bulk food intake but have lower basal metabolic rates and lower energy demands per unit of body mass, thus enabling a diet based on lower-density energy sources (Fleagle 1998) but precluding reliance on widely diffuse, small, high-energy resources such as insects. Moreover, since primates lack the enzyme cellulase, only large-bodied species are able to successfully exploit large quantities of foliage thanks to either complex foreguts or larger, more complex hindguts with long enough retention times to allow microbial fermentation and access to energy from digestible fiber (Chivers 1994). However, large-bodied primates lacking prey acquisition technology are virtually unable to consume large amounts of arthropods because of either the prohibitive pursuit-and-handling time involved in capturing widely dispersed small prey or anatomical and locomotor constraints on the kinematics of arthropod capture (Peres 1994a; Terborgh 1983). Body size constraints therefore appear to impose both upper limits on insectivory and lower limits on folivory, leading to the proposed dichotomy between frugivore-insectivores and frugivore-folivores (Rosenberger 1992) as predicted by Kay's threshold (Kay 1984).

The fundamental association between body size and basal metabolic rate led to the development of the Jarman-Bell principle, which has been used to interpret nutritional profiles in primates (Gaulin 1979). In general, primates fit the expected trend well, with small-bodied species focusing on rare, high-quality foods and large-bodied species focusing on more common, lower-quality foods. This is exemplified in the galagos, where the smallest-bodied species (*Galagoides* spp.) are more insectivorous, medium-sized species (*Galago* spp. and *Euoticus* spp.) consume more exudates, and the largest (*Sciurocheirus* spp. and *Otolemur* spp.) are more frugivorous (Nekaris and Bearder 2007). However, there are exceptions, particularly within the lemurs, where body size is not as useful in predicting diet. As noted in the sections above, the smallest lemurs (*Allocebus trichotis* and *Microcebus* spp.) are not strict insectivores, while the most folivorous species (*Lepilemur* spp.) are surprisingly small-bodied, and the bamboo lemurs (*Hapalemur* spp.) are extreme outliers, surviving on a diet of grasses or sedges. Such deviations from the expected body size–diet relationship are likely to be facilitated by behavioral or physiological adaptations, including torpor (Schmid and Speakman 2009), and many of the peculiarities observed in lemurs may be related to Madagascar's unique ecological conditions.

Exceptions to typical body size allometries are not, however, limited to Madagascar. For example, some of the largest-bodied primates, such as chimpanzees, gorillas, and orangutans, do not always consume low-quality foods as predicted (Boesch and Boesch 1989; Ganas and Robbins 2004; Hardus et al. 2012). To address the disparities observed, revisions to such relatively simplistic models are now needed, recognizing that body size is not the only important variable determining dietary choices. Understanding diet through models that assume the optimization of a single currency, such as energy or protein, is also increasingly considered insufficient, and recent methods take a more multivariate approach to consider how primate diets are adapted to balance multiple demands, including energy, protein, micronutrients, and toxins (Amato, Ulanov, et al. 2017; Ganzhorn, Arrigo-Nelson, Carrari, et al. 2017; C. A. Johnson, Raubenheimer, Chapman, et al. 2017; Righini 2017; Windley et al. 2022).

An appreciation of the synergies and trade-offs managed by primates when making foraging decisions that shape their dietary profile is provided by the nutritional geometry framework (Raubenheimer, Machovsky-Capuska, Chapman, et al. 2015; S. Simpson and Raubenheimer 1993; Raubenheimer, this volume). One of the nuances revealed by this approach is the degree of intraspecific variation in energy and macronutrient intake, particularly among age-sex classes; immature female orangutans had higher energy and protein intake than other age-sex classes, and lactating female sifakas had higher energy and macronutrient intake than males (Koch et al. 2017; E. Vogel et al. 2017). Such individual differences in energetic and nutritional requirements have long been described, particularly in relation to reproduction (e.g., M. J. S. Harrison 1983; P. Whitten 1983a), but intraspecific variation remains relatively underexplored and may represent one of the opportunities to deepen our understanding of primate feeding strategies.

Despite such complexities, there is perhaps still a complementary place for more simplistic models. While competition between multiple dietary currencies may provide more detailed mechanistic explanations to better understand feeding behavior, functional explanations such as optimal foraging theory can still be useful (S. Simpson and Raubenheimer 1993). Similarly, while it may be more realistic to value complex interactions between diet and multiple other traits including gut morphology, group size, and locomotion, the generalizations made from the relationship between mammalian diets and body size remain revealing (Pineda-Munoz et al. 2016; S. Price and Hopkins 2015). Indeed, examining the similarities and differences between models will potentially lead to new insights. Predictions from the Jarman-Bell principle and Kay's threshold can therefore still provide useful starting points to examine dietary profiles and have been best tested in the platyrrhine radiation, which, across their contemporary spectrum of body size, collectively spans a broad array of dietary classes (see below).

A CASE STUDY IN PLATYRRHINE FRUGIVORY

Acknowledging the diverse nature of primate diets as outlined in this chapter, it is clear that fruits are the dominant food source overall. Most primates consume at least some fruit, which in a large proportion of species represents at least half of observed diets or feeding time or both (Hohmann 2009). This is the case despite the patchy distribution of fruit resources in space and time (Fleming et al. 1987; Herrera 1998; Levey 1988), which necessitates the use of various alternative food items as dietary supplements. Neotropical primates provide a clear example of this, with all platyrrhine species frugivorous to at least some degree. The proportional balance between fruit and other food classes (notably invertebrates and leaves) is represented in the continuum from ripe fruit-pulp specialists, such as spider and woolly monkeys, to the highly insectivorous squirrel monkeys and the predominantly folivorous howler monkeys.

From an extensive literature review, Hawes and Peres (2014) compiled data from 290 individual studies spanning 42 years of research (1969–2011)—across 164 study sites in 17 Central and South American countries—to comprehensively describe the primate dietary spectrum, particularly its degree of frugivory. Frugivory among Neotropical primates was quantified using three complementary metrics: the total richness of fruit genera consumed, the number of fruit genera consumed per unit sampling effort, and the mean contribution of fruit as a proportion of the full diet (see Hawes and Peres [2014] for further details). This approach was adopted to account for the unequal distribution of sampling effort across the cumulative set of primate feeding observations, which show severe geographic and taxonomic biases in our knowledge, particularly toward large-bodied and widely distributed taxa (Hawes et al. 2013).

Intermediate fruit consumers such as opportunistic folivore-insectivores (e.g., titi and night monkeys) and frugivore-faunivores (e.g., capuchins) were generally represented along a three-dimensional folivore-frugivore-insectivore gradient, but any finer details were lost, particularly in relation to any additional food items such as consumption of seeds by pitheciines, exudates by marmosets and pygmy marmosets, and fungi by Goeldi's monkeys. Despite these fine-scale variations in diet across taxa, the results of this analysis corroborated the long-held hypotheses that broad-scale dietary preferences are closely related to body size as an overarching constraint (W. Calder 1996; Lindstedt and Boyce 1985; Peters 1986). Exudativory and insectivory are heavily represented in small-bodied primates, for which folivory is typically prohibitive due to their relatively short, simple guts (Chivers and Hladik 1980) and higher basal metabolic rates, which demand readily digestible and nutrient-rich food items (McNab 1986). Frugivory was revealed to be largely unimodal and gradually increasing toward a peak at intermediate body sizes before a decline in the two most folivorous large-bodied, prehensile-tailed atelines. Although this approach somewhat reduces the emphasis on rigid dietary classes, this particular frugivory-focused spectrum still represents a simplification of a reality that is frequently much more complex and is not entirely satisfactory in encompassing the full breadth of dietary strategies.

A GLOBAL REVIEW OF PRIMATE DIETS

The most complete multistudy primate dietary reviews currently conducted have focused either on distinct primate taxa, such as howler (Dias and Rangel-Negrín 2015) or spider monkeys (González-Zamora et al. 2009), or on individual food resources, such as flowers (Heymann 2011), foliage (Ganzhorn, Arrigo-Nelson, Carrari, et al. 2017), gums (L. Nash 1986; A. Smith 2010a), insects (Raubenheimer and Rothman 2013; Rothman, Raubenheimer, et al. 2014), or vertebrate prey (Butynski 1982). Collating a comprehensive collective database of all primate feeding ecology studies that includes all dietary classes is a hugely challenging task, considering the impressive volume of primatological field studies (Garber, Estrada, et al. 2009; Kappeler and Watts 2012). Seminal comparative analyses (e.g., Clutton-Brock and Harvey 1977) have perhaps been hampered by the fact that the cumulative efforts by independent researchers with differing objectives, amassed in a gradual and unguided fashion over decades, are inevitably concentrated on some geographic areas and study species, particularly large-bodied and widely distributed taxa (Hawes et al. 2013). Now that these taxonomic and geographic biases are better recognized, future efforts may be directed to focus further on those relatively understudied regions and species in order to complete our understanding of dietary patterns across all primates.

The continental-scale focus of the Neotropical frugivory review (Hawes and Peres 2014) highlighted the importance of such large-scale comparative analyses. Indeed, one step toward a global assessment would be to expand the frugivory review from the Neotropics (Hawes and Peres 2014) to explore the richness of plants exploited for fruit resources by primates worldwide. However, presenting a similarly quantitative review of global primate frugivory is a formidable undertaking, and so the focus here is less ambitious, providing only the relative contribution of each category to the diet, converted to mean values per primate genus. Proportional dietary data from 307 studies (the list of sources is available on request) were collated, covering 72 nonhuman primate genera from 16 families (table 2.1); the only genera without data were *Callibella*, *Mirza*, *Oreonax*, *Phaner*, and *Pseudopotto*. Dietary data were standardized into the following condensed categories: fruits (including ripe and unripe fruit, as well as seeds), foliage (including leaves, grass, and buds), flowers (including nectar), exudates (including gums and sap), animal prey (including invertebrates and vertebrates), and a single category for any other food sources. Finally, to assess the potential relationship between diet and body size, mean body mass was calculated per primate genus

Table 2.1. Mean proportional dietary composition and body mass (kg) for 72 extant nonhuman primate genera worldwide, excluding genera without dietary data (*Mirza, Phaner, Pseudopotto,* and *Oreonax*).

Family	Genus	Body mass (kg)[a]	Proportions of dietary classes[b]					
			Fruit/seeds	Flowers	Foliage	Exudates	Prey	Other
Cheirogaleidae	*Cheirogaleus*	0.28	68.5	23.0	0	1.5	7.0	0
	Microcebus	0.07	16.1	8.0	0	11.7	8.8	58.6
	Allocebus	0.09	14.0	0	0	45.0	41.0	0
Daubentoniidae	*Daubentonia*	2.56	47.4	8.0	0	0	29.4	15.2
Lemuridae	*Lemur*	2.21	70.0	5.0	25.0	0	0	0
	Eulemur	1.83	68.3	6.7	22.1	0	0	3.0
	Varecia	3.53	78.7	9.8	10.7	0	0	1.0
	Hapalemur	1.14	1.9	1.7	82.6	0	0	13.1
	Prolemur	1.73	0.6	4.4	92.7	0	0	2.2
Lepilemuridae	*Lepilemur*	0.76	0.6	4.4	90.1	0	0	2.5
Indriidae	*Indri*	6.34	20.7	6.7	73.7	0	0	0
	Avahi	0.99	0	7.2	92.9	0	0	0
	Propithecus	4.17	40.8	10.6	46.2	0	0	2.0
Lorisidae	*Arctocebus*	0.26	13.5	0	0	0	86.0	0
	Perodicticus	1.05	60.7	0	0	14.3	20.3	3.3
	Loris	0.23	2.1	0	0	0.9	96.7	0
	Nycticebus	0.61	31.2	10.6	0	35.5	21.5	1.3
Galagidae	*Otolemur*	0.96	41.5	0	0	31.0	27.5	0
	Euoticus	0.29	5.0	0	0	75.0	20.0	0
	Sciurocheirus	0.27	64.0	0	0	0	40.0	0
	Galago	0.22	0	0	0	48.0	52.0	0
	Galagoides	0.11	26.3	0	0	3.3	70.0	0
Tarsiidae	*Carlito*	0.13	0	0	0	0	100.0	0
	Cephalopachus	0.12	0	0	0	0	100.0	0
	Tarsius	0.11	0	0	0	0	100.0	0
Callitrichidae	*Cebuella*	0.12	0	0	0	76.7	23.3	0
	Mico	0.37	18.6	0	0	63.3	18.6	0
	Callithrix	0.37	17.9	0	0	48.4	22.8	11
	Callimico	0.53	29.0	0	0	7.3	25	38.3
	Saguinus	0.48	61.6	4.0	0.3	8.1	22.5	1.4
	Leontopithecus	0.58	76.1	10.8	0	3.5	9.5	0
Cebidae	*Cebus*	2.96	67.9	3.2	3.7	0	25.1	0
	Sapajus	2.96	48.5	3.6	16.2	0	22.5	2.8
	Saimiri	1.57	38.1	3.4	0	0	58.5	0
Aotidae	*Aotus*	0.93	76.5	14.0	2.3	0	7.2	0
Pitheciidae	*Callicebus*	1.04	70.8	2.4	21.1	0	3.7	2.0
	Cacajao	3.05	87.2	5.6	2.0	0	3.6	1.7

Family	Genus	Body mass (kg)[a]	Proportions of dietary classes[b]					
			Fruit/seeds	Flowers	Foliage	Exudates	Prey	Other
	Chiropotes	2.86	84.1	9.5	0.7	0	2.9	3
	Pithecia	2.31	85.0	4.2	9.1	0.1	0.8	0.6
Atelidae	Alouatta	6.32	35.1	7.2	54.7	0	0	2.9
	Ateles	8.56	78.3	4.4	12.0	0	0.3	5.1
	Brachyteles	8.84	42.6	9.3	45.5	0	0	2.5
	Lagothrix	8.10	73.4	2.5	12.2	1.5	8.6	2.0
Cercopithecidae	Allenopithecus	4.74	81.0	0	2.0	0	0	0
	Miopithecus	1.75	58.0	0	6.0	0	35.0	0
	Erythrocebus	9.45	5.9	6.6	2.9	39.4	43.1	2.4
	Chlorocebus	4.24	63.0	13.0	7.0	3.0	13.0	1.0
	Cercopithecus	4.38	63.7	4.5	10.4	0.4	13.2	5.6
	Macaca	8.89	43.4	3.7	26.0	0	16.6	10.4
	Lophocebus	6.97	60.4	5.6	5.1	1.2	24.6	3.2
	Rungwecebus		27.0	9.0	34.0	0	2.0	29.0
	Papio	16.27	38.5	7.6	21.0	0.1	3.2	24.5
	Theropithecus	15.35	5.5	0.3	81.5	0	0	12.8
	Cercocebus	7.78	79.5	0	2.0	0	13.0	6.4
	Mandrillus	19.08	82.7	0.4	5.8	0	6.3	4.8
	Colobus	9.10	35.9	3.8	56.8	0.2	1.3	2.3
	Piliocolobus	7.94	37.1	30.1	30.8	0	0	2.0
	Procolobus	4.45	19.5	7.2	73.4	0	0	0
	Trachypithecus	8.45	24.9	6.3	55.9	0	0.1	4.0
	Semnopithecus	12.53	23.5	12.7	55.9	0	0	7.8
	Presbytis	6.28	41.7	6.4	48.3	0	0	3.0
	Pygathrix	9.72	29.3	14.6	54.6	0	0	1.5
	Rhinopithecus	12.46	14.7	1.3	22.1	2.5	1.0	59.4
	Nasalis	15.11	24.3	7.6	62.4	0	0.2	6.0
	Simias	7.98	17.1	25.5	50.5	0	0	1.1
Hylobatidae	Hoolock	7.12	70.9	3.4	21.4	0	4.0	0.8
	Hylobates	5.78	65.2	8.7	17.5	0	8.3	0.4
	Symphalangus	11.30	61.0	1.0	17.0	0	21.0	0
	Nomascus	7.71	60.6	4.6	31.4	0	2.3	1.1
Hominidae	Pongo	56.95	63.6	0	18.3	7.0	6.4	3.7
	Gorilla	124.68	21.7	0.6	61.4	0	0.2	21.6
	Pan	43.50	74.5	2.7	16.4	0	2.3	4.0

a. Mean body mass per primate genus derived from R. Smith and Jungers (1997).
b. Condensed dietary classes show the proportional contributions of fruits (including ripe and unripe fruit pulp, and seeds), foliage (including leaves, grass, and buds), flowers (including nectar), exudates (including gums and sap), animal prey (including invertebrates and vertebrates), and a single category for any other food sources. The full list of references is available on request.

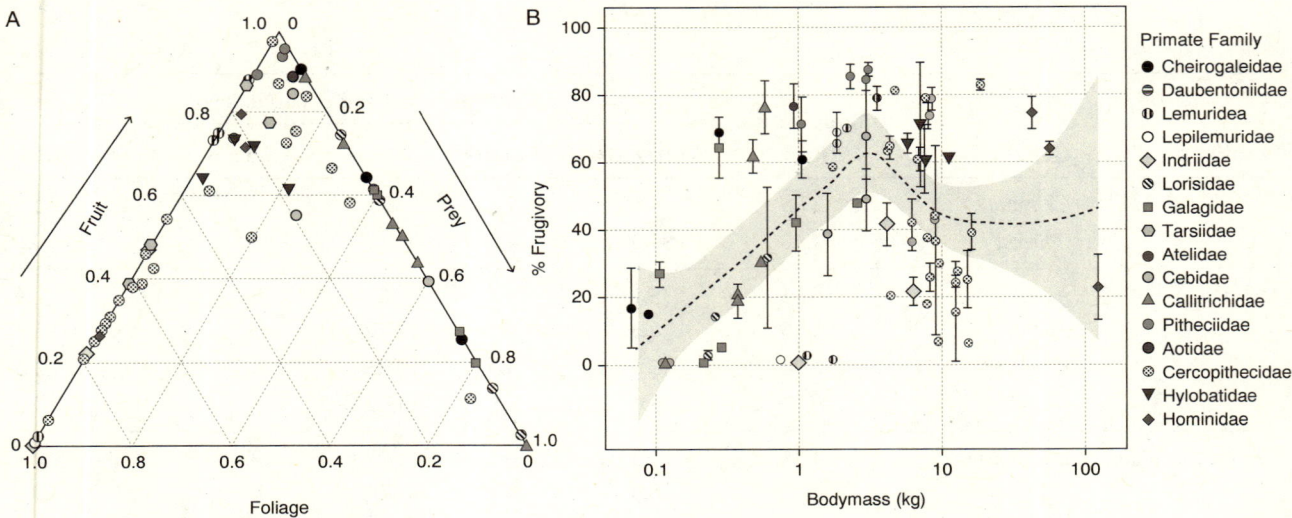

Figure 2.2. Summary of dietary differences across 72 primate genera worldwide. *A*, The relative three-way contributions of folivory, frugivory, and faunivory (excluding other dietary classes). *B*, The relationship between body size and frugivory (percentage of the overall diet including fruits and seeds, mean ± SE). Note that not all overlapping points are fully visible. The fill colors represent primate families, the dashed line represents the smoothed mean, and the gray shading represents 95% confidence intervals. Mean body mass values per genus are derived from R. Smith and Jungers (1997). Details of proportional contributions across all major dietary classes are presented in table 2.1.

(mean across sexes, each species weighted equally; R. Smith and Jungers 1997).

The results show the relative contributions of the three major dietary classes (fruit, foliage, and prey) as a ternary plot (fig. 2.2a) and the relationship between body size and the contribution of frugivory sensu lato (including fruit pulp and seeds) to the diet (fig. 2.2b). The figures were produced in R (R Development Core Team 2014), using the packages ggplot2 (Wickham 2009) and vcd (D. Meyer et al. 2016), and the Interactive Tree of Life (Letunic and Bork 2021), with minor adaptation of the consensus tree from 10kTrees (C. Arnold et al. 2010). These summary results show that while fruits often comprise a unifying resource for all extant primates worldwide, foliage and animal prey are often mutually exclusive (fig. 2.2a; fig. 2.3). They also strengthen our continental-scale review on Neotropical primates (Hawes and Peres 2014), demonstrating the remarkable diversity of primate diets worldwide and providing further support for a midrange peak in frugivory along the body size spectrum. This reinforces the notion that such compilations illustrate the limitations of oversimplified dietary categories and suggests that treating primate feeding strategies as points along a continuum may be more informative.

SUMMARY AND CONCLUSIONS

- The diets of nonhuman primates are diverse, and the relative contributions of different food resources vary spatially and temporally, both between and within species. With the resolution of dietary data from primate feeding studies continuing to improve, the boundaries between strictly defined dietary profiles are becoming less clear cut. While such classes are still useful generalizations, this chapter suggests a flexible continuum of dietary strategies that relates to multiple complex variables including cognitive psychology, digestive morphology and physiology, and locomotor and social behavior, in addition to gross anatomical constraints such as body size. The recent compilation of frugivory studies throughout the Neotropics and this preliminary global review across dietary classes should motivate similar regional assessments of primate diets elsewhere, eventually culminating in a worldwide synthesis and accompanying database of primate feeding studies.
- Primate diets are diverse and highly variable across families, with plant and animal food resources acquired from an array of fruits, seeds, leaves, flowers, exudates,

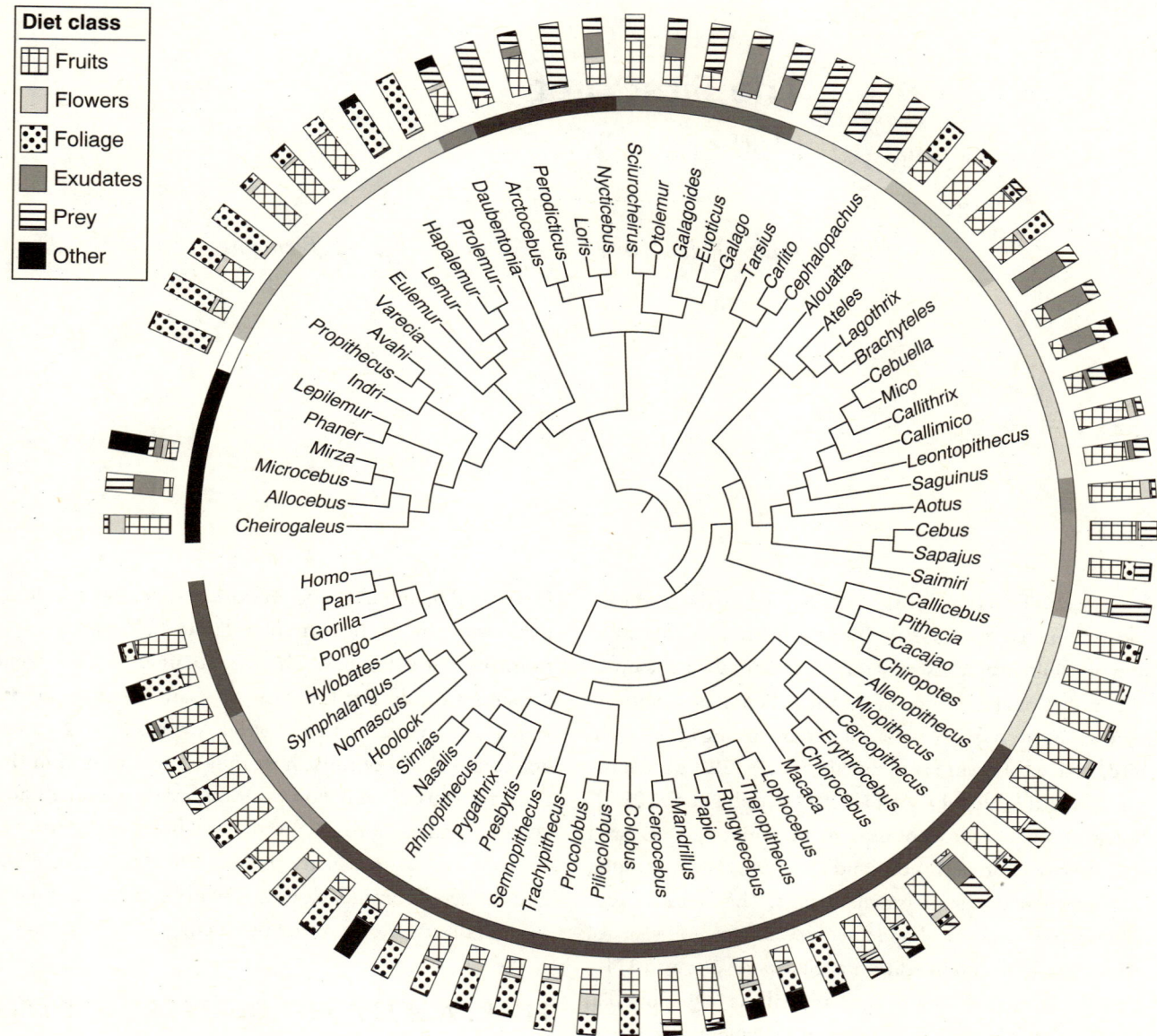

Figure 2.3. Phylogeny of primate genera worldwide (adapted from C. Arnold et al. 2010) showing the relative contributions of major dietary classes, as in table 2.1. The inner strip represents primate families, according to the legend in figure 2.2.

and vertebrate and invertebrate prey, as well as a selection of minor food items.
- Primate diets can be notoriously flexible within species and genera, likely in response to the combined influences of the patchy spatial and temporal distribution of resources and the multiple energetic and nutritional requirements that need to be balanced, in addition to differences among age-sex classes and individuals.
- The dietary flexibility in specialist and generalist feeders, demonstrated in both time and space, highlights the challenge of assigning primate species to fixed, discrete categories, and the concept of a continuum of dietary strategies offers the possibility of a more informative approach.
- The nutritional geometry framework provides a powerful means to explore the multidimensional demands that shape the continuum of dietary profiles observed across extant primates.
- With the improved resolution of cumulative primate dietary field studies, large-scale comparative analyses are increasingly useful tools to synthesize broad patterns and to test alternative models that seek to better understand the mechanisms behind dietary decisions.

3 The First Diet
Mother's Milk

Katie Hinde, Lauren A. Milligan, and Gregory E. Blomquist

Mother's milk is the first food consumed by the developing mammalian neonate. In this way, the nursing dynamic and physiological transfer of milk represent the behavioral and nutritional ecology of infancy. Parental secretions nourishing young can be found, rarely, among insect, fish, bird, and amphibian taxa (Buckley et al. 2010; M. Gillespie et al. 2012; Hosokawa et al. 2012; Kupfer et al. 2006); however, the synthesis of milk by mammary glands (and the consumption of this biofluid by young) is the defining characteristic of the mammalian class. This obligate synapomorphy has been shaped by over 200 million years of evolution. In the sixth edition of *Systema Naturæ* in 1748, Linnaeus described a group of animals as "Quadopedia, *corpus pilosum, pede quartuor, feminae viviparae, lactiferae*" (W. Gregory 1910). A decade and four editions later, this group of animals was identified as a class known as the Mammalia, so named for the specialized mammary glands. Mother's milk is an exquisite triumph of natural selection, affording mammals the ability to rear young in environments that are otherwise inhospitable and providing a context for the elaboration of nuanced social behavior and learning (Hinde and Milligan 2011; Pond 1977). This highly personalized fluid that is simultaneously food, medicine, and signal is essential to sustain and organize the developing neonate (Azad et al. 2021; de Weerth et al. 2022; Hinde and German 2012; Quinn 2021). Lactation is the most energetically costly phase of parental care for nearly all female mammals (Clutton-Brock 1991; Dufour and Sauther 2002; Gittleman and Thompson 1988; M. L. Power and Schulkin 2016). Importantly, energy transmitted to offspring in the form of milk must ultimately come from the maternal diet, whether obtained from current consumption during lactation or drawn from maternal somatic reserves. In this chapter, we review the nutrient composition of milk, how milk is synthesized in the mammary gland, milk composition across mammals and across primates more specifically, and how milk nourishes microbes that nourish the infant. We further highlight limitations of our understanding of primate milks and important directions for future research.

MILK: A NUTRIENT DELIVERY SYSTEM

Milk is food, medicine, and signal, but the nutritional component of milk is most explored across mammals (Akers 2002; Azad et al. 2021; Ballard and Morrow 2013; Donovan 2006; Donovan and Odle 1994; German et al. 2002; Hinde and Milligan 2011; Oftedal 1984; Skibiel et al. 2013). The energetic value of milk comes from fat, protein, and sugar concentrations, with the most energetically dense contribution from fat (Andreas et al. 2015; Hinde and Milligan 2011; Oftedal 1984, 2000). Fat, protein, and sugar concentrations in milk are necessarily interconnected (Oftedal 1984). Lactose, the principal carbohydrate in primate milks (Hinde and Milligan 2011; Oftedal 1984), is the primary osmole pulling water into the milk; lactose synthesis functionally dilutes milk while simultaneously increasing milk volume (Akers 2002;

Hinde 2009; Hinde and Milligan 2011; Hinde, Power, and Oftedal 2009). As such, milk synthesis generally involves a trade-off between milk energy density and milk yield.

Milk fats are used by the developing infant for brain growth and many other functions within the central nervous system (Brenna 2002; S. Carlson 1999; R. Gibson and Makrides 1999; Innis 2003) and, along with protein and carbohydrate, provide energy for behavioral activity (Hinde and Capitanio 2010). Fats are hydrophobic molecules, but milk is mostly water. To maintain a stable emulsion of fat in water required an evolutionarily novel delivery package known as the milk fat globule (MFG; Oftedal 2013; H. Singh and Gallier 2017). At the core of the MFG are the hydrophobic lipids, surrounded by three layers composed of phospholipids and glycoproteins (H. Singh and Gallier 2017) known as the milk fat globule membrane (MFGM). These layers stabilize fats as droplets in the aqueous environment of milk. Moreover, the MFGM, beyond the lipids in the core, may aid lipid digestion and provide antimicrobial functions in the neonate (Hernell et al. 2016; Rosqvist et al. 2015; H. Singh and Gallier 2017).

The synthesis of fatty acids, known as lipogenesis, by the mammary gland is influenced by the proportion and types of fatty acids or other precursors in the maternal diet and mobilization from maternal somatic lipid stores (R. Jensen et al. 1995; Koletzko et al. 1992; Oftedal and Iverson 1995; Rudolph et al. 2007), and the particular fatty acid profile can vary across species (Dils 1986). Fatty acids in milk may vary within and across species depending on the types and proportions of fats in the maternal diet and maternal body stores, as well as the evolutionary history of the species. The majority of fat in milk (approximately 98%) is triglycerols, three fatty acids on a glycerol backbone (R. Jensen et al. 1995; McManaman 2009). The remaining milk fats are in the form of free fatty acids, phospholipids, and cholesterol (R. Jensen et al. 1995). Free or glycerol-bound fatty acids originate from maternal plasma or are synthesized de novo by the mammary gland (Del Prado et al. 2001; Oftedal and Iverson 1995; Sauerwald et al. 2001). Fatty acids shorter than 16 carbons in length are generally uncommon in the diet, and their concentration in milk is directly associated with manufacture by the mammary gland. Conversely, the mammary gland cannot manufacture polyunsaturated fatty acids (PUFAs, such as 18:2n-6 [linoleic acid or LA] and 18:3n-3 [alpha-linolenic acid or ALA]), and the presence of these fatty acids in milk reflects their consumption in the maternal diet, either past or present (Brenna 2002; S. Carlson 1999, 2001; M. Huang and Brenna 2001; R. Jensen et al. 1995). Long-chain polyunsaturated fatty acids (LCPUFAs) in milk are supplied either directly from the maternal plasma (from the diet or depot fat stores) or as metabolites of precursor fatty acids (e.g., LA and ALA) in maternal plasma. Biosynthesis of LCPUFAs from PUFA precursors takes place in the liver and depends on the quantity of LA and ALA in the diet (Brenna 2002; S. Carlson 1999; R. Jensen et al. 1995), the ratio of omega 3 (n-3) to omega 6 (n-6) PUFAs in the diet (Brenna 2002; M. Huang and Brenna 2001), and the ability to convert n-3 and n-6 PUFAs into their longer-chain metabolites (C. Agostini et al. 2001; Brenna 2002; S. Carlson 2001). Thus, if species differ with respect to conversion efficiency of n-3 or n-6 PUFAs, milk LCPUFA proportions may differ, despite similar dietary intakes of n-3 and n-6 precursors. Conversion efficacy in both humans and baboons is relatively low (Brenna 2002; Burdge 2006; M. Huang and Brenna 2001), but data from other primates are not yet available.

The protein component of milk contains essential amino acids for tissue development and numerous immunofactors to protect the infant from pathogens (Hinde, Foster, et al. 2013; Lönnerdal 2003; Lönnerdal, Erdmann, et al. 2017). Aside from the MFGM, which contributes very little to the total protein, milk has two primary and unique classes of proteins: casein and whey (Lönnerdal, Erdmann, et al. 2017). Casein proteins are found in the pellet after milk has been centrifuged and are derived from the Golgi apparatus of mammary epithelial cells (Akers 2002; Lönnerdal 2003; Lönnerdal and Atkinson 1995). Serum proteins are more soluble and are generally the whey fraction of the milk (Lönnerdal and Atkinson 1995; Oftedal 1984). Milk's casein proteins have a unique structure known as micelles. Shaped like spheres, casein micelles have hydrophilic protein sections on the outside of the sphere and hydrophobic sections on the inside. Casein micelles are soluble in the aqueous solution of milk. However, once they reach the stomach, "one of the most ingenious events in nature takes place" (Lemay, Dillard, and German 2007): casein micelles are broken apart by digestive enzymes, and the hydrophobic subunits are exposed, become insoluble, and form a curd. Oftedal (2013) argues that the enzymatic mechanism that allows the conversion of liquid milk to a solid curd was a critical step in the evolution of efficient digestion among nursing offspring. Compared with whey proteins, whose amino acid building blocks appear quickly in the bloodstream after ingestion, casein curds slowly release amino acids (as well as protein-bound minerals such as calcium) into the infant's bloodstream. Whey proteins include antibodies, enzymes, antimicrobial factors, and α-lactalbumin, the latter of which is of high nutritional value and supplies many of the essential and branched-chain amino acids to infants (Kunz and Lönnerdal 1993; Lönnerdal 2003; Lönnerdal and Atkinson 1995). Colostrum, milk produced in primates during the immediate neonatal period, is the highest in protein concentration due to the high secretion of immune factors (Andreas et al. 2015; M. Barboza et al. 2012; Franca et al. 2011; J. Huang et al. 2015; Uruakpa et al. 2002). Protein concentration decreases over the first several weeks of lactation during a transitional phase and then is relatively stable in mature milk until weaning and involution (Kunz and Lönnerdal 1993; Lönnerdal and Atkinson 1995; Lönnerdal, Keen, et al. 1984; M. Power et al. 2002).

Lactose is the predominant carbohydrate of most mammalian milks, including those of human and nonhuman primates (Hinde and Milligan 2011; Oftedal and Iverson 1995), and is consistently the highest-concentrated macronutrient in human milk (Newberg and Neubauer 1995). Lactose, composed of one galactose and one glucose molecule, serves multiple functions within milk. Because it is a disaccharide, the osmolarity of lactose is less than that of two monosaccharides, and it would be less likely to cause osmotic stress in infants after they consume milk (Newberg and Neubauer 1995). The individual molecules—glucose and galactose—that make up lactose have important roles in energy supply to the infant's body and brain (Newberg and Neubauer 1995; Andreas et al. 2015). Other carbohydrates in milk include oligosaccharides, nucleotide sugars, glycolipids, glycoproteins, and mucins (Andreas et al. 2015). In eutherian mammals (including primates), milk oligosaccharides typically pass intact through low-acidity infant stomachs and function as ecosystem engineers in the infant lower intestinal tract (Urashima et al. 2022). Some milk oligosaccharide structures provide key nutritional substrates for beneficial commensal microbes or inhibit the colonization of pathogenic microbes (Allen-Blevins et al. 2015; Bode 2012; Goto et al. 2010; Hinde and Lewis 2015; Tao et al. 2011; Urashima et al. 2022).

In primate milk, the primary minerals are calcium, phosphorus, sodium, magnesium, and potassium (Oftedal 1984). Minerals in milk are used for skeletal growth and maintenance, muscle contractions, membrane fluidity, and protein synthesis, and as electrolytes (Atkinson et al. 1995; Hinde, Foster, et al. 2013; L. Klein, Breakey, et al. 2017; Lönnerdal 2000). The concentrations of phosphorus and calcium are positively correlated, and both are proportional to the amount of casein protein in milk, which binds these minerals (Atkinson et al. 1995; Hinde, Foster, et al. 2013; A. M. Prentice and Prentice 1995). Importantly, the ratio of ingested calcium and phosphorus may affect bone metabolism in the developing neonate (Hinde, Foster et al. 2013). Milk also contains iron, which is available in only trace amounts to human infants (0.2 to 0.8 mg/L; Casey et al. 1995). Reports on iron concentration in nonhuman primate milks are rare, but it is presumed to be comparably low as found in human milk due to similar milk protein composition and life history. Across mammals, approximately 30%–60% of iron is found in the casein fraction of milk, and thus milks with a higher casein concentration have more iron (Casey et al. 1995). Human infants have a compensatory mechanism for the low iron content of milk and are born with large iron stores in the liver and hemoglobin (Casey et al. 1995; Stini et al. 1980). The high bioavailability of iron in human milk suggests that it may play an important role for the infant, but its low, narrowly constrained concentration suggests that other mechanisms, perhaps competition with pathogens, may have selected for a conservative pattern of iron concentration (Casey et al. 1995; Quinn 2014; Stini et al. 1980).

THE SYNTHESIS OF MILK

Our understanding of the proximate mechanisms of milk synthesis derives primarily from research in rodent and dairy animal models and to a much lesser extent in humans (Hovey et al. 2002). For mothers to produce milk during lactation, morphological and functional development of the mammary gland must be established before parturition. Cells differentiate as mammary tissue in utero and continue to grow isometrically with the developing female until puberty, when mammary tissue grows allometrically (reviewed in Akers 2002; Hovey et al. 2002). Rapid somatic growth before puberty inhibits mammary development and has negative consequences for milk yield at reproductive maturity in dairy cows (Sejrsen and Purup 1997), but whether we can extrapolate to primates from dairy breeds under intense artificial selection for milk production remains unclear. Among rhesus macaques (*Macaca mulatta*), age of reproductive debut was associated with capacity to synthesize milk energy density and volume. Very young females produced lower milk volume per unit maternal mass at primiparity than did females who initiated reproduction at typical or older ages (Pittet et al. 2017). Rhesus primiparae who were older than typical produced lower milk energy density due to lower fat and protein concentrations (Pittet et al. 2017). These results suggest that there are trade-offs locally in the capacity of the mammary gland to synthesize milk and global somatic trade-offs that influence milk production in young rhesus macaques (Pittet et al. 2017). Functional development of the mammary precedes conception during the estrous cycle, and if conception occurs, the mammary continues to develop throughout pregnancy: mammary tissues undergo accelerated growth, differentiation, proliferation, and production (reviewed in Akers 2002; Hovey et al. 2002). This is orchestrated by precise up- and downregulation of endocrine hormones throughout pregnancy and lactation (Akers 2002; Hovey et al. 2002).

Active milk synthesis begins prior to parturition. During pregnancy, under the hormonal influence of progesterone, estrogen, prolactin, growth hormones, glucocorticoids, and insulin, mammary epithelial cells undergo secretory differentiation into lactocytes (or milk-secreting cells), which are capable of synthesizing lactose, casein protein micelles, and triglycerides (Pang and Hartmann 2007). Precursors for these milk ingredients (i.e., glucose, amino acids, and fatty acids) move into lactocytes across the basolateral membrane that separates the mammary epithelial cells from the maternal bloodstream, and they are synthesized by various organelles within the lactocyte. For example, amino acids are translated into milk-specific proteins at ribosomes located on the rough endoplasmic reticulum and are then moved to the cell's Golgi apparatus for further processing (e.g., casein micelle formation). The Golgi apparatus is also the site of lactose synthesis; glucose and galactose move into the cell's Golgi and, in the presence of α-lactalbumin, are bound together to form the disaccharide lactose. Secretory vesicles that bud off of the Golgi, containing both lactose and casein proteins, are transferred across the apical membrane of the lactocyte and release their contents, including water, into the lumen, in which milk accumulates until transfer to the young (Akers 2002).

Milk secretory cells are not activated until the final stages of pregnancy, when mammary glands begin to secrete colostrum, milk with relatively high carbohydrate and immunoprotein concentrations but low fat and nutritional protein concentrations (reviewed in Akers 2002; McManaman and Neville 2003; Truchet and Honvo-Houéto 2017). Humans may be rare among mammals in that secretory activation occurs postnatally, although many human mothers can express colostrum during the last weeks of pregnancy (Pang and Hartmann 2007). Colostrum is easily digestible for the infant intestinal tract, not yet colonized by commensal gut bacteria. Moreover, the high concentrations of oligosaccharides and immunoglobulins (e.g., sIgA) in colostrum facilitate the colonization of beneficial commensal bacteria, if present, while selectively inhibiting pathogenic bacteria (Hinde 2017; Hinde and Lewis 2015; Ward et al. 2006). Milk not only feeds intestinal microbes but also vertically seeds microbes into the primate neonatal intestinal tract (Dettmer et al. 2019; Groer et al. 2020; L. Jin et al. 2011; Muletz-Wolz et al. 2019; Petrullo, Jorgensen, et al. 2019) Additionally, milk includes enzymes that contribute to the nutrient bioavailability for the infant—milk in part digests itself to improve absorption and assimilation in the developing neonate (Dallas et al. 2014; Dallas and German 2017). Following birth, circulating prolactin and glucocorticoids remain elevated in the mother, but the delivery of the fetus and placenta causes downregulation of estrogen and progesterone. This combination stimulates the mammary to transition from colostrum production to more copious transitional milk production hours to days after parturition (Akers 2002; Neville and Morton 2001; Pang and Hartmann 2007). Specifically, without the opposing effects of progesterone, prolactin upregulates the production of α-lactalbumin. An important nutritional protein for infants, α-lactalbumin also acts as a rate-limiting enzyme in lactose production within the mammary cell's Golgi apparatus (Akers 2002; Oftedal 2013). Increased lactose synthesis functions to draw more water into the Golgi, which subsequently increases the volume of milk produced by the lactocytes. At the same time, infant suckling stimulates the pituitary gland to release oxytocin, the hormone responsible for promoting milk ejection, also referred to as milk letdown (Akers 2002; Pang and Hartmann 2007; Truchet and Honvo-Houéto 2017). The gene expression profiles of milk-producing cells in the mammary gland are appreciably different at each lactation stage (Lemay, Ballard, et al. 2013), The transcriptome of colostrum is characterized by immunofactors, whereas transcription in the mammary gland producing transitional milk and mature milk is characterized by protein synthesis and lipid synthesis, respectively (Lemay, Ballard, et al. 2013).

Milk synthesis and secretion increase toward peak lactation as the energetic needs of the young increase from growth and activity (Oftedal 1984; Riek 2008, 2021; Hinde and Capitanio 2010). Milk production during this time is under a positive feedback system in which mammary evacuation from infant suckling upregulates milk synthesis in the lactocytes as well as milk secretion from the lactocytes (Akers 2002; Neville, McFadden, and Forsyth 2002; Neville and Morton 2001; Wilde and Peaker 1990; Wilde et al. 1995). Involution occurs after peak lactation when significant decreases in the frequency of infant suckling increase milk stasis, downregulate lactose synthesis, and precipitate the onset of apoptosis in the mammary (Akers 2002). As lactose synthesis is downregulated during involution, less water is pulled into milk via osmotic pressure; as a consequence, milk energy density significantly increases while milk yield significantly decreases at the end of lactation (Akers 2002).

THE EVOLUTION OF MILK

Although the fossil record leaves little evidence of the soft-tissue structures of milk synthesis and transfer (Benoit et al. 2016), studies of fossils, genes, and modern-day mammals provide key opportunities to understand the evolution of lactation. All living mammal species are characterized by a complex system of lactation requiring the coordinated integration of morphology, physiology, and behavior in mother and young (Lefèvre et al. 2010). Transitional phases of less complex milk emerged among the synapsids 310 million years ago (mya) and continued into their descendants the therapsids (D. Blackburn et al. 1989; Lefèvre et al. 2010; Oftedal 2012). Morphological studies of fossils combined with molecular techniques indicate that modifications of the ear and brain and the emergence of hair and "mammary" tissue were evident by 240 million years ago in the stem taxa from which mammaliaforms arose (Benoit et al. 2017). The hatching of young without dentition indicates that milk of sufficient nutritive value to sustain the neonate had evolutionarily emerged just over 200 mya (Kielan-Jaworowska et al. 2005). Milk production by egg-laying monotremes provides strong evidence that the evolution of lactation preceded the emergence of viviparity in mammals (D. Blackburn et al. 1989; Pond 1977, 1984). Among extant mammals, there are many similarities in both mammary gland structure (D. Blackburn et al. 1989; Oftedal 2002a, 2002b) and milk composition (Oftedal and Iverson 1995). Genome-wide surveys of the presence, expression, and function of genes underlying lactation processes support this interpretation. Comparative analysis of disparate mammalian genomes—platypus, opossum, cow, human, dog, mouse, and rat—demonstrates that the secretory mechanisms of complexly structured milk evolved by 160 million years ago (Lemay, Lynn, et al. 2009). Taken together, we can infer that the ability to copiously synthesize milk was established before the present-day lineages of mammals diverged from their shared last common ancestor (D. Blackburn et al. 1989; Brawand et al. 2008; Long 1969; McClellan et al. 2008; Oftedal 2002a, 2002b). Indeed, among all descendants from synapsids, only the complex-milk-synthesizing mammals remain.

Despite the importance of milk as nourishment for offspring among extant mammals, the leading hypothesis on the origins of lactation suggests that the original functions of milk were not nutritional but rather to keep eggs hydrated and provide immunological factors to offspring (D. Blackburn et al. 1989; McClellan et al. 2008; Oftedal 2002a, 2012, 2013). Oftedal (2002a) posits that the mammary gland evolved from an ancestral apocrine-like gland associated with hair follicles and that the adaptive function of the original secretions was as a source of water and protective agents for the parchment-shelled eggs of synapsids (Oftedal 2002b). The hypothesis that the mammary gland was co-opted from existing structures and developmental pathways is supported by Lemay, Lynn, and colleague's (2009) identification of higher conservation of mammary genes relative to other genes among seven mammalian genomes spanning the monotremes, marsupials, and eutherian mammals. Further support for Oftedal's hypothesis comes from the identification of homologies between the structure and secretions of apocrine glands and the mammary gland, including structural similarity between lysozyme and α-lactalbumin (D. Blackburn 1991; D. Blackburn et al. 1989; Hayssen and Blackburn 1985; Oftedal 2012, 2013). Lysozyme, an integument-derived secretion found in tears, sweat, and milk, has antimicrobial properties, whereas α-lactalbumin is a nutritional protein that also plays a key enzymatic role in lactose synthesis within mammary cells. The first milk was not necessarily milk at all but what Oftedal (2013) refers to as "proto-lacteal secretions" that provided eggs with moisture, antimicrobial factors, and potentially some nutrients. Over the subsequent millennia, offspring ingestion of these secretions seemingly enhanced survival such that mutations that increased the nutritive content of milk were favored and the nutritional value of milk increased. In tandem, the physiology of milk synthesis, the anatomy of milk delivery (lumen, ducts, myoepithelial cells, and nipples) and ingestion, and the composition of milk likely coevolved in response to environmental and ecological pressures (Akers 2002; Oftedal 2002a, 2013; Pond 1977, 1984; Vorbach et al. 2006).

Lactation likely facilitated the evolution of several important mammalian characteristics and the ability to exploit nutrient-impoverished or rapidly changing environments (Pond 1977, 1984). By physiologically synthesizing milk, mothers were no longer required to forage for food to provision their young. Although the energetic costs of milk synthesis and provisioning food for the infant(s) can be equivalent for the mother, milk is more energetically efficient for the infant than are solid foods requiring mastication, detoxification, and digestion (Pond 1977). Indeed, maternal-origin proteases in milk contribute to the digestion of proteins within the infant's stomach; mother's milk contains factors that help the infant digest milk (Dallas et al. 2014; Dallas and German 2017). From the infant's perspective, surplus calories can be preferentially invested into other developmental priorities such as growth (Pond 1977) or behavioral activity (Hinde and Capitanio 2010). Lactation further enables mammals to reproduce successfully in environments that support the mother but would otherwise be inhospitable to infants (Pond 1984). Milk can also afford greater altriciality at birth (D. Blackburn et al. 1989) and an extended period of development and social learning (Bekoff 1974). Mammalian mothers may be further benefited by decoupling current dietary intake from reproductive effort, mobilizing maternal stores of lipids and minerals accumulated prior to lactation, and thereby buffering mothers and infants from current environmental conditions (Oftedal 1984; Pond 1977, 1984). Moreover, the behavioral biology of milk-feeding—the transfer of milk from mother to young—represents the first social interaction and social relationship. The neurobiological underpinnings for bonding, recognition, and social interaction between mother and young (and among littermates) are largely conserved across mammals (Jonas and Woodside 2016) and facilitated the emergence of greater social complexity in some mammalian lineages (Broad et al. 2006). These adaptations may have enabled mammalian ancestors to successfully reproduce in many types of environments, precipitating an adaptive radiation in the Cenozoic, when many other classes of animals, such as reptiles, were in decline (Pond 1977).

LACTATION STRATEGIES

The physiological synthesis of milk is embedded intimately with the other dimensions of the lactation strategy. The length of milk synthesis from parturition through weaning, the frequency and duration of nursing bouts, the litter size (and sex ratio), and whether milk synthesis is primarily supported by mobilization of maternal reserves or current dietary intake are expected to influence the volume and composition of milk that mothers produce (Hinde and Milligan 2011; M. L. Power and Schulkin 2016). Species are distributed among these interrelated dimensions of the lactation strategy, and individuals within those species exhibit variation within a narrower range along those dimensions.

Lactation Duration

Lactation begins with secretory activation and the transition to copious milk production (S. Anderson et al. 2007), increases toward peak lactation as the energetic needs of the young increase from growth and activity (Hinde and Capitanio 2010; Oftedal 1984; Riek 2008, 2011), and is then downregulated during the weaning process (Akers 2002). The range in lactation duration is highly variable among wild-living animals. On the shorter end are the mere 4 days of the hooded seal (W. Bowen et al. 1985) and 23 days in the house mouse (König and Markl 1987), producing 61.1% and 27% fat concentration in milk, respectively (Skibiel et al. 2013). On the other end of the distribution are the 4.75 years in the African elephant (P. Lee, Bussiére, et al. 2013) and up to 8.8 years among Bornean orangutans (T. M. Smith, Austin, Hinde, et al. 2017), producing 5% fat and 2.2% fat concentration in milk, respectively (Skibiel et al. 2013). Falling intermediately along this dimension is the sea otter lactation for 6 months (Reidman and Estes 1990), hyena for about 12 months (Holekamp et al. 1996), and Phayre's leaf monkey for about 18 months (Borries, Lu, et al. 2014), but milk composition values for these species have not yet been reported in the literature. These lactation lengths represent means at the population or species level, with important variation around that mean among individuals. Age at weaning is particularly challenging to interpret in part because weaning in many species, particularly primates, represents a complex biobehavioral process negotiated by both mother and infant (C. Austin et al. 2013 Borries, Lu, et al. 2014; P. Lee 1996; Rietsema et al. 2015), and behavioral observations of these dynamics have limited utility for understanding milk transfer from mother to infant (Cameron 1998; Cameron et al. 1999).

New methods for assessing milk ingestion are greatly improving our understanding of lactation duration and weaning dynamics in extant and fossil primates. Due to pioneering isotopic methods for assessing milk ingestion from primate infant feces, Reitsema (2012) has enabled detailed understanding of weaning dynamics in captive and wild living primates. Among captive rhesus macaques at the Yerkes National Primate Research Center, there is substantial variation in the timing and duration of weaning, with sons possibly weaning at younger ages (Rietsema et al. 2016). Bădescu, Watts, and colleagues (2022) found an opposite pattern among wild chimpanzees at Ngogo in Uganda, with sons consuming more milk for their age and for a longer period than did daughters. Analysis of isotopic signatures of milk ingestion recovered from infant hair among wild chacma baboons demonstrates that maternal social environment—in particular, low rank—is associated with accelerated weaning and possible nutritional stress (Carboni et al. 2022). Methods for investigating milk ingestion via teeth are particularly exciting, as mineralization during tooth development provides a longitudinal, daily record of lactation and weaning. These tooth records have revealed lactation curves in rhesus macaques with the weaning process typically unfolding from 6 to 9 months of infant age and cessation of milk transfer during the second trimester of the mother's subsequent pregnancy (C. Austin et al. 2013). Among orangutans, these milk ingestion signatures in teeth demonstrate a more variable, multiyear pattern characterized by periods of high and low milk ingestion that may correspond to seasonal variation in the availability of food items that young orangutans are able to access (T. M. Smith, Austin, Hinde, et al. 2017). In this way, mother's milk may represent the original fallback food for the primate weaning experiencing nutritional, immunological, and psychosocial challenges. Notably, these methods applied to fossil teeth of young Neandertals have demonstrated lactation curves of exclusive milk feeding for >6 months and cessation of lactation approaching 2 years of infant age, similar to humans (T. M. Smith, Austin, Green, et al. 2018).

The relationship between length of lactation and milk composition reflects the energetic requirements and developmental priorities of mammalian young (Barton and Capellini 2011; Riek 2021; Skibiel et al. 2013). For example, phylogenetic analyses using data from dozens of placental mammals have revealed that species with greater postnatal brain growth require longer lactations to support their extended period of neurodevelopment (Barton and Capellini 2011). Importantly, length of lactation also reflects challenges that mothers may face in synthesizing high-energy-density milk for months or years. An analysis of over 500 mammal species evaluated lactation length and weaning mass in relation to male contributions to rearing young, controlling for phylogeny (West and Capellini 2016). The length of lactation is shorter when males provide infant care by carrying the young, alleviating the combined energetic burden of lactation and increased locomotor costs on the mother (West and Capellini 2016). The weaning mass of young in species with male care of infants is the same on average as in species in which mothers provide all care over a longer lactation, but paternal care enabled more maternal energy toward supporting infant growth (West and Capellini 2016). These findings indicate that among species characterized by biparental care, females are able to synthesize richer or more milk to accelerate infant growth over a shorter lactation.

Traditionally, mammalian reproduction is dichotomized into gestation and lactation, but the emergence of the "first 1,000 days" public health messaging has been instrumental in anchoring concepts of early development more holistically (Geddes and Prescott 2013; Mahumud et al. 2022; Neville, Anderson, et al. 2012). Moreover, much as clinical management of pregnancy is arranged around trimesters, stages of lactation can be trifurcated. Milk composition is most frequently binned as colostrum, transitional, and mature milk, but public health recommendations and clinical management are anchored to the period of exclusive milk-feeding followed by a period of continued milk-feeding and complementary feeding (M. Martin 2017; Sellen 2009; Victora et al. 2016). On par with public health approaches, evolutionary lactation biologists have much to gain by addressing the interrelated durations of pregnancy, milk-only, and complementary feeding periods (Langer 2008). Controlling for phylogeny and body mass in dozens of placental mammal species, Langer (2008) discovered that rodents, insectivores, and carnivores had particularly long milk-only phases of maternal investment, with weaning occurring over a relatively shorter period of complementary feeding. In contrast, many ungulates had a proportionately longer "mixed feeding" phase compared to gestation and milk-only, possibly due to the complexity of digestion through fermentation requiring the development of commensal microbial communities for plant matter and upregulation of enzymatic capacity (Langer 2008). Among wild gelada (*Theropithecus gelada*), lengthy weaning transitions from primarily mother's milk to grazing are associated with substantial reorganization to an adult-type microbiome, a process that is especially protracted for the infants of primiparous mothers (Baniel et al. 2022). Among human subsistence populations, the cultural practice of pastoralism and the dietary use of animal milks are associated with a later age of introducing solids to infants but not with breastfeeding duration (M. Martin 2017; Sellen and Smay 2001).

Nursing Behavior

The length of lactation represents the aggregate of the shorter-time-scale negotiations of nursing behavior and the frequency and duration of suckling bouts. The opportunity to nurse is largely determined by the proximity of mother and young. Maternal foraging apart from young lengthens internursing intervals (Oftedal 1984). Many mammals, including most rodents, insectivores, rabbits, carnivores, and bats, cache young at a nest, burrow, den, or roost before they have achieved locomotor developmental milestones. Many of these same species are characterized by relatively long milk-only phases because their opportunity to begin exploiting adult food resources is limited by not foraging with the mother. Reduced gastric capacity combined with sustaining metabolic processes for longer internursing intervals requires relatively energetically dense milks high in fat and protein concentration (Ben Shaul 1963; Oftedal 1984; Oftedal and Iverson 1995; Tilden and Oftedal 1997). In contrast, mammalian young who locomote independently shortly after birth or are carried by their mother have more frequent access to suckling opportunities, and mothers produce more dilute milk that is lower in fat and protein (Oftedal 1984). Moreover, nursing patterns can transition across lactation. In the egg-laying short-beaked echidnas, the mother carries the newly hatched puggle in a pouch as she forages (Enjapoori et al. 2014; Rismiller and McKelvey 2009). When the puggle is too large for the pouch, it is cached in a burrow while the mother independently forages for up to three days (Enjapoori et al. 2014; Rismiller and McKelvey 2009). This three-day internursing interval requires milk to be especially energetically dense, with high fat and protein concentrations of 31% and 12% respectively, to sustain infants (M. Griffiths et al. 1984).

Litter Size and Sex Ratio of Young

Integral to the lactation strategy are the number and sex ratio of offspring that a mother is supporting via milk. This dimension, however, is complicated by high covariance among litter size, mass at birth, and altriciality at birth among mammalian species (Derrickson 1992; R. Martin 1984b; Oftedal 1984, 1985). Across mammals, neonate mass scales negatively allometrically with increasing maternal mass (Harvey and Clutton-Brock 1985; R. Smith and Leigh 1998). Small-bodied primates have relatively larger litter mass, though there is also a clear increase in relative litter mass from strepsirrhines to haplorrhines (Leutenegger 1979; R. Martin 1984b). Some of this variation may relate to maternal strategies, with relatively heavier litters reflecting greater prenatal investment.

Postnatal investment could then be negatively related, positively related, or unrelated to relative litter mass depending on the duration and intensity of nutrient transfer during lactation (P. Lee 1999).

Numerous hypotheses have been advanced to predict how mothers will differentially invest resources in sons and daughters (Cockburn et al. 2002), but few studies have addressed sex-differentiated milk synthesis (Bădescu et al. 2022; Hinde, Foster, et al. 2013). Among some mammals, there are suggestions of differentiated milk synthesis as a function of whether mothers gestated or reared sons or daughters (rhesus macaques: Hinde 2009; Hinde, Foster, et al. 2013; humans: C. Powe et al. 2010; Thakkar et al. 2013; Twigger et al. 2015; but see Quinn 2013; bovids: Hinde et al. 2014; cervids: Landete-Castillejos et al. 2005; marsupials: Quesnel et al. 2017; Robert and Braun 2012; pinnipeds: Piedrahita et al. 2014; with particularly elegant experimental research in rodents: Koskela et al. 2009). In these studies, researchers rarely control for offspring mass to disentangle what may be a by-product of the typically larger mass of sons (Hinde 2009). The pattern and directionality of sex-differentiated milk remain somewhat equivocal, but to date they suggest greater yield for daughters, whereas there may be energetically denser milk, through higher fat or protein concentrations (or both), for sons. Experimental manipulation has revealed that both gestational (Hinde et al. 2014) and postnatal (Koskela et al. 2009) factors may influence sex-differentiated milk synthesis. Sex-differentiated milk effects may be the greatest among primiparae, as cumulative changes in cellular architecture in the mammary gland across multiple pregnancies and lactations may diminish the magnitude of the effect of the current offspring sex (Hinde et al. 2014). Additionally, these sex-differentiated effects may be a by-product of the hormonal milieu of gestating daughters, as hormonal signals from fetal daughters crossing the placenta into maternal circulation may affect the functional development of the mammary gland during gestation (Hinde et al. 2014). Additionally, even when milk synthesis is the same for sons and daughters, sex-differentiated mechanisms for the assimilation of milk constituents may yield sex-differentiated outcomes (Badyaev 2002; Hinde 2009; Hinde, Foster, et al. 2013; Hinde et al. 2014). Future consideration of progeny-specific adaptations as well as differentiated maternal effort will contribute to a better understanding of the ontogeny of sexual dimorphism.

MILK COMPOSITION ACROSS MAMMALS

The extent to which the different dimensions of the lactation strategy exert a signature on milk composition varies. Systematic, comparative investigation of mammalian milk composition and energy production has explored the influences of phylogeny, ecology, behavioral biology, neurodevelopment, and life history (Barton and Capellini 2011; Ben Shaul 1963; Blomquist 2019; R. Martin 1984b; Oftedal 1984; Oftedal and Iverson 1995; Riek 2008, 2021; Skibiel et al. 2013). In the mid-20th century, "The Composition of the Milk of Wild Animals" presented a landmark consideration of the diversity of mammalian milks from a decade of systematic investigation by Devorah Ben Shaul (1963). Assessing concentrations of fat, protein, and carbohydrates from 101 species, Ben Shaul noted that "grizzly bear milk and kangaroo milk had virtually the same basic milk composition" (333), suggesting that degree of relatedness did not predict composition. Ben Shaul, and subsequently others, proposed that milk composition clustered in relation to the degree of maturity at birth, maternal attentiveness, nursing frequency, body mass, gastrointestinal capacity, exposure to marine ecosystems, ambient temperature, and aridity (Ben Shaul 1963; Oftedal 1984; Oftedal and Iverson 1995). Fifty years later, Skibiel and colleagues used data from 129 mammalian species, representing 51 families and 15 orders, to simultaneously investigate milk composition in relation to maternal body mass, adaptations to arid environment, maternal diet, length of lactation, altricial versus precocial young, aquatic versus terrestrial habitats, and total reproductive output (litter size and neonate size) while controlling for phylogeny using robust statistical models (Skibiel et al. 2013). Statistically controlling for phylogeny is important because closely related species have more similar genomes due to their recent common ancestor (Blomquist 2019; Nunn 2011), and there are genetic underpinnings to milk synthesis (Akers 2002; Lefèvre et al. 2010; Lemay, Lynn, et al. 2009). Despite Ben Shaul's prediction, phylogeny strongly predicts the gross composition of milk that a species produces, but maternal body mass, arid environment, reproductive output (litter size and mass), and precociality were not associated with milk gross composition across the mammalian class (Blomquist 2019; Skibiel et al. 2013). After controlling for phylogeny, two characteristics importantly predicted milk composition: maternal diet and duration of lactation. A species' diet had a strong influence on milk composition—carnivores synthesize milk with substantially higher concentrations of fat and protein than do herbivores and omnivores (Skibiel et al. 2013). The longer the length of lactation, the more dilute the milk a species produces; the shorter the lactation, the more energetically dense (Skibiel et al. 2013).

MILK COMPOSITION ACROSS PRIMATES

Primates serve as an ideal taxon for investigating patterns of milk composition due to the range of body sizes, locomotor habits, social systems, habitats, and diets. Compared to other mammals, primates produce relatively dilute milk, delivered to few infants over a long infancy (Hinde and Milligan 2011; Oftedal and Iverson 1995). With relatively large bodies and brain sizes, primate infants are very costly. Primate mothers appear to employ multiple strategies to meet the costs of lactation by increasing gross energy intake, reducing expenditures, or temporarily relying on stored reserves (cercopithecoids: J. Altmann 1983; J. Altmann and Samuels 1992; L. Barrett, Halliday, Henzi, et al. 2006; Dunbar and Dunbar 1988; A. Koenig, Borries, et al. 1997; hominoids: Bates and Byrne 2009; Lappan 2009; C. M. Murray, Lonsdorf, et al. 2009; Pontzer and Wrangham 2006; platyrrhines: Boinski 1988; Guedes et al. 2008; K. Miller et al. 2006; Nievergelt and Martin 1999; Rose 1994; Tardif 1994; strepsirrhines: Saito 1988; Sauther 1994; Vasey 2005; for reviews, see Dufour and Sauther 2002; Heldstab et al. 2017; P. Lee and Bowman 1995; Emery Thompson, this volume). The relatively slow somatic growth rates among primates and long female reproductive careers may allow for variation and flexibility in maternal investment strategies that are not possible in other mammalian orders (Pereira and Leigh 2003). From this perspective, Leigh (1994a) hypothesized that folivorous primates should have higher milk protein because they grow faster than other primates of comparable body size. He noted that the high protein content of structural plant parts could enable faster growth (J. Lambert 1998; Tanner 1990) and that rapid attainment of adult size and early maturation might reflect reduced feeding competition in folivores (Janson and van Schaik 1993). Alternatively, rapid growth might be facilitated by higher milk gross energy (kcal/g), which would be reflected primarily in milk fat concentration. Crudely, this would imply that a more faunivorous diet enables more rapid growth. A null alternative is that primate mothers produce essentially the same "recipe" of milk regardless of diet—leaving interspecific milk composition differences to be related to allometry, mother-infant contact, neonate or litter size, and other factors.

We explored milk synthesis, life history parameters, and nutritional ecology of $N = 36$ primate taxa, controlling for phylogeny, to evaluate the contributions of maternal diet to milk composition within the primate order (G. Blomquist et al. 2017). To date, milk composition analyses of relatively few primate taxa have been described in the literature, and most are from small samples of captive primates (Hinde and Milligan 2011). Notably absent are values from colobines, tarsiers, and many strepsirrhine clades (fig. 3.1). We incorporated adult female body mass, neonate mass, mean litter size, behavioral care (parking vs. carrying) (Kappeler 1998; Ross 2001), and diet (quantified as percent time foraging on reproductive plant parts, structural plant parts, and animal prey in the wild or for humans in a foraging context). We analyzed the relationship between milk composition and predictors by multiple regression of the species values (TIPS) and phylogenetic generalized least squares (Grafen 1989; E. Martins and Hansen 1997). Both methods allow simultaneously testing the effects of multiple predictors and controlling for correlations among them (Freckleton 2002). The phylogenetic generalized least squares model adjusted for nonindependence among the individual species arising from shared evolutionary history through a correlation matrix derived from an independently estimated phylogeny. Phylogenetic relationships among the primate species were taken from a consensus tree of version 3 of the 10,000 Trees Project (https://10ktrees.nunn-lab.org/; fig. 3.1; G. Blomquist et al. 2017). A strong phylogenetic signal was expected from a mammal-wide analysis of milk composition (Skibiel et al. 2013).

Primate milk composition reflected diet, maternal behavior, and maternal mass even after accounting for the phylogenetic signal in milk (figs. 3.2, 3.3; G. Blomquist et al. 2017). Greater consumption of structural plant parts or animal matter was positively associated with higher protein concentrations in milk. In contrast, sugar concentrations in milk were significantly reduced by greater

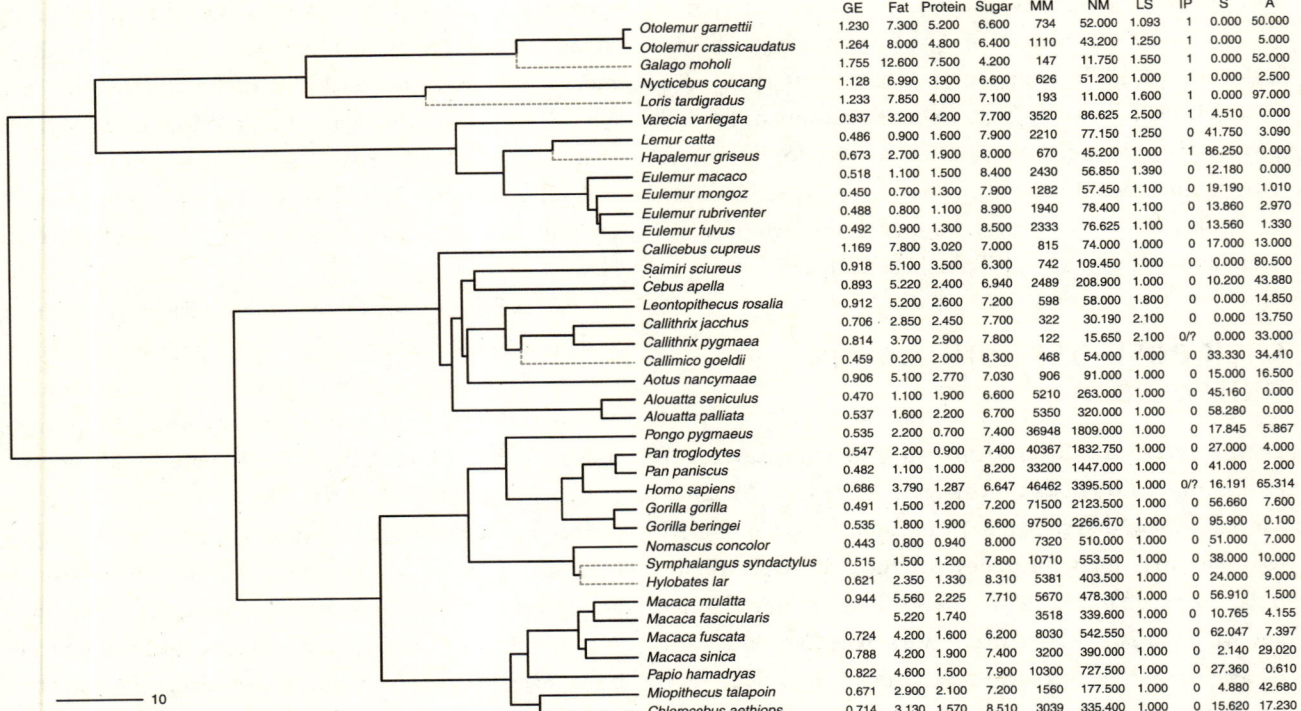

Figure 3.1. Phylogeny of primates from 10,000 Trees (version 3) and data set. The scale bar indicates 10 million years. Milk composition is reported as grams or kcal per 100 g of milk. Species with only a single milk sample are indicated in gray and are excluded from the reduced data set. Maternal (MM) and neonate mass (NM) are reported in grams. Litter size (LS) is the species mean, while infant parking (IP) is a binary code. Diet is indexed with the percentage of time spent feeding on structural plant parts (S) and animal material (A).

consumption of structural plant parts or animal matter across species. Importantly, diet was not associated with fat concentrations in milk composition across species. The association between wild diet and milk protein and sugar concentrations could be explained by several overlapping causes including the influence of species-typical foraging patterns, an independent correlation between diet and infant growth rates, variation among foods in the presence of nutrients that are costly to synthesize, and simple transmission of ingested nutrients. The first three of these we consider to be interrelated, plausible explanations that emphasize maternal energetics and can potentially be tested with additional data. Maternal mass, across primates, was negatively correlated with protein and fat concentrations, while litter size was positively correlated with fat concentration. Milk sugar concentration was not influenced by maternal or litter mass (G. Blomquist et al. 2017), effects strongly influenced by phylogeny (Hinde and Milligan 2011).

Our results provide general support for Leigh's (1994a) prediction that folivorous primate milk would be characterized by higher protein concentrations. However, because of the complicated network of possible causes of the milk composition–diet relationship and the paucity of folivores in our data set, we feel it is premature to assess its underlying mechanism or generality. Interestingly, *Callimico* and *Hapalemur*, both small-bodied primates with diets high in structural plant parts, were dramatic outliers, suggesting possible heterogeneity in milk composition among primate folivores (Leigh 1994a). We should be very cautious, however, since each of their values of milk composition derive from one individual. Aspects of infant and juvenile foraging and nutritional ecology, including relative acceleration or delay of digestive system maturation and availability of foods for complementary feeding (Charnov and Berrigan 1993; Godfrey, Samonds, et al. 2001; Janson and van Schaik 1993; Langer 2008, 2003; P. Lee 1999; Leigh 2004; R. Martin 1996), may further

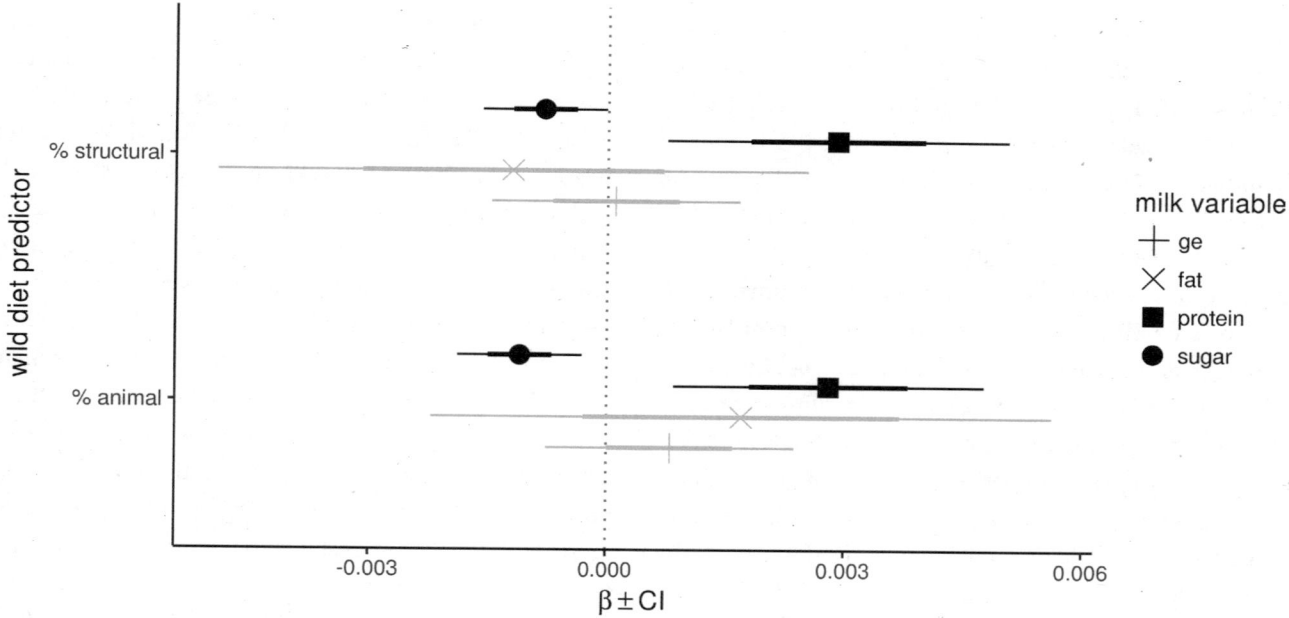

Figure 3.2. Regression coefficients (β) and confidence intervals for wild diet components predicting milk gross energy (ge) and percent fat, protein, and sugar in the phylogenetic generalized least squares analysis. The dashed vertical line indicates no effect of diet on the milk variable. An increase in either diet component significantly elevates milk protein content (positive coefficient) and reduces milk sugar concentration (negative coefficient). Neither dietary component has a significant effect on milk gross energy or fat content.

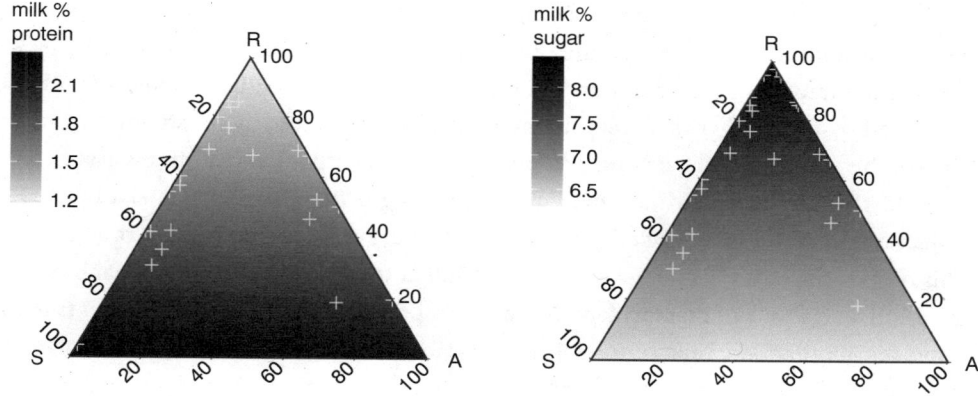

Figure 3.3. Triangle plots for the distribution of primate diets in the sample as percentages of time spent feeding on structural (S) or reproductive (R) plant parts and animal (A) material. Each species is a single plus sign within the triangle. The shading of each triangle indicates the predicted concentration of milk protein or sugar from the phylogenetic generalized least squares analysis (for a nonparking species of average maternal and litter mass). High milk protein concentrations are predicted for primates more reliant on structural plant or animal material. High milk sugar concentrations are predicted for diets rich in reproductive plant parts.

complicate our understanding of the milk-diet dynamic. Differentiating an offspring growth hypothesis from an explanation anchored to maternal energetics may amount to distinguishing between direct and indirect selection on mothers and offspring (S. Arnold 1994). Maternal dietary intake may directly influence the composition of milk by supplying specific nutrients that are transmitted to offspring or by affecting the energetics of milk synthesis. A strict interpretation of the relationship between consumed food and milk composition is complicated by the fact that milk is a synthesized product rather than a passive transfer mechanism for nutrients in the form they are encountered in foods. Body reserves are mobilized to varying degrees by mammalian mothers to support lactation (Dufour and Sauther 2002; Gittleman and Thompson 1988; Jönsson 1997; P. Stephens et al. 2009; Tardif, Power, Oftedal, et al. 2001). Fatty acids in milk are something of an exception in that they do often directly reflect consumed foods but may be obfuscated by total concentration of fats (L. Milligan and Bazinet 2008; L. Milligan et al. 2008). Further complicating a systematic understanding of primate milks is that the dietary categories of structural and reproductive plant parts and percent time feeding on them are often poor indicators of the nutritional composition of ingested foods (Chapman, Chapman, Rode, et al. 2003; Ganzhorn, Arrigo-Nelson, Boinski, et al. 2009; J. Lambert 2011; Oftedal 1991).

Infant parking was an important predictor of all aspects of proximate milk composition—elevating gross energy, fat concentration, and protein concentration while depressing sugar concentration. These findings confirm research among strepsirrhines (Tilden and Oftedal 1997) but refine and extend this phenomenon by demonstrating that the effect persists even when accounting for phylogeny, allometric effects of adult mass, litter mass, and diet. More precise data on the frequency and duration of internursing bouts caused by the mother-infant separation from parking, or possibly allomaternal carrying (Emery Thompson 2017; Heldstab et al. 2017; J. Mitani and Watts 1997; Ross and MacLarnon 2000), should clarify our understanding of this relationship.

Assessing gross concentrations of macrocategories in milk presents important limitations. Many proteins and sugars are not metabolized by the infant but rather facilitate development of the infant immune system (Chirico et al. 2008; Newburg and Walker 2007). Most notable are milk oligosaccharides, a complex sugar with lactose at the reducing end, which are found throughout monotremes, marsupials, and eutherians. Among eutherians, however, gastric enzymatic activity does not substantially cleave these glycans, and they pass intact into the infant's lower intestine, where they serve multiple functions influencing the colonization and maintenance of the infant's commensal gut microbes (Allen-Blevins et al. 2015; Bode 2012; Hinde and Lewis 2015; M. Martin and Sela 2013; Urashima et al. 2022). Microbes contribute to nutrient and caloric recovery from ingested foods, recruit and regulate the immune system, and influence neurobiology and mood (Allen-Blevins et al. 2015; Baniel et al. 2022; de Weerth et al. 2022; Hinde and Lewis 2015; M. Martin and Sela 2013; Lambert, Mutegeki, and Amato, this volume). Milk oligosaccharides partially engineer the community ecology of commensal microbes by inhibiting the establishment of some pathogenic microbes and providing a carbohydrate substrate for beneficial microbes, influencing the health, growth, and development of the neonate (Bode 2012; Charbonneau et al. 2016; Garrido et al. 2016; M. Martin and Sela 2013). Human milk oligosaccharides are substantially derived compared to those of the other primates studied to date, including all other great apes. Humans produce a greater diversity, abundance, and concentration of milk oligosaccharides, sometimes by an order of magnitude (Goto et al. 2010; Tao et al. 2011; Urashima et al. 2022). Oligosaccharides, although largely undigested by the primate infant, provide essential nutrients to the microbes that prevent diarrhea, improve growth, and serve other fitness-enhancing functions. In this way, maternal synthesis of these milk constituents is feeding "mother's littlest helpers" within the infant (Hinde and Lewis 2015). Moreover, recent research on captive vervet monkeys (*Chlorocebus pygerythrus*) further illuminates our understanding of milk complexities, as the milk produced by constrained primiparous mothers is enriched with microbial strains associated with improved infant growth (Petrullo, Baniel, et al. 2022). Additionally, mother's milk includes numerous hormones that shape infant temperament, cognition, and behavior (de Weerth et al. 2022). The myriad constituents in milk, their abundance, and their interactions demonstrate an exquisitely complex biofluid shaped by maternal constraints and capacities that in part orchestrates early life developmental trajectories of the primate infant (Allen-Blevins et al. 2015; Hinde, Foster, et al. 2013; Lu et al. 2019). In this way, mother's milk provides both the building blocks and to some extent the blueprints that organize infant trade-offs among physical growth, somatic development, immunological maturation, and behavioral activity.

MILK AND HUMAN EVOLUTION

From an anthropological perspective, data on primate lactation are critical to understanding evolved aspects of human life history in general and the human lactation strategy in particular. Over the last 20 years, our knowledge of primate milk composition, particularly nutritional composition, has greatly expanded (Cleland and Power 2022; M. Garcia et al. 2017; Hinde, Foster, et al. 2013; Hinde and Milligan 2011; Hinde, Power, and Oftedal 2009; L. Milligan and Bazinet 2008; L. Milligan et al. 2008; Petrullo, Baniel, et al. 2022; Petrullo, Jorgensen, et al. 2019; Pittet et al. 2017; M. L. Power, Oftedal, and Tardif 2002; Ziomkiewicz et al. 2021), allowing for a better understanding of how human milk composition compares with that of other primates. Human mothers produce slightly richer milks than do ape mothers; milk fat in humans provides roughly half of the energy, compared to approximately 25%–35% of energy in ape milks (Hinde and Milligan 2011). The human capacity to incorporate long-chain fatty acids into milk reflects the complexity of human nutritional ecology relative to other primates, as captive primates fed commercial diets produced in an industrialized food chain synthesize milk enriched with LCPUFAs (L. Milligan and Bazinet 2008). Quinn (2017) posits, however, that the human capacity to synthesize medium-chain PUFAs may be an important adaptation for protecting milk fat concentrations in the context of fluctuations in the dietary intake of mothers. Importantly, the total energy that infants derive from milk depends on both milk energy density and yield. Currently, there is a paucity of information about milk yield in primates (Hinde and Milligan 2011; Riek 2021; Rosetta et al. 2011) and none from nonhuman apes. Although it would be tempting to link the relatively larger adult brain size and more rapid postnatal brain growth rates of humans to their production of higher fat concentrations in milk, this hypothesis requires the inclusion of longitudinally collected data on ape milk production (or infant intake, similar to Wells, Jonsdottir, et al. 2012) and ape neurodevelopmental trajectories (Hinde and Milligan 2011). Milk proteomics comparisons between rhesus macaque and human milk reveal that humans produce many proteins in greater abundance than do rhesus macaques (K. Beck et al. 2015). Among the protein orthologs between the two species, over 90% were much more abundant in human milk and included proteins associated with the development of the gastrointestinal tract, the immune system, and the brain (K. Beck et al. 2015), features substantially derived in the hominin lineage.

Selection on milk composition and yield must also be considered in the context of other derived aspects of the human lactation strategy, specifically the shorter duration of lactation (Eerkens and Bartelink 2013; Greenwald 2017; Kennedy 2005; Sellen 2007). Richer milk, regardless of selection on increased yield, could be an important factor in enabling earlier weaning among human ancestors but is likely one of many factors contributing to this significant shift in life history. Indeed, this life history pattern of accelerated weaning is present in both humans and Neanderthals from tooth records of milk ingestion (C. Austin et al. 2013; T. M. Smith, Austin, Green, et al. 2018). Notably, elemental records of milk ingestion from fossil teeth of *Australopithecus africanus* demonstrate a much more chimpanzee-like duration of exclusive milk feeding for one year and cessation of milk transfer when young are four or more years old (Joannes-Boyau et al. 2019; Joannes-Boyau et al. 2020; T. M. Smith, Machanda, et al. 2013). The emergence of an extended childhood, when mother's milk is rarely provided, is facilitated by protection and provisioning from a network of allomaternal caregivers and maternal support networks (Bogin 1997; Hrdy 2007; Kaplan et al. 2000; M. Martin et al. 2022; Meehan et al. 2018; Scelza and Hinde 2019). Allomaternal care may even extend to allomaternal nursing, where other group members nurse a mother's offspring while babysitting, during maternal illness, or even in cases of maternal death (Bell et al. 2013; Hewlitt and Winn 2014). Allomaternal nursing is found across numerous traditional societies and contemporary foraging populations (Hewlett and Winn 2014) and is increasingly common among Westernized, educated, industrialized, rich, democratic populations (Palmquist 2020; Palmquist and Doehler 2016). Research into allomaternal nursing and milk synthesis among contemporary subsistence populations yields important insights into how allomothers' milk contributes to infant development and survival, possibly accelerates maternal postpartum recovery, and may have played an important role in the evolution of human life history (Bell et al. 2013; Hewlitt and Winn 2014; Hinde 2014; Meehan et al. 2018).

Among humans, milk composition varies in part as a function of local environmental adaptations and cultural ecology. Conventional wisdom from the public health sector at the end of the 20th century concluded that mean concentrations of fat, protein, and lactose were largely conserved across human populations, but ongoing research in recent decades has demonstrated substantial variation in many human milk constituents (Hinde and Milligan 2011; Quinn 2021). High-altitude adaptations among populations in the Himalayan and Peruvian Andes Mountains include a milk phenotype particularly enriched in fat concentration (Quinn et al. 2016; Schafrank et al. 2020). Recent analyses of human milk oligosaccharides among a global distribution of populations demonstrate significant variation in the presence and abundance of different oligosaccharides, with some variance a function of geographic regions, subsistence patterns, and genotype frequencies (Vinjamuri et al. 2022). Human milk oligosaccharides are important ecological engineers of infant microbial communities, and cultural patterns of breastfeeding are now implicated in the patterns of *Bifidobacterium* species colonization in infancy. Direct analyses of the milk microbiome among hunter-gatherers and horticulturist communities in the Central African Republic revealed some differences in microbial taxa among communities, but the size of a mother's social network and patterns of allomaternal care of infants were important predictors of microbial communities in human milk, indicating complex exchange of microbes between the mother-infant dyad embedded within a biosocial context (Meehan et al. 2018). Notably, concentrations of innate immune proteins in milk are similar across human populations, but adaptive immune protein concentrations in milk are more similar among populations in closer geographic proximity or with a shared subsistence pattern (L. Klein, Huang, et al. 2018). A comparison of mineral concentrations in breast milk samples across human populations in the United States, Poland, Argentina, and Namibia revealed that among Himba women the cultural practice of rubbing the skin with *otjize* (a traditional cosmetic paste made with red ochre high in iron ore) may increase the concentration of iron measured in milk from skin transfer (L. Klein, Breakey, et al. 2017). Taken together, the above findings suggest that cultural practices that influence breastfeeding initiation and duration, infant care networks, pathogen exposure, and diet and nutrition exert important influences on both milk composition and infant microbial communities, with important consequences for mothers and infants from both ultimate and proximate perspectives. Collectively, ongoing lactation research by primatologists and anthropologists highlights essential contributions of these fields to maternal and child health (M. Martin et al. 2022; Scelza and Hinde 2019).

SUMMARY AND CONCLUSIONS

The synthesis of milk by mammary glands is a key defining characteristic of our mammalian class and provides the first food of the mammalian neonate. The magnitude, sources, and consequences of variance in milk composition and yield across species, populations, and individuals, however, have not been significant targets of research efforts in anthropology, evolutionary biology, physiology, or nutrition (Hinde 2015). In the early 21st century, research in these topic areas has expanded and accelerated as next-generation tools and methods have provided new avenues for characterizing the library of constituents in mother's milk. A research agenda to systematically understand milk is "data hungry," requiring a high resolution of milk parameters, life history data, local ecological conditions, and nutritional composition of maternal diet both before and during lactation. Moreover, the dynamic aspects of milk synthesis across time are most effective where multiple assessments can be made (Mitoulas et al. 2002). Documenting ontogenetic variation among species in infant growth and key organ development and the continued study of the behavioral ecology of motherhood will better resolve the relative effects of costly nutrients, infant growth, and maternal foraging ecology on primate milk synthesis.

- Milk is a complex biofluid that functions as food, medicine, and signal. Mother's milk provides nutrients, immunofactors, and hormones that nourish, protect, and orchestrate the development of mammalian young.
- Milk composition and yield are two dimensions of a species' lactation strategy along with the length of lactation until the end of weaning, the frequency and duration of nursing bouts, and the number of offspring supported during lactation, with consideration of infant biological sex as well as infant care tactics (such as parking vs. carrying).

- The current understanding of milk composition across mammalian species broadly, including primates, is that the proportions of fat, protein, and sugar in milk reflect phylogeny, duration of lactation, and nutritional ecology, among other factors.
- Analysis of human milk reveals derived features of milk composition compared with that of other great apes. Particularly notable are a higher concentration of fat and a greater abundance, diversity, and complexity of milk oligosaccharides and other immunofactors in human milk.
- The complexity of milk composition both reflects maternal life history trade-offs and contributes to organizing trade-offs among developmental priorities in the primate infant.

4 Diet and the Energetics of Reproduction

Melissa Emery Thompson

Reproduction raises the energy requirements of primates, taxing their ability to maintain a healthy physical condition while successfully rearing young. These costs are central to shaping the evolution of species, as natural selection favors strategies that enable the most efficient conversion of energy into reproductive success. Part of this story involves developing a dietary strategy appropriate to the given environment. Layered on this are metabolic and life history strategies, such as whether to grow large or stay small, that determine the size of the energy budget and how much can be made available for reproduction. These issues are discussed elsewhere in this volume (see chapters by Hawes et al., DeCasien et al., and Borries and Koenig). However, animals also need to make proximate behavioral and physiological adjustments to coordinate these strategies. This chapter addresses how reproductive effort in primates affects nutritional needs and the strategies that primates use to accommodate these needs.

Given the tremendous diversity in primate dietary strategies, body size, social organization, and life history, primates experience and adapt to the costs of reproduction in a variety of ways. The field of primatology still has a limited grasp on this variation, given the lack of detailed data on the reproductive ecology of many primate species and populations. This is an area where captive and field studies can come together to provide increased resolution. Controlled captive feeding conditions and the ability to closely monitor changes in physical condition allow researchers to more clearly quantify the costs of various types of reproductive behavior. In turn, wild studies enable us to understand the ecological constraints on reproduction and the ways that primates respond to variation in energy availability. In practice, however, reproduction is well studied in a few, mostly small, species in captivity, while wild data are skewed toward larger, highly social species. Nevertheless, there appear to be broad differences in the reproductive ecology of large versus small primates. Not surprisingly, reproductive energetics is best characterized in humans. While the focus of this chapter is on nonhuman primates, I draw from the human reproductive ecology literature both as a point of comparison with other primates and to provide reference data that simply are not available for other species. One of the lessons that humans, who have been studied across vastly different ecological contexts, can provide is that even when offspring are very costly to produce, the strategies used to manage the costs of reproduction are highly flexible.

This chapter is organized around three central questions. First, what are the nutritional demands of reproduction, and how are they distributed across the reproductive effort? Second, what strategies do primates use to accommodate periods of higher cost? Third, what are the effects of nutritional shortcomings on reproductive outcomes at various stages of a reproductive bout? While the bulk of the chapter focuses on females and the associated costs of gestation and lactation, the chapter also presents a brief overview of emerging research on costs of reproduction in male primates.

ENERGETIC COSTS OF PREGNANCY AND LACTATION

The marked difference in reproductive investment between males and females lies at the heart of mammalian biology (Gittleman and Thompson 1988). Even in species with significant paternal care, females bear the obligate costs of pregnancy and lactation, subsidizing infant growth within their own energy budgets. Compared with other mammals, primate infants are relatively costly, as their large brains are metabolically expensive to run and take longer to develop (Harvey and Clutton-Brock 1985; Leigh 2004; R. Martin 1996). Thus, most primates can afford to produce only a single infant at a time, and primate birth rates and growth rates are substantially lower than those of comparably sized mammals (Charnov and Berrigan 1993; Dufour and Sauther 2002; Ulijaszek 2002). Slower growth rates correspond to reduced daily costs of lactation and pregnancy but longer periods of continuous investment in each offspring. For mothers, there is a reduced likelihood of energy shortfall each day. However, the burden of infant care, and thus the possibility of failure, looms longer. This pattern suggests that primate mothers need to adopt relatively complex strategies to manage the costs of reproduction over time.

The costs of pregnancy are not limited to the costs of producing infant mass. Mothers must produce a variety of supportive tissues, including the placenta and mammary apparatus. Mothers then pay the increased costs of maintenance and physical activity associated with a larger corpus (Dufour and Sauther 2002). Thus, while most neonatal primates weigh about 3%–10% of their mother's weight (Leutenegger 1979), nutritional needs during pregnancy exceed this. Increased demands on circulation and waste removal further tax maternal systems. Pregnant females may also suffer decreased mobility (Meder 1986; E. Price 1992a) or increased vulnerability to competition (Wasser and Starling 1988).

For most primates, the duration of lactation is longer than that of pregnancy (van Schaik 2000). Lactation is also costlier per day than pregnancy because the bodies and brains of nursing infants are larger than their fetal equivalents. Conversion of food energy into milk entails production costs and is therefore less efficient than is the placental transfer of energy. For humans, whose milk is not substantially different in composition from that of other primates (Hinde and Milligan 2011), the food-to-milk energy conversion efficiency is estimated at approximately 83% (A. M. Prentice and Prentice 1988). Still, primates produce milks that are very low in energy density; thus, the daily cost of milk production is estimated to be 4–15 times lower than in other mammals (A. M. Prentice and Prentice 1988). Unless other caregivers are available, lactating mothers pay the costs of transporting their growing infants. For baboons (*Papio cynocephalus*), this cost is estimated at 5%–10% above the mothers' solitary caloric requirements for the first six months of life (J. Altmann and Samuels 1992). However, when infants are nutritionally dependent, the cost of carrying can actually be less than if the infant moves independently because mothers would pay, through milk energy, for the less efficient locomotor patterns of infants (J. Altmann and Samuels 1992). Even after they are able to contribute to their own caloric needs, infants can reduce the foraging efficiency of their mothers (J. Altmann 1980). In some species, infant care also requires increased vigilance or social involvement, which create opportunity costs for foraging (J. Altmann 1983; L. Barrett, Dunbar, and Dunbar 1995).

While infant primates are likely to be most vulnerable to energy availability in the early postnatal period, mothers are increasingly energetically taxed as their infants grow larger, up until the infants can begin acquiring food for themselves. Therefore, infant weaning, or the transition from exclusive maternal nutritional support to completely independent offspring foraging, is theorized to begin when infant caloric needs are greater than the mother's ability to provide for them by increased intake or by drawing from diminishing reserves (J. Altmann 1983; P. Lee 1996; M. van Noordwijk, Willems, et al. 2013). In most primate species, this occurs when infants reach approximately four times their neonatal weight (P. Lee 1996; P. Lee, Majluf, and Gordon 1991). There can be considerable variation in when infants become too costly based on food availability and maternal condition, and this threshold can theoretically be reached in different ways. Mothers with abundant energetic resources can subsidize faster infant growth rates and reach an optimum weaning size relatively early. Mothers with poor resources may be forced to begin withdrawing investment early even if their infants are still small (Fairbanks and McGuire 1995; P. Lee, Majluf, and Gordon 1991). In smaller primates, which have lower energy budgets, weaning is often a relatively rapid process, as infants take over 100% of their needs relatively quickly. In contrast, larger species reach the "milk insufficiency point" early relative to the infant's ability to forage independently (M. van Noordwijk, Willems, et al. 2013). In these species, lactation may continue alongside independent foraging for multiple years. It is hypothesized that maternal costs remain near their plateau until infants are competent foragers, at which point the maternal contribution comprises only a fraction of the offspring's total budget (M. van Noordwijk, Willems, et al. 2013). However, seasonal fluctuations in resource availability experienced over multiyear lactation, and variation in the ability of infants to process particular foods, likely lead to peaks in the importance of maternal supplementation (L. Barrett, Henzi, and Lycett 2006; T. M. Smith, Austin, Hinde, et al. 2017). Final weaning in these species may be more strongly compelled by the resumption of maternal fertility than by specific landmarks in offspring development (Emery Thompson, Muller, Sabbi, et al. 2016; M. van Noordwijk, Willems, et al. 2013).

In addition to increased total energy needs, pregnant and lactating animals have stricter requirements for specific nutrients, such as protein (Dewey 1997; Oftedal et al. 1991; Riopelle, Hill, and Li 1975; Riopelle, Hill, et al. 1975), iron (Beard 2007; Coe et al. 2007; Golub, Hogrefe, et al. 2006), and zinc (Golub, Gershwin, et al. 1984). Despite a significantly increased need for the mobilization of calcium, experimental data from humans suggests that increased dietary intake of calcium is unnecessary (A. Prentice 1994; A. Prentice, Jarjou, et al. 1995). Manufacturing milk also requires water, which may be limiting in some wild diets (Gittleman and Thompson 1988; R. S. Nelson et al. 2022). Pregnant and lactating females may also be more vulnerable to plant secondary compounds (Sauther 1994).

Few quantitative data on the costs of pregnancy and lactation are available for nonhuman primates. In fact, there is often no practical way to calculate these, as both the costs and the strategies used to afford them vary with maternal ability to pay (G. Goldberg et al. 1991; Poppitt et al. 1993; Rosetta et al. 2011). For example, in captive baboon mothers, total energy expenditure during early lactation was positively predicted by maternal body mass and energy intake; those who could spend more on lactation, did (Rosetta et al. 2011). Some estimates put the peak costs of lactation for primates at about 50% over maternal baseline energy requirements, higher still for strepsirrhines and callitrichids (Key and Ross 1999; Portman 1970). Given the relatively low energy content of primate milks, it is more reasonable to assume that the costs for most primates fall at or below the low end of the range of 35%–149% cited for mammals (Gittleman and Thompson 1988, table 1). A variety of data support this. Captive baboon mothers granted free access to food increased their intake by 11% to 27% from the beginning to the peak of lactation (S. Roberts et al. 1985). These mothers also lost about 7% of their body weight before their infants were weaned, suggesting their caloric needs were elevated by slightly more than their intake reflected. Similar estimates of a 13%–24% increase in costs at peak lactation were obtained for wild baboons by estimation of daily energy expenditure (J. Altmann and Samuels 1992). Wild orangutan (*Pongo pygmaeus*) mothers were estimated to expend approximately 18%–25% above their baseline energy requirements at the point at which their infants began supplemental feeding (M. van Noordwijk, Willems, et al. 2013). These data suggest that the costs of lactation for most large-bodied primates may peak at 30% over the nonreproductive state.

While quantitative data on smaller primates is rare, the costs of lactation are likely to be higher in small than in large species, because neonates (and their daily growth requirements) are large relative to maternal weight (Leutenegger 1979). This effect is enhanced because many of these species, including the tamarins and marmosets (family Callitrichidae), some galagos (*Galago senegalensis* and *G. demidovii*), and gray mouse lemurs (*Microcebus murinus*), habitually give birth to two or three offspring at a time (G. Doyle, Andersson, and Bearder 1971; Eberle and Kappeler 2006; Goldizen 1987b), reaching a combined neonatal litter mass that is 14%–25% of maternal weight (versus 2%–14% in other primates; Leutenegger 1979). Despite this high demand on mothers, infants of small primate species maintain higher relative growth rates than those of larger species (Ross 1998; Tardif 1994). During pregnancy, callitrichid mothers appear to get by without substantially increasing their intake (Kirkwood and Underwood 1984; Nievergelt and Martin 1998), but a study of captive common marmosets (*Callithrix jacchus*) found that diet restriction to 75% of expected energy intake in midgestation led to fetal loss in every case (Tardif, Power, Layne, et al. 2004). Marmoset mothers consumed 50%–

102% more calories during lactation than when cycling but still lost 8% of their weight during only six weeks (Nievergelt and Martin 1998).

The energetic costs of mating activity are rarely addressed for females because it is assumed that they should be lower than the costs of pregnancy and lactation. For many species, the interval between lactation and mating is so short that it would be difficult to establish a baseline for comparison. However, mating behavior is likely to detract from foraging effort. In many species, mating activity involves larger or more cohesive groups, creating greater feeding competition, and disruptive behaviors such as male displays, mate guarding, and sexually coercive aggression. Female baboons reduce their time feeding when in consortships (Bercovitch 1983). In chimpanzees, females with sexual swellings often attract large parties with many males (Emery Thompson and Wrangham 2006; Hashimoto, Furuichi, and Tashiro 2001). Chimpanzee females feed less (Wrangham and Smuts 1980) and travel more (Matsumoto-Oda and Oda 1998) when they have sexual swellings than when they do not. Female chimpanzees experience lower energy balance when associating with large numbers of males (Emery Thompson, Muller, and Wrangham 2014). Chimpanzees and many other larger primate species cycle repeatedly over the course of several months before conceiving, so even if the energetic costs of mating activity are trivial, their recurrence may contribute to accumulating energetic stress.

HOW DO PRIMATES PAY THE COSTS OF REPRODUCTION?

Adjusting the Energy Budget

Increased costs of reproduction must be accommodated by making adjustments to the energy budget or to energetic allocations. Individuals can increase their caloric intake, draw from energy reserves, or decrease expenditures on physical activity or other metabolic processes. Generally speaking, primate mothers will make all of these adjustments at some point in their reproductive efforts, though how much they rely on each strategy (and their ability to do so) varies considerably. Ideally, we can come to understand whether the life history characteristics of different species predict or constrain their ability to adopt particular strategies. However, ecological variation within a species is expected to be a major mediating condition, and the data are as yet too sparsely distributed to justify structured conclusions about between- versus within-species variation.

Evidence from various species suggests that mothers are able to sustain pregnancy without increasing their intake (cotton-top tamarins, *Saguinus oedipus*, Kirkwood and Underwood 1984; white-faced capuchins, *Cebus capucinus*, McCabe and Fedigan 2007), and may not do so even when it is possible to (rhesus macaques, *Macaca mulatta*, Kemnitz et al. 1984). This suggests an increased energy efficiency during pregnancy. Humans with poor energy availability exhibit reduced basal metabolic rates (BMRs) during all but the latest weeks of pregnancy, suggesting that they are able to trim down enough of their maintenance energy budget to preserve the fetus (Poppitt et al. 1993; A. M. Prentice and Goldberg 2000). Doing this, however, requires drawing from energy stores or compromising activity budgets and somatic maintenance in the mother. It may also compromise fetal nutrition.

Changes in feeding behavior are more obvious during lactation (see also Hinde et al., this volume). For example, captive cotton-top tamarins did not increase their energy intake during pregnancy but nearly doubled it during lactation (Kirkwood and Underwood 1984). The most obvious way to increase intake is to devote more time to feeding, but this may not always be feasible if female activity budgets are stretched thin (J. Altmann 1983). Primates are often observed to forage more efficiently when lactating. For example, two studies of lactating female capuchins at Santa Rosa failed to find evidence that lactating females devoted more time to foraging than nonreproductive females. In one study, lactating females actually spent comparatively less time foraging but appeared to maximize caloric intake by focusing on foods with less complex processing requirements (Rose 1994). A later study found no difference in time spent feeding or in dietary composition but instead found that rates of food intake were much higher, equating to a 33% increase in protein and a doubling of fat intake (McCabe and Fedigan 2007). Lactating females of a number of species feed selectively on higher-quality foods (black howlers, Dias, Rangel-Negrin, and Canales-Espinosa 2011; sifakas, Hemingway 1999; vervets, *Chlorocebus aethiops*, P. Lee 1987; Assamese macaques, *Macaca assamensis*, Touitou et al. 2021; chimpanzees, C. M. Murray, Lonsdorf, et al. 2009).

Lactating female orangutans, for example, increased their consumption of fruit (E. Fox, van Schaik, et al. 2004). Guenon (*Cercopithecus* spp.) and red titi monkey (*Plecturocebus cupreus*) mothers ate more insects when lactating (Cords 1986; Dolotovskaya and Heymann 2020).

Mothers may not have the option to increase their intake or to find higher-quality foods. In fact, energy intake may fall because less food is available or because infants compromise foraging effort (J. Altmann 1980). For example, a study of time budgets in geladas found that infants interfered with foraging so much that their mothers conditioned them to nurse only during resting periods (L. Barrett, Dunbar, and Dunbar 1995). Lactating chacma baboon mothers failed to increase foraging effort because they increased time devoted to vigilance (L. Barrett, Halliday, and Henzi 2006). Siamang (*Symphalangus syndactylus*) mothers did not spend more time feeding or eat higher-quality foods during lactation but instead increased intake during and after infant weaning (Lappan 2009). At this time, males took over most of the carrying duties, and females used this respite to rebuild their condition. If intake cannot be raised during lactation, decreases in physical activity may be necessary. For example, lactating chacma baboon females reduced their activity levels when they could not increase feeding time (L. Barrett, Halliday, and Henzi 2006), while lactating vervet females had both decreased energy intake and expenditure (M. J. S. Harrison 1983). Well-nourished human mothers balanced their energy by a combination of increased intake and reduced expenditure, with individual variation in the reliance on each strategy (Butte and Hopkinson 1998; G. Goldberg et al. 1991; A. M. Prentice and Prentice 1988).

When lactation costs cannot be accommodated by shifts in diet or activity, mothers must draw from stored energy. Baboon mothers could withstand a reduction of their ad libitum intake by 20% without compromising milk output and without losing more weight than unrestricted mothers, but a reduction of 40% led to losses of about 3% of body weight per week (S. Roberts et al. 1985). Similarly, while relatively large common marmoset mothers could increase their intake sufficiently to rear twins and smaller mothers could raise singleton infants, small mothers with twins suffered significant weight loss (Tardif, Power, Oftedal, et al. 2001). Poorly nourished Gambian women reduced their energy expenditure during lactation, but when the ability to acquire more calories from their diet was limited (e.g., during the hunger season), they lost up to 2 kg per month (A. M. Prentice and Prentice 1988). This suggests that primate mothers have a variety of mechanisms to sustain lactation during energy shortages, but maternal and infant health may be compromised as a result.

Emerging evidence also suggests that maternal gastrointestinal microbiomes in human and nonhuman primates shift during both pregnancy and lactation (white-faced capuchins, Mallott, Garber, and Malhi 2018; humans, Sparvoli et al. 2020; Tibetan macaques, *Macaca thibetana*, B. Sun et al. 2021). In Phayre's leaf monkeys, *Trachypithecus phayrei* (Mallott, Borries, et al. 2020), these shifts were driven by shifts in reproductive hormones. It is not yet known how these changes affect maternal metabolism or nutrient availability.

Finally, changes in foraging during pregnancy and lactation may be constrained by the need to avoid particular foods, such as those containing secondary plant compounds. For example, Sauther (1994) noted that lactating ring-tailed lemur (*Lemur catta*) females avoided mature leaves, which contain higher concentrations of secondary plant compounds. In contrast, pregnant and lactating sifaka (*Propithecus verreauxi*) females consumed more tannin-rich plants than did males or nonreproductive females and were the only individuals seen to consume certain tannin-rich items (Carrai et al. 2003). Researchers speculated that despite negative impacts on protein absorption, moderate tannin consumption may have antiabortive or antihelminthic effects. Geophagy (soil eating) may help detoxify plant compounds and enhance mineral intake, functions that may explain its increased incidence during pregnancy in humans (S. Young et al. 2011). Limited evidence points to increased soil consumption during pregnancy in chacma baboons (*Papio ursinus*; Pebsworth, Bardi, and Huffman 2012) and chimpanzees (*Pan troglodytes*; Wrangham, Machanda, et al. 2014).

Seasonal Breeding

Timing is everything. Successful reproduction depends on the ability to coordinate energetic needs with supply over time. For many primates, the solution is to reproduce selectively during certain parts of the year, coordinating breeding seasons with changes in food abundance. The most common pattern is for the birth season to occur shortly before the peak in annual resource availability, thus aligning lactation's high costs with greater energy supply. This is often referred to as "income" breeding because most of the costs of reproduction will be paid with newly acquired energy (Stearns 1992). There is variation within this pattern as to whether peak resource availability corresponds to the peak periods of exclusive nursing or to the initiation of weaning, facilitating infant foraging success (Janson and Verdolin 2005). Other species adopt what has been termed a "capital" breeding strategy, with conceptions occurring during peaks in energy availability. These species rely on good maternal condition as a predictor of the ability to carry through the entirety of a reproductive event. Maternal energy reserves serve as capital on which to draw to buffer increased energy requirements. As detailed elsewhere, the income/capital breeding dichotomy is overly simplistic and conflates multiple, imperfectly correlated phenomena (Brockman and van Schaik 2005; Emery Thompson 2013; R. Lewis and Kappeler 2005). Primates rely on consistent daily energy intake to afford the costs of reproduction, and even under the best conditions, they may have to draw from energy reserves. While theoretically limited, the income and capital designations remain good rules of thumb for understanding how seasonal variation in diet covaries with reproduction.

Because income breeding anticipates future resource abundance, energy intake cannot directly drive the timing of reproduction. For births to coincide with the best resource availability, mating seasons and significant portions of pregnancy will tend to occur when there is relatively low resource availability. Most income-breeding species rely on indirect climatic cues, such as photoperiod or rainfall, to trigger fertile periods (Brockman and van Schaik 2005). In captivity, where resources are not seasonally variable, seasonal patterns persist, adjusting to the effect that latitude has on photoperiod (Di Bitetti and Janson 2000; Lindburg 1987). Even when conception seasons occur at low points in the annual cycle of resource availability, energetic factors can have an important impact on whether females reproduce at all (Brockman and van Schaik 2005). For example, lactation coincides with peak food availability in wild Assamese macaques, but those females in better energetic condition are most likely to conceive, and fertility rates are higher during years of high food availability (Heesen, Rogahn, et al. 2013).

For capital breeders, proximate changes in caloric intake and the resulting impacts on maternal energy balance are the primary, if not sole, cues for fertility (Brockman and van Schaik 2005). While this can result in a seasonal breeding pattern, it need not necessarily do so. It is debatable, therefore, whether any primates exhibiting the capital breeding characteristics are specifically adapted to breeding seasons or whether there is a more generalized sensitivity of conception to energetic condition that results in seasonal birth peaks in certain contexts. If so, there should be intraspecific variation in the tendency to breed seasonally. This is suggested by interpopulation variation in breeding seasonality among black-and-gold howlers (*Alouatta caraya*; Kowalewski and Zunino 2004) and Hanuman langurs (*Presbytis entellus*; Borries, Koenig, and Winkler 2001). In some environments, interannual variation in food availability may preclude having defined breeding seasons. For example, the propensity for capital breeding among frugivorous primates in Southeast Asia has been attributed to the mast fruiting phenomena that occur there (van Schaik and van Noordwijk 1985). Mast fruiting events are unpredictable from year to year yet are important times for females to capitalize on a superabundance of energy for reproduction.

There is a general trend in primates for larger species, which have longer periods of nursing, to conform to the capital breeding pattern, often without defined birth seasons (Brockman and van Schaik 2005; Di Bitetti and Janson 2000). For example, most Neotropical primates are seasonal breeders with births occurring just before the peak in fruit or insect abundance. Spider monkeys (*Ateles* spp.), muriquis (*Brachyteles arachnoides*), howler monkeys (*Alouatta* spp.), and woolly monkeys (*Lagothrix* spp.), four of the largest taxa, reverse the pattern, with more births during lean seasons and a weaker seasonal tendency (Di Bitetti and Janson 2000). Among catarrhines, the smaller guenons and some macaques have been classified as income breeders (Brockman and van Schaik 2005; Butynski 1988), whereas baboons (Alberts, Hollister-Smith, et al. 2005), Sanje mangabeys (*Cercocebus sanjei*; McCabe and Emery Thompson 2013), langurs (*Trachypithecus leucocephalus*, T. Jin et al. 2009; *Presbytis entellus*, A. Koenig, Borries, et al. 1997), and great apes (chimpanzees, Emery Thompson and Wrangham 2008; orangutans, Knott et al. 2009) conform to a capital breeding strategy, with peak energy availability or maternal condition predicting conception, often without defined birth seasons. Since the caloric requirements of larger animals are higher, these species often have more diverse diets, which may both reduce seasonal variation in caloric intake and make it more difficult to use climatic cues to forecast when the most food will be available (Alberts, Hollister-Smith, et al. 2005).

Infants of larger species are nutritionally dependent for longer than an annual breeding cycle allows, so the costliest stages of reproduction will necessarily extend over periods of both food abundance and food shortage (Brockman and van Schaik 2005). For larger species, the costs of failed reproductive efforts are likely to be greater than the costs of delayed reproduction. By contrast, small species may have little to lose by attempting reproduction under risky conditions if they are able to breed again in the next annual cycle. If female life histories depend on long life spans and long periods of infant dependency, as in larger species, the preservation of maternal physical condition during reproduction is of greater importance. Finally, since larger species experience less predation pressure, nutritional support of infants can be expected to play a relatively more important role in the probability of infant survival.

Cooperative Breeding

As noted above, small primate mothers face a particular challenge in paying the high costs of feeding and carrying infants that are proportionally large. Females often do not work alone but receive substantial amounts of caregiving assistance from other individuals. Tamarin and marmoset mothers habitually receive help from multiple caregivers, including fathers, secondary males, and older offspring (Goldizen 1990; Tardif 1994). In owl monkeys (*Aotus* spp.) and titi monkeys (*Callicebus* spp.), most of the help is provided by fathers, who carry infants even more than mothers do; older siblings and unrelated males will also contribute, particularly when fathers are not available (Fernandez-Duque, Juárez, and Di Fiore 2008; Fragaszy, Schwarz, and Shimosaka 1982). Without rearing assistance, it is unlikely that mothers could successful rear infants, particularly twins, in the wild (Terborgh and Goldizen 1985). The presence of more caregivers appears to enhance the probability of infant survival (Garber 1997; Heymann and Soini 1999). It is not clear that this is directly attributable to the infants' nutritional welfare, but there may be important trade-offs between infant care, foraging effort, and predator vigilance that are manageable only with contributions from multiple individuals (Savage et al. 1996). Aside from effects on infant survival, cooperative caregiving can facilitate higher reproductive rates. For example, male owl monkeys in an experimental paradigm were more likely to share food with lactating than with pregnant or cycling females—lactating females also begged for food more—and those females who received more food resumed cycling sooner (Wolovich, Evans, and French 2008). Comparative analyses indicate that nonhuman primate species with higher rates of nonmaternal care exhibit faster postnatal growth, earlier weaning, and faster rates of reproduction than do those with little allocare (Isler and van Schaik 2012; J. Mitani and Watts 1997; Ross and MacLarnon 2000).

A different form of cooperative breeding is observed in gray mouse lemurs. Groups of maternally related females nest communally, and while mothers appear to carry their own offspring exclusively, they nurse the offspring of relatives, particularly when the mother is away from the nest (Eberle and Kappeler 2006). In contrast to callitrichids, where the cooperative efforts of family groups are typically directed at one breeding female, mouse lemur care is truly communal, with multiple mothers breeding simultaneously and reciprocating care. This suggests that alloparenting in this species does not function as a way to subsidize the nutritional demands of lactation, per se, but may nevertheless be necessary to facilitate maternal energy acquisition, as communal care allows mothers to safely leave their infants in the nest rather than carrying them. Given a high risk of predation on mothers, related females were also ready to adopt orphans (Eberle and Kappeler 2006).

Even small amounts of alloparental care may provide energetic benefits to mothers. Among black-and-white colobus monkeys (*Colobus guereza*), during short infant-handling episodes (<5 min. per bout) by conspecifics, mothers prioritized feeding and resting and were able to forage more efficiently than when they held their own infants (Raboin et al. 2021).

Humans have also been described as cooperative breeders (Hrdy 2011; Kaplan et al. 2000; Kramer 2010). Extensive caregiving, and particularly nutritional provisioning, is provided by fathers (Marlowe 2000, 2001), grandparents (Hawkes, O'Connell, et al. 1998), older siblings (Kramer 2005), and a variety of other individuals (P. Hooper et al. 2015). In contrast to callitrichids, human infants are relatively small (Leutenegger 1979) and experience slow postnatal growth (Gurven and Walker 2006; Leigh 2001). However, the costs of feeding human infants are still unusually high because they must grow particularly large brains (Aiello and Wheeler 1995; Kuzawa et al. 2014). Humans in subsistence environments are unable to provide for all of their own nutritional needs until their teenage years, owing not only to their high caloric requirements but also to the complexity of the human diet (Kaplan et al. 2000). Therefore, while only one infant typically nurses at a time, parents in their peak reproductive years must provide nutritional inputs into several offspring of various ages (Gurven and Walker 2006). While fathers do not provide direct care in all human societies, biparental care may have been a necessity during our evolutionary past as hunter-gatherers (Lancaster and Lancaster 1983). Research on Hadza foragers indicates that fathers have the highest foraging returns when they have dependent offspring (Marlowe 2001). Cross-cultural data support the view that cooperative caregiving increases prospects for infant survival but reveal considerable variation in the importance of particular family members (e.g., grandparents vs. siblings; Sears and Mace 2008). It is less often appreciated that provisioning mothers may support their health and facilitate future reproduction as much as it provides nutritional support for existing children (Reiches et al. 2009).

Female Dominance

Nearly all lemurs share a social characteristic that is otherwise unusual among primates: females are consistently dominant to males in the context of feeding (A. Jolly 1984; A. Young et al. 1990). The most prominent hypothesis that has emerged to explain this characteristic is that lemurid females face atypically tight energetic constraints on reproduction, forcing males to cede foraging priority so that breeding can occur (A. Jolly 1984). Madagascar's island environment is both highly seasonal and unpredictable, and lemurs face prolonged periods of food shortage (P. Wright 1999). Female dominance might therefore be viewed as a part of the suite of adaptations for energy conservation observed in lemurs, including very conservative BMR (Müller 1985; Richard and Nicoll 1987) and the production of relatively small, altricial neonates (Kappeler 1996; A. Young et al. 1990). Raising the energy budget to accommodate the added costs of gestation and lactation may indeed be a significant challenge. A difficulty with this hypothesis is interpreting data on lorises, which are strepsirrhine primates that also have low metabolic rates and small litters but do not have female dominance. Evidence in support of the energy conservation hypothesis is that relative litter mass among strepsirrhines is correlated with BMR, suggesting that energy budgets constrain investment in gestation (A. Young et al. 1990). Female-dominant lemurs have relatively short gestation periods, resulting in significantly faster fetal growth for a given body size and BMR than in lorises, indicating that a larger portion of the energy budget is devoted to reproduction (A. Young et al. 1990). However, Kappeler (1996) has noted that the postnatal growth rates of lemurs and lorises do not differ and are uncorrelated to BMR, concluding that maternal costs at the height of investment are not sufficiently different to explain female dominance. Even if phylogenetic inertia can explain many of the metabolic differences between strepsirrhines and haplorrhines, it is still feasible that the harsh environments of Madagascar have forced the lemurs to adapt in novel ways. Torpor is also observed among some lemur species, such as the gray mouse lemur, and can be induced not only by low temperatures but also by restricting calories during gestation and lactation (C. Canale, Perret, and Henry 2012).

NUTRITION AND REPRODUCTIVE OUTCOMES

As outlined above, reproduction clearly necessitates increased energy availability, and primate females have varied and complex strategies to accomplish this. In the face of challenging environments, habitat change, and competition for resources, these strategies may not always be sufficient to sustain successful reproduction. This section reviews evidence for the effects of energy shortfall on reproductive success by examining key phases of the reproductive cycle.

Ovarian Function and Conception

Across many species, ovarian cycling is inhibited or impaired by food shortage. This effect is most obviously demonstrated through variation in conception success. In a number of primate populations, annual fecundity covaries with food availability, rainfall, or both (gray-cheeked mangabeys, *Lophocebus albigena*: Arlet et al. 2015; baboons: Beehner et al. 2006; Strum and Western 1982; red howlers, *Alouatta seniculus*: Crockett and Rudran 1987; white-capped capuchins, Fedigan, Carnegie, and Jack 2008; orangutans: Knott et al. 2009; Japanese macaques, *Macaca fuscata*: S. Suzuki, Noma, and Izawa 1998; long-tailed macaques, *M. fascicularis*: van Schaik and van Noordwijk 1985). Females in poor physical condition or those with low resource access are also less likely to conceive (chimpanzees: Emery Thompson, Kahlenberg, et al. 2007; Hanuman langurs: A. Koenig, Borries, et al. 1997; sifakas: Richard et al. 2000; long-tailed macaques: M. van Noordwijk and van Schaik 1999). Species that have lengthy birth intervals in the wild have significantly reduced birth intervals in the higher-resource conditions of captivity (baboons: C. Garcia et al. 2006; great apes: Knott 2001; tufted capuchins, *Cebus apella*: Wirz and Riviello 2008).

In species that experience repeated cycles before conception, fecundity may vary from cycle to cycle. Studies of wild Sanje mangabeys (McCabe et al. 2013), baboons (Lodge et al. 2013), chimpanzees (Emery Thompson, Kahlenberg, et al. 2007; Emery Thompson, Muller, and Wrangham 2014), orangutans (Knott et al. 2009), and humans (Bentley et al. 1998; Ellison 2003; Panter-Brick et al. 1993) have found that females produce higher amounts of the ovarian hormones estrogen and progesterone when dietary quality is better and when females have higher energy balance. Nutritional supplementation restores ovarian function after exercise-induced energetic stress in captive long-tailed macaques (N. Williams et al. 2001). In common marmosets, females have a higher number of ovulations per cycle when they are heavier (Tardif and Jaquish 1997). Golden lion tamarins (*Leontopithecus rosalia*) also produce more live births when heavier (Bales et al. 2001).

Fetal Development

The income-breeding patterns exhibited by many primates suggest that mothers habitually endure pregnancy during periods of relatively low food availability. Still, poor nutrition during pregnancy may be expected to compromise fetal development. Additionally, mothers faced with poor ecological conditions might be selected to abort fetuses before significant energy is invested (Wasser and Barash 1983). Undernourished Gambian women experience a high prevalence of low birth weights and perinatal mortality, both of which are substantially improved by nutritional supplementation during pregnancy (Ceesay et al. 1997). Supplementation was particularly effective at improving infant outcomes during the hunger season. Data on fetal loss and variation in fetal development are still very sparse for nonhuman primates. Studies of captive baboons found that restricting pregnant females to 70% of their ad libitum intake led to impaired placental function (C. Li et al. 2007) and cerebral development (Antonow-Schlorke et al. 2011). In the wild, baboons were more than twice as likely to experience fetal loss if they conceived after periods of drought (Beehner et al. 2006). Captive rhesus macaque mothers fed a low-protein diet had higher rates of fetal losses and neonatal deaths than did controls, despite an equivalent caloric intake (Kohrs et al. 1976; Riopelle, Hill, and Li 1975; Riopelle, Hill, et al. 1975). Facultative fetal loss in response to nutritional stress is also thought to explain the smaller litter sizes observed in wild versus captive common marmosets (Windle et al. 1999). Shorter gestation lengths have also been observed in response to nutritional stress during pregnancy (rhesus macaques, Riopelle and Hale 1975; baboons, Silk 1986).

Infant Mortality

As lactation is the costliest part of reproduction, it can be predicted that reproductive efforts are highly susceptible to failure during early infant development. Cause-specific infant mortality data are difficult to obtain for wild populations. However, nutritional factors are implicated in observations of higher infant mortality following drought in ring-tailed lemurs (L. Gould, Sussman, and Sauther 1999) and crested macaques (*Macaca nigra*; Kerhoas et al. 2014) and among siamangs in habitats disturbed by fire (O'Brien, Kinnaird, et al. 2003). In sifakas, tooth wear compromised the foraging ability of older mothers (S. King et al. 2005), and their infants experienced higher mortality rates than those of younger mothers (P. Wright et al. 2008). Across species, primates with income-breeding schedules appear to be more susceptible to poor infant outcomes because conceptions occur in anticipation of future resource abundance that may or may not materialize (Brockman and van Schaik 2005).

While these studies suggest that infant mortality is an important outcome of poor maternal nutrition, they often do not provide specific data on whether these effects are dependent on nutrition during lactation, during pregnancy, or both. A study of Sanje mangabeys found that maternal energy balance during pregnancy was a critical predictor of infant survival in the first year of life, while energy balance during lactation had no such effect (McCabe and Emery Thompson 2013). Among humans that survived the Dutch Hunger Winter, exposure to the famine during gestation, particularly the third trimester, strongly predicted infant mortality in the first year of life and could be linked to smaller birth weights, reduced maternal weight, and poor placental growth (Roseboom, van der Meulen, et al. 2001; Stein and Susser 1975). For some species, at least, the link between maternal nutrition and infant mortality appears to be mediated primarily through prenatal effects on offspring quality rather than through infant starvation, per se.

Offspring Growth and Development

Even if maternal reproductive adaptations minimize the probability of infant starvation, variation in maternal and infant nutrition can impact growth and development. When the caloric intake of pregnant mothers is sufficient to cause intrauterine growth restriction, this may have lifelong impacts on offspring health. For example, caloric restriction in pregnant baboons resulted in impaired cardiovascular health of their adult offspring (Kuo et al. 2017), mirroring a well-known effect in humans (Blackmore and Ozanne 2015). Caloric restriction during lactation in gray mouse lemur mothers resulted in partial (at 60% restriction) to complete (at 80% restriction) cessation of growth in pups (C. Canale, Perret, and Henry 2012). Similar effects can be observed when maternal condition varies naturally. For example, growth rates of ring-tailed lemurs slowed during the dry season when food was scarce (Pereira 1993). Among captive baboons, the offspring of heavier or higher-ranking mothers experienced faster growth (S. Roberts et al. 1985; Rosetta et al. 2011). Maternal rank also positively influenced growth in wild baboons (J. Altmann and Alberts 2005; S. E. Johnson 2003).

Milk quality is one important mediator of these effects (see Hinde et al., this volume). For example, milk output of captive baboons was resilient to a 20% reduction of caloric intake, but a 40% reduction led to lower milk output and reduced infant growth (S. Roberts et al. 1985). Among captive rhesus macaques, heavier mothers produced higher-quality milks, leading to faster growth, higher activity levels, and more confident temperament in their infants (Hinde 2009; Hinde and Capitanio 2010). Body weight also affected milk energy output and infant carrying frequency in common marmoset mothers (Tardif, Power, Oftedal, et al. 2001). Findings from humans are consistent with those of other primates. Undernourished Gambian women were able to produce the same quantity and quality of milk as women in the developed world, but only during the dry season, when their caloric intake was relatively high. During the hunger season, mothers produced less milk and milk with lower fat and energy content, and their neonates exhibited slower growth (A. M. Prentice, Whitehead, et al. 1981).

In some cases, reduced growth may be a temporary adaptation to accommodate temporal variation in food supply. For example, golden lion tamarin pups grew less well during the dry season than during the wet season, but individuals born during the dry season were no smaller as adults than those born during the wet season (Dietz, Baker, and Miglioretti 1994). However, long-lasting influences of nutritional variation have been documented in other species. In wild baboons, cumulative early life adversity, which includes nutritional predictors such as being born during poor feeding conditions or into a higher-density population, has been found to negatively impact longevity (Tung et al. 2016). The energy and protein content of infant baboon diets during their first year of life predicted survival to adulthood, adult reproductive life span, and the number of offspring produced (S. Altmann 1991). Additionally, baboon offspring in food-enhanced groups grew faster and reached maturity sooner (J. Altmann and Alberts 2005). In vervet populations with higher leptin, indicative of higher nutritional status, immature individuals experienced faster growth (P. Whitten and Turner 2009). Poor growth was associated with delayed reproductive maturation in males and accelerated maturation in females. Some of these correlations may reflect a more chronic effect of nutritional variation, in that high-status individuals may feed well as infants and continue to feed relatively well as they grow older. Data on humans, however, suggest that early nutritional insults can exert irreparable harm. In the Gambian population, the probability of premature death during adulthood was 3.65 times greater for individuals that were born during the hunger season (S. Moore et al. 1997). Individuals exposed prenatally to the Dutch famine exhibited a variety of chronic health deficits, including elevated rates of cardiovascular disease, schizophrenia, and breast cancer during adulthood (Roseboom, de Rooij, and Painter 2006; Roseboom, Painter, et al. 2011). There is now ample evidence from a variety of mammalian species that prenatal and neonatal nutrition exert epigenetic impacts and permanently alter pathways for the regulation of metabolism and stress, increasing susceptibility to chronic disease (Gluckman et al. 2008; M. Hanson and Gluckman 2014). For example, female baboons who experienced early life adversity had higher glucocorticoid production as adults and poorer offspring survival (S. Patterson et al. 2021; Rosenbaum et al. 2020).

Maternal Health and Reproductive Rate

While deficits in maternal energy may compromise offspring development, it is not necessarily the case that extra maternal energy will used to subsidize infant growth. Some of that energy may be used to invest in maternal condition and to increase future reproductive success. For example, heavier common marmoset mothers had faster-growing infants but were also less likely than small mothers to be ill and more likely to be fertile again in the year following lactation (Tardif, Power, Oftedal, et al. 2001). Similarly, captive baboon mothers with high energy reserves had both faster-growing infants and more rapid resumption of cycling than mothers with less energy (Rosetta et al. 2011). Providing nutritional supplementation to nursing Gambian women had relatively little effect on milk quality (A. M. Prentice, Roberts, et al. 1980), but it improved maternal health (A. Prentice, Lunn, et al. 1983) and reduced the duration of postpartum infertility (Lunn et al. 1984).

Under conditions of nutrient limitation experienced by wild primates, investments in fertility may have more obvious trade-offs with infant growth. Among wild, seasonally breeding vervets, those in habitats with better food availability were able to shift from a two-year breeding cycle to a one-year cycle, which engendered an increase in weaning conflict, as infants were rejected relatively early (Hauser and Fairbanks 1988). Among captive vervets, by contrast, the food supply was sufficient to support annual breeding without creating weaning conflict (Hauser and Fairbanks 1988). Wild chimpanzee mothers with better resource access and higher energy balance during lactation had shorter periods of postpartum infertility (Emery Thompson, Muller, and Wrangham 2012a), but the resultant short birth intervals corresponded with slower growth in their offspring (Emery Thompson, Muller, Sabbi, et al. 2016). Similarly, a technology that allowed women in rural Ethiopian villages to conserve energy led to a higher birth rate but poorer offspring growth (M. Gibson and Mace 2006).

On the other hand, maternal expenditures on reproduction may increase their own mortality. In ring-tailed lemurs, maternal mortality was 21%–30% following drought years relative to nondrought years (L. Gould, Sussman, and Sauther 1999). Protein deprivation during pregnancy increased the probability of maternal death in captive rhesus macaques (Kohrs et al. 1976). Female mortality peaked during the birth season in free-ranging, provisioned rhesus macaques (C. Hoffman, Ruiz-Lambides, et al. 2008). In baboons, lactating females healed from wounds more slowly than did cycling females (Archie et al. 2014).

COSTS OF REPRODUCTION FOR MALES

While male primates are not obligated to expend time and energy on infant care, many do (Fernandez-Duque, Valeggia, and Mendoza 2009; M. Muller and Emery Thompson 2012; P. Wright 1990). Males take over significant portions of the cost of infant carrying in owl monkeys (Dixson and Fleming 1981), siamangs (Lappan 2008), and callitrichids (Goldizen 1987a; Tardif 1994), particularly as infants grow larger. Male owl monkeys, titi monkeys, and callitrichids also contribute substantially to the provisioning of infants. Owl monkey fathers provision infants almost three times more than do mothers (Wolovich, Perea-Rodriguez, and Fernandez-Duque 2008). These behaviors not only cost males energy but reduce their ability to feed (Lappan 2009; E. Price 1992b). In captivity, male cotton-top tamarins can lose up to 10.8% of their body weight during the period of intense infant care, losing more when there are fewer alternative caregivers (Achenbach and Snowdon 2002; but see Nievergelt and Martin 1998). In wild siamangs, males carry infants almost exclusively after the first year of life, raising their travel costs by up to 23% (Lappan 2009). Males of many other species provide indirect care by maintaining vigilance against predators or infanticidal males, activities that have opportunity costs for foraging (M. Muller and Emery Thompson 2012).

Energy can constrain reproduction just as much, if not more so, for males in those species that provide little or no direct paternal care. Males of the most highly polygynous species invest in increased body mass, which increases their daily energy requirements for maintenance and physical activity. Key and Ross (1999) estimated that males of moderately dimorphic species, in which males are 30%–50% larger than females, pay as much for their added bulk as their female conspecifics pay for gestation and lactation. In highly dimorphic species, male costs are higher, even before they invest in active mating effort. In many primate species, males disperse between groups one or more times to seek new breeding opportunities, and they likely suffer energetically not only from needing to travel over long distances but also from a lack of knowledge of food resources in the new location (Isbell and Van Vuren 1996). Obtaining mates can involve costly searching, competition, and mate guarding. Reductions in male foraging effort during mating periods have been documented in a number of promiscuous species, including chimpanzees (Georgiev, Russell, et al. 2014), baboons (Alberts, Altmann, and Wilson 1996), rhesus macaques (Higham, Heistermann, and Maestripieri 2011), Japanese macaques (Matsubara 2003), long-tailed macaques (Girard-Buttoz, Heistermann, et al. 2014), and squirrel monkeys (*Siamiri sciureus*; A. Stone 2014).

If mating effort is energetically expensive yet compromises foraging, energy intake prior to mating seasons can have important impacts on male ability to breed successfully. For example, in rhesus macaques, only males with sufficient body fat at the beginning of the breeding season were successful in siring offspring (Bercovitch and Nürnberg 1996). For this species, aggressive competition over mates is relatively rare, and it is hypothesized that males engage in "endurance rivalry," with physical condition determining the ability to spend energy on successive consortships (Higham, Heistermann, and Maestripieri 2011; Higham and Maestripieri 2014). Dominant males had the highest energy balance during the birth season but the lowest levels at the end of the mating season.

Temporal variability in breeding opportunities can be expected to shape the influence of mating effort on male energetic condition. As the rhesus macaque example shows, extended periods of intense mating effort can be energetically exhausting for males. Even with provisioned food, males on Cayo Santiago lose weight over the mating season (Higham, Heistermann, and Maestripieri 2011), and this period corresponds to the peak in male mortality in this population (C. Hoffman, Ruiz-Lambides, et al. 2008). On the other hand, males of seasonally breeding species can avoid paying the costs of competition year-round. For example, squirrel monkeys and gray mouse lemurs produce sexually dimorphic body mass only for the breeding season (Schmid and Kappeler 1998; A. Stone 2014). For nonseasonally breeding males, having sporadic periods of mating activity could allow males to compensate for energy losses by increasing their intake on nonmating days. Available data suggest that this is not necessarily the case. Wild chimpanzee males, whose mating opportunities are spread over time, suffered energetic losses on mating days, with high copulation rates and increased aggression predicting low foraging effort (Georgiev, Russell, et al. 2014). Unlike their lactating female counterparts (C. M. Murray, Lonsdorf, et al. 2009), males did not compensate for reduced foraging effort by shifting to higher-quality foods (Georgiev, Russell, et al. 2014). However, even on nonmating days, high-ranking males spent the least time feeding. This suggests that the need for males to remain competitively viable for year-round breeding requires them to pay collateral costs of reproduction even when no immediate mating opportunities are available.

Seasonal variation in the costs of reproduction for males will depend on the breeding pattern of females. This means, effectively, that the costs of breeding for males and females will be asynchronous for many seasonally breeding species. In species where females are adapted to giving birth during food-rich seasons, males will experience their highest energetic investments during seasons of relative food scarcity. Unlike females, for which the initiation of a reproductive effort begins a lengthy energetic investment, males should have more flexibility in their allocation of reproductive effort in the face of poor energetic conditions.

This suggests that male resource access may be an important factor in the selection of alternative reproductive tactics and in determining reproductive skew. While theoretical models typically acknowledge that tactics should be selected based on costs relative to benefits, the majority of empirical work on these issues in primates has focused primarily on variation in reproductive benefits.

SUMMARY AND QUESTIONS FOR FUTURE RESEARCH

Both male and female primates face increased nutritional demands for reproduction. The need for increased energy for gestation and lactation has been widely recognized, and a growing literature documents how food availability and reproductive state interact to drive breeding seasonality and changes in foraging behavior, along with specialized adaptations such as cooperative breeding. Unfortunately, data on reproductive ecology are lacking for many of the world's wild primate species, while many others are represented only by basic data on reproductive rates and seasonality. Data for many species are extrapolated from one well-studied population, sometimes in captivity. Thus, we have a good grasp of how primates can meet the costs of reproduction but as yet very little resolution on how ecological and taxonomic variation affect the use of different strategies and to what extent resources, in comparison to other factors such as predation, constrain reproductive success. The effects of energetic constraints on male reproduction, and how that varies with breeding system, are only just beginning to be explored.

Primatology is experiencing a number of trends that promise to move this area of inquiry forward. One is the increasing focus on intraspecific diversity in foraging and reproductive ecology. A second is the cultivation of long-term field data sets, several generations deep, from a diverse selection of primate taxa that can document the ways that changes in ecology affect reproduction over time. Together with interspecific contrasts, these data will be vital grist for multilevel hypothesis testing. A third trend is the increasing use of physiological monitoring of hormones that reflect reproductive function as well as physiological and nutritional stress (Emery Thompson 2017). These methods allow for fine-grained analyses of the proximate effects of changing diet on reproductive function. Additionally, they can allow researchers to move from proxy measures, such as habitat-wide food availability, rainfall, and dominance rank, to assessments of how these variables affect the physical condition of individual animals. Key challenges remain. One major challenge is the ability to compare the quality of diets across different populations, particularly when it is not clear what nutritional factors are most limiting for reproduction. Emerging quantitative approaches in nutritional ecology, such as the geometric framework (Raubenheimer, Simpson, and Mayntz 2009; Raubenheimer, this volume), have a lot of promise for evaluating how dietary strategies change with reproduction (Felton, Felton, Lindenmayer, et al. 2009), but relevant data are only just emerging for primates (Dröscher et al. 2016).

As noted in this chapter, there is ample evidence for reproductive failure under conditions of nutritional stress. However, relatively little research has been done on the more moderate and long-term consequences. These include the influence of maternal nutrition on infant behavioral and motor development, juvenile mortality, social status, and long-term maternal health, as well as the influence of male nutrition on the ability to attain and maintain rank, selection and efficacy of alternative reproductive tactics, and survival. Similarly, increased research effort on these topics will facilitate comparative analyses of how different reproductive strategies influence the vulnerability of reproductive efforts to failure. For example, are capital breeders more susceptible to fetal loss than income breeders? Is maternal nutrition a stronger predictor of infant survival in larger species than smaller ones, and if so, does this affect the variance in investment by mothers? Are energy deficits during gestation more or less impactful than those during lactation, and does the answer vary by species? While answering these questions will require intensive, long-term research investment, this kind of research will be important for moving forward from an understanding of the energy requirements of reproduction toward a broader awareness of how nutrition structures primate life histories.

CONCLUSIONS

- Primate females produce relatively costly large-bodied, large-brained infants; thus, strategies to manage the costs of reproduction over time are central to their biology. Among these strategies are a lengthening of

pregnancy and lactation relative to other mammals, resulting in lower daily costs of reproduction.
- Lactation is the costliest part of the reproductive effort, raising energy requirements in most primate species by up to 30% by the time infants begin to forage independently. Despite increased energy needs, primate mothers face trade-offs between infant care and foraging efficiency.
- There is little evidence for changes in feeding behavior during pregnancy. During lactation, primate mothers often increase energy intake by spending more time feeding, increasing intake rates, or focusing on higher-quality foods. When this is not possible, mothers decrease the energy expended in travel or draw from energy reserves. Adaptations for seasonal breeding or for increased sensitivity of reproduction to physical condition help coordinate energy supply and energy need.
- Some primate species face unusually high reproductive costs (such as callitrichids and humans) or highly constrained environments (such as strepsirrhines). These species exhibit social adaptations, including cooperative breeding and female dominance, that defray the costs to mothers.
- Experimentally induced as well as seasonal or interindividual variation in energy availability can have consequences at various stages of reproduction, influencing rates of conception and fetal loss, infant mortality, growth and development, and maternal health. Data from two of the most well-studied species, baboons and humans, suggest that the influences of maternal nutrition on offspring health can persist into adulthood. Alongside its impacts on offspring development, energy availability affects the resumption of cycling in mothers, generating energy allocation trade-offs between infant growth and future reproduction that are just beginning to be empirically demonstrated.
- While female investment in producing offspring is the most direct and long-lasting, male primates often expend considerable energy on either paternal effort or mating competition. Their reproductive effort also produces foraging constraints that may be equal to or greater than those experienced by females. This emerging literature suggests a more important role for nutrition as a determinant of male reproductive success than has been recognized by socioecological theory.

5 Primate Energy Requirements
Brains, Babies, or Behavior?

Alex R. DeCasien, Mary H. Brown, Stephen R. Ross, and Herman Pontzer

All of life's tasks, from growth and reproduction to movement and maintenance, require energy. A basic accounting of any species' evolved strategies for survival and reproduction thus requires that we understand how much energy it needs each day and how that energy is allocated to competing tasks. The energy budgets of primates are of particular interest due to their remarkably slow life histories, large brains, and active lifestyles.

Historically, most of the work examining primate energetics has focused on basal metabolic rate (BMR, in kcal/d), the rate of energy expended while the organism is at rest, unfed, and unstressed. BMR is an important component of the daily energy budget, but the focus on BMR was largely a matter of methodological constraints. Prior to the development of doubly labeled water (DLW), there was no technique available to measure total energy expenditure (TEE, in kcal/d), the total energy expended over a 24-hour period. However, with the development of the DLW method and its application over the past three decades (Nagy et al. 1999; Speakman 1997), our understanding of the daily energy requirements of primates and other animals has grown. In this chapter, we examine the effects of ecology, anatomy, and life history on TEE among primate species.

BASAL METABOLIC RATE

Prior to 2000, nearly all the work in primate energetics focused on BMR. Measurements of BMR in humans and nonprimates date back to the late 1800s and early 1900s, the dawn of metabolic science and a period of rapid development in theory and methods (Pontzer 2017). Seminal work on nonhuman primates by Bruhn (Bruhn 1934; Bruhn and Benedict 1936) showed that primate BMRs were similar to those of other mammals, increasing in rough proportion to body surface area, or mass to the 0.66 power. Work by Kleiber (1932), beginning in the 1930s, would eventually overturn the notion that metabolic rates increase with surface area, showing instead that BMR increases with body mass to the 0.75 power, a relationship known as Kleiber's law. The early work on nonhuman primate BMR became part of a larger comparative data set that grew quickly through the mid-1900s, as respirometry equipment and methods became more commonplace and easier to use.

With large data sets and the focus on allometric analysis in the 1970s and 1980s, McNab and others sought to link primates' exceptionally slow life histories and arboreal lifestyle to variation in BMR (McNab 1980, 1986), largely without success. While McNab argued that primates evolved low BMR as a consequence of their arboreality, comparative analyses accounting for phylogenetic relatedness generally found no difference in BMR between primates and other placental mammals (Harvey et al. 1991). Others have suggested that parasite species richness or maintaining the immune system contribute to broad trends in BMR variation (Lochmiller and Deerenberg 2000; Morand and Harvey 2000). Strepsirrhine primates do have marginally lower BMR than monkeys and apes (Armstrong 1985; Müller 1985; Pontzer et al. 2014; Ross 1992; Snodgrass et al. 2007), but this only complicates McNab's proposed links to life history or arboreality, since both groups are arboreal. Within primates, nocturnality and folivory have also been linked to lower BMR (Dunbar and Shultz 2007; Ross 1992; but see Harvey et al. 1991; Ross 1992).

The most consistent determinant of variation in BMR is the size of metabolically costly organs, but even this effect is not always evident. Within species, studies have found seasonal differences in BMR that are attributable to variation in the sizes of metabolically active organs (e.g., tree sparrows, *Passer montanus*; J. Liu and Li 2006; Zheng et al. 2008), while others have shown no effect of organ mass (e.g., deer mice, *Peromyscus maniculatus*; Russell and Chappell 2007). Between species, relative BMR (i.e., residual BMR after accounting for body size) is positively correlated with both relative heart and kidney mass across birds (Daan et al. 1990). Relative BMR is positively correlated with relative brain size across mammals, accounting for up to 15% of brain size variation in mammals and up to 23% in primates (Isler and van Schaik 2006, 2009; but see Finarelli 2009). The relationship between brain size and BMR has also been replicated in other analyses limited to primates (Dunbar and Shultz 2007). Furthermore, there is some evidence that large-brained, altricial, polytocous species produce relatively heavier neonates than smaller-brained species, suggesting that they are able to disproportionately increase their metabolic rate during gestation (Isler and van Schaik 2009).

TOTAL ENERGY EXPENDITURE

Measurements of BMR are useful for understanding how energy is allocated to different organs and competing physiological tasks, but BMR is only one portion of the total energy budget, or TEE. Variation in BMR is not necessarily reflected in TEE. In fact, building on measurements of BMR to construct estimates of TEE and daily energy budgets has proved surprisingly difficult and unreliable.

In the 1970s and 1980s, researchers in primatology and in public health and nutrition developed methods for estimating TEE from an individual's body size and daily activity budget (Coelho 1974; Coelho et al. 1979; FAO 2004; James and Schofield 1990). This approach, sometimes called the "factorial method," assumes that variation in TEE is driven primarily by variation in activity levels, after accounting for effects of body size and BMR. While this approach continues to be used in public health (FAO 2004), growing evidence indicates that variation in TEE, both within and across species, is unrelated to behavioral differences.

The development of the DLW method has made it possible to accurately measure TEE during normal daily life without relying on activity-based estimations. DLW uses stable isotopes to track the body's production of carbon dioxide, a physiological measure of energy expenditure, over several days (typically one week; Nagy et al. 1999; Speakman 1997). The method has been widely validated in humans and other mammals (Speakman 1997) and has become the gold standard for measurements of TEE outside of a laboratory setting. Since the early 2000s, DLW studies have shown that TEE is not simply a function of body size and activity, as assumed by the factorial method.

DLW studies in humans and nonhuman primates have demonstrated that daily physical activity does not predict variation in TEE across populations or species. In nonhuman primates, captive and wild populations exhibit similar TEE despite obvious differences in lifestyle and activity levels (Pontzer et al. 2014), a surprising result that is evident in other mammalian groups as well (e.g., sheep, kangaroos, and pandas; Nie et al. 2015; Pontzer et al. 2014). In human studies, Luke, Ebersole, and colleagues demonstrated that daily energy expenditures were similar in relatively sedentary US women and relatively active Nigerian women (Ebersole et al. 2008). Adult Hadza hunter-gatherers and adults in the US and Europe have similar TEE in analyses controlling for body size (Pontzer et al. 2012), despite remarkably high levels of physical activity among the Hadza (Raichlen, Pontzer, et al. 2017). TEE is also similar among countries with high and low levels of socioeconomic development (Dugas et al. 2011). Urlacher and colleagues, studying energy expenditure among children in the Shuar population of Ecuador, found that daily physical activity levels and BMR were elevated among Shuar children in rural forager-horticultural communities, but TEE was similar to that of an age-matched cohort of Shuar children in more urban settings and to those of age-matched cohorts of children from the United States and United Kingdom (Urlacher,

Snodgrass, Dugas, Madimenos, et al. 2021; Urlacher, Snodgrass, Dugas, Sugiyama, et al. 2019). Adults among the Tsimane forager-horticulturalist community in Bolivia have somewhat elevated TEE, but this elevation is fully accounted for by their greater BMR, which is in turn related to elevated immune activity (Gurven et al. 2016). In laboratory studies of rodents and birds, experimentally induced increases in daily physical activity generally do not lead to corresponding increases in TEE (O'Neal et al. 2017; Pontzer 2015).

The lack of correspondence between TEE and activity is apparently due to evolved mechanisms that allow organisms to adapt dynamically to long-term changes in activity in order to maintain TEE within some relatively narrow range (Pontzer 2015, 2018). This constrained TEE model for daily energy expenditure is supported by cross-sectional studies of TEE and activity within humans (Pontzer, Brown, Raichlen, et al. 2016) as well as longitudinal exercise-intervention studies in humans reporting that the increase in TEE is less than expected from the imposed exercise workload and, in some cohorts, indistinguishable from baseline preintervention TEE (Donnelly et al. 2003; Goran and Poehlman 1992; C. Martin et al. 2019; Pontzer 2018; X. Wang et al. 2017; E. Willis et al. 2014). Similar results have been observed in experimental studies across a broad range of vertebrates, including birds and rodents (Pontzer 2015; Pontzer and McGrosky 2022), and are consistent with a broader literature on metabolic adaptation in public health (Ross and Janssen 2001; Thomas et al. 2012).

Rather than reflecting activity levels, DLW measurements suggest that TEE may be correlated with life history and other aspects of ecology (Nagy et al. 1999; Pontzer et al. 2014; Pontzer, Brown, Raichlen, et al. 2016; Pontzer and McGrosky 2022; Pontzer, Raichlen, et al. 2010). Across placental mammals, TEE is correlated with reproductive output (grams of offspring per year) but not growth or senescence, after controlling for body mass and phylogenetic relatedness (Pontzer et al. 2014). Primates' TEE is 50% lower than those of other placental mammals in analyses controlling for body size and phylogeny, and this drastic reduction in metabolic rate corresponds with primates' slow rates of growth, reproduction, and senescence (Pontzer et al. 2014). Among hominoids, humans exhibit an increased TEE and BMR, which is likely related to increased brain size and faster reproduction in the hominin lineage (Pontzer, Brown, Raichlen, et al. 2016). Orangutans have lower TEE relative to body mass than almost any primate or eutherian mammal ever measured (Pontzer, Raichlen, et al. 2010). Such an extremely low rate of energy use not only is consistent with their slow life histories but also likely reflects an additional evolutionary response to severe, unpredictable food shortages in their native Southeast Asian rainforests (Pontzer, Raichlen, et al. 2010).

TEE and BMR are only loosely correlated among mammals (Pontzer et al. 2014), suggesting they can and do evolve independently and may reflect different aspects of physiology and ecology. Primates provide a notable example; their low TEE is not evident in their BMR, which is similar to other mammals (or only marginally lower, in the case of strepsirrhines; Pontzer et al. 2014). However, while BMR has been the subject of many large comparative analyses over the past several decades, TEE has received relatively little attention. The relationship between life history and TEE has been examined across mammals, but not within primates. Apart from focused analyses of TEE among hominoids (Pontzer, Brown, Raichlen, et al. 2016), the effects of brain size, diet, and social ecology on variation in TEE among primate species is untested. These analyses are needed to reconstruct the broad framework of selection pressures shaping metabolic physiology, and therefore daily energy requirements, among primates.

DETERMINANTS OF TEE AMONG PRIMATES

We examined the effects of ecology and anatomy on TEE in primates. Specifically, we investigated the relationship between TEE and (1) brain size, (2) life history, (3) diet category, (4) social organization, and (5) physical activity among primate species. We also tested for sexual dimorphism in TEE, controlling for body size. We tested the hypotheses that TEE is positively related to variation in brain size, rates of reproduction and growth, and physical activity. We also tested the hypothesis that folivores follow an energy-minimizing strategy, with reduced TEE. To conduct these tests, we compiled TEE from a range of feeding studies and DLW studies. We therefore also tested for differences in TEE estimates between methods.

Our data set was assembled from 43 populations of 38 primate species (table 5.1; fig. 5.1). Measurements of

TEE done with the DLW method were taken from the literature (Edwards et al. 2017; Pontzer et al. 2014; Pontzer, Brown, Raichlen, et al. 2016). For humans, we used data from the Hadza hunter-gatherer population (Pontzer et al. 2012) and from a sedentary population (Pontzer, Brown, Raichlen, et al. 2016). For the great apes (orangutans, gorillas, chimpanzees, and bonobos), we used measurements from a recent analysis with larger sample sizes (Pontzer, Brown, Raichlen, et al. 2016) instead of values from initial reports.

We also included new measurements of two other hominoids, the siamang (*Symphalangus syndactylus*) and white-cheeked gibbon (*Nomascus leucogenys*). These species were measured ($N = 1$ for each species), with prior institutional approval, at the Gibbon Conservation Center (Santa Clarita, CA, USA) and Lincoln Park Zoo (Chicago, IL, USA), respectively. For these measurements, we followed the same protocol used previously for other apes (Pontzer, Brown, Raichlen, et al. 2016). Briefly, a dose of DLW (6% ^2H_2O, 10% $H_2^{18}O$) tailored to the subject's body mass (Speakman 1997) was mixed with fruit juice and given orally. Urine samples were collected via pipette from a completely dry floor, prior to dosing, 12 hours after dosing, and then on days 2, 4, 6, and 10. Urine samples were frozen at −5°C, shipped to the Pontzer Lab at Hunter College in New York, and then analyzed for isotope enrichment using cavity ring-down spectrometry (Picarro Inc.). We used the slope-intercept method to calculate dilution spaces for ^2H and ^{18}O. We calculated CO_2 production using equation 7.17 in Speakman (1997), assuming a dilution space ratio of 1.04 (measured space ratios were 1.04 for the gibbon and 1.03 for the siamang). TEE was calculated from CO_2 production using the modified Weir equation and a food quotient of 0.95 based on diet data from participating zoos. Body masses were measured using a digital scale.

We added estimates of TEE measured from careful studies of food intake, from both wild and captive populations. Wild populations were measured using long-term (several weeks to several months) focal observations paired with lab analyses for the nutrient content of the foods eaten (Conklin-Brittain, Knott, and Wrangham 2006; Felton, Felton, Raubenheimer, Simpson, Foley, et al. 2009; C. A. Johnson, Raubenheimer, Rothman, et al. 2013; Rothman, Dierenfeld, Hintz et al. 2008). Data for captive populations were taken primarily from entries for TEE in table 2.1 of *Nutrient Requirements of Nonhuman Primates* (NRC 2003).

Data on adult brain weights, life history variables (gestation length, litter size, interbirth interval, neonate mass, and maximum life span), and behavioral variables (home range, day range, diet category, group size, and mating system) were extracted from the published literature. Sources are listed in table 5.1. Additional variables were derived from these measures, including gestation growth rate (neonate mass / gestation length), litter mass (litter size × neonate mass), and reproductive output (litter mass / interbirth interval).

STATISTICAL ANALYSES

All statistical analyses were conducted in R 3.2.0 (R Development Core Team 2014). All continuous variables except gestation length and litter size were log transformed prior to analysis since these variables follow a lognormal distribution and transformation severely reduced skew.

Species represent nonindependent cases since they may share traits due to phylogenetic inertia, so we used phylogenetic mixed models implemented in the R package MCMCglmm (Hadfield 2010). This approach allowed us to incorporate population-level data while controlling for phylogenetic relatedness between species. The latter was accomplished by including the 10kTrees consensus phylogeny (C. Arnold et al. 2010) as a random effect. Across all models, the default prior for the mean and variance of fixed effects for Gaussian-family models in MCMCglmm and a relatively uninformative inverse-Wishart prior for the random effects and residual variances were used ($V = 1$; $v = 0.002$). Models were run for 1,000,000 iterations, sampling every 100 iterations with a burn-in of 500,000. The value of λ was estimated at each iteration, and for each model, we report the posterior mean for this term. We also report the deviance information criterion (DIC), a measure of model fit that is closely related to the Akaike information criterion (AIC). Markov chain Monte Carlo diagnostics were run using the R package coda (Plummer et al. 2006). Specifically, we ensured that proper mixing occurred by visually inspecting all trace and density plots, and we examined autocorrelation plots to confirm reduced correlation between successive samples. We confirmed that the effective sample sizes for all variables were greater than 1,000, and we ran each chain twice and confirmed convergence using the Gelman-Rubin statistic, with all models confirmed to have potential scaling reduction factors below 1.1 (Gelman and Rubin 1992).

We first tested whether there was a significant effect of body mass and methodology (DLW or food intake) on estimated TEE by modeling TEE (log) as a function of body mass (log) and method. In line with previous work, we found that body mass predicts TEE and that for every 1% increase in body mass, TEE increases 0.67% (fig. 5.2). We also found that studies using food intake and measures of metabolizable energy overestimate daily energy intake by approximately 31% compared to DLW (fig. 5.2), which is considered the gold standard for TEE measurement. The reasons that food intake studies tend to overestimate TEE are not immediately clear and may differ across studies. Intake studies may overestimate the amount of food consumed, its digestibility, or its energy content. Recent studies by Rothman and colleagues have worked to advance each of these areas (Rothman, Dierenfeld, Hintz, et al. 2008; C. A. Johnson, Raubenheimer, Rothman, et al. 2013), but more work is needed. Studies employing both methods (DLW and intake) in the same subjects would be particularly useful for unifying these approaches to estimating TEE.

To test whether any of our variables of interest explain additional variation in TEE after controlling for body mass and method, we created separate phylogenetic mixed models with TEE (log) as the dependent variable, body mass (log) and method as covariates, and each variable of interest as a predictor. We first tested for an effect of captivity to consider multiple populations in our analyses. In line with previous studies (e.g., Munn et al. 2011; Pontzer et al. 2014; Rezende et al. 2009; Stephenson et al. 1994), captivity did not have a significant effect on TEE after we controlled for body mass, method, and phylogeny (fig. 5.2). Consequently, we did not include captivity as a factor in models testing for effects of brain size, life history, and ecological variables.

Assumptions of linear modeling were tested and confirmed as follows. The linear relationship assumption was confirmed using plots of the dependent variable versus fitted values (even spread around a line with slope = 1) and of residuals versus fitted values (even spread around a line with an intercept and slope = 0). The exogeneity assumption was upheld since the mean of the error values was near zero and the errors were uncorrelated with the explanatory variables. The homoscedasticity assumption was upheld as the errors exhibited constant variance, which was confirmed using plots of residuals versus the dependent variable (constant variance around a line with intercept and slope = 0). The errors followed a normal distribution, confirmed using QQ plots and histograms of residual values. We calculated variance inflation factors using the car package (J. Fox et al. 2016). In most cases, these values did not exceed 3.3, indicating a lack of multicollinearity. Although there is a wide range of cutoff variance inflation factors in the literature (e.g., 10; Freund and Wilson 1998), a cutoff of 3.3 was employed here since it is a commonly used standard (e.g., Pettang 2016), indicating the point at which $R^2 = 0.70$ between variables. Some models exhibited high variance inflation factors (due to high correlations between the variable of interest and body mass), indicating that multicollinearity may affect results. In these cases, additional models were created in which these variables, rather than body mass, were included with method as predictor variables.

We also examined sex differences in a subset of species ($N = 11$) for which sex-specific information was available. In all cases, TEE was measured using the DLW method, so our phylogenetic mixed model included TEE (log) as the dependent variable and body mass (log) and sex as predictors. We also tested a model that included an interaction term between body mass and sex (to test for sex differences in slope).

PHYLOGENETIC EFFECTS

Posterior mean estimates for λ, a measure of the effect of phylogenetic relatedness, were near 0.30 across all models (fig. 5.2), indicating a moderate effect of phylogenetic relatedness on the interspecific patterning of relative TEE. These estimates were lower than previous analyses that included primates and other placental mammals, which reported λ values close to 1 (Pontzer et al. 2014). The lower value for λ here may be due to the narrower phylogenetic focus taken by the current study (as it excludes nonprimates).

Table 5.1. Total energy expenditure (TEE) and other variables.

Taxon	Body mass (kg)	TEE (kcal/day)	Captive (1, yes; 0, no)	Method	Brain size (g)	Gest. length (days)	Litter size	Inter-birth interval (days)	Neonate mass (g)	Litter mass† (g)	Gest. growth† (g/day)	Repro. output† (g/day)	Life span (months)	Home range (km²)	Day range (km)	Diet	Group size	Mating system§
Allenopithecus nigroviridis	7.90	524	1	DLW *a*	61.2 *d*	—	1.0 *e*	—	221.0 *e*	221.0	—	—	324 *e*	—	—	Omni. *d*	40.0 *d*	polyGA *d*
Alouatta caraya	8.11	402	1	intake *b*	52.7 *d*	187 *e*	1.0 *e*	338 *e*	187.5 *e*	187.5	1.00	0.55	389 *e*	0.03 *s*	0.40 *s*	Foli. *d*	6.7 *d*	polyGA *d*
Alouatta palliata	7.12	602	0	DLW *a*	50.9 *d*	186 *e*	1.0 *e*	384 *e*	409.0 *e*	409.0	2.20	1.07	300 *e*	0.18 *t*	0.30 *t*	Foli. *d*	13.1 *d*	polyGA *d*
Alouatta palliata	5.85	429	1	intake *b*	50.9 *d*	186 *e*	1.0 *e*	384 *e*	409.0 *e*	409.0	2.20	1.07	300 *t*	0.18 *t*	0.30 *t*	Foli. *d*	13.1 *d*	polyGA *d*
Alouatta sara	8.33	416	1	intake *b*	62.4 *d*	190 *e*	1.0 *e*	424 *e*	263.0 *e*	263.0	1.38	0.62	300 *e*	1.82 *u*	—	Foli. *d*	3.3 *d*	polyGA *d*
Ateles chamek	9.37	480	0	intake *b*	105.9 *d*	229 *e*	1.0 *i*	1072 *e*	480.0 *i*	480.0	2.10	0.45	552 *e*	2.00 *v*	—	Frug. *d*	23.7 *d*	polyGA *d*
Callithrix jacchus	0.47	52	1	DLW *a*	7.6 *d*	144 *e*	2.0 *e*	169 *e*	26.5 *e*	53.0	0.18	0.31	274 *e*	0.03 *w*	—	Frug. *d*	8.9 *d*	polyA *d*
Callithrix pygmaea	0.13	27	1	intake *b*	4.5 *d*	137 *n*	2.1 *i*	171 *e*	16.0 *i*	33.6	0.12	0.20	223 *e*	—	—	Omni. *d*	6.2 *d*	polyA *d*
Cebus apella	4.06	321	1	DLW *a*	68.7 *d*	158 *e*	1.5 *e*	671 *e*	239.7 *e*	359.6	1.52	0.54	552 *e*	0.80 *x*	2.07 *x*	Omni. *d*	12.1 *d*	polyGA *d*
Colobus guereza	10.90	542	1	intake *b*	77.4 *d*	175 *e*	1.0 *e*	365 *e*	397.8 *e*	397.8	2.27	1.09	420 *e*	0.10 *t*	0.26 *t*	Foli. *d*	8.6 *d*	HpolyG *d*
Daubentonia madagascariensis	2.46	260	1	intake *b*	46.3 *d*	167 *e*	1.0 *e*	912 *e*	109.0 *e*	109.0	0.65	0.12	292 *e*	1.03 *y*	1.72 *y*	Omni. *d*	1.8 *d*	sol *d*
Eulemur fulvus	1.84	146	0	DLW *a*	24.8 *d*	118 *e*	1.1 *e*	537 *e*	83.3 *e*	91.6	0.71	0.17	354 *e*	0.01 *t*	—	Fr./Fo. *d*	10.1 *d*	polyGA *d*
Gorilla beringei	200.0	9209	0	intake *a*	509.0 *d*	255 *f*	1.0 *e*	1494 *v*	1996.0 *q*	1996.0	7.83	1.34	720 *e*	5.00 *r*	0.40 *r*	Foli. *d*	13.0 *d*	HpolyG *d*
Gorilla gorilla	120.1	2830	1	DLW *c*	507.2 *d*	256 *e*	1.0 *e*	1397 *e*	2061.4 *e*	2061.4	8.05	1.48	665 *e*	8.00 *t*	0.47 *t*	Foli. *d*	9.0 *d*	HpolyG *d*
Homo sapiens (foragers)	46.63	2212	0	DLW *a*	1250 *d*	280 *e*	1.0 *e*	639 *e*	3312.5 *e*	3312.5	11.83	5.18	1470 *e*	—	—	Omni. *d*	150.0 *d*	—
Homo sapiens (sedentary)	76.95	2456	1	DLW *c*	1250 *d*	280 *e*	1.0 *e*	639 *e*	3312.5 *e*	3312.5	11.83	5.18	1470 *e*	—	—	Omni. *d*	150.0 *d*	—
Lemur catta	2.24	146	0	DLW *a*	23.2 *d*	135 *e*	1.1 *e*	365 *e*	70.6 *e*	77.7	0.52	0.21	448 *e*	0.06 *t*	0.75 *t*	Frug. *d*	15.8 *d*	polyGA *d*

(*continued*)

Table 5.1. (continued)

Taxon	Body mass (kg)	TEE (kcal/day)	Captive (1, yes; 0, no)	Method	Brain size (g)		Gest. length (days)		Litter size		Inter-birth interval (days)		Neonate mass (g)		Litter mass† (g)	Gest. growth† (g/day)	Repro. output† (g/day)	Life span (months)		Home range (km²)		Day range (km)		Diet		Group size		Mating system¶
Lemur catta	2.21	217	1	DLW a	23.2	d	135	e	1.1	e	365	e	70.6	e	77.7	0.52	0.21	448	e	0.06	t	0.75	t	Frug.	d	15.8	d	polyGA d
Leontopithecus rosalia	0.68	110	1	intake b	13.1	d	128	e	2.0	e	189	e	55.3	e	110.6	0.43	0.59	379	e	0.30	r	1.47	r	Omni.	d	5.4	d	mon d
Lepilemur ruficaudatus	0.77	121	0	DLW a	8.4	d	157	g	1.0	g	365	w	27.0	g	27.0	0.17	0.07	—		0.08	g	—		Foli.	d	1.0	d	sol d
Macaca fascicularis	5.70	627	1	intake b	66.5	d	165	e	1.0	e	431	e	320.0	e	320.0	1.94	0.74	468	e	1.13	r	1.90	cc	Frug.	d	27.5	d	polyGA d
Macaca mulatta	14.40	607	1	DLW a	88.9	d	165	e	1.0	e	444	e	464.0	e	464.0	2.81	1.05	480	e	2.30	t	1.40	t	Frug.	d	36.2	d	polyGA d
Macaca radiata	4.20	251	1	DLW a	79.5	d	161	e	1.0	e	365	e	394.0	e	394.0	2.45	1.08	360	e	2.30	t	0.88	t	Frug.	d	28.9	d	polyGA d
Microcebus murinus	0.06	28	0	DLW a	1.7	d	61	e	2.0	e	73	e	6.0	e	12.0	0.10	0.16	218	e	—		—		Omni.	d	5.1	d	sol d
Nasalis larvatus	12.00	1407	1	intake b	96.2	d	166	e	1.3	e	540	e	490.0	e	612.5	2.95	1.13	301	e	1.30	r	0.71	r	Foli.	d	14.5	d	HpolyG d
Nomascus leucogenys	8.10	786	1	DLW new	122.9	d‡	191	h	1.0	e	—		480.0	p	480.0	2.51	—	336	r	—		—		Frug.	d	3.7	d	mon d
Pan paniscus	38.32	1521	1	DLW c	346.2	d	232	e	1.0	e	1825	e	1331	e	1331.0	5.74	0.73	660	e	45.00	x	1.80	r	Frug.	d	67.5	d	polyGA d
Pan troglodytes	56.04	2047	1	DLW c	376.1	d	229	e	1.0	e	840	e	1821	e	1821.0	7.95	2.17	720	e	12.00	t	3.90	t	Frug.	d	28.8	d	polyGA d
Pan troglodytes	39.95	2022	0	intake b	376.1	d	229	e	1.0	e	840	e	1821	e	1821.0	7.95	2.17	720	e	12.00	t	3.90	t	Frug.	d	28.8	d	polyGA d
Papio anubis	16.18	832	1	DLW a	171.8	d	180	i	1.0	i	570	l	1068	i	1068.0	5.93	1.87	480	e	24.30	t	3.60	t	Frug.	d	46.0	d	polyGA d
Papio cynocephalus	12.00	813	0	DLW a	169.4	d	175	i	1.0	i	745	m	854.0	i	854.0	4.88	1.15	552	e	16.70	t	4.50	t	Frug.	d	48.2	d	polyGA d
Papio ursinus	14.80	940	0	intake b	195.4	d	171	e	1.0	e	568	e	814.0	e	814.0	4.76	1.43	540	e	14.70	t	4.80	t	Fr./Fo.	d	41.1	d	polyGA d
Pongo pygmaeus	67.45	1547	1	DLW c	382.5	d	249	e	1.0	e	1414	e	1736.5	e	1736.5	6.97	1.23	540	e	3.00	t	0.40	t	Frug.	d	1.6	d	sol d
Pongo pygmaeus	57.15	2866	0	intake a	382.5	d	249	e	1.0	e	1414	e	1736.5	e	1736.5	6.97	1.23	540	e	3.00	t	0.40	t	Frug.	d	1.6	d	sol d

Species																						
Propithecus diadema	4.90	346	0	DLW	a	41.2	d	157	e	1.0	e	569 z	145.0 e	145.0	0.92	0.25	—	0.25	t	—	Foli. d	5.6 d mon d
Pygathrix nemaeus	12.10	1563	1	intake	b	92.0	d	171	e	1.0	e	365 e	393.5 e	393.5	2.30	1.08	312	0.36	z	0.59 z	Foli. d	14.5 d HpolyG d
Saguinus fuscicollis	0.31	93	1	intake	b	8.1	d	147	e	1.8	e	202 e	39.9 e	71.8	0.27	0.36	299	0.30	x	1.22 x	Omni. d	5.3 d polyA d
Saguinus mystax	0.53	124	1	intake	b	11.5	d	145	e	2.0	e	168 e	46.9 e	93.8	0.32	0.56	240	0.30	r	1.85 r	Omni. d	5.3 d polyA d
Saguinus oedipus	0.47	101	1	intake	b	9.8	d	157	i	1.9	e	280 i	43.2 i	82.1	0.28	0.29	314	0.25	x	1.71 x	Omni. d	6.9 d polyA d
Saimiri sciureus	0.87	166	1	intake	b	24.1	d	178	e	1.9	e	244 e	39.4 e	74.8	0.22	0.31	363	2.50	x	1.50 cc	Omni. d	34.5 d HpolyG d
Trachypithecus francoisi	5.96	675	1	intake	b	94.4	d	—		1.0	k	489 n	457.0 e	457.0	0.00	0.93	315	1.57	aa	—	Foli. d	— HpolyG d
Varecia variegata	4.68	299	1	intake	b	32.6	d	99	e	2.2	e	365 e	87.2 e	191.8	0.88	0.53	432	0.20	r	1.35 r	Frug. d	5.9 d sol d
Symphalangus syndactylus	9.40	402	1	DLW	new	128.9	d	233	j	1.0	e	1369 o	536.9 p	536.9	2.31	0.39	516	0.23	bb	1.00 bb	Foli. d	3.8 d mon d

Sources: a. Pontzer et al. 2014; b. Knapka 2003; c. Pontzer, Durazo-Arvizu, et al. 2016; d. DeCasien et al. 2017; e. AnAge; f. M. Robbins 1999; g. Hilgartner et al. 2008; h. Rafacz et al. 2013; i. Harvey and Clutton-Brock 1985; j. Palombit 1995; k. Nadler et al. 2003; l. Smuts et al. 1987; m. R. Hill, Lycett, and Dunbar 2000; n. Borries et al. 2010; o. Lappan 2009; p. Geissmann and Orgeldinger 1995; q. Barrickman et al. 2008; r. Rowe 1996; s. Bicca Marques 2003; t. Clutton-Brock and Harvey 1977; u. Palacios and Rodriguez 2001; v. Mcfarland and Symington 1988; w. A. Mendes, Pontes, and Monteiro da Cruz 1995; x. Grant et al. 1992; y. Sterling 1993; z. Dlibarri 2013; aa. Kamilar and Paciulli 2008; bb. J. Mitani and Rodman 1979; cc. Wrangham, Gittleman, and Chapman 1993.

† Litter mass = neonate mass × litter size; gestation growth = neonate mass / gestation length; reproductive output = litter mass / interbirth interval.

¶ polyGA, polygynandry; polyA, polyandry; HpolyG, harem polygyny; sol, solitary; mon, monogamy.

‡ From closely related *N. gabriellae*.

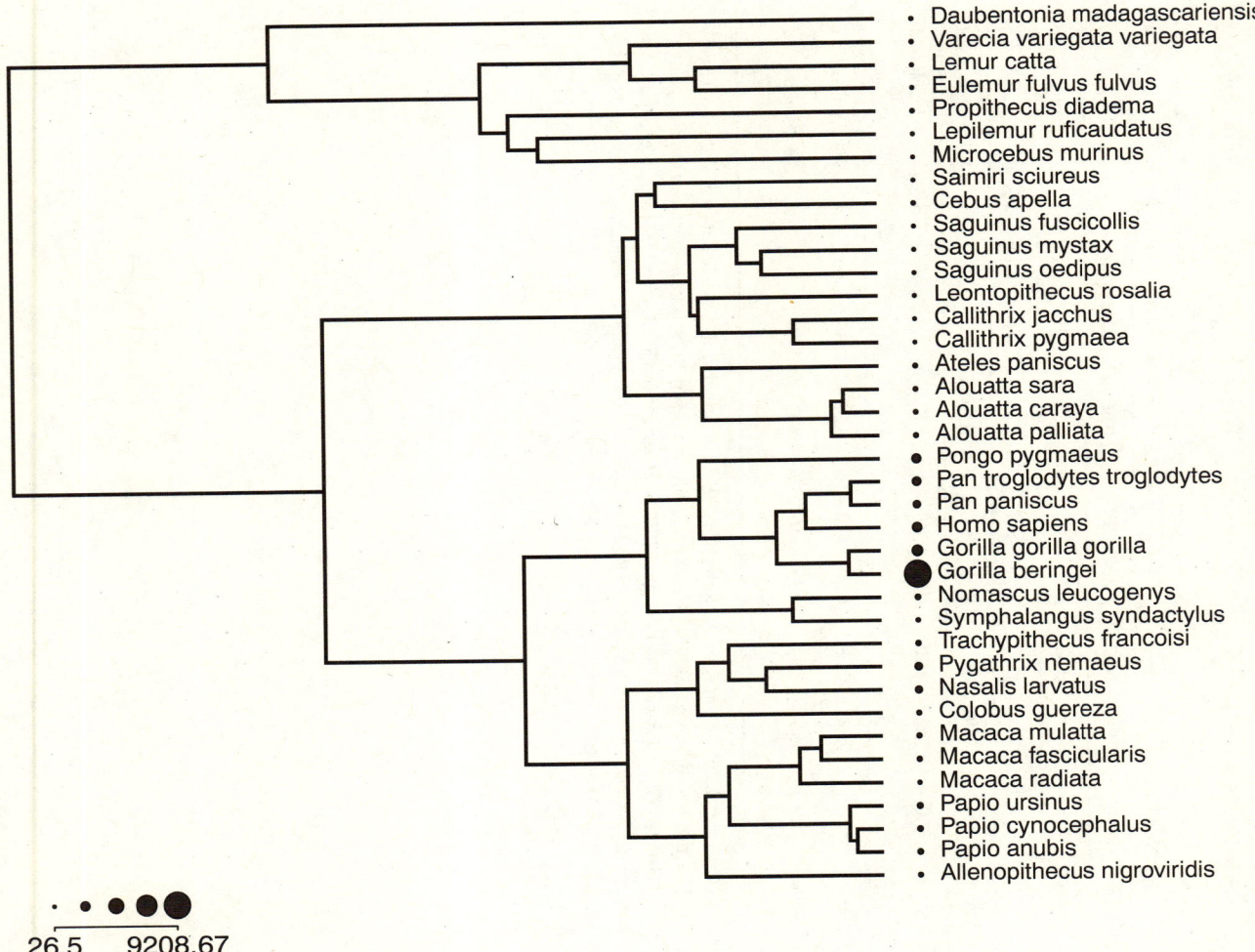

Figure 5.1. Phylogeny of species included in our data set. Species average TEE values (averaged across populations regardless of measurement method) are represented by the size of the circles.

BRAIN SIZE, LIFE HISTORY, AND ECOLOGY

In analyses controlling for the effects of body size, method, and phylogeny, there was no effect of brain size on TEE (fig. 5.2). These results run counter to previous studies reporting a positive effect of brain size on BMR (Isler and van Schaik 2006, 2009). Apparently, species that expend more energy to maintain a larger brain have corresponding reductions in energy expended on other tasks. These results underscore the difficulty in extrapolating from BMR to TEE and the importance of energy allocation.

We also found no correlation between TEE and various life history variables across primate species after controlling for body mass, method, and phylogenetic relatedness (fig. 5.2). These results are generally consistent with previous work that examined life history variables and TEE across a broader phylogenetic sample of primates and nonprimate placental mammals (Pontzer et al. 2014). Although previous work detected an effect of TEE on reproductive output, this inconsistency is likely to reflect differences across studies in species selection and statistical approaches.

The lack of correlation between TEE and life history in the present analysis is at odds with the correspondence between the remarkably slow life histories and low TEE in primates relative to other placental mammals (Pontzer et al. 2014). One possible explanation to reconcile these disparate results is that the correspondence between primate TEE and life history is spurious. We find this unlikely, given the quantitative agreement between TEE and life history. When plotted against TEE, primate growth

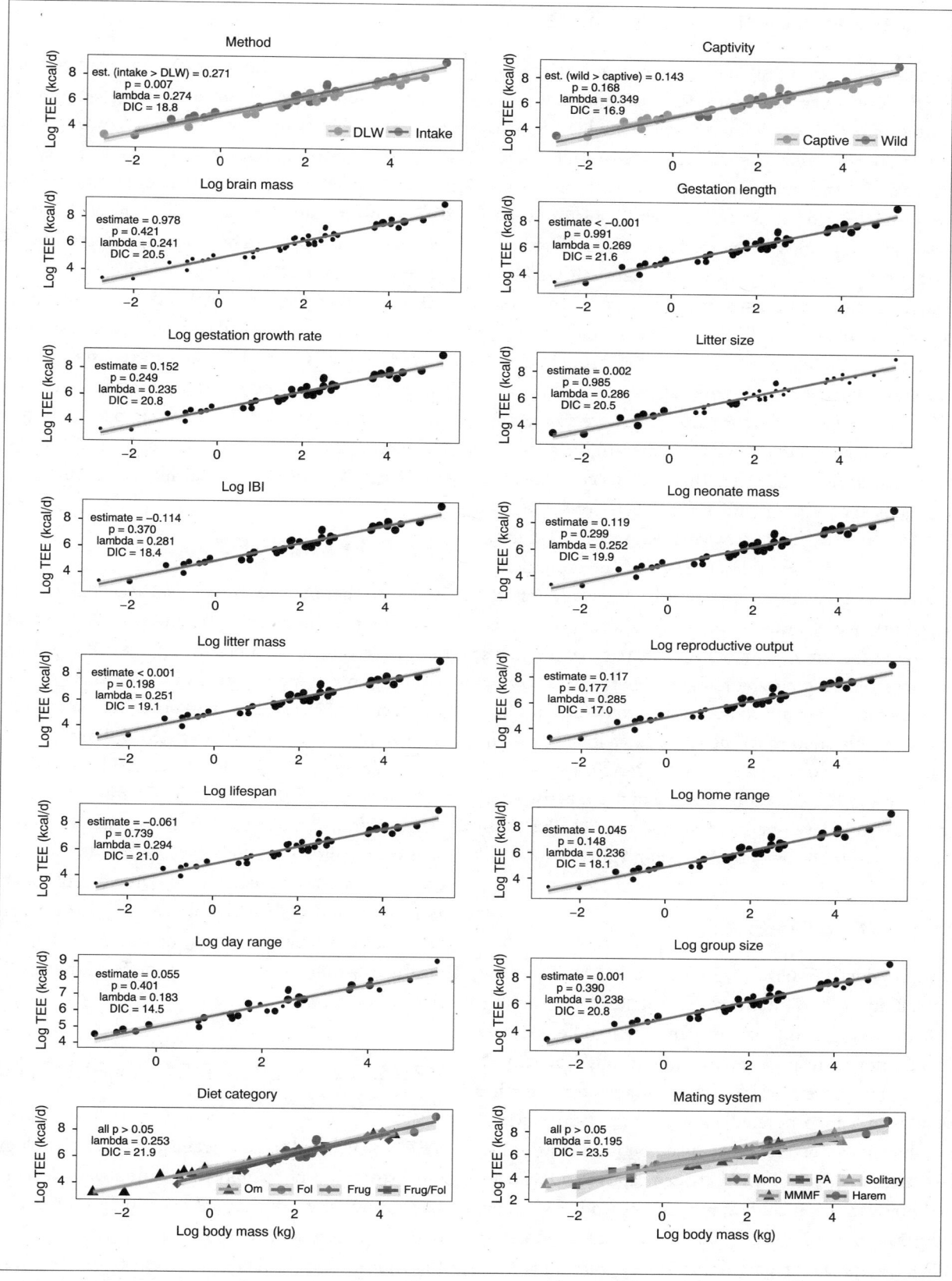

Figure 5.2. Scatterplots of TEE (log) against body mass (log) with other variables indicated by color scheme or size of data points. Linear regression lines and confidence intervals are shown (either for all species or within categories). Coefficient estimates, pMCMC values, DIC values, and mean lambda values are given for each variable from analyses controlling for body mass, method, and phylogeny. Estimates are not provided for categorical predictors with more than two levels given plotting space constraints and since these differences are not significant.

rates, reproductive outputs, and life expectancy fall in line with those of other placental mammals, indicating that primates' remarkable reduction of these life history rates matches the reduction in metabolic rate. An alternative hypothesis for the lack of correspondence between life history and TEE in this analysis is that life history variation within the primate order reflects evolved differences in energy allocation more than total energy throughput. Given the limited sample size and statistical power of our analysis, a small effect of TEE would be difficult to detect if variation in life history were more often brought about by evolved changes in energy allocation.

Diet category, home range size, day range size, group size, and mating system also had no discernible effect on TEE in our data set (fig. 5.2). These results contrast with the proposed energy-minimizing strategies of some folivores (Raño et al. 2016; Rosenberger and Strier 1989). As with life history variables, the lack of correspondence between these ecological traits and TEE likely reflects the importance of variation in energy allocation. That is, species may accommodate costly traits (e.g., larger brains and increased ranging) by reducing energy expenditure in other areas. The importance of such trade-offs in shaping species' physiology and behavior has a long history in evolutionary biology, dating back to Darwin (1861, 147) and Keith (1891), and more recently in the expensive tissue framework developed by Isler and van Schaik (2009, 2012) from seminal work by Aiello and Wheeler (1995).

CAPTIVITY EFFECTS

Captivity does not have a significant effect on TEE after controlling for body mass, method, and phylogeny (fig. 5.2), providing additional evidence that differences in TEE do not correspond to variation in physical activity. Consequently, we excluded this variable from further analyses and incorporated multiple populations where available. Previous work has also found similarities in TEE measured using DLW between captive and wild animal populations, including in red kangaroos (*Macropus rufus*; Munn et al. 2011), sheep (*Ovis aries*; Munn et al. 2011), two species of tenrec (*Microgale dobsoni* and *Microgale talazaci*; Stephenson et al. 1994), giant pandas (*Ailuropoda melanoleuca*; Nie et al. 2015), and deer mice (*Peromyscus maniculatus*) compared to control (warm climate) mice (Rezende et al. 2009).

The similarity in TEE in captive and wild populations is all the more surprising given that captive populations are presumably fatter. The strongest predictor of TEE within or between species is fat-free mass, because fat is much less metabolically active than other tissues (Pontzer 2015; Pontzer, Yamada, et al. 2021; Wang et al. 2001). For that reason, animals with higher body fat percentage (and hence lower fat-free mass) will usually exhibit reduced TEE when plotted against total body mass. The similarity in TEE between captive and wild populations adds weight to the hypothesis that TEE is an evolved, species-specific phenotype that the body works to maintain within a narrow physiological range, and not simply a reflection of body size and activity level (Pontzer 2015, 2018; Pontzer et al. 2012; Pontzer, Brown, Raichlen, et al. 2016).

SEX DIFFERENCES

We found that males and females have similar TEE after controlling for body mass (i.e., the sex effect and interaction effect were not significant; $p_{MCMC} > 0.05$; fig. 5.3). This finding is consistent with previous work in Hadza foragers (Pontzer et al. 2012, 2015) and other human populations (Pontzer, Brown, Raichlen, et al. 2016; Pontzer, Yamada, et al. 2021) reporting no difference in TEE between men and women, after controlling for fat-free mass. We note that very few of the females in these analyses were pregnant or nursing. Nonetheless, these results indicate that, outside of late-term pregnancy and nursing, physiological and behavioral differences between sexes (e.g., greater daily travel distances in males) do not generally result in measurable differences in TEE aside from those attributable to body size.

PRIMATE ENERGETICS: EXPENDITURE AND ALLOCATION

TEE is an evolved trait that reflects a species' ecology and life history; it is not simply the result of variation in body size and activity. Still, natural selection produces changes in both the size of the daily energy budget (i.e., TEE) and the allocation of energy to different tasks. As a result, it is nearly impossible to extrapolate evolved energy budgets from activity, the size of metabolically costly traits such as brain size, or life history alone. Measures of TEE and its components are needed to develop a coherent and

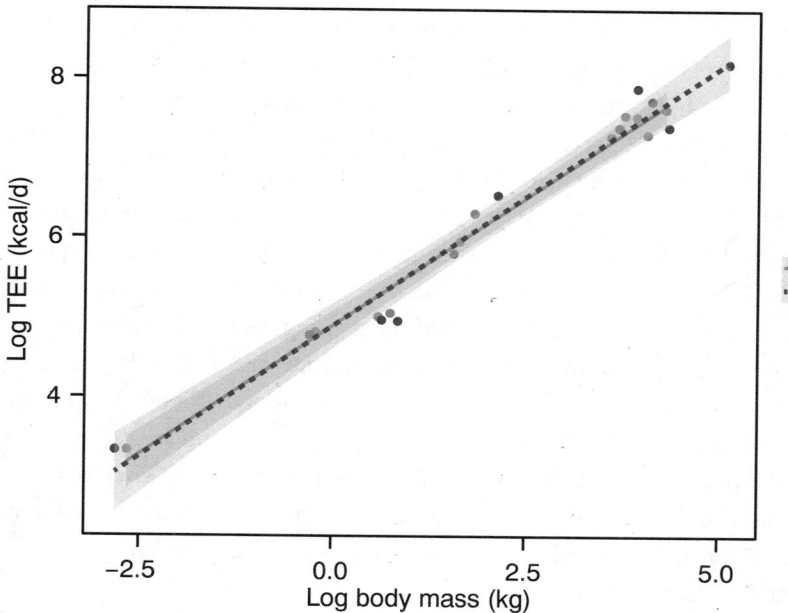

Figure 5.3. Scatterplots of sex-specific TEE (log) against body mass (log). Sex-specific linear regression lines and confidence intervals are shown.

comprehensive understanding of energy expenditure and allocation.

The similarity in TEE among captive and wild populations, across primates with different diets, brain sizes, and life histories, argues strongly against the notion that evolving costly traits necessarily results in increased daily energy requirements. More broadly, the results here and elsewhere caution against the use of activity budgets or other ecological, anatomical, or life history variables in estimating TEE. Such approaches, while intuitively appealing, are not supported by the available evidence. Instead, in broad comparisons across species, trade-offs in energy allocation apparently mask the effects of energetically costly traits on variation in TEE. The success of trade-off frameworks in predicting patterns of variation in primate brain size and life history (Isler and van Schaik 2009) is a testament to the importance of evolved energetic trade-offs.

Nonetheless, there is a growing body of evidence that evolutionary changes in BMR and TEE have played a critical role in primate evolution. Perhaps most notably, the entire primate order appears to have undergone a pronounced reduction in metabolic rate, corresponding with a marked reduction in the pace of growth, reproduction, and senescence (Pontzer et al. 2014). Changes in TEE have also figured prominently in specific lineages (Pontzer, Raichlen, et al. 2010; Pontzer, Brown, Raichlen, et al. 2016). Understanding evolutionary diversification among primates and other clades requires that we consider trade-offs in energy allocation as well as changes in size-adjusted metabolic rate. Such an integrated approach has been proposed by Isler and van Schaik (2009), and we have derived a similar model in our analysis of primate TEE and life history (Pontzer et al. 2014).

ENERGETICS IN HOMINOID EVOLUTION

Metabolic studies of hominoids, the most thoroughly studied primate clade, demonstrate the utility and importance of an integrative approach. Humans evince a suite of metabolically costly traits that define the lineage: a large, costly brain; greater reproductive output; longer life span; and increased daily travel distances and home range (Kraft et al. 2021; Pontzer, Brown, Raichlen, et al. 2016). Some of these increased metabolic costs are offset by the reduction in the size of the metabolically costly digestive tract and a more economical walking gait (Aiello and Wheeler 1995; Navarrete et al. 2011). However, these energy savings are insufficient to account for the complete suite of expensive human traits. Instead, the evolution of these metabolically costly traits was fueled by increases in both TEE and BMR that evolved in concert with decreases in the costs of walking and digestion (Kraft et al. 2021; Pontzer, Brown, Raichlen, et al. 2016).

In contrast, orangutans have evolved a substantially lower TEE and BMR, corresponding with a slower life history, apparently in response to the highly variable nature of food availability in Bornean and Sumatran rainforests, where crashes in ripe fruit, the staple of the orangutan diet, often occur (A. Marshall, Ancrenaz, et al. 2009). Low TEE reduces the probability of starvation, which may improve fitness in environments where food availability is extremely erratic and there are long periods of food shortage (McNab 1986; Pontzer and Kamilar 2009; Sibly and Brown 2007).

Metabolic evolution among hominoids also underscores the broad reach of energetics into other aspects of physiology and ecology. In humans, the evolution of increased TEE has coincided with a suite of behavioral and anatomical changes to increase energy intake and mitigate the risks of starvation. Hominins' dietary shift to more energy-dense foods, including meat, along with technologies for processing food both mechanically and through cooking, increased the net energy gained compared to other hominoids (Carmody et al. 2009; Kraft et al. 2021; Zink and Lieberman 2016). Food sharing, which is universal among human populations but relatively rare in other apes, reduces starvation risk, and humans also carry more body fat than other apes, providing an additional energy buffer (Pontzer, Brown, Raichlen, et al. 2016).

In orangutans, reduced energy availability and lower TEE appear to coincide with drastic changes in social organization, leading females to disperse more widely in order to avoid the foraging costs associated with group-living-induced food competition (M. Harrison and Chivers 2007; Wich, Buij, and van Schaik 2004). Reduced energy availability may have also affected orangutan ranging ecology, favoring more efficient travel (Thorpe et al. 2007).

SUMMARY AND CONCLUSIONS

Energetics is central to evolutionary biology, and interest in energy expenditure and allocation in evolution is as old as the field (Darwin 1861). Researchers have been measuring primate BMR for nearly a century, but measures of TEE have been elusive until recently. As a first pass, we can confidently say that the most important determinant of primate TEE and BMR is body mass and that both BMR and TEE increase with body mass in primates with a scaling exponent near Kleiber's 0.75. Moving beyond these allometric estimates to understand the effects of brain size, diet, or activity on BMR and TEE requires empirical measurements and comparison with closely related species. While the number of species and populations measured for TEE and BMR is growing, metabolic rates for most primate species remain unknown, and few data are available from the wild. Documenting variation in TEE and BMR among primate species and populations, as well as searching for patterns linking ecology to energetics and evolution, remains a promising and important area of research in primate biology. Here is what we know:

- Compared to other placental mammals, primates have TEEs that are 50% lower than expected for their body size. Primates' low TEE corresponds to their slow rates of growth, reproduction, and senescence.
- Anthropoid primates have BMRs that are similar to those of other placental mammals; strepsirrhines have marginally lower BMR.
- Metabolic methods matter. Estimates of TEE based on food intake or other estimates should be interpreted with appropriate consideration of potential measurement and prediction error. Where possible, direct physiological measures (DLW or respirometry) should be used to investigate daily energy throughput.
- Aside from body size, the factors shaping variation in TEE among individuals, populations, and species remain largely unknown. Traditional models implicating physical activity and body size as the primary determinants of TEE are not supported by the available data. Future studies should investigate the ecological, genetic, anatomical, and other determinants of TEE.
- Activity budget analyses are important for understanding energy and time allocation but should not be used to reconstruct TEE. For species or populations with similar body size, greater ranging does not necessarily indicate greater TEE. Instead, physical activity should be viewed as one component within an evolved and relatively inflexible energy budget.

6 Primate Senses
Finding and Evaluating Food

Amanda D. Melin and Carrie C. Veilleux

The study of primate foraging ecology has engaged scientists for decades and provides key insights into the ecology, adaptive radiation, and social organization of present and past species (Garber 1987; Janson 2000; McGraw and Daegling 2012). Sensory ecologists contribute to this area of research by investigating how animals perceive their world, use their sensory systems to find and evaluate foods, and interpret the stimuli emitted by vegetative foods and prey (Dangles et al. 2009; Veilleux et al. 2022). This relatively new area of research in primatology complements other ways of understanding foraging behavior and food selectivity—for example, approaches employing nutritional geometry or optimal foraging frameworks (Dangles et al. 2009; current volume)—and much progress has been made in understanding the evolution of primate senses in the last two decades (W. Allen et al. 2014; Dominy 2004; G. Jacobs 2009; Nevo, Heymann, et al. 2016; Veilleux et al. 2022). Five senses are commonly discussed in the context of primate foraging ecology: sight (vision), hearing (audition), smell (olfaction), taste (gustation), and touch (haptic sense) (Kawamura and Melin 2017). The focus of primate sensory ecology has been strongly biased toward researching vision (Heymann 2006), although promising trends in the last 15 years suggest that the relative amount of attention given to other senses may be increasing, despite visual ecology remaining an active field (fig. 6.1).

Revealing the sensory capacities and behaviors of foraging animals is important for understanding the evolutionary trajectories of the foods they consume. For example, in some ecological interactions, primates provide vital services for their foods, such as seed dispersal, which impacts the forest community structure (Beaudrot et al. 2013; Bufalo et al. 2016; Chapman, Bonnell, Gogoarte, et al. 2013; Guevara et al. 2016). Primates are highly effective seed dispersers known to enhance seed germination through spitting or defecating seeds away from the parental trees and by inducing favorable changes in the seeds during gut passage (Chapman 1995; Fuzessy et al. 2016; McConkey, Brockelman, and Saralamba 2014; Valenta, Burke, et al. 2013). Additionally, due to a long evolutionary history of mutualism, primate frugivores may have played an important role in the diversification and character evolution of angiosperm plants (Eriksson 2016; Fleming and Kress 2011; Guevara et al. 2016; Nevo et al. 2018a). A large amount of current research is now investigating the role that primates and other seed dispersers play in ecosystem conservation and reforestation/restoration projects (Carlo and Yang 2011; Link and Di Fiore 2006; McConkey, Prasad, et al. 2012); understanding the interactions between primate senses and fruit traits is critical for this process (Nevo, Heymann, et al. 2016; Nevo et al. 2018b; Valenta, Burke, et al. 2013; Valenta, Nevo, et al. 2017).

In circumstances where primates provide high-quality dispersal services, tree species may experience selective pressures to advertise their fruits to primates through vi-

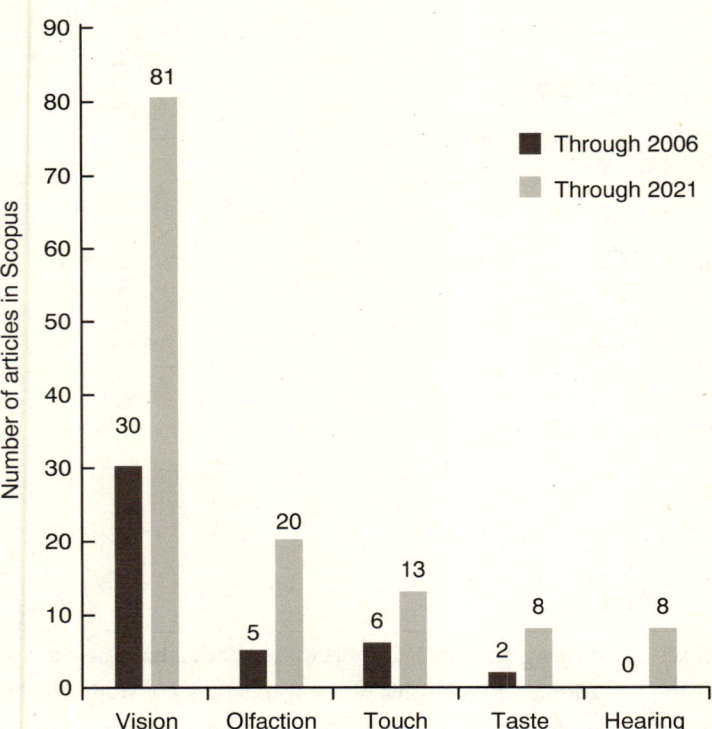

Figure 6.1. Number of papers identified in a Scopus query using the search terms "primate AND foraging AND vision OR visual OR olfaction OR smell OR haptic OR touch OR taste OR gustation OR audition OR hearing." Both the titles and the abstracts of research articles were included in the search. First, all records through 2021 were identified; then, the search was restricted to papers published in 2006 or earlier. Vision is topically dominating in both searches. However, the representation of papers in the topics of olfaction, taste, and hearing have increased in the past 15 years.

sual, olfactory, haptic, audible, or gustatory cues of edibility (Melin, Khetpal, et al. 2017; Nevo et al. 2018a; Veilleux et al. 2022). In turn, positive feedback may further shape primate sensory systems to be tuned and attracted to reliable cues of fruit nutritive reward. Such a relationship has been argued to result in fruit dispersal "syndromes" (Fleming and Kress 2011; Flörchinger et al. 2010; Janson 1983). This hypothesis posits coevolution between fruiting plants and their frugivore consumers—specifically, that certain fruit characteristics are adaptive for attracting different guilds of frugivores and that their consumers in turn provide high-quality seed dispersal and exert selective pressures back on the plant (Lomáscolo et al. 2008; Ridley 1930). While this is controversial, as coevolution is difficult to document (Fischer and Chapman 1993; Janzen 1980), robust differences in fruit traits of plant species that are primarily dispersed by birds versus mammals suggest that coadaptation has taken place (Janson 1983; Knight and Siegfried 1983; Link and Stevenson 2004; Nevo et al. 2018b; Willson et al. 1989). Bird-dispersed fruits tend to be small, red or black, nonodiferous, easy to process, and lipid rich, while mammal-dispersed fruits are larger, more odiferous, and "dull" in color (greenish-brownish; Gautier-Hion, Duplantier, et al. 1985; Willson et al. 1989). These differences can be understood when the sensory and masticatory systems of avian versus mammalian frugivores are examined. Compared to those of avian taxa, the olfactory systems of frugivorous mammals are often more sensitive to fruit-emitted odors, and mammalian jaws and dentition allow them to extract and chew the pulp or aril of larger, mechanically protected fruits (Janson 1983; Knight and Siegfried 1983; Link and Stevenson 2004; Valenta, Brown, et al. 2015). Some researchers have also argued for separate dispersal systems between primates and other mammals (Fleming and Kress 2011; Gautier-Hion, Duplantier, et al. 1985; Julliot 1996b; M. Leighton 1993; Terborgh 1983). In particular, the diurnal activity, manual dexterity, and derived color vision abilities of many primate species have been argued to have selected for large, colorful, thick-husked fruits, although not all of these traits are correlated or linked to primate feeding preference in each study. The strongest evidence for this comes from catarrhine primates in Asia, which are large, diurnal, and characterized by uniformly possessing a derived form of color vision (trichromacy) that allows differentiation between green leaves and yellowish-reddish fruits (Gautier-Hion, Duplantier, et al. 1985; G. Jacobs 1996; Kawamura et al. 2012; M. Leighton

1993). In this system, it is predicted that fruit whose seeds are dispersed primarily by nonprimate (often nocturnal) mammals remain green upon ripening and are easy to process and consume without extensive manipulation.

In other ecological interactions, primates are predators of seeds, leaves, or other animals. Here, sensory cues used by primates to locate prey (for example, tap scanning for embedded invertebrates, using color vision to identify reddish immature and easily processed leaves, or sniffing and probing manually for hidden prey) may induce selection on the prey to reduce their conspicuity or enhance their mechanical or chemical defenses (Dominy and Lucas 2001; Melin, Khetpal, et al. 2017; K. Phillips et al. 2003; Piep et al. 2008). As with the fruit dispersal syndromes discussed above, this scenario may shape sensory and cognitive abilities of the primate foragers as well as the traits of their foods (Bankoff et al. 2017; Melin, Young, et al. 2014). This question of reciprocal coevolution between primates and prey has received little attention, although there is evidence that adaptations for foraging on prey have shaped the evolution of primate sensory and communication systems (Siemers 2012). For example, auditory adaptations to detect reverberations created during percussive foraging may have impacted the evolution of vocalization frequencies in aye-ayes (Ramsier and Dominy 2012; Ramsier and Rauschecker 2017). Overall, the potential for tight coevolution between primates and their foods may be variable among systems and generally weak, given the influence of other selective pressures, such as the need to communicate with conspecifics or evade a variety of predators. However, in the cases of both predatory interactions and mutualism via seed dispersal, the foraging behaviors and sensory capacities of the primate consumers likely play a role in important ecological and evolutionary relationships that are worthy of investigation.

An exciting movement in the field of primate sensory ecology has been the recent upswing in studies investigating multiple sensory systems simultaneously and their integrated role in food localization and assessment. These studies, like those investigating multimodal foraging in other vertebrate and invertebrate species (e.g., Kulahci et al. 2008; Raguso and Willis 2005), demonstrate that animals are often better able to locate or select preferred foods when two or more types of sensory cues are available and used in concert (Kemp and Kaplan 2012; Rushmore et al. 2012; Siemers 2012; Valenta, Brown, et al. 2015). Multisensory assessment is especially important for foods that are novel to the foragers, and the use of nonvisual signals decreases as familiarity with food increases (Hegab et al. 2014; Laska, Freist, and Krause 2007). Furthermore, multisensory studies come far closer to describing how foraging unfolds in the natural world than do studies investigating one sense in isolation.

In this chapter, we discuss the sensory systems of primates and their involvement in finding and evaluating foods. Sensory challenges during foraging vary greatly with the type of food targeted. Thus, this chapter is divided into sections focusing on the major foraging stages ("Finding Food" and "Evaluating Foods"), with subsections reviewing the senses involved at each stage, and considering different food types (fruits, leaves, and invertebrates) separately when appropriate. The next section ("Sensory Evolution") summarizes current understanding of the evolution of primate senses with a focus on how comparative genetics and genomics can help reveal how diet and environment have contributed to the wide diversity of sensory systems seen intra- and interspecifically. The final section suggests ideas for future research directions and draws attention to the potential for interdisciplinary approaches in addressing complex systems.

FINDING FOOD

For wild primates, the search for food typically begins promptly after waking, and ranging choices are affected by numerous, interacting factors. Although not traditionally considered a "sense," cognition is pertinent to discussions of primate sensory ecology, as memory and problem-solving abilities work in conjunction with the senses to guide primates to fruit trees (Bicca-Marques and Garber 2004; Corlett 2011; Janmaat, Byrne, and Zuberbühler 2006a; Janson and Byrne 2007). Additionally, cognition may play an important role in localizing, assessing, and accessing foods that are hidden or well protected (DeCasien and Higham 2019; Melin, Young, et al. 2014). Much research has attended to the role of spatial cognition and memory in food search and (re)discovery by primates. For example, this topic has recently been explored in depth in a special issue of the *American Journal of Primatology* (Garber and Dolins 2014). A thorough review of this topic is beyond the scope of this chapter (but see Garber, this volume), although attention is drawn to the potential for interactions between cognition and sensation in primate foraging ecology. Whether engaged in directed travel to a recently visited food patch, monitoring patches visited in previous years or seasons, or searching the habitat for novel or ephemeral foods, the senses assist in the discovery or relocation of foods. Examples include finding small patches of colorful foods amid the background foliage, hearing the food-associated vocalizations of conspecific or allospecific foragers, or tracking an enticing odor (Hogan et al. 2018; Veilleux et al. 2022). Both the distance to the food patch and the nature of the environment (complexity, humidity, etc.) strongly impact the usefulness of the different senses and the types of information they can take in (Depasquale et al. 2022). In the sections below, the contributions of different senses to the localization of food sources by primates are reviewed and discussed in turn.

Hearing

Hearing is likely the farthest-reaching sense due to the extended propagation of sound waves and the high sensitivity of primate auditory systems (Dominy, Lucas, et al. 2001). The range over which sound carries decreases markedly in complex environments such as dense rainforests, where sound waves are scattered by tree trunks, leaves, and terrain (Brenowitz 1986). In particular, higher-frequency sounds drop off quickly, as they attenuate at a much faster rate than low-frequency sounds. These relationships have important consequences for the selective pressures shaping the structure of vocalizations in different environments (Brumm et al. 2004). Primates often respond to the calls of conspecifics given during foraging contexts, which may attract or repel others from clumped resources, depending on the social relationships within and between the two groups (Caine, Addington, and Windfelder 1995; Dittus 1984; Pollick et al. 2005).

Food-derived noises, such as fruit dropping to the forest floor, insects rustling in leaves, or noises made by vertebrate prey, are also potential foraging cues. For example, the fruit of *Sebastiana pavonia* eaten by capuchin monkeys (genus *Cebus*) in Sector Santa Rosa makes a popping noise when it dehisces, providing an auditory signal of its location (A. Melin, pers. obs.). Prey-derived sounds are especially important to chimpanzees, who will change their activities to hunt prey after hearing prey-based noises (Boesch and Boesch 1989). Nocturnal, insectivorous strepsirrhines and tarsiers are notably reliant on audition. A unique example is the aye-aye (*Daubentonia madagascariensis*), which uses a derived percussive foraging technique to locate hollow cavities caused by wood-boring invertebrates (Bankoff et al. 2017; Erickson et al. 1998). The large, mobile ears of tarsiers (genus *Tarsius*) are continually active during foraging and guide their prey capture (G. Doyle, Martin, and Niemitz 1979; Ramsier et al. 2012). Mouse lemurs (genus *Microcebus*) can successfully capture prey using only soft prey-derived noises in isolation from other sensory cues (Siemers 2012). However, hearing and vision often work together, and their evolution is correlated in small insectivorous mammals (Heffner 2004; Siemers 2012).

Olfaction

The sense of smell can guide animals to foods over long distances, as exemplified by migrating salmon and procellariform seabirds (DeBose and Nevitt 2008). It can also facilitate short-range food detection (Dominy, Lucas, et al. 2001). Odor molecules released from plants or animals are spread by moving air in plumes. As in the case of sound propagation, many environmental and climatic factors impact the spread of odor plumes in primate habitats, including vegetation structure and density, barometric pressure, temperature, humidity, and air currents (Depasquale et al. 2022; Müller-Schwarze 2006). Thus, the effective olfactory range for foraging primates will vary drastically between habitats (open canopy vs. close canopy, high- vs. low-altitude forests, forest edges, etc.) and within habitats (daily or seasonal temperature and humidity fluctuations, canopy levels, etc.). These environmental factors can also affect olfactory receptor functionality (Müller-Schwarze 2006). Thresholds for odorant detection are lower in humid than in dry conditions, for example, perhaps due to improved efficiency of the interactions between odorants and receptors of the olfactory mucosa, or because environmental humidity increases the capacity of air to carry odorants (Kuehn et al. 2008). Consequently, olfactory cues may be more useful for primates in humid conditions. Indeed, the role of olfaction in finding food may even vary seasonally for primates in deciduous environments (Veilleux et al. 2022).

In contrast to salmon and seabirds, long-distance odor tracking is likely more difficult for primates inhabiting forest environments due to the low speeds and shifting nature of forest air currents (Nevo and Heymann 2015; M. Willis 2008). However, olfaction may be a more reliable short-range foraging cue in forests, particularly at night when air is more stable (Depasquale et al. 2022; Müller-Schwarze 2006). Little is known about primates' abilities for chemotaxis (i.e., tracking a food source based on following an increasing odorant gradient), although one study found that humans can use odor to track chocolate dragged 10 m across a field (J. Porter et al. 2007), and a recent study found that ring-tailed lemurs (*Lemur catta*) can also track odor plumes (E. Cunningham et al. 2021). Nocturnal primates and other nocturnal mammals have been hypothesized to rely more heavily on odor cues than diurnal species during foraging because visual signals are more limited (Bolen and Green 1997; Kalko et al. 1996; Nevo and Heymann 2015). Bats, for example, have been found to use fruit and flower odors to find feeding sites (von Helversen et al. 2000; Hodgkison et al. 2007). In primates, experimental studies suggest that nocturnal strepsirrhines and platyrrhines also use olfactory cues for short-distance detection of fruits and insects (Bolen and Green 1997; Piep et al. 2008; Siemers et al. 2007). However, few studies have directly compared reliance on olfaction between sympatric diurnal and nocturnal primates. Captive free-ranging nocturnal owl monkeys (genus *Aotus*), for example, were able to locate baited containers of food using odor cues at a level significantly greater than chance, while diurnal capuchins (genus *Sapajus*) performed at chance levels (Bolen and Green 1997). In contrast, results from an experimental field study found that while nocturnal monkeys performed well using olfactory cues, some, but not all, diurnal monkeys did comparably well (Bicca-Marques and Garber 2004). Other studies have found that diurnal squirrel monkeys (genus *Saimiri*) and pigtail macaques (genus *Macaca*) can be trained to use olfactory cues to find food rewards in cups at very short range (Hübener and Laska 1998; Laska and Hudson 1993).

Olfactory cues are hypothesized to be particularly relevant for finding fruit and flower resources (Dominy, Lucas, et al. 2001; Nevo and Heymann 2015; Nevo and Valenta 2018). Among nonprimate mammals, field-based evidence suggests that diurnal coatis can use odor cues to find fruit, with vision potentially aiding the final stages of locating a resource initially detected by smell (Hirsch 2010). Within primates, wild diademed sifakas (genus *Propithecus*) have been documented using olfaction to forage for flowers among the leaf litter (Irwin, Raharison, Rakotoarimanana, et al. 2007). Valenta, Burke, and colleagues (2013) offered indirect evidence for the importance of odor in fruit foraging, finding that fruits dispersed by wild mouse lemurs had significantly higher odor intensities than nondispersed fruits. Experimental work suggests that mouse lemurs can use smell to detect both fruit and insects, but they appear to perform better with fruit odors (Piep et al. 2008; Siemers et al. 2007). In the context of hunting invertebrate or vertebrate prey, the limited research available suggests that the role of olfaction varies between species. Lorises (genus *Loris*) rarely use olfaction during hunting and were seen to use sniffing only to find strong-smelling invertebrates (Nekaris 2005). In contrast, Charles-Dominique (1977) describes angwantibos (genus *Arctocebus*) and pottos (genus *Perodicticus*) as using olfaction to detect most of their prey, at times at a distance of 1 m. Furthermore, Charles-Dominique wrote that needle-clawed bushbabies (genus *Euoticus*) use olfaction to localize gum exudates in lianas "within a range of 20–30 cm" (1977, 42). Although data indicate that odor cues may also be important during foraging in marsupial folivores (Stutz et al. 2017), there currently is no evidence of a role for olfaction in finding leaves among primates (Nevo and Heymann 2015).

Most studies of olfactory foraging in primates have used captive animals and experimental designs. The role of smell during long- and short-distance localization of fruits, prey, or exudates is at present quite difficult to assess in the wild (but see E. Cunningham et al. [2021] for a discussion of evidence from ring-tailed lemurs and Depasquale et al. [2022] for evidence from other animals). Emerging research also highlights use of primate scents as informative about the location of food sources (C. Thompson, et al. 2018). Further advances in this area may be enabled through advances in portable equipment for tracking ambient odorants (e.g., portable gas chromatography coupled with mass spectrometry [GC-MS]; Poirier et al. 2021; C. Thompson et al. 2020). Overall, studies of olfactory ecology are still very much in their infancy, and additional interspecific comparisons of diurnal and nocturnal foragers of different diets will shed light on the factors shaping primate olfaction in foraging contexts. In particular, studies where environments can be carefully manipulated and controlled, and animal responses monitored closely, may yield great advances in this area.

Vision

The visual range of food detection is limited by the primate's visual acuity, the size of the food target, and obstacles in the environment. Among mammals, primates have exceptional visual acuity, especially diurnal catarrhines—typically 30–60 cycles per degree or 20/40 to 20/20 vision (Melin, Kline, Hiramatsu, et al. 2016; Veilleux and Kirk 2014). However, in forested landscapes where most primates live, line-of-site visibility may be confined to distances of <20 m (Boesch and Boesch 1989). Thus, vision is typically limited to shorter distances than hearing or smell, which have a wider-reaching range—although, depending on the environmental conditions (e.g., wind), often a less precise one (Dominy, Lucas, et al. 2001).

Vision is useful for many different types of tasks, including detecting patterns, shape, size, movement, brightness, and color. Of these, detecting movement, shape, and brightness is more useful for invertebrate foraging, as many invertebrates consumed by primates are cryptically colored and use camouflage to hide against matching backgrounds. Luminance (brightness) contrast, rather than chromatic cues, may best indicate gum sources, which often stain darkly against the bark of trees that are gouged (Moritz 2015). Alternatively, food chroma (hue and saturation) is argued to be important for detecting the presences of fruits, flowers, or young leaves against mature foliage (Dominy 2004). The difference in color contrast between a fruit and background leaves affects how conspicuous a fruit is (Osorio et al. 2004). Certain fruit hues are more conspicuous than others, although the conspicuousness of a color contrast depends on the visual ecology of the perceiver. For example, red fruits against green leaves are conspicuous to birds and to primates with trichromatic vision (such as humans), but not to primates with dichromatic (red-green colorblind) vision (Melin, Chiou, et al. 2017; Osorio et al. 2004). However, blacks and dark purples are conspicuous to most animals against leaves because they differ in luminance, not hue, from the background. Fruit that are greenish are cryptic amid the background leaves to perceivers, regardless of their color vision.

Primate color vision is remarkably variable and has been the subject of decades of work. Trichromatic color vision is unique to primates among placental mammals and confers the ability to discriminate long-wavelength hues (i.e., greens, yellows, oranges, and reds). Catarrhine primates are nearly ubiquitously trichromatic, aside from relatively high rates of colorblindness in humans (Kawamura and Melin 2017; Munds et al., 2022). Nearly all species of platyrrhine monkeys, excluding monochromatic owl monkeys (genus *Aotus*) and routinely trichromatic howler monkeys (genus *Alouatta*), possess mixed populations of dichromatic and trichromatic individuals (Kawamura and Melin 2017). This polymorphism is due to allelic variation of an opsin gene coding mid-to-long-wavelength-sensitive photopigments (*OPN1LW*; G. Jacobs 2009; Kawamura et al. 2012). Remarkably, up to six alleles have been found in populations of wild primates (Corso et al. 2016; Goulart et al. 2017; Veilleux et al. 2021), and allelic variation shows signatures of being maintained by balancing selection (Hiwatashi et al. 2010). Although this form of color vision is also found in strepsirrhines, its distribution is currently believed to be more limited than that in platyrrhine monkeys, occurring in only a few genera (*Indri, Propithecus, Varecia*, and *Eulemur*; R. Jacobs et al. 2017, 2019; Y. Tan and Li 1999; Veilleux and Bolnick 2009).

The unique appearance of trichromacy in primates has been linked to their high levels of visual acuity, with a threshold of visual acuity perhaps around 8 cycles per degree (e.g., Caves et al. 2018; Melin, Kline, Hiramatsu, et al. 2016, Veilleux et al. 2022). Considerable discussion about the adaptive nature of color vision has been ongoing for decades (G. Jacobs 1996). Most prominent among these hypotheses are those linking color vision to foraging challenges, including finding reddish ripe fruit or young leaves, which are more nutritious and less tough than mature leaves (Dominy, Lucas, et al. 2001; Melin, Khetpal, et al. 2017). Trichromacy may be especially useful in helping foragers find small, unmemorable food patches (Bunce et al. 2011; Hogan et al. 2018; Melin, Hiramatsu, et al. 2014; Veilleux, Scarry, et al. 2016). Once they are in a given food patch, color vision may continue to help foragers find food over short distances—for example, to locate profitable clumps of foods and ideal "micropatches" (i.e., particular locations within a fruiting tree or other clumped patch), such as a branch with higher-than-average densities of ripe fruit, from which the primate bases its foraging activities.

The visual sense is useful for tasks not linked to color. For example, vision is indispensable to primates for detecting and tracking small prey such as insects and other invertebrates, as well as lizards and frogs at close distances. In fact, higher visual acuity is found in more predatory strepsirrhines (Veilleux and Kirk 2009) and more predatory mammals more broadly (Veilleux and Kirk 2014). In one of the few systematic studies on the sensory nature of prey capture, it was found that motion detection, together with auditory cues, was critical for the foraging success of mouse lemurs (Siemers 2012). Numerous other field studies have emphasized the importance of vision in detecting prey in nocturnal primates (Gursky 2003; Moritz, Ong, et al. 2017; Nekaris 2005; Veilleux 2020). Lorises, for example, switch from consuming slow-moving ants to more active insects with increasing nocturnal light levels (Bearder, Nekaris, and Buzzell 2002). These studies often identify a positive relationship between nocturnal light intensity and time spent foraging, suggesting an important role for vision among predatory nocturnal primates (Bearder, Nekaris, and Curtis 2006; R. Jacobs et al. 2019; L. Nash 2007; Valenta, Brown, et al. 2015). Indeed, prominent hypotheses of primate origins emphasize the importance of visual predation in driving the features of the visual systems seen in all primates, and predatory primates tend to have relatively large eyes with especially large binocular overlap (Cartmill 1992). Larger binocular overlap is associated with more stereoscopic vision and improved depth perception, which is especially important for detecting and grasping prey. Consequently, orbital convergence and binocularity are higher in more faunivorous mammals, including strepsirrhines (Heesy 2008). Binocularity also allows for a type of stereoscopic "X-ray" vision wherein primates can see through (around) obstacles in the environment due to the offset in overlapping visual fields from the left and right eyes. This property can help break the camouflage of prey through distance-based pop-out effects, as prey are always closer to the viewer than to their background (Changizi and Shimojo 2008). Unlike patterns seen in some mammalian species (including some bats and moles) in which the sense of sight is greatly reduced, the dim-light environments of cathemeral and nocturnal primates have not lessened the importance of vision, although certain aspects have been remodeled—see the section below on "Sensory Evolution." Rather, nocturnal primates have large eyes and rely on them heavily during foraging (Bearder, Nekaris, and Curtis 2006; Moritz, Melin, et al. 2014; Ross and Kirk 2007).

Touch

The range of the sense of touch is limited by the body size and arm/leg span of primates, and as such this is the closest-range sense of the five senses discussed in this chapter. During fruit and flower foraging, the food targets are exposed, and touch is primarily used in the context of food evaluation (discussed next). However, touch can play a role in food location via the search for hidden foods, such as small prey items under bark, on branches, or in crevices or tree holes. Capuchins, among other species, are known to thrust their fingers or hands into a potential prey hiding spot to feel around for and to grab any foods located there (Fragaszy, Visalberghi, and Fedigan 2004). In birds, species that use their bills to forage for prey hidden in the water have greater tactile sensitivity in their bills than visually foraging species (E. Schneider et al. 2014). The tactile pads in the fingers and hand must be sensitive enough to feel for prey items yet durable enough to withstand any bites, stings, or other defenses the prey employs, traits that should favor decreased sensitivity (H. Young et al. 2008) or at least decreased pain sensitivity. These opposing roles must act to shape the anatomy and haptic sensory systems of primate hands in contrasting ways and would be an intriguing topic for in-depth future study.

EVALUATING FOODS

Once a primate has located a potential food item, the senses are used to help assess edibility (Dominy, Lucas, et al. 2001). Food assessment presents numerous, diverse challenges that vary by food type. Foraging animals cannot know the nutritional quality of the foods they are eating, but they can rely on past experience with specific food characteristics as cues (Dominy, Lucas, et al. 2001; F. Huang et al. 2021). For example, chemical compounds are often perceived to have a distinctive taste or smell. Some, such as ethanol, can elicit a pleasant response among consumers and encourage feeding, while other compounds elicit an unpleasant response, ultimately leading to rejection (Breslin 2013; Gochman et al. 2016). For example, many compounds that are toxic to primates are perceived as bitter tasting and can therefore be deemed as unacceptable food items (Breslin 2013). Additionally, the physical characteristics of potential foods that can be seen (e.g., color and size) or felt (e.g., softness and texture) can also provide information about potential reward, effort, and risks (Melin et al. 2019; Valenta, Brown, et al. 2015; van Casteren and Lucas, this volume).

Fruits are perhaps the most ubiquitous type of food eaten by primates as well as the most studied food type (current volume). Due to the wealth of information on fruit assessment relative to other primate foods, the majority of the review below is focused on the sensory ecology of frugivory, treating each sense in turn. Although we address each sense independently, it is important to emphasize that fruit assessment is a multimodal process, and several senses are often used simultaneously or in rapid succession to assess fruits (Dominy, Yeakel, et al. 2016; Kawamura and Melin 2017).

Vision

Visual inspection of fruits at close range provides many potential cues of edibility, including size, chroma (hue and saturation), brightness, and texture. Primates can also examine fruits for signs of damage or spoiling such as spotting, holes, asymmetry, and chew marks (Dominy 2004; Hiramatsu, Goda, and Komatsu 2011; Melin, Fedigan, Hiramatsu, Hiwatashi, et al. 2009). Despite the unquestionable importance of vision, including color vision, to primates and its frequent use while foraging, it is perhaps the hardest sense to accurately measure in wild primates (Melin et al. 2018). This is especially true when primates are far away from observers or when fruits on the tree are numerous and small. Nevertheless, researchers studying vision have attempted to overcome these difficulties and assess the use of vision during food assessment in several ways.

Using computer simulations that approximate different forms of color vision ability together with digital images of natural foods, researchers have isolated the impact of color on food selection tasks (Bompas et al. 2013; Melin, Kline, Hickey, et al. 2013). Another approach to studying the impact of color vision on fruit assessment is to compare the foraging performances of members of the same species with different types of color vision (Hiramatsu, Melin, et al. 2008; Melin, Chiou, et al. 2017; Melin, Fedigan, Hiramatsu, Hiwatashi, et al. 2009; E. Vogel, Neitz, and Dominy 2007). The results from these studies are mixed, but at least two demonstrate that trichromatic color vision should confer an advantage for selecting ripe fruits from unripe ones (Melin, Chiou, et al. 2017; Melin, Kline, Hickey, et al. 2013). Other visual cues such as brightness can also be important over short distances (Hiramatsu, Melin, et al. 2008). Luminance changes may often be redundant with color; therefore, color vision might have the strongest effect while foraging on foods lacking achromatic cues (such as flowers) or under conditions when luminance is unreliable (Endler et al. 2005; Hogan et al. 2018). Despite our familiarity with the concept that fruits change color during ripening, color may not always be a particularly reliable cue of food quality, at least for some plant species. Recent studies have investigated how fruit size and color vary in relation to other sensory cues and measures of fruit quality within and between fruit species (Dominy, Yeakel, et al. 2016; Valenta, Brown, et al. 2015). A study published on foraging primates in Kibale, Uganda, found that color differences of greenish ripe figs were only weakly correlated with fructose concentration (red-green color information did not correlate; blue-yellowness correlated slightly but significantly, with $r^2 = 0.20$), making color an unreliable information source (Dominy, Yeakel, et al. 2016). Even in fruit species that have a "conspicuous" color change from greenish to yellowish or reddish, recent and ongoing work suggests that color change may happen well ahead of the fruits becoming ripe, as judged by softness and nutritional content (Melin, Shirasu, et al. 2015). In plant species where greenish-to-reddish color changes occur, however, this change allows individuals with trichromatic color vision to avoid clearly unripe fruits, and higher feeding efficiencies among individuals with trichromatic vision (versus group members with dichromatic vision) have been observed in the wild (Melin, Chiou, et al. 2017). In Verreaux's sifakas (fig. 6.2), this trichromatic advantage has been linked to better body condition, a trend for increased infant survival, and increased time feeding on fruit among trichromats (Veilleux, Scarry, et al. 2016). In the future, the use of slow-motion videography or other technology

Figure 6.2. A Verreaux's sifaka (*Propithecus verreauxi*) in the dry deciduous forest at Kirindy Mitea National Park in western Madagascar. Verreaux's sifakas are diurnal folivore-frugivores and exhibit allelic variation in the X-linked OPN1LW gene. Consequently, all males and homozygous females have dichromatic color vision, while heterozygous females have trichromatic color vision. A study of sifakas at this site (Veilleux, Scarry, et al. 2016) found that trichromat females and members of their social groups had feeding benefits compared to members of dichromat-only social groups.

Photo: Carrie Veilleux.

to measure the gaze of captive, semicaptive, or wild primates could broaden our understanding and provide new insights into the role of vision in feeding ecology (e.g., Ngo et al. 2022). Recent research on cone mosaics is adding new dimensions to color vision research in primatology (Munds et al. 2022) and will hopefully continue to be developed in the years to come (Munds et al., in review).

Touch

Once a potentially appealing fruit is spotted, it is often subjected to the other senses for further investigation, which may include touch, smell, taste, and auditory probing. Possessing sensitive Meissner's corpuscles and rich innervation in the fingertips of dexterous hands (Hoffmann et al. 2004; Verendeev et al. 2015) and in the maxillary area of the face (Muchlinski 2010), primates have a strongly developed sense of touch. Manual, lip, and incisal manipulation of fruits by primates to assesses the elastic modulus (squishiness) is a key source of foraging information (Dominy, Yeakel, et al. 2016). The sensitivity of the hands and face to haptic cues is hypothesized to be higher in frugivores and complex foragers than in folivores (Hoffmann et al. 2004; Muchlinski 2010; Winkelmann 1963). In a recent comparative study on sensory use during food investigation among three species of sympatric platyrrhines (fig. 6.3), the most frugivorous species (spider monkeys; *Ateles geoffroyi*) and the most dexterous

Figure 6.3. A wild Geoffroy's spider monkey (*left*; *Ateles geoffroyi*), white-faced capuchin (*middle*; *Cebus imitator*) and mantled howler monkey (*right*; *Alouatta palliata*) photographed in Sector Santa Rosa, northwestern Costa Rica. These three species live in sympatry and specialize in different dietary niches. Spider monkeys are ripe-fruit specialists, while capuchin monkeys are highly omnivorous, supplementing a primarily fruit diet with protein from invertebrate and vertebrate sources, including eggs, as imaged. Howler monkeys are the most folivorous species in Sector Santa Rosa but also eat fruits when they are abundant.
Photos: *Ateles geoffroyi*, Amanda D. Melin; *Cebus imitator*, Gabriel Benson; *Alouatta palliata*, Amanda D. Melin.

species (capuchins; *Cebus imitator*) used manual touch significantly more often than the more folivorous howler monkeys (*Alouatta palliata*; Melin et al. 2022). Touch and palpation are most useful for assessing fruits without hard carapaces, and in such species nutritional ripening is often strongly correlated with fruit softening (Dominy 2004). A study of lemur foraging ecology found softness to be the most reliable cue of fruit edibility (Valenta, Brown, et al. 2015). Pablo-Rodríguez and colleagues (2015) found that spider monkeys (genus *Ateles*) used touch and taste significantly more often when inspecting ripe fruit than unripe fruit (and olfaction more frequently for unripe fruits). The role of texture—a complex property of foods combining physical characteristics with their mechanical, visual, and auditory perception—in food acceptance or rejection is well known in humans but not yet well studied in other primates (Engelen and de Wijk 2012). The hands also often play an essential role in transporting food to the mouth, making it sometimes difficult to assess their role in food evaluation versus transportation. Still, observations of primates squeezing fruits and then leaving the fruits on the tree indicate the role of touch in food assessment. Mechanical probing using the mouth is also important, as many fruit are rejected after being bitten, although the impact of taste may also contribute at this stage (Ferrari and Lopes 2002). More force can be exerted by the jaws than by the hands, so biting might be especially important for assessing harder types of fruit, and fruit hardness is an important determinant of selectivity for some primates (Coiner-Collier et al. 2016).

Primate hand morphology varies greatly, with many species possessing limbs modified for suspensory locomotion, including the presence of vestigial thumbs in several species (e.g., spider monkeys, gibbons, and colobus monkeys). These morphological differences have the potential to affect manual dexterity and foraging behavior. Indeed, multiple studies have identified species differences in object manipulation ability and haptic sensitivity (Bishop 1962; Torigoe 1985). Using captive experiments, for example, Torigoe (1985) classified 74 primate species into three groups based on the number of manipulation patterns they were able to perform, with the least dexterous group including lemurs, marmosets, spider monkeys, and colobines and the most dexterous group including capuchins and apes. Field studies and observations from fieldworkers similarly suggest differences between species in the use of the hands during foraging, although these observations are limited. For example, spider monkeys often consume fruits directly from the branches of fruiting trees, whereas capuchins and squirrel monkeys typically use their hands to pluck fruits before consuming them (Laska, Freist, and Krause 2007; A. Melin, pers. obs.). Future studies investigating the interplay between sensory ecology, hand morphology, and feeding rates across sympatric species would be highly informative.

Olfaction

Following an increasing realization of the importance and nuance of the sense of smell to primates (Heymann 2011; Hoover 2010; Nevo and Heymann 2015; Nevo and Valenta 2018; Niimura et al. 2018) and an improvement in the technology and capacity for the study of volatile organic compounds under field conditions (e.g., Nevo, Heymann, et al. 2016; Valenta, Burke, et al. 2013), there has been a dramatic increase in the number of primate olfaction studies. Primates often sniff fruits before deciding to eat or reject them (Dominy 2004; Melin, Shirasu, et al. 2015), and wild capuchin and spider monkeys sniff visually "cryptic" (ever-greenish) or unripe fruits more often than fruits that change color with ripening (Hiramatsu, Melin, et al. 2009; Melin, Fedigan, Hiramatsu, Hiwatashi, et al. 2009; Pablo-Rodríguez et al. 2015). This may demonstrate increased reliance on the olfactory sense when other cues are not present, although it is possible that cryptic fruits also give off a stronger odor signal, which is consistent with the fruit syndrome hypothesis (Lomáscolo and Schaefer 2010). Additionally, a recent study on sympatric spider, capuchin, and howler monkeys suggests that more frugivorous species and those with larger olfactory bulbs are most likely to use sniffing behaviors during fruit assessment (Melin et al. 2022.) Fruits dispersed by mammals are often greenish and more odiferous than bird-dispersed fruits, suggesting a signal that evolved under natural selection to communicate information to foragers, rather than a plant by-product (Nevo, Orts Garri, et al. 2015). However, links between fruit odor and primate behavior may not be straightforward. In addition to cryptic figs, capuchin monkeys also sniffed fruits with high color contrast, showing signaling variation across plants. Brown lemurs (*Eulemur fulvus*) sniffed fruits with higher overall odorant emission less than fruits with lower overall odorant emission (Valenta, Brown, et al. 2015).

Primates can be highly sensitive to specific odors. Squirrel monkeys respond to extremely low concentrations of fruit-associated odorants, such as certain esters, aldehydes, and alcohols, that cannot be detected by "macrosomatic" mammals (e.g., dogs and rats) that are believed to have a good sense of smell (Laska and Seibt 2002). Spider monkeys can discern between odors of solutions mimicking unripe and ripe fruits, and controlled experiments suggest that the overall odor plume is more important than the amount of any specific compound (Nevo, Orts Garri, et al. 2015). Across 27 fruiting plant species in Madagascar, fruit sugar composition was correlated with scent chemical composition, suggesting that scent is an honest signal of sugar reward. Some compounds may be common to many fruits and therefore serve as a consistent guide of foraging behavior. For example, primates can smell and taste ethanol at low concentrations. Ethanol is known to act as a cue of fruit edibility, a source of nutrition, and likely an appetite stimulant (Dudley 2002). The importance of olfaction also varies among species according to their evolutionary history and ecological niches (E. Garrett and Steiper 2014). For example, frugivorous lemurs (ruffed lemurs) were more efficient at using odor cues than folivorous (sifaka) or generalist (ring-tailed) lemurs to select higher-quality foods in a captive experiment (Rushmore et al. 2012). Future work on the role of olfaction in evaluating fruit should explore interspecific differences in odor sensitivity and identify the important odor compounds of ripe and unripe fruit.

Audition

Literature discussing of the use of audition is essentially absent in primate foraging ecology. Anecdotal reports of primates tapping or knocking fruits indicate the potential for reverberating sounds to convey information about ripeness in fruits with thick exocarps. For example, humans are known to knock on watermelons and cantaloupes, and capuchins have been observed tapping the fruits of *Genipa americana* near their ears (A. Melin, pers.

obs.). Given the growing understanding of primate auditory processing, audiograms, and the nature of acoustic stimuli (Bankoff et al. 2017; M. Coleman 2009; Geerah et al. 2019; Ramsier and Dominy 2010), there will hopefully be more work on this topic in the future.

Taste

The final opportunity for assessing fruit prior to ingestion or rejection is by tasting it. The prolonged chewing process common to mammals exposes the tongue to compounds in the food (tastants) that stimulate taste receptors. Many fruits that appear ripe are rejected after being bitten by primates (Chapman 1995; McConkey 2000; Melin, Fedigan, Hiramatsu, Hiwatashi, et al. 2009), suggesting that buccal palpation or distaste led to the rejection of these fruits, yet the sensory basis for this is largely unknown. Our knowledge of the sense of taste among primates is still immature but growing rapidly. The sensitivity of primates to various compounds is correlated with the importance of consuming or avoiding them (Sugawara and Imai 2012). For example, many frugivorous primates are sensitive to small concentrations of sugars, especially fructose and ethanol (Dausch Ibañez et al. 2019; Dominy 2004; Dudley 2002; Remis 2006; Pereira et al. 2021; Sugawara and Imai 2012), and they are also sensitive to a wide variety of bitter compounds that may function as feeding deterrents (Itoigawa et al. 2021; Meyerhof et al. 2010; Purba et al. 2017; Sugawara et al. 2011; Widayati et al. 2019). Larger animals, who have a better metabolic ability to process tannins, may demonstrate tolerance to bitter compounds in their foods. For example, chimpanzees consume astringent fruits when they are also sweet (Remis 2006). Some plant species are hypothesized to build fruit using low-calorie protein "sugar mimics" rather than higher-calorie sugars in order to deceive primate frugivores (van der Wel et al. 1989). Primate species vary in whether they perceive these sugar mimics as sweet and whether they are "tricked" into dispersing these fruits (Guevara et al. 2016). Integrative research uniting behavioral ecology, genomics, and lab-based assays has revealed exciting evolutionary processes shaping taste sensitivities and changes over time in response to shifting diets (e.g., Toda et al. 2021; Chamoun et al. 2018) and promises to keep this field moving forward and reveal the mechanisms shaping the sense of taste within and between species.

OTHER FOODS

Flower foraging among primates is often overlooked, so although recent studies emphasize the importance that flowers have seasonally in primate diets (Heymann 2011; Hogan, Melin, et al. 2016), it is not surprising that we do not know how primates evaluate flower quality. Many flowering species specifically target mammals as pollinators by producing specific odors (S. D. Johnson, Burgoyne, et al. 2011), generating large volumes of pollen and nectar (S. D. Johnson, Pauw, and Midgley 2001), and tailoring the sugar composition of their nectar to a formula attractive to mammals (LaBare et al. 2000). Primates are well suited to evaluate many of these cues and can detect subtle changes in sugar ratios and minute concentrations of alcohol in fermented nectars (Gochman et al. 2016; Laska, Freist, and Krause 2007). To obtain nectar, many primates lick flowers, and they may use licking as an initial assessment of edibility (Hogan, Melin, et al. 2016; Torres de Assumpção 1981). Primates will also reject flowers after close-range visual inspection and may use visual cues as a measure of a flower's nectar content (Hogan, Melin, et al. 2016).

Tree gums are heavily consumed by primates including marmosets (genus *Callithrix*), pygmy marmosets (genus *Cebuella*), fork-marked lemurs (genus *Phaner*), and needle-clawed bushbabies (genus *Euoticus*), and these taxa have derived digestive systems facilitating microbial fermentation of gums. Other taxa, including tamarins (genus *Saguinus*) and mouse lemurs, among others, also feed on gum occasionally. Primates demonstrate selectivity in gummivory, feeding most often in a subset of tree species (C. Thompson et al. 2013). Yet the nutritional, mechanical, and chemical bases for gum preferences are poorly known. This topic is newly explored in marsupial gummivores (I. Wallis and Goldingay 2014) and is anticipated to provide insight into primate dietary ecology. The visual ecology of gummivory is also poorly known, but gum-feeding among nocturnal primates has been linked to the loss of functional color vision, as achromatic (brightness) cues—rather than chromatic cues—are likely to be more important in finding the darkly stained gum access points (Moritz 2015; Veilleux, Louis, and Bolnick 2013).

Leaves vary considerably in odor, color, toughness, and nutritional composition, both interspecifically and within plant species as they mature (Dominy and Lucas

2001; Rothman, Chapman, Struhsaker, et al. 2015; Rushmore et al. 2012). Among nonprimates, a recent study demonstrated that swamp wallabies (folivorous browsers) were able to differentiate between higher- and lower-quality *Eucalyptus* seedlings using odor cues alone (Stutz et al. 2017). Other studies have similarly revealed the importance of sensory cues in the feeding behavior of folivorous primates. For example, folivorous lemurs forage more efficiently when they can both see and smell leaves. However, changes in leaf odors are more subtle than those that occur in fruits, perhaps explaining why folivorous lemurs rely more on vision to assess food (Rushmore et al. 2012). Visual cues, including color, may also be important for evaluating leaves in some nocturnal folivores, such as woolly lemurs (genus *Avahi*; Veilleux, Jacobs, et al. 2014). Leaves and bark often contain distasteful compounds that may deter feeding by primates. Increased herbivory in primates appears to favor expansion of the ability to detect a different subset of bitter compounds, potentially to avoid the most toxic leaves (Hayakawa, Suzuki-Hashido, et al. 2014; Purba et al. 2017). The sensitivity threshold for individual bitter-tasting compounds also varies among primates and has been linked to both body size and diet. Primates that consume leaves and bark are less sensitive to some of the bitter dietary compounds present in those foods, an adaptation that would facilitate consumption of these items (Imai et al. 2012; Remis 2006). Leaf eating has also relaxed selection operating on the perception of sweet taste, although umami taste perception remains important (G. Liu et al. 2014).

Insects and other invertebrate prey can be generalized into two types of prey, and the senses used to assess each are very different. Foods that are gleaned opportunistically from surfaces of substrates are typically assessed visually by primates prior to pursuit. In particular, vision may be used to reject invertebrates with aposematic (warning) coloration. However, signal effectiveness depends on the visual system of the receiver (Fabricant and Herberstein 2015; Melin, Fedigan, Hiramatsu, Sendall, et al. 2007). Nonvisual senses are also important for the assessment of prey edibility. As with plant material, odor and taste receptors warn of noxious chemicals emitted by or within prey. Some invertebrates use mechanical defenses such as spines, stinging apparatus, or thick shells or embed themselves in branches or other substrates to deter predators, and proprioceptors in the mouth detect sharp or stinging objects and foods too hard to be masticated, thus preventing damage to teeth, soft tissues, or jaws (van der Bilt et al. 2006). A better understanding of the fracture mechanics of invertebrate exoskeletons coupled with dietary divergence within and between species would be a novel way to explore the sensory ecology of faunivory.

SENSORY EVOLUTION

An overarching goal of researchers studying foraging ecology is to link sensory genotype, phenotype, and fitness in natural populations (Bradley and Lawler 2011; Dangles et al. 2009; Endler et al. 2005). This section explores how foraging ecology has shaped the sensory systems of different species through natural selection, focusing on molecular ecology and comparative genomics as one way to understand evolutionary patterns among species. We summarize the state of knowledge in this area and highlight recent progress; where possible, we discuss the interplay among the senses along evolutionary timescales.

Color Vision

Color vision evolution among nocturnal, cathemeral, and diurnal primates has received a great deal of recent attention (recently reviewed by Moreira et al. 2019). Despite the presence of functional color vision and the use of color signals in nocturnal insects and reptiles (Kelber and Roth 2006), a prevailing belief until recently was that the presence of two intact opsin genes in nocturnal primates was a functionless vestige from a diurnal ancestor (Y. Tan et al. 2005). However, more recent studies on the opsin genes of some nocturnal primates (e.g., tarsiers and nocturnal lemurs) show evidence of purifying selection maintaining the function of these genes (Kawamura and Kubotera 2004; Melin, Matsushita, et al. 2013; Moritz, Ong, et al. 2017; G. Perry et al. 2007; Veilleux 2020; Veilleux, Louis, and Bolnick 2013). These and similar studies raise the exciting possibility that color vision is useful to nocturnal primates. Subsequent studies have documented that light levels in some nocturnal environments are bright enough to support mammalian color vision, that nocturnal primates inhabiting brighter habitats showed stronger evidence of purifying selection on opsin genes, and that color was important to the foraging ecology of nocturnal lemurs (Melin, Moritz, et al. 2012; Valenta,

Burke, et al. 2013; Veilleux, Jacobs, et al. 2014; Veilleux, Louis, and Bolnick 2013). In contrast, opsin genes may lose function in species that are active in scotopic (very dark) environments, such as nocturnal closed-canopy forests, or that are more heavily reliant on luminance rather than chromatic cues (Melin, Wells, et al. 2016; Veilleux, Louis, and Bolnick 2013). Under these conditions, color may be of little importance; if so, opsin genes would be released from purifying selection and would begin to evolve neutrally. Alternatively, natural selection may favor SWS opsin pseudogenization to improve motion detection and luminance contrast (reviewed in Regan et al. 2001). Tuning of opsins to different wavelengths of light also may impact the evolution of color vision. When patterns of dichromacy across all mammals are examined, the loss of the SWS opsin gene appears to be more common if the opsin is sensitive to blue light rather than to UV light (Melin, Wells, et al. 2016). This trend could be linked to visual acuity, signal parameters, or the spectral composition of light available, but it has not been systematically investigated. Nocturnal habitats are heterogeneous, and foods are variable, so species-specific foraging ecologies shape the loss or persistence of dichromacy (Dominy and Melin 2020). In sum, nocturnal activity seems necessary for the complete loss of color vision in mammals but not sufficient to predict it (G. Jacobs 2013). Once lost, color vision may be difficult to reclaim, as demonstrated by work on owl monkeys (Mundy et al. 2016). That said, dichromatic vision may hold advantages for some species: cathemeral *Eulemur rubriventer* are monomorphic for the L (red-shifted) opsin, and visual models suggest that luminance contrast is greatest with this phenotype (R. Jacobs et al. 2019).

Among day-active primates, the puzzle of polymorphic color vision continues to draw sustained interest and fuel new discoveries (e.g., Depasquale et al. 2021; Veilleux et al. 2021). A study of opsin gene diversity in platyrrhine monkeys offers strong support for balancing selection as a mechanism for maintaining the color vision variation (Hiwatashi et al. 2010). Yet, despite early evidence from captive experiments that trichromatic primates forage more efficiently for foods emulating ripe fruits (Caine et al. 2000; A. Smith et al. 2003), evidence of trichromatic advantage from the field has been scarce until recently. For the first time, three recent studies have provided evidence that trichromacy is advantageous for certain tasks under natural conditions: (1) members of sifaka groups containing at least one trichromatic female have better body condition, spend more of their time budget feeding on fruit, and engage in longer fruit-feeding bouts than members of groups of only dichromats (Veilleux, Scarry, et al. 2016); (2) trichromatic monkeys find a greater number of small, colorful flower patches than dichromats (Hogan et al. 2018); and (3) within fruit trees, trichromatic capuchins eat fruits at a faster rate once tree species and phenology are controlled for (an effect seen most strongly among juveniles; Melin, Chiou, et al. 2017). Earlier research has also documented foraging advantages to dichromatic monkeys for capturing camouflaged insects and other foods, particularly at low light intensity (Caine, Osorio, and Mundy 2010; Melin, Fedigan, Hiramatsu, Sendall, et al. 2007; A. Smith et al. 2012). These results, together with reports that dichromatic and trichromatic monkeys do not differ in net reproductive success, suggest that natural selection maintains color vision variation through some form of niche differentiation or mutual benefit of association (Fedigan, Melin, et al. 2014; G. Jacobs 1996).

All catarrhine primates and howler monkeys possess routine trichromacy due to the fixation of spectrally differentiated MWS and LWS opsin genes on the X chromosome (G. Jacobs et al. 1996). In the last decade, relatively little research has been done in this area. However, a recent study on howler monkey provides support for the hypothesis that seasonal dependence on reddish leaves favored the fixation of routine trichromacy in howlers, which may represent a convergent adaptation with catarrhine ancestors (Dominy and Lucas 2001; Melin, Khetpal, et al. 2017). In addition, novel work examining the impact of color vision on the detection of female sexual receptivity supports the hypothesis that primate trichromacy is well suited for interpreting social signals (Changizi et al. 2006; Hiramatsu, Melin, et al., "Experimental Evidence," 2017). Catarrhine primates show a wealth of colorful social signals (W. Allen et al. 2014; J. Setchell, Wickings, and Knapp 2006), and the radiation of these colorful signals was likely facilitated by the early fixation of routine trichromacy in catarrhines, following initial advantages in foraging (A. Fernandez and Morris 2007).

Chemosensation: Taste and Olfaction

Due to their dynamic evolution, olfactory receptors (ORs) and taste receptors (TRs) hold great promise for revealing patterns of adaptive radiation. As with the opsins underlying color vision, researchers seek to understand the evolution of olfactory and taste sensory systems by examining gene function, numbers, and "tuning" to sensory cues (e.g., Akhtar et al. 2022; Orkin et al. 2021; Trimmer et al. 2019). This task is considerably more difficult than for opsin genes for several reasons, including a greater number of OR and TR genes. Additionally, many of these genes are "orphaned," meaning we do not know the compounds to which they are sensitive or their roles in affecting phenotypes (Adipietro et al. 2012). Furthermore, while one chemosensory gene is expressed per sensory neuron (like opsins), chemosensory receptors have more complex "many-to-many" relationships with sensory stimuli than opsins. In other words, one receptor can respond to many compounds, and one compound can trigger many receptors. Particularly given the large number of functional OR genes, this relationship allows responses to huge numbers of odors (e.g., humans are estimated to sense more than one trillion odors; Bushdid et al. 2014) but complicates our understanding of olfactory function and evolution. Although much work lies ahead, the development of large-throughput technology and advances in comparative genomics, as well as the use of in vitro gene assays, have led to a dramatic increase in knowledge of olfactory genomics in the last 10 years (Adipietro et al. 2012; Hoover et al. 2015; Mainland et al. 2014; Meyerhof et al. 2010; Trimmer et al. 2019; Tsutsui et al. 2016).

A comparative analysis of the intact OR genes of 50 mammal species suggested a role of ecotype in shaping olfactory evolution (Hayden et al. 2010; but see Orkin et al. 2021). Ecotype (aquatic, semiaquatic, terrestrial, flying, etc.) was a more important predictor of OR gene similarity than was phylogenetic relatedness, and variation in OR genes among bats correlated with diet. OR orthologs, even among closely related species, can respond to the same ligand with very different efficacies, revealing another pathway for evolutionary adaptation (Adipietro et al. 2012). A comparative analysis of species-specific gene duplication in the OR repertoire across 94 mammalian species showed associations between OR families and dietary niches (G. Hughes et al. 2018). Looking at intraspecific patterns, humans demonstrate surprisingly large OR variation. Of 413 intact genes examined, 66% of ORs were affected by nonfunctional SNPs/indels or copy number variation (Olender et al. 2012). This variation has been linked to olfactory phenotypes and food experiences, in patterns that range from monogenic Mendelian patterns to more complex associations (Jaeger et al. 2013; McRae et al. 2013). A number of studies suggest that local environment and diet have influenced OR variation in modern and extinct Homo (Hoover et al. 2015; G. Hughes et al. 2014; Olender et al. 2012; Sorokowska et al. 2013; Veilleux et al. 2023), but more work in this area is needed to understand fundamental patterns of selection and to determine whether other primate populations show similar levels and causes of variation.

There are fewer TRs than ORs, and more progress has been made toward linking genotype to phenotype by associating TR responses with stimuli (tastants). Taste receptor genes are lumped into two broad classes: TAS1Rs and TAS2Rs. The TAS1R group consists of three genes in mammals and generates the perception of umami (TAS1R1 and TAS1R3), which is a "meaty" taste activated by amino acid L-glutamate and some ribonucleotides, and sweetness (TAS1R2 and TAS1R3). As with OR genes, dietary ecology appears to shape the tuning and function of TRs. For example, the giant panda, a specialized bamboo feeder descended from carnivorous ancestors, has lost the function of TAS1R1, presumably due to lower importance in detecting nucleic acids in their diet (Zhao et al. 2010). In contrast, some dedicated carnivores have lost the ability to taste sweetness through mutations in TAS1R2 (Jiang et al. 2012). This gene also appears to be under relaxed selection in the ancestral branch of colobine primates, which coincides with a shift from frugivory to specialization on leaves (G. Liu et al. 2014). A comparative study of TAS1R3 across primates suggests that primates and plants may also be coevolving in an "arms race" in regard to sweet-tasting protein sugar mimics. Although most catarrhine primates (and humans) perceive the sugar mimic brazzein as sweet, gorillas (genus *Gorilla*) have evolved an amino acid substitution in TAS1R3 that makes them insensitive to brazzein's sweetness, and they are reported to not consume fruit containing it (Guevara et al. 2016).

The genes underlying the second TR group, TAS2Rs, are more numerous (25–30 functional genes in primates) and diverse than *TAS1Rs*. Unlike the typical mammalian pattern for *ORs* and opsins, *TAS2Rs* are frequently coexpressed in the same cell, creating a dynamic sense that can recognize thousands of different bitter substances. Unsurprisingly, *TAS2Rs* have been well studied in the context of dietary ecology (Behrens and Meyerhof 2013; Meyerhof et al. 2010; Pereira et al. 2021; Purba et al. 2017; Widayati et al. 2019), yet the impact of diet on the molecular evolution of these genes is subject to debate. M. Campbell et al. (2012) suggest that ancient balancing selection maintained common haplotype variation in *TAS2R38* (the "PTC" gene) across global populations in humans and that recent selection pressures have yielded unusually high levels of rare variants in Africa, implying a complex model of selection. Furthermore, the distribution of common haplotypes of this gene was not correlated with diet, suggesting that allelic variation may be involved in nondietary processes. Alternatively, allelic diversity across *TAS2R* genes in a population of chimpanzees has been argued to indicate weak balancing selection though heterozygous advantage (Sugawara et al. 2011). Another chimpanzee population showed a different pattern of genetic diversity, which the authors suggested might reflect subspecies-specific dietary repertoires (Hayakawa, Sugawara, et al. 2012). Functional variation in *TAS2R38* has also been identified in one population of Japanese macaques (out of 17 sampled) and has been hypothesized to be an example of local adaptation in that population to local plant species (Suzuki-Hashido et al. 2015). Future work will shed insight into the complex pattern of functional and adaptive evolution in *TAS2Rs*. As more genomes become available and techniques for comparative genomics that explicitly include ecological parameters improve, our ability to detect meaningful differences in the responses of TRs and ORs within and between species will increase exponentially in the coming years (Kawamura and Melin 2017; Sugawara and Imai 2012; Tsutsui et al. 2016).

Mechanosensation: Audition and Touch

The sense of hearing in some mammals, especially bats, has been well studied. For example, a recent study reveals a fascinating arms race between predator and prey in shaping the acoustic sensitivity of echolocating species (Ter Hofstede and Ratcliffe 2016). Approximately 70 genes have been linked to the auditory sense in humans (Frenzel et al. 2012), but comparative genomics across primate species to examine the impact of ecological variables has not been attempted until recently, when seven orthologous genes underlying echolocation in whales and microbats were examined in aye-ayes and another lemur species without specialized hearing (Bankoff et al. 2017). Aye-ayes possess a derived auditory system, which they employ during percussive foraging (using a derived and highly irregular middle finger) to locate embedded invertebrate prey, and this study suggests that aye-aye tap-foraging auditory adaptations are not convergent with the genes of echolocating species but are rather a unique innovation. Future work linking genes underlying the acoustic sense (e.g., prestin; Shen et al. 2011) using audiograms, auditory reconstruction, and food-source noises across species will propel this area of sensory ecology forward (Quam et al. 2013; Ramsier and Dominy 2010).

Although the molecular basis of touch is still relatively unknown compared to that of other senses, several candidate genes have been identified (Poole et al. 2014; C. Walsh et al. 2015). Comparative genomic studies are currently still limited, and our understanding of evolutionary processes shaping the genes governing touch is lacking. Studies on birds and star-nosed moles, however, suggest that variation in the use of touch during foraging may be associated with changes in the levels of expression of candidate mechanoreceptor genes (Gerhold et al. 2013; E. Schneider et al. 2014). E. Schneider and colleagues (2014), for example, found that ducks, which use their bills for touch-based foraging, exhibit different patterns of gene expression in the neurons innervating their bills than visual-foraging chickens. Furthermore, at least one study suggests that some aspects of hearing and touch share a common genetic underpinning: people with hearing impairment also suffered from poor haptic acuity, which was linked to a common gene, *USH2A* (Frenzel et al. 2012). Both hearing and touch are based on the transformation of mechanical force (sound waves and physical objects, respectively) into electrical signals, so shared genetic mechanisms make intuitive sense. As discussed earlier in this chapter, comparative analysis of the haptic sense across species with varying morphologies and food palpation patterns would be a fascinating area for future work; identifying and examining candidate genes underlying this sense will fill an important piece of the puzzle.

FUTURE STUDIES

The future of primate sensory ecology is bright. Declining analytical costs and advances in portable field equipment are allowing us to investigate how wild primates and their natural foods interact in situ in ways that were not previously possible. In addition, studies that examine how, when, and why animals integrate information from multiple sensory systems to reach a decision will provide new insight into foraging ecology and sensory evolution. Currently, multisensory studies of primate foraging are in their infancy and have primarily been conducted on primates under captive or semicaptive experimental conditions. There is a clear need to increase the assessment of sensory integration in the wild. Ideally, experimental and field studies will be done in concert and build on each other. Critical to the process of understanding how and why primates use their senses to find and evaluate foods is the need to understand the physical and chemical characteristics of the foods and the information these properties might convey to foragers (chaps. 22–25, this volume). When designing studies, researchers should be mindful of collection methods and strive to sample foods in a standardized way to enhance abilities to draw comparisons across field sites. Primatology is becoming ever-more multidisciplinary, and the field of sensory ecology is well poised to benefit from this. To understand the importance of different sensory systems at proximate and evolutionary levels, we need to build teams with expertise in widely disparate areas. Figure 6.4 presents an example demonstrating the variety of diverse approaches that can contribute to a more comprehensive understanding of

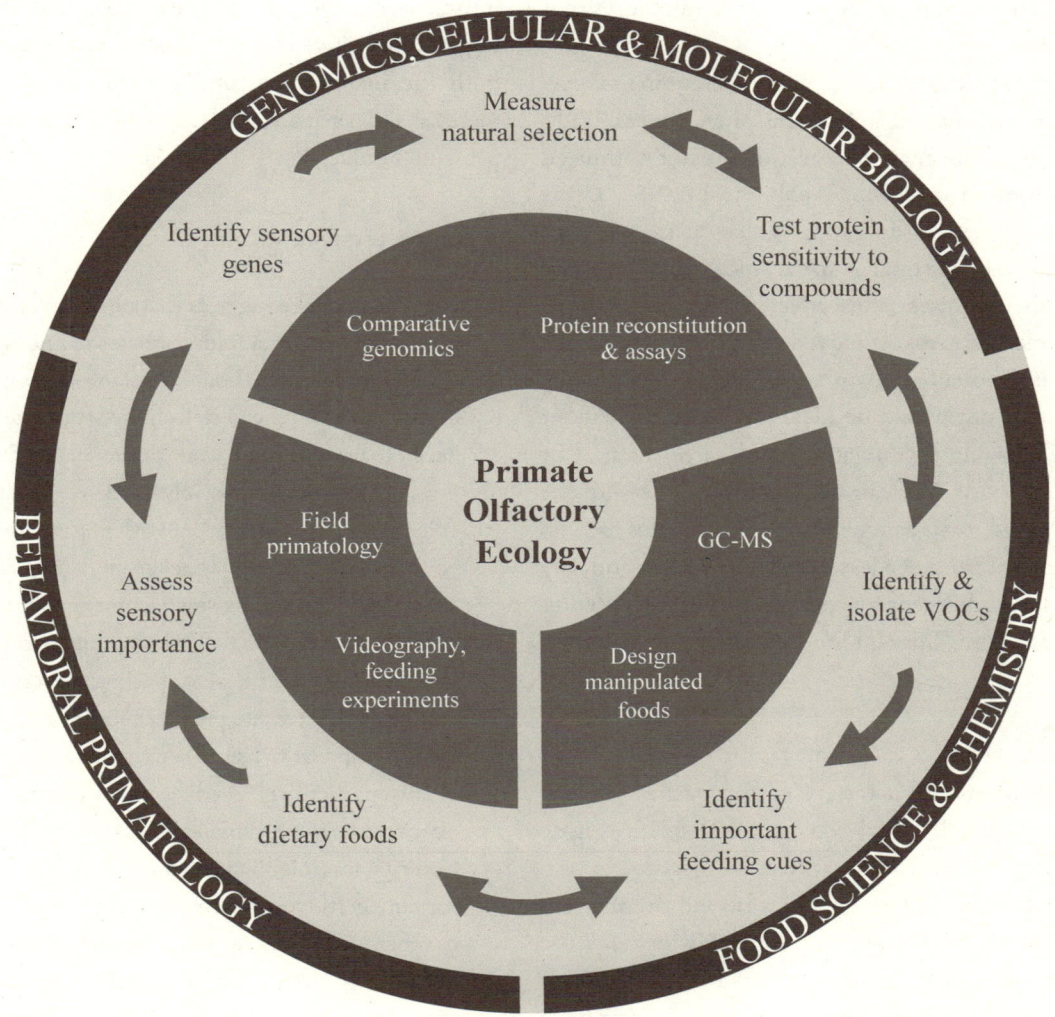

Figure 6.4. Multidisciplinary contributions toward an integrated understanding of primate olfactory evolution.

olfactory evolution. Similar schemes could be drawn for each of the senses discussed in this chapter. In color vision, there is a straightforward link between genotype and phenotype, but this is a greater challenge with the other senses, especially olfaction.

Advances in methods for obtaining high-quality DNA from feces, together with improved bioinformatic pipelines, allow for comparative genomics, even for rare, endangered, cryptic, or otherwise difficult-to-sample species (e.g., Snyder-Mackler, Majoros, et al. 2016; Orkin et al. 2021). These techniques have allowed a shift away from the biomedically important species, for which we currently have the best genomic data, and include taxa of key phylogenetic or ecological relevance to our study questions. Additionally, decreasing sequencing costs will facilitate population-level genotyping of study populations and augment the possibilities of linking genotypes to behavioral phenotypes in wild animals (Brent et al. 2014).

The goal of linking genes to functions requires achieving a better measure of behavioral phenotypes. Traditional methods of primate observation are not optimized for this (reviewed in Melin, Webb, et al. 2018). Using high-resolution, high-sensitivity video cameras with slow-motion replay, suitable for tropical fieldwork and captive study, promises future advancement (e.g., Ngo et al. 2022). Such cameras employed at field sites where primates are individually known and foraging experiments can be carried out would be ideal for dissecting subtler aspects of foraging investigation choices on natural or manipulated foods. We can also continue to refine the questions asked, such as whether some primates are less selective and eat fruits at a less ripe state, or when and why individuals with different sensory genotypes use different senses for assessing different foods.

SUMMARY

Sensory ecologists use a variety of approaches (experiments, genetics, behavioral ecology, etc.) to investigate how primates and other animals use their senses to perceive the world, including the roles played by different senses in finding and evaluating food. Historically, the visual sense has been prioritized in studies of primate sensory ecology, which is not surprising given the emphasis on visual adaptations throughout primate evolution (e.g., Cartmill 1992; G. Jacobs 2009; Ross and Kirk 2007). Although vision is important in primate foraging, recent work has revealed that other sensory modalities also play important roles. Indeed, evidence is growing that primates integrate cues from multiple senses to detect and evaluate foods. Furthermore, the roles played by different senses during foraging vary across primate species due to complex interacting factors such as food types, the time and light levels at which species are active, and the types of habitats and microhabitats preferred.

The next decade will likely be filled with advances in understanding how primates find and evaluate foods and how foraging pressures have contributed to primate adaptation and radiation. Exploring these relationships will provide new knowledge regarding how primates interact with their environments, how different diets within the order shape sensory systems, and how foraging decisions impact seed dispersal, parasite exposure, movement ecology, and cognitive ecology. These discoveries will inform both the management and enrichment of captive primates as well as decisions on how best to conserve natural primate biodiversity.

CONCLUSIONS

- Current evidence suggests that olfaction and vision are particularly important for primates in finding foods in the environment. Although tracking odors over long distances has not been well studied, primates are believed to use olfaction at shorter ranges to detect fruits, flowers, insects, and possibly exudates. Trichromatic color vision has been linked with finding fruits and red young leaves in a foliage background and may be especially useful for finding small, ephemeral food patches. Visual acuity and binocular vision, often in conjunction with hearing, are critical for finding and capturing invertebrate and vertebrate prey (especially at night).
- Primates appear to use all senses when evaluating food quality, particularly when feeding on fruit. Color vision, olfaction, taste, and smell can all be used to determine fruit ripeness, highlighting the role of multimodal sensory integration in food evaluation. Sensory evaluation of other food types has been less well studied than that of fruit. However, color, taste, and smell may all offer important cues for determining leaf and insect palatability (e.g., toxicity).

- Many studies have explored the evolution of primate color vision in relation to foraging ecology using analyses of opsin genes. Major questions investigated within this topic include factors influencing the retention or loss of dichromatic color vision among nocturnal primates, how polymorphic trichromacy has been maintained in platyrrhine monkeys and lemurs, and factors leading to the emergence of routine trichromacy in catarrhines and howler monkeys.
- The development of genetic and genomic technologies has facilitated a recent boom in studying the evolution of primate chemosensory genes. Comparative analyses suggest that diet may influence the diversity of olfactory receptor genes; however, much more work needs to be done in this topic. The evolution of taste receptor genes has yielded more results, with recent studies detecting changes in taste function associated with the evolution of dietary specializations in different lineages, as well as variation within species linked to local environmental conditions.
- Comparative genetic research exploring the evolution of mechanosensation (hearing and touch) is currently limited. Preliminary evidence indicates that there are likely shared genes between these two senses due to similar mechanisms of transforming mechanical forces (sound and pressure).

7 Seasonality in Food Availability and Energy Intake

Cheryl D. Knott and Andrea L. DiGiorgio

Primate diets are contingent on the availability of foods that often fluctuate dramatically in space and time (Hemingway and Pfannes 2005; van Schaik and Pfannes 2005; van Schaik, Terborgh, and Wright 1993). Understanding and measuring the ecological context of primate foraging has become increasingly important as caloric and nutrient intake have been found to be major determinants of both behavior and physiology, ultimately impacting evolutionary fitness. In this chapter, we first review the concept of seasonality as it applies to studies of wild primates. We focus on how changes in seasonality and food availability have been measured in primate studies and recommend the application of methods that will permit future comparisons between different primate populations and species. We then examine the concept of energy and nutrient intake, highlighting advances that allow us to more accurately estimate and understand changes in caloric and nutrient intake. The second part of the chapter describes the ways that primates respond to these seasonal changes in the foods that are available to them. Perhaps the majority of studies of wild primates contain at least some mention of changes in behavior due to diet and food availability. Thus, we do not attempt an exhaustive review of this literature. Instead, we synthesize the published data by presenting the theoretical approaches that have been used to understand primate responses to changes in food availability. Through providing specific examples from across the primate order, we hope to provide greater clarity on the complex ways that primates respond to changes in their environments in order to extract the energy and nutrients that sustain them.

WHAT IS SEASONALITY, AND HOW DO WE STUDY IT?

Defining Seasonality

When the first studies of wild primates were carried out, knowledge of changes in floristic patterns was limited (DeVore 1965; Rodman 1978). Although early primatologists were certainly aware that food availability changed and that this likely impacted caloric and nutritional intake (J. Altmann 1980), they usually were not equipped to systematically study these changes. It was the growth of the field of ecology, and the increasing recognition of the importance of understanding how an animal interacts with its environment, that led to the incorporation of phenological monitoring into primate studies. With the development of tropical research sites and an increased emphasis on ecology, many primatologists gained expertise in identifying and studying primate food plants and changes in floristic patterns (M. Leighton and Leighton 1983; Terborgh 1983; van Schaik 1986; van Schaik, Terborgh, and Wright 1993).

In primatology, we are particularly interested in seasonality as it relates to changes in food availability. In the temperate zone, seasonality follows relatively predictable cycles and is correlated with changes in temperature, rainfall, and plant productivity. However, most primates live in the tropics, where monthly temperature varies little (Whitmore 1986) and seasonality is usually defined by differences in rainfall (van Schaik and Pfannes 2005). This led early primatologists to use rainfall as an indicator of changes in food production (de Ruiter 1986; DeVore 1965; Goodall 1968). The daily rainfall was an easy measure to obtain, requiring a simple rain gauge. Yet, the relationship between rainfall and fruit (or food) production is not straightforward. Van Schaik and Pfannes (2005) in their analysis of the relationship between seasonality measures and phenology found that flowering and flushing (new leaf production) tended to occur at the start of the rainy season, especially in drier climates. However, rainfall patterns were *not* associated with fruit production (van Schaik and Pfannes 2005), except, perhaps, for the very driest sites in the Neotropics (Dezeure et al. 2021). Although rainfall might be associated with fruit production at a particular site, these results confirm that rainfall should not just be assumed to be an indicator of food availability.

Furthermore, the terms "seasonality" and "food availability" are often used as proxy measures for an animal's caloric intake. This is problematic. When overall forest fruit production is high, the actual caloric intake of an individual animal may not increase in response. There could be a variety of explanations for this, including that low-ranking animals may not be able to take advantage of high fruit availability because they are outcompeted. For example, in brown capuchins, high rates of aggression are associated with relatively scarce resources, and during a pronounced period of food scarcity, dominant brown capuchins consumed 20.5% more energy than did subordinates (Terborgh and Janson 1986). In gorillas, a species considered to have relatively weak dominance relationships, it was found that dominant females had higher intake rates than did subordinates in the same group (E. Wright et al. 2014). Likewise, a "low" fruit season may lead to high caloric intake if the particular plant species that are fruiting are of high caloric value. This was the case in orangutans at Gunung Palung National Park in Borneo, where, during a relatively low fruit season, orangutans (particularly males) ate *Neesia* seeds that were extremely high in lipids, and thus they had a high caloric intake during that period despite overall low fruit production in this forest (Knott 1998; van Schaik and Knott 2001).

An additional factor is heterogeneity in food availability within the study area. If food availability is assessed over an entire study area that consists of multiple habitat types, then primates may move to the more productive habitat, thus having a higher intake than would be expected; or, if their ranges are fixed or otherwise unable to change because of resource competition, they may be stuck in a lower-productivity area despite overall higher forest food production. In the Kanyawara study population in Kibale National Park, Uganda, female chimpanzees occupy "neighborhoods" that vary in their fruit productivity, and thus the chimpanzees vary in their fruit intake. This leads to profound consequences for the behavior and reproductive success of individuals occupying those areas (Emery Thompson, Kahlenberg, et al. 2007; Kahlenberg et al. 2008).

Thus, although food availability can correlate with food intake, this cannot be assumed and may not hold for every individual in the population. Indeed, Hemingway and Bynum (2005) found in their analysis of 157 studies of primate diet that fruit availability *did not* correlate with time spent eating fruit. The exception was Malagasy primates, where availability explained 45% of the variation. Similarly, leaf flush was a good predictor of leaf feeding in Madagascar but not in Asia (sample sizes were too small to test in Africa and South America). Thus, although phenological measures are extremely useful indicators of what is theoretically available for a primate, they may be poor measures of actual caloric intake.

One of the most important features of primate food availability, and one that structures primate behavioral responses, is the degree to which food is temporally and spatially predictable. In some locations, seasonality connotes temporal predictability, with the same plants fruiting at the same time every year. However, at other sites, such as the lowland dipterocarp forests of the Southeast Asian tropics, forest fruit production varies tremendously but is decidedly unpredictable (M. Leighton and Leighton 1983; van Schaik 1986). These forests are characterized by mast fruiting, a phenomenon that happens every 2–10 years in which a majority of canopy trees fruit in synchrony (Ashton et al. 1988). This is argued to be a strategy to swamp out seed predators (Ashton et al. 1988; Janzen 1974) and seems to be tied to the El Niño weather pattern (Curran 2000; F. Ng 1977; Wich and van Schaik 2000). Norden et al. (2007) demonstrate that mast fruit-

Figure 7.1. Female orangutan (*Pongo pygmaeus wurmbii*) and seven-month-old infant feeding on fruit. Gunung Palung National Park, West Kalimantan, Borneo, Indonesia.
Photo © Tim Laman.

ing also occurs in South American rainforests, and they argue that masting may be a more common plant strategy than previously supposed. Thus, in many primate habitats there are no "seasons" per se, and it is more appropriate to refer to changes in food availability rather than food seasonality.

Measuring Seasonality in Primate Habitats

Quantifying changes in food availability presents a number of challenges (Heymann, this volume; A. Marshall and Wich 2013). Is it appropriate to measure the production of all fruits in the forest or only those that your species of interest actually feeds on? If you decide to measure the fruits and plants that your species consumes, do you monitor all food species that could possibly be consumed by that primate taxon, only the foods consumed at your site, or the most commonly consumed foods? In sites that constitute a mosaic of habitats, food availability may vary by habitat type. In such cases, a given individual may have a home range that does not encompass the high-fruit-availability area. Thus, one site-wide measure of food availability may not apply to animals that use different portions of that habitat.

Comparisons of the relative degree of seasonality between different study populations and different species are very problematic because of the lack of comparable measures. Many primatologists use measures of food availability that are applicable only to their particular study site. One common practice is to estimate the percentage of individual tree crowns that are flowering or fruiting. This index measure does allow for comparisons of food availability between different years and months for that site. However, this method does not account for differences between trees or between species in crown size, and it does not allow for comparisons between study sites (Knott 2005; van Schaik and Pfannes 2005). For example, many small trees could have 100% full crowns of fruit, but because of the size of the trees, this could represent much lower food availability than another sample of much larger trees with only a fraction of full crowns. Thus, primate ecologists have argued that the best phenological measures involve estimates of crop size, the standard methodology of rainforest ecologists (M. Leighton 1993). This provides a quantifiable measure of the actual amount of food that is available for primate consumption. These studies typically estimate flower and fruit crops in categories of exponentially increasing size (A. Marshall and Wich 2013). Another method of quantifying fruit production is to use the diameter at breast height of the tree. This method has been shown to reflect crop size and fruit biomass fairly well (Chapman, Chapman, Wrangham, et al. 1992). However, diameter at breast height is not a useful measurement for lianas or figs (*Ficus*), which are often important primate food sources (A. Marshall, Ancrenaz et al. 2009).

Furthermore, primate foods differ vastly in size, energy, and nutrient content (Chapman, Chapman, Rode, et al. 2003; Conklin-Brittain, Wrangham, and Hunt 1998; Génard and Bruchou 1992; Georgiadis and McNaughton 1990; Houle, Chapman, and Vickery 2010; Norconk, Grafton, and Conklin-Brittain 1998; Felton and Lambert, this volume). Thus, measures of fruit availability should consider the energy and nutrient content, as well as the quantity, of fruit available. A full canopy of small, low-calorie fruits presents a different resource than a full canopy of large, high-calorie fruits. Ideally, fruit availability measures should incorporate the size and the caloric content of the fruit, in addition to the number of fruits per crown (Knott 2005). Where feasible, "kcals of primate food available/hectare" should be calculated as a measure that is comparable between study sites and between species (Knott 2005).

Research on arboreal, frugivorous primates focuses on phenological assessments of fruit and flowers from trees, lianas, and figs. However, terrestrial primates may be accessing fallen fruits and the seeds they contain (McGraw, Vick, and Daegling 2014a). For these species, using ground phenology measures may enhance our ability to estimate food availability. In a study of Tana River mangabeys, seeds from three out of seven highly consumed food species remained on the ground well after the canopy was bare of these foods (Wieczkowski 2013). Wieczkowski (2013) describes combining a measure of ground seed availability with canopy phenology and discusses for which mangabey foods each measure was the most effective.

Primates also eat many nonreproductive plant parts such as leaves, herbaceous plants, underground storage organs (USOs), exudates (gums and saps), and bark. A fair number of studies include leaf phenology measurements (Agetsuma 1995; Campera et al. 2021; Ganzhorn 2002; Hemingway 1998; Overdorff 1993a; Stanford 1991). Leaf flushing may even be more seasonal than fruit production (Hemingway 1998) and is strongly related to climate, whereas fruiting patterns are not (van Schaik and Pfannes 2005). Some studies have measured the availability of terrestrial herbaceous vegetation (Malenky and Stiles 1991; Malenky and Wrangham 1994; Malenky et al. 1993; Wrangham, Rogers, and Basuta 1993) or the presence of parasitic and epiphytic plants (Le et al. 2019). Arboreal herbaceous plants or epiphytes present an interesting challenge, as they often grow above and along large branches, invisible from the ground (for method recommendations, see A. Marshall and Wich 2013). In addition, they can be completely consumed or destroyed by a primate or group of primates in a single feeding bout. Terrestrial primates may access USOs, such as roots, tubers, rhizomes, and corms. Thus, researchers have also used soil cores to sample the availability of these foods underground (Jarvey et al. 2018). Exudates can be important food sources for a number of primates, especially for callitrichines (Burrows and Nash 2010). Availability can be difficult to assess, but a number of studies have used natural or experimental monitoring of individual trees to measure exudate production (Garber and Porter 2010; Génin et al. 2010; Isbell 1998) as well as the role of consumers in stimulating production (C. P. Jackson and Reichard 2021). Bark is also an important primate food source, but we are aware of no studies that assess bark (or, more specifically, cambium) availability. We do not expect seasonal differences in bark availability, but the nutrient composition of the bark that primates consume may change seasonally.

Many primates include insects and even small vertebrates in their diets, either seasonally or year-round. Some studies have sampled insects and other prey items, demonstrating that they do vary spatially and temporally (Janson and Emmons 1990). However, measuring insect and small-vertebrate abundance is difficult and rarely attempted (see Heymann, this volume, for a review of studies). Studies of tarsiers, the most insectivorous primates, call for an assessment of insect "availability" beyond assuming seasonal variation (Gursky 2000). For research on primates consuming flying and terrestrial insects, malaise traps, composite traps, and pitfall traps are common methods of assessing prey abundance (Gursky 2000; Mallott, Garber, and Malhi 2017; Muirhead-Thompson 2012; Skvarla 2015), although other methods have also been used (Boinski and Fowler 1989; Janson and Emmons 1990). Termites, ants, and other nest-dwelling and colonial insects present a major challenge in assessing food abundance, especially when these insects inhabit the canopy and interior of trees. Without knowledge of seasonal changes in insect and larva abundance, it is assumed that these resources are constant. However, insect abundance likely also fluctuates in response to changes in rainfall, temperature, and plant flowering and fruiting patterns.

Our ability to assess primate diet and seasonality in terms of food availability has become much more refined, thanks to advances from primatologists, ecologists, and botanists alike. Yet, to facilitate intersite comparisons, more application of absolute measures rather than relative indices is needed. While many primates specialize on fruit and flowers, most consume other foods as well. Especially when primate nutrition is viewed from the perspective of seasonality or fallback foods, the need for methodologies to assess mature and new leaf, bark (tree and liana), USO, exudate, and other nonreproductive plant parts, as well as insect availability, becomes crucial. Thus, as we become more precise in our assessment of primate feeding ecology, a more comprehensive approach to overall food availability and abundance is warranted.

Assessing Seasonal Caloric and Nutrient Intake

In parallel with more sophisticated studies of primate habitats and food availability, the study of caloric and nutrient intake has become increasingly precise as well. The assessment of energy intake has evolved over the past 50 years from studies of simple feeding time to measured estimates of energy and nutrient intake. This reflects improved technology and resources as well as theoretical advances in our understanding of primate physiology, nutrient requirements, and feeding behavior. The first studies of primate feeding behavior focused on quantifying time spent feeding, with the assumption that animals that fed longer were also consuming more energy. Although this is an obvious first step in understanding energy intake, increased feeding time does not necessarily mean increased caloric intake (Aristizabal et al. 2017). Some foods are much more calorie-dense, commonly called *high-quality* foods, and others, with lower caloric content, are considered *low-quality* foods—with the caveat that food "quality" may be determined by multiple nutritional components (Felton and Lambert, this volume; DiGiorgio et al. 2021). During low-food-availability periods, primates may actually be feeding longer but on poorer-quality foods (foods from which primates can extract less energy) and thus obtaining less energy overall. For example, *Macaca fuscata yukui* in the very seasonal Yakushima Island habitat of Japan ate a higher proportion of leaves with increased feeding time (Agetsuma 1995). When temperatures were coldest, the macaques conserved energy by moving around less, likely to maintain body temperature, and by eating leaves that were more concentrated in distribution than fruit. In this case, temperature and energy expenditure were important factors in feeding decisions. Thus, feeding time alone is not a reliable measure of caloric intake.

Beyond ascertaining what primates eat and for how long, more precise measures analyze the nutrient content of wild primate foods (M. Leighton 1993; Milton 1979, 1981c; Rogers, Maisels, et al. 1990; Williamson et al. 1990; Wrangham, Conklin, Chapman, et al. 1991). The first studies used bomb calorimetry, in which the overall energy content of the sample was determined (Nagy and Milton 1979b). Later studies measured the macronutrient components separately to determine both the nutrient composition of the diet and how much energy was actually metabolically available to the animal (Conklin-Brittain, Wrangham, and Hunt 1998; Rogers, Maisels, et al. 1990; Sterling et al. 1994). Eventually, actual daily or hourly caloric and nutrient intake could be estimated (M. E. Harrison, Morrogh-Bernard, and Chivers 2010; C. A. Johnson, Raubenheimer, Rothman, et al. 2013; Knott 1998; Masi, Mundry, et al. 2015; Rothman, Dierenfeld, Hintz, et al. 2008). Measuring energy intake requires collecting detailed data on feeding behavior, weighing and processing food samples, and subsequently conducting laboratory analyses of the nutrient and caloric content of those samples (see Conklin-Brittain, this volume).

Other factors that relate to seasonal changes in the diet must also be considered when reporting either the number of kilocalories or the nutrients consumed. Many primates rely on high-fiber foods during periods of low fruit availability, with some species obtaining energy from fiber through hindgut fermentation (Chang et al. 2016). However, quantifying the amount of energy an animal obtains from fiber is particularly problematic. Although most fiber is not digested, the hemicellulose portion of neutral detergent fiber (NDF) is partially fermentable through the action of anaerobic gut microbes (Conklin-Brittain, Knott, and Wrangham 2006). The percentage of fiber that is digested has been determined for a small number of primates (de Andrade Carneiro et al. 2021; M. Edwards and Ullrey 1999a, 1999b; Milton and Demment 1988; Remis and Dierenfeld 2004; D. Schmidt et al. 2005; Takahashi et al. 2019). Another complication of fiber digestion is that lignin, an additional component of the NDF portion of fiber, is not fermentable. One approach to account for this, taken by Masi, Mundry, et al. (2015), is to first subtract the lignin from the NDF and then use the experimentally determined digestion coefficient. This allows the lignin component of the fiber to be considered separately for each food.

The intake of high-fiber foods may also impact the digestibility of other foods. During low-food periods, some primates may consume a large volume of low-quality, high-fiber foods (Conklin-Brittain, Knott, and Wrangham 2006). The sheer bulk of the diet can then lead to faster gut passage rates, which lowers the digestion coefficients of all the macronutrients (Milton and Demment 1988; Righini et al. 2017; van Soest 1994). However, this may vary by species. Chang et al. (2016) found that gut retention time was longer for captive gibbons and orangutans fed a high-fiber diet than those fed a low-fiber diet, but this was not the case for two species of macaques (*Macaca cyclopis* and *Macaca fascicularis*). This was attributed to the apes having the ability to retain food longer or having more extensive fermentation capabilities than the monkeys. Howler monkeys fed diets higher in NDF showed reduced digestibility of crude protein (de Andrade Carneiro et al. 2021). The lignin portion of the fiber can also bind with protein, which then makes this bound protein unavailable for digestion (Conklin-Brittain, Dierenfeld, et al. 1999; Rothman, Chapman, and Pell 2008). Thus, as diets change in fiber, specifically lignin content, this may affect the amount of protein that is metabolized from food, which may impact foraging decisions. For example, the feeding time for the folivorous lemur *Avahi meridionalis* is positively correlated to nitrogen (protein) content and negatively correlated to polyphenol (plant chemicals that bind to proteins such as lignin) during the food-rich season, suggesting that when seasonal conditions allow, these strepsirrhines avoid fibrous foods and instead seek out more available protein (Campera et al. 2021).

As our understanding of primate nutritional chemistry becomes more sophisticated, we will better be able to appreciate how these seasonal changes in the types and amounts of foods primates consume affect their digestibility. Although studies of seasonality and primate diet have gained finer detail over time, research in primate nutritional ecology highlights the need for assessing the nutritional content of primate foods at an even finer level (e.g., between fruits on the same stem or between stems of the same species; Chapman, Chapman, Rode, et al. 2003; Houle et al. 2010; K. Potts, Chapman, and Lwanga 2009). Technologies such as near-infrared reflectance spectroscopy (Rothman, Chapman, Hansen, et al. 2009; Conklin-Brittain, this volume) may provide finer resolution to these issues. Additionally, techniques such as stable isotope analysis may give us new insights into primate diets and their incorporation into body tissues (Oelze, Percher, et al. 2020).

PRIMATE BEHAVIORAL RESPONSES TO SEASONALITY: THEORETICAL APPROACHES

Studies of primate diet and responses to seasonal changes are built on theoretical frameworks including optimal foraging theory and the geometric framework of nutrition (also see the discussion of these frameworks in Felton and Lambert, this volume). Each of these provide different understandings of primate foraging goals and thus are important in a discussion of how seasonality impacts primate diet. Here, we discuss these theoretical frameworks as they relate to primate responses to seasonality.

Optimal Foraging Theory

Most research in primate and mammalian nutritional ecology is built on the framework of optimal foraging

Figure 7.2. Spectral tarsier (*Tarsius tarsier*) eating a cockroach at night. Tangkoko-Duasudara Nature Reserve, Sulawesi, Indonesia. Photo © Tim Laman.

theory (OFT; Pyke et al. 1977; D. Stephens and Krebs 1986). OFT assumes that foraging has evolved via natural selection to optimize feeding in order to maximize fitness (Pyke et al. 1977). In primate studies, OFT-based categorizations include energy maximizers, protein maximizers, and toxin (plant secondary compounds, indigestible fiber, etc.) minimizers, though in most cases, energy maximization (calories per unit time) is thought to be the most important fitness "currency" (Pyke et al. 1977; Schoener 1971; Westoby 1974). These approaches recognize factors, such as predation, macro- and micronutrient requirements, plant toxins, aggressive interactions, and searching for mates, that may limit or constrain an animal's ability to achieve the "optimal" diet and conceptualize these as "constraints" (Pyke et al. 1977).

One of the most important constraints on achieving an optimal diet is the spatiotemporal distribution of food, meaning that seasonality often plays an important role in OFT. Achieving an optimal diet is likely easier during seasons when food resources are more varied and abundant. As primate research developed within the OFT paradigm, resources were categorized as preferred or fallback (discussed below), and the intrinsic quality of a resource could be described in relation to the primate's primary foraging goal (e.g., fruit is a high-quality resource for an energy-maximizing species.)

OFT and its extensions have been and continue to be an important part of primate nutritional studies, from Stuart Altmann's extensive work on the foraging strategies of baboons at Amboseli (S. Altmann 1998, 2009) to research on species ranging from Campbell's monkeys (Zausa et al. 2018) to gibbons (Grether et al. 1992) to spatial decision making in lemurs (Teichroeb and Vining 2019). Larger-scale group movement and foraging decisions have also been understood through the lens of OFT (G. Davis et al. 2022). Even human foraging studies have benefited from an OFT approach (B. Griffiths et al. 2022; Raichlen, Wood, et al. 2014; Venkataraman 2017).

Geometric Framework

Another approach to studying nutrient intake and nutrient balancing is nutritional geometry, sometimes referred to as the geometric framework of nutrition (Felton, Felton, Lindenmayer, et al. 2009; Felton, Felton, Raubenheimer, Simpson, Foley, et al. 2009; C. A. Johnson, Raubenheimer, Chapman, et al. 2017; Rothman, Raubenheimer, and Chapman 2011; S. Simpson and Raubenheimer 2012; Raubenheimer, this volume). This framework allows the examination of how the intake of nutritional components (and biologically relevant ratios of those components) varies with food selection and whether the intake of a particular nutrient stays at a target level or within a given range while other nutrients vary. The intake target is the optimal nutrient ratio or requirement of an animal (S. Simpson and Raubenheimer 2012). Modeling primate diet using the geometric framework adds dimensions to OFT by looking at the entire content and nutrient balance of an animal's diet and how the nutrient and caloric elements interact. Described as the "packaging problem" (S. Altmann 2009), nutrients in foods are combined in varying quantities and ratios, and foods often have both desired nutrients and toxins that protect them. No food is perfect. There are several ways an animal can reach the appropriate nutrient targets and ratios. For example, Rothman, Raubenheimer, and Chapman (2011), using the geometric framework, found that mountain gorillas (*Gorilla beringei*) of Uganda shifted their diet based on seasonal food availability to maintain their nutrient balance. Specifically, gorillas prioritized nonprotein energy (i.e., energy from carbohydrates and fats) over different seasons and different levels of food availability, while intake of kilocalories from macronutrients other than protein did not change. When primarily leaves were available, gorillas dramatically increased their intake of leaves, even though this meant having a much higher protein intake. This was necessary to maintain adequate intake of nonprotein energy (Rothman, Raubenheimer, and Chapman 2011). In contrast, Felton, Felton, Raubenheimer, Simpson, Foley, et al. (2009) found that spider monkey intake of nonprotein energy was significantly related to the availability of ripe fruit, but protein intake remained stable regardless of season (Felton, Felton, Lindenmayer, et al. 2009; Felton, Felton, Raubenheimer, Simpson, Foley, et al. 2009).

Preferred versus Fallback Foods

The terms "fallback foods" and "preferred foods" have been used to conceptualize how dietary choices change with differences in availability caused by seasonality. A. Marshall and Wrangham (2007) operationalize *preferred foods* as food items that are overselected, or selected disproportionately, based on their relative abundance within the habitat. Thus, preferred foods are defined by both rarity and frequent utilization. Since this definition relies on selectivity and abundance, which are both difficult to define in the complex generalist diets of primates, methodologically determining preference is often challenging. To overcome this, A. Marshall and Wrangham (2007) suggest a detailed framework to address preferred and fallback foods and their adaptive significances. They suggest that *preferred foods* should be associated with adaptations for harvesting. In other words, if a food is high quality and rare, procurement of that food (searching for it or detecting foods within the habitat) should be a driver for adaptations to reduce search efforts. Harvesting adaptations expected to arise from seeking preferred plant foods include general cognitive abilities, spatial navigation, locomotor efficiencies, visual acuity, and improved olfaction. Noted examples of primate harvesting adaptations that fit this model include the locomotor adaptations allowing gibbons to travel efficiently between small and dispersed patches of fruits that are not accessible by sympatric macaques (Cannon and Leighton 1996; Cannon et al. 1994).

In contrast to preferred foods, A. Marshall and Wrangham (2007) define *fallback foods* as those food items that are abundant and easy to locate but hard to process. Noting that the term "fallback foods" is often used to indicate a food of relatively poor nutritional quality and high abundance that is utilized particularly during periods of preferred fruit scarcity, they operationalize the definition of fallback foods as "foods whose use is negatively correlated with the availability of preferred foods" (A. Marshall and Wrangham 2007, 1220). This definition thus suggests a distinction between preference (dietary choice) and importance (dietary composition). Preferred foods are often of high quality and can be either a large part of the diet during some seasons (high importance) or rarely consumed (low importance). Low-preference foods are of low quality but can form a major part of the diet, as do fallback foods (high importance), or they can be of low quality and rarely consumed (low importance). Overall, then, according to A. Marshall and Wrangham (2007), fallback foods would provide lower rates of energy gain than preferred foods. Adaptations to fallback foods would be expected to reduce processing effort (i.e., in procurement, ingestion, or digestion). In this framework, suggested adaptations to fallback foods include food-processing skills and morphologies, dental topographies (sheering crests, grinding cusps, etc.), enamel thickness, digestive adaptations (gut length and morphology of the digestive system), body size, and tool use (A. Marshall and Wrangham 2007).

Recognizing the high degree of variability with which certain fallback foods are utilized, A. Marshall and Wrangham (2007) also classify fallback foods into "staple" and "filler" fallback foods. The hallmark of this distinction is whether a food is ever used by a population or species for 100% of the diet (staple) or not (filler). Staple fallback foods are capable of sustaining a species seasonally, even if this yields a cessation in reproductive functioning. Staple fallback foods are thus expected to exhibit low seasonality and a more uniform spatial distribution. In contrast, filler fallback foods are food items that are likely not capable of providing complete sustenance over a season and that would fluctuate in availability and be patchily distributed.

J. Lambert (2007) proposes another framework with which to view the evolutionary importance of fallback foods. Drawing from optimal foraging theories, she suggests that animals should prefer foods that as a whole are the most energetically profitable, being nutrient dense, easy to find (low search time), and easy to access (low handling time). Lambert suggests two categories of fallback foods: (1) those of lower nutritional density that are often more abundant and thus require less search time but more processing or handling time, and (2) those of higher nutritional density that are often mechanically protected and difficult to find. In addition, Lambert (2007) suggests two fallback strategies that an organism can use to cope with seasonal food scarcity. One strategy is to fall back on more abundant foods that may require dental or digestive processing because they are less nutrient dense, higher in secondary or toxic compounds, or mechanically defended. The second strategy is to switch to foods that are either less nutrient dense or harder to find but that do *not* reduce the overall quality of the diet because they are being added to an already high-quality diet, and thus the profitability of these foods is still positive. Examples of these two strategies are found in the sympatric monkeys and chimpanzees of Kibale National Park: monkeys fall back on foods that are higher in toxins and digestion inhibitors, thus decreasing their overall quality, while chimpanzees fall back on foods that are of high quality, such as pith (Conklin-Brittain, Wrangham, and Hunt 1998; Wrangham, Conklin-Brittain, and Hunt 1998). J. Lambert (2007) thus suggests that we see the evolution of anatomical, physiological, and behavioral solutions more markedly if a species falls back on foods that have significantly different chemical, mechanical, or ecological characteristics than their preferred foods. This yields a continuum of adaptations in relation to fallback foods.

More recently, J. Lambert and Rothman (2015) have proposed that the use of the terms "fallback" and "preferred" foods is not particularly useful when we consider daily nutrient balancing. For example, we (DiGiorgio et al. 2022) have evidence that wild Bornean orangutans leave available fruit crops to consume other types of food, such as leaves, which normally would be seen as a fallback food. Thus, in this situation, orangutans are turning to less-energy-dense foods, possibly to obtain a particular protein balance. Within the geometric framework, primates are seen as foraging in a "nutritional landscape" of variably available nutrients and variable needs (J. Lambert and Rothman 2015). When viewed from this perspective, fallback foods and high-quality foods will take on new characteristics. Seasonality will impact what a fallback food is. For example, if an animal prioritizes protein (i.e., attempts to get a certain amount of protein, within a range, each day) during high-fruit periods, young leaves will become a critical and high-quality resource, even though leaves are traditionally classified as a fallback food. This is because young leaves are often a good source of protein in primate diets but a poor source of carbohydrates and a high source of indigestible fiber. During periods when young leaves are available and fruit is not, protein is easier to come by, and carbohydrates and energy are in limited supply. In many tropical forests, protein may be abundant enough that it is not a limiting resource for folivorous primates (Ganzhorn, Arrigo-Nelson, Carrari, et al. 2017).

BEHAVIORAL RESPONSES: CHANGES IN DIET, FORAGING PATTERNS, AND HABITAT USAGE

As the quantity, quality, and types of food vary seasonally, primates employ different strategies to cope with these changes (van Schaik, Terborgh, and Wright 1993). Natural selection favors individuals that are able to obtain the energy and nutrients needed to grow to reproductive age, survive long enough to reproduce (ideally repeatedly), and then, in some cases, provide adequate energy for offspring survival. Fluctuations in the food supply can present challenges to each of these goals. In figure 7.3, we present a schematic of the different strategies that primates use to cope with fluctuating food availability. We break these responses into behavioral, physiological, and social responses.

Below, we focus primarily on the behavioral and physiological responses to seasonal changes in energy intake. Many excellent reviews have been published that examine the diversity of primate responses to food seasonality (Hemingway and Pfannes 2005; van Schaik and Pfannes 2005; van Schaik, Terborgh, and Wright 1993). Here, we highlight specific examples that illustrate the range of responses that are found across the primate order, particularly focusing on research that uses the newest methodologies and that examines actual energy or nutrient intake. We also refer to our own work on wild orangutans. Because orangutans live in one of the most unpredictably fluctuating habitats, they provide particularly clear examples of some of these behavioral, physiological, and morphological responses.

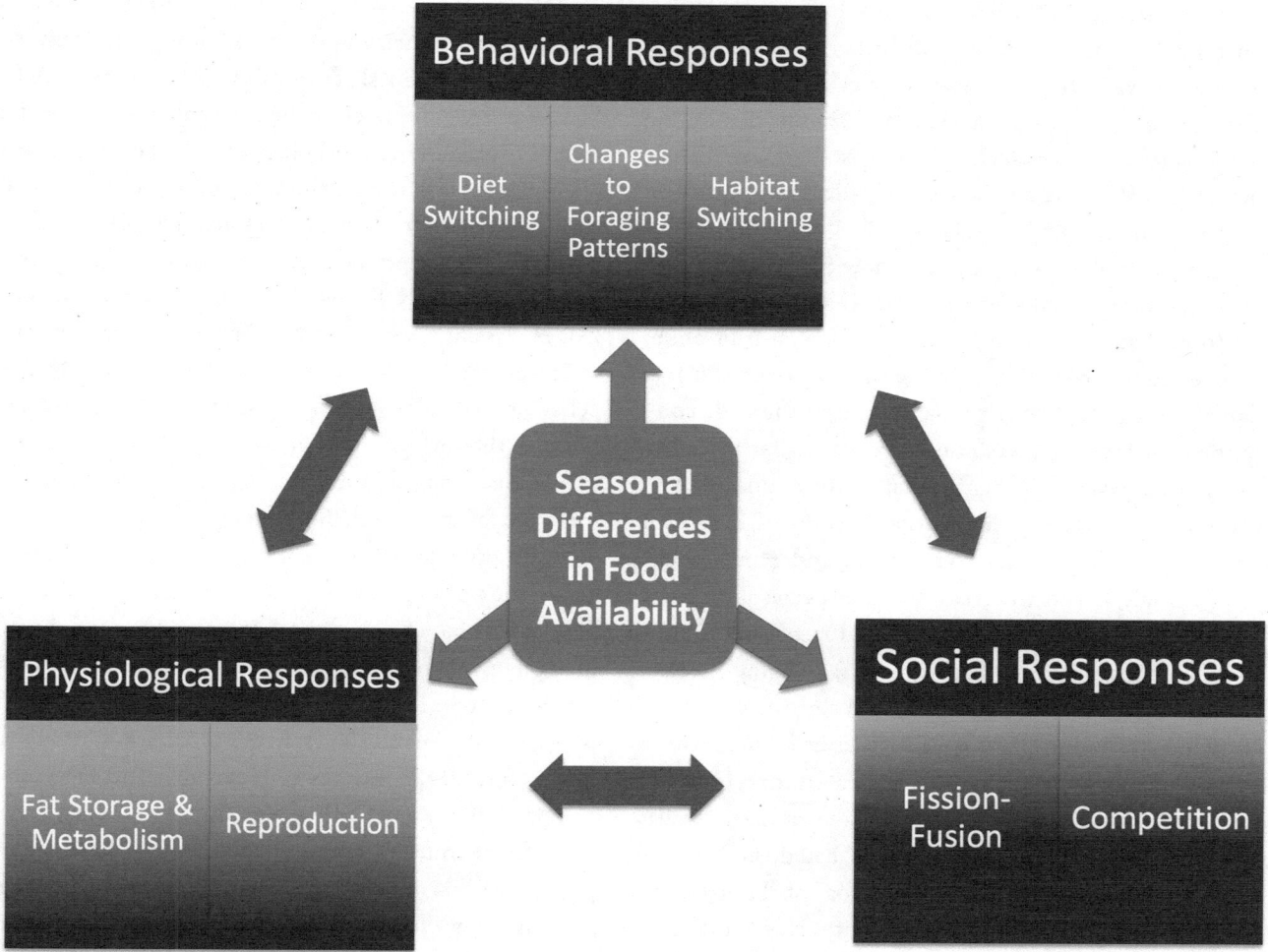

Figure 7.3. Schematic diagram of primate responses to seasonality in food availability and the interrelation between each.

Dietary Switching

Perhaps the most obvious behavioral strategy to cope with food seasonality is to change the composition of the diet (van Schaik, Terborgh, and Wright 1993) with considerations for maintaining an adequate caloric intake, acquiring an appropriate balance of nutrients, and avoiding toxins. As preferred foods become less abundant, primates may increase the diversity of items in their diet and change the relative proportions of items eaten (Colquhoun 1993; M. Leighton 1993). For example, the two *Eulemur macaco* subspecies experience marked dry and wet seasons (Colquhoun 1993). During the wet season, these lemurs consume fruit supplemented with mushrooms and millipedes. In contrast, during the dry season, they switch to a diet of flowers, nectar, seedpods, and some leaves. Leaf monkeys (*Presbytis rubicunda rubida*), so named for the importance of leaves in their diet, increase seed consumption with higher fruit availability, similar to sympatric gibbons (Clink et al. 2017). The folivorous lemur *Lepilemur fleuretae* switches its diet during periods of low leaf availability to consume more mature leaves (73.5% of diet) than during the flushing season (13.5% of diet; Campera et al. 2021).

Some studies have looked at how primate nutrition changes during food switching due to seasonal variation in food availability. During periods of high fruit abundance, chimpanzees (*Pan troglodytes schweinfurthii*) in Kibale National Park, Uganda, increased the fruit component of their diet, resulting in diets higher in carbohydrates and lower in lignin; that is, they consumed an overall higher-quality diet during periods of high fruit intake (Conklin-Brittain, Wrangham, and Hunt 1998). Surprisingly, the three cercopithecine species monitored in the same forest (redtail monkey, *Cercopothecus ascanius*; blue monkey, *Cercopothecus mitis*; and gray-cheeked mangabey, *Lophocebus albigena*) did *not* increase their fruit intake during periods of high fruit abundance. In fact, blue monkeys significantly increased their intake of *leaves* when fruit was abundant. All three monkey species, due to their greater leaf consumption, had diets that were significantly higher in protein than those of chimpanzees. Protein intake showed little seasonality, except that blue monkeys consumed diets higher in protein during ripe-fruit abundance. For the monkeys, dietary fiber (NDF) varied little with season, except for one group of mangabeys who had a *higher* lignin intake during the high-fruit period. Finally, all four primate species consumed more lipids in their diet during the high-fruit season and thus may have had increased energy intake during this period, although this was not directly measured.

The Taihangshan macaque (*Macaca mulatta tcheliensis*) inhabits a harsh high-altitude mountainous region of China and exhibits large shifts in diet over the seasons. In the Taihang Mountains, seeds are abundant in the fall and winter but scarce in the spring and summer. During seed-rich periods, 68% of the macaques' diet consists of seeds, with herbs contributing 19%. However, in the spring and summer, seed consumption drops off, and more than 60% of the diet comes from leaves, with an additional 32% coming from herbs (Cui, Shao, et al. 2019). Young bark and twigs are available year-round but are consumed only during the winter. Although the makeup and nutrient balance of the diet changes by season, pregnant and lactating mothers are able to maintain their caloric intake throughout seasons if seed availability persists (Cui, Wang, Zhang, et al. 2020). Macaques are considered ecological generalists adapted to survive extreme variances in habitat, and the Taihangshan macaques in particular exhibit tremendous flexibility in dietary switching in response to seasonal changes that allows for survival and reproduction (Cui, Shao, et al. 2019; Sayers 2013). Similarly, the western Chinese rhesus macaque (*M. m. lasiotis*), which also inhabits high-altitude areas of Yajiang County, China, exhibits differences in feeding behaviors over four seasons, consuming primarily young leaves in the spring; young and mature leaves in the summer; roots, seeds, and mature leaves in the autumn; and roots and fallen leaves in the winter (K. Zhang et al. 2022).

Even primates who live in or interact with anthropogenic contexts experience seasonality related to diet. Robust capuchins (*Sapajus* spp.) are omnivorous primates that often inhabit fragmented forests that are in close proximity to human environments or are visited by tourists (Gonçalves et al. 2022). In the urban forest of Foz do Iguaçu, Brazil, the months of November and December comprise the season of higher food availability, when more fruits and arthropods are abundant. While the capuchins of this forest do not alter feeding time between seasons, they do consume more plants during the low season than during the high season, and they even diet-switch to consume more anthropogenic foods during the low season (Gonçalves et al. 2022).

Whether primates switch their diet in response to seasonality may also interact with other variables such as group size and predator avoidance. In three groups of sympatric wedge-capped capuchin monkeys (*Cebus olivaceus*) in Venezuela, group size influenced seasonal dietary switching (de Ruiter 1986). During the dry season, when food availability was low, members of the two larger groups switched to eating palm leaf pith and snails, two foods that were typically avoided. The smaller group did not do this. De Ruiter attributed this to smaller groups having less within-group feeding competition due to their fewer group members. Additionally, snails are found on the ground, where predation risk is greater, which could deter smaller groups, who are more vulnerable to predators. Thus, even within a single population of primates, responses to seasonality may vary depending on other factors.

Foraging Patterns: Time Spent Feeding, Foraging, and Moving

Along with changes in diet selection may come adjustments in *foraging patterns*, which we define here as including both the time and travel distance spent obtaining food. We find it helpful to view the behavioral responses of primates to seasonal changes in food availability as occurring along three continua: lower to higher energy intake, reduced to increased movement, and reduced to increased feeding time (fig. 7.4). Some primates, deemed *energy maximizers*, commonly live in highly fluctuating environments and need to take advantage of periods when food is abundant to increase their caloric intake (Shoener 1971), often storing excess calories as fat (Knott 1998). When high-energy foods are available, these primates increase their energy expenditure by traveling more to find food and often feed for longer. This increased energy expenditure is compensated for by an increased energy intake from these high-energy foods (Dasilva 1992). Some orangutan populations offer a good example of this energy-maximizing strategy, as they take advantage of mast fruiting and other fruit peaks to maximize their caloric intake and store that extra energy as fat reserves, which can be relied on later when fruit is scarce (Knott 1998; M. Leighton 1993; Wheatley 1982, 1987). Animals that employ an energy-maximizing strategy must not be overly compromised, though, by the excess body weight that increased caloric intake brings.

The opposite response is to take advantage of the availability of high-calorie or nutrient-dense foods to reduce time spent foraging, called a *time-minimizing* strategy. Rothman, Dierenfeld, Hintz, et al. (2008) found that mountain gorillas (*Gorilla beringei*) in Bwindi, Uganda, were able to maintain a stable caloric intake across seasons by shifting to different foods and spending more time feeding during periods of poor food availability. Masi, Mundry, et al. (2015), studying seasonal differences in diet and caloric intake in western lowland gorillas (*Gorilla gorilla*) in Bai-Hokou, Central African Republic, found that in this forest fruits and leaves did not differ in the amount of digestible energy. However, during the high-fruit period, the foods eaten by gorillas (including fruit and other types of vegetation) had a higher caloric content than foods eaten during the low-fruit period. In response to eating higher-quality foods, during the high-fruit season the gorillas spent less time feeding and more time traveling (Masi, Cipolletta, and Robbins 2009). Thus, in these examples, energy intake remained relatively constant, but the time spent feeding varied. As no seasonal differences were found in overall energy *intake*, Masi, Mundry, et al. (2015) argue that these gorillas are adopting a time-minimizing strategy.

Variable responses also occur during periods of low food availability. Some primates respond to lower food availability by conserving energy through reducing their travel time and spending more time resting. For example, the southern woolly lemur, *Avahi meridionalis*, reduces feeding time and daily travel distance during lean periods (Campera et al. 2021). Western black-and-white colobus (*Colobus polykomos*) were found to make adjustments along all three continua, decreasing their feeding time, travel time, and energy intake during low-food periods (Dasilva 1992). This change in activity pattern may also lead to increased gut retention time and thus more efficient digestion of the food that is consumed (Dasilva 1992). Martin's bare-faced tamarin (*Saguinus martinsi martinsi*) in the Brazilian rainforest maintained a highly frugivorous diet through high- and low-food seasons and spent the same amount of time feeding. However, these callitrichids visited fewer trees during the low-fruit season and spent more time, on average, in each tree (L. Silva et al. 2021). This was due to their heavy reliance on two fruit species during the low-fruit season, thus also leading to lower dietary diversity.

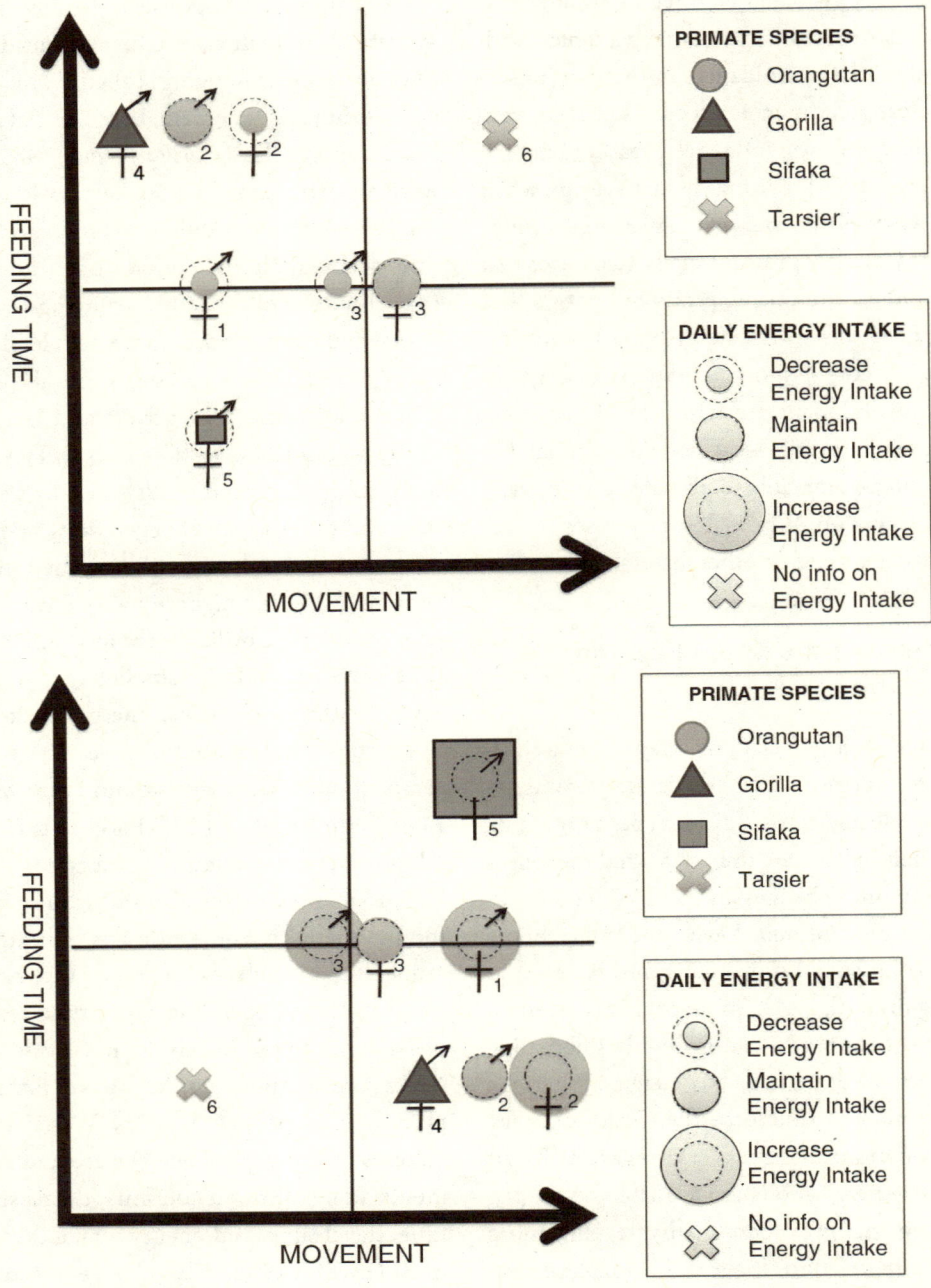

Figure 7.4. Changes in movement, feeding time, and energy intake in response to (*top*) low food availability and (*bottom*) high food availability—examples from the literature. Most data are from full-day follows or extrapolated from partial-day follows, and similar behavioral and nutrient methodologies were employed in each study. When available, data from each sex of a taxon are shown. [1]Knott 1998, 1999b (*Pongo pygmaeus wurmbii*); [2]E. Vogel, Alavi, et al. 2016 (*Pongo pygmaeus wurmbii*); [3]M. E. Harrison 2009; M. E. Harrison, Morrogh-Bernard, and Chivers 2010 (*Pongo pygmaeus wurmbii*); [4]Masi, Cipolletta, and Robbins 2009; Masi, Mundry, et al. 2015 (*Gorilla gorilla*); [5]Irwin 2008; Irwin, Raharison, Raubenheimer, et al. 2014, 2015 (*Propithecus diadema*); [6]Gursky 2000 (*Tarsius spectrum*).

In contrast, other primates may increase their day range (Barton, Whiten, et al. 1992; Overdorff 1993b) or their foraging time (Gursky 2000; J. G. Robinson 1986) during low-food periods to find adequate resources. The Cat Ba langurs (*Trachypithecus poliocephalus*) increase their foraging time but not travel time during periods when preferred foods are not available (Hendershott et al. 2016). Squirrel monkeys (*Saimiri vanzolinii*) in the floodplain forests of Central Amazonia increase travel time but not foraging time during food-poor periods (Paim et al. 2017). During periods of low food availability, red colobus monkeys (*Piliocolobus tephrosceles*) range farther (C. Marsh 1981), and guereza (*Colobus guereza*) diversify their diet, increase their daily path length, spend more time feeding, and visit more patches per day (T. Harris et al. 2009). Furthermore, both male and female spectral tarsiers (*Tarsius spectrum*) modify their activity budgets during times of low resource availability (the dry season) by spending more time traveling and foraging than during the wet season (Gursky 2000). Primates that increase their day range in response to low food availability must live in environments where increased foraging leads to enough of a net caloric gain to compensate for the increase in energy expenditure required to obtain that food. Which strategy a species adopts likely reflects the types and distribution of foods that are eaten during food-scarce periods and whether increased foraging time will be compensated by increased energy intake.

Often, there are interpopulation and interspecies differences in these responses. For example, one population of *Lagothrix lagotricha poeppigii* increased foraging time during periods of higher ripe-fruit availability (Di Fiore and Rodman 2001). Other populations of *L. l. poeppigii* ranged further, spent more time moving, or moved faster during periods of high fruit abundance, but in relation to insect prey availability, not to fruit abundance (Di Fiore 2003). Gibbons also show population-level and perhaps species-level differences in responses to seasonality. *Hylobates albibarbis* did not change daily ranging patterns in response to a 50% seasonal reduction in food availability at Tuanan in Borneo but did shift to eating unripe fruit, lianas, and figs (E. Vogel, Haag, et al. 2009). In contrast, black crested gibbons in China (*Nomascus concolor*) constricted their range use in response to reduced food availability and switched to eating more leaves (P. Fan and Jiang 2008; Ning et al. 2019). A number of factors could account for the differences in these ranging responses including study length, topography, habitat differences in the foods available, food distribution, and taxonomic differences.

Feeding time also seems to vary along a continuum associated with resource availability. Sifakas (*Propithecus diadema*) in five groups living in habitats with varying degrees of disturbance increased feeding time during the fruit-abundant season, with groups in the least disturbed forests showing the biggest increase in energy intake (Irwin, Raharison, Raubenheimer, et al. 2014). Groups in the most disturbed habitats spent more time feeding, but this did result in greater daily intake. During the low-fruit season, sifakas decreased their feeding time and travel time, leading to a decrease in energy intake (Irwin 2008; Irwin, Raharison, Raubenheimer, et al. 2014, 2015). Irwin and colleagues hypothesize that sifakas did not increase their feeding time, despite having low energy intake, in order to avoid toxicity from secondary plant compounds found in these foods.

Additional factors that may impact the diet seasonally are the rate of travel or food ingestion and sex differences in seasonal responses. E. Vogel, Alavi, et al. (2016) report that at Tuanan in Borneo, most age-sex classes of orangutans spent more time traveling and had longer day journey lengths with higher food availability. However, flanged males did *not* spend more time traveling, though they *did* have longer day journey lengths; thus, they must have increased their rate of travel. Rothman, Dierenfeld, Hintz, et al. (2008) also report that, overall, gorilla males at Bwindi spent less time foraging but consumed more food than did females. They attribute this to eating larger food items, eating faster, or staying longer in feeding patches, leading to increased intake per unit time.

Studies that go beyond recording feeding, foraging, and travel time to measuring *net energy* or *nutrient intake* allow us to evaluate the physiological impact of these responses. For example, N. Dunham and Rodriguez-Saona (2018) studied nutrient and energy intake in three groups of colobus monkeys (*Colobus angolensis*) and found that despite significant differences in plant species consumed (N. Dunham 2017), all three groups were able to maintain consistent macronutrient intake across seasons. Contrary to expectation, females in early lactation showed the lowest intake of metabolizable energy. In contrast, sifakas had higher energy and macronutrient intake during the wet season than during the dry season, despite no difference in the amount of food eaten (Koch et al. 2017).

Orangutans are one of the few species where daily caloric intake has been studied in several wild populations, and these studies reveal considerable variation between populations, sexes, and individuals in the interaction between seasonality, feeding time, and energy intake (M. E. Harrison, Morrogh-Bernard, and Chivers 2010; Knott 1998, 1999b; E. Vogel, Alavi, et al. 2016). As noted in figure 7.4, all age-sex classes of orangutans in masting forests increase their caloric intake during periods of high fruit abundance (Knott 1998). However, in peat swamp forests, which do not have mast fruiting and where fluctuations in fruit availability are not as dramatic, some studies have found that only flanged males (M. E. Harrison, Morrogh-Bernard, and Chivers 2010) or only adult females (E. Vogel, Alavi, et al. 2016) increase their caloric intake with overall higher forest fruit availability. Whether increases in energy intake are accompanied by changes in feeding time shows considerable variation as well (M. E. Harrison, Morrogh-Bernard, and Chivers 2010; Knott 1998, 1999b; E. Vogel, Alavi, et al. 2016). Rather than attribute these varied responses to intrinsic differences between these populations, we interpret this as facultative adjustment of daily ranging, feeding time, and diet in order to maintain adequate caloric and nutrient intake and to increase energy intake when possible. More detailed comparisons that consider additional factors such as social context and the life history status of individuals, as well as the distribution and caloric content of the foods at the time of study, are warranted to determine how foraging strategies may differ within and between individuals and populations.

Habitat Switching

Habitat shifting is a common strategy in many migratory taxa, notably including ungulates, birds, and insects. This is less common in primates, where habitat shifting often describes the opportunistic use of different habitat types or microhabitats during periods of food scarcity (Hemingway and Bynum 2005). Habitat shifting will likely co-occur with increased day range, increased feeding time, or both. In an analysis of 329 distinct primate responses to seasonality (drawn from 157 records), habitat shifting occurred in only 10% of the records, and small primates rarely shifted habitats in response to food scarcity (Hemingway and Bynum 2005). However, examples do occur in all the major primate groups. One example in strepsirrhines is rufous lemurs (*Eulemur fulvus rufus*), who, in one study, left their home range area for six weeks during reduced food availability, possibly moving to fruiting *Psidium* up to 6 km away (Overdorff 1993a). Erhart et al. (2018) also found that habitat switching was the main response of *Eulemur rufifrons* to dietary stress, with groups migrating up to 5 km when there were two or fewer fruiting species in their home range.

Interestingly, Neotropical primates seem more likely to switch habitats than to switch diet categories (i.e., between leaves and fruit; Hongo et al. 2018). This may be because fruit and new leaf availability tend to coincide more in the Neotropics, and thus there is less of an option to switch to leaves when fruit are not available (van Schaik and Pfannes 2005). Hemingway and Bynum (2005) suggest that platyrrhines switch habitats because they have a limited ability to increase their dietary diversity. For example, red howler monkeys in a Colombian rainforest shifted to flooded forest habitats (edge habitats with high productivity and variability) during periods of fruit scarcity (Palacios and Rodriguez 2001). Peres (1994b) found a variety of responses in the different primates of the Urucu River region of the western Brazilian Amazon, including habitat shifting to creek-side forests in *Callicebus torquatus purinus* and to non-Igapó habitats in *Cebus albifrons unicolor*.

There are also examples of Old World primates shifting habitats in response to food reduction. Mandrills (*Mandrillus sphinx*) showed seasonal range expansion as well as more uniform habitat use during fruit-scarce periods (Hongo et al. 2018). Chimpanzee nest surveys suggested that, during low fruit seasons, chimps switched from mixed mature forest habitat to *Musanga*-dominated secondary forest (Furuichi, Hashimoto, and Tashiro 2001). Chimpanzees living in Nyungwe National Park, Rwanda, at the altitudinal limit of their species distribution, shifted their home ranges along altitudinal gradients depending on which species were fruiting (Green et al. 2020). Orangutans, with access to multiple habitats, show habitat shifting in response to changes in fruit abundance (M. Leighton and Leighton 1983; A. Marshall, Beaudrot, and Wittmer 2014). Overall, while there are not many accounts of migration in primates, it is important to con-

Figure 7.5. Pygmy marmoset (*Callithrix pygmaea*) feeding on sap. Yasuni National Park, Ecuador.
Photo © Tim Laman.

sider habitat shifting as a strategy in times of low food availability.

Changes in Social Structure

Primates also exhibit a suite of social flexibilities to deal with seasonality in food availability and energy intake. Historically, the most commonly cited examples are the fission-fusion social structures of the *Pan* (chimpanzee; e.g., Boesch and Boesch-Achermann 2000; J. Mitani et al. 2002; Nishida and Hiraiwa-Hasegawa 1987) and *Ateles* (spider monkey) genera (e.g., Chapman, Wrangham, and Chapman 1995b; Pinacho-Guendulain and Ramos-Fernandez 2017; Symington 1990). However, the primate literature contains other examples of primate species either reducing or increasing group size and altering social behavior in response to seasonal changes in their environment. A synthesis of fission-fusion primate literature (Aureli et al. 2008) demonstrates the breadth of social responses in the primate order and suggests that the term "fission-fusion" be viewed not as a modal social structure but rather as dynamic, encompassing a continuum of variation in both spatial cohesion and individual membership within a group over time. We concur with this viewpoint, and here we briefly present examples of primates changing their social structure in response to seasonal changes in energy availability.

The fission-fusion dynamic is a social response to seasonality and energy intake (among other factors) in which animals change their group size (decrease or *fission* into smaller subgroups or parties, or increase or *fuse* into larger groups and parties) commonly based on the availability, distribution, and heterogeneity of resources as well as activity patterns (Aureli et al. 2008; Goodall 1986; Kummer 1971; Nishida 1968; Pinacho-Guendulain and Ramos-Fernandez 2017). These relationships may relate to overall food abundance but may be particularly driven by the availability and distribution of key species (Doran 1997; Pinacho-Guendulain and Ramos-Fernandez 2017). Modern humans are also thought to exhibit a fission-fusion social structure (Aureli et al. 2008; Marlowe 2005; Rodseth et al. 1991). The fission-fusion dynamic varies by species, ecology, and population (Aureli et al. 2008).

However, examples of primate fission-fusion responses to food availability are not limited to the haplorrhines or, more specifically, the anthropoid taxa. An excellent example of this adaptive social structure comes from ruffed lemurs (*Varecia variegata*; Erhart et al. 2018; S. Holmes et al. 2016). When fruit is available from several sources, black-and-white ruffed lemurs often fission into smaller groups to avoid intraspecific competition. These lemurs remain in or fuse into larger groups, however, in response to rare food sources when only a few resources are fruiting. Ruffed lemur social structure and group size also vary intraspecifically between sites based on several factors including fruit patch size, tree size, and site-wide flower availability (S. Holmes et al. 2016). Several studies have focused on populations of these animals within areas where cyclones strike frequently. After cyclone Gretelle struck in 1997, destroying more than 50% of the preferred food trees, black-and-white ruffed lemurs in Manombo reduced group sizes to two or three animals and often foraged alone (Ratsimbazafy 2002).

Fission-fusion social dynamics are not limited to fruit, flower, and leaf consumers. The spectral tarsier (*Tarsius spectrum*) provides an example of an insectivorous primate that changes its social behavior in response to seasonality and availability of food resources. These tarsiers alter their diet and activity budgets as a result of differing resource availability—with more time spent traveling and foraging during the dry season (low resource availability) than during the wet season (high resource availability). Socially, spectral tarsiers were more likely to be involved in territory disputes during the dry season when there was lower resource availability. Intragroup encounters decreased during the dry season as well (Gursky 2000).

Within the Asian catarrhines, Yunnan snub-nosed monkeys (*Rhinopithecus bieti*) of China also alter their sociality apparently in response to seasonal differences in food and energy availability. These monkeys fissioned during June and July, when bamboo shoots (a high-protein resource) formed an important part of the diet. This decrease in group size also occurred at the end of the birth season, when females were lactating and required increased energy for their offspring (Ren et al. 2012). Orangutans (*Pongo*) also show a fission-fusion dynamic response to changes in food availability. Although orangutans are often in a fission state, when fruit is abundant, they commonly become more social and eat and travel in subgroups (van Schaik 1999).

PHYSIOLOGICAL RESPONSES: FAT STORAGE, METABOLISM, AND REPRODUCTION

Although many studies have documented changes in diet composition, how those changes affect the caloric and nutrient adequacy of the diet is often unclear. Are the compensatory mechanisms used enough to maintain diet adequacy, or do these primates suffer physiological costs? Such knowledge can be gained by comparing seasonal changes in calorie or nutrient intake to direct physiological measures.

Fat Storage and Metabolism

Primates exhibit varied physiological adaptations to cope with seasonal changes in food availability and food type. One major form of managing energy levels within a seasonal environment is fat storage. The strepsirrhines provide examples of a wide breadth of fat storage adaptations, starting with the ability of several species of lemurs to store fat, often in their tails. The fat-tailed dwarf lemur (*Cheirogaleus medius*) enters a period of hibernation for at least six months. During this time, the lemur uses fat stored in its tail and under its skin to survive an inactive period. Fat-tailed dwarf lemur tails averaged 42 cm^3 prior to hibernation and 15 cm^3 after hibernation, an overall body mass reduction of approximately 35% (C. Hladik et al. 1980). The gray mouse lemur (*Microcebus murinus*) also stores fat in its tail. The hairy-eared dwarf lemur (*Allocebus trichotis*) stores fat throughout its body to rely on during low-food seasons (Meier and Albignac 1991). In this species, males are more active during periods of low availability, whereas females are largely inactive (Rasoazanabary 2004). Within the genus *Microcebus*, we see a wide swath of physiological responses to food scarcity—from fat storage and torpor (Atsalis 1999) to daily torpor with no overall changes in activity level (Radespiel et al. 2003; Randrianambinina et al. 2003; Reimann and Zimmermann 2002).

The African lesser galago (*Galago moholi*) provides an example of an adaptive and flexible response to seasonality. Recent research has demonstrated that these strepsirrhines, like the lemurs, are capable of torpor (Nowack, Mzilikazi, and Dausmann 2010). *Galago moholi*, however, enters into torpor only under emergency situations of cold and low food and water intake (Nowack, Mzilikazi, and Dausmann 2010; Nowack, Wippich, et al. 2013). During the cold and dry winter months, *Galago moholi* increases huddling and use of insulated and enclosed sleeping sites and reduces nighttime activity instead of utilizing torpor. In addition, this primate alters its diet by reducing insect hunting and increasing intake of gum, a resource found to be higher in gross energy content during the winter (Nowack, Wippich, et al. 2013). Nowack and colleagues suggest that this primate is able to avoid torpor by reducing activity and consuming higher-quality food (though in lower quantities) during the winter, and it resorts to the use of torpor only in an emergency.

Periods of fat storage in response to seasonality have been inferred in many anthropoid species of primates (K. Miller et al. 2006) but are difficult to positively determine in the wild without invasive capture. Among these taxa, fat storage is perhaps most well documented in orangutans (Knott 1998; M. Leighton 1993; MacKinnon 1974; Wheatley 1982, 1987). Orangutans take advantage of periods of high food availability, particularly during masts, to significantly increase their caloric intake as shown through calculations of energy balance and measurements of C-peptide in urine, indicating increased energetic stores (Emery Thompson and Knott 2008; Knott 1999b). During extended periods of low food availability, they have been shown, through the detection of ketones in urine, to rely on these fat reserves as a source of energy (Knott 1998) and to go into negative energy balance (Emery Thompson and Knott 2008). During some low-fruit periods, orangutans even resort to muscle catabolism (E. Vogel, Knott, et al. 2012; C. O'Connell et al. 2021).

Reproductive Responses

Seasonality in energy intake plays an important role in primate reproduction through its influence on maternal energy intake and energy balance (see also Emery Thompson, this volume). Nutrition has been shown to be a key factor regulating the variance in reproductive success in primates, affecting birth season, age at menarche, interbirth intervals, and number of live births (J. Altmann 1980, 1983; J. Altmann, Altmann, and Hausfater 1978; Cheney et al. 1986; Dunbar and Dunbar 1988; Gaulin and Konner 1977; Knott 2001; P. Lee 1987; Lindburg 1987; Sadleir 1969; Strum and Western 1982; van Schaik and van Noordwijk 1985; P. Whitten 1982; Emery Thompson, this volume; Borries and Koenig, this volume). Increased energy intake may be inadequate in many cases to overcome the costs of reproduction, leading to maternal energetic depletion, typically at the end of lactation (Brockman and van Schaik 2005; Emery Thompson 2017). Seasonal differences in energy intake may also impact male primate reproductive effort (Emery Thompson 2017). Ultimately, the degree of resource predictability may be a significant selective force leading to differences in life history between closely related species (Knott and Harwell 2020; Lodwick and Salmi 2019; M. Robbins, Gray, Kagoda, et al. 2009; Stoinski et al. 2013).

In seasonal breeders, food availability may influence the timing of conception, the timing of weaning (when offspring need to forage on their own), or both. For example, Verreaux's sifaka (*Propithecus verreauxi*), in the Kirindy Forest of Madagascar, mate at the end of the high-food-availability period, thus allowing the females to recover from lower food availability during the dry season. The mothers give birth to infants and lactate during periods of low food availability, and weaning is timed with the following wet season and high food availability (R. Lewis and Kappeler 2005). Sifaka infant growth is stalled, and these infants generally lose as much as 10% of their body weight during the dry season (R. Lewis and Kappeler 2005). Similarly, in squirrel monkeys (*Saimiri*), weaning across several species coincides with the period of maximum fruit availability (Boinski 1987; A. Stone and Ruivo 2020). Studies of saddleback tamarins (*Saguinus fuscicollis*) at Cocha Cashu found that births were timed so that lactation and weaning occurred when food was abundant (Goldizen et al. 1988). Dramatic decreases in food availability may have drastic impacts on reproduction. A population of black-and-white ruffed lemurs in a habitat severely impacted by a cyclone ceased reproduction for five years, likely as an energetic response to decreased food availability (more than half of the food trees had been destroyed; Ratsimbazafy 2002). In golden lion tamarins (*Leontopithecus rosalia*), during the dry season and prior to conception, reproductive females were found to ingest 3.2 times the energy taken in by nonreproductive females. During the wet season, when these females were gestating, lactating, and carrying infants, they spent most of their time sleeping or being stationary, interpreted as an energy-conserving strategy (K. Miller et al. 2006).

Primates that are large-bodied, have long interbirth intervals, and live where food availability is more unpredictable are likely to be nonseasonal breeders. Instead, conception is usually tied to periods of high energy balance, whenever that may occur, enabling them to begin a new bout of reproductive investment while in the best energetic condition (Knott 2001). Van Schaik and van Noordwijk (1985), studying long-tailed macaques in Sumatra, discovered that conception was much more likely during food-rich periods. A similar pattern has been found for wild Sanje mangabeys (*Cercocebus sanjei*) where hormones were directly measured (McCabe and Emery Thompson 2013; McCabe et al. 2013). Emery Thompson and colleagues have shown that wild female chimpanzees in the Kibale Forest of Uganda who lived in neighborhoods with more preferred fruit trees had higher estrogen and progesterone levels, shorter interbirth intervals, and higher infant survivorship than females living in adjacent neighborhoods that were lower in quality (Emery Thompson 2005; Emery Thompson and Wrangham 2008; Emery Thompson, Kahlenberg, et al. 2007). In our own work, we have shown that female orangutans in Gunung Palung National Park, Borneo, have significantly higher levels of estrogen during periods of positive energy balance, measured through caloric intake and energy expenditure, and that these periods are associated with higher conception rates, leading to birth clustering (Knott 1999b; Knott et al. 2009). Further studies are needed that obtain detailed measures of both endocrinology and nutritional intake on more primate species in order to fully appreciate the impact of seasonal differences in nutrient intake on reproduction.

APPLICATIONS TO CONSERVATION

The importance of seasonality in affecting primate diets and foraging strategies needs to be considered in our efforts to protect and conserve primate populations. Research into how seasonality impacts primate foraging and energy intake also provides evidence for how these species may react to anthropogenic disturbance and climate change, as these inevitably cause changes in food availability and distribution. Increasingly, studies are examining the intersection between seasonality and foraging in secondary forests, forest fragments, and human-modified landscapes (Abwe et al. 2020; Souza-Alves, Chagas, et al. 2021). It cannot be assumed that disturbed habitats provide less food for primates, as they may have increased numbers of exotic, invasive, and pioneer plant species (N. Dunham 2017). Thus, food availability needs to be directly assessed to account for the particular history and degree of human impact at each site. Of course, even if there is adequate food, anthropogenic disturbance brings additional sets of risks to wild primates from contact with humans. Nevertheless, even very disturbed habitats can provide refuge for wild primates, an important point to consider in conservation planning.

Currently, when suitable habitat is being selected for conservation concessions, rehabilitation, and reintroduction, the availability of fruit and "preferred" feeding resources is assessed. However, because fruiting is seasonal, even areas that do have a high density of preferred fruit trees will have periods when those fruits are not available. The highest levels of primate mortality are experienced during periods of low food availability. Thus, assessing the adequacy of nonfruit and fallback foods, in addition to fruit, is critically important. While fruit and preferred food re-

Figure 7.6. Red howler monkey (*Alouatta seniculus*) feeding on leaves. Yasuní National Park, Ecuador. Photo © Tim Laman.

sources are very important in allowing primates to build up energy reserves and reproduce, evaluating the availability of less preferred and fallback foods is also critical to finding suitable habitat that can sustain the animals, both calorically and in nutrient balance, through seasonal fluctuations. Recent advances in understanding primate nutrient balance highlight that primates cannot survive on preferred foods alone and even during times of high fruit availability may still need to incorporate what are typically considered less preferred food items in their diet (J. Lambert and Rothman 2015). For example, our research suggests that Bornean orangutans commonly eat leaves of trees from the genera *Durio* and *Xanthophyllum*, several epiphytes, and the pith of *Pandanus* and *Callamus* (Rotan) instead of fruit (DiGiorgio et al. 2023). Thus, we recommend that conservation and habitat selection strategies involve assessing the seasonal availability of not only fruits but also other common food items such as leaves, cambiums, piths, and nonreproductive plant parts that can serve as sustenance.

Similarly, we suggest that the feeding recommendations and foraging training of captive and rehabilitant animals include a seasonal component. Observation of captive and orphaned orangutans demonstrates that they gravitate toward fruit when given the choice, and leafy greens and nonreproductive plant parts are passed over and often wilt before they are consumed (Karmele Sanchez and Gail Campbell-Smith, pers. comm.). This seems to occur at a higher rate in young, orphaned animals who have not foraged in the wild. Such foraging behaviors make it difficult to teach orangutans the strategies required to find, and especially process, nonreproductive plant parts during seasons of low fruit and flower availability. These findings may apply to other formerly captive primates released into the wild. Thus, future research should assess different feeding regimens that allow captive and rehabilitant animals to become familiar with subsisting on fallback and nonpreferred foods to a great extent in anticipation of seasonality.

Another conservation application of primate diet seasonality research is that low population densities of endangered primates could be alleviated through enrichment planting of preferred species as well as those that fruit when other species do not (M. Leighton 1993; A. Marshall, Salas, et al. 2007). Thus, tropical rainforest restoration and the establishment of corridors and "fruit fences" should aim to be more like the original habitat, including not only preferred fruit trees but also the range of other foods that will be available when those preferred trees are not fruiting. Dillis et al. (2015), who found that gibbons relied most often on fruit that reproduced asynchronously, emphasize that forest restoration should include asynchronous fruit genera, such as *Ficus* (fig) and other trees and lianas, that may be particularly important food resources during periods of low food availability.

Finally, as we aim to conserve and restore primate populations, the link between seasonal food availability, nutrition, and reproduction must be considered. Although some primate populations may survive in degraded habitats where food availability is poor, individuals may have difficulty conceiving and obtaining enough energy for lactation if caloric and nutrient intake is inadequate.

SUMMARY AND CONCLUSIONS

Several important points emerge from this review. First, as the number and diversity of primate studies continue to increase, so too does the recognition that the responses seen in any one population may not be representative of the entire species. Substantial interpopulation, intrapopulation, and even intraindividual responses exist, even within a single species, largely dependent on local ecology. Second, primates show a large suite of behavioral responses and physiological adaptations to cope with both the challenges and opportunities presented by fluctuating environments. Third, the strategies employed by some primates may allow them to adequately respond to fluctuations in their food supply; however, with increased environmental variability or habitat degradation brought about by anthropogenic actions, these responses may not be adequate for successful survival and reproduction. Furthermore, we make the following recommendations and conclusions:

- Primatologists should follow standard ecological methods to measure seasonality and food availability, such as using crop size and biomass measures that allow for comparisons between sites.
- Measurement of seasonality and food availability should include all relevant plant parts, such as leaves, bark, and terrestrial herbaceous vegetation.
- If the caloric values of available foods are known, we suggest calculating the kilocalories of primate food available per hectare (Knott 2005) as a comparative measure.
- More studies are needed that measure daily caloric and nutrient intake to establish the actual physiological impact of primate foraging and to compare between populations.
- Primates as a clade exhibit a full spectrum of strategies to cope with seasonality in food availability including changes in diet, changes in feeding and foraging time, changes in travel and range use, alterations in social structure, seasonal and aseasonal breeding strategies, and adaptations for fat storage and physiological energy reduction. Further research will highlight the evolutionary, community, and conservation implications of these strategies.
- Significant species-level and population-level differences in primate foraging and energy strategies highlight the importance of seasonality in the lives of these taxa and the flexibility of primates to cope with seasonality. Further research focusing on the interaction between seasonality and these strategies will continue to illuminate the importance of the environment in the evolution and behavioral ecology of primates.

PART II

Nutrients, Nutrition, and Food Processing

The chapters in this section highlight the work that must go on in the body to overcome the chemical constraints imposed by a diet before accessing and using nutrients and energy. This includes food processing by the consumer itself (autoenzymatic digestion) as well as by the symbiont microbes (fermentation and alloenzymatic digestion) hosted in various regions of the digestive tract (Lambert, Mutegeki, and Amato). Regardless of their packaging, most foods present mechanical and especially chemical challenges in the form of plant secondary metabolites (PSMs) that must be solved upon consumption (Stalenberg, Ganzhorn, and Foley); as with the breakdown and use of the nutritive elements of foods, breakdown of the nonnutritive aspects of diet involves action both by the consumer and by symbiont microbes. A subset of these PSMs have important consequences for endocrinology, neurotransmission, and behavior (Wasserman, Després-Einspenner, Mutegeki, and Steiniche). Once processed, foods also have important implications for immunoresponse, with complex interactions (some synergistic and others antagonistic) among total energy intake, macronutrients, and immune function (Vogel, Zulfa, Utami Atmoko, and Moldawer) and, ultimately, life history (Borries and Koenig).

8 Enzymes and Microbes of the Mammalian Gut
Toward an Integrated Understanding of Digestion

Joanna E. Lambert, Richard Mutegeki, and Katherine R. Amato

Animals must procure a predictable supply of nutrient- and energy-yielding organic molecules from their habitats. Put more simply: they need to eat. This means not only finding and procuring food but also processing that food into a form that is available for metabolism and cellular processes (Hartenstein and Martinez 2019). This processing takes place through digestion. Beyond just needing to ingest food, animals are further distinguished based on the evolution of the primary anatomy of digestion—the gastrointestinal tract (see table 8.1 for definitions). In one form or another, gastrointestinal tracts—in which the ingestion of food occurs at the proximal region of the tract and elimination at the terminal—have existed for at least 558 million years and have evolved twice in the Bilatera: once in the protostomes and once in the deuterostomes (Erwin et al. 2002; Ruppert et al. 2004). The derivation of the gut from germ layers differs in these two clades. Embryologically, the mouth of protostomes develops first, and among deuterostomes, second—after the anus. Along with echinoderms, hemichordates, xenacoelomorpha, and all chordate animals, primates are deuterostomes (Bourlat et al. 2006), the defining characteristic of which is that the mouth develops at the opposite region of the developing embryo's anus and that a tract (alimentary canal) connects the two (Martín-Durán et al. 2012).

Digestion is typically understood in terms of both mechanical digestion, in which food is physically broken down via oral mastication and gastrointestinal peristalsis into smaller units, and chemical digestion, in which enzymes further process foods into molecules that are used directly by the body (Voet et al. 2016). However, this classical division of mechanical and chemical digestion fails to capture the significance of microbial fermentation in the use of food—largely carbohydrates—by the consumer. Thus, here we distinguish between autoenzymatic chemical digestion, in which endogenous enzymes break down foods, and alloenzymatic, in which symbiont microbes metabolize complex carbohydrates through a process known as fermentation, breaking them down into units that can be used by the body as energy.

In this chapter, we provide a review of digestion in mammals, with an emphasis on primates and with reference to humans. We start with a brief review of the basics of digestion and the accessory organs of digestion. Overall, our emphasis will be on chemical digestion rather than either mechanical digestion or anatomy, as several exhaustive reviews of primate and mammal digestive morphology exist with little change in our understanding since they were published (Chivers and Hladik 1980, 1984; Chivers and Langer 1994; Clemens 1980; J. Lambert 1998; Stevens and Hume 1995). We discuss chemical digestion in terms of autoenzymatic (digestion via endogenous enzymes in the gut) and alloenzymatic (microbial fermentation) processes. We present details of the enzymes involved with a focus on the particular challenges of carbohydrates and conclude by evaluating the centrality of symbiont microbes in all aspects of primate diet, nutrition, and host health.

Table 8.1. Key terms related to food, digestion, and fermentation in animals.

Key term	Definition
alloenzymatic chemical digestion	The process in which symbiont microbes ferment plant fibers, breaking them down into units that can be used by the body as energy.
autoenzymatic chemical digestion	The process in which endogenous enzymes break down foods.
alimentary canal	An alternative term for "gastrointestinal tract" or "digestive tract." Refers to the extent of the passage through which food is passed, processed, and absorbed from mouth to anus.
amylase	An enzyme found in saliva and the pancreas that catalyzes starches into monosaccharides.
Bilatera	Animals with bilateral symmetry and three germ layers.
carbohydrase	A class of enzymes that catalyze carbohydrates.
catabolic process (catabolism)	A sequence of enzyme-induced reactions in which larger molecules are broken down into smaller units, resulting in a release of energy.
cellulase	A class of enzymes found in fungi, bacteria, and protozoans that catalyze cellulose and other polysaccharides into monosaccharides. Not found in vertebrates.
cephalic phase of digestion	The phase that occurs before food enters the body. After receiving external cues such as odor, neural signaling (in the amygdala and hypothalamus) transmitted by the vagus nerve prepares the body for food consumption (e.g., salivation).
chemical digestion	Processes by which enzymes further process foods into molecules used directly by the body.
chitin	A fibrous polysaccharide component of arthropod exoskeletons and fungi cell walls that is catalyzed by chitinase.
deuterostomes	Bilaterally symmetrical animals in which the anus is formed first, embryologically, from the blastopore.
digestion	A catabolic process in which large molecules of insoluble foodstuff are broken down into small, water-soluble molecules that can be used by the consumer by direct delivery to the blood plasma.
disaccharide	A food substance comprising two molecules of simple sugars (monosaccharides). Examples include sucrose and lactose.
enzymes	Proteins that catalyze chemical reactions in the body.
foregut fermenter	An animal that houses symbiotic, fermenting microbes in a specialized alkaline chamber of the stomach. In this system, microbes alloenzymatically break down carbohydrates before foodstuff enters the enzyme-producing regions of the gut, where it is further processed as energy. In primates, this system is found only among the subfamily Colobinae.
gastric phase of digestion	Mechanical (peristalsis) and chemical (enzyme) digestion of food that occurs in the stomach, stimulated by the distension of the stomach by the food bolus.
gastrointestinal tract	The continuous tract running through the ventral surface of the body through which food is passed from mouth to anus. Often referred to as the alimentary canal.
heterotrophy	The requirement to consume foods external to the body, in contrast to autotrophy, in which the organism produces its own food.

Key term	Definition
hindgut fermenter	An animal with a simple acid stomach in which most digestion takes place. Plant materials (e.g., complex carbohydrates) that are refractory to autoenzymatic processing are fermented in expanded regions of the large intestine—typically either an enlarged colon or a specialized caecum. This is the ancestral condition of primates and is most common.
intestinal phase of digestion	The arrival of foodstuff to the duodenum (anterior portion of small intestine) stimulates nervous and hormonal responses (largely, the secretion of gastrin), thereby activating stomach activity.
lipolytic esterase (lipase)	A class of enzymes that catalyze the hydrolysis of dietary fats (lipids).
macronutrients	The three primary classes of nutrients required by consumers, including carbohydrates, lipids, and proteins.
mechanical digestion	Processes by which food is physically broken down, via oral mastication and gastrointestinal peristalsis, into smaller units.
microbial fermentation	A metabolic process by which the enzymes of symbiont microbes in the gut extract energy from carbohydrates.
monosaccharides	The smallest unit of carbohydrate; serve as the constituent units of disaccharides and polysaccharides. Also known as simple sugars. Examples include glucose and fructose.
neurogenic	Originating in and controlled by the nervous system.
oligosaccharide	A carbohydrate comprising 3 to 10 monosaccharide units. An example is raffinose, found in legumes.
omnivore	A consumer that eats from two or more trophic levels; omnivory is the consumption of both plants and animals.
peristalsis	Involuntary wavelike constriction and relaxation of smooth muscle in the gastrointestinal tract; serves to move the food bolus.
polysaccharides	Complex carbohydrates comprising long chains (polymers) of simple sugars. Include both storage carbohydrates and structural polysaccharides. Examples include cellulose and lignin.
proteases	A class of enzymes that catalyze proteins.
primary consumer	A consumer that eats from the first trophic level. Also known as an herbivore.
protostomes	Bilaterally symmetrical animals in which the mouth is formed first, embryologically, from the blastopore.
secondary consumer	A consumer that eats primary consumers. Also known as a carnivore.
short-chain fatty acid	A fatty acid with fewer than six carbon atoms. A metabolite by-product of microbial fermentation in the gut that is used directly by host cells for energy.
vagus nerve	The tenth cranial nerve, representing the main component of the parasympathetic nervous system and the primary modulator of the gut-brain axis.

SOME BASICS OF DIGESTION AND DIGESTIVE ANATOMY

Digestion is a catabolic process in which large molecules of insoluble foodstuff are broken down into small, water-soluble molecules that can be used by the consumer through direct delivery to the blood plasma. It occurs in three phases. The cephalic phase is neurogenic and occurs before food has entered the body. Signaling that results in salivation and gastric secretion originates in the appetite regions of the amygdala and hypothalamus (after receiving stimuli such as food odors) and are transmitted to the vagus nerve to the stomach. The gastric phase is stimulated by distension of the stomach by food. As digesta leaves the stomach via the duodenum, the intestinal phase is initiated, and gastrin is released (Cheeke and Dierenfeld 2010; Rees and Turnberg 1981).

The primary organ of digestion is the gastrointestinal tract, often referred to as the alimentary canal (Chivers and Langer 1994, and references therein; Karasov and Martinez del Rio 2007; Stevens and Hume 1995). The alimentary canal comprises four layers. The most interior layer is known as the mucosa and is the absorptive layer, made up of epithelial cells and connective tissue; the surface area of the interior mucosal wall is greatly increased by the presence of villi and microvilli comprising what is generally referred to as the brush border. Adjacent to the mucosal layer is the submucosa, a thick layer that captures absorbed materials. The muscularis layer is made up of a circular layer of smooth muscle and a longitudinal muscle layer—together, these muscles (circular and longitudinal) are responsible for the movement of foodstuff through the gut via contractions and peristalsis. The final, most external layer of the gastrointestinal tract is the serosa, a protective layer comprising connective and epithelial tissue (Chivers and Langer 1994, and references therein; Karasov and Martinez del Rio 2007; Stevens and Hume 1995).

The major readily identified components of the gastrointestinal tract will be familiar to all readers and include the mouth (oral cavity), in which mechanical and some chemical digestion is initiated; the pharynx and esophagus, through which digesta is conveyed to more distal regions; a thick-walled stomach, in which food is either mixed with acids and enzymes (in the case of hindgut fermenters) or fermented via symbiont microbes (in the case of foregut fermenters); the small intestine, where absorption occurs; and the large intestine, where fermentation of complex carbohydrates occurs along with water uptake and, finally, processing of waste material into feces.

Accessory—or secondary—organs of digestion include the teeth and tongue, which initiate mechanical digestion through the mastication of food into smaller particles that are more readily chemically digested and fermented (P. Barboza et al. 2009; Cheeke and Dierenfeld 2010; Janiak 2016; Karasov and Martinez del Rio 2007; NRC 2003; C. Stephens and Hume 1995). The parotid, submandibular, and sublingual glands of the exocrine system yield saliva, which contains mucus to lubricate food, salivary amylase that initiates the digestion of starch (more on this below), and hydrogen carbonate, which creates an alkaline environment for amylase to work. The liver plays diverse roles including, among many others, breaking down toxins (e.g., alkaloids in seeds), synthesizing plasma protein, and regulating glycogen storage. For the purposes of digestion, its main function is in the production of bile, critical for the emulsification of dietary fats; much of this bile is stored in another accessory organ, the gallbladder. The pancreas is part of both the exocrine and endocrine systems and is responsible for the secretion of numerous digestive enzymes (e.g., lipase and amylase) and regulation of various hormones that are directly or indirectly involved with digestion (e.g., somatostatin, glucagon, and insulin; P. Barboza et al. 2009; Cheeke and Dierenfeld 2010; Janiak 2016; Karasov and Martinez del Rio 2007; NRC 2003; C. Stephens and Hume 1995).

ENDOGENOUS ENZYMES

Enzymes are proteins that catalyze chemical reactions in the body (Cheeke and Dierenfeld 2010; Janiak 2016; Karasov and Martinez del Rio 2007). Thousands of enzymes occur in the bodies of mammals, and traditionally these proteins were named simply by adding "ase" to the ending of whatever substrate that they are specialized to catalyze (e.g., lactose—lactase). A more current nomenclature classifies enzymes according to the form of chemical reaction they initiate and thus their function in the body (e.g., oxidoreductase, transferase, and isomerase; Cheeke and Dierenfeld 2010; Karasov and Martinez del Rio 2007).

Enzymatic catalyzation occurs throughout the body, but all enzymes involved in digestion are hydrolases, which require water to break chemical bonds and split large molecules into smaller ones, yielding molecules that can be absorbed through the intestinal wall (see Janiak 2016 for a recent review of primate digestive enzymes). Digestive enzymes can be further classified into three broad categories, each of which comprises multiple specific enzymes depending on the substrate: proteases (for protein digestion), esterases (for lipid digestion), and carbohydrases (for carbohydrates—discussed in the next section). Note that until fairly recently, "esterases" as a class of enzymes was often used interchangeably with "lipases." But the nomenclature of enzymes continues to be revised, and esterases comprise both lipolytic esterases (acting on lipids) and nonlipolytic esterases (not acting on lipids; Ali et al. 2012).

Proteases operate by hydrolyzing the peptide bonds between amino acids in a process known as proteolysis: the hydrolytic catabolism that cleaves the linkages between amino acids in a polypeptide chain. The catalyzation of proteins is initiated in the stomach through secretions of pepsin. However, this results in only partial digestion in that pepsin reduces proteins into shorter chains known as peptides. It is in the small intestine where peptides are broken down into their constituent amino acids via the action of pancreatic enzymes (Cheeke and Dierenfeld 2010; Janiak 2016; Karasov and Martinez del Rio 2007).

The digestion of lipids is quite different from the digestion of proteins and carbohydrates in that lipids are hydrophobic. In the lumen, lipids form globules that are suspended in water; these droplets must be emulsified before further digestion can take place. Bile salts, secreted by the liver along with colipase, serve to greatly reduce the size of the fat globules into droplets, providing surface area for lipolytic digestion (e.g., from lipase) to occur. As lipase is water-soluble, its action can take place only at the surface of lipid droplets. Emulsification occurs in the duodenum. The effect of emulsification is the creation extremely small units called micelles that are small enough (approximately 200 times smaller than the droplets produced by emulsion) to then be hydrolyzed either in solution or at the lipid-water surface, resulting in monoglycerides and fatty acids that can be absorbed in the small intestine (Karasov and Martinez del Rio 2007).

A few digestive enzymes are secreted in the mouth (amylase) and stomach (pepsin), but the vast majority are secreted by either the pancreas or the small intestine. A subset of digestive enzymes are secreted directly into the lumen of the gut, resulting in luminal digestion. Another set of digestive enzymes remain in the brush border of the mucosal lining of the gut and engage in what has been called "membrane digestion" (Karasov and Martinez del Rio 2007). Some substrates (e.g., sucrose) are digested entirely by membrane digestion, while others (e.g., starch) require luminal digestion first, followed by a second step of membrane digestion (more on starch below) (Cheeke and Dierenfeld 2010; Janiak 2016; Karasov and Martinez del Rio 2007).

An enzyme of interest to discussions related to primate feeding is chitinase, which is of particular relevance in the context of recent discussion of the centrality of insectivory in the primate lineage and in human evolution (Janiak 2016; Raubenheimer and Rothman 2012; Rothman et al. 2014). Chitin is a derivative of glucose and is very similar in its chemical structure to cellulose, made up of nitrogen-containing glucose amines with additional hydrogen bonding, which makes it stiffer and stronger than other polysaccharides. Chitin is extremely abundant in distribution; it is found in the cell walls of fungi and in the exoskeletons of insects, crustaceans, and other arthropods. A high diversity of primates in all taxonomic groups consume arthropods, particularly insects, whose exoskeletons comprise upward of 85% chitin, which means that primates and other mammals must have some means to gain access to protein- and lipid-rich insect tissue (Janiak 2016). Consumers must manually or dentally remove this cuticle or break it down digestively in some fashion. Because it is a polysaccharide, chitin may also be broken down via microbial fermentation, which would explain why the virtually completely faunivorous tarsiers, as well as other primarily insectivorous prosimians, have extremely large caeca. However, there has also been a long-standing question as to whether mammals, including primates, have the enzyme—chitinase—to break this polysaccharide down (Kay and Sheine 1979). It has long been assumed that mammals do not have this endogenously produced enzyme. However, most recent analyses suggest that in fact chitinase occurs in multiple vertebrate taxa, including humans (Tang et al. 2015). Humans have two chitinase types (AMCase and chtotriosidase), each of which operates in a different pH optimum—AMCase performs best under acidic conditions and is found in abundance in the stomach and lungs of both humans and mice (Krykbaev et al. 2010). The functions of these chitinases are not clear, although there are some compelling suggestions that they may be involved in immunoresponse and protect humans from asthma (Krykbaev et al. 2010; Tang et al. 2015). These recent results also point to an evolutionary origin of insectivory among the earliest primates. Janiak et al. (2018) have documented the gene(s) responsible for encoding AMCase (CHIA). Though most extant primates have only one or perhaps two genes, comparative analyses demonstrate a high probability that ancestral primates had three, indicating the central role of insects in early primate diets (Janiak et al. 2018).

THE SPECIAL CASE OF CARBOHYDRATE DIGESTION

Solar energy is converted into carbohydrate energy via photosynthesis by plant producers. Plants comprise the most biomass on the planet (80%), and indeed life on Earth is founded on carbohydrate energy, which is the most abundant macronutrient on Earth. As heterotrophs, primates can occupy multiple levels in food webs, including as primary consumers (herbivores), secondary consumers (carnivores), or both (omnivores). With a few exceptions, however (e.g., tarsiers), for most primate species—including humans—most calories are netted directly from carbohydrates.

Carbohydrates (or saccharides) are classified on the basis of the number of carbons in the carbohydrate molecule and take the form of either monomers (monosaccharides) or polymers (disaccharides or polysaccharides) of hexose sugar molecules (e.g., glucose, fructose, or galactose; NRC 2003). Monosaccharides (simple carbohydrates) such as glucose or fructose are readily absorbed by the body. Disaccharides such as sucrose or lactose must be hydrolyzed into simple sugars in the small intestine before they can be absorbed and used. Polysaccharides include starches (polymers of single glucose units), which are the plant's stored energy (storage carbohydrates), and the saccharides of the plant wall, which include cellulose, hemicellulose, lignin, and pectin—often referred to as the structural carbohydrates or structural polysaccharides. Monosaccharides, disaccharides, and storage polysaccharides are broken down autoenzymatically. Structural polysaccharides (especially cellulose) require fermentation (NRC 2003).

The diversity of carbohydrate forms and the centrality of plants in the diets of primates means that primate solve the challenge of digesting them by different means. Some carbohydrates require only the action of enzymes; some of these enzymatically digested carbohydrates require only membrane digestion (simple sugars, or monosaccharides), while others must first be partially digested via luminal digestion and then undergo membrane digestion (disaccharides and oligosaccharides). Others still (polysaccharides) require further digestion because of the complexity of their carbon chains. These latter carbohydrates require alloenzymatic chemical digestion—otherwise known as microbial fermentation, described below (NRC 2003).

Carbohydrases comprise a group of endogenously secreted enzymes that catalyze the breakdown of carbohydrate types into simple sugars that can be directly absorbed by the brush border of the intestinal wall. As with all enzymes, there has been considerable discussion of the classification of carbohydrases, but generally enzymes are required for all carbohydrate types (mono-, di-, oligo-, and polysaccharides), and a classic classification has been to designate those enzymes that hydrolyze di- and oligosaccharides as "glycosidases" and those that hydrolyze polysaccharides as "polysaccharidases" (Pigman 1943).

Carbohydrases are secreted from two glandular sources in the body, including the salivary glands (the parotid and submaxillary) and the pancreas (Cheeke and Dierenfeld 2010; Janiak 2016; Karasov and Martinez del Rio 2007). The pancreas consists of both endocrine and exocrine gland; the exocrine cells produce digestive enzymes, and the endocrine cells release digestive hormones such as insulin and glucagon into the circulatory system. The pancreas is by far the more significant source of digestive enzymes and is directly connected to the small intestine via the pancreatic duct. The most abundant enzyme produced in the mouth is amylase, which catalyzes the hydrolysis of the polysaccharide starch into its constituent monosaccharide (glucose) and disaccharide (maltose). The glucose product of amylolysis can be used immediately by the body, as it is a monosaccharide, while maltose requires hydrolysis into its constituent molecules of glucose + glucose. Other enzymes, intestinal mucosal enzymes produced by the pancreas, are implemented primarily in the most proximal portion of the small intestine (duodenum) and complete the digestion of disaccharides and some fractions of oligosaccharides and polysaccharides (e.g., the starches). Included among them are maltase (yielding glucose), sucrase (yielding glucose and fructose), and lactase (yielding glucose and galactose; Cheeke and Dierenfeld 2010; Janiak 2016; Karasov and Martinez del Rio 2007; NRC 2003).

Starch is an exceptionally important source of carbohydrate energy—both for the plant and for the consumer. Starch, like all carbohydrates, is an end product of photosynthesis, in which plants convert light energy to produce glucose from carbon dioxide. Starch is synthesized in leaves during daytime hours, stored as granulates in amyloplasts, and used at night or—in deciduous species—during those seasons when leaves are shed. In humans and some nonhuman primates, α-amylase is produced both in the mouth (by the parotid or submaxillary gland or both) and in the pancreas. The genes responsible for salivary α-amylase (AM1) and pancreatic α-amylase (AM2) have been investigated and thus far have been found only among Catarrhini, but not Platyrrhini; no work to date has investigated the presence of AM1/α-amylase in Strepsirrhini (Mau et al. 2010; Pajic et al. 2019; G. Perry et al. 2007). It is not clear whether salivary α-amylase is primitive to Catarrhini or whether it evolved independently in Cercopithecoidea and Hominoidea (Mau et al. 2010). Interestingly, salivary α-amylase has evolved independently in other taxa, including murid and arvicoline rodents, some chiropterans, and pigs (Redondo and Santos 2006). There is considerable diversity both within and between species in the number of gene copies of AM1 and the resulting levels of salivary α-amylase (Mau et al. 2010; G. Perry et al. 2007). With multiple copies of AM1, higher levels of salivary α-amylase are produced; a selective advantage is presumed in those populations and species with greater numbers of gene copies. Humans have multiple copies of AM1, although this varies by population and by the historical context, in which traditional diets were either low or high in starch (G. Perry et al. 2007). As brains run on glucose, starch consumption is viewed to have played a particularly important role in human brain evolution (G. Perry et al. 2007). Modern humans are indeed extremely efficient starch digesters, with high uptake of starch polymers in the small intestine and a gut microbiome that is highly responsive to the total loads of dietary starch (Walter and Ley 2011). The cercopithecine cheek pouch is so high in α-amylase that approximately 50% of starch is digested within minutes of its ingestion, but the production of salivary α-amylase in the cheek pouch differs between cercopithecine species (Jacobsen 1970; Mau et al. 2010; McGeachin and Akin 1982; Rahaman et al. 1975). Hominoids, too, show both intra- and interspecific variation in gene copies of AM1 and production of salivary α-amylase. Gorillas and orangutans (*Pongo* spp.) have similar levels of salivary α-amylase, similar numbers of AM1 copies, and overall more of both than the two *Pan* species (Behringer, Borchers, et al. 2013).

HOW TO USE STRUCTURAL POLYSACCHARIDES—MICROBES AND FERMENTATION

Although carbohydrates are the most abundant macronutrient on Earth and cellulose the most abundant carbohydrate, no vertebrate has the enzyme (cellulase) to access the energy locked up in cell walls. For this, mammal consumers rely on symbiont microbes and microbial fermentation. All mammals possess a complex microbial community in the gastrointestinal tract composed of bacteria, archaea, anaerobic fungi, protozoa, and viruses (Mackie 2002).

Mammals (including primates) can be classified into two large functional groups based on where microbial fermentation occurs in the gastrointestinal tract—either in a modified stomach (foregut fermenters) or in an expanded region of the large intestine (hindgut or caeco-colic fermenters; fig. 8.1). Hindgut fermenters first autoenzymatically digest plant foods in the stomach and small intestine before the food bolus enters the large intestine; fermenting microbes reside in a modified caecum or enlarged colon or both (Chivers and Hladik 1980; Kay and Davies 1994; Parra 1978; Penry 1993). So-called foregut fermenters have evolved an alloenzymatic system facilitating microbial action on plant foods before they enter acid- and enzyme-producing regions of the gut; microbes reside in one to several sections of a sacculated stomach (Clauss et al. 2007; Parra 1978; Penry 1993). Among primates, hindgut fermentation is the ancestral state. Foregut fermentation without rumination has evolved only among the Colobinae (Kay and Davies 1994; Matsuda et al. 2011;

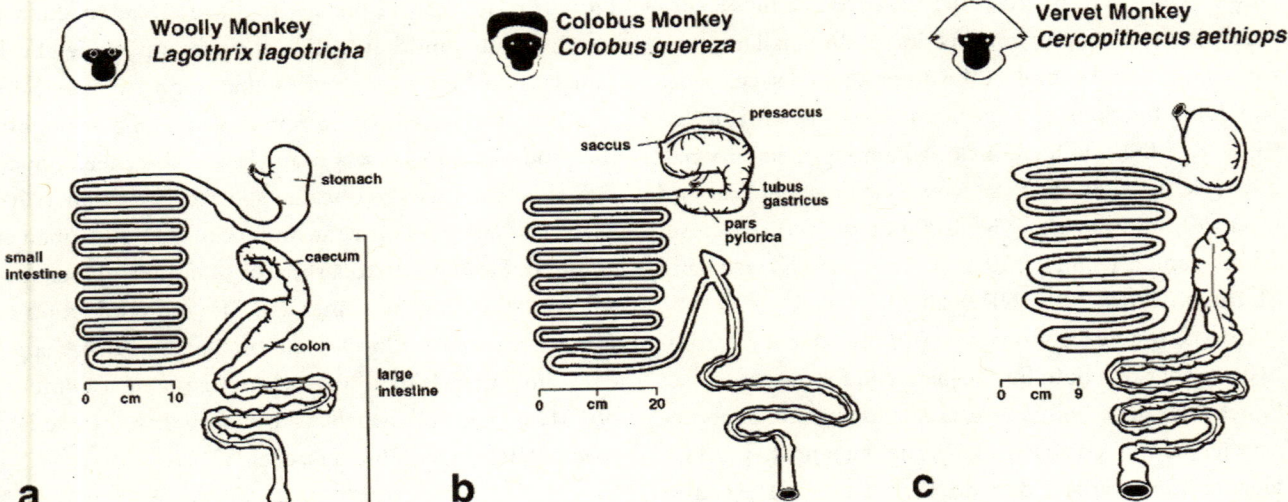

Figure 8.1. Gastrointestinal tracts of woolly monkey (*a*; *Lagothrix* sp.), black-and-white colobus (*b*; *Colobus guereza*), and vervet monkey (*c*; *Chlorocebus pygerythrus*). This graphic depicts major regions of the ancestral mammalian gut, including the stomach, small intestine, and large intestine (colon and caecum) as well as specialized regions (presaccus, sacks, tutus gastric, and pars pylorica) found only in Colobinae. Reprinted with permission from J. Lambert 1998.

Milton 1998). At least one colobine species (*Nasalis larvatus*) has been reported to regurgitate and remasticate plant foods in a manner converging with ruminant artiodactyls (Matsuda et al. 2011).

In fermenting regions of the gut, microbes release volatile fatty acids, also known as short-chain fatty acids (SCFAs), as major by-products of fermentation (J. Lambert and Fellner 2011). These acids (e.g., acetic, propionic, and butyric) are used by the consumer as readily available energy or for glucose storage in the liver (J. Lambert 1998). Fermentation also produces CO_2 and methane, which are eliminated. Additionally, as symbiotic bacteria die off, lysozymes (digestive enzymes) break down the bacterial cell walls, reducing the microbes to a form that is usable for protein by the host, particularly in foregut fermenters.

Because approximately 90% of gut microbes are bacteria (Riedel et al. 2014; Sender et al. 2016), most examinations of the gut microbiota focus on bacteria. In fact, it is estimated that the human body houses roughly the same number of human cells and bacterial cells ($\sim 3.9 \times 10^{13}$ bacterial cells; Sender et al. 2016), emphasizing the potential of gut bacteria to influence host physiology. Nonetheless, recent work highlights the importance of microbial eukaryotes and viruses in the gut as well (Arrieta et al. 2017; Beghini et al. 2017; Minot, Bryson, et al. 2013; Minot, Sinha, et al. 2011; Mirzaei et al. 2020; Nishijima et al. 2022; Parfrey, Walters, and Knight 2011; Parfrey, Walters, et al. 2014; Scanlan et al. 2014).

Microbes can be found throughout the mammalian gastrointestinal tract, but in hindgut fermenters, the colon contains more than 70% of all of the microbes present in the body (Sekirov et al. 2010). In addition, the composition of microbial communities varies along the gastrointestinal tract, with diversity generally increasing from the stomach (~100 phylotypes) to the caecum and colon (~1,000 phylotypes; Riedel et al. 2014). Due to the complexity of the colonic microbiota and the utility of fecal samples for noninvasive sampling of the colon, the majority of gut microbiota research targets this community. As a result, we know that the microbiota of the colon is generally dominated by obligate anaerobes, and approximately 80% of all colon microbes (henceforth gut microbes) belong to the phyla Firmicutes and Bacteroidetes (Ley, Hamady, et al. 2008). Other major phyla include Proteobacteria, Actinobacteria, and Verrucomicrobia (Ley, Hamady, et al. 2008). Some of the most common genera of microbes in the colon include *Bacteroides* (phylum Bacteroidetes), *Prevotella* (phylum Bacteroidetes), and *Clostridium* (phylum Firmicutes; Riedel et al. 2014).

However, the relative abundances of these microbes vary among individuals and within individuals across time.

The gut microbiota begins to establish in mammals at birth when infants are inoculated by their mothers' vaginal microbiota (but see Aagaard et al. 2014; J. Koenig et al. 2011). Throughout development, individuals continue to acquire microbes through contact with conspecifics and the environment (Ferretti et al. 2018; Lane et al. 2019; Mackie et al. 1999; Yatsunenko et al. 2012). However, there is a particularly strong maternal influence during the neonatal period, since the majority of social contact is between mother and offspring. For mice fed the same diet, gut microbiota composition is more similar between mother and offspring than between unrelated individuals, regardless of genetic similarities shared by unrelated individuals (Ley, Backhed, et al. 2005). In addition, breast milk has been identified as a source of gut microbes for newborns (Cabrera-Rubio et al. 2012; Donnet-Hughes et al. 2010; Hinde et al., this volume), and it contains oligosaccharides that stimulate the growth of beneficial gut microbes (Marcobal and Sonnenburg 2012; Zivkovic et al. 2011).

The gut microbiota of infant mammals is relatively simple—mainly composed of lactic acid bacteria, enterobacteria, and streptococci in humans (J. Koenig et al. 2011). However, its composition varies greatly among individuals and across time due to environmental factors such as diet and contact with conspecifics as well as stochastic processes (Mackie et al. 1999; Sekirov et al. 2010; Spor et al. 2011). As individuals mature, the gut microbiota develops in a successional manner. Its composition stabilizes, increases in diversity, and begins to include higher numbers of obligate anaerobes (J. Koenig et al. 2011; Mackie et al. 1999). The most dramatic shift toward an adult gut microbiota occurs with the introduction of solid foods into the diet (J. Koenig et al. 2011).

While host-associated evolutionary traits and early life influences have the potential to influence gut microbiota composition and function in adult mammals, a range of other factors can also affect the adult gut microbiota and incite shifts in its composition and function. Therefore, although adult mammals possess a relatively stable gut microbiota, temporal changes in its composition and function do occur. Below we discuss several traits and factors that impact the gut microbiota as well as some of the mechanisms via which the gut microbiota shapes host nutrition and health.

EFFECT OF DIET ON MICROBE COMMUNITY COMPOSITION AND FUNCTION

Because different microbes utilize different substrates, host diet creates strong selective pressure in the gut that shapes the structure of the gut microbial community (Duncan et al. 2003; A. Johnson et al. 2019; Turnbaugh et al. 2009; G. Wu et al. 2011). Since microbes have a short generation time, these shifts in community structure can occur rapidly. For example, changing the diet of 340 mice from a low-fat diet rich in plant polysaccharides to a high-fat, high-sugar diet results in a dramatic increase in the abundance of several classes of bacteria in the Firmicutes phylum over the course of one day (Turnbaugh et al. 2009). Long-term diet patterns have also been associated with differences in the mammalian gut microbiota. For example, a study of humans detected more genes related to amino acid breakdown, vitamin synthesis, xenobiotic breakdown, and bile salt metabolism in the gut microbiota of adults from the US than in Amerindians and Malawians, who possessed more genes for starch breakdown (Yatsunenko et al. 2012). These differences appeared to be the result of a diet dominated by meat and fat in the US versus a diet dominated by corn in Malawi and Amazon Venezuela (Yatsunenko et al. 2012). Similarly, studies of howler monkeys (*Alouatta* spp.) demonstrate distinct gut microbial communities for the same species inhabiting different types of forest with distinct plant communities and therefore distinct diets (Amato, Martinez-Mota, et al. 2016).

Because diet specializations in mammals represent diet patterns that span evolutionary timescales, it is perhaps not surprising that they have also been associated with distinct gut microbial characteristics. In broad terms, herbivores have been reported to have higher gut microbial diversity than omnivores, and carnivores have the lowest gut microbial diversity (Ley, Hamady, et al. 2008). Gut microbiota composition and function also differ between these diet groups (Ley, Hamady, et al. 2008; Muegge et al. 2011; Youngblut et al. 2019). For example, carnivores exhibit more genes associated with protein degradation, while herbivores possess more genes for amino acid biosynthesis (Muegge et al. 2011). More extreme diet specializations are also associated with microbial signatures. Myrmecophages have a distinct gut microbiota enriched in *Streptococcus* compared to closely related nonmyrmecophages, and gut microbiota of the aardwolf (*Proteles cristata*) and the sloth bear (*Melursus ursinus*) resemble those of other myrmecophages more closely than those of other species of Carnivora (Delsuc et al. 2014). Baleen whales differ from terrestrial mammals in that their gut microbiota is enriched for chitin degradation (Sanders, Beichman, et al. 2015), and woodrats (*Neotoma lepida*) that specialize on toxic creosote bush are unable to consume the plant without certain gut microbes (Kohl et al. 2014).

Although most mammal species are adapted to exploit certain food resources (e.g., primates; Chivers and Hladik 1980; J. Lambert 2011; Norconk, Wright, and Conklin-Brittain 2009), shifts in diet are a regular occurrence for wild animals. Changes in abiotic factors such as climate across habitats and seasons, as well as anthropogenic disturbance, lead to spatial and temporal variation in food availability (Albon and Langvatn 1992; Chapman, Chapman, Rode, et al. 2003; van Schaik, Terborgh, and Wright 1993), and as food availability changes, many animals respond by altering their diet (Andelt et al. 1987; Cantu-Salazar et al. 2005; Cerling and Viehl 2004; Milton 1980; van Schaik, Terborgh, and Wright 1993; Knott and DiGiorgio, this volume; Chapman et al., this volume; Hardie and Strier, this volume). For example, mantled howler monkeys (*A. palliata*) on Barro Colorado Island, Panama, spend 46% of feeding time consuming fruits during months when fruit availability is high but spend 85% of feeding time consuming leaves during months when fruit availability is low (Milton 1980). These seasonal diet changes can impact the gut microbiota. A study of wild black howler monkeys directly correlates differences in gut microbial community composition across four habitats with differences in diet composition (Spearman's ρ = 0.82; Amato, Yeoman, et al. 2013), and a related study demonstrates that changes in the relative abundances of individual bacterial genera such as *Acetivibrio* and *Butyricicoccus* are correlated with shifts in diet composition across time (Amato, Leigh, et al. 2015). A study of Hokkaido native horses also reveals changes in gut microbial community composition across seasons (Kobayashi et al. 2006), which may be related to diet shifts, although diet was not measured.

Despite the clear evidence that diet affects the mammalian gut microbiota, current research that includes quantitative diet data is limited to a handful of host species, especially in the wild. Additional research that describes which microbes vary in abundance with host diet on different timescales will provide important information regarding host plasticity and the response of mammals to variable environments. If the abundance of a subset of gut microbes can change in response to short-term changes in host diet across months or long-term changes in diet incurred by habitat alteration, it may allow hosts to adapt digestively to fulfill nutritional demands while maintaining activity patterns and life history processes despite changes in the types and amounts of food items available. These dynamics may also guide our understanding of how diet specializations evolve, since shifts in the gut microbiota that allow hosts to better utilize certain food resources are likely to occur before physiological adaptations emerge.

EFFECT OF PHYLOGENY AND HOST ON MICROBE COMMUNITY COMPOSITION AND FUNCTION

Within mammals, host phylogeny has also been shown to strongly determine the composition of the gut microbiota. One study examining 60 species of mammals reported that individuals within the same taxonomic order had more similar gut microbiota than individuals in different taxonomic orders (Ley, Hamady, et al. 2008), and subsequent studies have demonstrated similar patterns (Delsuc et al. 2014; Youngblut et al. 2019). Likewise, within primates, host phylogeny can be reproduced using gut microbiome taxonomic composition data. Humans appear to break this pattern, since the composition and function of their gut microbiomes converge with those of baboons (Amato, Mallot, et al. 2019; Amato, Sanders, et al. 2018). However, this convergence appears to be a result of a shared ecology across evolutionary timescales that led to digestive adaptations in humans that also break phylogenetic patterns among apes.

While phylosymbioses—similar patterns of phylogenetic branching observed in hosts and their gut microbial communities—may indicate coevolution of host and gut microbiota, they may also reflect codiversification over evolutionary time or simply be a product of similar selective environments for gut microbes in related hosts (Brucker and Bordenstein 2012; Sanders, Powell, et al. 2014). Continued research is necessary to distinguish between these alternatives and more accurately situate gut microbes within the evolutionary history of their hosts. Additionally, while current research suggests that phylosymbiosis is more prevalent in mammals due to physiological and behavioral factors (e.g., vaginal birth, nursing, and immune complexity), the microbiomes of other vertebrates are understudied (Mallot and Amato 2021). Continuing to build knowledge in these areas may shed additional light on the mechanisms driving the evolution of host diet, life history, behavior, and additional factors.

Host physiology also has strong effects on the gut microbiota. Therefore, like diet, host physiology can result in differences in gut microbiota composition and function among different hosts and also within hosts across time. To begin with, gut morphology and physiology affect the selective environment occupied by gut microbes. Variation in retention time across hosts determines how long gut microbes have access to host diet components, and properties of the mucus layer covering the intestinal epithelial cell layer regulate the adhesion of microbes. For example, different host species may possess distinct glycans that provide nutrients to specific bacteria and allow adhesion of these bacteria via lectins and glycosidases (Johansson et al. 2011). These digestive features are likely to be driving microbiome convergence between humans and cercopithecines, but more data on gut physiology and anatomy are necessary to test this hypothesis (Amato, Mallot, et al. 2019). Furthermore, specialized gastrointestinal chambers house microbial communities that may be distinct from those found in the colon and alter the process of microbial fermentation. For instance, while the same phyla of bacteria dominate the rumen and colon microbiota, cellulolytic activity is higher in rumen bacteria than in colon bacteria (Flint, Bayer, et al. 2008; M. Kim et al. 2011). Additionally, fungi and protozoa are thought to play a more prominent role in fiber breakdown in the rumen than in the colon (M. Kim et al. 2011). Dead microbes from the rumen also pass through the rest of the gastrointestinal tract, allowing hosts to digest them to obtain protein and other nutrients. These nutritional resources are lost to hindgut fermenters. While less is known about the microbial communities of nonrumen foreguts, some descriptive studies suggest unique microbial signatures in the wallaby (*Macropus eugenii*) foregut and the multichambered stomachs of baleen whales (Sanders, Beichman, et al. 2015). Foregut-fermenting primates have microbiomes that appear to converge with those of ruminants, and these primates exhibit similar microbiomes regardless of geographical location. Foregut-fermenting primates also exhibit substantial microbial fermentation in the hindgut, which may be a result of inefficient foregut digestion in these species (Amato, Clayton, et al. 2022; Amato, Sanders, et al. 2018; R. Liu et al. 2022). In comparison, almost nothing is known about the microbial community of the caecum.

Independently of gut morphology, the host immune system is constantly monitoring the gut microbiota and has a variety of effects on its composition via the excretion of several compounds (Artis 2008; L. Hooper et al. 2012; Vogel et al., this volume). Immunoglobulin A, secreted by B cells in the gut, can differentially target gut microbes (L. Hooper et al. 2012), and gut epithelial cells release antibacterial peptides that have been shown to impact gut microbiota composition (Salzman et al. 2003). Genetically impairing the immune system by knocking out genes that are involved in the host inflammatory response or cell signaling mechanisms leads to changes in the gut microbiota (Elinav et al. 2011; W. Garrett et al. 2007; Vijay-Kumar et al. 2010). On a broader scale, immune complexity appears to be positively associated with gut microbial diversity, and the immune features that evolved in placental mammals to protect the fetus have been hypothesized to increase host microbiota specificity in mammals (Mallott and Amato 2021; Woodhams et al. 2020).

Furthermore, neurohormones such as dopamine and norepinephrine as well as other hormones such as estrogens and glucocorticoids have been shown to impact gut microbiota composition and function (Neuman et al. 2015). For example, increases in plasma cortisol in rhesus macaques (*Macaca mulatta*) are associated with a reduction in fecal lactobacilli (M. Bailey and Coe 1999), and rats and chicks exposed to stress from heat and crowding possess distinct gut microbiota compared to individuals not exposed to these stressors (K. Suzuki et al. 1983). Elevated plasma norepinephrine associated with stress increases the virulence of gastrointestinal pathogens such as *E. coli* and *Salmonella enterica*

(Freestone et al. 2007; Pullinger et al. 2010), and mouse models of depression, associated with altered levels of neurohormones, exhibit changes in the gut microbiota (Park et al. 2013).

In addition to the interspecific differences in the gut microbiota that these processes are likely to cause, individual variations within host species in physiology associated with the gut, the immune system, the endocrine system, and even the nervous system are likely to lead to intraspecific differences in gut microbiota composition and function. Much remains to be learned about these interactions, but weak effects of host genotype on the composition of the gut microbiota suggest that interindividual variation is not solely a product of host environmental factors acting directly on the gut microbiota (Goodrich et al. 2014; Grieneisen et al. 2021; Zoetendal et al. 2001)

EFFECT OF COMMUNITY ON HOST NUTRITION

While a variety of host-associated factors affect gut microbiota composition and function, the gut microbiota also has a strong effect on host nutrition, immune function, and behavior. The effects of the gut microbiota on nutrition are the most well studied. First, as discussed earlier, the gut microbiota converts these indigestible plant fractions into SCFAs, thereby providing hosts with upward of 70% of their daily energy needs (P. Barboza et al. 2009; Flint and Bayer 2008; Mackie 2002; Stevens and Hume 1995). SCFAs can also act as signaling molecules that alter host metabolic programming, as well as other body system functions (Visconti et al. 2019). Gut microbiota also reduce the pH of the intestinal lumen to facilitate nutrient absorption and prevent the accumulation of toxic metabolic by-products (Neish 2009; Sekirov et al. 2010). Germ-free rats have reduced intestinal levels of SCFAs (Hoverstad and Midtvedt 1986). They excrete twice as many calories in urine and feces as conventional rats fed the same diet and must compensate for the lack of energy-rich SCFAs by increasing their food intake (Wostmann et al. 1983). Interactions between microbes in the gut can also affect SCFA production (Duncan et al. 2003; Flint, Duncan, et al. 2007; Samuel and Gordon 2006). For example, the fermentation of dietary fructans increases when germ-free mice are colonized with both *Bacteroides thetaiotaomicron* and *Methanobrevibacter smithii* since *M. smithii* uses formate for methanogenesis, and *B. thetaiotaomicron* produces more acetate and formate in its presence (Samuel and Gordon 2006). These interactions promote more efficient fermentation and energy production in the gut, and co-colonized mice exhibit increased adiposity compared with mice colonized with only *B. thetaiotaomicron* (Samuel and Gordon 2006).

In addition to producing SCFAs, the gut microbiota affects host nutrition by regulating xenobiotic metabolism (Bjorkholm et al. 2009) and producing vitamins (M. Hill 1997). For example, the health benefits of soya—such as improvements in vasomotor symptoms, osteoporosis, prostate cancer, and cardiovascular disease—have been attributed to (S)-equol produced by gut bacteria (R. Jackson et al. 2011). It has also been suggested that the production of folic acid by the gut microbiota benefits women and female nonhuman primates during reproduction (Amato, Leigh, et al. 2014; Santacruz et al. 2010). Differences in the relative abundances of bacterial taxa in the gut therefore may influence host nutrition by affecting the production of these compounds. A study of seven individuals from a Chinese family provided evidence for this process by demonstrating that relative abundances of *Bacteroides*, *Clostridium*, and *Bifidobacteria* were correlated with the concentrations of urinary metabolites, and variation in the relative abundance of *Faecalibacterium prausnitzii* was associated with changes in eight urinary metabolites (M. Li et al. 2008). Germ-free rats have also been shown to become anemic when fed a low-iron diet and exhibit increased fecal iron content compared to rats with a healthy microbiota (Reddy et al. 1972).

EFFECT OF COMMUNITY ON HOST HEALTH

The gut microbiota also affects host health via nonnutritional pathways (Cryan et al. 2019; Leshem et al. 2020; Nabhani and Eberl 2020; Pronovost and Hsiao 2019). Metabolites produced by the gut microbiota have a range of effects on hosts. For example, SCFAs help exclude potential pathogens from the gut by improving gut epithelial barrier function, and the production of acetate by *Bifidobacterium longum* appears to improve rodents' ability to survive infection by a lethal *E. coli* strain (Fukuda et al. 2011). Antioxidant metabolites such as indole-3-propionic acid have been associated with the gut microbiota, and gut microbes have been shown to produce a variety of neurohormones such as serotonin, dopamine, norepinephrine, and gamma-aminobutyric acid (H. Li et al. 2008; Lyte et al. 2010; Wikoff et al. 2009). In particular, *Lactobacillus rhamnosus* has been shown to alter gamma-aminobutyric acid receptor mRNA levels in the mouse brain and reduce corticosterone and anxiety- and depression-related behaviors in maze and forced swim tests (Bravo et al. 2011; Gareau et al. 2007).

The gut microbiota also interacts directly with the immune system and the nervous system (see Vogel et al., this volume, for further detail; see also Leshem et al. 2020; Nabhani and Eberl 2020; Pronovost and Hsiao 2019). In the gut, regulatory T cells, which suppress unwanted immune reactions, and Th-17 cells, which stimulate the epithelium to produce antimicrobial proteins, are partially regulated by the gut microbial community (Atarashi et al. 2008; Gaboriau-Routhiau et al. 2009; Hall et al. 2008; Ivanov et al. 2009; Neiss et al. 2008; O'Mahony et al. 2008; Round and Mazmanian 2010; Wen et al. 2008). *Bacteroides fragilis* influences the levels of helper T cells in the spleen and systemic antibodies (E. Bauer et al. 2006; C. Hansen et al. 2012; Mazmanian et al. 2005; Noverr and Huffnagle 2004). Additionally, mice treated with antibiotics and then infected with an influenza virus have reduced immunoglobulin and T-cell responses (Ichinohe et al. 2011), indicating reduced immune function.

Exposure to gut microbes is critical for normal development of the immune system. Colonizing germ-free mice with segmented filamentous bacteria leads to a 24%–63% increase in immunoglobulin A (IgA) production, which is used by the innate immune system to tag potential pathogenic invaders and prevent them from entering the body (Talham et al. 1999). Similarly, the development of gut-associated lymphoid tissue, which is used by the intestinal mucosa for defense, depends on the gut microbial community (L. Hooper et al. 2012; Neish 2009; Pirarat et al. 2006), and germ-free mice have fewer Toll-like receptors and class II major histocompatibility complex molecules (Lundin et al. 2008; Matsumoto et al. 1992), which are involved in pathogen detection, on epithelial cells in the gut.

Analogously, there is evidence that the gut microbiota impacts brain development and function by altering gene expression as well as neuronal circuits involved in motor control and anxiety (Buffington et al. 2016; Forsythe et al. 2010; Foster and McVey Neufeld 2013; Gehrig et al. 2019; X. Liu, Li, et al. 2021; Y. Silva et al. 2020). These impacts are likely to influence host behavior. In fact, research suggests that germ-free mice are more susceptible to stress when physically restrained than specific pathogen-free mice (Sudo 2006), and the administration of certain bacterial strains (such as *Lactobacillus* and *Bifidobacterium*) has been shown to reduce anxiety and cortisol in rats and humans (Messaoudi et al. 2010).

As a result, it is perhaps not surprising that many diseases have been associated with shifts in the gut microbial community (Foster and McVey Neufeld 2013; Krajmalnik-Brown et al. 2015; Larsen et al. 2010; Severance et al. 2014). However, many perspectives on mammalian nutrition and health do not incorporate gut microbe–host interactions. Incorporating these processes into nutrition research will provide insight into gut microbial relationships with diet and nutrition and may also provide new mechanisms via which host diet affects host health.

SUMMARY AND CONCLUSIONS

- Like all animals generally and deuterostome chordates more specifically, primates must consume nutrient- and energy-yielding organic substrates from their habitat and digest that food in a gastrointestinal tract. Knowing the processes by which food is digested in the gastrointestinal tract is fundamental to understanding the biology of consumers.
- After entering the body, food is broken down into smaller units either autoenzymatically by endogenous enzymes or alloenzymatically by symbiont gut microbiota.
- Carbohydrates comprise the most abundantly distributed macronutrient on the planet, but most of the biomass of carbohydrates consists of structural polysaccharides, which require fermentation by symbiont microbes found either in a specialized stomach or in the large intestine of primates.
- A consumer's gut microbiota is influenced by many factors including individual, population, and species history, phylogeny, and diet.
- Differences in the community composition of an animal's gut microbiome have important nutrition and health consequences.

9 Secondary Compounds in Primate Foods

Time for New Approaches

Eleanor M. Stalenberg, Jörg U. Ganzhorn, and William J. Foley

Plant secondary metabolites (PSMs) are ubiquitous in woody plants and have long been believed to be important determinants of feeding and habitat suitability for vertebrate herbivores. Extensive evidence for the importance of particular PSMs in influencing vertebrate feeding behavior has accumulated for many different species, with key examples arising from studies of mammalian specialists that feed on potentially toxic foods such as woodrats (*Neotoma* spp.; Sorensen et al. 2005), snowshoe hares (*Lepus americanus*; Bryant et al. 1983; Reichardt et al. 1984), and the marsupial folivores (*Phascolarctos cinereus*, *Pseudocheirus* spp., *Trichosurus* spp., and *Petauroides volans*; Lawler, Foley, et al. 1998; B. Moore and Foley 2005; Pass et al. 1998).

Although primatologists were among the first to test early theories about the role of PSMs in feeding ecology (Freeland and Janzen 1974; C. Hladik 1977a; Milton 1979; Oates, Swain, and Zantovska 1977), that early promise has delivered few clear examples where PSMs have been shown to structure feeding or define habitat quality for any group of primates. More importantly, most studies on PSMs and primate feeding have, with few exceptions, focused on crude analyses of poorly defined chemical groupings. Such an approach has not delivered significant insights in other mammal-plant systems, and the shortcomings have been documented many times (Dearing et al. 2005; W. Foley and Moore 2005; Torregrossa and Dearing 2009; I. Wallis et al. 2012).

Our aim in this chapter is to review previous studies of the role of secondary metabolites in primate feeding and to discuss the possible reasons why the links between PSMs and feeding ecology, niche partitioning, and reproductive success emerging from primatology are not as apparent as they are in other plant-vertebrate systems. We suggest potential ways to overcome some of these difficulties in the future. We start by conducting a review of previous primate field studies to evaluate the evidence for the effects of some groups of secondary metabolites on diet choice in primates. We go on to identify the particular strengths and weaknesses of primate dietary ecology studies and contrast these with other vertebrate-plant systems where secondary metabolites are prominent. We emphasize the importance of identifying the chemical principles that underlie variations in food choice and describe how different approaches have been used to identify specific PSMs that play a strong role in mediating feeding in other mammals, which are likely to be equally applicable in primates.

WHAT ARE PLANT SECONDARY METABOLITES?

Plant secondary metabolites can be defined as all compounds produced by plants that are not required for basic metabolism (Glander 1982; Masi, Gustafsson, et al. 2012) and are thus distinct from primary metabolites (see box 9.1). PSMs are ubiquitous across the plant kingdom and comprise a vast variety of structures and functions. Tens of thousands of PSMs have been

BOX 9.1 LIMITATIONS OF CHEMICAL ANALYSIS OF SECONDARY COMPOUNDS IN FEEDING STUDIES.

Limitations of chemical analysis using standardized tests are widely realized and have been described often by primatologists (e.g., Lucas et al. 2003; Waterman 1984). However, these limitations are not always acknowledged, and the same tests have been used for almost 40 years. We summarize some of the issues briefly below.

The measurement of **tannins** in any plant material is difficult, and attributing effects of tannins on animals based on simple assays of chemical groups is even more fraught. This is well known among those working in plant-animal interactions and specifically in primatology (Rothman, Dusinberre, and Pell 2009; Yeager et al. 1997). For example, Rothman, Dusinberre, and Pell (2009) confirmed that using a single standard for the chemical assay of tannins in multiple species was likely to introduce significant errors. Since the vast majority of primate feeding studies focus on many different plant species, this is a major issue. Significant improvements in analytical techniques for tannins are being made (e.g., Engström et al. 2019; Salminen and Karonen 2011), and these have already changed our understanding of tannins in insect-plant interactions. Nonetheless, the current approach has not yielded strong patterns, and continuing to use simple chemical assays of tannins is unlikely to be successful.

An alternative approach is the use of tannin-binding chemicals such as polyethylene glycol and polyvinyl pyrrolidine. These approaches have been widely applied in agricultural work and in studies of folivorous marsupials to understand how tannins might affect the intake and digestibility of food. In some situations, this approach might be useful to primatologists, although we (Stalenberg, Ganzhorn, and Foley, unpubl.) have found that (not surprisingly!) it is much harder to use tannin-binding agents under field conditions than in the laboratory. We suggest other alternative approaches to estimating the intake of bioactive tannins in free-ranging primates elsewhere in this paper (see "Countermeasures against Plant Secondary Metabolites").

Alkaloids are the second most tested group of secondary metabolites in primate foods. Most commonly, alkaloids have been identified using spot tests such as Dragendorff's, Marquis's, and Libermann's reagents. The limitations of these approaches are also well recognized among primatologists (e.g., Lucas et al. 2003), and the disconnect between the assay and inferences about bioactivity in animals is even greater than it is with tannins. These tests provide little reliable information about the types and quantities of alkaloids present and nothing about their likely impacts on animals. Certainly, the presence of some specific alkaloids can be inferred from knowledge of the plants being eaten (e.g., indole alkaloids in *Rauvolfia vomitoria* leaves at Douala-Edea; McKey, Gartlan, et al. 1981; Waterman 1984). However, we suggest that the use of spot tests to infer a relationship between alkaloids and food selection in primates be discontinued except perhaps as an adjunct to selection of plants for more detailed analysis.

Cyanogens are known in some primate foods, but for this group, the spot test methods (Feigl-Anger test) are a more reliable guide to compounds likely to have a biological action against herbivores. However, spot tests for cyanogens must be followed up with a reliable quantitative assay.

Surprisingly, little attention has been focused on **terpenoids**, and we are aware of few studies that have quantified terpenes in primate foods (Welker et al. 2007). Methods for extraction, identification, and quantification of terpenes tend to be more robust than those for other secondary metabolites.

---

identified in plants, and new compounds are continuously being discovered; however, our knowledge of how PSMs influence herbivore feeding is limited to a very small number of these compounds, and relatively little is known about the function of these compounds in plant–mammalian herbivore interactions. Any classification of the structural aspects or mode of action of PSMs is limited by the multiplicity of compounds but nevertheless helps organize understanding of this extraordinary natural diversity (Wink 2008). PSMs fall into three broad categories based on their chemical structure: phenolics (including tannins), terpenoids, and nitrogen-containing compounds (such as alkaloids and cyanogenic glycosides). PSMs are also often classified by vertebrate ecologists according to their mode of action as either toxic or digestibility-reducing (Dearing et al. 2005), although the more we learn, the more blurred these distinctions become. Importantly, not all PSMs have negative impacts on consumers, and some are known to have medicinal properties as antiparasitic agents and as antioxidants (Forbey, Harvey, et al. 2009).

In 1974, Freeland and Janzen published a seminal review on the ecological significance of PSMs for vertebrate herbivores, stating that "the ubiquitous nature of these compounds would make herbivory impossible unless animals had mechanisms for degrading and excreting them" (Freeland and Janzen 1974). They predicted that the physiological limits to PSM detoxification force mammals to prefer foods that contain small amounts of secondary compounds, to sample new foods with caution, and to choose foods from a variety of sources containing different toxins metabolized by different detoxification pathways. Over the subsequent 40 years, much progress has been made in testing these ideas (e.g., K. Marsh 2006) and in linking them to broader influences on the distribution and abundance of mammals, but few of these insights have been made in studies with primates. There are many reasons for this, but we argue below that one of the main impediments has been the difficulty of conducting manipulative experimental work with primates.

PREVIOUS WORK WITH PRIMATES AND PSMS: A REVIEW

We undertook a review of studies examining the hypothesis that PSMs (mainly condensed tannins, total phenolics, or unspecified alkaloids) in plant parts act as feeding deterrents for primates. We restricted our review to published studies that quantified feeding preferences for individual primate species, such as eaten versus not-eaten items, a feeding preference rank of plant items, or the amount of time primates spent feeding on plant items. We did not consider studies that only compared preferences between primate species or those that compared between major plant parts without a measure of primate feeding preference of the different items. Many studies examined the preferences of multiple primate species, and so where possible, we report the findings for each primate species in each study separately in table 9.1. This review is expanded elsewhere into a formal meta-analysis of the role of PSMs in primate diet selection (Windley et al. 2022).

We examined 73 peer-reviewed manuscripts and split these into 84 separate examinations of the effects of PSMs on the feeding preferences of individual primate species (table 9.1). Of these individual primate species studies, 73 used statistics to test the effect of PSMs on primate feeding preferences, and the remaining 11 looked at patterns in feeding without statistical analysis. At least one group of PSMs was found to be a statistically significant feeding deterrent for primates in half of the studies reviewed (34 of the 73 studies that used statistics), and the remaining studies found no evidence that PSMs significantly influence primate feeding. When examining the influence of individual PSMs, we found that condensed tannins were significant deterrents in 22 of the 68 statistical studies of tannins in the diets of individual primate species, total phenolics were significant deterrents in 14 of the 41 studies of phenolics, and alkaloids were significant deterrents in 6 of the 31 studies of alkaloids (plus 6 more studies that indicated a nonstatistical aversion to unspecified alkaloids). More than half of the reviewed studies examining the effect of tannins and phenolics on colobine feeding preferences found a significant deterrence of both these PSMs (table 9.1).

This review of previous work suggests there is little consistent evidence to either reject or accept the hypothesis that PSMs in general act as feeding deterrents for primates. However, caution must be taken in drawing conclusions from our literature review alone, as we have drawn together many studies that used different field, laboratory, and statistical approaches and many that found mixed results depending on the hypotheses (i.e., different patterns for different plant parts or for different primate populations of the same species). Accordingly, we undertook a formal meta-analysis of primate feeding studies (Windley et al. 2022) that accounts for different field and statistical approaches and effect size. The results of the meta-analysis revealed that PSMs do affect diet selection in wild primates and that this is largely driven by a significant deterrent effect of condensed tannins on the diet choice of colobine primates. No other PSMs were found to influence diet selection in colobine primates, and there was no significant effect of any PSM on noncolobine primates when considered as a single group (Windley et al. 2022).

Below, we summarize much of the work on primate feeding selectivity in relation to PSMs among the major studied groups of primates, including the lemurs, New

Table 9.1. Summary of studies examining the influence of plant secondary metabolites (PSMs; mainly tannins, phenolics, and alkaloids) on primate feeding preferences and the statistics used, if any.

Primate group	Species	Location	Tannins	Phenolics	Alkaloids	Statistics	Comment	Reference
Lemurs								
Avahi	*Avahi laniger*	Andasibe, Madagascar	− L		+ L	Compared food and nonfood items using a Mann-Whitney *U* test.	Significant effect of alkaloids. Leaves eaten did not have alkaloids.	(Ganzhorn 1988; Ganzhorn, Abraham, and Razanahoera-Rakotomalala 1985)
	Avahi occidentalis	Andasibe, Madagascar	− L		+ L	Compared food and nonfood items using a Mann-Whitney *U* test.	Significant effect of alkaloids. Leaves eaten did not have alkaloids.	(Ganzhorn 1988)
	Avahi meridonalis	Saint Luce, SE Madagascar	− L Fl	+ L Fl	− L Fl	F and *t*-tests to compare food and nonfood and regressions with feeding time.	Significant effect of phenolics. No difference between food and nonfood, but feeding time decreased with increasing phenolics.	(Norscia, Ramanamanjato, and Ganzhorn 2012)
Indri	*Indri indri*	Perinet, Andasibe, Madagascar	− L Fl	− L / (+) Fl	+ L	Compared food and nonfood items using a Mann-Whitney *U* test.	Significant effect of alkaloids. Leaves eaten did not have alkaloids.	(Ganzhorn 1988)
	Indri indri	Perinet, Andasibe, Madagascar	− L		− L / + Fr	Spearman rank correlation of leaf chemistry and time spent feeding.	Significant effect of alkaloids. No correlation between time feeding and tannins, phenolics, or alkaloids, but fruit eaten had very low phenolics and no alkaloids.	(Powzyk and Mowry 2003)
Propithecus	*Propithecus coquereli*	Andasibe, Madagascar	− L		− L	Compared eaten and not eaten using a Wilcoxon matched-pair rank test.	No significant effect of PSMs.	(Ganzhorn and Abraham 1991)
	Propithecus verreauxi	Berenty, Madagascar		− L	− L	Compared eaten with representative subset by a chi-squared test.	No significant effect of PSMs.	(Simmen, Hladik, et al. 1999)
	Propithecus diadema diadema	Perinet, Andasibe, Madagascar	− L Fr	− L Fr	− L Fr	Spearman rank correlation of leaf chemistry and time spent feeding.	No significant effect of PSMs.	(Powzyk and Mowry 2003)

(continued)

Table 9.1. (continued)

Primate group	Species	Location	Tannins	Phenolics	Alkaloids	Statistics	Comment	Reference
	Propithecus verreauxi	Kirindy, Madagascar	+**L**			Spearman rank correlations of tannins and time feeding.	Significant effect of tannins.	(Norscia, Carrai, and Borgogini-Tarli 2006)
	Propithecus verreauxi	Beza Mahafaly, Madagascar	(−) L	(−) L		No statistics of feeding preferences.	No statistics. Tannins and phenolics are present in most foods and correlated with nutrients, suggesting no effect.	(Yamashita 2008)
	Propithecus verreauxi	Berenty, Madagascar	− L	− L	− L	Logistic regression: probability of plants being either major or minor food items.	No significant effect of PSMs. Major foods had higher concentrations of tannins than minor foods.	(Simmen, Tarnaud, Marez, et al. 2014)
Lepilemur	Lepilemur mustelinus	Andasibe, Madagascar	+**L**		− L	Compared food and nonfood items using a Mann-Whitney U test.	Significant effect of tannins.	(Ganzhorn 1988)
	Lepilemur edwardsii	Andasibe, Madagascar	+**L**		− L	Compared food and nonfood items using a Mann-Whitney U test.	Significant effect of tannins.	(Ganzhorn 1988)
Cheirogaleus	Cheirogaleus major	Andasibe, Madagascar	− L Fr		− L Fr	Compared food and nonfood items using a Mann-Whitney U test.	No significant effect of PSMs.	(Ganzhorn 1988)
Eulemur	Eulemur fulvus	Andasibe, Madagascar	− L Fr		− L Fr	Compared food and nonfood items using a Mann-Whitney U test.	No significant effect of PSMs.	(Ganzhorn 1988)
	Eulemur fulvus	Berenty, Madagascar		− L	− L	Compared eaten with representative subset by a chi-squared test.	No significant effect of PSMs.	(Simmen, Hladik, et al. 1999)
	Eulemur fulvus	Saziley, Mayotte	+**L**	− L	− L	Compared eaten with uneaten using a Mann-Whitney U test.	Significant effect of tannins.	(Simmen, Tarnaud, Bayart, et al. 2005)
	Eulemur macaco	Ampasikely, Madagascar	−yL	+**yL**	−yL	Compared eaten with uneaten using a Mann-Whitney U test.	Significant effect of phenolics.	(Simmen, Tarnaud, Bayart, et al. 2005)
	Eulemur collaris	Mandena and Saint Luce, Madagascar	− Fr	+/− **Fr**		Compared marginal and primary fruits using one-way ANOVA.	Mixed effect of PSMs. One group preferred fruits with lower phenolics; other groups showed no preference. No effect of tannins.	(Donati et al. 2011)

Genus	Species	Location				Methods	Findings	Reference
Lemur	Lemur catta	Berenty, Madagascar		+L	−L	Compared eaten with representative subset using a chi-squared test.	Significant effect of phenolics.	(Simmen, Hladik, et al. 1999)
	Lemur catta	Berenty, Madagascar	−L	−L	−L	Compared eaten and not eaten using a Mann-Whitney U test.	No significant effect of PSMs. Sometimes selected foods very high in alkaloids.	(Simmen, Peronny, et al. 2006)
	Lemur catta	Beza Mahafaly, Madagascar	(+) L Fl Fr	(+) L Fl Fr		No statistics of feeding preferences.	No statistics. Tannins and phenolics not present in foods and so potentially avoided.	(Yamashita 2008)
	Lemur catta	Berenty, Madagascar	−L Fl Fr			Compared major foods with minor foods using a Mann-Whitney U test.	No significant effect of tannins.	(L. Gould, Constabel, et al. 2009)
Hapalemur	Hapalemur griseus	Andasibe, Madagascar	+L		(+) L	Compared major foods with minor foods using a Mann-Whitney U test.	Significant effect of tannins.	(Ganzhorn 1988, 1989b)
	Hapalemur aureus	Mandena, Madagascar	−L Sh	−L Sh		Compared primary, secondary, and marginal foods using a Kruskall-Wallis ANOVA.	No significant effect of PSMs.	(Eppley et al. 2011)
Microcebus	Microcebus rufus	Andasibe, Madagascar	+L		(+) L	Compared major foods with minor foods using a Mann Whitney U test.	Significant effect of tannins.	(Ganzhorn 1988, 1989b)
Atelinae	Alouatta palliata	Barro Colorado Forest, Panama	(−) L	(−) L		No statistics.	No statistics. Preferred young leaves, which had higher tannin and phenolic concentrations.	(Milton 1979)
	Alouatta palliata	Hacienda La Pacifica, Costa Rica	(+) L Fl		(+) L Fl	No statistics on PSM content. Did not quantify PSMs.	No statistics, mixed influence of PSMs. Avoided trees with high concentrations of tannins and alkaloids in some species.	(Glander 1981)

(continued)

Table 9.1. (continued)

Primate group	Species	Location	Tannins	Phenolics	Alkaloids	Statistics	Comment	Reference
	Alouatta palliata	Guanacasta Province, Costa Rica	−/+ **L**	− L		Compared eaten vs. uneaten plants and correlated feeding time with PSMs.	Mixed effects of PSMs. No significant difference between eaten and uneaten items, but preferred lower tannins in protein-rich species.	(Bilenger 1995)
	Alouatta palliata	Guanacasta Province, Costa Rica				Compared terpene concentrations of eaten vs. uneaten items.	Significant effect of terpenes. Selection of particular plants likely due to sesquiterpene concentrations.	(Welker et al. 2007)
	Alouatta pigra	El Tormento, Mexico	(−) L Fl			No statistics, did not quantify PSMs.	No statistics. Species with high tannins were consumed frequently.	(Righini et al. 2015)
Pitheciidae	*Pithecia pithecia*	Venezuela	(+) S			No statistics.	No statistics, but uneaten seeds had higher tannin concentrations than eaten seeds.	(Kinzey and Norconk 1993)
	Pithecia pithecia	Venezuela	− L Fr			Compared eaten vs. not eaten and used multiple regressions of time feeding.	No significant effect of tannins, although preferred foods (yL) tended to have lower tannins.	(Norconk and Conklin-Brittain 2004)
	Callicebus p. melanochir	Eastern Brazil	− L F S	− L F S		Pearson correlation between time feeding and chemical composition.	No significant effect of PSMs.	(Heiduck 1997)

Old World monkeys

Colobinae

| Colobus | *Colobus guereza* | Kibale, Uganda | (+) L | | − L | Correlation with frequency of consumption, compared food with nonfoods. | Possibly deterred by PSMs. Preferred foods (yL) tended to have lower tannins. | (Oates 1977; Oates, Swain, and Zantovska 1977) |
| | *Colobus satanus* | Douala-Edea, Cameroon | + **L S** | + **L S** | − L | Spearman rank correlation with food selection ratio. | Found a weak negative correlation between feeding and tannins and phenolics. | (McKey 1978; McKey, Gartlan, et al. 1981; McKey, Waterman, et al. 1978) |

Species	Location				Method	Result	Reference
Procolobus verus	Tiwai Is, Sierra Leone	+LS			Compared eaten with uneaten and correlation of feeding.	Significant effect of tannins.	(Oates 1988)
Piliocolobus tholloni	Salonga NP, DRC	−L/+S	−LS/+Fr		Compared eaten and not eaten (combined items with *C. a. angolensis*).	Mixed effect of PSMs. Preferred fruits with lower total phenolics and seeds with lower tannins.	(Maisels et al. 1994)
Colobus angolensis angolensis	Salonga NP, DRC	−L/+S	−LS/+Fr		Compared eaten and not eaten (combined items with *p. badius*).	Mixed effect of PSMs. Preferred fruits with lower total phenolics and seeds with lower tannins.	(Maisels et al. 1994)
Piliocolobus rufomitratus	Tana River, Kenya	−L	(+)L		Pearson correlations of percentage feeding.	No significant effect of PSMs. Potentially avoided hydrolyzable tannins and alkaloids.	(Mowry et al. 1996)
Colobus guereza and Piliocolobus tephrosceles (combined)	Kibale, Uganda	−L	−L		Linear and logistic regression of feeding effort.	No significant effect of PSMs.	(M. A. Burgess and Chapman 2005)
Piliocolobus tephrosceles	Kibale, Uganda	−L	−L		Compared foods and nonfoods and regression with foraging effort.	No significant effect of PSMs. Saponins and cyanogenic glycosides also had no effect on foraging.	(Chapman and Chapman 2002)
Colobus guereza	Kakamega, Kenya	+L		Alkaloids not found in plants	Compared eaten and not eaten and regression of feeding selection.	Significant effect of tannins.	(Fashing et al. 2007)
Colobus guereza	Kibale, Uganda	−M			Linear regression of tannins in eaten items with patch occupancy time and time spent feeding in a patch.	No significant effect of tannins.	(C. A. Johnson, Raubenheimer, Chapman, et al. 2017)
Langurs (leaf monkeys)							
Trachypithecus johnii	Kakachi, India	−L	−L		Correlation between food selection and PSMs.	No significant effect of PSMs, but highly preferred items had low tannins (not significant).	(Oates, Waterman, and Choo 1980)
Presbytis rubicunda	Sepilok, Malaysia	−yLS	−yLS		Compared eaten and uneaten using a Mann-Whitney test and Spearman correlation.	No significant effect of PSMs. Preferred seeds with higher tannins.	(A. Davies, Bennett, and Waterman 1988)

(continued)

Table 9.1. (continued)

Primate group	Species	Location	Tannins	Phenolics	Alkaloids	Statistics	Comment	Reference
	Presbytis melalophos	Kuala Lompat, Malaysia	– L	– L		Compared eaten and uneaten using a Mann-Whitney test and Spearman correlation.	No significant effect of PSMs.	(A. Davies, Bennett, and Waterman 1988)
	Trachypithecus auratus sondaicus	Pangandaran NR, West Java	– L Fr	– L Fr		Compared eaten with not eaten using a Mann-Whitney U test.	No significant effect of PSMs.	(Kool 1992)
	Nasalis larvatus	Kinabatangan Borneo	– L			General linear model of leaf preference.	No significant effect of tannins.	(Matsuda et al. 2013)
	Presbytis rubicunda	Danum Valley, Borneo	– L			Compared preferences of PSMs to those of N. larvatus using a chi-squared test.	No significant effect of tannins. The ratio of consumed plant species with tannins was the same as for N. larvartus.	(Matsuda et al. 2013)
Odd-nosed monkeys (Rhinopithecus)	Rhinopithecus bieti	Mt. Longma, Himalayas		– L		Compared food and nonfood using a Mann-Whitney U test.	No significant effect of phenolics.	(Z. Huang et al. 2010)
	Rhinopithecus roxellana	Shennongjia, China	– L Li	– L Li		Compared food and nonfood using a Mann-Whitney U test.	No significant effect of PSMs.	(X. Liu, Stanford, et al. 2013)
	Rhinopithecus roxellana	Qinling Mountains, China	+ M			Correlation between percent contribution of species to diet and PSMs.	Negative correlation between tannins and feeding preference.	(Zhao et al. 2020)
Cercopithecinae								
Baboons	Papio ursinus	Drakensberg Mts S. Africa	– M	+ M	+ M	Multivariate models to compare foods vs. nonfoods.	Significant effect of phenolics and alkaloids but not tannins.	(Whiten, Byrne, et al. 1991)
	Papio anubis	Laikipia Plateau, Kenya	+ M	– M	– M	Multivariate models to compare foods vs. nonfoods.	Significant effect of tannins.	(Whiten, Byrne, et al. 1991)
	Papio anubis	Laikipia Plateau, Kenya	– M	+ M		Compared foods and nonfoods and multiple regressions of time feeding.	Significant effect of phenolics; foods tended to be lower in tannins (not significant).	(Barton and Whiten 1994)

	Species	Location			Method	Result	Reference
	Papio anubis	Kibale, Uganda	− M		Compared eaten vs. discarded foods.	No significant effect of tannins, but eaten foods tended to be lower than uneaten (not significant).	(C. A. Johnson, Swedell, and Rothman 2012)
Macaques	*Macaca mulatta*	Murree Hills, Northern Pakistan	(+) L Fr *Hydrolyzable tannins	+ **L Fl Fr** − L Fl Fr	Pearson correlation between feeding and chemical contents.	Significant effect of phenolics, astringency, and hydrolyzable tannins but not condensed tannins or alkaloids.	(Marks et al. 1988)
	Macaca fuscata	Yakushima, Japan	+/− **mL**		Logistic regression of food vs. nonfood leaves.	Mixed effect of tannins. Tannins negatively correlated with selection in coniferous forest but not in coast forest; no effect of hydrolyzable tannins.	(Hanya, Kiyono, Takafumi, et al. 2007)
	Macaca fuscata	Yakushima, Japan	+ **mL**		Compared food vs. nonfood leaves using a *t*-test and linear models.	Significant effect of condensed tannins and hydrolyzable tannins.	(Hanya, Ménard, et al. 2011)
	Macaca sylvanus	Moyen Atlas, Morocco	− mL		Compared food vs. nonfood using a *t*-test and linear models.	No significant effect of tannins.	(Hanya, Ménard, et al. 2011)
	Macaca assamensis	Guangxi, China	+ yL		Compared food and nonfood and relationship with feeding time using GLMM and ANOVA.	Significant effect of tannins when considering a single species (*Bonia saxatilis*), but no overall effect of tannins.	(Y. Li et al. 2020)
Mangabeys	*Cercocebus albigena Lophocebus albigena*			(+) Fr	No statistics.	No statistics; possible deterrence of alkaloids.	(Waser 1977)
		Mabira and Lwamunda Forest Reserves, Uganda	+ **S Fr**		Logistic regression eaten vs. avoided and ANOVA.	Significant effect of tannins. Reduced consumption with increasing tannin concentrations.	(Masette, Isabirye-Basuta, Baranga, Chapman, et al. 2015; Masette, Isabirye-Basuta, Baranga, and Chemurot 2015)
Vervets	*Chlorocebus pygerythrus*	Amboseli NP, Kenya	+ **L Fr Fl G**	+ **L Fr Fl G**	Inconclusive	Correlation between PSMs and time feeding. Selection among two *Acacia* spp.	Significant effect of tannins and phenolics when feeding on *Acacia xanthophloea* and *A. tortilis*.	(Wrangham and Waterman 1981)

(*continued*)

Table 9.1. *(continued)*

Primate group	Species	Location	Tannins	Phenolics	Alkaloids	Statistics	Comment	Reference
Anthropoids								
Chimpanzee	*Pan troglodytes*	Gabon and Gombe			(−) L S	No statistics.	No statistics. Alkaloids present in about 15% of species eaten.	(A. Hladik and Hladik 1977; C. Hladik 1977a)
	Pan troglodytes	Gombe, Tanzania	(+) Fr			No statistics.	No statistics. Avoided high-tannin unripe fruit.	(Waterman 1984; Wrangham and Waterman 1983)
	Pan troglodytes	Kibale, Tanzania	(−) Fr			No statistics.	No statistics; prefer fruits that are higher in tannins. Chimps have a lower intake of tannins than mangabeys.	(Wrangham, Conklin-Brittain, and Hunt 1998)
	Pan troglodytes schweinfurthii	Budongo, Uganda	− L Fr B mix			*t*-test to compare eaten and not-eaten items.	No significant effect of tannins.	(Reynolds, Plumptre, et al. 1998)
	Pan troglodytes schweinfurthii	Bulindi, Uganda	− Fr	− Fr		Compared eaten and not eaten fruits using Mann-Whitney tests with Holm-Bonferroni adjustment.	No significant differences in PSMs and only small concentrations of tannins in fruits.	(McLennan and Ganzhorn 2017)
	Pan troglodytes schweinfurthii	Kalinzu forst, SW Uganda	* Fr	(+) Fr		Person correlation and multiple regression models on contribution of fruit species to diet.	Significant preference for fruits with higher tannin concentrations and nonsignificant preference for lower total phenolics.	(Kagoro-Rugunda 2020)
	Pan troglodytes	Bossou, Guinea	+ **mL**			Two-tailed Mann-Whitney *U* test (eaten vs. not) and Spearman rank correlation.	Significant effect of tannins. Avoided mature leaves with higher tannins.	(Takemoto 2003)
	Pan troglodytes	Gashaka, Nigeria	− Fr	− Fr		Compared food and nonfood using ANOVA.	No significant effect of PSMs, but tended to choose fruits with lower phenolics (not significant).	(Hohmann, Fowler, et al. 2006)
	Pan paniscus	Salonga, DRC	− Fr	− Fr		Compared food and nonfood using ANOVA.	No significant effect of PSMs.	(Hohmann, Fowler, et al. 2006)

	Pan t. troglodytes, P. t vellerosus, P. t. schweinfurthii, and P. paniscus (combined)	Four sites in West and Central Africa	− L Fr	− L Fr	Two-way ANOVA and PCA comparing food vs. nonfood.	No significant effect of PSMs, but foods tended to contain lower tannins and phenolics than nonfood (not significant).	(Hohmann, Potts, et al. 2010)
Gorillas	Gorilla g gorilla	Campo, Cameroon	− L Fr Sh	− L Fr Sh	Welch t-test between high and low preference and regression.	No significant effect of PSMs.	(Calvert 1985)
	Gorilla g gorilla	Lope, Gabon	− L Fr S	− **L S** / + **Fr**	Wilcoxon-Mann-Whitney test to compare eaten and not-eaten items.	Mixed effects of PSMs. Eaten fruits (ripe) contained higher tannins and lower phenolics than uneaten fruits (unripe) for a particular species.	(Rogers, Maisels, et al. 1990)
	Gorilla gorilla	Bai Hokou, C.A.R	− Fr	(+) L Fr	Mann-Whitney U test to compare important vs. less important foods.	No significant effect of PSMs, but only two eaten fruits and no leaves had alkaloids.	(Remis et al. 2001)
	Gorilla gorilla	Captive study	+ **Fr**		Paired t-test comparing water with experimental solution.	Significant effect of tannins.	(Remis and Kerr 2002)
	Gorilla beringei	Bwindi, Uganda	(−) L B Sh		No statistics.	~35% of plants eaten contained condensed tannins and two foods contained cyanogenic glycosides.	(Rothman, Dierenfeld, Molina, et al. 2006; Rothman, Dusinberre, and Pell. 2009)
	Gorilla beringei	Bwindi, Uganda	− L Fr	− L Fr	PCA.	No significant effect of PSMs; preferred items with higher phenolics, and one group preferred fruit with high tannins.	(Ganas et al. 2008)
	Gorilla beringei	Bwindi, Uganda	+ **H Fr**	+ **H**	Multiple regression and PCA with feeding preference.	Significant effect of PSMs. Preferred items with lower tannins and phenolics.	(Ganas et al. 2009)
Orangutans	Pongo pygmaeus pygmaeus	Kutai National Park, Borneo	+ **P S**	+ **Fr**	Multiple regression with factor analysis with feeding preference.	Significant effect of PSMs. Preferred items with lower tannins and phenolics.	(M. Leighton 1993)

(continued)

Table 9.1. (continued)

Primate group	Species	Location	Tannins	Phenolics	Alkaloids	Statistics	Comment	Reference
Gibbons	*Nomascus nasutus*	Jingxi, China	– L Fr Fl			Compared important, less-important, and nonfoods using a Kruskal-Wallis test.	No significant effect of tannins.	(Ma et al. 2016)

Summary of statistical findings

	Tannins	Phenolics	Alkaloids
No. of studies that found PSMs significantly deterred feeding	22	14	6
No. of studies that tested for statistical significance and found that PSMs had no significant effect of deterrence	46	27	25

Only studies with a measure of feeding preference for individual primate species (e.g., eaten or not eaten, or time spent feeding) were included. Studies finding a statistically significant avoidance of PSMs are in boldface. Tannins are condensed tannin concentrations unless otherwise specified as hydrolyzable tannin concentrations, phenolics are total phenolic concentrations, and alkaloids are the presence of unspecified alkaloids.

+: statistically significant avoidance of PSMs, boldface.
(+): pattern of avoidance, but not statistically significant.
–: not statistically significant—i.e., no avoidance of PSMs.
(–): no statistics, but no other evidence or pattern of PSM avoidance.
*: statistically significant preference for higher PSM concentrations.
L, leaves; S, seeds; Fr, fruit; Fl, flowers; B, bark; H, herbs; yL, young leaves; mL, mature leaves; Sh, shoots; Li, lichen; G, gum; M, mixed.

World monkeys, colobine and other Old World monkeys, and apes.

Lemurs

Ganzhorn studied the feeding preferences of a lemur community in the eastern rainforest of Madagascar and found that each species had a unique response to different PSMs. Alkaloids but not tannins were found to be important feeding deterrents for *Avahi* and *Indri* species (*Avahi laniger*: Ganzhorn 1988; Ganzhorn, Abraham, and Razanahoera-Rakotomalala 1985; *A. occidentalis* and *Indri indri*: Ganzhorn 1988; Ganzhorn, Abraham, and Razanahoera-Rakotomalala 1985) but did not affect *Propithecus coquereli* (Ganzhorn 1989b). *Hapalemur griseus* were found to avoid both tannins and alkaloids, and *Lepilemur mustelinus* chose leaves with lower tannin concentrations but were not affected by unspecified alkaloids (Ganzhorn 1989b). These diverse results led the authors to propose that distinct tolerances to PSMs by individual primate species allows for niche separation in sympatric primate communities (Ganzhorn 1989b).

Subsequent researchers have also predominantly found that sifakas (*Propithecus*) were unaffected by PSMs: tannins, phenolics, and unspecified alkaloids. For example, *Propithecus diadema diadema* consumed fruits containing alkaloids and also those with high phenolics, whereas *I. indri* did not feed on any fruits containing alkaloids or phenolics (Powzyk and Mowry 2003). *Propithecus verreauxi* at Berenty in southeast Madagascar were also found to be unaffected by alkaloids and frequently fed on the tannin-rich leaves of *Vernonia pectoralis* (Asteraceae; Simmen, Hladik, et al. 1999). There was no difference in condensed-tannin concentrations in items eaten compared with those not eaten by *Propithecus coquereli* in western Madagascar (Ganzhorn and Abraham 1991). Indeed, the tannin intake of *Propithecus verrauxi* was found to be higher in pregnant and lactating females than in nonreproductive females (Carrai et al. 2003), suggesting that sifakas may increase tannin intake for antihelminthic or other self-medicative properties. Conversely, at the same site, Norscia, Carrai, and Borgogini-Tarli (2006) found that feeding preference (measured as time spent feeding) of sifakas was negatively correlated with condensed-tannin concentrations in their food plants during the lean dry season. The authors also found that, contrary to earlier studies, *Avahi meridonalis* were unaffected by alkaloids and decreased feeding as phenolic concentrations increased (Norscia, Ramanamanjato, and Ganzhorn 2012).

Hapalemur aureus, *H. griseus*, and *Prolemur simus* specialize on the cyanogenic shoots of giant bamboo (*Cathariostachys madagascariensis*), consuming high doses of cyanide that would be lethal to humans (Ballhorn et al. 2009; Glander, Wright, Siegler, et al. 1989). Ballhorn and colleagues (2009) found that cyanide concentrations in *C. madagascariensis* were closely correlated with protein, particularly in new bamboo shoots, which are most preferred by *Hapalemur*. Eppley, Tan, and colleagues (2017) suggest that bamboo lemurs offset the negative effects of the toxin with protein, consuming up to twice the amount of protein when feeding on bamboo compared to lemurs in areas without bamboo and sympatric lemur species that feed on leaves of trees. It is still unclear how these species cope with such high concentrations of toxins in their diet (Jeannoda et al. 2003). The diet of the southern gentle lemur (*Hapalemur meridionalis*) in southeast Madagascar is distinct from those of other *Hapalemur* species, as it does not feed on bamboo. Eppley, Verjans, and Donati (2011) found that there was no relationship between time spent feeding and concentrations of total phenolics and condensed tannins in plant items.

Ring-tailed lemurs (*Lemur catta*) and *Eulemur* spp. in Madagascar do not appear to be dissuaded by tannin concentrations or by unspecified alkaloids in their food, but they are potentially affected by phenolics (Donati et al. 2011; Simmen, Peronny, et al. 2006; Simmen, Tarnaud, Bayart, et al. 2005). Ring-tailed lemurs at Berenty chose leaves with lower concentrations of phenolics compared with a representative sample of leaves from their habitat (Simmen, Hladik, et al. 1999); however, later studies at the same site found that leaves selected as foods by *L. catta* did not differ significantly in PSMs (tannins, total phenolics, or alkaloids) from leaves not eaten or a random sample; and in some cases, ring-tailed lemurs were found to select leaves with very high amounts of putative alkaloids and phenolics (L. Gould, Constabel, et al. 2009; Simmen, Peronny, et al. 2006; Simmen, Sauther, et al. 2006).

New World Monkeys

Studies on New World monkeys generally support the statement by Milton that there is "little direct evidence toxic secondary compounds are strongly implicated in patterns of primate feeding" (Milton 1998, 534). Following extensive studies on the diet and nutrient balance of mantled howler monkeys (*Alouatta palliata*) in Panama, Milton suggested that higher total phenolic concentrations may have deterred howlers from feeding on particular plant species but did not influence the choice between mature and young leaves of the same species (Milton 1979, 1980). Glander (1981) suggested that howler monkeys (*A. palliata*) generally avoided leaves with alkaloids and tended to prefer items without condensed tannins, but this was not true for all plant species. Selectivity was more closely associated with other constituents such as fiber, amino acid, and total protein concentrations than with PSMs. Some researchers have found that howlers prefer items with higher concentrations of condensed tannins and frequently feed on plant species with high tannins (Righini et al. 2015). For example, Bilgener (1995) found that feeding time and preferences of *A. palliata* were positively correlated with tannin concentrations; however, when howlers fed on protein-rich plants, they preferred those with lower tannin concentrations. Welker and colleagues (2007) suggested that PSMs influence fine-scale intraspecific plant selection rather than interspecific choices of howlers. They reported that *A. palliata* selected particular leaf stages of *Hymenaea courbaril* with lower condensed tannins and total phenolics compared to other leaf stages of the same plant that were not selected, and that preferences for particular trees were related to variability in the sesquiterpene α-copaene.

Masked titi monkeys (*Callicebus personatus melanochir*) in Brazil and white-faced saki (*Pithecia pithecia*) in Venezuela were not deterred by PSMs (Heiduck 1997; Norconk and Conklin-Brittain 2004). It has been suggested that saki monkeys might be able to cope with high tannins in foods if they are also high in lipids (Kinzey and Norconk 1993; Norconk and Conklin-Brittain 2004), but the mechanisms of such a connection are unclear.

Colobines and Other Old World Monkeys

Our review of studies of colobine monkeys suggests that PSMs are likely to be significant feeding deterrents for this primate group, with over half of the reviewed studies of colobus species (excluding studies of leaf and odd-nosed monkeys) finding a significant effect of both tannins and phenolics on feeding preferences (table 9.1). This hypothesis is supported and expanded by the meta-analysis by Windley et al. (2022). For example, Oates and colleagues (Oates 1977; Oates, Swain, and Zantovska 1977) investigated the feeding behavior of the black-and-white colobus (*Colobus guereza*) at Kibale, Uganda, and found that they preferred to eat plant items with lower condensed-tannin concentrations. They proposed that tannins, but not alkaloids, were effective feeding deterrents for colobine monkeys below a threshold concentration (Oates 1977; Oates, Swain, and Zantovska 1977). Subsequent research also found that black-and-white colobus at Kibale (Fashing et al. 2007), black colobus (*Colobus satanas*) at Douala-Edea in Cameroon (Choo et al. 1981; McKey, Gartlan, et al. 1981; McKey, Waterman, et al. 1978), and olive colobus (*Procolobus verus*) in west Africa (Oates 1988) feeding on leaves was negatively related to condensed tannins and total phenolics, but not alkaloids. Furthermore, Maisels and colleagues (1994) found that red colobus (*Piliocolobus badius*) and black-and-white colobus (*Colobus angolensis angolensis*) in the Democratic Republic of the Congo preferred young leaves, which were high in condensed tannins, but that the seeds and fruit eaten had significantly lower phenolics than those that were avoided.

Despite the overall trend of tannin avoidance among colobine primates, there is certainly diversity among different colobine species (Rothman, DePasquale, et al. 2022). Red colobus monkeys (*Piliocolobus tephrosceles*) at Kibale showed no preference for lower condensed tannins or total phenolics, or for the presence or absence of alkaloids or saponins (M. A. Burgess and Chapman 2005; Chapman and Chapman 2002). Furthermore, a highly preferred species at Kibale (*Prunus africana*) contained the highest concentrations of cyanogenic glycosides (Chapman and Chapman 2002), and another preferred species (*Albizia grandibracteata*) contained high amounts of unspecified saponins. C. A. Johnson and colleagues (C. A. Johnson et al. 2015; C. A. Johnson, Raubenheimer, Chapman, et al. 2017) found no relationship between time spent feeding or patch departure time and condensed-tannin concentrations in items eaten by *Colobus guereza*. Condensed tannins were not correlated with feeding preferences of Tana River red colobus monkeys (*Piliocolobus rufomitratus*) in Kenya, although two plant species with high concentrations of hydrolyzable tannins and one with high concentrations of naphthoquinones were avoided, suggesting that PSMs other than condensed tannins may be important (Mowry et al. 1996). There is no evidence that phenolics or alkaloids are feeding deterrents for Asian colobine monkeys such as langurs (*Trachypithecus auratus sondaicus* [Kool 1992], *Presbytis johnii* [Oates, Waterman, and Choo 1980], *Presbytis rubicunda* and *P. melalophos* [A. Davies, Bennett, and Waterman 1988], and *Nasalis larvatus* [Matsuda et al. 2013]) and snub-nosed monkeys (*Rhinopithecus roxellana* [X. Liu, Stanford, et al. 2013] and *Rhinopithecus bieti* [Z. Huang et al. 2010]).

Our review of research on baboons and macaques suggests that the Cercopithecinae are likely to be deterred by tannins and phenolics in plants, and six of the seven studies we reviewed found a significant deterrence of these PSMs. Selection of leaves by chacma baboons (*Papio ursinus*) in South Africa (Tutin, Fernandez, et al. 1991; Whiten, Byrne, et al. 1991) and olive baboons (*P. anubis*) in Kenya (Barton and Whiten 1994) was negatively related to total phenolics, tannins, and alkaloids. Condensed tannins were present in a higher (but not statistically significant) proportion of discarded items compared with those items eaten by olive baboons at Kibale (C. A. Johnson, Swedell, and Rothman 2012). Marks and colleagues (1988) found that feeding by rhesus monkeys (*Macaca mulatta*) in northern Pakistan was negatively correlated with total phenolics, hydrolyzable tannins, ellagitannins, and the protein-precipitating capacity of extractable tannins, but not with condensed tannins. Hanya, Ménard, and colleagues (2011) showed that condensed tannins were negatively correlated with leaf selection by Japanese macaques (*M. fuscata*) but not by Barbary macaques (*M. sylvanus*) in North Africa and that selection by macaques varied depending on the habitat quality (i.e., the chemistry of plant items available; Hanya, Kiyono, Takafumi, et al. 2007).

Apes

Extensive studies have been undertaken on the feeding preferences of apes that have advanced our knowledge of this diverse group. Wrangham and colleagues (Wrangham, Conklin, Etot, et al. 1993; Wrangham et al. 1983) suggested that higher tannin concentrations make fruit unpalatable to chimpanzees. Conversely, chimpanzees at Kanyawara, Budongo, and Kalinzu Forest tend to prefer ripe fruits that are higher in tannins than other ripe fruits (Kagoro-Rugunda 2020; Reynolds, Plumptre, et al. 1998). Hohmann, Fowler, and colleagues (2006) did not find evidence that condensed tannins deter feeding by bonobos in the Democratic Republic of the Congo and chimpanzees in Nigeria, but other authors have found that foods selected by chimpanzees had lower tannin and phenolic concentrations than nonfoods (Hohmann, Potts, et al. 2010; M. Leighton 1993; Takemoto 2003).

Studies on the feeding of gorillas suggest that neither tannins nor total phenolics are significant feeding deterrents for this group (e.g., Calvert 1985; Remis 2002; Remis et al. 2001; Rothman, Dierenfeld, Molina, et al. 2006); however, unspecified alkaloids were implicated as deterrents by Remis and colleagues (2001). Fruits and seeds chosen by gorillas as food were higher in both sugars and condensed tannins than those not eaten; however, gorillas preferred ripe fruits with lower total phenolics over unripe fruits for a given species (Rogers, Maisels, et al. 1990). Gorillas at Bwindi, Uganda, regularly ate foods with very high tannin concentrations (20%), although many of the staple foods were low in tannins, suggesting that they may switch between foods to dilute the negative effects of PSMs (Rothman, Dusinberre, and Pell 2009; Rothman, Pell, and Bowman 2009).

In a captive study, Remis and Kerr (2002) found that captive gorillas reduced consumption of tannin as concentrations increased, although their intake of toxins interacted with sugars, as the animals consumed higher concentrations of tannins if sugar concentrations were also increased. Ganas and colleagues (2008) undertook a principal component analysis on the chemical characteristics of plants eaten and not eaten by gorillas at Bwindi in an attempt to incorporate the trade-offs between nutrients and toxins in plant feeding. Gorillas were deterred by a combination of higher fiber and condensed-tannin concentrations in leaves but also preferred leaves with higher sugars, and one group preferred fruit from a particular plant with very high tannin concentrations. A later study (Ganas et al. 2009) found that gorilla choice of fruit and herbs was negatively influenced by condensed tannins. These seemingly conflicting results are perhaps a consequence of varying tannin-protein interactions (I. Wallis et al. 2012) and the behavioral trade-offs that individual herbivores make between the many diverse compounds in plants to achieve a dietary balance (Raubenheimer, this volume; e.g., Rothman, Raubenheimer, and Chapman 2011). Studying the interactions directly and using targeted approaches may help untangle the question of the influence of PSMs in primate feeding.

NUTRITIONAL ECOLOGY OF SECONDARY METABOLITES IN PRIMATOLOGY: STRENGTHS AND WEAKNESSES

Interactions among Nutrients and Secondary Metabolites Are Key

The review above and the expanded formal meta-analysis (Windley et al. 2022) suggest that there is little consistent evidence to either support or refute the hypothesis that PSMs in general act as feeding deterrents for primates, apart from condensed tannins acting as deterrents for colobine primates. However, as many of the studies use different approaches, caution must be taken in rejecting the hypothesis without further consideration. The vast majority of studies on primate feeding ecology attempt to correlate primate feeding preferences (usually a measure of time spent feeding) with concentrations of PSM complexes (primarily condensed tannins, though also total phenolics, alkaloids, hydrolyzable tannins, and saponins) or test for plant-chemical differences between plant items grouped by primate feeding preferences (e.g., primary, marginal, eaten, or avoided) of the primate under study.

A number of problems arise when compiling a review of these approaches. The first is the lack of specificity in most of the chemical assays (see box 9.1), which means that the chemistry of plants in different studies is unlikely to be comparable (Rothman, Dusinberre, and Pell 2009). Second, it is not possible to definitively identify foods that are avoided by primates or to determine whether the apparent avoidance is due to chance. It is also difficult to know whether feeding on a plant item for a longer time actually means that the item is preferred or whether it is simply a reflection of processing difficulties, such as leaf toughness (Teaford, Lucas, et al. 2006). As a consequence, many studies report only anecdotal or qualitative data, do not test hypotheses statistically, or do not provide the full details of statistical tests such as sample numbers and effect sizes to allow for adequate comparisons between studies needed for meta-analysis. Third, the complex trade-offs that wild animals make between different nutrients and potentially toxic compounds suggest that we should not expect simple relationships between PSMs and feeding (Raubenheimer, this volume; Felton and Lambert, this volume). Rather, future work could benefit from using better statistical approaches that cope with more complex interactions such as quantile regressions to estimate functional relationships between food selection and chemical composition (DeGabriel, Moore, Felton, et al. 2014).

An alternative approach is to use targeted chemical assays that attempt to estimate the bioactivity and structure-activity of PSMs in specific plant samples. In an early study of primate feeding, Ganzhorn (1988, 1989b) measured the protein-precipitating effect of tannins and fiber (measured as "extractable protein") in plant samples chosen by seven lemur species. All lemur species except *L. mustelinus* selected for leaves with higher concentrations of extractable protein (i.e., not bound by tannins or fiber) compared with leaves they did not choose; however, only one lemur species selected for leaves with higher total protein. Ganzhorn (1992) later showed that the leaf selection of 11 species of folivorous lemurs was related to concentrations of extractable protein, whereas condensed tannins significantly explained the feeding preferences of only one species (Ganzhorn 1988, 1992). This suggests that it is the interaction of tannins and protein that is likely to be important for lemurs, rather than the concentrations of protein or phenolics in isolation. In vitro studies suggest that tannin-protein binding occurs in many lemur foods during digestion (I. Wallis et al. 2012), but in vivo confirmation is lacking.

Elsewhere, studies of other vertebrate folivores such as marsupials have developed new approaches to overcome problems of directly measuring tannins. DeGabriel, Wallis, et al. (2008) drew on agricultural studies to develop a simple in vitro assay to measure the digestibility-reducing effect of tannins and fiber on protein. The outcome, "available protein," is a measure of the proportion of the total foliar crude protein that herbivores can obtain from their foods and may be a more appropriate measure to explain the densities and population dynamics of animals eating tannin-rich diets (DeGabriel, Moore, Foley, et al. 2009;

DeGabriel, Moore, Shipley, et al. 2009; McArt, Spalinger, Kennish, et al. 2006; I. Wallis et al. 2012).

When applying this measure to diets of Peruvian spider monkeys (*Ateles chamek*), Felton, Felton, Wood, Foley et al. (2009) found that available protein, but not crude protein or condensed tannins, explained the preference for unripe fruits over ripe fruits. Importantly, the daily intake of crude protein by spider monkeys fluctuated by 75% across all three seasons, whereas available protein intake remained steady regardless of diet composition and food availability. Felton, Felton, Raubenheimer, Simpson, Foley et al. (2009) further analyzed the patterns of nutritional prioritization of *A. chamek* using the geometric framework (S. Simpson and Raubenheimer 1993; Raubenheimer, this volume) and found that spider monkeys prioritize the balanced intake of available protein over the intake of nonprotein energy (energy from nonstructural carbohydrates, fat, and digestible fiber). This pattern of protein leverage by spider monkeys would not have been apparent if crude protein had been used; it was apparent only when protein intake was expressed as available protein.

Using similar methods, Dröscher and colleagues (2016) found that *Lepilemur leucopus* in southern Madagascar also prioritize available protein in relation to nonprotein energy, whereas mountain gorillas (*G. beringei*) show the reverse pattern and overconsume available protein to meet a nonprotein energy target (Rothman, Raubenheimer, and Chapman 2011). In a study of the feeding preferences of red leaf monkeys (*Presbytis rubicunda*), Hanya and Bernard (2015) found that leaves chosen as food had significantly higher available protein contents than leaves not chosen, though selection for total protein (when tannins were not considered) was not significant. K. Evans and colleagues (2021) used the available protein assay to show that while tannins reduced the digestibility of protein in colobine diets by up to 50%, concentrations of available protein exceeded primate protein requirements at all sites. A meta-analysis of studies of the role of protein in the feeding choices of wild primates showed that soluble protein and acid detergent fiber were limiting factors in primate feeding choice, but crude protein was not (Ganzhorn, Arrigo-Nelson, Carrari, et al. 2017). Furthermore, the authors found that selection for soluble protein was reinforced in forests where protein was less available in the habitat (Ganzhorn, Arrigo-Nelson, Carrari, et al. 2017).

These recent studies suggest that it is likely that tannins have always played a role in primate nutrition, particularly in habitats where tannins are abundant and where protein is limiting (I. Wallis et al. 2012), but that by measuring tannins and crude protein in isolation, we have not captured the ecological interactions of this PSM group and have provided potentially misleading evidence of the actual nutrient goals and feeding preferences of primates.

Lack of Experimental Feeding Manipulations Impacts Hypothesis Generation

Wild systems are infinitely complex, so controlled in vivo studies with captive animals are often necessary to demonstrate the effects of particular PSMs on diet selection and physiology (Dearing et al. 2005). These captive studies have been essential for generating specific hypotheses that researchers can then test under field conditions. Studies with captive vertebrate herbivores have found that animals will reduce their intake of or avoid feeding on plants or plant parts with increasing concentrations of particular PSMs, extend their feeding bouts or increase time between feeding bouts to allow for detoxification (Wiggins, McArthur, and Davies 2006; Wiggins, McArthur, et al. 2003), and mix food sources to control for detoxification limitations (K. Marsh, Wallis, Andrew, et al. 2006). Although there have been a number of excellent captive studies on primate nutritional ecology (e.g., digesta passage), many zoo-based studies of primates can work with only a limited range of diets.

A strength of primatology is the success of field studies in obtaining highly detailed and long-term data on the behavior and ecology of individual animals and groups, though conversely, ethical and conservation considerations particular to primate research have largely prevented the study of diet selection and digestive physiology of captive primates (for exceptions, see Espinosa-Gómez, Gómez-Rosales, et al. 2013; Glander and Rabin 1983; Milton, Van Soest, and Robertson 1980). Restricting investigations to observational field studies restricts studies to investigations of correlations between preferences and broad groups of chemical constituents, rather than examining the effects of individual PSMs or demonstrating predictable patterns, causal links, or mechanisms driving feeding choice.

Two studies of the trade-off between predation risk and dietary PSMs show the possibility of manipulative experiments with free-ranging primates. Emersen and colleagues (2015) conducted giving-up-density experiments with wild samango monkeys (*Cercopithecus* [*nictitans*] *mitis erythrarchus*) in South Africa to test whether toxins added to experimental feeding stations reduced the time animals were willing to spend at each station and whether there was an interaction between toxin concentration in foods and perceived predation risk. The authors found that while tannins deterred feeding (monkeys gave up feeding sooner with increasing concentrations of tannins), oxalates had a greater impact on feeding, particularly if the feeding station was in a "risky" location. A similar result was found in wild bushbabies (*Otolemur crassicaudatus*) with the monoterpene 1,8-cineole added to foods at risky and nonrisky locations (McArthur, Orlando, et al. 2012). These studies show that negative effects of PSMs may compound with other ecological risk factors such as predation to create "hazard zones" for primates in their natural habitat (Emerson and Brown 2015).

Experimental studies of taste discrimination in primates have helped reveal interesting interactions between plant compounds and behavior. However, although taste may play some role in diet selection (Melin and Veilleux, this volume), the mechanisms underlying feeding choices and the consequences for wild animals remain unclear. For example, Simmen and colleagues showed that *Lemur catta* are very sensitive to sweet and tannic tastes—that is, they respond to (accept) low concentrations of fructose and reject low concentrations of tannic acid—compared to other primate species (Simmen, Peronny, et al. 2006; Simmen, Sauther, et al. 2006). In contrast, the lemurs were less sensitive to bitter tastes than other primates and rejected quinine (a bitter-tasting compound) at relatively high concentrations. The authors found no difference in PSM concentrations (tannins, phenolics, or alkaloids) between eaten and uneaten items and also found that alkaloids were widespread at the site and in the lemurs' diet. The authors hypothesize that a low bitter taste sensitivity reflects an adaptation of *L. catta* that allows them to ingest food with bitter qualities (such as alkaloids and some cyanogenic glucosides or saponosides) in quantities that may be toxic to other species and that this may assist lemurs in coping with the marked seasonal variations in food availability and nutritional quality seen throughout southern Madagascar (Simmen, Peronny, et al. 2006). However, bitterness has been shown to correlate poorly with toxicity (Glendinning 1994), and preferences for bitter compounds are easily learned in the absence of negative feedback such as toxicity (Launchbaugh et al. 1993).

Interestingly, the strength of primate field studies but the paucity of captive studies places primatologists in a converse situation from other fields of vertebrate nutritional ecology, where the number of captive studies tends to far outweigh the number and quality of field studies (DeGabriel, Moore, Felton, et al. 2014). The detailed field methods used in primatology and the extensive baseline data on primates set an example to which nutritional studies of other free-living vertebrates can aspire. The extensive data available on primate natural history and behavior are not equaled by other fields of animal ecology and mean that primatologists can achieve a much more detailed level of inquiry into free-living animal nutrition than in any other vertebrate group.

Identifying the Chemistry behind Ecological Observations of Feeding

Careful observations of feeding preferences of animals in the wild are the foundation on which ecologically significant chemistry can be discovered. When clear preferences against certain species or individual plants are noted, be they perennial or seasonal, the presence of deterrent PSMs should be considered. However, in any natural system, where there are many possible food choices, a plant may not be eaten for a variety of reasons. Eliminating these possibilities and determining whether PSMs are responsible requires careful evaluation, and the first step is often to reduce the choices available through captive feeding.

The plant-herbivore systems where PSMs have been demonstrated to influence the feeding ecology of wild vertebrate herbivores tend to have one thing in common: they are focused on animal species where captive feeding experiments that manipulate the diet of animals are possible. Examples include studies of snowshoe hares in Alaska that identified pinosylvin and its methyl ether as the key compound in the defense of juvenile green alder and papyriferic acid in paper willow (Bryant et al. 1983; Clausen et al. 1986; Reichardt et al. 1984), studies of marsupial folivores that identified formylated phloroglucinol compounds (e.g., macrocarpal G and sideroxylonal A) as the elusive key defense compounds in the majority of *Eucalyptus* (Lawler, Foley, et al. 1998; B. Moore, Wallis, Pala-Paul, et al. 2004; Pass et al. 1998), and studies of ruffed grouse (*Bonasa umbellus*) where coniferyl benzoate was shown to strongly influence the consumption of aspen buds (Jakubas and Gullion 1990).

In all examples given above, the compounds were identified by bioassay-guided fractionation, wherein sequential extracts of plant parts that are eaten or not eaten are fed to captive animals and their preferences noted until such time as a specific compound can be purified and its antifeedant actions confirmed. Isolation and purification of single compounds by repeated fractionation is the classical approach of natural products chemistry and is a very powerful approach to identify antifeedant compounds. At each step, we ask animals to express their preferences by measuring, for example, how much food is consumed in either choice or no-choice tests. Often, only a few solvent partitioning steps are required to identify likely deterrent principles.

When specific compounds are identified as being deterrents to feeding, it opens the door to a wide range of follow-up studies including structure-function assays (Lawler, Eschler, et al. 1999), an exploration of how the intake of compounds is regulated physiologically (Lawler, Foley, et al. 1998; K. Marsh, Wallis, McLean, et al. 2006; Torregrossa and Dearing 2009) and the specific effects of compounds on different groups of animals (L. Jensen, Wallis, Marsh, et al. 2014).

Of course, this approach is not always successful in identifying specific candidate compounds, and there may be many issues to consider, including:

1. Compounds that are unstable in the solvents and are thus lost from extracts.
2. Systems where deterrent and nondeterrent compounds are correlated phenotypically. For example, in *Eucalyptus*, the foliar concentration of the monoterpene 1,8-cineole is correlated with the concentration of formylated phloroglucinol compounds (Lawler, Eschler, et al. 1999; Pass et al. 1998). This may be partly because of colocation of biosynthetic pathways in the genome or shared elements of biosynthesis (Henery et al. 2007). Formylated phloroglucinol compounds are deterrents to marsupials at low concentrations, whereas cineole is not a deterrent at even artificially high concentrations (Lawler, Stapley, et al. 1999).
3. The context of the captive feeding experiments, where nutritional factors such as protein concentration and extrinsic influences including temperature may be poorly controlled (Dearing 2013).
4. The loss of complex interactions between multiple compounds that may interact additively or synergistically in the bioassay (L. Jensen, Wallis, and Foley 2015).

The last point is perhaps the most important limitation of the fractionation approach. Since most of the sample is discarded, important data on the biological complexity of the actual mixture are lost. These types of effects are particularly well demonstrated in animal pheromones (Linn et al. 1986).

METABOLOMICS—PLAYING TO THE STRENGTHS OF PRIMATE FIELD STUDIES

In contrast to the classical fractionation approach, metabolomics describes a more comprehensive chemical fingerprinting that can provide significant chemical information and even identify specific antifeedant structures (Robinette et al. 2011). In this protocol, the complexity of the mixture is maintained rather than being discarded as is the case with the fractionation approach.

Metabolomics (see O. Jones et al. [2013] for a clarification of the sometimes confusing terminology used in the field) is a broad name given to comprehensive chemical profiling of samples of plants or animals based typically on nuclear magnetic resonance (NMR) spectra or gas or liquid chromatography coupled with mass spectrometry (GC-MS or LC-MS). Although this is often purveyed as a new branch of science, it is essentially the high-throughput application of standard chemical techniques to large numbers of samples combined with common statistical methods (e.g., principal component analysis and partial least squares regression) that allow the complexity of the spectral or chromatographic information to be reduced so that similarities and differences can be identified.

The application of metabolomics to ecology is expanding, and we refer readers to several reviews, which cover the topic in more depth than we are able to in this chapter

(O. Jones et al. 2013; Kuhlisch and Pohnert 2015; Macel et al. 2010; Sardans et al. 2011). Kuhlisch and Pohnert (2015) describe three broad approaches that can be used in studies of ecological chemistry: the comparative metabolomics approach, the systems biology approach, and the comprehensive metabolome approach.

The comparative metabolomics approach, wherein samples that are eaten or rejected by animals are compared, is by far the best suited for studying chemically mediated interactions among species such as between primates and their food plants (Kuhlisch and Pohnert 2015; Sardans et al. 2011). The stronger the ecological basis of the comparison, the more likely that the comparative metabolomics approach will be successful. The comparative approach allows the tissues and targets to be refined via prior knowledge, but it follows that in this approach efforts should be made to ensure that other potential influences are controlled (e.g., sun vs. shade leaves, maturity of the plant, and time of collection). In contrast to the comparative metabolomics model, the systems biology approach aims to comprehensively investigate the metabolome along with the transcriptome or genome to get the best possible understanding of the complex phenome. The comprehensive metabolome approach is more focused on a broad understanding of biochemical regulation, particularly in model organisms (Kuhlisch and Pohnert 2015)

Many of the metabolomic techniques used in ecology have been derived from other areas of science, including human medicine and plant science, and have focused on genome-phenome studies of crops and model organisms. Although the standardization of analytical techniques is essential (e.g., H. Kim et al. 2010), it may not always be possible to adhere to the ideal sample collection conditions when working with forest trees or in remote locations. For example, a strict collection protocol was drawn up for us by a metabolomics expert who was used to working with *Arabidopsis*. This involved freezing intact leaves in liquid nitrogen before they were cut from a tree and sampling all trees at the same time of day. This is clearly impractical at almost all field sites where primates are studied. It is simply not possible to climb multiple trees at the same time of day with a flask of liquid nitrogen!

Whether deviation from these conditions is important depends entirely on the aim of the study, and we believe that few if any metabolomic studies of primate foods in the immediate future would suffer. For example, if the aim were to understand whether contrasts between foods that are eaten and not eaten are associated with particular PSMs, then collecting freshly cut leaves from branches within a few minutes would be acceptable. Similarly, in many instances, well-preserved dried tissue would be adequate for studying these kinds of questions. However, if the question focused on diurnal variations in particular sugars in a leaf, then more stringent protocols would be needed.

While we do not intend to describe protocols for all purposes, ecologists should not be deterred because of seemingly impossible published sampling protocols. Nonetheless, it is essential to describe exactly how the samples were prepared and to be aware of possible effects of sampling protocols in interpreting results.

The merits of different metabolomic methods have been debated in the literature. Approaches based on NMR spectroscopy and on GC-MS or LC-MS all have advantages and disadvantages.

NMR experiments provide data on the structure of compounds and can allow unknown compounds to be identified from first principles. The major disadvantage is their low sensitivity, though this can be overcome with modern high-field NMR instruments, which are capable of discriminating small concentrations of many compounds and can be automated for high throughput. In contrast, GC-MS and LC-MS for metabolomics has traditionally relied on spectral libraries to identify structures, but this has limited utility in diverse forests.

However, recent advances in bioinformatics have allowed plant metabolomes (based on NMR, LC-MS, or LC-MS/MS [tandem mass spectrometry] data) to be assembled into molecular networks based on their chemical structural similarity (Richards et al. 2015; Wang et al. 2016). These approaches have been highly productive in quantifying chemical similarity and identifying chemical differences between plants in complex communities even though relatively few compounds can be definitely identified (Sedio 2017). Metabolomics combined with molecular networks is answering long-standing questions about the chemical ecology of forest communities (e.g., Massad et al. 2022; Sedio et al. 2019), and we believe that these approaches could also be applied to questions in primate feeding.

Metabolomic methods may work best when there is a clear contrast to test (box 9.2). In controlled experiments, this might be different treatments that are applied, different

BOX 9.2 EXAMPLE OF A SIMPLE, UNTARGETED METABOLOMICS APPROACH TO IDENTIFYING ECOLOGICALLY RELEVANT CHEMISTRY IN A MAMMALIAN ARBOREAL FOLIVORE.

The metabolomics approach has been used to successfully identify plant antifeedant compounds in studies of marsupial folivore consumption of *Eucalyptus*. K. Marsh, Foley, and colleagues (2003) and later B. Moore, I. Wallis, K. Marsh, and W. Foley (2004a) summarized data that showed that some species of marsupial fed little from one of the major groups of *Eucalyptus*, the subgenus *Monocalyptus* (= subgenus *Eucalyptus*), whereas other species fed predominantly from this group. Tucker and colleagues (2010) hypothesized that this was because of the presence of an unknown antifeedant in leaves of monocalypts.

NMR spectra collected from different subgenera of *Eucalyptus* revealed clear chemical differences. The NMR spectra showed that a major signal in the dichloromethane extract of monocalypts was absent from the symphyomyrtles. Additional 2D NMR experiments suggested that these were likely to be flavanones with unsubstituted B-rings. This group of compounds are known to be biologically active in several other species of animals (e.g., Napal et al. 2010), and subsequent bioassays with captive common brushtail possums (K. Marsh, Yin, et al. 2015) demonstrated that this class of compound has significant antifeedant effects in common brushtail possums that feed sparingly on the *Monocalyptus* subgenus of *Eucalyptus*. Importantly, knowing the relevant chemistry allows variants to be explored and the chemical basis of deterrence to be defined. In this case (fig. A), it is the lack of substitution on the B-ring of the flavanone structure that is deterrent, whereas compounds with substitutions on the B-ring or unsubstituted flavones are not.

Had this comparative metabolomics approach been applied to eucalypts many years previously, it would also have readily identified the FPC compounds that are the most important in the other major group of eucalypts, the subgenus *Symphyomyrtus* (B. Moore, I. Wallis, J. Pala-Paul, J. Brophy, R. Willis, and W. Foley 2004). Bioassays consistently showed strong preferences for feeding between different individual trees of the same species, but the chemical principles were unknown. The role of FPC compounds as feeding deterrents was instead discovered following extensive bioassay-guided fractionation. The implications of the strong phylogenetic differences between the occurrences of different classes of vertebrate antifeedant compounds would have driven hypotheses about niche separation much earlier had the metabolomic approach been used from the outset.

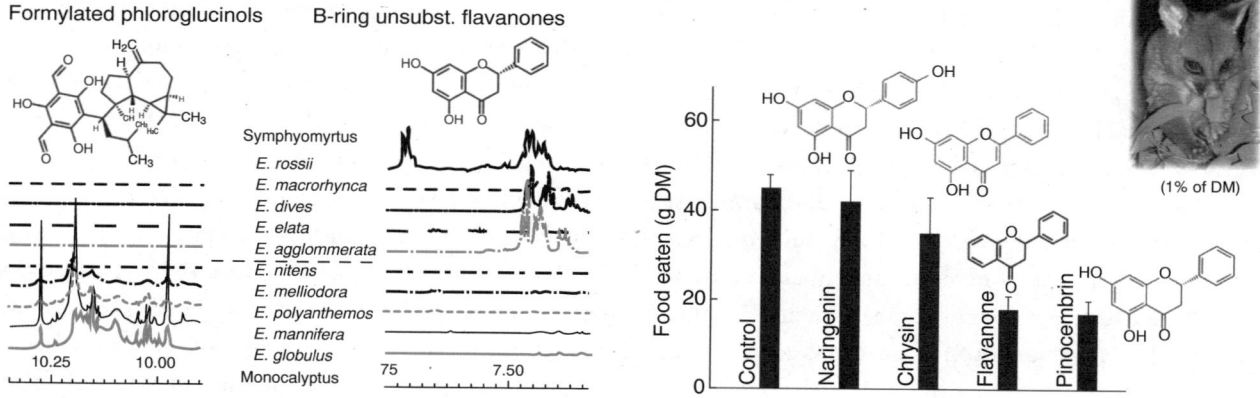

Figure A. Detection and bioassay of B-ring unsubstituted flavanones as feeding deterrents against common brushtail possums. *Left*, ¹H NMR spectra of 10 *Eucalyptus* species, five from each of the subgenera *Symphyomyrtus* and *Eucalyptus* (*Monocalyptus*); scale 1:0.025. Compounds that are characteristic of these signals (i.e., the formylated phloroglucinol compound macrocarpal G and the B-ring unsubstituted flavanone pinocembrin) are shown. *Right*, summary of results of feeding common brushtail possums on a basal diet supplemented with 1% DM of narigenein (substituted B-ring substituted flavanone), chrysin (B-ring unsubstituted flavone), and flavanone and pinocembrin (B-ring unsubstituted flavanones). Data derived from Tucker et al. (2010) and K. Marsh, Yin, et al. (2015).

disease states, or significantly different phenotypes. However, in studies of wild species, we have to rely on good observations of animal feeding to provide the contrasts. This is a strength of primate field studies where feeding data are collected meticulously, over multiple seasons, and in multiple groups. Therefore, contrasts can be made between plant parts that are eaten and not eaten or between species or individual trees that are used and never used as food. Undoubtedly, many of these contrasts may be due not to distinct chemical differences but perhaps to nutritional variables or local ecological effects (predators, physical defenses, or effects of conspecifics). However, a more holistic approach to chemical characterization must be the aim of future research.

METABOLOMICS USING SPECTRAL DATA ALONE

We have argued that a major limitation to primate nutritional ecology studies has been a lack of knowledge of specific small molecules that appear to modify feeding by wild primates. We have described metabolomic approaches to doing so and shown a simple example where an antifeedant compound was identified from untargeted metabolomics. However, there are cases where metabolomics may be useful without analysis of the underlying chemistry.

All who have worked in the field of primate dietary ecology know that one of the most important determinants of whether a food will be extensively eaten is the interaction between various nutrients and various secondary metabolites. But rarely, if ever, can we measure all these different components accurately, and even then, it is difficult to know how to combine them. Elsewhere, we have described some of the problems with the ratio methods that have often been used to characterize primate foods (I. Wallis et al. 2012). These include protein-to-fiber ratios (Chapman, Chapman, Naughton-Treves, et al. 2004) and the ratio of protein to some combination of fiber and tannins (Oates, Whitesides, et al. 1990).

An alternative approach is to use the metabolomic data (as coded in spectra) of different foods in place of analyses of each component individually. Above, we have described NMR spectra of primate foods, but other spectroscopic methods such as near-infrared reflectance spectroscopy (NIRS) are also useful. NIRS has been used in the ecology of plant-herbivore interactions for many years (DeGabriel, Moore, Felton, et al. 2014; W. Foley et al. 1998; Landau et al. 2006) and is really just another metabolomic method with a strong statistical underpinning. NIRS is an indirect method of analysis that relies on establishing the relationship between the near-infrared spectrum (~700–2,500 nm) of a set of plant tissues and a set of laboratory reference values (e.g., protein concentration). This "calibration equation" is then used to predict the reference values from the spectrum of additional "unknown" samples. That these relationships can be derived for multiple chemical constituents (both primary and secondary metabolites) implies that the near-infrared spectra contain information about the composition of the plant. Several studies have shown that the near-infrared spectrum of primate foods is an extremely valuable way of characterizing their nutritional quality (Rothman, Chapman, Hansen, et al. 2009).

The same is true in studies of marsupials (McIlwee et al. 2001), but B. Moore and colleagues (2010) also showed that the NIR spectrum was better able to predict a standardized measure of feeding rates in koalas than multiple measurements of chemical composition. Thus, the NIR spectrum captures data about multiple chemical components and interactions between them. Whether the same would be achievable with NMR or LC-MS spectra has not been tested, but, in our opinion, this is highly likely. NIRS has the advantage of speed, lack of sample preparation, and low cost, but a combination of different spectral approaches could be very powerful. Knowledge of the underlying chemistry does provide novel insights, but using spectral data without necessarily linking directly to chemistry may be a useful approach when the questions are broad, as they often are in ecology. A recent review of NIRS in wildlife and biodiversity research is a useful starting point for those considering such an approach (Vance et al. 2016).

BEYOND METABOLOMICS

Other "omics" approaches might be able to be combined with metabolomics to help understand patterns of diet selection in primates. The increasing availability of genomic data for a wide range of species offers the possibility that comparative genomics may be able to provide insight into the mechanisms of detoxification of PSMs in primates.

Consider, for example, the well-known human enzyme cytochrome P450 (CYP) 2D6 (CYP2D6). This enzyme has a very strong affinity for alkaloids (Fonne-Pfister and Meyer 1988) and metabolizes a large proportion of human drugs. Variation in the gene coding for CYP2D6 has been extensively studied in humans, and four phenotypes are recognized: poor (PM), intermediate (IM), efficient (EM), and ultra-rapid (UM) metabolizers. The reasons for these differences are a combination of allelic differences that result in PMs having no functional CYP2D6 enzyme and UMs having up to 13 copies of the genes, allowing them to rapidly metabolize alkaloids and related medicines (Yasukochi and Satta 2015).

The evolution of CYP2D6 has been studied in 14 nonhuman primates (Yasukochi and Satta 2015). These authors found that purifying selection and gene conversion have affected genes paralogous to human CYP2D6 in primates. They further suggest that the number of copies of CYP2D6 in macaques (*Macaca mulatta* and *M. fascicularis*) and white tufted-ear marmoset (*Callithrix jacchus*) suggest that these primates likely metabolize toxins more efficiently than do humans. This example illustrates the insights that are possible from a comparative genomics approach and how this approach is likely to prove more useful as sequence data from more wild primates are made available in future. A genomic approach might provide confirmatory data on the apparent indifference of some species of lemurs to plants rich in alkaloids (Ganzhorn 1988). While metabolomic approaches are increasingly being applied to primate feeding (Amato, Ulanov, et al. 2017; Garber et al. 2019), there has been no specific focus on PSMs to date.

COUNTERMEASURES AGAINST PLANT SECONDARY METABOLITES

If PSMs are common in the diets of many primates, then we might expect animals to have a series of countermeasures that enable limited ingestion without significant deleterious consequences. All mammals have the capacity to detoxify (in the broadest sense of the word) significant xenobiotics whether or not these are regularly encountered, and there is no reason to think that the methods that primates use to do so are any different. Many specific mechanisms have been suggested and are undoubtedly involved in mediating toxic effects, but evidence is currently limited. For example, the forestomach of colobine monkeys is widely regarded as a site where metabolic changes can occur that reduce the toxicity of ingested PSMs. Evidence for a similar function of the ruminant forestomach is equivocal (Dearing et al. 2005), but advances being made in functional metagenomics may soon provide a way to test these ideas (e.g., Kohl and Dearing 2012). Kohl and Dearing (2016) also demonstrated that the detoxification function of the foregut of woodrats (*Neotoma* sp.) was transferable between congeners and even across distantly related species. Similarly, although drug metabolism has been extensively studied in biomedical primate models, we are not aware of examples where these concepts have been transferred to wild primates to understand how, for example, limits to detoxification might constrain feeding.

TANNIN-BINDING SALIVARY PROTEINS

One countermeasure that has received significant attention is the presence of tannin-binding salivary proteins (TBSPs). TBSPs are a group of proteins produced in the saliva of most mammals that can be effective precipitators of some tannins (McArthur, Sanson, and Beal 1995; Shimada, Saitoh, et al. 2006).

The most common components of TBSPs are proline-rich proteins (PRPs). TBSPs have been detected in many species of wild mammals, including several species of primates—howler monkeys (*Alouatta palliata*; Espinosa-Gómez, García, et al. 2015), macaques (*Macaca fascicularis*; Ann and Lin 1993; Bennick 2002; Sabatini et al. 1989), and Hamadryas baboons (*Papio hamadryas*; Mau, de Almeida, et al. 2011; Mau, Südekum, et al. 2009)—but relatively few studies have demonstrated functional significance in the nutritional ecology of the animals (McArthur, Sanson, and Beal 1995; Shimada, Saitoh, et al. 2006).

The assays that are used to detect TBSPs are usually based on the disappearance of protein bands from gels after mixing with a tannic acid solution (P. Austin et al. 1989). Although this can reveal proteins with affinities for the substrate in the assay (usually tannic acid), this does not necessarily predict binding by the types of tannins naturally present in food plants (Hagerman and Robbins 1993). Mole and colleagues (1990) concluded that several species in their survey produced TBSPs that are inactive against the tannins found in representative food plants of those species. This was very much unexpected.

A second issue is that for TBSPs to be functionally significant, they need to be secreted at a high rate and not just be shown to be present in high concentrations (McArthur, Sanson, and Beal 1995). Although the identification of TBSPs is a necessary first step, it is not sufficient in itself to conclude that TBSPs serve a major ecological role because it captures data from a single moment in time. Measuring secretion rates of salivary proteins is difficult to do with wild species (but see Espinosa-Gómez, Serio-Silva, et al. 2018).

Although we believe that it is likely that TBSPs do provide some degree of first-line defense against deleterious effects of dietary tannins in at least some primates, the evidence for a significant relationship with diet remains weak. This is largely because of the difficulties of obtaining the necessary data on saliva composition and production rates. However, an alternative, noninvasive approach suitable to field studies was developed for use with forest rodents (Shimada, Nishii, and Saitoh 2011), which may be better for primatological research.

Complexes between tannins and proline-rich proteins are stable in the gut and excreted in the feces (Skopec et al. 2004). Shimada, Nishii, and Saitoh (2011) were able to show a strong correlation between dietary tannin intake and fecal proline using a simple ninhydrin assay and could use this to estimate tannin intake in the field. Although Shimada and colleagues discuss several limitations and caveats of the technique, it seems likely that this type of approach combined with observations of feeding would allow the collection of longitudinal data on the importance of TBSPs in primates in the field. In many instances, feces are readily collected from focal individuals (e.g., T. Gillespie 2006; Rangel-Negrín et al. 2009), and the assays of proline richness are relatively simple. We believe that fecal proline excretion could prove a better index of variations in TBSPs and their correlation with different diets than continuing to measure putative concentrations of TPSPs directly from very limited saliva samples of primates.

BENEFICIAL IMPACTS OF PLANT SECONDARY METABOLITES— SELF-MEDICATION AND BEYOND

Plant secondary compounds encompass a huge variety of structures, activities, and effects on animals. The potential for harmful effects on animals is by far the major focus of most scientists—that is, animals should limit their ingestion of PSMs to avoid a wide range of negative effects. However, the long history of the human use of plants for medicinal or other indirect therapeutic effects attests to the potential beneficial effects of some PSMs. The "dose makes the poison" (Stumpf 2006), and few if any compounds are invariably toxic or invariably beneficial (S. Simpson and Raubenheimer 2012). A good example is the occurrence of PSMs that can bind with estrogen receptors and thus produce effects similar to those of steroids produced by the endocrine system. These are commonly called phytoestrogens. Wasserman, Milton, and Chapman (2013; see also Wasserman et al., this volume) have recently reviewed the literature on the potential interactions of primates with phytoestrogens. Wasserman, Milton, and Chapman (2013) highlight the two opposing views that phytoestrogens provide health benefits or that they can act as endocrine disruptors. Field studies correlating behavior with the consumption of plants having estrogenic activities (e.g., Wasserman, Chapman, Milton, et al. 2012) suggest that these compounds may have significant effects on primate behaviors, but clearly there is more to discover.

Primatology has contributed significantly to concepts of self-medication in wild mammals, and the topic has been well reviewed in several places in the past few years (Forbey, Harvey, et al. 2009; Huffman 1997; Villalba and Provenza 2007); we refer readers to these sources. Although experimental tests of possible self-medication in primates are difficult, the accumulated observations together with long-term cultural transmission of these behaviors argue strongly for a therapeutic function, particularly in the case of apes (Huffman 2015), but, as Huffman (1997) cautions, greater efforts need to be made in experimental testing. Certainly, where specific secondary metabolites have been isolated and shown to have appropriate dose-related effects, the evidence is stronger. Related behaviors such as geophagy or the consumption of charcoal may also have medicative functions (e.g., Pebsworth, Hillier, Wendler, et al. 2019b; Ta et al. 2018), but these need not be exclusive, and several reviews demonstrate the multifactorial nature of ingesting soil (Krishnamani and Mahaney 2000; Pebsworth, Huffman, Lambert, et al. 2019a).

The challenge for nutritional ecologists, whether the

focus is primatology or other systems, is to start to bring our existing concepts of potentially toxic compounds, medicines, and nutrients into a single framework. As Raubenheimer and Simpson (2009) point out, the distinctions between these categories are vague and poorly defined, and the strong interdependence between these different parts of foods argues for a more integrative treatment (the need for an improved framework is fully characterized for primate nutrition in general in Raubenheimer's chapter in this volume). For example, several studies show that the consumption of some secondary metabolites is enhanced when animals have access to adequate protein in the diet (O. Jones et al. 2013; Villalba and Provenza 2005). Thus, is protein a nutrient or a medicine? Villalba and Provenza (2007) argue that nutrients and medicines are both means to the same end—"staying well"—and their ingestion is a natural part of regulated homeostatic behavior in animals. Animals learn either through experience or from peers how to incorporate diverse foods into their diets to serve all these functions. Integrating these concepts is not easy, but where potentially toxic compounds and medicines are involved, understanding the chemical structures and their concentrations and bioavailability is critical to determining their roles.

SUMMARY AND CONCLUSIONS

- This chapter presents a broad view of secondary metabolites in primate diets with examples and suggestions for future directions drawn from studies of marsupials and rodents, as well as primates.
- We reviewed previous published literature of primate nutrition, focusing on studies that statistically evaluated the influence of PSMs on feeding preferences of individual primate species. We found that half of the statistical studies detected a significant influence of PSM concentrations (either condensed tannins, total phenolics, or unspecified alkaloids) in plant parts on the feeding preferences of individual primate species. We also found that the effects of different PSMs on feeding varied among primate groups and species, when primates feed on different plant parts and plant species, and that preferences even varied for different local populations of the same primate species.
- We highlight reasons why we think primate studies have not found a consistent relationship between PSMs and primate feeding behaviors and ecology. In particular, we consider that various field, laboratory, and statistical approaches used by different authors prevent us from drawing general conclusions from the studies and that the paucity of studies of primate feeding under controlled captive conditions has hampered our ability to generate testable hypotheses for individual species.
- We suggest alternative approaches that will allow the identification of relevant chemistry, which builds on a renewed interest in the plant metabolome and a broader appreciation of the beneficial effects of diets rich in some groups of PSMs. We believe that there are many opportunities for primatologists to engage in productive collaborations in this area.
- We hope to see renewed interest in a broader approach to questions about the role of PSMs in primate nutritional ecology that draws on studies in agriculture and other mammal-plant systems and harnesses the new technologies of metabolomics and ultimately genomics. Cross-fertilization of ideas between primatology and other systems can be hugely beneficial, but we believe that this has not been as strong in recent years as it once was. As an example, consider the impact that the 1978 book *The Ecology of Arboreal Folivores* (Montgomery 1978) had on studies of primates and other folivores and the fact that it continues to be regularly cited 40 years later.

10 Hormonally Active Phytochemicals in Primate Diets
Prevalence across the Order

Michael D. Wasserman, Marie-Lyne Després-Einspenner, Richard Mutegeki, and Tessa Steiniche

Food matters. Nutrition matters. However, how much of what you put into your mouth is simply for nutritional content? What about coffee, tea, beer, wine, and spices? At a planetary scale, how much land is used to grow plants not for the protein, energy, fats, minerals, or vitamins they contain but rather for nonnutritive molecules that affect physiology? As humans, we seek out these nonnutritive phytochemicals for their ability to change physiology or behavior, as medicine or for recreational or mind-altering effects, but what about other primates?

Most primate species are highly dependent on plant foods for meeting energetic and nutritional demands critical to survival, growth, development, and reproduction (Chapman, Chapman, Naughton-Treves, et al. 2004; Irwin, Raharison, Chapman, et al. 2017; Milton 1979, 1993, 2003b; Rothman, Raubenheimer, and Chapman 2011). However, plants are chemical powerhouses producing at least 100,000 other plant secondary metabolites (PSMs), chemicals produced by plants that have no known role in their growth and development and are often utilized as a defense against herbivory (Harborne 1988, 2000; Lambers et al. 1998). From the primate consumer's perspective, these PSMs, although ingested, may or may not have significant biological effects at the doses consumed and may or may not be avoided or preferred based on their costs and benefits. Due to the complexities of PSM effects on primate biology, including cost-benefit trade-offs of ingestion, interactions with nutrients and other nonnutritive molecules, and the effects of dose, concentration, and timing of ingestion in determining whether a molecule is toxic or beneficial, the traditional dichotomy of nutrients and PSMs oversimplifies the food-medicine continuum (Forbey and Foley 2009; T. Johns 1999; Raubenheimer and Simpson 2009). All molecules can be toxic at high enough doses, while only a few may be beneficial, including all classic nutrients and a subset of the thousands of PSMs. As the classic saying goes, the dose makes the poison (or drug).

Many PSMs can trigger direct changes in the physiology of the consumer. For primates, PSMs with known biological effects include alkaloids, tannins, cyanogenic glycosides, saponins, and phytosterols, among many other categories of plant compounds (Chapman and Chapman 2002; Harborne 2001; Wasserman, Milton, and Chapman 2013). Ingestion of these compounds comes with trade-offs, such as detoxification costs when dealing with plant defenses and self-medication benefits when dealing with parasites, and has been examined across many species, including lemurs, colobine monkeys, and chimpanzees (Chapman and Chapman 2002; W. Foley and Moore 2005; Forbey, Dearing, et al. 2013; Freeland and Janzen 1974; Glander, Wright, Siegler, et al. 1989; Huffman 2003; Krief, Huffman, et al. 2006; Oates, Waterman, and Choo 1980; Rothman, Dusinberre, and Pell 2009; Waterman, Ross, et al. 1988; Wrangham, Conklin-Brittain, and Hunt 1998).

In this chapter, we focus on a specific subset of PSMs: hormonally active phytochemicals, or naturally occurring plant compounds that alter physiology or behavior via interactions with the neuroendocrine system. The vertebrate neuroendocrine system, which is highly conserved across fish, amphibians, reptiles, birds, and mammals, modulates physiology and behavior by releasing endogenous hormones and neurotransmitters that interact with receptors across the body and brain (Hadley 2000; R. J. Nelson 2011). Hormonally active phytochemicals can mimic or block the activity of these hormones and neurotransmitters upon ingestion, thus causing physiological and behavioral changes through hormone-dependent gene transcription and translation, nongenomic pathways, or neuronal activity.

To summarize what is currently known about primate interactions with these hormonally active phytochemicals and the significance of these interactions for primate ecology, evolution, and conservation, we address three main questions:

1. How prevalent are hormonally active phytochemicals across primate diets?
2. What are the physiological and behavioral effects of consuming various hormonally active phytochemicals?
3. How do primate interactions with hormonally active phytochemicals vary by phylogeny, geography, and dietary niche?

We address question 1 by examining the diversity of known hormonally active phytochemicals and their presence across various plant families commonly consumed by primates. We address question 2 by exploring the range of mechanisms by which these compounds interact with the primate endocrine system, the effects of these interactions on primate physiology and behavior, and the overall significance of hormonally active phytochemicals in primate feeding ecology. We address question 3 by reviewing the primate literature for studies of hormonally active phytochemicals (table 10.1) and present recent data from our research on the presence of such compounds in the diets of the primate community in Kibale National Park, Uganda (table 10.2). We further assess the potential for phytochemical consumption in other primate species based on an analysis of the proportion of dietary items from each plant family across primates (tables 10.3 and 10.4). We end the chapter by presenting future research directions that will connect the influence of hormonally active phytochemicals on primate physiology and behavior with anthropogenic chemical pollutants, landscape analyses of land use, and the microbiome.

CATEGORIES OF HORMONALLY ACTIVE PHYTOCHEMICALS

Hormonally active phytochemicals have been identified across more than 300 plant species in 32 plant families (Dixon 2004; Reynaud et al. 2005). Known hormonally active phytochemicals include those that can directly interact with the thyroid hormone axis, having implications for growth and metabolism; others that act as steroid hormones (i.e., estrogens, androgens, progestins, and glucocorticoids), thus affecting the hormonal regulation of reproduction, behavior, and the stress response; and those that interfere with neurohormones such as serotonin, dopamine, acetylcholine, and endocannabinoids.

Among the most studied hormonally active phytochemicals that primates encounter in their diets are phytosteroids. Phytosteroids are naturally occurring plant molecules that, upon ingestion, can either promote or block the activity of endogenous vertebrate steroid hormones, such as estradiol, cortisol, testosterone, or progesterone. Bioactivity can occur if the phytosteroid directly binds to hormone receptors or affects hormone synthesis, metabolism, availability, or activity through effects on enzymes or other cellular processes (V. Beck et al. 2003; Benie and Thieulant 2003; Rosenberg et al. 1998). Usually, phytosteroids have vertebrate hormone activity because they are similar in structure and able to bind to hormone receptors. For example, phytoestrogens are plant compounds that enhance or block estrogenic activity in vertebrates, often by binding to estrogen receptors. Vertebrate estrogens play a major role in female growth and development as well as reproductive physiology and behavior, but they serve important physiological and behavioral functions in both males and females, including fat deposition, fertility, sexual behavior, and aggression. Similarly, phytoprogestins are plant compounds with progestogenic activity, often acting on the progesterone receptor. Vertebrate progesterone plays a main role in preparing the reproductive tract for implantation and maintaining pregnancy, while its role in males includes sperm production (Mannowetz et al. 2017). Phytoandrogens, plant compounds with androgenic activity in vertebrate organisms, often exhibit bi-

Table 10.1. Summary of field studies that either examined or suggested the possibility of phytosteroid-containing plants to influence primate ecology (updated from Wasserman, Milton, and Chapman 2013).

Primate species	Plant species	Evidence
Modern human (*Homo sapiens*) (Chester et al. 2020; P. Whitten and Patisaul 2001; K. Wynne-Edwards 2001)	Many, e.g., soy (*Glycine max*; Fabaceae)	Ethnobotanical, hormonal (serum), in vitro and in vivo assays, phytochemical, epidemiological, clinical
Common chimpanzee (*Pan troglodytes*) (Emery Thompson, Wilson, et al. 2008; McCarthy et al. 2017; J. Wallis 1997)	*Vitex fisheri* fruit (Lamiaceae)	Hormonal (urinary), behavioral, ethnobotanical for phytoprogesterones
Mountain gorilla (*Gorilla beringei*) (Wasserman, Taylor-Futt, et al. 2012)	*Ipomoea involucrata* leaves (Convolvulaceae)	Behavioral, transfection assay for plant estrogenic activity
Olive baboon (*Papio anubis*) (Higham, Ross, et al. 2007; Lodge et al. 2013)	*Vitex doniana* ripe fruit and young leaves (Lamiaceae)	Hormonal (fecal), behavioral, morphological, immunoassay for phytoprogesterones, ethnobotanical
Vervet (*Chlorocebus aethiops*) (Garey 1993; P. Whitten 1983b)	*Acacia elatior* flowers (Fabaceae)	Behavioral, in vitro bioassay for plant estrogenic activity
Red colobus (*Procolobus rufomitratus*) (Aronsen et al. 2015; Wasserman, Chapman, Milton, Gogarten, et al. 2012; Wasserman, Taylor-Futt, et al. 2012)	Many, but especially *Millettia dura* young leaves (Fabaceae)	Hormonal (fecal), behavioral, transfection assay for plant estrogenic activity, ethnobotanical
Phayre's leaf monkey (*Trachypithecus phayrei*) (Lu, Beehner, et al. 2011)	*Vitex* spp. fruit and leaves (Lamiaceae)	Hormonal (fecal), behavioral, ethnobotanical for phytoprogesterones
Muriquis (*Brachyteles arachnoides*) (Strier 1993)	*Enterolobium contortisiliquum* fruit (Fabaceae)	Behavioral, phytochemical for presence of stigmasterol
Kenyan galago (*Galago senegalensis*) (L. Nash and Whitten 1989)	*Acacia drepanolobium* gum (Fabaceae)	Behavioral, phytochemical for presence of flavonoids with weak estrogenic activity and antiestrogenic activity
Black and white colobus (*Colobus guereza*) (Benavidez et al. 2015)	Many, incl. *Erythrina abyssinica* young leaves and flowers (Fabaceae) and *Ficus sansibarica* unripe fruit and young and mature leaves	Behavioral, transfection assay for plant estrogenic activity

ological effects by either directly binding to androgen receptors or interfering with the functioning of aromatase, the enzyme that converts testosterone into estrogen. Vertebrate androgens, including testosterone, play a major role in male growth and development, including growth, the development of secondary sexual characteristics, and the initiation of sperm production, but they also serve important physiological and behavioral functions in both females and males, including hair growth, sexual motivation, aggression, and muscle growth. Several phytochemicals have been recognized for their antiandrogenic effects, including licorice (*Glycyrrhiza glabra*), red reishi (*Ganoderma lucidum*), and Chinese peony (*Paeonia lactiflora*), which achieve endocrine effects by reducing testosterone levels, reducing the enzymes that convert testosterone to its more biologically active form of dihydrotestosterone, and promoting the aromatization of testosterone into estrogen, respectively (Grant and Ramasamy 2012). Finally, phytoglucocorticoids are plant compounds with glucocorticoid activity in vertebrate organisms. Vertebrate glucocorticoids, including cortisol, regulate blood glucose by increasing gluconeogenesis, lipolysis, and protein catabolism, all of which increase blood glucose levels, and in doing so play a major role in the stress response by making energy available to deal with stressors (see further discussion in relation to the immune response in Vogel et al., this volume). For further details on the specific functions of steroid hormones, see Hadley (2000) and R. J. Nelson (2011).

PHYTOESTROGENS

As studies on phytochemicals increase, the number of compounds known to have endocrine effects continues to grow. In particular, estrogenic activity has been documented in more than 160 different plant compounds (Dixon 2004; Reynaud et al. 2005). Due to the major focus of research on phytoestrogens, this chapter focuses on this group of phytosterols. Classes of phytochemicals that have at least some specific molecules with demonstrated estrogenic activity include flavonoids, lignans, phytosterols, terpenoids, and the bioactive metabolites of fungi and bacteria (Dixon 2004; Harborne 1988). Many of these estrogenic compounds are phenolics, including flavonoids and lignans, which are ubiquitous across plants and may have important roles in allelopathy, such as the inhibition of seed germination of grasses and herbs, as well as herbivore defense (Lambers et al. 1998), while also acting as important antioxidants for delaying food spoilage and providing human health benefits (Shahidi and Ambigaipalan 2015). Phenolics are structurally defined as organic compounds consisting of an aromatic hydrocarbon ring with at least one hydroxyl group directly bonded to the carbon ring. Polyphenols are molecules consisting of two or more of these aromatic hydrocarbon ring–hydroxyl structures bonded together. There are many types of phenolic compounds, including tannins, which precipitate protein and interfere with food digestion and are thus considered one of most important feeding deterrents for primates, though this varies across primate taxa (Espinosa-Gomez, Serio-Silva, et al. 208; Lambers et al. 1998; Rothman, Dusinberre, and Pell 2009).

One particularly important set of phenolic compounds with estrogenic activity are the flavonoids. Flavonoids are ubiquitous in angiosperms, gymnosperms, and ferns, with over 4,000 identified compounds playing roles in plants such as nodule formation with nitrogen-fixing bacteria, pathogen defense, flower coloration, and deterrence of herbivore feeding (Harborne 1988; Lambers et al. 1998). Regarding human consumption, they are found in many whole grains (e.g., barley, millet, and rye), fruits (e.g., citrus, berries, apples, and grapes), vegetables (e.g., onions, broccoli, and celery), herbs (e.g., parsley and thyme), flowers (e.g., hops), seeds (e.g., cocoa), and beverages (e.g., wine and tea; Beecher 2003; Dykes and Rooney 2006). The main group of flavonoids with estrogenic activity are the isoflavones (Bucar 2013). Isoflavones include genistein, daidzein, biochanin A, and formononetin, are commonly found in legumes, and are well studied due to their presence in soy, clover, and alfalfa, thus having significance for human and livestock health and reproduction (N. R. Adams 1990; P. Whitten and Patisaul 2001). Despite evidence of negative effects on fertility in livestock, isoflavones have long been viewed as beneficial to human health through potential reduction of sex-hormone-dependent cancers, such as estrogen-dependent breast cancer and androgen-dependent prostate cancer, due to their hormonal activity (Lambers et al. 1998). Other well-studied estrogenic flavonoids include coumestrol, a coumestan also found in legumes such as alfalfa and clover, which shows greater estrogenic activity than the isoflavones through estrogen receptor binding but also exerts androgenic effects through the disruption of steroid-converting enzymes (C. Blomquist et al. 2005; Konar 2013). Furthermore, the prenylated flavonoids, found in hops (Cannabaceae), contain one of the most estrogenic phytochemicals consumed by humans, 8-prenylnaringenin (S. Milligan et al. 1999; Stevens and Page 2004).

There are also several documented nonflavonoid phytoestrogens, including lignans, phytosterols, terpenoids, and mycoestrogens. Lignans are compounds associated with plant cell walls and fiber and therefore mostly found in fiber-rich foods (Peterson et al. 2010). Lignan precursors, although not estrogenic when consumed, are found in the highest amounts in seeds (e.g., flaxseed, sesame seed, peanut, hazelnut, and walnut) but also in some whole grains (e.g., wheat, oat, and corn), fruits (e.g., apricot), vegetables (e.g., garlic, carrot, and cabbage), and beverages (e.g., coffee, tea, and wine; Meagher and Beecher 2000). When plant lignan precursors are ingested, their bioactivity is altered by intestinal bacteria that create the hormonally active enterlignan compounds enterodiol and enterolactone, which bind to estrogen receptors (Carreau et al. 2008; Landete 2012; Rowland et al. 2000). Terpenoids are cyclic unsaturated hydrocarbons with various numbers of oxygen atoms and carbon rings synthesized by all plants that provide both coloration and scent as well as defensive protection against herbivory, especially the toxic cardiac glycosides and saponins (Harborne 1988; Langenheim 1994). The category of terpenoids that are estrogenic are the phytosterols, which are structurally and functionally similar to animal sterols (e.g., cholesterol; Harborne 1988; Ling and Jones 1995), and although they are ubiquitous in plants, they occur in the highest quantity in nuts, seeds, cereal grains, and certain fruits (e.g., olives; Ostlund 2002). Their effects on estrogenic activity, as well as that of other steroids, likely arise from their ability to block cholesterol absorption from the gut, as cholesterol is the precursor molecule for all vertebrate steroid hormones (Bradford and Awad 2007; Hadley 2000; Ostlund 2002), along with direct estrogenic activity, as seen in the phytosterol miroestrol of *Pueraria mirifica* (Fabaceae; Harborne 1988). In addition to estrogenic phytochemicals, the vertebrate endocrine system can be affected by mycoestrogens, the estrogenic molecules found in fungi. For example, *Fusarium* can infest storage crops, such as maize and hay, and produces the metabolite zearalenone, a molecule with demonstrated estrogenic effects in livestock species (Metzler et al. 2010).

MECHANISMS OF ACTION OF HORMONALLY ACTIVE PHYTOCHEMICALS

The vertebrate neuroendocrine system plays a central role in modulating physiological and behavioral responses to development and life history, as well as changes in the external environment (ecological or social; see further discussion of interactions among the neuroendocrine system, immune function, and nutrition in Vogel et al., this volume). This system consists of a series of ductless glands found throughout the body (e.g., thyroid, adrenals, and gonads), the hypothalamus and pituitary in the brain, and the central nervous system. These glands are usually organized along a hypothalamic-pituitary-peripheral gland axis in which signals from the central nervous system trigger chemicals (i.e., hormones) to be secreted from the hypothalamus and transported directly to the pituitary via the hypophyseal portal system, blood vessels that connect the hypothalamus and pituitary. These hypothalamus hormones trigger the production and secretion of pituitary hormones that then enter the general circulatory system to be carried to target sites throughout the body. When they encounter these target sites, in the form of protein or glycoprotein hormone receptors, a hormone-receptor complex is formed, which triggers changes in cellular activity or genomic transcription and translation. For steroid hormones, after binding to their receptor, they move to the nucleus and bind to a response element (promoter region), which is responsible for up- or downregulating the transcription of downstream genes.

Ingestion and absorption of hormonally active phytochemicals into the bloodstream can bypass this pathway and interact directly with the endocrine system, thus influencing rates of gene transcription and translation through the same receptor-response element pathway. As a result, phytochemicals can alter physiological and behavioral endpoints through changes in the production of proteins, which can occur via a number of mechanisms. Some phytochemicals share a similar molecular structure to endogenous vertebrate hormones and can compete for receptor sites. The occurrence of structurally similar molecules in plants and animals that have hormonal activity may simply be due to a biochemical coincidence, as both steroid hormones and phytosteroids are simple molecules. For example, all vertebrate steroid hormones (e.g., cortisol, estradiol, testosterone, and progesterone) are formed from cholesterol and then metabolized into various forms of a molecule with four carbon rings via enzymatic activity. Likewise, in plants, most phytoestrogens (such as the isoflavones genistein and daidzein) have a molecular structure consisting of two to four carbon rings, which allows them to mimic the structure of endogenous vertebrate estradiol inside the vertebrate body. Of particular importance to the ability to bind to estrogen receptors is the presence of a phenolic and hydroxyl moiety in the molecular structure, similar to that found in the estradiol molecule.

When a phytochemical binds to a receptor, it can mimic the actions of endogenous hormones by activating a response (i.e., chemical agonists) or block the receptor without activating a response, thereby inhibiting endogenous hormone function (i.e., chemical antagonists). Additionally, phytochemicals can interfere with hormone synthesis by altering hormone precursors (e.g., cholesterol) or enzymes responsible for steroid conversion (Kao et al. 1998). For example, the flavonoid phytochemicals chrysin and quercetin can inhibit the actions of aromatase, the enzyme responsible for converting testosterone into estradiol (Oberdorster et al. 2001). Alternatively, phytochemicals can affect the rate of hormone metabolism by stimulating or inhibiting enzymes responsible for hormone degradation (Clotfelter et al. 2004; P. Whitten and Patisaul 2001). Moreover, phytochemicals can interfere with hormone transport by stimulating or suppressing the synthesis of sex-hormone-binding globulins, used to move hormones throughout the body (C. Hobbs et al. 1992).

Phytochemical interference with estrogen signaling pathways, or the entire estrogen hormone axis from regulation by the hypothalamus to synthesis and secretion from endocrine cells (e.g., ovaries) to receptor binding and subsequent activity within the cell via genomic (e.g., estrogen response element) and nongenomic (e.g., protein-kinase cascades) pathways, is most well understood via competition between the phytoestrogen and endogenous vertebrate estrogens for binding to nuclear estrogen receptors (P. Whitten and Patisaul 2001). Nuclear estrogen receptors (ERs) have two distinct subtypes, ERα and the more recently evolved ERβ, which vary in distribution across and within tissues and between developmental stages in all vertebrates (Kuiper et al. 1998). The location of ER subtypes is also expected to influence different aspects of behavior. For example, in laboratory rat models, ERα has been shown to mediate sexual behavior, whereas ERβ plays a role in mediating aggression and anxiety (Patisaul and Bateman 2008; Patisaul et al. 2004). Phytoestrogens have demonstrated an affinity for ERβ. Recent work has begun to demonstrate adaptations in vertebrates for reducing the effect of the external noise of phytoestrogens on ERβ signaling. For example, phytoestrogen activation of ERs between two closely related species of rhinoceros indicate adaptation to these compounds in species with greater exposure in their natural diets (Tubbs, Hartig, et al. 2012). Similarly, in primates, differences in ERβ interactions may also suggest variation in adaptations to diets high in estrogen-active phytochemicals (Weckle et al. 2012).

EFFECTS OF HORMONALLY ACTIVE PHYTOCHEMICALS ON PRIMATE PHYSIOLOGY, BEHAVIOR, AND REPRODUCTION

Given the various ways hormonally active phytochemicals interact with the primate endocrine system, the consumption (or avoidance) of plants containing these molecules has the potential to play important roles in primate ecology and evolution. Consumption may create fertility costs or benefits through alterations in hormonal signaling key to gamete production, implantation, or pregnancy; changes in behaviors important to survival or reproduction; and self-medication. Many laboratory-, agricultural-, and zoo-based studies have shown that hormonally active phytochemicals, particularly phytoestrogens, decrease fertility and fecundity (N. R. Adams 1990; Cederroth, Auger, et al. 2010; Cederroth, Zimmerman, et al. 2010; Cline and Wood 2009; Tubbs, Moley, et al. 2016). Alternatively, it has also been suggested that phytoestrogens may be selected for medicinal purposes (Forbey, Harvey, et al. 2009; Glander 1980; Huffman 1997; Leopold et al. 1976; Tham et al. 1998). Interpretations of phytochemical-induced hormonal effects are limited by the fact that most studies on phytochemical consumption have been conducted in evolutionarily novel contexts rather than in a natural environment. By incorporating ecological and evolutionary perspectives, we can examine the effects of hormonally active phytochemicals on primate fitness.

One hypothesis for the relationship between primates and hormonally active phytochemicals is that plants produce these compounds as a feeding deterrent that negatively affects herbivore reproduction (Wasserman, Milton, and Chapman 2013). However, only a few primate studies have provided direct evidence of hormonally active phytochemical effects on hormones or behaviors important to reproduction (table 10.1). Higham, Ross, et al. (2007) showed that during periods of consumption of African black plum (*Vitex doniana*), a plant exhibiting progesterone-like compounds, baboons exhibited higher levels of fecal progesterone and decreases in sexual swellings. Similarly, increased progesterone levels in wild female Phayre's leaf monkeys and wild female chimpanzees have also been associated with seasonal availability and consumption of *Vitex*, respectively (Emery Thompson,

Wilson, et al. 2008; Lu, Beehner, et al. 2011). Moreover, the consumption of phytoestrogens in wild red colobus males correlated with elevated fecal estradiol and cortisol levels, increased aggressive and mating behaviors, and decreased social grooming behaviors (Wasserman, Chapman, Milton, Gogarten, et al. 2012). Understanding the significance of consumption of these hormonally active phytochemicals is complicated based on timing and dose of exposure, age, sex, and reproductive state, as well as whether changes in hormone levels or behaviors are significant enough to affect reproduction. In captive female cynomolgus monkeys (*Macaca fascicularis*), those fed low and moderate doses of estrogenic *Pueraria mirifica* had increased estrous cycle length, while those fed the highest dose ceased ovulation all together (Trisomboon et al. 2005), a clear indication of direct effects on reproduction and fertility, with potential implications for effects on lifetime fitness if a similar diet were consumed on a regular basis. Furthermore, captive male cynomolgus monkeys fed a soy-based high-isoflavone diet were more aggressive and less affiliative (Simon et al. 2004). More substantial lifetime effects on reproduction are likely from exposure to phytochemicals in utero and during early stages of development, as this has the most distinct influence on endocrine system organization and activation later in life. For example, the consumption of the phytoestrogen genistein in pregnant females has been shown to influence fetal estradiol levels by 70% in rhesus macaques (*Macaca mulatta*; R. M. Harrison et al. 1999) and testosterone levels in captive newborn male common marmoset monkeys (*Callithrix jacchus*; Sharpe et al. 2002). Furthermore, neonatal phytoestrogen exposure in rodent models is associated with alteration in the development of sex-typical organs (including the reproductive tract and external genitalia), disruption of estrous cycles and sociosexual behavior, and increases in embryo loss (Jefferson et al. 2012).

Alternatively, hormonally active phytochemicals may provide functional environmental cues, serving as proximate mediators of reproduction that are beneficial to the primate. For several decades, scholars speculated whether phytochemicals can influence various reproductive parameters in primates, such as the regulation of estrus and conception, timing of birth, and infant sex ratio (Garey 1984; Glander 1980; Strier 1993; J. Wallis 1997). Such effects could allow for timing reproduction to only those times when sufficient resources are available to support pregnancy, lactation, and weaning, similar to what was reported for the consumption of estrogenic clover by California quail (Leopold et al. 1976). While a few primate studies have suggested both physiological and social mechanisms of altered reproduction, the potential functional and adaptive value of such effects are still not well understood.

Endocrine effects from phytochemicals may also influence primate health through positive changes in overall physiological function. The consumption of phytochemicals has been associated with antiviral, antibacterial, anti-inflammatory, antidiabetic, neuroprotective, and cardioprotective medicinal benefits (Briskin 2000). Many phytochemicals, especially phytoestrogens, have also been suggested to protect against steroid-dependent cancers. In humans, evidence that phytoestrogens may contain some cancer-protective properties arose from studies indicating an increased prevalence of estrogen-dependent cancers since the agricultural revolution (Crespi and Summers 2005). Specifically, it has been proposed that a diet low in plant consumption, especially those with phytoestrogens, contributes to increased rates of estrogen-dependent breast and androgen-dependent prostate cancers (e.g., Adlercreutz 1995). For example, Japanese men have been found to have ten times the amount of plasma phytoestrogens genistein and daidzein and a lower incidence of prostate cancer compared to British men (Guerrero-Bosagna and Skinner 2014), and the overall prevalence of prostate cancer is ten times lower in Asia than in the United States (Coffey 2001). However, these associations have been shown to dramatically differ between regions and populations: the risk of developing breast cancer has been shown to decrease with increased soy intake in Asian but not Western women (A. Wu et al. 2008). Paradoxically, studies from the same regions or populations can also indicate opposing relationships between phytoestrogen consumption and estrogen-dependent cancer (e.g., Trock et al. 2006). This lack of consensus may be influenced by variation in dietary phytoestrogen levels based on modern food-processing techniques (C. J. Jackson et al. 2002).

Moreover, the microbiome may play a significant role in how phytochemicals impact physiology. Gastrointestinal microbiota can form symbiotic relationships with hosts to help shape the immune and endocrine system (Kau et al. 2011; Y. Lee and Mazmanian 2010; Neuman et al. 2015). As such, the microbes that metabolize phytochemicals in the gut may also play a role in understanding cancer-protective properties of hormonally active phytochemicals, as they often alter the bioactivity of the ingested compound. For example, certain microbes convert daidzein, a phytoestrogen found in soy and many other legumes, to the more estrogenic compound equol. This occurs in only some individuals, with 30%–50% of the human population estimated to be equol producers (Atkinson et al. 2005). These equol producers are thought to be the ones benefiting from the possible cancer-protective properties of some phytoestrogens. Differences in equol production are also hypothesized to explain differences in rates of prostate and breast cancer between Western and Asian populations, with 25%–35% of Western adults compared to 50%–60% of Asian adults being equol producers. Becoming an equol producer appears to be due to the gut microbiome and diet early in life, as early exposure to high levels of phytoestrogens and equol-producing bacteria results in a higher likelihood of becoming an equol producer (N. Brown et al. 2014).

Similar to human studies, a cross-species review of cancer prevalence in wild and domesticated mammals documented 48 species of mammals to develop cancer, including African lions, meerkats, black rhinoceros, Asian elephant, and Nubian ibex, but no nonhuman primates (Nagy et al. 2007). Clearly, more research on cancer prevalence in primates other than humans is needed. Nonetheless, approximately 20%–30% of humans suffer from cancer at some point in their lives, while only 1.5%–4.5% of captive and free-ranging wildlife have been found to develop cancer (Nagy et al. 2007). There are many factors that explain these differences, including extended life history and increased exposure to carcinogens, but it has also been suggested that chimpanzees may be particularly resistant to steroid-dependent cancers due to a combined effect of plant-based diets, especially tropical fruits, leaves, flowers, and seeds, and microbial metabolism of phytoestrogens from that plant-based diet in the gut (Musey et al. 1995). Since altered estrogen:androgen levels influence incidences of steroid-dependent cancers (Coffey 2001), it is possible that variation in phytoestrogen consumption over primate evolutionary history and across human cultures at least partially explains variation in the prevalence of these types of cancers.

Another hypothesis for the association between primates and hormonally active phytochemicals is that these plant molecules have no significant biological costs or benefits for primates and they interact with the vertebrate endocrine system due to their molecular structure and shared biochemical pathways between animals and plants (e.g., cholesterol). Phytochemicals play numerous essential roles for the plants that produce them. Many compounds serve as biochemical adaptations, including roles to deter herbivores, attract pollinators, defend against pathogens, attract mutualistic microbes, provide UV protection, act as precursors for pigment synthesis, and regulate overall plant growth and physiology (R. Bennett and Wallsgrove 1994; Glick 1995; Harborne 1988; Winkel-Shirley 2002). As a result, quantities of phytochemicals can vary in response to changing biotic and abiotic conditions. For example, environmental stressors can cause plants to increase concentrations of phytochemical expression (Shulaev et al. 2008), which can change during a matter of seconds to years within individual plants (Liechti and Farmer 2002). Since exposure to phytochemicals in primate foods can fluctuate over space and time due to biotic and abiotic stressors on plants, if primates do incur costs or benefits from consumption, then wild primates are likely making foraging decisions (i.e., selection or avoidance) partially based on hormonally active compounds in their wild plant foods. However, if no significant biological effects occur from the ingestion of phytoestrogens, then exposure to phytoestrogens is simply a coincidence of seeking nutritionally rich or readily available foods that just happen to be estrogenic.

In some cases, potential costs of phytochemical consumption (such as the disruption of normal reproductive function) may be offset by energetic or medicinal benefits of the plants. The establishment of longitudinal data on phytochemical consumption can help discern potential trade-offs between short-term negative effects on reproductive function and long-term reproductive success. Given the demonstrated endocrine effects of many phytochemicals and their potential impacts on primate ecology, evolution, and conservation, further examination of the role of phytochemicals in primate ecology and evolution is warranted.

MEASURING HORMONALLY ACTIVE PHYTOCHEMICALS IN PRIMATE FOODS

Current, relatively simple methods exist to detect steroidal activity in primate plant foods using reporter assays, described here (Chester et al. 2020). Fresh plant material of known primate foods is collected in the field and dried at low temperatures out of direct sunlight at the field station. Plant samples are then shipped to a laboratory and ground to a constant particle size using a Wiley mill (Rothman, Chapman, and Van Soest 2012). Any potential steroidal compounds are extracted from the dried, ground samples using methanol, and the plant extracts are dissolved in dimethyl sulfoxide as a noncytotoxic assay medium. Small quantities of the dimethyl sulfoxide extracts are added to a transfected cell line that has had the luciferase gene, responsible for the light produced by fireflies, attached to the steroid response element within its genome using electrophoresis. If a phytochemical with steroidal activity is present, the cells emit light, which can be measured using a luminometer. Relative light units of all plant samples are compared to a negative control where no estrogenic compounds are present and a positive control of cells with pure estradiol added, resulting in large amounts of light produced (Vivar et al. 2010; Wasserman, Taylor-Gutt, et al. 2012). Samples significantly different from the blank and approaching levels of light produced in the positive control are identified as having steroid-active phytochemicals. While this assay provides a simple test for the presence of steroidal phytochemicals, it does not quantify the amount of steroid-active phytochemical present or the actual specific compound responsible for the activity.

Alternatively, steroid enzyme immunoassays or radioimmunoassays can be used to look for compounds that have similar structure to endogenous hormones through their binding to antibodies. The immunoassay method utilizes competitive binding to an antibody to quantify an unknown sample based on the steroid hormones (or structural mimics) in that sample competing with a known amount of a labeled hormone added to the well. Based on the amount of labeled hormone remaining after incubation and washing of the assay plate (only hormones bound to antibodies in the well remain) and a comparison to a standard curve of known amounts of hormone, the amount of the steroid hormone (or structural mimic) can be determined. However, steroid enzyme immunoassays or radioimmunoassays only provide quantification of phytochemicals with structural similarities to vertebrate hormones; they cannot provide evidence for actual steroid activity from the structural mimic. If a sample is found to contain structurally similar compounds or possesses known phytoestrogens, compound-specific immunoassays can be used to measure the plant directly or the presence of compounds in the animal consumer via urine or feces (Lapcik et al. 2004; M. Uehara et al. 2000).

Determining specific compounds responsible for either activity or structural similarity and quantifying the amounts of these compounds present in different primate foods requires more sophisticated chemical analyses, including gas or liquid chromatography coupled with mass spectrometry (GS-MS or LC-MS). Chromatography is a technique used for separating the components of a mixture based on physical and chemical properties. Briefly, solvents (or mobile phases) get passed through a medium (or stationary phase) where the components move at different rates, allowing for the separation of specific analytes (Tsao and Deng 2004). High-performance liquid chromatography (HPLC) utilizes a pressurized pump to pass liquid solvents through columns, allowing for higher resolving power when separating mixtures. Analytes can then be identified and quantified using ultraviolet-visible spectroscopy (e.g., Andlauer et al. 1999). More recent methodologies of LC-MS and GC-MS combine separation techniques of HPLC with mass spectrometry, allowing for even greater sensitivity in compound detection. For more information about specific methodological protocols for these techniques, see Lopez-Gutierrez et al. (2014). Because these methods allow for high specificity and simultaneous measures of many compounds, they are ideal for detecting and quantifying phytochemicals of interest in primate foods. However, bioassays are still necessary to determine the estrogenic activity of those compounds, especially those not yet classified.

CASE STUDY 1: DIETARY PHYTOESTROGENS IN A PRIMATE COMMUNITY

Significant attention has been given to the study of phytoestrogens across disciplines, accumulating over 10,000 peer-reviewed articles on the topic (Wasserman, Milton, and Chapman 2013). Many of these studies suggest that phytoestrogens have important effects on health and reproduction and therefore have the potential to act as a selective force on primate ecology and evolution. Despite widespread interest in phytoestrogens, the relationship between nonhuman primates and estrogenic foods in their wild habitats has been largely ignored. Do primates actively select or avoid estrogenic foods? What are the effects of phytoestrogen consumption on physiology and behavior, and are these effects enough to affect lifetime fitness?

To begin addressing these questions, it is first necessary to understand which primate foods contain phytoestrogens. We examined the prevalence of phytoestrogens in the diets of the primate community in Kibale National Park, Uganda. Using reporter assays, we screened 45 plant food items for estrogenic activity at ERβ. Samples included foods of the red colobus monkey (*Piliocolobus tephrosceles*, fig. 10.2), black-and-white colobus (*Colobus guereza*), redtail monkey (*Cercopithecus ascanius*), blue monkey (*Cercopithecus mitis*), gray-cheeked mangabey (*Lophocebus albigena*), and chimpanzee (*Pan troglodytes*, fig. 10.3) as determined by long-term dietary data (Chapman, Struhsaker, et al. 2010; T. R. Harris and Chapman 2007; K. Potts, Watts, and Wrangham 2011; Rothman, Plumptre, et al. 2007; Wasserman, Taylor-Gutt, et al. 2012).

From these assays, we found that 5 of the 45 items tested were estrogenic, all of which were in the genus *Ficus* (fig. 10.1). Combined with results from previous studies, we have now screened a total of 165 plant food items from 94 species throughout the forests of Kibale and Bwindi Impenetrable National Parks, Uganda (Wasserman, Taylor-Gutt, et al. 2012) and have found 23 estrogenic items from 13 species (table 10.2). Many of these estrogenic species were found in the Moraceae (i.e., six *Ficus* species) and Fabaceae (i.e., two legume species from the Faboideae/Papilionoideae subfamily) plant families. For these estrogenic species, all plant parts tested except for ripe fruits have been found to have estrogenic activity. Within the primate community, it appears that only the highly folivorous species regularly consume estrogenic plant items for more than 1% of their overall diet (table 10.2). We found that 10% of the red colobus diet consisted of three estrogenic staple foods (Wasserman, Taylor-Gutt, et al. 2012), 1%–6% of the black-and-white colobus diet consisted of five estrogenic staples (Benavidez et al. 2015), and 9% of the mountain gorilla diet consisted of one estrogenic staple food (Wasserman, Taylor-Gutt, et al. 2012). However, this difference between more folivorous and less folivorous species should be interpreted with caution, as we have screened only about a third to a half of the less folivorous primate species' diets thus far.

Our results thus far indicate three interesting trends. First, primates do consume estrogenic plants in their diets, especially from legumes and figs. Since these two plant families are commonly consumed across the tropics, this suggests many primates are likely consuming estrogenic plants. Second, estrogenic phytochemicals are found in leaves, seeds, unripe fruits, pods, and bark, but not in ripe fruits of the plant species examined here, thus explaining why more folivorous species are consuming estrogenic foods, but more frugivorous species are not. Third, consuming estrogenic plants appears to affect physiology and behavior in more folivorous species, as significant relationships have been found between the consumption of young leaves of the estrogenic legume *Millettia dura* and red colobus fecal estradiol and cortisol levels, along with rates of grooming, aggression, and mating (Wasserman, Chapman, Milton, Gogarten, et al. 2012).

CASE STUDY 2: DIETARY PHYTOESTROGENS ACROSS THE PRIMATE PHYLOGENY

Given the significant effects of phytoestrogens on physiology and behavior and their presence in tropical plant-based diets of more folivorous primate species of Uganda, we explored the potential for the presence of estrogenic plants in diets of primates from across the order. It is expected that most, if not all, primate species are exposed to phytoestrogens in their natural environment based on their presence in the Fabaceae and Moraceae plant families. Fabaceae is the third-largest angiosperm family, with more than 700 genera and 20,000 species (J. Doyle and Luckow 2003). A striking adaptation found among some members of this family is their symbiosis with nitrogen-fixing bacteria, which has permitted them to flourish across a wide range of habitats, from mountains to lowland forests and even aquatic environments (J. Doyle and Luckow 2003). As a result of their ability

Table 10.2. Plant items from Kibale National Park and Bwindi Impenetrable National Park, Uganda, that have thus far been identified as having estrogenic activity based on transfection assays.

Plant species	Plant part	Plant family	RC	BWC	RT	Blue	Mang	Chimp	Gorilla
Ipomoea involucrata	Leaves	Colvolvulaceae							X
Macaranga sp.	Mature leaves	Euphorbiceae							
Erythrina abyssinica	Young leaves	Fabaceae (Papilionoideae)		X					
Erythrina abyssinica	Mature leaves	Fabaceae (Papilionoideae)		X					
Erythrina abyssinica	Flowers	Fabaceae (Papilionoideae)		X					
Millettia dura	Pods	Fabaceae (Papilionoideae)							
Millettia dura	Seeds	Fabaceae (Papilionoideae)							
Millettia dura	Mature leaves	Fabaceae (Papilionoideae)							
Millettia dura	Young leaves	Fabaceae (Papilionoideae)	X						
Xymalos monospora	Bark	Monimiaceae							
Ficus brachypoda	Unripe fruit	Moraceae							
Ficus capensis	Unripe fruit	Moraceae							
Ficus mucuso	Mature leaves	Moraceae							
Ficus mucuso	Young leaves	Moraceae							
Ficus natalensis	Young leaves	Moraceae	X						
Ficus natalensis	Mature leaves	Moraceae							
Ficus natalensis	Unripe fruit	Moraceae							
Ficus sansibarica	Unripe fruit	Moraceae		X					
Ficus sansibarica	Young leaves	Moraceae		X					
Ficus sansibarica	Mature leaves	Moraceae		X					
Ficus stipulifera	Mature leaves	Moraceae							
Eucalyptus grandis	Bark	Myrtaceae	X	X					
Olea welwitschii	Mature leaves	Oleaceae							

The primate species known to feed on each part for at least 1% of the diet are indicated by an X from the following literature: Wasserman, Chapman, Milton et al. 2012 (RC, red colobus), T. R. Harris and Chapman 2007 (BWC, black-and-white colobus), Chapman et al. 2010 (RT, redtail monkey; Blue, blue monkey; Mang, mangabey), K. Potts, Watts, and Wrangham 2011 (Chimp, chimpanzee), and Rothman, Plumptre, et al. 2007 (Gorilla).

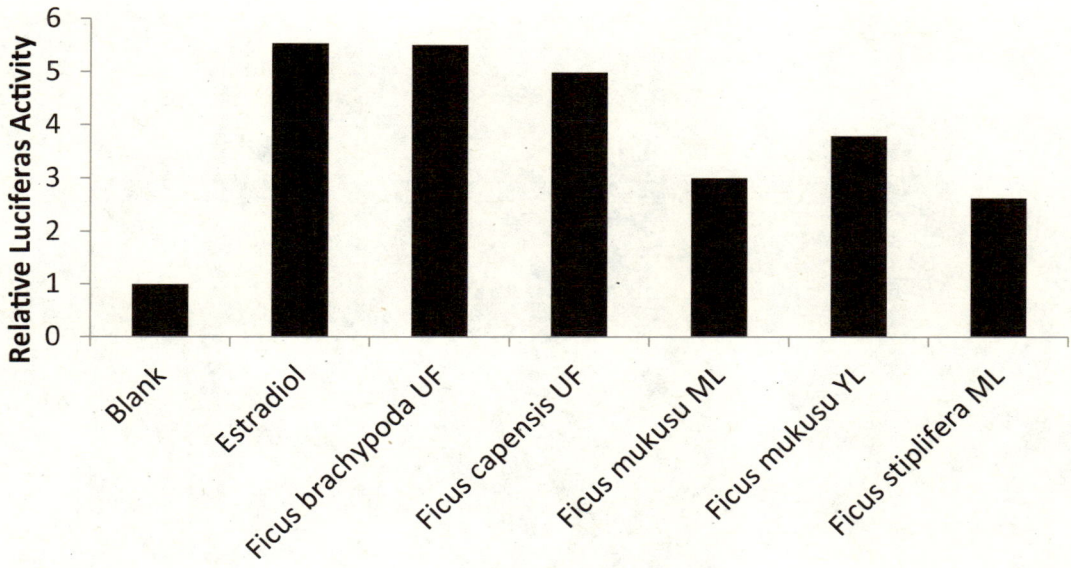

Figure 10.1. Transfection assay results showing five estrogenic plant items from Kibale National Park, Uganda. Any samples expressing at least a twofold increase in relative light units compared to a blank of no estrogenic chemicals added to the cells are considered to have estrogenic activity. Estradiol is the positive control with maximum relative light units. The results thus far indicate that numerous *Ficus* species and parts are estrogenic. See Wasserman et al. (2012b) and Chester et al. (2020) for descriptions of the transfection assay.

Figure 10.2. Red colobus monkey (*Piliocolobus tephrosceles*) in Kibale National Park, Uganda.

Figure 10.3. Chimpanzee (*Pan troglodytes*) in Kibale National Park, Uganda.

to fix nitrogen, Fabaceae plants tend to have high levels of protein, and high-protein food items are often, but not always, prioritized by primates (Chapman, Chapman, Bjorndale, et al. 2002; Ganzhorn, Arrigo-Nelson, Carrari, et al. 2017; Milton 1979; Rothman, Raubenheimer, and Chapman 2011; Wasserman and Chapman 2003). Several plants from this family, in particular those from the Faboideae subfamily, also contain high levels of phytoestrogens (N. R. Adams 1990, 1995; Wasserman, Chapman, Milton, Gogarten, et al. 2012; Wasserman, Milton, and Chapman 2013; Wasserman, Taylor-Gutt, et al. 2012; P. Whitten and Patisaul 2001).

Species of Moraceae, the fig family, are found in many tropical habitats, and their fruits and leaves are consumed by a wide variety of primate species, including more frugivorous species such as spider monkeys, chimpanzees, and gibbons as well as more folivorous species such as howler monkeys, colobus monkeys, and gorillas (Dominy, Yeakel, et al. 2016; Janmaat, Polansky, et al. 2014; Milton 1981b; Rothman, Dierenfeld, Molina, et al. 2006; Shanahan et al. 2001; Wasserman, Taylor-Gutt, et al. 2012). Of special interest in the Moraceae family are *Ficus* species, some of which have been shown to contain phytoestrogens (Greenham et al. 2007; Reynaud et al. 2005; Wasserman, Taylor-Gutt, et al. 2012). *Ficus* is one of the most speciose genera of angiosperm, with an estimated 750 species, 500 of which reside in the Asian-Australasian region (Ronsted et al. 2008; Shanahan et al. 2001). Adding to their ubiquitous availability, young leaves and ripe fruits of *Ficus* are available throughout the year since they are asynchronous in phenology, thus offering a constant food supply for primates to meet their basic nutrient needs, including high levels of calcium (Conklin and Wrangham 1994; Milton 1991; O'Brien, Kinnaird, and Dierenfeld 1998; Serio-Silva, Rico-Gray, et al. 2002; Wrangham, Conklin, Etot, et al. 1993). Despite the presence of phytoestrogens in some *Ficus* species and in some parts of those estrogenic species, ripe fruits have only recently been identified as having estrogenic activity from some species, so actual estrogenic parts of these species may not be consumed by more frugivorous primates. Many species do, however, consume the leaves and unripe fruit of *Ficus* species, which have been shown to be estrogenic across more plant species (Milton 1979, 1981c; Rothman, Dierenfeld, Molina, et al. 2006; Wasserman, Chapman, Milton, Gogarten, et al. 2012; Wasserman, Taylor-Futt, et al. 2012). Therefore, despite an expected widespread distribution of *Ficus* in primate diets, it is still unknown which species of primates are eating estrogenic species or the parts containing phytoestrogens. Does dietary niche influence phytoestrogen consumption based on differences in presence between leaves and fruits? Do primates in certain geographic areas consume more estrogenic plants than those in other areas based on the distribution and abundance of estrogenic plant species? Are there differences in consumption between apes and monkeys based on differences in digestive or detoxification adaptations?

To begin addressing these questions, we collected data from primary literature ($N = 50$) and calculated the relative percentages of total diet from all Moraceae and specifically *Ficus* species and all Fabaceae and specifically the subfamilies Caesalpinioideae, Faboideae, and Mimosoideae. These families, subfamilies, and genera were selected based on previous research indicating their production of phytoestrogens. To allow for a high consistency across dietary data, we selected papers based on the following two conditions: (1) the data had to be collected via behavioral methods (focal animal or scan sampling) and (2) the study had to be conducted for a period of at least nine months. Two of the studies were less than nine months long (*Pan paniscus*, four months [Georgiev, Thompson, et al. 2011], and *Otolemur crassicaudatus*, eight months [Ejidike and Okosodo 2007]) but were considered in order to cover a wider range of primate taxa.

We found that Moraceae and Fabaceae were commonly consumed across the primate order, making up an average of 23% of the total diet across the 50 primate species (tables 10.3 and 10.4). Although much variation exists across species, all geographic (Africa, Asia, Latin America, and Madagascar), phylogenetic (apes, colobines, cercopithecines, prosimians, and Ceboidea), and dietary niche (frugivore, folivore, omnivore, and gummivore) categories were found to consume these two families (table 10.4). All primate species included in the analysis, except *Gorilla* spp. and *Rhinopithecus bieti*, fed on Moraceae or Fabaceae plants or both (table 10.3).

Other studies have shown that gorillas and snub-nosed monkeys consume *Ficus* (Rothman, Dierenfeld, Molina, et al. 2006; P. Zhang et al. 2016). Relative percentages show that *Ficus* spp. account for at least 50% of Moraceae consumption (8.19% out of an average of 11.7%), except for the colobine category, while among Fabaceae plants, the most consumed subfamily varied across all categories (table 10.4).

As a whole, based on our comparative analysis of the primate diet (table 10.2), we have shown that Fabaceae and Moraceae are the two most consumed plant families across the primate phylogeny and are consumed across all categories of geographic location, dietary niche, and phylogeny (table 10.3). This further supports the role of *Ficus* as keystone species in tropical habitats and suggests that the importance of legumes should be further studied (Janzen 1979a; F. Lambert and Marshall 1991; Peres 2000b; Shanahan et al. 2001; Tello 2003; Wrangham, Conklin, Etot, et al. 1993). Given the occurrence of phytoestrogens in *Ficus* and Fabaceae, most primates are probably eating estrogenic plant species. However, based on our results from the primate community at Kibale, some highly frugivorous, insectivorous, or omnivorous primates may be avoiding the estrogenic parts of these plant species. Even primate species with a more folivorous diet may limit consumption of phytoestrogens in *Ficus* through selection of ripe fruits over leaves; for example, howler monkeys at Agaltepec, Mexico, and on Barro Colorado Island, Panama, tend to be frugivorous with a focus on *Ficus* fruits (e.g., 64% of the Agaltepec diet came from *Ficus*; Milton 1979; Serio-Silva, Rico-Gray, et al. 2002).

Screening a number of ripe fruits, especially figs, for estrogenic activity will be especially useful for understanding the distribution of phytoestrogens within estrogenic plant species and helpful in clarifying the ecological role of these compounds in primate-plant relationships. Whether they serve a role as a primate feeding deterrent for plants (Harborne 1988) or a form of self-medication for primates (Forbey, Harvey, et al. 2009; Glander 1980; Huffman 1997; Leopold et al. 1976; Tham et al. 1998) is yet to be demonstrated. Current evidence suggests that more highly folivorous species are likely interacting the most with estrogenic plant foods, while other primates may be avoiding such plant items. It may be that folivorous primates have adapted to the presence of phytoestro-

Table 10.3. Comparative analysis of primate diet based on published primate field studies with a focus on the percentage of diet from known estrogenic plant families of Fabaceae and Moraceae.

Phylogenetic group	Species	Study	Duration of study	Dietary niche	Percentage of total diet reported (%)	Percent *Ficus* spp. (%)	Percent Moraceae (including *Ficus* spp.) (%)	Percent Fabaceae (total) (%)	Percent Fabaceae (Ceasalpinoideae) (%)	Percent Fabaceae (Faboideae) (%)	Percent Fabaceae (Mimosoideae) (%)
Africa											
Hominoidea	*Pan troglodytes schweinfurthii*	K. Potts, Watts, and Wrangham (2011)	19	Fr	92.3–99.5	45.9	48.4	0.0	0.0	0.0	0.0
Hominoidea	*Pan paniscus*	Georgiev et al. (2012)	4	Fr	95.3	1.3	1.3	47.9	46.8	0.0	1.1
Hominoidea	*Gorilla beringei*	D. Watts (1984)	17	Fo	99.1–100	0.0	0.0	0.0	0.0	0.0	0.0
Hominoidea	*Gorilla gorilla gorilla*	Magliocca and Gauthier-Hion (2002)	10	Fo	100	0.0	0.0	0.0	0.0	0.0	0.0
Colobinae	*Colobus guereza*	Fashing (2001)	12	Fo	91.5–94.7	10.8	35.5	4.1	0.0	0.0	4.1
Colobinae	*Piliocolobus tephrosceles*	Wasserman, Chapman, Milton, et al. (2012)	11	Fo	79.3	1.5	10.8	25.9	0.0	5.1	20.8
Colobinae	*Procolobus verus*	Oates (1988)	9	Fo	54.8	0.0	0.0	15.2	4.6	2.3	8.3
Papionini	*Lophocebus albigena*	Poulsen, Clark, and Smith (2001)	11	Fr	59.2	0.0	0.0	23.1	21.0	0.0	2.1
Papionini	*Papio hamadryas*	Swedell, Hailemeskel, and Schreier (2008)	12	Om	99	0.0	0.0	18.6	0.0	0.0	18.6
Papionini	*Papio cynocephalus*	Bentley-Condit (2009)	55	Om	99.7	1.1	1.1	2.7	0.1	0.8	1.8

	Species	Reference	N	Diet							
Cercopithecini	*Erythrocebus patas*	Isbell (1998)	17	Om	91.9	0.0	0.0	85.3	0.0	0.0	85.3
Cercopithecini	*Cercopithecus doggetti*	Kaplin et al. (1998)	10	Fr	83.9	8.3	8.3	0.0	0.0	0.0	0.0
Cercopithecini	*Cercopithecus aethiops*	Wrangham and Waterman (1981)	11	Om	84.8	0.0	0.0	59.1	0.0	0.0	59.1
Cercopithecini	*Chlorocebus sabaeus*	M. J. S. Harrison (1984)	14	Om	68.2	9.1	10.2	10.6	2.0	7.8	0.7
Galagonidae	*Otolemur crassicaudatus*	Ejidike and Okosodo (2007)	8	Gum	100	0.0	1.5	6.0	3.0	3.0	0.0
Asia											
Hominoidea	*Pongo pygmaeus*	Kanamori et al. (2010)	27	Fr	66.5	21.1	21.1	18.6	0.0	18.6	0.0
Hominoidea	*Hylobates moloch*	S. Kim et al. (2012)	12	Fr	58–72.5	35.5	36.6	0.0	0.0	0.0	0.0
Hominoidea	*Nomascus concolor*	P. Fan, Ni, et al. (2009)	14	Fr	97.9	24.4	24.4	0.0	0.0	0.0	0.0
Colobinae	*Rhinopithecus roxellana*	Yiming (2006)	14	Fo	100	0.0	2.9	0.0	0.0	0.0	0.0
Colobinae	*Presbytis entellus*	Newton (1992)	12	Fo	93.4	8.3	8.3	32.0	9.3	22.2	0.5
Colobinae	*Trachypithecus auratus sondaicus*	Kool (1993)	13	Fo	80.3–83.1	18.8	18.8	8.4	1.8	5.5	1.1
Colobinae	*Trachypithecus vetulus*	Dela (2007)	32	Fo	72.7–82.1	0.3	39.6	4.2	3.3	0.9	0.0
Colobinae	*Presbytis pileata*	Craig (1991)	12	Fo	78.1	0.0	0.0	10.4	7.4	3.0	0.0
Colobinae	*Rhinopithecus bieti*	Grueter, Li, et al., "Dietary Profile" (2009)	20	Fo	91.7	0.0	0.0	0.0	0.0	0.0	0.0

(continued)

Table 10.3. (continued)

Phylogenetic group	Species	Study	Duration of study	Dietary niche	Percentage of total diet reported (%)	Percent Ficus spp. (%)	Percent Moraceae (including Ficus spp.) (%)	Percent Fabaceae (total) (%)	Percent Fabaceae (Ceasalpinoideae) (%)	Percent Fabaceae (Faboideae) (%)	Percent Fabaceae (Mimosoideae) (%)
Colobinae	Nasalis larvatus	Yeager (1989)	12	Fo	100	0.0	0.0	2.0	2.0	0.0	0.0
Papionini	Macaca fascicularis	Ungar (1995)	12	Fr	100	21.1	21.1	1.0	0.0	0.0	1.0
Papionini	Macaca fuscata	Hanya (2004)	12	Om	98.5	0.8	0.8	0.0	0.0	0.0	0.0
Papionini	Macaca assamensis	Zhou, Wei, Huang, et al. (2011)	12	Om	100	8.1	8.5	0.3	0.0	0.0	0.3
Lorisidae	Nycticebus coucang	F. Wiens et al. (2006)	49	Gum	99.2	6.5	6.5	0.0	0.0	0.0	0.0

Madagascar

Phylogenetic group	Species	Study	Duration of study	Dietary niche	Percentage of total diet reported (%)	Percent Ficus spp. (%)	Percent Moraceae (including Ficus spp.) (%)	Percent Fabaceae (total) (%)	Percent Fabaceae (Ceasalpinoideae) (%)	Percent Fabaceae (Faboideae) (%)	Percent Fabaceae (Mimosoideae) (%)
Lemuriformes	Indri indri	Britt et al. (2002)	12	Fo	92.1	0.0	3.8	4.7	4.7	0.0	0.0
Lemuriformes	Eulemur fulvus	Overdorff (1993b)	12	Fo	63.1	10.7	10.7	2.9			
Lemuriformes	Hapalemur griseus	Grassi (2006)	12	Fo	88.7–97.1	12.6	14.5	0.0	0.0	0.0	0.0
Lemuriformes	Avahi occidentalis	Thalmann (2001)	12	Fo	88.3	2.4	11.0	2.4	0.0	2.4	0.0
Lemuriformes	Lepilemur edwardsi	Thalmann (2001)	24	Fo	55.3	0.0	0.0	6.7	0.5	6.2	0.0
Lemuriformes	Propithecus diadema edwardsi	Hemingway (1999)	13	Fo	59.0–67.0	0.0	8.7	0.9	0.0	0.9	0.0
Lemuriformes	Eulemur rubriventer	Overdorff (1993b)	12	Fr	57.5	6.6	6.6	0.0	0.0	0.0	0.0

Family	Species	Reference	N	Dietary niche							
Lemuriformes	*Microcebus murinus*	Thorén et al. (2011)	11	Gum	81.0–100	0.0	0.0	27.0	13.0	14.0	0.0
Lemuriformes	*Microcebus ravelobensis*	Thorén et al. (2011)	11	Gum	45.0–80.0	0.0	0.0	10.0	10.0	0.0	0.0
Latin America											
Atelidae	*Alouatta palliata*	Estrada, Juan-Solano, et al. (1999)	12	Fo	100	55.1	78.8	3.7	0.2	1.7	1.8
Atelidae	*Alouatta pigra*	Silver et al. (1998)	11	Fo	88	29.4	33.1	28.3	0.0	6.8	21.5
Atelidae	*Ateles chamek*	R. Wallace (2005)	11	Fr	81.1	10.7	28.1	1.6	0.0	0.0	1.6
Cebidae	*Cebus apella*	Galetti and Pedroni (1994)	44	Fr	100	4.1	4.6	20.5	11.2	1.1	8.2
Cebidae	*Saguinus fuscicollis*	Peres (1993)	13	Fr	82.9	9.5	15.0	16.9	0.0	0.0	16.9
Cebidae	*Saguinus mystax*	Peres (1993)	13	Fr	82.4	7.9	13.9	17.2	0.0	0.0	17.2
Cebidae	*Callithrix flaviceps*	Hilário and Ferrari (2010)	11	Gum	94.8	0.0	0.0	4.0	2.4	0.0	1.6
Cebidae	*Saimiri sciureus*	A. Stone (2007)	12	Om	56.0–92.0	0.0	0.0	18.4	0.0	0.0	18.4
Pitheciinae	*Callicebus coimbrai*	Souza-Alves, Fontes, et al. (2011)	12	Fr	99.5	0.0	0.0	6.4	3.0	1.0	2.4
Pitheciinae	*Callicebus nigrifrons*	Caselli and Setz (2011)	9	Fr	83.5	6.2	6.2	4.5	0.0	4.2	0.3
Pitheciinae	*Cacajao calvus ucayalii*	Bowler and Bodmer (2011)	25	Gran	86.6	5.1	7.6				

Dietary niche: Fr, frugivore; Fo, folivore; Om, omnivore; Gum, gummivore; Gran, granivore. Percentage of total diet reported: a range indicates that the study reported data for more than one group or according to different time periods. When the percentage is not listed, specific data were not reported in the article.

Table 10.4. Percentage of diet from known estrogenic plant families for primate species based on geographic location, dietary niche, and phylogeny.

Category	Percent *Ficus* spp. (%)	Percent Moraceae (including *Ficus* spp.) (%)	Percent Fabaceae (total) (%)	Percent Fabaceae (Ceasalpinoideae) (%)	Percent Fabaceae (Faboideae) (%)	Percent Fabaceae (Mimosoideae) (%)
Geography						
Africa	5.2 (11.9)	7.8 (14.6)	19.9 (25.5)	5.2 (12.7)	1.3 (2.4)	13.5 (25.3)
Asia	11.2 (11.7)	15.2 (15.2)	5.2 (9.2)	1.6 (2.9)	3.3 (7.1)	0.2 (0.4)
Madagascar	3.6 (5.1)	6.1 (5.5)	6.1 (8.5)	3.5 (5.2)	2.9 (5.0)	0
Latin America	12.3 (17.4)	16.8 (23.4)	11.7 (8.8)	1.7 (3.5)	1.5 (2.3)	9.0 (8.5)
Total	8.2 (12.5)	11.7 (16.3)	11.2 (16.6)	3.1 (7.7)	2.2 (4.7)	6.1 (15.5)
Phylogeny						
Hominoidea	18.9 (17.2)	21.8 (19.8)	8.4 (17.2)	6.0 (16.5)	2.3 (6.6)	0.1 (0.4)
Colobinae	4.0 (6.5)	11.6 (15.0)	10.2 (11.1)	2.8 (3.3)	3.9 (6.8)	3.5 (6.7)
Cercopithecinae	4.9 (6.9)	5.0 (7.0)	20.0 (29.4)	2.3 (6.6)	0.9 (2.5)	16.9 (30.3)
Strepsirrhini	3.5 (4.8)	5.5 (5.1)	5.5 (7.8)	3.1 (4.8)	2.6 (4.5)	0
Platyrrhini	12.3 (17.4)	16.8 (23.4)	11.7 (8.8)	1.7 (3.5)	1.5 (2.3)	9.0 (8.5)
Total	8.2 (12.5)	11.7 (16.3)	11.2 (16.6)	3.1 (7.7)	2.2 (4.7)	6.1 (15.5)
Dietary niche						
Frugivore	13.5 (13.5)	15.7 (14.3)	10.5 (13.6)	5.5 (12.9)	1.7 (4.8)	3.4 (5.9)
Folivore	7.5 (13.8)	13.8 (19.7)	7.6 (10.0)	1.8 (2.8)	3.0 (5.2)	3.1 (6.7)
Omnivore	2.4 (3.9)	2.6 (4.2)	24.3 (31.3)	0.3 (0.7)	1.1 (2.7)	23.0 (32.1)
Gummivore	1.3 (2.9)	1.0 (1.6)	9.4 (10.5)	5.7 (5.5)	3.4 (6.1)	0.3 (0.7)
Total	7.9 (12.4)	11.2 (15.9)	11.5 (16.9)	3.1 (7.8)	2.3 (4.8)	6.3 (15.7)

Standard deviations are provided in parentheses. Total is from all species considered, which varies for dietary niche because granivore was excluded from the analysis due to a sample size of one.

gens physiologically, such as increasing endogenous hormone levels to dilute any effect of exogenous hormones (K. Wynne-Edwards 2001), or behaviorally by limiting ingestion below a threshold that would result in negative effects (Forbey, Dearing, et al. 2013). It may even be possible that highly folivorous primates have developed a need for a certain level of phytoestrogens to assist with their gut microbiota, similar to how legumes use these plant compounds to communicate with mutualistic soil microbes (J. Fox et al. 2004). Furthermore, if folivores have evolved to better cope with higher phytoestrogen loads, then more frugivorous, insectivorous, or omnivorous primate species may be more susceptible to the effects of phytoestrogens (and possibly other anthropogenic estrogen-active compounds) due to little exposure to such compounds in their diets. Future studies should focus on whether these trends are simply coincidental or due to active selection for such plants by colobines and other folivores and avoidance by other primates.

FUTURE INVESTIGATIONS: ANTHROPOGENIC ENDOCRINE-DISRUPTING CHEMICALS, LANDSCAPE MODELS OF CHEMICAL EXPOSURE, AND THE MICROBIOME

In addition to phytochemicals in the diet, nonhuman primate populations may encounter numerous anthropogenic chemicals that can directly interfere with endocrine function (i.e., endocrine-disrupting chemicals [EDCs]) through exposure to pollutants from agriculture, urbanization, or pest control. Concern over the effects of synthetic chemicals became a major public issue when Rachel Carson (1962) warned of the health effects of pesticides in *Silent Spring*. Years later, Colborn et al. (1993) proposed an "endocrine disruption hypothesis," outlining how some synthetic chemicals alter physiology and behavior through direct interactions with the endocrine system. Recently, endocrine disruption has been the focus of much investigation and debate within environmental science and public health (A. Gore et al. 2015; T. Hayes et al. 2011; Vandenberg et al. 2013), including work on estrogenic bisphenol A, a component of many plastics present in the urine of approximately 93% of the American population (Calafat et al. 2008; Krishnan et al. 1993).

While synthetic EDCs can act on endocrine pathways, following similar mechanisms of action to hormonally active phytochemicals, they differ from these phytochemicals in several key ways. Synthetic EDCs can be more persistent, often exhibiting longer biological half-lives in the environment as well as longer periods of biological activity within organisms (Patisaul and Adewale 2009). Additionally, the context in which organisms are exposed to EDCs and phytochemicals has ecological and evolutionary significance. While geographic and temporal exposure to phytochemicals is expected to reflect adaptive environmental cues, the distribution of pollutants and exposure to synthetic EDCs occur in novel contexts (McLachlan 2001). Finally, exposure to synthetic EDCs may have more pronounced and adverse outcomes than exposure to naturally occurring phytochemicals. Persistent EDCs can bioaccumulate within bodily systems and biomagnify within food webs (Mackay and Fraser 2000). In addition to adverse effects of direct exposure, EDCs can indirectly influence primate feeding ecology through the disruption of plant-hormone signaling pathways. For example, EDCs known to interfere with estrogen receptors in humans and wildlife have also demonstrated the ability to disrupt phytoestrogen signaling pathways used by leguminous plants for nitrogen fixation (J. Fox et al. 2004). Consequently, EDCs can affect the phytochemical and other nutritional aspects of primate plant foods.

Given the proximity of many human and nonhuman primate populations, EDCs are expected to be prevalent in many natural habitats, even in protected areas that are assumed to be buffered from human disturbance (S. Wang, Steiniche, Romanak, et al. 2019). Although certain human activities such as hunting and logging may be suppressed in protected areas, other indirect effects such as climate change and pollution have no boundaries. While evidence for primate exposure to EDCs is currently limited (Brockman et al. 2009; Krief, Iglesias-González, et al. 2022; Serio-Silva, Olguin, et al. 2015; S. Wang, Steiniche, Rothman, et al. 2020), the extent to which nonhuman primates are directly exposed to or indirectly affected by EDCs may have major effects on the ability of primate species to maintain populations. Examining the diversity of anthropogenic pollutants that wild primates are exposed to through water, soil, diet, and air is an area of research that deserves much more attention than it currently receives (Chapman, Steiniche, et al. 2022). Because they are our closest-living relatives, we have used captive nonhuman primates to advance biomedical research for decades. By broadening our scope to noninvasively monitoring strepsirrhines, monkeys, and apes for negative effects of synthetic compounds, we will provide two important benefits: (1) increasing the likelihood of conserving threatened primate populations by monitoring a direct threat on survival and reproduction and (2) developing a systematic warning system for human health by watching for cases of unusual health or reproductive patterns in primates and investigating the synthetic chemical load of these troubled populations. Utilizing an EcoHealth perspective for investigations of interactions between endocrine-active compounds and primates will be necessary as more and more species and populations are exposed to urbanization, roads, and agriculture (fig. 10.4).

Landscape models can be valuable tools for understanding spatial and temporal dynamics of primate interactions with exogenous chemicals. Spatial analyses of the density and distribution of plant species that express hormonally active phytochemicals can shed light on the subsequent density, distribution, and behavior of primates. Furthermore, spatial variation in anthropogenic land use can influence nonhuman primate exposure to synthetic EDCs, as seen in other wildlife populations. The use of landscape models to understand how EDCs move through the environment can help determine which species or populations may be most vulnerable to EDC effects. Landscape models can also be useful for evaluating temporal variation of exposure to exogenous chemicals. The chemical landscape is not static but constantly

Figure 10.4. Agricultural land adjacent to Kibale National Park, Uganda.

changing in response to trophic interactions and feedback processes. For example, feeding pressure from herbivores induces changes in plant chemistry, which in turn affect patterns of plant consumption. Such interactions can create temporal variation in the nutritional value of primate foods (Chapman, Chapman, Rode, et al. 2003). Novel methods in remote sensing, imaging spectroscopy, and GIS may be especially useful for characterizing plant chemistry and understanding how primates navigate the chemical landscape (Asner et al. 2011; Kokaly et al. 2009; Rothman, Chapman, Hansen, et al. 2009).

Finally, a factor expected to have an important influence on the biological activity of phytochemicals is gut morphology and the gut microbiome. Differences in gut morphology and gut microbial communities can influence the metabolism of phytochemicals and determine the type and amount of compound that gets passed from the gastrointestinal tract into the blood (N. Brown et al. 2014; K. Setchell and Clerici 2010a, 2010b). Consequently, interspecific and interindividual variation in phytochemical metabolism can influence downstream endocrine effects. Furthermore, the endocrine system and gut microbiota exhibit a bidirectional feedback relationship (Clavel, Fallani, et al. 2005; Clavel, Henderson, et al. 2005), raising questions as to how these interactions influence feeding behavior and ecology. As such, incorporating measures of gut microbiota into phytochemical studies can help illuminate the evolutionary role gut microbes may have played in establishing dietary niches (Amato 2016; Amato, Martinez-Mota, et al. 2016; Amato, Ulanov, et al. 2017).

SUMMARY AND CONCLUSIONS

Plant foods provide most of the required nutrients to most primate species, but the other molecules ingested along with these nutrients also play important roles in influencing physiology, behavior, health, and reproduction (see also Stalenberg et al., this volume). Here, we have reviewed what is known about one subset of these other chemicals that can affect primate biology through direct interactions with the endocrine system: hormonally active phytochemicals. Despite theoretical and empirical evidence of their importance, much more research is needed to document the prevalence of hormonally active chemicals in primate diets and environments, both natural and synthetic, and the effects of their presence and consumption on primates.

Moving forward, by conducting comparative studies of primate diets and chemical analyses of their plant foods, we will be able to understand the relative importance of dietary niche, phylogeny, and geographic location to variation in exposure to both anthropogenic pollutants and hormonally active phytochemicals. Through environmental sampling and noninvasive biomonitoring, we can better understand the presence of hormonally active synthetic chemicals in primate habitats and their exposure to these chemicals, allowing us to develop a landscape of chemical-primate interactions. By incorporating a landscape-level perspective, we will be able to map relative chemical risk to threatened populations and species (see Chapman et al., this volume). Additionally, since much of the metabolizing of chemicals is done by microbial organisms, advances in methods for documenting variation in the microbiome across soil, vegetation, and primate digestive systems will provide novel insights into how plant and synthetic chemicals are altered in the external environment and internally in the gut (Caporaso et al. 2011; Costello et al. 2009; Fierer and Jackson 2006; Lau and Lennon 2011). Studies of microbe-chemical interactions will be critical for clarifying the importance of hormonally active phytochemicals to primate ecological interactions and evolutionary processes (see Lambert, Mutegeki, and Amato, this volume).

- Hormonally active phytochemicals, such as phytoestrogens and other phytosteroids, occur in common primate plant foods, having the potential to influence the physiology and behavior of many species of primates.
- The influence of hormonally active phytochemicals on primate ecology and evolution is likely a combination of costs to reproduction and benefits to survival, but data are currently lacking to determine any overall net effect on fitness.
- Of 94 plant species tested from Kibale and Bwindi Impenetrable National Parks in Uganda, 13 were identified as having estrogenic activity at ERβ, and most of these species were members of the Moraceae and Fabaceae plant families. For these estrogenic plant species, all parts tested except for ripe fruits were identified as being estrogenic; thus, the highly folivorous primates (i.e., red colobus, black-and-white colobus, and mountain gorillas) were found to regularly ingest phytoestrogens, while other primates may avoid these compounds.
- Based on a meta-analysis of 50 primate species, Fabaceae and Moraceae are the two most consumed plant families across the primate order.
- Due to the prevalence of phytoestrogens in Fabaceae and Moraceae and their importance to primate diets, most primate species are likely consuming estrogenic plant species, although some may avoid estrogenic parts of those species based on their dietary niche.
- The consumption of hormonally active phytochemicals and synthetic chemicals should be considered by primate field researchers, especially when measuring hormone levels, as this may be an important source of variation in data sets.

11 Nutrition and Immune Function in Primates

Erin R. Vogel, Astri Zulfa, Sri Suci Utami Atmoko, and Lyle L. Moldawer

Diet and nutrition are critical factors influencing the health of humans and other primates. Primates are known to exhibit a diverse array of behavioral and physiological strategies to maintain adequate energy and nutrient intake. There is a large body of literature focusing on how their diets vary in response to seasonal changes in overall food availability (reviewed by Hemingway and Bynum 2005). These seasonal changes also have consequences for reproduction, survival, and health of primates—patterns that have been well documented in longitudinal studies within populations and comparative studies across taxa. This has led many to assume a causative relationship between primate nutrition and susceptibility to infectious disease, and experimental research with model organisms has begun to reveal some of the underlying mechanisms (reviewed by Ponton, Wilson, Holmes, et al. 2013). Due to a great void of studies on this topic in nonhuman primates, we still lack a basic understanding of how variation in nutritional intake modulates health and immune function in wild nonhuman primates, as most of what we do know comes from captive primates, human studies, or research on other model organisms (see table 11.1).

The objectives of this chapter are to focus on how nutritional status and dietary intake influence the inflammatory response and how host immunity mediates dietary needs. We first examine the concept of allostasis as it relates to nutrition to support physiological maintenance in nonhuman primates. The complex relationships among total energy intake, macronutrients, and immune function are assessed, and in addition, we survey how micronutrients (including vitamins and trace minerals) interact with each other to modulate the immune response. Some vitamins and minerals have a synergistic effect on immune response, while other combinations have antagonistic effects. While studies of trace mineral and vitamin intake in wild primates remain limited, we explore future avenues of exploration based on human studies. We conclude this chapter with a summary of noninvasive methods that can be applied to studying immune function in the context of nutrition.

BASIC PRINCIPLES OF NUTRITIONAL IMMUNOLOGY

The immune response is divided into two integrated systems: innate (natural) immunity and adaptive (acquired) immunity. Innate immunity typically provides an immediate first line of defense against invading microorganisms and occurs in both vertebrates and invertebrates, while adaptive immunity provides the host with a more delayed but ultimately more effective immune response upon the host's exposure to microorganisms. Although innate immunity can be "trained" and therefore has some rudimentary memory (Nica et al. 2022), it is primarily the adaptive immune response that possesses memory and therefore adapts through lifetime exposure to infectious agents. Despite their fundamental differences, both innate and adaptive immune responses are essential for

Table 11.1. Summary table of research on the relationship between nutrition and immune function in nonhuman primates.

Species	Wild/captive	Nutritional status	Pathology	Physiological/biological data	Results/significance	Reference
Alouatta palliata	Wild	Natural	Atherosclerosis	N/A: plant nutrient analysis	Low-grade atherosclerosis exists in wild populations of howler monkeys, but dietary intake of fatty acids in the wild was less than in an atherotic human diet. It remains unclear if howler monkeys are fatty acid sensitive or if other nondietary sources induced atherosclerosis.	J. Chamberlain et al. (1993)
Callithrix jacchus	Captive	Vitamin D restriction	Vitamin D deficiency; calcium insufficiency	Serum calcium; vitamin D; bone mineral density	Colitis was associated with malabsorption of nutrients and interferes with micronutrient absorption. In a calcium- and vitamin-D-deficient state, marmosets had lower bone mineral density. Captive animals may be more prone to this if they do not get adequate sunlight.	Jarcho et al. (2013)
Callithrix jacchus, Saguinus oedipus	Captive	Rice and wheat introduced to diet	Colitis; wasting syndrome	Immunoglobin groups (IgG, IgA)	Both wheat- and rice-based food products cause gastrointestinal inflammation in tamarin and marmoset populations.	M. Gore et al. (2001)
Cercopithecus aethiops	Captive	Saturated, monosaturated, polyunsaturated fat	Atherosclerosis	Coronary artery atherosclerosis; cholesteryl ester accumulation; low- and high-density lipoprotein cholesterol	Saturated, monosaturated, and polyunsaturated fats caused atherosclerosis, but polyunsaturated fats had the slowest rates of increase. Particle size and composition of low-density lipoproteins may be more important for inducing atherosclerosis than low- and high-density lipoprotein ratios.	Rudel et al. (1995)

(continued)

Table 11.1. (continued)

Species	Wild/captive	Nutritional status	Pathology	Physiological/biological data	Results/significance	Reference
Macaca fascicularis and *Papio* sp.	Captive	Dietary protein restriction	Protein-energy malnutrition	Plasma/serum proteins; erythrocyte rosettes (T-cells)	Severe immunosuppression occurs in hosts experiencing protein-energy malnutrition, both before and without infection.	Qazzaz et al. (1981)
Macaca fuscata	Captive	High-fat diet (HFD)	Endothelial dysfunction; cardiovascular disease	Insulin; triglycerides (TG); cytokines (VEGF, TNFα, and ICAM-1); intima thickness—vascular morphology	Infants who nursed from mothers on a HFD had an increase in proinflammatory cytokines associated with endothelial dysfunction, increased intima thickness, and decreased aorta dilation capacity. Infants then switched to a control diet showed nonsignificant reductions in proinflammatory cytokines and only minimal improvement.	L. Fan et al. (2013)
Macaca fuscata	Captive	HFD	Metabolic disease; pancreatic inflammation; glucose dysregulation	Cytokines (IL-6, IL-1β); insulin	Juvenile Japanese macaques fed chronic HFD had higher levels of pancreatic inflammatory markers before the prevalence of obesity and glucose dysregulation relative to controls.	Nicol et al. (2013)
Macaca fuscata	Captive	HFD	Placental inflammation; placental blood flow; stillbirths	Uterine and placental hemodynamics; cytokines (includes IL-1β, MCP-1, TLR4); insulin	Pregnant females fed a HFD were associated with placental dysfunction, decreased uterine blood flow, and increased risk of stillbirth. Maternal plasma cytokines did not increase due to HFD, but cytokines increased in the placenta.	Frias et al. (2011)

Macaca fuscata	Captive	HFD	Fetal nonalcoholic fatty liver disease (NAFLD)	Glycerol; hepatic TG; serum insulin; glucose; gluconeogenic enzyme; cytokines; stress-activated protein kinase (JNK-1)	HFDs in pregnant females were associated with a large suite of metabolic disorders in fetal offspring. Liver TG levels increased during NAFLD. Fetal livers had premature activation of the gluconeogenic pathway, lipid accumulation, increased proinflammatory cytokines, and activation of oxidative stress pathways.	McCurdy et al. (2009)
Macaca mulatta	Captive	HFD	Cardiovascular disease; insulin resistance	Truncal fat; carotid intima-media thickness; plasma inflammatory biomarkers (adiponectin, RANTES, MCP-1, CRP, IL-18, sV-CAM, IL-8, vWF); carotid P-selectin; vascular cell adhesion molecule-1; glucose; lipid profiles	Endothelial activation coincided with insulin resistance, but the severity of insulin resistance did not determine endothelial activation; rather, the duration of insulin resistance impacted the severity of endothelial activation.	Chadderdon et al. (2014)
Macaca mulatta	Captive	Had access to feed freely on commercial monkey chow	Periodontitis; metabolic syndrome	C-reactive protein; microRNAs (gingival tissues)	Metabolic syndrome was associated with spontaneous periodontitis.	H. Sun et al. (2014)
Macaca mulatta	Captive	HFD	Metabolic disease; ARH melanocortin system dysregulation	Insulin; leptin; melancortin system; cytokines (IL-1β, IL-1R1, IL-1F6, IL-1F7, eotaxin, eotaxin-R, MIP-3, MCP-1); cortisol	Elevated maternal cortisol levels were also found in fetal serum; activation in neural and fetal hypothalamus inflammation was associated with abnormalities of the melancortin system.	Grayson et al. (2010)

(continued)

Table 11.1. (continued)

Species	Wild/captive	Nutritional status	Pathology	Physiological/biological data	Results/significance	Reference
Macaca mulatta	Captive	Calorie restriction	Periodontitis	Plaque; probing pocket depth; bleeding on probing; gingival index; calculated attachment level	Periodontal destruction occurred at slower rates in calorie-restricted macaques, suggesting that calorie restriction buffers the inflammatory response or offers an anti-inflammatory effect.	Branch-Mays et al. (2008)
Papio cynocephalus	Wild	Natural	White monkey syndrome; zinc toxicity due to copper deficiency	Prenatal copper deficiency; copper content in breast milk; trace mineral profile	Despite the benefits of zinc, zinc toxicity occurred due to copper deficiency or zinc inhibiting copper uptake. Infant baboons that nursed from mothers with high zinc-to-copper ratios in their system experienced symptoms of white monkey syndrome.	Markham, Gesquiere, et al. (2011)
Papio cynocephalus	Wild	Natural and garbage-feeding	Body mass; body fat; energy expenditure	Morphometric techniques; skin-fold measures; isotope-labeled water; estimated BMI	Wild baboons feeding from garbage were more sedentary and had larger body masses and percent body fat, but were similar in body length compared to wild-feeding baboons. Potential model for energy expenditure and nutrient intake in humans.	J. Altmann et al. (1993)
Papio hamadryas	Captive	High-fat, high-cholesterol diet; vitamin E and coenzyme Q10 (Co Q10) supplements	Vascular disease	Cytokine level (CRP)	Vitamin E supplementation was correlated with an anti-inflammatory effect on circulating proinflammatory biomarkers (CRP), and cosupplementation of Co Q10 further enhanced these effects.	X. L. Wang et al. (2004)

"Natural" indicates only wild food sources and no provisioning unless noted.

host survival responses required for resisting infection, recovering from injuries, and maintaining overall homeostasis (Batool et al. 2015; McEwen and Wingfield 2003). The allostatic response to inflammation—that is, the response necessary to achieve homeostasis or stability through environmental change—is highly dependent on the nutritional state of the host, with the large majority of studies showing that protein and energy malnutrition lead to greater disease susceptibility in the host (McEwen and Wingfield 2003). The main mediators of allostasis that support achieving the homeostatic state of an organism include the hormones of the hypothalamic-pituitary-adrenal axis, catecholamines, and cytokines (McEwen and Wingfield 2003). During allostasis, the activity levels of these mediators are altered in response to a changing physical and social environment such that there may be inadequate production of some with concurrently excess production of others, resulting in an imbalance.

These main mediators of allostasis are intricately linked. Cytokines are low-molecular-weight proteins that provide communication among different tissues and cells, activate and regulate inflammatory responses, direct leucocyte movement, modulate the balance between humoral and cell-based adaptive immune responses, and stimulate the production of blood cellular components (Delano and Moldawer 2006; Elenkov and Chrousos 2002; Kubena and McMurray 1996; Zimmerman et al. 2014). Homeostasis within the immune system is dependent to a large extent on the balance of cytokines and other inflammatory mediators, as they are the predominant signaling molecules of the immune system among itself and with parenchymal tissues. The regulation and function of cytokines has been well studied in vertebrates, making them excellent markers for the study of nutritional immunology in primates (Banuls et al. 2019; Higham, Kraus, et al. 2015; C. Hoffman, Higham, et al. 2011; Zimmerman et al. 2014). For example, human studies have shown that cytokine production and regulation are profoundly affected by nutrient balance and that malnutrition impairs immune competence (Iyer et al. 2012; Kudsk et al. 2000; Michael et al. 2022; Sävendahl and Underwood 1997; F. Wallace et al. 2001; Wan et al. 1989). Elenkov and colleagues (2005) reviewed cytokine production in humans and mice and found that overall, dysregulation of cytokine signaling leads to greater susceptibility to pathogens and infection. One major cause of cytokine dysregulation is nutritional deficits, as the interactions of immune cells, phagocytes, and the release of cytokines are costly in terms of energy and protein requirements (Chandra 1983; Dos Santos et al. 2017; Wan et al. 1989). It is well known that diet-derived nutrients affect immune function at several levels (e.g., thymus, spleen, lymph nodes, and circulating immune cells; Cunningham-Rundles 2002; Cunningham-Rundles et al. 2005; Dos Santos et al. 2017; Stelmasiak et al. 2021). Proteins, specifically branched-chain amino acids, arginine, and glutamine, are integral for launching the host immune response and have been shown to improve nitrogen balance and increase whole-body protein synthesis (Cruzat et al. 2014; Keenan et al. 1982; Negro et al. 2008; Wan et al. 1989). Several cytokines (e.g., IL-1, TNFα, and IL-6) are involved in triggering the transcription and synthesis of acute-phase proteins used to support metabolic function by infected organisms (Aliyu et al. 2022; Wan et al. 1989; Wolf et al. 2014).

While data relating cytokine production to nutrition are available for humans and some other model organisms (e.g., rodents and insects), to our knowledge no studies have examined these relationships in wild nonhuman primates, although cytokines have been used as markers of health and immune function in free-ranging primates (C. Hoffman, Higham, et al. 2011). In captive macaques, infants born to mothers fed high-calorie, high-fat diets had localized increases in proinflammatory cytokines in the hypothalamus, and macaques fed a high-fat diet exhibited an amplified acute inflammatory response (Frias et al. 2011; Grayson et al. 2010). While these primate-model studies have relevance for understanding the prevalence of obesity and its health consequences in humans, wild primate diets are not typically high in fat (Janson and Chapman 1999). However, future studies focusing on energy or protein intake combined with noninvasive markers of inflammation such as cytokines will be particularly relevant to understanding how nutrition modulates immune function and the evolution of the immune response in humans.

The immune system and hypothalamic-pituitary-adrenal axis are intricately linked, and both play major roles in the regulation of homeostasis. Perturbations, such as nutrient deficits, to either system lead to an increase in allostatic load, ultimately increasing an organism's susceptibility to infection and disease (McEwen and Wingfield 2003). Studies on mice, humans, and nonhuman primates have shown simultaneous elevations in both glucocorticoids (GCs) and proinflammatory cytokines during periods of environmental stress, be it trauma, infection, or psychological stress (Altemus et al. 2001; Avitsur et al. 2006; Christian et al. 2006; Higham, Kraus, et al. 2015; C. Hoffman, Higham, et al. 2011; Reyes and Coe 1998; Rozlog et al. 1999). Elevations in GCs can inhibit the production of certain proinflammatory cytokines (e.g., IL-12, IFNγ, and TNFα) and stimulate the production of anti-inflammatory cytokines (IL-4 and IL-10), providing a negative feedback mechanism that protects the host (reviewed in Elenkov and Chrousos 2002; C. Hoffman, Higham, et al. 2011). In the long term, this can lead to reduced immune function, impaired healing, and increased susceptibility to disease (Elenkov and Chrousos 2002; Philip et al. 2012).

Many of the macroendocrine and inflammatory cytokine responses to infection and injury are evolved mechanisms to support short-term host protective immunity at the expense of long-term nutritional status (Cunningham-Rundles 2002; Michael et al. 2022). For example, the early corticosteroid and catecholamine response to inflammation increases glucose availability for inflammatory and wound cell populations by mobilizing glycogen reserves and increasing protein catabolism for gluconeogenesis (McKay and Cidlowski 2003). Simultaneously, increased production of inflammatory cytokines, such as TNFα, IL-1, and IL-6, promotes amino acid release from muscles, skin, and the gut (L. Moldawer and Copeland 1997; Roh et al. 1986), which are directed to the liver to support both gluconeogenesis and the synthesis of acute-phase proteins such as haptoglobin, C-reactive protein, complement, and opsonins (Koj 1970; Sganga et al. 1985; fig. 11.1). In the short term, these responses benefit the host, but the continued loss of skeletal protein, skin, and gut results in reduced physical activity and decreased barrier function, which are detrimental. Continued infections, coupled with inadequate dietary protein intake, lead to a form of protein-calorie malnutrition that ultimately leads to a loss of host protective immunity and then death. Studies have shown that in children with this kwashiorkor form of protein-calorie malnutrition have inadequate quantities of host proteins to mobilize in support of host protective immunity (Ibrahim et al. 2017; Jahoor et al. 2008). While these relationships are well documented for humans (Elenkov and Chrousos 2002), fewer studies have examined how stress mediates immune function in nonhuman primates (but see Georgiev, Muehlenbein, et al. 2015a; Higham, Kraus, et al. 2015; Higham, Vitale, et al. 2010; C. Hoffman, Higham, et al. 2011; Muehlenbein 2015). High GC levels as well as nutritional imbalance can lead to greater immunosuppression and susceptibility to disease, which can in turn affect energy and reproduction (Chapman, Schoof, et al. 2015; Muehlenbein et al. 2003). For instance, a study of wild chimpanzees showed that GCs were positively associated with parasite prevalence and richness, indicating a relationship among stress, immunity, and health (Muehlenbein 2006; Muehlenbein and Watts 2010).

Most of what we know about the relationship between nutrition and host immunity comes from either studies on model organisms that are more easily manipulated in a laboratory setting or human studies on populations that have experienced trauma, infection, or environmental catastrophes. From these studies, we know that dietary intake, inflammation, and protective host immunity are inextricably linked, and when an organism is malnourished, host protective immunity is significantly impaired (Cunningham-Rundles 2002; Lazzaro and Little 2009; Michael et al. 2022). Because mounting an immune response is energetically costly (Bonneaud et al. 2003; Muehlenbein 2008; K. Wilson and Cotter 2013), the outcome of host fitness in response to infection is greatly influenced by host nutrition. The relationship between nutrition and immune response is bidirectional and complex, with nutrition affecting the host's immune response both directly and indirectly and the host's immune response affecting its ability to acquire and assimilate environmental nutrients as well as the utilization of endogenous reserves (Cotter et al. 2011; Cunningham-Rundles et al. 2005; Lazzaro and Little 2009). For example, chronic protein and calorie insufficiency is commonly associated with declines in protective immunity, increased susceptibility to and persistence of infections, and ultimately the development of protein-calorie malnutrition (G. Blackburn and Bistrian 1976; Dewan et al. 2009; Moreira, Burghi, and Manzanares 2018; Wan et al. 1989). Chronic infections also induce anorexia, lean-tissue wasting, loss of visceral proteins, and the development of protein-calorie malnutrition (G. Fernandes 2008; Peck, Babcock, and Alexander 1992; Rincon et al. 2022). Chronic infections are frequently associated with "sickness syndromes" that reduce spontaneous nutrient intake, reduce nutrient utilization, and depress host pro-

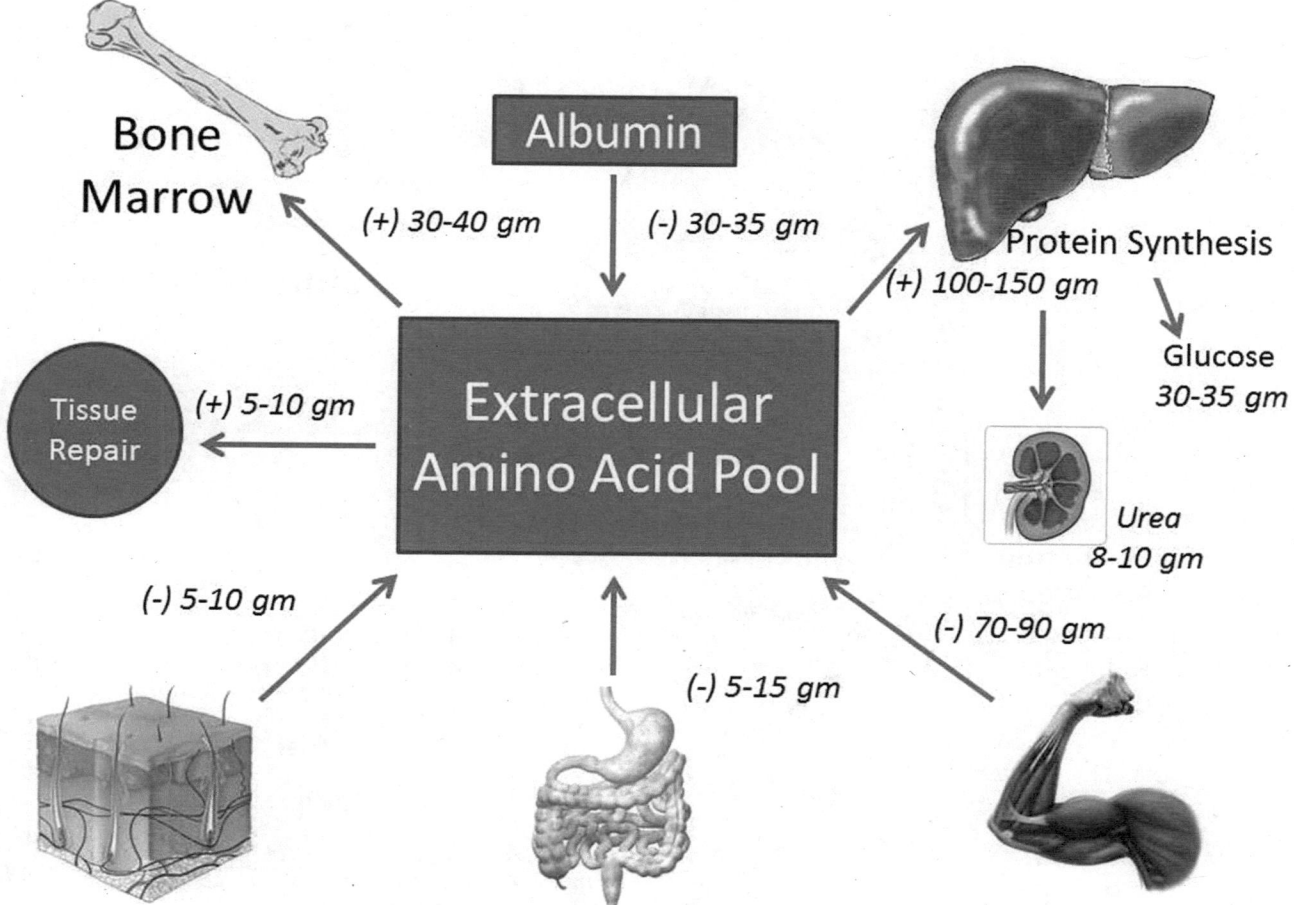

Figure 11.1. Functional redistribution of the body cell mass. During periods of infection and reduced dietary protein intake, the host relies to a greater extent on free amino acids released from skeletal muscle, skin, and gut tissue net catabolism to provide the substrate for the increased anabolic needs. The increased anabolic response is primarily in tissues and organs involved in host protective immunity and wound healing, and it involves the synthesis of acute phase proteins and complement in the liver, bone-marrow-derived new leukocyte populations, and healing tissues. Figure by L. Moldawer.

tective immunity. This repetitive incidence of nutritional insufficiency coupled with endemic and opportunistic infections is a vicious cycle that ultimately leads to reduced survival from the failure of host protective immunity (Rincon et al. 2022; fig. 11. 2).

However, the relationships between nutrient availability, dietary intake, decreased protective immunity, inflammation, and protein-calorie malnutrition depend on a large number of independent variables. For example, the age of the host, its prior nutritional status, its current energy demands, and external physical and psychological cues all influence the dynamic interactions between nutrient availability, protective immunity, inflammation, and nutritional status (Shepherd 2008, 2009; Soultoukis and Partridge 2016). Humans, the most studied primate in the field of nutritional immunology, have adapted a significant number of mechanisms to deal with variations in nutrient availability. Although first-world living and a shifting diet have made many of these adaptations either irrelevant or actually harmful, humans are remarkably resistant to transient nutrient deficiencies, and they invoke both metabolic and immunologic responses meant to maintain homeostasis in the face of dramatic changes in nutrient availability (M. L. Power and Schulkin 2009).

Whereas nutritional status and the host response to inflammation and protective immunity are reasonably well understood in humans, how dietary intake per se influences inflammation and protective immunity is

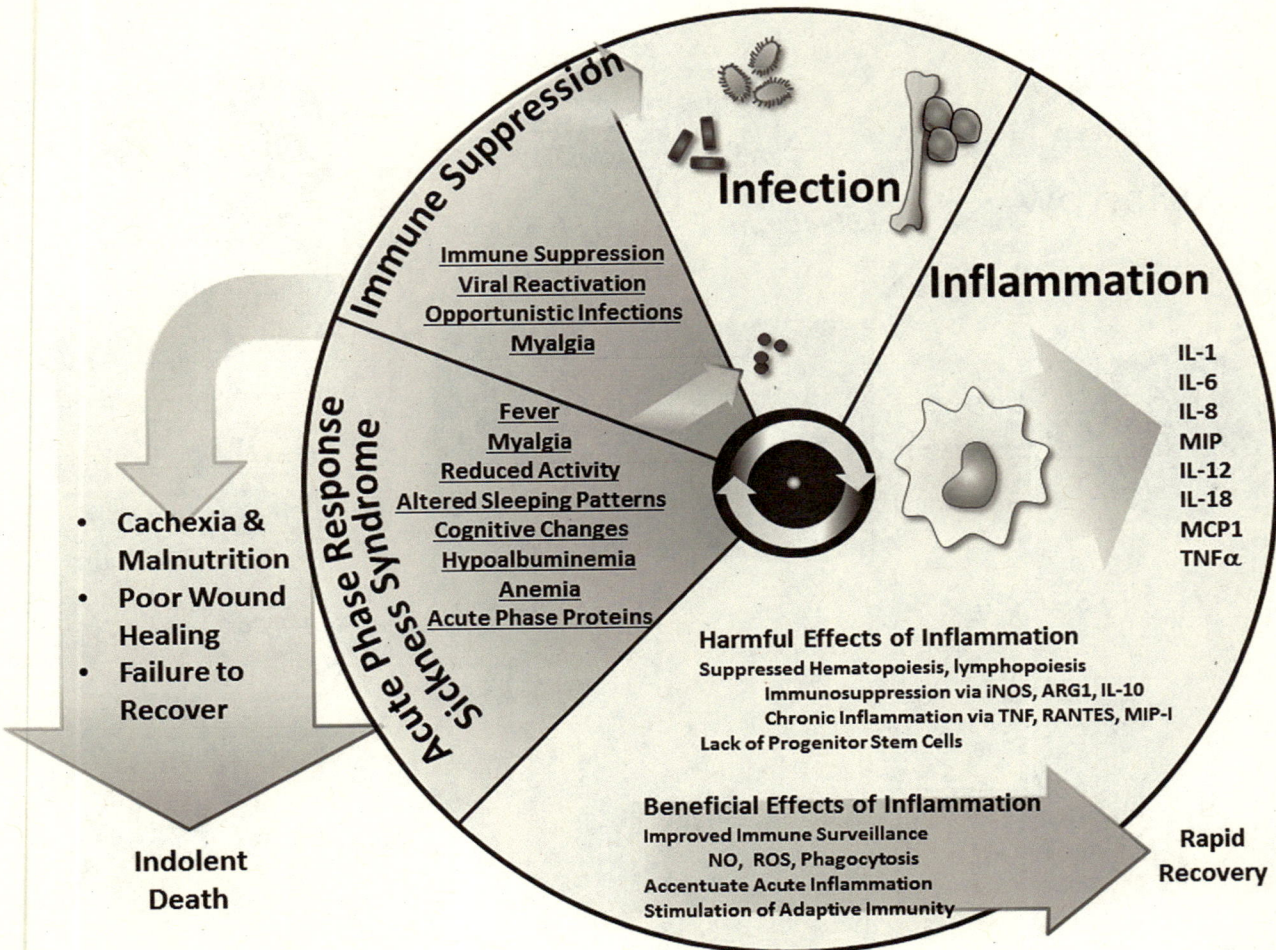

Figure 11.2. Cycle linking nutritional status, immunological function, and incidence of infection. Infection results in the activation of innate immunity, inflammation, and "sickness syndrome." This often leads to reduced food intake, including dietary protein. Persistent reductions in dietary protein and inflammation promote losses of both protein reserves and host protective immunity. The more susceptible organism has a greater risk of additional infections, which repeats the cycle. The end result is protein calorie malnutrition and reduced protective immunity, which, if severe, inevitably leads to death. Figure by L. Moldawer.

considerably less well understood. Most of the data we have come either from healthy individuals fed modified diets for extended periods or from hospitalized patients fed enterally or parenterally. In the latter case, the effect of diet on host inflammation or protective immunity is generally evaluated either in the context of a chronic inflammatory process, such as cancer, or after trauma or acute infection. Most studies of malnutrition that have focused on nonhuman primates have used them as models to better understand human syndromes of malnutrition (Qazzaz et al. 1981). We emphasize that we have much to learn about nutritional immunology in wild, nonhuman primates. In the following sections, we provide details on what is known about the influence of nutrition on immune function in model organisms, including nonhuman primates. We end this chapter with a synthesis of noninvasive field methods that can be used to study immunological responses to nutrition in nonhuman primates.

ENERGY DEFICITS AND IMMUNE FUNCTION

Energy deficits, or in the extreme case, starvation, represent perhaps the greatest challenge to human and nonhuman primates; yet, there are adaptive mechanisms that preserve essential protein stores and functions for extended periods across mammals. Glycogen reserves are

Table 11.2. Anabolic processes during early innate immune and inflammatory responses.

Organ or tissue	Function
Bone marrow	Increased hematopoietic stem cell proliferation
	Increased mesenchymal stem cell proliferation
	Increased myelopoiesis
	Differentiation of myeloid progenitors
Spleen	Extramedullary hematopoiesis
	Expansion of stem cell populations
	T-cell clonal expansion
Lymph node	T-cell clonal expansion
	Antibody production
Gut	Antibody production
Liver	Acute-phase protein synthesis
	Increased synthesis of coagulation cascade proteins
	Complement and opsonin production
	Glucose production for obligatory requirements
Wounds and damaged tissue	Cell proliferation
	Collagen deposition
	Cytokine and inflammatory mediator production
	Expansion and maturation of inflammatory cell populations

generally exhausted within 48 hours, but the host quickly transitions to a predominantly lipid-based fuel source (Bassili and Deitel 1981) (Table 11.2). Obligatory glucose requirements are dramatically reduced and generally met by gluconeogenesis, or they are replaced with beta oxidation of fatty acids and ketone bodies (Bassili and Deitel 1981). The previously well-nourished host has substantial amounts of protein reserves, not only for gluconeogenesis but also for providing substrates for essential host functions, such as protective immunity (table 11.3). During gluconeogenesis, reduced insulin and elevated glucagon concentrations favor protein catabolism of skeletal muscle, skin, and gut to provide the substrate (amino acids) for more vital functions, such as red blood cell and leukocyte production, as well as hepatic protein synthesis of opsonins and other proteins involved in innate immunity and pathogen recognition (Bistrian, Blackburn, et al. 1975; Flatt and Blackburn 1974; Keusch 2003; Rui 2014; table 11.2). This "redistribution of the body cell mass," as it was first described, is the primary host mechanism to maintain vital functions in the face of total inanition (G. Blackburn and Bistrian 1976; Flatt and Blackburn 1974; Rincon et al. 2022; fig. 11.1). There is convincing evidence that modest periods of total starvation have only minimal impact on both innate and adaptive immunity, and in the absence of infection or injury, humans can survive up to 70 days (C. J. Murray et al. 2015). Studies have demonstrated that two weeks of fasting have variable adaptive immune responses (Galland and Polk 1982; Walrand et al. 2001), and, in some cases, protective immunity is actually increased (Lin et al. 1998). Furthermore, T-cell proliferative responses to mitogens and antigens were found to be only modestly affected by fasting. For example, Bistrian and colleagues showed in the early 1980s that short-term fasting was not associated with any significant reduction in the acute phase response to yellow fever vaccination (Bistrian, George, et al. 1981; Bistrian, Winterer, et al. 1977). For human populations, however, reduced nutrient availability and inanition do not exist in a void but rather occur in environments where endemic infections are common, and infections or injury represent a predominant cause of death.

Adaptations to maintain homeostasis during periods of reduced nutrient availability, increased starvation,

Table 11.3. Energy requirements and sources during inflammation and injury.

Tissue or diet	Nutrient type	Response
Dietary intake	Protein and calories	Significantly decreased due to anorexia and immobility.
Skeletal muscle	Protein and energy sources	Increased early glycogenolysis for muscle use. Release of amino acids from protein degradation with oxidation of branched-chain amino acids for energy, and release of increased amounts of alanine and glutamine for gluconeogenesis. Beta-oxidation of free fatty acids and ketone bodies.
Liver	Protein, free amino acids, glycogen, and glucose	Increased glycogenolysis early and release of glucose; later, gluconeogenesis of alanine and other glucogenic amino acids; beta oxidation of free fatty acids and ketone body generation.
Kidney	Free amino acids	Gluconeogenesis from glutamine and to a limited extent from other sources.
Brain	Glucose	An obligatory requirement for glucose as a fuel; alternative fuel sources limited.
Wound tissue	Glucose	Granulating tissue utilizes predominantly glucose.

and thus increased allostatic load are often disrupted by inflammation and infection (Scrimshaw et al. 1959; Solomons 2007). As shown in table 11.3, the host response to infection or inflammation places immediate energy and protein demands on the host, which take priority over adaptive mechanisms meant to spare essential body proteins during inanition and reduced nutrient availability. The term "acute phase response" to inflammation, of either infectious or traumatic origin, reflects the changes in host metabolism meant to support antimicrobial and host survival functions on a short-term basis (Kapetanovic and Cavaillon 2007; Wan et al. 1989). These short-term responses can be considered beneficial to the host but become pathologic when they are prolonged or exaggerated (R. Hawkins et al. 2018; Rincon et al. 2022). Indeed, the inflammatory response presents a paradox, as many of our body's most vital responses are expressed as primary effects necessary for survival but become detrimental to our overall health when the response is sustained over the long term (McEwen 2000).

The immediate demands on the infected or injured host are for substrates for glucose and new protein synthesis, which are essential for wound healing, reducing the replication and dissemination of the invading pathogen, and preventing secondary opportunistic infections (Beisel 1975). The quantitative protein demands on the host are not trivial. Blood neutrophils, the primary first cell population at the site of infection and injury, have a half-life of only 4–12 hours, and to maintain a neutrophil count of 15,000 cells/mm^3 requires 20–30 g of amino acids per day for new protein synthesis in the bone marrow (Beisel 1975; fig. 11.1). Hepatic protein synthesis dramatically changes as the synthesis of albumin is replaced and expanded by the required synthesis of multiple acute-phase proteins that are both antimicrobial and reparative in nature (Bostian et al. 1976; Sehgal 1990). Increased immunoglobulin synthesis is required for an adaptive protein response (Tayek and Blackburn 1984). Equally important, many of these inflammatory cells and granulating wound tissue require glucose almost exclusively as a fuel, increasing its requirement by gluconeogenesis from amino acids. Finally, the reduced energy expenditure often seen as an adaptive response to starvation or reduced nutrient intake gives way to increased energy demands as the host increases net protein synthesis to generate the products required for host protective immunity (Clowes et al. 1976; N. Ryan 1976).

Persistent inflammation in the presence of reduced nutrient intake can have devastating results, including muscle wasting, hypoalbuminemia, anemia, lymphopenia, and attrition of both primary and secondary lymphoid organs. In their most severe forms, these alterations lead to the loss of both innate and adaptive immunity, making the host more susceptible to opportunistic infections and increasing their severity (R. Hawkins et al. 2018). The net result is a vicious, destructive cycle, termed the "persistent, inflammatory, immunosuppressive and protein catabolic syndrome" (Darden et al. 2021; R. Hawkins et al. 2018). This destructive cycle is more pronounced depending on the age of the organism, as age can also have a dramatic effect on the relationship between nutritional status and inflammatory challenges (Mankowski et al. 2022). For example, the aged frequently have age-associated sarcopenia, reduced adaptive immunity, higher rates of mortality due to trauma and infections, and frequently, a chronic low-grade inflammatory state (M. Bauer and Fuente 2016; Hazeldine et al. 2015; Nacionales, Gentile, et al. 2014; Nacionales, Szpila, et al. 2015). These studies along with studies on rhesus macaques (C. Hoffman, Higham, et al. 2011), Barbary macaques (N. Muller et al. 2017), and chimpanzees (Negrey, Behringer, et al. 2021) provide increasing data to suggest that increasing age is commonly associated with a chronic inflammatory status. The causes are likely multifactorial, but the general consensus is that chronic inflammation is secondary to long-term and accumulated oxidative injury (M. Bauer and Fuente 2016).

Like the aged, neonates also exhibit some of the highest mortality rates to infection and injury (Rodriguez Cervilla et al. 1998; J. L. Wynn et al. 2015) and are poorly prepared for periods of nutrient insufficiency, especially in the presence of infection or injury-associated inflammation. In newborns and especially the premature, there are few if any protein or lipid substrate reserves, and thus, chronic or persistent inflammation rapidly exhausts the host nutrients required to mount an effective immune response (J. Cunningham 1995). Combined with an incomplete adaptive and defective innate immune response, it is not surprising that morbidity and mortality in humans are so high where infections are endemic. Thus, as in many organisms (including nonhuman primates), the aged and newborn individuals within a population experience the most dramatic response to nutritional imbalance when faced with immunological challenges.

Extreme and chronic starvation appear to promote susceptibility to disease by impairing the development and expression of the immune response (Cunningham-Rundles et al. 2005). Indeed, several experimental studies on insects have shown a direct link between both short- and long-term starvation and decreased immune response (J. Ayres and Schneider 2009; De Block and Stoks 2008; Moret and Schmid-Hempel 2000; Siva-Jothy and Thompson 2002). Evidence from humans, nonhuman primates, and mice has shown that prolonged starvation or inanition leads to increased susceptibility to disease, inflammation, and mortality (Loy 1970; Sävendahl and Underwood 1997). Due to ethical concerns, however, the large majority of experimental studies on human and nonhuman primates have focused on caloric restriction (versus starvation) and have yielded mixed results (Nikolich-Žugich and Messaoudi 2005). For example, caloric restriction has been associated with reduced inflammation in captive rhesus macaques (*Macaca mulatta*; Branch-Mays et al. 2008; Willette et al. 2010). However, calorie-restricted bonobos showed evidence of the breakdown of tissues and body energy reserves (Deschner, Fuller, Oelze, Ortmann, et al. 2010), which has been associated with the loss of both innate and adaptive immunity in humans (Cunningham-Rundles et al. 2005). Mattison and colleagues (2012) found no effect of caloric restriction on survival outcomes in rhesus macaques. Similar conflicting results have been obtained in caloric restriction experiments and starvation studies in humans (Genton et al. 1998). These varying results may be due to differences in the ratio of macronutrients in the diet (Kapahi et al. 2010; Sohal and Forster 2014).

ENERGY EXCESS AND IMMUNE FUNCTION

Nutrient excess can also have a significant impact on host immunity and inflammation. This may be a unique problem in first-world countries where obesity is rampant, with approximately 30% of the population considered obese (Ogden et al. 2007). The available data on human obesity and host protective immunity are modest and often conflicting. There appears to be a general consensus that obesity is associated with a chronic inflammatory state (Bastard et al. 2006; Frydrych et al. 2018). The argument is that adipose tissue contains a number of different cell populations, including adipocytes, fibroblasts, pluripotent stem cells, and other myeloid populations. Many of these cell populations are capable of producing inflammatory cytokines, such as TNFα, and other members of the IL-6 superfamily. In addition, several of the "adipokines," such as leptin and ghrelin, may have proinflammatory and energy-regulating properties of their own. Individuals with increased adipose tissue often have increased basal production of these inflammatory mediators, and as such, increased chronic inflammation (Booth et al. 2016; Lackey and Olefsky 2016).

We do know that obese individuals who experience trauma or septicemia have higher rates of secondary infections, although mortality is not necessarily increased (Winfield 2014). However, it has been difficult to ascribe this increased frequency of secondary infections to obesity and chronic inflammation per se, since management of these patients is frequently more difficult and more invasive (Winfield 2014). Obese patients are less likely to ambulate and often have considerably more comorbidities, such as diabetes and cardiovascular disease (Winfield 2014). Some studies have actually shown an attenuated inflammatory response to trauma (Winfield et al. 2012).

For nonhuman primates living in their natural environments without provisions from humans, obesity has yet to be documented, although it certainly is a problem in captivity and zoological settings. Indeed, finding enough food for survival is likely one of the greatest challenges for nonhuman primates. Primates have evolved a suite of adaptions for coping with seasonality in fruit availability (Hemingway and Bynum 2005). For example, Knott (1998) found that orangutans deposit fat reserves during periods of high fruit availability and catabolize these reserves when fruit is scarce and caloric intake is low, which, combined with a uniquely slow metabolism (Pontzer, Durazo-Arvizu, et al. 2016; Pontzer, Raichlen, et al. 2010), may help explain why captive orangutans are especially prone to obesity and diabetes relative to other great apes (Gresl et al. 2000; Zihlman et al. 2011). This variation in caloric intake (M. E. Harrison, Morrogh-Bernard, and Chivers 2010; Knott 1998; E. Vogel, Harrison, et al. 2015), which often leads to a negative energy balance (Emery Thompson and Knott 2008; M. E. Harrison, Morrogh-Bernard, and Chivers 2010; Knott 1998), has led researchers to proposed that the exceptionally low energy expenditure and metabolic rate of orangutans relative to other hominids are adaptations to their nutritionally variable habitats (Pontzer, Raichlen, et al. 2010).

Recent studies on diet selection in nonhuman primates have focused on calories obtained from the intake of protein and nonprotein energy (Cui, Wang, Shao, et al. 2018; Felton, Felton, Raubenheimer, Simpson, Foley, et al. 2009; Hou, Chapman, Jay, et al. 2020; Irwin, Raharison, Raubenheimer, et al. 2015; C. A. Johnson, Raubenheimer, Rothman, et al. 2013; Raubenheimer, Machovsky-Capuska, Chapman, et al. 2015; Rothman, Raubenheimer, and Chapman 2011). Caloric intake plays an important role in nutrient utilization and therefore in the long-term effects of diet on host immunity (Afacan et al. 2012). Dietary protein utilization is highly dependent on meeting the energy needs of the host from nonprotein energy sources. The source of nonprotein calories, whether carbohydrate or lipid, is less important in their protein-sparing effect than the total quantity of calories consumed (Bark et al. 1976; B. Wolfe et al. 1977). Infection, inflammation, and trauma produce anorexia and reduced calorie intake, which are known as the classic components of sickness syndromes. The survival benefit of this response has always been questioned, since energy requirements are often increased after trauma and infections (R. Wolfe et al. 1983). Surprisingly, however, human studies of hospitalized patients have shown that administration of calories in excess of energy requirements can be detrimental to the recovery of the patient (M. Berger 2014; Schetz et al. 2013). Common hospital practice is to deliver calories at energy requirements but to increase dietary protein intakes often to as high as 1.5–2.0 g per kg of body weight per day (M. Berger 2014). Indeed, several studies on model organisms have found that limiting individual macronutrients may have more salient effects on immunity than restricting overall caloric intake. For example, Mair and colleagues (2005) found that *Drosophila* subjected to dietary protein and lipid restrictions had higher rates of mortality than calorie-restricted *Drosophila*.

PROTEIN INTAKE AND HOST INFLAMMATORY RESPONSE

While it is generally accepted that chronic protein insufficiency can have a dramatic effect on host adaptive and innate immunity, short-term dietary protein intake appears to have only a modest impact on immediate inflammatory and immunological responses in humans and mice (Cunningham-Rundles 2002; Peck, Babcock, and Alexander 1992). Rather, the effects of protein and calorie restriction on host immunity generally require days or weeks to manifest. This is not surprising when one considers that dietary protein intake is only a small component of whole-body protein turnover rates in humans. While dietary protein intakes range from 0.6 to 1.5

g of protein per kg of body weight per day (50–150 g/day), whole-body protein breakdown and synthesis range from 3.0 to 4.0 g of protein per kg of body weight per day (200–300 g/day; Fukatsu and Kudsk 2011). Thus, dietary protein intake represents only 25%–50% of whole-body protein breakdown and synthesis daily, and acute changes in dietary protein intake likely have only modest impacts on whole-body protein metabolism, at least in the short term.

There are, however, a number of caveats to this conclusion. First, dietary protein often has a different fate than protein catabolized to free amino acids in peripheral tissues. Dietary protein and its constituent amino acids come into contact first with luminal cell populations including both epithelial and lymphoid/myeloid cells (e.g., Peyer's patch and gut-associated lymphoid tissues). Approximately 60% of the host immune system is in the gut and splanchnic region (Conroy and Walker 2008), and they are preferentially exposed to dietary constituents. As a result, acute and chronic changes in oral nutrient availability often have a weighted effect on gut immunological function.

This effect of protein and its constituent amino acids is most greatly seen during periods of injury or infection when inflammation is at its greatest (fig. 11.3). Mucosal atrophy depends primarily on the quantity of dietary protein intake (Siddiqui et al. 2020; Steffee et al. 1976) and is characterized by diminished intestinal function as well as morphological changes including decreased villous height, crypt depth, surface area, and epithelial cell numbers. Animal studies suggest incremental relief of atrophy with progressively greater protein intake and that morpho-

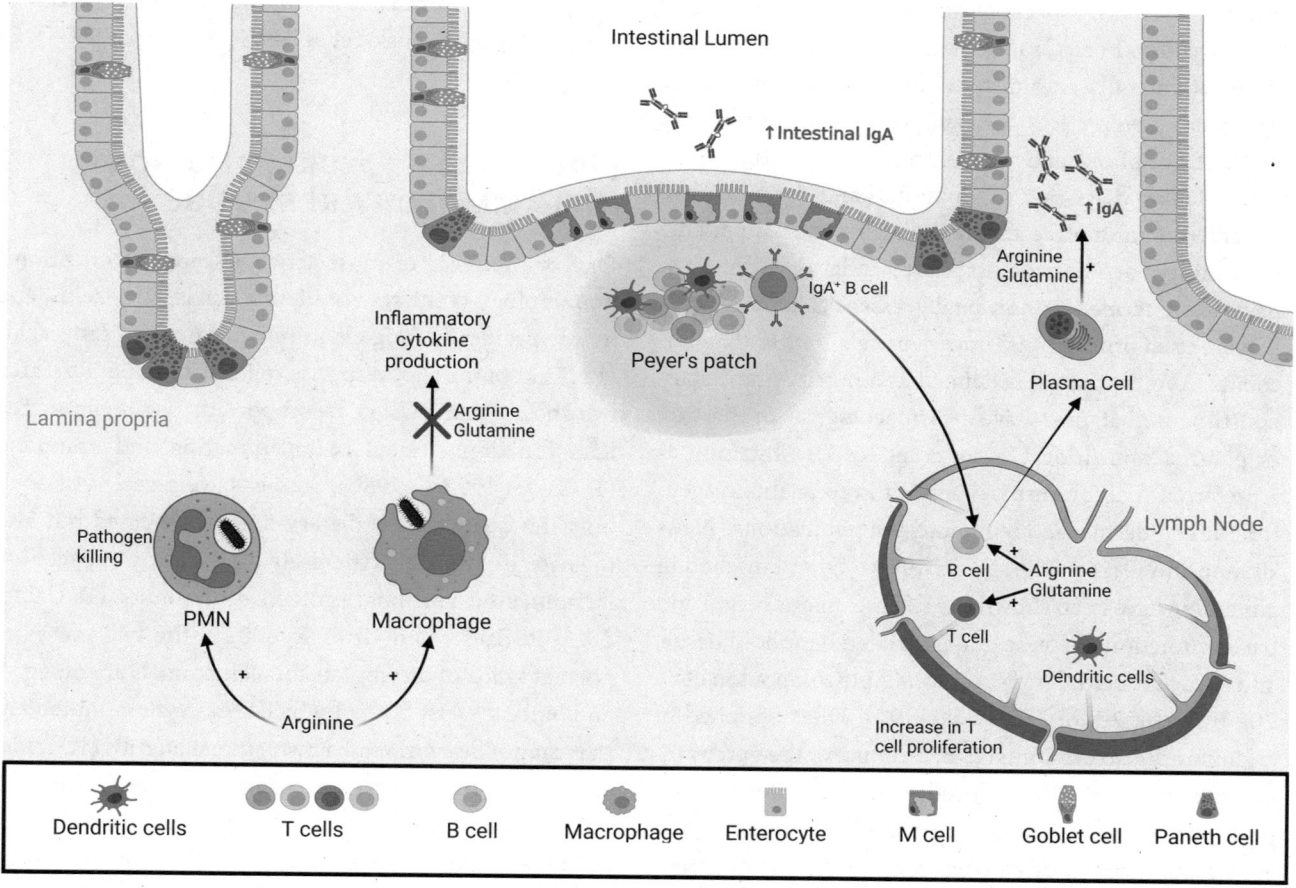

Figure 11.3. Amino acid and protein utilization by gut lymphoid populations during infection. Glutamine and arginine in particular are utilized at enhanced rates by T-cell populations, including effector T cells and gamma-delta T cells, as well as by innate lymphoid populations during trauma and infection. Both glutamine and arginine are required for T-cell proliferation, and arginine is both a precursor for nitric oxide production and a Krebs cycle intermediate. These amino acids become conditionally essential in severe trauma and infection. Figure created with BioRender by L. Moldawer.

logic atrophy is most evident at the villous tip (Kang and Kudsk 2007). Protein intake also has a dramatic impact on lymphocytes and in the gut (Vidueiros et al. 2008). This is critically important to the prevention of bacterial translocation from the gut to mesenteric lymph nodes, the liver, and distant organs (fig. 11.3). The mucosal immune system produces about 7% of the antibodies made by the body and produces specific antibodies against intraluminal bacteria antibody in the form of secretory IgA (sIgA), which functions not through inflammation but rather through adhesion and bacterial exclusion (C. D. Johnson et al. 2003; Kang and Kudsk 2007). In both experimental and clinical work, alterations in diet and injury dramatically affect this coordinated system of acquired immunity in gastrointestinal tracts (Kang and Kudsk 2007).

Alterations in the amino acid patterns of dietary protein affect both epithelial barrier function and host immune systems, especially during periods of injury or inflammation. In fact, modifying the amino acid composition of enterally administered proteins or amino acids has been attempted as a means to support enterocyte and gut lymphoid and myeloid function (Latifi 2011). In particular, diets rich in glutamine and arginine have become commonplace for treating hospitalized patients with injury- or infection-associated inflammation. Both have been termed conditionally essential amino acids and are used primarily as "nutraceuticals" rather than nutrients. A nutraceutical is defined as a nutrient or dietary constituent that may have pharmacological properties, or pharmaconutrition (Pierre et al. 2013). Glutamine is a preferred fuel for enterocytes and is an anabolic agent for both epidermal and lymphoid cell populations (Newsholme and Carrie 1994; Sacks 1999). Diets enriched in glutamine have been shown to preserve villous height and barrier function, increase gut-associated lymphoid tissue, and reduce bacterial translocation (Apostolopoulou et al. 2020; Kudsk 2006; J. Li et al. 1997). Diets enriched in arginine are also presumed to be trophic for the gut, by increasing nitric oxide (NO) production and enterohepatic perfusion. Arginine is also required by T-lymphocytes for proliferation and is often consumed in the gut microenvironment for NO production (C. Morris et al. 2017; fig. 11.3). In other words, certain amino acids improve protective immunity in humans, yet we still know little about variation in amino acid consumption in wild primate diets and its implications for primate health.

Longer-term protein deficits have greater effects on immune function by negatively impacting the epithelial barrier, immature T-cells, cytokine production, and phagocytic function (Cooper et al. 1974; Cunningham-Rundles 2002; Peck, Babcock, and Alexander 1992). Experimental research has shown that disease-challenged caterpillars have lower mortality on a high-protein diet, and these individuals preferentially select a diet with a higher protein-to-carbon ratio (K. Lee, Cory, et al. 2006). These results provide strong evidence that (1) the host immune response is more tightly regulated by protein than by total energy during a viral infection, (2) there are high protein costs associated with immune defenses, and (3) the costs may be offset by increasing protein intake. Surprisingly, there are no studies focusing on the effect of dietary protein on immune function in nonhuman primates, even though protein is thought to be a limiting nutrient for primates in tropical habitats (Ganzhorn, Arrigo-Nelson, Boinski, et al. 2009; Mattson 1980; T. C. White 1993).

LIPIDS, FATTY ACIDS, AND HOST INFLAMMATORY RESPONSE

One of the areas of most active research in nutritional immunology is the role of dietary lipids, fatty acids, and more specifically long-chain polyunsaturated fatty acids (PUFAs) in the host response to inflammation. Research has shown that PUFAs are important regulators of cellular functions related to inflammation and immunity (P. Calder 1998a, 1998b). Studies have revealed that not only the quantities of dietary lipids consumed but also the nature of the lipids themselves can clearly impact both inflammation and host protective immunity (P. Calder 2013; P. Calder and Grimble 2002). The two most important kinds of dietary fats for all mammals are omega-6 and omega-3 fats, both of which are polyunsaturated fats that cannot be synthesized by mammalian cells (P. Calder and Field 2002). The large majority of laboratory studies have found that high-fat diets, particularly those rich in linoleic acid, an omega-6 fatty acid, result in reduced innate immune response (P. Calder 1998b), but there is a threshold, as other studies on rodents have found that diets deficient in omega-6 or omega-3 fatty acids reduce immune response (D. Kelley and Daudu 1993). Kelly and colleagues (1992, 2005) found that humans fed a diet where fat contributed 25% of the daily energy intake had enhanced lymphocyte proliferation, but high-fat diets were associated with suppressed T-cell proliferation. However, another study found that essential fatty-acid deficiency was associated with suppressed cell-mediated immune responses (D. Kelley and Daudu 1993). In other words, both fatty acid excess and deficiency have similar immunological effects in laboratory studies, but the type of fatty acid (e.g., those high in linoleic acids verses high-saturated-fat diets) also influences the outcomes (D. Kelley and Daudu 1993).

For humans, saturated animal fats and oils from monocot seeds make up the majority of fatty acids in the diet. Omega-6 fatty acids are the precursors of prostaglandins and leukotrienes that in general have proinflammatory properties (Whelan 2008; table 11.4). Western diets rich in omega-6 have been suggested to be "pro-inflammatory" because increased prostaglandin synthesis during injury or infection yields increased amounts of lipoxins, prostacyclins, and thromboxanes based on eicanoic acid (Whelan 2008). These prostaglandins and lipoxins generally have both an immunosuppressive and inflammatory phenotype (Blok et al. 1996). In contrast, oils based on increased levels of omega-3 fatty acids are thought to be less proinflammatory, since the same cyclooxygenases generate prostaglandins and lipoxins based on eicosapentanoic acids that have been shown to possess different inflammatory and immunosuppressive phenotypes (P. Calder 1997). The average Western diet contains between 10:1 and 30:1 omega-6 to omega-3 fatty acids (Nettleton et al. 2013); a healthier diet recommends a 1:1 ratio. Some have even speculated that the increased incidence of chronic inflammatory diseases in the West, such as cancer, coronary artery disease, arthritis, asthma, and psoriasis, are exacerbated by the high-fat, high-omega-6-fatty-acid diets we consume (Manzel et al. 2014).

Primates in the wild consume polyunsaturated and saturated fats in relatively equal proportions (J. Chamberlain et al. 1993; Milton 2000a), with animal matter and some fatty fruits acting as the primary fat sources in their diets (Rothman, Vogel, and Blumenthal 2013). However, the caloric contribution of fat to the primate diet relative to carbohydrates and protein is low (J. Lambert and Rothman 2015; Milton 2000a). For example, dietary fats contribute only 17% of daily caloric intake for howler monkeys, comparable to Western humans (J. Chamberlain et al. 1993). Intake of fat by primates has been explored by examining how different fatty acid types (e.g., monounsaturated, polyunsaturated, and saturated fats) affect primate immune response. Juvenile Japanese macaques (*Macaca fuscata*) fed a high-fat diet (HFD) showed evidence of increased fasting insulin levels and increased cytokine

Table 11.4. Lipid precursors and their inflammatory products.

Lipid precursor	Product	Biological effect(s)	Selected products
Omega-3 fatty acids	Alpha-linolenic 18:3 Octadecatetraenoic 18:4 Eicosatetraenoic 20:4 Eicosapentaenoic 20:5 Docasapentanaenoic 22:5 Tetracosapentanaenoic 24:5 Tetrahexaenoic 24:6 Docosahexaenoic 22:6	Pro-inflammatory Immunosuppressive Fails to resolve	Prostaglandin E1 (PGE1) PGE2 Thromboxane A2 (TBA2) Leukotriene B4 (LTB4) Lipoxins
Omega-6 fatty acids	Linoleic 18:2 Gamma-linoleic 18:3 Dihomogamma-linoleic 20:3 Arachidonic acid 20:4 Adrenic 22:4 Tetracosatetraenoic 24:4 Tetracosapentaenoic 24:5 Docosapentaenoic 22:5	Less inflammatory Promotes resolution of inflammation	PGE3 TBA3 LTB5 Resolvins D, E Neuroprotectin D1

levels associated with obesity-related inflammation (e.g., IL-6, IL1β, TNFα, CRP, and IL10) relative to controls, suggesting increased pancreatic inflammation and the onset of type 2 diabetes (Nicol et al. 2013).

Because of the high prevalence of obesity in humans, and particularly women and children (Swinburn et al. 2011), studies using nonhuman primate models have focused on the influence of a HFD consumed by pregnant females on the health and morbidity of offspring (E. Sullivan, Nousen, and Chamlou 2014; E. Sullivan, Smith, and Grove 2011). Nutrition and obesity in pregnant females is of interest because the interaction of nutrition and inflammation in the mother can directly affect the immune function of both the mother and her offspring. Japanese macaques fed a HFD while pregnant showed evidence of acute inflammatory responses and a lack of adequate blood flow to and from the placenta (Frias et al. 2011). The HFD female subjects experienced an increase in the number of stillborn infants, particularly for females with preexisting diet sensitivities (Frias et al. 2011).

Because mother's milk supports an infant's immunological, hormonal, somatic, and neural development (Hinde and Milligan 2011), an infant's immune function is ultimately affected by its mother's nutrition. Infants born to Japanese macaque mothers on a HFD showed signs of lipotoxicity, cellular dysfunction, and apoptosis; conversely, obese macaques removed from a HFD and switched to a controlled diet had infants with lower signs of lipotoxicity (McCurdy et al. 2009). In another study, infants born to female rhesus macaques fed a HFD during pregnancy had increased levels proinflammatory cytokines in the hypothalamus relative to control subjects (Grayson et al. 2010). These elevated proinflammatory cytokines in the hypothalamus during fetal development can make the infant more vulnerable to metabolic disease later in life (Grayson et al. 2010). Nonhuman primate models have also been utilized to test whether infants born to mothers fed a HFD can later stabilize and return to control levels when switched to a recommended diet. Japanese macaque infants nursing from mothers fed a HFD were observed to have increased production of cytokines (e.g., VEGF and TNFα) in their serum (L. Fan et al. 2013). After a sample of these infants were weaned from their mothers and placed on a controlled diet, these abnormalities could be partially reversed (L. Fan et al. 2013). Thus, both early programing and postweaning diet contribute to immune function in infants. Although these studies are critical for understanding metabolic pathways, obesity, and inflammation in humans, the utility of these studies for understanding nonhuman primate nutritional immunology may be limited because it is rare for unprovisioned, wild primates to consume high-fat diets.

MICRONUTRIENT INTAKE AND HOST INFLAMMATORY RESPONSE

A complete description of dietary micronutrients and the inflammatory response is well beyond the limits of this review. Micronutrient deficiencies and the diseases or syndromes they produce have been well described in the literature and clearly have significant implications for both inflammation and host protective immunity. Because diseases associated with deficiencies in micronutrients are still prevalent in many parts of the world, interest has increased in the effect of relative insufficiencies and excesses in micronutrient intake on inflammation and host immunity. In fact, whole industries have arisen and are focused on the supplemental use of several micronutrients to minimize inflammation and oxidative injury. Probably, the most prevalent and least understood are the roles of dietary micronutrients and supplements in oxidative injury and inflammation. It is generally accepted that aging and chronic oxidative injury are inevitable consequences of living in an oxygen-rich environment and using oxygen as the final proton acceptor in the electron transport chain (K. Jacob et al. 2013). It is the great paradox of life that oxygen is essential but at the same time destructive to proteins, nucleic acids, and lipids. Several vitamins and trace minerals have antioxidant properties, including beta-carotene, vitamin A, vitamin C, vitamin D, vitamin E, and selenium, although much of the immunological literature focuses on vitamins A and D (Mora et al. 2008; Wintergerst et al. 2007). However, their antioxidant properties are not necessarily their essential host function, and their importance as antioxidants under normal dietary conditions is still controversial (D. Hughes 2002).

There have been several randomized controlled trials examining antioxidant supplements in patients with oxidant injury. Beta-carotene, vitamin A, and vitamin E supplementation were given to patients with cancer or coronary artery disease, both known chronic inflammatory diseases, with no beneficial effect and with the possibility that it may have harmed individuals (Tinkel et al. 2012). Vitamin C supplementation was evaluated in smokers and during exercise, again with no substantial benefit (Nam et al. 2003). However, vitamin A supplementation reduced mortality in children with acute measles (Ogaro et al. 1993), *P. falciparum* malaria (Friis et al. 1997), HIV (Coutsoudis et al. 1992), and diarrheal diseases (A. Sommer et al. 1984) but had no effect on mortality in acute lower respiratory infection (Quinlan and Hayani 1996).

Milton (2003b) brought attention to the lack of data on the vitamin content of foods consumed by wild primates, yet there still remains a paucity of studies focusing on vitamins in primate diets (Felton, Felton, Raubenheimer, Simpson, Foley, et al. 2009). Thus, we know very little about the functional relationship between vitamin intake and immunity in nonhuman primates, other than a few studies that have been conducted on captive primates (see below).

Vitamin E plays a major role in maintaining immune cell function (D. Hughes 2002), having both antioxidant (Rimm et al. 1993) and anti-inflammatory properties (U. Singh et al. 2005). Vitamin E deficiency is associated with a suite of pathologies, with anemia being the most common. Evidence for anemia, a lack of red blood cells or hemoglobin, has been assessed in captive primates using blood smears and observation for weight loss and behavioral changes, including lack of appetite and lethargy (Ausman and Hayes 1974; K. Hayes 1974). In a study on baboons (*Papio hamadryas*) fed a high-fat diet, oxidative stress was reduced when study subjects were administered vitamin E supplements (X. L. Wang et al. 2004). It is clear that vitamin deficiencies have similar health consequences in nonhuman primates and humans, but there has been very little research on micronutrient intake in wild primates. Milton (2003b) reviewed the available data focusing on vitamin and mineral composition of wild primate plant foods. She found that, overall, primate foods (e.g., fruits and leaves) are higher in vitamin C than those consumed by human populations but stated that we know very little about vitamin intake and requirements in nonhuman primates (Milton 2003b). Furthermore, primates may appear to take in excess micronutrients in some cases, but it may be that the bioavailability of micronutrients in primate plant foods is low; that is, they may assimilate only a small proportion of the micronutrients consumed.

In terms of mineral intake and immune function, there has been considerable evidence to suggest that dietary iron and copper can play a significant role in inflammation. In mammals, both metals rarely travel in the bloodstream in an unbound form; iron travels bound to transferrin and copper to ceruloplasmin (Fraser et al. 1989). During inflammation, iron concentrations decline and copper concentrations increase, and these two responses are known to have survival benefit (Besold et al. 2016; Neyrolles et al. 2015). Iron can play a significant role in the pathogenicity of several bacterial species, and in response to infection or inflammation, there is strong competition between the host and pathogenic bacteria for free iron. Pathogenic bacteria release siderophores to trap and capture free iron in the circulation. In hospitalized patients receiving enteral or parenteral diets, free iron is generally not administered to infected patients because of its ability to enhance the pathogenicity of bacteria (Pieracci and Barie 2005; Rech et al. 2014). In mammalian hosts, iron is tightly bound to proteins such as hemoglobin, transferrin, lactoferrin, and ferritin. Inflammation is associated with a marked increase in the synthesis of ferritin and lactoferrin (Bornman et al. 1999; Palma et al. 1992). Surprisingly, ceruloplasmin, the primary copper-carrying protein in mammalian systems, also plays a crucial role in iron metabolism. This enzyme reduces iron by transferring an electron between iron and its own copper, thereby making the iron ion more readily captured by transferrin. Ceruloplasmin expression is dramatically increased in inflammation (Mayer et al. 1991).

As with vitamins, there has been little research focusing on mineral intake and immune function in wild and captive primates. Milton (2003b) found great variation in mineral availability in wild primate plant foods across continents, but her data on howler monkeys suggest that, as with vitamins, wild primates ingest higher levels of minerals relative to the recommended human requirements. However, while intake may be higher, the assimilation of these micronutrients, such as calcium, magnesium, phosphorous, and iron, may actually be quite low (Nagy and Milton 1979b). The combined effects of zinc and copper have been examined in terms of heath and survivorship in wild baboons (*Papio cynocephalus*; Markham, Gesquiere, et al. 2011). Copper and zinc deficiencies are known to impair immune function (Bhaskaram 2001; Bo et al. 2008), and zinc has been shown to reduce the bioavailablity of copper (L. McDowell 2003). Markham, Gesquiere, and colleagues (2011) found that Zn/Cu ratios were higher in females that had given birth to at least one infant with white monkey syndrome, suggesting a link between the disease and copper deficiency induced by zinc toxicity. Copper intake has also been implicated in variation in redtail monkey (*Cercopithecus ascanius*) population density in logged and unlogged forests (Rode, Chapman, McDowell, et al. 2006). Rode and colleagues (2006) found that monkey groups with greater intake of absorbable copper and a greater ratio of copper intake to caloric intake had higher population density, suggesting a link between copper deficiencies and health in these monkeys.

NUTRITIONAL IMMUNOLOGY AND NUTRITIONAL GEOMETRY: A NEW APPROACH TO EXAMINING NUTRIENT INTAKE AND HEALTH

The geometric framework of nutrition (GF) describes how animals move across a nutritional landscape as they encounter foods within a "nutrient space," which is represented by the area inside a Cartesian plot (S. Simpson and Raubenheimer 2012; Raubenheimer, this volume). These foods have different nutrient compositions, and foraging animals will consume a suite of foods within their landscape, typically consuming differing amounts of foods that vary in their nutrient compositions. The balance (or ratio) of nutrient constituents in a food item is represented by the slope of the line. The "nutrient target," which can vary depending on the reproductive state or age class of the animal, represents the optimal nutrient requirement of the foraging animal. The foraging animal can regulate its nutritional intake to reach its nutritional target by selecting foods exhibiting a similar balance of nutrients to the target or by mixing food items that are nutritionally complementary. The latter is what would be expected for primates with broad and varied diets. Feeding is represented in GF models by the change in nutritional state of the animal as it forages and is plotted as a trajectory through nutritional space (S. Simpson and Raubenheimer 2012). Plots of daily nutrient intake in two or more dimensions allow for a novel comparison of the nutrient balance of animals as they consume foods over time.

The GF has been used in experimental studies ranging from insects to humans and has recently been applied to a variety of wild primates including sifakas (Irwin, Raharison, Raubenheimer, et al. 2015), baboons (C. A. Johnson, Raubenheimer, Rothman, et al. 2013), spider monkeys (Felton, Felton, Raubenheimer, Simpson, Foley, et al. 2009), gorillas (Rothman, Raubenheimer, and Chapman 2011), and black-and-white colobus monkeys (C. A. Johnson, Raubenheimer, Chapman, et al. 2017). The GF is also a means of assessing the "prioritization" of nutrients, or regulation of nutrients when faced with nutritional constraints due to changes in the spatial and temporal availability of food; and it can be used to decipher intricate interactions among nutrient availability, intake, and physiological state (K. Lee, Simpson, Clissold, et al. 2008; K. Lee, Simpson, and Wilson 2008; Ponton, Wilson, Cotter, et al. 2011; Povey et al. 2009). While the majority of studies have focused on the effects of individual macro- or micronutrients on immune response (Cunningham-Rundles 2002; but see Kubena and McMurray 1996), the GF offers a new perspective to examine nutritional immunology by focusing on the interactive effects of dietary nutrients on immune function (Cotter et al. 2011; K. Lee, Behmer, et al. 2002; K. Lee, Cory, et al. 2006; Ponton, Wilson, Cotter, et al. 2011; Povey et al. 2009; Raubenheimer and Simpson 2003; Raubenheimer, Simpson, and Mayntz 2009; S. Simpson and Raubenheimer 2009).

Ponton, Wilson, Holmes, and colleagues (2013) provide several examples that demonstrate how immune function in insects is influenced more by nutrient ratios than by overall energy intake. The use of GF to study nutritional immunology has revealed that the balance of protein to carbohydrates (P:C)—not total caloric intake—influences immune response and overall health in insects (Cotter et al. 2011; K. Lee, Cory, et al. 2006; Povey et al. 2009) and mice (X. Huang et al. 2013; Le Couteur, Solon-Biet, McMahon, et al. 2014; Le Couteur, Tay, et al. 2015; Solon-Biet et al. 2014). In contrast to K. Lee, Cory, and colleagues (2006), mice fed relatively high-P:C diets showed mixed results in terms of longevity and immune function, whereas mice fed a low-P:C diet had greater longevity, similar to studies on caloric restriction (Le Couteur, Tay, et al. 2015). Thus, while there is clear evidence that nutritional balancing influences immune function and health in some model organisms, we still know very little about how the balance of nutrients influences health in primates. The use of GF to study nutritional immunology in primates will improve our understanding of how and why primates select their diets and how these decisions ultimately influence health and fitness.

METHODOLOGIES FOR STUDYING NUTRITIONAL IMMUNOLOGY IN WILD PRIMATES

Romero et al. (2009), following McEwen and Wingfield (2003), posit that researchers in field settings should strive for a top-down approach—in other words, examine how an animal responds to "stressors" in the short term and then identify how these responses can become pathological if sustained in the long term. In their model, they first identify the discrete conditions and physiological mediators of reactive homeostasis during both short-term and long-term bouts of allostatic loading, then identify how particular mechanisms influence the animal during this period (Romero et al. 2009). This option in the field may be limited because the opportunities to quantify discrete forms of allostatic load are unpredictable, and while noninvasive sampling techniques have improved greatly, the remote setting that characterizes most primate field sites create challenges for collecting physiological samples.

Methods for studying primate nutritional ecology, specifically feeding observations, plant sample collection and preparation, and nutrient intake calculations, are detailed elsewhere (Conklin-Brittain, Knott, and Wrangham 2006; Ortmann et al. 2006; Rothman, Chapman, and Van Soest 2012; Conklin-Brittain, this volume). Here, we focus on noninvasive methods to monitor immune activation and inflammation in wild nonhuman primates. Over the past few decades, field methods to examine physiological indicators of energetic and immune stress have been developed, enabling researchers to gain a deeper understanding of the effects of allostatic load on behavior, life-history traits, and fitness in nonhuman primates (Bergstrom, Kalbitzer, et al. 2020; Emery Thompson 2017; Emery Thompson, Fox, et al. 2020; Emery Thompson and Knott 2008; Gonzalez et al. 2020; Heistermann and Higham 2015; Higham, Kraus, et al. 2015; C. Hoffman, Higham, et al. 2011; Negrey, Reddy, et al. 2019; Sherry and Ellison 2007; E. Vogel, Crowley, et al. 2012). The most common types of noninvasively collected samples for studies of immune function in wild nonhuman primates have been urine (Behringer, Deimel, et al. 2021; Behringer, Preis, et al. 2020; Behringer, Stevens, Leendertz, et al. 2017; Behringer, Stevens, Wittig, et al. 2019; Bergstrom, Kalbitzer, et al. 2020; Emery Thompson 2017; Emery Thompson, Fox, et al. 2020; Emery Thompson, Machanda, et al. 2020; Emery Thompson, Muller, and Wrangham 2012b; Heistermann and Higham 2015; Higham, Kraus, et al. 2015; Higham, Stahl-Hennig, and Heistermann 2020; C. Hoffman, Higham, et al. 2011; Lohrich et al. 2018; A. Moldawer et al. 2014; Negrey, Reddy, et al. 2019; E. Vogel, Rothman, et al. 2015; D. Wu et al. 2018) and feces (Behringer, Muller-Klein, et al. 2021; Dibakou et al. 2019; Muehlenbein 2006, 2015; Muehlenbein and Watts 2010).

Sample collection methods for urine and feces are fairly straightforward and have been outlined in detail elsewhere (Hodges and Heistermann 2011; Knott 1997; Nguyen 2013). For most types of samples for studies of inflammation, it is important to (1) note the time of day, as some markers of inflammation follow circadian rhythms, and thus collecting the first morning void is best for most assays; (2) collect multiple samples from each individual across time; (3) collect additional data with each sample including weather (e.g., rain), behavioral data, and where the sample was collected from (e.g., leaves or plastic); and (4) for urine, note whether there is feces or dirt in the sample. If urine or fecal samples cannot be processed immediately, we recommend bringing a thermos with ice and keeping the sample in a collection vial in the thermos during the day. While many field stations are remote and do not have electricity, running a small freezer on solar panels has worked very well for our project in Indonesia (Tuanan Orangutan Research Program) for six years. We have found this method sufficient to keep urine and fecal samples at −16°C until they can be transported each month. Samples should be transported on ice, dry ice, or liquid nitrogen to prevent freeze/thaw cycles. Sample preservation methods for both urine and fecal samples can vary depending on the marker of inflammation of interest (reviewed in Muehlenbein and Lewis 2013). Thus, careful consideration must be made before the start of the study, and researchers should validate each assay in the laboratory before the start of a study.

Several different physiological parameters have been measured to examine immune function in wild nonhuman primates. Urinalysis chemical test strips most often used in hospitals and veterinarian clinics have been used to obtain simple quantitative information on nonhuman primates (e.g., Roche Diagnostics Chemstrip 10 UA; Knott 1997, 1998). Researchers have found some success measuring ketones and pH as indicators of nutritional state (MacIntosh et al. 2012), but these strips are more used for preliminary diagnostics (i.e., presence or absence or some basic categories) and are thus limited. Higham, Kraus, and colleagues (2015) examined a number of markers of inflammation from urine samples and tested their utility for studies of inflammation. They found that of the three biomarkers of inflammation examined (C-reactive protein [CRP], haptoglobin, and neopterin), only urinary neopterin concentrations were correlated with serum concentrations, whereas fecal and urinary neopterin concentrations were not correlated. They also found that urinary neopterin is a reliable noninvasive marker of inflammation in rhesus macaques, while CRP and haptoglobin may also be useful. However, they found that for both CRP and haptoglobin, fecal and urinary measures did not correlate with serum values (Higham, Kraus, et al. 2015). Neopterin has been shown to be highly stable under various conditions in the field (Behringer, Stevens, Leendertz, et al. 2017; Heistermann and Higham 2015; Higham, Stahl-Hennig, and Heistermann 2020) and provides general information about T-helper-cell-derived cellular immune activation and oxidative stress elicited by the immune system (Murr et al. 2002). Because of its stability in remote field settings, urinary neopterin as a marker of immune activation has been successfully quantified in a variety of primate species in relationship to energetic stress, life history, injuries, and disease (Ballare 2021; Behringer, Deimel, et al. 2021; Behringer, Preis, et al. 2020; Behringer, Stevens, Wittig, et al. 2019; Danish et al. 2015; Gonzalez et al. 2020; Lohrich et al. 2018; Lucore et al. 2022; N. Muller et al. 2017; Muller-Klein et al. 2019; Negrey, Behringer, et al. 2021; E. Vogel, Naumenko, et al. 2019; D. Wu et al. 2018).

Because spot urine and feces collections are most common in field settings, it is frequently required to standardize measurements based on some physiological standard. Dilution of urinary samples by the hydration status of the animal or contamination with water from environmental sources makes absolute concentrations frequently unreliable. As a result, it is essential that such measurements be standardized with an auxiliary compound whose output is constant. Urinary creatinine is often used; its urinary excretion is both relatively constant and directly related to skeletal muscle mass, it is stable in a field setting, and it is easily measured. Urinary analytes are usually reported per unit of creatinine. Recent studies have also recommended the use of specific gravity (SG) as a means to standardize urine samples for water content (Emery Thompson, Muller, and Wrangham 2012b; R. Miller et al. 2004), which can be quantified easily in the field using a handheld refractometer (Atago PAL-10S). The advantage of using SG instead of creatinine is that SG is not affected by muscle mass (R. Miller et al. 2004). Emery Thompson, Muller, and Wrangham (2012b) validated the use of urinary SG combined with creatinine to estimate lean body mass in wild nonhuman primates. This method has since been used to estimate changes in muscle mass in relationship to food availability in both capuchin monkeys (Bergstrom, Thompson, et al. 2017) and orangutans (C. O'Connell et al. 2021).

Another promising method to examine immune function in relation to nutrition in nonhuman primates is to evaluate urinary cytokine concentrations. C. Hoffman, Higham, and colleagues (2011) measured three proinflammatory cytokines (IL-6, IL-8, and IL-ra) from plasma samples in free-ranging rhesus macaques and found positive correlations for IL-6 and IL-ra but not IL-8. Maestripieri and Georgiev (2016) also examined cytokine concentrations and found a significant correlation between IL-6 and fecal glucocorticoids. We validated the use of six cytokines involved in both the adaptive and innate immune responses (GCS-F, MCP-1, TNF-α, IL-RA, IL-8, and IL-10) on orangutan blood and urine samples (A. Moldawer et al. 2014). We used enzyme-linked immunosorbent assays (ELISA) to quantify these cytokines using Millipore kit PRCYTOMAG-40K-06, following the kit instructions, and then analyzed the results on a Luminex MagPix with Analyst software. Samples were run without dilution, as validation experiments found the highest recovery (85%) with undiluted samples. We found that in both urine and blood samples, there was a positive correlation among the majority of these markers of inflammation (A. Moldawer et al. 2014). Ballare (2021) measured and compared cytokines in wild and rehabilitant orangutans (*Pongo pygmaeus wurmbii*) and found that cytokine values were more variable and higher in the wild orangutan population, suggesting greater immune activation in response to a more variable environment and diet. With the rapid development of new methods for noninvasively collected samples, we expect this list of markers to continue to grow (see table 11.5 for a summary of markers of inflammation and immune stress).

As the forested habitats of primate species around the world continue to suffer from rapid deforestation and human encroachment (Estrada, Garber, et al. 2017), the

Table 11.5. Examples of markers of immune stress, their tissue source, and methods of analysis.

	Mediator class	Mediator	Tissue source	Information gained	Method of analysis	References
Proteins	Cytokine	IL-6, IL-8, IL-1ra, other cytokines	Plasma, urine, feces	Inflammation	Immunoassays	(Ballare 2021; Kamakoti et al. 2018; Maestripieri and Georgiev 2016; Rivera et al. 2001)
	Acute-phase reactant	C-reactive protein, haptoglobin	Plasma, urine	Acute-phase protein response	Immunoassays, electrochemical impedence spectroscopy	(Georgiev, Muehlenbein, et al. 2015b; Higham, Kraus, et al. 2015; Kamakoti et al. 2018; Vandeleest et al. 2016)
	Kidney function	IL-18, lipocalin-2[1]	Urine	Kidney function, inflammation	Immunoassays, chemoluminescence	(Y. Li et al. 2013; Reiter et al. 2018; Siew et al. 2010; Singer et al. 2013)
Nucleic acids	Cell-free DNA	Eukaryotic, mitochondrial DNA, CpG	Plasma, urine, feces (both eukaryotic and prokaryotic)	Organ injury, source of microbial infections	PCR technologies, NanoString	(Burnham et al. 2018; Stewart et al. 2018; Streleckiene et al. 2018)
	RNA	RNA transcripts	Plasma, urine, feces	Transcriptomics	RT-PCR, microarray, NanoString	(Y. Fan et al. 2014)
	MicroRNA	MicroRNA	Plasma, urine, feces	Transcriptomics	RT-PCR, microarray, NanoString	(Cardenas-Gonzalez et al. 2017)
	Pteridines	Neopterin	Plasma, feces, urine	DNA replication, cell proliferation, reactive oxygen injury		(Behringer, Deimel, et al. 2021; Behringer, Muller-Klein, et al. 2021; Behringer, Stevens, Leendertz, et al. 2017; Behringer, Stevens, Wittig, et al. 2019; Bendlin et al. 2011; Bergstrom, Thompson, et al. 2017; Danish et al. 2015; Dibakou et al. 2019; Georgiev, Muehlenbein, et al. 2015a; Gonzalez et al. 2020; Higham, Kraus, et al. 2015; Lohrich et al. 2018; Maestripieri and Georgiev 2016; N. Muller et al. 2017; Muller-Klein et al. 2019; Murr et al. 2002; Negrey, Behringer, et al. 2021; O'Connell et al. 2021; D. Wu et al. 2018)

(continued)

Table 11.5. (continued)

Mediator class		Mediator	Tissue source	Information gained	Method of analysis	References
Lipids	Prostaglandins	PGE2	Plasma, urine	Inflammatory mediators	Immunoassay	(Colas et al. 2018; Das 2011; Sasaki et al. 2015)
	Leukotrienes	LTB4, LTC4, LTD4, LXA5	Plasma, urine	Inflammatory and anti-inflammatory mediators	Gas chromatography, mass spectrometry (GC-MS)	
	Resolvins, marestins, protectins	Resolvins D (RvD1-5), E (RvE1-3, 18s-RvE)	Plasma, urine	Anti-inflammatory mediators	Liquid chromatography and GC-MS	
Leukocytes		Total WBCs, PMNs, T cells	Plasma, urine	Inflammation	Flow cytometry, immunoassay, leukocyte esterase/catalase	(Langlois et al. 1999; Pels et al. 1989)
Hormones		Cortisol, corticosterone, catecholamines	Plasma, urine, feces	Stress response	Electrochemoluminescence, immunoassay, HPLC	(Hostinar et al. 2015; Mesa et al. 2014)

field of nutritional immunology will become increasingly important for making informed conservation decisions regarding species survival plans. Numerous studies have found that human pathogen strains have been or can be transmitted to wild primates and that transmission is more common closer to human contact zones (Carne et al. 2013, 2014; Formenty et al. 1999; Grutzmacher et al. 2018; Kaur et al. 2009; Nunn et al. 2008; Patrono et al. 2018; Wyers et al. 1999). Given the strong relationship between nutrition, health, and immune function in primates, detailed planning regarding the availability of nutrients in the habitat, the borders surrounding the habitat, the amount of contact between wild nonhuman primates and humans, and the risk of disease transmission should be considered for future relocation and primate protection programs.

SUMMARY AND CONCLUSIONS

Nutrition is critical to immunity and pathogen resistance in humans and other animals, yet we lack a basic understanding of how nutrient availability varies in natural habitats, how this variation shapes primate nutritional strategies, and how nutritional strategies affect the energetics and health of wild primates. The use of noninvasive urine sampling provides an innovative approach to understanding how nutritional strategy can modulate immune function in response to natural variation in nutrient availability in nonhuman primates. There is strong experimental evidence that the balance of nutrients is the driving force in the nutrition-behavior-health link, as demonstrated in a diverse array of nonprimate taxa. While some studies of primates in their natural habitats have examined the relationship between nutrient balance and foraging strategy, none have extended this to include the implications for immune function and health. Studying nutritional immunology in wild primates offers a unique opportunity to integrate metabolic physiology and immunity with foraging in an ecological context, providing a natural experiment to examine the multidimensional relationships of nutrition and health in our closest living relatives.

- Very little is known about nutritional immunology in wild primates, making this an open area for exploratory research.
- Studies on model organisms have demonstrated that immune function is tightly linked to the balance of protein and nonprotein energy in the diet.
- Nutritional status and resistance to infection are inextricably linked, and improved understanding of how diet regulates host protective immunity can provide insights into the overall health and well-being of nonhuman primates both in captivity and in the field.
- Newly developed noninvasive methods will enable us to gain significant insights into how ecological variation and consequent nutritional strategies regulate the immune system in our closet living relatives.
- Studies on primate nutritional immunology can inform decisions that influence the conservation status of primates by examining how land use and wildlife reserve design influence nutrition and immune function in nonhuman primates.

12 Nutrition and Primate Life History

Carola Borries and Andreas Koenig

When we initiated our study of a population of Nepal gray langurs (*Semnopithecus schistaceus*) in the 1990s, we had a set of expectations based on prior studies of gray langur populations in India (plains gray langurs, *Semnopithecus entellus*; Arekar et al. 2021). For example, we expected that females would form age-inversed dominance hierarchies (Hrdy and Hrdy 1976) and that immature females would begin to rise within the adult hierarchy at about two years of age (Borries, Sommer, and Srivastava 1991). We could confirm the age-inversed hierarchy for Nepal gray langurs, but documenting the process of integration into the adult female hierarchy met with some obstacles (Lu, Borries, et al. 2013). We followed immature females in Nepal as soon as they had completed their second year of life. To our surprise, it took an additional three years before the youngest, still immature female in Nepal began to ascend into the adult female hierarchy (Apelt 1995). We also determined a matching, significantly older age at first parturition in this population (mean 6.7 versus 3.5 years; Borries, Koenig, and Winkler 2001). Growth, maturation, and, with it, social integration of females were all significantly delayed at the site in Nepal. Populations studied earlier in India had all been crop-raiding, or they were provisioned by people with human-made food, which was much higher in nutrient content than the natural vegetation in their habitats and led to a higher caloric intake (*Semnopithecus* spp.; A. Koenig and Borries 2001; Schuelke 1997). When we were finally able to compare growth and reproduction in the wild with those in the provisioned population, all variables were significantly accelerated in the provisioned one (Borries, Koenig, and Winkler 2001).

The above examples illustrate the extent of phenotypic plasticity in primate life history and the strong impact that nutrient availability in particular and ecology in general can have on the speed of growth and reproduction in primates. In this chapter, we explore several dimensions of this impact and how they relate to primate life history. We begin with a brief review of the principal life-history variables, which helps in selecting the variables most relevant in the context of nutrient availability. Next, we summarize studies illustrating nutritional effects on primate life history and those that do not seem to fit the general trend. Based on well-documented example species, we visualize the potential effect of nutrient availability to examine whether the effect is similar for different life-history variables and whether its relative effect is proportional to the duration of the variable considered. We close with a brief review of potential caveats and conclusions.

LIFE-HISTORY(-RELATED) VARIABLES: A BRIEF OVERVIEW

Life history is of central importance for our understanding of nonhuman and human primate biology and paleontology. Because it encompasses all developmental changes and states of an individual from conception through death (Lande 1982), it sets the stage and boundaries for individual tactics and helps in understanding energy allocations and trade-offs and their evolution (Leigh and Blomquist 2011; A. van Noordwijk and de Jong 1986). Through correlations with well-preserving hard tissues such as bones and teeth, some life-history traits can even be estimated for extinct species (Dean 2010; J. Kelley and Schwartz 2012; T. M. Smith, Tafforeau, et al. 2007).

Generally, life history captures the speed of growth and the speed of reproduction (Lande 1982; Skinner and Wood 2006; see also chapter 6). Growth processes are estimated via variables such as gestation length and developmental landmarks such as age at weaning and age at first parturition. In addition, the life-history variable litter size helps further characterize the energy allocated to each reproductive event, while age at death and maximum life span limit the time available for reproduction (Leigh and Blomquist 2011). In the past, the length of the menstrual cycle has sometimes been mistaken for a life-history variable (Harvey and Clutton-Brock 1985; Skinner and Wood 2006), even though it does not capture the speed of growth or the speed of reproduction. Not surprisingly, cycle length produced confusing results in past analyses (e.g., Bromage et al. 2012; Harvey and Clutton-Brock 1985).

In addition to life-history variables, Skinner and Wood (2006), among others, identified "life history related variables," which include tooth formation times, dental eruption schedules, brain mass, and body mass. Life-history traits are allometrically related to the latter two variables (e.g., Isler et al. 2008; R. Martin 1984a), which means that they scale exponentially with adult female body mass and with brain mass (R. Martin 1982; van Schaik, Barrickman, et al. 2006). It makes intuitive sense that larger species (i.e., higher mass) require more time to form the larger amount of tissue that constitutes their bodies and consequently also reproduce more slowly and—as a trade-off—for longer (Ross 1998).

LIFE HISTORY VARIATION: ADAPTATION VERSUS PHENOTYPIC PLASTICITY

Arguably, the factor molding life-history evolution the most is mortality (Harvey, Promislow, and Read 1989; Roff 1992). For example, if predation pressure on immatures is high, this developmental period will often be short. Consequently, adulthood, and with it reproduction, will begin sooner, thus accelerating life history (Michod 1979). These effects have an adaptive, genetic basis that can be experimentally reenacted (Reznick et al. 2001), at least in guppies. In primates, patas monkeys (*Erythrocebus patas*) are a good example of early maturation likely fueled by high adult mortality (Isbell, Young, et al. 2009). The opposite extreme is realized in stable populations with very low mortality where life histories can be much slower than predicted by adult female body mass. A good primate example likely are blue monkeys (*Cercopithecus mitis stuhlmanni*) which begin to reproduce late, at a low rate, and for an extended period relative to body mass (Bronikowski, Altmann, et al. 2011; Bronikowski, Cords, et al. 2016; Cords and Chowdhury 2010). Generally, these different life-history trajectories are characterized by trade-offs, such as a faster reproductive rate and a shortened reproductive life span under high mortality, and the opposite when mortality is low (Charnov 2001).

In contrast to mortality, the effects of nutrient availability on life history are largely an expression of phenotypic plasticity; they are somatic reactions to specific ecological conditions (Ricklefs and Wikelski 2002; Wells and Stock 2011). When conditions worsen, the speed of growth and of reproduction slows down quickly, and body mass may be lost, while the effects are reversed with improving conditions (olive baboons, *Papio anubis*, Strum and Western 1982; Japanese macaques, *Macaca fuscata*, Sugiyama and Ohsawa 1982). This flexible reaction to the current environmental conditions does not necessarily entail trade-off costs. If the conditions improve, all aspects of life history are improved, which highlights the great advantages connected with such a flexible reaction to changing ecological conditions (Alberts and Altmann 2006). Ecological flexibility of life history is known to occur in mammals in general (Bronson 1989; Gilmore and Cook 1981; Sadleir 1969) and in primates in particular (Asquith 1989; Hendrickx and Dukelow 1995; P. Lee and Kappeler 2003). Individuals in provisioned populations of Japanese macaques were significantly younger at first parturition and had a faster reproductive rate than those from wild-feeding populations (Sugiyama and Ohsawa 1982; Takahata, Suzuki, et al. 1998). Similarly, in yellow baboons (*Papio cynocephalus*), individuals from a group that regularly fed at a local dumpster were significantly heavier (J. Altmann, Schoeller, Altmann, Muruthi and Sapolsky 1993), and females had a younger age at menarche and a faster reproductive rate (J. Altmann and Alberts 2003) than individuals in the same population that did not supplement their diet with human-made food. Increased nutrient availability led to significantly faster growth and reproduction—that is, increased nutrient availability accelerated life history. We recently compiled additional examples of this effect from the primate literature (Borries, Lodwick, et al. 2022).

An effect of differential access to and intake of food on life history can also be found independently of nutrient supplementation in wild-feeding primates as a rank effect within groups. This assumes that dominance rank, especially in females, provides nutritional advantages to higher-ranking individuals, which has been shown for several wild-feeding primates (e.g., brown capuchin monkeys, *Cebus apella*, Janson 1985; Nepal gray langurs, A. Koenig 2000; long-tailed macaques, *Macaca fascicularis*, M. van Noordwijk and van Schaik 1987; vervet monkeys, *Cercopithecus aethiops*, P. Whitten 1983a). Improved nutrient availability reduced the age at maturity and accelerated reproductive rates in high-ranking females (yellow baboons, J. Altmann and Alberts 2003; long-tailed macaques, M. van Noordwijk and van Schaik 1999). Similarly, high-ranking female gray-cheeked mangabeys (*Lophocebus albigena*) had their first infant sooner than lower-ranking females (Arlet et al. 2015). In chacma baboons (*Papio ursinus*), infant body mass gain was positively related to maternal rank (S. E. Johnson 2003). Similar effects can also be manifested between groups in wild primate populations. Slower maturation, slower infant growth rates, and reduced reproductive rates in larger groups were interpreted as a consequence of feeding competition increasing with group size (Phayre's leaf monkeys, *Trachypithecus phayrei crepusculus*, Borries, Larney, et al. 2008; olive baboons, Packer et al. 2000; long-tailed macaques, M. van Noordwijk and van Schaik 1999).

THE ORANGUTAN PUZZLE HAS BEEN SOLVED

While most results support an accelerated life history with improved nutrient availability, some studies did not seem to easily fit the paradigm. One such case were wild orangutans, the primates with the slowest life history (*Pongo abelii* and *Pongo pygmaeus*; Wich, Utami Atmoko, Mitra Setia, Rijksen, et al. 2004). The two species distinguished until recently (there are new arguments for a third species; Nater et al. 2017) live on two different Indonesian islands, Borneo and Sumatra (Mittermeier, Rylands and Wilson 2013). Borneo has the poorer habitat with a lower and less reliable fruit availability (A. Marshall, Ancrenaz, et al. 2009) and consequently a higher percentage of feeding on the less nutritious fallback food inner bark (Wich, De Vries, et al. 2009). Despite these seemingly clear ecological differences, the life histories as well as mortality patterns of both species were very similar, with the exception of female reproductive rates. Reproductive rates were faster on Borneo, the island with the lower nutrient availability, which is the opposite of what would be expected. This exception was interpreted as a genetic difference between the two species, with orangutans in the better habitats having adapted to lower reproductive rates because of habitat saturation (Wich, Utami Atmoko, et al. 2004). However, it is now clear that this exception was driven by a sample size issue. A recent analysis based on a larger sample size found no difference in interbirth interval between the orangutan populations (M. van Noordwijk, Utami Atmoko, et al. 2018).

LIFE HISTORY AND ECOLOGICAL INSTABILITY

The previous struggle to explain potential orangutan life history differences highlights another aspect of food availability—namely, ecological instability as manifested in food distribution in space and especially in time. Habitats on Sumatra are characterized by smaller annual variation of fruits in the diet, much longer fruiting periods, stronger mast fruiting peaks, and an overall higher fruit density, while on Borneo the periods with low fruit abundance are longer (A. Marshall, Ancrenaz, et al. 2009). In addition, ecological differences may relate to fig availability, the density of less preferred dipterocarp trees, general soil fertility, and predator abundance (van Schaik, Marshall, and Wich 2009).

On a more general level, primary production in a habitat can vary tremendously, leading to temporal food availability varying randomly or seasonally (van Schaik and Brockman 2005). This variation leads to adjusted growth periods, speed of growth, and timing of reproduction to the current ecological conditions (Ricklefs and Wikelski 2002). Where primary productivity is low and seasonality is strong, one would expect slow growth, smaller body mass, and slow reproduction, especially during lean seasons. The opposite is expected if primary productivity is high (S. Ferguson and McLoughlin 2000; Varpe 2017). The effect of seasonality on reproductive function in primates is discussed elsewhere in this volume (chap. 7). Seasonal effects on growth and maturation have been found in some primates (e.g., golden lion tamarins, *Leontopithecus rosalia*, Dietz, Baker, and Miglioretti 1994; Japanese macaques, Hamada et al. 1999), but overall, this topic has not yet been well researched.

WHEN NUTRIENT AVAILABILITY IS NOT CONSIDERED

In light of the fact that primate phenotypic plasticity facilitates the effects of nutrient availability on life-history variables, the question arises about the extent of these effects. We start addressing this with the following two examples, which illustrate that not considering nutrient availability in the context of primate life history can in fact lead to questionable results.

Example 1: Life-History Analyses May Be Meaningless

In a previous comparison of Asian colobines and Asian macaques (Borries, Gordon, and Koenig 2013; Borries, Lu, et al. 2011), we ran two phylogenetic generalized least squares models (a-models and b-models in the following) each for gestation length, age at first parturition, and interbirth interval (IBI). A-models included adult female body mass, taxon (here Asian colobines versus Asian macaques), and nutrition, a binary variable indicating either access to human-made food or no such access. Each b-model differed from the a-model by not considering nutrition.

Gestation length in the two taxa was only minimally altered by the inclusion of nutrition ($P = 0.067$). Both, a-model as well as b-model, accounted for more than 90% of the variance, with both adult female body mass and taxon as significant factors. For age at first parturition, nutrition was a significant factor ($P < 0.001$), which is mirrored in the fact that the a-model accounted for 42.0% of the variance while the b-model accounted for only 0.3% of the variance. In the b-model, adult female body mass no longer had an effect, and the trend for taxon had disappeared. For IBI, nutrition was a significant factor as well ($P < 0.001$). Accordingly, the a-model accounted for 65% of the variance, while the b-model accounted for 24% of the variance. In the a- and b-models, the effects of adult female body mass and taxon were reversed. Adult female body mass was significant ($P = 0.009$) only in the a-model, and taxon revealed a trend only in the b-model. Basically, the a-model and b-model contradicted each other.

Taken together, the variance accounted for was always higher when nutritional conditions were considered. In the two more flexible traits examined, age at first parturition and IBI, the basic, allometric relationship with adult female body mass that is widely established for life-history traits (W. Calder 1996; Roff 1992) held only if nutrient availability (a-models) was considered. The simple coding of nutrient availability as a binary variable improved all three analyses, which emphasizes the pivotal importance of considering nutrition in these comparisons even if in very basic, binary terms.

Example 2: Incorrect Conclusions May Be Drawn Based on Body Mass

The second example relates to the suggestion that human neonates (*Homo sapiens*) are heavy compared to apes and especially compared to chimpanzees (*Pan troglodytes*; DeSilva 2011). These heavier babies would impact maternal energetics and locomotion as well as the social system, and they are even linked to the evolution of communal infant care in humans (Hrdy 2016). A comparison was designed to identify when in human evolution this shift to heavier babies occurred (DeSilva 2011). Part of the analysis was based on 18 human and 3 chimpanzee populations, all based on very large sample sizes (DeSilva 2011). In contrast to earlier studies using population means, mothers' body mass was directly matched with their own neonates' mass. Across the populations considered, neonate mass in chimpanzees averaged 3.3% (range: 3.3%–3.7%) of maternal mass, and in humans it averaged 5.7% (range: 4.8%–6.5%). It is, however, possible that this significant difference was partly driven by high body mass values of the adult female chimpanzees, which were housed in research centers. At 53.6 kg, the average chimpanzee mother was just 3.5 kg lighter than the average human mother (57.1 kg) in this comparison. Because wild adult female chimpanzees weigh only about 31.3 kg (Pusey et al. 2005) or up to 45.8 kg, depending on the subspecies (R. Smith and Jungers 1997), the chimpanzees in DeSilva's (2011) study were much heavier than their wild counterparts. Excess body mass, mainly composed of fat, can cause an allometric problem because, at least in humans, maternal obesity did not increase neonate mass (Ode et al. 2012). Consequently, the percentage of neonate on maternal mass dropped to 3.6% in obese mothers (Ode et al. 2012) because of the increased maternal mass. This lower percentage value for human neonates relative to their obese mothers falls within the range for captive chimpanzees (DeSilva 2011, as cited above). Assuming the same obesity effects in chimpanzees, neonatal mass in captivity and in the wild would be very similar; only maternal mass would differ. Consequently, if body mass in the captive chimpanzee mothers in DeSilva's sample had been as low as in some wild populations (31.3 kg), neonates in chimpanzees and humans would have a very similar relative mass (chimpanzees: 5.6%, DeSilva 2011; humans: 5.9%, Robson et al. 2006), and human mothers would not be exceptionally burdened. However, these percentages will change with changes in female body mass depending on the chimpanzee subspecies (not known for the DeSilva samples; Cofran 2018) as well as the respective human populations considered.

This example highlights how differential effects of nutrient availability on neonatal versus adult body mass may drive results with far-reaching consequences for interpretations. A trait considered unique to our species—that is, heavy neonate mass—may not exist, illustrating that nutrient availability can have a distorting impact on comparative analyses. However, mass values for wild neonate chimpanzees are needed to increase confidence in our conclusion.

EXPLORING AND VISUALIZING THE EFFECTS OF NUTRIENT AVAILABILITY

The above two examples illustrate how variation in nutrient availability and the corresponding plasticity of life-history values can distort comparative analyses, with the latter being only as relevant as the data used (Borries, Sandel, et al. 2016; T. Parker et al. 2016). More generally, while many researchers seem to be aware of the impact of diet and nutrient availability on life history, the exact nature of this effect and how systematically it acts is unknown.

In the primatological literature, nutrient availability at the respective study site is rarely mentioned, let alone quantified (Sugiyama 2015). In the following overview, we therefore distinguish only two major nutritional regimes: wild (unprovisioned) if the study animals had no access to human-made food; all other conditions are considered provisioned (see also Strandin et al. 2018). The provisioned category includes animals in captivity as well as free-ranging populations as long as they have access to human-made food via crop raiding, feeding by people, or other means. This very crude "measure" of nutrient availability has proved to be a meaningful distinction in prior life-history analyses (e.g., example 1 above; for caveats on this approach, see below). The question can still be asked whether the simple dichotomy used here (wild versus provisioned) indeed resembles differences in nutrient availability. After all, in most cases, the calories individuals consumed are unknown. If indeed provisioning resembles higher nutrient availability and intake, provisioned individuals should be heavier than wild ones, which has been confirmed repeatedly (e.g., J. Altmann and Alberts 2005; J. Altmann et al. 1993; Leigh 1994b; Strum 1991; Stulp et al. 2015).

Other life-history-related variables, such as dental schedules, have rarely been compared between captive and wild populations and have produced mixed results (accelerated in captivity and dumpster-feeding individuals in yellow baboons, Galbany, Tung, et al. 2015; Phillips-Conroy and Jolly 1988; no difference in wild versus captive chimpanzees, T. M. Smith, Machanda, et al. 2013). With respect to the brain, there is limited evidence that captivity, with its higher nutrient availability, does not impact brain volume in primates (17 species; Isler et al. 2008), a finding recently confirmed for a larger sample of chimpanzees (Cofran 2018). These two life-history-related variables are therefore not considered in the following.

The data are visualized as classic double-log-transformed plots, with values for wild populations on the x-axis and those for provisioned ones on the y-axis (e.g., Leigh 1994b). The transformation accounts for the presumed underlying allometric relationship. Bold, solid lines ($y = x$) indicate identical measures for the two conditions or no effect of nutrient availability. The majority of our examples stem from Old World primates. Unfortunately, we could not consider chimpanzees even though extensive data for wild and provisioned populations are available because clear subspecies assignment was missing for most captive populations. The three chimpanzee subspecies differ noticeably in mean adult female body mass (33.7–45.8 kg; R. Smith and Jungers 1997, 547), and life-history traits strongly relate to this measure (see also above and below).

Assuming a simple, direct relationship and all else being equal, the impact of nutrition on the different life-history variables should be stronger the longer the variable lasts simply because there is more time available for somatic reactions. Under this hypothetical assumption, effects would be smallest for gestation lengths (up to nine months in primates), moderate for interbirth intervals (mostly at least one year, often longer), and strongest for female age at first parturition (several years). For body mass, we expect improved nutrient conditions to lead to heavier individuals.

PROVISIONING, NUTRIENT AVAILABILITY, AND ADULT FEMALE BODY MASS

In mammals, body mass and size are subject to nutrient availability, and classic effects, such as the island rule, Bergmann's rule, Cope's rule, and Dehnel's phenomenon, all relate back to nutrient availability; therefore, it has been suggested to name the underlying general pattern the "resource rule" (McNab 2010). On the basis of the individual, it probably does not come as a surprise that the more we eat, the heavier we are—all else being equal. Research on humans suggests that in order to lose weight, reducing the intake of nutrients, as opposed to increasing physical activity, is most effective (Pontzer, Durazo-Avizu, et al. 2016). If food is nutritious and abundant, it often leads to a heavier adult body mass, even in wild primate populations (yellow baboons, J. Altmann and Alberts 2005; vervet monkeys, Turner et al. 2016). In his classic comparison of 53 nonhuman primates, Leigh (1994b) documented a higher mass for captive individuals of both sexes than for wild ones. He also found a high correlation between values from captivity and from the wild, suggesting that nutrient availability alters body mass in a systematic, predictable way. To avoid a simple repetition of Leigh's work, who compiled data from studbooks, we mostly chose mass values from other studies (table 12.1).

With one exception (black-and-white ruffed lemurs, *Varecia variegata*), the provisioned animals were heavier than wild ones, indicated by the positive percentage value, which averaged +29.6% across species ($N = 12$; range = −0.3% to +70.1%). The data highlight two basic reactions to provisioning: either no or a small mass gain (−0.3% to +14.3%; $N = 6$; mean = 5.3%) or a very strong one (+32.9% to +70.1%; $N = 6$; mean = 53.9%). Species seem to differ in their susceptibility to mass gain and obesity, as already suggested by Leigh (1994b). When the allometric relationship with body mass is considered as a double-log-transformed plot (fig. 12.1), the bold, solid diagonal representing identical values for wild and provisioned populations falls close to the confidence limits of the values compiled. Whether there is indeed a significant difference between the two regimes awaits a larger sample as well as phylogenetic testing.

Surprisingly, the lightest species on our list, *Cheirogaleus medius*, had one of the highest percentages of mass gain when provisioned. This is likely a reflection of specific adaptations to fat storage and mass gain in connection with hibernation in these dwarf lemurs. Prior to hibernation, wild *Cheirogaleus medius* were able to add 88% to their body mass in just four months (Fietz and Ganzhorn 1999). Orangutans are another species with advanced capabilities to store body fat: provisioned females were 70.1% heavier than their wild counterparts (table 12.1). In a recent analysis of hominoids, captive female orangutans had the highest body fat percentages of all apes examined (Pontzer, Brown, Raichlen, et al. 2016). Overall, however, wild primates generally have a very low percentage of body fat (1.9% in yellow baboons, J. Altmann et al. 1993; 2.1% in toque macaques, *Macaca sinica*, Dittus 2013), likely because, in stable populations, individuals live at carrying capacity so that there is minimal surplus energy to be stored as fat.

The examples above hint at the complex relationship between nutrient availability and body mass, but they also indicate that provisioned primate populations have more surplus nutrients available, which are often stored as body fat. We interpret this as support for the dichotomous approach to nutrient availability employed in this chapter.

Besides body mass, increased food intake can also increase adult body length (also called structural size),

Table 12.1. Examples of adult body mass in relation to nutrient availability.

		Intake of human-made food						
		None (wild)				At least some (provisioned)		
Species	Sex	Reference	N	Mass (g)	[%]	Mass (g)	N	Reference
Cheirogaleus medius	F	R. Smith and Jungers (1997)	6	172	+64.0	282	25	Kappeler (1991)
Saimiri sciureus	F	R. Smith and Jungers (1997)	40	662	+5.6	699	58	R. Smith and Jungers (1997)
Aotus lemurinus	F	R. Smith and Jungers (1997)	12	874	+1.7	889	101	R. Smith and Jungers (1997)
Eulemur macaco	F	Terranova and Coffman (1997); R. Smith and Jungers (1997)	5	1,760	+42.6	2,510	41	Terranova and Coffman (1997); R. Smith and Jungers (1997)
Sapajus spp.	F	Fragaszy, Izar, et al. (2016)	7	2,100	+14.3	2,400	13	Fragaszy and Bard (1997)
Eulemur fulvus rufus	M, F	Glander, Wright, Daniels, et al. (1992)	33	2,215	+2.1	2,261	41	Terranova and Coffman (1997)
Chlorocebus aethiops sabaeus	F	Turner et al. (2016)	24	3,020	+44.4	4,360	34	Turner et al. (2016)
Varecia variegata	F	Terranova and Coffman (1997); R. Smith and Jungers (1997)	5	3,520	−0.3	3,510	35	Kappeler (1991)
Macaca mulatta	F	Gordon (2006)	6	4,940	+69.2	8,360	25	S. Schwartz and Kemnitz (1992)
Trachypithecus cristatus	F	Bolter (2004)	20+	5,800	+8.6	6,300	22	Shelmidine et al. (2009)
Pongo pygmaeus wurmbii	F	R. Smith and Jungers (1997)	13	35,800	+70.1	60,900	?	Leigh (1994b)
Gorilla gorilla	F	Jungers and Susman (1984)	3	71,500	+32.9	95,000	27	Hoellein Less (2012)

Whenever possible, adult female mass was selected. The data are arranged in ascending order for values under wild conditions. [%] indicates differences between the two nutritional regimes expressed as percentage points: $(T_p - T_w) \times 100 / T_w$, where T_p = trait when provisioned, T_w = trait when wild. A positive value indicates a higher mass under provisioning, and a negative value indicates a lower mass.

as has been shown, for example, in African elephants (*Loxodonta africana*; Chiyo et al. 2011), humans (Eveleth and Tanner 1990; Stulp et al. 2015), and several smaller, terrestrial mammals (review in Yom-Tov and Geffen 2011). There is also some evidence for nonhuman primates (Barbary macaques, *Macaca sylvanus*, Borg et al. 2014; Japanese macaques, Kimura and Hamada 1995, 1996; vervet monkeys, Turner et al. 2016). It is, however, likely that improved nutrient availability increases adult body mass much more than body length (e.g., yellow baboons, J. Altmann et al. 1993; alpine marmots, *Marmota marmota*, C. Canale, Ozgula, et al. 2016; male white-tailed deer, *Odocoileus virginianus*, Michel et al. 2016).

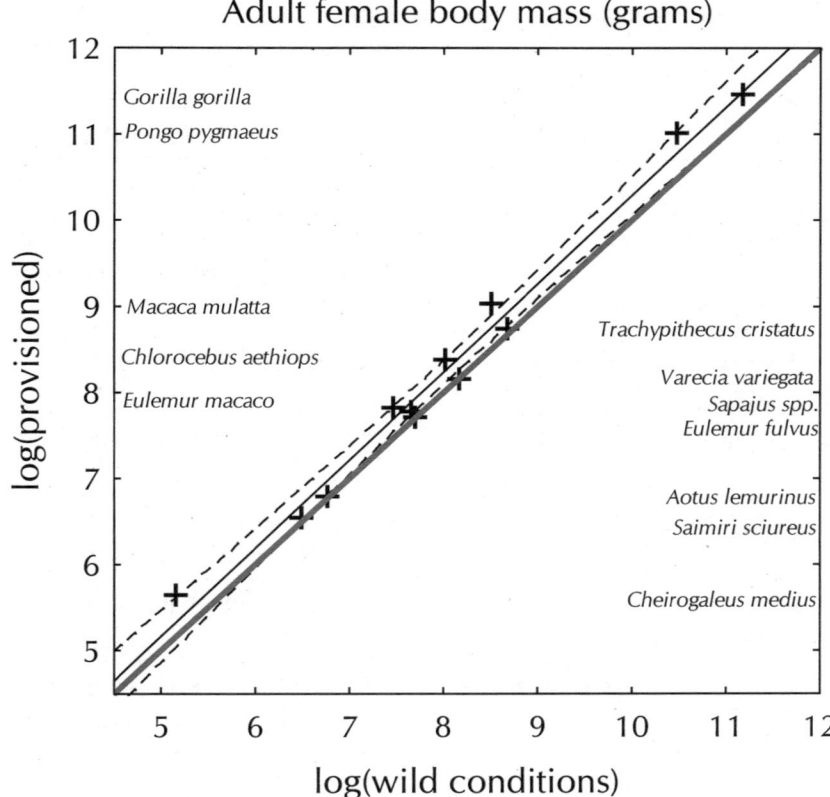

Figure 12.1. Examples of adult female body mass (in grams) in wild versus provisioned populations. The bold, solid diagonal indicates the null hypothesis of identical values (i.e., no effect of nutrient availability); the thin, solid, line is the regression line (standard, not phylogenetic); and the dotted lines are the 95% confidence limits. Data in table 12.1.

NUTRIENT AVAILABILITY AND LIFE-HISTORY VARIABLES

Gestation Length

The first readily available measure of growth in an individual's life is gestation length, which encompasses the entire period of prenatal growth. Gestation length is usually considered a conservative trait unaffected by environmental conditions and mainly determined by adult female body mass and phylogeny (Clutton-Brock 2016; Harvey and Clutton-Brock 1985; P. Lee 2012; R. Martin and MacLarnon 1985). However, farmers have long known that, in livestock, reduced food intake during gestation reduces neonatal mass and increases gestation length, or it reduces gestation length, depending on when during gestation the food shortage occurred (overview in Gilmore and Cook 1981; Sadleir 1969). An early study on wild yellow baboons documented longer gestation periods when maternal feeding time declined throughout gestation compared to when it increased (Silk 1986). Overall, however, ecological effects on primate gestation length have rarely been examined, perhaps because no relevant impact is expected.

In general, gestation length in primates is characterized by pronounced grade shifts, with gestation length being, for example, shorter than predicted by adult female body mass in species with altricial neonates and longer in species with a diapause (R. Martin 2007). Our current overview therefore does not consider such species. We compiled data for gestation length if conceptions (i.e., the beginning of gestation) were determined with reliable methods such as changing hormone levels or single days of housing with a male and when parturition dates were known to within a few days (R. Martin 2007).

In keeping with the general perception of a very stable, species-specific gestation length, we found only a small average difference of −2.7% between the provisioned and the wild conditions in our sample (range = −0.2% to −5.4%; $N = 8$; table 12.2). The double-log transformation (fig. 12.2) emphasizes the relative similarity between the two nutritional regimes, as parts of the bold, solid diagonal fall within the confidence limits

Table 12.2. Examples of gestation length (in days) in relation to nutrient availability.

Species	Intake of human-made food						
	None (wild)				At least some (provisioned)		
	Reference: study site	N	Mean	[%]	Mean	N	Reference: facility/site
Macaca fascicularis	Engelhardt et al. (2006): Ketambe, Gunung Leuser, Indonesia	6	163.0	−0.2	162.7	10	MacDonald (1971): New England Regional Primate Research Center, USA
Macaca sylvanus	C. Young et al. (2013): Middle Atlas, Morocco	?	170.0	−4.0	163.2	56	A. Paul and Kuester (1987): Affenberg Salem, Germany
Macaca fuscata	Fujita et al. (2004): Kinkazan, Japan	9	176.3	−1.9	173.0	17	Nigi (1976): Japan Monkey Center, Japan
Papio cynocephalus	Beehner et al. (2006): Amboseli, Kenya	590	177	−1.1	175	>50	Kriewaldt and Hendrickx (1968): Southwest Foundation, USA
Theropithecus gelada	E. Roberts et al. (2017): Simian Mountains, Ethiopia	6	182.7	−0.5	181.8	3	McCann (1995): Bronx Zoo, USA
Papio anubis	Higham, Warren, et al. (2009): Gashaka-Gumti, Nigeria	4	185.3	−2.9	180	?	van Calsteren et al. (2009): Institute of Primate Research, Kenya
Cercocebus albigena	S. Wallis (1983): Ngogo, Uganda	3	186.0	−5.4	176.0	3	Rowell and Chalmers (1970): Makerere University College, Uganda
Semnopithecus spp.	Ziegler et al. (2000): Ramnagar, Nepal	6	211.6	−5.3	200.3	31	V. Sommer et al. (1992): Jodhpur, India

The data are arranged in ascending order for values under wild conditions. [%] indicates differences between the two nutritional regimes expressed as percentage points: $(T_p - T_w) \times 100 / T_w$, where T_p = trait when provisioned, T_w = trait when wild. A positive value indicates a longer duration under provisioning, and a negative value indicates a shorter duration.

of the regression line (see also example 1 above). Still, all species in our sample had a shorter gestation length under improved nutritional conditions. Furthermore, in the one species for which individual data were available for testing, the gestation length under provisioning was significantly accelerated, although it amounted to only −5.3% (*Semnopithecus* spp., Borries, Koenig, and Winkler 2001; table 12.2). Nevertheless, the overall effect of provisioning on gestation length and thus prenatal growth in primates was small, which could be related to effective maternal buffering mechanisms (Schlabritz-Loutsevitch et al. 2007).

In humans, pregnancy is often shorter and neonate mass lower in undernourished than in well-nourished women (Woods 1989), and additional effects on mother and child have been described, depending on the kind of maternal malnutrition (review in Papathakis et al. 2016). For example, if food was scarce during pregnancy, neonate mass, body length, and head circumference were all lower than under a sufficient food supply (Black et al. 2008; Papathakis et al. 2016; Woods 1989). It is not entirely clear why poor nutrient availability shortens pregnancy in humans but lengthens gestation in nonhuman primates. Perhaps the mechanisms inducing birth differ. In humans, parturition is induced when the fetal energy requirements are no longer met (Dunsworth and Eccleston 2015; Ellison 2001), a condition that might not occur in nonhuman primates.

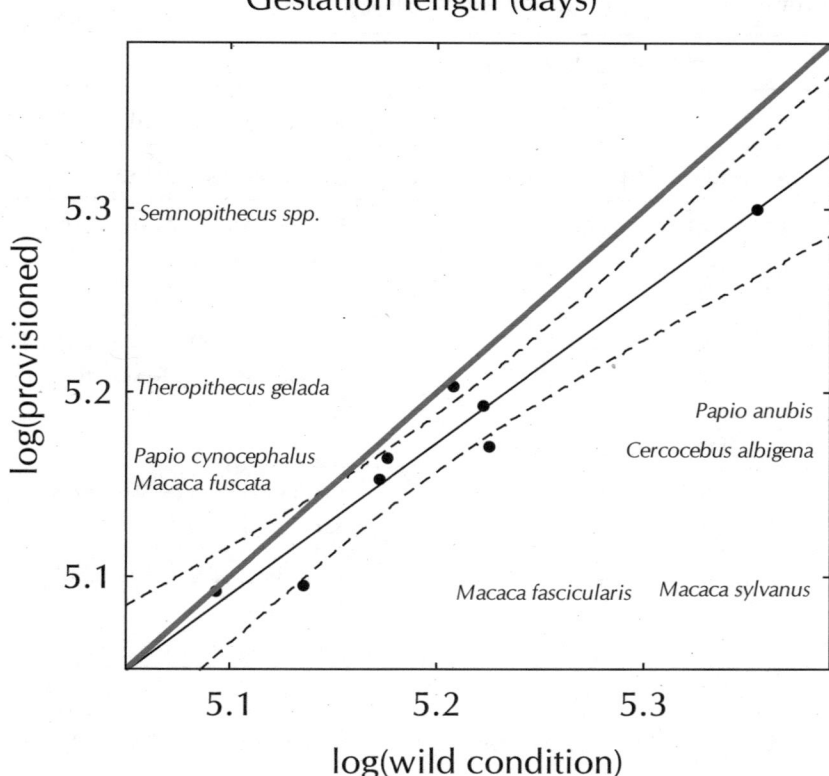

Figure 12.2. Examples of gestation length (in days) in wild versus provisioned populations. The bold, solid diagonal indicates the null hypothesis of identical values (i.e., no effect of nutrient availability); the thin, solid line is the regression line (standard, not phylogenetic); and the dotted lines are the 95% confidence limits. Data in table 12.2.

Age at First Parturition

The second measure for growth that is relatively frequently available for primates is the age at cessation of somatic growth captured by the attainment of adult head-body length (i.e., the cessation of structural growth). In female primates, this threshold is often approximated by the age at first parturition, which is much easier to determine in wild populations. We consider the age at first parturition to be a good though not perfect estimate for the turning point from investment in somatic growth toward investment in reproduction (see also K. Hill and Hurtado 1996 for humans; Lu, Bergman, et al. 2016 for geladas, *Theropithecus gelada*), even though females may continue to grow for some time after starting to reproduce (e.g., several colobines, Bolter 2004; gray langurs, Borries, pers. obs.; toque macaques, Cheverud et al. 1992).

All species in our sample had a younger age at first parturition under improved nutritional conditions, averaging −27.2% (range = −15.3% to −47.8%; N = 8; table 12.3). The effect was rather uniform across species (fig. 12.3), resulting in a regression line almost parallel to the bold, solid diagonal. The bold, solid diagonal also fell mostly outside of the confidence limits of the regression line. In the most extreme case, already mentioned at the beginning of this chapter, provisioned females had their first infant when they were only about half as old as their wild counterparts (*Semnopithecus* spp.; table 12.3). In *Pongo*, although the difference between mean values was not extreme, the difference in minimum age at first parturition was large (−46.2%; table 12.3).

In humans, the onset of reproduction is usually captured as the age at menarche (first menstruation) rather than first parturition because the latter depends on factors unrelated to somatic capabilities (Morabia et al. 1998). Initially, it was assumed that a particular threshold body mass triggers menarche in girls (Frisch and Revelle 1971), but this was soon broadened to a body mass index, considering height in addition to mass (Frisch et al. 1973). Today, however, the bi-iliac diameter as a proxy for the pelvis diameter is considered as the most important factor (Ellison 1990). Girls attain the adult pelvis shape late, likely because the final modeling is influenced by increasing estrogen levels, which also terminate structural growth (Ellison 2001).

Table 12.3. Examples of female age at first parturition (in years) in relation to nutrient availability.

Species	Intake of human-made food						
	None (wild)				At least some (provisioned)		
	Reference: study site	N	Mean	[%]	Mean	N	Reference: facility/site
*Papio cynocephalus**	J. Altmann and Alberts (2005): Amboseli, Kenya	56	4.7	−27.7	3.4	14	J. Altmann and Alberts (2005): Amboseli, Kenya
Macaca fascicularis	M. van Noordwijk and van Schaik (1999): Ketambe, Gunung Leuser, Indonesia	22	5.2	−25.0	3.9	252	Petto et al. (1995): New England Regional Primate Research Center, USA
Macaca silenus	Kumar (1987): Varagaliyar, India	5	6.6	−25.8	4.9	39	Lindburg et al. (1989): North America studbook
Semnopithecus spp.	Borries, Koenig, and Winkler (2001): Ramnagar, Nepal	26	6.7	−47.8	3.5	12	V. Sommer et al. (1992): Jodhpur, India
Semnopithecus spp.	As above	1	6.0	**−48.3**	3.1	1	as above
Macaca fuscata	Takahata, Suzuki, et al. (1998): Kinkazan, Japan	20	7.1	−23.9	5.4	182	Koyama et al. (1992): Arashiyama, Japan
Hylobates lar	U. Reichard et al. (2012): Khao Yai, Thailand	5	10.5	−24.8	7.9	4	calculated for individuals of known age from Geissmann (1991): several zoos
Pongo abelii	Wich et al. (2009): Ketambe, Gunung Leuser, Indonesia	9	15.2	−27.0	11.1	13	Kuze et al. (2012): Bukit Lawang Rehabilitation, Indonesia
Pongo pygmaeus	Galdikas and Ashbury (2013): Tanjung Puting, Indonesia	3	15.7	−15.3	13.3	19	Galdikas and Ashbury (2013): Camp Leakey, Indonesia
Pongo	As above	1	13.0	**−46.2**	7.0	1	Shumaker, Wich, and Perkins (2008)

Some select youngest ages are given to highlight the maximum acceleration realized with the difference highlighted in boldface. * = menarche age. The data are arranged in ascending order for values under wild conditions. [%] indicates differences between the two nutritional regimes expressed as percentage points: $(T_p - T_w) \times 100 / T_w$, where T_p = trait when provisioned, T_w = trait when wild. A positive value indicates a longer duration under provisioning, and a negative value indicates a shorter duration.

These increasing estrogen levels furthermore induce and modify fat depositions, which might be the reason why, in past studies, body fat and body mass index correlated with menarche age in humans. In most nonhuman primates, however, the bi-iliac diameter is unlikely to restrict parturition because the newborn's head is smaller than the pelvic diameter (Leutenegger 1970; Wells, deSilva, and Stock 2012). Only humans are faced with the obstetric dilemma (Leutenegger 1982; Haeusler et al. 2021; but see Dunsworth et al. 2012). Regarding our main question, menarche age in humans is influenced not only by nutrient availability but also by health factors (Karapanou

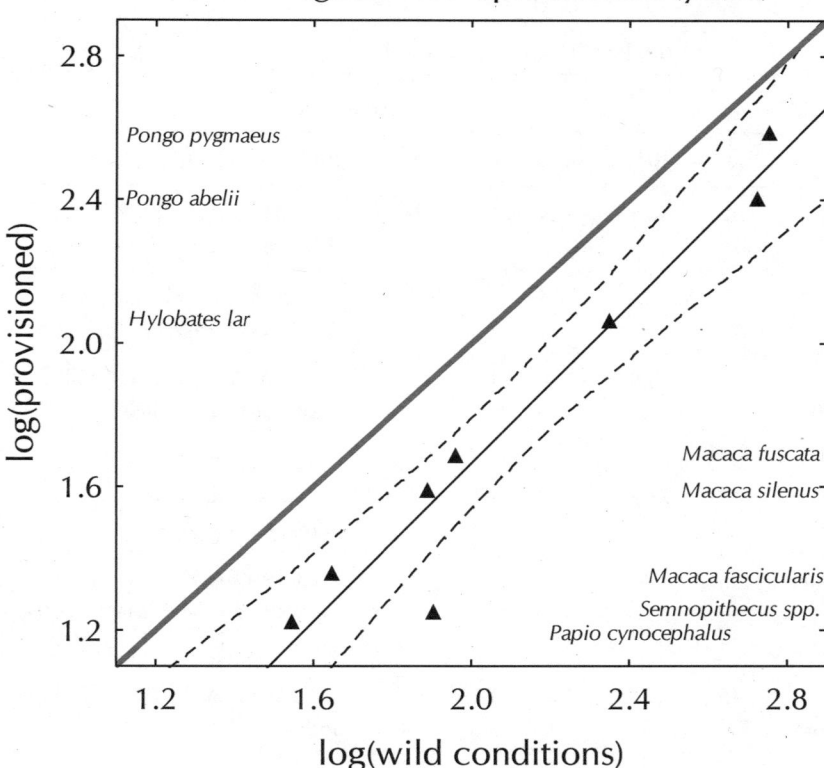

Figure 12.3. Examples of female age at first parturition (in years) in wild versus provisioned populations. The bold, solid diagonal indicates the null hypothesis of identical values (i.e., no effect of nutrient availability); the thin, solid line is the regression line (standard, not phylogenetic); and the dotted lines are the 95% confidence limits. Data in table 12.3.

and Papadimitriou 2010; Simondon et al. 1997; Thomas et al. 2001), with malnutrition delaying menarche (A. Riley 1994; Simondon et al. 1997) and modern environmental pollutants or hormones added during food production accelerating it (Steingraber 2007). Furthermore, during the past two centuries, a secular trend reduced the average menarche age by more than three years (Lehmann and Scheffler 2016). Improved economic status and nutrition are the strongest correlates of this trend (Berkey et al. 2000; Wronka and Pawlinska-Chmara 2005). In line with the presumed mechanisms regulating menarche (described above), obesity (i.e., excess fat tissue) alone had only a limited effect, reducing menarche age by another five months on average (Bau et al. 2009).

Reproductive Rate (the Interbirth Interval)

Besides the age at first parturition (e.g., J. Altmann, Hausfater and Altmann 1988), female lifetime reproductive output is mainly influenced by the speed of reproduction modified by mortality patterns and reproductive life span (Pusey 2012; Tung et al. 2016). Because premature loss of a dependent offspring reduces the IBI in most primates (van Schaik 2000; exceptions are seasonal breeders reproducing every year), we considered only mean intervals following the birth of a surviving infant. This standardization removed the effect of local mortality patterns.

All species in our sample had a shorter IBI under provisioning. The acceleration averaged −39.8% (range = −27.3% to −56.3%; $N = 9$; table 12.4). The bold, solid diagonal fell outside of the confidence limits of the regression line (fig. 12.4). With almost 40% acceleration under provisioning, this life-history trait seems to be the most impacted by nutrient availability of the four traits considered here.

Compared to nonhuman primates, humans have a short IBI (Robson et al. 2006), especially in relation to adult female body mass (Emery Thompson 2013b). It is assumed that the maternal recovery period in humans is so short relative to that in nonhuman primates because of the extensive communal care, with several individuals providing food and other support (Hrdy 1999), including immature household members (Kramer 2005). What has turned out to be an advantage for our lineage may be a disadvantage for nonhuman primates. It is theoretically

Table 12.4. Examples of interbirth intervals (IBIs) after a surviving offspring (in months) in relation to nutrient availability.

	Intake of human-made food						
	None (wild)				At least some (provisioned)		
Species	Reference: study site	N	IBI	[%]	IBI	N	Reference: facility/site
Papio hamadryas	Swedell, Leedom, et al. (2014): Filoha, Ethiopia	51	22.4	−38.4	13.8	18	Kaumanns et al. (1989): German Primate Center, Germany
Papio cynocephalus	J. Altmann and Alberts (2003): Amboseli, Kenya		~22	−27.3	~16		J. Altmann and Alberts (2003): Amboseli, Kenya
Papio anubis	Higham, Warren, et al. (2009): Gashaka-Gumti, Nigeria	2	25.0	−36.0	16.0	4	Higham, Warren, et al. (2009): Gashaka-Gumti, Nigeria
Macaca fuscata	Takahata, Suzuki, et al. (1998): Kinkazan, Japan	49	28.4	−38.4	17.5	770	Koyama et al. (1992): Arashi-yama, Japan
Macaca fascicularis	M. van Noordwijk and van Schaik (1999): Ketambe, Gunung Leuser, Indonesia	33	29.3	−56.3	12.8	22	Hadidian and Bernstein (1979): Yerkes Regional Primate Center, USA
Semnopithecus spp.	Borries and Koenig (2000): Ramnagar, Nepal	45	32.4	−46.9	17.2	82	V. Sommer et al. (1992): Jodhpur, India
Hylobates spp.	U. Reichard et al. (2012): Khao Yai, Thailand	22	40.8	−32.4	27.6		Hodgkiss et al. (2010): Howletts Wild Animal Park, England
Pongo pygmaeus	M. van Noordwijk, Willems, et al. (2013): Tuanan, Indonesia		~84	−31.2	57.8	249	H. Anderson et al. (2008): International studbook
Pongo abelii	Wich et al. (2004): Ketambe, Gunung Leuser, Indonesia	19	111.1	−51.7	53.7	308	H. Anderson et al. (2008): International studbook

Offspring sex not considered. The data are arranged in ascending order for values under wild conditions. [%] indicates differences between the two nutritional regimes expressed as percentage points: $(T_p - T_w) \times 100 / T_w$, where T_p = trait when provisioned, T_w = trait when wild. A positive value indicates a longer duration under provisioning, and a negative value indicates a shorter duration.

possible that nonhuman primate females under excellent nutrient availability recover faster than their infants are able to grow and gain independence. Under such conditions, infants could suffer higher mortality, as females would likely wean them prematurely.

Other Life-History Variables (Less Well Studied)

The age at nutritional independence of an offspring is another variable to estimate growth and thus describe the life-history pattern of a population. The variable is, however, problematic because it has been measured in many different ways, such as the first intake of solid food or the period of peak lactation (Borries, Lu, et al. 2014; see discussion in, e.g., P. Lee 1996). Despite much improved methods (Bădescu, Katzenberg, et al. 2017; Reitsema and Muir 2015), it is still difficult to pinpoint the moment when an immature primate gains nutritional independence. At present, we can only provide comparative data from our own research on Asian colobines for which we measured the age at nutritional independence via the age at the cessation of nipple contact. This measure was closely related to the age when immatures were

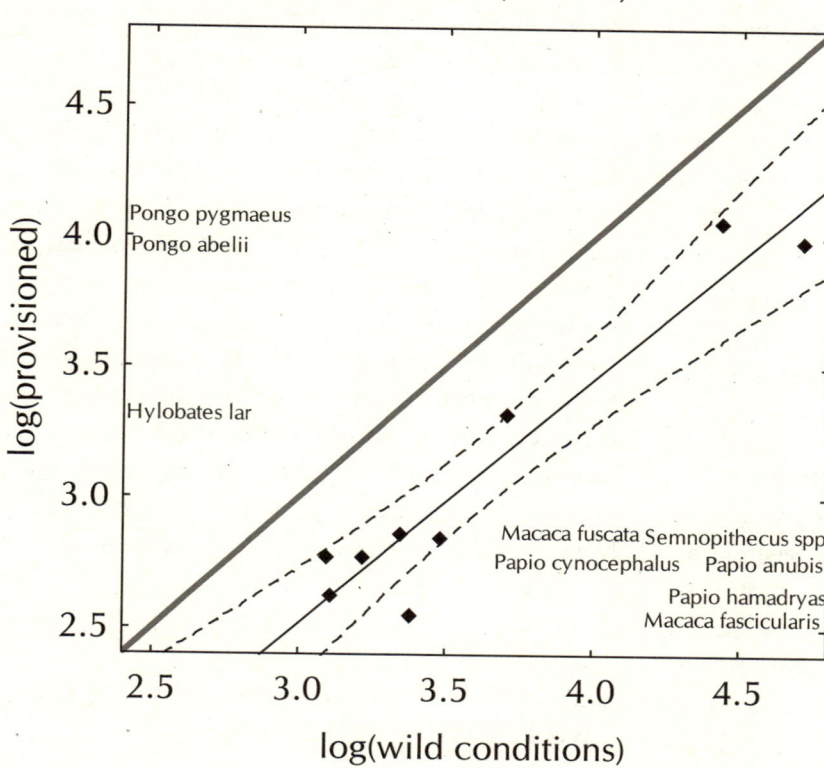

Figure 12.4. Examples of interbirth intervals after a surviving offspring (in months) in wild versus provisioned populations. The bold, solid diagonal indicates the null hypothesis of identical values (i.e., no effect of nutrient availability); the thin, solid line is the regression line (standard, not phylogenetic); and the dotted lines are the 95% confidence limits. Data in table 12.4.

able to survive the loss of their mother (Phayre's leaf monkeys; Borries, Lu, et al. 2014). The impact of nutrient availability on the age at nipple contact cessation was strong. In provisioned *Semnopithecus* spp., nipple contact terminated at an average age of 12.8 months (Rajpurohit and Mohnot 1991) compared to 28.9 months at the wild forest site, a significant difference (Borries, Koenig, and Winkler 2001). Similarly, comparing across species, provisioned silvered leaf monkeys, *Trachypithecus cristatus*, at the Bronx Zoo had an average age at last nipple contact of 12.1 months, whereas wild Phayre's leaf monkeys, the sister taxon, ceased to have nipple contact at 19.0 months on average (Borries, Lu, et al. 2014). With −55.7% and −36.3%, respectively, the magnitude of the difference between the two nutritional regimes in these age-at-last-nipple-contact examples resembles that of the IBI comparisons above. A close connection between the two parameters is indeed very likely because, in many primates, the current offspring is weaned right before the next sibling is born (Borries, Lu, et al. 2014), emphasizing that maternal recovery and reconception regulate investment in the current offspring. We elaborate on the potential problems this disconnect can cause below when comparing the relative effects of nutrient availability on several life-history variables.

Litter size is another life-history variable to be considered in estimating reproductive effort. Most primate species bear singletons, but some taxa such as ruffed lemurs, *Varecia* spp., bear multiple young. Arguably the best-studied litter-bearing primate taxon are the Callitrichidae, although the majority of studies are from captivity, allowing only for a few comparisons with wild conditions. For example, common marmosets (*Callithrix jacchus*) had an average litter size of 2.27 in captivity (Tardif, Smucny, et al. 2003) and 1.77 in the wild (calculated from Digby 1995), but the sample size for the wild population was small. Similarly, in golden lion tamarins, litter size in captivity averaged 2.14 (A. Baker and Woods 1992) versus 1.80 in the wild, with a higher proportion of twins and fewer triplets in the wild (Dietz, Baker, and Miglioretti 1994). In addition, several colonies report an increasing number of larger litters with time (e.g., Hiddleston 1978;

Rothe et al. 1987; Tardif, Smucny, et al. 2003). In contrast to these results for platyrrhine species, litter size in gray mouse lemurs (*Microcebus murinus*) did not differ, averaging 2.0 both in captivity and in the wild (Eberle and Kappeler 2004; Wrogemann et al. 2001). Thus, whether litter size in primates is generally sensitive to nutrient availability awaits further study.

IS THE RELATIVE EFFECT OF NUTRIENT AVAILABILITY PROPORTIONAL TO THE DURATION OF THE VARIABLE?

Based on the above explorations, the answer to this question is no. First of all, the variables examined above changed in the predicted direction under provisioning: body mass increased, indicating surplus energy; and gestation length was shorter, age at first parturition decreased, and the IBI was reduced—that is, the speed of reproduction was accelerated. However, contrary to our expectation—which assumed all else being equal—the relative effect was not directly proportional to the duration of the respective variable. Of the three life-history variables examined, gestation had the shortest duration, followed by the IBI and then age at first parturition. However, the largest relative effect was found for the IBI, which did not have the longest duration. This suggests a fundamentally different underlying mechanism for the IBI compared to the other two life-history variables of gestation length and age at first parturition.

Gestation length and age at first parturition both capture the speed of somatic, structural growth. At first glance, the IBI is also composed of two growth periods: fetal growth in the form of gestation length and subsequent postnatal infant growth. However, the length of the IBI is de facto determined much earlier, at the time when the next conception occurs. Hence, the time from parturition until subsequent conception is decisive for the length of the IBI. This fact moves the focus from the infant to the mother, to the time needed for maternal energy replenishment after parturition. Once a positive energy balance is achieved and sustained, the next reproductive event can be initiated (Ellison 2001). This can happen very quickly if nutrient availability is high and there are no seasonal constraints on reproduction (as in several primate species; see chap. 7). Because maternal energy balance triggers the onset of the next reproductive event, communal breeding, which saves maternal energy, can accelerate the frequency of reproduction even further. For example, humans, with their extensive food-provisioning schemes, increased their reproductive rate to the extent that it became largely decoupled from infant growth and development, resulting in multiple dependent offspring (e.g., Kramer 2005).

Our results indicate that high nutrient availability facilitates energy replenishment for mothers rather fast and exponentially, which is also reflected as body mass gains. We would like to caution, however, that there could be gross differences in the susceptibility of species to pack on pounds when food is plentiful, as has previously been suggested (Leigh 1994b). This adds yet another argument in favor of using body mass data from wild animals rather than provisioned ones when accounting for the underlying allometric relationship in comparative life-history analyses.

BEYOND PROVISIONING: FOOD AVAILABILITY AND OTHER CAVEATS

Variation in Food Availability and Its Measurements

Our above exploration is only preliminary, because sample sizes are small and taxonomically biased. A thorough, statistical analysis based on a much larger sample and controlled for phylogeny is urgently needed. Such a comparison is, however, hampered not only by the lack of a vetted compilation of primate life-history data (Borries, Sandel, et al. 2016) but also by the lack of fine-grained, quantitative assessments of food availability and, more importantly, energy intake and expenditure. This is why we coded food availability very broadly and crudely as a dichotomous variable. Obviously, this has several disadvantages.

Provisioned primate populations can be subject to vast differences in food availability and energy intake. In a study of provisioned gray langurs, individual gross energy intake varied by a factor of seven from lowest to highest, while energy expenditure was rather similar (factor 1.2; calculated from fig. 1 in Schuelke 2001). Strong differences in energy intake also occur across populations. In our sample, populations were labeled as provisioned when they had access to human-made food during just a few weeks each year as well as when they were in captivity and entirely dependent on provided food. Similarly, we assumed constant and optimal conditions in captivity even though primate husbandry has improved in the past few decades. Captivity today and 50 years ago can entail very different conditions, rendering direct comparisons of data collected in different decades problematic. Even more recently, changes in housing conditions can have strong effects on life-history variables (e.g., IBIs in baboons; Cary et al. 2003). While we would argue that the strong and uniform effect of provisioning on body mass and life history justifies our approach, a more detailed quantification of nutrition in provisioned and captive populations would certainly be helpful.

In addition, natural, undisturbed habitats differ in the number of primate and nonprimate food competitors (e.g., Elder 2013; Fleagle et al. 1999), which can impact food availability. Consequently, measuring food availability in the habitat may not reflect its availability for a particular group or species. Difficulties in quantifying food availability were recently demonstrated in a large-scale study of IBIs in yellow baboons, in which two different measures of food availability—namely, rainfall and a habitat quality index—showed inverse associations with gestation length (Gesquiere et al. 2018). Across populations, natural habitats also differ widely in their general productivity and food availability, which translates into different reproductive rates (e.g., baboons; L. Barrett, Henzi, and Lycett 2006) or even life expectancies (Wood et al. 2017). Populations may also not be at equilibrium, nor do the animals necessarily have an optimal diet, especially when inhabiting a marginal habitat. One solution to this problem might be to characterize populations as expanding or stable because life history changes with population dynamics in a predictable manner (Charnov 2009). However, to determine such broad population trends requires reliable and extensive data, which are rarely available.

A striking example of local differences in food availability is provided by the chimpanzees of the Kibale forest in Uganda (see also Borries, Lodwick, et al. 2022). There, two populations live just 10 km apart, one at Kanyawara, the other at Ngogo, and each has been studied for several decades (D. Watts 2012b; Wood et al. 2017). Fruits, the main food source for chimpanzees (Balcomb et al. 2000), are much more abundant and less seasonal at Ngogo than at Kanyawara (K. Potts, Baken, Ortmann, et al. 2015). At Ngogo, the chimpanzee density is three times higher, and community size is still increasing (from about 144 in the late 1990s to 204 in May 2016; Wood et al. 2017), indicating a population not yet at carrying capacity. The energy gain per minute feeding time is much higher (71.88 kcal at Ngogo versus 44.97 kcal at Kanyawara; K. Potts, Baken, Ortmann, et al. 2015), and IBIs after surviving infants are 23% shorter at Ngogo (62.9 months, $N = 20$, uncensored) than at Kanyawara (81.8 months, $N = 31$; Emery Thompson, Kahlenberg, et al. 2007). The documented differences in food abundance match these differences in life history, yet both populations are wild. In fact, the lower food density at Kanyawara potentially resulted from its past logging history. If true, this would be an example of reduced rather than improved food availability caused by prior human impacts on the habitat. This adds yet another dimension to food availability in relation to human influences.

For the purpose of this overview, we have considered human impact as a factor improving nutrient availability, which holds true for primate colonies, many free-ranging but provisioned populations, and those raiding crops (e.g., Barbary macaques; Borg et al. 2014). However, the opposite may occur as well, as documented for western jackdaws (*Corvus monedula*; Meyrier et al. 2017). Negative impacts on food availability can also be related to habitat degradation and fragmentation (A. D. Johns 1986; L. Marsh and Chapman 2013). As a scientific community, we need to develop better and standardized descriptions of human impacts on the system under study (Sugiyama 2015). More generally, the local nutritional conditions, be it in the wild, at the dumpster, or in captivity, need to be quantified comprehensively, for the entire system, to allow for meaningful future data matching, comparisons, and analyses. How to meaningfully quantify habitat quality and nutrient availability across populations and study sites will remain a challenge.

What's Wrong with Ape Life-History Data?

In addition to the problems mentioned above, our compilation must remain preliminary because of the taxonomic imbalance of the sample. For example, the sample for gestation length consisted exclusively of Cercopithecoidea. To some degree, this relates to the stronger focus of behavioral studies on this taxon. However, it relates also to the tricky nature of the available data for several species, particularly apes.

Apes have the slowest life histories of all nonhuman primates, which in itself is a problem because it requires decades to accumulate sufficiently large sample sizes. This is exaggerated by their small group sizes (*Pongo* is solitary or semisolitary, and in *Gorilla*, small one-male groups predominate) or small party sizes in a fission-fusion system (both *Pan* species; D. Watts 2012a). Consequently, at any point in time, one can collect data on only the few individuals nearby, and large data sets accumulate painfully slowly.

In addition, existing ape life-history(-related) data have been plagued with multiple shortfalls, one of which is body mass assessment, a crucial measure in the context of life history. Perhaps the most prominent example is adult female body mass in the genus *Gorilla* (recently summarized in Borries, Lodwick, et al. 2022). R. Smith and Jungers (1997) gave 97.5 kg ($N = 1$) for wild mountain gorillas (*Gorilla beringei*) and 71.5 kg for wild western gorillas (*Gorilla gorilla*, $N = 3$), while a more recent compilation (Gordon 2006) did not distinguish the two gorilla species and gave a mean of 75.7 kg ($N = 6$), citing R. Smith and Jungers (1997). Based on new morphometric data, adult female body mass in mountain gorillas has been determined as 66.3 kg ($N = 3$, M. L. Burgess et al. 2018) and thus 31.2 kg or 32% lighter than the value used thus far. As it stands right now, mountain gorilla females are in fact lighter, not heavier, than western gorilla females; it therefore makes sense that mountain gorillas have a slightly faster life history than western gorillas, who also live in more seasonal habitats (Doran-Sheehy, Mongo, Lodwick, Salmi, et al. 2012; Stoinski et al. 2013). Further back in time, life-history values for wild mountain gorillas were compared with those for captive western gorillas because no data from captivity were available for mountain gorillas and no data from the wild were available for western gorillas. For additional misunderstandings in the interpretation of variation in gorilla life history, see Borries, Lodwick, et al. (2022). In this paper, we propose that the most parsimonious explanation for the variation is phenotypic plasticity driven by differences in nutrient availability and energy budgets.

Some of the core problems with life-history data for chimpanzees also revolve around adult female body mass. The three subspecies distinguished differ noticeably in this measure, with females in two subspecies (*P. t. troglodytes* and *P. t. verus*) heavier (45.8 kg, $N = 4$ and 41.6 kg, $N = 3$, respectively) than those in the third (*P. t. schweinfurthii*, 33.7 kg, $N = 26$; R. Smith and Jungers 1997). Unfortunately, at the longest-running study site, Gombe (subspecies *P. t. schweinfurthii*), food was provided until July 2000, which increased mass values by up to 10% (Pusey et al. 2005), although, at 31.3 kg, the median adult female body mass was even slightly lower than the values from the literature. Comparing wild populations with captive colonies touches on another problem because the subspecies of colony founders was often unknown and, through the years, hybridization with other subspecies may have occurred. When mitochondrial DNA for more than 200 founder individuals was tested, *P. t. schweinfurthii* individuals were rarely represented, and most individuals belonged to *P. t. verus* (Ely et al. 2005). For this review, our solution has been to exclude the species with its subspecies.

For the other *Pan*, the bonobo (*P. paniscus*), there are still very few life-history data available from wild populations (Kano 1996; Robson et al. 2006; Sugiyama 2015). Together, the data for most apes, which represent a crucial extreme of the primate life-history spectrum, are currently not suitable for certain life-history comparisons.

SUMMARY AND CONCLUSIONS

With respect to potential environmental effects on primate life history, research is still in its infancy. Most of the published compilations of primate life-history data combine values from different populations and thus different nutritional regimes under the same label (see, e.g., the comparison in Borries, Gordon, and Koenig 2013). To make primate life-history data suitable for comparative analyses, it will be necessary to settle on basic, comparative measures of nutrient availability and report those in future publications in a standardized manner. Based on what we know thus far (above) and given the time-consuming collection of precise and comparable ecological data, identifying access to any kind of human-made food can already be a useful step in the right direction (Sugiyama 2015).

Based on data compiled from the literature, we explored the potential impact of nutrient availability on primate life history. Nutrient availability was rated binary as good under provisioning (any consumption of human-made food) and less favorable in the wild, where provisioning was absent. Although the sample sizes of our compilations were naturally small, the following patterns emerged:

- Adult female body mass was larger in provisioned populations, which supports the assumption of improved nutrient availability under this regime. Some species were more likely than others to increase mass under favorable nutritional conditions.
- Gestation length was consistently shorter but only marginally affected by provisioning.
- Provisioning strongly accelerated the age at first parturition and the interbirth interval.
- There are indications for a fundamentally different and potentially conflicting mechanism regulating female reproductive rate and offspring growth that require further studies.
- Except for gestation length, human life history reacts in a similar fashion to differences in nutrient availability as nonhuman primate life history.
- Our preliminary and simple (binary) comparisons demonstrate the high plasticity of primate life-history variables, which—if unaccounted for—may distort the results of comparative studies.
- As a community, we need to develop standards to determine the relative food availability for our study systems and how to code it in comparative analyses.

PART III

Food Acquisition and Nutrition in Social Environments

It is well understood that primates are among the most social mammals—with upward of 70%–80% of all species living in stable, socially cohesive groups—as well as among the most encephalized of mammalian orders. As examined in detail in this section, social living and cognition interact at many scales, with consequences for virtually all aspects of behavior, including locating and procuring sufficient energy and nutrition. The evolutionary and ecological significance of being a social forager has occupied a central position in the theoretical conceptualization of primate feeding and socioecology (Janson). Overarchingly, life as a social animal both hinders and facilitates the meeting of nutritional goals (Bryer and Uwimbabazi). Consuming diverse food types varying in spatial and temporal availability and simultaneously solving the challenges of tracking social relationships and neighborhoods are no doubt significant contributors to primate cognitive abilities in general (Garber) and are clearly implicated in tool use (Pruetz, Badji, Bogart, Lindshield, Ndiaye, and Walkup), cooperative hunting (Watts), and decisions regarding how and when to move (Crofoot and Alavi). Moreover, as prey species, primates forage in landscapes of fear (Hill), which has significant ecological and evolutionary implications and which shapes both intra- and interspecific interactions. The costs of predation and fear, however, are buffered somewhat by an overall high degree of phenotypic plasticity in foraging—arguably another hallmark of being a primate (Hardie and Strier).

13 Social Food Competition, Then and Now

Charles H. Janson

Food competition has long held a central place in the conceptual theory of primate socioecology (e.g., Alexander 1974; Clutton-Brock and Janson 2012; Isbell 1991; Janson 1988b; Janson and Goldsmith 1995; A. Koenig 2002; Pruetz and Isbell 1999; Sterck 1997; Terborgh 1983; van Schaik 1989). Food competition has also proved to be of enduring empirical importance, with hundreds of papers that have focused on the existence, strength, and mechanisms of food competition in particular primate species or populations. Recent meta-analyses based on these population studies have shown that the expected negative effects of food competition on fitness (measured primarily as reproductive success) are common across primate populations, although with several interesting exceptions (Majolo, Lehmann, et al. 2012; Majolo, Vizioli, and Schino 2008).

The overarching goal of this chapter is to review the development of socioecological theory applied to primates, highlight some unresolved questions about the conceptual basis of social food competition in primates, present a more explicit and comprehensive conceptual theory of how competition affects primate group size and structure, and discuss how spoken and unspoken assumptions of the theory have been tested. This review is not comprehensive; some topics such as interspecies competition for food are not discussed, even though we have learned a tremendous amount about this in the past two decades. I apologize in advance to the authors of a considerable number of excellent empirical studies on food competition whose work I cannot include here.

My specific aims here are to (1) review the development of the four main types of social food competition in primates; (2) discuss the common perception that scramble (indirect) and contest (direct) competition are independent of each other; (3) examine the mechanisms that translate food distribution into food competition, including interindividual spacing, knowledge of food patch location and condition, and time constraints; (4) relate food competition to group size and structure; and (5) consider some of the tests of assumptions concerning the causes of food competition and its consequences for social living.

THE MAJOR CATEGORIES OF SOCIAL FOOD COMPETITION

It is hard to convey the state of ignorance and confusion that reigned in socioecology in the 1970s. There were no agreed-upon mechanisms that related particular aspects of ecology to social structure, despite several correlational studies relating social group size and stability to broad ecological patterns (e.g., diet, activity, and habitat; Crook 1964; Crook and Gartlan 1966; E. Wilson 1975) and suggestions about the likely selective factors that might generate them (e.g., food abundance and distribution, predation, and between-group competition; Alexander 1974; Clutton-Brock 1977b; Clutton-Brock and Harvey 1976, 1977; A. Jolly 1972; E. Wilson 1975; Wrangham

1979, 1980). The limited size of most social groups was counterintuitive given the widely held dogma that population sizes were determined by both population density and habitat area. Unless something was limiting the area a group occupied, there should be no natural limit to its size; indeed, some bird flocks reach numbers in the millions. In this context, what seem like simple intuitions today came as significant moments of inspiration, sometimes following on the heels of intense discussions into the early morning hours with colleagues at a conference. The clues we had were a set of common observations, based almost entirely on diurnal primates: (1) aggression was fairly common, at least in fruit-eating or omnivorous species; (2) aggression often occurred in the context of feeding; and (3) most fruit-eating primates spent a large majority of their time, and sometimes all of it, feeding or foraging. Thus, it seemed that food was worth fighting over, that group members were willing to stay close enough to each other while feeding to allow aggression between them, and that it was not always easy for primates to get enough to eat, despite the prevailing myth of tropical superabundance of food. Two major insights helped bring these ideas together: (1) Trivers (1972) argued that for females, food should be the main limiting resource for fitness, whereas for males it should be access to fertile females, and (2) fruit often occurs in discrete limited patches that may be scarce (e.g., Terborgh 1983). The idea emerged that the willingness of females to tolerate other females might be a function of the amount of food available in each patch (M. Leighton and Leighton 1982), while males largely competed against other males over access to the female group (e.g., Andelman 1986; Dunbar 1984). Limited patch size could act as the spatial constraint that limits group size. If so, this insight opened a new question: if females in a group compete over food, why bother living in groups at all—what selective forces favor grouping in female primates (van Schaik 1983)? This question received several answers. First, larger groups of a species might win encounters over valuable food patches often enough that their group members actually had higher food intake than individuals in smaller groups (Wrangham 1980); in practice, food "patches" may equate to entire territories (Cheney 1981), although the two scales of patchiness have different implications. Second, females might cluster around a particular strong male either to avoid infanticide (Wrangham 1979) or harassment (Clutton-Brock et al. 1992) or to improve access to food (T. R. Harris 2006; Wrangham 1980). Third, groups might form to reduce each individual's risk of predation through a variety of distinct mechanisms (van Schaik 1983; review in Janson 1998b), similar to long-standing ideas and evidence from bird flocks.

Because behavioral competition over food was either observed or suspected in most primates (Terborgh 1986), we focused efforts on understanding this common constraint. Socioecologists observed that animals possessed two common tactics to obtain food—by being the first to reach a feeding site and ingest the food, or by being strong enough to displace or chase others from good feeding sites (e.g., Murton et al. 1971; Pulliam and Caraco 1984). Indirect or scramble competition is the reduction in food intake to group members that occurs because some animals arrive earlier to find and ingest food items; implicit in scramble competition is the condition that food density recovers slowly enough that later arrivals experience reduced food availability. Direct or contest competition is the loss of food intake when one animal displaces another from its food, whether accompanied by agonism or not (Janson and van Schaik 1988; van Schaik and van Noordwijk 1988). Both kinds of competition can occur within a social group and between distinct groups, resulting in four main categories of social food competition (fig. 13.1, after Janson and van Schaik 1988): within-group scramble (WGS), within-group contest (WGC), between-group scramble (BGS), and between group contest (BGC). BGC is the only form of social food competition that might favor larger groups; the remaining all favor reduced group size either by imposing foraging costs on all group members (WGS and BGS) or by skewing the costs toward particular individuals, who should therefore be more likely to leave the group (WGC; e.g., Giraldeau and Caraco 2000; Vehrencamp 1983).

In the nearly 40 years since the description of these forms of competition, direct or proxy measures of food competition have been made in many primate species for all competitive categories except BGS, which is thought unlikely to have a strong effect on group size since it should affect all groups equally. However, it is premature to dismiss BGS: the extent of social food competition will depend on the degree to which food is limiting to the population overall (fig. 13.1e). If disease, predation, or hunting reduces the population density well below its carrying capacity, all forms of social food competition will be reduced, including BGS. Similarly, if food production temporarily exceeds the needs of a population, all forms of competition will be reduced or absent (e.g., Janson 1984a; Terborgh 1983). Conversely, if predation is eliminated, a population may grow "unnaturally" dense, and the increased density may intensify and skew the importance of different forms of food competition. Indeed, in the second major iteration of the socioecological model of primate social structure (Sterck et al. 1997), habitat disturbance was added to help explain some of the variation in sociality among populations. One factor opposing food competition is foraging facilitation (fig. 13.1f; Rodman 1988), defined as the ability of larger groups to find or capture more food than smaller groups. In its simplest form, foraging facilitation occurs because individuals spread apart from each other and so have at least partly nonoverlapping search fields. In some cases, such as searching for camouflaged or hidden insects, the distance between foraging group members may be larger than the distance at which they can detect such cryptic food items. In this case, total foraging effort effectively scales in proportion to group size, and it completely cancels out the scramble competition caused by the increasing nutrient demands of a larger group (Janson 1988b); each forager's success is independent of the number of group members. Although each monkey's consumption of an insect deprives another monkey the chance to find and eat it, the effect of WGS at the group level is zero in this situation. The extent to which sociality affects an individual's food intake is the net effect of all four forms of social food competition, plus facilitation. For instance, if both WGS and BGC are present, then foraging efficiency may be "bowl-shaped"—higher at the smallest and largest group sizes (Grueter, Robbins, et al. 2018).

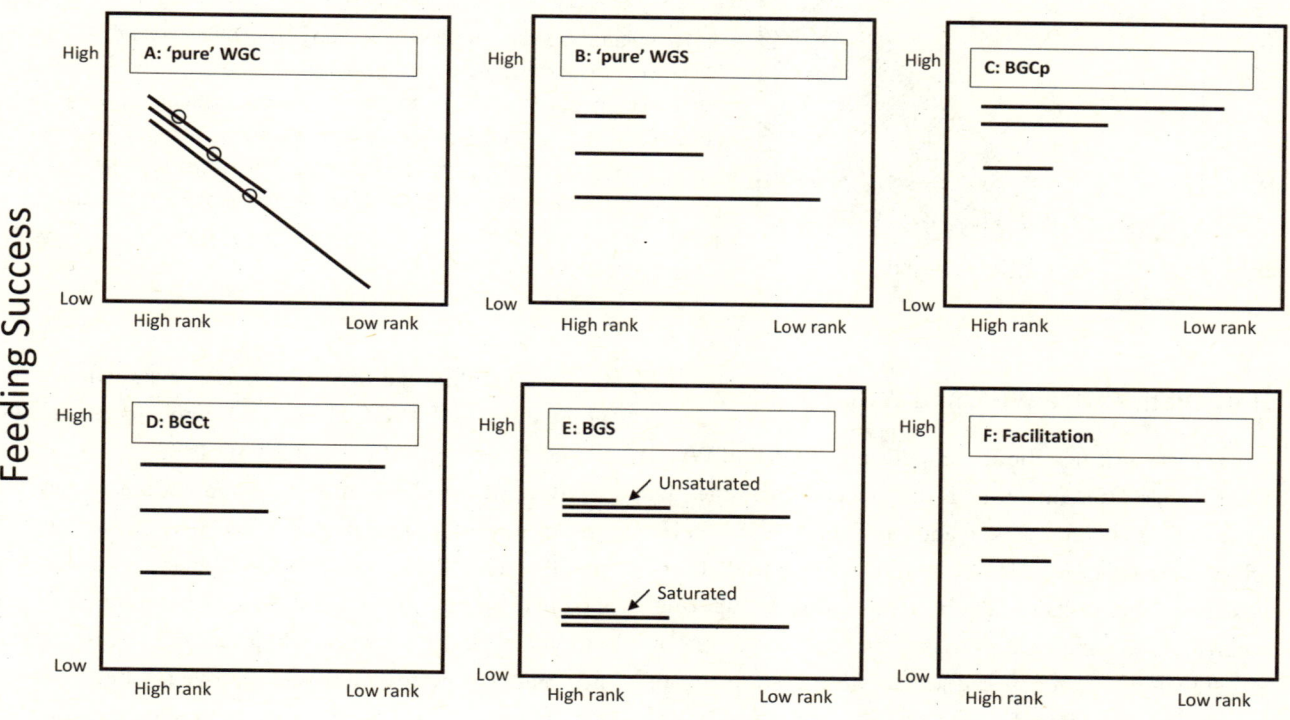

Figure 13.1. Schematic relationships between individual feeding success, social rank, and group size (longer lines within a plot denote larger social groups). Each subgraph denotes a distinct component of social food competition or facilitation: A, within-group contest; B, within-group scramble; C, between-group contest over resource patches; D, between-group contest over territories; E, between-group scramble; F, group facilitation.

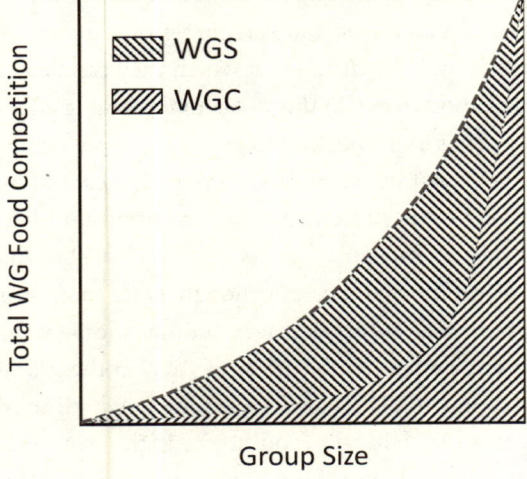

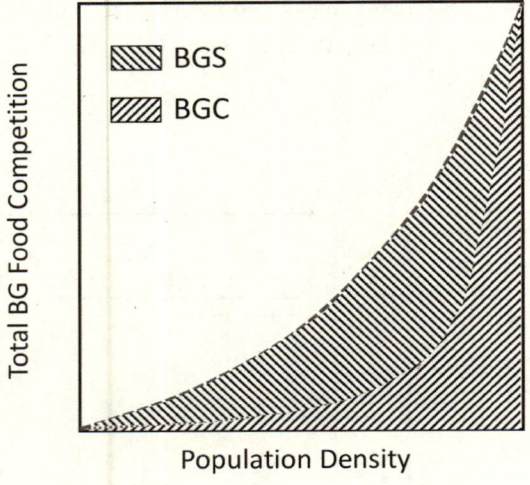

Figure 13.2. Illustration of how scramble and contest components of social food competition may vary as a function of factors that determine the total competitive environment: group size for within-group competition and population density for between-group competition. Note that scramble and contest competition are only different behavioral expressions of the total competitive environment: when total competition is low, neither component can be large. For a given competitive situation (point on the x-axis), increasing one expression of competition necessarily reduces the other. In general, the fraction of total competition that is expressed via contest increases with population numbers (group size and population density).

How Distinct Are WGS and WGC?

In the early writings about competition, WGS and WGC were discussed as conceptually distinct and separable (fig. 13.1a and 13.1b; after Janson and van Schaik 1988): WGS affects the average food intake in different groups that vary in size or other traits such as between-individual spacing, whereas WGC affects the variation in food intake among individuals of different social status within a group. In particular, fig. 13.1b was described as an example of "pure" WGS, although a considerable part of the text of the original article was devoted to disentangling the confounding effects of WGS, BGC, and facilitation on individual food intake in different groups (the y-axis in this graph). The labeling of this graph and similar ones might lead researchers to treat WGS and WGC as though they were independent, although most discussions acknowledge that populations with WGC always have some degree of WGS as well (A. Koenig 2002; van Schaik 1989).

Beyond the observation that populations with WGC are known to have detectable WGS, it is not logically possible for WGS and WGC to be independent (fig. 13.2). Why not? If a food item that is removed by one individual decreases the feeding rate of another, it does not matter how the item was removed, whether by intimidation (WGC) or by being consumed first (WGS). They are merely two mechanisms by which a food item ingested by one individual can affect the feeding rate of another. If ingestion does not affect the feeding rate of another individual, then neither WGS nor WGC exists, but if either exists, then the potential for the other also exists. If individuals rarely overlap in the same resource long enough for behavioral interactions to occur, then WGS may the primary mode of competition, yet when conditions change to increase the overlap of individuals at feeding sites, agonistic interactions may increase to the point that WGC is common (Isbell and Young 2002). In short, it is possible to measure WGS without WGC, but if WGC occurs, it substitutes for potential WGS; the common observation that subordinate individuals race in front of their group to be the first to enter a food patch is simply a method for them to "force" WGS on dominant group members. What happens when food is locally superabundant (in the sense that there is more food than the group can eat) but patchy, as is often the case in captivity or in naturally superproductive fruit trees (Janson 1988a)? In this case, patch depletion does not occur, so WGS is absent by definition, yet agonism may still occur, suggesting the presence of WGC. In such cases, it is important to measure the total food intake of subordinates, allowing for the fact that they may find another feeding place of equal value to the one they gave up, or they may return to the patch after the dominant has left. In cases in which such measurements have been performed carefully, the subordinate may suffer no detectable loss of feeding efficiency (Post et al. 1980) or may have only a modest (ca. 10%) decline relative to dominants (Janson 1985). In such cases, researchers have suggested that agonism benefits dominants not by increasing their food intake but by reinforcing their dominance rank (Shopland 1987) or suppressing reproduction by subordinates (Silk 2002).

If WGC cannot occur without WGS being at least possible, is the pure WGC case depicted in figure 13.1a incorrect? What is confusing here is that figure 13.1a is referred to as "pure" WGC even though it does include WGS. To see this fact, note that group average food intake declines as group size increases, indicated by the circles near the center of each line. If there were no WGC, the slope of each line would be zero (horizontal), but the mean food intake would still be at the level indicated by the circles, which is exactly what is depicted in figure 13.1b for pure WGS. In other words, individuals of lower social rank will have lower feeding success only if food overall is limiting, in which case WGS must exist.

The relationship between WGC and BGC might appear to be simpler. Like WGC, BGC requires that a resource be limiting. In this case, however, it matters if the resource is a relatively small food patch or is a territory (van Schaik 1989). If BGC occurs over food patches (BGCp), then it should occur only when there is an advantage to gaining access to the patch—the patch provides notably more or better food than alternative patches, and feeding bouts are long enough that groups are likely to overlap. These are the same conditions that favor WGC, which requires that patch depletion can occur and thus that WGS is present. In the case that BGC is over territories (BGCt), the situation is more complex. While food may be limiting at the scale of the population and thus worth defending, it may not be distributed in a way that promotes WGS, or at least it is likely that the net effects of WGS and facilitation cancel out. For instance, scattered insects are the main diet of many birds during the breeding season; even though birds defend the territory, they rarely fight over individual insects inside the territory. Although both forms of BGC favor larger groups, it is more difficult for subordinate groups to find alternative resources when competition is over territories than when it is over patches. When competition is over patches, middle-sized groups may be able to achieve feeding success close to that of larger (dominant) groups by using contested resources at other times or choosing alternative resources (fig. 13.1c).

Relating Food Competition to Food Abundance and Distribution

Once the various forms of social food competition and facilitation were described, we had to figure out how these were related to food abundance and distribution. Accepting Terborgh's (1983) premise that food patch size was a simple measure of food distribution, how should it relate to the various kinds of food competition and facilitation? The correlation between patch size and fruit tree diameter in Terborgh's study suggested that the primary effect was through WGS—the more food there was in a patch, the less each group member's food intake was depressed by the feeding of others. However, it also seemed likely that patch size would relate to WGC, since a variety of studies in captivity had shown that aggression increases in frequency as animals become crowded (Mathy and Isbell 2001). Because primate groups often have stable social and foraging group sizes (with some notable exceptions), one might expect that aggression would increase when the foraging group size exceeds patch capacity (Symington 1986). Patch size might also affect BGC because larger fruit patches are generally more attractive to a group and the larger amount of fruit allows them to stay longer, both factors making it more likely that groups will meet at a patch as well as increasing the value of defending the patch. Thus, the larger the patches used by groups in a population, the larger groups should be (because of both reduced WGS and increased BGC) and the less often group members should fight over access to food. Conversely, if small patches predominate, groups should be small and within-patch aggression more common. These predictions work across a modest range of patch and group sizes (Terborgh 1983), but they break down because of the other major ecological factor that I have not yet discussed: patch density.

Patch density affects all three major forms of food competition as well as facilitation. The greater the density of food patches, the shorter the distance between them and therefore the less "patchy" the food availability (all else equal). As interpatch distances become smaller, the cost of finding the next patch becomes lower (Chapman and Chapman 2000), and this applies equally to entire groups (hence lowering WGS) and to subordinate individuals within a group that cannot access the current patch (hence reducing WGC). More generally, the willingness of any social forager to contest access to a feeding place is a function of the "opportunity cost" of not gaining access to it (E. Vogel and Janson 2007). The opportunity cost is defined by what an animal gives up by choosing or accepting one outcome rather than another and is measured by the difference between the fitness consequences of the two (or more) outcomes. When patches are far apart, the opportunity cost (food intake in the current patch minus the food intake if you are excluded from it) is large, so individuals should be willing to expend considerable energy or effort to retain or gain access to the current patch, especially if it is richer than alternative patches. This situation leads to high WGC. Conversely, when patches are overall common or the current patch is not very rich compared to other patches, an individual would be better off leaving to find another patch than risking energy or injury gaining access to the current patch (E. Vogel and Janson 2007, 2011). Thus, high patch density leads to low WGC. Similarly, BGC should decline as patch density increases because the chances that any two groups will meet at a given patch becomes smaller (they have more patches to choose from), and the value of defending (or challenging for access to) a patch is lower since there are more alternative patches to feed in.

Because increasing patch density is expected to reduce all three major forms of competition, group size might not change dramatically as patch density increases. The relaxed competition due to higher food availability and lower aggression within groups could allow groups to become larger, but this trend may be counterbalanced by the reduced benefit of larger groups due to lower BGC. Facilitation is also affected by patch density. When food items do not run away, facilitation takes the form of increased detection of food items. When items or patches are very common, each group member can detect its own food item or patch with little interference from others, as described above for insects. When patches are very scarce, larger groups could, in principle, find them faster than smaller groups because the larger group has more eyes (or other senses) for detection. However, group facilitation for finding scarce patches is not likely to be a large effect. First, it is unlikely that a larger group of N individuals could detect patches N times faster than a solitary animal, because the search fields of group members are likely to overlap (S. Altmann 1974). Second, many primates that use sparse patches remember the location, the quality, and even the ripening state of productive patches (Janson 2016) and therefore do not "find" them. In this case, larger groups do not benefit from facilitation in food finding, or at best they benefit from minor facilitation in finding a patch for the first time before they begin to use it repeatedly. Primates do not generally remember the locations of common patches because they are easy enough to find without remembering their locations—it pays to remember the location of rare patches because it would take considerably more effort to detect them (Janson 1998a).

What happens when we combine all forms of competition and facilitation as patch density increases? Both WGS and WGC should be relaxed as patch density increases (it is easier to find alternative good resources), and facilitation would also increase with patch density, both trends favoring larger groups. BGC for patches (BGCp) should decline as patch density increases, perhaps favoring smaller groups, although, in practice, the intensity of BGCp may be small relative to the effects of WGC or WGS (Janson 1985, 1988a). BGC for territories

(BGCt) should increase with patch density, since home ranges can be smaller and thus easier to defend. However, because territory size increases with group size, BGCt should limit group size once the costs of territory defense increase faster than the increased territory quality that accrues to the larger group (e.g., J. Brown 1982). Thus, for both forms of BGC, increasing patch density should have a relatively small impact on the relative foraging success of different-sized groups. Summing across all forms of food competition, we would expect group size to increase with increasing patch density and with increasing patch size. If patch density and patch size were often or usually independent of each other, we could expect to test the preceding predictions for the effects of patch size and density on group size, all without much fuss. Alas, nature is not so simple.

A common feature of plant size and density is that the larger the mature plant of a given species, the less common are individuals of that species (e.g., Harper 1977). In any plant community, smaller plants in general have higher density than larger plants simply because the former take up less space. In forests with low diversity such as high-latitude coniferous forests, a few large tree species may dominate the canopy and be common enough to form a near-continuous matrix, but this pattern is not typical of the tropical forests where most primates live. Therefore, it is usually the case that small food patches are relatively common, and large patches are rare. This sets up a complex interaction of foraging advantages and disadvantages as patch size and density covary. Small patches should favor small groups, but small patches are often common and may be common enough that patch density outweighs WGS and allows large, sometimes very large, groups to form. Examples include primates that consume a large fraction of insects in the diet, such as *Saimiri* (Boinski, Sughrue, et al. 2002), or that use extremely common vegetal materials, such as the snub-nosed monkeys that depend largely on lichen (Grueter, Li, et al., "Dietary Profile," 2009; Kirkpatrick et al. 1998; Ren et al. 2012). If small patches are less common and are worth remembering, then both patch size and lack of foraging facilitation combine to favor small group sizes, as are found in tamarins (Terborgh 1987). Large patches are generally scarce in tropical forests, so they are worth remembering and thus will not favor larger groups by increasing food encounter rates (e.g., Janson 1988a). However, many individuals can feed together in a large patch without appreciable interference, up to the number that exhausts the patch's productivity. This constraint may account for both the modest size of groups of rainforest frugivores and the correlation between group size and patch size noted by Terborgh (1983). This descriptive model is summarized in figure 13.3.

What happens if large patches are common? This combination was once thought to occur for leaf-eating primates, and it was the basis for the "folivore paradox": the observation that leaf-eating primates often lived in groups of relatively small size despite feeding on what appeared to be superabundant resources (Janson and Goldsmith 1995; Steenbeek and van Schaik 2001). There are two accepted resolutions to this paradox: (1) selection pressures other than food competition, such as avoidance of infanticide (Crockett and Janson 2000; Steenbeek and van Schaik 2001), might be more important to females in leaf-eating species (see below); and (2) the apparent superabundance of leaves as a resource was simply wrong. Indeed, it has become understood that most leaves are either poisonous or very difficult to digest (Milton 1981b), and they often vary in quality within a species or over time seasonally or even within a day (e.g., Ganzhorn 1989a; Ganzhorn and Wright 1994). In addition, many leaf-eating primates also consume immature fruits and seeds (e.g., MacKinnon and MacKinnon 1980). As more primatologists studied folivorous primates, it became clear that their diets could be patchy enough to engender aggression over food. Indeed, the old idea that leaf-eating species were rarely aggressive enough to allow or favor consistent dominance hierarchies has been disproved in several species (e.g., A. Koenig 2000; E. Wright et al. 2014). Conversely, particular habitats (such as early successional areas and swamps; Janson and Emmons 1990) can be dominated by a few common fruit-bearing species. In most cases, the one or few common species will ripen fruit within a restricted season, during which fruit production may exceed the ability of frugivores to consume it (Janson and Emmons 1990). At these times, food competition should be relaxed, and many of the predictions of conventional theory will be hard to document. However, every good time has its price, and in these low-diversity habitats, the price is that for much of the year, there is little to no fruit production. Primates that use these habitats have to either find food elsewhere or subsist on generally

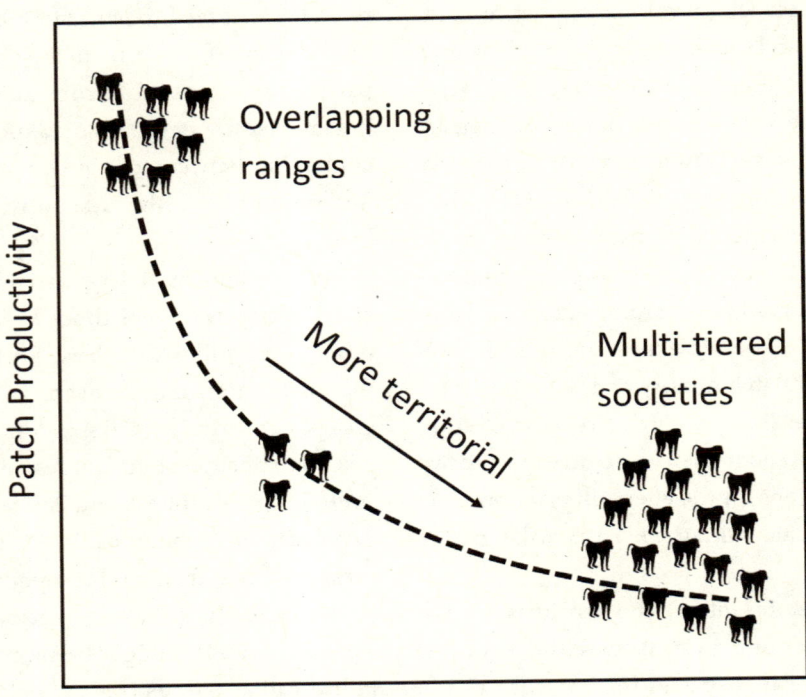

Figure 13.3. Expected relationships of group size (roughly indicated by the relative numbers of monkey silhouettes grouped together) to the combinations of patch productivity ("size") and density, assuming that food competition places the upper limit on group size. The dotted line shows the trade-off expected between patch size and density for a community of a given overall productivity and species diversity. Communities cannot occur above this line without increasing overall productivity. Communities may exist below the line, but where patch productivity and density are both low, primates are likely to be scarce. Groups are smallest where patch density and productivity are both low or moderate, leading to intense within-group scramble and contest competition. When patch density is low, increases in patch productivity allow increased group size, as argued by Terborgh (1983). When patch productivity is low, increased patch density can lead to rapid increases in group size as it becomes easier for each group member to forage with little or no overlap with others, thus increasing group facilitation of foraging.

less nutritious fallback foods (A. Marshall and Wrangham 2007; Terborgh 1983). Group sizes should then equilibrate at the level determined by food production during the least favorable time of year (Janson 1984a), unless fallback foods are evenly dispersed or relatively common and so do not place a strong constraint on group size (Wrangham 1986; "staple foods," A. Marshall and Wrangham 2007).

What if resources are truly not patchy yet still limited? In such cases, it may pay an animal or group to defend a territory to limit access by competitors; this is the main case in which social behavior directly impacts BGS as well as BGC. Because territorial defense is costly, territories are typically small relative to the day range of a group (J. Mitani and Rodman 1979). Limited territory size in turn restricts group size (Isbell 1991), even though the relatively uniform dispersion of the food items could allow large groups to form. The net result is that territorial species tend to be insectivorous (often nocturnal) or folivorous, live in small to moderate group sizes, and yet can achieve high population densities. There are many exceptions, including numerous guenon species that are territorial despite being frugivorous and some frugivorous species that maintain very large territories by cooperative male defense (e.g., chimpanzees and spider monkeys). When groups become extremely large, they are very sensitive to seasonal changes in food abundance and distribution (in effect, moving away from the right side of fig. 13.3 toward the middle) and so may break apart (fission) or adopt multitiered societies in which small foraging units can be independent but gather under favorable conditions (Kirkpatrick et al. 1998).

Synthesis

The precepts and empirical generalizations of primate social food competition can be summarized in a few sentences. The patchier the resources used by a population, the more motivated individual foragers are to crowd into resource concentrations. The more individuals crowd together, the more likely they are to fight over feeding spaces or items. To reduce the stress of frequent fighting, evolution favors behaviors that allow individuals to recognize each other and to quickly establish consistent behavioral signals to decide the outcome of an encounter without injury; such signals lead to predictable, often-linear dominance hierarchies. More successful individuals in a hierarchy leave more offspring (for diverse reasons). In some cases, coalitions of individuals may outcompete single individuals or smaller coalitions, thereby favoring the formation of groups that defend resource patches or clusters of potential mates. Group formation may be favored by factors other than defense of food patches, such as mutual reduction in predation risk, risk of infanticide, or harassment, or by high fitness costs of dispersal (Isbell and Young 2002; Port, Hildenbrandt, et al. 2020). If individuals in larger groups have higher fitness than those in smaller groups, evolution may favor philopatry so that the benefits of group formation are shared with kin. Although the theory is clear, and in many cases it is supported by empirical studies, each sentence in the preceding list hides an abundance of mechanistic assumptions and nuanced variations between populations that have been the subject of much of the development of social food competition studies during the past 30 years. The remainder of this chapter surveys some of these developments.

TESTING THE ASSUMPTIONS AND MECHANISMS OF SOCIAL FOOD COMPETITION

There have been several major advances in our understanding of social food competition since the 1990s. I group these into six categories: (1) integrating social food competition into the theory of games, producing the field of social foraging theory; (2) learning what primates know about their food items, especially fruit trees, and how that knowledge affects food competition; (3) challenging assumptions about how diet relates to the ecological causes of competition; (4) revealing the scale and sources of spatial and temporal variation in food quality; (5) questioning the causal relationship between food competition and social structure; and (6) relating food competition to physiological measures of stress and health.

Integrating Individual Foraging Strategies into Food Competition

To move beyond the traditional view that treated entire foraging groups as single economic units, it was necessary to view each individual in a group as a competitor with potentially distinct foraging strategies, costs, and benefits. The earliest model of such a process was the "producer-scrounger" game (Pulliam and Caraco 1984), in which individuals adopt one of two pure strategies, producing (finding their own food) or scrounging (joining producers at food sources they find). Like most evolutionary games, the population (or group) may equilibrate at one pure strategy or a mixture of strategies. This basic model soon spawned a cottage industry of variants, including ones with more strategies such as "opportunists," who can both produce and scrounge, albeit with a cost in performance to each (Vickery et al. 1991); models with differences in individual ability to displace others from food sources (although hierarchies and aggression were not modeled explicitly); and spatially explicit models, in which individuals using different strategies occupied different spatial positions in a group (review in Giraldeau and Caraco 2000). How beneficial it was to be a scrounger depended a lot on how much food was left in a patch at the time the scrounger joined the producer, or conversely how much food the finder had consumed before the arrival of the scrounger, termed the "finder's advantage." The concept of the finder's advantage was implicit in early work on spatially structured foraging in primate groups (Janson 1990b) but developed explicitly in subsequent work (Di Bitetti and Janson 2001). Independently, Isbell and Young (2002) focused on the time it took to deplete a patch as a critical parameter linking foraging to social behavior, primarily by affecting whether two individuals would meet up at a patch and interact. The finder's advantage and patch depletion time are clearly linked to each other and to spatial structure: the farther apart individuals forage, the less likely a scrounger is to join a producer at a patch before the patch is depleted, thus increasing the finder's advantage. Recent field experiments show that subordinate foragers make different spatial decisions about feeding sites depending on the presence and proximity of dominant group members as well as on the required feeding time at the site (Arseneau-Robar et al. 2022). The social and ecological factors that determine between-individual spacing are generally understood (e.g., Janson 1990a), but few comparative studies have focused on this critical aspect of social living (Heesen, Macdonald, et al. 2015). More recent work has instead focused on group cohesion and the mechanisms that promote communication about and coordination of group movements (e.g., Farine, Strandburg-Peshkin, et al. 2016; A. King and Cowlishaw 2009; Strandburg-Peshkin, Farine, Couzin, et al. 2015a).

Ecological Cognition of Primates

Primatologists have long assumed that their study animals know a lot about their ecological world and perceive it largely as we human observers do. Indeed, the complexity of a primate's ecological world has prompted the hypothesis that the large brains of primates must be able to solve complex ecological problems (e.g., E. Menzel 1973; but see Janson 2014) or in fact evolved specifically to allow primates to cope with the complex challenges of harvesting sparse and seasonal fruit patches in the tropics (Milton 1981a). The general premise for all this work is that primates appear to be highly goal-oriented and often move in apparently deliberate and quite straight lines from one food patch to the next (Byrne, Noser, et al. 2009; Garber, this volume). Demonstrating that such movements are in fact not "random" has spawned a diverse set of methods, including observational protocols, statistical models, and movement simulations, to distinguish apparently goal-directed movement from random (e.g., Janson 2007, 2012). There is now evidence that most primates know and remember the locations of major food patches (Janmaat, Byrne, and Zuberbühler 2006a; Janson 1998a; review in Trapanese et al. 2019), sometimes use the degree of synchrony of ripening as a cue to find other maturing fruit patches (Janmaat, Chapman, et al. 2012; C. Menzel 1991), and in most cases also anticipate the amount of food typically available at the patch (Ban et al. 2014; Janmaat, Polansky, et al. 2014).

At least part of their ability has a neural basis, as primate hippocampal neurons are specialized to encode an "allocentric" view of their environment and specifically where the individual is looking, unlike rodents, whose hippocampal neurons respond to where the individual is located (Rolls and Wirth 2018). How primates manage to anticipate the reward in a fruit patch is somewhat mysterious, as the amount of ripe fruit in a patch varies over time within the fruiting season and as a function of how much time the patch has not been visited since it was last depleted of ripe fruit. Recent evidence suggests that primates avoid revisiting renewing food patches by tracking the interval since they last visited each patch and integrating that information with spatial and productivity information to predictably choose the most profitable patches available at any given time (Janson 2016). The apparent ability of primates to track many of the details of the production and location of their fruit sources has two implications for the importance of social food competition. First, for food patches that are shared by most or all members of the group, location and condition should be known to all group members. If so, there is no group facilitation in finding such patches, and thus a major benefit countering within-patch food competition is absent (Janson 1988b). Second, some renewing food patches may not be productive enough to attract the entire group and thus may be visited by only a subset of the group, even just a single individual. In such a case, the knowledge of the location and condition of these patches is not shared by the whole group and may allow individuals to reduce WGS and WGC by visiting such patches when they are excluded from feeding with the rest of the group. Because many early studies of primate behavior and ecology were likely biased toward observations in or near the main concentration of group members (simply because it is difficult to follow isolated primates for an appreciable distance), it remains generally unknown to what extent subordinates can compensate for within-group competition by using ecological knowledge of alternative resources. The ability of subordinates to employ alternative foraging strategies is likely a function of the patchiness, predictability, and abundance of resources, key aspects of diet that have long been thought to be the true causes of the relationship between diet and food competition.

Diet and Food Competition

How (if at all) does diet relate to ecological parameters that should impact social food competition? As mentioned above, the early view that leaves were ubiquitous (nonpatchy) and of low nutritional value led to expectations that leaf-eating species should not live in large groups or show much aggression over food (A. Jolly 1972). Although some early studies of folivores (e.g., howler monkeys and guerezas) seemed to confirm both expectations, other studies quickly challenged one or the other of these predictions: red colobus were found to live in groups as large as those of many savanna baboons (Struhsaker 1969), while hanuman langurs were discovered to be frequently aggressive, sometimes over food (Hrdy 1977). As the number of long-term studies of leaf-eating primates has increased, their social diversity has proved to be very large, including (1) some of the smallest social groups (indri) and largest (snub-nosed monkeys) of all the primates; (2) mating systems including monogamy, harem, multimale, and multitiered flexible associations; (3) social organization including primarily male or female dispersal; and (4) home ranges that are defended or overlap completely.

Among the folivores best studied both socially and ecologically are the red colobus, hanuman langur, and gorilla. Each has contributed particular insights, several of which have challenged conventional wisdom. The hanuman langur did not fit the expectation that females of leaf-eating species would rarely be aggressive, yet detailed study showed that they fit the major predictions of social ecology quite well. Despite being folivores, they showed predictable patterns of aggression both within and between groups, sorting into matrilines of related females, among which there was a marked dominance hierarchy (Borries, Sommer, and Srivastava 1991). Such a pattern would be consistent with the use of patchy, sparse resources; indeed, careful study of the resources in one population showed that there was considerable variation in the amount and quality of leafy resources consumed by females, such that the best foods were scarce and patchy (A. Koenig 2000). While demonstrating that at least this population of hanuman langurs fit the predictions of socioecological models, this series of studies also convincingly showed that leaves cannot be stereotyped as uniform, dense, low-quality foods. Gorillas have similarly shown considerable flexibility in diet across populations, with some showing at least seasonal reliance on fruits while others depend almost exclusively on vegetation. Interestingly, however, both frugivorous and folivorous populations show similar social structure centered on one or two large adult males, a few females, and their offspring (Harcourt and Stewart 2007). Although females do not interact aggressively very often and appear not to form matrilines or form coalitions consistently with relatives, there is a discernible hierarchy that can remain stable over years (M. Robbins, Robbins, et al. 2005). Thus, gorillas present the paradox that diet can vary considerably without a marked social response and that consistent dominance hierarchies can exist even when females do not assist relatives and when aggression is rare overall. Red colobus would seem to be a type example of folivores as predicted by socioecological models for species with low WGS: aggression is rare, groups are large, and daily path lengths are variable but not correlated with group size. Because their groups are large, they do not fit the folivore paradox discussed earlier, yet detailed study demonstrated that the population at Kibale showed direct evidence of scramble competition (Snaith and Chapman 2005): reduced feeding rates and increased movement with increased feeding time in a patch, and reduced patch feeding times as the number of competitors increased. As such, red colobus present a double paradox—not only do they show evidence of WGS despite an essentially leaf-based diet, but in this case, they ought to live in small groups and show increases in daily travel distance with group size, yet they do not do so (Janson and Goldsmith 1995; but see T. Gillespie and Chapman 2001). This challenge to orthodoxy may have either a simple or a complex resolution. The simple resolution is that the Kibale population or study period may be unusual; studies of other populations of red colobus have failed to find evidence of WGS (Isbell 2012). The complex resolution is that WGS is present but is often confounded with differences between groups in food availability; in fact, when food density in distinct home ranges was controlled statistically, there was a significant positive correlation between group size and daily path length among Kibale red colobus (Snaith and Chapman 2008). This result suggests that variation in food density between home ranges can mask the effects of scramble competition: groups in better home ranges can become larger at no additional cost of WGS. Certainly, differences in food density and quality between home ranges should be assessed and included in any analysis of socioecological patterns when study groups do not occupy largely overlapping home ranges (e.g., T. R. Harris 2006) or core areas (Scarry 2013).

Describing Spatial and Temporal Variation in Food Quality and Patchiness

We now recognize that diet categories are not robust predictors of competitive regime (A. Koenig and Borries 2006). Further evidence suggests that within a plant species there is a lot of variation between and even within individual trees in food quality and other foraging-related parameters. Advanced analytic methods (Conklin-Brittain, this volume; Rothman, Chapman, Hansen, et al. 2009) have allowed researchers to describe the nutrient contents of many different feeding patches of a single species (Chapman, Chapman, Rode, et al. 2003; Rothman, Chapman, Hansen, et al. 2009) rather than characterizing the nutritional composition of a species based on a single sample from an individual tree or a mixture of several trees (e.g., Janson 1985). The implications of this variation, which is not readily apparent to human senses, are potentially profound. First, much variation in the choice of individual trees of a given food species, currently unexplained, may be due to such variation in food nutrient quality (Felton and Lambert, this volume) or levels of secondary compounds (Glander 1982; Stalenberg et al., this volume). Second, variation in food quality across space and time may promote within-group competition even when there appears at first to be little reason to expect any (e.g., T. Harris et al. 2009; A. Koenig, Beise, et al. 1998), especially for folivores, for which the quality of food ingested, rather than its quantity, may be critical to fitness (e.g., Chapman and Chapman 2002; Milton 1980; Rothman, Raubenheimer, and Chapman 2011). Optimal foraging rules designed for patch-foraging animals may not apply well to leaf-eating primates that must often balance intake of a variety of nutrients and plant poisons (Felton and Lambert, this volume; Raubenheimer, this volume; e.g., Irwin, Raharison, Raubenheimer, et al. 2015; J. Johnson et al. 2015). Subtle but predictable variation in food quality is rampant when it is looked for—food items may be more nutritious as the day progresses (B. Carlson et al. 2013; Ganzhorn and Wright 1994) or in upper versus lower layers of a tree (Houle, Chapman, and Vickery 2007). Food quality may differ markedly in adjacent home ranges in ways that are not apparent to casual observation (e.g., T. R. Harris 2006). Indeed, given the documentation of widespread variation in food quality, it is a bit amazing that "old-fashioned" primate ecology ever managed to document any relationships between feeding ecology and social structure!

Relating Emergent Properties of Social Structure to Food Competition

Socioecological models of primate social systems (Sterck et al. 1997; van Schaik 1989) posit that there is a natural and strong relationship between the abundance and distribution of food and the intensity of the various forms of social food competition. The premise appears simple (although it was far from obvious to primatologists in the mid-1980s): if food patches are sparse and worth defending, then WGC will occur. If WGC is common, then predictable hierarchies of aggressive success will be evident among individuals. If larger groups or coalitions (including subgroups within a social group) can defend a patch more easily than smaller groups, then the benefit of increased access to food should be preferentially shared with kin. Having kin to share with implies philopatry, so the sex for which access to food is most important should be philopatric. Although early data appeared to support this model in a general way and sometimes in specific contrasts among closely related species (e.g., Barton, Byrne, and Whiten 1996; Mitchell et al. 1991), increasingly detailed studies often failed to agree with the original model. Several reviews of the socioecological model summarize many of these apparent exceptions (Clutton-Brock and Janson 2012; Isbell and Young 2002; A. Koenig 2002; A. Koenig and Borries 2006; A. Koenig, Scarry, et al. 2013). There are at least four categories of exceptions: (1) constraints of phylogeny, (2) effects of male harassment and infanticide, (3) other sources of selection on philopatry, and (4) revised expectations about how nepotism and dominance hierarchies are related.

Although phylogeny is a fundamental aspect of understanding variation in traits among organisms, its use has been more often limited to morphological traits. It was (and still is) often assumed that behavioral traits, including social structure, would be able to adjust rapidly to prevailing ecological selection pressures. However, pioneering analyses by Di Fiore and Rendall (1994) showed that there was a strong phylogenetic signal in some aspects of social behavior, such as the existence of matrilineal social structure. Similar inertia in the response of social traits to ecological conditions has been emphasized in macaques by Thierry and colleagues (2000). It is now clear that the lemurs, while they may respond to food competition in broadly similar ways as other primates, possess a suite of morphological and physiological traits that lead to distinctively different social systems (Kappeler 1999). Are these examples idiosyncratic, the result of historical artifacts of genetic structure that dictate which routes of evolutionary change are more likely, or are there patterns? It appears that, in many cases, inertia is most pronounced in the structure of dominance interactions: whether females favor kin, whether fights typically escalate or are avoided, and whether females are mostly dominant to males or vice versa. Might there be reasons that dominance interactions, once evolved in a particular form, are less flexible than other socioecological traits? Given the focus of this chapter on food competition, I can only raise the question here. However, it is worth considering the impacts of possible social inertia on the intensity and expression of food competition, a topic that has received relatively little attention.

Male harassment and infanticide may induce the formation of, place upper limits on, or force the breakup of female groups for reasons that have little to do with food competition (e.g., Crockett and Janson 2000; Janson, Baldovino, and Di Bitetti 2012; Steenbeek and van Schaik 2001). In such cases, food competition may appear to be absent or weak, even though the basic mechanisms linking food discovery and consumption are the same as in species in which food competition is more apparent. Conversely, male coalitions may defend food patches to the benefit of females and offspring (Scarry 2013), as originally posited by Wrangham (1980). In this case, males may benefit from tolerance and individualistic social relationships more typically described for females (Janson 1986b; van Hooff and van Schaik 1994).

Philopatry in the socioecological model is assumed to be a consequence of a need for (or benefit of) nepotism in the defense of food (van Schaik 1989). However, others have argued that philopatry benefits females in acquiring food when patterns of food availability vary over space and time, so that long-term knowledge (presumably involving memory and cognition) of a home range increases foraging success (Isbell and Van Vuren 1996). That group foraging patterns may differ between established and novel ranging areas is supported by some case studies (Janmaat and Chancellor 2010). More generally, female dispersal may be so costly that it benefits mothers to tolerate daughters in the natal home range as long as the costs to the mother are modest (Isbell 2004). In this case, females might be philopatric in general, with nepotism as a side effect. A somewhat similar argument is made by Clutton-Brock and Janson (2012) that philopatry is generally beneficial, regardless of the sex, but, to avoid inbreeding, one sex or the other (or both) must disperse to breed. Which sex disperses may be a direct consequence of the degree to which a dominant individual of one sex can best monopolize breeding opportunities. If that individual is male, then males should disperse, but if that individual is female, then females should disperse. Again, nepotism is a side effect of philopatry in this case. More generally, models of social foraging (Giraldeau and Caraco 2000) show that nepotism is beneficial only when social groups have an aggregation economy—one in which fitness benefits increase with group size—whereas nepotism is disadvantageous when social groups have a dispersion economy (benefits decline with group size). The application of this model to primate groups is not trivial, as it is not always clear what defines a "group" (is it a matriline or a whole group?) or what the benefits are. In van Schaik's 1989 model, the benefits are restricted to the acquisition of food or safety from predators, whereas there is no a priori reason that safety from infanticide or parasites might not be equally important. Conversely, if increasing these sources of mortality has direct negative impacts on primate density (as is sometimes the case at least for predation; Karpanty 2006; Stanford 1995), they could limit population densities to levels low enough that both WGS and WGC are markedly reduced and hard to detect (fig. 13.2, left side).

Finally, recent challenges have arisen to the assumption that the existence of statistically definable dominance hierarchies requires either nepotism or philopatry or a fruit-based diet (A. Koenig, Scarry, et al. 2013; Wheeler et al. 2013). In many cases, relatively rare agonistic interactions limit the ability to detect significant hierarchies, yet with sufficient sample size, such hierarchies may emerge.

In some cases (such as gorillas; M. Robbins 2008), the scarcity of agonism over food may indicate an overall lack of local food depletion (thus limiting both WGS and WGC), even though occasional patches favor competition. In other cases (baboons and some macaques), intense fights over status occur mostly outside of the context of food, while interactions at food sources appear to be mediated by avoidance and subtle displacement (Shopland 1987). Even if interactions over food are subtle and uncommon, there may still be a significant relationship between dominance rank and food intake (e.g., gorillas; M. Robbins, Robbins, et al. 2005). In species with egalitarian relationships, dominance status may be transitory (e.g., related to age; Tombak et al. 2019), so that hierarchies are evident in the short term, but long-term fitness consequences among females are slight. Even if there is no short-term gain in access to food, it is possible that females might benefit from costly "random" aggression toward subordinates as part of a long-term strategy of reproductive suppression (Silk 2002). Finally, well-supported hierarchies with frequent aggression may contain complex triadic relationships as well as a notable fraction of wins "against the hierarchy" due to polyadic or coalition-based aggression (A. Koenig and Borries 2006). Thus, the existence of a simple agonistic hierarchy among females in a group is neither sufficient nor necessary to demonstrate food competition. If you want to know how much food competition there is, you have to document either reduced food intake or a physiological consequence of this reduction (the subject of the next section).

Consequences of Food Competition for Individual Physiology: Stress, Metabolism, and Parasite Resistance

Ideally, food competition would be assessed by its effects on both intake (thereby confirming the assumptions of the model) and fitness (testing the importance of food competition), although we have data on both aspects for relatively few species. Even if we have both forms of evidence, however, the exact mechanism by which reduced food intake is translated into lower fitness is not necessarily obvious—fecundity can decrease because a female has less energy to devote to offspring, because low intake leads to increased stress, or because low nutritional condition results in higher parasite loads (e.g., I. Agostini 2017). The advent of noninvasive methods to obtain highly sensitive measures of physiological indicators of stress (cortisol by-products in feces and urine), metabolites (ketone bodies, creatinine, and C-peptide of insulin), and parasitism (egg counts in feces) allows us to evaluate the diverse mechanisms by which food intake affects fitness. Studies of ring-tailed lemurs show that females in larger groups may experience higher stress (Pride 2005), which accords with demographic evidence (A. Jolly, Dobson, et al. 2002), whereas in red colobus, levels of stress hormones do not increase with group size (Snaith et al. 2008) even though there is evidence for within-group scramble competition in the same population (Snaith and Chapman 2005). Several studies have been able to show strong correlations between changes in food availability and seasonal changes in fat catabolism (e.g., Knott 1998), relative muscle mass (Bergstrom, Emery Thompson, et al. 2017), or metabolic proxies of energy throughput (Emery Thompson and Knott 2008; Emery Thompson, Muller, Wrangham, et al. 2009). Some studies that used C-peptides as a proxy for energy status have found the expected positive correlations between male rank and C-peptide levels at least some of the time (Surbeck et al. 2015), but sometimes the opposite pattern (Emery Thompson, Muller, Wrangham, et al. 2009) or complex changes with rank during the mating season (Higham, Heistermann, and Maestripieri 2011) have been found. Among female capuchin monkeys, C-peptide levels correlated with estimated energy balance, but neither depended on dominance rank despite considerable variation in food competition (Bergstrom, Kalbitzer, et al. 2020). The richness of these patterns suggests that these integrative and indirect measures of nutritional status may reveal hitherto-unexpected dynamic patterns in food competition.

Given the widespread effects of group size and social rank on food intake, plus the demonstrable and reciprocal effects of nutritional status on parasitism (review in Nunn and Altizer 2006), one might expect that these social variables would also correlate with parasitism levels or diversity. While there are now plenty of studies that examine these correlations, there is no simple picture that emerges. Parasite egg counts in feces sometimes decrease with group size (Snaith et al. 2008), sometimes increase (Freeland 1979b), and sometimes show no clear trend (review in Côté and Poulin 1995). Likewise, parasite levels show diverse patterns with respect to social rank—sometimes levels are higher in high-ranking animals (e.g., male baboons; Hausfater and Watson 1976), sometimes they are lower (Foerster, Kithome, et al. 2015), or they may show no pattern even in species with notable hierarchies (e.g., in female baboons; Hausfater and Watson 1976). In a recent meta-analysis (Habig et al. 2018), parasitism was generally found to increase with social rank and thus may represent a cost of high rank, particularly in males. This could make sense in the context of several studies showing significant energetic costs of maintenance of high rank in males, both within a single breeding season (Higham, Heistermann, and Maestripieri 2011) and in the long term (Emery Thompson, Muller, Wrangham, et al. 2009). In short, direct measures of fitness correlates, such as nutritional status, metabolism, and parasite loads, may support or change our perspectives on food competition based on behavioral observations alone. In cases in which they conflict, we must be prepared to throw out old assumptions as well as ask if one or the other measure properly assesses the relationship—the data could be biased, based on short-term sampling, or confounded by additional variables not measured in the original study.

WHAT'S NEXT?

What are the major unanswered questions in primate social food competition? There are certainly dozens of relationships among individual traits, diet, food distribution, phylogeny, predation, and human disturbance that remain to be documented or could be documented better (fig. 13.4). Several of these have been the subjects of the preceding text. It would take another chapter to explore the many relationships that have only barely been addressed, including the question of how human-caused environmental changes may affect food competition and resulting social structure (see Sterck 1998). However, I want to point out one broad gap in most contemporary approaches to fitting social food competition into other aspects of primate biology: the problem of confusing current conditions, which maintain a trait or suite of traits, with ancestral conditions that favored the origin or initial spread of such traits. As in the case of predation (Janson 1998b), primatologists have tried to relate observed rates of a putative selection pressure (intensities of modes of food competition) to the social behaviors thought to respond to these adaptively (e.g., Janson 1990b; Mitchell et al. 1991). Rarely, however, have we considered how changing social behaviors end up affecting the intensity of food competition (Janson 1986b; C. Richter et al. 2015). Depending on other selection pressures, competition may be reduced to the point that it is negligible or only weakly related to the potential food competition that might have existed during the early evolution of sociality in primates. To cite a specific possible example, between-group competition is found to be important in affecting feeding success and fitness in some territorial species (Cheney and Seyfarth 1987; T. R. Harris 2006), but it appears to be weak in some nonterritorial species that compete over shared food patches (e.g., Janson 1985; but see Scarry 2013). Even if it appears to be weak under current conditions, however, it would not be logically correct to conclude that between-group competition was not important in the evolution of social groups. Why? The intensity of between-group competition in a population is proportional to the rate of between-group encounters, and this rate increases rapidly as group size declines (assuming constant population density and group movement speed). For instance, the observed rate of about one between-group encounter per week in brown capuchins in Manu National Park (Janson 1985) occurs in a population with a current median group size of 10. If the same density of monkeys were foraging solitarily, "group" density would be 10-fold higher, and the encounter rate between "groups" would increase 100-fold (10 × 10). Thus, during the initial evolution of sociality, a solitarily foraging capuchin-like ancestor could have experienced encounters with competing foragers about once per hour! At these high encounter rates, if there were even a modest net advantage for groups of two in winning encounters with solitary foragers, group foraging should have been strongly favored. In short, the competitive conditions that might have favored the origin of sociality may no longer exist under current conditions and may or may not be important in maintaining sociality now; a similar argument has been given about the origin versus maintenance of group territoriality in primates (Port, Schülke, and Ostner 2017). It may turn out that current social structure is better related to the potential for food competition, as indexed by the spacing and nutritional productivity of food patches (E. Vogel and Janson 2011), than to measured current levels of the different kinds of social food competition. As more data on food distribution, nutritional

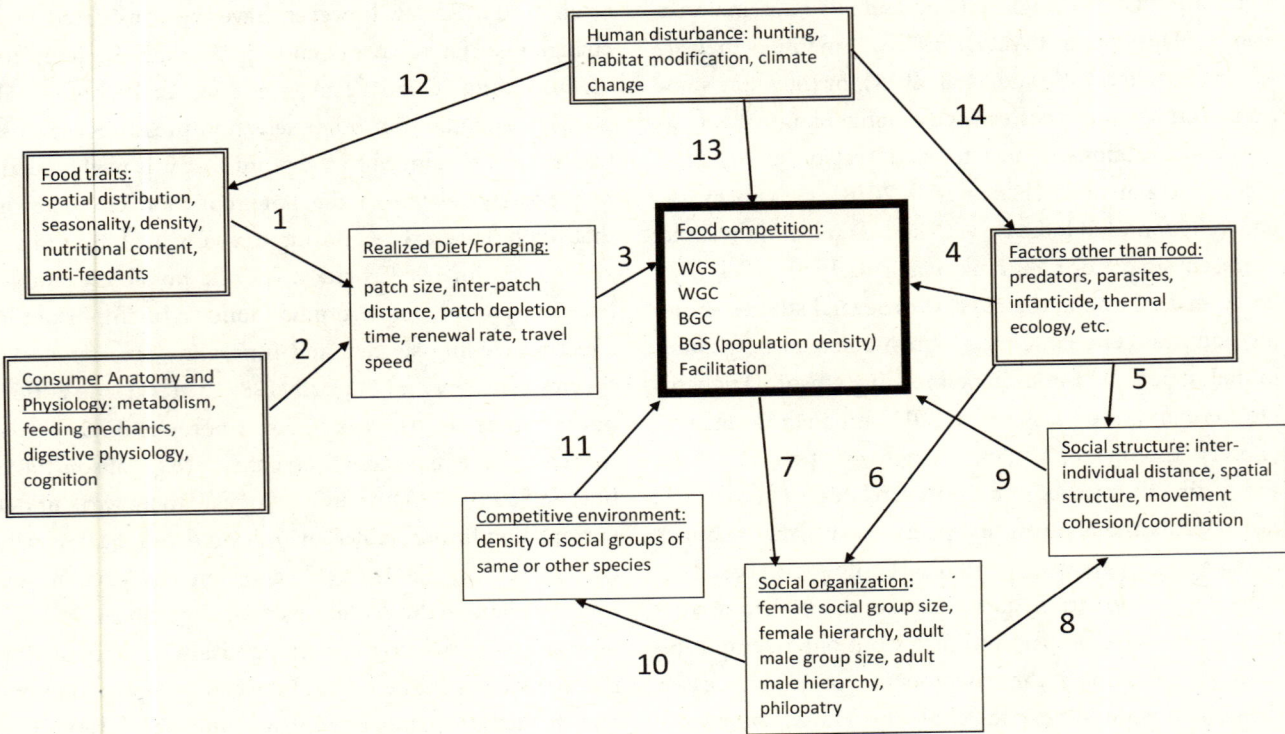

Figure 13.4. Box diagram of ecological and social components that affect social food competition and vice versa. Several of these components (e.g., arrows 12 and 14) are outside of the scope of this chapter but are included for completeness. Note that arrows often create feedback loops from food competition to social traits and back to food competition.

values, and digestive constraints accumulate, it should be possible to relate parameters of social structure to these underlying foraging constraints in the context of other ecological and social selection pressures.

SUMMARY

Competition for food is a common feature of primate populations. It can occur within and between groups. The overall level of food competition in a population depends on the food available per individual. Food may not be limiting to individuals for two general reasons: short-term superabundance of food or a depressed population size of the consumer. Examples of short-term superabundance include seasonal overproduction, extremely productive food patches, and large between-year variation in total production. Factors that may depress primate populations below their potential full population density include various sources of mortality (predation and hunting, parasitism, and infanticide). For a given level of overall food competition, how it affects individuals depends on the use of two behavioral mechanisms. The first is removing a food item before someone else can obtain it; this mechanism results in indirect or scramble competition and can be used by individuals of any social rank. The second is using force (or the threat of force) to prevent others from accessing a resource or to make them give up the resource; this mechanism results in direct or contest competition and is more available to those with more social status or power.

The extent to which scramble or contest competition affects individual food intake depends not only on the characteristics of the individual but also on the dispersion of food. Small, common, evenly distributed foods are fast to consume and rarely worth defending, thus favoring scramble competition. In contrast, the more food occurs in less common and larger patches, the more individuals will overlap in food sources and the easier it is for some individuals to intimidate others, thereby favoring contest competition. Contest competition may favor a variety of social strategies to enhance access to food, either by enhancing or entrenching individual competitive ability

Figure 13.5. *a–c*, Capuchin monkeys (*Sapajus nigritus*) in Argentina. Photos: Charles Janson.

(e.g., by the use of dominance hierarchies and ritualized signaling of submission) or by promoting the use of coalitions to enhance the chance of winning or defending a resource. In contests between groups, larger groups or groups with more adult males may have an advantage over other groups.

Beyond the preceding generalizations, trying to relate various forms of food competition to other aspects of social organization has proved problematic or at least filled with apparent exceptions. Early theories suggested that contest competition would favor nepotism so that the benefits of greater food acquisition would accrue to relatives, and thus nepotism would favor philopatry of the sex that defends food patches (or territories). However, more recent work suggests that philopatry is primarily a function of sexual competition (the sex that can most easily monopolize breeding can favor or force emigration of subordinate would-be breeders of that sex); in this case, nepotism is a secondary consequence of philopatry. Similarly, attempts to link types of food competition with broad diet or food categories have inevitably run into well-supported exceptions—not because the principles are wrong but because the "contest potential" of any diet or food type varies a lot within the category, so that some contest competition is often present even in situations when food is small, common, and relatively evenly distributed (as leaves were once thought to be). In addition, social selection pressures (including infanticide and social intimidation) can modify or even reverse the expected patterns of fertility and survival otherwise expected from food competition. Finally, phylogenetic history can play a major role by predisposing some lineages of primates to social organizations that vary within narrow confines or are biased toward certain patterns that are otherwise rare among primates (e.g., female dominance in lemurs).

CONCLUSIONS

- Social food competition is measured by the extent to which the foraging success of one individual is depressed by the presence of other individuals in the same group (within-group) or other social groups (between-group). Two distinct behaviors are used to reduce access to food of other group members: eating it before the other arrives (scramble) or agonistically displacing or chasing the other (contest). The two forms of social food competition are complementary, but both depend on the overall level of food competition—for instance, when overall competition is low, neither scramble nor contest will be strong. Overall competition should be low when habitat-wide food production notably exceeds the metabolic needs of the primate group or population.
- Whether scramble or contest strategies are favored in a given situation depends on the size and abundance of food patches. Small items (or patches) are quickly exhausted by whichever individual arrives first, leaving little or nothing to fight over when later individuals arrive. A high density of food items does not favor fighting, as each item is easily replaced. Thus, a high density of small items favors scramble competition, whereas agonistic contests are more likely when using a low density of large patches. When groups do not need to be very cohesive, individuals can avoid fighting by spreading apart while foraging, forming either widely dispersed or fission-fusion societies, which gather in large groups when needed or when possible but otherwise forage in small subgroups.
- Each individual in a group has distinct foraging needs depending mostly on its sex, reproductive status, and fighting ability. These combine to determine when that individual can use scramble or contest competition to acquire food. The group, then, is a sum of individuals, each carrying out a somewhat distinct set of foraging strategies, usually within the context of a predictable social setting. Although individuals are in principle free to leave a group, the fitness costs of doing so are often severe. A current challenge is to understand how the distinct needs of group members are accommodated in decisions of when and where to move.
- All primates know something about where to find their main food sources. In some cases, they may remember where individual feeding trees are relative to other trees or landmarks, as well as how productive each tree is, when it has edible food, and how soon a food patch is worth visiting again. This knowledge can be very important to social food competition, as scrambling is much easier when the location and qualities of food sources are known ahead of time instead of having to be discovered each time.
- Diet alone is not a robust predictor of the major form of social food competition in a given population. While ripe fruits often occur in large and sparse patches, they can be seasonally superabundant and thus produce little overall competition. Conversely, mature leaves seem dense or even superabundant, but leaves vary tremendously in quality so that the choicest leaves may be sparse and patchy. Even when contests are relatively uncommon, dominance hierarchies can be present and stable.
- Philopatry and nepotism often co-occur with high levels of contest competition. Why this is the case is a matter of debate. Early arguments favored the logic of food contests → nepotism → philopatry, but others have argued that philopatry is intrinsically beneficial, so the only reason to leave a natal territory is to seek reproduction elsewhere. This argument yields the logic of sexual competition → emigration → absence of nepotism. A complicating factor is incest avoidance—if one sex is philopatric, offspring of the other sex must usually emigrate to find reproductive opportunities.
- Factors other than food (phylogenetic history, level of human disturbance, and social pressures such as infanticide and sexual competition) can modify, dilute, or reverse patterns expected purely from food competition.
- The rapid increase in the precision and kinds of noninvasive measures of an animal's physiological condition and genetics promises a future of far greater mechanistic understanding of both the behavioral and genetic mechanisms and the physiological and fitness consequences of social food competition.

14 Applying a Framework of Social Nutrition to Primate Behavioral Ecology

Margaret A. H. Bryer and Moreen Uwimbabazi

Sociality is critical to both helping individuals reach their nutritional goals and hindering them in doing so, has fitness consequences, and likely played an important role in primate (including human) evolution. Nutritional ecology can be used to examine links between sociality and ecology in primates by explicitly testing socioecological theory (Sterck et al. 1997; van Schaik 1989; Wrangham 1980). An animal moves through its environment seeking and avoiding a diverse array of nutritional and antifeedant components: seeking macronutrients (fat, nonstructural carbohydrates, available protein, and digestible fiber), energy, and micronutrients (S. Altmann 1998; Cheeke and Dierenfeld 2010; NRC 2003; Felton and Lambert, this volume) and avoiding indigestible fiber (Milton and Demment 1988; Oates, Waterman, and Choo 1980; Van Soest 1978) and diverse plant secondary compounds (Dearing et al. 2005; Freeland and Janzen 1974; Glander 1982; Stalenberg et al., this volume). "Social nutrition" (sensu Lihoreau, Buhl, et al. 2015; S. Simpson and Raubenheimer 2012; S. Simpson, Clissold, et al. 2015) quantifies the complex interplay among an animal's physiology, behavior, and environment, an environment that includes other foraging animals in addition to diverse nutrient and antifeedant sources (Lihoreau, Charleston, et al. 2017; Lihoreau et al. 2018; S. Simpson, Clissold, et al. 2015; fig. 14.1). Social nutrition specifically involves testing aspects of social foraging theory with multivariate nutritional ecology. Additionally, social effects on nutrition are not confined to conspecifics; the presence of other species can also alter individual nutritional intake in complex ways (Goodale, Beauchamp, and Ruxton 2017; Houle, Conklin-Brittain, and Wrangham 2014; D. Thompson and Barnard 1984).

In this review, we discuss the effects of sociality with conspecifics and heterospecifics on nutrient access and intake. Specifically, we review (1) social foraging theory, (2) social nutrition via empirical studies in nonprimate taxa and theoretical modeling, (3) primate socioecological models (see Janson, this volume, for further discussion), (4) the ecology of primate polyspecific associations, and (5) the application and future directions of social nutrition in primate conspecific and polyspecific interactions.

SOCIAL FORAGING THEORY

Social foraging theory broadly states that animal social behavior—cooperative or competitive, with conspecifics or with other sympatric species—influences individual feeding behavior (Giraldeau and Caraco 2000). Social foraging involves the complexity of multiple players influencing each other's feeding decisions (Clark and Mangel 1986; Giraldeau and Caraco 2000). At a basic level, the presence of an individual (or individuals) at a feeding site attracts other individuals (Chapman, Snaith, and Gogarten 2014; Judd and Sherman 1996; J. Krebs et al. 1972; Lihoreau and Rivault 2011). Multiple factors then influence which individuals in a social group gain access to preferred foods, and skewed food access leads to variation

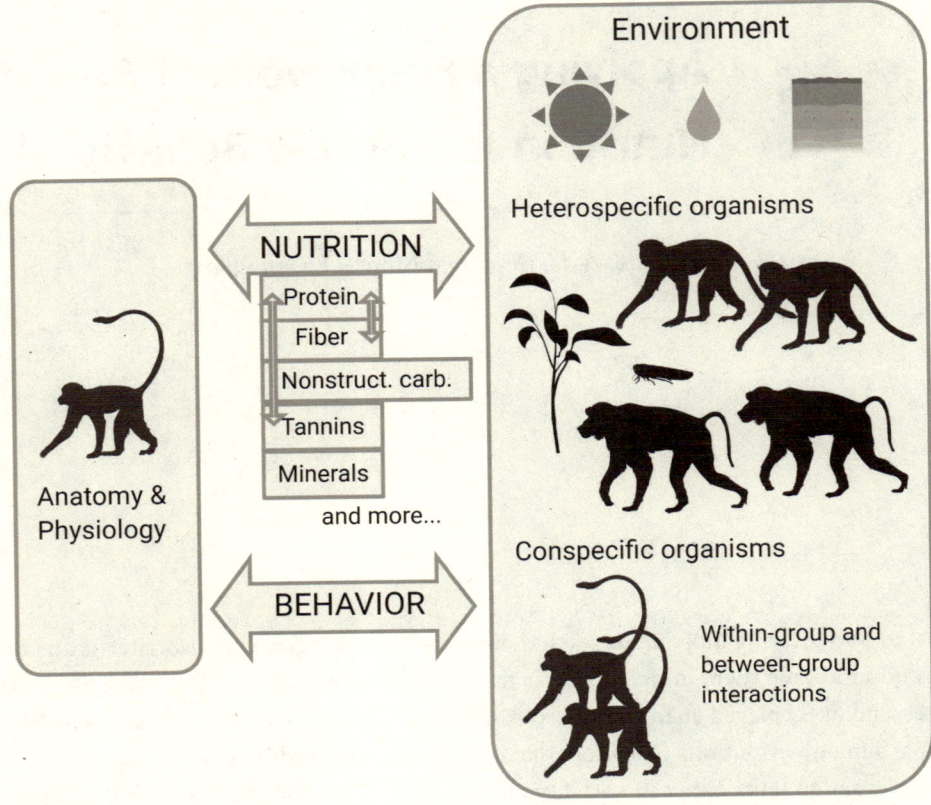

Figure 14.1. An individual primate moves through an environment that includes, in addition to abiotic factors, diverse heterospecific and conspecific organisms. This primate acquires diverse nutritional components from plants and insects (components that also interact with one another, affecting nutrient availability to the primate, as indicated for protein and fiber and protein and condensed tannins). The environment in which the primate acquires nutritional components also includes other conspecifics in its group or in other neighboring groups whose nutritional goals may be similar to or different from the individual's due to age, sex, rank, group, and so on. The environment also includes heterospecifics of other primate species whose nutritional goals may be similar to or different from the individual's goals. Though only two primate polyspecific associate species are presented in the figure, the composition of polyspecific groups can involve more participant species and multiple groups of each species over different timescales. Additional heterospecific organisms in the primate's environment (not included in the figure) that also interact with nutrition include predators and microorganisms. This figure was made with BioRender by Margaret Bryer.

in fitness within the population (Caraco and Giraldeau 1991; Giraldeau and Dubois 2008). The costs and benefits of social foraging fall under the costs and benefits of participating in social groups generally: though foraging with other individuals can aid in locating and accessing food, leading to increased individual fitness, foraging with others can also cause feeding competition, leading to decreased individual fitness (Giraldeau and Caraco 2000; Janson, this volume).

Producer-scrounger decision-making addresses the widespread phenomenon of food patch discovery by certain individuals ("producers") and exploitation of that discovery by other individuals ("scroungers"; Barnard and Sibly 1981; Di Bitetti and Janson 2001; Giraldeau and Caraco 2000; A. King, Isaac, and Cowlishaw 2009; Janson, this volume). Depending on food distribution, individuals in a group who find food may or may not differ from members who exploit food. For example, subordinate chacma baboons (*Papio ursinus*) were able to scrounge dominant baboons at large food patches but not at smaller clumps of food ("sub-patches") that could be monopolized by dominant individuals (A. Lee and Cowlishaw 2017). Social patch models consider more than one scale of competition over a food patch in a frequency-dependent framework: (1) models assuming one resource patch per individual, so competition is over access to a patch, and (2) models assuming that one patch is exploited by multiple individuals at the same time, so competition is over resources within a patch (Janson 1988b; Livoreil and Giraldeau 1997). Scenario 2 is complicated by social factors including how many and when individuals discover the patch (producers in the producer-scrounger framework) and how many and when others join in the patch (scroungers; Vickery et al. 1991), as well as interactions between individuals in the patch that interfere with individual food intake (contest; Giraldeau and Caraco 2000). Producers and scroungers affect each other's energetic gains, and individuals may switch between foraging tactics due to intrinsic or extrinsic factors (Fülöp et al. 2019; Harten et al. 2018; Lendvai et al. 2004; Morand-Ferron et al. 2011; Vickery et al. 1991).

Despite the prominence of characterizing food distribution in the ecological literature using terms such as "patch," "patchiness," and "clumped," definitions are often inconsistent (Giraldeau and Caraco 2000; Isbell, Pruetz, and Young 1998; Searle, Thompson Hobbs, and Shipley 2005; E. Vogel and Janson 2011; J. Wiens 1976). Trying to form a universal definition of a food patch or a "clumped" resource is problematic because how, when, and at what scale a particular animal interacts with components of its heterogeneous environment determine what is clumped or patchy to the animal (Isbell, Pruetz, and Young 1998; J. Wiens 1976) and how this affects interactions with other individuals. Landscape ecologists propose delineating patches hierarchically (Girvetz and Greco 2007; Kotliar and Wiens 1990), with animals responding to multiple scales of resource heterogeneity within the environment; however, empirical testing of these models is usually restricted to two levels (A. Lee and Cowlishaw 2017). Alternatively, food site depletion time (Isbell, Pruetz, and Young 1998) or food site residence time (Chancellor and Isbell 2009) may be more informative than characterizing spatial characteristics of the food patch. Rather than how "clumped" a food is, whether or not the food is "usurpable" or "monopolizable" has also been presented as a more biologically meaningful variable (Isbell, Pruetz, and Young 1998), though effectively measuring this variable can be difficult (E. Vogel and Janson 2011). These ambiguities are important to clarify and define in studies examining how individual animals influence each other's feeding behavior in a complex, heterogeneous food environment.

SOCIAL NUTRITION: EMPIRICAL NONPRIMATE STUDIES

Empirical research on "social nutrition" (sensu Lihoreau, Buhl, et al. 2015; S. Simpson and Raubenheimer 2012; S. Simpson, Clissold, et al. 2015), or socially mediated multivariate nutrition, has been done predominantly in insects (S. Cook et al. 2010; Dussutour and Simpson 2009; Grover et al. 2007; Lihoreau, Clarke, et al. 2016; S. Simpson and Raubenheimer 2012) because entomology allows for direct testing of fitness effects of nutrition and for determining adaptive intake targets (Eggert et al. 2008; Salomon et al. 2008). Directly linking intake and balance of multiple nutritional components to fitness is more challenging in longer-lived organisms, with much left to explore. Simulation models in conjunction with nutritional geometry (Raubenheimer, Simpson, and Mayntz 2009; S. Simpson and Raubenheimer 2012; Raubenheimer, this volume), discussed further below, offer predictions of how social behavior, including, for example, contest competition, may affect the nutritional ecology of social vertebrates (Lihoreau, Buhl, et al. 2015).

Eusocial insects provide an opportunity to examine intraspecific social behaviors in relation to nutrition. Producer-scrounger decision-making in eusocial insects incorporates both colony overall nutrition and individual insect nutrition (Mayack and Naug 2013; Toth and Robinson 2005). In eusocial bees, producers seek out resources and use the waggle dance to convey information about resources, while scroungers exploit this information. When individual-honeybee or whole-colony nutritional state was high (defined as an individual fed sucrose and a fully populated colony, respectively), honeybees explored (produced) more and exploited (scrounged) less (Katz and Naug 2016). However, different nutritional pathways likely regulate the behavior of individuals engaged in these two foraging strategies (Ament et al. 2010), as demonstrated by differential sensitivity to individual and colony nutritional states, with producers more sensitive to colony nutritional status over individual nutritional status if there was a mismatch (Katz and Naug 2016). Individual eusocial insects make strategic feeding decisions for the benefit of the colony for multiple nutrients: honeybees experiencing an experimental deficit in one amino acid subsequently chose food options that addressed this nutritional deficit (Hendriksma and Shafir 2016). Similarly, worker ants, both in an experimental laboratory setting (*Solenopsis invicta*; S. Cook et al. 2010) and in the wild (*Ectatomma ruidum*; S. Cook and Behmer 2010), prioritized carbohydrates over protein while collecting food for all members of the colony (in which workers need a high-carbohydrate diet and larvae and the queen need a high-protein diet).

Social nutrition also provides a framework for investigating how insects navigate a nutritional landscape in the context of interspecific interactions, specifically mutualism or parasitism (Crumière et al. 2020). Argentine ants (*Linepithema humile*) rely on carbohydrate-rich honeydew from aphid mutualists, particularly when establishing colonies, in addition to protein-rich insect prey (Shik and Silverman 2013). Species of ants (Shik et al. 2018) and termites (da Costa et al. 2019) rely on fungal symbionts for carbohydrates. Ants (*Sericomyrmex amabilis*) that cultivate fungus maintained a protein:carbohydrate balance of 3:4, but when *S. amabilis* also contended with the nutritional demands of a parasitic ant species (*Megalomyrmex symmetochus*), the protein:carbohydrate balance was adjusted to 7:4 (Shik et al. 2018). Though the underlying mechanism of this shift is unclear (Shik et al. 2018), it indicates dynamic nutritional intake influenced by heterospecific interactions.

Intraspecific social nutrition studies in nonprimate vertebrate taxa have also addressed social foraging questions of patch use. In field-based feeding trials, insectivorous ibis (*Threskiornis moluccus*) chose higher-lipid food as group size increased at a clumped resource, which may indicate switching to a less preferred food or less strategic feeding under scramble or contest competition (Coogan, Machovsky-Capuska, et al. 2017). Antifeedant plant secondary compounds affect food patch use in marsupials, partially driving the heterogeneity of the feeding environment and potentially population density (W. Foley and Moore 2005; B. Moore and Foley 2005; Stalenberg et al., this volume).

Social nutrition has also been linked to the gut microbiome (Di Stefano et al. 2019; Pasquaretta et al. 2018; Wong et al. 2015) and immune function (Cotter et al. 2019; Ponton, Morimoto, et al. 2019; Ponton, Wilson, Holmes, et al. 2013), indicating diverse avenues for interdisciplinary social nutrition research connecting nutrition, health, and social behavior.

SOCIAL NUTRITION: MODELS

To test complex hypotheses related to social nutrition, researchers have recently combined state-space models of nutritional geometry (S. Simpson and Raubenheimer 2012; Raubenheimer, this volume) with other models, including agent-based models (Hosking et al. 2019; Senior, Charleston, et al. 2015; Senior, Lihoreau, Charleston, et al. 2016; S. Simpson et al. 2010), social network analysis (Senior, Lihoreau, Buhl, et al. 2016), phase change models (Lihoreau, Charleston, et al. 2017), and landscape ecology models (S. Simpson et al. 2010). Using these hybrid models, researchers can test hypotheses about the relationships among social and nutritional variables that are difficult to test in wild vertebrates in complex environments.

Several model studies have examined individual nutritional strategies in a heterogenous landscape in a group context to assess patch use. Lihoreau, Charleston, et al. (2017) created a model based on phase change models in physics (Vicsek et al. 2009) to test predictions about the nutritional intake of social animals in a heterogenous food landscape. The model included a nutritional environment with foods with different features in terms of protein-to-carbohydrate content (though this can be generalized to any food components that affect fitness), overall abundance ("proportion of cells with that food"), and patchiness ("fractal dimension of all cells containing that food"; low is isolated large patches, and high is evenly distributed small patches). In this model, overlaying the nutritional space is the individual's nutritional state (location in the nutrient space in terms of nutrient intake), intake target (coordinates representing the goal balance of nutrients that maximizes fitness), and social variables indicating spatial alignment with or repulsion from other individuals. When they ran simulations of groups of 50 individuals for 10,000 time steps across multiple food environments, they found that social interactions improved individual nutrient balancing performance (efficiency of reaching the intake target) in environments where foods are rare, distributed in clumps, and contain different though complementary carbohydrate:protein ratios (i.e., they can be mixed to reach the intake target).

Lihoreau, Buhl, and colleagues (2015) predicted, through a series of simulation models, differential nutrient balancing strategies in the context of competition in social vertebrates. Individuals better able to reach or get close to their nutritional intake target are better able to win contest competitions, gaining "winner's advantage," which leads to improved balance (ratios) of limiting nutritional components (Lihoreau, Buhl, et al. 2015). As a result of competition, individual variation in intake and balance of nutritional components is predicted to emerge based on these models. Senior, Lihoreau, Buhl, and colleagues (2016) used nutritional geometry with agent-based models and social network analysis to further examine how nutritional state affects dominance relationships. They generated dominance networks over time in three different nutritional environments containing three foods (with variation in food abundance) and found that when food was more abundant, fewer contests occurred and the network was minimally connected over time. Intuitively, if food was less abundant, the dominance networks were more connected due to more contest competition, leading to increased variation in fitness across group members. Senior, Lihoreau, Buhl, and colleagues' (2016) model also indicated that network centrality variables are good indicators of future fitness, a finding that suggests that collecting certain more limited social network centrality measures in wild populations can be as effective in linking behavior to fitness as more complex social network analyses.

The nutritional components that can be examined in relation to social behavior in these models are diverse: examining individual amino acids and fatty acids, instead of protein and fat, requires expanding the state-space model of nutritional geometry to fully characterize and visualize these relationships in nutritional performance landscapes. Methods have been proposed such as measuring angles and distances between peaks in nutritional performance landscapes (Bunning et al. 2015), using a vector-based approach (Morimoto and Lihoreau 2019), or using a trigonometric approach (Morimoto 2022) to better characterize the peaks (best diets) and valleys (worst diets) of performance landscapes.

Though modeling approaches, including agent-based models and social network analyses, have been used to examine aspects of primate socioecology (Bonnell, Chapman, and Sengupta 2016; Bonnell, Henzi, and Barrett 2019; Bonnell, Sengupta, et al. 2010; Boyer et al. 2006; Ramos-Fernández et al. 2006), multivariate nutritional landscapes and intake have not been explicitly integrated into these models, though some models include estimates of energy or macronutrient requirements (Sueur, Deneubourg, and Petit 2012; Sueur, Deneubourg, et al. 2010; Sueur, MacIntosh, et al. 2013).

PRIMATE SOCIOECOLOGICAL MODELS

Primate socioecological models have been built on the assumption that primate sociality is influenced by (1) food distribution in the environment (how frequently an individual interacts with a finite amount of food in the environment), (2) food density (the density of finite food resources in the environment), and (3) food "quality"—a term used with varying levels of ambiguity, meaning how much of a nutritional payoff the individual will get from a given finite food resource (Isbell 1991; Janson, this volume; A. Koenig and Borries 2009; Sterck et al. 1997; van Schaik 1989; Wrangham 1980). Food distribution in the environment and food quality dictate the types of food competition that primates experience: within-group scramble, within-group contest, between-group scramble, and between-group contest (van Schaik 1989; see Janson, this volume, for further discussion).

Primate intraspecific variation in social structure includes variation in group size, cohesion, and membership at multiple timescales, referred to, especially at short timescales, as fission-fusion dynamics (Aureli et al. 2008). Reduction of feeding competition is frequently the driver of primate group fission (Asensio, Korstjens, et al. 2009; Baden et al. 2016; Hardie and Strier, this volume; Shaffer 2013a; Symington 1988), though food abundance and distribution measures do not always explain subgroup size (D. P. Anderson et al. 2002; Chapman, Wrangham, and Chapman 1995b; Newton-Fisher et al. 2000). The nutritional composition of foods (Busia, Schaffner, et al. 2016) and the macronutrient and energy intake of an individual are potential mechanisms of fission-or-fusion decision-making to explore.

ECOLOGY OF PRIMATE POLYSPECIFIC ASSOCIATIONS

Diverse taxa form mixed-species groups (birds: Dolby and Grubb 1998; D. Morse 1970; Sridar et al. 2009; marine mammals: Cords and Wursing 2014; Frantzis and Herzing 2002; ungulates: FitzGibbon 1990; Kiffner et al. 2014; Pays et al. 2014; primates: Heymann and Buchanan-Smith 2000; Noë and Bshary 1997; Terborgh 1990), and mixed-species groups may not be restricted to species of just one order (French and Smith 2005; Heymann and Hsia 2015; Minta et al. 1992). Examining social foraging in a mixed-species group of any combination of species is more complex than conspecific social foraging in that subtle costs and benefits of grouping vary not only over time and space and by individual forager, but also by the number of and roles of the species participating (Farine, Aplin, et al. 2015; Farine, Garroway, et al. 2012; Goodale, Beauchamp, and Ruxton 2017; Meise et al. 2019). Additionally, conspecific competitive dynamics can be influenced by heterospecific competitive dynamics (Farine, Garroway, et al. 2012; Goodale, Sridhar, et al. 2020). The focus here is specifically on mixed-species associations between or among primate species. Mixed-species association or polyspecific association occurs in many primates, especially those living in forest habitats, though the duration and stability of these groupings vary (Platyrrhini: Garber 1988a; Heymann and Buchanan-Smith 2000; Norconk 1990; Peres 1992b; Cercopithecidae: Chapman and Chapman 2000a; Gautier-Hion, Ouris, and Gautier-Hion 1983; Struhsaker 1981; Strepsirrhini: Eppley, Hall, et al. 2015; Freed 2006). Polyspecific association among forest primate species has been defined in various ways. Recent comprehensive reviews of mixed-species association across animal taxa (Goodale, Beauchamp, and Ruxton 2017; Goodale, Sridhar, et al. 2020) define mixed-species association in more general terms previously proposed by Powell (1985) and others: "moving animal groups that owe their existence to social interactions between species" (Goodale, Sridhar, et al. 2020, 3). Goodale, Sridhar, and colleagues (2020) distinguish between mixed-species aggregations (stationary at a resource) and mixed-species associations (moving through the environment).

As with other taxa, mixed-species or polyspecific association in primates is driven by a combination of foraging and predation protection benefits, with implications for individual nutritional intake within and across species. The food and antipredator benefits of polyspecific association are not mutually exclusive, as the landscapes of foods and predators influence one another at the individual scale of a potential trade-off between vigilance and feeding, as well as at the larger scale of niche construction in relation to navigating predation risk and food availability (Gil et al. 2017; Hill, this volume). If dietary overlap between two primate species (or among more than two primate species) in association is low, then species in the group can gain predator protection without the potential costs of feeding competition (Oates and Whitesides 1990). If dietary overlap is high, the question of whether dietary niche overlap increases or decreases among participants is a central ecological question in the primate polyspecific association literature. Increased diet overlap among primate species in polyspecific association has been observed at multiple sites (Chapman and Chapman 2000a; Cords 1987; Gautier-Hion, Ouris, and Gautier-Hion 1983; Struhsaker 1981), with associates converging on preferred food resources, thereby temporarily increasing their diet overlap. During periods of low fruit availability, dietary overlap among three associating guenon species (*Cercopithecus nictitans*, *C. pogonias*, and *C. cephus*) decreased, but the species remained in polyspecific association as frequently as in high-fruit-availability periods (Gautier-Hion 1980, 1988). Maintaining polyspecific association regardless of fruit availability suggests the antipredator value of polyspecific groups compared to single-species groups. Increased travel by participants in a polyspecific association compared to those in conspecific groups suggests a potential energetic cost to dietary overlap (Chapman and Chapman 2000a; Cords 1987; Gautier-Hion 1988).

Whether interspecific contest competition occurs is dependent on multiple factors, including the distribution of food in the habitat, group spread among individuals of different species, which species joins the mixed association when, and the satiation of one species before another (sensu Janson 1988b for conspecifics). Vertical stratification of species in the canopy or wider group spread help minimize interspecific contest competition (Garber 1988a; Gautier-Hion, Ouris, and Gautier-Hion 1983; Heymann and Buchanan-Smith 2000; Norconk 1990). This kind of spatial stratification can facilitate and even improve access to a variety of food types; for example, different spatial positions and insect-capture behaviors by two species of tamarins meant that during insect feeding moustached tamarins (*Saguinus mystax pileatu*) flushed insects from foliage that was then eaten by saddleback tamarins (*S. fuscicollis avilapiresi*), with both species maintaining similar insect capture rates (Peres 1992b). The foraging benefit of flushing (sometimes referred to in the literature as beating or disturbing) food to the benefit of another species has been observed in a variety of animal taxa during mixed-species associations (Au and Pitman 1986; Courser and Dinsmore 1975; Källander 2005; Sazima et al. 2007; E. Willis and Oniki 1978). Flushing or beating benefits of mixed-species association are frequently observed between species that are not closely related (for example, birds following ungulates; E. Fernandez et al. 2014), including in primates and nonprimate species associations (Heymann and Hsia 2015).

Patch use and scale in mixed-species groups are complicated by spatial stratification of species who forage on similar foods (Farine and Milburn 2013) and by merging (Powell 1989) or diverging (Alatalo et al. 1986) shifts in foraging niche during mixed-species association. In birds, heterospecific groups often lead to competition over food, with interspecific dominance hierarchies forming and dominant bird species benefiting the most from the mixed-species group (Cimprich and Grubb 1994). Interspecific dominance hierarchies determined by species body size occur in some primate mixed-species associations (Bicca-Marques and Garber 2003; Cords 1990; Houle et al. 2010; Struhsaker 1981). Dominant species may exclude subordinate species when food is usurpable (Houle et al. 2010; Houle, Conklin-Brittain, and Wrangham 2014; Peres 1996). In the same two *Saguinus* species referenced above, both species had long food patch residence time when in large food patches ("defined as trees >10cm in DBH and lianas and epiphytes >10m in height"), whereas the dominant species excluded the subordinate species from small patches ("understory trees and treelets <10cm in DBH or other plants <10m in height") with clumped food ("food items in a highly clustered fashion"). This exclusion of the subordinate tamarin species usually took the form of making the subordinate wait outside the patch, though contest competition over the patch did also occur less frequently (Peres 1996). Aggressive intergroup encounters between polyspecific groups have also been observed, indicating cases of a polyspecific group participating in between-group contest competition in defense of preferred resources (Garber 1988a; Gautier-Hion, Ouris, and Gautier-Hion 1983; Peres 1992a).

Even if contest competition occurs, polyspecific association persists in many primate systems due to mixed-species grouping conferring greater protection from predators than conspecific groups. Protection from predation through mixed-species grouping follows the same reasoning as in conspecific groups, except that complementary alarm systems (Goodale and Kotagama 2005), division of vigilance roles among species (Munn 1986), and differential levels of predation risk (Goodale, Beauchamp, and Ruxton 2017) affect the types of species that make beneficial associates. In primates, reduced predation risk via polyspecific association may be due to increased group size compared to conspecific groups, one species responding aggressively to a particular predator (Arlet and Isbell 2009; Gautier-Hion 1988; Stanford 1995), or spreading the burden and benefits of vigilance and alarm calling across species (Gautier-Hion, Quris, and Gautier 1983; Heymann and Buchanan-Smith 2000; Noë and Bshary 1997). Arboreal guenon species in polyspecific association have a similar vocal repertoire, which allows efficient interspecific alarm-call communication (Cords 1987; Gautier-Hion 1988; Marler 1973).

The ecological and social implications of polyspecific association are increasingly complex as more species participate and more groups of a given species participate, especially as variation in the comingling of groups may occur daily, seasonally, or interannually (Gautier-Hion 1988). The following polyspecific association scenarios for one group of redtail monkeys (*Cercopithecus ascanius*) at Kibale National Park, Uganda, have potentially different social and ecological implications for participants: the time the redtail monkey group spends in polyspecific association with (1) any group of blue monkeys at the site; (2) a specific group of blue monkeys (group A); (3) another specific group of blue monkeys (group B); (4) any group of blue monkeys and any group of gray-cheeked mangabeys simultaneously; (5) blue monkey group A and gray-cheeked mangabey group A simultaneously, and so on; or (6) any combination of these scenarios in succession—over the course of one day in some cases. As a result, the data collection workload of fully characterizing and quantifying the ecological implications of these associations can become extremely complex. Social network analyses of mixed-species bird flocks (Farine, Aplin, et al. 2015; Farine, Garroway, et al. 2012; Farine and Milburn 2013) point to bottom-up ways to quantify this mixed-species group complexity in primates.

SOCIAL NUTRITION IN PRIMATES

The relationship between social behavior and diet has been extensively explored in testing primate socioecological models (Barton and Whiten 1993; Chancellor and Isbell 2009; C. M. Murray, Eberly, and Pusey 2006; M. van Noordwijk and van Schaik 1987; P. Whitten 1983a; Wittig and Boesch 2003; Hardie and Strier, this volume; Janson, this volume), with food quality characterized in a variety of ways. However, spatiotemporally dynamic nutritional and antifeedant components and an animal's ability to digest these components determine the quality of food (B. Carlson et al. 2013; Chapman, Chapman, Rode, et al. 2003; Ganzhorn 1995b; Hohmann, Potts, et al. 2010; Houle, Conklin-Brittain, and Wrangham 2014; Masette et al. 2014; C. Robbins 1987; Rothman, Chapman, and Pell 2008; S. Simpson and Raubenheimer 2012; Worman and Chapman 2005). Therefore, few primate socioecological studies explicitly explore the intake of energy or specific macronutrients in relation to social behavior (*Cebus apella*, Janson 1985; *Cebus capucinus*, E. Vogel 2005; *Cercopithecus mitis*, Takahashi 2018). Some efforts have also been made to quantify the nutritional costs of primate within-group scramble competition (Grueter, Robbins, et al. 2018; Stacey 1986).

In the context of contest competition, dominant individuals will prevent subordinate individuals from entering a food resource, eject them from the food resource, or force them to move to a less desirable portion of the food resource patch (Janson 1988b; Peres 1996); these behaviors by dominant individuals may result in nutritional costs for subordinate individuals. With increased rates of aggression over a food resource, dominant capuchin monkey individuals skewed energy intake in their own favor (*Cebus apella*, Janson 1985; *Cebus capucinus*, E. Vogel 2005). When minimal aggression occurred over a resource, however, rank effects were not observed in energy intake, indicating that aggression is the tool that dominant individuals use to skew nutritional intake in their favor. In contrast, dominance rank did not affect energy intake rates in Assamese macaques (*Macaca assamensis*; Heesen, Rogahn, et al. 2013) or mountain gorillas (Grueter et al. 2016), even in food distribution contexts in which contest competition would be expected and where aggression occurred. Similarly, though high-ranking blue monkey (*Cercopithecus mitis*) females gained preferential access to fruits and had lower glucocorticoid levels than subordinates (Foerster, Cords, and Monfort 2011), rank was unrelated to protein intake and the ratio of nonprotein energy to protein (Takahashi 2018). However, higher-ranking blue monkey females fed from fewer unique foods daily (Takahashi 2018), and when fruit availability was low, low-ranking females spent more time feeding (Pazol and Cords 2005), which arguably both indicate subtle reduction in foraging efficiency by subordinates.

Much is left to be explored in the nutrition of primate fission-fusion dynamics. Spider monkey (*Ateles geoffroyi*) mean daily subgroup size was linked to the crude protein content of food patches visited, while no such relationship was found for nonprotein energy content (Busia, Schaffner, et al. 2016). Other mechanisms of spider monkey fission-fusion dynamics that may help explain this result include social preferences (Busia, Schaffner, and Aureli 2017) and predation risk; additionally, intake data would be needed to assess whether nutrition state affects fission-or-fusion decision-making. Models suggesting that nutritional state or differential nutritional requirements across individuals in a group can predict who leads the group in movement (Sueur, Deneubourg, et al. 2010; Sueur, MacIntosh, et al. 2013) also point to questions of if and how nutritional intake influences leadership (including in fission-fusion contexts) in primate social groups. In a similar vein, Lihoreau, Buhl, and colleagues (2015) predicted via simulation models that individuals in a group that deviate the most from their nutritional intake target (sensu S. Simpson and Raubenheimer 2012; Raubenheimer, this volume) would lead in switching to a different food.

The use of the social nutrition framework (Lihoreau, Buhl, et al. 2015; Lihoreau et al. 2018) and the multidimensional nutritional niche framework (Machovsky-Capuska, Senior, et al. 2016) holds promise in exploring the effects of primate sociality on nutritional landscapes, including nutritional niche shifts in conspecific and polyspecific contexts. The social nutrition framework expands on nutritional geometry (Raubenheimer, this volume) to incorporate the social environment. In nutritional space (sensu S. Simpson and Raubenheimer 2012; Raubenheimer, this volume), individuals of a social group are each represented and are moving toward either the same intake target or different intake targets (if they have different nutritional needs). The shape and amplitude of variation in nutritional states and intake targets of indi-

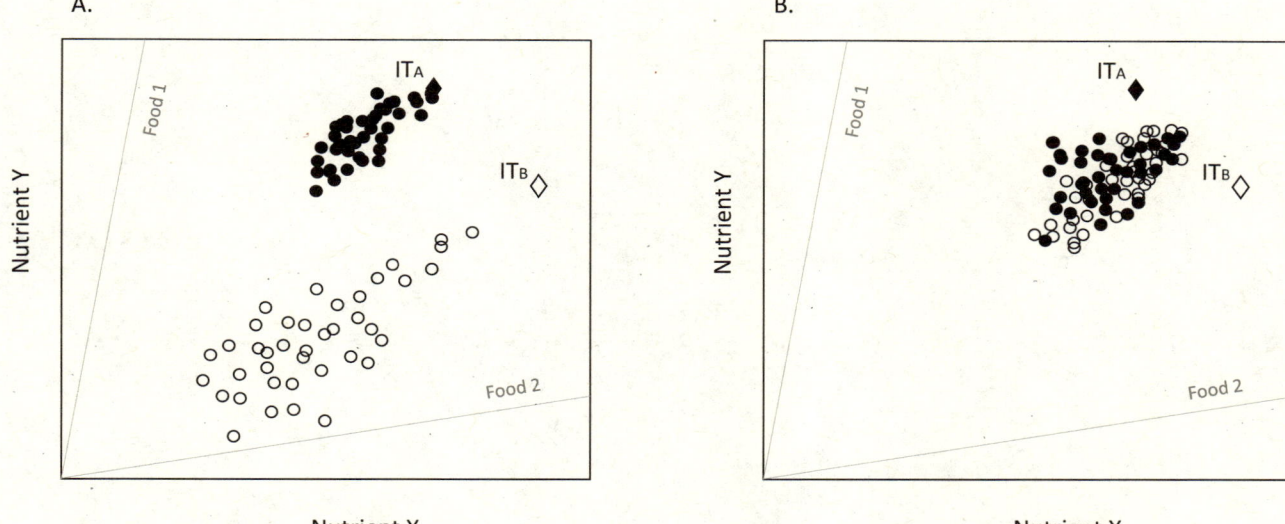

Figure 14.2. Two hypothetical scenarios for how the intake and balance of two nutrients are influenced by intraspecific competition and polyspecific association for two primate species, with different intake targets (IT_A and IT_B) for those nutrients, in polyspecific association with two complementary foods (Food 1 and Food 2). In scenario A, the hypothetical data for individual intake show intraspecific competition among individuals of species B leading to a wide variation in intake while maintaining balance, with no species B individuals reaching the intake target. All individuals in species A are able to get closer to or reach the intake target, demonstrating that species A experiences lower intraspecific competition than species B. In scenario B, neither polyspecific associate reaches their respective intake targets; instead, they converge on a compromise in nutrient space.

viduals in a group show how nutrition can mediate social structure and vice versa in contexts such as competition, collective movement, and temporary roles (Lihoreau, Buhl, et al. 2015). The social nutrition framework can be similarly applied to nutritional states and intake targets of individuals from multiple species in a polyspecific association (fig. 14.2; fig.14.3).

The multidimensional nutritional niche framework enables first the assessment of diversity of foods in terms of macronutrient composition: "food composition generalists" pull their diet from foods of diverse macronutrient composition (*Ursus arctos*, Coogan, Raubenheimer, et al. 2018; *Morus serrator*, Machovsky-Capuska, Miller, et al. 2018; *Sus scrofa*, Senior, Grueber, Mackovsky-Capuska, et al. 2016; *Cercopithecus mitis*, Takahashi et al. 2019). Though foods may be diverse in nutritional composition (making the animal a "food composition generalist"), the amount and ratio of key macronutrients ingested may be similar ("macronutrient specialist") regardless of the foods ingested (Takahashi et al. 2019), indicating maintenance of a macronutritional strategy via flexible feeding. This multilevel characterization of nutritional niche may vary by extrinsic factors, including interactions with conspecifics and heterospecifics, in conjunction with predation risk. In the context of polyspecific association, do shifts in nutritional niche overlap occur among participating species even if they are feeding from the same food type? Nutritional costs to polyspecific association despite feeding on the same food type have been inferred for some primate species. For example, redtail monkeys (*Cercopithecus ascanius*), the smallest-bodied diurnal frugivores (or omnivores) at Kanyawara site in Kibale National Park, Uganda, fed lower in feeding trees and altered their feeding rates when larger-bodied blue monkeys or gray-cheeked mangabeys were present, regardless of whether overt interspecific aggression occurred (Houle et al. 2010), and the upper crowns of four species of feeding trees in Kibale had higher fruit density, larger fruit crops, and higher sugar concentration than lower strata, which may result in interspecific contest competition over higher-strata fruits (Houle, Conklin-Brittain, and Wrangham 2014). Possible nutritional mechanisms of coexistence in this study include larger-bodied frugivores consum-

Figure 14.3. A gray-cheeked mangabey (*Lophocebus albigena*) forages in the same tree as blue monkeys (*Cercopithecus mitis*) in Kibale National Park, Uganda.
Photo: Margaret Bryer.

ing higher-energy foods more selectively while redtails switch to the remaining medium-to-low-quality food left behind.

Multidimensional nutritional niche analyses could also illuminate how the number and kind of participants in a primate polyspecific association affect nutritional niche. Feeding with some polyspecific partners may be more costly than feeding with others. This complexity is well illustrated by studies of mixed-species association in birds (Barnard et al. 1982), including those that use social network analysis (Farine, Aplin, et al. 2015; Farine and Milburn 2013). Individual food intake ("net rate of energy intake" based on intake rates of worms of estimated size) was affected by flock composition and number of species in a mixed-species flock of birds including some combination of lapwings, black-headed gulls, and golden plovers (Barnard et al. 1982). By examining the number of birds and the presence and absence of species in relation to individual intake, Barnard and colleagues found that when gulls were present, lapwing net rate of energy intake decreased, whereas, without gulls, even if still in a mixed flock with golden plovers, lapwing net rate of energy intake was affected only by conspecific number. With the inclusion of one particular species, intraspecific competition was drowned out by interspecific competition. Golder plover net rate of energy intake was affected by the number of lapwings in the flock, unless gulls arrived, in which case the lapwings had no effect on the golden plovers (Barnard et al. 1982). Studies using social network analysis of mixed-species bird flocks (Farine, Aplin, et al. 2015; Farine, Garroway, et al. 2012; Farine and Milburn 2013; Farine et al. 2014) point to how the complexity of relationships and nutritional effects of primate polyspecific associations could be more fully characterized via social network and nutritional geometry (including multidimensional nutritional niche) analyses.

This discussion of multivariate nutrition and sociality has been focused on macronutrients and energy, for which there has been more empirical research (S. Cook et al. 2010; Dussutour and Simpson 2009; Grover et al. 2007; Lihoreau, Clarke, et al. 2016; Janson 1985; Takahashi 2018; Uwimbabazi et al. 2021). The biological importance of micronutrients indicates the need to also examine mineral intake in relation to primate conspecific and heterospecific sociality. The availability (bioavailability) of minerals in wild animal foods and which minerals are limiting are less clear than for macronutrients, except for sodium (Belovsky 1978; R. L. Jones and Hanson 1985; Kaspari 2020; Oates 1978; Reynolds, Lloyd, et al. 2009; Rothman, Van Soest, and Pell 2006; Tracy and McNaughton 1995; Venable et al. 2020). High-sodium decaying wood elicits aggressive within-group contest competition in gray-cheeked mangabeys (Chancellor and Isbell 2009) and mountain gorillas (Rothman, Van Soest, and Pell 2006; E. Wright and Robbins 2014) and is an example of the mineral composition of a food affecting conspecific social interactions. Investigating micronutrient strategies in the nutritional geometry framework is more challenging than for macronutrients (but see Nie et al. 2015), and much is left to be explored.

SUMMARY AND CONCLUSIONS

In this chapter, we reviewed the effects of social behavior on nutrient access and intake by first reviewing social foraging theory, empirical studies on social nutrition in nonprimate taxa, and theoretical modeling of social nutrition. We then reviewed the rich primatological literature on socioecological models and the ecology of polyspecific associations. Finally, we discussed the application and potential of using a modern social nutrition framework in testing socioecological questions in conspecific and polyspecific contexts.

There are rich research avenues in primate social nutrition:

- In addition to macronutrients and energy, micronutrients (specifically limiting minerals) should also be examined in relation to social patch use and social dynamics.
- Explicit testing of predictions made in social nutrition and nutritional geometry models (i.e., Lihoreau, Buhl, et al. 2015) is needed to examine differential macronutrient intake and balance in relation to within- and between-group contest competition as well as subgroup dynamics in fission-fusion systems.
- The social nutrition framework (Lihoreau, Buhl, et al. 2015; Lihoreau et al. 2018) in conjunction with the multidimensional nutritional niche framework (Machovsky-Capuska, Senior, et al. 2016) enables primate polyspecific association research to go beyond niche partitioning as variation in food type and examine potential multidimensional nutritional niche shifts. Cross-site comparisons in which certain species are missing or play different roles in polyspecific association will be informative in understanding nutritional niche construction.
- Anthropogenic change may affect species' roles or participation in polyspecific association, which, in turn, has implications for shifts in nutritional intake and balance.
- A social nutrition framework enables the examination of complex interactions between primate nutritional landscapes and landscapes of fear (sensu Hill, this volume) in conspecific and polyspecific contexts.

15 Primate Cognitive Ecology
Challenges and Solutions to Locating and Acquiring Resources in Social Foragers

Paul A. Garber

> Most recorded observations of primate innovation are in the foraging context.
>
> READER AND MACDONALD (2003, 94)

Compared to many groups of mammals, primates are characterized by large brain size, increased neuron packing density (especially in the primary sensory cortex), and complex cognitive skills that influence the ability of individuals to store, retrieve, and integrate different types of social and ecological information (Byrne 2000; Garber, Bicca-Marques, and Azevedo-Lopes 2009; M. Hoffman et al. 2009; Kaas 2013; Tomasello and Call 1997; van Schaik, Isler, and Burkart 2012). Relatedly, both juvenile and adult primates frequently engage in exploratory and innovative behaviors in the context of searching for, acquiring, manipulating, and processing food, as well as during bouts of play and in the development of social traditions (K. Gibson 1986; Kummer and Goodall 1985; McGrew 1992; F. Mendes et al. 2015; S. Perry and Manson 2003; Reader and MacDonald 2003; van Schaik, van Noordwijk, and Wich 2006; van Schaik, Deaner, and Merrill 1999; Visalberghi, Fragaszy, Ottoni, et al. 2007). Although an increase in total brain volume or in the relative proportions of particular regions of the brain (e.g., neocortex and hippocampus) is not a direct index of cognitive ability (van Schaik, Isler, and Burkart 2012), in the case of both avians and mammals, there is evidence that, when body mass is controlled for, larger-brained species (especially those with an expanded forebrain) are significantly more successful than smaller-brained species in their ability to engage in innovative behavior and to invade and establish themselves when introduced into new environments (Lefebvre et al. 2004; Sol, Bacher, et al. 2008; Sol, Duncan, et al. 2005).

A link between brain size, cognition, social learning, planning, problem-solving, tool use, and innovation has been proposed for primates, including humans (*Homo sapiens*; Reader et al. 2011; van Schaik, Damerius, and Isler 2013; van Schaik, Isler, and Burkart 2012). Innovation is defined as the ability to apply a new behavior to solve a previously encountered social or ecological problem or to modify an existing behavior to more efficiently solve a social or ecological problem encountered in a novel context (Kummer and Goodall 1985; P. Lee 2003; Reader et al. 2011). For example, observations that some capuchins of the genus *Sapajus* use stones as tools to crack open hard palm nuts, which appears to represent the modification of a more common capuchin behavior (present in both *Cebus* and *Sapajus*) of accessing difficult-to-open fruits or nuts by pounding them on hard surfaces, is best described as innovative behavior (Izawa and Mizuno 1977; F. Mendes et al. 2015; Visalberghi, Fragaszy, Ottoni, et al. 2007). This change in behavior may have occurred by chance, error, or insight. Nevertheless, its persistence highlights the ability of tufted capuchins to use a set of cause-and-effect relationships between the properties of a food item, a hard surface, and the mechanical act of pounding using a tool. The fact that an innovative behavior can spread among group members or between populations is consistent with the assumption that social learning plays a critical role in the transmission of problem-solving and technological information (van Schaik, Deaner, and Merrill 1999).

Similarly, Byrne, Hobaiter, and Klailova (2011, 683) compared the ability of wild (*Gorilla beringei*) and captive gorillas (*Gorilla gorilla*) to manipulate and ingest stinging nettles. These authors argued that although many of the complex motor and cognitive steps involved in nettle processing were similar across these gorilla populations, "inter-site differences in nettle-eating techniques are best explained as a consequence of social transmission" resulting in shared knowledge among group members. In this regard, there is evidence that the distinctiveness of an event (e.g., an observed or engaged-in behavioral sequence that contains novel or unexpected elements or outcomes) can improve recall of both short-term and long-term information (Beran 2011; Martin-Ordas et al. 2013). This may offer a mechanism to explain the transmission of novel or innovative behavioral solutions to social and ecological problems within a group or local population.

Planning also appears to represent a hallmark of primate intelligence and decision-making. Planning can be defined as the ability to integrate disparate types of information to anticipate or predict outcomes or future changes in the social and ecological environment (van Schaik, Damerius, and Isler 2013). In contrast, the concept of intelligence refers to the ability of an individual to solve a problem; intelligence, however, does not specify the cognitive processes involved in solving that problem. In this chapter, I examine primate cognitive ecology and the ability of individuals to integrate social and ecological information in decision-making. Specifically, I focus on the kinds of information primates use in making foraging decisions, evidence for planning a route of travel, and the ability of prosimians, monkeys, and apes to form an internal representation (e.g., mental map) integrating spatial, temporal, and quantity information to predict the present and future availability and spatial distribution of feeding sites within their home range. Although some aspects of primate foraging and traveling are likely to represent opportunistic or random movements (A. Schreier and Grove 2014; Shaffer 2014), there is overwhelming evidence that primates encode and recall spatial information in the form of egocentric (representing the location of an object relative to the individual) and allocentric (representing the location of an object relative to the locations of other objects) frames of reference and that foraging is principally goal oriented (Dolins and Mitchell 2010; Garber and Porter 2014; Janmaat, Ban, and Boesch 2013a; Janmaat, Boesch, et al. 2016; table 15.1). What remains less clear is the specific form or forms of mental map individual primates use, how spatial information is represented, how detailed these representations are, what information is contained in these representations, how long this information may be retained, and the degree to which different primate taxa differ in computational and learning abilities. As suggested by Kummer and Goodall (1985, 203), whereas "instinct is like a key fitting a single lock," learning and the ability to integrate and apply past and current information to generate an effective solution represent a master key that opens many doors.

PRIMATE INTELLIGENCE

Four main theories have been proposed to explain the evolution of encephalization and complex decision-making in primates. First, it has been argued that challenges associated with exploiting a complex resource base (food patches that vary in time, space, and quantity, methods of resource acquisition, and nutrient balancing) have been the main driver of primate cognitive abilities (e.g., the ecological intelligence hypothesis; Clutton-Brock and Harvey 1980; K. Gibson 1986; Janmaat, Boesch, et al. 2016; Milton 1981a, 2000b; R. Potts 2004). This includes species that forage for embedded plant tissues or animal prey, species that use tools to access food items, and species that exploit environments in which the ability to accurately track resource availability and distribution is critical to foraging success. In this regard, primates that exploit resources that vary markedly in their spatiotemporal distribution and production schedules or that exploit extensive home ranges (e.g., several square kilometers, such as chimpanzees, *Pan troglodytes*; golden snub-nosed monkeys, *Rhinopithecus roxellana*; hamadryas baboons, *Papio hamadryas*; chacma baboons, *P. ursinus*; and gorillas, *Gorilla gorilla*) are likely to encounter very different problems associated with finding food and returning to previously visited feeding sites compared to primates that exploit a very small home range or resources that exhibit a more clumped distribution and a relatively stable production schedule (e.g., <1–10 hectares, such as pygmy marmosets, *Cebuella pygmaea*; common marmosets, *Callithrix jacchus*; howler monkeys, *Alouatta* spp.; and mouse lemurs, *Microcebus* spp.). A second possibility is that challenges associated with living in a complex and changing social environment and the ability to track and evaluate social relationships (e.g., dominance hierarchies, partner competency, and tactical deception) across days, weeks, months, and possibly years have played a primary role in primate cognitive abilities (e.g., the social intelligence hypothesis: Dunbar 1992, 1995, 1998, 2003; Humphrey 1976; Reader and Laland 2002; Machiavellian intelligence: Byrne and Whiten 1988; Seyfarth and Cheney 2012). With increasing group size, alternative patterns of group cohesion and social organization, and changes in group membership or dominance status, strategies required to track and score dyadic and social network relationships are likely to increase in complexity. A third hypothesis, the technical intelligence hypothesis, attempts to explain cognitive differences between monkeys and apes by arguing that in apes "the ability to build hierarchical programs of action," as evidenced by tool manufacture and tool use, the construction of night nests, theory of mind, and mirror recognition, "evolved in the technical sphere" to enhance foraging efficiency (Byrne 1997, 306). Finally, the innovation and exploration hypothesis argues that conditions favoring exploration, possibly associated with an extended juvenile period, in which individuals obtain positive feedback and adapt new behaviors to solve current problems, drive primate cognitive abilities (Kerr and Feldman 2003; Sol, Duncan, et al. 2005). In this regard, it has been argued that selection for complex learning, innovation, and planning may follow the Goldilocks principle. That is, when the amount of environmental or social variability is "just right" (not too low to foster minimal benefits associated with change and not so high that outcomes have low predictability), the advantages of learning to accurately predict changes in contingent relationships and develop new or more effective solutions to a problem are maximized (Kerr and Feldman 2003). However, given that virtually all primates are social foragers and live in groups containing individuals of both sexes and all age categories, including kin and nonkin group members, individuals commonly encounter problems whose effective solutions require the ability to integrate social and ecological information (Afshar and Giraldeau 2014; De Petrillo et al. 2022; Garber, Bicca-Marques, and Azevedo-Lopes 2009). This includes decisions such as whom to follow, whom to avoid, whom to form long-term partnerships with, and which nearby or distant feeding sites are likely to contain a sufficient food reward, as well as situations favoring social contagion in which the collective actions of several individuals offer advantages over the lone action of a single individual (Milton 2000b; Sussman and Garber 2011; Videan et al. 2005).

Although any one of these four theories may provide a productive framework to explain cognitive complexity for a single or small set of species, hypotheses that integrate information processing across both social and ecological domains are likely to offer the most comprehensive explanation of decision-making among a broad range of primate taxa. Moreover, the demands of social group living in primates appear to have selected for enhanced analytical skills that enable individuals to assess trade-offs associated with the potential short-term costs of cofeeding or sharing a food patch with conspecifics versus the long-term benefits that accrue to individuals who act as alliance partners and engage in reciprocal, mutualistic, and cooperative behaviors to acquire and maintain access to resources and mates (De la Fuente et al. 2022; Sussman and Garber 2011; van Schaik, Isler, and Burkart 2012). In this regard, Barraclough et al. (2004, 406) have identified the existence of neurons in the prefrontal cortex of the rhesus macaque (*Macaca mulatta*) that actively update the value of a given behavioral strategy "according to the animal's decisions and the outcomes of those decisions." Combined with evidence of hippocampal place and grid cells that link locations, routes, and experiences (Moser et al. 2008), this provides a critical neurobehavioral substrate for primate flexible learning and the ability of individuals to adopt and refine a set of search strategies under changing social, spatial, temporal, and ecological conditions.

ECOLOGICAL INFORMATION

Wild primates exploit a wide range of food types that differ in nutrient and energy content (e.g., young leaves, mature leaves, ripe fruits, unripe fruit, fungi, lichen, grasses, flowers, floral nectar, bark, exudates, and invertebrate and vertebrate prey; Garber 1987; Janmart et al. 2013a; J. Lambert 2011; Rothman et al. 2006; Rothman, Chapman, and Pell 2008). These resources also vary in abundance and distribution across temporal scales of hours in the day, days of the week, and seasons of the year (Chapman, Chapman, Naughton-Treves, et al. 2004; Garber 2000; Janmaat, Boesch, et al. 2016). Some tree species produce edible food items synchronously, with most trees fruiting, flowering, or flushing leaves during the same one-to-two-month period. This presents an element of predictability to foragers who can encode and integrate plant-specific spatial, temporal, and phenological information to develop a foraging pattern that may include visiting several trees of the same species during the same day or across sequential days (referred to as traplining, Garber 1988b; or a synchrony-based inspection strategy, Janmaat, Chapman, et al. 2012). In contrast, other tree species may fruit or flush leaves asynchronously, in a piecemeal fashion, or once every few years, and a "knowledgeable" forager could use this information to periodically monitor particular trees across its range to determine individual fruiting, flowering, and leafing schedules. Finally, individual trees of the same species, perhaps due to age or microhabitat conditions, may vary predictably in patterns of food production (e.g., the amount of food produced each day or the variance in the amount of food produced across days [Garber 1988b; Garber and Porter 2010], seasons, or years [Chapman, Chapman, Naughton-Treves, et al. 2004; Janmaat, Ban, and Boesch 2013b]) or nutritional content (Houle, Conklin-Brittain, and Wrangham 2014), such that foragers who inhabit the same home range over long periods may benefit from remembering individual tree production histories and "monitor a small number of trees that have a high probability of producing large fruit crops" (Janmaat, Boesch, et al. 2016, 15). Resource monitoring, which is likely to represent a critical and regular component of the foraging strategies of many primate species, has received only limited attention and empirical study. Under conditions of resource monitoring, foragers are expected to maximize opportunities to update information on forest productivity rather than minimize the distance traveled between sequential feeding sites (Garber and Porter 2014; Janmaat, Boesch, et al. 2016; L. Porter, Garber, et al. 2021). This information could then be used to plan future foraging routes. In the case of Weddell's saddleback tamarins (*Leontocebus weddelli*, formerly *Saguinus fuscicollis weddelli*), for example, the use of less direct (e.g., non-straight-line) travel routes enabled individuals to monitor the availability of potential insect and fruit feeding sites that were then exploited during the following one-to-two-week period (Garber and Porter 2014). In the case of female chimpanzees (*Pan troglodytes verus*), on days on which solitary individuals traveled across an unusually large area of their home range (i.e., expedition days), they moved in broad arcs and rarely backtracked (L. Porter, Garber, et al. 2021). The authors speculate that "chimpanzees may monitor trees through expeditions in less familiar areas of their home range but rely on long-term spatial memory to relocate productive feeding sites in more familiar areas, a possibility that requires further study" (L. Porter, Garber, et al. 2021, 241).

SPATIAL MAP FORMATION IN PRIMATES

There is considerable evidence that wild primates rely on a range of spatial strategies, sensory cues, and past experiences to navigate between nearby and distant feeding and refuge sites across their home range (Garber and Dolins 2014). This might include olfactory information (the smell of ripening or rotting fruit, floral nectar, or scent marks left by conspecifics; Garber and Hannon 1993; Siemers et al. 2007), auditory information (the sound of birds feeding nearby or conspecific calls; van Schaik, Damerius, and Isler 2013), social information (following, avoiding, or joining other group members; Afshar and Giraldeau 2014; Garber, Bicca-Marques, and Azevedo-lopes 2009; Giraldeau and Caraco 2000), or visual information (sighting nearby landmarks or following natural features of the environment to return to a feeding site, or using color cues to differentiate ripe from unripe fruits or young leaves from mature leaves; Di Fiore and Suarez 2007; Dominy, Lucas, et al. 2001; Garber and Porter 2014). Figure 15.1 highlights the challenges that different species of arboreal primates face in using visual cues to directly sight to distant feeding or resting sites. Although a primate's field of view is likely to vary across different locations of the same forest, across forests of different canopy structure and tree density, and in deciduous versus evergreen forests, in most instances, individuals are likely to have unobstructed views of targets or goals within a distance of only 10 to 30 m (Garber and Porter 2014; L. Porter and Garber 2013).

Figure 15.1. A primate view of the forest highlighting the challenges associated with using nearby and distant visual information to navigate to previously visited feeding and sleeping sites and to monitor future feeding sites. *From left to right*: (*top*) mantled howler monkey (*Alouatta palliata*), François' langur (*Trachypithecus francoisi*), common marmoset (*Callithrix jacchus*); (*middle*) white-headed black langur (*Trachypithecus leucocephalus*), Geoffroy's spider monkey (*Ateles geoffroyi*), Tibetan macaque (*Macaca thibetana*); (*bottom*) Geoffroy's saddleback tamarin (*Leontocebus nigrifrons*), golden snub-nosed monkey (*Rhinopithecus roxellana*), and black-and-white snub-nosed monkey (*Rhinopithecus bieti*).
Photos: Paul A. Garber.

Field studies of primate navigation have tended to focus on the use of spatial information and the ability of foragers to take relatively straight-line, direct, and least-distance routes between both distant and nearby feeding sites. Although there are elements of primate ranging that are opportunistic (i.e., Brownian walks or Lévy walks; see A. Schreier and Grove 2014; Shaffer 2014), computer simulations of random movement are generally not consistent with the actual travel routes taken by wild primates, who are best described as goal-directed foragers (Asensio, Brockelman, et al. 2011; Garber 2000; Garber and Porter 2014; Janmaat, Boesch, et al. 2016; Janson 1998a; Suarez et al. 2014). What remains less clear, however, is whether primates (a) remember the temporal availability and distribution of tens or hundreds of individual trees in their home range; (b) encode spatial information in a Euclidean spatial representation (i.e., coordinate or metric representation in which salient points are stored as angles and distances) and compute direct and novel paths to reach distant targets; (c) encode spatial information in a topological representation in which individuals remember the location, connectivity, and geometric arrangements of a defined set of routes that are associated with particular landmarks or features of the environment and search for resources adjacent to these routes (i.e., a network or route-based map that contains "a mental representation which accurately preserves only topological connectedness (the order of locations and turns)" rather than true distances and angles; Byrne 1979, 153); (d) plan foraging movements one, two, or more steps ahead (e.g., multistep foraging); and (e) internally represent spatial information differently in small-scale versus large-scale space.

MENTAL MAPS IN LARGE- AND SMALL-SCALE SPACE

Poucet (1993) proposed a model of primate spatial memory in which in large-scale space (i.e., when traveling between distant feeding sites that lie outside the forager's field of view), individuals are expected to rely on a topological internal spatial representation, also referred to as a route-based or network map, whereas in small-scale space (i.e., a localized area in which individuals can obtain views of the same set of salient landmarks from multiple perspectives), individuals have the opportunity to encode spatial information in a Euclidean or metric-based internal spatial representation in which "spatial relationships between distinct places are encoded in polar coordinates as vectors, that is, pairing of information about distances and directions" (Poucet 1993, 169). A forager using a route-based spatial representation is expected to restrict travel to a defined set of reused "paths" or route segments that intersect at a limited number of landmarks or choice points (Byrne 2000; Garber 2000). These choice points link multiple paths and serve to redirect or reorient travel (L. Porter and Garber 2013). A forager using a topological representation is expected to reuse the same route or a small number of alternative routes when revisiting a feeding site. Some of these routes may be fairly direct or straight-line, as the forager may select among the shortest routes or the route with the fewest choice points (turns or segments) to navigate from its present position to its goal. However, in some cases, routes used may be constrained by topographical features of the environment (e.g., following ridges or watercourses, varying based on the location and availability of suitable weight-bearing supports in a tree crown or adjacent tree crowns, or deviating to avoid a perceived barrier such as a large open area in the forest or an area of high predation risk), guided by orienting to a single landmark (beacon), or constrained by following a sequence of individual landmarks that are used as stepping stones to reach the general area of the target (Garber and Dolins 2014). In some primates, such as spider monkeys and woolly monkeys, the same set of routes may be reused across years (Di Fiore and Suarez 2007), whereas in other species, such as Weddell's saddleback tamarins and West African chimpanzees, new routes become incorporated and old routes abandoned over a more limited period (L. Porter, Garber, et al. 2021; table 15.1).

In contrast, a Euclidean or metric spatial representation is similar to a "view from above" in which salient features of the environment are encoded as X and Y coordinates. Using this frame of reference or mental grid, a forager is expected to calculate a relatively direct route or novel shortcut from its present location to both nearby and distant feeding sites. In the absence of observing the group continuously or introducing a set of field experiments in which provisioned foods are positioned in new locations, it is not possible to be certain if a wild primate is using a truly novel shortcut, and caution must be exercised because what might be considered a novel shortcut also could represent long-term memory for a rarely used travel route located in a peripheral or infrequently visited part of an individual's home range. In practice, however, it is generally assumed that, after some extended period of observation (i.e., months), the use of not previously observed shortcuts and a pattern of relatively straight-line travel between distant sequential feeding sites is consistent with the expectations of a Euclidean spatial representation.

One measure that is commonly used to evaluate directness of travel in primate field studies is the circuity index (CI). The CI is the actual distance traveled divided by the straight-line distance between a starting point and an ending point (Garber and Hannon 1993; L. Porter and Garber 2013). A CI of 1.0 or slightly greater indicates that the forager traveled in a relatively straight line to reach its target and is consistent with the expectations of a Euclidean-based spatial representation. A CI close to 1.0 also is consistent with a topological map that contains a highly interconnected set of routes such that a forager could reach its goal by moving through a single choice point connecting two relatively straight-line travel routes. However, a forager using a Euclidean-based spatial representation is expected to exhibit both a lower mean CI and less variance in their CI than a forager using a topological representation.

A CI of 1.3, for example, indicates that the forager has traveled 30% further than the straight-line distance. This is consistent with the expectations of a route-based mental map or may be indicative of a pattern of travel in which resource monitoring is valued over minimizing the distance between sequential feeding sites. Whether members of a given species are able to construct a Euclidean-based spatial representation or a route-based spatial representation in small- or large-scale space likely depends on species differences in cognitive architecture and sensory abilities, as well as ecological conditions that permit an individual to obtain multiple views of the same set of landmarks or targets from alternative spatial perspectives. The ability to form a Euclidean spatial representation also may require that individuals encode the relative spatial positions of multiple landmarks simultaneously to form a landmark array or a geometric configuration that can be mentally rotated such that the forager can arrive at its goal from any number of cardinal directions (Garber and Dolins 2014). There is evidence that primates such as humans, chimpanzees, baboons, tamarins, and capuchin monkeys are capable of mentally rotating the spatial configurations of landmarks to locate a food reward (Dolins 2009; Garber and Brown 2006; Garber and Porter 2014; Tomasello and Call 1997).

Table 15.1 summarizes data on spatial strategies used by wild prosimians, monkeys, and apes to locate feeding sites. There is evidence from several species that individuals encode information on the spatial location of multiple feeding sites; target particular plant species, especially those characterized by intraspecific synchronous fruiting; monitor patch productivity (crop size), phenology (stage of fruit maturation), and variance in daily food reward of individual trees; preferentially select a nearby tree of a targeted plant species; and track the period of time since a food patch was last visited in deciding when to return (Garber 1988, 1989, 2000; Janmaat, Ban, and Boesch 2013a, 2013b; Janmaat, Boesch, et al. 2016; Janson 1998a, 2014; Janmaat, Byrne, and Zuberbühler 2006a; L. Porter and Garber 2013; L. Porter, Garber, et al. 2021). Thus, primates are capable of integrating several types of information in deciding where and when to forage (see fig. 15.2). Two major questions associated with these forag-

ing decisions are (1) how many steps ahead do primates plan (e.g., is there evidence that individuals of some species are capable of planning a foraging route for the entire day, also known as the traveling salesperson problem), and (2) when navigating in large-scale space, do primates form a Euclidean or route-based spatial representation?

PRIMATES RARELY ACT AS TRAVELING SALESPERSONS

Janson (2014) tested the traveling salesperson model and found very limited empirical evidence that primates plan a route of travel more than one or two steps ahead. In this regard, Janson (2014, 410) argued that primate foragers consistently employ a rule "that mentally sums spatial information from all unused resources in a given trial into a single 'gravity' measure that guides movements to one destination at a time." Such a rule is cognitively simpler than a three-or-greater look-ahead rule. Moreover, planning too far ahead might limit the ability of a forager to correct or alter its itinerary in the face of new or updated information or unanticipated changes in energetic or nutritional demands that can be better fulfilled by visiting other feeding sites (Janson 2014; Noser and Byrne 2014).

MODELS OF OPTIMAL FORAGING AND MENTAL MAPS IN LARGE-SCALE SPACE

Models of optimal foraging generally assume that individuals act to minimize time and energy spent in locating or traveling to feeding sites while maximizing energy intake at a feeding site. This also is consistent with the marginal value theorem that argues that a forager should remain in its current food patch until resources in that patch fall below the average value of food patches in the environment (Charnov 1976). However, there is considerable evidence that primate foraging goals are less dependent on minimizing travel distance per se (Sigg and Stolba 1981; see CIs listed in table 15.1) and more dependent on taking advantage of opportunities for resource monitoring and revisiting individual trees of target species or microhabitats that offer highly productive, predictable, or nutritionally complementary foods. This makes considerable theoretical and empirical sense—traveling an additional 100, 200, or even 500 meters over the course of a day ac-

Table 15.1. Evidence of cognitive map formation in wild prosimians, monkeys, and apes in small-scale and large-scale space.

Species and references	Large-scale space (LSS)	Small-scale space (SSS)	Planning	Resource monitoring/ecological information
Prosimians				
Eulemur fulvus rufus (Erhart and Overdorff 2008)	Network map—reuse travel routes, exploit clumped feeding sites.		Planning	
Microcebus murinus (Joly and Zimmermann 2011; Luhrs et al. 2009)	Network map—reuse routes and route segments, change points as landmarks. The authors argue that "a network of common routes and olfactory or visual landmarks on a large scale is combined with a more detailed representation on smaller scales, requiring storage of detailed place representations only for essential resources within the animal's home range (e.g., feeding and sleeping sites)" (Luhrs et al. 2009, 607).	Experimental field study in SSS (seven baited platforms located 15 m apart). Four of the seven platforms contained a food reward. CI values ranged from 1.03 to 1.30, although the authors suggest that the mouse lemurs probably traveled further than indicated by the CI. Employ a highly detailed form of a route-based map. Also possible lemurs used path integration to navigate (Luhrs et al. 2009).	Planning	
Propithecus edwardsi (Erhart and Overdorff 2008)	Network map—reuse travel routes.		Planning	
New World monkeys				
Alouatta palliata (Milton 1981a, 2000b)	Network map—central place forager, use traditional routes to move between areas of their home range. Routes persist over long periods (years).			
Alouatta palliata (Garber and Jelinek 2006)		Network map—reuse of route segments, travel between nearby feeding sites (<50 m). CI = 1.05, home range = 3.9 ha.		
Aotus nigriceps (Bicca-Marques and Garber 2004)		Experimental field study to locate food rewards in SSS. In the absence of visual, spatial, and olfactory cues, the monkeys did not score better than chance. With spatial information only, wild night monkeys correctly located feeding sites significantly above c		

(continued)

Table 15.1. (continued)

Species and references	Large-scale space (LSS)	Small-scale space (SSS)	Planning	Resource monitoring/ecological information
		hance. With visual information only (sight of real vs. plastic banana), night monkeys scored above chance but were less efficient than wild tamarins and titi monkeys. When presented with olfactory information only, night monkeys scored significantly above chance and generally performed better than diurnal primates.		
Ateles belzebul (Di Fiore and Suarez 2007; Suarez et al. 2014)	Network map—reuse travel routes that followed topological features of the habitat. Routes persist over long periods (years). CI = 1.4.			Resource monitoring
Ateles geoffroyi (Valero and Byrne 2007)	Do not use habitual routes. Route segments averaged 149.8 m. 78% of routes had a CI < 1.29, and 22% had a CI > 1.65.		Planning > 1 step ahead	
Callicebus cupreus (Bicca-Marques and Garber 2004)		Experimental field study to locate food rewards in SSS. In the absence of visual, spatial, and olfactory cues, the monkeys did not score better than chance. With spatial information only or visual information only (sight of real vs. plastic banana), wild titi monkeys correctly located feeding sites significantly above chance. When presented with olfactory information only, titi monkeys were unable to locate food rewards above chance levels.		
Callithrix jacchus (Abreu et al. 2021; De la Fuente et al. 2019; De la Fuente et al. 2022)	Network map—reuse of route segments, and in some cases entirely daily routes were repeated. Over the 46-day study period, the monkeys used 43 route segments that were reused 20–43 times. The most frequently used route segments were adjacent to productive feeding sites and distant to sleeping sites. Group members reoriented travel at 56 node clusters, which were often located near	Experimental field study examining the use of social and ecological information in selecting feeding sites in SSS. Four groups of common marmosets were each presented with access to a feeding station composed of four identical wooden platforms in their home range containing accessible or inaccessible food rewards. The amount of accessible food varied systematically and was visually concealed. The results indicated that the single		

	breeding female in each group had priority access to feeding sites through behaviors associated with social tolerance, contest competition, and scramble competition. Individuals who acted as finders had higher feeding success than joiners. Across most experimental conditions, juveniles and adults obtained a similar level of feeding success.	
	feeding and resting sites. The group's home range was 7.31 ha. Overall, the CI was 1.9. However, path straightness increased with more frequent visits to a feeding site. A small number of exudate trees were revisited daily during the entire study period.	
Callithrix penicillata (Sacramento and Bicca-Marques 2022)	Experimental field study examining the use of social and ecological information in selecting feeding sites in SSS. Three groups of black-tufted marmosets were each presented with access to a feeding station composed of 12 or 18 identical wooden platforms in their home range containing accessible or inaccessible food rewards. The area occupied by the platforms was 24.5 m^2. The amount of accessible food varied systematically and was visually concealed. The marmosets were presented with two conditions: a small number of rich food patches or a large number of poor food patches. The marmosets were found to consume a higher food reward if they acted as scroungers at productive feeding sites. However, given high levels of social tolerance on reward platforms, many individuals continued to act as scroungers under conditions of low food availability.	
Cebus capucinus (Garber 2000; Garber and Brown 2006; Garber and Paciulli 1997)	Experimental field study shows use of multiple landmarks simultaneously and the ability to rotate spatial information in the form of landmark arrays to predict the location of baited feeding sites.	Planning
Cebus capucinus (Urbani 2009)	Network map—reuse travel routes, use landmarks and topological features of the environment to navigate. Mean distance between sequential feeding sites was 112 m. CI = 1.42; home range = 40 ha.	Planning

Resource monitoring and selected feeding sites using temporal and quantity information

Focused on trees of targeted species

(*continued*)

Table 15.1. (continued)

Species and references	Large-scale space (LSS)	Small-scale space (SSS)	Planning	Resource monitoring/ecological information
Chiropotes satanus (Shaffer 2014)	Network map—however, evidence of random movement in infrequently used areas of home range. Mean distance between food patches was 77 m, CI = 1.21. Mean distance to high-quality food patches was 236 m, CI = 1.37. Deviation from direct travel often occurred during insect foraging. Home range = 1,000 ha.		Planning	Focused on trees of targeted species
Lagothrix poeppigii (Di Fiore and Suarez 2007)	Network map—reuse travel routes that followed topological features of the habitat. Routes persist over long periods (years).			Resource monitoring
Pithecia Pithecia (E. Cunningham and Janson 2013)	Mean distance from start to goal = 242 m; CI = 1.68. Visited the same feeding site from multiple directions. Passed by nearer feeding sites to visit more distant feeding sites.		Planning often selected sites based on distance + patch productivity	Resource monitoring Revisit highly productive trees that were visited earlier in the same week
Saguinus imperator (Bicca-Marques and Garber 2005)		Experimental field study to locate food rewards in SSS. In the absence of visual, spatial, and olfactory cues, the monkeys did not score better than chance. With spatial information only or visual information only (sight of real vs. plastic banana), wild emperor tamarins correctly located feeding sites significantly above chance. When presented with olfactory information only, individuals from one group were able and from another group were not able to locate food rewards above chance levels.		

Saguinus mystax (Garber 1988b, 1989; Garber and Dolins 1996; Garber and Hannon 1993)	Network map—often reuse the same set of travel routes. Mean distance between sequential feeding sites was 107 m (Garber 1988b) and 92 m in a second study (Garber 1989). However, the tamarins revisited feeding sites from different directions and distances. When visiting non-nearest neighbor trees of a target species, the monkeys used information on patch productivity and patch predictability (consistent food reward across visits) to select feeding sites. Monkeys often visited the same trees on consecutive days. Home range = 40 ha.	In an experimental field study, individuals recalled the locations of 16 platforms located across a distance of 200 m. These platforms differed in the amount of food during different periods of the day (a.m. vs. p.m.).	Planning at least one step ahead. Individuals selected the nearest tree of a target species 70% of the time. Several individual trees of two or three target tree species were the focus of daily ranging activities.	Resource monitoring Used temporal, productivity, and place information to select feeding sites, as well as visual associative cues (landmarks) in SSS
Saguinus weddelli (Leontocebus weddelli) (L. Porter and Garber 2013)	Network map—over the course of two months, habitually reused 29 route segments (mean length of route segment was 155 m) and 9 major switch point clusters (landmark arrays) to reach feeding sites. Mean distance between sequential feeding sites was 129 m. CI = 1.43. No change in CI when traveling to nearer or further-out-of-sight goals. Routes and switch points change across years.		Planning —several individual trees of two or three target tree species were the focus of daily ranging activities	Resource monitoring Used temporal, productivity, and place information to select feeding sites

(continued)

Table 15.1. (continued)

Species and references	Large-scale space (LSS)	Small-scale space (SSS)	Planning	Resource monitoring/ ecological information
Saguinus weddelli (*Leontocebus weddelli*) (Garber and Porter 2014)		Use multiple spatial strategies. SSS defined as navigating within an area of 60 m. In 31% of cases, travel was route based. In 69% of cases, tamarins appeared to attend to near-to-site landmarks to revisit feeding site. Reached feeding sites from several different directions and arboreal pathways. CI = 1.30. No reduction of CI distances with repeated visits to the same feeding site. In the vicinity of a feeding site, the tamarins appeared to rely on multiple landmarks or the area bounded by these landmarks to compute a search trajectory to relocate feeding sites.		Resource monitoring—monkeys searched for arthropods at sites closer to currently used feeding trees than to random sites. 30%–78% of feeding trees fed in during a given week were located within 60 m of a tree fed in during the previous week.
Saguinus weddelli (*Leontocebus weddelli*) (Bicca-Marques and Garber 2003, 2005)		Experimental field study to locate food rewards in SSS. In the absence of visual, spatial, and olfactory cues, the monkeys did not score better than chance. With spatial information only or visual information only (sight of real vs. plastic banana), wild saddleback tamarins correctly located feeding sites significantly above chance. When presented with olfactory information only, the tamarins were unable to locate food rewards above chance levels.		
Sapajus nigritus (Presotto and Izar 2010)	Multiple spatial strategies—reuse route segments, resulting in a network of repeated routes as well as visited feeding sites from different starting points, using different paths and routes. Traveled through habitual routes <30%. Revisited the same feeding site from multiple directions and routes. Evidence of a network of repeatedly used routes in the core area of the group's range,		Planning	

	not in the periphery. The authors suggest the possibility of a Euclidean map; however, I would argue the data are inconclusive.		
Sapajus nigritus (Janson 1998a)	Experimental field study—15 feeding platforms each located 200 m apart. The author assumed a detection field of 82 m. Search was nonrandom. Approximately 50% of CIs were between 1.0 and 1.1	Planning one step ahead	Used information on the location and productivity of feeding sites
Old World monkeys			
Cercocebus atys (Janmaat et al. 2006)	Travel faster when at a distance of 100–150 m to trees that contain fruit. Speed of travel when approaching a feeding site correlates with the value of the food patch. 700–800 ha home range.	Planning	Remember previous feeding experiences in a particular tree and distinguish spatial locations of trees that recently had and had not carried fruit. Maintain knowledge of the fruiting state of individual target trees of a given species. Distinguish between different types of fruit-bearing trees.
Lophocebus albigena (Janmaat, Byrne, and Zuberbühler 2006a; Janmaat, Chapman, et al. 2012)	Travel faster when at a distance of 100–150 m to trees that contain fruit. Speed of travel when approaching a feeding site correlates with the value of the food patch. 300–600 ha home range. Employed a tree species "synchrony based inspection strategy" when exploiting fruit patches.	Planning	Remember previous feeding experiences in a particular tree and distinguish spatial locations of trees that recently had and had not carried fruit. Maintain knowledge of the fruiting state of individual target trees of a given species. Distinguish between different types of fruit-bearing trees.

(*continued*)

Table 15.1. (continued)

Species and references	Large-scale space (LSS)	Small-scale space (SSS)	Planning	Resource monitoring/ ecological information
Papio hamadryas (A. Schreier and Grove 2010, 2014)	Network map—route-based travel. Daily path length = 8.3 km.	Evidence of Brownian random movement in SSS or during area-restricted search (intrapatch search), which may be pervasive in *Acacia* patch.	Planning	
Papio hamadryas (Sigg and Stolba 1981)	Network map—reuse route segments of 500 m as well as topological features of the environment and landmarks to navigate. Reuse familiar routes. Home range = 28 km2. Daily path length = 8.6–10.4 km.		Planning	Resource monitoring
Papio ursinus (Noser and Byrne 2007, 2010, 2014)	Network map—composed of route segments and change points. Reuse familiar routes. Anticipate location, distance, and features of feeding site (travel faster when near a target or begin traveling earlier when moving to a more distant target). Goal-directed travel associated with a network of familiar, memorized routes. Navigation includes the use of single landmarks and possibly a sequence of landmarks. Bypass closer food patches to travel to more distant and more productive feeding sites.		Planning one step ahead	

Apes

Hylobates lar (Asensio, Brockelman, et al. 2011)	Network map—with routes and switch points that change during different periods of the year. Route use in LSS overlapped by 31%. Mean distance between change points was 165 m. Not a static topological map but a changing topological.		Planning one step ahead	Used temporal, spatial, and quantity information. Change points often located in the near preferred feeding trees.

Gorilla gorilla (Salmi et al. 2020)	Dense network map or, as proposed by the authors, "some findings potentially suggesting Euclidean mapping ability," warranting further investigation." The gorillas did not travel in straight lines to reach feeding and resting sites. CI = 1.36 when traveling between current location and goal. Twenty-two percent of travel steps had a CI ≤1.1. Travel speed and travel distance were greatest when swamp herbs and fruits were the foraging targets. The initial direction taken when traveling to the next feeding site closely matched the straight-line trajectory to that site. Gorillas maintained a detailed spatial representation of the distribution and availability of resources across their 9.04 km² home range.	Planning: gorillas travel in the direction of their next feeding site and rarely re-orient travel
Pan paniscus (C. Menzel et al. 2002)	A language-trained bonobo, Kanzi, was presented with a symbol for the location of a food reward. In 99 of 127 trials in a heavily forested area of 20 ha, Kanzi went directly to the goal location (distance between starting point and goal ranged from 21 to 170 m). The second experiment had 12 control trials in which Kanzi began at different starting points in the forest. Although he successfully navigated to the goal, he rarely took the most direct route with a CI = 1.46. In some cases, Kanzi did take novel shortcuts.	

(*continued*)

Table 15.1. (continued)

Species and references	Large-scale space (LSS)	Small-scale space (SSS)	Planning	Resource monitoring/ecological information
Pan troglodytes (Janmaat, Ban, and Boesch 2013a; Normand and Boesch 2009; Normand et al. 2010)	Euclidean map—distance between starting point and goal =294 m. CI = 1.04. Returned to a food resource from many different directions rather than repeatedly using the same paths. Initial direction taken to reach a food source corresponded closely to the direction needed to reach that resource, suggesting that they did not navigate using landmarks. Direct travel occurred in both the core and periphery. Home range = 21.4 km^2.	Chimpanzees may combine Euclidian map with landmarks to orient in SSS.	Planning: chimps base expectations of finding fruit on previous rates of discovery and recognizing phenological patterns of individual fruit species. Sighting or taste of a particular type of fruit triggered chimps to visit other trees of that species.	Resource monitoring—synchrony in fruit production of encountered trees was a strong predictor of inspection activity. Use quantity, synchrony, and fruit type in selecting feeding sites. Associate information about previous feeding times in the particular trees.
Pan troglodytes (Janmaat, Boesch, et al. 2016)				Remember the location, phenology, and fruit production of individual fruiting trees across years and regulate return times.

Pan troglodytes (Boesch and Boesch 1984)		Transport of over 500 hammers (stone and wood) in small-scale space (5–50 m) to nut-cracking sites. The authors hypothesize that the chimpanzees maintain a mental map of the locations of the potential tools and trees and minimize the distance and weight of tools transported. This is consistent with a Euclidean spatial representation.
Pongo abelii (Van Schaik, Damerius, and Isler 2013)		Planning: flanged adult males were found to give long calls that predict and advertise to conspecifics the direction of future travel.
Pongo pygmaeus (McKinnon 1974)	Network map—reuse routes or particular arboreal highways. Often routes followed topological features of the habitat.	
Pongo pygmaeus morio (Bebko 2018)	Network map—characterized by the reuse of travel routes that connected at switch points and were located in the vicinity of major fruit feeding trees. In some parts of their range, travel between adjacent tree	Planning: orangutans were found to alter their behavior (travel faster, increase time spent traveling, and travel to areas often avoided)

(continued)

Table 15.1. (*continued*)

Species and references	Large-scale space (LSS)	Small-scale space (SSS)	Planning	Resource monitoring/ ecological information
	crowns was constrained by the absence of lianas and large tree branches (limited crown connectivity). Orangutan travel routes contained a significantly greater number of large trees than control or randomly selected travel routes.		hours prior to successfully escaping human observers. This may represent a form of planned travel and avoidance behavior.	

Included in the table is information on the linearity of travel. This is defined in terms of a circuity index (CI), which represents the actual distance traveled by a forager to reach a feeding site divided by the straight-line distance to that feeding site. A CI of 1.0 indicates direct or straight-line travel.

counts for less than 1% of a primate's total daily energy expenditure (Steudel 2000). Moreover, most primate species can cover such distances in minutes, and any increase in travel time or travel energy is likely compensated for by additional feeding opportunities (but see Janmaat, Boesch, et al. [2016] for an alternative explanation of foraging decisions in chimpanzees). In this regard, tracking and monitoring resource availability and productivity and obtaining nutritionally complementary resources to achieve a balanced diet within the same day or across days appear to represent high-priority goals for many primate foragers (C. A. Johnson, Raubenheimer, Rothman, et al. 2013; Righini et al. 2017; Rothman, Dierenfeld, Molina et al. 2006; Rothman, Chapman, and Pell 2008). Thus, which feeding site is targeted or preferred next is likely to be more dependent on a primates' current perception of value (e.g., a calculation using information on distance, location, productivity, predictability, nutrient quality, nutrient intake earlier in the day or on previous days, an estimate of predation risk, and the likelihood of conspecific aggression at that feeding site) and knowledge of local resource availability and distribution than on simply minimizing travel distance or travel time to the next available feeding site (fig. 15.2).

Based on current empirical evidence (table 15.1), the movement and ranging patterns of virtually all wild primates are most consistent with a route- or network-based (topological) mental map in large-scale space. This is true for prosimians, New World monkeys, Old World monkeys, gibbons (*Hylobates lar*), and great apes. It is true for primates that have a home range of several square kilometers and for primates that have a home range of less than 10 hectares. Route-based travel may offer several advantages, including the ability to reuse travel paths that cross microhabitats that consistently contain available feeding sites; access to natural, unchanging, and human-made topological features of the environment (ridges, ravines, rivers and streams, emergent trees, forest margins, areas of habitat disturbance, and roads) as orientation points; reusing familiar and tested arboreal pathways or areas of the forest that offer reduced predation risk (at present there are virtually no studies that have examined the degree to which reusing travel paths or core areas of a group's home range affects predation risk); and increasing opportunities for monitoring the presence and availability of nearby feeding sites. Assuming that all or most group members contain similar information in their mental representation (e.g., public information; Milton 1981a) and use a similar set of decision-making rules in selecting feeding sites (see below), a route-based or network map may serve to promote group cohesion and advantages associated with shared predator vigilance during traveling and foraging (Di Fiore and Suarez 2007).

Two important questions related to route-based maps are (1) how many routes or choice points are located on the map, and (2) how often are new routes added or old routes abandoned? Here, I describe two field studies that examined these questions. Di Fiore and Suarez (2007) studied the ranging patterns of sympatric white-bellied spider monkeys (*Ateles belzebuth*) and red woolly monkeys (*Lagothrix lagotricha poeppigii*) in Ecuador over an eight-year period. Although these primate species are characterized by slight differences in diet (the white-bellied spider monkey diet consists of 87% fruits/seeds, 9% leaves, buds, and shoots, 1% decayed wood, 1% flowers, nectar, and pollen, and 0.7% insects, whereas that of red woolly monkeys consists of 73% fruits/seeds, 10% leaves, buds, and shoots, 6% insect prey, and 5% flowers, nectar, and pollen; Dew 2005) and forage at different heights in the canopy, individuals of both species often used the same arboreal travel routes, which generally followed natural topographical features of the area. Di Fiore and Suarez (2007) report that, for spider monkeys, 68% of over 1,000 feeding trees were located within 25 m of an established route. In the case of woolly monkeys, 69% of 989 trees were located within 25 m of an established route. These results suggest that, given the local topography and distribution of preferred feeding sites, these primate foragers were able to navigate effectively using a mental map that included a limited number of travel routes.

A second study examined travel routes and feeding sites used by the same group of saddleback tamarins (*Leontocebus weddelli*) during two successive dry seasons in northern Bolivia (L. Porter and Garber 2013; L. Porter, Garber, et al. 2021). During each study season, the tamarins used a limited number of travel routes ($N = 29$ and $N = 26$; the mean distance of travel routes was 155 m and 146 m, and the daily path length was 1,718 and 1,821) and switch points (50 and 42) to navigate within a 30–35 ha home range. Although the tamarin dietary pattern (based on time spent feeding) was virtually identical

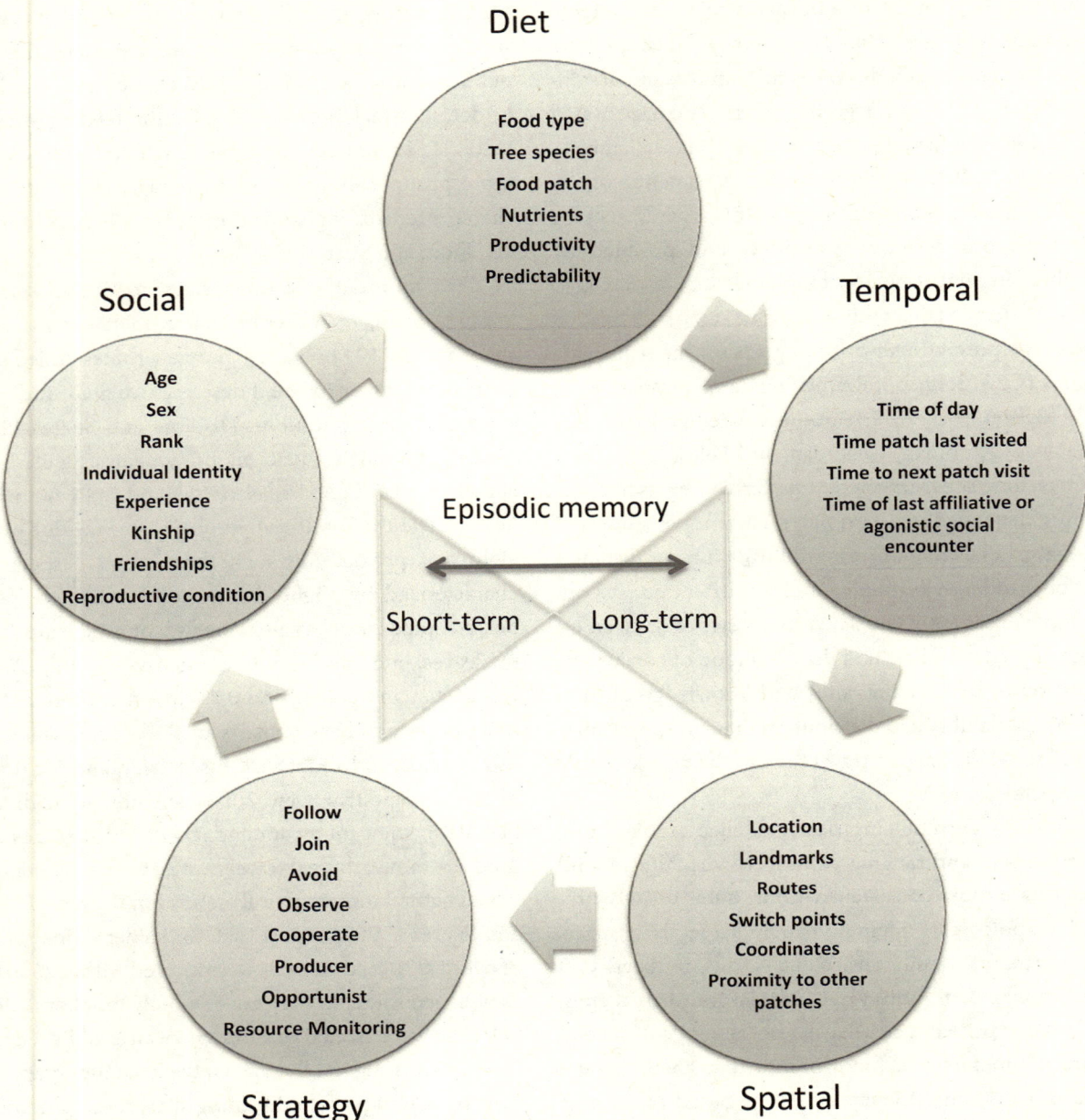

Figure 15.2. Hypothesis-based foraging. This model assumes that primate foragers integrate ecological, temporal, spatial, social, and strategic information using both short-term and long-term episodic-like event memory to generate a set of hypotheses or expectations on which to base decision-making. The saliency of particular types of information may vary within or between species based on age, sex, dominance status, experience, sensory acuity, current ecological and social conditions, and differences in an individual's capacity to store, recall, and integrate alternative types of information.

during each dry season (60%–61% ripe fruits, 20%–22% arthropods, 5%–11% exudates, and 4%–6% nectar) and the number of feeding trees exploited per season ranged from 82 to 109, only 8 travel routes and 7 individual feeding trees were reused across years. Thus, unlike spider monkeys and woolly monkeys, saddleback tamarins were found to change travel routes as they incorporated new feeding sites into their foraging itinerary. L. Porter, Garber, et al. (2021) report that on days on which travel distances exceeded one standard deviation of daily path length, the tamarins were found to monitor a significantly larger number of future feeding sites (i.e., feeding sites that were visited for the first time during the next week) than on days on which travel distance was shorter. In the case of saddleback tamarins, this suggests that the added information obtained through increased time and energy devoted to travel and resource monitoring was stored as short-term episodic-like or event memory (i.e., the ability to integrate and recall information on time, location, context, and experience related to a specific event that one observed or participated in; Rubin and Umanath 2015; van Schaik, Damerius, and Isler 2013) and used to identify future feeding sites or nearby clusters of feeding sites. The fact that tamarins incorporated new routes into their travel itinerary more commonly than spider monkeys and woolly monkeys is unlikely to reflect greater cognitive abilities or memory capacity in tamarins. Rather, it is likely a consequence of different social and ecological challenges faced by individual species or different groups of the same species in exploiting local habitats.

Two striking aspects of table 15.1 are the absence of quantitative field studies of spatial memory and navigation in wild Asian catarrhines, and the fact that virtually all species of wild nonhuman primates are reported to navigate in large-scale space using a topological or route-based map. This is true for prosimians, New World monkeys, Old World monkeys, and apes. In most cases the CI for individual species varied from 1.2 to 1.6. The only species for which there is evidence pointing to the possibility of a Euclidean mental map in large-scale space are West African chimpanzees and possibly western gorillas. Normand and Boesch (2009) and Normand and Boesch (2010) report that when traveling to distant feeding sites located an average of 300 m apart, West African chimpanzees (*Pan troglodytes verus*) traveled only 3%–4% farther (CI = 1.03–1.04) than the straight-line or most direct distance between their starting point and goal. Moreover, the chimpanzees could reach these distant sites from any number of different directions, rarely reused the same paths, and only occasionally had to adjust or change their direction of travel. This occurred despite the fact that these Taï Forest chimpanzees are reported to visit hundreds of individual feeding trees each year across their 21 km² home range (Janmaat, Boesch, et al. 2016). Direct travel to feeding sites occurred in both the core and peripheral areas of their range (Normand and Boesch 2009). Although these results also may be consistent with the expected travel paths followed by a forager using a highly interconnected route-based spatial representation, such a mental map would appear to be considerably more complex (contain a greater number of choice points, route segments, and remembered feeding sites) than reported for other primates.

In a related set of detailed field studies, Janmaat and colleagues (Janmaat, Ban, and Boesch 2013a, 2013b; Janmaat, Boesch, et al. 2016) and L. Porter, Garber, et al. (2021) report that, across years, female chimpanzees in the Taï Forest rarely reused the same individual feeding trees or travel routes. These chimpanzee foragers appeared to integrate information on food quantity, resource synchrony, and fruit type in selecting feeding sites, and to recall and use information concerning previous visits and feeding times in a particular tree in making foraging decisions. Given their large home range, individual chimpanzees travel through only a small portion of this area during any month or season. Therefore, it appears that rather than utilizing a ranging and foraging pattern focused on monitoring resource availability across their entire range, the chimpanzees relied on opportunistic local resource monitoring plus long-term episodic-like memory and knowledge of the location and fruiting patterns of a targeted group of highly productive fruit trees to select current feeding sites (Janmaat, Ban, and Boesch 2013a, 2013b; L. Porter, Garber, et al. 2021). This is supported by 3–20 months of phenological data from three chimpanzee field sites (Kibale National Park, Uganda; Loango National Park, Gabon; and Taï National Park, Côte d'Ivoire; Janmaat, Boesch, et al. 2016). At these sites, on average only 3.4%–10.9% of trees contained ripe fruit in monthly samples; however, during some years, this dropped to only 0.3%–1.1%. And, at the site of Kanyawara, in Kibale National Park, "in the most fruit-scarce month they found one ripe fruit-bearing chimpanzee food tree every 1730 m" (Janmaat, Boesch, et al. 2016, 631). Moreover, only between 0% and 7% of the 118–268 fruiting trees visited by a chimpanzee in a given season were revisited by that same chimpanzee in subsequent fruiting seasons (Janmaat, Boesch, et al. 2016). Given chimpanzee ranging patterns and the limited temporal and spatial availability of productive fruiting trees across their home range, it appears that individuals store and integrate information on specific tree locations, tree fruiting histories, and their personal feeding experiences obtained over the course of several years with current ecological and social information in deciding where to travel, whom to travel with, and which feeding sites to visit (Janmaat, Boesch, et al. 2016). Finally, a study of captive chimpanzees navigating in virtual reality indicated that subjects learned to recognize individual objects as landmarks and to search for these landmarks and reorient travel when these landmarks were obscured (Allritz et al. 2022). In this regard, it is possible that chimpanzee mental maps differ in detail (storage capacity or information contained) or computational abilities from those of other primate species studied. Clearly, chimpanzees are presented with a set of unique dietary challenges in terms of their large body size and the fact that they aggregate in relatively large communities (25–150 members), range over large areas (5–20 km^2), require a large ripe fruit crop, and exploit otherwise difficult-to-access resources through the manufacture and use of tools (e.g., for purposes of nut cracking, termite fishing, honey gathering, water dipping, spearing of solitary prey, and obtaining algae; Boesch, Kalan, et al. 2016; Pruetz and Bertolani 2007; Sanz and Morgan 2009). In addition, individuals within communities benefit dietarily from advantages associated with cooperative hunting (J. Mitani and Watts 2001).

A recent study of spatial memory in wild western gorillas (*Gorilla gorilla*) suggests that individuals navigate using either "a dense network map" (i.e., a large number of remembered routes located across their entire home range) or possibly a Euclidean spatial representation (Salmi et al. 2020). The diet of western gorillas includes the exploitation of terrestrial herbaceous vegetation and leaves, two food types that are readily available throughout the year; swamp herbs, which are an abundant but spatially restricted resource; and fruit trees, which are extremely patchy in time and space (Salmi et al. 2020). Overall, the gorillas' CI averaged 1.36, with 22% of steps approximating straight-line travel (CI ≤ 1.1). The average distance traveled from a starting location to a target location was 216 m. Travel to swamp herbs (CI = 1.11) and fruit trees CI = 1.27) was more direct than travel to terrestrial herbaceous vegetation and leaves (CI = 1.34 and CI = 1.37, respectively). These authors found that the first step taken by gorillas when leaving a feeding site was oriented in the straight-line direction of their next feeding site. Overall, western gorilla movement patterns were consistent with the ability to remember the location of previously visited feeding sites and to distinguish feeding sites based on food type. In addition, their pattern of travel speed and direction indicated spatial knowledge of resources in both the core and periphery of their home range. However, since Salmi et al. (2020) did not examine evidence of route networks, choice points, or the ability of western gorillas to take novel shortcuts in large-scale space, it remains unclear whether gorilla spatial representation is most consistent with a network map or a Euclidean map. In this regard, Barks et al. (2015) report that western gorillas have a significantly larger hippocampus and cerebellum than eastern gorillas (*Gorilla beringei*). These authors offer the possibility that differences in the sizes of these brain structures may reflect species-specific ecological challenges faced by western gorillas, which include an increased dependence on ripe fruit, a larger home range, and increased arboreal travel compared to eastern gorillas (Barks et al. 2015).

Given that ripe fruits represent an ephemeral resource in time and space, it has been hypothesized that large-bodied primate frugivores face foraging challenges that require more detailed and longer-term spatial memory. However, relatively large-bodied primates such as orangutans (*Pongo* spp.), baboons (*Papio* spp.), and mangabeys (*Lophocebus albigena* and *Cercocebus atys*) also consume a large proportion of fruit in their diet, and there is very limited evidence from the wild to support their reliance on either a Euclidean mental representation or a highly integrated and complex route-based map to navigate and forage in large-scale space (table 15.1). In the case of wild Sumatran orangutans (*Pongo abelii*), van Schaik, Damerius, and Isler (2013) have argued that flanged males plan their direction of travel as much as a day in advance and communicate both travel direction and changes in travel direction via loud calls. Although this may enable fertile females or other flanged males to seek out or avoid the caller, it offers little insight into how orangutans represent spatial information. A study of wild Bornean orangutans (*P. pygmaeus*) conducted by MacKinnon (1974) indicated that travel was most consistent with a route-based spatial representation and that travel paths followed topological features of the environment. Bebko (2018) also studied spatial navigation in Bornean orangutans. He found that navigation in large-scale space was most consistent with a network map characterized by the reuse of travel routes that were connected to choice points located in the vicinity of major fruit feeding trees. Similarly, a study of one language-trained bonobo (*Pan paniscus*) navigating within a 20 ha wooded area in Georgia (C. Menzel et al. 2002) did not provide strong support for a Euclidean representation of space. Unfortunately, there are no published studies examining mental map formation in wild eastern gorillas or bonobos. Such studies are critical for evaluating the full range of cognitive abilities and spatial strategies that characterize large-bodied fruit-eating primates.

MENTAL MAP FORMATION IN SMALL-SCALE SPACE

In contrast to the numerous studies of primate navigation in large-scale space, there are only a small number of field studies examining wild primate navigation in small-scale space. For example, in a series of experimental field studies that controlled for the effects of odor cues influencing food choice, white-faced capuchins (Garber and Brown and Garber 2006) and saddleback tamarins (Garber and Porter 2014) used the spatial configuration of landmark arrays, along with a route-based map and possibly some form of Euclidean spatial representation, to locate nearby feeding sites in small-scale space (see also Bicca-Marques and Garber 2003, 2004, 2005; Garber 2000). When presented with multiple views of the same set of targets and landmarks from alternative visual perspectives, a broader set of primate species may compute novel shortcuts and possibly encode spatial information in a Euclidean-based framework. Taï Forest chimpanzees also may use a Euclidean mental representation in small-scale space when transporting stone tools to crack open hard nuts. Boesch and Boesch (1984, 160) argue that, in small-scale space, chimpanzees remember the locations and properties of a large number of hammer stones and are able to "choose the stone with the shortest distance to a goal tree; to correctly locate a new stone location with reference to different trees; and to change their reference point so as to measure the distance to each Panda tree from any stone location." New innovative research based on experimental field studies, natural field studies, captive studies, and virtual reality is needed to more fully examine evidence for species differences in the ability to encode single landmarks and multiple landmarks and to calculate novel shortcuts in small-scale space (see Allritz et al. 2022; Dolins et al. 2014).

SOCIAL FORAGING THEORY

Finally, since virtually all primates are social foragers, individuals are likely to account for the actions of more dominant, more experienced, or more knowledgeable group members in deciding where to feed (Garber, Bicca-Marques, and Azevedo-Lopes 2009; Milton 1981a). One aspect of social foraging theory (Afshar and Giraldeau 2014; Giraldeau and Caraco 2000; Sacramento and Bicca-Marques 2022) highlights three basic foraging strategies: searcher, joiner, and opportunist. The effectiveness of each strategy is determined, in part, by the productivity of the food patch and the number and identity of other individuals adopting a similar or different strategy. In this model, individuals may act as searchers (also referred to as producers) and use self-generated or learned ecological information to locate a new or previously visited feeding site. Depending on the productivity of the food patch, searchers may benefit through a "finder's advantage" and consume most or all food items in the patch. Searchers, however, may face increased foraging costs and predation risk, as they are the first individual to encounter a feeding site. Alternatively, individuals may act as joiners (also referred to as scroungers) and monitor the behavior of searchers (use social information) to avoid, cofeed at, or usurp a feeding site. At productive feeding sites, the finder's advantage is relatively small (there are sufficient resources for several group members to feed; Garber, Bicca-Marques, and Azevedo-Lopes 2009), and joiners can take advantage of the foraging efforts of searchers (time, energy, and increased predation risk) to obtain food rewards (Sacramento and Bicca-Marques 2022). Finally, given the ability to assess and update how changes in social and ecological conditions may affect foraging success, some individuals are expected to adopt an opportunist strategy and sometimes search and other times join in deciding where to forage, whom to follow, and whom to avoid. Decisions made by social foragers are frequency dependent, and therefore payoffs are expected to be contingent on the number of other group members adopting the same and different strategies (Afshar and Giraldeau 2014; but see De la Fuente 2019; Sacramento and Bicca-Marques 2022). For example, in an experimental field study of social foraging strategies in wild saddleback tamarins and emperor tamarins (*Saguinus imperator*), Garber, Bicca-Marques, and Azevedo-Lopes (2009) found that the same individual adopted a searcher, joiner, or opportunist foraging strategy depending on the number of nearby feeding sites, expectations regarding the amount of food at a feeding site, and the rank and degree of social tolerance of other group members currently feeding at the site. Although an opportunist foraging strategy was the strategy most commonly employed across experimental conditions (44% in emperor tamarins and 57% in saddleback tamarins), when food availability was high, differences in feeding success across alternative strategies were minimal. In contrast, "when the potential for a finder's advantage was high (small and/or monopolizable feeding sites), searchers experienced greater feeding success than joiners or opportunists" (Garber, Bicca-Marques, and Azevedo-Lopes 2009, 382). Finally, rank alone was not a reliable predictor of feeding success across experimental conditions (high, medium, and low food rewards; see also De la Fuente et al. 2019; Sacramento and Bicca-Marques 2022). Overall, the results indicated that tamarins flexibly altered their foraging strategies and preferentially cofed with particular group members under conditions of changing food availability. In this regard, producer-scrounger-opportunist and other game theory models offer strong insight into decision rules that form the basis of primate foraging strategies (Afshar and Giraldeau 2014; Sacramento and Bicca-Marques 2022).

HYPOTHESIS-BASED FORAGING

Next presented is a framework for envisioning how individuals of the same primate species or individuals of different primate species might encode, integrate, and assess spatial, temporal, quantity, ecological, and social information in decision-making (fig. 15.2). This model is predicated on data from experimental and natural field studies that indicate that many species of nonhuman primates forage socially, engage in goal-directed behavior, may plan or make decisions at least one step ahead, update information on patch size and patch predictability, distinguish between feeding sites that recently contained or did not contain a food reward, preferentially select and return to more productive feeding sites over less productive feeding sites, and encode information on the spatial relationships of nearby and distant food patches in deciding where to feed next (Asensio, Brockelman, et al. 2011; Garber 1988b, 1989; Jaanmat et al. 2006; Jaanmat et al. 2012; table 15.1). In addition, it appears that the ability to learn and understand contingent relationships, not only from personal experience but also by observing the actions of others and evaluating the consequences of the behaviors of conspecifics, plays a critical role in primate problem-solving, innovation, and planning. In this regard, Tomasello and Call (1997, 370) have argued that primates "understand the interactions and relationships that other objects and individuals have with one another," including "relations among third parties in which the observer is not directly involved" (see also Rubin and Umanath 2015). In figure 15.2, it is assumed that primates initiate their decision-making process by identifying, matching, and prioritizing common elements, motivations, and information based on personal experiences (perhaps more recent experiences are prioritized over less recent experiences, or more predictable variables such as place are prioritized over less predictable variables such as patch productivity, or the expected response of a long-term partner is discounted relative to the expected response of a less familiar partner) and on the observed experiences of others (e.g., female D observed that female A was threatened when she fed next to female B in a small-crowned fruit tree but that female A was not threatened when she fed next to female C in a small-crowned tree). This information is then evaluated to generate a set of working hypotheses or strategies (see Afshar and Girldeau 2014) to guide decision-making. In many cases, these hypotheses are likely to be relatively simple: for example, if a nearby feeding site visited yesterday contained a large fruit crop, then returning to that site the next day also will likely result in a large food reward (win-return); or, if an individual has limited information regarding the location of nearby feeding sites, it may be advantageous for that animal to monitor the actions of a more dominant group member in order to locate a food reward (follow the leader). In other cases, these hypotheses may be more complex. For example, if over the past several days, female D successfully joined females A and B when they were feeding on fruits in a small-crowned tree, then female D will continue to join females A and B in other small food patches unless threatened (adopt a joiner strategy). Based on past experiences and expected outcomes, two individuals might make the same decision, make different decisions, switch to a new hypothesis as social and ecological information is updated, or vary in how persistent each is in continuing to make the same decision in the face of failure or limited success (Garber, Gomes, and Bicca-Marques 2012). In those cases in which an individual solved a problem by adopting a novel or innovative solution, the rate at which this behavior may spread through the group likely depends on group cohesion and social tolerance, the frequency with which the problem is encountered, the effectiveness of the solution, the rank or age of the innovator, and the cognitive capacity of conspecifics to understand cause-and-effect relationships by observing others (van Schaik, Deaner, and Merrill 1999). Given numerous examples of planning and innovation in wild primates, a model of decision-making predicated on hypothesis-based foraging offers a productive framework for identifying individually based and species-specific differences in cognitive ability.

Finally, researchers have access to a range of innovative methodological tools to examine the kinds of information and strategies wild primates use in decision-making. For example, the use of GPS technology has greatly expanded the ability to accurately monitor short-term changes in forager movement patterns (see Crofoot and Alavi, this volume), travel paths, and the location and availability of feeding sites to test and refine theories of primate spatial memory. Similarly, trap cameras offer an opportunity to document, at close range, individual behaviors, decision-making, and social interactions without any potential bias of observer effect. Although there is no substitute for long-term field studies designed to examine the specific social and ecological problems that individuals of different ages, sexes, and species face in acquiring resources and building strong partner preferences, carefully designed experimental field studies add an additional level of control required to test hypotheses related to social learning, problem-solving, planning, innovation, rule-based foraging, and species differences in cognitive ability (Garber 2000).

SUMMARY AND CONCLUSIONS

Compared to many groups of mammals, primates are characterized by large brain size and complex cognitive skills that influence the ability of individuals to store, retrieve, and integrate different types of information. Four main theories have been proposed to explain the evolution of encephalization and decision-making in wild primates. These are the ecological hypothesis, the social brain or Machiavellian intelligence hypothesis, the technical intelligence hypothesis, and the innovation and exploration hypothesis. Although each hypothesis tends to prioritize either ecological or social factors in driving primate intelligence, given that virtually all species of primates are social foragers and live in groups containing individuals of both sexes and all age categories, group members commonly encounter problems whose effective solutions require the ability to integrate and evaluate both social and ecological information. The information used by prosimians, monkeys, and apes to navigate across their home range, monitor resources, and revisit productive feeding sites in both large-scale and small-scale space was examined. The currently available data indicate that virtually all primate species, with the possible exception of West African chimpanzees and perhaps western gorillas, represent spatial information in a route-based mental map (also referred to as a network map) in large-scale space. In small-scale space, however, there is evidence that chimpanzees, capuchins, tamarins, baboons, and perhaps other primates use a range of spatial strategies, including a Euclidean mental representation and the ability to rotate landmark arrays to reach their goal. Exploiting episodic-like memory and a hypothesis-testing cognitive framework, many primate species dynamically integrate spatial, temporal, quantity, ecological, and social information in deciding where to forage and whom to forage with. How this information is hierarchically organized or internally represented is likely to differ across taxa and may explain proposed differences in monkey and ape intelligence.

- Wild nonhuman primates commonly integrate a complex set of social and ecological information in decision-making.
- In large-scale space (when moving between out-of-sight food patches), virtually all primate species appear to rely on a route- or network-based mental representation. The possible exceptions are West African chimpanzees and potentially western gorillas, but the currently available data are contradictory.
- In small-scale space (localized areas in which individuals can obtain views of the same set of salient landmarks from multiple perspectives), chimpanzee, baboon, tamarin, and capuchin monkey movement patterns are consistent with the ability to mentally rotate the spatial configurations of landmarks to locate a food reward.

- This is suggestive of a Euclidean-based spatial representation but will require further verification.
- Although models focused on the energetics of primate foraging assume that selection should favor taking the most direct route of travel (the least-distance principle), there is evidence that resource monitoring represents a critical and regular component of primate travel. Under conditions of resource monitoring, foragers are expected to maximize opportunities to update information on forest productivity rather than minimize the distance traveled between sequential feeding sites.
- The ability to learn and understand contingent relationships, not only from personal experience but also by observing the actions of others and evaluating the consequences of the behaviors of conspecifics, appears to play a critical role in primate problem-solving, innovation, and planning.

16 Feeding-Related Tool Use in Primates
A Systematic Overview

Jill D. Pruetz, Landing Badji, Stephanie L. Bogart, Stacy M. Lindshield, Papa Ibnou Ndiaye, and Kristina R. Walkup

Tools provide a means to acquire foods that are usually rich in nutrients and difficult to obtain. However, only certain nonhuman primate species have been recorded to use tool-assisted foraging and feeding (TAF) behaviors. Foraging is defined as searching for and extracting food, whereas feeding is the consumption of food. These terms are often linked due to the overlapping nature of the behaviors; for example, a chimpanzee (*Pan troglodytes*) may be eating termites (feeding) while she is also extracting the termites from a mound (foraging). In addition to reviewing records of such tool behavior, this chapter considers possible reasons for differences found between species and populations, including biology, ecology, and culture. Furthermore, this review highlights the ecological context and nutritional contributions of TAF as well as proto-tool use (p-TAF). Following Shumaker, Walkup, and Beck (2011, 5), tool use is defined as

> the external employment of an unattached or manipulable attached environmental object to alter more efficiently the form, position, or condition of another object, another organism, or the user itself, when the user holds and directly manipulates the tool during or prior to use and is responsible for the proper and effective orientation of the tool.

While Shumaker, Walkup, and Beck (2011) reviewed all cases of tool use, we focus on foraging behaviors in nonhuman primates (from here on "primates") and the ecological ramifications of such behavior. Tool mode categories follow Shumaker, Walkup, and Beck (2011). We deconstruct these 22 categories into more specific behaviors and compare them across species for TAF frequency and diversity. We include proto-tools (p-TAF) in this chapter because of their similarity to tools and their importance to foraging. Proto-tools are "functionally analogous to tools but are not held and directly manipulated during or prior to use" (Shumaker, Walkup, and Beck 2011, 6). Examples of p-TAF include fixed anvil use to open nuts (without hammers) by some capuchin monkeys (*Cebus capucinus*; Freese 1977) and hard-shelled fruits (fig. 16.1) or small termite mounds by chimpanzees (*Pan troglodytes*; Gašperšič and Pruetz 2008; Hicks et al. 2019; Marchant and McGrew 2005; Reynolds and Gruber, pers. comm.). In such cases, the user does not hold or manipulate the anvil, and the anvil is not modified; however, the result of acquiring the food is the same as in true tool use, and the nature of the food and anvil properties must also be considered or at least associated. As was argued by Shumaker, Walkup, and Beck (2011), there continues to be no evidence to support the notion that proto-tool use is more primitive, less cognitively demanding, or less ecologically important than tool use. Furthermore, p-TAF implies problem-solving skills that would likely be important for the evolution of tool use. In the general discussion of TAF and p-TAF below, "TAF" is in reference to both types of tool behavior unless the behavior in question is specifically a proto-tool one.

Figure 16.1. Adult male Fongoli chimpanzee eats baobab (*Adansonia digitata*) fruit after using a stone anvil (right, foreground) to open fruit (FSCP photo).

Our chapter focuses on wild (excluding captives, ex-captives, semiwild individuals, and rehabilitants) foraging behaviors in relation to tool-related behaviors (see also Haslam et al. 2013). Although primate tool behavior under other conditions is illuminating from a cognitive stance, it is less revealing regarding the ecological importance of the behavior in the wild. Foraging tools are often the focus in examining the evolution of human tool use for several reasons: these tools (1) contribute to fitness, (2) are extensively manufactured and used in great apes and extinct hominins, (3) make up the majority of hominin tools found in the fossil record, and (4) are more likely to require modification (J. Lambert 2007; van Schaik, Deaner, and Merrill 1999). Furthermore, foraging skills are an important component in the evolution of cognition (S. Parker and Gibson 1977; Reader et al. 2011). Thus, examining TAF in nonhuman primates stands to follow this reasoning and can allow for referential and conceptual modeling of related aspects of human evolution.

We highlight primate TAF from the perspective of general animal tool use and synthesize the available information regarding this behavior in nonhuman primates. Furthermore, we review various hypotheses and conditions set forth to explain TAF (e.g., E. Fox, Sitompul, and van Schaik 1999; van Schaik, Deaner, and Merrill 1999) and discuss them in light of this synthesis. We also address questions such as why TAF exists in some primates and not others, and we conclude with a discussion of the significance of TAF to understanding the evolution of hominin tool use. Tool and proto-tool use in feeding and foraging includes a variety of food species, unsurprisingly, given the geographically broad and environmentally diverse range of tool-using primate species. The fact that dipping for honey is the most common TAF behavior, and that honey is one of the most energetically valuable foods eaten by primates, indicates that nutritional data especially should be considered when testing hypotheses related to explanations of TAF in nonhuman primates in the future.

TOOL-ASSISTED FORAGING IN NON-PRIMATE ANIMALS

In addition to primates, other animals such as birds, fish, and even invertebrates have been documented to use tools, but fewer nonprimate mammals have been recorded to do so (Deecke 2012; see Shumaker, Walkup, and Beck 2011 for review). Tool use in general has thus far been cited in less than 1% of all known animal genera (Biro, Haslam, and Rutz 2013). Following primates, birds are the taxon exhibiting the highest frequency and diversity of tool behavior (Shumaker, Walkup, and Beck 2011). Although the New Caledonian crow (*Corvus moneduloides*) is cited as rivaling primate tool users (e.g., G. Hunt and Gray 2004), both Shumaker, Walkup, and Beck (2011) and McGrew (2013) found this claim unsubstantiated in comparing them with chimpanzees. Of the possible modes and functions (not limited to TAF), wild chimpanzees used 20 of 22 tool modes (the Cut and Hang modes are recorded with certainty only in captivity, but see Koops, McGrew, and Matsuzawa 2010) whereas wild New Caledonian crows used 2–4 modes depending on classification (Insert and Probe, Jab, Reach, and Dig). In terms of function, chimpanzees showed six of seven functions (missing only Hide/Camouflage), while crows showed just one (Extend Reach). Regarding tool manufacture, the species were similar, with wild chimpanzees using all four modes and wild crows using three (Detach, Subtract, and Reshape). Furthermore, chimpanzees routinely use Tool Crafting in the wild, while wild crows have crafted tools only in a designed experiment (Gašperšič and Pruetz 2008; G. Hunt and Gray 2004). Finally, and perhaps most importantly in regard to the cognitive aspects of tool use, wild chimpanzees unambiguously used three forms of Associative Tool Use in the wild (Tool Set, Tool Composite, and Multifunction Tool), whereas wild New Caledonian crows did not use any of these behaviors. However, Rutz, Klump, et al. (2016) suggest that the Hawaiian crow (*C. hawaiiensis*) possesses naturally occurring tool use, even though all individuals of this species are in captivity. The fact that juvenile Hawaiian crows develop tool use without training and that proficient tool use characterizes most individuals of this species led these authors to suggest that ecological factors such as island habitats influenced convergent tool use in New Caledonian and Hawaiian crows (Rutz, Klump, et al. 2016). They maintain that these crow species possess a genetic predisposition for functional tool use and suggest that similar experiments and analyses be conducted for nonhuman primates (Rutz, Klump, et al. 2016).

In the few mammals that do exhibit TAF, variation occurs across and within populations. Wild bottlenose dolphin (*Tursiops* cf. *aduncus*) females in Shark Bay, Australia, use marine sponges while foraging, with certain individuals spending up to 96% of their foraging time doing so (Mann et al. 2008; E. Patterson et al. 2016). However, a portion of the same population does not forage with sponges (Mann et al. 2008). The use of hammer and anvils to open hard-shelled prey in a population of eight sea otters (*Enhydra lutris*) studied over the course of 17 years differed greatly in prevalence (10%–93% of individuals), but variation in diet explained most of the difference (Fuiji et al. 2015). Sea otter tool use is conspicuous, and accounts of vertebrate tool use indicate that observational bias may explain the lack of TAF for at least some species (e.g., most marine animals; Mann and Patterson 2013). In the first confirmed account of tool use by wild bears, a brown bear (*Ursus arctos*) was observed to manipulate barnacle-encrusted rocks that it then used to rub its face and neck in what was interpreted as efforts to remove food debris or otherwise groom itself (Deecke 2012). For other nonprimate vertebrates, only a few accounts exist (Shumaker, Walkup, and Beck 2011). Recently, crocodilians have been observed using sticks to lure nesting birds within reach (Dinets et al. 2013). In certain crocodilian populations, the behavior is seasonal, coinciding with the nesting behavior of the bird species using rookeries in their habitats (Dinets et al. 2013). These new accounts of tool use in various species are intriguing and suggest that more remains to be discovered, but many of these are anecdotal and likely to reflect behaviors that are "hardwired" within a species (Biro, Haslam, and Rutz 2013).

The frequency of TAF in animals also depends on the definition of tool use applied. A number of species engage in p-TAF. Mongoose species have long been known to smash snails and other food items against hard substrates (Eisner and Davis 1967; Rasa 1973). Proto-tool use requires some knowledge or association of the properties of an object (e.g., hardness in the case of anvil use).

TOOL-ASSISTED FORAGING IN PRIMATES

Primates continue to exhibit the highest diversity of tool use among nonhumans and, arguably, the most sophisticated tool use outside of our own species. Table 16.1 provides an overview of TAF and p-TAF in nonhuman primates. In general, great apes are the most prolific tool-using foragers among primates, although not all apes use tools to the same degree. Several monkey species also use tools to forage, but only an anecdotal account of strepsirrhine tool use exists (Shumaker, Walkup, and Beck 2011). A wild aye-aye (*Daubentonia madagascariensis*) repositioned an attached liana so that it could be used to access food (Sterling and Povinelli 1999). Our review also considers accounts from populations exhibiting diverse and frequent TAF and p-TAF (defined as exhibiting two or more modes of TAF or p-TAF and where at least one of the modes is considered habitual or customary, following Whiten, Goodall, et al. 1999). Honey dip ($N = 11$) followed by anvil use (p-TAF; $N = 9$), sponge or dip for liquid ($N = 9$), and nest/hive open ($N = 9$) were the most common types of TAF exhibited across primate species and populations (table 16.2). Chimpanzees in Central (*Pan troglodytes troglodytes*) and West Africa (*P. t. verus*) exhibit a greater frequency and diversity of TAF, followed by East African chimpanzees (*P. t. schweinfurthii*), bearded capuchins (*Sapajus libidinosus*), and then Sumatran orangutans (*Pongo abelii*). Below, these accounts are discussed in more detail according to taxonomic divisions with specific reference to TAF.

TAF in Monkeys

Of the platyrrhine primates, capuchins (*Cebus* and *Sapajus* spp.) are the most prolific in terms of TAF, with the most expansive monkey Tool Kit (table 16.2). Habitual or even anecdotal TAF is notably absent in other monkey species of the Americas. Bearded capuchins frequently Hammer with stones and sometimes use sticks and occasionally other nuts to Pound open palm nuts at certain sites (Ottoni and Izar 2008; Struhsaker and Leland 1977; Urbani 1998). Other resources such as hard fruits, cactus, seeds, tubers, subterranean arthropod nests, caiman eggs, snails, crabs, and oysters are also accessed by capuchin monkeys using stone or branch hammers through Pounding, Digging, and in some cases even Cutting the resource open with a stone (Falótico, Siqueira, and Ottoni 2017; Mannu and Ottoni 2009; Torralvo et al. 2017; reviewed in Shumaker, Walkup, and Beck 2011). Palm nut kernels would be unobtainable without the use of such tools (Visalberghi, Sabbatini, et al. 2008). As in chimpanzee populations (see below), bearded capuchins demonstrate foresight in their activities by transporting both nuts and hammer stones to anvil sites (Visalberghi, Fragaszy, Ottoni, et al. 2007). In addition to hammering, capuchin monkeys demonstrate a wide range of other TAF behaviors, using a variety of tools (twigs, vines, branches, etc.) to Probe into various crevices (for lizards, rodents, etc.) and dip for food (arthropods, honey, etc.) or water (Castro et al. 2017; Chevalier-Skolnikoff 1990; Mannu and Ottoni 2009; Moura and Lee 2004; Souto et al. 2011). These tools are often manufactured prior to use, with the tool Detached from the substrate and leaves and stems Subtracted (Moura and Lee 2004; Shumaker, Walkup, and Beck 2011).

Tool manufacture for TAF has been reported only for wild bearded capuchin monkeys—namely, in terms of Detaching twigs or vines from a substrate and modifying them further in many cases by Subtraction of leaves or stems (Moura and Lee 2004). Associative Tool Use has also been documented for wild capuchins (Shumaker, Walkup, and Beck 2011). Mannu and Ottoni (2009) describe three cases of capuchins using smaller stones to dislodge other stones before using the latter for nut cracking, which Shumaker, Walkup, and Beck (2011) classify as Sequential Tool Use (using a tool to gain a tool) and Mannu and Ottoni (2009) call Secondary Tool Use (using a tool to make another tool). Mannu and Ottoni (2009) also documented capuchins using Tool Sets, using a stone to hammer on a tree cavity followed by probing in the cavity with a stick (Mannu and Ottoni 2009). Additionally, capuchins have been observed on several instances using the same tool for multiple TAF functions (Multifunction

Table 16.1. Accounts of TAF in nonhuman primates.

Species	Absorb	Affix, Apply, Drape	Brandish, Wave, Shake	Contain	Cut	Dig	Insert and Probe	Jab, Stab, Penetrate	Pound, Hammer	Prop and Climb, Bridge, Reposition	Pry, Apply Leverage	Reach	Wipe	Ref.
Daubentonia madagascarensis										Reposition—food access*				1
Sapajus libidinosus				Leaf cup—water	Tuber, cactus	Tuber, root, arthropod	Insect, lizard, wax, honey, water		Hammer—nut, fruit, seed, oyster, etc.					2–6
S. flavius							Insect							7
S. xanthosternos									Hammer—nut					8
S. macrocephalus						Eggs								9
Cebus capucinus		Wrap—insect, fruit			Tuber		Insect				Fruit, insect			10
C. albifrons	Sponge—water													11
Macaca fascicularis									Hammer—shellfish			Dip—water	Insect	12–15
M. sylvanus									Hammer—insect*					16
M. silenus	Sponge—water	Drape—insect					Insect*							17–18
Papio hamadryas sspp.						Clay	Insect*		Hammer—fruit*					19–21

(continued)

Table 16.1. (continued)

Species	Absorb	Affix, Apply, Drape	Brandish, Wave, Shake	Contain	Cut	Dig	Insert and Probe	Jab, Stab, Penetrate	Pound, Hammer	Prop and Climb, Bridge, Reposition	Pry, Apply Leverage	Reach	Wipe	Ref.
Chlorocebus aethiops	Sponge—water, sap*													22
Pongo abelii	Sponge—water	Leaf glove/cushion—insect, spines	Wave—insect				Insect, seed, honey	Jab—insect nest	Hammer—insect nest		Pry—seeds	Dip—water	Fruit	23–26
Pongo pygmaeus	Sponge—water	Leaf glove—spines, insect	Wave—insect											27
Gorilla gorilla sspp.		Stabilize								Reposition—food access*		Dip—insect*		28–29
Pan troglodytes	Sponge—water, algae, wine			Leaf spoon—water	Fruit?	Tuber, water, insect	Insect, honey, water	Stab—Galago, squirrel	Hammer and anvil, nut, pith, juice		Pry—nest	Dip—water, insect	??	30–55
Pan paniscus	Sponge—water													56

Categories follow Shumaker, Walkup, and Beck 2011.

(1) Sterling and Povinelli 1999; (2) Visalberghi, Fragaszy, Izar, et al. 2006; (3) Waga et al. 2006; (4) G. Canale et al. 2009; (5) Fragaszy, Visalberghi, and Fedigan 2004; (6) Mannu and Ottoni 2009; (7) Souto et al. 2011; (8) Chevalier-Skolnikoff 1990; (9) Torralvo et al. 2017; (10) Panger et al. 2002; (11) K. Phillips 1998; (12) Malaivijitnond et al. 2007; (13) Gumert and Malaivijitnond 2012; (14) A. Tan, Tan, et al. 2015; (15) Wheatley 1988; (16) C. Hladik 1973 in Shumaker et al. 2001; (17) Hohmann 1988; (18) Bhat 1990; (19) Oyen 1979; (20) Marais 1969; (21) Broda cited in Shumaker, Walkup, and Beck 2011; (22) Hauser 1988; (23) van Schaik, Ancrenaz, Borgen, et al. 2003; (24) van Schaik, van Noordwijk, and Wich 2006; (25) E. Fox et al. 1999; (26) van Schaik and Knott 2001; (27) Rijksen 1978; (28) Kinani and Zimmerman 2015; (29) Breuer et al. 2005; (30) Hernandez-Aguilar et al. 2007; (31) Koops, Schöning, et al. 2015; (32) Matsusaka et al. 2006; (33) Huffman and Kalunde 1993; (34) Pruetz and Bertolani 2007; (35) Sugiyama 1989; (36) Nakamura and Itoh 2008; (37) Pruetz et al. 2015; (38) Kortlandt and Holzhaus 1987; (39) Boesch 1978; (40) Goodall 1963; (41) Goodall 1964; (42) van Lawick-Goodall 1968; (43) Goodall 1986; (44) Sanz and Morgan 2007; (45) D. Watts 2008; (46) Humle 1999; (47) McGrew 1974; (48) McGrew 1992; (49) McGrew and Collins 1985; (50) K. Hunt and McGrew 2002; (51) Fay and Carroll 1994; (52) Lanjouw 2002; (53) Matsuzawa 1991; (54) Bogart and Pruetz 2011; (55) Boesch et al. 2017; (56) Hohmann and Fruth 2003.

*Anecdotal (see "Present" in Whiten, Goodall, et al. 2001).

Table 16.2. Patterns of TAF in primate species and populations exhibiting frequent and diverse TAF and p-TAF.

Species	Site	Liquid and/or algae — Fluid: Sponge, leaf scoop	dip, branch scoop	Algae scoop	Fruit, nuts, and/or pith: Hammer	Anvil (fruit)	Pith pound	Anvil prop	Cut extract	Seed, extract	Insects and/or honey: Branch as swatter	Honey dip	Ant dip	Ant fish/arboreal insect	Ter-mite fish	Hive nest open	Nest perfo-rate	Hive pound	Anvil (insect)	Vertebrates: Leaf glove (insect)	Spear hunt	Mar-row extract	Other (tubers, spines): Tuber, insect dig	Leaf glove (spine)
Pan troglodytes verus	Fongoli, Senegal	F				F*						R		R	F						F			
	Bossou, Guinea	F	R		F	F*	F	F				R	F	R	R	F?	R						F	
	Taï, Côte d'Ivoire	R	F		F	F*	F	F	R?	F		F	F		F?	R		R*			F		
P. t. troglodytes	Goualougo, Triangle, Congo					F*				R		F	F	F	F	F	F					R		
P. t. schweinfurthii	Gombe, Tanzania	F				F*						F?	F	F	F?	R								
	Mahale, Tanzania	F										F	F	F	F	R?				R				
	Kalinzu, Uganda	F											F		R									
	Kanyawara, Uganda	F	R									E		F	R?									
	Ngogo, Uganda	F										R		R	R?									
	Budongo, Uganda	F				F*						E							R*					
Pongo abelii	Ketambe, Sumatra	R									F	F		F	R	F	F			F			F	
	Suaq Balimbing, Sumatra	F							F		F	F		F		F								
Pongo pygmaeus wurmbii	Gunung Palung, Borneo	R									R													
	Tanjung Puting, Borneo										R									R			R	
	Sabangau, Borneo	F									R													
P. p. morio	Lower Kinabatangan, Borneo										F									R			R	

(continued)

Table 16.2. (continued)

		Liquid and/or algae			Fruit, nuts, and/or pith					Insects and/or honey									Vertebrates			Other (tubers, spines)	
		Sponge, leaf scoop	Fluid dip, branch scoop	Algae scoop	Hammer	Anvil (fruit) pound	Pith prop	Anvil Cut extract	Seed extract	Branch as swatter	Honey dip	Ant dip	Ant fish/ arboreal insect fish	Termite nest fish	Hive, Nest open	Nest perforate	Hive pound (insect)	Anvil (insect)	Leaf glove (insect)	Spear hunt	Marrow extract	Tuber dig	Leaf glove (spine)
Species	Site																						
Cebus capucinus	Lomas Barbudal				F	F*																	
Sapajus libidinosus	Fazenda Boa Vista, Brazil				F	F*																	
	Serra da Capivara, Brazil			?	F	F*					F?									R		R?	

*Proto-tool use.

This review considers habituated study populations exhibiting at least two modes of TAF where records are not anecdotal in nature. Values for multiple study groups are combined here for each site. The coding system is from Whiten, Goodall, et al. (1999) and van Schaik, van Noordwijk, and Wich (2006) with modifications by the authors to rank subjects at each study site according to their use of TAF. "F" denotes "frequent" where Whiten, Goodall, et al. (1999) used "customary" (C; where the behavior occurs in all or most able-bodied members of at least one age-sex class) or "habitual" (H; where the behavior has occurred repeatedly in several individuals). Van Schaik's (2006) category of "rare" (R) is used to indicate that the behavior is present but only a few cases exist, and the data are indirect or anecdotal evidence in nature. "E" indicates that the behavior was elicited via experimental setup.

References: Multiple sites (Nishida, Wrangham, Goodall, et al. 1983; Shumaker, Walkup, and Beck 2011; Whiten, Goodall, et al. 1999, 2001); Fongoli (Bogart and Pruetz 2011; Gašperšič 2008; Pruetz and Bertolani 2007; Pruetz, unpublished data); Bossou (Carvalho 2011; Humle 1999, 2011; Humle and Matsuzawa 2002; Matsuzawa 1999; Sugiyama and Koman 1979; Yamakoshi 1998); Taï (Boesch and Boesch 1982, 1990; Boesch and Boesch-Achermann 2000; Boesch, Marchesi, et al. 1994; Luncz and Boesch 2015; Luncz, pers. comm., as cited in Hicks 2010); Goualougo (Sanz and Morgan 2007, 2013; Sanz, Morgan, and Gulick 2004; Sanz and Morgan, pers. comm. to SML); Gombe (O'Malley et al. 2012; van Lawick-Goodall 1968); Mahale (Huffman and Kalunde 1993; Matsusaka et al. 2006; Nakamura and Itoh 2008; Nishida, Matsusaka, and McGrew 2009; S. Uehara 1982); Kalinzu (Hashimoto, Isaji, et al. 2015; Koops, Schöning, et al. 2015); Kanyawara (Whiten, Goodall, et al. 1999); Ngogo (D. Watts 2008; Sherrow 2005); Budongo (Gruber, Potts, et al. 2012; Hobaiter, Poisot, et al. 2014; Reynolds 2005; Reynolds and Gruber, pers. comm. to SML); Ketambe (van Schaik, van Noordwijk, and Wich et al. 2006; van Schaik, Ancrenaz, Djojoasmoro, et al. 2009); Suaq Balimbing (van Schaik, Ancrenaz, Djojoasmoro, et al. 2009); Gunung Palung (van Schaik, Barrickman, et al. 2006; van Schaik, van Noordwijk, and Wich 2006); Tanjung Putting (van Schaik et al. 2009); Sabangau (van Schaik, Ancrenaz, Djojoasmoro, et al. 2009); Kutai (van Schaik, Ancrenaz, Djojoasmoro, et al. 2009); Lower Kinabatangan (van Schaik et al. 2009); Fazenda Boa Vista (Spagnoletti, Visalberghi, Ottoni, et al. 2011); Serra da Capivera (Ottoni 2015; Ottoni and Izar 2008; Ottoni, pers. comm.).

Tool), such as using the same tool at different sites to dig and also pound (Mannu and Ottoni 2009). In a yearlong study of wild bearded capuchins at Fazenda Boa Vista in Brazil, Spagnoletti, Visalberghi, Ottoni, et al. (2011) reported that stone tool use was habitual (viz McGrew 1992) or customary (viz Whiten, Goodall, et al. 1999) in that almost all of the two study group members engaged in this behavior.

For water extraction, leaves have been utilized by wild capuchins as "cups" or crumpled and used to Absorb the water (*Cebus albifrons*; K. Phillips 1998). At some sites, capuchins also use leaves to wrap around *Automeris* caterpillars and *Sloanea terniflora* fruits, prior to rubbing the objects on a substrate presumably to remove the harmful substances covering each while protecting their hands (*Cebus capucinus*; Panger et al. 2002).

Our review and that of Shumaker, Walkup, and Beck (2011) do not report any habitual cases of tool manufacture or Associative Tool Use by wild catarrhine monkeys. Some Afro-Eurasian monkeys use TAF, although only one species approaches the regularity of bearded capuchins. Thai long-tailed macaques (*Macaca fascicularis*) use and transport stone hammers to access over 47 different food species that include mostly marine invertebrates, such as rock oysters (*Saccostrea cucullata*), nerite snails (*Nerita* spp.), drills (*Thais bitubercularis*), and gastropods (*Thais* spp.), but also sea almonds (*Terminalia catappa*; Falótico, Siqueira, and Ottoni 2017; Gumert and Malaivijitnond 2012; Malaivijitnond et al. 2007; A. Tan, Tan, et al. 2015). The Nicobar long-tailed macaque (*M. f. umbrosus*) has been recorded to wrap and wipe using leaves during extractive foraging (Pal et al. 2017) and use scraping tools in foraging. Lion-tailed macaques (*M. silenus*) have been documented on four occasions covering caterpillar chrysalises with leaves and rubbing them on a substrate before consuming the insect within (Hohmann 1988).

Other cercopithecine TAF and p-TAF are largely restricted to anecdotal accounts. Shumaker, Walkup, and Beck (2011) summarize these, including the use of stones to access tough-skinned fruits by chacma baboons (*Papio hamadryas ursinus*); scorpion immobilization with stones by Barbary macaques (*M. sylvanus*) and baboons (unknown species); digging or reaching with sticks to access insect nests by colobus (unknown species), mangabeys (unknown species), and baboons (unknown species); and a yellow baboon (*P. h. cynocephalus*) using a stick to probe into a termite nest. Hauser (1988) reported wild African green monkeys (*Chlorocebus aethiops*) absorbing water and *Acacia* exudate using pods. The absence of more numerous and recent reports in the literature indicates a lack of customary or habitual tool use in most Afro-Eurasian monkey species, with the exception of long-tailed macaques.

TAF in Apes

There have not yet been any cases of TAF reported in gibbons (*Hylobates* spp.) or siamangs (*Symphalangus* spp.), and tool use in general is limited in the lesser apes. Regarding great apes, only limited TAF examples exist for bonobos and gorillas. Hohmann and Fruth (2003) documented juvenile and female wild bonobos at Lomako (Democratic Republic of the Congo) Absorbing water with moss sponges. At present, only circumstantial evidence exists for termite fishing, with broken termite mounds and presumed extraction tools (sticks and stems) documented from two bonobo research sites, Lomako (Badrian et al. 1981) and Salonga (Bila-Isa 2003). Only a few studies pertaining to TAF in gorillas have been published, with both cases being anecdotal (Kinani and Zimmerman 2015). A young female mountain gorilla (*Gorilla beringei*) used a stick tool to ant dip after first using her arm and after observing an adult male using his arm (Kinani and Zimmerman 2015). In another example, a wild female gorilla (*G. gorilla*) at Mbeli Bai (Congo) Repositioned a dead shrub trunk in a swampy habitat and then used it to stabilize her position while she foraged, dredging aquatic herbs out of the swamp (Breuer et al. 2005). Orangutans and chimpanzees more consistently engage in TAF, although there is substantial intraspecific variation in such behavior.

TAF in Orangutans

Analysis of TAF across orangutans comes from six research sites across Borneo (*Pongo pygmaeus wurmbii* and *P. p. morio*) and two sites in Sumatra (*Pongo abelii*). Noted differences exist in TAF between Bornean and Sumatran orangutans, with the former exhibiting a smaller Tool Kit and only 4 different patterns of TAF but the latter showing 11 different TAF types (Table 16.2). Van Schaik, Laland, and Galef (2009) summarized cultural behaviors in orangutans including four that are utilized for TAF. The first of these, Leaf gloves/cushion, is a customary behavior at Ketambe in Sumatra and has been documented (although rarely) on Borneo (E. Fox and bin'Muhammad 2002; Rijksen 1978; van Schaik, Laland and Galef 2009). The user manufactures the tool by Detaching leaves and Adding and Combining them together and Applies it to protect the hands while handling spiny foods or vegetation. Alternatively, cushions of leaves may be employed for sitting while the ape forages or sits in a spiny tree.

Tree-hole tool use, in which a tool is used to probe into a tree hole for insects (mostly stingless bees but also termites and ants) or honey, is customary at Suaq Balimbing in Sumatra but is absent at all other research sites (E. Fox, Sitompul, and van Schaik 1999; van Schaik, Ancrenaz, Djojoasmoro, et al. 2009). This behavior encompasses a number of modes of tool use and tool manufacture. Regarding Tool Crafting, tools are manufactured specific to the target insect, with fresh branches Detached from nearby and twigs, leaves, and often bark Subtracted. The end of the tool may be chewed to fray or split open. Ant tools are the thinnest and nearly always bark-stripped, termite tools are the widest and often not bark-stripped, and stingless bee tools are of a medium width and often bark-stripped (E. Fox, Sitompul, and van Schaik 1999). These tools are then used to Hammer into insect nests, Jab at the nest for further opening, push occupants out, and finally Insert and Probe to extract insects, eggs, or honey. Multiple tools were often used per session, with the original tool either being replaced due to wear or discarded and replaced for presumed inadequacy (E. Fox, Sitompul, and van Schaik 1999).

Seed-extraction tool use is also customary at Suaq Balimbing but is absent elsewhere. The fruit (*Neesia* spp.) is difficult to open without a tool, especially due to the presence of irritating hairs protecting the seeds. These tools are also crafted; the resulting tools are thin and short, with bark often stripped and the end of the tool frayed. The tool is held via the mouth and Inserted into the cracked fruit to remove the hairs and then Pry the seeds from the fruit. These tools are often reused and carried to different feeding sites (E. Fox, Sitompul, and van Schaik 1999).

Branch scoop, using a leafy branch to Reach and Absorb water from a tree hole, is habitual across two orangutan populations, Suaq Balimbing and Sabangau (Borneo), but is marked as absent elsewhere for nonecological reasons (van Schaik, Laland, and Galef 2009). Crumpling leaves (Reshape manufacture) and using them to Absorb water is also a rare behavior but is present at Ketambe and Sabangau (van Schaik, Ancrenaz, Borgen, et al. 2003; van Schaik, Laland and Galef 2009). Knott (1999a) documented Gunung Palung (Borneo) orangutans using leaves as drinking tools (Contain or Absorb), noting that the presence of such behaviors, although rarer than in chimpanzees, supports the hypothesis that the capacity is present in wild orangutans.

TAF in Chimpanzees

A total of 10 long-term chimpanzee study sites with direct observational data are included in this TAF analysis. Chimpanzees are arguably the most prolific nonhuman tool users. Considering Associative Tool Use, wild chimpanzees often use Tool Sets, Multifunction Tools, and Tool Composites, while these forms are mainly absent in the other wild apes (an exception being the anecdotal case of Multifunction Tool described for gorillas above). Currently, almost every chimpanzee population studied uses tools habitually (Whiten, Goodall, et al. 2001 and this chapter). However, chimpanzee communities vary in the degree to which they include TAF and whether or not this is habitual or customary (Whiten, Goodall, et al. 2001). As Spagnoletti, Visalberghi, Ottoni, et al. (2011) note, habitual tool use suggests that this behavior is significant in the ecology of a population rather than an incidental activity. Most chimpanzee TAF fits this description.

Chimpanzees consume a variety of different resources with the assistance of tools (tables 16.2 and 16.3). Of 39 tool behaviors that were considered to vary culturally across nine chimpanzee study sites, approximately 80% ($N = 31$) were TAF behaviors (calculated using Whiten, Goodall, et al. 2001). The only novel mode of TAF that should be added to the Whiten, Goodall, et al. (2001) database is tool-assisted hunting (Pruetz et al. 2015; fig. 16.2), but our understanding of TAF is enhanced with new study sites and observations continually providing updated or novel data. When novel behavioral patterns are found, those linked to food resources are generally more likely to be maintained in a population as a tradition (Nishida, Matsusaka, and McGrew 2009). Nishida, Matsusaka, and McGrew (2009) examined their 43-year

Figure 16.2. Adult female Fongoli chimpanzee uses a tool to hunt *Galago senegalensis* as her juvenile daughter looks on. Photo: McKensey Miller.

study of Mahale chimpanzees (*P. t. schweinfurthii*) in Tanzania and found eight novel behaviors that had emerged during the study, two of which involved TAF (leaf sponge and leaf spoon; Nishida, Matsusaka, and McGrew 2009).

Chimpanzees use tools to extract a variety of different resources and use many different modes of TAF to accomplish this (tables 16.2 and 16.3), with many tool types comprising the species' Tool Kit. Due to the expansive nature of this Tool Kit, readers are referred to Shumaker et al. (2011) for a complete literature review. TAF behaviors in chimpanzees, with specific reference to manufacture and Associative Tool Use (notably Tool Sets, Multifunction Tools, Metatools, and Composite Tools), are summarized below.

Table 16.3. Types of foods extracted or processed with the aid of tools or proto-tools given by genus or species.

Food type	Taxa (refs. in parentheses)			Food type total (percentage of overall total)
	Cebus spp. and *Sapajus* spp. (1–12)	*Macaca fascicularis* (13–14)	*Pan troglodytes* (15–37)	
Fruit	*Acacia* spp.[P]	*Cocos nucifera*	*Adansonia digitata*[P]	
	Anacardium sp.	*Opuntia* sp.[P]	*Balsamocitrus dawei*[P]	
	Annona reticulata[P]	*Pandanus tectorius*	*Calancoba glauca*[P]	
	Apeiba tibourbou[P]	*Terminalia catappa*	*Ceiba pentandra*	
	Astrocarpum campestre		*Coula edulis*	
	Astrocaryum chambira*		*Detarium senegalensis*	
	Attalea spp.		*Eleais guinensis*	
	Bactris minor[P]		*Panda oleosa*	
	Cassia grandis[P]		*Parinari excels*	
	Cecropia peltata[P]		*Saba comorensis*[P]	
	Cnidoscolus sp.		*Sacoglottis gabonensis*	
	Cocos nucifera*		*Strychnos* sp.*	
	Genipa americana[P]		*Treculia africana*	
	Hymenaea courbaril			
	Jacaranda sp.			
	Luehea candida[P]			
	Mangifera indica[P]			
	Manihot dichotoma			
	Manilkara chicle[P]			
	Orbignya sp.			
	Pithecellobium dulces[P]			
	Quercus spp.[P]			
	Randia armata[P]			
	Sideroxylon spp.[P]			
	Sloanea terniflora*			
	Stemmedenia donnell-smithii[P]			
	Sterculia apetala[P]			
	Syagrus cearensis			
	Tabebuia ochracea[P]			
	Terminalia catappa			
Total (*N*)	31	4	13	48 (39.7%)
Underground storage organ (USO)	*Astronium* sp.		*Brachystegia bussei*	
	Combretum sp.		*Dolichus kilimandscharicus*	
	Manihot spp.		*Fadogia quarrei*	
	Spondias tuberosa		*Raphionacme welwitschii*	
	Thiloa glaucocarpa		*Smilax* sp.	
			Tacca leontopetaloides	
Total (*N*)	5	0	6	11 (9.3%)

Food type	Cebus spp. and Sapajus spp. (1–12)	Macaca fascicularis (13–14)	Pan troglodytes (15–37)	Food type total (percentage of overall total)
Stem, pith, leaf (shoots)	*Opuntia* sp. *Pilosocereus piauhyensis*	*Rhizophora apiculata*	*Eleais guinensis*	
Total (*N*)	2	1	1	4 (3.4%)
Gum			*Carapa procera*	
Total (*N*)	0	0	1	1 (0.8%)
Palm wine			*Eleais guinensis* *Raphia* sp.	
Total (*N*)	0	0	2	2 (1.7%)
Algae			*Spirogyra* sp.	
Total (*N*)	0	0	1	1 (0.8%)
Insect, insect product (honey, bee bread)	*Automeris* spp. *Merobruchus columbinus* *Nasutitermes* sp.		*Apis mellifera* *Camponotus* spp. *Crematogaster* spp. *Cubitermes* sp.[P] *Dorylus* spp. *Hypotrigona* spp. *Macrotermes* spp. *Meliponula* spp. *Pachycondyla* spp. *Pseudocanthotermes* sp. *Thoracotermes* sp.[P] *Trinervitermes* spp. *Xylocopa* sp.	
Total (*N*)	3	0	13	16 (13.6%)

(continued)

Table 16.3. (continued)

Food type	Cebus spp. and Sapajus spp. (1–12)	Macaca fascicularis (13–14)	Pan troglodytes (15–37)	Food type total (percentage of overall total)
Mollusks, crustaceans, sea cucumber	Crassostrea rhizophorae Gecarcinus quadratus	Asaphis violascens Balanus sp. Cellana radiata Chicoreus spp. Clypeomorus bifasciatus Ellobium spp. Euraphia sp. Gafrarium spp. Grapsus albolineatus Holothuria leucospilota Laevistrombus canarium Lambis lambis Littoraria spp. Marcia marmorata Mitra scutulata Monodonta labio Morula spp. Myomenippe hardwickii Nerita spp. Perna viridis Pinna bicolor Planaxis sulcatus Pugilina cochlidium Ruditapes sp. Saccostrea cucullata Scylla olivacea Thais spp. Thalamita sp. Trochus maculatus Vasticardium flavum	Arachatina marginata [p]	
Total (N)	2	28	1	31 (25.6%)
Reptile	Melanoschucus niger	Hemidactylus sp.	Kinixys spp.[p]	
Total (N)	1	1	1	3 (2.5%)
Bird	Dendrocygna autumnalis[p]			
Total (N)	1	0	0	1 (0.8%)

	Taxa (refs. in parentheses)			
Food type	*Cebus* spp. and *Sapajus* spp. (1–12)	*Macaca fascicularis* (13–14)	*Pan troglodytes* (15–37)	Food type total (percentage of overall total)
Mammal			*Colobus* spp. *Galago senegalensis* *Heliosciurus rufobrachium*	
Total (*N*)	0	0	3	3 (2.5%)
Total per consumer taxon	45	34	42	121 (100%)

Includes primate taxa that use TAF or p-TAF with multiple food species. Several foods were identified to the genus level only (sp.). Species pluralis (spp.) is used to indicate that there are multiple known or suspected food species from a single genus. *Cebus* and *Sapajus* sister taxa are lumped together because they highly overlap in (p-)TAF food types. Reports of single food items identified to the genus or species level for other primate taxa include gum from *Acacia* sp. for *Chlorocebus aethiops* (Hauser 1988), insect from *Dorylus* sp. for *Gorilla beringei* (Kinani and Zimmerman 2015), and fruit from *Neesia* sp. (van Schaik, Fox, and Sitompul 1996) and *Durio* sp. (van Schaik, Ancrenaz, Djojoasmoro, et al. 2009) for *Pongo abelii*. Food types procured with TAF but not identified to genus or species (refs. in parentheses) include honey/bee (3), wasp (4), gastropod (4), hermit crab and marine snail (12), and lizard (7) by *Cebus* or *Sapajus*; fish (13) and caterpillar and worm (Wheatley 1988) by *Macaca fascicularis*; caterpillar by *Macaca silenus* (Hohmann 1988); scorpion by *Macaca sylvanus* (C. Hladik 1973 in Shumaker et al. 2001); termite by *Papio hamadryas cynocephalus* (Broda cited in Shumaker, Walkup, and Beck 2011); fruit by *Papio hamadryas ursinus* (Marais 1969); and ant, termite, and honey/bee (van Schaik, Fox, and Sitompul 1996) by *Pongo* spp.

(1) Spagnoletti, Visalberghi, Verderane, et al. 2012; (2) Visalberghi, Sabbatini, et al. 2008; (3) Mannu and Ottoni 2009; (4) Panger 1997; (5) Panger et al. 2002; (6) Struhsaker and Leland 1977; (7) Moura and Lee 2004; (8) M. Fernandes 1991; (9) Souto et al. 2011; (10) G. Canale et al. 2009; (11) Torralvo et al. 2017; (12) B. Barrett et al. 2018; (13) Gumert and Malaivijitnond 2012; (14) A. Tan, Luncz, et al. 2016; (15) Alp 1997; (16) Boesch, Marchesi, et al. 1994; (17) Carvalho 2011; (18) Marchant and McGrew 2005; (19) Koops, McGrew, and Matsuzawa 2010; (20) van Lawick-Goodall 1968; (21) Wrangham 1977; (22) Sugiyama and Koman 1979; (23) Sugiyama 1994; (24) Hernandez-Aguilar et al. 2007; (25) Boesch and Boesch 1990; (26) Nakamura and Itoh 2008; (27) Pruetz et al. 2015; (28) Pika et al. 2019; (29) Schöning et al. 2008; (30) Yamamoto et al. 2008; (31) Boesch 2003; (32) Bogart and Pruetz 2008; (33) Sanz and Morgan 2009; (34) Humle 2003; (35) Humle 2011; (36) Boesch, Kalan, et al. 2016; (37) V. Reynolds and T. Gruber, pers. comm. to SML.

*Species associated with TAF and p-TAF.

Foods procured with proto-tools are denoted with superscript [p].

Sponge and Fluid Dip

Using tools to sponge or dip for liquid is nearly universal at all long-term chimpanzee study sites to date (Whiten, Goodall, et al. 2001; table 16.2). In sponging, vegetation such as leaves, bark, stems, or moss are Detached from the substrate and then in some cases Reshaped by wadging, typically for Absorbing water from tree hollows, holes, and open waterways (such as streams; Goodall 1964, 1986). Ohashi (2006) documented wild Bossou chimpanzees using this technique to extract palm wine that local humans were collecting in bottles from the crown of the oil palm tree. Fluid Dipping involves using a modified stick or stem (Detached from substrate, leaves Subtracted) to extract fluid (honey or water) via Inserting and Probing into a cavity (Lanjouw 2002; Whiten, Goodall, et al. 2001). Some chimpanzee populations have been recorded to use Reach tools to obtain algae (Boesch et al. 2017). Honey dipping in table 16.2 is classified in the insect resource category, as bees are sometimes consumed during this process.

Insect Extraction

Chimpanzees at most study sites use tools to extract termites or ants with the use of manufactured tools (Detached, Subtracted, and in some cases Reshaped; Whiten, Goodall, et al. 1999, 2001; table 16.2; fig. 16.3). A number of different tool modes are associated with insect extraction, including Digging with stout sticks to expose underground nests, Prying with sticks to access nests, Inserting and Probing into hives or mounds to extract insects or honey, and Reaching with sticks to capture army ants or other prey. The use of modified sticks or other vegetation to extract ants occurs at all West and Central African communities in table 16.2, along with Gombe, Mahale, Ngogo, and Kalinzu in East Africa. TAF for ants is conducted in two main ways: ant fishing includes the Insert and Probe tool use mode to extract ants from arboreal tunnels, while ant dipping involves a probe in the Reach tool use mode to collect ants on a surface or within a terrestrial nest (Shumaker, Walkup, and Beck 2011; Whiten, Goodall, et al. 2001). Using modified probes to extract termites from tunnels in mounds (Whiten, Goodall, et al. 2001) occurs at five of the ten sites in table 16.2. Honey dipping is found at all chimpanzee sites listed in table 16.2 except Kalinzu. This behavior can be relatively simple, by modifying (Detach and Subtract) twigs or branches to Insert into a beehive to Fluid Dip, or it can be one of the most complex types of TAF seen in chimpanzees. Chimpanzees at Goualougo, Congo (*P. t. troglodytes*; Sanz, Call, and Morgan 2009), and at Fongoli, Senegal (*P. t. verus*; Gašperšič and Pruetz, in prep.; fig. 16.3), employ Tool Crafting to manufacture (Reshape) a brush tip at the end of termite fishing probes, which Sanz, Call, and Morgan (2009) showed is more effective in gathering insects based on experiments they conducted. Sherrow (2005) also observed five individual chimpanzees (*P. t. schweinfurthii*) at Ngogo, Uganda, making brush tools to extract insects from holes in fallen dead trees.

Regarding Associative Tool Use for Insect Extraction, at several sites, termite-fishing tools are Multifunction tools, used for both Digging and Probing into a termite mound (Sugiyama 1997). S. Suzuki, Kuroda, and Nishihara (1995) found that Ndoki chimpanzees (*P. t. troglodytes*) use small, sturdy sticks to perforate deep, narrow holes into termite mounds, followed by using termite fishing probes, thus making a Tool Set. A similar perforating tool has also been recorded in the chimpanzees (*P. t. troglodytes*) at the Goualougo Triangle site in the Republic of Congo, where they use a Tool Set to obtain termites and honey, in that they sequentially use multiple tools to achieve one goal (Sanz, Call, and Morgan 2009; Sanz, Morgan, and Gulick 2004). Furthermore, these apes use a puncturing stick, a stout tool pushed (often with their foot) into a subterranean termite nest to make a hole, and then use a probe tool to extract the termites (Sanz, Morgan, and Gulick 2004). At Goualougo, apes (*P. t. troglodytes*) use a Tool Set to Pound, Pry Open, and then Insert and Probe to remove honey from stingless sweat-bee hives (Sanz and Morgan 2009). In Gabon, Boesch, Head, and Robbins (2009) found evidence that chimpanzees (*P. t. troglodytes*) use from three to five steps to acquire honey from different types of beehives, including Penetrate underground bee nests, Insert and Probe for honey extraction, and a Contain tool (called a swabber) to spoon the honey out. Koops, Schöning, et al. (2015) recently proposed that Seringbara, Guinea, chimpanzees' (*P. t. verus*) use of an elevated perch to escape army ant aggression during army ant dipping is an example of Composite Tool use, where two or more tool types are used simultaneously and complementarily. They used indirect evidence of tool remains and perch modification in this unhabituated community of apes and found that

Figure 16.3. Adult male Fongoli chimpanzee uses a tool to termite fish.
Photo: Stephanie Bogart.

Composite Tools (i.e., perches) were found at 40% of the dipping sites recorded (Koops, Schöning, et al. 2015). This same behavior occurs at Fongoli, Senegal (Pruetz, unpubl. data).

Tool-Assisted Vertebrate Hunting

Chimpanzees at Fongoli systematically craft tools for hunting bushbaby prey (*Galago senegalensis*; Pruetz and Bertolani 2007; Pruetz et al. 2015; fig. 16.2). Chimpanzees Detach live branches, Subtract leaves and side branches, and often break off the terminal end of the branch; some individuals Reshape the tip via trimming with their teeth before jabbing this tool into cavities used by bushbaby prey (Pruetz and Bertolani 2007; Pruetz et al. 2015). This behavior is otherwise known only from anecdotal reports of two squirrel (*Heliosciurus rufobrachium*) hunts and one hyrax (*Heterohyrax brucei*) hunt by Mahale, Tanzania, chimpanzees (Huffman and Kalunde 1993; Nakamura and Itoh 2008), with only the former being successful. Tortoise predation via percussive technology was also reported in Gabon (Pika et al. 2019).

Other TAF Behaviors

Nut cracking is another well-known TAF behavior but is found only in West African chimpanzees (*P. t. verus*). Records stem from Liberia (Kortlandt and Holzhaus 1987; Ohashi 2015), Bossou in Guinea (Sugiyama and Koman 1979), and the Taï Forest of Côte d'Ivoire (Boesch 1978). Archaeological analysis indicates that at least at the Taï Forest site, this behavior has been in practice for at least 4,300 years (Mercader, Barton, et al. 2007; Mercader, Panger, and Boesch 2002). In nut cracking, using a hammer and anvil as a Tool Composite is a complex tool-using skill that requires social learning, including imitation and possibly direct teaching (Biro, Inoue-Nakamura, et al. 2003; Boesch 1991b). Like other primates (see above overview of capuchins), chimpanzees exhibit forethought with tool transport (Byrne, Sanz, and Morgan 2013; Osvath and Osvath 2008). Chimpanzees have been observed to obtain materials to make tools at a later place and time (Byrne, Sanz, and Morgan 2013; Goodall 1964; McGrew 1974). For example, Boesch and Boesch (1984) found that, to crack the harder *Panda oleosa* nuts at Taï Forest, chimpanzees transport heavy stone hammers over a longer distance in comparison to a shorter transport distance of lighter (and more abundant) wooden hammers used to crack the softer *Coala edulis* nut. At Bossou, nut cracking also includes Metatool use, involving using a small stone as a wedge to stabilize the anvil, termed Anvil Prop, and results in a Tool Composite (Matsuzawa 1994).

Chimpanzees also engage in a variety of other Pounding behaviors to extract food. Some of these behaviors were described under "Insect Extraction" above. Pestle-Pounding, in which the ape Detaches a hard leaf petiole and then uses it to Pound into the crown of an oil palm tree to extract juice and vegetation, is a habitual behavior at Bossou (Yamakoshi and Sugiyama 1995). This behavior has recently been documented in a neighboring Liberian forest, indicating cultural transmission from emigrant chimpanzees (Ohashi 2015).

WHY USE TOOLS DURING FEEDING?

Tool-assisted and proto-tool-assisted foraging by primates has been examined on various levels. At the proximate level, tools may be used by primates to overcome the mechanical, behavioral, and chemical antipredator strategies found in several plant and animal resource (i.e., prey) species. These non-mutually-exclusive strategies can prevent consumers (i.e., predators) from accessing resource species for a variety of reasons. Mechanical defenses, such as a hard and relatively indigestible nutshell encasing a seed, might make the seed completely or partially inaccessible to a consumer with the use of brute force (e.g., prying or biting) alone. Primates can use several approaches to overcome this problem, including the use of anvil or hammer tools to fracture these protective structures and probing or jabbing tools to access resources located in deep chambers of their habitats, such as in termitaria and tree cavities (table 16.4). Tools can be used to overcome the behavior-based antipredator strategies of prey species as well. For instance, ant dipping tools can minimize painful bites and stings, and rousing tools may flush mammals out of their hiding spaces. Chemical defense strategies involve the synthesis and discharge of noxious substances by prey species to deter their predators. Moreover, prey can use these poisons in combination with physical structures, such as urticating hairs that contain toxins, as well as bristles or spines. Primates can remove or avoid these hairs with the use of tools (see below).

While tools and proto-tools are a means to accessing many food resources, they are not a prerequisite in all instances. Take the example of chimpanzees feeding on *Macrotermes* termites. The hardness of *Macrotermes* termitaria ranges from very soft areas that can easily be torn apart by chimpanzees to areas of concrete-like hardness. In some cases, chimpanzees can crumble apart softer sections of mounds to quickly capture fleeing termites. However, the use of a probing or fishing tool likely increases termite intake rate and feeding bout length. Accessing foods and increasing foraging efficiency are therefore key benefits of tool use (S. Parker and Gibson 1977). Ultimately, in an evolutionary sense, selective pressures influencing TAF include social, ecological, and cognitive explanations.

Van Schaik, Deaner, and Merrill (1999) proposed several conditions that allow for the evolution of TAF in primates. First, a population is characterized by an ecological niche that allows the use of tools to extract a resource. They are also characterized by the appropriate anatomy or manipulative abilities (e.g., motor control) necessary to modify and use tools. Next, cognition also plays a significant role, as the animal needs to be able to invent or learn the tool-using skill. Finally, sociality will determine the mode and likelihood of transmission, or social learning, of the behavior. In an examination of TAF in chimpanzees, orangutans, and capuchins, Koops, Visalberghi, and van Schaik (2014) propose a three-factor model of primate material culture. They propose that cognition, sociality, and environment are the main factors influencing primate material culture to varying degrees, but each depends on opportunity (Koops, Visalberghi, and van Schaik 2014). Ecological opportunities include resource density and degree of terrestriality (e.g., in bearded capuchins); cognitive opportunities include social and individual learning; and social opportunities include tolerance, gregariousness, and material artifacts (Koops, Visalberghi, and van Schaik 2014). These factors are discussed below as well as major hypotheses traditionally used to explain TAF in wild primates (e.g., social and ecological). Here, the focus is on the most prolific tool users among primates in terms of both frequency and diversity of TAF. In addition, a greater emphasis on understanding nutritional factors influencing primate TAF is highlighted as an area for future research.

NUTRITIONAL COMPONENTS OF TAF

Primates primarily use foraging tools to access high-quality foods, or those foods that tend to be easily digestible and excellent sources of macro- or micronutrients (see other chapters in this volume for discussions of the definition of food "quality"). Plant foods comprise more than one-half of the food species ingested with the aid of (proto-)tools (56%; N = 59), followed by coastal marine invertebrates (27%; N = 29), insects (12%; N = 13), and vertebrate prey (5%; N = 5; table 16.3). Long-tailed macaques occupying coastal forests in Thailand allocate most of their (p-)TAF efforts to marine animal foods (fig. 16.4; table 16.3). In contrast, capuchins (*Cebus* and *Sapajus*) and chimpanzees mainly focus on plant food species, except for *Cebus* that use coastal resources (B. Barrett et al. 2018). Capuchins primarily use TAF and p-TAF to access fruits (70% of food species; table 16.3), whereas chimpanzees access a wider array of plant foods with tools, and insect prey or products comprise almost one-third of their (proto-)TAF food species (fig. 16.4; table 16.3). Such foods tend to be concentrated packages of macronutrients (fat, protein, and carbohydrate) and calories. For instance, invertebrates and vertebrates are high in protein (J. Lambert and Rothman 2015), and honey is one of the most energetically concentrated wild foods (O'Malley and Power 2012). A review of the nutritional composition of p-TAF and TAF foods is provided in table 16.4. While plant and animal foods that are relatively high in proteins, sugars, or fats are routinely acquired without tools or proto-tools, TAF and p-TAF rarely involve food items with lower to intermediate concentrations of macronutrients, suggesting that, in general, the relative profitability hypothesis (Rutz and St. Clair 2012) to explain TAF should be further investigated (fig. 16.4; table 16.4). Foraging and feeding with tools or proto-tools can substantially contribute to the amount of food mass, calories, and nutrients that an individual primate ingests within a single day, particularly when the energy intake rate is relatively high. This nutritional payoff is especially pronounced when tools or proto-tools are used to access an abundant supply of fruit seeds or pulp (Boesch and Boesch 1982; Izar et al. 2022; Lindshield et al. 2021; N'Guessan et al. 2009; but

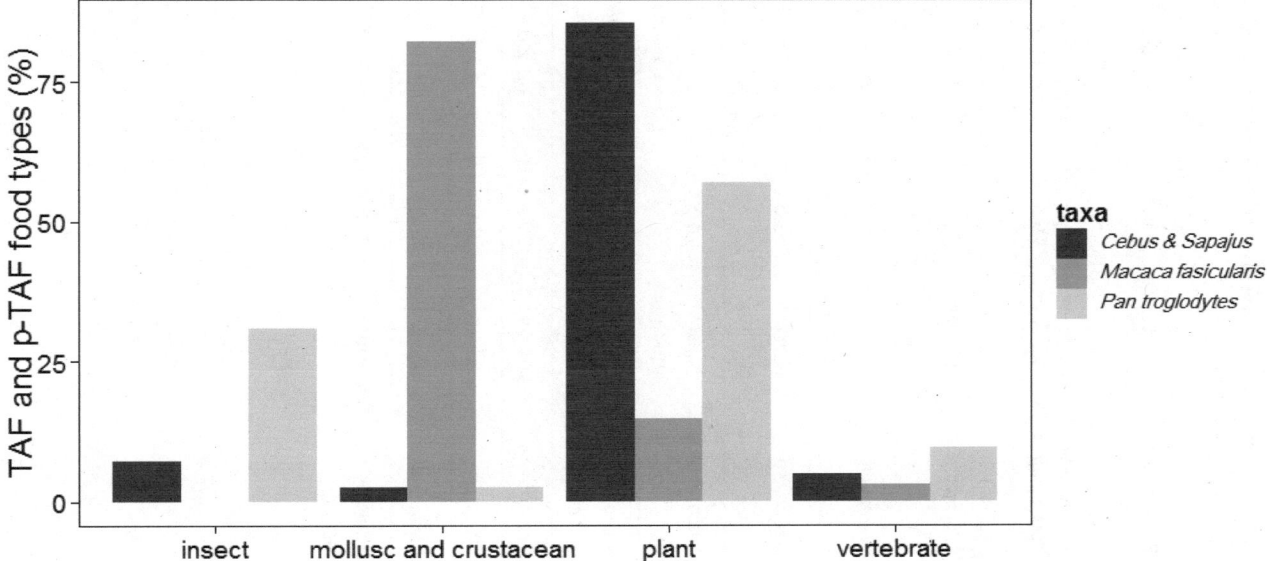

Figure 16.4. A comparison of the food types (insects, other invertebrates, plants, and vertebrates) targeted by capuchins, macaques, and chimpanzees with (proto)tools shows taxon-level differences. The complete list of foods to the species or genus level is given in table 16.3.

Table 16.4. Proximate composition (moisture, protein, carbohydrate, lipid, fiber, and ash) and energetic value of foods acquired with the aid of tools or proto-tools.

Consumer	Food species	Part	Moisture (%)*	Protein (%)	Carbohydrate (%)	Lipid (%)	Fiber (%)	Crude Ash (%)	Energy (kcal/g)	Refs.
Animal foods										
T	*Apis mellifera*	Workers	79.5	11.1^{A*} 10.3^{B*}	na	1.9^{R*}	2.5*	1.4*	1.1^{N*} 0.6^{O*}	1
T	*Apis mellifera*	Honey	na	0.3^{A*}	na	0^{R*}	na	na	3.4^{N*} 3.3^{O*}	2
T	*Camponotus* sp.	Caste mixture	75.0 ± 4.1	15.9 ± 4.2^A 12.3 ± 3.9^B	na	2.2 ± 0.9^{R*}	5.4 ± 1.0*	1.0 ± 0.3*	1.4 ± 0.3^{N*} 0.7 ± 0.2^{O*}	1
T	*Dorylus* sp.	Caste mixture	62.6 ± 3.7	52.4 ± 33.3^{A*} 40.0 ± 24.1^{B*}	na	4.2 ± 3.9^{R*}	26.8 ± 25.7^{J*}	1.6 ± 0.5*	1.8^{N*} 1.7 ± 0.9^{O*}	1, 3
T	*Macrotermes* sp.	Caste mixture	74.5 ± 4.5	52.9 ± 20.2^A 41.1 ± 16.9^B	na	3.1 ± 2.0^R	25.4 ± 15.8^J	5.2 ± 3.8	1.2 ± 0.4^{N*} 1.0 ± 0.1^{O*}	1, 3–4
T	*Pseudacanthotermes* sp.	Caste mixture	70.6 ± 1.6	20 ± 1.7^{A*} 21.0^{B*}	na	4.0^{R*}	5.3^{J*}	5.8 ± 4.5	1.6 ± 0.5^{N*} 1.4^{O*}	1
S	*Crassostrea* sp.**	Meat	79.7 ± 0.6	8.1 ± 0.4^{A*}	8.6 ± 1.1^{T*}	1.4 ± 0.2^{S*}	na	2.1 ± 0.2*	na	5
M	*Scylla* sp.**	Meat	81.0 ± 1.9	14.1 ± 1.2^{A*}	na	1.0 ± 0.04^{Q*}	na	2.2 ± 0.2*	na	6
Plant foods										
T	*Adansonia digitata*	Ripe fruit pulp	17.6 ± 10.3	3.3 ± 0.7^A 3.2 ± 1.0^B	54.2 ± 5.2^D 3.8 ± 2.9^F 3.5 ± 1.2^G 0.2 ± 0.1^H 32.4 ± 10.7^I	0.9 ± 0.5^R	13.3 ± 1.8^K	4.7 ± 2.8	3.6 ± 0.1^O	7
T	*Adansonia digitata*	Mature seed kernel	30.1 ± 16.7	38.6 ± 2.2^A 36.1 ± 4.4^B	10.8 ± 6.3^D 0.2^F 0^G 0^H 0.6^I	29.2 ± 4.1^R	22.0 ± 3.9^K	11.2	4.7 ± 0.2^O	7
M	*Cocos nucifera*	Seed kernel	7.51 ± 0.13	10.57 ± 0.63^A	32.84 ± 0.26^C	47.80 ± 0.38^R	7.70 ± 0.18^L	1.02 ± 0.02	5.5^P	8

Consumer	Species	Part								Ref
T	*Coula edulis*	Seed	na	5.3[P]*	1.2[P]*	13.4[P]*	1.9[P]*	1.1*	3.6[P]*	9
T	*Detarium senegalensis*	Seed	na	7.2[P]*	0.51[P]*	4.7[P]*	1.8[P]*	1.1[P]*	2.7[P]*	9
P	*Durio* sp.	Pulp/seed	na	16.2[A]	40.2[C]	2	41.9[K]	na	2.7[O]	10
T	*Eleais guinensis*	Seed	4.4 ± 0.7	10.9 ± 0.1[A]	35.1[C]	32.6 ± 0.8[R]	15.6 ± 1.9[L]	1.7 ± 0.1	na	11
S	*Manihot* sp.**	Seed	na	na	na	na	na	na	2.5[O]	12
P	*Neesia* sp.	Seed	na	12.4[C]	31.7[C]	46[R]	9.8[K]	na	6.0[O]	10
S	*Orbignya* sp.	Seed	na	na	na	na	na	na	4.2[O]	13
T	*Panda oleosa*	Seed	na	17.8[P]*	0.87[P]*	17.8[P]*	5.8[P]*	2[P]*	4.1[P]*	9
T	*Parinari excelsa*	Seed	na	8.7[P]*	0.97[P]*	48.4[P]*	5.2[P]*	1.8*	5.4[P]*	9
T	*Spirogyra* sp.**	Algae	na	16.7 ± 1.5[A]	55.7 ± 2.4[D]	18.1 ± 0.7[Q]	na	na	na	14
S	*Syagrus* sp.	Seed	na	na	na	na	na	na	5.3[O]	15
M	*Terminalia catappa*	Seed kernel	6.23 ± 0.09	17.66 ± 0.13[A]	1.36 ± 0.16[E] 7.68 ± 0.06[C] 1.22 ± 0.15[F] 1.95 ± 0.99[G] 2.23 ± 1.89[H] 95.82 ± 0.90[I]	54.68 ± 014[R]	9.97 ± 0.08[M]	3.78 ± 0.04	5.9[O]	16
T	*Treculia africana*	Seed	7.8 ± 0.02	13.4 ± 1.7[A]	58.1 ± 1.2[C]	18.9 ± 0.3[R]	1.4 ± 0.1[L]	2.1 ± 0.03	4.6[P]	17

Consumer species: M, *Macaca fascicularis*; P, *Pongo* sp.; S, *Sapajus* sp.; T, *Pan troglodytes*.

Each estimate is based on dry weight except for wet weight values marked with an asterisk (*).

na, not applicable.

Methods of analysis include the following: [A] crude protein; [B] available protein; [C] TNC; [D] water-soluble carbohydrates; [E] reducing sugars; [F] starch; [G] glucose; [H] fructose; [I] sucrose; [J] acid detergent fiber, a proxy for chitin; [K] neutral detergent fiber; [L] crude fiber; [M] total dietary fiber; [N] gross energy; [O] metabolizable energy; [P] method not specified; [Q] chloroform and methanol extraction; [R] petroleum ether extraction; [S] dichloromethane extraction; [T] glycogen.

(1) O'Malley and Power 2012; (2) Lakmanji et al. 2008 as cited in O'Malley and Power 2012; (3) Deblauwe and Janssens 2008; (4) C. Hladik 1977a; (5) Oliveira et al. 2006; (6) Sreelakshmi et al. 2016; (7) Lindshield et al., in prep; (8) Obasi et al. 2012; (9) Boesch and Boesch 1982; (10) Knott 1998; (11) Bora et al. 2003; (12) Martins et al. 2007 as cited in Emidio and Ferreira 2012; (13) Izar et al. 2022; (14) Tipnee et al. 2015; (15) Crepaldi et al. 2001 as cited in Emidio and Ferreira 2012; (16) S. Ng et al. 2015; (17) Edet et al. 1985.

**Values from congeners were substituted (Boesch, Kalan, et al. 2016; Emidio and Ferreira 2012).

see Emidio and Ferreira 2012). For instance, the food-pounding behavior (p-TAF) exhibited by chimpanzees of the Fongoli community is a core feature of their foraging strategy during the baobab (*Adansonia digitata*) fruiting season, where individuals crack open these fruits on anvils or other hard substrates (Gašperšič 2008). This behavior enables them to access the nutritious fruit pulp and seeds (table 16.4). Baobab feeding comprises about two-thirds of adult male subjects' feeding time during the baobab season and accounts for a similar proportion of daily energy intake (Lindshield et al. 2017). Although individuals at Fongoli do not use anvils to open every fruit, as they occasionally use their anterior teeth to open the husk, baobab cracking is likely the more efficient means of accessing the edible interior of this fruit and, furthermore, could reduce jaw strain and tooth wear (Boesch and Boesch 1982).

Recent research on the nutritional geometry of female chimpanzees provides an example of an additional potential framework in which to examine TAF and p-TAF in primates. Uwimbabazi et al. (2021) concluded that female chimpanzees (*P. t. schweinfurthii*) at Kanyawara, Uganda, prioritized available protein relative to nonprotein energy. Given that most TAF and p-TAF foods are high in protein, Uwimbabazi and colleagues' (2020) use of a geometric framework of nutrition (i.e., modeling an animal's nutrition in a multidimensional context; Raubenheimer, Machovsky-Capuska, Chapman, et al. 2015) is relevant to better understanding TAF and p-TAF in primates. However, micronutrient studies should be included in such a framework as well (see chap. 1 this volume).

The nutritional benefits of tool use can extend beyond the traditional view of energy or macronutrient gains, and a closer look at these foods reveals that each one is characterized by myriad nutrients that could motivate foraging tool use. For instance, oil palm kernels contain essential amino acids in addition to saturated fats (Bora et al. 2003), meat is an excellent source of fat as well as essential vitamins and minerals (Tennie, Gilby, and Mundry 2009), and some insects are richer in minerals than other foods (Rothman, Raubenheimer, et al. 2014). Emidio and Ferreira (2012) found that while the nuts cracked open with stone tools and anvils by *Sapajus* spp. had unexpectedly low net energy gains, these fruits were excellent sources of vitamins such as beta-carotene or ascorbic acid. Similarly, foraging for termites with tools may result in unimpressive macronutrient gains because such extractive foraging is tedious and termites are very small food packages, where each bite usually consists of only a few individuals (O'Malley and Power 2012). However, the average yield from a *Macrotermes subhyalinus* fishing bout contributed a large or entire portion of the estimated daily recommended intakes for iron, copper, and manganese for the Kasakela chimpanzee community at Gombe (O'Malley and Power 2014). These findings reveal the nutritional complexity of subsistence tool use behavior, and, given that few studies have addressed this issue, primatology will benefit from more nutritional studies of foods acquired with tools.

SOCIAL, CULTURAL, AND ECOLOGICAL ASPECTS OF TAF

Predominantly, variation in tool behaviors among primates has been interpreted as a result of cultural traditions (van Schaik and Pradhan 2003; Whiten, Goodall, et al. 1999, 2001): behaviors are passed down generations through social learning and are spread throughout a community but may vary between different groups of the same species (Whiten et al. 2011). Social intelligence—the ability to apply complex information in a society—has been linked to an animal's ability to display varied cultural behaviors (Whiten and van Schaik 2007). At some sites where related hypotheses have been tested, social explanations find more support than ecological explanations. In a study of three chimpanzee (*P. t. verus*) communities at Taï, Côte d'Ivoire, Luncz and Boesch (2015) found that differences in tool use to obtain insect prey could not be explained by seasonal availability or by the availability of foods eaten without the use of tools for each of the groups studied. These authors attributed the differences to localized diversification and learning (Luncz and Boesch 2015). The prevalence of tool-assisted drinking at Mahale, Tanzania, by immature chimpanzees was similarly interpreted as an example of social learning given that the behavior was unessential and the variations at the site could not be explained by local ecological conditions (Matsusaka et al. 2006). This is likely to be the case for the use of water-dipping sticks by immature chimpanzees at Fongoli, Senegal (Pruetz, unpubl. data). Records of daily presence/absence of TAF by chimpanzees at Fongoli, Senegal, revealed similarities in dry (64.2% of days) versus wet (67.9% of days) season values, also suggesting that perhaps factors other than ecology influence TAF, even in apes living in highly seasonal, open environments, and tool-assisted hunting in particular showed no pattern of correlation with the period of ripe fruit scarcity (Pruetz et al. 2020). Finally, Hobaiter, Poisot, et al. (2014) used social network analyses to support the hypothesis that one of two novel tool-use variants on liquid sponging was better explained by social learning than by individual learning in chimpanzees (*P. t. schweinfurthii*) at Budongo, Uganda. However, most of these studies equated what some scholars term the "necessity hypothesis" with a general ecological hypothesis (e.g., Hobaiter, Poisot, et al. 2014). Additionally, social explanations of TAF on a proximal level do not exclude ecological explanations of TAF ultimately.

Earlier studies of cultural behavior in primates used an exclusionary method to determine whether ecological or genetic explanations might otherwise explain candidate behaviors, including TAF (Whiten, Goodall, et al. 1999, 2001). This lent initial support for the idea of culture in nonhuman species. However, excluding environmental impacts and the impacts of other factors, such as genetics, on tool use has been criticized (Laland and Janik 2006; Laland et al. 2009; Schöning et al. 2008; van Schaik 2009), and research recognizes the complexity of dynamics among ecology, culture, and biology (Koops, Visalberghi, and van Schaik 2014). Some studies suggest that TAF may be important for primates experiencing periods of unreliable resources, such as fruit for chimpanzees at Bossou, Guinea (J. Lambert 2007; Yamakoshi 1998). J. Lambert (2007) points out that the isolated nature of the Bossou community prevents these chimpanzees from expanding their day and home ranges in search of food, a pressure that could have brought about the innovation of various TAF behaviors. This point is relevant to the fact that this particular ape community exhibits the highest level of variation in TAF behaviors in our review (table 16.2).

Fallback foods, resources exploited during periods when highly preferred foods are scarce (A. Marshall and Wrangham 2007), play an important role in the fitness of a population, impacting anatomy and behavior (S. Altmann 2009; J. Lambert 2007; A. Marshall and Wrangham 2007; A. Marshall, Boyko et al. 2009; Melin, Young et al. 2014; K. Phillips and Hopkins 2007). The foods consumed when preferred foods are unavailable are often (although not always) difficult to obtain, such as insects or hard nuts, and need to be extracted by using a tool. The opportunity to use tools can depend on the presence of a fallback food, the opportunity to acquire the food, and the need for the nutritional component of the resource (S. Altmann 2009; A. Marshall and Wrangham 2007). Thus, ecology has an important role in explaining variation within tool-using primate species (Koops, McGrew, and Matsuzawa 2013; Koops, Visalberghi, and van Schaik 2014) and appears to explain much of the TAF in several species exhibiting prolific tool use.

EXPLANATIONS FOR TAF IN PROLIFIC TOOL-USING PRIMATE SPECIES

Capuchins

Ottoni and Izar (2008) noted that for savanna-dwelling *Sapajus*, using tools is the norm rather than the exception. They interpreted tool use as a socially learned behavioral tradition that is primarily associated with more terrestrial habits (Ottoni and Izar 2008). Other authors note the need to consider alternative hypotheses to social explanations as new data emerge (Souto et al. 2011). Moura and Lee (2004) also stress the terrestrial aspect of capuchin tool use at their Serra da Capivara, Brazil, study site. They, along with others, stress the energy bottlenecks associated with the dry and more open sites in the Caatinga and Cerrado biomes in Brazil (G. Canale et al. 2009; Moura and Lee 2004). In particular, G. Canale and colleagues (2009) surveyed *S. xanthosternos* and *S. libidinosus* across biomes and found evidence for stone tool use only in drier environments. From the hypotheses set out by E. Fox, Sitompul, and van Schaik (1999), much capuchin TAF can be explained by the opportunity hypothesis. Some studies show no support for a link between food availability and capuchin tool use, which is the main assumption of the necessity hypothesis (Spagnoletti, Visalberghi, Verderane, et al. 2012). However, most nut cracking was found when most nuts were more available, supporting the ecological opportunity hypothesis (Spagnoletti, Visalberghi, Verderane, et al. 2012). Koops, Visalberghi, and van Schaik (2014) also examined capuchin, as well as chimpanzee, TAF relative to the opportunity and necessity hypotheses. They concluded that data on insectivory in chimpanzees and nut cracking by chimpanzees and bearded capuchins also supported the opportunity hypothesis (Koops, Visalberghi, and van Schaik 2014).

Orangutans

The preponderance of tool use in Sumatran orangutans compared to their Bornean congenerics was originally explained via the higher density and more social nature of the former species, which provides for the subsequent maintenance of TAF (van Schaik and Knott 2001). Sumatran forests have higher food productivity than Bornean forests, resulting in higher densities of orangutans (Russon, Wich, et al. 2009; van Schaik, Priatna, and Priatna 1995). Higher densities are believed to contribute to more social interactions and could allow for greater opportunities for tool behavior transmission (Knott 1999a). On both islands, food can shift dramatically by season, with mast fruiting and El Niño events linked to food shortages (Russon, Wich, et al. 2009). The resulting variability in food availability contributes to a large diet breadth that is comparable to or possibly even exceeds what is found in the African apes (Russon, Wich, et al. 2009). However, neither geographic variation nor ecological differences fully explained the use of TAF to process *Neesia* fruits and to obtain insects or honey from tree holes by orangutans at Suaq Balimbing in Sumatra or the lack of this behavior in orangutans at Gunung Palung, Borneo (van Schaik and Knott 2001). Van Schaik and Knott (2001) concluded that since nonhuman primate tool use is characterized as a largely learned behavior, the less social nature of Bornean orangutans provides fewer opportunities for learning and passing on innovations (van Schaik and Knott 2001). However, van Schaik et al. (2009) note that seed-extraction tool use is absent for a clear ecological reason (absence of the fruit) at some sites. The absence of this behavior at sites where the seeds are eaten (without tools, such as at Gunung Palung, Borneo) may be explained by the lower population density of orangutans than at Suaq Balimbing, lessening opportunities for social transmission of the behavior (Knott 1999a; E. Fox, van Schaik, Laland and Galef 2004). In addition, the absence of tool use for the opening of *Neesia* fruits at sites with low *Neesia* tree densities is in line with limited ecological opportunities for invention and transmission of the tool use behavior (Koops, Visalberghi, and van Schaik 2014). Specifically, the availability of tree holes at Ketambe and Suaq relative to orangutan density, *Neesia* fruit tool use, and *Neesia* availability at several study sites supported the opportunity hypothesis (Koops, Visalberghi, and van Schaik 2014). Conversely, a lack of support for the necessity hypothesis stemmed from examining orangutan tree-hole tool use relative to fruit scarcity and overall variation in tool use relative to feeding on cambium, a fallback food for orangutans (Koops, Visalberghi, and van Schaik 2014).

Chimpanzees

Chimpanzees use TAF at almost every site where they have been studied over the long term. This allows us to examine multiple tool modalities and multiple sites, continuing the trend by Whiten, Goodall, et al. (1999) to examine putative cultural behaviors across chimpanzee populations. Using 17 cases of TAF in chimpanzees at eight study sites, Sanz and Morgan (2013) examined the necessity, opportunity, and invention hypotheses set out by E. Fox, Sitompul, and van Schaik (1999), in addition to the relative profitability hypothesis posed by Rutz and St. Clair (2012). They assigned hypotheses to explain these cases post hoc and concluded that over half of all TAF types could be explained by the opportunity hypothesis, which states that repeated exposure to appropriate conditions results in tool use, while 20% of tool use types were better explained by the necessity hypothesis, which maintains that resource scarcity prompts tool use (E. Fox, Sitompul, and van Schaik 1999; Sanz and Morgan 2013). The invention hypothesis, which focuses on social factors that maintain tool use, better explained 16% of the tool types, and the profitability hypothesis, where TAF is more profitable than nontool foraging, accounted for 8% of the tool types examined (E. Fox, Sitompul, and van Schaik 1999; Sanz and Morgan 2013). These cases were not tested explicitly, however. Additionally, to varying degrees, the hypotheses are not incompatible with explanations that look to culture or tradition as a means to explain variation. Using indirect evidence of TAF in ant dipping at Seringbara, Guinea, Koops, Schöning, et al. (2015) came to the same conclusion for chimpanzees in West Africa (see also Koops, McGrew, and Matsuzawa 2013)—namely, that the opportunity hypothesis best explained this behavior at this site. Therefore, many factors potentially contribute to whether primates will use tools in their natural habitat, including (but not limited to) genetics, development, ecology, learning in a social group, and cultural traditions. In a field experiment, researchers at Sonso, Budongo, in Uganda examined wild chimpanzees' (*P. t. schweinfurthii*) interest in a novel foraging problem (concealed honey) within the context of energy expenditure (previous distance traveled) and previous feeding time. They concluded that necessity influenced the chimpanzees' likelihood of engaging in an opportunity (using tools to access honey) and suggest that the necessity and opportunity hypotheses have a scaffolding effect in explaining TAF in chimpanzees (Grund et al. 2019). However, the contribution of macronutrient and micronutrient contents of foods acquired with tools to explanations for this behavior is relatively overlooked but provides another level of explanation.

WHY DOES TAF OCCUR IN SOME PRIMATES BUT NOT OTHERS?

Tool use allows an individual to solve a task that cannot be accomplished with one's own body (Goodall 1964; Shumaker, Walkup, and Beck 2011) or allows an individual to more efficiently exploit resources. Some evidence suggests that the most prolific primate tool users also have relatively large encephalization quotients and neocortex sizes (i.e., *Homo*, *Pan*, and *Cebus/Sapajus*; Panger 2007). However, discussing brain size (or neocortex size) can be problematic in terms of varying measures and data acquired (Reader et al. 2011). Regardless, primate tool use assumes some level of cognitive capacity, since it is usually the result of a combination of learning factors, individual or social, rather than hardwired (Biro, Haslam, and Rutz 2013). TAF relies on motor skills and sensory perceptions, which have been used to define cognitive development in primates, termed "sensorimotor intelligence" (S. Parker and Gibson 1977). Such sensorimotor intelligence, including extractive foraging and feeding adaptions, is described as "intelligent" tool use. Van Schaik, Deaner, and Merrill (1999) demonstrate that the use of feeding tools evolved independently in capuchins, cercopithecines, and the great apes. Complex object manipulation (manipulation of a detached object relative to another object, involving a change of state for one of the objects—e.g., hitting one object with another object) is a major component of tertiary circular sensorimotor intelligence as well (Piaget 1952). Again, capuchins, cercopithecines, and the great apes reveal more varied and complex object manipulation than all other primates (Torigoe 1985). Reader and colleagues (2011) conclude that intelligence includes many domains within society, environment, and technological abilities (see Seyfarth and Cheney [2015] for a review of social cognition). High levels of general intelligence correlate with the most prolific tool use (Panger 2007; Reader et al. 2011). Reader et al. (2011) suggest that convergent evolution in high levels of general intelligence emerged in *Cebus*, *Papio*, *Macaca*, and the great apes, and larger brains facilitated greater behavioral flexibility rather than just innovation, social learning, and extractive foraging.

If closely related taxa exhibit similar cognitive and sensory motor abilities, then how is significant variation in TAF explained? The opportunity hypothesis could, at least in part, explain differences between groups, subspecies, and species regarding TAF. However, genetic differences may also factor into the presence and frequency of TAF in some primates. All primates have various degrees of enhanced visual and grasping abilities, including some opposability of digits, in contrast to taxa such as cetaceans (Mann and Patterson 2013). Troscianko et al. (2012) cite the lack of digit opposability in most nonhumans as partial explanation for why relatively few animal species exhibit tool use (see also Fragaszy 1998). Regardless of similar traits related to grasping and vision, however, biological differences among primate species may still contribute to the frequency and diversity of TAF in closely related taxa. In contrast to chimpanzees, bonobos (*Pan paniscus*) have not been observed to use TAF in the wild save for using leaf sponges to extract water from tree holes at Lomako (Furuichi, Sanz, et al. 2015; Hohmann and Fruth 2003; Ingmanson 1996; Koops, Furuichi, and Hashimoto 2015), even though bonobos have the capability of using tools, both physically and cognitively (Boose et al. 2013; Gruber, Clay, and Zuberbühler 2010). In a comparison of chimpanzees (*P. t. schweinfurthii*) at Kalinzu, Uganda, and bonobos at Wamba, Democratic Republic of the Congo, Koops, Furuichi, and Hashimoto (2015) concluded that extrinsic factors (ecology and sociality) did not successfully explain species differences between these apes. Bonobos have the ecological means to use tools to extract termites and ants at Wamba, as well as the social requirements to learn tool-related behaviors, yet they do not exhibit TAF (Koops, Furuichi, and Hashimoto 2015). Koops, Furuichi, and Hashimoto (2015) found that chimpanzees at Kalinzu had higher rates of object manipulation and play than bonobos, indicating that intrinsic factors (i.e., predisposition) explain species differences in tool use. Similar results were found in corvid species (Kenward et al. 2011). Finally, recent analyses have suggested that genetic factors play a significant role in tool use among chimpanzees (W. Hopkins et al. 2015).

Tools may give users an advantage over interspecific competitors, particularly sympatric nonhuman primates, or enable prolific tool-using primates to coexist with these competitors by way of niche partitioning, or both. Using the premise that certain foods are exceedingly difficult to access and that some primates have overcome this problem by incorporating tools into their foraging repertoires, it follows that closely related species that do not use tools are mostly unable to exploit such foods. Furthermore, it is expected that prolific tool users are mainly allopatric, because sympatry may lead to increasing niche overlap unless subsistence tools were used by each species to acquire different foods. In this view, tool use could be an adaptation that enables species to coexist through reducing niche overlap. In support of this idea, there is very little overlap in species ranges among the most prolific tool-using primates, the robust (*Sapajus* spp.) and gracile (*Cebus* spp.) capuchins, chimpanzees, orangutans, and long-tailed macaques (table 16.3). There are two important exceptions, however. First, the orangutan range is nested within the geographical distribution of long-tailed macaques in Borneo and Sumatra. Although both of these primates are frugivores, most information on TAF in *M. fascicularis* comes from Thailand and involves stone tools to acquire marine invertebrates (Gumert and Malaivijitnond 2012). This region falls outside of the *Pongo* geographical range, and, furthermore, wild orangutans mainly apply their TAF skills to fruits and social insects in the forest canopy (E. Fox, Sitompul, and van Schaik 1999), not intertidal environments (table 16.1). Second, while gracile and robust capuchins are mostly allopatric, some overlap occurs in Amazonia. In their review of capuchin species ranges, Fragaszy, Visalberghi, and Fedigan (2004) found strong support for niche partitioning between *Cebus* and *Sapajus* in areas of sympatry, particularly concerning habitat use (Defler 1985; Janson 1986a; Mittermeier and van Roosmalen 1981). While more research on how foraging strategies contribute to niche partitioning is needed, the geographical distribution of these primates largely supports the notion that tool use plays a role in constructing primate communities.

In addition, it is expected that when tool-using primates are sympatric with competitor species, particularly those with similar phylogenetic constraints (e.g., daily nutritional requirements and gut physiology), diets might significantly diverge where tool use is concerned. In other words, the tool user would consume foods that the non–tool user ignores or rarely consumes. This dietary divergence is apparent with the insect prey of chimpanzees (*P. t. troglodytes*) and gorillas (*G. g. gorilla*) at La Belgique in Cameroon, where Deblauwe and Janssens (2008) found little overlap in the species of ants and termites targeted by each species. Moreover, while each African ape targeted termites, termite fishing for *Macrotermes* was pronounced in chimpanzees, while gorillas used brute force to break or push over *Cubitermes* or *Thoracotermes* mounds (Deblauwe and Janssens 2008). Similarly, gorillas (*G. g. gorilla*) in the Central African Republic broke open termite mounds and mouthed termites off these fragments or further pounded these fragments on open hands and subsequently consumed the insects (Cipolletta et al. 2007). Additionally, gorillas (*G. g. beringei*) typically pull open army ant nests instead of using tools (Kinani and Zimmerman 2015). In sympatric populations of lowland gorillas and chimpanzees, where the same resources are available to both species and they consume similar foods, TAF has been found for only the latter (Tutin and Fernandez 1992). However, in a more recent account of wild gorilla tool use, an immature mountain gorilla (*G. g. beringei*) female used a tool to dip for ants after observing an adult male forage for these insects using only his hands (Kinani and Zimmerman 2015).

How often TAF is used depends to some extent on the species, but differences also exist depending on the community or population being studied (table 16.3). In *Pan troglodytes*, some communities use TAF frequently, while others do not. Luncz and Boesch (2015) quantified tool use in three chimpanzee communities in Taï Forest, Côte d'Ivoire, over the course of 22, 12, and 2 years, respectively. They reported daily tool use in one group, twice-weekly tool use in the second, and weekly tool use in the third community studied (Luncz and Boesch 2015). In cases where absence is more remarkable than presence, such as in the Budongo, Kanyawara, and Ngogo chimpanzee communities in Uganda, the explanation is unclear (McGrew 2013). However, given the preponderance of cases suggesting that various modes of TAF can be explained by the opportunity hypothesis (Koops, Visalberghi, and van Schaik 2014; Sanz and Morgan 2013) and that differing local traditions can also explain variation in TAF (Luncz and Boesch 2015), similar explanations may be used for chimpanzee communities where such behavior is notably absent. The same may be said for bearded capuchins, where significant variation in TAF between sites exists (Ottoni and Izar 2008). Finally, Gruber (2013) stresses the importance of considering temporal changes in ecology, especially those brought about by humans.

ORIGINS OF TAF

Some of the first steps in tool evolution were likely the breaking of hard fruits on rocks or trees or digging in the ground (Ambrose 2001; Hernandez-Aguilar et al. 2007; Marchant and McGrew 2005), similar to what we found in our overview of p-TAF. Once an individual recognizes that a resource is concealed, the thought process must then be how to get that resource out. Trial and error may be used. Primates use a variety of ways to get at unattainable resources, from the base of a tree used as an anvil to modified tools that are transported (see above and tables 16.1 and 16.2). Modifying materials to be used as tools is considered to be the more derived condition, but this does not mean it is always the best solution for a species or even an individual within a social group. For example, gorillas can use their strength to get into insect nests (Cipolletta et al. 2007) or, in the case of the only TAF reported for wild gorillas, are not deterred by biting army ants and do not need to modify a tool to get their desired resource. An immature female (*G. g. beringei*), after observing an adult male use his arm to mop ants and after trying this method herself, used a stick tool to feed on these aggressively biting ants (Kinani and Zimmerman 2015). Using primate models of TAF can help inform hypotheses of what may have been the impetus for TAF in early hominins.

It has been suggested that tool making became more complex in the genus *Homo* with the reduction of canine size and dimorphism and an increase in social tolerance (Plavcan and van Schaik 1997; van Schaik, Deaner, and Merrill 1999). These factors contribute to the higher level of intelligence associated with increased social gregariousness and tolerance, particularly in hunting and food sharing (Stanford 1996), linked with later hominins. Although it has been shown that many animals use tools (Shumaker, Walkup, and Beck 2011), more recent cases of tool use, such as reports from bears and crocodiles, demonstrate how much is unknown regarding animal tool behavior. A yearlong study of bearded capuchins at Fazenda Boa Vista in Brazil revealed that individuals of two social groups exhibited stone tool use at a rate of 0.92 episodes per hour per group (Spagnoletti, Visalberghi, Ottoni, et al. 2011). Moura and Lee (2004) found digging, cracking, or probing TAF more than five times per day on average in bearded capuchins at the Serra da Capivara, Brazil, site, with overall TAF occurring at a rate of 0.43 instances per hour. At Fongoli, Senegal, daily records of tool use on 433 observational days over the course of 3.5 years revealed that these chimpanzees used TAF (tool-assisted hunting, termite fishing, ant dipping, and water dipping) or p-TAF (baobab fruit cracking on anvils) on 66.5% of days overall (Pruetz et al. 2020). At Taï, Côte d'Ivoire, three neighboring chimpanzee communities used tools daily, weekly, and every other week, respectively (Luncz and Boesch 2015). As Biro, Haslam, and Rutz (2013) note, few studies have attempted to assess the effects of TAF on fitness, except for bottlenose dolphins (Mann et al. 2008). To fully explore TAF as an adaptation, studies of energetic costs and benefits, as have been conducted with bird species (Rutz, Bluff, et al. 2010; Tebbich et al. 2002), are a step toward this goal. Taking into account details of nutritional benefits associated with TAF that are now available make this possible.

SUMMARY AND CONCLUSIONS

This review of TAF in nonhuman primates confirms the mammalian pattern that relatively few species exhibit TAF.

Certain species exhibit frequent or diverse TAF:

- Chimpanzees at all long-term study sites engage in some type of TAF, and usually in tool manufacture as well, for a wide variety of foods.
- Capuchins, especially some *Sapajus* species and groups,

show extensive TAF in terms of frequency and diversity, mostly for fruits.
- Orangutans show diverse and frequent TAF but with a bias toward the more social Sumatran species.
- Thai long-tailed macaques show extensive reliance on hammer and anvil use to obtain coastal marine foods but exhibit less diversity in tool use modes than chimpanzees, orangutans, and bearded capuchin monkeys.

Among the primate species that use TAF frequently and in diverse ways, several different modes of tool use stand out. These modes cross taxonomic lines, occurring in apes and in capuchin monkeys (table 16.4):

- Dipping or sponging for liquid of some sort, opening insect hives or nests, and opening hard fruits or nuts are the most common forms of TAF or p-TAF across species.
- Honey dip is the most common type of TAF, occurring at 11 sites, including at least three species and five subspecies. The high quality of this food for primates supports adopting a nutritional geometry approach to studies of primates in helping us better understand TAF and p-TAF in the context of specific study groups.
- Anvil use (a form of p-TAF), using tools to sponge or dip for liquid, and using tools to open hives or nests are the most common types of TAF/p-TAF after honey dip, each occurring at nine different sites.

This chapter reviewed hypotheses to explain TAF and addressed associated questions, such as why TAF is used by some species and not others:

- Ecological factors explain many of the differences in TAF within species and even within populations, especially for apes, but thus far the necessity hypothesis has less support than other explanatory hypotheses, such as the opportunity hypothesis.
- In addition to ecological opportunities (including resource density and terrestriality), social opportunities (including social tolerance) and individual and social learning abilities influence TAF in primates.

17 Hunting by Primates

David Watts

Nearly all primates eat food from two or more trophic levels, even if their diets comprise mostly plant-source foods (PSFs; Milton 2003a, 2003b; cf. R. A. Harding 1981). Animal-source foods (ASFs) are the most common non-plant diet components (Milton 2003a, 2003b). Most are invertebrates, especially arthropods (McGrew 2014), but many primates eat vertebrates (R. A. Harding 1981; Butynski 1982). In contrast to PSFs, ASFs contain all the amino acids needed for protein synthesis. Moreover, animal protein is more bioavailable than plant protein (K. Carpenter 1994), so primates could meet protein requirements with smaller quantities of ASFs than PSFs (Milton 2003a). ASFs also are better sources of some micronutrients than are most PSFs (see below).

Small-bodied primates generally meet more of their protein needs from ASFs than do larger species (Kay 1984). Body size variation accounts for only a limited amount of interspecific variation in faunivory, however. At one extreme, tarsiers (*Tarsius* spp. and *Carlito* spp.), with female body masses of less than 150 g, are exclusively faunivorous (Gursky 2007; Niemitz 1984). In contrast, sportive lemurs (*Lepilemur* spp.), with body masses less than 1 kg, consume only PSFs (Dröscher et al. 2016). ASFs also contribute little to the diets of most colobines, in contrast to similar-sized cercopithecines; colobines are foregut fermenters that obtain most protein from leaves and also obtain amino acids from gut microbes (Sterck 2012). Limits on prey-harvesting rates constrain the ability of relatively large-bodied primates to rely on invertebrates (Kay 1984), although chimpanzees (McGrew 1992, 2014; O'Malley and Power 2014) and some Sumatran orangutans (*Pongo abelii*; E. Fox, van Schaik, et al. 2004) partly overcome such constraints by using tools (Pruetz et al., this volume). Still, chimpanzees acquire nutrients much more efficiently from vertebrate prey than from insects (Tennie, O'Malley, and Gilby 2014). Body size and other aspects of morphology likewise constrain the ability to prey on vertebrates, in part because nonhuman primates rarely use tools in the context of hunting and meat eating. The only known example of regular tool-assisted predation on vertebrates by nonhuman primates is the use of sticks by chimpanzees at Fongoli to extract galagos (*Galago senegalensis*) from tree holes (Pruetz and Bertolani 2007; Pruetz et al. 2015; Pruetz et al., this volume), although a few other cases are known for chimpanzees (D. Watts 2020; Takenaka et al. 2022), and Russon, Compost, et al. (2014) described a few cases of tool-assisted fishing by rehabilitant orangutans. The near absence of tool use helps explain why all nonhuman primates other than tarsiers that prey on vertebrates only take prey smaller than themselves (Butynski 1982).

D. Watts (2020) recently thoroughly reviewed meat eating, defined as the ingestion of vertebrate tissue (muscle, viscera, brains, etc.), by nonhuman primates. In this chapter, I summarize and update that review's findings, expand on its themes, and briefly consider the importance of hunting and meat eating in human evolution. Active pursuit and capture of invertebrates (e.g., capture of crabs by chimpanzees [*Pan troglodytes*] in the Nimba Mountains, Guinea; Koops, Richard, et al. 2019) could be construed as hunting, and the dietary importance of invertebrates deserves more attention (McGrew 2014). However, I define hunting operationally as the capture of vertebrate prey, even though not all meat eating involves active prey pursuit. For example, predation on the eggs of multiple bird species by white-faced capuchins (*Cebus capucinus*; Riehl and Jara 2009; Rose et al. 2003) qualifies as "prey pursuit" in terms of optimal foraging theory, although the prey does not flee. I restrict discussion mostly to the capture of terrestrial vertebrates but briefly summarize data on fish eating and note that the use and nutritional importance of both invertebrate and vertebrate aquatic resources warrant more attention (Gumert and Malaivijitnond 2012; Russon, Compost, et al. 2014), especially given the putative importance of such resources in the evolution of *Homo sapiens* (Marean 2011, 2014, 2015). I address the following major questions: (1) How widespread is hunting and meat eating among nonhuman primates? (2) What is the nutritional importance of meat? (3) Do nonhuman primates hunt cooperatively? (4) When did hunting assume major importance in hominin behavioral ecology, and how did it influence human social and life-history evolution?

Predation on vertebrates by nonhuman primates is most common and best described in chimpanzees, baboons (*Papio* spp.), and capuchins (*Cebus* spp. and *Sapajus* spp.). The literature on hunting and meat eating in chimpanzees is especially extensive, and chimpanzees stand out regarding the dietary contribution of meat despite considerable variation in hunting frequency and meat consumption among populations. For these reasons, I give particular attention to these taxa.

HOW WIDESPREAD IS HUNTING AMONG NONHUMAN PRIMATES?

Overall Incidence of Vertebrate Predation by Primates

D. Watts (2020) plus one omitted paper from that review (Rodel et al. 2002) and subsequent publications (Filho et al. 2020; Garbino et al. 2020; H. Klein et al. 2021; Lile et al. 2020; Lüffe et al. 2018; McLester 2022) document vertebrate predation by 96 nonhuman primate species (and by *Callithrix jacchus* × *C. pencillata* hybrids; Begotti and Landesmann 2008) belonging to 11 families and 38 genera (table 17.1). This undoubtedly underestimates the overall prevalence of meat eating, given that few or no relevant data are available for many primate species.

Taxonomic Distribution of Prey

Prey taxa and size cover a wide spectrum that includes teleost fish; anuran eggs and adults; lizards, snakes, and tortoises; bird eggs, nestlings, fledglings, and adults; and immatures and adults of a wide range of small and medium-sized mammals. Russon, Compost, et al. (2014) reviewed reports of fish eating by wild nonhuman primates and concluded that firm evidence exists for Bornean tarsiers (*Tarsius bancanus*), Philippine tarsiers (*Carlito syrichta*), black-capped capuchins (*Sapajus apella*), chacma and olive baboons (*Papio ursinus* and *P. anubis*), Japanese and long-tailed macaques (*Macaca fuscata* and *M. fascicularis*), and Allen's swamp monkey (*Allenopithecus nigriviridus*). Chimpanzees (Sugiyama and Koman 1987) and De Brazza's monkey (*Cercopithecus neglectus*; Zeeve 1993) should be added to this list (table 17.1; see D. Watts 2020). Prey include eight genera and species belonging to five families (Russon, Compost, et al. 2014; Takahata, Hayashi et al., 2022; D. Watts 2020).

Forty-two primate species for which data are available prey on amphibians, and 49 prey on reptiles (table 17.1). Species from 12 anuran genera and 18 genera of Squamata and Testudines are included in food lists, but many reports simply state that study taxa ate "frogs, unknown species" or "lizards, unknown species," which prevents accurate estimation of the numbers of prey species. Predation on frogs and lizards, which comprise most prey taxa (D. Watts 2020), is undoubtedly far more widespread, especially given that both are commonly included in the diets of those few members of several species-rich genera for which data are available (notably *Saguinus* and *Callithrix*; table 17.1; D. Watts 2020).

The diets of 56 of the primate species listed in table 17.1 include bird eggs, nestlings, fledglings, or adults, making birds the most common prey category. Prey lists include 94 avian species belonging to 85 genera. Again, though, many reports do not give information on prey taxa (D. Watts 2020), so the actual numbers are certainly higher.

Table 17.1. Consumption of vertebrate prey by nonhuman primates.

Predator	Prey class				
	Teleostei	Amphibia	Reptilia	Aves	Mammalia
Strepsirrhini Lorisiformes Lorisidae					
Euoticus elegantulus				X	
Galago alleni		X			
Galago moholi				X	
Loris lydekkerianus			X		
Loris tardigradus			X		
Nycticebus coucang				X	
Otolemur crassicaudatus			X	X	Tooler
Perodicticus potto				X	CH
Lemuriformes Lemuridae					
Cheirogaleus medius			X		
Eulemur fulvus				X	PR
Microcebus berthae			X		
Microcebus murinus		X	X		
Mirza coquereli		X	X	X	PR,R
Haplorrhini Tarsiiformes Tarsiidae					
Carlito syrichta	X				
Tarsius bancanus	X		X	X	CH
Anthropoidea Platyrrhini Atelidae					
Alouatta caraya				X	
Alouatta palliata				X	
Alouatta seniculus			X		
Lagothrix flavicauda			X		

(continued)

Table 17.1. (continued)

Predator	Prey class				
	Teleostei	Amphibia	Reptilia	Aves	Mammalia
Lagothrix lagotricha				X	R
Callitrichidae					
Callimico goeldii		X	X	X	
Callithrix aurita				X	
Callithrix flaviceps		X	X	X	
Callithrix geoffroyi		X	X		
Callithrix humeralifer		X	X	X	
Callithrix jacchus		X	X	X	D, R
Callithrix jacchus-pencillata					X
Callithrix kuhli		X	X	X	
Callithrix penicillata		X	X	X	D, R
Cebuella pygmaea		X	X		
Leontopithecus chrysomelas			X	X	X
Leontopithecus chrysopygus			X	X	
Leontopithecus nigrifrons		X			
Leontopithecus rosalia		X	X	X	
Saguinus fuscicollis		X	X	X	
Saguinus imperator		X	X		
Saguinus mystax		X	X		
Saguinus oedipus		X	X		
Cebidae					
Cebus albifrons		X	X	X	DR
Cebus capucinus		X	X	X	CA, CH, R
Cebus cay				X	R
Cebus olivaceous		X	X	X	R
Saimiri boliviensis		X	X	X	

	Prey class				
Predator	Teleostei	Amphibia	Reptilia	Aves	Mammalia
Saimiri cassaqueriensis[a]		? X	? X	? X	CH
Saimiri oerstedi		X			
Sapajus apella		X	X	X	CH, D, PR, R
Sapajus flavius			X		
Sapajus libidinosus			X		R
Sapajus nigritus				X	
Sapajus xanthosternos					PR
Pitheciidae					
Cacajao calvus			X		
Cacajao melanocephala		X			
Cacajao ouakary			X		
Callicebus coimbrai				X	
Cercopithecoidea					
Cercopithecidae					
Cercopithecinae					
Allenopithecus nigroviridis	X				
Cercocebus agilis		X			A
Cercocebus atys		X			R
Cercocebus chrysogaster					A
Cercocebus galeritus		X			
Cercocebus sanjei			X		
Cercopithecus ascanius			X		
Cercopithecus campbelli		X			
Cercopithecus cephus			X		

(continued)

Table 17.1. (*continued*)

Predator	Prey class				
	Teleostei	Amphibia	Reptilia	Aves	Mammalia
Cercopithecus petaurista		X			
Cercopithecus diana		X			
Cercopithecus l'hoesti				X	
Cercopithecus mitis			X	X	PR, R
Cercopithecus neglectus	X				
Chlorocebus pygerythrus				X	
Chlorocebus sabaeus			X	X	L, R
Erthyrocebus patas			X		
Macaca assamensis		X	X	X	X
Macaca fascicularis	X	X			
Macaca fuscata	X	X	X	X	
Macaca leonina				X	R
Macaca nemestrina[b]		?	?	?	R
Macaca nigra		X	X	X	CH
Macaca radiata			X		
Macaca silenus					R
Macaca sinica		X	X		X
Mandrillus leucophaeus		X			R
Mandrillus sphinx		X	X		A, E, R
Miopithecus talapoensis				X	
Papio anubis	X	X		X	A, CH, L, R
Papio cynocephalus		X	X	X	A, L, PR
Papio hamadryas			X	X	A, H, L
Papio papio				X	A, L
Papio ursinus	X		X	X	A, L, PR, R

	Prey class				
Predator	Teleostei	Amphibia	Reptilia	Aves	Mammalia
Colobinae					
Rhinopithecus bieti				X	R
Hominoidea					
Hylobatidae					
Hoolock leuconedys				X	
Hylobates lar				X	
Nomascus concolor				X	
Pongidae					
Pongo abelii					PR
Pongo pygmaeus	X				R
Hominidae					
Pan paniscus				X	A, CH, E, PR, R
Pan troglodytes	X	X	X	X	A, CA, M, PH, PR, R

For fish, amphibians, reptiles, and birds, X indicates that the row species preys on ≥1 species of column vertebrate class. Letters other than X under Mammalia indicate orders: A, Artiodactyla; CA, Carnivora; CH, Chiroptera; D, Didelphimorphia; E, Eulipotyphla; H, Hyracoidea; L, Lagomorpha; M, Macroscelidea; PH, Pholidota; PR, Primates; R, Rodentia; X, prey taxa not identified. Predation on Amphibia includes consumption of eggs and adults; predation on birds includes eggs, nestlings, fledglings, and adults. For complete information concerning families, genera, and species on which each primate species preyed, see D. Watts (2020) and also Rodel (2002), Lüffe et al. (2018), Filho et al. (2020), Garbino et al. (2020), H. Klein et al. (2021), Lile et al. (2020), and McLester (2022).

[a] *Saimiri cassaqueriensis* hunts unspecified "small vertebrates" (Janson and Boinski 1992).
[b] *Macaca nemestrina* also preys on unspecified "small vertebrates" (Caldecott 1986).

Mammalian prey contribute to the diets of 43 of the primate species listed in table 17.1. Prey includes rodents, artiodactyls, chiropterans, lagomorphs, hyracoidea, didelphids, pholidota, eulipotyphla, macroscelidea, carnivores, and primates, although most prey taxa are other primates (17 genera, 37 species), artiodactyls (17 genera, 27 species), and rodents (19 genera, 22 species). Hunting of primates by primates includes predation by dwarf lemurs (*Mirza coquereli*) on mouse lemurs (*Microcebus murinus*); by yellow and anubis baboons (*Papio cynocephalus* and *P. ursinus*) on vervet monkeys (*Chlorocebus aethiops* and *C. pygerethrus*) and galagos (*Galago senegalensis*); by capuchins on several haplorrhine species (below); by orangutans (*Pongo pygmaeus*) on slow lorises (*Nycticebus coucang*); by bonobos (*Pan paniscus*) on several cercopithecoid species; and, most notably, by chimpanzees on many cercopithecoids, especially red colobus (*Piliocolobus* spp. and *Procolobus* spp.), and on galagos (*Galago senegalensis*, *G. galago*, and *Galagoides demidovii*).

Prey Choice and Hunting Frequency

Most (67/97) taxa listed in table 17.1 hunt prey from only one or two vertebrate classes. Twelve take prey from all four classes of terrestrial vertebrates; these include one lemur (*Mirza coquereli*), two callitrichids (*Callithrix jacchus* and *C. pencillata*), five cebids (*Cebus albifrons*, *C. capucinus*, *C. olivaceous*, *Saimiri oerstedi*, and *Sapajus apella*), three cercopithecines (*Macaca assamensis*, *M. nigra* and *Papio cynocephalus*), and chimpanzees (although only one case of chimpanzee predation on fish [Sugiyama and Koman 1987] and three on frogs [Hosaka et al. 2020] are known). Four species (*Tarsier bancanus*, *Macaca fuscata*, *Papio anubis*, and *P. ursinus*) prey on three such classes and occasionally eat fish. Another 12 (*Otolemur crassicaudatus*, *Callimico goeldii*, *Callithrix flaviceps*, *C. humeralifer*, *C. kuhli*, *Leontopithecus chrysomelas*, *L. rosalia*, *Saguinus fuscicollis*, *Saimiri boliviensis*, *Macaca sinica*, *Mandrillus sphinx*, and *Papio hamadryas*) prey on three classes of terrestrial vertebrates. The high representation of callitrichids and capuchins in these lists indicates that other, less well-studied species in these taxa may also prey on more than two vertebrate classes. Most small-bodied primates (including callitrichids) prey only on anurans, lizards, or birds (table 17.1), but a few eat small mammals (e.g., *Saimiri oerstedi* prey on tent-making bats [*Artibeus watsoni*]; Boinski and Timm 1985). All tarsier species for which data are available apparently are exclusively faunivorous, but not all hunt by the definition used here. Notably, *Tarsius bancanus* hunts a variety of small vertebrates (Gursky 2007; Niemitz 1984; table 17.1), while *T. tarsier* apparently prey only on invertebrates (Gursky 2007). Captive but wild-caught *Carlito syrichta* readily eat fish (N. Cook 1939), and both Kempf (2009) and Russon, Compost, et al. (2014) list them as among those few primates that eat fish in the wild.

Hunting is uncommon or rare in most taxa listed in table 17.1, although few quantitative data on hunting and prey capture rates and on meat intake are available for species other than chimpanzees and, to some extent, capuchins and baboons (below; reviewed in D. Watts 2020). For example, Lüffe et al. (2018) reported that mean rates at which *Saguinus mystax* at Estación Biológica Quebrada Blanco (Peru) captured frogs ranged from 0 to 0.32/individual/month. Meat eating accounted for less than 1% of feeding time for *Macaca fuscata* on Yakushima Island (Hanya 2004) and for *Saguinus fuscicollis* at San Sebastian, Bolivia (L. Porter 2001), although lizards accounted for as much as 9% of monthly feeding time for *Callimico goeldii* at the same site (ibid.).

Chimpanzees hunt more often and eat more meat than all other nonhuman primates. The known chimpanzee prey repertoire (reviewed in D. Watts 2020; see also H. Klein et al. 2021) includes at least 71 species: 1 tortoise, at least 1 frog, at least 18 birds (birds are often recorded simply as "unidentified species"), 7 artiodactyls, 2 suids, 3 viverrid carnivores, 1 macroscelid, 1 Pholidota, 28 primates, and 9 rodents. Prey choice and hunting frequency vary considerably among populations, among communities in the same population, and within communities over time (table 17.2; reviewed in D. Watts 2020; see also H. Klein et al. 2021), although comparison of hunting rates is not straightforward, because some reports give only absolute numbers of hunts without providing observation time. Also, some published data pertain only to hunts of red colobus monkeys (table 17.2), the main prey of chimpanzees at nearly all sites where the two are sympatric (table 17.3). Moreover, researchers are likely to miss some hunts because the high fission-fusion dynamics of chimpanzee communities can make it impossible to monitor all community members on any given day. Rates can be quite high, however, with all but one com-

Table 17.2. Hunting rates and predation rates for chimpanzees.

Site	Hunt rate	Capture rate	Reference
Sonso[1]	0.99/month	0.89/month	Hobaiter, Samuni, et al. (2017)
Kanyawara[2,3]	0.45/100 h		Gilby and Wrangham (2007)
Kanyawara[4]	0.3/100 h (1.1/month)	0.8/month	Gilby, Machanda, Mjungu, et al. (2015)
Kanyawara[2]	2.75/month	4.75/month	Gilby, Machanda, O'Malley, et al. (2017)
Kahuzi-Biega	2.5/month	≥ 2.5/month	Basabose and Yamagiwa (1997)
Fongoli[5]		c. 2.5/month	Pruetz et al. (2015)
Taï[6]	10/month	2.79–6.21/month	Boesch and Boesch (1989)
Taï East Group[6]	4.7/100 h	2.50/100 h	Samuni et al. (2018)
Taï South Group[6]	1.6/100 h	0.98/100 h	Samuni et al. (2018)
Kasekela[7]		1.41/month	Wrangham and Bergmann-Riss (1990)
Kahama[8]		0.79/month	Wrangham and Bergmann-Riss (1990)
Kasekela[9]	8/month	6.48/month	Stanford, Wallis, Matema, et al. (1994)
Kasekela[10]	2.5/month	4.75/month	Gilby, Machanda, O'Malley, et al. (2017)
Loango	0.83/100 h (2.70/month)	0.6/100 h (2.0/month)	H. Klein et al. (2021)
Mahale M Group[11]	1.56/month		Takahata, Hasegawa, and Nishida (1984)
Mahale M Group[12]		5/month	S. Uehara et al. (1992)
Mahale M Group[13,14]	3.88/month	6.67/month	Hosaka et al. (2001)
Mahale M Group[15]	1.91/month	1.44/month	Hosaka et al. (2020)
Mitumba[15]	1.11/100 h (4.0/month)	2.8/month	Gilby, Machanda, Mjungu, et al. (2015)
Ngogo[16]	4.9/month (0–21)	7.6/month (0–48)	Gilby, Machanda, Mjungu, et al. (2015)
	1.8/100 h		unpublished data

Data from Sonso, Kahuzi-Biega, Fongoli, Mahale, and Ngogo include all prey species; data from other sites as noted.
Values in parentheses are ranges among months.

[1] 1999–2017; Hobaiter, Samuni, et al. (2017) stress that these are rough estimates and actual hunting rates were almost certainly higher, given that unsuccessful hunts often went unrecorded.
[2] Kanyawara data are for red colobus only.
[3] Minimum estimate based on length of study.
[4] 1995–2014; calculated from data in Gilby, Machanda, Mjungu, et al. (2015).
[5] Estimated from data in Pruetz et al. (2015).
[6] Data from Boesch and Boesch (1989) cover 22 months in 1984–86; hunting rates are corrected for observation time, while the kill rate includes only documented kills (a similar correction would double this value). Data in Samuni et al. (2018) cover 18 months in 2013–15.
[7] 1973–75.
[8] Data on males only.
[9] Red colobus only; 1989.
[10] Red colobus only; 1976–2013.
[11] 1982–92.
[12] 1979–82.
[13] 1981–90.
[14] Corrected for observation days.
[15] 1965–2010.
[16] 2000–2014; calculated from data in Gilby, Machanda, Mjungu, et al. (2015).
[16] Values in parentheses are ranges among months.

munity for which relevant data are available (Sonso; Hobaiter, Samuni, et al. 2017) showing an estimated rate of more than one hunt per month over the long term (table 17.2). For example, during a 22-month period in 1984–86, chimpanzees in the Taï National Park, Ivory Coast, hunted red colobus (*Piliocolobus badius*) about 10 times per month; less often, they also hunted other primate and nonprimate species (Boesch and Boesch 1989; Boesch and Boesch-Achermann 2000). Kasekela data collected between 1976 and 2013 (Gilby, Machanda, O'Malley, et al. 2017) included 2,167 prey captures during 1,254 successful and unsuccessful hunts; the authors did not report total observation time, but this gives minimum values of 2.75 hunts per month, or about 1 per 11 days, and 4.75 prey captures per month, or about 1 per 6.3 days. Chimpanzees at Ngogo sometimes go for up to two months without hunting but have hunted as many as 21 times in a single month (table 17.2; D. Watts and Mitani 2015). During a seven-month period in 2002, they captured at least 225 prey, including at least 186 red colobus. Several hunting "binges" occurred during this time, including one in which the chimpanzees hunted 22 times in 21 days. Similar binges have also been reported at other sites (e.g., Kasekela: Stanford, Wallis, Matema, et al. 1994; Mahale: Hosaka et al. 2020).

Red colobus (*Piliocolobus* spp.) account for most hunts and most prey captures at all sites where they are sympatric with chimpanzees (table 17.3) other than Sebitoli (below). Only 13% of hunts by Mahale M Group were of red colobus between 1965 and 1982, but the rate of red colobus hunting subsequently increased greatly, for uncertain reasons (although it varied considerably from year to year), and red colobus accounted for 72.1% of all hunts and 68.1% of prey captures between 1965 and 2010 (table 17.3; Hosaka et al. 2020). In contrast, the rate of red colobus (*P. tephrosceles*) hunting at Ngogo has decreased since the start of long-term research there, because severe predation pressure on red colobus at Ngogo has caused a major decline in the local population (Teelen 2007, 2008; D. Watts and Amsler 2013; D. Watts and Mitani 2015). Chimpanzees hunt red colobus disproportionately in relation to their abundance. Those at Ngogo hunt all seven species of sympatric cercopithecoid primates at least occasionally, but red colobus accounted for 64% of all hunts and 83% of all prey captures over a 20-year period, whereas redtail monkeys (*Cercopithecus ascanius*) and gray-cheeked mangabeys (*Lophocebus albigena*) accounted for only 8.1% and 3.6% of hunts, respectively, and 3.8% and 1.5% of kills, despite being far more abundant (D. Watts and Mitani 2015). Boesch and Boesch (1989; Boesch and Boesch-Achermann 2000) documented predation on seven of the eight other diurnal primates at Taï (only spot-nosed guenons, *Cercopithecus nictitans*, were not included) and on pottos (*Perodicticus potto*), but most prey captures were of red colobus (see also Samuni et al. 2018). Data from Gombe (Stanford 1998; Gilby, Eberle, et al. 2006), Kanyawara (Gilby and Wrangham 2007), and Mahale (Hosaka et al. 2020) confirm selective predation on red colobus (reviewed in Bugir et al. 2021). Two factors that contribute to disproportionate predation on red colobus where they are sympatric with chimpanzees are that (1) at least in Kibale, red colobus take longer to prepare for leaps between tree canopies than do sympatric cercopithecines and guerezas and therefore are slower to flee from chimpanzees (Stern and Goldstone 2005); and (2) red colobus groups often hold their ground rather than fleeing and rely on cooperative defense by adult males to drive off attacking chimpanzees, but chimpanzees can overcome the monkeys' defenses by hunting in groups, especially if these contain many adult males (J. Mitani and Watts 1999). Black-and-white colobus are the second-most common primate prey after red colobus at both Ngogo (*Colobus guereza*; D. Watts and Mitani 2015) and Taï (*C. polykomos*; Samuni et al. 2018), and either those species or other black-and-white colobus (*C. angolensis*) are the most common prey at several sites where red colobus are absent (e.g., Sonso: Hobaiter, Samuni, et al. 2017; Makamba: H. Klein et al. 2021).

Chimpanzees hunt multiple ungulate species (reviewed in D. Watts 2020; see also H. Klein et al. 2021), most commonly duiker. Chimpanzees prey on at least eight bovid species in various parts of their range. These include at least four duiker: red duiker (*Cephalophus natalensis*) at Ngogo (J. Mitani and Watts 1999; D. Watts and Mitani 2002b, 2015) and at Sonso and Waibira in Budongo (Hobaiter, Samuni, et al. 2017); blue duiker (*Philantomba monticola*) at these sites, at Mahale (Hosaka et al. 2020; Nishida and Uehara 1983; S. Uehara 1997; S. Uehara et al. 1992), and at Rekambo in Loango (H. Klein et al. 2021); and bay (*Cephalophus dorsalis*) and yellow-backed duiker (*Cephalophus silvicultor*) at Rekambo (H. Klein et al. 2021). Predation on unknown

Table 17.3. Chimpanzee hunting success rates.

Site	Prey	% Hunts successful	% Hunts of RC	% Kills of RC	Reference
Kanyawara	RC	61.3	n.d.	n.d.	Gilby, Machanda, Mjungu, et al. (2015)
Kasakela	RC	62.3	n.d.	82.2	Gilby, Machanda, Mjungu, et al. (2015); Stanford, Wallis, Matema, et al. (1994)
Rekambo	All	62.3	n.a.	n.a.	Klein et al. (2019)
Mahale	All	69.7	n.d.		Hosaka et al. (2020)
Mahale	RC	64.2	72.1	68.1	Hosaka et al. (2020)
Mitumba	RC	62.3	n.d.	n.d	Gilby, Machanda, Mjungu, et al. (2015)
Ngogo	All	72.1			J. Mitani and Watts (2015)
Ngogo	RC	85.3	64.0	83.0	J. Mitani and Watts (2015)
Sonso	All	42.0[a]	n.a.	n.a.	Hobaiter, Samuni, et al. (2017)
Taï	Allk	57.0	n.d.	n.d.	Boesch and Boesch (1989)
Taï	RC	54.0	81.0	77	Boesch and Boesch (1989)
Taï South	All	61.0		> 72.5	Samuni et al. (2018)
Taï East	All	54.0[b]			Samuni et al. (2018)

RC, red colobus. "Hunts of RC" is the percentage of all known hunts in which red colobus were targets. n.d., data not provided; n.a., red colobus not present. "Kills of RC" is the percentage of all recorded kills in which prey were red colobus. The Taï data are from 1984–86; those from Taï South and Taï East are from 2014–16.

[a] The Sonso value is based on only 19 hunts in 2010–11, when researchers consistently documented unsuccessful hunts during a 22-month interval. In earlier data, 173 of 184 reported hunts (94%) were successful, but this was an overestimate because unsuccessful hunts were underreported (Hobaiter, Samuni, et al. 2017). However, the 2010–11 value could be an underestimate, given the small sample size.

[b] The Taï value from Samuni et al. (2018) is approximate, because the authors did not gather precise capture data for each prey species, but red colobus hunts comprised 72.5% of successful hunts for both communities combined, and the chimpanzees captured only one prey item in 84% of successful hunts.

duiker species has also been documented in the Ugalla region (J. Moore, Black, Hernandez-Aguilar at al. 2017), at Ndoki (Republic of Congo; Kuroda et al. 1996) and at Outamba-Kilimi (Sierra Leone; Alp 1993). Duiker are absent at Gombe, but chimpanzees there hunt infant bushbuck (*Tragelaphus scriptus*) relatively often (Goodall 1986; Stanford 1998). Infant bushbuck are also occasional prey at Ngogo (D. Watts and Mitani 2015), Mahale (Hosaka et al. 2020; Nishida and Uehara 1983), and Fongoli (Bogart and Pruetz 2008; Pruetz et al. 2015), and A. Suzuki (1975) documented predation on bushbuck in the Kasakati Basin in Tanzania.

Chimpanzees are remarkably successful hunters. Those in Taï, Gombe, Kibale, and Mahale capture prey in the majority of hunts (table 17.3). Ngogo stands out: hunting rates there have not been especially high compared to Taï or Gombe, but success rates have been much higher, with over 80% of red colobus hunts and more than half those of black-and-white colobus, redtail monkeys, mangabeys, and blue monkeys leading to kills (table 17.3; J. Mitani and Watts 1999; D. Watts and Mitani 2002a, 2015). Chimpanzees commonly capture multiple individuals per hunt of red colobus. This has been the case at Kasekela, in Gombe, and Kanyawara, Kibale (Gilby, Machanda, Mjungu, et al. 2015); for Mahale M Group (Hosaka et al. 2020); and at Ngogo, where multiple kills are most common, and the long-term mean of kills per successful hunt has exceeded three (D. Watts and Mitani 2015).

Notable contrasts in chimpanzee hunting behavior exist among and within populations, not all of them due to variation in prey availability. Hunting rates seem to be relatively low for chimpanzees in savanna-woodland habitats that have long dry seasons, low primate species diversity, and low chimpanzee population densities, including woodland-savanna-farmland mosaics in West Africa (Senegal: McGrew et al. 1988; Pruetz and Bertolani 2007; Pruetz et al. 2015; Guinea-Bissau: Bessa et al. 2015; Ugalla, Tanzania: J. Moore et al. 2017). On the basis of comparative data from fecal samples, J. Moore et al. (2017) questioned whether "savanna" chimpanzees really hunt less often than those in forest habitats on average, although they noted that the frequencies of prey remains in feces were the lowest for the two savanna sites with by far the largest sample sizes (Ugalla: J. Moore et al. 2017; Fongoli: Pruetz 2006) and were among the lowest reported. Regardless, J. Moore et al. (2017) noted that savanna chimpanzees seem to prey less often on other primates than do those in forest habitats and mostly to take smaller prey. Tool-assisted hunting of galagos at Fongoli (Pruetz et al. 2015), which is relatively common, is a partial exception.

Hunting is also uncommon in montane forest habitats (e.g., Kahuzi-Biega: Basabose and Yamagiwa 1997; Nyungwe: Matthews et al. 2019). However, hunting frequencies can be relatively low in rainforest habitats at lower altitudes, especially if red colobus monkeys are absent (e.g., Sonso; Hobaiter, Samuni, et al. 2017; Newton-Fisher et al. 2002). Chimpanzee demographic variation, particularly variation in the number of adult males per community, explains some of the variation (J. Mitani and Watts 1999).

Intriguing contrasts occur among communities in Kibale. Red colobus are the main prey at Kanyawara and Ngogo, but hunting frequency has been much lower at Kanyawara (especially prior to the red colobus population decline at Ngogo) despite higher red colobus population density there (Gilby, Eberle, and Wrangham 2008; Gilby, Machanda, Mjungu, et al. 2015; Gilby, Machanda, O'Malley, et al. 2017; Gilby and Wrangham 2007; D. Watts and Amsler 2013; D. Watts and Mitani 2002a, 2015). Unusually large average hunting party size at Ngogo, which reduces the risks of hunting and increases the probability of prey captures compared to Kanyawara and other communities with fewer adult males, presumably helps explain this difference (Gilby, Machanda, Mjungu, et al. 2015; J. Mitani and Watts 1999; D. Watts and Mitani 2002a, 2015). Chimpanzees at Ngogo hunt duiker relatively often (D. Watts and Mitani 2015), but only one case of feeding on red duiker (probably scavenging) is known from Kanyawara, and no predation on blue duiker has been seen there (Gilby, Machanda, O'Malley, et al. 2017). Chimpanzees at a third Kibale site (Sebitoli) hunt guerezas (*Colobus guereza*), the second-most common prey at Ngogo and Kanyawara, but are not known to hunt red colobus despite the high local population density of this species (S. Krief, pers. comm.) and the presence of 15 adult males in the chimpanzee community (Krief, Cibol, et al. 2014). Sebitoli is the only study site where chimpanzees and red colobus are sympatric but predation on red colobus is unknown.

Prey choices also seem to differ between two chimpanzee communities in Budongo (Hobaiter, Samuni, et al. 2017). Guerezas accounted for 74.4% of hunts and 75.3% of prey consumption events at Sonso, but only 36.1% and 23.3%, respectively, at Waibira despite similar guereza population densities at both sites and the presence of more adult males in the Waibira community. Red duiker accounted for 30% of prey captures at Waibira, but researchers saw only one red duiker hunt in 18 years at Sonso, although the abundance of this species appears similar for the two sites. However, the Sonso sample included 202 hunts plus meat-eating events following unobserved hunts, whereas that for Waibira was only 36 hunts, and the Waibira community was not fully habituated, which might have deterred the chimpanzees from hunting when humans were present (ibid.). Kibale data show that within-population variation in hunting frequency and prey choice can exist independently of variation in prey densities and in chimpanzee demography. Budongo data point in this direction, but the small Waibira sample prevents claims about socially mediated variation in prey preference, as Hobaiter, Samuni, et al. (2017) acknowledge.

Duiker hunting is unknown at Taï despite the presence of six species there (Boesch and Boesch 1989; Boesch and Boesch-Achermann 2000). Only chimpanzees at Loango are known to hunt tortoises (Pika et al. 2019), although this behavior is also strongly suspected in the Bili region (Hicks et al. 2019). Finally, temporal changes in hunting have been documented at Mahale, where hunting of red colobus was initially rare but has since greatly increased in frequency for reasons that are not entirely clear (Hosaka at al. 2020); at Ngogo, where the frequency of red colobus hunting has decreased and that of several other primate species has increased because overhunting by the chimpanzees has caused a red colobus population decline (D. Watts and Amsler 2013; D. Watts and Mitani 2015); and at Sonso, where hunting of black-and-white colobus seems to have increased over time (although the apparent increase might have been at least partly an artifact of habituation; Hobaiter, Samuni, et al. 2017).

Bonobos (*Pan paniscus*) generally hunt less than chimpanzees. As in chimpanzees, their prey choices vary. Those at Lomako (Badrian and Malenky 1984; Hohmann and Fruth 2002), Lui Kotale (Hohmann and Surbeck 2008; Surbeck et al. 2009), Lilongo (Bermejo, Illera, and Pi 1994), and Iyondji (Sakamaki et al. 2016) hunt duiker, but duiker predation is unknown at Wamba. Predation on bats is known only at Wamba (Hirata et al. 2010; Ihobe 1992) and Lilongo (Bermejo, Illera, and Pi 1994), and predation on squirrels is known only from Iyondji (Sakamaki et al. 2016). Bonobos rarely hunt other primates, but predation on four cercopithecoid taxa and on *Galagoides demidovii* is known at Lui Kotale (Hohmann and Fruth 2008; Surbeck and Hohmann 2008; Surbeck et al. 2009), and bonobos at Isonji hunt redtail monkeys (Sakamaki et al. 2016).

Multiple populations and species of capuchins hunt a variety of vertebrates (reviewed in D. Watts 2020). However, hunting is infrequent except for predation by white-faced capuchins (*Cebus capucinus*) on coati (*Nasua narica*) pups during the coati birth season in Costa Rica (S. Perry 2008; Rose 1997, 2001; Rose et al. 2003) and raids of nests of seasonally breeding birds (e.g., Menezes and Martini 2017; J. G. Robinson 1986). Capuchins obtain vertebrate ASFs mostly by raiding nests made by birds, squirrels, and coatis and seizing bird nestlings and eggs and squirrel and coati pups. White-faced capuchins actively pursue juvenile and adult squirrels (S. Perry 2008; Rose 1997, 2001; Rose et al. 2003). Species of *Cebus* and *Sapajus* occasionally pursue small rodents, frogs, and lizards (S. Perry 2008; Rose 1997, 2001; Rose et al. 2003; Terborgh 1985; table 17.1). Tufted capuchins (*Sapajus* spp.) occasionally hunt other primates, including predation by *Sapajus xanthosternos* on common marmosets (*Callithrix jacchus*; Sousa et al. 2013, cited in Hilário and Ferrari 2015) and by *Sapajus apella* on brown titi monkeys (*Callicebus brunneus*; Lawrence 2003; Sempaio and Ferrari 2005), dusky titi monkeys (*Callicebus molloch*; Lawrence 2003), and owl monkeys (*Aotus brumbacki*; Carretero-Pinzon et al. 2003). Common marmosets in a Brazilian Atlantic coast forest fragment showed typical antipredation vigilance or flight in response to blonde capuchin (*Sapajus flavius*) long calls, to playbacks of those calls, and to approaching capuchin groups (Bastos et al. 2018). Rose (1997) estimated that white-faced capuchins at Santa Rosa captured one bird prey item per group per 4.2 days and one mammalian prey item per 3.1 days; these values, and success rates of 50% (Rose 1997) and 81% (Fedigan 1990) for squirrel hunts, were similar to those of chimpanzees.

Baboons hunt a wide range of prey, and hunting frequencies and prey sets vary among populations, partly due to habitat variation (reviewed in D. Watts 2020). Fawns of Thomson's gazelle (*Gazella thomsonii*) and impala (*Aepyceros melampus*) were among the most common prey at Gilgil (R. A. Harding 1975; Strum 1981) and Amboseli (S. Altmann and Altmann 1970; Hausfater 1976) in Kenya. Neither species occurs at Moremi (Botswana), where Cape hares (*Lepus capensis*) were the most common mammalian prey (Hamilton and Busse 1978, 1982). Rhine et al. (1986) estimated that on average, individual yellow baboons (*P. cynocephalus*) at Mikumi (Tanzania) captured only 4.34 vertebrate items, including birds' eggs and lizards, per year. Meat eating was similarly infrequent at sites in Kenya, Uganda, Botswana, and South Africa (Rhine et al. 1986). Hunting frequencies were higher at Gilgil than at other East African sites, perhaps because of the near absence of, and hence low competition with, large mammalian carnivores (Strum 1981).

HOW MUCH MEAT DO NONHUMAN PRIMATES EAT?

In the first review of data on meat eating by nonhuman primates, Butynski (1982; cf. K. Hill 1981) estimated that meat accounted for at most 1% of food intake, on average, for baboons and chimpanzees and was almost certainly lower for other species, with the possible exception of *Tarsius bancanus*. Subsequent data generally support this conclusion (above; D. Watts 2020). For example, meat eating accounted for only 1% of overall feeding time for white-faced capuchins at Lomas Barbudal (S. Perry

2008), although values were considerably higher during the coati birth season (cf. Rose 2001). This was far less than the 47% of feeding time devoted to foraging for invertebrates (S. Perry 2008), although biomass intake rates from invertebrates can be low (Terborgh 1985) and feeding time data may not accurately reflect nutritional and caloric intake. Rose (1999) estimated that meat eating accounted for only 1%–3% of feeding time for white-faced capuchins at Lomas Barbudal. Likewise, most individual baboons eat little meat (S. Altmann and Altmann 1970; Hamilton and Busse 1978, 1982; R. A. Harding 1975; Hausfater 1976; Rhine et al. 1986; Strum 1981). Based on their review of data from Mikumi and other sites, Rhine et al. (1986) concluded that vertebrate meat is a minor component of baboon diets in terms of energy and macronutrients. However, olive baboons captured about 15 mammal prey per month during R. A. Harding's (1973, 1975) and Strum's (1991) research at Gilgil, a rate that exceeded means for most chimpanzee communities (table 17.2). Notably, large carnivores were rare at Gilgil, so the baboons faced little competition for gazelle and impala fawns with other predators (cf. R. Taylor et al. 2016). However, meat intake can be highly skewed among individuals in baboon groups. For example, five males accounted for 73% of meat consumption during Hausfater's (1976) study of yellow baboons in Amboseli, and males there and at Moremi (Hamilton and Busse 1978), Mikumi (Rhine et al. 1986), and Gilgil (Strum 1981) sometimes gained enough meat by capturing or stealing prey to feed to satiation.

Meat eating also accounts for only small proportions of chimpanzee feeding time. However, considerable variation exists among sites, and some individuals eat substantial amounts of meat (reviewed in D. Watts 2020). For example, quantitative data on meat intake are not available from Mahale, but much higher hunting rates for M Group than for K Group indicate that mean per capita meat availability was lower for K group (Hosaka et al. 2020). Likewise, mean per capita meat intake in Kibale has been considerably lower at Kanyawara and Sebitoli than at Ngogo, where hunting rates have been much higher (above). Low hunting frequencies or smaller average prey size or both presumably lead to relatively low per capita meat intake at Sonso (Hobaiter, Samuni, et al. 2017; Newton-Fisher et al. 2002), in savanna-woodland habitats (Bessa et al. 2015; McGrew et al. 1988; J. Moore et al. 2017; Pruetz and Bertolani 2007; Pruetz et al. 2015), and in montane forests (Basabose and Yamagiwa 1997).

Regardless of variation in hunting rates, the total amount of meat that chimpanzees eat at a successful hunt can be high. Per capita meat availability can also be high, and some individuals eat large amounts of meat. The mean amount of meat obtained per successful hunt at Taï was positively associated with the number of hunters and varied from about 0.8 kg when single males hunted to 9.3 kg when six or more males hunted (Boesch 1994). Between 1995 and 2001, the mean prey biomass harvested per successful red colobus hunt at Ngogo was 15.6 kg, the maximum total meat availability from a single hunt exceeded 50 kg, and the estimated per capita meat available to males present at meat-eating sessions averaged nearly 1 kg and reached as high as 3 kg (D. Watts and Mitani 2002a).

Sex differences in meat intake exist in chimpanzees and baboons. Male baboons are larger than females and better able to capture prey. Also, they are likely to steal all but small prey items from females, while females are unlikely to acquire meat from males unless the males discard some of it (Hamilton and Busse 1982; Hausfater 1976). Male chimpanzees hunt more often than females do, and they especially hunt monkeys more often (Boesch 1994; Boesch and Boesch 1989; Boesch and Boesch-Achermann 2000; Gilby 2006; Gilby, Eberle, and Wrangham 2008; Gilby, Machanda, Mjungu, et al. 2015; Gilby, Machanda, O'Malley, et al. 2017; Hosaka et al. 2020; Stanford 1998; D. Watts and Mitani 2002a, 2002b). Data from the Kasekela and Mitumba communities at Gombe and from Kanyawara offer some support for three mutually compatible hypotheses proposed to explain this difference, although none enjoyed unequivocal support: females encounter the most sought-after prey type (red colobus) less often than males, they are more risk averse than males, and they are more likely to lose prey they capture to theft (Gilby, Machanda, O'Malley, et al. 2017). Boesch and Boesch-Achermann (2000) estimated that on average, males at Taï ate 186 grams of meat per day (or 68.9 kg annually) during their study, almost an order of magnitude more than females (25 grams per day, or 9.1 kg annually). Adult male meat intake also considerably exceeded that by adult females and by immature individuals at Ngogo (D. Watts and Mitani 2002a), Kasekela (Gilby 2006; Gilby, Machanda, O'Malley, et al. 2017; Stanford

1998), Kanyawara (Gilby, Machanda, O'Malley, et al. 2017), and Mahale (S. Uehara 1997). Sex differences may be less pronounced or even absent in chimpanzee populations that hunt infrequently. Initial data seemed to indicate that galago predation at Fongoli was predominantly a female activity; because the Fongoli chimpanzees rarely hunt other species, this raised the possibility that females ate more meat than males (Pruetz and Bertolani 2007). However, subsequent data showed no obvious sex difference and implied that the initial result was a sample-size artifact (Pruetz et al. 2015).

Despite the general sex difference, variation in chimpanzee male access to meat varies considerably at all sites for which data exist, largely due to variation in hunting skill and prey capture success (Boesch 1994; Boesch and Boesch-Achermann 2000; Fahy et al. 2013; Gilby 2006; Gilby and Wrangham 2008; L. Klein, Huang, et al. 2018; J. Mitani and Watts 1999; Stanford 1998; Tennie, Gilby, and Mundry 2009; Wrangham 1977; Sponheimer and Crowley, this volume). For example, between 1995 and 2001, the average annual amount of meat available per capita to males at Ngogo from all prey species was 15–20 kg, but the most successful males harvested an estimated 40–60 kg of red colobus per year (D. Watts and Mitani 2002a). Chimpanzees typically consume all edible parts of prey carcasses, but these values overestimate the amount of edible matter, and actual meat availability was lower (ibid.). Also, meat transfers reduced some of the disparity in meat availability among males and between males and females (below). Nevertheless, some males at Ngogo ate large amounts of meat annually, and individuals often fed to satiation on medium-to-large carcasses or parts of these. Similar variation characterized males at Gombe (Gilby 2006), Kanyawara (Gilby, Eberle, and Wrangham 2008), and Taï (Boesch 1994; Boesch and Boesch-Achermann 2000). Boesch (1994) reported that meat intake per hunt at Taï was highest for males who captured prey, intermediate for noncaptors who actively pursued prey, and lowest for individuals who did not participate in hunts but begged for meat. Samuni et al. (2018), analyzing later data from Taï, found that captors always obtained meat, although individuals who solicited meat from possessors gained shares in 80% of cases, regardless of whether they had actively pursued prey.

The sex difference in meat intake is the opposite for bonobos, at least for predation on duiker, in keeping with chimpanzee-bonobo differences in the relative power of males and females. Females control access to most duiker carcasses, and while males sometimes try to steal carcasses from females (Wakefield et al. 2019), they mostly obtain meat by successfully soliciting it from females (Fruth and Hohmann 2018; Hohmann and Fruth 1994; Wakefield et al. 2019).

Secondary meat acquisition reduces interindividual skew in intake in capuchins, baboons, chimpanzees, and bonobos. Individuals other than captors may gain meat by aggressive theft, by retrieving dropped or discarded items, by successfully begging for (or "soliciting") carcass portions from meat possessors, or by receiving unsolicited portions freely offered by possessors (Jaeggi and Gurven 2013; Silk et al. 2013). Noncaptors sometimes acquire meat via secondary transfers—that is, from others who did not capture the prey themselves but instead obtained meat from the captors via one of these modes. Transfers in response to solicitation may or may not be accompanied by aggression by the possessor or the soliciting individual, and transfers in response to solicitations can be passive (possessor allows solicitor to take portions) or active (possessor gives portions to solicitor). Unsolicited donations are active by definition.

Retrieval of dropped or discarded items seems to be the most common means of secondary acquisition in capuchins, although both theft and peaceful transfers in response to solicitations occur occasionally. In one study, 104 squirrel and coati carcasses consumed by white-faced capuchins had means of 1.68 and 2.97 "owners" at Santa Rosa and Lomas Barbudal, respectively, and other individuals got meat by retrieving fallen pieces (Rose et al. 2003). Theft and retrieval of dropped or discarded items are the main modes of secondary acquisition in baboons (R. A. Harding 1975; Strum 1981). At Gilgil, from 1 to 16 individuals obtained meat from single carcasses, with up to 7 feeding directly on a single carcass (often after one or more others had discarded it) and up to 11 eating fallen scraps; 51% of those who ate meat only obtained scraps (Strum 1981). Between two and five individuals fed on infant impala carcasses at Moremi; one carcass transfer apparently involved theft by the alpha male, while the other six involved lower-ranking males retrieving carcasses dropped by higher-ranking males who had already eaten much of the meat (Hamilton and Busse 1982). Other prey items (e.g., juvenile vervets) also had two or more possessors; in all cases ($N = 18$), secondary users acquired meat by appropriating it from captors or after captors discarded the items (ibid.).

All four modes occur in chimpanzees. Theft is uncommon, although males more often steal meat from females than from each other (Gilby, Machanda, O'Malley, et al. 2017; Watts and Mitani, unpubl. Ngogo data). Passive transfer following solicitation is the most common mode of secondary acquisition (Boesch 1994; Boesch and Boesch 1989; Gilby 2006; J. Mitani and Watts 2001, unpubl. Ngogo data), although active transfers are also common at Ngogo (Watts and Mitani, unpubl. data) and individuals sometimes acquire meat by retrieving scraps dropped by others (Samuni et al. 2018; Tennie et al. 2008). The number of individuals who obtained meat per hunt increased with hunting party size at Kanyawara (Gilby, Eberle, and Wrangham 2008) and Ngogo and was positively associated with carcass size and total meat availability at Ngogo (D. Watts and Mitani 2002a, unpubl. Ngogo data) and Taï (Samuni et al. 2018). Multiple factors influence meat transfer in chimpanzees (Gilby 2006; J. Mitani and Watts 2001; Silk et al. 2013; D. Watts and Mitani 2002b). How much of it represents tolerated theft in response to harassment, rather than sharing, has been debated (Gilby 2006; J. Mitani and Watts 2001). Regardless, secondary consumers sometimes acquire substantial meat (e.g., entire limbs of adult red colobus; pers. obs.). At Ngogo (J. Mitani and Watts 2001, unpubl. data) and Taï during the 1980s (Boesch 1994), adult males most often controlled access to meat, and most transfers were between males. However, Taï data from 2014 to 2016 showed no sex difference in success at soliciting meat from possessors (Samuni et al. 2018), and transfers from adult males to adult females and to immature individuals and from females to their offspring are also common at other sites (e.g., Kasekela; Gilby, Emery Thompson, et al. 2010; Stanford et al. 1996). In contrast, bonobo meat transfers are most common between adult females and between females and offspring; transfers between females and males and between males are uncommon (Hohmann and Fruth 1994; Wakefield et al. 2019; but see Fruth and Hohmann 2018).

WHY EAT MEAT?

Hamilton et al. (1978) asserted that baboons prefer ASFs to PSFs and reported that chacma baboons at Moremi devoted considerable foraging effort to animal prey, but these mostly comprised invertebrates (see also Rhine et al. [1986] for yellow baboons at Mikumi and S. Altmann and Altmann [1970] for yellow baboons in Amboseli). White-faced capuchins eagerly pursue and consume vertebrate prey, as S. Perry (2008) vividly described, despite its small overall dietary contribution. Likewise, chimpanzees eagerly pursue prey and consume meat, and they assiduously and persistently solicit meat from others who possess it. Such behavior raises fundamental questions about the nutritional value of hunting and meat eating. Three major hypotheses have been proposed as answers; these are not mutually exclusive and could apply differently to different cases:

1. Individuals eat meat to gain calories (Boesch 1994; Hamilton and Busse 1978). If the value of meat depends on its energy content, net energy gain should be higher for meat than for alternative foods. The more specific "energy shortfall hypothesis" proposes that energy from meat is particularly important when other sources of easily assimilated energy (notably drupaceous fruit) are scarce and predicts that hunting frequency and meat intake should vary inversely with the availability of such foods.
2. Nonhuman primates eat meat to obtain protein. Meat is a valuable source of protein, particularly because protein in vertebrate tissues is usually nutritionally complete, unlike that in most PSFs (Goodall 1986; Hamilton and Busse 1978; Milton 2003a, 2003b; J. Mitani and Watts 2001; Takahata, Hasegawa, and Nishida 1984; Teleki 1975; Tennie, Gilby, and Mundry 2009).
3. Meat eating occurs primarily because meat is an important source of micronutrients, especially those scarce in many PSFs (Hamilton and Busse 1978; Milton 2003a, 2003b; J. Mitani and Watts 2001; Tennie et al. 2008).

A strict version of the energy hypothesis—that meat is primarily important as a calorie source—is unlikely to apply. Some small-bodied primates meet large proportions of their energy needs from faunivory but mostly or entirely by eating invertebrates, which are also likely to be the main sources of dietary protein and important sources of micronutrients. Hamilton and Busse (1978) proposed that faunivory by nonhuman primates is generally most important because of its energy returns but that these come mostly from foraging for invertebrates. They and others who have analyzed data on vertebrate predation by baboons (Rhine et al. 1986) or capuchins (Rose et al. 2003) concur that meat eating is probably not an important long-term source of energy. However, this does not exclude the possibility of occasional considerable short-term net energy gains, although whether these exceed gains available from alternative foods is an open question.

Boesch (1994) used behavioral data, including estimated meat consumption rates, plus assumptions about the costs of pursuing prey and the caloric returns of meat eating to estimate the net energy gains of hunting red colobus for chimpanzees at Taï and Kasekela. Despite the unknown magnitude of error associated with such estimates (e.g., the caloric values of red colobus tissues have not been directly measured; Tennie, O'Malley, and Gilby 2014), the major result probably held at least qualitatively: per capita gains were positive for all hunting party sizes but varied in association with the number of males who participated in hunts. Gains were highest for hunting parties of four males at Taï but for single hunters at Kasekela. For these party sizes, males potentially gained net benefits of 5,166 kJ/h at Taï and 4,245 kJ/h at Kasekela; estimated net gains at other group sizes ranged from 1,250 (two hunters) to 5,020 kJ/h (more than six hunters) at Taï and 861 (six hunters) to 3,201 kJ/h (two hunters) at Kasekela. Comparative data were not available for fruit. Tool-assisted nut cracking provides as much as 3,450 kcal, or 14,441 kJ, per day at Taï (Boesch and Boesch-Achermann 2000), although such high returns are available only seasonally. Depending on how much time the chimpanzees require to extract this much energy from nuts, their expected energy return rates from nut cracking might exceed those from hunting.

Gilby (2006) estimated that at Kasekela, chimpanzees who possessed red colobus meat consumed it at mean rates of 1.16 kg per meat-eating bout and 1.9 kg/h and ate up to 2.5 kg of meat in single bouts. If we assume that the percent dry matter (DM) and the energy content of red colobus tissues are similar to those of vertebrates subject to predation by wolves (Bosch et al. 2015) and various African ungulates (L. Hoffman 2008; L. Hoffman and Ferreira 2004; Kryiakou et al. 2016; Ntiamoa-Baidu 1997) and apply the mean of these values (38.6% DM), 2,244 kJ/100 g DM yields values of 14,451 kJ/h (or 3,454 kcal/h) and 10,041 kJ (2,400 kcal/h) per average meat-eating bout in Gilby's (2005) sample (D. Watts 2020). These estimates considerably exceed Boesch's (1994) for the same community, although Boesch's were of net gains, while Gilby's were of gross energy intake. In comparison, O'Malley and Power (2012) estimated that the maximum caloric return from a bout of ant-dipping at Kasekela was only 59 kcal (247.2 kJ), more than an order of magnitude less than the average gain from eating meat. The estimated energy returns from meat eating are undoubtedly too high, partly because of limits on the ability to chew and digest meat (Ragir et al. 2000), but they support Tennie, O'Malley, and Gilby's (2014) contention that the estimated 433 kJ (103 kcal) available from a typical small scrap of meat (about 50 g; Gilby 2006) would surpass that available from ant-dipping. However, ant-dipping is a less risky foraging choice, given that not all hunts succeed, pursuing prey carries a risk of injury, and not all individuals present at meat-eating sessions necessarily obtain meat (McGrew 1979; Tennie, O'Malley, and Gilby 2014).

Still, chimpanzees obtain most of their energy from PSFs, not insects. Information on the energy and macronutrient content of major chimpanzee plant foods are available from Kanyawara (Uwimbabazi et al. 2019; Wrangham, Conklin, Chapman, et al. 1991; Wrangham, Conklin, Etot, et al. 1993; Wrangham, Conklin-Brittain, and Hunt 1998), Gashaka and Taï (Hohmann, Potts, et al. 2010), and Ngogo (Hohmann, Potts, et al. 2010; K. Potts, unpubl. data). Values are on a DM basis, and data on water content have not been published; this plus variation in the digestibility of different plant parts and food fractions, the potential to derive energy from hindgut fermentation, and other factors makes directly comparing nutritional value between meat and PSFs difficult. Also, intake rate data are generally unavailable for PSFs. However, Wrangham, Conklin, Chapman, et al. (1991) gave DM estimates of 26.7% for fruit and 24.8% for leaves at Kanyawara, while Potts (unpubl. data) estimated that nonfig fruit eaten by chimpanzees at Ngogo provided a mean of 190 kJ/100 g DM and figs a mean of 185 kJ/100 g DM. If energy digestibility is the same for meat and PSFs, chimpanzees would require 2.23 kg of nonfig fruit (wet weight) or 2.29 kg of figs to acquire the energy potentially available from the average meat meal at Gombe (D. Watts 2020). Much smaller quantities (96.2 g of nonfig fruit and 98.8 g of figs) would supply the calories available from a 50 g meat scrap (ibid.). These are only rough estimates, but they imply that while ingesting relatively large quantities of meat brings major energetic payoffs, energy pursuit is not the main motivator of attempts to get scraps (cf. Tennie, O'Malley, and Gilby 2014).

The more general energy shortfall hypothesis has not been explicitly investigated in baboons, but available data do not obviously support it (J. Mitani and Watts 2005b). Rose (1997) addressed it qualitatively for capuchins at Santa Rosa and noted that fruit—the main energy source—was relatively scarce during dry seasons, whereas meat eating was most common then because of the availability of coati pups. However, meat intake was probably too low to provide substantial energy returns (ibid.). Boesch and Boesch (1989; see also Boesch and Boesch-Achermann 2000) addressed the hypothesis indirectly for chimpanzees at Taï, where red colobus hunts were most common during the rainy season; this is the red colobus birth season, but also capturing prey might be easier then because slippery conditions in the forest canopy reduce red colobus mobility. They did not provide data on temporal variation in fruit abundance, but subsequent analyses showed that no clear relationship between rainfall and fruit availability existed at Taï and that, if anything, fruit abundance was relatively high during months transitional between rainy and dry seasons (Janmaat, Boesch, et al. 2016; Polansky and Boesch 2013).

Data on the relationship between hunting frequency and fruit availability from three other chimpanzee sites do not support the energy shortfall hypothesis. Gilby and Wrangham (2007) found that chimpanzees at Kanyawara were more likely to hunt red colobus on encounter during "non-fig fruit seasons" (periods of at least two consecutive months when fruit other than figs accounted for at least 40% of feeding time, an animal-centered measure of fruit availability) than when nonfig fruit was less abundant. High fruit abundance should have meant that the chimpanzees could easily meet their caloric needs and did not need to rely on meat for energy. The effect of fruit abundance was independent of a significant positive relationship between fruit abundance and male chimpanzee party size. Gilby and Wrangham (2007) argued that the chimpanzees could afford to engage in risk-prone foraging (hunting) to gain nutritional benefits other than energy when they could easily obtain energy from fruit. Long-term data from Ngogo showed that monthly hunting rates increased significantly in association with another animal-centered measure of fruit abundance, the proportion of monthly feeding time devoted to the top fruit species in the chimpanzees' diet (D. Watts and Mitani 2015). This relationship was driven by a positive association between the rate of red colobus hunts and fruit intake; the rate at which the chimpanzees hunted other prey species was independent of fruit intake. Notably, most red colobus hunts occurred during "hunting patrols," when the chimpanzees spent up to several hours searching for red colobus groups and during which they fed little on plant foods (cf. J. Mitani and Watts 1999; D. Watts and Mitani 2002a). Foregoing immediate opportunities to harvest fruit under such conditions to expend energy searching for and hunting red colobus, activities with uncertain returns, should not impose net costs, because the chimpanzees gain energy from any meat they harvest and can easily gain large amounts of energy from fruit before and after patrols and hunts (D. Watts and Mitani 2002a, 2015). Gilby and Wrangham (2007) dubbed this the "energy surplus hypothesis" and distinguished it from their "abundance/risk" hypothesis, but the two are essentially the same. Finally, chimpanzees at Rekambo (Loango, Gabon) also hunted more often when fruit availability was high, and researchers cited this as support for the energy surplus hypothesis (H. Klein et al. 2021).

Meat is an excellent protein source and can provide considerable protein in the short term, especially to those individuals most successful at capturing relatively large prey. Assuming again that values for the nutritional composition of wolf prey species apply approximately to chimpanzee prey species, applying the mean value of 68.2% protein by DM to Gilby's (2006) intake estimates yields values of 493 g protein/h, 173 g protein/average meat-eating bout, and 13 g protein per average meat scrap. Leaves eaten by chimpanzees are generally higher in protein than fruit (Hohmann, Potts, et al. 2010). Potts (unpubl. data) calculated a mean of 26% protein by DM in a sample of leaf foods from Ngogo. Making the same assumptions as above, about 4.7 kg of leaves by wet weight provide the protein content of the average meat meal at Gombe, and 200 g would equal the content of a 50 g meat scrap. Error in these estimates could be large; still, a chimpanzee would need to eat a large volume of leaves to acquire as much gross protein as it would from eating a juvenile red colobus or the limb of an adult. Moreover, the meat protein would be nutritionally complete. Gaining the gross protein equivalent of a meat scrap from leaves would be less challenging.

Alternatively, or besides providing occasional protein and energy bonanzas, meat and other components of vertebrate carcasses could be particularly important sources of lipids and of vitamins, minerals, and other micronutrients. If data on the nutritional content of different tissues available for other wild mammals (e.g., L. Hoffman 2008; Kyriacou et al. 2016) generally apply to chimpanzee prey, liver and viscera (but not muscle tissue) are good sources of fat. Brain tissue also is high in lipids and can be a particularly valuable source of both the long-chain polyunsaturated fatty acids necessary for brain growth and metabolism and for retinal development and the omega-3 and omega-6 precursors from which these can be synthesized, as can liver (Kyriakou et al. 2016). Chimpanzees at Ngogo eagerly consume viscera, including the stomach and intestinal contents of colobines, which they seem to prefer to other carcass parts (pers. obs.), and similar behavior occurs at other sites (e.g., Kasekela; Gilby, Machanda, O'Malley, et al. 2017). The importance of long-chain polyunsaturated fatty acids could explain why Gombe chimpanzees preferentially consume the brains of infant red colobus, to which they can gain access by biting directly into the skulls, before other portions of carcasses (Gilby and Wawrzyniak 2018). Micronutrients are important for the maintenance of immune function, for brain growth and cognitive development, for cell division, and for myriad other physiological processes (M. Bailey et al. 2015; Stammer et al. 2015). Some are widely available in PSFs in primate diets, and estimated intakes can equal or greatly exceed recommended daily allowances for humans and nonhuman primates (Cancelliere et al. 2014; Milton 2003b). Because some plant parts typically contain more of certain micronutrients than others and the micronutrient content of specific plant parts varies among species, obtaining adequate and balanced intake of all micronutrients requires a diverse array of foods (Cancelliere et al. 2014; Milton 2003b). ASFs are rich in micronutrients that are often not abundant in plants generally or at least in certain plant parts. Variation in the bioavailability of different micronutrients in PSFs also may complicate micronutrient access via a plant-based diet, at least for some micronutrients. Notable examples of micronutrients in which ASFs are rich include vitamins A, K, B_6, and B_{12}; calcium, phosphorus, sodium, and potassium; and iron, selenium, and zinc. As for protein and lipids, mineral and micronutrient content varies among animal tissues; for example, copper content in liver exceeded that in muscle in all six wild southern African ungulate species for which Kyriakou et al. (2016) had data, while iron content was higher in liver in four of the six. Tennie, Gilby, and Mundry (2009), invoking earlier suggestions regarding the value of meat as a source of micronutrients (Hamilton and Busse 1983; Milton 2003a, 2003b; Teleki 1975), proposed that the need for vitamins, minerals, and trace elements explained why chimpanzees are so eager to obtain even small amounts that would yield little energy or protein. Data showing increases in the number of individuals who obtain at least small amounts of meat in association with increases in hunting party size, but no increase in per capita meat availability, at Kanyawara (Gilby, Eberle, and Wrangham 2008) and Ngogo (D. Watts and Mitani 2002a) are consistent with this "meat scrap hypothesis." The hypothesis could apply equally well to other nonhuman primates that occasionally eat meat and that also show great eagerness to obtain meat from others.

COOPERATIVE HUNTING BY NONHUMAN PRIMATES

Addressing the question of whether nonhuman primates hunt cooperatively is crucial for assessing the relevance of meat eating by other primates to understanding its importance in human evolution and to the emergence of obligatorily cooperative foraging in the ancestors of *Homo sapiens*. It also addresses the issue of what energy or nutritional gains nonhumans make from hunting. From an ecological perspective, cooperative hunting can be identified operationally when individuals make higher net energy returns by hunting in groups, at least up to some optimal group size, than by hunting alone (Creel and Creel 1995). Hunting by nonhuman primates in this sense has received some attention, but most discussions of whether hunting and meat eating in nonhuman primates is cooperative have centered on the extent to which multiple individuals coordinate their actions while pursuing prey—hence on cooperation in a behavioral sense—and on meat transfers. Unsurprisingly, chimpanzees have been the main subjects of discussion.

Boesch (1994) argued that red colobus hunts at Taï mostly met the ecological definition of cooperation because the relationship between net individual energy intake and hunting group size had an inverted-U shape: up to groups of four or perhaps somewhat larger, individuals achieved higher energy gains by hunting in groups than by hunting alone (like the inverted-U-shaped relationship of net energy return to hunting group size reported for African wild dogs in the Selous; Creel and Creel 1995). In contrast, Boesch (1994) estimated that per capita gains were highest for solitary hunters at Kasekela and for Mahale M Group. Average values masked variation among individuals, and per-hunt payoffs were highest for males who actually caught prey. However, Boesch (1994; see also Boesch and Boesch-Achermann 2000) estimated that males who actively hunted, but did not capture prey, and bystanders who did not actively hunt also achieved their highest gains when hunting parties had four or five members. He claimed that a "behavioral rule" discriminating against bystanders prevented them from receiving meat, but this was belied by his own data showing that they often obtained meat. Later Taï data also showed that most individuals present at meat-eating sessions obtained meat, regardless of their behavior during hunts (Samuni et al. 2018).

Other Kasekela data and data from Ngogo and Kanyawara showed no clear relationship between per capita energy gain (estimated as per capita meat availability) and hunting group size, although comparisons with Boesch's (1994) findings must be qualified because the observers did not estimate energy expenditure during hunts and, for Ngogo, considered all males present to be potential hunters instead of distinguishing hunters from bystanders. The probability of success and the mean biomass of prey harvested in red colobus hunts at Ngogo increased significantly with hunting party size, but estimated per capita meat availability was independent of hunting party size (D. Watts and Mitani 2002a). Kasekela data also showed that the probability of capturing at least one red colobus increased with hunting party size, but per capita meat availability significantly decreased as hunting group size increased (Gilby 2006). At Kanyawara, the likelihood that individuals obtained some meat increased with hunting party size, but the probability that individual males pursued prey reached an asymptote at large hunting party sizes (Gilby, Eberle, and Wrangham 2008). Gilby, Eberle, and Wrangham (2008) did not estimate total meat availability, but they suggested that when relatively many males hunt, the high probability of success and correspondingly high average total meat availability means that many individuals switch from hunting to soliciting meat once other individuals have some (similar behavior occurs at Ngogo; pers. obs.). This limits synergistic effects on meat availability and thus limits per capita meat availability.

Nevertheless, the probability that an individual ate some meat increased with hunting party size at Kasekela (Tennie et al. 2014) and Kanyawara (Gilby, Eberle, and Wrangham 2008). This is consistent with the meat scrap hypothesis. Also, if the main benefit of hunting is the provision of micronutrients, it fits the definition of cooperation as behavior by one individual that provides benefits to one or more others, independently of whether the behavior imposes direct costs on the actor or provides immediate net benefits. If the meat scrap hypothesis is the main explanation for why chimpanzees hunt, any net caloric gains and protein gains from hunting may be by-products of behavior motivated by the possibility of acquiring micronutrients (Tennie, Gilby, and Mundry 2009).

With regard to behavioral definitions of cooperation, Boesch and Boesch (1989) proposed that four tactics used in hunts of red colobus at Taï potentially qualify:

1. Similarity: Multiple individuals pursue the same prey or group of prey, but their actions are not synchronized or aimed at the same individual prey.
2. Synchrony: Multiple individuals simultaneously pursue the same group of prey.
3. Coordination: Multiple individuals simultaneously pursue the same individual prey.
4. Collaboration: Multiple individuals take mutual, complementary roles during simultaneous pursuit of the same prey and thereby increase the probability that at least one hunter captures prey compared to probabilities associated with the first three tactics.

I. Bailey et al. (2013) added a fifth category of "passive cooperation": predators hunt in the same location as others, and the presence of multiple hunters somehow increases individual success, even though they do not pay attention to each other's actions. Hunters share focal prey only by chance and rarely share meat. They added that hunters synchronize their activities to prey in similarity but also to each other's behavior in synchrony, and that the positioning of hunters relative each other must not be incidental for behavior to count as coordination. Finally, hunters' timing and behavior must be much more strongly based on each other's behavior than on that of the prey, and role differentiation must be clear, for behavior to qualify as collaboration. In reviewing the literature on group hunting by carnivores, I. Bailey et al. (2013) concluded that 8 of 21 carnivore species for which adequate data existed sometimes collaborate and that collaboration possibly occurred in 6 others. Examples included group hunting by lionesses in the Etosha Pan, Namibia (Stander 1992). They also concluded that simpler mechanisms (e.g., associative learning) could explain collaborative hunting and that shared intentionality is not necessary.

Albiach-Serrano (2015) also pointed out that collaboration could just require understanding that others have intentions plus some understanding of the importance of their attentional states, contrary to Boesch and Boesch's (1989) argument that it requires shared intentionality. Like I. Bailey et al. (2013), she noted that Boesch and Boesch's (1989) scheme had no clear criteria for distinguishing synchrony from coordination and did not specify whether individuals adjusted their behavior (a) to the mere presence of conspecifics, (b) to whether conspecifics were engaged in some form of action, (c) to the kind of actions in which conspecifics were engaged, or (d) to specific acts by individual conspecifics. Her alternative scheme included five categories:

1. Independent cooperation: Individuals perform actions independently, without adjusting to each other (the equivalent of "similarity").
2. Presence-dependent cooperation: Individuals are more likely to act when conspecifics are present.
3. Action-dependent cooperation: Individuals are more likely to act when others perform a particular action (e.g., climb toward prey).
4. Form-dependent cooperation/coordination: Individuals adjust their behavior to that of others in space, time, or both. Albiach-Serrano saw this as equivalent to "coordination," and it presumably applies to Stander's (1992) description of role-taking during lion hunts.
5. Intentional cooperation: Individuals attend to others' actions and flexibly adjust their behavior based on their understanding of others' roles in achieving shared goals.

They might understand only that others have goals and that their attentional states are relevant to these, or—in true collaboration—they share attention and intentions.

Boesch and Boesch (1989) classified most hunts (77%) at Taï as involving collaboration by their definition, and Boesch (1994) argued that collaboration was much more common there than at Kasekela or Mahale (cf. Boesch 2004; Boesch and Boesch-Achermann 2000). However, Albiach-Serrano (2015) pointed out that controlled experiments with captive chimpanzees provide compelling evidence for presence-, action-, and form-dependent cooperation but not for intentional cooperation. Gilby, Eberle, and Wrangham (2008) reported that collaboration by Boesch and Boesch's (1989) definition was uncommon at Kanyawara. D. Watts and Mitani (2002a) noted that it also seemed to be uncommon at Ngogo, although they qualified this by noting that constraints on visibility made accurate assessment of tactics at many hunts impossible, and they provided anecdotal evidence for form-dependent cooperation. In the end, we should be cautious regarding claims that hunt participants impute mental states to others (e.g., desires to capture prey or knowledge of how to maximize capture probabilities), although this does not necessarily mean that chimpanzees do not engage in coordination during hunts.

Relevant here is that hunting tactics that qualify as cooperative in that behavior by one individual provides benefits to others may ultimately be examples of by-product mutualism (Gilby and Connor 2010; cf. Gilby, Eberle, and Wrangham 2008). That is, hunters try to maximize their own chances of capturing prey and thereby take advantage of how the actions of other hunters affect prey behavior and incidentally create situations that others can exploit. Exploitable opportunities are most likely to arise when particularly skilled or motivated hunters ("impact hunters") are present (Gilby, Machanda, Mjungu, et al. 2015). Chimpanzees are adept at exploiting such effects, and behavior that looks like coordination could result from independent efforts by individuals to maximize their own prey capture probabilities. This contrasts with the claim that in most red colobus hunts at Taï, some hunters tried to force the prey to move toward other chimpanzees or to cut off prey escape routes but did not try to capture prey (Boesch and Boesch 1989; Boesch and Boesch-Achermann 2000). Alternatively, however, "blockers" could have been trying to place themselves in positions where they had good chances of catching fleeing monkeys, similar to the way in which males at Ngogo sometimes follow the progress of hunts from the ground and can be in good positions to capture monkeys that fall or are isolated in vulnerable positions in the canopy (D. Watts and Mitani 1999).

Debates about cooperative hunting highlight the need to consider observational constraints, particularly the difficulty that single observers face in trying to monitor the behavior of all chimpanzees present during group hunts of arboreal monkeys consistently and precisely (e.g., Boesch 1994). This is possible in some contexts—for example, when the canopy is low and broken, when few individuals pursue prey, and when the prey group spread is sufficiently narrow—but can be impossible when many chimpanzees are simultaneously active in the canopy or moving on the ground and the prey are widely dispersed (D. Watts and Mitani 1999; pers. obs.). Consistently accurate data collection can be impossible even when multiple observers are present. For example, Gilby, Machanda, O'Malley, et al. (2017) used data on the identities of prey captors at hunts of arboreal monkeys (defined as prey encounters in which at least one chimpanzee climbed toward the monkeys) to test the hypothesis that females with young offspring are more sensitive to risks and thus pursue arboreal monkeys less often than do males or females without young offspring. They stated that their data, which came from Kasekela, Mitumba, and Kanyawara, were collected by teams of two field assistants, supplemented with focal-sampling data that K. Walker collected at Kasekela. They did not find strong support for the hypothesis but qualified their findings by noting that females with young infants might leave these behind when they hunt or might pursue prey less intensively and abandon pursuits more readily than females without such offspring. Essentially, this means that the observer teams could not consistently collect the data needed for strong tests of hypotheses about behavioral cooperation or for estimating energy returns.

Cooperation and possible synergistic effects of hunting by capuchins and baboons have received less attention. Baboon hunts sometimes involve synchrony and even coordination (Hamilton and Busse 1978; Hausfater 1975; Strum 1981), but whether this increases per capita meat availability or meat scrap availability is unknown. Given the extent to which individuals, especially high-ranking males, monopolize carcasses (above), the amount of meat actually available per capita might not increase even if captures are more likely when multiple individuals chase prey. Multiple individuals pursued squirrels during most hunts by white-faced capuchins at Santa Rosa (Rose 2001), where the monkeys sometimes actively searched for prey (behavior not seen at Lomas Barbudal or Palo Verde) and where hunting tactics possibly included collaboration (e.g., one individual pursued a squirrel while two others blocked its escape routes; Rose 2001; Rose et al. 2003). However, synergistic effects on meat availability seem unlikely because captors monopolized carcasses (Rose 2001; Rose et al. 2003).

HUNTING AND HUMAN EVOLUTION

Even in chimpanzee populations where hunting is most frequent, meat intake is far less than documented among recent and extant human hunter-gatherer societies (Wood and Gilby 2017; K. Hill 1982). In contrast to the uncertain nutritional importance of hunting for chimpanzees, baboons, and capuchins, hunting and meat eating clearly have had great importance in human evolution, although heated debates exist about when they became major components of hominin behavioral ecology and how much they have influenced human social and life-history evolution.

Kelly's (2007) review of the ethnographic literature revealed that values for the percentage of the diet contributed by meat varied from 10% to 90% in 123 recent and contemporary hunter-gatherer societies. He distinguished hunting of terrestrial vertebrates from fishing, broadly construed to include collection of shellfish and hunting of marine mammals, which in turn contributed from 0% to 80% of the diets. Hunting and fishing together ostensibly contributed from 15% to 100%, although it seems implausible that any hunter-gatherers had diets devoid of PSFs. Kelly's analysis confirmed that geographic variation explains much of the variation in the overall importance of ASFs versus PSFs, with hunting and fishing contributing more to the diets of temperate-zone and Arctic hunter-gatherers than those in the tropics. Cordain et al. (2000) obtained similar results from a larger literature survey that included 220 societies.

Such broad samples necessarily included data of widely varying quantitative precision; indeed, in some cases only qualitative data were available. Kaplan et al. (2000) summarized detailed quantitative data on food harvesting and intake in nine hunter-gatherer societies and used a series of assumptions about the caloric value and protein content of different food types to estimate that meat provided an average of 52.7% of all food calories (range = 25.8%–78.8%). They noted that this might have been an underestimate, given that two studies of the !Kung yielded quite different estimates of meat intake, although only one allowed estimation of total calorie intake. They also estimated that individual meat intake varied from about 270 g to 1,400 g per day. Invertebrates also contributed substantially to people's energy intake in several cases (Anbarra, Nukak, and Ache). Also, qualitative reports indicate that foraging for invertebrates, including insects, was common in many human societies (McGrew 2014). By implication, meat also contributed the great majority of dietary protein, on average, in the societies included in Kaplan and colleagues' (2000) study. The precision of their estimates is uncertain, but even the lowest—260 g meat/person/day and 29% of total calories from meat (one study of the !Kung)—are much higher than the maximum estimates for individual chimpanzees (above).

Given that both chimpanzees and bonobos—the closest living relatives of humans—hunt and eat meat, the assumption that these were characteristics of the last common ancestor of these taxa and humans is reasonable. The earliest uncontested evidence for a change in the use of vertebrate source foods coincides with the earliest evidence of Oldowan stone tool technology, in the form of bones bearing stone-tool cut marks and signs of fracturing by hammer stones at sites in the Bouri formation in Ethiopia (de Heinzelin et al. 1999) and of cut-marked and fragmented bones unequivocally associated with Oldowan flakes, cores, and debitage at several sites in the nearby Gona region (Semaw et al. 2003). Both sets of sites are dated to about 2.6 million years ago (mya). Most remains are those of equids and, more commonly, bovids, some of them large. The samples are too small to determine whether the hominins who modified the bones had early access to intact or nearly intact carcasses that they obtained either by hunting or by power scavenging (aggressively driving carnivores from fresh kills), or only engaged in passive scavenging of mostly defleshed carcasses. Dominguez-Rodrigo, Pickering, et al. (2005) argued that the presence of cut marks on the midshafts of upper and intermediate limb bones in these assemblages is at least consistent with early access. This would open the possibility that whichever hominins made the tools and modified the bones sometimes acquired large amounts of meat from animals larger than themselves; such behavior is absent from the chimpanzee repertoire but is a hallmark of the human predatory patter (J. Thompson et al. 2019). More cautiously, H. Bunn (2007) argued that the small Bouri and Gona samples do not allow the kinds of the inferences about carcass access possible with much larger bone assemblages at later Oldowan sites but added that these samples at least show that some hominins tried to obtain meat and bone marrow from large mammals. Besides increasing the quantitative importance of ASFs as sources of lipids and calories, the use of stone tools to fracture bones from scavenged carcasses of medium- to large-sized bovids and thereby gain access to marrow quite likely marked the first qualitative difference in the use of ASFs between at least some hominins and the last common ancestor with chimpanzees (J. Thompson et al. 2019). Which hominin taxon might have been responsible for the earliest evidence of large-carcass processing is uncertain; a common assumption is that it was *Homo*, but the presence of *Australopithecus ghari* at Bouri (de Heinzelin et al. 1999) leaves the question open. Moreover, isotopic evidence from Sterkfontein and Swartkrans (South Africa) suggests that the diets of both *Australopithecus africanus* and *Paranthropus robustus* might have included meat (Lee-Thorp, Sponheimer, et al. 2010).

Multiple lines of inquiry contribute to the debate about whether evidence from Oldowan and early Acheulean sites indicates that hominins had regular access to largely intact carcasses of medium-to-large bovids and other vertebrate prey or that they mostly obtained small amounts of meat via passive scavenging. These include taphonomic analysis; assessment of whether the representation of skeletal parts showing signs of butchery is biased toward meat-rich elements and how it compares to the sequence with which extant carnivora typically consume carcasses; enumeration of where stone-tool cut marks and carnivore tooth marks, when present, occur on the bones and claims about how well these distinguish between the removal of large amounts of meat by hominins and the removal of small amounts remaining after carnivore ravaging; assessment of whether bones modified by both hominins and carnivores were handled first by one or the other; and prey mortality profiles (reviewed in H. Bunn 2007; Dominguez-Rodrigo 2012; Lupo 2012; Pickering and Bunn 2012). No consensus exists about how hominins contributed to accumulations of skeletal remains at sites such as FK22 Zinj (Olduvai; 1.86 mya), FxJj 50 (Koobi Fora, Kenya), and the ST site complex (Peninj, Tanzania), but a compelling case exists that some Oldowan and early Acheulean hominins (presumably members of *Homo*) had regular access to intact carcasses of medium-to-large ungulates, butchered meat-rich skeletal elements, and transported substantial quantities of meat from butchery sites to places where they could consume it with minimal risk from large carnivores (H. Bunn 2001, 2007; Dominguez-Rodrigo 2012; Dominguez-Rodrigo and Barba 2006; Pickering and Bunn 2012). Cobo-Sanchez et al. (2022) recently strengthened this case by using deep learning and computer vision to analyze pits and scores that lions and spotted hyenas made on equid long bones and applying the resulting models to faunal remains from DS Bed 1 at Olduvai, dated to 1.84 mya and penecontemporary with FLK Zinj. The models distinguished marks made by lions from the effects of hyenas with high accuracy. Applying them to cut-marked bones from DS Bed 1 showed that nearly all signs of carnivore damage represented hyena ravaging of bones defleshed by hominins who had access to intact carcasses, consistent with analysis of the distribution of cut marks (Dominguez-Rodrigo, Baquedano, et al. 2021).

Proponents of regular early access concede that data on skeletal element representation and the distribution of cut marks cannot distinguish between hunting and power scavenging when assemblages also show damage inflicted by carnivores. Also, the question of how Oldowan hominins could have killed large mammals remains open, although H. Bunn and Pickering (2010; see also Pickering and Bunn 2012) proposed that stalking and ambushing prey—including ambushes from trees—would have enabled hominins who knew how fashion wooden spears using stone tools could have hunted prey like those at the sites cited above. The two-million-year-old site of Kanjera (Kenya) is particularly important for the question of when *Homo* entered the carnivore guild, because it has provided abundant remains of medium-sized bovids that show signs of hominin butchery consistent with early access but few or no signs of damage by carnivores; this clearly indicates that at least one hominin taxon regularly hunted and engaged in tool-assisted carnivory (Ferraro et al. 2013). Moreover, use-wear analysis of Oldowan tools from Kanjera indicates that some were used to process carcasses, while others were used to scrape wood; woodworking could have included the fashioning of spears (Lemorini et al. 2014).

Later BK sites in Olduvai dated to 1.34 mya provide abundant evidence for systematic butchery of complete or nearly complete carcasses of ungulates, including megafauna, ranging in size from gazelle to elephants and including the extinct giant buffalo *Pelorovis* (Dominguez-Rodrigo, Bunn, et al. 2014). The total amount of meat represented at these sites would have surpassed that from all known earlier East African sites combined, and the larger carcasses would have provided meat bonanzas to large hominin groups (ibid.). The inference that *Homo erectus* (or *H. ergaster*) was procuring meat at these sites (Dominguez-Rodrigo, Bunn, et al. 2014) is reasonable, as is the inference that *H. erectus* was responsible for processing carcasses at DS Bed 1, FLK Zinj, and other sites dated to about 1.8 mya (Cobo-Sanchez et al. 2022) and even at Kanjera.

Hominins dispersed to largely temperate habitats in Eurasia by about 1.85 mya (Gabunia et al. 2000) and to temperate habitats in southwestern France by about 1.23 mya (Huguet et al. 2013). They successfully maintained a presence in warm temperate southern Europe and, during warmer periods, in more northern areas in habitats in which PSFs would have been seasonally scarce. This would have been impossible without reliance on a year-round supply of meat that included consumption of small, medium, and large mammals and preferential targeting of herd-forming species such as bison and horses (Landeck and Garriga 2016). Evidence of regular butchery of carcasses of large mammals (e.g., *Cervus*, *Equus*, and *Bison* spp.) to which the hominins had early access at Untermassfeld (Germany; ibid.), Sima del Elefante at Gran Dolina (Spain; Huguet et al. 2013), and multiple sites in southwestern France (ibid.), among others, attest to the ability of early *Homo* to disperse widely by engaging in flexible and opportunistic predation (ibid.).

This conclusion is reinforced by isotopic analysis of Neanderthal teeth and skeletal remains that, in combination with analysis of fauna found at archaeological sites, shows them to have relied heavily on meat, their main source of dietary protein (Bocherens 2009). Neanderthals had a wide geographic distribution characterized by considerable ecological diversity and by climate-driven contractions and expansions. Much of their range was highly productive "mammoth steppe" where the biomass of medium and large grazing mammals was high, and they also extensively used temperate woodlands. Prey choice, and the relative dietary contributions of meat and PSFs to Neanderthal diets, varied temporally and geographically in association with variation in climates, degree of forest cover and plant species diversity, and the composition of faunal communities (Bocherens 2009; Fiorenza et al. 2014; Henry, Brooks, and Piperno 2014; Roebroeks and Verpoorte 2009). PSFs presumably contributed more to diets in the southern, circum-Mediterranean part of the Neanderthal range than in western and northern Europe and Eurasia (Henry, Brooks, and Piperno 2014; Weyrich et al. 2017). Neanderthals also relied on small vertebrate prey and invertebrates to varying extents (Henry, Brooks, and Piperno 2014; Weyrich et al. 2017) and might sometimes have overexploited large prey (Speth 2013). Overall, though, both PSFs and meat and bone marrow from large mammals contributed to Neanderthal diets throughout their range, and the contribution of ASFs was substantial and persisted throughout the year, sometimes with seasonal variation in prey choice. For example, an archaeological record that spans 60,000 years at Peche de L'Azé, in southwest France, shows a shift from year-round occupation during relatively mild temperature regimes and concentration on nonmigratory cervids (roe deer, *Capreolous capreolus*, and red deer, *Cervus elaphus*) to seasonal occupation during colder conditions and concentration on reindeer (*Rangifer tarandus*), which would have been locally abundant while migrating in the spring (Niven 2013). Many Neanderthal archaeological sites show evidence of specialization on single prey species, and some evidence points to game drives that must have required collaboration and possibly meat storage (reviewed in Fiorenza et al. 2014).

Evidence for group hunts during which hominins killed large ungulates (and sometimes made multiple kills) and that presumably involved collaboration also comes from several European Middle Paleolithic sites that pre-date Neanderthals. Among others, these include TD6-2 in Gran Dolina, Atapuerca, dated to about 730 thousand years ago (kya; Rodríguez-Hidalgo et al. 2017); TD10-2 at Sima de los Huesos at Atapuerca, where a large bone assemblage dominated by bison is associated with *Homo antecessor* (ibid.); and the "Horse Butchery" site at Schöningen, dated to about 300 kya (Conard et al. 2015). Schöningen is particularly notable for the remarkable preservation of organic material, including multiple throwing spears that show aerodynamically sophisticated design and required considerable investments of labor to produce and a presumed thrusting spear. The spears were associated with the remains of dozens of horses, presumably killed during repeated, preplanned ambush attacks by groups of hominins, acting in well-coordinated fashion, on horse bands along a lake margin (ibid.). As Conard et al. (2015) argued, envisioning how this could have happened without the use of language is difficult. In any case, coordinated attacks on large prey, planned in advance, have no equivalent in the behavior of nonhuman primates, nor does seasonal tracking of prey movements.

MEAT EATING AND HUMAN LIFE-HISTORY EVOLUTION

Human life histories have several notable features not found in other primates (Borries and Koenig, this volume). These include "childhood," during which somatic growth is relatively slow but steady, brain growth is rapid, and immature individuals are weaned but still depend nutritionally on their mothers and on other, older individuals who more competent foragers or food producers (Bogin 1997); late age at maturity for females, but then reproductive rates anomalously high relative to body and brain size (K. Hill et al. 2009; Kaplan et al. 2000); and extended female postreproductive life spans (Hawkes, O'Connell, and Blurton-Jones 1997; Hawkes, O'Connell, et al. 1998). Evolution in the genus *Homo* has also involved a tremendous increase in the size and complexity of the brain—an organ that requires a constant supply of energy vastly greater than its proportional contribution to overall body mass—and tremendously increased reliance on socially transmitted knowledge. A combination of extensive allocaretaking, especially provisioning, that subsidized the energetic demands of reproduction, with "skill-intensive foraging," notably involving the acquisition of large, high-quality resource packages, facilitated the evolution of unexpectedly fast human female reproductive rates (Kaplan et al. 2000). Kaplan et al. (2000) proposed that hunting was the crucial component of skill-intensive foraging and meat the crucial type of high-quality resource package. They developed a feedback model in which skill-intensive foraging that generated surplus calories subsidized female reproduction and reduced infant and juvenile mortality, in turn allowing females to reproduce at shorter intervals without imposing costs on their current offspring and allowing those offspring to invest in social learning while remaining nutritionally dependent on others. Payoffs from skill-intensive foraging provided a selective advantage to anything that favored increased longevity and the concomitantly longer periods during which adults provided energy surpluses from which others benefited, with the benefits including the ability to meet the energy demands of larger brains and investing more in social learning. Their review of quantitative data on hunter-gatherer diets showed that in most cases, most calories came from men's activities—overwhelmingly from hunting—and that both men and women produced energy surpluses as adults, with men starting to produce surpluses at later ages but then achieving surpluses far in excess of those from women's activities and maintaining these until late in life. They took this as support for their "embodied capital" model of human life-history evolution. Gurven et al. (2006) added support to this in a study of the Tsimane, an indigenous Bolivian group who have a mixed economy but still rely importantly on hunting wild game. Tsimane hunting return rates continue to increase until men are in their forties, long past the ages at which they reach peak physical strength and attain maximum archery skills. The most difficult aspect of hunting is integrating archery skills with knowledge of prey natural history (e.g., recognizing signs of current and recent prey presence) and predicting the likelihood that pursuing prey encountered directly or indirectly will lead to captures.

Hawkes and colleagues (Hawkes, O'Connell, and Blurton-Jones 1997; Hawkes, O'Connell, et al. 1998; see also Hawkes 2001, 2003) proposed that instead of hunting and meat sharing, the potential for older females to obtain fitness gains by provisioning adult daughters accounted for evolutionary increases in the pace of reproduction and in longevity and for the evolution of menopause. They also appealed to ethnographic data, especially from Hadza hunter-gatherers in Tanzania, to support their argument that grandmothers provided substantial energy supplements to their daughters and grandchildren. Hawkes (2001; see also Hawkes and Bliege Bird 2003) further argued that meat provided by men did not necessarily benefit their own families disproportionately; they proposed instead that displays of hunting prowess and widespread provision of benefits available from meat have positive effects on male reproductive success that better explain big-game hunting. Some ethnographic data support this "showing off" hypothesis (Hawkes and Bliege Bird 2003). Likewise, demographic data from contemporary small-scale human societies provide some support for the "grandmother" hypothesis (Sears and Mace 2008), and the existence of fitness benefits from grandmaternal effects would be compatible with the "embodied capital" model (K. Hill et al. 2009). However, detailed data on male foraging and meat sharing among the Hadza, who provided the main inspiration for both of these hypotheses, contradict the "showing off hypothesis": men give disproportionately large shares of meat to their own families and increase their foraging effort and caloric returns when they have infants and children, although they do so mainly by harvesting more honey (Wood and Marlowe 2013).

When the fully modern human life history emerged has been a highly contentious question. Posing the question as "when does modern morphology appear" implies that the answer is 300,000 years ago, given evidence for a fully modern pattern of dental development, thus of overall growth and development, in the early anatomically modern human remains from Djebel Irhoud, Morocco (T. M. Smith, Tafforeau, et al. 2007), dated to between 277 and 354 kya (D. Richter et al. 2017). The paleoarchaeological record yields no evidence of major shifts in hunting technology and meat acquisition strategies around this time. However, evidence of reliance on coastal resources, especially shellfish and fish harvested in the intertidal zone, starts to appear in southern Africa at around 165 kya and is abundant by about 110 kya (Marean 2010). This pre-dates evidence of advanced projectile weapons from the same region, dated to about 71 kya and including microlithic technology and bone projectile points, that would have allowed hunters to kill prey at greater distances (K. Brown et al. 2012; Henshilwood et al. 2001). Marean (2010, 2014, 2015, 2016) in particular has argued that reliance on densely distributed and more or less constantly available resources, like those available from the intertidal zone and from geophytic plants endemic to the Cape Floral Region, was crucial for the ability of *Homo sapiens* to survive periods of dry climates in southern Africa refuges during marine isotope age 6 (ca. 195–25 kya), and that "coastal adaptation" was also crucial for subsequent dispersal out of Africa and through Eurasia, although hunting large and medium-sized mammals retained great importance (Cowling et al. 2020; Marean 2015) and early *Homo sapiens* in Africa would have had diverse diets that included a broad range of vertebrate prey.

Reviewing the dietary contribution of hunting following the dispersal of modern humans from Africa, including the debate about what explained the expansion of prey choice to include a wider range of smaller vertebrates (Stiner 2013; Zeder 2012), is beyond the scope of this chapter. However, the importance of culturally grounded niche construction for transforming subsistence strategies, including those for vertebrate prey (Zeder 2012), points to another fundamental difference between humans and nonhuman primates. Less risky and more efficient hunting made possible by more sophisticated technology, plus increased ability to acquire and transmit knowledge of prey ecology and of ecological opportunities generally, would have provided survival and reproductive benefits that contributed to the successful dispersal of modern humans and to their replacement of other hominins (K. Brown et al. 2012; Marean 2015, 2016). Arguably, the fully modern complex cognition and hyper-prosociality and obligatory cooperation in subsistence and infant and childcare envisioned by the "embodied capital" hypothesis of Kaplan et al. (2000) emerged in concert with these fundamental shifts in foraging strategies and with concomitant shifts in between-group resource competition (K. Brown et al. 2012; Marean 2014, 2015).

OPEN QUESTIONS AND FUTURE RESEARCH DIRECTIONS

Identifying the exact nutritional importance of hunting and meat eating for most nonhuman primate taxa in which such behavior occurs will be difficult, given its low frequency, although continued collection of data on incidental meat-eating events will be useful, especially for small-bodied primates for which vertebrate prey, when eaten, probably constitutes rich packages of energy and nutrients. Continued collection of detailed data on the nutritional composition of PSFs eaten by chimpanzees (e.g., Hohmann, Potts, et al. 2010) holds promise for identifying specific ways in which ASFs, including but not restricted to meat, have supplementary or complementary nutritional value. Better estimates of meat intake would be helpful. These have sometimes (e.g., D. Watts and Mitani 2002a) been based on limited data on prey body masses without correction for the proportion of edible tissue. The results are global estimates at best, but they can provide relative measures of individual variation in intake when combined with behavioral data. Boesch (1994) and Gilby (2005) tried to measure intake rates directly; the measurement error is unknown, and similar efforts by other researchers would be worthwhile. Further use of isotopic analysis (Fahy et al. 2013) would provide more data on variation in relative meat consumption. Data on variation in the nutritional composition of prey tissues would also be valuable. For example, viscera and skeletal muscle presumably do not provide the same mix of nutrients.

Albiach-Serrano's (2015) revised categorization of behavioral cooperation combined with a review of the experimental literature on cooperation by chimpanzees in captivity would be a good starting point for rethinking questions about cooperative hunting in the wild. However, field researchers still face the problem that constraints on visibility limit the ability to apply her scheme in an unbiased fashion. Moreover, experiments provide the only way to test hypotheses about mental state attribution, although these must be grounded in observations made in the wild to have ecological validity.

Finally, despite decades of attention, many questions about the importance of hunting and meat eating in human evolution remain open. Research on these questions is thriving, thanks to isotopic analysis, refinements to methods used to identify causes of modifications to animal bones, and other methodological developments. Compelling evidence supports the hypotheses that meat eating assumed an importance for *Homo* far surpassing its importance for any nonhuman primate not long after the genus emerged and that it subsequently profoundly influenced human social and life-history evolution, but disagreement still exists regarding whether, in H. Bunn's (2007) words, "meat made us human." Exciting potential to resolve this disagreement exists.

SUMMARY AND CONCLUSIONS

Plant source foods dominate the diets of most primates, but nearly all consume animal source foods, which contribute substantially to nutrient intake in some taxa. Many nonhuman primate species occasionally hunt (capture terrestrial vertebrate prey) and eat meat (ingest vertebrate tissues), although usually only in small quantities. Hunting and meat eating are most common in capuchins, baboons, and, especially, chimpanzees, but meat contributes much less even to chimpanzee diets than it does or did to those of human hunter-gatherers. Chimpanzees sometimes obtain large amounts of calories and protein from meat, and individuals of other nonhuman primate species occasionally gain substantial calories and protein from meat. However, for many or most nonhuman primates that eat meat, it may be most important as a source of micronutrients, especially those scarce in most plant source foods, and eating even small amounts could bring important micronutritional payoffs.

In theory, individuals gain greater energetic payoffs by hunting cooperatively than by hunting alone. Discussion of cooperative hunting in nonhuman primates focuses mostly on chimpanzees, but whether they hunt cooperatively in this sense is an open question, as is the extent to which individuals coordinate their behavior during group hunts, as opposed to independently pursuing strategies that maximize the probability of capturing at least one prey item. Meat transfers, which are particularly common following successful hunts by chimpanzees, reduce skew in individual meat intake; these occur for multiple reasons, and most involve meat possessors allowing others to take portions of meat. Active transfers qualify as cooperation in that meat possessors behave in ways that benefit others.

By at least two million years ago, meat eating became much more important for one or more hominin taxa than for any nonhuman primates. Considerable evidence indicates that Oldowan and early Acheulean hominins had regular access to intact or nearly intact animal carcasses that served as sources of meat and bone marrow and that they obtained many of these by hunting, and thus that the human predatory pattern, characterized by regular hunting of prey larger than the hunters (J. Thompson et al. 2019), had been established. The initial dispersal of hominins into temperate habitats would have been impossible without extensive reliance on meat eating. Evidence for coordinated group hunts of large prey exists in the Middle Pleistocene European archaeological record, and Neanderthals have been called "top predators" on the basis of isotopic data. The major nutritional payoffs available from skilled hunting, especially of large mammals, probably had an important role in the evolution of modern human life histories.

- Many nonhuman primates at least occasionally capture and consume vertebrate prey, but in most cases the dietary contribution of meat (vertebrate tissues) is quantitatively minor.
- Capuchins, baboons, and especially chimpanzees hunt more often and eat more meat than any other nonhuman primates, although even chimpanzees eat far less meat than typical for human hunter-gatherers.
- Meat is a valuable source of protein and, when available in relatively large quantities, of energy. However, it may be most important as a source of micronutrients that are uncommon in foods of plant origin; this makes access to even small amounts nutritionally valuable.
- Disagreement exists about whether group hunts by chimpanzees yield net energetic payoffs to all participants and whether group hunting tactics include active collaboration or just by-product mutualism. Meat transfers following successful hunts occur for multiple reasons, although disagreement also exists about the extent to which they involve cooperation.
- By at least two million years ago, the nutritional importance of hunting and meat eating for at least some hominins far exceeded its importance for any nonhuman primates. Technology allowed hominins to obtain prey larger than themselves and to process carcasses in ways not possible for nonhuman primates. The initial Pleistocene dispersal of *Homo* into temperate habitats depended on year-round meat eating, and strong archaeological evidence exists for collaborative group hunts of large mammals by the middle Pleistocene.
- The cognitive demands of skilled hunting and the high nutritional returns from successful predation on large mammals quite likely had crucially important roles in human life-history evolution and in the unique human mode of biocultural niche construction.

18 Movement Ecology and Feeding Neighborhoods

Margaret C. Crofoot and Shauhin E. Alavi

The search for food drives many movement decisions of primates. Whether it is baboons marching across the savanna or gibbons swinging through the trees, to remain healthy, primates must not only locate sufficient calories but also find the necessary balance of micro- and macronutrients (Chapman, Rothman, and Lambert 2012; Felton, Felton, Raubenheimer, et al. 2009) while avoiding antifeedants that are toxic or inhibit digestion (Wrangham, Conklin-Brittain, and Hunt 1998). In some cases, such as forays to clay licks or salt flats, movement decisions have obvious goals that can be clearly linked to dietary requirements (e.g., Fashing et al. 2007). More often, however, primates' ranging patterns are part of an inscrutable dietary selection process—why does a group pass by one fruit-laden tree to feed in another of a different species nearby?

To understand the movement strategies underlying primates' foraging behavior, it is critical to both recognize key motivations—what are individuals trying to accomplish as they move through their habitat?—and be able to assess performance—how efficiently do they achieve these goals? Most studies of primate ranging behavior, however, focus on emergent patterns of space use (e.g., home range size, day range length, and degree of range overlap), which are poorly suited to answering either of these questions. It is the dynamic decision-making process underlying these larger-scale patterns that can best shed light on how evolution has shaped primates' movement strategies to meet their physiological needs. A major challenge in primatology is thus to move beyond monitoring where our study animals go and investigate, at a mechanistic level, how they get there and why.

FROM PATTERN TO PROCESS: PRIMATE NUTRITION WITHIN A MOVEMENT ECOLOGY FRAMEWORK

The emerging field of movement ecology provides a conceptual framework for linking patterns of primate space use to the movement processes that drive them (Nathan et al. 2008). This framework focuses on four fundamental drivers of animal movement—the internal state, motion capacity, and navigation capacity of individuals, and the external features of the environment that influence them—and seeks to explain how these factors interact to shape patterns of movement across the range of scales from specific displacement decisions to lifetime tracks (fig. 18.1; Nathan et al. 2008).

Internal State

What motivates an animal to move? The answer to this question lies in an individual's physiological and, in some cases, psychological state, which spurs it to accomplish one or more goals. Nutritional considerations have a major impact on animal movement, influencing decisions about whether to move as well as shaping the way in which individuals travel through their habitat (McIn-

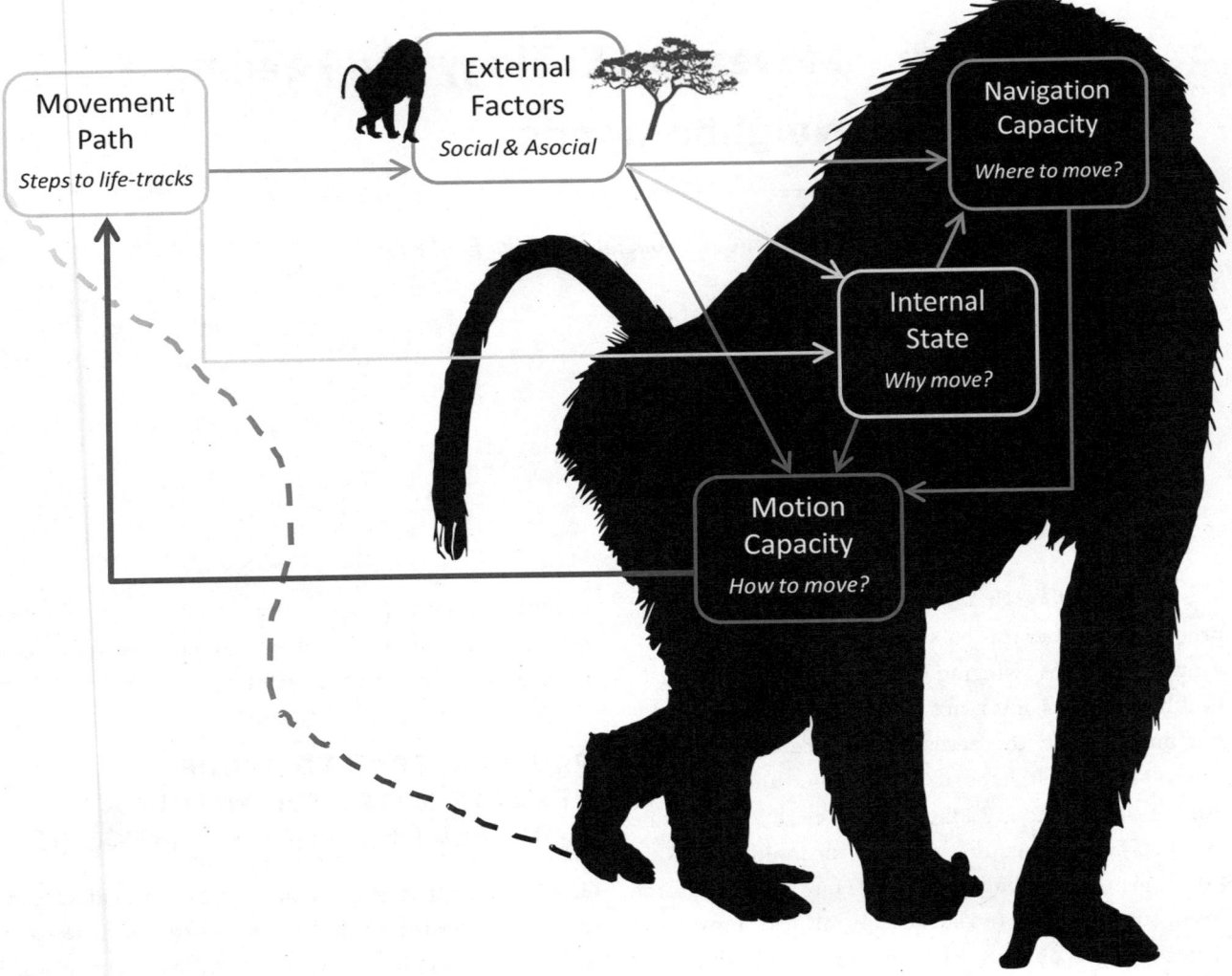

Figure 18.1. The movement ecology framework (redrawn from Nathan et al. 2008).

tyre and Wiens 1999; Spiegel et al. 2013). For example, Spiegel and colleagues (2013) show that important aspects of vulture ranging behavior, including how far and how high they flew, changed depending on the amount of time that had passed since they last fed. Energetic status has also been shown to have important implications for social movement, with hungry individuals acting as group leaders because of slight changes in their movement tendencies, including their travel speed, and the range at which they show attraction to other members of their group (Conradt et al. 2009). In primates, evidence suggests that satiation may influence both patch choice and the duration of visits to food sources (Plante et al. 2014). However, rigorously testing how aspects of primates' nutritional state influence their movement (or how movement decisions change internal state) is complicated by the fact that such states are hidden from observers and generally cannot be experimentally manipulated in studies of free-ranging individuals. It may be possible to accurately infer hunger/satiety based on past behavior (e.g., time since food was last consumed), and, with sampling conducted at appropriate scales, noninvasive methods for assessing energy balance (e.g., C-peptide of insulin measured in urine; Emery Thompson and Knott 2008; Emery Thompson, Muller, Wrangham, et al. 2009) may provide even greater insight into the interaction between energetic status and movement in free-ranging primates (Bergstom 2015; Girard-Buttoz, Higham, et al. 2011; Grueter, Deschner, et al. 2014). However, more nuanced nutritional considerations (e.g., macro- or micronutrient balancing) may also be crucial drivers of primate movement but are largely inaccessible to observers without more invasive physiological monitoring (but see E. Vogel, Crowley, et al. 2012).

Motion Capacity

Biomechanical adaptations determine how far, how fast, and through which environments (or strata of the environment; Harel, Alavi, et al. 2022) animals can move. This capacity for motion has important nutritional consequences because it constrains the set of resources individuals are able to exploit. For example, a shoulder morphology specialized for brachiation makes it possible for gibbons to cross larger gaps in the forest canopy than sympatric macaques that engage in quadrupedal locomotion (Cannon and Leighton 1994) and thus to minimize travel costs by using more direct routes between resources (Cannon and Leighton 1996). Whereas long-tailed macaques ranging in a Bornean rainforest face a highly fragmented environment, gibbons moving through the same area experience a functionally continuous canopy—a difference that allows them to visit more fruit patches per day (Cannon and Leighton 1996) and possibly maintain a higher-quality diet (A. Marshall, Cannon, and Leighton 2009). While motion capacity might appear to be a static characteristic of individuals or species and thus relevant only to comparative studies of movement ecology, it is important to remember that primates often use distinct modes of locomotion depending on environmental conditions (Furnell et al. 2015; Guillot 2011; Schoonaert et al. 2016; Shapiro et al. 2016) or their own internal state. Because gaits differ in their energetic implications (e.g., O'Neill 2012), biomechanics—how, in a mechanistic sense, individuals get from point A to point B—is key to understanding the efficiency of primate foraging decisions. However, a fundamental mismatch in the spatial scales at which the biomechanics and ecology of movement are studied impedes our ability to understand how context-dependent changes in locomotor mode shape larger-scale patterns of primate foraging behavior. The use of biotelemetry—namely, three-dimensional accelerometers—to remotely monitor modes of locomotion in free-ranging individuals at large spatial scales (D. Brown et al. 2013; Harel, Carter Loftus, and Crofoot 2021; Sellers and Crompton 2004) may make it possible to overcome this obstacle and is a promising avenue for future research in primate movement ecology.

Navigation Capacity

What animals know about their environment is crucial for understanding observed patterns of movement and space use (Fagan et al. 2013). At the same time, movement itself shapes the information that animals acquire due to the effects of habitat sampling, speed-accuracy trade-offs, and cognitive limitations (Spiegel and Crofoot 2016). To meet their nutritional requirements, all animals must use information acquired via their senses or from their memory about what food resources exist in the environment, where they are located, and when they are available to time and orient their travel toward appropriate goals. Some evidence suggests that there are differences in the movement patterns of primates, compared to other animals that rely on similar types of foods, that can be explained by differences in their sensory systems, travel costs, and patch exploitation dynamics (Hirsch et al. 2013). However, Zuberbühler and Janmaat (2010) have argued that primates possess a suite of cognitive adaptations that allow them to integrate what/when/where information in surprisingly complex ways (e.g., Asensio, Brockelman, et al. 2011; E. Cunningham and Janson 2007; Erhart and Overdorff 2008; Janmaat, Byrne, and Zuberbühler 2006a, 2006b; Janmaat and Chancellor 2010; Janson 1998a, 2007; Normand et al. 2009; Normand and Boesch 2009; Presotto and Izar 2010; Valero and Byrne 2007) and thus to exploit food resources more efficiently than sympatric resource competitors. Wild capuchins, for example, appear capable of path integration, choosing multisite foraging routes that maximize energy intake over visits to several different resources (Janson 2007). Gray-cheeked mangabeys seem to take past weather conditions and fruit ripening dynamics into account when searching for food, behavior consistent with the capacity for episodic memory (i.e., mental "time-travel"; Janmaat, Byrne, and Zuberbühler 2006b). Some primates also appear to use Euclidean mental maps (where relationships between known locations are encoded as a set of angles and distances), rather than cognitively simpler (and, in some instances, less efficient) route-based maps, to navigate in their environment (Luhrs et al. 2009; Normand and Boesch 2009; Presotto and Izar 2010). Interestingly, primates appear to vary the degree to which they rely on spatial memory in their foraging decisions depending on the distribution and abundance of food resources in the habitat. White-faced sakis show clear evidence of memory-based foraging during times of resource abundance when foods are dense and patchily distributed but not during periods of low food availability (E. Cunningham and Janson 2013). Unfortunately, relatively few studies have explored the ecological cognition of primate and nonprimate frugivores in comparable ways (Corlett 2011), and thus the degree to which navigation capacity varies from species to species and the impact it has on movement patterns and foraging success remain largely unknown (Corlett 2011; Janmaat, de Guinea, et al. 2021; Zuberbühler and Janmaat 2010).

Environment

Observed patterns of animal movement emerge from an interaction between the three attributes of moving individuals, discussed above, and features of the environment they are traveling through. For example, for many primates, predation risk varies depending on the vegetation structure of the habitat, creating dynamic "landscapes of fear" that impact movement behavior (Willems and Hill 2009). Asocial aspects of the environment, including topography (Di Fiore and Suarez 2007; T. Gregory et al. 2014; Howard et al. 2015), vegetation type and density (Buchin et al. 2015; Strandburg-Peshkin, Farine, Crofoot, et al. 2017; Willems and Hill 2009), canopy structure (McLean et al. 2016; Palminteri, Powell, Asner, et al. 2012), and food type and location (Asensio, Brockelman, et al. 2014; Plante et al. 2014; Suarez 2014) have well-documented impacts on primate movement decisions and patterns of space use. However, social factors can also modulate individual movement. A rapidly expanding body of research focuses on understanding how group-living animals influence one another's behavior to produce coordinated patterns of movement (Conradt and List 2009; Conradt and Roper 2003; Couzin and Krause 2003; Couzin, Krause, et al. 2005). Research on collective movement in primates has shed light on how groups of individuals reach consensus about travel direction (e.g., A. King, Douglas, et al. 2008; Strandburg-Peshkin, Farine, Couzin, et al. 2015a) and the timing of group travel (A. King, Sueur, et al. 2011; Meunier et al. 2006; Petit et al. 2009; Stueckle and Zinner 2008; Sueur 2011; Sueur, Petit, and Deneubourg 2009). However, few studies of any species have explicitly investigated how social interactions impact movement via their influence on individuals' internal states, motion capacity, or navigation capacity (but see Bazazi et al. 2011) or explored how social and asocial features of the environment interact to shape patterns of movement (Basille et al. 2015; Latombe et al. 2014; Strandburg-Peshkin, Farine, Crofoot, et al. 2017). Such an integration of collective movement into the movement ecology paradigm is a promising avenue for future research (Bonnell, Campenni, et al. 2013; Strandburg-Peshkin, Farine, Crofoot, et al. 2017; H. Williams and Safi 2021).

IMPLEMENTING A MOVEMENT ECOLOGY FRAMEWORK IN STUDIES OF PRIMATE DIET AND NUTRITION: OBSTACLES AND OPPORTUNITIES

Compared to many mammalian species, primates are quite amenable to study within a movement ecology framework because they can be directly observed. Most species are diurnal and habituate quickly to the presence of human observers, allowing researchers to directly record data on their behavior, interactions, and movements. While this makes it possible to collect extremely detailed information about primate diets and foraging behavior (Rothman, Chapman, and Van Soest 2012), a number of challenges remain when it comes to linking the nutritional causes and consequences of primate movement.

1. A fundamental mismatch often exists between the spatial and temporal scales at which animals make decisions about foraging movements and the scale at which researchers collect movement data. Because many of their food resources are found in relatively small, discrete patches (e.g., ripe fruit, young leaves, and larvae of social insects), primates tend to have relatively short patch residence times and move frequently between food patches. However, movement data have traditionally been collected at much coarser spatial and temporal scales (e.g., 50 m × 50 m grid cells within

the study site; one location estimate recorded every 30 minutes) that cannot capture these dynamics. The improved performance of Global Positioning Systems (GPS) in forested environments (Crofoot 2021), as well as its increasing adoption in field-based primatology, provides an opportunity to overcome this limitation. Whether integrated into tags attached to study animals (e.g., Buchin et al. 2015; Farine, Strandburg-Peshkin, et al. 2016; Markham and Altmann 2008; Markham, Alberts, and Altmann 2012; Markham, Guttal, et al. 2013; Strandburg-Peshkin, Farine, Couzin, et al. 2015a; Strandburg-Peshkin, Farine, Crofoot, et al. 2017) or handheld units carried by researchers, GPS tracking technology can produce the high-resolution movement data needed to explore primate foraging decisions at appropriate spatiotemporal scales.

2. Many key drivers of behavior within the movement ecology framework are not observable. Hunger, thirst, fatigue, health—all these dimensions of a primate's internal state are expected to impact their movement decisions. While some can be inferred based on current or past behavior, others cannot, and the accuracy of such inferences is generally unknown. Although rarely used in field-based primatology, biotelemetry provides another avenue for monitoring internal states. Small, energy-efficient sensors can be fit to study animals and used to log important aspects of behavior and physiology (D. Brown et al. 2013; R. Wilson et al. 2008). For example, sensors that record body temperature can be used to track the incidence of fevers (Adelman et al. 2014) and may be key to understanding behavioral responses to disease (McFarland, Henzi, et al. 2021). Similarly, heart-rate monitors can provide otherwise-unobtainable insight into both levels of arousal (A. Davies et al. 2014) and energetics (Tomlinson et al. 2014) and how these impact the foraging behavior and movement decisions of free-ranging primates.

3. Due to the diversity of their diets, fine-scale mapping of the environment is required to capture the full range of foraging options available to most primate species. Navigation capacity—what individuals know about their habitat and how they use this information to orient travel—remains a black box (Janson and Byrne 2007). The kinds of carefully controlled experiments needed to rigorously test hypotheses about cognition are difficult, if not impossible, to conduct in the wild (but see Janson 1996, 1998a, 2007), while replicating ecologically relevant conditions in captive settings is often infeasible. However, combining movement data with detailed information about the distribution of resources in the habitat can provide important insight into what animals know about their habitat and how they use this information (Asensio, Brockelman, et al. 2011; Ban et al. 2014; Di Fiore and Suarez 2007; Janmaat, Byrne, and Zuberbühler 2006a, 2006b; Janmaat, Polansky, et al. 2014). For species with small home ranges or specialized diets, it is sometimes possible to identify and map all possible food sources by hand (black-and-white colobus, T. R. Harris 2006; white-handed gibbon, Asensio, Brockelman, et al. 2011; common marmoset, C. Thompson et al. 2013). In many cases, however, this is not logistically feasible because primates tend to live in large home ranges (tens to hundreds of hectares) and consume hundreds of species of plants and animals (e.g., Quéméré et al. 2013; Crofoot, unpubl. data). The increasingly widespread use of unmanned aerial vehicles (i.e., drones) in field research (Ivosevic et al. 2015; J. Zhang et al. 2016) may facilitate this kind of data collection. With drone-based aerial photography, it is possible to map the distribution of certain visually distinctive tree species at much larger spatial scales than is possible using ground-based surveys (Caillaud et al. 2010; Jansen et al. 2008), while hyperspectral imaging holds promise for our ability to locate and identify a wide range of important food species via remote sensing (Asner and Martin 2009; Feret and Asner 2011).

4. Traditional methods of behavioral observation are inherently unsuited to answering questions about the dynamics of social foraging. In gregarious species, group-mates play a central role in defining the environment an individual inhabits. While observational data collection has allowed primatologists to collect rich, highly detailed data on the social interactions of wild primates as they go about their daily lives, these traditional methods are limited in their usefulness for studying collective phenomena such as group foraging because it is usually impossible to collect data on more than a few individuals at a time. Group-mates may differ dramatically in their preferences about what to eat, where to search, and when to leave one patch to travel to the next, and how these differences of opinion are

integrated to yield a single, group-level decision is a key but to date largely unstudied aspect of primate movement ecology (but see A. King, Douglas, et al. 2008). As Byrne observed in his review of collective movement in primates, "It is an observational task of daunting dimensions to attempt to record the actions of many potential decision-maker animals at once" (2000, 501). Remote-tracking technology, however, can be used to simultaneously monitor the movements and behaviors of entire groups, providing new and exciting opportunities to tackle previously intractable questions about collective movement (Farine, Strandburg-Peshkin, et al. 2016; Kays, Tilak, et al. 2011; Strandburg-Peshkin, Farine, Couzin, et al. 2015a) and social foraging in primates.

DIVERSE GROUPS, DIVERSE NEEDS: CONFLICTS OF INTEREST AND SOCIAL FORAGING IN GROUP-LIVING PRIMATES

Most primates are gregarious, and thus their movements, rather than simply reflecting the choices of a single individual, instead emerge from the integration of and negotiation between the needs and capabilities of multiple group-mates. Primate groups tend to be heterogeneous, with individuals who differ in age, sex, size, reproductive status, and rank ranging together in search of food. The dietary requirements of these classes of individuals differ (NRC 2003), as do their ability to extract resources from the habitat (S. Altmann 1998; Gunst et al. 2008; S. Perry 2009) and the costs they pay to travel between resources (J. Altmann and Samuels 1992; Caperos et al. 2012; Pontzer and Wrangham 2006; Wunderlich et al. 2011), making conflicts of interest over when, where, and what to eat inevitable. The social dynamics of primate groups exaggerate such conflicts by creating disparities in the nutritional landscapes that group members face. For example, when food distribution is heterogeneous, high-ranking individuals are often able to monopolize the best feeding sites (Janson 1985). This priority access profoundly alters their foraging experience compared to that of subordinate group-mates (H. Marshall et al. 2015) and creates a situation in which the best interests of high- and low-ranking group members diverge (A. King, Douglas, et al. 2008).

Furthermore, differences in experience, memory, or sensory capacity will lead group-mates to differ in what they know about the location and state of food resources in the habitat and thus in their preferences about when, where, and how their group should forage.

Social formulations of well-known optimal foraging theory models (Giraldeau and Caraco 2000) provide novel insights into the dynamics of groups of foraging animals and how the interests of coforagers may diverge (G. Davis et al. 2022). Charnov's (1976) marginal value theorem, for example, can be extended to predict the optimal patch departure time for socially foraging individuals as well as for their group as a whole (fig. 18.2; Giraldeau and Caraco 2000; Livoreil and Giraldeau 1997). This model clearly illustrates how differences in individual behavior—food-processing ability or intake rate, for example—can create conflicts of interest over collective foraging decisions (G. Davis et al. 2022). However, social foraging theory models generally assume fission-fusion social dynamics, where individuals are free to leave if their interests do not align with those of other group members. With a few exceptions, primates are social foragers that travel from food patch to food patch as a cohesive unit despite conflicts of interest that may be at play in these foraging decisions.

The need to reach consensus imposes a constraint on primate behavior that has yet to be formally incorporated into existing models of social foraging but must be taken into account when attempting to understand how individuals' dietary needs are linked to patterns of group movement and space use. In stable social groups, the process by which individuals reach consensus (i.e., who has influence over collective decisions) will not only shape what group members eat, and when and where they eat it, but also determine which individuals bear the costs of compromise (consensus costs; Conradt and Roper 2005). Shared decision-making where influence is evenly distributed among all group members will, under many conditions, minimize consensus costs for the group as a whole because democratic decisions tend to be less extreme than the decisions of despotic leaders. Unshared (or partially shared) decisions, while beneficial to one or a few leaders, typically impose significant consensus costs on the other members of the group who must compromise their own preferred patterns of behavior to maintain group cohesion

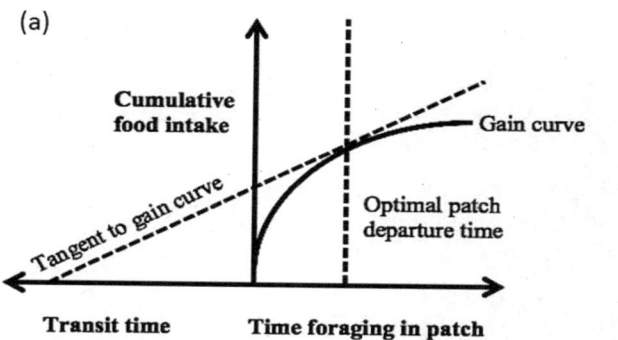
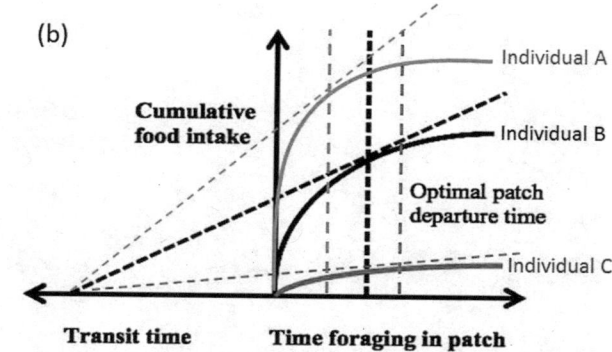

Figure 18.2. When consuming a depleting resource, individuals must make a choice about when to stop feeding in one patch and move to the next (J. Brown 1988). Such decisions in solitary foragers can be predicted based on the decline in their food intake rate within a patch, illustrated as a "gain curve"—the total amount of food obtained as a function of the time spent foraging in a given patch (a; Charnov 1976; D. Stephens and Krebs 1986). In socially foraging species, within-group feeding competition or interindividual variation in feeding behavior may lead to differences in the shape of each group member's gain curve and thus their optimal patch departure times (b; Olsson et al. 2001). The cost an individual incurs by leaving a patch before its preferred departure time to maintain group cohesion (its consensus cost) can be quantified by estimating the discrepancy between the individual's optimal departure time and the group's observed departure time from the patch.

(Conradt and Roper 2007, 2009). Because in all group-living species, individuals have an interest in minimizing their own consensus costs, two interconnected questions must be addressed to understand primate foraging decisions. First, studies need to examine how the dynamics of social foraging shape individuals' preferences about where to go and what to eat and quantify the resulting conflicts of interest among group-mates. Second, it is critical to recognize that observed patterns of foraging behavior are the result of a consensus decision-making process, and how individuals' preferences are aggregated to yield these group-level decisions will determine whose nutritional requirements are reflected in observed patterns of movement (fig. 18.3).

SUMMARY

A suite of empirical and theoretical advances are paving the way for a new and exciting epoch in research on the connection between the nutritional and movement ecology of wild primate populations. Remote sensing technology provides a welcome addition to the primatological toolkit, complementing the fine-scale dietary information that can be collected only via direct observation of habituated animals (Rothman, Chapman, and Van Soest 2012) with previously inaccessible information about the internal state of foragers (e.g., McFarland, Barrett, Fuller, et al. 2017), the small- and large-scale movement decisions of individuals and groups (Strandburg-Peshkin, Farine, Couzin, et al. 2015a; Strandburg-Peshkin, Farine, Crofoot, et al. 2017), and a detailed understanding of the energetic landscape that primate groups navigate (Caillaud et al. 2010). A merging of theoretical models of the behavior of socially foraging individuals (Giraldeau and Caraco 2000) with the rapidly growing body of literature on group decision-making (Bazazi et al. 2011; Conradt and Roper 2009; Couzin, Ioannou, et al. 2011; Gavrilets et al. 2016) promises to yield a suite of new predictions about how differences in the needs and capabilities of group-mates will influence patterns of foraging behavior in group-living species. However, there are well-known limitations to the optimal foraging theory framework (e.g., its focus on energy maximization), and the incorporation of more nuanced assumptions about the nutritional goals of foragers (e.g., nutrient balancing; Raubenheimer, Simpson, and Mayntz 2009) into social foraging models is a promising area for future research (J. Johnson et al. 2015).

Figure 18.3. Capuchin monkeys moving through their environment and foraging, including with tool use (*bottom panel*).

Photos: *Top, middle,* Christian Ziegler; *bottom,* courtesy of Max Planck Institute of Animal Behavior.

Wild primates navigate a complex environment replete with predators and competitors, and nutritional concerns will rarely be the sole or perhaps even the primary factor driving their movement decisions (McArthur, Orlando, et al. 2012; Willems and Hill 2009). Foraging is a multidimensional problem, and behaviors taken in pursuit of one goal may negatively impact another. Experimental approaches will likely be useful for explicitly addressing the trade-offs that free-ranging animals face—for example, between their nutritional requirements and their need to avoid predators (Emerson and Brown 2012, 2013; McArthur, Orlando, et al. 2012). Agent-based foraging models provide a flexible platform for exploring how animals balance multiple needs and can generate nonintuitive predictions about the dynamics of social foraging (Bonnell, Campenni, et al. 2013) that serve as inspiration for further field-based research.

CONCLUSIONS

- A new urgency is driving efforts to understand the mechanistic underpinnings of animals' movement patterns due to the threats posed to so many populations by human-induced rapid environmental change.
- While technological advances are making it possible to monitor the movements of many species in unprecedented detail, at scales both small and large, direct observation of behavior remains key to understanding why animals move when and where they do.
- For this reason, primates are uniquely suited for addressing questions about how nutritional considerations impact ranging patterns and space use and vice versa.
- Exciting opportunities exist in research that merges the theoretical frameworks and empirical toolkits of nutritional ecology, collective behavior, and movement ecology.

19 Foraging in a Landscape of Fear

Russell Hill

Fear, n.: The emotion of pain or uneasiness caused by the sense of impending danger, or by the prospect of some possible evil.

OXFORD ENGLISH DICTIONARY

The incorporation of fear into ecology is a relatively new concept (Laundré, Hernández, and Ripple 2010). Initially introduced by J. Brown, Laundré, and Gurung (1999) as the "ecology of fear" and applied to traditional ecological predator-prey models, the concept of fear was also proposed at a similar time by Laundré, Hernandez, and Altendorf (2001) and Altendorf et al. (2001) as useful in explaining foraging patterns of animals. In doing so, they introduced the term "landscape of fear" as a visual model to help explain how fear could alter an animal's use of an area as it tries to reduce its vulnerability to predation. Conceptually, the landscape of fear represents relative levels of predation risk as peaks and valleys that denote the fear of predation a prey perceives in different parts of its home range (Laundré, Hernández, and Ripple 2010). This landscape can be considered alongside the others that an animal inhabits, such as vegetation structure or food availability. Nevertheless, the most important landscape to an animal is often considered to be its landscape of fear (J. Brown and Kotler 2004). Despite this, the concept of the landscape of fear (table 19.1) has found little traction in primatology, though many primate species and primatological methods have the potential to place the discipline at the forefront of the field. To illustrate this fact, this chapter reviews the ecological literature on the landscape of fear, focusing initially on methods for measuring it in foraging contexts. These methods range from behavioral to experimental approaches and have all been employed in primatology to some extent, although there is considerable scope for further work. The various anthropogenic impacts on the landscape of fear are then explored.

In evaluating the importance of predation on prey species, the distinction between the direct or consumptive effects of predation (predation rate) and the indirect or nonconsumptive effects (predation risk) is critically important (Creel 2011; Creel and Christianson 2008; Gaynor, Brown, et al. 2019; R. Hill and Dunbar 1998; Laundré, Hernandez, et al. 2014; Matassa and Trussell 2011; Peacor et al. 2020). While predators affect prey demography and social structure through direct predation, risk effects can influence behavior through changes in habitat use (Creel, Winnie, et al. 2005; Valeix et al. 2009), vigilance (Childress and Lung 2003; L. Hunter and Skinner 1998), foraging (S. Lima and Bednekoff 1999), aggregation (Boesch 1991a; R. Hill and Lee 1998), and movement patterns (Fortin et al. 2005). In turn, the costs of these responses can reduce survival, growth, or reproduction (Creel, Christianson, et al. 2007; Ruxton and Lima 1997). Changes in the landscape of fear can alter predator-prey relationships and so produce cascading effects at a variety of ecological levels (Laundré, Hernández, and Ripple 2010). Empirical research has shown that risk

Table 19.1. Key terms used in fear ecology.

Term	Definition
Landscape of fear	"Represents relative levels of predation risk as peaks and valleys that reflect the level of fear of predation a prey experiences in different parts of its area of use" (Laundré, Hernández, and Ripple 2010). "The spatially explicit distribution of perceived predation risk" (Bleicher 2017). "The spatial variation in prey perception of predation risk" (Gaynor, Brown, et al. 2019).
Predation risk	"The probability of being killed by a predator" (S. Lima and Dill 1900). "The likelihood of a prey animal being killed by a predator. Predation risk varies at multiple scales in space and time and among individual prey" (Gaynor, Brown, et al. 2019).
Indirect/risk effects	"The costs of antipredator behaviours that reduce direct predation but carry fitness costs, such as vigilance and retreat to safe habitats" (Creel and Christianson 2008). "The nonlethal or non-consumptive effects of predators on a prey population, brought about by costly antipredator behavior that affects survival and reproduction" (Gaynor, Brown, et al. 2019).
Predation rate	"The annual mortality within a population directly attributable to predation; it represents the level of successful predator attacks that the animals are unable to control after they have implemented their antipredation strategies" (R. Hill and Dunbar 1998). (Analogous to direct predation effects [Creel and Christianson 2008]).
Intrinsic risk	"The expected predation rate of a prey population holding all anti-predator strategies (both behavioral and morphological) constant" (Janson 1998b).

effects on prey dynamics can be as large as direct effects, or even larger (E. Nelson et al. 2004; Preisser, Bolnick, and Benard 2005), highlighting the importance of quantifying this landscape in studies of foraging ecology.

Spatial variation in predation risk is one of the most important drivers of the nonlethal effects of predation (Cresswell and Quinn 2013). Implicit in the concept of the landscape of fear, therefore, is that animals can learn to differentiate dangerous versus safe habitats before they are killed (Laundré, Hernández, and Ripple 2010). Evidence from predator eradications and reintroductions suggests that animals do have the ability to learn and respond to spatial changes in predation risk (Laundré, Hernandez, and Altendorf 2001; Ripple and Beschta 2003, 2004). As a consequence, if predation risk varies physically over space and time, we should be able to actually measure that risk relative to the specific prey-and-predator system we are considering (Laundré, Hernández, and Ripple 2010). Of course, these risk effects can be among the most difficult to quantify (Creel and Christianson 2008), and predation risk, the landscape of fear, and prey antipredator responses my map imperfectly onto each other (Gaynor, Brown, et al. 2019).

Traditionally, foraging studies have considered the impacts of a number of key aspects of predation risk. The most notable aspects are the likelihood of detecting predators before attack and the subsequent probability of escape, and landscape features influence them both (S. Lima and Dill 1990). Prey species should thus alter their use of space and their foraging activity with regard to landscape features (Ripple and Beschta 2003). A series of primate studies have exploited this fact to explore the impact of predation on foraging decisions. Cowlishaw (1997) quantified the food availability and predation risk (based on habitat visibility and refuge distance) of four habitats used by chacma baboons (*Papio ursinus*) at Tasobis, Namibia. The baboons spent less time feeding in the high-risk, food-rich habitat but more time feeding in the low-risk, relatively food-poor habitat, while resting and grooming were confined to the safest but most food-poor habitat. Comparable results were reported by R. Hill and Weingrill (2007) for a South African population. Similarly, blue monkeys (*Cercopithecus mitis stuhlmanni*) emptied their cheek pouches in safer habitats where they were less exposed (L. Smith et al. 2008). Terrestrial foraging behavior of golden-backed uacaris (*Cacajao melanocephalus ouakary*) increased in larger food patches that were close to arboreal refuges and distant from dense ground-based vegetation (Barnett et al. 2012), while Thomas langurs (*Presbytis thomasi*) were cautious and foraged in larger groups when exploiting food patches on the ground, which were assumed to be riskier (Sterck 2002). White-bellied spider monkeys (*Ateles belzebuth*) responded to high levels of predation risk in areas immediately surrounding mineral licks by maximizing subgroup size during lick visitation and with higher fusion rates as animals approached the lick (Link and Di Fiore 2013).

Such studies generally rely on broad classifications of risky and safe habitats or areas, but since landscape features are not uniform, an animal's perception of safe and unsafe areas may change over very short distances. For example, the feeding intensity of elk (*Cervus canadensis*) can drop dramatically in the space of a few meters as they move closer to features such as riverbanks and gullies that can impede escape (Ripple and Beschta 2003). As a consequence, while the landscape-of-fear concept offers significant potential for understanding the trade-off between food availability and predation risk, methods are required that allow assessments of fear and risk on refined spatial scales that go beyond broad classifications of habitat types.

MEASURING THE LANDSCAPE OF FEAR

Internally, fear can be measured via changes in corticosteroid levels, which increase with risk and anxiety (Anson et al. 2013; Aronsen et al. 2015; Clinchy et al. 2013; Creel, Fox, et al. 2002; B. Jones et al. 2016; LaBarge, Allan, Berman, Hill, et al. 2022; Sheriff et al. 2011; Tkaczynski et al. 2014; Wasserman, Chapman, Milton, Goldberg, et al. 2013). Externally, the landscape of fear adds an additional layer to the environmental factors traditionally considered in studies of foraging ecology such as landscapes of productivity, vegetation structure, food availability, soil and physical features, and climate (J. Brown and Kotler 2004). Many of these other landscapes can be measured directly in the field, sometimes with the use of satellite imaging and other remote sensing techniques (Campos, Bergstrom, et al. 2014; Palminteri, Powell, Asner, et al. 2012; Peck, Thorn, et al. 2011; Willems et al. 2009). A variety of methods also exist for quantifying the landscape of fear.

Traditionally, studies exploring the landscape of fear have suggested that it is best assessed through behavioral titrations that allow animals to reveal their foraging cost of predation (J. Brown and Kotler 2004). A range of alternative approaches to these giving-up density methods also exist, however (Laundré, Hernández, and Ripple 2010; Moll et al. 2017). A series of studies have explored the landscape of fear through vigilance observations in a range of taxa, while primates offer additional opportunities through their acoustically distinct alarm calls (B. Coleman and Hill 2014; Willems and Hill 2009). For rodents, frequency of trapping has been used to assess risk landscapes (Kotler 1985; M. Price et al. 1984), while foraging surveys of plants are another alternative (Schmitz, Hamback, and Beckerman 2000; Schmitz, Krivan, and Ovadia 2004) that could offer significant potential for primatology. For example, vegetation changes in Yellowstone National Park documented after wolf reintroduction were suggested to result from changes in the way elk browsed and an avoidance of certain areas due to their fear of wolves (*Canis lupus*; Ripple and Beschta, 2003, 2004). Similarly, C. White et al. (2003) suggested that the distribution patterns of aspen (*Populus tremuloides*) could be explained by predation-driven foraging patterns of elk. As a consequence, the amount of browsing on preferred feeding plants of a prey species warrants investigation in primatology since it could be a valuable method of mapping the landscape of fear. This review, however, concentrates on primate-focused methods to estimate the level of risk perceived by primate prey and so map the landscape of fear. Primatology is at the forefront of some of these approaches such that there is significant potential for greatly increasing our understanding of the role of predation in primate foraging decisions.

ALARM CALL DISTRIBUTIONS

As first described in the classic studies of vervet monkeys (*Chlorocebus pygerythrus*; Seyfarth et al. 1980a, 1980b), a range of primate species have been reported to give acoustically distinct alarm calls to different predator guilds (K. Arnold and Zuberbühler 2006; Fichtel et al. 2005; Kirchhof and Hammerschmidt 2006; Murphy et al. 2013; T. Price et al. 2015; M. Seiler, Schwitzer, Gamba, et al. 2013; Zuberbühler 2000b, 2001). While there is some debate about whether these represent functionally referential signals (Wheeler and Fischer 2012), these distinct and easily recognizable alarm calls and responses to different predator guilds allow perceived predation risk to be estimated relatively easily in some nonhuman primates (Campos and Fedigan 2014; B. Coleman and Hill 2014; Willems and Hill 2009). The need to identify risk from different predator guilds is critical since when prey are subject to attack from several predators that represent different types of risk, the appropriate antipredator responses differ among predators (B. Coleman and Hill 2014; Cresswell and Quinn 2013; Preisser, Orrock, and Schmitz 2007; Shultz et al. 2004). Acoustically distinct alarm calls allow the threats from different predators to be recorded in space and time, and as a consequence they represent a unique opportunity to place primatology at the forefront of studies using behavioral measures to investigate the landscape of fear.

Willems and Hill (2009) first developed this novel approach for creating predator-specific landscapes of fear, allowing the impact of predation risk and the distribution of resources on primate range use to be assessed within a single spatial model. Vervet monkeys in the Soutpansberg Mountains, South Africa, were exposed to predation pressure from leopards (*Panthera pardus*), chacma baboons, raptors (crowned eagle, *Stephanoaetus coronatus*, and African black eagle, *Aquila verreauxii*), and African rock python (*Python sebae*). Through mapping out the locations of the alarm calls to each of these predator types, Willems and Hill (2009) created predator-specific landscapes of fear (signifying the probability of an alarm response occurring at any location within the home range per unit time spent there; fig. 19.1). These landscapes were then

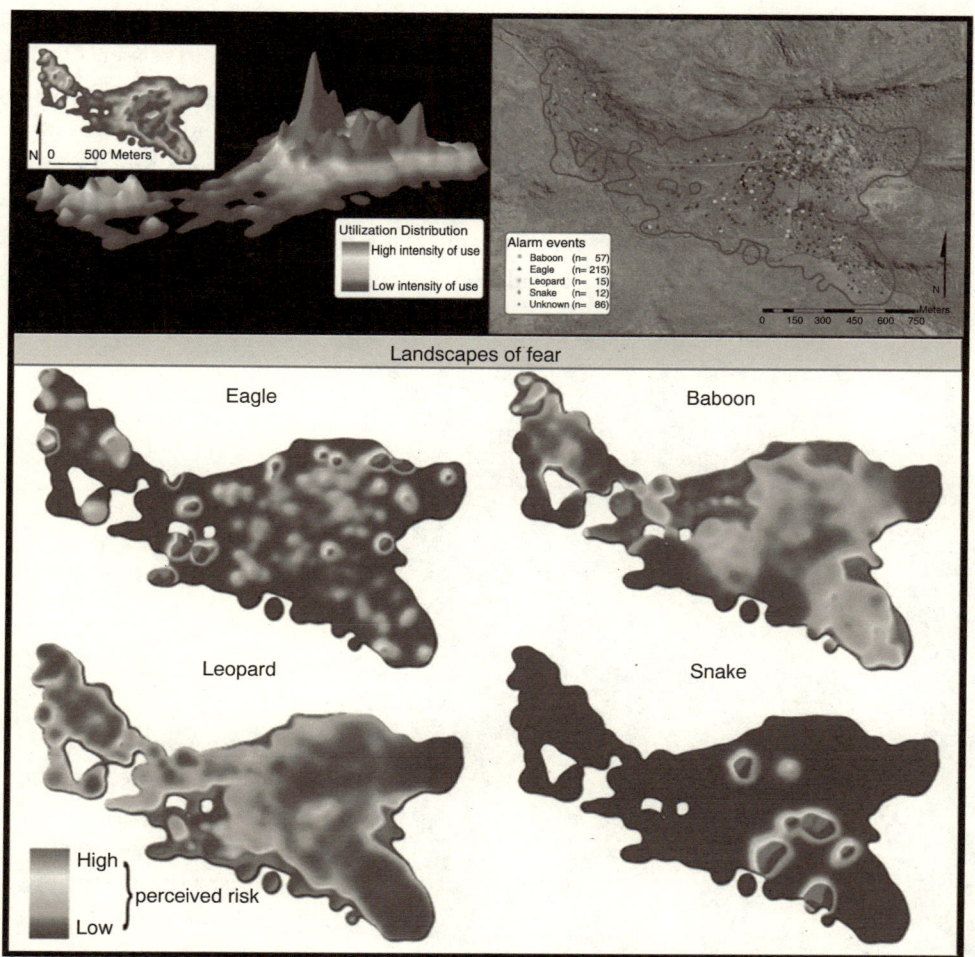

Figure 19.1. Utilization distribution of a group of vervet monkeys in the Soutpansberg Mountains, South Africa. Locations of acoustically distinct alarms calls were used to create probabilistic predator-specific landscapes of fear, based on the probability of an alarm response occurring at each location within the home range per unit time the monkeys spent there (adapted from Willems and Hill 2009).

assessed alongside habitat type (food availability) and distance to key resources such as water and sleeping sites, to determine which had the greatest impact on ranging patterns. The results of this spatial analysis demonstrated that ranging behavior could be interpreted as an adaptive response to perceived risk of predation by some (but not all) predators; leopards and baboons were a significant determinant of vervet monkey space use, but eagles and snakes had no significant effect on ranging, despite eagles soliciting the most alarm calls during the study (Willems and Hill 2009). Most importantly, however, the statistical effects of fear exceeded those of resource distribution in explaining vervet range use, suggesting that the landscape of fear may indeed be the most important landscape experienced by primates.

Willems and Hill (2009) highlighted that their framework could be extended to incorporate the effects of additional factors potentially shaping animal range use. The approach also allowed an assessment of whether primate species inhabiting the same environment, exposed to the same predator guild, and accessing the same resources responded to the risk of predation in the same way. Samango monkeys (*Cercopithecus albogularis schwarzi*) inhabiting the same multipredator environment in the Soutpansberg Mountains as in the original vervet study (fig. 19.2) are reported to give acoustically distinct alarm calls to different predator types (Papworth, Bose, et al. 2008). B. Coleman and Hill (2014) used the locations of these calls to create predator-specific landscapes of fear for the study group and showed that the landscape of fear from eagles (the only discernible risk landscape) was the most significant determinant of samango range use. There was no significant effect of resource availability. Instead, the monkeys also selected areas of their range with higher canopies and higher understory visibility, factors consistent with reducing predation risk from terrestrial predators. The monkeys visited high-risk areas only to forage (E. Parker et al. 2022), and to mitigate this risk, the samangos increased cohesion between group members (LaBarge, Allan, Berman, Margulis, et al. 2020; E. Parker et al. 2022). While reinforcing the conclusion that the effects of predation outweigh those of resource availability in determining primate habitat use, the results indicate that the landscapes of fear experienced by samango monkeys and vervet monkeys appeared to differ despite exposure to identical predator guilds at the site (B. Coleman and Hill 2014; Willems and Hill 2009). This further emphasizes the importance of distinguishing between the risk effects of different predators in understanding prey ecology and behavior (Cresswell and Quinn 2013; Preisser, Orrock, and Schmitz 2007; Shultz et al. 2004) and shows that closely related prey species may respond to local predator guilds in different ways. As a consequence, broad assumptions about the likely importance of specific predators or specific landscape features in different environments could lead to erroneous conclusions.

Figure 19.2. Samango monkeys (*Cercopithecus albogularis schwarzi*) in the western Soutpansberg Mountains, South Africa, show strong responses to the landscape of fear. Photo: Russell Hill.

Studies of wild groups of white-faced capuchins (*Cebus capucinus*) in Costa Rica further highlight the importance of distinguishing between predator types in understanding the importance of predation risk in different habitat types (Campos and Fedigan 2014). Relative risk maps based on alarm calls and thus the locations of potential predator encounters revealed that high-risk areas for raptors and for all predator guilds combined consisted of more mature forest, whereas relatively younger forest appeared lower risk. Nevertheless, relationships for different predator guilds were inconsistent such that habitat type was not a particularly accurate cue on which to base behavioral decisions for reducing predation risk (Campos and Fedigan 2014). These findings highlight the misconceptions that may arise from using simple habitat classifications as proxies for predation risk and show that detailed ecological characterization of risk zones is needed to prevent drawing misleading conclusions when exploring primate foraging behavior.

Isbell and Bidner (2016) cautioned against alarm calls to assess spatial variation in risk since their data suggested that while alarm calls may be good indicators of leopard presence, they are not necessarily good indicators of risk because they acted as a predator deterrent. They argued that leopards could approach closely only when vervets did not give alarm calls, so that alarm call location could lead to the illogical conclusion that vervets are at greater risk when leopards are farther away (Isbell and Bidner 2016). Such an interpretation conflates many of the different stages of predation (such as the probability of encounter and probability of attack; see R. Hill and Weingrill 2007; S. Lima and Dill 1990) and fails to distinguish between the intrinsic risks associated with locations (risky places) and the immediate risks (risky times) once predators are detected in the area (Basille et al. 2015; Creel, Schuette, and Christianson 2014; Moll et al. 2017; Periquet et al. 2012; Valeix et al. 2009). In that regard, false alarms may be equally informative about the perceived risk in an area as those given in direct response to predator presence (Willems and Hill 2009), since prey will hardly ever know when a predator is nearby and so should maintain a background level of fear consistent with their perceived risk of attack (J. Brown, Laundré, and Gurung 1999). Indeed, Campos and Fedigan (2014) noted that despite the frequent false alarms at nonpredatory birds, antipredatory behavior was more attuned to the perceived risk landscape for birds than for other predators. Since raptors are the primary predators of many Neotropical primates (Hart 2007), these false alarms undoubtedly reflect perceived risks of attack in certain areas (risky places). Thus, while others have also suggested that alarm calls and observations of predator encounters are not reliable indicators of predation risk (Janson 1998b), such caution is likely to be misguided, since primate alarm calls may actually help us understand the mismatch between fear and risk (see Gaynor, Brown, et al. 2019), particularly if we can account for the underlying utilization distributions of both predator and prey. As a consequence, the location of predator-specific alarm calls represents a powerful method for determining spatial variation in perceived predation risk (Campos and Fedigan 2014; B. Coleman and Hill 2014; Willems and Hill 2009) and thus assessing the impact of the landscape of fear on primate foraging decisions. Given the prevalence of acoustically distinct alarm calls for different predator guilds across the primate order, this method has the potential to place primatology at the forefront of developments in this field of foraging ecology.

VIGILANCE BEHAVIOR

A series of studies have demonstrated that vigilance increases as mammals become more fearful (Childress and Lung 2003; L. Hunter and Skinner 1998; Kuijper et al. 2015; Quenette 1990; Sherwen et al. 2015; Treves et al. 2003; Wolff and Van Horn 2003) such that fear can be measured by levels of vigilance (Laundré, Hernández, and Ripple 2010; Welp et al. 2004). Lesser rhea (*Rhea pennata pennata*) were more vigilant in areas with low visibility that hindered escape by running, as well as in areas exposed to hunting (Barri et al. 2012). Nubian ibex (*Capra nubiana*) are more apprehensive and have higher vigilance levels at greater distances from cliff refuges (Hochman and Kotler 2007), while mule deer (*Odocoileus hemionus*) increase vigilance behavior at patch edges when in open microhabitats or forest interiors (Altendorf et al. 2001). Elk and bison (*Bison bison*) showed increased vigilance after the reintroduction of wolves into Yellowstone National Park (Laundré, Hernandez, and Altendorf 2001).

Similar relationships have been reported for primates (Allan and Hill 2018). Individuals in a mixed-species group of saddleback (*Saguinus fuscicollis*) and moustached (*S. mystax*) tamarins were more vigilant at lower heights in the forest (A. Smith, Kelez, and Buchanan-Smith 2004), and increases in routine and induced vigilance were greatest at lower canopy heights in ursine colobus monkeys (*Colobus vellerosus*; Teichroeb and Sicotte 2012). In squirrel monkeys (*Saimiri* spp.), preemptive vigilance increased in open habitats to scan for the approach of avian predators, while reactive vigilance predominated in areas where raptors could make effective ambush attacks from forest canopy cover (Boinski, Kauffman, et al. 2003). Habitat visibility (Chapman 1985; Cowlishaw 1998), refuge proximity (Cowlishaw 1998; R. Hill and Cowlishaw 2002), and proximity to waterholes (Rose and Fedigan 1995) have also been reported to influence vigilance and so identify key risk parameters. In a comprehensive analysis, Gaynor and Cords (2012) found vigilance to be explained by predation risk factors (recency of an antipredator event, height in canopy, and position in forest [edge/interior]) as well as by extra-group and within-group monitoring. While this highlights the importance of social factors, which are particularly important for primates (Allan and Hill 2018), the model relating vigilance to antipredator functions was superior to the models relating vigilance to conspecific monitoring and social risk. As a consequence, vigilance patterns can serve as a reliable predictor of perceived predation risk in primates, and spatial variation in vigilance may thus allow an alternative behavioral titration of the landscape of fear. Vigilance definitions and methods require careful consideration, however, for this approach to be applied successfully (Allan and Hill 2018, 2021)

Campos and Fedigan (2014) conducted a spatially explicit investigation into the vigilance behavior of white-faced capuchins that integrated landscapes of fear based on alarm call distributions as predictor variables. Their models highlighted that capuchins reduced vigilance and so perceived lower predation risk in the high and middle forest layers but that they adjusted their vigilance behavior to small-scale spatial variation in perceived risk. The alarm call landscapes of fear were thus predictive of perceived predation risk, with the combined risk model for all predators having considerably more empirical support than any of the guild-specific risk models (Campos and Fedigan 2014). This suggests that the perceived risk landscape integrates information about risk from multiple predator guilds (Campos and Fedigan 2014; B. Coleman and Hill 2014; Willems and Hill 2009) but that vigilance patterns can detect the subtle variation in risk arising from multiple environmental sources. Detailed vigilance studies are thus able to investigate the impact of spatial variation in predation risk on primate foraging behavior and in turn allow the landscape of fear to be mapped for primate species.

EXPERIMENTAL TITRATIONS OF RISK

In 1988, Joel S. Brown introduced an elegant experimental and mathematical approach to assess an animal's foraging decisions in natural environments based on artificial food patches. This "giving-up density" (GUD) framework (J. Brown 1988) drew on Charnov's (1976) marginal value theorem, where food patches were viewed as a depletable resource that foragers exploited differentially to maximize fitness. The amount of food that foragers left in a foraging patch (i.e., the giving-up density) should therefore reflect the perceived cost of foraging in that patch. A high GUD would indicate high net costs of foraging in that area or habitat (high predation risk), whereas low GUDs imply comparatively low foraging costs and a relatively safe environment. Subsequent publications have advocated that the landscape of fear can be best assessed through these behavioral titrations that allow animals to reveal their foraging cost of predation (J. Brown 2000; J. Brown and Kotler 2004). To date, however, the approach has been little used in primatology (Emerson et al. 2011).

The GUD framework can be envisioned as a simple rule for patch departure (J. Brown 2000). The patch should be exploited until the marginal cost of foraging equals the marginal benefits of exploitation (J. Brown 1988, 1992). For a forager in a risky environment, this can be expressed as $H = C + P + MOC$ (J. Brown 1988): a food patch should be left when the benefits of the harvest rate (H) no longer exceed the sum of the energetic (C), predation (P), and missed opportunity costs of foraging (MOC—which emerge from not being able to forage elsewhere or engage in other fitness-enhancing activities while at the patch; J. Brown 1988, 1992). J. Brown (1988) demonstrated that P could be titrated by measuring the GUDs of standardized depletable food sources, although experiments may be designed to examine any of the parameters in the equation (Bedoya-Perez, Carthey, et al. 2013). Nevertheless, the predation cost is generally deemed the most significant foraging cost (J. Brown 2000; J. Brown and Kotler 2004).

Since their inception, GUD experiments have been employed in a large number of studies across a variety of taxa (Bedoya-Perez, Carthey, et al. 2013; Bleicher 2017) including invertebrates (Korb and Linsenmair 2002), fish (Alofs and Polivka 2004; K. Hedges and Abrahams 2015; Persson and Stenberg 2006), birds (Tang and Schwarzkopf 2013; Tsurim et al. 2010; van Gils and Tijsen 2007), lagomorphs (Abu Baker et al. 2015; D. Morris 2005), hyraxes (Druce et al. 2006; Kotler, Brown, and Knight 1999), and ungulates (Abu Baker and Brown 2014; Altendorf et al. 2001; Hochman and Kotler 2007; Rieucau et al. 2009; Shrader, Brown, et al. 2008; Stears and Shrader 2015), although the vast majority of research has focused on rodents (Abu Baker and Brown 2010; Carthey and Banks 2015; Embar et al. 2011; Falcy and Danielson 2013; J. Jacob and Brown 2000; Kotler, Brown, et al. 2010; M. van der Merwe and Brown 2008). Measuring perceived predation risk remains the focus of the majority of the research using the GUD framework, accounting for approximately 50% of the nearly 200 GUD papers published to date (Bedoya-Perez, Carthey, et al. 2013). Surprisingly, therefore, given the scope of primatology and the interest in exploring trade-offs between foraging and predation risk, the use of GUDs has been limited to just a few primate species (galagos [McArthur, Orlando, et al. 2012] and guenons [Emerson and Brown 2012, 2013, 2015; Emerson et al. 2011; Houle, Vickery, and Chapman 2006; Makin et al. 2012; Nowak, le Roux, et al. 2014]), largely from a single research site in South Africa. Nevertheless, the diversity of questions and approaches tackled for nonprimate taxa suggests significant potential in the GUD framework for future studies of primate ecology, even though practical challenges inevitably exist in implementing this apparently simple technique—not least the effects of group foraging (Bedoya-Perez, Carthey, et al. 2013; Carthey and Banks 2015; Emerson and Brown 2013; Makin et al. 2012), which are important given the diversity of social organization in primates. At the same time, however, the arboreal nature of most primates offers the opportunity to assess the foraging costs of predation risk in three dimensions (Emerson et al. 2011; Makin et al. 2012), something that has received only limited attention in other taxa (Carrascal and Alonso 2006; Holbrook and Schmitt 1988).

GUD experiments are based on creating a depletable food patch that is appropriate for the focal species. Diminishing returns are created by mixing a small (known) amount of food into a large, fixed volume of inedible substrate (such as sand or sawdust), so that, as each piece of food is located, the next piece becomes progressively harder to find than the last (Emerson et al. 2011). The patches must provide a rich enough food source to attract the forager to it while being small enough to avoid satiation in the foraging animal. Furthermore, the diminishing returns must be severe enough to cause the forager to abandon the patch before all of the food items have been harvested. Such requirements often need a significant period of refinement, initially through the assessment of preferred dietary items and establishing an appropriate food and substrate volume (Emerson and Brown 2012). For samango monkeys, Emerson et al. (2011) developed food patches consisting of plastic tubs (15 cm high and 45 cm diameter) filled with 4 liters of untreated pine sawdust and a fixed amount of raw, dried peanut halves (normally 25) thoroughly mixed into the substrate. Artificial food patches were then placed at multiple heights in the canopy to titrate the vertical axis of fear (Emerson et al. 2011; see also Nowak, le Roux, et al. 2014; fig. 19.3). It was predicted for this arboreal study species that GUDs would decrease with greater height in the canopy, increase on the ground when sightlines were blocked (increased ambush probability), decrease in patches near trees (refuges), and increase in patches near shrubs (predator concealment). Along the vertical axis, GUDs were found to decrease by 15% from 0.1 m to 2 m and by 10% from 2 m to 5 m (Emerson et al. 2011), a result that was broadly replicated by Nowak, le Roux, et al. (2014) across a greater range of heights. Along the horizontal axis, GUDs were affected by curtain-blocked sightlines (approximately 5% higher with curtains), highlighting the effectiveness of these titrations of risk in quantifying subtle variation in risk perceptions. Distance to vegetation did not influence GUDs, however, suggesting that the vertical landscape gradient is more meaningful to samango monkeys than the terrestrial horizontal axis (Emerson et al. 2011). For more terrestrial vervet monkeys, however, GUD experiments have indicated strong vertical and horizontal effects of predation risk from both aerial and terrestrial predators, revealing a three-dimensional landscape of fear (Makin et al. 2012). Differences in substrate preferences and exposure to dif-

Figure 19.3. Experimental giving-up density (GUD) setup with artificial food patches hung at four different heights (0.1 m, 2.5 m, 5 m, and 7.5 m) to assess the samango monkeys' vertical axis of fear (Nowak, le Roux, et al. 2014). Basins are filled with 4 L of sawdust and 25 shelled peanut halves, with ropes crossed over the top to restrict access and slow foraging rates to yield more reliable GUDs. Camera traps captured monkeys' visits and behavior at trees to confirm that the focal species is the one exploiting the experimental food patches.

ferent predator guilds may thus influence responses to the landscape of fear (B. Coleman and Hill 2014). Interestingly, in a follow-up study, Emerson and Brown (2013) found that forest height did not significantly impact samango monkey foraging despite the strong vertical axis of fear; when monkeys were given feeding opportunities in fragments of tall forest and in bordering secondary-growth short forest, GUDs were similar despite a predicted habitat preference for tall forest. Such results are in contrast to conclusions made on the basis of ranging data (B. Coleman and Hill 2014) but have been supported by similar fine-scale assessments from other populations (Nowak, Wimberger, et al. 2017). The results also suggested that spatial cohesion during group movement could be an equally important predictor of habitat use since the benefits of staying with the group may be more significant than habitat type in driving individual foraging decisions (Emerson and Brown 2013). Collectively, these studies highlight the precision to which the predation costs on primates' foraging can be assessed, particularly at small spatial scales.

The GUD framework may also be used to explore the importance of food chemistry in the trade-off between resource acquisition and predation risk (McArthur, Banks, et al. 2014). Initially studied in rodents (Fedriani and Boulay 2006; K. Schmidt 2000; Tuen and Brown 1996), ungulates (Abu Baker 2015; Shrader, Kotler, et al. 2008), and marsupials (Bedoya-Perez, Issa, et al. 2014; Kirmani et al. 2010; Mella, Banks, and McArthur 2014; Mella, Ward, et al. 2015; Nersesian et al. 2011), the techniques have also been applied to understanding the interactions between food chemistry and safety with primates (Emerson and Brown 2015; McArthur, Orlando, et al. 2012). McArthur, Orlando, et al. (2012) illustrated that in thick-tailed bushbabies (*Otolemur crassicaudatus*) the use of artificial food patches reflected the interplay between toxin concentration in food and patch safety. Specifically, GUDs increased as the concentration of both cineole (a terpene toxin) and gallic acid (a phenolic toxin) increased, reflecting the increased foraging costs. In the cineole experiment, the bushbabies also perceived the toxin-free ground feeders as riskier than the toxin-free tree feeders. However, while increased toxicity in tree feeders resulted in reduced foraging (higher GUDs) at safe heights within the canopy, as well as temporal changes in foraging patterns throughout the night, it did not lead to increased harvesting (lower GUD) at the risky ground feeders (McArthur, Orlando, et al. 2012). Instead, the bushbabies sought out natural food patches with lower net foraging costs, despite the high-quality food in the ground feeders. In a similar experiment, Emerson and Brown (2015) showed that GUDs were higher in artificial patches containing food treated with secondary compounds (oxalic acid) than in controls, but there was also an interaction with the costs of predation; GUDs for secondary compounds were higher relative to control patches in the risky microhabitat (i.e., ground level). In natural situations, high toxin concentrations in plants located in safe habitats could make food in risky locations relatively less costly and hence encourage animals to forage in higher-risk locations. For frugivorous primates, such shifts in foraging could occur when the availability of low-toxin (often ripe) fruit in low-risk habitats declines toward the end of the fruiting season or when the ripe fruits have been consumed. Such trade-offs might be difficult to detect from behavioral observations, but experimental approaches based on the GUD framework provide the opportunity to titrate the relative costs of toxins and fear and pinpoint where these costs are equivalent (McArthur, Orlando, et al. 2012).

Although GUD experiments have traditionally quantified the amount of food left by a forager at artificial feeding patches, one notable exception has been the measurement of the ripe fruit left unpicked in a tree by blue and redtail (*Cercopithecus ascanius*) monkeys in Kibale National Park, Uganda (Houle, Vickery, and Chapman 2006). GUDs were estimated by climbing trees after foraging bouts to count the volume of ripe and unripe fruit left by the primate groups. The results highlighted the mechanism of potential coexistence between the species. Blue monkeys were aggressive toward redtail monkeys and could displace them from fruiting trees; they were thus able to coexist with redtail monkeys through aggressive interference (Houle, Vickery, and Chapman 2006). In contrast, blue monkeys were less efficient at exploiting fruiting trees (higher GUDs), allowing redtail monkeys to coexist by exploiting these trees more thoroughly (lower GUDs) at fruit densities below those at which blue monkeys give up (Houle, Vickery, and Chapman 2006). Similar approaches have been used on a range of other species (J. Brown, Kotler, and Mitchell 1994; Guerra and Vickery 1998; M. Jones et al. 2001; Kotler and Brown 1999; Yunger et al. 2002), suggesting a profitable approach for future studies of coexistence in primate communities. Collectively, therefore, the GUD framework opens up a number of exciting potential research avenues in primatology.

TELEMETRY

Although radiotelemetry has been used to assist with the tracking of primates in a wide variety of species (A. Campbell and Sussman 1994; Fedigan, Fedigan, et al. 1988; Ganzhorn, Pietsch, et al. 2004; R. Martin and Bearder 1979; Neri-Arboleda et al. 2002; Paola Juarez et al. 2011; Pimley et al. 2005; M. Seiler, Holderied, and Schwitzer 2014; Trayford and Farmer 2012), it generally provides data sets with location points distributed at irregular time intervals that can reveal only the most general patterns of range space use (Cagnacci et al. 2010). Furthermore, the manual nature of data collection means that it is no more efficient than following primates for observational data collection and is therefore of greatest benefit for tracking species that are difficult to observe (particularly nocturnal species). While significant improvements have been made to this approach, particularly with the innovative system on Barro Colorado Island (Kays, Tilak, et al. 2011) that led to some novel outputs (Alba-Mejia et al. 2013; Crofoot 2013; Crofoot et al. 2010), there are limits to the range and accuracy of the system. In contrast, GPS-enabled collars generate volumes of precise data that exceed those achievable via traditional methods (Bridge et al. 2011; Dore et al. 2020; Kays, Crofoot, et al. 2015; Urbano et al. 2010). To date, however, their use has largely been restricted to large-bodied species such as baboons inhabiting open habitats (Ayers et al. 2020; Fehlmann, O'Riain, Kerr-Smith, et al. 2017; Markham and Altmann 2008; Markham, Guttal, et al. 2013; Pebsworth, MacIntosh, et al. 2012; Pebsworth, Morgan, and Huffman 2012; Strandburg-Peshkin, Farine, Couzin, et al. 2015a, 2015b), although there have been some reports for other primates (Dore et al. 2020; Isbell and Bidner 2016; Qi et al. 2014; Sprague et al. 2004; Takenoshita et al. 2005). Nevertheless, while telemetry studies on multiple groups have led to some valuable insights on the determinants of primate range use (Alba-Mejia et al. 2013; Crofoot 2013; Markham, Guttal, et al. 2013), the significant benefits of simultaneous tracking of multiple animals from different species, particularly in the context of species interactions and predator-prey interactions (Kays, Crofoot, et al. 2015), have been generally overlooked in studies of primate foraging and ecology (Isbell and Bidner 2016). Indeed, studies integrating the behavior of primates and their main predators have largely been restricted to systems where the predators themselves are also a primate species (Hausfater 1976; Noë and Bshary 1997; Stanford 1998; Treves 1999; D. Watts and Amsler 2013).

Studies of a number of carnivore species have used the clustering of GPS points from collar data to estimate kill rates (C. Anderson and Lindzey 2003; Cavalcanti and Gese 2010; Krofel et al. 2013; Q. Martins et al. 2011; Sand et al. 2005; Tambling et al. 2010), giving more precise estimates of predation rates in these systems. Some of these have revealed high levels of predation on primates (Jooste et al. 2012). It is the application of these techniques in understanding spatial variation in risk, however, that offers the greatest potential for understanding the influence of predation on foraging decisions (Arias-Del Razo et al. 2012; Lone, Loe, Gobakken, et al. 2014; Oriol-Cotterill et al. 2015; Valeix et al. 2009).

Basille et al. (2015) explored the dynamic antipredator tactics of boreal caribou (*Rangifer tarandus*) in response to predator proximity by wolves using spatially explicit data on predator and prey movements from GPS collars. Female caribou avoided open areas and deciduous forests and crossed the riskiest areas at relatively high speed to minimize the risks of encounter when wolves were in relatively close proximity (within 2.5 km). When risk became more acute, however, with wolves closer than 1 km, caribou switched to a strong avoidance of preferred food-rich foraging areas in favor of areas providing a denser cover. Similar differences between the effects of long-term risk (predator utilization distribution) and short-term risk (predator presence in the last 24 hours) were detected for multiple herbivores exposed to the risk of predation by lions in Hwange National Park, Zimbabwe (Valeix et al. 2009). The long-term risk of predation by lions appeared to influence only the distribution of browsers, with no effects on grazers, indicating that species with different ecological constraints may be influenced at different spatial and temporal scales by long-term risk from particular predators. Nevertheless, all herbivores used more open habitats preferentially when lions were in their vicinity to reduce risk from this ambush predator (Valeix et al. 2009). Collectively, these results highlight the importance of understanding predator location and behavior in exploring risk-sensitive foraging in prey species.

In a pioneering study, Isbell, Bidner, and colleagues (2018) used high-resolution GPS tracking of leopards, vervet monkeys, and olive baboons (*Papio anubis*) to explore the interaction between multiple groups of these primates and their main predator on the Laikipia Plateau of central Kenya. Over the 14-month study, the data revealed that leopards and primates encountered each other relatively rarely, with encounters between leopards and vervets more common than leopards and baboons, despite the greater ranging distances of baboons. Leopards encountered vervets more often during the day, while encounters between leopards and baboons were distributed more evenly between day and night. Leopard encounters with baboons were longer than with vervets, particularly at night, with leopards generally responsible for initiating encounters. Based on movement patterns, the primates also often failed to detect leopards during encounters, with baboons appearing to detect leopards less often than vervets during the day. Collectively, these results highlight the richness of information on predator-prey encounters that can be derived from telemetry data, and similar studies over the next few years will undoubtedly significantly increase our understanding of how predators influence primate foraging and ranging behavior.

While such techniques offer significant potential in primatology, there are important methodological considerations that need to be addressed to implement these methods effectively. For example, if GPS fixes are not collected frequently or only a small proportion of the predator population is collared, encounter rates may substantially underestimate the nonconsumptive effects of predators on prey, potentially by a factor of 10 (Creel, Winnie, and Christianson 2013). Traditionally, such requirements relating to the frequency of GPS fixes rendered these approaches appropriate to only larger species that could carry appropriate collars within the recommended weight guidelines (Casper 2009). But significant advances in technology each year, with trends for GPS receivers to be smaller, operate at lower voltages, consume less power, and have reduced times to obtain fixes, open up the technology and methods to smaller species (Tomkiewicz et al. 2010). Indeed, a series of studies using small, lightweight collars in recent years point to the increasing accessibility of this approach for many primate species (A. L. Adams et al. 2013; T. Dennis et al. 2010; Dore et al. 2020; Forin-Wiart et al. 2015; Havmøller et al. 2021; Recio et al. 2011; Sánchez-Giraldo and Daza 2019). Furthermore, recent advances in collar GPS telemetry for raptors make them appropriate for monitoring the key predators of many primate species (Lebeau et al. 2015; Nadjafzadeh et al. 2016), and proximity collars may allow interesting opportunities for determining where predator-prey interactions occur (M. Davis et al. 2013; Ellis et al. 2015; Prange et al. 2006; Ralls et al. 2013). An increasing number of studies are also illustrating how behavior can be inferred from accelerometers (Fehlmann, O'Riain, Hopkins, et al. 2017; Kooros et al. 2022; Nekaris, Campera, et al. 2022). Given the critical importance of understanding the effects of predator behavior on primate foraging ecology, telemetry technology offers enormous untapped potential (Dore et al. 2020), although performance validation will be required for arboreal species to assess the impact of vegetation structure and canopy height on collar performance (A. L. Adams et al. 2013; T. Dennis et al. 2010).

ANTHROPOGENIC IMPACTS ON THE LANDSCAPE OF FEAR

Humans can profoundly affect the ways in which wild animals assess risk (J. Berger 2007; Ciuti et al. 2012; A. Coleman et al. 2008; Geffroy et al. 2015; LaBarge, Hill, et al. 2020; Nowak, le Roux, et al. 2014), although the nature of the effects can vary dramatically and have both positive and negative consequences. In some instances, the negative consequences of human activity can exceed those of natural predators and habitat type on foraging and vigilance behavior (Ciuti et al. 2012). In contrast, opportunistic mammals such as primates may be attracted into human-occupied areas due to the potential resources they offer (Fehlmann, O'Riain, Kerr-Smith, et al. 2017; T. Hoffman and O'Riain 2012; Krief, Cibol, et al. 2014; McKinney 2011; Priston et al. 2012; E. Riley et al. 2013; Saraswat et al. 2015; Strum 2010; G. Wallace and Hill 2012; Y. Warren et al. 2011). Studies have also illustrated that human infrastructure can act as a shield providing safety from natural predators (Atickem et al. 2014; J. Berger 2007; Nowak, le Roux, et al. 2014). The strength of an animal's behavioral response to human presence can also be influenced by its condition (Beale and Monaghan 2004), and as risk of starvation increases, animals will select more hazardous foraging sites and engage in riskier behavior (Sih 1980; Verdolin 2006). Thus, while interaction between humans and wildlife generally leads to flight (Stankowich 2008), alarm calls and threat displays (Soltis et al. 2014), active avoidance (Tadesse and Kotler 2012), and stress (Creel, Fox, et al. 2002), the relationships are complex and can significantly alter the nature of predator-prey interactions.

RISK DISTURBANCE HYPOTHESIS

According to the risk disturbance hypothesis, animals respond to human disturbance in a way that is analogous to their response to natural predators (Frid and Dill 2002). As a consequence, high perceived risks of negative interactions with humans in certain locations can lead to animals spending less time than expected in these areas despite high resource abundance (Frid and Dill 2002). For example, the efficiency with which pink-footed geese (*Anser brachyrhynchus*) exploited arable fields declined with increased disturbance from roads (Gill et al. 1996), and elk also decreased their feeding time when closer to roads (Ciuti et al. 2012). Similarly, foraging by ibex in a national park was reduced in proximity to humans and during weekends when human visitation levels were high (Tadesse and Kotler 2012). In turn, these effects may have similar consequences for a species as habitat loss or degradation due to reduced access to high-quality resources (Gill et al. 1996), such that restricted distributions, reduced reproductive output, and population declines can result (Benitez-Lopez et al. 2010; Gill et al. 1996; Leblond et al. 2013; Paudel and Kindlmann 2012).

In a formal test of the risk disturbance hypothesis, E. Hawkins and Papworth (2022) found little support from pygmy marmosets (*Cebuella niveiventris*) using a playback experiment with anthropogenic noise (human speech and motorboats) and avian predators. Despite evidence from an earlier study (Sheehan and Papworth 2019), there was no behavior change by the marmosets in response to playbacks of human speech, although they increased scanning during playbacks of cicadas and predators compared to baseline. While noise can thus change the behavior of pygmy marmosets, it was not in a manner consistent with the risk disturbance hypothesis. Similarly, Reisland and Lambert (2016) examined whether Javan gibbons (*Hylobates moloch*) avoided areas visited by humans in a manner consistent with responses to perceived predation risk. In the nature reserve and sacred forest of Cagar Alam Leuweung Sancang, West Java, Indonesian spiritual tourists entered the forest seeking a change in luck, while villagers entered for resource extraction (e.g., fishing, hunting, and timber extraction) or to act as guides, porters, or kuncen (spirit masters) for the spiritual tourists. As a consequence, human activity was concentrated in sacred areas. One gibbon group significantly avoided these sacred areas, and locations of high human use did not overlap with areas of high gibbon use (Reisland and Lambert 2016). In contrast, a second group continued to use areas that were heavily visited by humans, although a sampling bias to more accessible areas with high human activity and increased levels of habituation for this group may offer partial explanations (Reisland and Lambert 2016). Nevertheless, if areas of high resource abundance coincide with the locations of sacred sites and high human density, gibbons may be unable to avoid foraging in these areas despite the perceived risk from humans. While the study provides only partial support for the risk disturbance hypothesis, it does confirm that areas of heavy human disturbance may be perceived as risky by gibbons, resulting in a reduction in use of such areas relative to other parts of the home range. It is important that the potential implications of human use of the landscape be formally assessed in future studies of primate foraging ecology.

Classical foraging theory predicts that, if risk of starvation is high, animals will make high-risk decisions to obtain food and avoid death (Abrahams and Dill 1989; J. Brown and Kotler 2007; Dill and Fraser 1984). During the food-scarce winter, samango monkeys at Hogsback, South Africa, foraged outside indigenous forest and entered gardens where they fed on exotic species such as fallen acorns (*Quercus* spp.) on the ground, despite potential threats from humans and domestic dogs (Nowak, Wimberger, et al. 2017; Wimberger et al. 2017). Experimental manipulations to change this economic calculation showed that the provision of equal feeding opportunities (in the form of artificial food patches) in both gardens and forest resulted in the monkeys reverting to a preference for foraging in the safer forest habitat, with higher visitation rates to experimental food patches in the forest relative to those in gardens (Nowak, Hill, et al. 2016; Nowak, Wimberger, et al. 2017). This supports the risk disturbance hypothesis and highlights that monkeys demonstrate highly flexible, responsive, risk-sensitive foraging. Nevertheless, once inside gardens, monkeys depleted patches in an identical fashion to those inside indigenous forest, showing high sensitivity to risk at lower strata in both habitats, despite their extensive use of the ground when foraging in gardens in winter (Nowak, Wimberger, et al. 2017). The monkeys thus appear to respond to human disturbance in a way that is analogous to responses to natural predators, consistent with the predictions of the risk disturbance hypothesis (Frid and Dill 2002). The findings also confirm one of the central tenets of optimal foraging theory in that risk of starvation (during the cold winters) can be as important a driver of behavior as risk of predation (Nowak, Wimberger, et al. 2017). The utility of the risk disturbance hypotheses thus appears context specific.

Numerous other interactions between humans and primates may also have an impact on risk perceptions and therefore behavior, although the effects may often be specific to the context, species, or study location. Many of these human-primate interactions emerge directly from research practices. Live-capturing or darting of animals is used for a variety of research purposes, including to enable observers to distinguish individuals (Glander, Fedigan, et al. 1991; Rocha et al. 2007; A. Stone et al. 2015), to collect biological samples (Fietz 2003), to deploy GPS or VHF collars to study movement patterns (Moehrenschlager et al. 2003), or to remove snares or provide other medical interventions (Sleeman et al. 2000). Such activities may have short-term or long-term influences on behavior and thus foraging patterns through altering risk perceptions. Red colobus monkeys (*Procolobus rufomitratus*) respond similarly to darting and collaring as to a predatory attack by chimpanzees (*Pan troglodytes*) with an acute but short-term stress response (Wasserman, Chapman, Milton, Goldberg, et al. 2013). Such responses would be anticipated to influence subsequent foraging behavior with a shift to lower-risk areas. Stress responses to live capture have also been reported for nonprimate species (Delehanty and Boonstra 2009; Fletcher and Boonstra 2006). Rhesus macaque (*Macaca mulatta*) mothers that have experienced an extended period of trapping on Cayo Santiago were more likely to maintain closer proximity to their infants and less likely to encourage independence or reject infants (Berman 1989). In contrast, no long-term effect of trapping was found on the stress physiology of mouse lemurs (*Microcebus murinus*), which appear to readily habituate to trapping and are easily retrapped (Haemaelaeinen et al. 2014). Similarly, a study of the effects of trapping on baboons (*Papio hamadryas*) and vervet monkeys found no obvious effects on individual or group behavior, nor did animals become more wary of traps following previous capture (Brett et al. 1982). Despite samango monkeys becoming incredibly trap-shy following a capture event, such that animals cannot be recaptured following a single exposure, live-capture led to no perceptible changes in the use of space, vigilance, or exploitation of experimental food patches (Nowak, Richards, et al. 2016). Length, frequency, and method of capture may thus all influence response type and magnitude, as does the species under study. Critically, it cannot be assumed that research practices such as darting and trapping are entirely neutral, and for some species they may result in changes of behavior consistent with the risk disturbance hypothesis. Nevertheless, these effects may be species- or location-specific, requiring researchers to properly assess the effects of live capture, darting, and associated handling on their study animals to ensure that the behavior is not significantly altered by the intervention.

While primates may learn the association between researchers and perceived danger and so habituate to their presence, even if practices such as live capture are used, it remains unclear how nonlethal human "predators" can influence the perceived risk and therefore foraging costs of wild animals. Human hunters are expected to adopt hunting tactics that increase the encounter rate and hunting success of target prey, but their activity may also be limited by time constraints or light conditions or possibly by legislation, traditions, or equipment (Norum et al. 2015). As a consequence, their impacts may differ fundamentally from those of "natural" predators (Gaynor, McInturff, and Brashares 2022). However, a series of studies from nonprimate taxa suggest that animals such as ungulates may not readily distinguish hunting from other human activities. For example, red deer (*Cervus elephus*) respond to both recreational park users and hunters with increased vigilance, although vigilance levels are higher overall in the hunting season (Jayakody et al. 2008). Roe deer (*Capreolus capreolus*; Benhaiem et al. 2008) and mountain gazelle (*Gazella gazelle*; Manor and Saltz 2003) become more vigilant when and where they are hunted or exposed to human nuisance disturbance. Similarly, while carnivores (Cristescu et al. 2013; Newby et al. 2013) and ungulates (Lone, Loe, Gobakken, et al. 2014; Lone, Loe, Meisingset, et al. 2015; Manor and Saltz 2005; Norum et al. 2015; Proffitt, Grigg, et al. 2009; Proffitt, Gude, et al. 2013; Theuerkauf and Rouys 2008) are known to shift ranging patterns in response to human hunters, Nubian ibex do the same in response to tourists (Tadesse and Kotler 2012). Identifying the precise risks from humans appears difficult for many mammals, despite the fact that many of these studies are conducted in regions with hunting legislation and seasons that can make predation risk from humans more predictable in space and time.

Although human predation has rarely been included in studies of primate antipredator behavior (Dooley and Judge 2015; Papworth, Milner-Gulland, and Slocombe 2013), evidence also suggests inconsistent responses. While Bshary (2001) found that Diana monkeys (*Cercopithecus diana*) responded to humans cryptically, Zuberbühler (2000a) reported Diana monkeys giving vocalizations in response to human models. Similarly, K. Arnold, Pohlner, and Zuberbühler (2008) also suggested inconsistent reactions to human presence in putty-nose monkeys (*Cercopithecus nititans*) in Nigeria. Nevertheless, where woolly monkeys (*Lagothrix poeppigii*) were hunted, Papworth, Milner-Gulland, and Slocombe (2013) showed that they learn to distinguish among three types of humans (hunters, gatherers, and researchers), responding most strongly to hunters. Vervet monkeys could also distinguish between researchers and Maasai herdsmen in Amboseli, Kenya (Isbell 1994b). Kloss gibbons (*Hylobates klossii*) reduced the risk of being stalked by hunters by singing less often during daylight hours and by leaving the location of male predawn singing before full light in the Mentawai Islands, Indonesia (Dooley and Judge 2015). This suggests that primates are acutely aware of the risks from hunters and that their behavioral responses could impact subsequent foraging and range use. In a comparison between an area where hunting is prohibited and a nearby forest where hunting pressure was moderate but spatially variable in southwestern Gabon, Croes et al. (2007) found that monkeys become more secretive when hunted, commencing alarm calls only when at a certain distance (typically >50 m) from humans. Monkeys in the no-hunting area displayed similar responses to human observers as to their natural predators (in line with the risk disturbance hypothesis), but in hunted areas the monkeys never approached observers and usually fled on being encountered (Croes et al. 2007). The monkeys thus seemed to distinguish the increased risks arising from the presence of hunters in their environment, although researchers conducting surveys elicited these same responses. Humans thus present nonhuman primates with complex problems in terms of estimating predation risk, since different classes of humans represent different orders of magnitude of risks, with hunters placing significant constraints on ranging behavior and activity. Despite the strong selective pressure of long-term human predation for multiple primate species, studies have been slow to explore the importance of human predation in driving antipredator behavior (Dooley and Judge 2015; Papworth, Milner-Gulland, and Slocombe 2013). As a consequence, the significance of human activity in limiting foraging options in primates is likely greatly underestimated.

HUMAN SHIELDS

While humans can undoubtedly pose risks to primates, particularly in the context of hunting, humans and human infrastructure can also represent a potential source of safety for wild animals, adding another layer of complexity to animals' perceptions of risk and their subsequent foraging decisions. Some mammals may indirectly benefit from human presence by reducing their exposure to predation risk (P. Leighton et al. 2010; Muhly et al. 2011). For example, while both wolves and elk avoid areas within 50 m of human trails in two national parks in Canada, elk will approach trails to distances of 50 m, which wolves continue to avoid (not approaching human trails within 400 m; Rogala et al. 2011). As a result, the areas between 50 m and 400 m of the trails become elk predation refugia, altering underlying trophic interactions (Rogala et al. 2011), and prey benefit through increased feeding and reduced vigilance in these predator shelters (Shannon et al. 2014). Similarly, female moose (*Alces alces*) in Yellowstone National Park move closer to paved roads to give birth to protect their newborns from bears (*Ursus arctos*; J. Berger 2007). In the Ethiopian highlands, mountain nyala (*Tragelaphus buxtoni*) overnight near human settlements to avoid predation by hyenas (*Crocuta crocuta*), although there is some individual heterogeneity in their use of humans as shields (Atickem et al. 2014). Human shields may also provide protection from sexually selected infanticide: successful mothers in Scandinavian brown bears were more likely to use humans as protective associates than unsuccessful mothers were, benefiting from the fact that potentially infanticidal adult males generally avoided humans (Steyaert et al. 2016). For many mammalian species, therefore, humans and human infrastructure may reduce predation risk for prey species in certain habitats or parts of their range.

Within primates, vervet monkeys in Amboseli National Park, Kenya, were subject to increased levels of predation by leopards when people were away from the field site, with animals 3.6 times more likely to disappear during observer absence than when researchers were present (Isbell and Young 1993). This classic example was informally labeled "the Nairobi effect" by Dorothy Cheney and Robert Seyfarth (Isbell 1994a) and suggested that researchers shielded monkeys from terrestrial predators simply by being present. A similar effect inhibiting leopard predation was produced by a ranger station on the edge of the study area, with losses from vervet groups declining the closer they lived to the station (Isbell and Young 1993). At the time, Isbell (1994b) also noted that unhabituated vervets spent more time scanning for predators than habituated monkeys but commented that it was not possible to assess the scanning rate for the habituated animals when researchers were not present. An increasing body of literature suggests that habituation may never result in observer presence becoming truly "neutral" to study animals, however (Allan, Bailey and Hill 2020, 2021; Allan, White and Hill 2022; Crofoot et al. 2010; Jack et al. 2008; LaBarge, Allan, Berman, Hill, et al. 2022; McDougall 2012; Nowak, le Roux, et al. 2014), and given the prominence of habituation in studies of primate behavior (Aguiar and Moro-Rios 2009; Hammond et al. 2022; Williamson and Feistner 2011), innovative methods are required to determine whether this influences the data collected, particularly in the context of predator-prey interactions.

Crofoot et al. (2010) exploited an automated radiotelemetry system that remotely monitored the movement and activity of radio-collared white-faced capuchins to examine the impact of observers. They found no evidence for observer presence impacting the ranging behavior or activity of the monkeys, although the study did not explore how researchers might influence responses to predation risk and the frequency of predator-prey interactions. In contrast, Nowak, le Roux, et al. (2014) quantified the magnitude of the "human shield effect" experimentally using giving-up densities for two groups of samango monkeys. The study reported a strong vertical axis of fear, with these arboreal monkeys preferring to feed higher in the canopy. When human followers were present, however, giving-up densities were reduced at all heights, with the vertical axis of fear disappearing entirely in one of the focal groups in the presence of researchers (Nowak, le Roux, et al. 2014). This suggests that observers reduce the monkeys' perceived risk of terrestrial predators and so affect their foraging decisions at or near ground level. Indeed, LaBarge, Allan, Berman, Hill, et al. (2022) demonstrated that while fecal cortisol metabolite levels were higher on the day following potential predator encounters (but not competitive interactions) and increased in relation to the number of likely predator encounters, as observer numbers increased the responses to predators flattened to the extent that when three or four researchers were present, there was no discernable impact of predator encounters on fecal cortisol metabolite level. The presence of several human observers thus appears to influence the samango monkeys' perception of danger such that it modulates the physiological response to predators. The anthropogenic alteration of risk-taking behavior has rarely been acknowledged or quantified, particularly in behavioral ecological studies reliant on habituated animals, but these findings have significant implications for future studies of foraging responses to predation risk that use habituation and observational methods. Further work to replicate these methods and explore the generality of these findings is also urgently required.

While interactions between humans and primates are generally perceived as negative or risky by primates, it is clear that monkeys may come to associate humans or their infrastructure as beneficial if these act as predator refugia (P. Leighton et al. 2010; Muhly et al. 2011). In turn, commensalism could develop from this human-predator refugia association or the eventual result of the habituation process (Nowak, Hill, et al. 2016) as primates associate humans with sources of food. For opportunistic mammals such as baboons (*Papio* spp.), animals may subsequently be further attracted into human-occupied areas due to the potential resources they offer (Strum 2010), and significant investment may then be needed to mitigate conflict (T. Hoffman and O'Riain 2012). Indeed, the substantial nutritional benefits from exploiting crops and other human foods (C. Hill 2000; Y. Warren et al. 2011) may outweigh the high risks and increased mortality associated with raiding (Strum 2010). Therefore, where human shield effects lead to increased proximity between humans and primates, the landscape of fear may not easily be reestablished. Investigations into the sophisticated and flexible responses of primates to humans and their infrastructure are likely to be a profitable direction of future research since the risk disturbance hypothesis is not a blanket term describing human impacts on primate behavior, and primates may benefit significantly from foraging in close proximity to humans.

SUMMARY AND CONCLUSIONS

While predation risk has long been assumed to be an important selective force on primate behavior and sociality (Alexander 1974), the concept of the landscape of fear, well established in the animal behavior and ecology literature, has received comparatively little attention in primatology. Nevertheless, evidence from primates suggests that the effects of fear outweigh those of food availability in determining primate range use, highlighting its importance for future studies. Furthermore, a range of methods exist for measuring the landscape of fear and risk effects more generally, many of which can be easily adapted to primates. The rapid increase in technology is also increasing the options available. Nevertheless, we need to be mindful of our presence as observers and researchers, since humans can have a series of impacts on risk perception by primates. As the impact of human activities on natural habitats continues to rise, it will become increasingly complex to understand the role of predator-prey interactions in shaping primate foraging behavior.

- Studies of predation in primates must distinguish between the consumptive (direct) effects of predators and the nonconsumptive (indirect) risk effects, with the landscape of fear providing a useful framework for exploring how risk perceptions vary spatially.
- Predator-specific alarm calls provide a powerful method for assessing the landscape of fear, offering the potential to place primatology at the forefront of this field of animal behavior and ecology.
- Classic experimental approaches to quantifying predation, such as giving-up densities (GUDs), provide opportunities for teasing apart different elements of the foraging cost of predation.
- New technologies, such as GPS collars, offer significant potential for mapping predator-prey interactions and quantifying the landscape of fear.
- Primates may respond to human disturbance in a way that is analogous to their response to natural predators, reducing foraging efficiency in areas where the probability of encounter with humans is high.
- In contrast, researchers may reduce primates' perceived risk of predation and so influence their landscape of fear. This has significant implications for future studies of primate foraging responses to predation risk that use habituation and observational methods.

20 Behavioral Flexibility and Diet

AJ Hardie and Karen B. Strier

The ability to alter feeding, ranging, or social patterns in response to temporal variation in ecological or demographic conditions is an essential part of primate adaptation (Strier 2017). One component of primate behavioral flexibility is the capacity that many primates exhibit to switch food types along different spatiotemporal scales (e.g., between feeding patches, between forest fragments, seasonally, or interannually because of El Niño–Southern Oscillation [ENSO] events or climate change). However, dietary flexibility is one of many behavioral strategies primates employ to cope with fluctuations in the availability of preferred foods.

In this chapter, we begin with a brief overview of theoretical models predicting primate behavior and food choice, as well as factors that impose limits on the extent to which primates can shift behavior and diet to cope with food scarcity. We next review the ways in which variation in the spatial and temporal distribution of primate foods affect primate distribution and behavior. We specifically consider the dietary and behavioral flexibility of primate populations living in marginal habitats, as the ability to survive extreme climatic, latitudinal, or altitudinal conditions can provide important insights into the potential of primates to adjust to anthropogenic pressures including global climate change. In our final section, we provide a schematic model for integrating the dynamic interactions among behavioral adjustments in feeding, ranging, and grouping patterns in wild primates.

THEORETICAL PERSPECTIVES ON DRIVERS OF FOOD CHOICE AND BEHAVIORAL FLEXIBILITY

No perfect food exists for any primate, so individuals are faced with negotiating a series of trade-offs to effectively balance the energy expended in obtaining food with the energy and nutrients gained from that food. The amount of food an individual can consume is limited by the time available for foraging, body size, and intestinal morphology, among other factors. Thus, the chemical composition and abundance of the foods an individual selects are two inherent limitations in the quest to obtain sufficient nutrients for maintenance, growth, and reproduction (Borries and Koenig, this volume). Primate food choice is explained as a function of this balancing act between energy expenditure and gain, as well as between the consumption of different macronutrients (i.e., carbohydrates, lipids, and protein) and micronutrients (i.e., vitamins and minerals; Felton and Lambert, this volume).

Models of foraging theory (e.g., the optimal diet model, J. Krebs 1978; the patch depletion model, J. Krebs 1978; D. Stephens and Krebs 1986) predict foraging decisions based on the trade-offs between maximizing energy input and minimizing energy output such that energy is optimized in foraging choices (D. Stephens and Krebs 1986). For example, according to the patch depletion model (or marginal value theorem), foragers should vacate patches when it becomes more profitable to expend energy searching for a new patch than to remain in a patch with rapidly diminishing energetic returns (J. Krebs 1978; D. Stephens and Krebs 1986; Ydenberg et al. 2007). In most applications of foraging theory, energy is considered the fundamental currency. Consistent with this focus on energy, primate species have previously been classified as energy maximizers or minimizers, a distinction based on whether they seek to maximize energy intake or minimizing energy expenditure (Milton 1980; Schoener 1971). Among the Atelidae, for example, *Alouatta* are distinguished by following a behavioral and dietary strategy that minimizes energy expenditure (Milton 1980; Strier 1992), while *Ateles*, *Lagothrix*, and *Brachyteles* maximize energy intake by consuming a diet higher in easily digested, high-energy foods such as fruits (Strier 1992).

Protein is also thought to influence primate food choice. To acquire essential amino acids, frugivorous primates include some leaves, insects, or other proteinaceous source in their diets (Hawes et al., this volume). Therefore, protein is often talked about as a limiting factor in food choice for primates (Ganzhorn et al. 2017; Mattson 1980). Protein intake is most frequently explored as a driver of food selection, and especially leaf selection in folivorous primates (reviewed in Ganzhorn et al. 2017), but is rarely included in maximization or optimization models as energy is (Felton, Felton, Lindenmayer, et al. 2009). Indeed, part of the difficulty in considering protein is that it is often difficult to disentangle what drives protein consumption, especially in folivorous primates. This raises the question of whether high protein consumption is indicative of higher protein requirements or merely a consequence of the consumption of protein-rich leaves in the quest to obtain sufficient energy (e.g., K. Evans et al. 2021; Rothman, Raubenheimer, and Chapman 2011; Thurau et al. 2021).

Models of primate food choice are most often expressed in economic terms of expenditure and gain. In many instances, these models discriminate between energy (primarily derived from carbohydrates, though also from fat) and protein as the explanatory variables driving primate food choice. The earliest models of foraging theory (e.g., J. Krebs 1978; D. Stephens and Krebs 1986) were most frequently based on one currency (i.e., energy or protein). Yet, despite their utility for formulating general hypotheses, models based on the maximization or minimization of one particular currency such as energy or protein are difficult to apply to primates that have notably complex nutritional requirements (Felton, Felton, Lindenmayer, et al. 2009).

Unidimensional frameworks have ceded primacy in primate nutritional and foraging ecology to multidimensional models (e.g., nutritional geometry; S. Simpson and Raubenheimer 1993, 2012; Raubenheimer, this volume) that consider multiple nutritional components and biologically meaningful ratios between them to determine which nutritional components and interactions among components influence primate foraging behavior (Raubenheimer, Machovsky-Capuska, Chapman, et al. 2015; S. Simpson and Raubenheimer 2012). For example, the ratio of nonprotein energy (NPE, defined as the intake of energy from nonstructural carbohydrates, digestible fiber, and fat) to protein (NPE:P) in animal diets has been found to be a predictor of fitness-related variables (Cui, Wang, Zhang, et al. 2020; K. Lee, Simpson, Clissold, et al. 2008; Lodwick and Salmi 2019; Ponton, Wilson, Cotter, et al. 2011; S. Simpson and Raubenheimer 2009; Solon-Biet et al. 2014). Highly frugivorous spider monkeys (*Ateles chamek*) managed nutrients in their diet by maintaining a balance of NPE:P of 8:1 kcal; as preferred food availability varied, nonprotein energy fluctuated while protein was regulated (Felton, Felton, Raubenheimer, Simpson, Foley, et al. 2009). Similarly, frugivorous chimpanzees (*Pan troglodytes*) maintained an NPE:P of 7:1 kcal, prioritizing protein intake while NPE intake varied with daily diet (Uwimbabazi et al. 2021). Unlike spider monkeys and chimpanzees, folivorous gorillas prioritized nonprotein energy when constrained by food availability, overeating protein and maintaining nonprotein energy, indicating a different NPE:P strategy under environmental constraints (Rothman, Raubenheimer, and Chapman 2011).

Another aspect of this multidimensional approach is consideration of micronutrients and the often-related investigation of the importance of foods that may be eaten infrequently or in small quantities. For example, decaying wood eaten by mountain gorillas in Uganda was revealed to provide 95% of the gorillas' dietary sodium despite being only 3.9% of their wet food intake (Rothman, Van Soest, and Pell 2006). Decaying wood is otherwise a very poor food source, low in protein and sugars, high in fiber, and containing higher amounts of digestion-inhibiting compounds such as lignin (Rothman, Van Soest, and Pell 2006).

The complexity of the chemical and ecological factors that impact primate food choices further compounds the challenges an individual must navigate to find and consume enough food to survive, grow, and reproduce. These challenges can be thought of as variables that can negatively impact an individual's ability to maintain a positive balance between energy intake and expenditure. When an individual fails to maintain a positive energy intake-expenditure balance, negative physiological consequences can seriously impact behavior, reproduction, and survival (Cui, Wang, Zhang, et al. 2020; Emery Thompson 2017; J. Lambert and Rothman 2015; E. Vogel, Knott. et al. 2012). Not acquiring enough energy or protein for prolonged periods can have significant physiological consequences, such as decreased reproductive success; therefore, primates exhibit numerous behavioral and morphological adaptations to ensure that energy and protein needs can be met during periods of food shortage (Emery Thompson 2017; Simmen and Rasamimanana 2018; E. Vogel, Knott, et al. 2012).

Changes in foraging behavior can lead to different diets or result from an individual trying to maintain a similar diet despite variation in the availability and distribution of these foods. Shifts in both may reflect behavioral strategies to reduce energy expenditure in times of food shortage. Often, as primates increase time spent looking for food or processing difficult-to-access foods, time devoted to other behaviors such as grooming or resting may decrease (van Schaik and Brockman 2005). Furthermore, spatiotemporal differences in food abundance have been used to explain the significant variation in primate social relationships (especially among female primates) and grouping systems (e.g., Wrangham 1980). However, many other factors such as demography and phylogeny play significant roles in determining the range of behavioral responses available for a species (Chapman and Rothman 2009; Kamilar and Baden 2014; A. Koenig, Scarry, et al. 2013).

CONSTRAINTS ON BEHAVIORAL FLEXIBILITY AND DIET

How animals meet their energetic and nutritional demands varies depending on evolutionary history, life-history stage, and behavior. While foraging theory predicts that individuals should behave in such a way to maximize intake while minimizing expenditure, there are numerous constraints on an individual's ability to obtain the ideal mix of nutrients they need to fuel maintenance and growth. Exogenous and endogenous constraints on an individual's ability to alter behavior or diet interact with spatiotemporal variation in food abundance to influence which behavioral strategies (e.g., changing ranging or grouping patterns, increasing foraging time during periods of food shortage, or increasing reliance on lower-quality food items) a species may adopt. The degree to which animals can alter their behavior in response to the decreased availability of preferred foods is dynamic, influenced by phylogeny, demography, physiology, and morphology.

Behavioral shifts, such as changes in group size or day range length, can occur within feeding bouts, between seasons, or interannually as food availability varies (Chapman, Rothman, and Lambert 2012; J. Lambert and Rothman 2015; Strier 2017; van Schaik, Terborgh, and Wright 1993). Dietary shifts, which are both a component and an outcome of behavioral flexibility, also occur on variable timescales and are often short-term strategies of expedience for primates faced with nutritional challenges (Strier 2017). Spatiotemporal variation in food abundance (e.g., seasonal shifts, or variation in habitat quality between feeding patches) will result in behavioral changes that reflect a species' adaptations and ability to adjust behavior in response to external stimuli.

COMPETITION

Resource scarcity, whether the result of periods of food shortage or decreased food abundance in low-quality habitats, can lead to increased inter- and intraspecific competition. Demographic factors, such as the population densities and group sizes of both conspecifics and sympatric species with overlapping diets, vary over individual lifetimes as well as across and within populations. Competitive abilities, dominance rank (for those species in which it occurs), and cooperative alliances that affect access to food can also change during a lifetime, with consequences for diets at both individual and interspecific levels.

Levels of interspecific feeding competition depend on the degree of dietary overlap between sympatric species

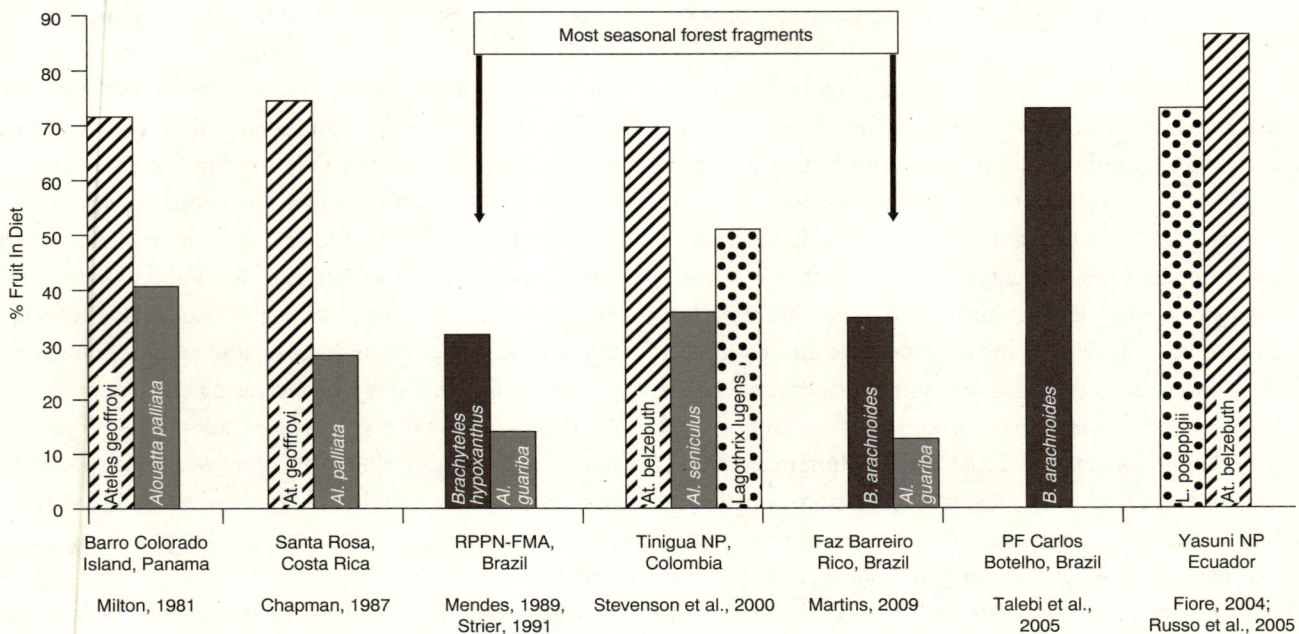

Figure 20.1. Sympatric and intergeneric comparisons of Atelidae diet. Both *Brachyteles* and *Alouatta* are less frugivorous in the most seasonal, fragmented forests. Nonetheless, where dietary data are available for both genera (Reserva Particular do Patrimônio Natural-Feliciano Miguel Abdala), *Brachyteles* resembles *Ateles* elsewhere in being proportionately more frugivorous relative to sympatric *Alouatta*. (Updated from Strier 2021b, online supplement.)

and how that overlap varies in accordance with spatio-temporal shifts in food availability (Campera et al. 2021; Neha et al. 2021; Ruslin et al. 2019). Strategies by which primates partition dietary niches include varying feeding strata, food types, or food species. Extreme cases of one or more of these kinds of niche divergence can be seen in polyspecific associations. For example, differing rates of association between Diana monkeys (*Cercopithecus diana*) and other guenon species in the Taï forest in Côte d'Ivoire did not influence the food type consumed; indeed, food type consumed during polyspecific associations overlapped by 90.8% to 98.1% between guenons (Kane and McGraw 2017). However, the proportion of different food species consumed varied between Diana monkey groups. These results indicate that, while guenons maintained a similar level of frugivory, the species consumed varied in such a way that competition was effectively reduced between groups.

Comparisons of closely related species of sympatric Atelidae help illustrate how variation in local ecological conditions, such as rainfall and food availability, may shape both intra- and interspecific competition in diets (fig. 20.1). In general, atelines are more frugivorous than *Alouatta*, though the proportion of fruits in their respective diets varies due to ecological factors such as seasonality. Specifically, both *Brachyteles* and *Alouatta* are less frugivorous in the most seasonal fragmented forests than in continuous forests, but *Brachyteles* is more frugivorous than sympatric *Alouatta* (expanded from Strier 1992). These comparisons highlight the role of dietary niche partitioning within primate communities and its importance for understanding both dietary competition and variation within and among species. They also illustrate the importance of intraspecific comparisons for understanding the degree to which primates can shift their diets to mitigate the impacts of interspecific competition.

Niche partitioning or differentiation is not always evident among sympatric species. For example, black-and-gold howler monkeys (*Alouatta caraya*) and brown howler monkeys (*Alouatta guariba clamitans*) in northern Argentina exhibited different behavioral responses to seasonal changes in fruit availability yet had largely similar diets (I. Agostini et al. 2010, 2012). Black-and-gold howlers, which lived in larger groups than brown howlers, adopted a clear energy minimization strategy during periods of decreased fruit availability by spending significantly less time traveling and moving, while brown howlers did not significantly alter their behavior in relation to food availability (I. Agostini et al. 2010, 2012).

PHYSIOLOGY AND MORPHOLOGY

The ability to change behavior or switch between different food types in response to resource scarcity is further constrained by physiology and morphology. The nutritional requirements of an individual depend on activity, body size, age, and reproductive state. Thus, responses to food abundance can change at many points throughout an individual's life concurrent with life-history stages. Body mass typically accounts for most variation in absolute nutrient intake, but scaling for body mass (i.e., $m^{0.75}$) reveals that females have higher relative nutritional demands than males (as a result of reproduction and lactation) and that juveniles have high relative nutritional demands to fuel growth and development (Key and Ross 1999; Oftedal 1991). For example, despite similar absolute energy and macronutrient intake patterns across age-sex classes, when intake was scaled to body mass, age and sex were found to influence energy and macronutrient intake in Bornean orangutans (*Pongo pygmaeus wurmbii*; E. Vogel, Alavi, et al. 2016).

Digestive morphology and food passage rates play significant roles in an individual's ability to switch between food types. For the most part, protein digestion occurs in the stomach, lipid digestion occurs in the small intestine, and carbohydrate digestion occurs in the stomach and large intestine; as a result, most digestive specializations seen in primates are seen in the stomach and large intestine (J. Lambert 1998). Foods like leaves require more time in the gut to maximize nutrient absorption, while foods like fruits can pass through relatively quickly because they are more easily broken down.

For primarily folivorous species such as howler monkeys (*Alouatta* spp.), enlarged hindguts (the caecum and colon) combined with slow food passage rates enable them to extract a significant proportion of the nutrients available in leaves (J. Lambert 1998; R. Liu et al. 2022; McGrosky et al. 2019; Milton 1986, 1997). The intestinal morphology of howler monkeys is a significant component of their ability to switch between numerous food types and species, allowing substantial dietary flexibility within and between species in the genus (de Andrade Carneiro et al. 2021; Espinosa-Gómez, Gómez-Rosales, et al. 2013; Garber, Righini, and Kowalewski 2015). However, the same mechanisms that facilitate food processing and nutrient absorption can impose limits on digestive efficiency: an individual needs to balance the amount of food consumed (volume) with nutrient extraction (time) in the gut (Milton 1986). Digestive capabilities can impose limits on diet selection and thus constrain an individual's capacity for diet switching in the face of preferred-food shortages (Penry 1993).

Additional processing before digestion may address some of the limitations imposed by intestinal morphology. Many primates exhibit dental morphology (e.g., molar enamel thickness and occlusal surface complexity or topography) associated with processing fallback foods, indicating a critical function of this morphology in breaking down foods that are more difficult to access or process (McGraw and Daegling 2020; Rosenberger and Kinzey 1976; Rosenberger and Strier 1989; E. Vogel, Haag, et al. 2009). For example, primarily frugivorous Bornean orangutans possess molars with thick enamel and crenulations on the occlusal surface that are adaptations for processing tough foods, such as bark (Ungar 2007b; E. Vogel, Haag, et al. 2009). The interaction of morphology and body size imposes limits on the speed and efficiency with which an individual can digest and absorb nutrients (Chapman, Rothman, and Lambert 2012; J. Lambert 1998; Strier 1992). For instance, leaves require more time to be fully digested than fruit pulp or insects because most of the energy contained in leaves is in the form of structural carbohydrates (i.e., cellulose and hemicellulose) that can be broken down only via microbial action in the gut (Karasov and Martinez del Rio 2007; Kay 1984). Thus, below a body size of approximately 700 g, a folivorous diet is difficult to maintain because of the imbalance between the energy required to digest leaves and the energy obtained from leaf material (Kay 1984).

Smaller primates that require more energy per unit of body mass are restricted to energy-dense foods such as insects. Small-bodied primates are able to obtain sufficient energy from insects due to highly specialized morphology and, for some species, endogenous production of the enzyme chitinase that breaks down the structural carbohydrate chitin, which forms a large part of insect exoskeletons (Janiak, Chaney, and Tosi 2017). For the most part, larger primates cannot subsist on insect-based diets due to the energy that must be expended in obtaining a sufficient volume of food. However, social insects such as ants or termites are important food resources for large-bodied primates such as chimpanzees (*Pan troglodytes*), though usually extractive foraging aided by tools is necessary to exploit these resources (Isbell 1998; Raubenheimer and Rothman 2013). Body size, morphology, and physiology interact to inform an individual's nutritional requirements. How an animal meets those requirements is in turn limited by competition with conspecifics and sympatric species. The limits that these variables impose integrate with evolutionary history (i.e., phylogeny) to form the basis for responses to food scarcity.

HABITAT HETEROGENEITY, RESOURCE DISTRIBUTION, AND SOCIAL BEHAVIOR

Derived from foraging theory, the ideal free model posits that, given certain assumptions about movement and resource distribution, foragers should be distributed in such a way that energy gain is highest—they should occupy the feeding patches where food quality or potential energy gain is highest (Fretwell 1972; Fretwell and Lucas 1970). However, interpretations of the ideal free model are complicated by the significant role that competition plays in shaping foraging decisions for most primates and by the complexity of primate diets.

High heterogeneity, typically reflecting local legacies of human occupation or use, causes significant variation in forest type and habitat quality (e.g., nutrient content of food items, tree species diversity, and density of top food species) within forest fragments (Chapman, Chapman, Rode, et al. 2003; L. Jung et al. 2015; Chapman et al., this volume). Ecological variation within mosaic habitats often leads to increased use of areas with high plant species diversity or a higher density of food species (Sha et al. 2017). Densities of top food species, in particular, were correlated with habitat use in both mantled howler monkeys (*Alouatta palliata*) in Costa Rica (Stoner 1996) and black howler monkeys (*Alouatta pigra*) in Belize (Ostro, Silver, Koontz, and Young 2000). However, at higher population densities in highly heterogeneous habitats, groups may occupy lower-quality areas than conspecifics and be faced with additional nutritional or energetic costs. In the Reserva Particular do Patrimônio Natural-Feliciano Miguel Abdala in Minas Gerais, Brazil, two groups of brown howler monkeys were shown to have significantly different diets and behavioral patterns that correlated with local ecology (L. Jung et al. 2015). A group in a higher-quality area (valley with evergreen forest) spent less time traveling and had a smaller home range, in addition to a diet higher in fruits and lower in mature leaves, when compared to a group in a lower-quality area (hillside with secondary, mostly deciduous forest that was previously a coffee plantation) of the forest fragment (L. Jung et al. 2015).

In some instances, primate distribution or abundance may be shaped not by energy gain but by characteristics of plant foods such as the protein-to-fiber ratio of leaves (but see I. Wallis et al. 2012; Chapman et al., this volume) or the macro- and micronutrient content of available foods. Furthermore, food types (i.e., flower, fruit, leaf, etc.) vary naturally according to plant species characteristics and developmental stage. It is generally accepted that foods like fruits are more patchily distributed than items like leaves. While all leaves are certainly not equal, their abundance in relation to fruits generally means that even if they occur in discrete patches, those patches are still more abundant or contain more food items. Yet, leaves can vary significantly in quality based on phenophase, plant species, soil composition, or sun exposure, among other factors, and folivorous primates often choose leaves that have higher protein and lower fiber contents and fewer secondary metabolites or toxins (Chapman, Chapman, Rode, et al. 2003; Ganzhorn 1995b; Glander 1978); this variation can impact primate behavior and distribution. Population density of redtail monkeys (*Cercopithecus ascanius*) was significantly related to the amount of copper consumed per calorie of food, but not the basal area of food trees or density of food trees per hectare, suggesting that copper may be limiting for this primate (Rode, Chapman, McDowell, et al. 2006), though further information is needed on copper availability in the environment to better assess this.

Feeding patch size can impose an upper limit on the number of individuals that can feed in one location at any time (Strier 2016). Intragroup competition increases with the number of animals foraging within a finite patch, and, to compensate for increased intragroup feeding competition, animals must either move more frequently to find feeding patches (thus expending more energy) or adjust feeding group size (Aureli et al. 2008; A. Koenig, Scarry, et al. 2013; Strier 2016). Perhaps the most well-known examples of this strategy are chimpanzees (*Pan troglodytes*) and spider monkeys (*Ateles* spp.) that routinely fission into smaller feeding parties when preferred fruit patches are small and form larger parties when these patches can accommodate more individuals (Aguilar-Melo et al. 2020; Chapman, Chapman, and Wrangham 1995b; Hartwell et al. 2021). For female spider monkeys (*Ateles geoffroyi*) in Costa Rica, decreased fruit abundance was correlated with smaller subgroup sizes but not with fecal glucocorticoids, suggesting that smaller subgroups provide an effective buffer against physiological stress caused by fluctuations in food abundance (M. Rodrigues 2017). Muriquis (*Brachyteles* spp.) are also known to adjust party sizes in response to fruit availability, though the same pattern does not hold for abundance of leaves (Moraes et al. 1998; Strier 1989).

The separation of a larger group into smaller subgroups—which persist for several hours, several days, or longer—is a strategy seen to differing degrees in many primate taxa as a mechanism for coping with changes in resource abundance (Aureli et al. 2008). The emergence of these fission-fusion dynamics is often attributed to changes in fruit abundance, as is seen in chimpanzees and spider monkeys (e.g., Chapman, Chapman, and Wrangham 1995b). However, many other taxa exhibit flexible social systems in response to resource scarcity. For instance, red-capped mangabeys (*Cercocebus torquatus*) in Gabon were observed to forage in decreased party sizes when fruit availability increased (Dolado et al. 2016), whereas black capuchin monkeys (*Sapajus nigritus*) increased group spread when feeding on fruit, adjusting cohesion dynamically in accordance with activity (Luccas and Izar 2021). Other taxa, such as ruffed lemurs (*Varecia* spp.), exhibit significant degrees of intraspecific variation in the degree of social flexibility, ranging from cohesive to highly fluid (Vasey 2006). One population of black-and-white ruffed lemurs (*Varecia variegata*) at Ranomafana National Park, Madagascar, exhibited a high degree of fission-fusion dynamics, with subgroup transitions occurring every 90 minutes on average (Baden et al. 2016). Furthermore, smaller and less cohesive subgroups were observed during cooler and wetter months when fruit availability was lower, indicating that flexible social dynamics in black-and-white ruffed lemurs serve to reduce intragroup feeding competition (Baden et al. 2016). The degree to which primates can adjust group size is contingent on phylogenetic constraints, local ecology, and the complex interplay of spatial and temporal changes in food abundance.

HABITAT DISTURBANCE, DIET, AND BEHAVIOR

Variation in habitat quality, especially forest contiguity and floristic structure, can have significant impacts on food availability. Habitat quality is often measured using variables such as tree species diversity or observation of anthropogenic disturbances such as logging (E. Vogel and Dominy 2011). In general, habitats that are considered to be of higher quality have greater tree species diversity, higher densities of top food species, and lower levels of disturbance, though many of these ecological variables are site specific. Primates living in lower-quality habitats often have reduced home range sizes, decreased dietary diversity, and increased competition with both conspecifics and sympatric species (Strier 2009).

Residency in lower-quality habitats can result in significant changes to primate behavior and physiology. For instance, Sulawesi macaques (*Macaca tonkeana*) living in lower-quality habitats relied more heavily on alternative food items (e.g., insects or fungi) and spent more time foraging, less time moving, and more time resting than a group in a higher-quality habitat (E. Riley 2007). In addition, Ménard et al. (2013, 2014) documented significant differences in dietary composition and behavior between Barbary macaques (*Macaca sylvanus*) living in habitats with differing degrees of anthropogenic disturbance. Macaques in lower-quality habitats where livestock grazing was more common fed more on underground resources, shrubs, and acorns and also had increased day range lengths and spent more time foraging and moving than conspecifics in higher-quality habitats (Ménard et al. 2013, 2014). Life in lower-quality habitats often results in increased energy expenditure and reduced diet quality, both of which are associated with reduced fitness, though the effects of anthropogenic disturbance on diet are varied and complex (Chapman et al., this volume).

Degradation can alter the floristic composition of primate habitats, in turn causing changes to diet and behavior. In some instances, fragmentation or degradation decreases the availability of preferred foods and increases reliance on fallback foods (Onderdonk and Chapman 2000). Geladas (*Theropithecus gelada*) are graminivorous throughout the year but rely on underground storage organs in the dry season when green grasses are less abundant and may, in fact, rely more on underground foods such as corms or rhizomes in anthropogenically altered landscapes (Jarvey et al. 2018; Kifle and Bekele 2021). Other primates, such as spider monkeys, that may alter group size or cohesion to cope with fluctuations in preferred food availability may also rely on fallback foods in degraded habitats. For example, brown spider monkeys (*Ateles hybridus*) living in a 65 ha forest fragment in northern Colombia consumed significantly more leaves and decaying wood than expected (de Luna et al. 2017). Thus, habitat degradation imposes additional significant constraints on dietary and behavioral flexibility.

Habitat degradation and fragmentation can also push primates to use novel or exotic food resources, including invasive species or cultivated crops. Crop feeding has been reported across the order Primates, and dietary changes in response to the introduction of cultivated foods have been recorded in many taxa. For example, brown howler monkeys in Rio Grande do Sul state, Brazil, were observed feeding from six species of cultivated fruits, accounting for <1% to 18% of annual feeding (Chaves and Bicca-Marques 2017). Consumption of cultivated fruits was in proportion to their availability; however, cultivated fruit feeding was unrelated to the availability of wild fruits in the area (Chaves and Bicca-Marques 2017). Similar exploitation of nutrient- and energy-dense cultivated fruits outside of periods of decreased fruit availability has been observed in great apes (Hockings and McLennan 2012; Naughton-Treves 1998; N. Seiler and Robbins 2016). Of course, increased reliance on cultivated crops also increases the likelihood of negative human-nonhuman primate interactions, which can have significant conservation implications (Hill 2017).

Investigations of dietary diversity for primates in disturbed habitats indicate the inclusion of exotic or novel plant species as an additional strategy for coping with decreased food availability. The diet of colobus monkeys (*Colobus angolensis palliatus*) in Kenya's Diani Forest consisted of up to 40.3% exotic (nonindigenous) plant species across home ranges with varying levels of disturbance (N. Dunham 2017). The inclusion of novel food sources is seen in lemuroids, as well: folivorous southern bamboo lemurs (*Hapalemur meridionalis*) included a higher proportion of exotic and pioneer species in their diets than did sympatric, frugivorous collared brown lemurs (*Eulemur collaris*; Eppley, Balestri, et al. 2017). Folivorous primates such as southern bamboo lemurs or brown howler monkeys may be less severely impacted by habitat degradation due to higher levels of dietary flexibility than frugivorous primates.

TEMPORAL SHIFTS IN FOOD AVAILABILITY, BEHAVIORAL FLEXIBILITY, AND DIET

Variation in the size and distribution of food patches can occur over time as well as across space. There are often trade-offs between behavioral responses that result in flexible, fission-fusion grouping patterns and the inclusion of abundant food sources where patchy foods are scarce. Female-bonded species, such as baboons and macaques, are more likely to shift their diets, while male-bonded species, such as chimpanzees and spider monkeys, are more likely to adjust their party sizes (Wrangham 1980).

Food availability varies along multiple timescales in response to shifting ecological conditions. The tropical forests where nearly all primate species are found do not typically experience high degrees of seasonality in temperature or day length; however, many tropical forests are characterized by significant rainfall seasonality (van Schaik, Terborgh, and Wright 1993). Seasonal patterns of rainfall vary interannually and can be impacted by climatic events such as ENSO—with resulting interannual changes in, for example, fruit availability (Chapman, Omeja, et al. 2018)—and global climate change (Rothman, Chapman, Struhsaker, et al. 2015). Additionally, in some regions, changes in climate and weather patterns contribute to peaks in fruit production known as masting events that occur every 2–10 years (Ashton 1988; Knott 1998). The abundance of preferred foods shifts weekly based on phenophase, seasonally with rainfall, and interannually with larger climatic patterns, but because responses to ecological variation may be constrained by physiology or morphology, behavioral responses to ecological shifts are highly variable (van Schaik and Brockman 2005).

For species that live in highly seasonal habitats, behavioral flexibility provides a key advantage in adapting to fluctuations in food abundance. Indeed, most primates living in highly seasonal habitats cope with intra-annual shifts in resource availability by altering dietary or behavioral patterns throughout the year (Hemingway and Bynum 2005). Seasonal shifts in behavior are often the result of increased competition over scarce resources or adjustments to cope with reduced energy intake. Typical behavioral responses to decreases in the availability of preferred foods include changing ranging patterns or altering activity patterns to conserve energy (J. Lambert and Rothman 2015; van Schaik, Terborgh, and Wright 1993). However, it is important to note that the rainy season and dry season may have different meanings in different habitats. For example, in Kibale National Park, Uganda, fruiting peaks tend to occur between the end of the first rainy season and the beginning of the dry season (Chapman, Wrangham, et al. 1999), whereas in Caratinga, Minas Gerais, Brazil, fruiting peaks during the rainy season (Strier 1991; Strier and Boubli 2006).

Howler monkeys are often characterized as behavioral or facultative folivores because they typically consume fruit in proportion to its spatiotemporal availability (Rosenberger, Halenar, and Cooke 2011; Strier 1992). Black-and-gold howlers and brown howlers in highly seasonal habitats at the southernmost edge of the genus's distribution (southern Brazil and northern Argentina) generally consume a more folivorous diet than Central American species (Garber, Righini, and Kowalewski 2015). Across study sites, brown howlers consume at least 60% leaves throughout the year but consume fruit in proportion to its availability in their habitats (Aguiar et al. 2003; Chaves and Bicca-Marques 2016; S. Mendes 1989; Strier 1992; but see Santos et al. 2013). For instance, S. Mendes (1989) found that, during the wet season (a period of increased fruit abundance), brown howler monkeys at the Reserva Particular do Patrimônio Natural-Feliciano Miguel Abdala in Minas Gerais, Brazil, spent 14.5% less time feeding on leaves but 28.4% more time feeding on fruits. Additional studies with brown howler monkeys report high year-round consumption of leaves with fruit consumption typically varying seasonally and with local ecological characteristics (I. Agostini et al. 2010; Aguiar et al. 2003; Chiarello 1993, 1994; Miranda and Passos 2004). Brown howler monkeys can be thought of as shifting diets along a folivory-frugivory spectrum as resource availability shifts in their habitats (S. Mendes 1989; Strier 1992).

Dietary flexibility in brown howler monkeys is associated with spatial and temporal fluctuations in the abundance of high-quality foods such as fruits; however, relationships between diet and behavior are more difficult to discern. Both S. Mendes (1989) and I. Agostini et al. (2012) reported that resting time did not change between rainy and dry seasons in southeastern Brazil and northeastern Argentina, respectively. However, Chiarello (1993) reported that, in a population of brown howler monkeys in São Paulo state, resting time was significantly higher in the summer (rainy season) than in the winter (dry season). Increased resting time among howler monkeys compared to other primates is generally attributed to their highly folivorous diet because leaves are more difficult to digest than fruits (Milton 1980). Resting time, it seems, should be significantly related to the degree of leaves consumed, yet the conflicting findings from multiple studies with brown howlers indicate that additional factors beyond diet may influence the amount of time spent resting.

Furthermore, the results of the few yearlong studies conducted with brown howler monkeys indicate that the proportions of time dedicated to socializing, traveling, and feeding also show differential patterns of shifting between rainy and dry seasons. Brown howler monkeys at the Reserva Particular do Patrimônio Natural-Feliciano Miguel Abdala increased time spent traveling and decreased time spent feeding during the rainy season (S. Mendes 1989).

Brown howlers in São Paulo state also spent significantly less time feeding during the rainy season and less time in social interactions but did not shift time spent traveling between the rainy and dry seasons (Chiarello 1993, 1995). Similarly to Chiarello (1993, 1995), I. Agostini and colleagues (2012) found that brown howlers in northeastern Argentina did not significantly shift ranging patterns between the rainy and dry seasons; however, I. Agostini et al. (2012) observed significantly more social interactions during the rainy season. Indeed, for howler monkeys, their high degree of behavioral flexibility appears to enable multiple behavioral adjustments to cope with seasonal shifts in the availability of high-quality foods.

Supra-annual changes in fruit availability, such as fruit masting in Southeast Asian forests, also have significant impacts on behavior and diet. During fruit masting events, the diet of orangutans in Gunung Palung National Park, Borneo, consisted of up to 100% fruits, yet during periods of low fruit availability, orangutans consumed a diet consisting of as little as 21% fruit and up to 37% tree bark (Knott 1998). The presence of ketones in orangutan urine analyzed from the low-fruit period indicates that these orangutans were under energetic and nutritional stress during severe food scarcity, despite spending the same amount of time foraging, and relied partly on catabolization of body fat for survival (Knott 1998). Leaf monkeys (*Presbytis rubicunda rubida*) and gibbons (*Hylobates albibarbis*), also in Gunung Palung National Park, show differing responses to fruit masting and periods of low fruit availability (Clink et al. 2017). Leaf monkeys had higher dietary diversity and richness than gibbons (likely due to higher consumption of nonfruit items), but both species increased their fruit and seed consumption when fruits became more abundant and increased the amount of leaves and figs in their diets during periods of decreased fruit abundance (Clink et al. 2017).

In the short term, seasonal changes in the availability of preferred foods or high-quality foods can put significant energetic or nutritional stress on primates, which must be addressed via changes in behavior. Prolonged periods of nutritional or energetic limitation can have significant physiological consequences for primates (e.g., Knott 1998). Behavioral flexibility is one of the principal mechanisms for primates coping with temporal variation in food abundance. Conserving energy by reducing costly behaviors such as traveling while increasing behaviors such as resting can have a significant impact on an individual's ability to maintain a positive or neutral energy balance.

BEHAVIOR AND DIET IN MARGINAL HABITATS

Most primates live in tropical regions. A limited number of species occupy latitudes significantly farther north or south of the equator, and life at the extremes can create considerable challenges for these animals. Distance from the equator is positively correlated with increased seasonality and greater changes in day length (hours of sunlight) between seasons. Seasonal fluctuations in temperature and rainfall have significant effects on primate behavior, imposing constraints via fluctuations in the availability of preferred foods that can impact an individual's ability obtain sufficient energy (reviewed above). Seasonal shifts in day length can impose additional constraints by limiting the amount of time individuals can allocate to various behaviors such as foraging (B. Coleman et al. 2021; R. Hill et al. 2003; Van Doorn et al. 2010).

Primates that live at high altitudes on Madagascar and in Asia, Africa, and Central and South America exhibit a wide range of adaptations to cope with the unique physiological challenges that altitude can bring (Grow, Gursky-Doyen, and Krzton 2014; Hou, Chapman, Jay, et al. 2020). A comparison of macaques and colobines in temperate Asian forests revealed that both species adjusted diet as latitude and altitude increased; however, the effect of altitude on diet was stronger for colobines, which live as high as 4,500 m above sea level in China (Kirkpatrick and Grueter 2010; Tsuji et al. 2013). Asian colobines (e.g., *Rhinopithecus* spp.) typically exhibit pronounced seasonal variation in diet and rely on lichen as a principal component of diet for up to 10 months of the year at some sites (Grueter et al. 2010; Hou, He, et al. 2018). For snub-nosed monkeys in particular, altitude and temperature are correlated with increased time spent feeding (Kraus and Strier 2022). Indeed, altitude and latitude better predicted feeding effort of black-and-white snub-nosed monkeys (*Rhinopithecus bieti*) in Yunnan Province, China, than forest area, home range, group size, or population density (Z. Huang et al. 2017).

Extreme weather and deviation from average climate conditions also impact primate diet and behavior. There are numerous examples of phenomena such as hurricanes or cyclones impacting primate populations. Analysis of hair cortisol concentrations and body weight in a population of ring-tailed lemurs (*Lemur catta*) at Beza Mahafaly Special Reserve in Madagascar revealed significant variation associated with drought and cyclonic activity (Fardi et al. 2017). Both hair cortisol concentration and body weight responses to drought and a cyclone varied by age-sex class, indicating that an individual's sex and life stage can significantly influence their ability to adjust behavior in response to extreme weather events. In general, female ring-tailed lemurs had higher hair cortisol concentration as a result of drought, while males exhibited higher hair cortisol concentration during the postcyclone period (both of which were periods of reduced food availability; Fardi et al. 2017). The impacts of events such as cyclones likely extend many months past the initial occurrence due to defoliation and tree damage reducing the availability or diversity of foods, but behavioral and dietary flexibility may lessen the impacts of these events on some primate populations (Dinsmore et al. 2021).

Semiregular weather patterns such as ENSO can cause significantly different weather conditions in primate habitats, which in turn can impact food availability and behavior. For instance, during an extreme drought that occurred in an El Niño year, gray-cheeked mangabeys (*Lophocebus albigena*) frequently fed on bark, a very low-quality food resource (J. Lambert et al. 2004). ENSO events have also been associated with reduced fecundity in Milne-Edwards's sifaka, likely as a result of decreased rainfall and food abundance (*Propithecus edwardsi*; A. Dunham, Erhart, Overdorff, et al. 2008). Similarly, ENSO conditions in three years preceding population censuses were associated with decreased reproductive output and increased offspring mortality in white-faced capuchins (*Cebus capucinus*) in Costa Rica (Campos, Jack, and Fedigan 2015). ENSO events can increase or decrease rainfall or temperature based on geography; therefore, the effects of ENSO on primate behavior and diet can vary significantly within and between species and geographic regions.

There is emerging evidence that over longer timescales (i.e., decades), global climate change is also having a significant impact on plant species and thus primate food availability. A short-term comparison of leaf chemistry at Kibale National Park revealed that variation among individual trees was greater than variation over the study periods of August 1998–June 1999 and July 1999–May 2000 (Chapman, Chapman, Rode, et al. 2003). However, analysis of long-term data on leaf chemistry also conducted at Kibale revealed that leaf quality declined significantly when the same trees were reassessed after 15 and 30 years (Rothman, Chapman, Struhsaker, et al. 2015). Specifically, the fiber content of leaves from the same trees increased over 15 and 30 years, while protein content declined over 15 years (Rothman, Chapman, Struhsaker, et al. 2015). Indeed, greenhouse experiments support the conclusion that global climate change may have a significant negative impact on leaf quality: in multiple experiments, nitrogen (a common proxy for protein) declined as greenhouse gas levels were artificially increased (Stiling and Cornelissen 2007; Zvereva and Kozlov 2006). For numerous folivorous species, leaf quality, especially protein-to-fiber ratio, is positively correlated with habitat use and group distribution (Ganzhorn 1992; but see I. Wallis et al. 2012). Declining leaf quality could have significant impacts on primate communities as the planet's climate continues to change.

The effects of climate change can compound as increasingly extreme weather patterns impact the phenology and productivity of forests (Butt et al. 2015), thus reducing food availability for consumers at different trophic levels. While it can be difficult to assess the impacts of climatic variation on primate communities without long-term studies, data from several projects of at least 20 years are beginning to show the potential impact of large-scale climatic variation on primate behavior, ecology, and demography (Campos, Morris, et al. 2017). For instance, in their analysis of seven primate species for which long-term data are available, Campos, Morris, and colleagues (2017) found that for seasonally breeding species (the northern muriqui, *Brachyteles hypoxanthus*; Milne-Edwards's sifaka, *Propithecus edwardsi*; and blue monkeys, *Cercopithecus mitis stuhlmanni*), there was strong evidence of decreased fertility as a result of climatic variables such as warmer temperatures. In many cases, primates are also at a heightened risk of extinction due to the compounding effects of global climate change and anthropogenic habitat alteration, hunting, and disease (Estrada, Garber, et al. 2017).

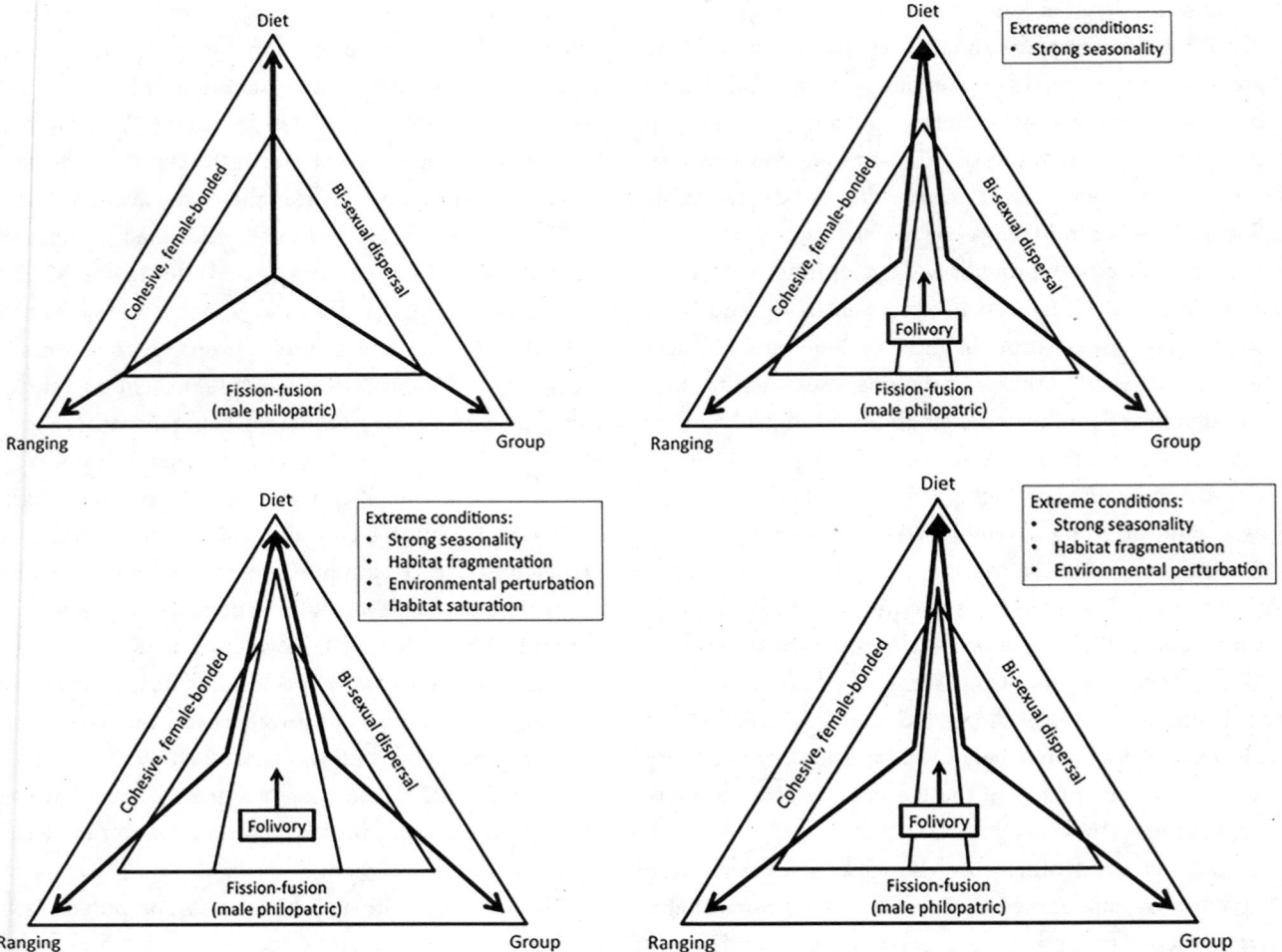

Figure 20.2. Schematic model of behavioral responses to food availability. Behavioral adjustments to changes in food availability include changes in diet, grouping, and ranging patterns (*upper left*). The way that primates respond to changes in food availability also reflects patterns of dispersal and social systems, following Strier (2009). Female-bonded species are most likely to change ranging and diet; fission-fusion (male-philopatric) species (see Strier, Possamai, and Mendes 2015) are most likely to change ranging and grouping; and species with bisexual dispersal regimes are most likely to change grouping patterns and diet, as originally described by Strier (2009). Extreme ecological conditions, such as those shown in the sequential panels (*clockwise from upper right*), can impact the ability of a species to respond to food availability, which can in turn interact with constraints on a species' ability to subsist on leaves and other low-quality foods (Strier 2021b).

DYNAMIC TRADE-OFFS BETWEEN DIETARY AND BEHAVIORAL FLEXIBILITY

The concept of flexibility has become increasingly important in comparative models of primate behavior as our knowledge of behavioral variation within and among species has grown (Chapman and Rothman 2009; Strier 2009, 2017). Dietary flexibility, which represents a subset of behavioral flexibility, has long been associated with dietary shifts documented to occur in response to fluctuations in food supplies due to seasonality and extreme climatic events such as extended droughts (e.g., van Schaik and Brockman 2005). The dichotomy between the ability to shift to alternative foods and shifting other aspects of behavior, such as ranging and grouping patterns, has also been a key distinguishing feature among some of the early predictive models of primate socioecology (e.g., Isbell 1991; Sterck et al. 1997; van Schaik 1989; Wrangham 1980). Indeed, it appears that

Figure 20.3. A female brown howler monkey feeds on *Cecropia glaziovii* leaves while her infant clings ventrally. Cecropia trees are characteristic of disturbed forests and secondary growth but are an important source of both fruit and leaves for howler monkeys throughout their range in the Americas.

Photo © AJ Hardie/Projeto Muriqui de Caratinga.

the tendency of a species to respond to scarcities in the availability of preferred foods may be related, at least in part, to its dispersal regime and corresponding social systems (fig. 20.2). Thus, female-bonded primates living in cohesive groups are more likely to shift their diet and ranging patterns in response to declines in preferred food availability, whereas primates living in fission-fusion societies with male philopatry are more likely to shift their grouping and ranging patterns (fig. 20.3). However, extreme conditions related to strong seasonality, habitat fragmentation, perturbations, and saturation can force increasingly greater shifts toward folivory that may be beyond the physiological limits for some species (e.g., Korstjens et al. 2010).

Because of their relatively high energy requirements, extended exposure to extreme conditions that decrease the availability of energy-rich foods can result in fitness consequences such as decreased fecundity and increased mortality. Constraints on adjusting dispersal regimes, which are phylogenetically constrained (Strier, Lee, and Ives 2014), require behavioral and dietary flexibility that may negatively impact the resilience of primate populations over time (figs. 20.4, 20.5; Bicca-Marques et al. 2020; Strier 2021a).

Figure 20.4. A female brown howler monkey feeds on paineira-rosa flowers (*Ceiba* sp.) with her young infant clinging ventromedially. While brown howler monkeys are mostly folivorous throughout the year, seasonally abundant foods such as flowers are an important supplement to their diets.

Photo © AJ Hardie/Projeto Muriqui de Caratinga.

SUMMARY AND CONCLUSIONS

Despite its relatively advanced empirical and theoretical foundation, there are still many gaps in our understanding of primate dietary and behavioral flexibility. For example, the relatively short duration of many studies makes it difficult to distinguish behavioral and dietary variation from flexibility (Strier 2017). These distinctions are confounded by the challenges of obtaining reliable data on physiological regulating mechanisms, which underlie, for example, nutrient balancing, or the relative trade-offs between energy stress and protein stress in wild primates (e.g., E. Vogel, Knott, et al. 2012). Nonetheless, major advances have occurred in the development of noninvasive, affordable assessments of primate physiology, including fecal steroid assays to measure levels of glucocorticoids and urinary C-peptide assays, which now make it possible to evaluate levels of social and energetic stress in wild primates (e.g., Chen et al. 2021; Emery Thompson 2017; Sacco et al. 2021). Ongoing refinements, including species-specific validations of these assays, will permit their application across a wider diversity of species, generating the comparative data necessary to detect patterns in physiological mechanisms that may underlie species differences in behavioral flexibility.

- Physiology (e.g., nutritional requirements) and morphology (e.g., dentition or body size) impose significant constraints on dietary and behavioral flexibility in pri-

Figure 20.5. A muriqui eats coquinho fruit from a palm tree. Muriquis at the highly seasonal Reserva Particular do Patrimônio Natural-Feliciano Miguel Abdala are less frugivorous than at other, less seasonal sites.
Photo © AJ Hardie/Muriqui Project of Caratinga.

mates. These limitations are compounded by ecological, demographic, and behavioral characteristics that may change significantly within individual lifetimes.
- The abundance of high-quality (energy-dense) foods fluctuates across numerous spatial and temporal scales. Assessments considering scale (e.g., within one forest fragment, L. Jung et al. 2015; seasonal, I. Agostini et al. 2012; interannual, Campos, Morris, et al. 2017) are necessary to fully understanding the degree to which behavior can vary within and among species.
- Understanding behavioral and dietary flexibility is increasingly important to conservation and management programs on behalf of populations of endangered species that persist in severely altered habitats on suboptimal diets (Korstjens et al. 2010; A. Meyer and Pie 2022; McLennan et al. 2017; Strier 2021a). Incorporating the variation in primate behavioral flexibility into the assessments of their conservation status can provide additional insights into their ecological resilience and adaptive potential.

PART IV

Methods, Practice, and Application

This section explores how nutrition science continues to expand methodologically and be applied through space and time and across systems. As discussed by David Chivers and Kim McConkey at the outset of this volume, we have come a long way as scientists in the acquisition of samples in the field to measure diet (Heymann) and the associated laboratory methods to assay nutrition (Conklin-Brittain), stable isotopes (Sponheimer and Crowley), and mechanical properties of those foods (van Casteren and Lucas). We have also moved vastly beyond considerations of just energy (as in optimal foraging models of the last century) to integrating detailed information on macro- and micronutrients into our models (Raubenheimer). We can use these updated methods and nuanced understanding of food at both dietary (the foods) and nutritional scales (macro and micro) in application to the fossil record (Ungar) and to test hypotheses regarding primate population carrying capacity (Marshall). Most urgent, however, is the application of our knowledge to on-the-ground exigencies of the impacts of climate change (Rothman, Makombo, and Irwin) and anthropogenic conversion of habitat (Chapman, Valenta, Espinosa-Gómez, Corriveau, and Bortolamiol) on primate food nutritional content and availability.

21 Measuring Food in the Field

Eckhard W. Heymann

Acquiring sufficient amounts of macro- and micronutrients is essential for maintaining the structure and metabolism of any organism, for growth and reproduction, and thus ultimately for gaining evolutionary fitness (Felton and Lambert, this volume). Feeding—the uptake of living or dead organic matter and of inorganic minerals—provides primates and other animals with these nutrients. Depending on the type of food through which the majority of energy is acquired, primates are coarsely categorized as frugivores, folivores, insectivores/faunivores, exudativores/gummivores, or a combination of these (Hawes et al., this volume). These categorizations are based on field observations and the quantification of feeding behavior. Therefore, the simplest reason to measure food in the field is to obtain information on the dietary strategies of primates, which is essentially basic primate autecology.

However, there are more complex reasons for measuring food in the field. Before feeding can take place, primates must find food (searching phase; Dominy et al. 2006). This requires different sensory capacities—for example, the ability to visually detect fruit or leaves against a background or interpret the camouflage of prey. Once food has been detected, animals must make feeding choices (e.g., between ripe and unripe fruits or between mature and young leaves) that can be based on visual, olfactory, or haptic inspection (Dominy et al. 2006). Then, a final decision is made on whether a food item is actually consumed or rejected. In this phase, additional sensory input on the mechanical (e.g., hardness; van Casteren and Lucas, this volume) and chemical (e.g., taste) properties can be involved (Dominy et al. 2006). This entire process of finding and selecting food is part of the inquiry into primate sensory ecology (Dominy, Lucas, et al. 2001; Melin and Veilleux, this volume).

Measuring the amount of food ingested is but the first step in primate nutritional ecology. Combining the amount with data on the chemical composition of food allows estimating the uptake of energy and specific nutrients (Conklin-Brittain, this volume) but also of secondary plant compounds (Stalenberg et al., this volume).

Since the amount of food available in the habitat or in a food patch can be limited and food may vary in quality, primates may compete over food within and between groups of the same species (Janson, this volume; Hardie and Strier, this volume). The social structure of primate groups can influence access to and the amount of food consumed by any individual, which may ultimately affect reproductive success. Therefore, pertinent measures of food availability are part of the inquiry into primate socioecology (Houle et al. 2010; Janson 1988a; P. Stevenson et al. 1998).

Competition over food can also take place between different primate species and between primates and non-primates. Measuring food by recording the frequency of use of different food items or food species or by quantifying the amount of food items ingested can be used to estimate the dietary overlap between species and to evaluate the potential for interspecific competition; this method is therefore also a tool in studies of community ecology.

As feeding actually implies predation sensu lato on another organism—be it plants, fungi, lichens, or animals—measuring food can also provide information on the potential impact of primate feeding ecology on populations of their food species. This also falls into the field of community ecology. So far, there are few examples of such impacts—for example, of flower feeding on fruit set (Riba-Hernández and Stoner 2005) or of predation on prey populations (Lwanga et al. 2011; Teelen 2008)—but these ecological roles and impacts of primates are likely to become a more important field of inquiry in the future (Chapman, Bonnell, Gogarten, et al. 2013). While the impact on affected plant populations of seed predation by granivorous primates has not been addressed at all, Souza-Alves, Barbosa, and Hilário (2020) recently showed that gouging by *Callithrix jacchus* may affect tree survival and turnover rates. In contrast to potentially negative fitness effects on individuals of food species (be it plants, fungi, lichens, or animals), feeding may also have positive effects. When nectar is consumed and reproductive flower parts remain intact, primates can act as pollinators (Heymann 2011). The number of subsequent visits to different flowers (i.e., nectar feeding bouts) and to different flowering individuals of the same plant species influence the probability of pollination (Janson, Terborgh, and Emmons 1981). When seeds are swallowed along with pulp and voided intact with the feces or are spit out at some distance from source trees, primates function as seed dispersers, which is probably one of the major ecological functions of primates (Chapman and Russo 2007; Corlett and Primack 2011; J. Lambert and Garber 1998). The number of visits to food plants and the number of seeds dispersed per visit are important quantitative parameters of seed dispersal effectiveness (Schupp et al. 2010). Thus, with regard to both pollination and seed dispersal, pertinent measures of food are relevant for the analyses of primate ecological functions.

Any animal can only feed on what is available in its habitat, and differential availability can influence feeding strategies. Therefore, measuring the abundance of (potential) food items in the habitat is also key for understanding primate dietary strategies and feeding ecology—for example, how they respond to seasonal food shortages.

Finally, measuring food in the wild can also inform studies on functional morphology and physiology. Over the course of primate evolution, the challenges provided by the chemical and physical properties of food—for procurement, mastication, and digestion—may have resulted in specific adaptations in sensory systems, dentition, and the gastrointestinal tract. To interpret these adaptions as a consequence of continuous challenges provided by the bulk of the diet or of the exploitation of minor resources that are critical for survival during periods of food scarcity (J. Lambert 2007; Rosenberger 1992) requires quantitative information on feeding behavior in the wild.

A requisite for measuring abundance and consumption is the proper taxonomic identification of food items. This is relatively straightforward for most plant food through determination in situ with the help of pertinent identification keys or through sampling reproductive and vegetative plant parts and comparing them with herbarium specimens. However, it creates major challenges with regard to prey items, particularly invertebrate prey, because it is difficult to identify invertebrates from a distance and invertebrates can be difficult to catch. Proper taxonomic identification is essential for understanding not only population ecology but also (and perhaps even more so) community ecology—for example, how different primate species or primate and nonprimate species with similar and overlapping ecological requirements coexist or how primates interact with and impact their food.

In this chapter, I use the term "measuring food" in a wider sense. I do not refer to the intrinsic characteristics of food such as chemical composition or physical properties, as they are addressed elsewhere (Conklin-Brittain, this volume; van Casteren and Lucas, this volume). Rather, I address questions of how to quantify feeding behavior and the availability of food in the wild. These include the definition and levels of resolutions of food; behavioral methods to record feeding behavior; ecological indices to characterize the diversity, preferences, and overlap of food; and methods to quantify the availability of food in the habitat. For the latter, I emphasize food types (e.g., arthropods and exudates) that have received less attention in primate ecological studies than fruits, leaves, and flowers. The chapter aims to provide an overview of methods for measuring food (in the sense defined above). Apart from providing an overview of standard methods, I give special emphasis to food types that have traditionally received less attention in primate research and to new tools for and approaches to measuring food. The chapter ends with a look at a new approach ("foodscape") that might become a useful concept in primate ecological studies in the future.

QUANTIFYING FEEDING BEHAVIOR

Definition of Food Types and Levels of Resolution

Before feeding behavior can be quantified, it is necessary to define the food types to be recorded. Different levels of resolution can be applied, and the degree of resolution obviously depends on the scope of the study. A coarse categorization into fruit, leaf, exudate, and animal prey will usually be sufficient to examine primate dietary strategies. However, depending on the primate species under study, additional food types that are relevant for nutrition (e.g., flower/nectar, fungus, lichen, or bark) need to be recognized and recorded separately (L. Porter and Garber 2010; Sussman and Tattersall 1976; Wich, Utami-Atmoko, Mitra Setia, Djoyosudharmo, et al. 2006; Xiang et al. 2007; fig. 21.1).

In studies of nutritional ecology and community ecology (particularly ecological functions of primates), it is important to specify the part of the food item that is consumed (unless the entire item is ingested). For fruits, a distinction can be made between the ingestion of entire fruits including husk, pulp, and seeds; of pulp and seeds with husks dropped (unless the husk itself is edible); of pulp only with seeds being spit out (fig. 21.2); or of seeds only, as these parts differ in their chemical composition

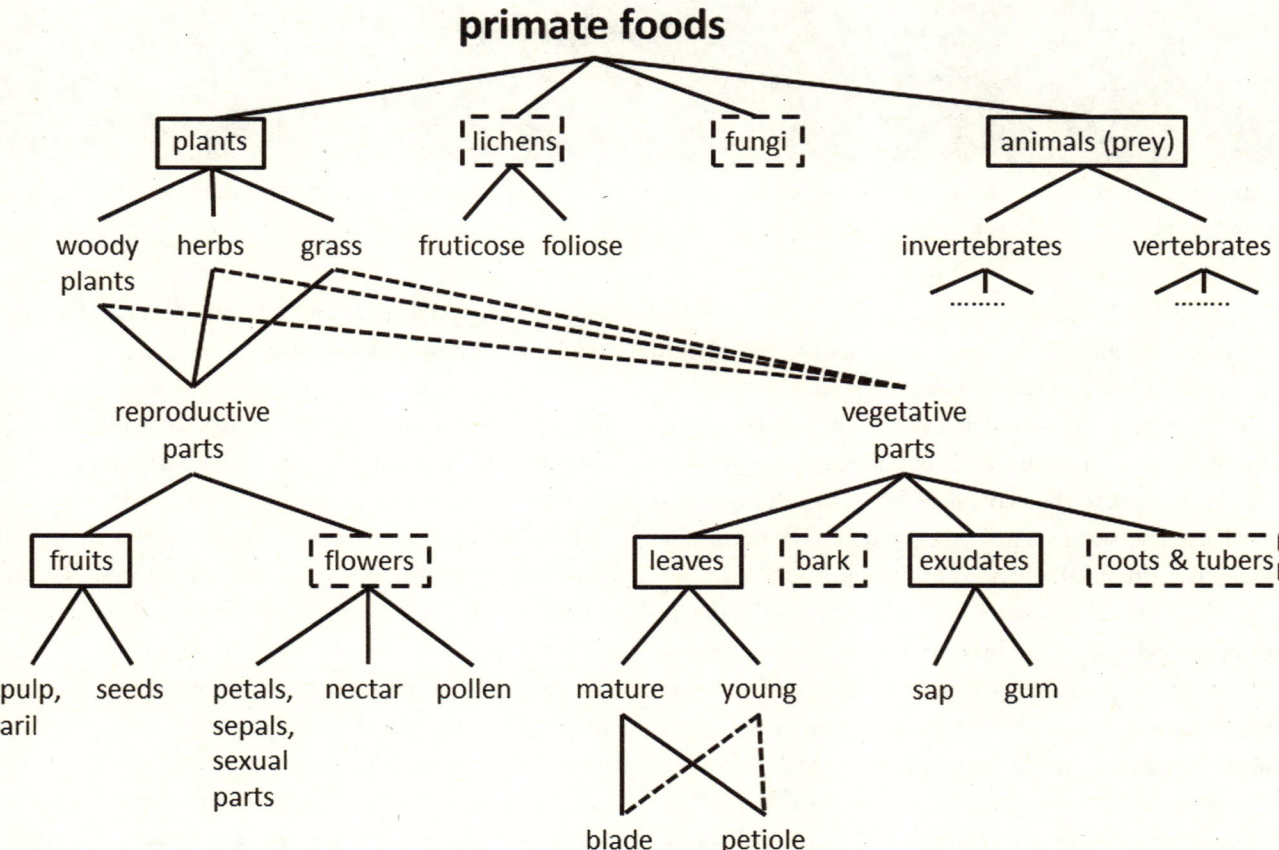

Figure 21.1. Food types of primates with different levels of resolution. This figure is an extension of the scheme provided by Richard (1985). Solid boxes indicate the food categories classically included in most studies of primate feeding ecology; hatched boxes indicate additional types required for a number of primate species. For animals, particularly for invertebrates, further distinctions can be made on whether the entire animal, parts of its body, or secretions (Corbin and Schmid 1995; Dammhahn and Kappeler 2008) are consumed.

Figure 21.2. A coppery titi monkey, *Plecturocebus cupreus*, spitting the seed of a shimbillo, *Inga edulis*, after removal of the pulp. Photo © Fabian Nummert (used with permission).

(Wagner et al. 2015) and physical properties. When entire fruits or pulp and seeds are ingested, fecal samples should be examined to determine whether seeds are passed intact through the gastrointestinal tract or whether seeds are masticated. In addition to nutritional consequences, this is relevant to the ecological function of primates as seed dispersers or seed predators (Norconk, Grafton, and Conklin-Brittain 1998; Russo and Chapman 2010).

For leaves, distinction can be made regarding the developmental stage (leaf buds, young leaves, and mature leaves; fig. 21.3) and the part that is consumed (entire leaf including petiole, leaf blade, apical or basal parts of leaf blade, or petiole), as the chemical and physical properties may vary between developmental stages and leaf parts (A. Davies, Bennett, and Waterman 1988; Teaford, Lucas, et al. 2006). For primates in open habitats that include plants from the ground in their diet, a distinction between dicotyledonous plants and grasses may be relevant, as these differ drastically in the amount of silica (Russel 1961, quoted in Massey et al. 2006) and in the relative proportions of stable carbon isotopes (^{12}C and ^{13}C; Sponheimer and Crowley, this volume).

Similarly, when flowers are exploited, nutritional ecology and community ecology studies demand a distinction between the parts of the flower consumed: entire flowers; petals, sepals, and reproductive parts; nectar; or pollen. As with fruits and leaves, these parts differ in their chemical composition and physical properties. The consumption of entire flowers, of the reproductive parts, or of pollen reduces the reproductive potential of a plant, while the consumption of only nectar might even be beneficial for plants, as it may result in pollination as a by-product of primate feeding activity (Heymann 2011).

"Exudate" is a collective term for substances emerging from plant tissue, usually as a response to damage (F. Smith and Montgomery 1959). Exudates consumed by primates include phloem sap and gums and possibly latex, while resins are not included in the diet of any primate species (L. Nash 1986; M. L. Power 2010). Phloem sap is a watery solution of mainly simple carbohydrates and pro-

Figure 21.3. A coppery titi monkey, *Plecturocebus cupreus*, feeding on a young leaf of an unidentified liana.

Photo © Fabian Nummert (used with permission).

teins produced in photosynthetically active plant organs and transported to other plant parts. Gums are chemically complex, including water-soluble polysaccharides, simple sugars, and secondary compounds (D. M. W. Anderson et al. 1990; Dewi et al. 2022; Hegnauer 1994; F. Smith and Montgomery 1959). They are liquid to viscous when oozing out and may harden in contact with air; time since gum was exuded (gum age) has previously been linked to intraspecific variation in digestibility (L. Nash 1986). Often produced in specialized cells, their function mainly consists in sealing wounds (Kozlowski et al. 1991; F. Smith and Montgomery 1959). The distinction between sap and gum may not always be easy in the field, but it is essential in light of their different chemical properties. In the primate ecology literature, "resin" is occasionally used to refer to gums or to exudates in general (e.g., Buchanan-Smith 1991; Santhosh et al. 2015). However, given their chemical nature (high concentrations of terpenoids and phenolics; Langenheim 2003), it is extremely unlikely that any primates consume resins at all.

Fungi consist of vegetative mycelia and reproductive sporocarps ("fruiting bodies"). While A. Hanson and colleagues (A. Hanson, Hall, et al. 2006; A. Hanson, Hodge, and Porter 2003) report the consumption of sporocarps in *Callimico goeldii*, the part consumed by other primates is not reported in other studies (see A. Hanson, Hodge, and Porter 2003). Sporocarps are mainly composed of structural carbohydrates, and although they are rich in nitrogen, this is largely associated with cell and spore walls and thus unavailable to digestion (Cork and Kenagy 1989; A. Hanson, Hall, et al. 2006).

Lichens are composite organisms formed through the mutualistic interaction between fungi and photosynthetic algae or cyanobacteria (T. Nash 2008). Depending on the growth form of the thallus, lichens are distinguished into various types, including fruticose, foliose, crustose, and others (T. Nash 2008). *Rhinopithecus bieti* has been reported to feed on fruticose and foliose lichens (Grueter, Li, et al., "Dietary Profile," 2009). Lichens contain polysaccharides such as lichenin or ß-glucan that require microbial fermentation and are rich in secondary compounds (Bissell 2014; Svihus and Holand 2000).

Taxonomic Resolution of Primate Foods

Taxonomic identification of food is crucial, especially when measuring the availability of food or when assessing dietary overlap and thus the potential for feeding competition between syntopic or sympatric species. Identifying plants is generally not a major problem. Florulas for many tropical and subtropical regions and the expert knowledge of local or national botanists generally allow for a good taxonomic resolution, and many publications in primate ecology provide exhaustive lists of plant food species. On a cautionary note, taxonomic revisions (which now increase through the availability of molecular ecology techniques) or different taxonomic approaches (lumping vs. splitting) by botanical experts may introduce variance or inconsistencies between studies, particularly when these have been executed at different times. This may create a problem for comparative analyses or for analyses of dietary overlap. Consulting large internet databases such as Tropicos.org or theplantlist.org that provide accepted names and synonyms helps alleviate this problem.

In stark contrast to the detailed taxonomic list of plant foods is the low resolution when it comes to the arthropod prey of primates. It has been (and still mainly is) common practice in primate field studies to classify the arthropod prey into coarse taxonomic categories (classes, orders, or at best families; e.g., Gautier-Hion 1980; E. Lima and Ferrari 2003; Terborgh 1983). This is the almost unavoidable consequence of prey being eaten completely, prey residuals being difficult to recover, or the difficulty of identification on a lower taxonomic level (genus or species).

This difficulty results from the high levels of training and specialization (compared to plant identification) required for identifying arthropods where sometimes microscopic differences may be critical for species assignation (e.g., structure of genitalia) and the lack or scarcity of field identification guides that are feasible to nonexperts.

A high taxonomic resolution of prey, comparable to the resolution achieved for plant food, is important for intraspecific (variation between primate groups and populations) and interspecific comparisons (differences or overlap between syntopic or sympatric primate species). The following two examples illustrate this point. Chimpanzee, *Pan troglodytes*, populations vary in tool use to gather termites (Sanz, Deblauwe, et al. 2014). Only the species-level identification of termites allowed excluding that this is simply a consequence of different termite species being harvested. For the two tamarin species *Saguinus mystax* and *Leontocebus nigrifrons*, Orthoptera are the major prey type (fig. 21.4), and dietary overlap would be very high if an ordinal level of identification were employed. Only a higher level of taxonomic resolution revealed that the two species differ strongly in the spectrum of orthopteran prey as a consequence of vertical segregation and different strategies of prey search and capture (Nickle and Heymann 1996; A. Smith 2000).

It is not surprising that for socially living prey like most ants and termites, taxonomic identification is easier than for solitary prey (O'Malley and Power 2012; J. G. Robinson 1986). After the end of a feeding bout, individuals of socially living prey can be collected from the nest, mound, or column for identification, which is not possible for solitary prey. In this case, only dropped residuals, such as tegmina (i.e., the modified forewings of Orthoptera, Blattodea, and other insect orders) and hindwings, are available for identification (Nickle and Heymann 1996). If prey items are ingested completely, their identification may become impossible. There are two potential solutions: identification of prey residuals in feces and metabarcoding approaches. Identification of prey residuals from feces will allow for a high taxonomic resolution only when even small remains (e.g., head capsules), provide taxonomically informative clues that allow identification to the genus level, as is the case in ants (Manfred Verhaagh, pers. comm.). The metabarcoding approach is recently emerging as a tool in primate ecology. Hofreiter et al. (2010) amplified vertebrate mitochondrial DNA

Figure 21.4. A black-fronted saddleback tamarin, *Leontocebus nigrifrons*, feeding on a phaneropterine katydid, probably *Titanacris olfersii*. Photo © Mojca Stojan-Dolar (used with permission).

(mtDNA) segments from feces of bonobos, *Pan paniscus*, and western gorillas, *Gorilla gorilla*, to examine whether they prey on vertebrates, as direct observations of vertebrate consumption are extremely rare from bonobos and lacking from gorillas. Metabarcoding was employed for a comparative study on the arthropod prey of Amazonian primates and for analyses of invertebrate foraging strategies in wild white-faced capuchins, *Cebus capucinus* (Mallott, Garber, and Malhi 2017; Pickett et al. 2012). Metabarcoding resulted in the detection of arthropod consumption in lemur species where this had not been previously known (Rowe et al. 2021), both in a broader taxonomic coverage and in resolution of arthropod prey consumed by vervet monkeys, *Chlorocebus pygerythrus* (Brun et al. 2022). The paucity of published DNA sequences from arthropods (in relation to the huge number of arthropod, particularly insect, species) may in most cases currently restrict the taxonomic resolution to the family or genus level, but with the increasing use of metabarcoding in biodiversity and ecological research, this is likely to change in the future. Another limitation of metabarcoding is that some prey taxa may be too degraded to identify. Metabarcoding can also be employed for prey residuals that do not provide taxonomically informative morphological characteristics (Lyke et al. 2019). Again, the resolution will depend on the availability of published DNA sequences.

Metabarcoding has also been used in a few studies to examine the plant portion of the diet in *Gorilla gorilla*, *Colobus guereza*, *Cebus capucinus*, and *Macaca arctoides* (Bradley et al. 2007; Mallott, Garber, and Malhi 2018; Osman et al. 2020). The use of the chloroplast marker trnL as a marker instead of rbcLis has been recommended by Mallott, Garber, and Malhi (2018).

Observational Methods for Quantifying Diet Composition and Food Intake

Various observational methods are available to quantify diet composition and food intake. The choice of method employed in any study depends on the specific research questions, the observational conditions (e.g., dense rainforest or open savanna), the degree of habituation of study subjects, and whether or not they can be individually identified. It may be imperative to combine different methods to obtain the optimal set and amount of data. An exhaustive review is far beyond the scope of this chapter, and readers are advised to consult J. Altmann (1974) and P. Martin and Bateson (2021) for concise introductions to observational methods.

Instantaneous scan sampling allows quantifying the basic dietary strategy (diet composition) of a primate species, as well as activity patterns, habitat use (including vertical use of space), and other ecological data. It is suitable to record the feeding behavior of entire groups or subgroups. For this method, the observation period (e.g., an observation day) is subdivided into intervals of fixed length (e.g., 30 minutes). At the end of each interval, the behavior of all visible group members is recorded. When an individual is feeding, the food type (with the resolution needed for the study; see above) is specified. Additional information, such as the height of feeding, the position in a tree crown, and distance to other group members, can be recorded. Calculating the percentage of records for any food type in relation to the total number of all feeding records gives an estimate of the relative contribution of fruit, leaves, and so on to the diet (i.e., the time spent feeding on different food types). One limitation of scan sampling in characterizing primate feeding is that rarely eaten food items may be overemphasized when calculated as contribution by number of records rather than as percentage of time, for example. Including taxonomic identification of the food, this method also provides a ranking of food species.

Ideally, the behavior of all group members would be recorded at the same moment, but this is practically impossible. Therefore, the observer has to scan through the group. This requires fixing a scan length—that is, the time allowed for concluding a scan. Any individual that comes into sight during this time is included in the sample, but any individual that comes into sight later is excluded. The probability that an animal comes into sight is influenced by the conspicuousness of its (feeding) activity. This may introduce a bias toward more conspicuous activities. To reduce this bias, observers may record the activity only if it is sustained for a certain time (e.g., 5 seconds) or record the activity that the animal performs at a fixed time after it came into view.

The scan length should be short in relation to the interval length but long enough to allow scanning through a group and recording the behaviors of a substantial fraction of the group. Obviously, for small groups or for groups spread over a small area, the scan length can be shorter than for large groups spread over a large area. If groups of different size or cohesion are observed within the same study, the interval and scan lengths must be the same for all observed groups to guarantee comparability of the data.

What is an advantage of instantaneous scan sampling on one hand—simultaneous recording of many individuals—can create a bias on the other hand: food in large patches that accommodate many or all group members at the same time (e.g., trees and lianas with large crowns) might be overrepresented, while food in small patches that are visited by only one or a few animals (e.g., trees and lianas with small crowns) or that are exploited on an individual level (solitary prey) will tend to be underestimated. Group size and spread are variables that affect the degree of this bias. Another bias can be introduced by the length of the sampling interval. If the interval length is large (e.g., 30 or 60 minutes), foods that are exploited in short feeding bouts tend to be underrepresented, even though their consumption may account for a considerable proportion of time.

Such biases can be reduced by employing focal animal sampling with continuous recording (P. Martin and Bateson 2021). This method requires that individuals be reliably identified. Focal animals are observed for a predetermined period, and the onset and end of feeding bouts are timed. As the feeding behavior of the focal animal is recorded independent of whether it is feeding alone or in a group, different food types or food in patches of different sizes are equally likely to be recorded. Focal animal sampling can be difficult under conditions of restricted visibility, as in dense tropical forests. When the focal animal goes out of sight, time-out must be taken. This can be cumbersome and error-prone under circumstances where animals continuously come into and get out of sight. When an animal is out of sight for a prolonged period, the observation should be terminated; a time criterion should be fixed for this.

When activities change too rapidly to time them accurately, focal animal sampling can also be combined with instantaneous recording. For this, the focal sample period is divided into short intervals.

Ideally, a randomized order of focal animals should be predetermined for every observation period (day, week, or month), and the same individual should not be selected as a focal animal subsequently or within a short period. If a focal animal cannot be found or if the observation has to be terminated because the animal moved out of view for too long, the next animal on the predetermined list should be selected, and the missed individual should be observed at the next opportunity.

In the field, this ideal procedure can be unrealistic, and focal animals need to be selected opportunistically. Nevertheless, the same animals must not be observed in subsequent focal samples or within a short period. The number of focal observations per individual should be balanced between group members and over time (i.e., evenly spread out over the day and over the study period).

A third potential method is behavior sampling (P. Martin and Bateson 2021). Whenever feeding by one or more individuals is observed, the type (and species) of food, the number of individuals involved, and any other relevant details are recorded. This method is also prone to biases—for example, feeding bouts involving several individuals are more likely to be detected than feeding bouts of a single individual. While it is not recommended to employ this method alone for quantifying feeding behavior, it may complement instantaneous scan sampling (particularly when intervals are long) to detect short feeding bouts on rare food types or species.

When analyzing data to determine the time spent feeding (calculated from the proportion of scan sampling points where this activity is recorded) or the number and length of feeding bouts, it must be acknowledged that none of the abovementioned observational methods provides accurate information on the ingested amount of food. In a classical study on the feeding ecology of leaf monkeys, C. Hladik (1977b) showed that ripe fruits of one plant species accounted for 28% of feeding time but 77.3% of fresh mass intake, while all other fruit species accounted for 46% of feeding time but only 16.6% of fresh mass intake. Similarly, in Hamadryas baboons, feeding time and the number of feeding bouts could explain only 30% and 40%, respectively, of the variation in the ingested amount of food, and no relationship existed between the length of feeding bouts and the ingested amount of food (Zinner 1999). Although this study was conducted on captive animals, it is plausible that this applies under field conditions as well.

Another bias can occur when comparing the diet composition in terms of time spent feeding on different dietary items between individuals, age-sex classes, seasons, populations, or species. The bias may result from different sample sizes and variable length of focal animal observations, termed "individual follows" in M. E. Harrison, Vogel, et al. (2009). These authors recommended empirically determining a minimal length of individual follows and only including data in the analyses that comply with this length.

Focal animal sampling provides the opportunity for more detailed measurements of food intake. In combination with nutritional analyses of food, this allows for estimating the uptake of energy and specific nutrients (Conklin-Brittain, this volume). Feeding rates (the number of food items ingested per time unit) or bite rates (the number of bites of a specific food item per time unit) are suitable measures as long as the size of items or of bites is relatively uniform. Bite size may vary due to multiple factors, including body size (Ross, Washington, et al. 2009), but may be recorded as an additional measure. For example, the area bitten from leaves or the volume bitten from fruits can be measured from dropped leaves and fruits, and units of the same size can then be cut from leaves or fruits that match the consumed items as closely as possible. The mass of these samples is then determined before and after drying.

There is no single ideal method or combination of methods suited for all studies of feeding ecology. The method used will depend on the question of interest for the researcher. For each study, the optimal method or combination should be identified and tried out through preliminary observations before the onset of the study.

MEASURING FOOD DIVERSITY, PREFERENCES, SIMILARITY AND OVERLAP

When a sufficient taxonomic resolution in the identification of food items is achieved, the diversity of and preferences for food species and the overlap between sympatric primate species can be estimated. The measures that can be applied are those that are also used in biodiversity research. For a full treatment of these measures, see C. Krebs (1999); all formulas shown here are adopted from this source.

Food Diversity

The simplest measure of food diversity is species richness—that is, the number of species in the diet. This measure does not consider differential representation of food species in the diet. As the number of food species is likely to increase with the length of the study period and approaches an asymptote at some point, it is important to obtain an estimate of the proportion of species that has been recorded with a given length of the study. Therefore, in a rarefaction analysis, the expected number of species is plotted against the time units (the number of hours or days of observation).

Generally, different food species contribute differently to the diet. This heterogeneity can be expressed with Simpson's index or the Shannon-Wiener index. Simpson's index puts more weight on common species in the diet, the Shannon-Wiener index on rare species. There are different variants of and thus different formulas for the calculation of Simpson's index (C. Krebs 1999). As the number of food species is finite, the most appropriate formula is

$$1 - D = \sum_{i=1}^{s}[n_i \times (n_i - 1)/N \times (N - 1)],$$

where $1 - D$ is Simpson's index, s the total number of food species, n_i the number of feeding records for food species i, and N the total number of feeding records. The index ranges from 0 (practically never realized, as this would mean feeding on only one food species) to almost 1.

The Shannon-Wiener index, H', is a measure derived from information theory and is calculated as

$$H' = \sum_{i=1}^{s}(p_i \times \log p_i),$$

where s is the total number of food species and p_i the proportion of feeding records for food species i. Theoretically, it ranges from 0 (again, practically never realized, as this would mean feeding on only one food species) to infinity (also not realized as the number of food species is finite).

A complementary but conceptually different measure is evenness. Evenness indices quantify how equally or unequally food species are represented in the diet. In its simplest form, evenness, E, is expressed as

$$E = I / I_{max},$$

where I is a diversity index and I_{max} the maximum possible value of this index, given the number of food species and feeding records. A shortcoming of this index is its sensitivity to species richness, and the use of other indices that are independent of species richness has been suggested (C. Krebs 1999).

Food Preferences

As different food species are not generally equally abundant, either primates can consume different species according to abundance or they can exert preferences. A simple measure of preference feasible under a wide range of conditions is the forage ratio or selection index

$$S = d_i / e_i,$$

where d_i is the proportion of a food species in the diet and e_i the proportion of the food species available in the habitat. Distinct statistical tests have to be employed depending on whether the available resources have been censused completely or estimated from sampling (C. Krebs 1999). Given the complexity of most primate habitats and the diversity of foods they consume, the former situation (complete census) is unlikely to apply in primate field studies.

Another more complex measure for preferences is Manly's α. This measure is based on the probability of a consumer feeding on a resource upon encounter (C. Krebs 1999). Therefore, a distinction has to be made between different situations: (a) the fraction of a certain resource (e.g., fruits of a given plant species) that is consumed is negligible in relation to its overall abundance, or renewal rates are high; or (b) a large portion of the resource is consumed, or renewal rates are low. In the latter case, the changing availability needs to be taken into account when calculating Manly's α (C. Krebs 1999). While both situations require detailed information on resource abundance, the latter obviously creates larger demands, as sampling of resource abundance has to take place at much shorter intervals. For a detailed statistical treatment of preference measures and selection indices, see Manly and coworkers (2007).

Dietary Similarity and Overlap

Most primates live in complex tropical ecosystems where they have to share food resources with other primate or nonprimate species (Cowlishaw and Dunbar 2000; Waser 1987). Depending on resource abundance, this creates a potential for competition, and measures of dietary similarity and overlap are a means of estimating this potential. Various indices of different complexity are available for similarity and overlap. A simple one is the binary Jaccard's coefficient

$$J = a / (a + b + c),$$

where a is the number of food species consumed by primate species A and B, b the number of food species consumed by primate species A but not B, and c the number of food species consumed by primate species B but not A. The Sorenson coefficient is a modification where a is multiplied by 2, thus weighting matches more strongly than mismatches.

The Renkonen index or percentage similarity / percentage overlap (also called the Schoener overlap index) is based on relative abundances of food species that sum up to 100%. For each food species i consumed by primate species A and B, the lower percentage value (from either A or B) is taken, and the values are summed up over all food species:

$$R = \sum_i \text{minimum}\,(p_{A,i}, p_{B,i}),$$

where $p_{A,i}$ is the percentage or proportion of food species i in the diet of primate species A, and $p_{B,i}$ the percentage or proportion of food species i in the diet of primate species B. The Renkonen index is relatively insensitive to sample size and food species diversity (C. Krebs 1999). A value of 0 indicates no similarity, and a value of 1 or 100 indicates complete similarity, depending on whether proportions or percentages are used.

Other, more complex indices are Pianka's measure, Morisita's measure, and the simplified Morisita index (also called the Morisita-Horn index). Examples of the use of Pianka's measure include analyses of the dietary overlap between sympatric guenons (Gautier-Hion 1980), sympatric mouse lemurs (Dammhahn and Kappeler 2008), sympatric langurs (Hadi et al. 2012), and sympatric macaques (Zhou, Wei, Tang, et al. 2014). Morisita's measure has been employed to examine diet overlap in cercopithecine polyspecific associations (M. Mitani 1991), sympatric howler and spider monkeys (Stoner et al. 2005), and sympatric macaques and langurs (M. Singh et al. 2011). P. Stevenson and colleagues (2000) and Nadjafzadeh and Heymann (2008) used the simplified Morisita index to analyze overlap in diet and prey foraging strategies, respectively, between sympatric New World monkeys. These indices can also be instrumental for comparisons within species (e.g., the interannual overlap of diets; Ni et al. 2014), for determining the consistency of data on fruit consumption obtained through different methods (Chaves, Stoner, Arroyo-Rodrigues, et al. 2011), and for the overlap in other ecological variables (e.g., vertical strata use; Norconk 1986). The reader is referred to C. Krebs (1999) for the respective formulas and for a discussion of the pros and cons of each index.

Potential Problems and Error Estimates

All indices presented above are derived from biodiversity and ecological research. They are based on measures of abundance, biomass, and so on and assume independence of samples. However, feeding records from group-living primates are rarely if ever independent samples. Therefore, their applicability in studies of primate feeding ecology may be debated. It remains a task for expert statisticians and modelers to test the applicability with real and simulated data sets of primate feeding.

Most often, no or very few replicates are available in primate ecological studies, so no standard errors or confidence intervals can be calculated, and thus no tests for statistical significance are possible. Therefore, resampling procedures such as the jackknife method and bootstrapping have to be employed (C. Krebs 1999).

Other Measures

Insectivorous or faunivorous primates may spend considerable amounts of time searching for prey. As a measure of the success and efficiency of prey foraging, an index of foraging success can be calculated as the number of records of prey feeding per foraging effort (time spent foraging). E. Lima and Ferrari (2003) did this with data from instantaneous scan sampling, while Kupsch and colleagues (2014) used data from focal animal sampling. Given the potential for underestimating the amount of prey consumption with instantaneous scan sampling (see above), focal animal sampling is suggested when foraging success or efficiency are to be estimated.

MEASURING FOOD ABUNDANCE

The tropics exhibit minor variation in day length and ambient temperature but vary considerably in the amount and temporal distribution of rainfall (Kricher 2011). The seasonality of rainfall affects the phenology of plants, although flushing, flowering, and fruiting are affected to different degrees (van Schaik and Pfannes 2005). There is also geographic variation in rainfall, which together with variation in soil properties (Ghazoul and Sheil 2010) influences vegetation composition and productivity on local, regional, or continent-wide scales (Arago et al. 2009; Gentry and Emmons 1987; Honorio Coronado et al. 2009; ter Steege et al. 2006). Primates are affected by this variation directly as primary consumers (frugivores, folivores, etc.) and indirectly as secondary consumers (insectivores/faunivores). Studies of primate feeding and nutritional ecology therefore need to consider variation in food availability for a proper interpretation of behavioral, morphological, and physiological adaptations—that is, they must include information on the temporal and spatial variation in food availability (i.e., phenological patterns).

This section only briefly addresses the measurement of fruit, leaf, and flower availability. Rather, I focus on insects and exudates, food types whose availability to primates and overall abundance in the environment are more challenging to measure.

Fruits, Leaves, and Flowers

Primatologists have used a variety of methods to quantify the availability of fruits, leaves, and flowers, and a number of overviews and methodological comparisons are available (e.g., Chapman, Chapman, Wrangham, et al. 1992; Chapman, Wrangham, and Chapman 1994; Ganzhorn, Rakotondranary, and Ratovomanana 2011; A. Marshall and Wich 2013; Parrado-Rosselli et al. 2006; P. Stevenson 2004; S. Zhang and Wang 1995).

Phenological measures are made along transects, in plots, or in a combination (plots systematically placed along transects). Quantification of the availability of fruits, leaves, and flowers can take numerous forms. The simplest is to record the presence or absence of fruits, leaves, and flowers (or finer levels of resolution—e.g., ripe and unripe fruits or mature and young leaves) and calculate the proportion of plants with presence in relation to the total number of plants examined. Abundance is more commonly estimated in a semiquantitative way. Categorical scores (e.g., 0: none, 1: few, 2: moderate, 3: many items) estimate the abundance of items in relation to the size of the plant (Chapman, Wrangham, and Chapman 1994; Peres 1994b). Since plants vary in their reproductive strategies (e.g., the number and size of fruits and seeds; Willson 1983), this estimate requires some familiarity with the amount of flowers and fruits a plant can produce. Presence/absence records or categorical scores should be weighted with the diameter at breast height or the crown volume of the monitored plants; otherwise, small plants will bias the abundance estimates (e.g., a small tree or liana, for which many items might mean less than one hundred fruits, receives the same score as a large tree or liana, where many means several thousand). Logarithmic scores (0: none, 1: 1–10, 2: 10–100, 3: 100–1,000 items, etc.) avoid this problem (Janson and Chapman 1999). Since counting the number of fruits in a large crown can be quite cumbersome, an alternative is to select a number of 1 m^3 sections within the crown, count fruits or other items within these sections, and then extrapolate to the entire crown (Chapman, Chapman, Wrangham, et al. 1992). Some studies have separately examined abundance in canopy and understory plants, using categorical scores for the canopy and presence/absence scores for the understory (e.g., Peres 1994b).

Phenology can be measured generally for all tree and woody liana species (beyond a size that makes them potentially suitable as primate food plants) to reflect overall habitat productivity and its temporal dynamics. This is the logical approach if information on habitat productivity is needed or if at the beginning of a study no or only limited knowledge is available on the plant species exploited by the primate species under study. If such knowledge is available, the phenological sampling can be restricted to those species that form part of the diet of the primate to provide an estimate of food availability and its temporal dynamics (Ganzhorn, Rakotondranary, and Ratovomanana 2011). In this case or after sufficient information on exploited plant species has accumulated, so-called phenology trails (or "fruit trails") are an additional option for phenological sampling (Chapman, Wrangham, and Chapman 1994). Without fixing any transects or plots, relevant trees and woody lianas are visited on a regular schedule, and phenological measures (presence/absence, categorical scores, counts, etc.) are taken. Fruit traps are receptacles (e.g., plastic bags) with openings of a defined area (e.g., 1 m^2) placed below the crown of trees or lianas to collect falling fruits or leaves, to subsequently measure their fresh and dry mass. However, fruit traps do not measure overall phenology as reliably as the other methods outlined above do (Chapman, Wrangham, and Chapman 1994). They can be useful for comparing production (and consumption) rates between trees or lianas of the same or different species.

Phenological patterns exist on different temporal scales. Food availability can change weekly (for example, the flowers of some species may bloom for very short periods), monthly, and seasonally but may also vary between years. In masting years—on a community level as in Southeast Asian dipterocarp forests or on a taxon level as in Neotropical Lecythidaceae—fruit availability is strongly increased (Sabatier 1985; Sakai 2002). Furthermore, some plant taxa, notoriously species of the genus *Ficus*, show unpredictable phenological patterns (Gautier-Hion and Michaloud 1989). Finally, global climate change creates long-term alterations of phenological patterns (Chapman et al. 2005; Visser and Both 2005). Disentangling changes on different temporal scales is instrumental to understanding the response of consumers to these changes. Recently, Polansky and Robbins (2013) employed generalized additive models and generalized additive mixed models to determine how fruit availability changes on different temporal scales. This approach is particularly useful because it accommodates different types of data (presence/absence, counts, etc.) as are collected in primate ecological studies.

Insects and Other Arthropods

All primates except for the completely faunivorous tarsiers include variable proportions of prey, particularly insects and other arthropods, in their diet. For a number of species, searching for and consuming prey account for a major fraction of the time and food budget, respectively (Charles-Dominique 1977; Nekaris and Rasmussen 2003; Terborgh 1983). Substantial amounts of nutrients can be obtained from prey (Bergstrom et al. 2019; Bryer, Chapman, et al. 2015; O'Malley and Power 2014). As there is ample evidence for seasonal variation in arthropod abundance (D. Pearson and Derr 1986; L. Richards and Windsor 2007; Smythe 1982; Wolda 1980), information on the availability of prey can be as important as information on the availability of plant foods for the interpretation of primate feeding strategies and adaptations.

Insects and other arthropods are generally mobile animals. Upon approach, they may escape, but they can also be cryptic or hidden within substrates (crevices, rolled-up dead leaves, etc.) and thus difficult to detect (e.g., Nickle and Castner 1995 for katydids). Noninvasive monitoring of their abundance is thus generally difficult, if not impossible. Rather, trapping remains the only means to quantify the abundance of most insects and other arthropods. However, trapping potentially creates problems itself. As Ferrari (1988, 81) already pointed out in a study on *Callithrix flaviceps*, "insect trapping, in contrast to the more passive observation of plants, is open to a wide range of possible biases and random effects . . . animals are killed during capture, effectively reducing overall abundance. This can, in turn, have a direct and disproportionate influence on measured abundance in subsequent months." Even though not all trapping methods involve killing, Ferrari's warning should be kept in mind, for scientific but also for ethical reasons.

More previous knowledge is required before measuring arthropod abundance than before measuring plant food abundance. It must be known where arthropods are searched for and captured (ground and forest strata, specific habitats and microhabitats, etc.), whether mobile or sedentary arthropods or both are preyed on, and whether the prey moves mainly by flying, walking, or jumping. This previous knowledge is essential, as different trapping methods are suitable for different types of arthropods and have different degrees of specificity (Leather 2008).

In some studies, the lack of previous knowledge was a major reason for the decision not to monitor arthropod abundance (Barnett, Ronchi-Teles, Almeida, et al. 2013). Another aspect is the timing of trapping. Ozanne et al. (2011) recommended that the timing of trapping coincide with primate activity periods. However, this may not be appropriate for all prey types. For example, most katydids, the major group of insect prey for tamarins (Peres 1993; Terborgh 1983), are actually nocturnal and spend the day resting, either hidden or exposed but camouflaged through cryptic coloration and positioning (Belwood 1990; Nickle and Castner 1995); therefore, they may not be captured by traps operating during the day. Instead, it is recommended to set and empty traps (except light traps that are operational at night only) at sunrise and sunset to capture arthropods with different activity periods that may or may not coincide with primate activity periods.

An exhaustive treatment of arthropod trapping methods is provided by Leather (2008). Ozanne and colleagues (2011) give a useful overview of arthropod trapping methods for primate ecological studies, including comments on the pros and cons of each method.

Probably the simplest method is sweep netting, where a net with a round opening is swept repeatedly through the vegetation. Ferrari (1988) and L. Porter (2001) used sweep netting along trails or transects. A problem is that the net can be caught by vegetation (Ferrari 1988). The operational height is restricted to about 2 m above ground. It requires some practice to standardize the speed, width, and angle of sweeps; otherwise, biases will be introduced.

Pitfall traps are the trap type most often employed in primate ecological studies (table 21.1). They are made of circular containers partially filled with water, detergent, or a preservation medium. The diameter of the container will determine the maximum size of arthropods that can be trapped. Pitfall traps are usually embedded in the soil and thus capture ground-living insects. Thus, they are employed mainly if the primate species under study also forage for prey on the ground (Nowack, Wippich, et al. 2013; L. Porter 2001). In her study on *Saimiri sciureus*, A. Stone (2007) placed pitfall traps on platforms at 1.5 m height. The lack of seasonal variation in arthropod abundance in her study, despite the strong variation shown in many studies (see above), casts some doubt on the validity of this approach.

Malaise traps are suitable for flying insects. They are basically tentlike and can be set in flight lines at different heights. Insects fly into the trap and are then routed to a collection container at the top of the trap. Like pitfall traps, they can be operated day and night.

Light traps are suitable for trapping nocturnally active insects that are attracted by light. Flying insects are collected in a container below the light source, or they are killed on contact with a high-voltage cord from which they are then collected.

Clipping branches and shaking arthropods into bags is a method restricted to arthropods that to not escape when their substrate is moved. Similarly, shaking trees, bushes, or branches will only record insects that drop down rather than flying away.

To quantify the abundance and spatial distribution within trees of domatia-dwelling ants, Isbell and Young (2007) sealed the exit holes of swollen thorns, clipped the thorns, and then identified the ant species, numbers, and life stages of ants found in the thorns.

To estimate the abundance of katydids, Kupsch et al. (2014) used ultrasound detection to count the number of singing individuals at night. While this method can only estimate the number of singing males, it provided information on the relative abundance of katydids in primary and secondary forest. It is also unspecific, as songs of different species cannot be differentiated with ultrasound detectors, but in this respect does not perform worse than trapping methods unless the trapped arthropods are identified on a taxonomic level lower than family.

Apart from the seasonal variation in the abundance of arthropods, information on the vertical stratification of arthropod communities (Basset et al. 1992; Wardhaugh 2014) can be highly relevant in primate ecological studies. This is the case for analyzing and interpreting niche differentiation between strongly insectivorous sympatric primate species (Terborgh 1983; Yoneda 1984). However, to my knowledge, pertinent data collections have not been undertaken in primate ecological studies.

In summary, various methods have been employed for measuring arthropod abundance in primate ecological studies (table 21.1). For the moment, it remains unclear how reliably they reflect the abundance. Seasonal variation for entire orders of arthropods (the usual level of resolution) may or may not reflect variation in the abundance of those species that are actually preyed on by pri-

Table 21.1. Trapping methods employed for estimating arthropod abundance in primate field studies.

Primate species	Trapping method	Source
Microcebus murinus	Light traps, tree shaking	Corbin and Schmid 1995
Microcebus berthae, Microcebus murinus	Malaise traps, light traps, pitfall traps	Dammhahn and Kappeler 2008
Galago moholi	Pitfall traps, light traps	Nowack, Wippich, et al. 2013
Tarsius spectrum	Malaise traps, pitfall traps	Gursky 2000
Tarsius pumilus	Malaise traps	Grow, Gursky, and Duma 2013
Cercopithecus ascanius	Light traps, sweep netting	Gathua 2000
Callithrix flaviceps	Pitfall (water) traps, sweep netting	Ferrari 1988
Callimico goeldii, Saguinus labiatus, Leontocebus weddelli*	Sweep netting	L. Porter 2001
Cebus capucinus	Malaise traps (canopy, terrestrial), pans, frass traps	Mosdossy et al. 2015
Saimiri sciureus	Pitfall (water) traps	E. Lima and Ferrari 2003; A. Stone 2007
Saimiri oerstedi	Sweep netting, dead foliage bagging	Boinski 1988
Lagothrix lagotricha	Branch bagging and clipping	Fonseca Aldana 2019
Multiple platyrrhine spp.	Light traps	Janson and Emmons 1990

*Previously *Saguinus fuscicollis weddelli*; see Rylands et al. (2016).

mates. Ideally, primatologists would combine efforts with entomologists to evaluate the relevance of the abundance measures obtained with various standard methods.

Exudates

Quite a number of primate species include plant exudates in their diet, and for some species and populations, exudates represent the major proportion of dietary items, either overall or seasonally. This is the case for members of the family Galagidae, Bengal slow lorises (*Nycticebus bengalensis*), fork-marked lemurs (*Phaner* spp.), and marmosets (genera *Callithrix*, *Mico*, and *Cebuella*; A. Smith 2010b). Exudativorous primate specialists have evolved adaptations in the dentition, the gastrointestinal tract, or both to procure and process exudates (see L. Nash [1986] and chapters in Burrows and Nash [2010]).

Measuring the availability of phloem sap and gum provides a considerable challenge, and up to now no standardized methods are available (L. Nash and Burrows 2010). To the best of my knowledge, no study has attempted to measure the availability of phloem sap, and it can be questioned whether measuring this may make sense at all. At least in evergreen tropical forests, phloem sap is likely to be present year round. There might be seasonal and diurnal variation in relation to fluctuating photosynthetic activity in leaves (Bi et al. 2015; Goulden et al. 2004), but this may result in variation of the chemical or nutritional content rather than availability. In contrast, in woody plants that shed their leaves seasonally (e.g., in semideciduous and deciduous forests), phloem sap is strongly reduced when leaves have been shed. In such forests, a simple and obvious measure for the availability of sap would therefore be the number of trees or woody lianas per area that bear leaves from plant species whose sap is exploited. As the amount of phloem sap is likely to correlate with the size of a woody plant, diameter at breast height or crown diameter or volume (as a proxy for the quantity of photosynthetically active tissue—i.e., leaves) could be factored in.

The availability of gum has been measured in a few studies in various ways (Garber and Porter 2010; Génin 2008; Isbell 1998; Joly-Radko and Zimmerman 2010). Isbell (1998) counted the number of gum-producing sites, their height on trees, and the total height of gum-producing trees in 50 m × 5 m transects and obtained visual estimates of the surface area of gum sites or the volume in the case of globular gum. Garber and Porter (2010) collected and weighed gums exuding from natural holes (gouged by *Cebuella niveiventris* or by wood-boring insects) and from experimental holes. The latter were produced by boring or cutting with a machete into the bark. This allowed them to detect three different patterns of gum production: immediate and short-term, delayed and continuous, and immediate and continuous. They also detected seasonal variation in gum production with a trend for lower production in the drier period of the year when considering three plant species together, but no clear pattern when examining each plant species separately. Similarly, Génin et al. (2010) used experimental cuts and weighed gum drops over seven days at the end of the rainy season and the end of the dry season. They found no clear seasonal variation in gum production. C. P. Jackson and Reichard (2021) sampled and weighed the gum from randomly selected holes gouged by *C. niveiventris*. After removing all gum from the selected holes, they sampled the newly exuding gum every hour for five hours or once after five hours. Gum production was higher in the former treatment, suggesting that the feeding activity of *C. niveiventris* could stimulate gum production, further complicating the objective measurement of gum availability.

A potential drawback of experimental cutting to stimulate gum flow is that it might affect the feeding behavior. Tamarins in northeastern Peru rapidly started to exploit gum exuding from wounds inflicted by climbing spurs used in the collection of botanical samples (pers. obs.). It is therefore advisable to exclude primate access to experimentally induced gum flow to avoid biasing their feeding behavior.

Like for arthropod abundance, there is not yet an optimal solution for estimating the availability of exudates. Obviously, for primates that actively stimulate exudate flow, the approach must be different from (and will be more complicated than) the one for primates that rely on flow caused by damage inflicted by wood-boring insects or other agents. In the former case, it might even be debated whether measuring availability is theoretically plausible, as the primates influence availability through their gouging activity (see above). At the current state of knowledge, it might be more appropriate to first find out about the selection criteria employed by specialized exudativores for identifying the trunks that they gouge to stimulate the flow of phloem sap and gum. Do they select trunks according to size, spatial distribution, or properties of the bark (C. Thompson et al. 2013), or do they probe several trunks and then select the one that is the most productive? C. Thompson et al. (2013) observed higher gouging intensity in trees with a larger circumference, and Francisco et al. (2014) found a weak positive correlation between the number of gouge holes and height as well as diameter at breast height of *Anadenanthera peregrina* trees exploited by *Callithrix* hybrids. However, this might reflect a simple geometric relationship—larger trees provide larger trunk surface areas that can be gouged—rather than a specific selection process. Once the selection criteria are better known, it will be more feasible to conceive measures for the availability of exudates.

Other Food Types

Fungi are important components of the diet of several primate species (A. Hanson, Hodge, and Porter 2003). For *Callimico goeldii*, fungi comprise the major portion of the diet in certain months (L. Porter 2001). Information on their abundance is therefore important to evaluate fungivory as a dietary strategy. L. Porter (2001) counted the number of fungi on dead and fallen trees and on bamboo stalks and branches along transects that were also employed for estimating the abundance of fruits and flowers.

Some Malagasy primates feed on insect secretions. For *Microcebus berthae*, such secretions account for more than 80% of the overall diet (Dammhahn and Kappeler 2008). Corbin and Schmid (1995) estimated the abundance of this resource by counting the number of patches with secreting insects (homopterans) and measuring the surface area covered by these insects in 25 m × 2 m plots. No other studies have followed up on this so far, despite the importance of insect secretions in the diet of some Malagasy primates and the influence on their ranging patterns that the spatial distribution of these secretions potentially has (Corbin and Schmid 1995).

Predation on vertebrates has been observed in many primate species (Butynski 1982). The proportion of vertebrates in the diet is generally low but may vary between seasons, species, populations, groups, and age-sex classes (Chapman and Fedigan 1990; Fedigan 1990; Heymann et al. 2000; Lüffe et al. 2018; J. Mitani and Watts 2005c; Pruetz 2006). Despite vertebrates making up a low proportion of the diet, information on the abundance of vertebrate prey may be desirable in some cases—for example, for understanding the impact of primate predation on prey species (Teelen 2008) or seasonal patterns and episodic outbreaks of vertebrate hunting (Stanford 1998). In the case of large, diurnal vertebrate prey such as red colobus monkeys (hunted by chimpanzees), estimates of prey availability (population density) can be available from transect censuses (Lwanga et al. 2011). However, with smaller prey such as frogs and lizards, determination of population densities is challenging due to small body size, secretive behavior, or immobility (Smolensky and Fitzgerald 2010). A combination of visual and acoustic surveys along rectangular transects is a feasible method for amphibians in tropical habitats (Rödel and Ernst 2011). For reptiles, transect census can be employed but may underestimate densities (Smolensky and Fitzgerald 2010). Given the specialized effort that has to be made, it is unlikely that small vertebrate prey abundance will be estimated in many primate field studies. However, at field sites where not only primatologists but also herpetologists are working, combining efforts might render some new insights, particularly with regard to the impact of primate predation.

"FOODSCAPES"—A NEW APPROACH FOR THE FUTURE?

While the measures described in the previous sections can provide reasonably good estimates of food abundance, diversity, and preferences and of similarities or differences between individuals, populations, and species, they may not necessarily reflect the perspective of the consumers. This is particularly obvious in socioecological studies—for example, when examining the effects of varying availability on feeding competition. Therefore, E. Vogel and Janson (2007, 2011) devised a "focal tree method" that takes the perspective of the primate consumers. For this, the food (fruit) abundance both in trees visited by a primate group and in the neighborhood is estimated within 48 hours after a visit. The size (area) of the neighborhood is determined by the average group spread of the primate species under study. By including information on the nutritional composition, this approach allows researchers to measure the supposed value of a tree (or food plant in general) and relate this to behavioral patterns and strategies, such as competitive interactions (E. Vogel, Munch, and Janson 2007).

Conceptually similar is the "foodscape" approach, which integrates information on the spatial distribution of resources, their nutritional properties, and the "context"—that is, the plants growing in the surroundings (neighborhood)—to test predictions on foraging decisions by consumers (K. Marsh, Moore, et al. 2014; B. Moore, Lawler, et al. 2010; Searle, Hobbs, and Gordon 2007). This approach has been primarily employed for folivores with relatively narrow diets (e.g., koalas) and in habitats with low complexity. It remains to be determined whether this approach is feasible in primate ecology. For instance, when a primate must decide whether to feed on fruit or to search for and feed on prey, plant-based measures will not suffice. Moving from the habitat or home-range scale, on which food is usually measured, to the local scale will increase the level of analytical resolution but will also strongly increase the effort required for data collection in the field. In any case, a foodscape approach will require a strong integration of different but complementary approaches, from behavioral ecology through nutritional ecology to sensory ecology (just to mention major pillars), and technical tools that facilitate the measurement of food availability and quality. But even then, further integration, such as including the "landscape of fear" (Laundré, Hernández, and Ripple 2010; McArthur, Banks, et al. 2014) or the "social landscape," will be necessary for a full understanding of the foraging decisions and ecological strategies of primates (fig. 21.5). Taking these other influencing factors into consideration could then eventually lead to a theoretical distinction between the "fundamental foodscape" (what the habitat offers on a local scale) and the "realized foodscape" (what animals actually make out of it).

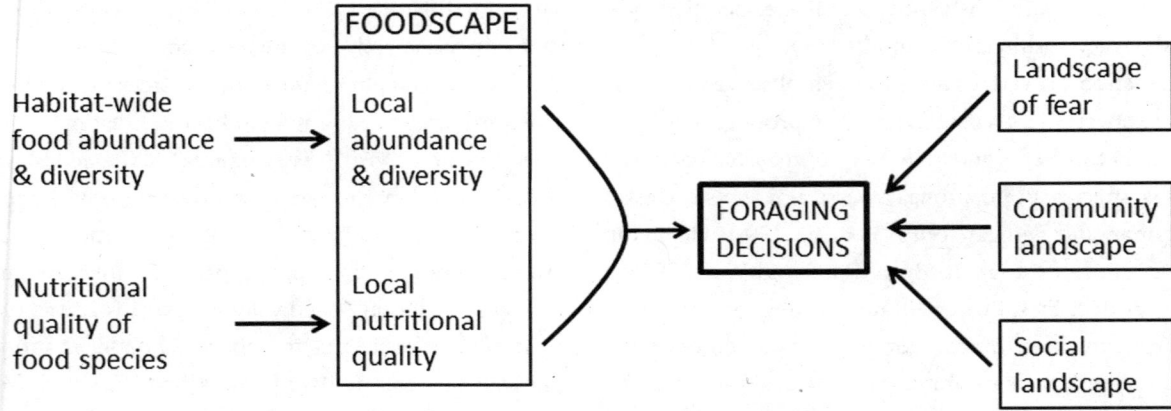

Figure 21.5. A simple visualization of the "foodscape" together with other major factors that can influence foraging decisions. The "landscape of fear" refers to the spatial distribution of predation risk (Laundré, Hernández, and Ripple 2010; Hill, this volume). The "community landscape" means the local abundance, distribution of, and interactions with heterospecific members of the ecological community. The "social landscape" refers to the number and age-sex composition of the social unit or subunit in which an animal is foraging and the social interactions (e.g., competition) with conspecifics. The scheme does not include potential interactions and feedback loops. For instance, foraging decisions will feed back on the local abundance of food, local food abundance could influence predators' decisions on where to search for prey and thus modify the landscape of fear, and the social unit or subunit can influence the predation risk through vigilance, dilution effect, and other mechanisms. It also ignores the potential influence of internal factors (e.g., physiological status and motivation).

SUMMARY AND CONCLUSIONS

There are many reasons to measure food in the field and many ways in which to do it. What to measure and how to measure depend on the specific research questions. Practicability under the field conditions of a specific project may often dominate over theoretical demands. With the limitations of time and funding that projects are generally confronted with, compromises between theoretical demands and practicability have to be made. But even when optimal measures are possible, we need to be aware that in the complex ecosystems that most primate species live in, any measures are an approximation but never a replica of ecological reality. Nevertheless, in a rapidly changing world where global climate change and encroachment on tropical habitats are increasingly affecting primate populations (A. Dunham, Erhart, and Wright 2011; Estrada, Garber, et al. 2017; Graham et al. 2016; Luo et al. 2015; Chapman et al., this volume) measuring food in the wide sense will be essential for understanding whether and how primates will be able to adapt to the challenges of survival.

- Measuring primate foods needs to be expanded beyond fruits, leaves, and flowers, given that many primates include substantial proportions of prey and other plant materials in their diets.
- Genetic tools (metagenomics) will become increasingly instrumental for higher resolution in the taxonomic identification of prey.
- New theoretical approaches (such as the "foodscape" concept) to measuring food could be implemented in primate ecological studies, but their feasibility and heuristic value remain to be tested.
- Measuring food will be an important component of understanding whether and how primates will adapt to the challenges imposed by global climate change and increasing habitat destruction.

22 Wild Plant Food Chemistry

Nancy Lou Conklin-Brittain

In the following I describe how to add a nutritional ecology element to your overall primate ecology program. If you search "proximate analysis of food" online, you will find many books and articles about food chemistry. This chapter is distinct by providing the reader a wild plant food perspective and a primatology angle. The information here will help you talk to laboratory managers in the United States and around the world to assure them and yourself that standard livestock or human nutrition laboratories can analyze wild plant foods.

To nutritionally characterize your field site, you will want to first perform "crude" or "proximate" determinations of the macronutrient contents of the wild foods at your site. The word "crude" is used to acknowledge that each of these nutrient assays will include some impurities—for example, crude lipid includes waxes as well as nutritional fat. Another example is that crude protein includes nonprotein nitrogen as well as the nitrogen in proteins.

Most published nutrition work in primatology to date is in these early stages of the process, but a few sites have moved into a second level of food chemistry by, for example, considering how available to digestion and absorption the macronutrients are. For example, the crude protein (CP) values can be corrected by determining the acid-detergent insoluble crude protein (ADI-CP) and subtracting that value from the CP value (Conklin-Brittain, Dierenfeld, et al. 1999; Prichard and Van Soest 1977; Rothman, Chapman, and Pell 2008) to give you available protein (AP). This example and others are explained below. A glossary of abbreviations relevant to nutritional chemistry analyses is provided (table 22.1).

As you are planning a research project that will include the collection and processing of the plant foods seen eaten by the primate species at your field site, and in anticipation of taking samples to a laboratory, you will need to follow a systemic series of activities. I will assume you have been awarded a grant that will fund your efforts in the field and in the laboratory. Before leaving for your field site and if you will not be doing the lab analyses in-country, make sure you have permits from the country of origin allowing you to export your samples, as well as permits for the country where the lab is, to import the samples. In addition, all of the permits need to be country specific, and the rules can change from year to year, so make sure you have the most up-to-date permits. Rothman, Chapman, and Van Soest (2012) discuss this in detail using Uganda as an example.

COLLECTING WILD PLANT FOOD SAMPLES

To collect wild foods, you first need to answer a few questions. For example, where do your animals feed—in the canopy or on the ground? If in the canopy, ideally you will plan on climbing the trees that the monkeys or apes are climbing. However, if this is impractical, careful observations and collections off the ground are acceptable.

Table 22.1. Glossary of relevant abbreviations.

Abbreviation	Definition
ADF	acid-detergent fiber
ADIN	acid-detergent insoluble nitrogen
ADI-CP	acid-detergent insoluble crude protein
ADL	acid-detergent lignin
AEW	as-eaten weight
AP	available protein
CP	crude protein
Cs	sulfuric acid cellulose
DM	100% dry matter
EE	crude lipids or fat (ether extract)
FDW	field dry weight
FSS	free simple sugars
FW	fresh weight
Ls	sulfuric acid lignin
NDF	neutral detergent fiber
NPN	nonprotein nitrogen
PE	petroleum ether
TDF	total dietary fiber (both soluble and insoluble fibers)
TDIF	total dietary insoluble fiber
TDSF	total dietary soluble fiber
TNC	total nonstructural carbohydrates
WW	wet weight

Wildlife moving in the trees dislodge good, fresh fruits that fall to the ground, and you can collect them when newly fallen. But not all the fruit on the ground represents items rejected by arboreal animals. Alternatively, if your research questions are very specific and need precise collections, but the canopy is very high, it is possible to hire people trained in climbing. For example, Cornell Tree Climbing is part of the Cornell Outdoor Education Program; you can hire their graduates, and they can train field assistants at your field site.

A second question is, how do the animals process their food items? Do they use their hands or their mouth, do they take bites, or do they pop the whole food item in the mouth? Do they chew a lot? Do they wadge (make a quid) to spit out? Do they reject and drop parts of the food item? Collect and examine the rejected parts. You might keep them for analysis, depending on your research questions, but, if not, at least describe them. You will also want to collect feces and describe them to determine whether the fruit seeds are digested, are somewhat digested, or pass through whole (Rothman, Chapman, and Van Soest 2012).

The third question is, how many samples do you need to collect? Keep in mind when you are collecting samples that it will be difficult for you to analyze more than 400 individual samples in a timely manner in a wet chemistry laboratory. If your site has a near-infrared reflectance spectroscopy (NIRS) machine calibrated for the kinds of samples you are collecting and for the habitat your animals are in, potentially you can process many hundreds of samples per project. However, if your field site does not have a NIRS machine, and you plan to buy one, be warned that you will need to go through the wet chemistry process first, analyzing about 400 representative samples before you can calibrate and use the NIRS machine. This chapter is about the wet chemistry process, which is time consuming and labor intensive, and your 400 samples might take a year to finish. Putting a positive spin to this, you can learn a lot about these plant foods as you handle them and process them in the lab, and remember that future nutritional monitoring of your site will be easier once you have done this groundwork (Chapman, Chapman, Rode, et al. 2003; Rothman, Dusinberre, and Pell 2009).

Another collection question is, should you collect only from the trees used by the animals you are observing? Ideally, yes, but some trees are small, and perhaps the arboreal and terrestrial communities have depleted the patch. It is a judgment call whether you should collect the leftovers or get some from other trees. Sometimes you have no choice. Rothman, Chapman, and Van Soest (2012) describe a detailed collecting process if you are ready to move beyond the basics.

You will need plastic bags for each different plant part collected and for each plant species, to not lose any moisture. The process of transpiration will still be going on in all the fresh samples, releasing water into the air. Start with preweighed plastic bags and write the weight on the bag with a permanent black ink pen. Keep ripe fruits in separate bags from unripe fruit, separate from young leaves, separate from mature leaves, and so on.

How much sample do you need to collect of fresh weight? Ideally, the more the better—you are sampling a rainforest, and it is hard to draw grand conclusions with only ten grams of the rainforest. In addition, the animals are your guides regarding which foods are more important. You should collect at least five to ten separate samples per food eaten in quantity by the animals, and from several preferred trees. If you will be shipping all the samples to another country, you might want to limit the weight to 30 g per dry sample to control shipping costs. From the point of view of the laboratory, to do the basics—ash, lipids, protein, total fiber, and nothing else—your minimum fresh weight should be 50–100 grams fresh, estimating a yield of ~10 grams of dry weight and assuming no accidents in the lab. Rothman, Chapman, and Van Soest (2012) recommend larger quantities; after all, you are characterizing a whole forest.

Create a collection data checklist including species collected, plant part, stage of maturity, wet/fresh weight, date collected, date dried and packed, field dry weight, and average dimensions of fruit sizes or leaf sizes. This will be useful for calculating food intake. Use the following procedure for each food sample:

1. Place a labeled tag inside the sample bag or envelope.
2. Write on the outside of the bag or envelope itself (customs might want to see it).
3. Give each bag a unique identifying number or code.
4. Handwrite everything in a notebook.
5. Enter everything in your computer. Redundancy is good.

INITIAL SAMPLE PROCESSING IN THE FIELD

Imagine you are in camp, standing at a table covered with carefully collected, fresh wild plant food items in plastic bags. Next you need a portable weighing balance (aka scales) that ideally has at least one decimal place, hopefully two. It is best not to move the balance too much, and do not take it out in the field unless you have at least one or two spare balances at camp. Even those labeled "field" balances have short lives if they bounce around in a backpack.

Your final results are only as good as your initial processing of your samples. However, and arguably, your job is to be watching the animals being active in the wild. You might want to hire field assistants to specialize on collecting and processing as follows:

1. Weigh each bag with its content of fresh food, usually referred to as "wet weight" (WW) or "fresh weight" (FW), but some foods might be eaten somewhat dry and can be labeled "as-eaten weight" (AEW). This weight minus the weight of the preweighed bag equals the whole sample weight. In addition, weigh several whole fruits or whole leaves individually; this will be used in step 4 below. Do this as soon as possible after collection.
2. Quickly begin processing fresh samples according to how you saw the items being processed by the primate while eating. A common one is to separate pulp and seeds and weigh them separately. In addition, if a fruit has a thick outer skin or husk, this should be separated from the pulp. Check what is routinely spat out or in the feces to decide whether the seeds or the hard or leathery skins are actually being digested. Separate petiole from leaf blade from veins only if necessary; bark can have layers that need to be separated, and so on. Consult Rothman, Chapman, and Van Soest (2012) for additional examples.
3. Weigh the seeds instead of the pulp because the pulp often loses or evaporates off moisture during processing, seeds less so. The weight of the fresh, whole fruit minus the weight of the recently processed seeds can give you a more accurate fresh pulp weight.
4. For the fruits that were also weighed individually, calculate individual fresh fruit weight minus fresh seed weight; this is the fresh pulp weight per fruit. You will need individual weights when you are calculating grams of intake.
5. Special processing tip for figs: Cut figs into at least halves (small figs) or maybe up to 8 pieces (large figs). Let them dry (see next section) without removing the seeds. After field drying, gently scrape out the seeds. Weigh the seeds, and calculate pulp weight as in step 3 or 4 above. Scraping fresh figs removes too much non-seed, pulpy material.

There are micronutrients, such as vitamins and fatty acids, as well as some secondary plant compounds that should be assayed with fresh, not dried, samples. This chapter does not cover procedures that require fresh or frozen samples (but see examples in Dierenfeld et al. 1995; Milton and Jenness 1987; J. Sullivan 1973). This is to alert you to the fact that you will need to have special arrangements organized in the field and lab before you collect for these types of assays. You should not do these assays in the field, as some of the procedures use dangerous chemicals, and you need the safety features that a real laboratory gives you.

FIELD DRYING

Your goal is to dry your samples as quickly as possible without damaging them. Field drying under cover or shade with a good draft of warm air is usually the least expensive and most efficient method to prevent mold or fermentation while drying fresh/wet samples (Adeyemi et al. 2014; Harborne 1984, 4; Mudau and Ngezimana 2014). Simple heat sources, such as kerosene lamps, can help the drying process during the night without needing electricity. Direct sun exposure damages tannins and various other secondary plant compounds (D. Bernard et al. 2014). It also accelerates plant transpiration initially, causing some small loss of volatile carbohydrates. Microwave drying should not be used, even if only as a pretreatment to stop plant transpiration; it degrades secondary plant compounds and creates Maillard product, which then contaminates and elevates your fiber measurement (Darrah et al. 1977; Van Soest and Mason 1991; Youssef and Mokhtar 2014).

Ideal ambient conditions cannot be assumed every day or at all field sites, so the drying process often needs to be helped. If electricity is available, start a drying oven or food dehydrator at 35°C–45°C (A. Hanson, Hall, et al. 2006). If you do not see significant signs of drying within 24 h, you can start increasing the temperature to a maximum of 55°C to speed up the drying process.

Livestock forages are often dried at 60°C to quickly dry enough to shut down the enzymes of plant transpiration (J. Sullivan 1973). However, these temperatures can damage tannins and create Maillard product (Van Soest 1994, and see above), especially if left at 60°C for 24 h (Goering et al. 1973). Maillard product is the result of simple sugars and proteins (especially free amino acids) in the food permanently bonding together due to heat. This reduces the measurable amount of sugar and protein and increases the apparent fiber measurement (Dzowela et al. 1995; Van Soest 1994). Macronutrients, like those assayed by the proximate analyses, are not damaged by the ambient air or warm-air drying processes up to 55°C (Palmer et al. 2000).

To dry large numbers of samples at ambient temperatures, you can build drying racks with fine mesh wire fencing or chicken wire, in an airy space with air circulation, out of the sun and under cover. Consider using a pop-up canopy. Make aluminum foil trays to hold each sample as they dry. Monitor the temperature on the racks. Monitor the drying samples to make sure none are molding or fermenting. If they are, you need to throw them out and collect new samples. If they look and smell good, weigh them for a couple consecutive days, and if they seem dry, package them in paper envelopes, put them in a basket, and hang the basket from the ceiling, assuming that is the warmest and driest place in camp. If your site is extremely humid and that plan would cause them to reabsorb moisture, put a source of heat, such as kerosene lamps, semipermanently under the drying rack. When the samples are as dry as possible, put them in plastic bags and seal the bags to keep out the humidity. Then keep your fingers crossed that they do not mold despite your efforts. You can do the following additional things to prevent mold: (1) cut the sample (often fruit) up into smaller pieces to dry more quickly; (2) separate pieces so they do not touch each other; (3) spread out leaves only one leaf deep; (4) split or slightly crush stems, petioles, or veins if thick or dense; and (5) be sure to have more than one sample per food.

Bottling fresh samples in alcohol is rarely recommended as a preservation method in nutrition. It immediately stops transpiration (Harborne 1984), which is good, but introduces an organic solvent, alcohol, which has consequences for laboratory analysis methods, as explained below. If you have no other method possible, cut up and stuff the sample into the smallest jar available and then add a minimum of boiling alcohol to cover the sample and cap the jar very tightly.

Liquid nitrogen is the most expensive preservation option. It is recommended for micronutrients such as vitamins and fatty acids, as well as some secondary plant compounds and metabolites such as *n*-alkanes. Liquid N is not necessary for the macronutrients in the proximate analysis.

Investigate the possible use of a food chemistry laboratory in the country where your field site is located. Most countries have a human food lab or livestock lab set up in or near the country capital. Some are locally funded; others have been set up or are supported by donor agencies. The labs might be at the national university or agricultural university or in a government department or ministry.

SHIPPING

As previously mentioned, if you have made arrangements with a lab in a different country, have your permits prepared for removing the samples from the country where you are working and entering the country where your laboratory is. Carry the permits in your carry-on and put copies inside the sample trunk or box.

I do not recommend silica gel when shipping. However, if you want to use it, never pour loose granular silica gel into the bags with the loose food sample. You have just added a contaminant that is very difficult to pick out. The silica gel must be contained in cloth bags with a very tight weave that will keep all silica particles from contaminating the food samples. Natural biogenic silica can be an antifeedant, so do not add more silica as a contaminant (McNaughton et al. 1985; Van Soest and Jones 1968).

If your plant samples are part of your PhD thesis work, it is best to make sure you, personally, take your precious samples out of the field site and all the way to the laboratory you will be using. The only other people who should be allowed to carry your samples are your parents and your committee chair. Leaving your samples behind for someone else to mail them is absolutely not recommended. Professional couriers have been known to lose packages but rarely, so I recommend them for faculty research.

IN THE LABORATORY

The lab procedures we use in the Nutritional Ecology Laboratory at Harvard University have other alternatives in apparatus and methods that give equivalent results (Rothman, Chapman, and Van Soest 2012). The following discussion focuses on the Harvard laboratory methods, but I also mention other options.

You need to decide which laboratory you will use if it is not your home institution. In some countries, you do not have a choice because there is only one food laboratory. Nevertheless, you need to visit the lab, ask for a tour, ask for a copy of the lab manual, ask where the equipment came from, and ask what the costs will be. It is important that you include in your reports or thesis what equipment was used (model and manufacturer), the list of chemicals used, and details of the procedures used. This helps you and any reader compare your results to those from other labs. Ask permission to watch them performing the lab procedures if lab technicians will be doing the work rather than you. Before you start giving them samples, find out if any machines are not currently working and what it will take to get it fixed so you can budget your time. The proximate analyses are very basic, and any food lab should be able to perform them. However, if you have any misgivings, make sure you have permits for sending your samples home.

Assume that you have chosen a lab and that the lab manager will let you process some of your samples, with the help of technicians. Ideally, you should first equilibrate all your field-dry weights when you get to a lab by putting the samples in a forced-air drying oven set at an average ambient daytime temperature in the lab. Leave them overnight at the day temperature and then reweigh them in the morning and calculate by the equation below. This becomes the official "field dry weight." Store your samples at room temperature in the lab. Store them off the floor, in cabinets or on shelves away from lights or sun, in a dry room. If you must store your samples in, for example, a walk-in cooler, take them out in advance to equilibrate to room temperature before weighing out samples for the assays. However, you might not be able to control all this, depending on whether you or only the laboratory employees are allowed to do the assays. You can use the dry weights you recorded in the field; they are not as exact but fine for the proximate procedures.

Water is the nutrient that is being measured during the drying process (R. Bailey 1973). The water or moisture content of foods for primates is important because available water is scarce in tree canopies. The field dry weight is a percentage of fresh weight. The water is also a percentage of fresh weight but is determined by subtraction.

$$\%\text{water} = 100 - (\text{dry weight} / \text{fresh weight} \times 100);$$

that is, %water = 100 − %dry matter.

PREPARING SAMPLES FOR ANALYSIS THAT DO NOT ARRIVE DRY

The field-dried samples are immediately ready for grinding and analysis. If you preserved samples in alcohol in a jar, before you can grind the samples you need to evaporate off all of the alcohol. Pour out all the solid sample pieces and all the liquid into a shallow glass dish big enough to hold all the jar contents and rinsings; use the smallest shallow dish possible. You need to scrape the insides of the jar and rinse the jar with alcohol. Unfortunately, the alcohol has extracted an unknown-but-significant amount of plant cell solubles, especially sugars. You want to retrieve those solubles. Then put the dish into a laboratory hood and close the window to increase the current (Remis et al. 2001). Monitor progress every several hours; stir every time you check. Ideally, drying should be complete within 24 to 36 h. If the sample is not drying noticeably after 12 to 18 h, put the sample in a forced-air drying oven at 40°C–55°C. Once it is dry, scrape off the evaporated alcohol residue in the dish, mix it with the solid pieces, and store it in a plastic bag. Our nutrition methods are quantitative, so you need to recover as much of the sample as possible. Be careful if you used methanol, as it is toxic and absorbs through your skin.

If you used the liquid-nitrogen method, you need a freeze-drying machine in the lab. Once the samples are freeze-dried, they should be thought of as equivalent to field-dried samples (J. Sullivan 1973) and stored in a plastic bag. This method is not worth the cost for a preliminary, proximate food survey but is highly recommended if your research questions are about micronutrients. You might also consider looking for an in-country lab to analyze fresh samples for vitamins and other micronutrients or certain phytochemicals (secondary plant compounds).

GRINDING OR MILLING

If you brought more than 400 samples to the lab, prioritize which ones will be the 400 to be analyzed in the lab by wet chemistry. Only grind the initial 400; whole samples suffer less from oxidation damage and have a longer shelf life than do milled samples. Keep in mind also that 400 samples might not be enough to calibrate your NIRS machine, and you might have to process more eventually.

Grinding approximates the effect of chewing, and, just as chewing breaks down food particles so the enzymes can digest the food better, grinding allows the chemicals used in the lab to break down the food during analysis. There are different types of mills; common in labs are cutting mills (good for fibrous plant material) and hammer mills (good for seeds or grains). Either will work; we use a cutting mill.

The type of mill is not as important as using the right size screen to have a standardized fineness to the milled product. The detergent system of fiber analysis and the total dietary fiber analysis methods recommend #20 or #30 screens (Van Soest 2015, 41). Screens are made of woven wire mesh, and the holes come in various sizes—for example, the #10 is 2.00 mm, #20 is 0.841 mm, and #30 is 0.595. The nutrition literature commonly refers to 1 mm mesh (Mowlana et al. 1994), which technically is a #18 designation but in practice is often the #20 mesh. Whatever size you use, be sure to state it in your report. These numbers stand for guaranteed mesh hole sizes and are important for keeping the results repeatable.

We have standardized all analyses in the Harvard lab to #20 mesh. Using a smaller mesh size makes grinding very slow and laborious. The #20 mesh optimizes ease of grinding and accuracy in analysis. The #40 or #60 designations are commonly recommended for secondary plant compounds, but #60 has only 0.250 mm holes, making grinding very slow and difficult. The range of food particle sizes found in the feces is very wide, but 1 mm seems reasonable for nonruminant species (Fritz et al. 2009; C. Phillips and McGrew 2014) given their various dentition and body sizes (Sheine and Kay 1977; Uden and Van Soest 1982).

Mortar and pestle, blenders, or most coffee grinders are acceptable for qualitative analyses but not for quantitative assays. All assays here are quantitative.

I have the following tips for grinding with a cutting mill, especially for fruit pulp high in sugar and seeds high in fat. We start by grinding through a #20 mesh screen. The majority of samples will go through without problems. Use garden snippers to cut large chunks of sample into pieces that fit into the mill hopper. Occasionally, you might need to switch to a #10 screen (2 mm mesh). Once the sample has gone through #10, regrind it with the #20 screen (Van Soest 2015, 41).

If you suspect high fat because the contents of the

grinding chamber look like butter and nothing is coming through the screen, grind only as much as will fit in the grinding chamber and turn it off when it just barely turns into a smooth butter. Then scrape that amount out and put it in a brown glass jar. High-fat samples slowly dissolve plastic bags, so it is best to put them directly into glass. Then put more sample into the hopper and repeat. It is easy to overgrind, so do not grind for too long but also check for big chunks to regrind as you empty the chamber. It might help to switch to #10, and you probably will not need to regrind with #20. Clean the screens with soapy water and the chamber with ethanol and cotton-tipped sticks. If you have access to a hammer mill, you might want to try it for high-fat samples. In addition, liquid N might help; see step 2 below.

If you suspect high fat but the sample turns into stiff toffee, not butter, you probably have high sugar. Proceed as follows:

1. First, try dry ice (CO_2; Van Soest 2015, 43). The dry ice will keep the chamber cool and will not add water. You do not want the grinding chamber heating up.
   a. Put the dry ice in a plastic bag and tap the ice lumps with a hammer to make smaller lumps that fit in the mill hopper.
   b. Grind up a couple of cups of dry ice by running it through the grinder by itself to chill the chamber and to get the dry ice into a smaller particle size, usually #10.
   c. Mix the sample you want to grind with some of the milled dry ice and start grinding at #20. To keep the chamber cool, keep grinding the mixture of sample plus dry ice.
   d. After you are done grinding, spread the sample-plus-ice mixture on a tray to evaporate overnight at room temperature.
   e. High-sugar samples can usually be stored in plastic bags.
2. If dry ice isn't enough, use liquid nitrogen:
   a. Use a large, fine metal-mesh tea ball. Put as much sample you want to grind into the tea ball as will fit and immerse it in a 1 or 2 liter Dewar flask (special for liquid N) that is ¾ to ⅔ full of liquid N.
   b. While the sample is freezing, chill the grinder with dry ice, as described above (pouring liquid nitrogen in might crack the mill). Then start grinding the N-frozen sample with the milled dry ice, as described above.
   c. Repeat until all the sample is milled.
   d. Evaporate the mixture in a tray.
3. It might help to first grind a sample through a #10 mesh screen with dry ice and then mix the #10 milled sample with more dry ice and run that mixture through the #20 mesh.
   a. Or put the #10 milled sample into the tea ball again and into the liquid nitrogen again. Check that premilled sample is not escaping from the tea ball. Then grind it with dry ice through #20.
4. Plants containing latex (for example, figs) generally grind up without any special procedures. If you have problems, try any of the above suggestions.

The following is the recommended order of procedures for the basic plant analyses:

1. 100% dry matter (DM) coefficient
2. Total ash (aka total minerals)
3. Crude lipids (aka EE for ether extract)
4. Crude protein (CP)
5. Detergent system of fiber analysis (NDF) to measure insoluble, structural fibers
6. The calculation of total nonstructural carbohydrates (TNC)

$$100\% - Ash\% - CP\% - EE\% - Fiber\% = TNC$$

TNC was historically called "nitrogen-free extract" (NFE), and you might see the older term in the literature.

Why perform these analyses in a particular order? This is an attempt to increase efficiency in the lab. The EE and CP do not compete for the same equipment. However, it is useful to know how much ash is in a sample before you do a Kjeldahl nitrogen assay. Samples with high ash content (>25% of DM) can cause moderate explosions during the Kjeldahl distillation step. Tree barks and some seeds can be high in total ash. It might be necessary to use 0.4 g or 0.3 g of sample instead of 0.5 g per test tube for samples with high ash. In addition, high lipid levels can start a fire in the Dumas method, so you should be prepared.

Another issue with lipids is that during fiber analyses of samples with >10%–15% lipid, the excess fat might not rinse out and will stay as a contaminant in the NDF. Samples containing more than 10%–15% lipid should be defatted first, and then the fiber methods can be done using fat-free sample. To create fat-free sample, see below and http://dx.doi.org/10.13140/RG.2.2.26387.25120

Most of the procedures in the proximate analyses of food are gravimetric, meaning that the weight of the product or residue is the measurement to be reported. In the past, the weighing process was cumbersome and slow, using a drying oven, a desiccator, and silica gel. Now the process is faster by using a hot weighing technique. Each crucible is weighed straight out of the forced-air oven without needing a desiccator with silica gel. A detailed, step-by-step description of how to do this is at http://dx.doi.org/10.13140/RG.2.2.26387.25120. This method is also more accurate than using a desiccator, mainly because it is quicker (Van Soest 2015, 57–60). The balance used for weighing should have four decimal places.

We also use a quality-control measure. All samples are analyzed in duplicate, and you subtract the two values obtained; if they are less than 2 units different, they are accepted and averaged, and the mean is considered the official value for that sample in any statistical use. If the difference is more than 2, you need to reanalyze the sample, in duplicate again, more carefully, until you get a pair that are ≤2 units different. A difference of 2 is for procedures that do not have many steps—for example, the ash, protein, and lipid procedures have fewer steps than the fiber and carbohydrate procedures do; the sugar or starch duplicates can be allowed to differ by 3 to 4 units. Where did the number 2 come from? Tradition and an estimate of reasonable challenge; some labs use as much as 5.

100% DM COEFFICIENT

Gravimetric procedures start with pre-hot-weighed, empty crucibles; record the tare weights. Then, using the field-dried, milled sample, you take subsamples of about 0.3 g, record all four decimal places, place them into each of two temperature-proof porcelain crucibles in a heat-proof tray, and then place them in a forced-air oven at 105°C. After drying overnight, hot-weigh the crucibles again and record the weights. This drying of the subsamples evaporates off the last amount of water, and the resulting dry matter is officially called "100% dry" (Association of Official Analytical Chemists 1990). The calculations are shown at http://dx.doi.org/10.13140/RG.2.2.26387.25120. Samples that were freeze-dried need to be treated this way also. Never dry any whole sample at 105°C or higher— that would cause a great deal of Millard product and, as discussed above, irreversible damage to your samples.

The 100%DM coefficient will be used to correct all subsequent nutrient values from the procedures below. All the samples are analyzed in the field-dry-matter state. The equations correcting the values to be percentages of 100%DM are at http://dx.doi.org/10.13140/RG.2.2.26387.25120.

TOTAL ASH DETERMINATION

After hot-weighing the crucibles plus subsamples, put them into a muffle furnace to burn them at 500°C to 550°C overnight (maximum 15 hours). Ideally, the furnace is in some kind of fume hood; there will be a lot of smoke. The upper limit of 550°C is because Pyrex or Kimex glass starts to melt above that temperature (Van Soest 2015, 52). The porcelain crucibles tolerate 1,000°C, but we keep our furnace set at 520°C because I have seen glass filters and crucibles deform when hotter than that.

Do not put the tray into the furnace; put the crucibles on the floor of the furnace. In the morning, turn off the furnace, wait several hours for it to cool off before opening the door, and remove the crucibles with tongs or wearing heat-tolerant gloves. Put the crucibles in the forced-air drying oven set for 105°C and equilibrate to 105°C overnight before hot-weighing them. The contents of the crucibles are the total ash; report the weights. When you tour the laboratory you plan on using, look for a forced-air (or forced-draft) drying oven and a muffle furnace.

As previously stated, these beginning analyses can also be referred to as "crude," implying in this case that the ash can be contaminated with something, mostly soil or airborne dust. If you collected all your samples from the ground, there might be dirt on the fruits or herbaceous plants. You can brush this dirt off when you pick samples up and before you bag them. Do not wash with water, because you do not want added water complicating the fresh weight. If you are in a dry site, airborne dust might be part of the consumer's diet, and you can leave it alone. However, if there are windstorms, they can be a problem. In addition to shaking and brushing, you can mathematically remove the mineral contamination; this is explained at http://dx.doi.org/10.13140/RG.2.2.26387.25120, where the templates for calculations are also discussed. Whatever you decide to do, be sure to describe it in your report.

The other issue regarding ash is the bioavailability of the minerals. This is not a topic for the basic proximate analysis, but think about finding a lab that can determine individual minerals and do bioavailability tests eventually.

CRUDE LIPIDS

"Lipid" is an umbrella term for all the hydrophobic compounds such as nutritional fats (triglycerides), sterols, phospholipids, and things generally indigestible such as waxes (cutins; Engels 1996) and antifeedants such as terpenoids and latexes (Mazoir 2008; Wrangham, Conklin-Brittain, and Hunt 1998).

"Soxhlet" is the name of a common apparatus that extracts lipid from foods. The Golfisch Fat Extractor is another type of fat extractor. These machines are designed for safety, but both use boiling-hot petroleum ether (or some other powerful organic solvent) to achieve the extraction relatively quickly (within a few hours). The petroleum or ethyl ethers are toxic, and ethyl ether is carcinogenic. You should wear a lab coat and lab glasses, and, regarding lab gloves, nothing will keep the ethers from penetrating through to your skin. Fortunately, they do not do it quickly. You have about 15 minutes to carefully take off and throw out the glove that contacted the ether and put on a new glove. Work slowly and calmly.

Most nutrition laboratories have a Soxhlet machine of some kind to measure lipid content, except ours. A previous lab manager of the Harvard lab was focused on phytochemicals instead of nutrients, and he wanted to extract lipids gently for the phytochemical methods he was using. He extracted lipids at room temperature, in petroleum ether (PE), for five to seven days. The idea came from Harborne (1984, 6, 154, 161). When I became lab manager, an undergraduate did a project to examine how many days of room-temperature extraction time were needed to get the equivalent amount of lipid extract obtained by a Soxhlet machine in a matter of hours. Using soybeans and the Soxhlet machine in the Bronx Zoo Nutrition Laboratory, he found that three days (72 h) of extraction, sitting in a fume hood at room temperature, extracted as much crude lipid as the Soxhlet following its standard procedure (Scot Zens, unpubl. data). The extracted amount leveled off at the start of the fourth day; essentially, no more lipid was extracted from the soybean meal at that point.

We were also investigating terpenoids and wanted gentle extraction of the lipid, so we continued with the room-temperature extractions (Wrangham, Conklin-Brittain, and Hunt 1998). We have been extracting lipid at room temperature with a four-day minimum schedule ever since. We use two dozen glass Erlenmeyer flasks with glass stoppers, and the whole process is done in a hood. Our assumption has been that soybeans are good substitutes for wild plant material. This might not be true, which explains the four-day schedule instead of three-day. This would make an excellent project for undergraduates to test the rate of extraction and amount of crude lipid extracted using various wild plant foods compared to a Soxhlet.

Another assumption with our method has been that extracting at room temperature prevents extracting non-nutritional long-chain waxes and other indigestible, hydrophobic, large compounds such as latexes, which are probably extracted by the high temperatures used in the Soxhlet process. The cuticle waxes of plants are universal in leaf coatings to protect the plant's water balance (Harborne 1984, 160) but are assumed to be nonnutritional to most consumers. We assume that our extractions favor nutritional fats over nonnutritional "contaminations," which also needs to be tested.

Regarding your project, you should check whether there is a lipid extractor of some kind in the laboratory and a working fume hood whether you use an extractor or do room-temperature extractions. The procedures for the room-temperature extractions are at http://dx.doi.org/10.13140/RG.2.2.26387.25120.

If there is only one fume hood, you will need to share it between the organic solvent procedures (e.g., lipid) and the acidic procedures (e.g., Kjeldahl). The different apparatus and reagents for one category should not be together in the same hood or storage cabinets at the same time. This is awkward, but for safety it is worth moving them out of the hood and storing them in separate cabinets.

Regardless of what method you use, you should save the residue left after the extraction, especially if the results show that there is more than 10% fat in the sample. The residue left after removing the fat will be the fat-free residue of the sample that you should use for various carbohydrate assays. If the amount of fat-free residue is not enough for carbohydrate assays, you can do more ether extractions.

Nutritional fats and specifically their building blocks, fatty acids, are subjected to breakdown over time that causes the whole sample to become rancid. Storing them in a refrigerator will slow the inevitable. The structures of the fatty acids change, and special ones (such as the omegas) lose their special value. You can still perform a crude lipid measure because the breakdown products are usually still hydrophobic and can be extracted by organic solvents, but do not try to find individual fatty acids, because they have changed. There are interesting questions to study with fatty acids (J. Chamberlain et al. 1993; Reiner et al. 2014), but in keeping with the preliminary-analyses status of this review, I present only crude fat, aka ether extract (EE).

CRUDE PROTEIN ANALYSES

There are two methods commonly used for measuring the total amount of nitrogen in a food sample: the Kjeldahl procedure (Sáez-Plaza et al. 2013a; Sáez-Plaza et al. 2013b) and the Dumas method, also called the combustion analyzer (Chang 2010). The methods give comparable results (S. Jung et al. 2003; Wiles et al. 1998). We use Kjeldahl because it has decades of testing and works well. The Dumas method also has decades of development but has only relatively recently been adapted to fit on a laboratory counter. There are very few problems with either method: high-fat samples, such as nuts, might start a fire in the Dumas apparatus (Chang 2010), but on the other side, samples containing >20% ash can cause the Kjeldahl distilling step to explode. Some tree barks contain high ash levels (Conklin-Brittain, pers. obs.). You should determine both the lipid percentage and the ash percentage of your samples in advance and reduce the size of the subsamples analyzed for those plant parts. Also, be sure the safety shields on the Kjeldahl apparatus are down.

In the Kjeldahl method, there are choices depending on what type of food you are analyzing—for example, plant material versus meats versus dairy. Wild plant materials are similar to vegetables, fruit, and livestock forages, which are relatively simple to analyze, using the standard Kjeldahl method without modifications. Our chemical digestion step uses concentrated sulfuric acid, which burns off organic matter by boiling the sample with a catalyst; we use copper. You are left with N in a clear, turquoise, acidic liquid, which is neutralized with sodium hydroxide (turning brown or muddy blue). The next step is distillation, where we distill the N liquid into weak boric acid with a color indicator (purple turns to green). In the third step, we titrate to a cobalt blue using weak hydrochloric acid. Our step-by-step procedure is online at http://dx.doi.org/10.13140/RG.2.2.26387.25120.

The apparatus for the Dumas method uses combustion to oxidize the sample into simple molecules that are then detected using thermal conductivity or infrared spectroscopy. Nitrogen is one of the elements that can be measured by this procedure. The Dumas method is quicker than Kjeldahl, so you can process more samples per hour. When you tour the laboratory, you can ask which apparatus is used. You need to include this information in your report.

Either of these methods gives an accurate measure of N, and then to convert the N values into crude protein values, you need to use a conversion factor. The most used factor is 6.25, which is derived from the observation that the protein in meat in general contains 16% N (the rest is carbon, hydrogen, and oxygen) and that 100% divided by 16% = 6.25. Consequently, this factor is applied to the N measure from Kjeldahl or Dumas results, and the product of total N times 6.25 is called the percent crude protein of a food. However, there is a controversy about the 6.25 factor because it represents the percentage of protein in meats more than it reflects the protein in plant foods, and that tends to overestimate protein content.

I do not recommend that the 6.25-versus-4.3 discussion (the range goes as low as 3.9) be continued by primatologists, unless they are interested in the details of food chemistry and plant physiology. This chapter is a basic nutrition document. The 6.25 factor is supported by the weight of history. It has been used for decades (Krul 2019), but I am not going to talk about determining the "best" conversion factors.

Fortunately, we can follow an easier path. Figure 22.1 gives an overview of an accessible understanding of crude protein. Instead of talking about individual amino acids, I will talk about categories of available proteins within crude protein.

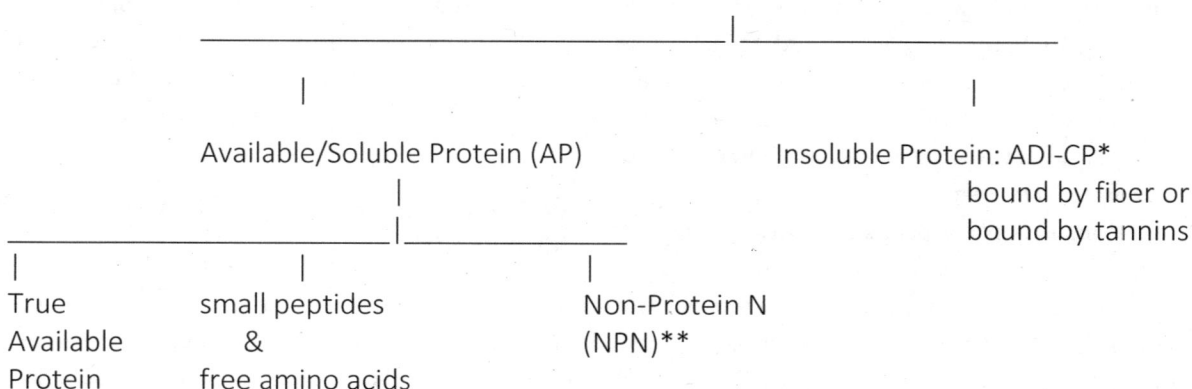

Figure 22.1. A breakdown of crude protein fractions.

Plants produce compounds called lignin and cellulose while they grow. Lignin is a structural fiber that makes it possible for plants to stand up, and it is a major component of insoluble fiber, which is discussed in the next section of this chapter. These two compounds, especially lignin, are not nutritional; in fact, they are antinutritional structures that bind up nutrients, especially long-strand proteins, and that protein becomes unavailable to the consumer to eat or digest (fig. 22.1). Tannins can also be present and are chemically similar to lignin by binding up specifically long strands of protein, but tannins are not present in all plant species.

To measure unavailable or insoluble protein versus available or soluble protein in plant food samples, follow this procedure, which determines how much unavailable protein is in a plant food and, by subtraction, how much is nutritionally available. First, you perform the basic N assay (Kjeldahl or Dumas) on a subsample of a food and calculated the crude protein (CP) using factor 6.25, as described above. Second, you perform the acid-detergent (AD) procedure on another subsample of the same food as described in the "Detergent System of Fiber Analysis" section below and in detail at http://dx.doi.org/10.13140/RG.2.2.26387.25120. This gives you the AD residue, which consists of cellulose and lignin. Third, this AD residue is subjected to the N assay, Kjeldahl or Dumas (again). The result of that procedure is the acid-detergent insoluble nitrogen (ADIN), which is then multiplied by 6.25 and becomes ADI-CP (acid-detergent insoluble crude protein). Finally, CP minus ADI-CD equals available protein (AP; Conklin-Brittain, Dierenfeld, et al. 1999; Pichard and Van Soest 1977; Rothman, Chapman, and Pell 2008). The CP value has lost a substantial portion of the undigestible part (fig. 22.1). You should report both CP and AP because you will find both or either in the literature.

The usefulness of the AP method is reduced if fresh samples are dried above 60°C, especially fruit samples (I recommend 50°C for fruit). As discussed before, higher temperatures create Maillard product, which is a bonding of molecules of protein and carbohydrates, and this means that some of the naturally available protein will unfortunately end up in the insoluble protein category (see fig. 22.1) due to human error.

You might be wondering about the nonprotein nitrogen (NPN) on figure 22.1. It is a nonnutritional contaminant in the AP. Lignin, cellulose, and tannins do not generally bind with small compounds, so NPNs are not taken up with the ADI-CP; they stay with the AP by default. In figure 22.1, there is a list of the nonnutritional NPN (Imafidon and Sosulski 1990; Milton and Dintzis 1981). This contaminant is what keeps AP as a version, but cleaner, of the crude protein category.

In addition, some plant species have small but important amounts of nonnutritious phytochemicals (aka secondary plant compounds), some of which are toxic—for example, alkaloids or toxic nonprotein amino acids. These sources of NPN need to be included in the NPN count (D'mello 1992) in the AP. On the other hand, most of the tannin-bound N probably goes into the ADI-CP (Rothman, Chapman, and Pell 2008; Waterman and McKey 1989), so it probably does not contaminate the AP. Finally, most peptides too small to be precipitated, as well as free amino acids (fig. 22.1), are nutritional but unfortunately are extracted in the NPN category.

The following are some further tips for protein analyses:

1. If you are monitoring for the basic nutritional value of a habitat or just monitoring nitrogen, pay for the crude protein procedure.
2. If your questions are somewhat more complicated—for example, about nutritional ecology or optimal foraging—pay for the available protein procedure.
3. If you are monitoring a small, protected reserve and the resident primates are not thriving or not reproducing, then you need to examine the quality of the protein in the foods available to the primates more carefully. Then, you should look for a lab with a gas chromatography machine designed to identify individual amino acids and quantify all the amino acids available to the consumers. Get a budget estimate first.
4. If you want to compare your data with a published data set calculated using 4.3 (for example) instead of 6.25, divide the CP values by 4.3 to get back to the original %N, and then multiply all the %N values by 6.25.

Finally, use the following protein safety precautions. The Kjeldahl apparatus digestion step needs to be in a hood with a scrubber to remove the boiling H_2SO_4 fumes. The distiller and titration setup can be on a lab bench, but if you have a large fume hood, put the digester at one end and the distiller at the other. Wear standard lab gloves, lab glasses, and a lab coat. Remember that sulfuric acid (H_2SO_4) dissolves cotton and hydrochloric acid (HCl) dissolves polyester fabric; a blended fabric will protect your skin better through the Kjeldahl procedure.

THE DETERGENT SYSTEM OF FIBER ANALYSIS

The detergent system of fiber analysis is the method that gives you neutral-detergent fiber (NDF), acid-detergent fiber (ADF), and acid-detergent lignin (ADL), first published by Goering and Van Soest (1970). You will find that most animal science departments in the world have apparatus to do this system. There are at least four types of apparatus: Fibertec from FOSS, ANKOM's fiber analyzer, LABCONCO reflux apparatus, and the FIWE instrument-VELP Scientifica. I have had experience on LABCONCO, which is completely manual, and FOSS, which is automated to different levels, depending on which model you buy.

The detergent system has been designed to quantify the three insoluble fibers that make up the plant cell walls—hemicellulose, lignin, and cellulose—and remove all the soluble cell contents. The acronyms "NDF" and "ADF" are not for fiber types; they are for the dominant reagent (detergent) and the pH (neutral or acid). Hence, the neutral-detergent fiber is NDF and acid-detergent fiber is ADF. It was somewhat less confusing when the alternate names were used: neutral detergent residue, or NDR, and acid detergent residue, or ADR. You will see them in the literature sometimes. The residue is the substance left on the bottom of the crucible used to filter off the soluble cell contents and leave the insoluble fibers (residues) in the crucible. But NDF and ADF are more commonly used.

The NDF procedure removes everything in the plant cell that is soluble and retains all three of the principal insoluble fibers: hemicellulose (HC), cellulose (Cs), and lignin (ADL or Ls). The next step is the ADF procedure, which removes hemicellulose and retains cellulose and lignin. Then, the ADL procedure removes cellulose and retains lignin. The actual carbohydrate fibers, hemicellulose and cellulose, are determined by subtraction: NDF − ADF = HC, and ADF − ADL = Cs. The ADL is not a carbohydrate-based fiber; it is a phenol-based compound. The acronym "ADL" refers to the acid-detergent lignin procedure, which is performed sequentially on the residue left after the ADF procedure is performed. The ADL is also written as "Ls" because sulfuric acid was used. The "Cs" abbreviation indicates that the cellulose was derived from the Ls procedure. The step-by-step lab procedures are posted at http://dx.doi.org/10.13140/RG.2.2.26387.25120.

If you are interested in quantifying cutin, the waxy cover on most leaves and fruit, you can also use the detergent system. You treat the ADF first with a potassium permanganate reagent, which removes the lignin but keeps the cellulose, with "contaminants," in this case cutin. Second, you treat the cellulose residue with the 72% sulfuric acid reagent, dissolving the cellulose, and the remaining residue is cutin. I found that the fruit in Kibale National Park contain a surprisingly high amount of lignin, which is usually an antifeedant (Van Soest 1994). When I tested for cutin, I found that the lignin was at least half cutin. Being a wax, soft and chewy, it might be less of an antifeedant than is sharp, harsh lignin, even though it is equally indigestible (Conklin-Brittain, unpubl. data).

For additional information, theoretical and applied, consult the 2015 book, *The Detergent System for Analysis of Foods and Feeds*, by Peter Van Soest. The word "feeds" refers to what domestic livestock are fed; "foods" refers to what all other animals, including humans, eat.

The commercial reflux apparatus by LABCONCO was originally used to determine "crude fiber" (CF), an old and discredited technique (Van Soest 1994), but the apparatus can still be used for the NDF system. If a lab offers you the crude fiber procedure, do not accept it, but ask them for the NDF procedures. Some veterinarian labs still do the crude fiber method. The CF situation can be somewhat confusing because commercial bags of animal food, such as monkey biscuits or pet chow, still list CF on their "Guaranteed Analysis" tags. This reflects very old government policies that have not been changed, so CF is still required by law on commercial pet and livestock food/feed.

If the food you want to analyze is insect, you need to determine what percentage of the insect is chitin by performing the ADF procedure on a sample of milled insect (M. E. Allen 1989). The regular plant fibers are not present, but chitin, the exoskeleton of insects, is essentially cellulose with a nitrogen moiety attached to each glucose sugar, and the AD reagent dissolves everything except the chitin. Then you apply a Kjeldahl procedure to the chitin residue and determine how much N is in the chitin and therefore unavailable to the consumer. Then you subtract the bound N (times 6.25) from the crude protein value of the insect and calculate available protein (AP), the same way you do for plants; see http://dx.doi.org/10.13140/RG.2.2.26387.25120.

The detergent system is comparatively safe. Nevertheless, you should wear a lab coat, glasses, and gloves. The neutral reagent is a neutral pH, and the acid-detergent reagent contains diluted sulfuric acid. However, the ADL reagent is 72% sulfuric acid, so you need to treat it carefully. Remember, sulfuric acid dissolves cotton and burns skin. The room should be well vented and the ADL done in a fume hood, but the acid fumes are not as strong as those in the Kjeldahl procedure.

TOTAL NONSTRUCTURAL CARBOHYDRATES—DIGESTIBLE CARBOHYDRATES, SOLUBLE FIBERS AND PREPARATIONS FOR ENERGY CALCULATIONS

The TNC is the last element in the traditional proximate analysis process and is calculated based on the other measurements. Arguably, you can stop analyzing things now. As discussed above, and ideally, the TNC is only sugars and starch, and you are now ready to apply the physiological fuel values (PFVs) and calculate the energy content of your food.

Unfortunately, because TNC is a calculation, not a direct measurement, the TNC is affected by the contaminants in the NDF, CP, EE, and ash. For example, our proximate method to measure ash content can contain nonnutritional ash from soil contamination. The crude lipid can contain nonnutritional wax or latex. Available protein can contain NPN, and the crude protein can contain ADI-CP. This means that the NDF value is also contaminated by ADI-CP. The NDF can also be contaminated by some of the fat in high-fat foods. Because of this, the TNC values will probably be slightly lower than what is true. Occasionally, the TNC is zero—for example, if the sample is high in fat and high in fiber, like some seeds. Nevertheless, this simple equation has been used for decades, and it is a good point to start with when you are trying to estimate the nutritional value of the habitats at your field site.

The classical, basic equation is:

$$TNC = 100 - ash - CP - EE - NDF$$

The TNC contains various soluble carbohydrates: sugars, starch, and soluble fibers. It is very useful to know what the ratios of these three categories are, but this is not necessary to calculate the "crude" energy content.

ENERGY CALCULATIONS

You are now ready to apply the PFVs (Merrill and Watt 1955) to the nutrient measurements you obtained from the assays above for EE, CP, and TNC values and to finally calculate energy estimates for all your food samples. The PFVs are 9 kcal/g EE, 4 kcal/g CP, and 4 kcal/g TNC. There have been discussions in the literature about potential PFVs for soluble fibers in the diet. I have not included a section on soluble fibers in this chapter because it is like the search for the perfect correction factors in the protein section—long and arduous. I have kept this chapter focused on the basics and the "crude" first steps of food chemistry.

$$9 \text{ kcal/g EE}, 4 \text{ kcal/g CP, and } 4 \text{ kcal/g TNC}$$
$$= \text{Metabolic Energy (ME)}$$

You might be surprised that protein (CP) is given a PFV even though it does not have an energetic job in the body; it has a metabolic and construction job. However, eventually a protein will be broken down, and the individual amino acids will be broken down. The nitrogen will go out in the urine, and the sugar backbone of the amino acid will become CO_2, H_2O, and several molecules of ATP, in other words, it is metabolized as a source of energy. However, if you are comparing protein and energy as opposite factors in a food choice equation, protein should not be in the energy calculation. You need them to be independent of each other.

SUMMARY AND CONCLUSIONS

This is a comprehensive beginner's program for adding a nutritional ecology element to your overall primate ecology program. These proximate methods will also help you monitor your field site's nutritional value over the years, especially correlating changes in the weather that might influence the nutrient value of the site into the future. To perform this type of survey efficiently, I recommend the traditional proximate analyses: 100%DM, water, ash, EE, AP, NDF, calculated TNC, and calculated energy content. The AP method uses the same equipment and reagents as the CP, so not causing excessive additional cost compared to CP.

The nutritional hypotheses that are increasingly tested in primatology use additional, more complicated, and more costly procedures. Nevertheless, the procedures that I think will help advance nutritional ecology can be found at http://dx.doi.org/10.13140/RG.2.2.26387.25120, along with the following:

1. A phenol-sulfuric acid assay to measure free simple sugars
2. The Megazyme total starch assay
3. The total dietary fiber assay
4. A ninhydrin assay to measure total protein in the presence of tannins

And finally, the following are topics that I do not have any expertise with, but someone needs to fix them:

- Resolve whether room-temperature extractions of fat contain fewer nonnutritional hydrophobic compounds in your lipid measure than the Soxhlet method.
- Measure the tannin activity and content levels of your food samples. This will help you determine whether enzymatic assays are working properly.
- Once we have characterized the nutrient content of the foods available in our various field sites, then we need to work on the fact that the accuracy of the calculated energetic equation given in this document needs to be, and can be, improved a great deal.

23 Evaluating Primate Diets with Stable Isotopes

Matt Sponheimer and Brooke Crowley

Stable isotope studies of primates have become common of late; most of these focus on diet, but some are more concerned with environment or aspects of physiology. While all of these studies are question driven, in a great many cases the questions are primarily methodological. Specifically, researchers have typically investigated the degree to which stable isotope compositions of primate tissues allow us to understand things such as habitat use or niche partitioning, to distinguish between the consumption of certain classes of foods, or to track group membership and dispersal. This is wholly appropriate, as proper interpretation of stable isotope data requires a solid understanding of the way stable isotopes are distributed in plants, the individual primates that eat them, and the communities in which these individuals are embedded. It is also fair to say that it is only through such studies that stable isotope practitioners will continue to find new ways to add value to traditional studies of primate diet.

However, this tendency toward methodologically driven research also makes it less clear in what areas stable isotopes have made novel contributions and where they are poised to make the largest contributions to our understanding of primate dietary ecology. Our goal in this review is to address this gap and more. We begin with a brief discussion of terminology and the isotopic systems of greatest relevance to current isotopic studies of primate diets. The meat of this review, however, is an extended personal reflection on the state of the discipline. This starts with an analysis of questions and types of research where stable isotopes have, beyond any reasonable doubt, taught us something novel about primate diets. From there we try to derive general lessons about where stable isotopes are especially well positioned to contribute to our understanding of primate diets in the future. But first a few caveats. This is not intended to be a general review of stable isotope uses and methods in primatological research. We direct those who want such reviews to several papers along these lines that have been published in recent years (e.g., Ben-David and Flaherty 2012; Crowley 2012; Reitsema 2015; Sandberg et al. 2012; Schoeninger 2010). Also, given the nature of this volume, we focus primarily on diet rather than other aspects of primate ecology, such as habitat use or physiology. It is also important to acknowledge that stable isotope work with primates will be worthwhile in many contexts not discussed herein. However, we hope that any deficits accrued by a lack of breadth in this contribution will be repaid by our, admittedly idiosyncratic, attempt to step back from recent contributions, take stock, and focus on issues of broad importance.

A BRIEF PRIMER

All isotopes of an element have the same number of protons and electrons but differ in the number of neutrons that bind the nuclei. For example, the stable isotopes of carbon are ^{12}C (six protons and six neutrons) and ^{13}C (six protons and seven neutrons). An element's chemical behavior depends on its electron configuration, and because isotopes of an element do not differ in the number of electrons, they have the same chemical properties. However, different numbers of neutrons lead to differences in mass between isotopes of the same element. These mass differences affect the physical properties and reaction rates of molecules (especially light elements such as C and N), which result in enrichment or depletion of the heavy isotope relative to the light isotope during physical and chemical reactions such as tissue synthesis (anabolism) and metabolic processes. This change in the relative proportion of heavy and light isotopes, called isotopic fractionation, leads to considerable variation in the stable isotope compositions of plants and the animals that eat them.

We express isotopic compositions using δ-notation: $\delta^{13}C$ or $\delta^{15}N = (R_{sample} / R_{standard} - 1) \times 1{,}000$. R_{sample} and $R_{standard}$ are the $^{13}C/^{12}C$ or $^{15}N/^{14}N$ ratios in the sample and an international standard, respectively. These ratios are expressed in parts per thousand (‰; see Sharp [2006] for a review of relevant terminology).

Multiple body tissues and by-products can be employed for isotopic studies of primate diets. Bone and enamel are typically used for paleoprimatological applications for the obvious reason that they are usually the only materials preserved (e.g., Lee-Thorp, van der Merwe, and Brain 1994; S. Nelson 2007). Hair and feces are most common for studies of living primates, as they can be obtained easily and relatively noninvasively from a large number of individuals (e.g., Blumenthal, Chritz, et al. 2012; D. Codron et al. 2006; T. O'Connell and Hedges 1999; Schoeninger, Iwaniec, and Glander 1997). However, blood, breath, and a variety of other tissues can be employed depending on availability and the question of interest. Each material for analysis differs to some extent in either turnover (the speed at which the isotopic signal changes) or the component of diet it principally reflects. On the fast-turnover end of the spectrum, breath CO_2, dissolved inorganic carbon in blood, and to a lesser extent feces, urine, and blood plasma reflect diet at the scale of hours to a week or more (reviewed in Dalerum and Angerbjorn 2005). At the other end of the spectrum, we have bone, which, depending on the species, the element, and the manner in which it is sampled, can integrate dietary signals over months or years (R. Hedges et al. 2007; Sealy et al. 1995). Enamel and hair (or fur) keratin, which are target tissues for many paleoanthropological and primatological studies, respectively, do not turn over and will reflect diet over the period of formation, which in practice is typically weeks to months or even years (Ayliffe et al. 2004; Balasse 2002). The temporal span of some materials, as well as the discontinuity between material types, offers the possibility of addressing questions of diet from multiple time periods even if material is collected only once. This is an unusual property of stable isotope approaches to dietary research and one that, arguably, has been underappreciated by primatologists.

Body tissues and by-products also differ considerably in which portions of the diet they reflect. For instance, the stable isotope composition of feces largely reflects the undigested portion of diet (R. J. Jones et al. 1979; Reed, Crowley, and Haupt 2023), while tissues reflect digested foods (see below). Differences are also present among tissues/materials. Hair, plasma, bone collagen, and other organic materials preferentially (though not exclusively) reflect the isotopic composition of the protein component of the diet (Ambrose and Norr 1993; Ambrose et al. 1997; Jim et al. 2004; Tieszen and Fagre 1993). Inorganics, such as breath CO_2 and the mineral component of bone or enamel (bioapatite), tend to better reflect the isotopic composition of the whole diet, including other macronutrients such as carbohydrates and lipids (Ambrose and Norr 1993; Ambrose et al. 1997; Jim et al. 2004; see Passey et al. [2005] for relationships between breath CO_2 and enamel in multiple herbivores). While this may complicate interpretation of stable isotope data from multiple tissues in practice, it also opens the door to asking new questions. For instance, a multiple-tissue study might allow us to explore the possibility of a primate deriving most of its energy from one dietary source (e.g., fruit) while getting most of its protein from an uncommon but high-protein component of its diet (e.g., animal food).

It is also worth noting that the isotopic composition of the component of the diet that is reflected in a tissue is not typically transferred directly because of isotopic fractionation (as referenced above). This fractionation depends on the sampled material, the consumed diet, and the animal's physiology and metabolism. For most species, the fractionation between proteinaceous tissues and diet is between 2‰ and 5‰ for both carbon and nitrogen (reviewed in Crowley 2012). If we assume a fractionation of about +3‰ between hair keratin and diet for both carbon and nitrogen (Nakashita et al. 2013), then if a monkey consumes plants with $\delta^{13}C$ and $\delta^{15}N$ values of −27‰ and 3‰, respectively, its hair $\delta^{13}C$ and $\delta^{15}N$ values will be about −24‰ and 6‰. In contrast, there is a much greater fractionation in the carbon isotopes of bioapatite in bone and tooth enamel relative to diet (there is no N in the mineral itself). It is about +12‰ to 14‰ for most large-bodied herbivores and lower (as low as 10‰) for many smaller animals and those that produce very little ^{13}C-depleted methane (Ambrose et al. 1997; Cerling, Bernasconi, et al. 2021; Cerling, Hart, and Hart 2004; Krueger and Sullivan 1984; Lee-Thorp et al. 1989; Passey et al. 2005). Assuming an offset of about 12‰, the folivorous monkey above with a diet of −27‰ would have an enamel mineral $\delta^{13}C$ value of about −15‰ (Lee-Thorp et al. 1989; Malone et al. 2021).

Virtually all existing primate isotope studies are focused on carbon isotopes, nitrogen isotopes, or both. Other isotopic systems are sometimes employed, most notably oxygen isotopes ($\delta^{18}O$; $^{18}O/^{16}O$), but as such studies are few and far between, we do not discuss them in depth here (but some discussion of oxygen and nontraditional isotopes is found below). Carbon isotopes can be used to distinguish plants that use the C_3 (most trees, shrubs, herbs, and forbs), C_4 (most tropical grasses and many sedges), and CAM (crassulacean acid metabolism; succulents) photosynthetic pathways. The carbon isotopic compositions of C_3 and C_4 plants are highly distinct and nonoverlapping, while CAM plants can have intermediate compositions depending on their physiology and the local climate conditions (fig. 23.1; Kluge et al. 2001; O'Leary 1981; B. Smith and Epstein 1971; Winter 1979). Thus, primates that consume different amounts of C_3 and C_4 foods, such as colobus monkeys (e.g., *Colobus guereza*) and many savanna baboons (*Papio*), have very different carbon isotope ratios (Ambrose and DeNiro 1986; Cerling, Hart, and Hart 2004; D. Codron et al. 2006; Krigbaum et al. 2013; Levin et al. 2008; Thackeray et al. 1996). There are also smaller differences in the carbon isotope compositions of co-occurring plant species that use the same photosynthetic pathway, and even within plants themselves. For instance, fleshy fruits and seeds typically have higher $\delta^{13}C$ values than do leaves from the same tree (e.g., Cernusak et al. 2009; J. Codron, Codron, Lee-Thorp, Sponheimer, Bond, et al. 2005), and these differences may be passed on to primate consumers (fig. 23.2).

Carbon isotope distributions also vary across environments. The $\delta^{13}C$ values of plants (and the primates that eat them) are correlated with rainfall and, to a lesser extent, temperature (Cerling and Harris 1999; J. Codron, Codron, Lee-Thorp, Sponheimer, Bond, et al. 2005; Crowley, Thorén, et al. 2011; Loudon et al. 2016; Oelze, Fahey, et al. 2016; Schoeninger, Most, et al. 2016). C_3 plants have higher $\delta^{13}C$ values under xeric conditions because partial stomatal closure prevents water loss, and this results in less isotopic discrimination during photosynthesis (reviewed in Farquhar et al. 1989). In contrast, C_4 plants are much less affected by climate (e.g., Swap et al. 2004). Additionally, C_3 plants in open environments have higher $\delta^{13}C$ values than those in closed environments, largely because CO_2 is recycled, moisture is higher, and light levels are lower in the latter (Ehleringer et al. 1986; Medina and Minchin 1980). Such carbon isotopic zonation is also manifested vertically in forests and is frequently referred to as the canopy effect. Plants on the forest floor have the lowest $\delta^{13}C$ values because the available CO_2 there has low $\delta^{13}C$ values (due to respiration by soil microbes) and because there is very little light. Values are higher in the subcanopy and highest in the emergent canopy, where air moves freely (CO_2 is not trapped and recycled to the same extent) and light levels are high (Medina and Minchen 1980; fig. 23.2). Thus, carbon isotopes offer the possibility of tracking spatial segregation of sympatric primates (Carter and Bradbury 2016; Cerling, Hart, and Hart 2004; Krigbaum et al. 2013; S. Nelson 2007).

Nitrogen isotopes have often been used as tools for distinguishing trophic level given the well-known stepwise enrichment of about 3‰ between trophic levels (Schoeninger and DeNiro 1984; fig. 23.2). All else being equal, within any given ecosystem we would expect a primate that eats nothing but leaves to have tissue $\delta^{15}N$ values that are about 3‰ higher than the leaves it consumes and 3‰ lower than a carnivore that eats nothing but leaf-eating primates. Thus, nitrogen isotopes may be particularly powerful for studying omnivorous primates that are believed to change their diet over time or space, niche partitioning among species, or differences in meat consumption between sexes. Nitrogen isotopes are also useful in distinguishing the consumption of legumes from nonleguminous plants (at least in certain environments; e.g., Schoeninger, Iwaniec, and Nash 1998). This is possible because legumes, unlike most other plants, can derive nitrogen from symbiotic bacteria that directly fix atmo-

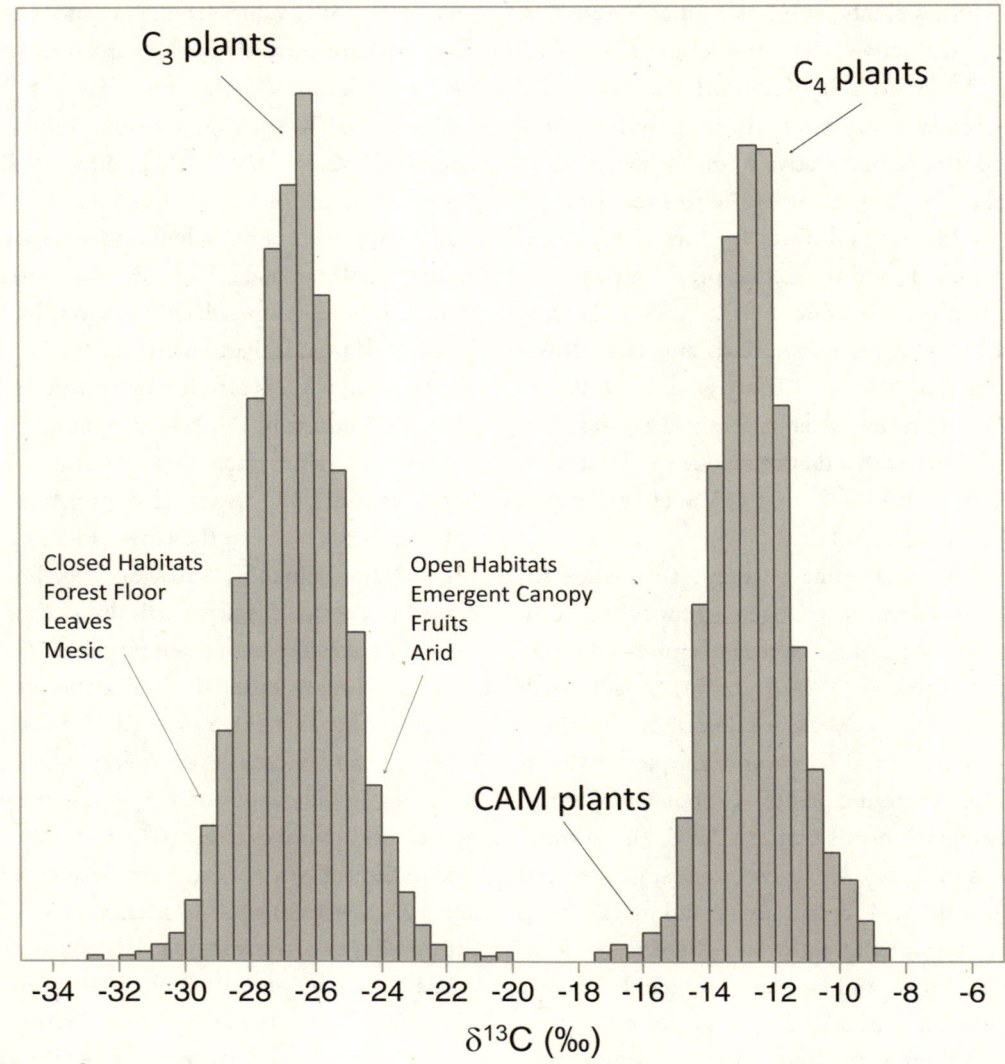

Figure 23.1. $\delta^{13}C$ values of C_3, C_4, and CAM plants from Kruger National Park, South Africa (N = 2,729; data from J. Codron, Lee-Thorp, et al. 2013). CAM plants appear as a shoulder at the lower end of the distribution for C_4 plants. Trends toward higher (e.g., open habitats, fruit) or lower (e.g., closed habitats, leaves) $\delta^{13}C$ values are noted for the C_3 plants.

spheric nitrogen and therefore can have $^{15}N/^{14}N$ values close to those in air (0‰; Virginia and Delwiche 1982). Nitrogen isotopes are also employed as environmental indicators because $\delta^{15}N$ values of plants (and the animals that eat them) are typically lower in cool and moist environments than they are in hot and dry ones (Amundson et al. 2003; Craine, Elmore, Aidar, et al. 2009; Handley et al. 1999). Other uses of nitrogen isotopes include looking for evidence of consumption of human foods (Loudon, Sandberg, et al. 2014; Loudon et al. 2016; Schurr et al. 2012), addressing questions about physiological condition (e.g., starvation; Mekota et al. 2006; E. Vogel, Knott, et al. 2012; reproductive status; Oelze, Douglas, et al. 2016), and determining the timing of weaning (Bădescu, Katzenberg, et al. 2017; Rietsema et al. 2016).

LOW-HANGING FRUIT: EXAMPLES WHERE STABLE ISOTOPE STUDIES EXCEL

Stable isotope approaches are one of the only available methods to study individuals, groups, or species that are no longer living. So long as there is material such as hair or tooth enamel preserved, we can learn at least something

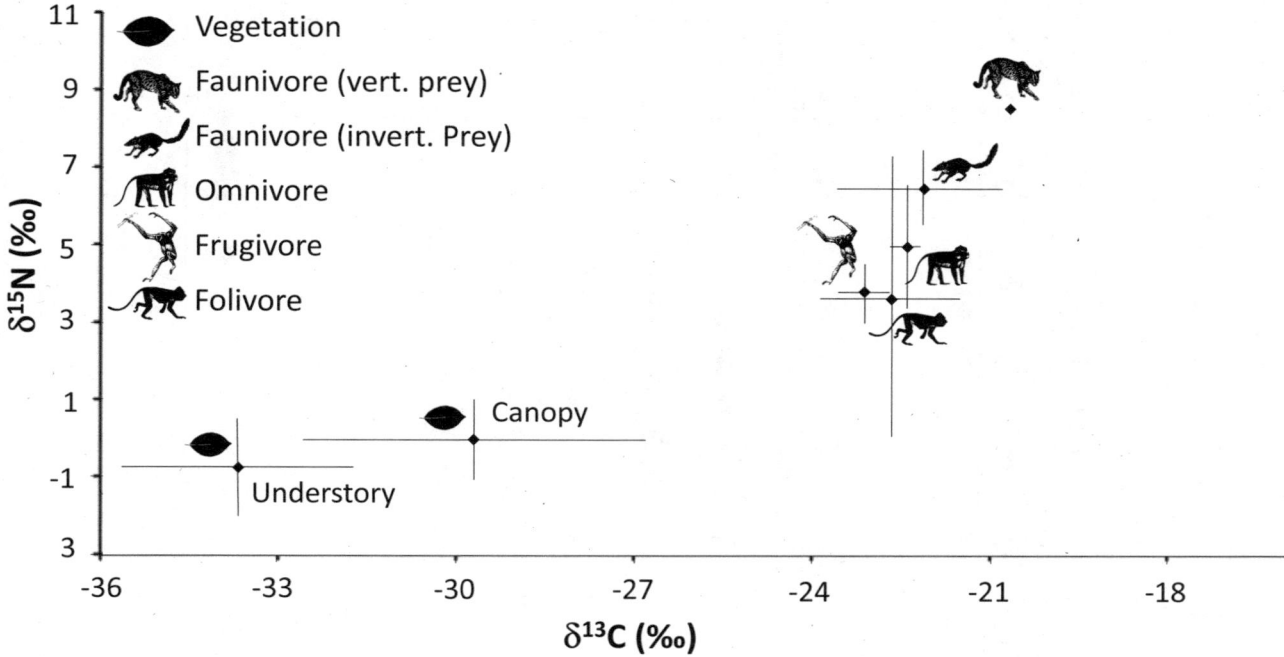

Figure 23.2. δ¹³C values and δ¹⁵N values of selected plants and mammals from Borneo. Note the trophic step in δ¹⁵N values from plants to frugivores to faunivores. Figure modified after Melin et al. (2014).

about the diets of deceased individuals. Given this, many of the most obvious applications for stable isotopes are in paleoprimatology, most notably with early hominins. There have long been questions about whether early hominins primarily ate fleshy fruits and leafy material as do modern chimpanzees (Grine and Kay 1988) or whether they had begun to consume savanna resources such as C_4 grasses or animals that ate grass (C. Jolly 1970a). Stable isotope data are consistent with both being true. Carbon isotope data indicate that, like most living primates, the earliest hominins (prior to about four million years ago) ate C_3 vegetation (e.g., trees, shrubs, and their various parts with mean δ¹³C values of about −27‰; Cerling, Manthi, et al. 2013), which is consistent with other evidence (e.g., microwear) for a diet of fleshy fruits and leaves (Grine et al. 2006; fig. 23.3). On the other hand, many later taxa ate C_4 vegetation (e.g., tropical grasses and some sedges with mean δ¹³C values of about −12‰) to greater or lesser extents (Sponheimer, Alemseged, et al. 2013). Most notably, *Paranthropus boisei* had a diet of about 80% C_4 (or CAM) plants, which is completely unlike that of any living hominoid (Cerling, Mbua, et al. 2011; fig. 23.3). Moreover, there is a correlation between cheek tooth size and C_4/CAM consumption among the australopiths (Sponheimer, Alemseged, et al. 2013), which suggests that this C_4/CAM resource, whatever it was, had an important influence on the highly derived masticatory package of the australopiths. This is a nice story that allows us to look at the interconnections among environment, diet, and morphological adaptations. Moreover, it is a story that could not have been told without an isotopic approach.

And stable isotopes have made similar contributions far afield from the Hominini. One conspicuous example is the curious case of *Hadropithecus stenognathus* (e.g., Godfrey, Crowley, and Dumont 2011; Godfrey, Crowley, et al. 2016). *Hadropithecus* is an extinct strepsirrhine from Madagascar that has long been known to have morphological convergences with early hominins, especially the robust australopiths, such as small incisors, robust mandibular corpora, and enlarged cheek teeth (C. Jolly 1970b). These features led some researchers to assume that *Hadropithecus* ate relatively hard foods, as "Nutcracker Man," *P. boisei*, was supposed to have done (Rafferty et al. 2002; J. R. Scott et al. 2009). However, *Hadropithecus* molars have relatively thin, weakly decussated enamel and seem poorly suited to resist fractures from the consumption of hard items

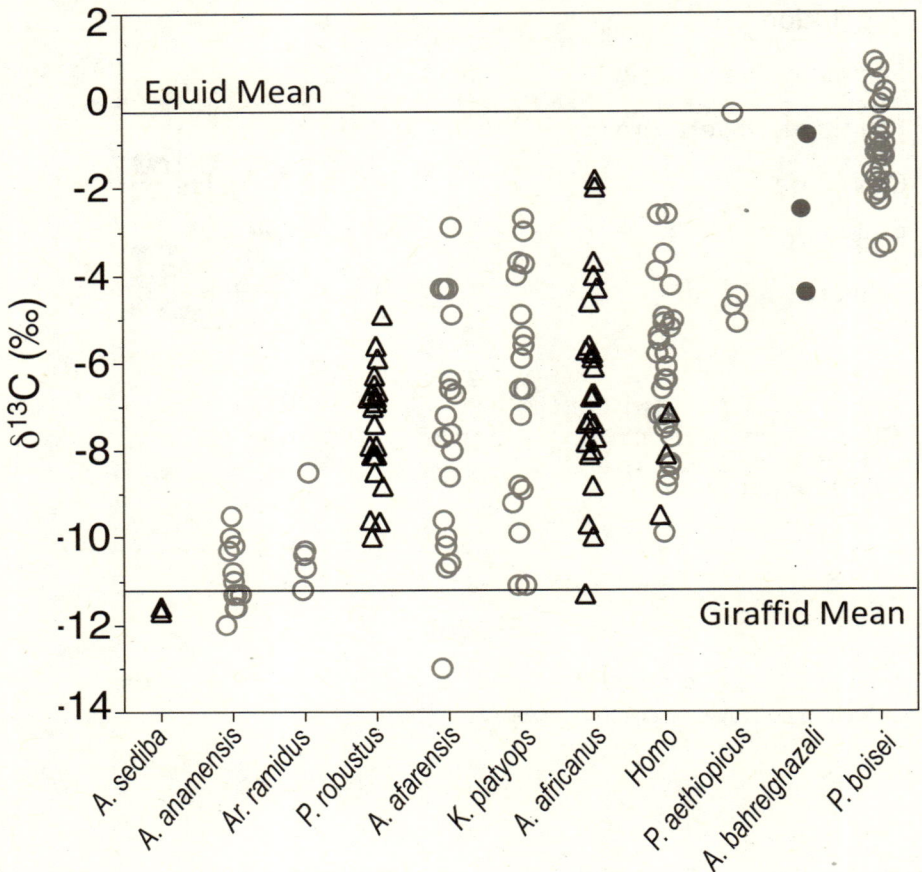

Figure 23.3. δ¹³C values of Plio-Pleistocene hominins from Africa. Triangles, open circles, and filled circles indicate South African, eastern African, and central African specimens, respectively. The means for equids and giraffids (from multiple sites in southern and eastern Africa) are shown to provide rough endmembers for C_4 and C_3 consumers, respectively. Figure modified after Sponheimer, Alemseged, et al. (2013).

(Dumont et al. 2011). Thus, the diet of this taxon was up in the air. Here again, stable isotopes were able to make an important contribution. The δ¹³C values preserved in bone collagen of subfossil *Hadropithecus* are consistent with the consumption of large quantities of non-C_3 plants (Godfrey, Crowley, and Dumont 2011; Godfrey, Crowley, et al. 2016; fig. 23.4). This suggests that, like *P. boisei*, *Hadropithecus* may have eaten large quantities of C_4 grasses or sedges. However, nitrogen isotope data suggest that CAM succulents, particularly from an endemic superfamily of spiny species called Didierioideae, which is very common in the spiny forests of southwestern Madagascar where *Hadropithecus* lived, are a much more likely candidate (Crowley and Godfrey 2013; Godfrey, Crowley, et al. 2016).

So what do these two examples have in common? For one, they are both about taxa for which relatively little dietary information was available, primarily because direct observation of extinct taxa is impossible. A second commonality is that, in both instances, species consumed significant quantities of an isotopically distinct food (i.e., vegetation that did not use the C_3 photosynthetic pathway). We discuss each of these commonalities in more detail below.

WHEN LITTLE DIETARY INFORMATION IS AVAILABLE

Stable isotope analysis will always have a place at the table when it comes to explicating the diets of fossil taxa. When there are few sources of information available, one will usually accept any help that one can get. It is worth emphasizing, however, that the same benefits apply to living taxa that are difficult or impossible to observe. For

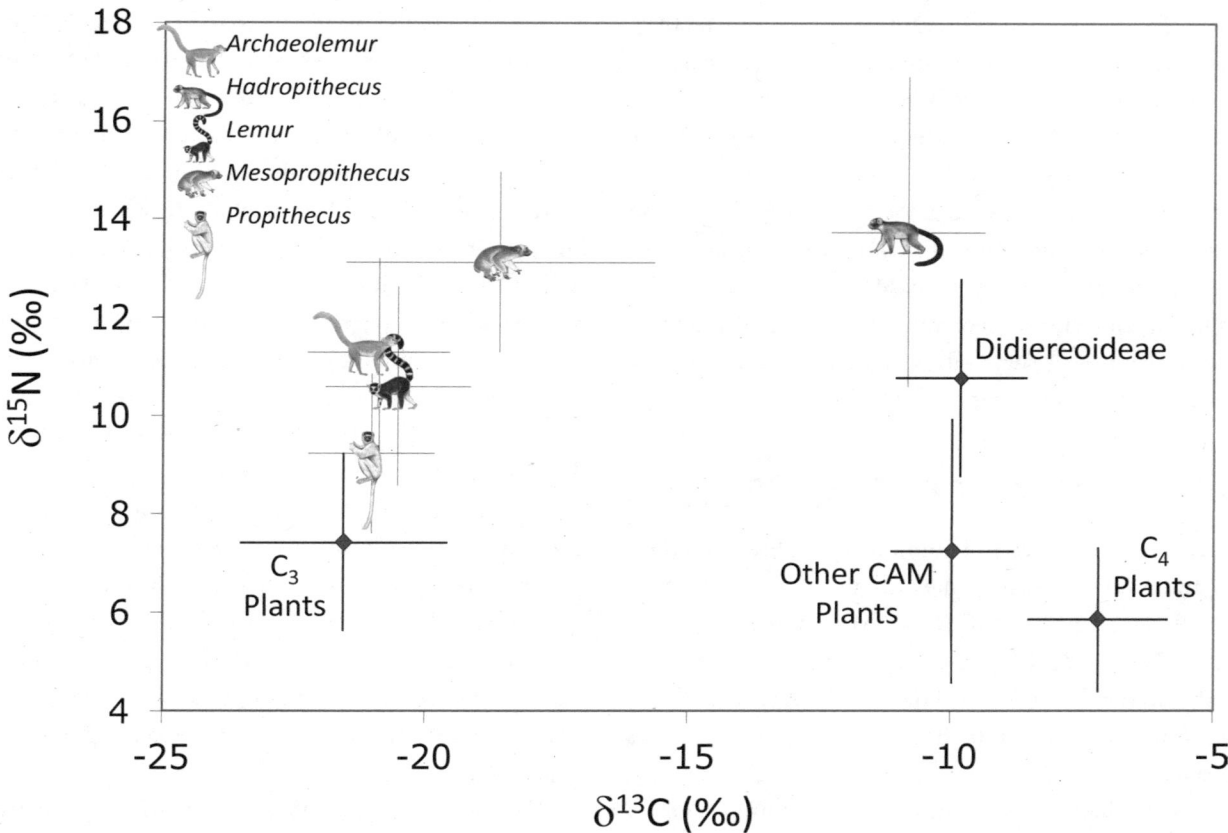

Figure 23.4. δ¹³C values and δ¹⁵N values for subfossil lemurs and modern C_3, C_4, and CAM plants from southwestern Madagascar. While most lemurs consume C_3 plants, carbon isotopes indicate that *Hadropithecus* clearly relied heavily on CAM or C_4 resources. Nitrogen isotope data suggest that plants belonging to the spiny superfamily Didiereoideae were the most likely food. Figure modified from Crowley and Godfrey (2013).

instance, carbon and nitrogen isotopes have been used to verify niche partitioning between sympatric small-bodied nocturnal lemurs in Madagascar (e.g., Crowley, Blanco, et al. 2013; Dammhahn and Kappeler 2014; Heck et al. 2016; Rakotondranary et al. 2011) as well as tarsiers and owls in Borneo and the Philippines (Moritz, Melin, et al. 2014).

Similarly, if one is interested in working out the impacts of ecological change or the increasing influence of humans on primate diets, one can readily trace a community's dietary evolution back through time (so long as time-transgressive samples are available). There have been few such attempts with primates (see Crowley, Godfrey, Bankoff, et al. 2017; Crowley, Godfrey, Guilderson, et al. 2012; L. Gibson 2011). However, quite a few studies have looked at how diets have changed over time within other living taxa, including California condors, gray foxes, and elephants (e.g., C. Chamberlain et al. 2005; J. Codron, Codron, Sponheimer, et al. 2012; Hofman et al. 2016).

Along these lines, there is also much promise in looking at community-level changes over time. From a primatological perspective, for instance, one might use stable isotopes to look at how the recent extinction of the large Malagasy subfossil lemurs impacted the diets and niche breadth of still-living taxa (see Crowley, Godfrey, Guilderson, et al. 2012). Despite considerable research, there is still much we do not know about the recent history of primates on this island. Crowley, Godfrey, Guilderson, et al. (2012) examined changes in the δ¹³C and δ¹⁵N values for four extant lemur species in southwestern Madagascar following (1) the extinction of large-bodied taxa and (2) extensive anthropogenic habitat modification. They concluded that the living lemur species have not filled niches occupied by the extinct taxa. Instead, modern individuals appear to be living in habitats that are moister than those inhabited by the same species in the past, perhaps because these habitats have been protected from logging and hunting. Similar questions could be asked in other settings where there is sufficient time-transgressive fossil or historical material available. We are only just beginning to address such questions, at least partially because getting sufficient samples of subfossil and fossil material remains difficult. But the potential to use isotopes to track change over time is very real and arguably one of the most powerful aspects of isotopic approaches.

ISOTOPICALLY DISTINCT FOODS

Returning to the second commonality of the *Paranthropus* and *Hadropithecus* studies discussed above, in both cases, large quantities of C_4 or CAM resources were consumed. Why does this matter? We would not be so readily able to detect the unusual diets of these species if they consumed only C_3 plants. Because C_3 vegetation differs considerably from C_4 (and much CAM) vegetation, and because C_3 and C_4 plants have nonoverlapping $^{13}C/^{12}C$ ratios (B. Smith and Epstein 1971; fig. 23.1), it can be very easy to detect the presence of C_4/CAM vegetation in a primate's diet using stable isotopes. Indeed, the first application of stable isotope paleodietary analysis in anthropological circles was to look for the influx of maize (a C_4 tropical grass) in the diets of Native Americans in the northeast of what is now the United States, where virtually all other vegetation uses C_3 photosynthesis (J. Vogel and van der Merwe 1977). This was no accident; these authors realized that the consumption of C_4 vegetation in an otherwise C_3-dominated environment sticks out like a sore thumb. The takeaway point here is that C_3/C_4 or CAM questions are particularly amenable to isotopic approaches because the basic signal is so strong. Most primates, in contrast, consume almost exclusively C_3 vegetation, with the result that the isotopic differences between the food classes of interest (e.g., fruits vs. leaves) are usually three to ten times smaller (see Cernusak et al. 2009), making them much more difficult to distinguish in practice (fig. 23.2). This is particularly problematic when comparing individuals from different localities or times, as climate, habitat, and even physiological variables can make the interpretation of small isotopic differences complex.

But there is also an upside to the C_3-dominated diets of most primates—namely, that it makes diet provisioning and raiding of human C_4 crops such as maize, sugarcane, sorghum, and millet immediately apparent in the tissues of the raiders (e.g., Loudon et al. 2014; Loudon, Sandberg, et al. 2016; Schurr et al. 2012). For instance, Loudon et al. (2014) found that a vervet troop from Oribi Gorge, South Africa, had fur $\delta^{13}C$ values indicative of large-scale C_4 plant consumption, which was unanticipated since initial observations revealed no evidence of such dietary items (fig. 23.5). Only after the isotopic data illuminated this possibility did observations confirm that the monkeys cryptically raided a sugarcane field on the reserve's border. More recently, Loudon, Sandberg, et al. (2016) found that chimpanzees with regular access to human resources also had higher hair $\delta^{13}C$ values than those that did not access human foods (fig. 23.5). Such studies could have wide-scale relevance given the increasing importance of human and nonhuman primate interactions from both ecological and conservation perspectives (Fuentes 2012).

Fortunately, C_3/C_4 differences are not the only isotopic pattern of sufficient magnitude for ready application in studies of primate diet. There is also the well-known trophic-level shift in nitrogen isotopes (described in the primer section above): plants have lower nitrogen isotope ratios than do herbivores, which in turn have lower $\delta^{15}N$ values than do carnivores or insectivores from the same locality (e.g., fig. 23.2). Perhaps most notably, nitrogen has been used to investigate the diets of Neanderthals (Bocherens et al. 2005; M. Richards, Pettitt, et al. 2000). These studies revealed that the $\delta^{15}N$ values of Neanderthal bone collagen were more similar to those of contemporaneous carnivores such as wolves than to those of herbivores such as reindeer. While this result is hardly shocking given the strong associations between Neanderthal remains and butchered animal bones (Mellars 1996; Speth and Clark 2006), it nonetheless provides another form of evidence for the importance of animal products in meeting Neanderthal dietary protein requirements. Alas, this method has not been applied to the likes of early *Homo*, where hypotheses of links between animal food consumption and encephalization abound (e.g., Aiello and Wheeler 1995; Dart 1957; Milton 1999), for the simple reasons that collagen does not survive so deep into the past and tooth enamel contains negligible nitrogen

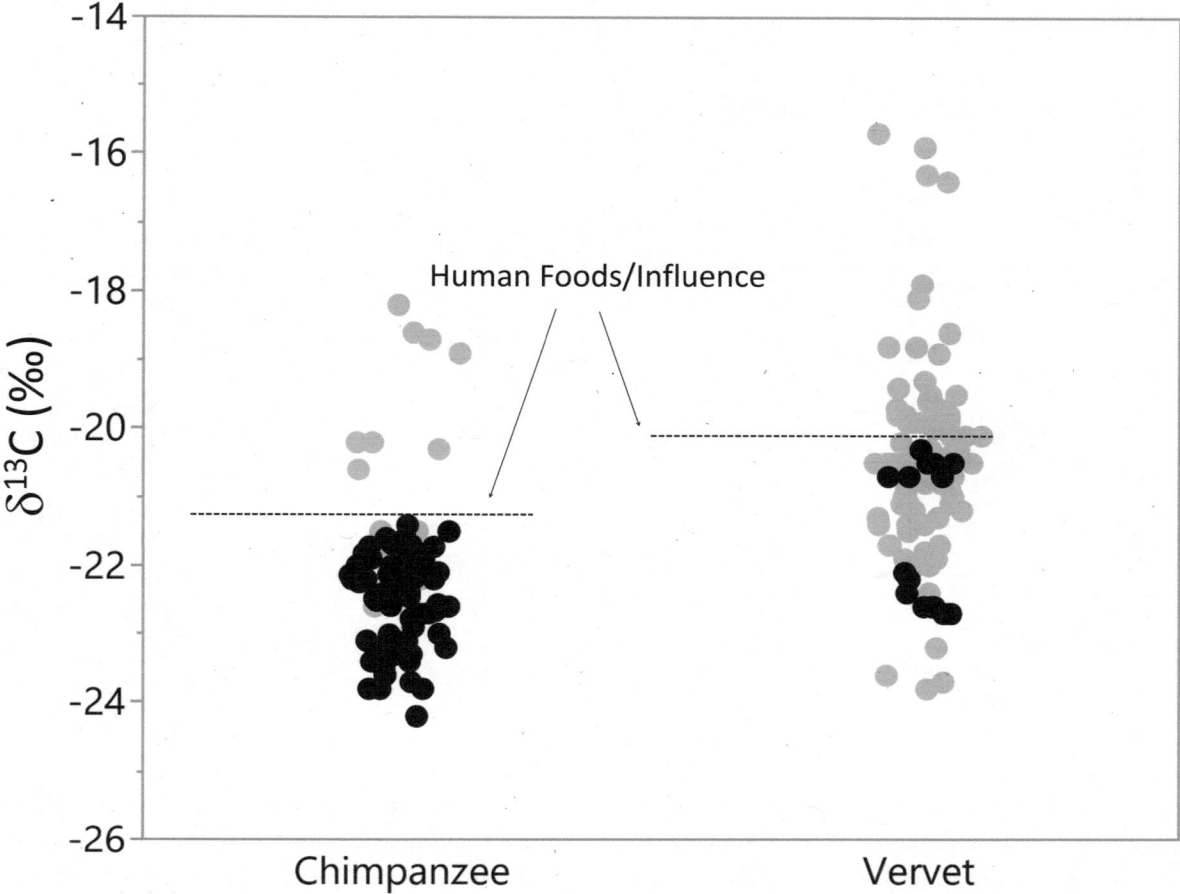

Figure 23.5. $\delta^{13}C$ values for chimpanzee hair and vervet monkey fur from multiple African sites. Gray circles represent high levels of access to human resources, and black circles represent modest or no access to human resources (data from Loudon et al. 2014; Loudon, Sandberg, et al. 2016). There is some overlap between high- and modest/no-access sites for both species; however, $\delta^{13}C$ values above about −21.5‰ and −20.5‰ for chimpanzees and vervets, respectively, are always from individuals that have been strongly impacted by humans.

(although we soon may be able to measure $\delta^{15}N$ values in enamel, as in Leichliter et al. 2021).

There are, however, abundant examples of modern primates where questions about animal food consumption are highly relevant. For instance, Crowley, Blanco, and colleagues (2013) used carbon and nitrogen isotope values in fur from mouse lemurs (*Microcebus*) and dwarf lemurs (*Cheirogaleus*) from continuous and fragmented forest in eastern Madagascar to investigate whether lemur diets differ in fragments. After controlling for baseline isotopic differences in vegetation among localities, they found that all three species have higher $\delta^{15}N$ values in disturbed forest than in undisturbed forest, which suggests that individuals consume more arthropods in fragments. In a separate study, Crowley, Rasoazanabary, and Godfrey (2014) showed that in three habitats in arid southwestern Madagascar, mouse lemurs have higher nitrogen isotope values in the rainy season than in the dry season, suggesting greater arthropod consumption during the former. Their data further suggest greater arthropod consumption by males than by females, which is consistent with behavioral observations. In a similar vein, Fahy et al. (2013) showed that male chimpanzees in the Taï Forest had higher nitrogen isotope compositions than did females (presumably indicating greater animal food consumption) and that highly successful hunters had higher $\delta^{15}N$ values than did others (fig. 23.6). Again, these results were consonant with observations of meat consumption within the community.

It should be emphasized that stable isotope approaches can do more than just identify the consumption of C_3 and C_4 foods and distinguish between herbivo-

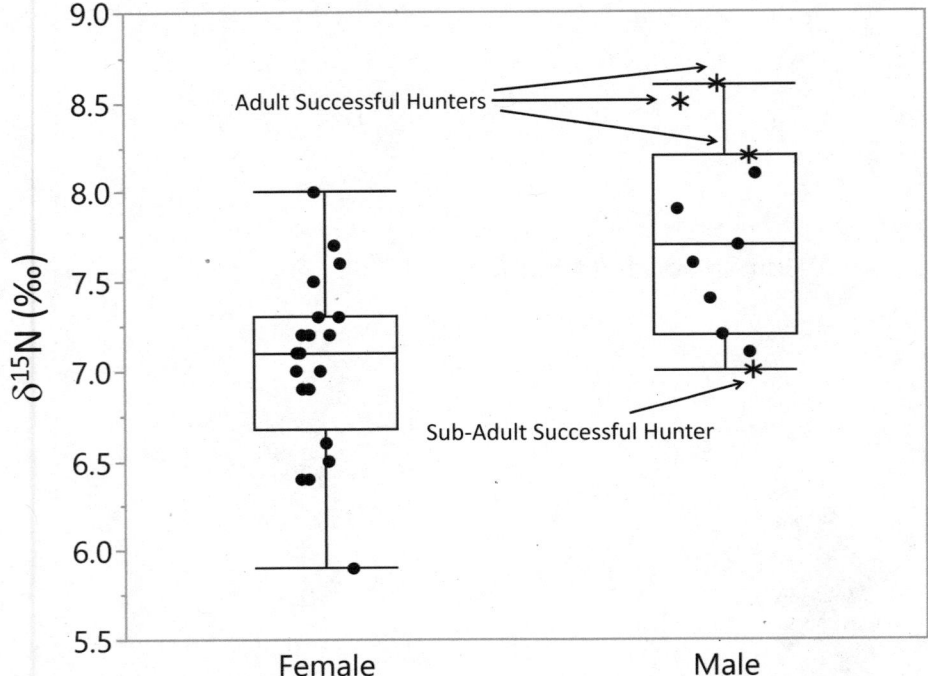

Figure 23.6. $\delta^{15}N$ values of chimpanzees from Taï National Forest, Cote d'Ivoire. Males, which tend to consume more meat, have higher $\delta^{15}N$ values. Among males, highly successful hunters have higher $\delta^{15}N$ values than do other males. The one individual (Fitz) that does not fit this pattern was not fully adult during the growth period of the sampled hair. Data from Fahy et al. (2013).

rous and faunivorous diets. They can potentially provide highly resolved information about diets in simple systems where only a few major foods (or broad classes of foods) are consumed. For example, stable isotopes have proved efficacious for distinguishing marine and terrestrial diets generally (Schoeninger, DeNiro, and Tauber 1983; Stapp 2002), and this has recently been confirmed in nonhuman primates by M. Lewis et al. (2018), who showed that chacma baboon (*Papio ursinus*) fecal $\delta^{15}N$ values are correlated with measures of observed marine food consumption. The isotopic data also suggested that few if any marine resources were consumed by these baboons in the spring and summer despite their local abundance. Other primate species are known to consume marine resources (e.g., *Papio cynocephalus* [Messeri 1978] and *Macaca fascicularis* [Son 2003]), so there exists considerable potential for using isotopes to trace marine protein inputs to primate diets in the future.

ISOTOPES ACROSS SPACE AND TIME

This brings us to another area where stable isotopes may be uniquely poised to contribute to discussions about primate ecology: variation across space and time. It can be quite challenging to illuminate the diet of a taxon across large areas to investigate questions such as, "How is diet impacted by habitat?" This would be very time consuming if observations were required, but hair (from nests, etc.) or feces can be collected fairly quickly and easily from across a wide geographic range, and the isotopic analysis of these materials is straightforward and relatively inexpensive. There have been relatively few attempts to do this with primates, although the practice is growing. Thackeray et al. (1996) examined collagen for *Papio* species from across southern Africa and found that within C_4 biomes, $\delta^{13}C$ values could vary by more than 4‰, indicating large differences in the amount of C_4 or CAM plants consumed by various populations. D. Codron et al. (2006) examined spatial isotopic variability in baboon feces in Kruger National Park and the Free State, South Africa; they documented significant differences in CAM/C_4 consumption across space. Spatial variability in primate fur and hair has been investigated for mouse lemurs (Crowley, Thorén, et al. 2011; Heck et al. 2016; Rakotondranary et al. 2011), macaques (O'Regan et al. 2008; Schillaci et al. 2013), and African great apes (Loudon, Sandberg, et al. 2016; Oelze, Fahy, et al. 2016; Schoeninger, Most, et al. 2016). Other primate studies have focused on change over time rather than across space, and such research may be very promising for documenting shifts in diet related to seasonality, migration, habitat, or body condition (e.g., Blumenthal, Chritz, et al. 2012; Mekota et al. 2006; Oelze, Douglas, et al. 2016; Oelze, Fuller, et al. 2011). Nevertheless, it is probably fair to say that while the studies focused on space or time have provided novel data regarding primate diets (or at least their stable isotope compositions), they have tended toward the descriptive, largely confirmed what we already knew, and typically focused on factors other than potential drivers of dietary change.

More sophisticated analyses have been carried out for nonprimates. For instance, $\delta^{13}C$ values for more than 1,500 elephant fecal samples from across Kruger National Park, South Africa, were able to document strong seasonal differences in C_3/C_4 consumption; differences between regions of the park with differing habitats; positive correlations between rainfall, crude protein, and grass consumption; and a negative correlation between tree diversity and grass consumption (J. Codron, Codron, Lee-Thorp, Sponheimer, Kirkman, et al. 2011). A follow-up study coupling these stable isotope data with satellite imagery has revealed a positive correlation between grass consumption by elephants and grass abundance in the southern portion of the park, but only during the dry season (Marston et al. 2020). Such studies allow us to ask much more nuanced questions about diet choice and would be virtually impossible, or at least extremely time consuming, without isotopic methods. An added advantage of isotopic studies using feces is that they can be readily supplemented by traditional fecal analysis, which, while having limitations, still provides us with some of the best diet data available for unhabituated primate populations (e.g., Yoshikawa and Ogawa 2015).

Of course, the above study on elephants was primarily useful because changes in C_3 browse and C_4 grass consumption are fundamental to our understanding of elephant diets. This is less relevant for most primate species, except for *Papio*, some *Theropithecus* populations, and a few other cercopithecines. Researchers have sought to test the idea that chimpanzees in relatively open "savanna" habitats eat C_4 foods in significant quantities (Schoeninger, Moore, and Sept 1999; Sponheimer, Loudon, et al. 2006), but these studies have not been able to document C_4 resource consumption among chimpanzees. To be honest, however, this was hardly an unanticipated insight. Chimpanzees have certainly been observed to eat C_4 foods, but the relative contribution of such foods is so small that we might not expect it to be readily identifiable using isotopes (e.g., Itoh and Nakamura 2005; A. Suzuki 1969; but see Wrangham, Conklin, Chapman, et al. 1991), particularly in organic tissues such as hair or bone collagen that primarily reflect dietary protein (Ambrose and Norr 1993; Tieszen and Fagre 1993). A possible exception would be in situations where high-protein foods with C_4-derived carbon are consumed (e.g., eating insects that eat C_4 grasses).

OTHER ISOTOPIC APPLICATIONS

The ability of stable isotopes to stretch temporal and spatial windows relative to observational studies also makes them well suited to testing questions about the impact of changes in physical or biotic conditions on diet at both the individual and aggregate levels, and, with complementary data sets, testing the fitness consequences of such behavioral changes. For instance, a carbon and nitrogen isotope study of living black-tailed deer and black-tailed deer carcasses from British Columbia showed that individuals that foraged in some hemlock- and cedar-dominated habitats (as evidenced by their stable isotope compositions) were more likely to be preyed on by wolves than individuals in other environments (Darimont et al. 2007). It was further argued that these dangerous habitats provided higher-protein forage than others, and since protein is crucial for survival and reproduction in mammalian herbivores, dealing with increased predation might be an acceptable trade-off. This study is an excellent example of what stable isotopes can bring to the table. Both the spatial scope (~1,500 km²) and the need to glean data from previously killed individuals would have made such a study impractical or impossible using other methods.

Similarly, L. G. Adams et al. (2010) showed that some inland wolves in Denali National Park, Alaska, used salmon from rivers (determined via stable isotope ratios in the bone collagen of dead individuals) and were therefore able to maintain comparable population densities to wolves in other areas of the park despite substantially (six times) lower ungulate biomass. Furthermore, given the relatively high ratio of wolves to ungulates in this area of the park and consequent high levels of ungulate predation (which was possible only because of the subsidy provided by allochthonous salmon), the authors surmised that the presence of salmon is indirectly responsible for the low ungulate biomass. While one might argue about the direction of the arrow of causality here, it is easy to see how stable isotopes, once again, allowed investigation of individual and community dynamics that would have been impossible otherwise. In this case, dead individuals from over a wide spatial area were used to determine the amount of salmon consumed. When coupled with population density data for wolves and prey species, the isotopic data allowed investigation of environmental impacts on wolves and the broader communities. The capacity of stable isotopes to allow analysis of variation over multiple spatial and temporal scales, coupled with data on the physical and biotic environments (including intraspecies density differences over space), allows one to answer otherwise unassailable questions.

A FEW LIMITATIONS AND PRACTICAL CONCERNS

Despite all of this promise, there remain some persistent and important limitations of isotopic applications. First, one needs to have good control over the isotopic composition of dietary items. In some regards, this has become much easier over the past few decades as analyses have become more automated and less expensive, and we can now perform hundreds of analyses in the time it took to run only a few when the field was developing. However, getting baseline diet data is still very difficult in practice. For example, one ideally should sample dietary items from the time they were consumed to properly interpret tissue isotopic compositions, as the isotopic compositions of foods are not so much discrete as moving targets. Carbon and nitrogen isotope compositions of plants can vary between seasons and potentially among years (e.g., Bump et al. 2007; J. Codron, Codron, Lee-Thorp, Sponheimer, Bond, et al. 2005), although the degree to which these fluctuations occur depends on location. Moreover, as mentioned earlier, there is variation in individual specimens related to organ (e.g., fruit, leaf, or root; J. Codron, Codron, Lee-Thorp, Sponheimer, Bond, et al. 2005), tree height (e.g., N. McDowell et al. 2011), or species of plant (e.g., J. Codron, Lee-Thorp, et al. 2013; Ruiz-Navarro et al. 2016). In a perfect world, one would also sample the exact food items that were partially consumed by primates to establish whether there are systematic isotopic differences between consumed and avoided items within any particular food type.

The upshot of this is that it is difficult to get a good handle on the isotopic composition of the diet of a generalist primate without collecting many samples or making many assumptions. The degree to which this matters will depend a great deal on the questions being asked. For instance, if one is trying to distinguish between three foods, all of which are quite distinct in carbon and nitrogen isotope compositions, the problems are not severe. Additionally, if one is only interested in assessing differences in trophic level among populations living in different habitats, it may be sufficient to establish baseline isotopic variability among the localities of interest by measuring isotope values in a variety of plants. However, as one tries to distinguish between more food items, the likelihood of the dietary values isotopically overlapping increases, necessitating even better control over the dietary isotopic compositions. If analyzing feces, collecting plant samples at the same time as fecal samples will not be difficult, but if analyzing hair, one ideally needs to assess isotope values for the food groups in question over the period of tissue formation. Once again, the degree to which this will be a problem will depend on the kind of question being asked, as well as the system of interest. An annual average value is more than sufficient if one is just trying to get a sense of things like C_3 versus C_4 consumption (even the absence of vegetation data is not a disaster in such cases), but if the goal is to distinguish between the values of fruits of two species, or even between leaves and fruit within a habitat, one will likely want to have better control over dietary values. Since so many primates consume a large number of species, and this may vary among seasons, sample sizes can very quickly become unwieldy (e.g., species × plant part × season).

One solution to this is to try to limit oneself to asking questions that do not demand such resolution. This will also help ameliorate uncertainty about other factors affecting mammalian stable isotope compositions. For instance, there are good reasons to believe that diet-to-tissue offsets differ between food types (e.g., C. Robbins, Felicetti, and Sponheimer 2005) and body states (e.g., negative energy balance or negative protein balance; Mekota et al. 2006; Reitsema et al. 2016). In most cases, these differences are likely to be small, on the order of 1‰ or so; however, larger differences have been reported. For example, for a broad suite of ungulates, high-protein diets lead to much higher tissue $\delta^{15}N$ values than do low-protein diets (Spon-

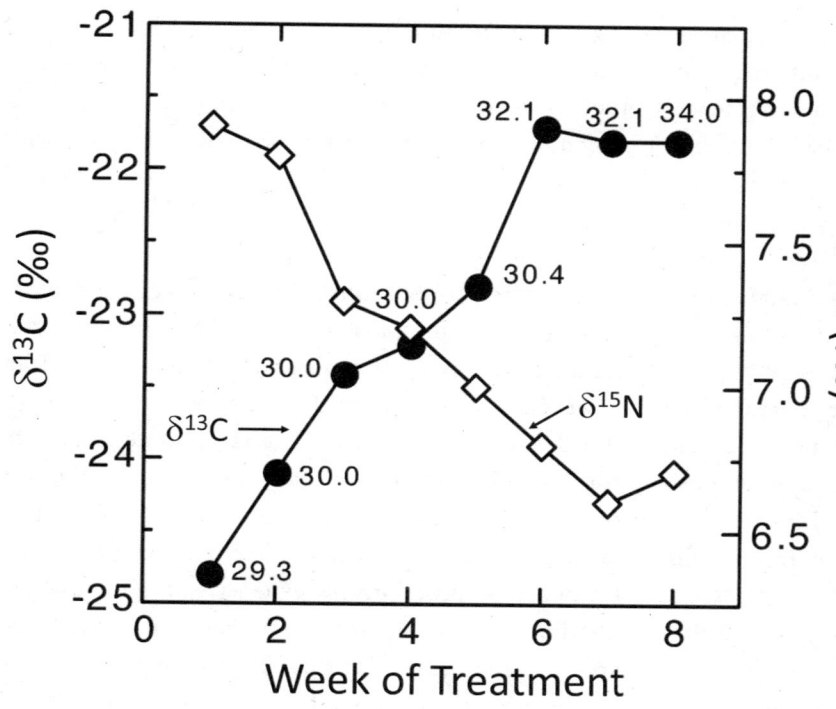

Figure 23.7. Incremental $\delta^{13}C$ (circles) and $\delta^{15}N$ (diamonds) data for hair from a 25-year-old human female who was suffering from anorexia nervosa. The numbers next to each datapoint are the patient's weight in kilograms during eight weeks of treatment. Hair $\delta^{15}N$ values decreased and $\delta^{13}C$ values increased systematically after she was placed on a nutritionally adequate diet. Data are from Mekota et al. (2006).

heimer, Robinson, et al. 2003), but in nectarivorous bats, the opposite pattern is observed (Mirón et al. 2006). An advantage of this is that these potentially confounding factors may allow us to address questions beyond diet that are also of interest, such as the degree to which an individual is suffering from nutritional stress (Deschner, Fuller, Oelze, Boesch, et al. 2012; Mekota et al. 2006; E. Vogel, Knott, et al. 2012) or, as shown in figure 23.7, when they are recovering from such deficiencies.

Another potential solution is to use a stable isotope mixing model, such as SIAR, which uses Bayesian methods to produce posterior probability distributions of food source proportions while accounting for potential variability in food isotopic compositions, fractionations, and even elemental concentrations (Parnell et al. 2010). Such models represent a major advance but are far from catholicons. They do not, for instance, account for isotopic routing (e.g., preferential routing of carbon from dietary protein to hair; Ambrose and Norr 1993; Tieszen and Fagre 1993), which is known to occur and is of greater importance for omnivorous primates than for most purely carnivorous or herbivorous taxa. Moreover, as the creators of the SIAR model emphasize, its dietary proportion estimates can be "highly uncertain" and may include large ranges. Consequently, using single summary output values (e.g., mode leaf consumption = 50%) should be done with caution or perhaps avoided completely (Parnell et al. 2010).

OTHER ISOTOPES TO THE RESCUE!

To this point, we have focused entirely on stable carbon and nitrogen isotopes. It is important, however, to acknowledge that there are a host of other isotopic systems that either independently or in tandem with C, N, or other isotopes (e.g., oxygen and sulfur) might reveal much about the diets of extant and extinct primates. Oxygen isotopes ($\delta^{18}O$) have received the most attention. Like carbon and nitrogen, $\delta^{18}O$ values are impacted by climate, and they typically are higher in hot and dry habitats than in cool, moist ones (G. Bowen, Wassenaar, and Hobson 2005; Dansgaard 1964; Luz et al. 1990). There are also differences in the oxygen isotope compositions of plants in closed (lower $\delta^{18}O$ values) and open environments (higher $\delta^{18}O$ values), and $\delta^{18}O$ values in plants increase with canopy height (Ometto et al. 2005; Sternberg et al. 1989), reflecting degrees of evaporative enrichment of ^{18}O in plant water. Additionally, leaves typically have higher $\delta^{18}O$ values than fruits because they are photosynthetically active (which results in some water loss) and have greater surface area per unit volume than other plant tissues (reviewed in Barbour 2007). These differences in plant tissues are passed on to animal consumers. Generally, $\delta^{18}O$ values are higher in individuals that spend time in the canopy or forest clearings or gaps rather than forest subcanopies or understories because evapotranspiration is generally higher in more open microhabitats (Carter and Bradbury 2015; Cerling, Hart, and Hart 2004; Crowley 2014; Crowley, Melin, et al. 2015; Krigbaum et al. 2013; T. D. White et al. 2009).

Mammalian oxygen isotope ratios are governed by a complex interaction between drinking behavior, diet, and physiology (e.g., Kohn 1996). Oxygen isotope compositions tend to differ between animals that drink regularly and those that get most of their water from food (Kohn et al. 1996; Sponheimer and Lee-Thorp 1999) because water in plants (particularly leaves) tends to be evaporatively enriched (Barbour 2007; Yakir 1992). Oxygen isotope ratios tend to be higher in folivores than in frugivores, and herbivores typically have higher $\delta^{18}O$ values than omnivorous and faunivorous taxa (Crowley, Melin, et al. 2015; Sponheimer and Lee-Thorp 2001). Given this, it might appear that oxygen isotopes are exquisitely sensitive diet and habitat indicators. But this sensitivity is mired in a sea of equifinality, meaning that various combinations of the above factors can result in the same oxygen isotope composition. For example, the relatively low oxygen isotope ratios of some early hominins (e.g., Cerling, Mbua, et al. 2011; Lee-Thorp and Sponheimer 2006) may indicate a tendency toward omnivory or frugivory, but they could just as easily indicate a strong dependence on free water from less evaporatively enriched sources such as rivers, a tendency to occupy the most heavily wooded portions of landscapes, or both (as might be found in gallery forests). Thus, without other evidence, one will almost always be able to offer several plausible interpretations of such data. The same holds for modern taxa. Where several factors affect $\delta^{18}O$ values in the same manner (e.g., folivory, upper canopy use, and open environments), it is nearly impossible (unless data from other isotopic systems are available) to interpret oxygen isotopic data without data from behavioral studies—in which case the stable isotope data provide a useful complement but do not teach us much that we did not already know. Of course, when stable isotope data defy expectations, they may precipitate the reinterrogation of existing behavioral data, which can be very important. Moreover, it is possible that with various isotopic systems in tandem, one can get better discriminatory power, but there is reason to believe that even with three isotopes (C, N, and O), resolution may be limited (e.g., Cerling, Hart, and Hart 2004; Crowley, Melin, et al. 2015; Krigbaum et al. 2013).

There are unquestionably many other isotopic systems that, in some capacity or another, can prove useful for addressing questions about diet. It is not clear, however, how broadly applicable these will be for primates. New systems are introduced, sometimes with considerable fanfare, but often make only a limited contribution (see below). Sometimes this is because making the measurements is complex (e.g., calcium isotopes) or prohibitively expensive, or because the distributions of the isotopes in modern ecosystems, and the mechanisms for their partitioning therein, are poorly understood. Regarding this latter point, J. Martin, Vance, and Balter (2015) recently published a paper showing that $^{26}Mg/^{24}Mg$ ($\delta^{26}Mg$) is generally higher in modern carnivores and omnivores than in herbivores of La Lope National Park, Gabon. Accordingly, magnesium isotopes could hold some promise for addressing questions about faunivory in modern primates (as discussed above), and they could be especially important for addressing questions about early meat consumption in our lineage, on which there is a large body of empirical archaeological work (e.g., Binford 1981; H. Bunn and Ezzo 1993) and a similarly well-developed body of theory (e.g., Aiello and Wheeler 1995; Milton 1999). This has occasioned excitement in some corners of the paleoanthropological and biogeochemical communities, at least by the metric of people contacting one of us (MS) about exploring research along these lines. But how promising are magnesium isotopes? Our reading of the existing data leaves us unexcited. Why? Because although the method shows some gross differences between carnivores and herbivores en masse, there is weak to little support for differences among individual carnivorous and herbivorous taxa. For instance, the La Lope study mentioned above reveals no statistical difference between genets, which are categorized as carnivores (although some eat considerable quantities of leaves and fruit), and herbivorous buffalo (data in J. Martin, Vance, and Balter 2015). Moreover, a previous study, which measured the δ^{26}Mg values of mammals in Kruger National Park, South Africa, could not reliably differentiate lions (or hyenas) and grazing herbivores such as tsessebe and roan antelope (data in J. Martin, Vance, and Balter 2014). Given this, considerable ground-truthing will be required if we are to accept the reliability of magnesium isotopes or their superiority to the long-established Ba/Ca and Sr/Ca ratios as trophic-level indicators (e.g., Balter 2004). Nevertheless, this isotopic system shows promise as a trophic-level indicator, as do calcium, hydrogen, and zinc isotopes (G. Bowen, James, et al. 2009; Clementz et al. 2003; DePaolo 2004; Jaouen, Beasley, et al. 2016; Melin, Crowley, et al. 2014; Reynard and Hedges 2008; Reynard et al. 2010). From the perspective of early hominin ecology, calcium isotopes are of particular interest because *P. boisei*'s $\delta^{44/42}$Ca values are much higher than those of other hominins and co-occurring mammals (J. Martin, Tacail, et al. 2020). This probably indicates herbivory, as carnivores tend to have low $\delta^{44/42}$Ca values, but beyond that, these results are uninterpretable given our poor understanding of the mechanisms that drive variation in plant calcium isotope compositions. Lastly, iron and copper isotopes may also prove useful for sexing unknown individuals (Jaouen, Balter, et al. 2012; Walczyk and von Blanckenburg 2002).

SUMMARY AND CONCLUSIONS

Stable isotopes have made important contributions to the field we call "anthropology" in the United States, and there is little doubt that they will continue to do so for the foreseeable future, particularly with anticipated technical innovations that will improve analytical possibilities. That said, we do feel it is an appropriate time to step back and critically assess what we have accomplished lest the direction of our field gets caught up in a wave of enthusiasm that is warranted but perhaps not appropriately or sufficiently directed or validated. On the one hand, we should acknowledge that stable isotope approaches have taught us far more about ancient specimens than about modern ones, while on the other we should recognize that the confirmatory and methodological research of the past decade has provided much firmer footing for stable isotope studies in primatology.

What we have not seen much of in modern primatology are applications that are directed toward questions for which stable carbon and nitrogen isotopes are particularly well suited. For instance, there is a relative dearth of studies on modern baboons (*Papio* spp.), vervets (*Chlorocebus pygerythrus*), blue monkeys (*Cercopithecus mitis*), and macaques (*Macaca* spp.), all of which live in C_4 biomes (though not exclusively) and all of which come in regular contact with human foods. Other taxa of particular interest are (1) many of the lemurs of Madagascar, which, like the above, are often in mixed C_3/CAM environments and frequently have omnivorous diets, or (2) omnivorous New World primates, such as squirrel monkeys or capuchins. Places where both modern and historical or fossil materials are available are also particularly attractive for isotopic research. Studies that leverage the power of stable isotopes to provide diachronic dietary data by using multiple tissues or incremental sampling (e.g., Oelze 2016; C. Smith et al. 2010) or that use stable isotopes to investigate the impacts of the biotic and physical environment (as in L. G. Adams et al. 2010; Darimont et al. 2007; E. Vogel, Knott, et al. 2012) on individual fitness and community structure and function would also be welcome. This is far from an exhaustive list of potential high-priority applications, but it is meant to emphasize some of the main points we discussed above.

- Stable isotope approaches have taught us far more about ancient specimens than about modern ones. Most work with extant primates has been either confirmatory or methodological in nature.
- Stable isotopes are at their most powerful when they are used to answer questions that cannot be answered with other methods (and sometimes when coupled with other data sources).
- Stable isotopes have the most discriminatory power when distinguishing between diets or habitats that have disparate isotopic compositions.
- Stable isotope studies of primates have barely begun to explore the impacts of diet and environment on individual fitness or community structure and function.
- Stable isotopes will become increasingly crucial tools in primate evolutionary biology, ecology, and conservation research, although we must remain mindful of the strengths and limitations of isotopic approaches.

24 Mechanical Properties of Primate Foods

Adam van Casteren and Peter Lucas

Technical measurements in the context of primate fieldwork are always going to be troublesome because of the remote locations where most primatologists work. However, field methods for both physical and chemical tests (P. Lucas, Copes, et al. 2011; Lucas, Corlett, et al. 2011) are gradually becoming more tractable as a result of simplifications in the equipment needed to obtain results. Even field laboratories are now perfectly feasible in many locations, extending the kit (incorporating mechanical testing machines, low-cost spectrometers for food color and chemistry, and global positioning units) introduced by Lucas et al. (2001).

In this chapter, we deal with the mechanical aspects of foods that primates consume. It is important to measure foods as close to the time of collection as possible because the physical properties of foods deteriorate quickly. The exceptions are mineral components, either intrinsic or associated, and woody tissues, which are durable. Here, we describe recent advances that make measurements feasible almost anywhere and indicate the possible significance of these measurements for understanding the acquisition and oral processing of foods by primates.

There is a long tradition of simple mechanical field measurement, particularly in studies of plant-animal interactions. The oldest devices are the penetrometers, wherein a flat-ended cylindrical probe is pushed completely through a leaf to measure the force required to do so. There is an extensive history of the use of these in ecology (e.g., Coley 1983; Feeny 1970; Raupp 1985; Sanson et al. 2001; L. Williams 1954), but very few studies have used penetrometers in primatology. Little or no expertise or design standardization has been reported for these devices, with guidelines being set only quite recently (Onoda et al. 2011). To a lesser extent, using a protocol derived from agricultural research (Magness and Taylor 1925), penetrometers have also been used to test fruits where a cylinder, with either a flat or a hemispherical tip, is pushed partially into fruit tissue—beginning with the pioneering work of Kinzey and Norconk (1990). Even though these "fruit pressure testers" are described as assessing firmness, they really measure the maximum load needed to produce some type of fracture. Variants of such devices, including durometers, have been used by primatologists more often for testing fruits or bulk food items than for leaves (e.g., J. Lambert et al. 2004; Pampush et al. 2011; Wieczkowski 2009; Yamashita 1998). However, all these devices tend to produce results unique to the device employed and its dimensions. As currently configured, results are not generalizable because they do not measure an actual mechanical property, a simple list of which is given in table 24.1. However, there is scope to improve this technique (Lucas, in prep.).

Fundamental properties such as the elastic modulus, the degree of plastic response, and the fracture toughness are essential for trying to characterize foods objectively on a continuous scale that are otherwise described qualitatively as some combination of "soft-hard" or "brittle-tough" dichotomies (Thiery et al. 2017). In contrast, there has been a recent trend toward the creation of what are essentially small-scale universal testing machines with the overt aim of measuring such mechanical properties. Originally, a full-scale testing machine was taken to the field (Strait and Overdorff 1996), which is perfectly feasible for sites with full laboratory facilities. However, miniaturized versions have since been custom-made to perform a variety of possible tests via the attachment of different jigs (Darvell et al. 1996), and although there may be differences in testing methods (see the section "Actual Field Measurements" for details), the properties being measured are fundamental and therefore comparable across biological materials. The difference between these approaches reflects a simple choice that can be made for any kind of physical or chemical test. Should you make a unique measurement dependent on the exact equipment and conditions involved in the test or instead try to measure something more fundamental that underpins physical or chemical behavior in any circumstance? The former option tends to ignore concepts such as concentration—for example, the concentration of ions involved in a chemical reaction or the concentration of the force in a mechanical test. The latter choice offers the potential for prediction via the creation of a formula governing the behavior concerned. Predictions require reproducible estimates of fundamental properties (table 24.1). However, the tests designed for that purpose have often been very complicated.

Here, we try to bring the two approaches together and show how a mechanical tester equipped with the capacity for a variety of jigs (including the equivalent of penetrometer probes) can be used to obtain much more information about primate foods than instruments that have a single function. In contrast to other accounts, we indicate

Table 24.1. Common mechanical properties that are pertinent to understanding the physics of primate feeding.

Property	Formula	Description
Stress	$\sigma = \dfrac{F}{A}$	The force (F) per unit area of a material (A).
Strain	$\varepsilon = \dfrac{dL}{L}$	The measure of how much a material stretches or compresses relative to original length. Calculated when strains are small by dividing change in length (dL) by original length (L).
Fracture strength	$\sigma_{max} = \dfrac{F_{max}}{A}$	The maximum stress a given material can withstand before fracturing, calculated at small strain by dividing maximum force (F_{max}) by an original cross-sectional area (A) the force acts on.
Yield stress	$Y = \dfrac{F_{yield}}{A}$	The stress at which a material stops behaving elastically and begins to behave plastically, calculated by dividing the maximum force at yield (F_{yield}) by original cross-sectional area (A).
Elastic (Young's) modulus	$E = \dfrac{d\sigma}{d\varepsilon}$	This is the material stiffness: its resistance to elastic deformation. It can be calculated from the slope of the initial linear region of a stress/strain graph.
Fracture toughness	$R = \dfrac{e}{A}$	The energy needed to propagate a crack through a given material. It can be calculated from the energy needed to produce the crack (e) divided by the area of one of the sides of the crack (A) that develops.
Poisson's ratio	$\nu = \dfrac{\varepsilon_{lateral}}{\varepsilon_{axial}}$	A measure of how much a material expands in compression or narrows in tension: the ratio of lateral to axial strain, where axial strain is that in the direction of the applied force.
Hardness	$H = \dfrac{F_{max}}{A_p}$	Not really a mechanical property as such, but results from an indentation test that give Y indirectly. It is essentially material resistance to permanent plastic deformation, measured at a variety of scales and defined as the maximum force (F_{max}) divided by the projected area of contact (A_p).

here the implications of our measurements in terms of the understanding of food acquisition and oral processing in primates. Are the mechanical properties of foods a factor in dietary selection by primates? What might the consequences be of consuming large quantities of foods to which the anatomy and physiology of the mouth of any particular primate species (i.e., the mechanical apparatus of that animal) is poorly adapted?

DEFINITIONS AND TERMINOLOGY

Terms used to describe mechanical properties have been complicated by a recent debate contrasting the language that primatologists and biologists, including the current authors, have been using to definitions that engineers have adopted (Berthaume 2016). Actually, conflict between these usages has not really centered on definitions or on how to measure the parameters described in those definitions. Instead, the focus of the debate has basically been about labels, particularly the meaning of the terms "hardness" and "toughness." As we define properties, we will indicate where the disagreement seems to lie.

Pertinent food properties to measure are those that resist the force imposed on a food object and that could be incorporated into a formula from which the force could be predicted. Resistance to breakdown is a function of external surface attributes (such as object size, shape, adhesiveness, and roughness) and internal mechanical properties (such as the elastic modulus, yield stress, and fracture toughness). Although this chapter is concerned mostly with internal mechanical properties, which we describe next, it is important to realize that these have to be combined with other properties (such as those at the food surface, which have a critical effect on friction) in order to properly understand the conditions under which foods break down.

Elastic Modulus

A force on a food object causes it to deform, where deformation is the movement of any given point within the object acted on, measured in the direction of the force. To predict the intensity of the disturbance to the food object, both force and displacement have to be normalized to the object's dimensions. The quantities of stress and strain achieve this (table 24.1). Stress is the concentration of a force, obtained by dividing the force by the area over which it acts. Strain is the ratio of the displacement to the object's linear dimension in the direction of the force. At small strains in plant tissues, the response is largely elastic, and stress and strain are often approximately linearly related. Accordingly, resistance at small deformations can be described by the ratio of the stress to the strain and termed (in engineering and materials science) the elastic or Young's modulus (symbol E; table 24.1). Strains are dimensionless, so the elastic modulus has the dimensions of stress. The Pascal (Pa) is an engineering unit designed for stress calculations, where 1 Pa equals 1 Newton per square meter. This is a very small unit, and so megaPascals (1 MPa = 10^6 Pa) or gigaPascals (1 GPa = 10^9 Pa) are employed to describe the modulus. In many plant tissues, although the stress/strain ratio is more or less linear at small deformations, there is an element of time dependence, meaning that a different value is to be expected if a test is conducted at another rate. This dependence can easily be exaggerated, but some estimate of how much this affects measured values is important.

Force can be applied directly onto an object, being tensile if it pulls it apart (direct tension being relevant to the use of the front teeth of colobines in "stripping" leaves) and compressive if it presses on the object (which is more relevant inside the mouth). The stress can also be oblique, as in sliding across an object; this is referred to as a shear stress. Virtually all loads induced by the teeth within the mouth will involve compression and shear. However, compression also results in indirect tension at right angles to the load because the food object (usually with a fluid content) also strains in this direction. In fact, there are very few loadings conceivable in the mouth that do not combine tension, compression, and shear. Normally, the elastic modulus is defined from tensile or compression tests.

Yield Stress

At a given stress, most structures fail to resist in a recoverable elastic manner and instead start to develop permanent distortions. The stress at which this happens is called the yield stress (Y; table 24.1). This stress is often difficult to measure directly and is inferred from indentation tests, wherein a small indenter is positioned over a flat surface of a specimen and impressed into it.

Indentation is one of the oldest mechanical tests and has been the subject of intense development over the past 25 years. Engineers call an indentation test a "hardness" test, and herein lies part of the controversy. We digress to address this.

Engineers have proceeded, consciously or not, by assigning terms that have been derived from everyday usage to describe the results of experimental tests designed to estimate individual properties. Some terms have a long tradition of specialized usage, such as, indeed, "hardness" (table 24.1). Usually, "hardness" to an engineer means "resistance to plastic deformation," but the term is actually linked to some variant of an indentation test, which can involve purely elastic deformation (such as with rubber polymers) or time-dependent elasticity. Though never overtly stated, such tests will make sense to a budding materials scientist or engineer in that the use of the fingers to squeeze an object is presumed to give rise to the sensation of pressure by which anyone, scientist or not, could decide whether that object is "hard" or not. Of course, the range of hardness that the fingers can determine is limited because the soft tissues of the fingers are also deforming, as much as or more so than the object under evaluation (Peleg 1980). A mechanical indentation test uses very hard machinery to indent a specimen and find out accurately the force that produces a particular surface area of deformed material lying beneath it (where this deformed area is measured in the plane of the surface). Hardness is simply the force divided by surface area; that is, it is a pressure (table 24.1). However, hardness is not really a property in itself, instead being shorthand for the yield stress, something that can be evaluated directly in tensile tests as the stress at which a material changes from elastic behavior to permanent (plastic) deformation. The hardness H is usually a small multiple of the yield stress, with $H \approx 3Y$ being a common approximation. This relationship was first worked out by Tabor (1950), a century or more after the indentation test was in routine service, and can be generalized to most engineering materials, as explained in Atkins (1982) or in textbooks such as Ashby and Jones (2006, 2011) and Gibson and Ashby (1999). In some cellular materials, it is thought that the yield stress and hardness can be more or less equal (i.e., $H \approx Y$) due to their compressibility (Wilsea et al. 1975), so the relation $H \approx 3Y$ is not necessarily generalizable to a lot of plant tissues.

There is, however, another relevant scientific tradition for investigating the use of terms to deform and break materials, associated with food scientists. Sensory assessments in feeding have been subjected to psychophysical investigation, with results being analyzed statistically. What does "hardness" mean during biting with the teeth rather than when deformed with the fingers? Evidence suggests strongly that hardness evaluated by subjects in experiments refers to fracture, in particular to the critical stress intensity factor, K_c (E. Kim et al. 2012; Vincent et al. 2002). This K_c parameter is also not in itself a property but a combination of the elastic modulus E and toughness R (defined below), where $K_c \approx (ER)^{0.5}$. Is it permissible to talk of "hardness" in two different ways? We say certainly, provided that it is clear in what sense the word is being used. No one is asking engineers to drop their traditions, but it would be difficult to change the minds of food scientists about their terminology, only recently so persuaded via experiments, without doing further studies.

Toughness

Another term to cause disagreement is "toughness." In its modern form, in materials science, engineering, and food science, this refers to fracture. The fundamental concept of fracture is now understood in energetic terms as the energy consumed in making a given surface area of a crack, symbolized as R. The units are Joules per meter squared, or $J\ m^{-2}$, with toughness values often being given in $kJ\ m^{-2}$ (i.e., $10^3\ J\ m^{-2}$). The understanding that energy is required for surface formation goes back a very long way in physics to the concept of surface tension of liquids (even if quoted as "force per length," it is actually understood physically as an "energy per unit area") or in chemistry to the surface energy of a solid. Unfortunately, there are two older usages of the term "toughness" that have nothing to do with fracture. One refers to the area under a stress-strain curve, which survives in some engineering textbooks, and the other to the area under a force-displacement curve (in dentistry; Darvell 2009). These usages should be replaced by the plain descriptive terms "strain energy density" and "strain energy," respectively. Stored elastic energy provides the drive for cracks, but a valid test needs to isolate the energy used in the fracture process from that which is producing deformation remote to the crack. To avoid confusion, we suggest referring to "fracture toughness" instead of just "toughness."

Unfortunately, this is not the end of it. In many parts of the world, materials scientists have decided that "toughness" should actually be defined as R, while in others, particularly the US, they have decided that "toughness" is actually K_c. K_c is the value of the (tensile) stress applied to a cracked specimen multiplied by the square root of the length of that crack at the critical moment when the crack starts to grow further. Its use requires linear elastic conditions, which is limiting, but as stated above, it is really two properties combined. The higher the value of E in a material, the tougher that material seems to be. Defined in terms of K_c, glass is about as tough as a common polymer such as nylon, when clearly their fracture behavior is very different.

Lucas, Turner, et al. (2000) and Lucas (2004) use the term "toughness" in a biological context to refer to a combination of properties, specifically $(R/E)^{0.5}$, rather than just to R. The reason for this is that any test of toughness in terms of R, the energy required to make a given crack area, is not going to be what any animal will sense or "measure" with their nervous system as a food fractures, because an animal will always also sense the excess energy used up in elastic deformation that toughness tests cleverly manage to exclude. The terms "hardness" as $(ER)^{0.5}$ and "toughness" as $(R/E)^{0.5}$ define two different conditions under which the fracture behavior of a food material could be detected in the mouth. "Hardness" refers here to the resistance of a food when the stress is limiting, while "toughness" defines a situation where the displacement rather than the stress is the limiting factor. These property groups emerge as that combination of properties critical for a food particle in resisting fracture under either limited stress or limited displacement regardless of the form of loading.

Is this extension of terminology allowable? We follow Popperian reasoning here, explained lucidly in the short biography of the philosopher Karl Popper by Magee (1973), which states that scientific hypotheses have to be testable and falsifiable in order to be valid. Furthermore, it is not always wise or productive in science to linger on the exact definitions of many fundamental concepts. The concept of mechanical "force" is a classic example of this. We see no reason for obsession over the "toughness" dispute provided that it is clear what is being measured and how properties are being related to the problem at hand—that is, how foods are broken by the teeth. Everyone is clear about what the elastic modulus means, understood at the atomic level, but no one is really sure about the deepest meaning of fundamental terms such as "force" or "energy." Therefore, adding "hardness" and "toughness" to that list seems acceptable. The important thing is to be clear about what is being measured and to demonstrate some value for this usage (ideally as a potentially falsifiable theory). We base our faith in our interest in understanding what induces animals to feed or deters them from doing so—that is, in terms of what they sense.

FUNDAMENTAL FOOD PHYSICS

Appropriate mechanical techniques for field use will depend very much on the questions being asked. We will try to derive a relevant strategy from observations that a fieldworker might make, formulating these observations into measurable concepts. Very commonly, fieldworkers may notice a food or foods that seem difficult for a primate to process. The deduction may be that such processing requires a lot of effort. Effort might equate here to "high forces" or to "lots of work," either of which will have to be estimated indirectly (just as in laboratory studies of humans and other primates). In lay terms, the food may also be said to be "sliced," "sheared," "crushed," or "ground." These are among a huge family of terms in English (with parallels in other languages) that connote fracture, but what they have in common is that they fail to denote (specify) the conditions involved and so are not useful for an analysis. It has been suggested that these terms relate to what are called "modes of fracture" (explained in Lucas 2004). When homogeneous materials are loaded, the resulting fractures can be defined in terms of crack path with respect to the direction of the load. There are three geometrical possibilities that lead to crack growth, one of which is tensile, while the other two involve shear. However, modes likely explain little for fracturing biological tissues because their complex three-dimensional microstructure constantly changes the mode. Instead of referring to modes as such, we lump them into one term, the "loading pattern," which we consider below.

If we take "force" as the governing criterion above, then in any mechanical activity, we might state that the force a primate applies will be resisted by some combination of event geometry, a group of material properties of the food object and the factors involved in resisting loading and consequent fracture:

Force ∝ geometry × material properties × loading pattern

Included as a resisting category, "geometry" would include surface attributes such as the extent of the contacting surfaces and the shape and size of the contacting bodies (e.g., tooth and food object). Under the category of "material properties" comes a small suite of properties that define resistance to deformation and fracture. "Loading pattern" refers to the direction of loading, such as compression or tension, or to the path of fracture as the food object breaks.

This may seem like a very difficult situation, but there are ways to analyze it. The geometry of loading is going to be apparent to an observer only during ingestion. If one particular type of feeding is the focus, then it may be worth designing equipment or a specific jig useful for analyzing this event. However, imitation will not by itself provide any basic understanding because it does not allow extrapolation from any one such event to another. For that purpose, material properties may be a vital element of testing. These properties come completely to the fore during mastication because the movement of particles around the mouth and their continual size reduction mean that geometrical parameters cannot be specified. In this case, the critical relation is with the properties involved, which, as we shall show, can be estimated in the field.

Putting this together so far, an attempt to get at the forces involved when a food particle is fractured is probably going to depend on measuring material properties of foods coupled with an estimate of the size of the event. However, this leaves the phrase "loading pattern" to be explained. In a material without any complex microstructure—something like a crystal or a nonoriented polymer—it makes little difference to its deformation whether it is compressed or subjected to tension. It can certainly make a difference to biological tissues because their internal structure is paramount in controlling their behavior. A simple example is the way in which compression pushes cells together such that in the end it is no longer individual cells bounded by their walls that matter but just the properties of the walls as they get pressed tightly together. The seeds in van Casteren et al. (2016) provide an example of this. Therefore, materials testing must try to use the same general direction of loading as a primate. It is sufficient to describe conditions in the mouth for this purpose as "compressive."

As an extra catch to all this, foods that primates eat contain a lot of water (so much so that some primates rarely drink), often having additional air spaces in their structure. In other words, foods contain gas and liquid, which are viscous and are likely to change property estimates with the rate of loading. How much so? Our first method (below) estimates this.

ACTUAL FIELD MEASUREMENTS

Elastic (Young's) Modulus

Until recently, it has been difficult to estimate the elastic moduli of food objects very effectively. This is because specimens generally need to be in the form of cylinders, cubes, or elongated strips of tissue for bending, compression, or tensile tests (fig. 24.1a). Descriptions of these tests can be found in Lucas (2004) with implementation in primatology illustrated by Yamashita (2003) and Dominy, Yeakel, et al. (2016). However, most food objects have to be shaped for this purpose, and doing this in the field can be difficult. There are prescribed limits on specimen shape for any test, significant departures from which will result in error. In addition, measurements are often thwarted by the fact that tissue relaxes significantly between making the specimen and performing the test. Such viscoelastic behavior is exhibited by most moisture-laden biological tissues, with ripening fruit flesh being a conspicuous example. Finally, a decision has to be made as to what region of the stress-strain graph to use for estimating the modulus. For a relatively linear stress-strain relationship (fig. 24.1b), it is important to specify the strain at which the modulus is estimated.

There is now an effective solution to these issues in terms of penetrometry. However, rather than using a probe to fracture a specimen, as is the norm in ecology or agriculture, the aim is to use very small displacements with a blunt hemispherical probe in order to measure the elastic modulus. The test is actually a version of an indentation test, now a very popular technique at a variety of scales in materials science and one rapidly gaining ground in biology. As described here, this blunt indentation test can not only provide an estimate of the elastic modulus but also indicate the extent of viscoelastic behavior in a succinct manner (Chua and Oyen 2009; Oyen et al. 2008). What we mean by this is that biological tissues, containing water and sometimes air, tend both to flow (particularly obvious when loaded very slowly) and to display elastic recoil. The resulting combination in a viscoelastic solid can investigated by a test that loads a specimen of tissue, holds that load at a specified deformation for a specified time, and then examines how rapidly the load declines. Operated like this as a load-relaxation test, blunt indentation offers three important measurements: (1) an instantaneous elastic modulus, called E_0—the predicted result if you could load a material in an instant, such as when teeth interact with food; (2) an infinite (fully relaxed) elastic modulus, E_∞, which is an estimate of elastic behavior under an infinitely slow loading regime; and (3) the average relaxation ratio, or the ratio of the above moduli (E_∞/E_0), which indicates how elastic the specimen is and effectively how rate-sensitive the food tissue is likely to be. This ratio allows an experimenter to investigate how a food may behave at a range of mastication speeds or ingestive behaviors.

There are two versions of the blunt indentation test, one for relatively thick blocks of tissue, such as fruit flesh, and the other for thin specimens, such as leaves. In both, the probe has a hemispherical head, but its radius will vary according to the specimen dimensions. If the tissue is thick, then the only necessary prerequisites for a test are a flat specimen surface normal to the load and a measurement of specimen thickness. The radius of the probe will ideally be <50% of specimen thickness, and probe movement in the test will be limited to about 10% of the same (Galli and Oyen 2008). At the start of the test, the probe is positioned as in figure 24.1c, left. The probe is advanced into the tissue for a short time (e.g., 10 seconds) and then stopped (fig. 24.1c, middle). Recording now continues with the displacement fixed but the load allowed to relax. After a defined period of such relaxation (e.g., 1–2 minutes), the force-time graph will appear as in figure 24.1c, right. A curve is then fitted to the relaxation part of the graph (shaded in fig. 24.1c, right). From this, E_∞, E_0, and their ratio can be obtained (Chua and Oyen 2009).

The other version of the test is designed for thin sheets, such as leaf specimens. A leaf specimen is held between two plates, just as in most penetrometer designs, with a circular disc of tissue exposed between aligned circular holes in the plates. The exposed disc is thus clamped at its circumference. The center of the disc is then contacted by a hemispherical probe, the radius of which must be less than ⅛ of the radius of the disc itself (fig. 24.1d). The test is operated similarly to that for bulk tissue, with similar curve-fitting for the relaxing tissue. However, there are two possible modes of deformation. The laminae of stiffer mature leaves seem to bend like plates, while those of younger ones may deform like a membrane in which the entire leaf is stretched (Talebi et al. 2016). The results can reveal whether the leaf specimen is behaving as a membrane or as a plate. As is the case for fruits, the analysis can give the two moduli, E_∞ and E_0, most accurately if the leaf behaves as a membrane.

What are the implications of modulus measurement for ecological and morphological studies in primates? The deformability of foods is crucial mechanical knowledge because of the way in which it influences the distribution of the load of a specimen over the dentition. This is shown schematically for a single tooth in figure 24.2. A food particle with a high elastic modulus, such as a seed, contacts a very small part of a tooth's surface (fig. 24.2a). As it breaks down, this area of contact will change, depending very much on the properties of the inside of the seed, but each individual contact will remain small compared to that of a particle with a much lower modulus, which would smother the tooth crown surface (fig. 24.2b). Dependent on the loads involved, fractures can be generated in the enamel near the point of contact in figure 24.2a. There are two possibilities: a crack may pass from the outside of the tooth inward, usually causing a chip, or inside-outward (Lee et al. 2011). The latter starts near the enamel-dentin junction under the point of contact. Neither type of fracture can result with the pliant food shown in figure 24.2b because the deforming food compresses the tooth crown, preventing such cracks from opening. However, the tooth is still vulnerable to fracture lower down the crown, away from the contact. These cracks are often seen in human teeth and result from the bulging of the interior of the tooth, similar to the way in which a barrel can break laterally by being vertically compressed (Chai et al. 2009).

Fracture Toughness

To measure the energy needed to propagate a crack through a material, it is necessary to fracture the material of interest, ensuring that no excess energy is included in

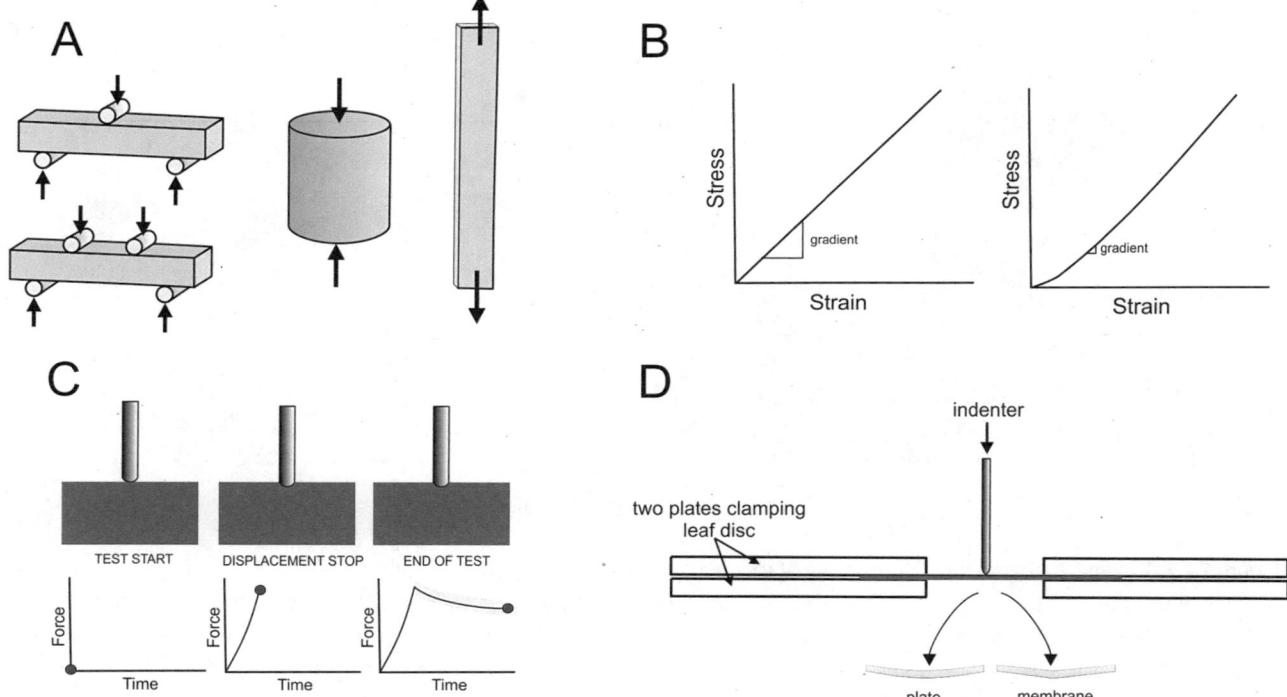

Figure 24.1. Tests to establish deformation behavior and the elastic modulus. *A*, Types of standard tests for obtaining stress-strain behavior in materials. *B*, Many biological materials do not have completely linear stress-strain curves. *C*, An alternative to standard tests in *A* is to do a load relaxation test with a blunt indenter. This test involves driving a probe into a sample for a given period (usually 10 seconds) and recoding the resultant force. Displacement is then stopped for a longer period (1 to 2 minutes), and the relaxation of the material is measured. Fitting a curve to this relaxation behavior allows the calculation of E_0 and E_∞. *D*, The variant of blunt indentation that can be used for a leaf specimen.

the calculation. This can be done in one of two ways: a crack can run freely through a specimen, or a controlled crack can be produced (van Casteren et al. 2016). Free-running cracks can be produced by artificially inducing a flaw, such as a notch, with a crack propagated from the notch tip by applying controlled loads. Bending tests (shown in fig. 24.1a) can be adapted to this end, with a very common type of such test for fracture shown in figure 24.3b,c. While championed by some (Berthaume 2016), this method has drawbacks. The crack will always take the path of least resistance through a sample, which may not be within the tissue of interest or a particularly useful analogy for primate feeding, where the teeth often dictate the path of a crack. The method often requires substantial sample preparation, which can be tricky when handling primate foods. Samples can be rare, present themselves as small food objects, and be moisture-laden, making them difficult to clamp or place in such bending tests (Atkins 2009; van Casteren et al. 2016). Alternatively, bladed cracking tests can be a much more useful testing scenario whereby blades or wedges extend cracks through a material, measuring the force and displacement needed to do so. A second pass of the blade or wedge is needed to account for frictional effects and any unused elastic energy, which otherwise could lead to an overestimation of toughness. Once the work used during fracture is isolated, dividing this by the crack area gives an estimate of toughness (Lucas, Osorio, et al. 2011). Most tests in the current literature of fracture toughness in primate diets utilize these controlled crack methods, as they require relatively little sample preparation and are very suitable for in-field tests of primate foods. For bulk material, such as fruit flesh, a narrow wedge (with 15° included angle) is used to produce a crack (Lucas, Osorio, et al. 2011). For foods in the form of rods or sheets, a single blade or two blades that pass each other can be used (Ang et al. 2008; Lucas, Osorio, et al. 2011).

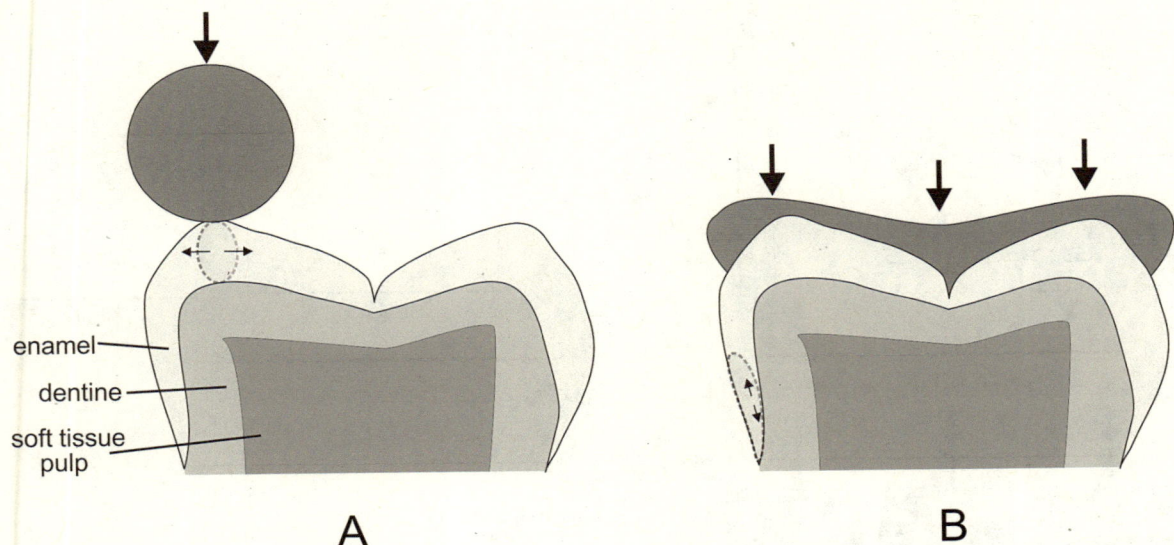

Figure 24.2. The influence that the deformability of foods may have on tooth form. *A*, A stiff food (shown as a circle) will have a small contact with a tooth, causing indirect and locally high tensile stresses in the enamel (in the dotted region, directed as the arrows) as it bends on the dentin. Radial cracks may start deep in the enamel at the enamel-dentin junction passing vertically up toward the food particle. This form of failure is resisted by thick enamel and a broad tooth crown, for which the tooth form shown is well adapted (J. Lee et al. 2011). *B*, A pliant food may deform over the tooth surface, spreading a highly compressive stress field that suppresses any local tensile stresses that could lead to cracking but leaves the cervical margins of the enamel susceptible (e.g., in the dotted region shown). The cause of margin fractures is the incompressibility of the pulp that presses the dentin outward, producing "hoop stresses." The tooth form shown is poorly adapted to resist such margin cracks, for which a cingulum (bulging of the tooth in the cervical region, totally lacking in the tooth shown) is superior (Lucas, Constantino, et al. 2008).

However, recently these types of test have garnered criticism (Berthaume 2016), focused on two main points: (1) that the apparent different modes of fracture involved in different bladed tests make them not comparable and (2) that toughness measured like this is not a material property, as it depends on specimen thickness.

To address point 1, while it is true that in homogenous engineering materials fracture toughness often varies with different modes of fracture (i.e., with the geometry of crack), this effect is difficult to justify in biological materials as one has to consider a three-dimensional cellular framework, such as those found in plant tissues. As a crack progresses through such a structure, the fracture path will constantly vary, so inducing different modes of fracture at the crack tip. Up to the level of a two-dimensional mat, the influence of the mode of fracture can be analyzed (e.g., Lucas and Pereira 1991), but for three-dimensional structures such as plants (the staple foods of primates), it makes little sense. We can think of this with respect to the defense of a plant: why would a plant allow its defenses to be weak with respect to one direction of attack? It is fracture mechanisms, not modes, that predominate in plant structures.

The principal mechanism that matters is whether cracks pass through cells or around them. Figure 24.3a gives an overview of this with respect to the volume fraction of the tissue occupied by the cell wall (V_c). Plant tissues with primary cell walls are the principal foods for primates. With generally low V_c, these tissues almost always fracture within cells (intracellular fracture), probably because intercellular fracture paths are generally more expensive, as shown in figure 24.3a. Cells with secondary cell walls are "woody" tissues that act as mechanical supports or defenses—in both respects, they are deterrents to feeding. Primates must break through such tissues in a variety of situations—for example, when an aye-aye breaks through bark to obtain larvae, when seed-eaters break through a woody shell to eat the endosperm or cotyledons, or when any leaf-eating primate eats all but the very youngest leaves, needing to contend with woody tis-

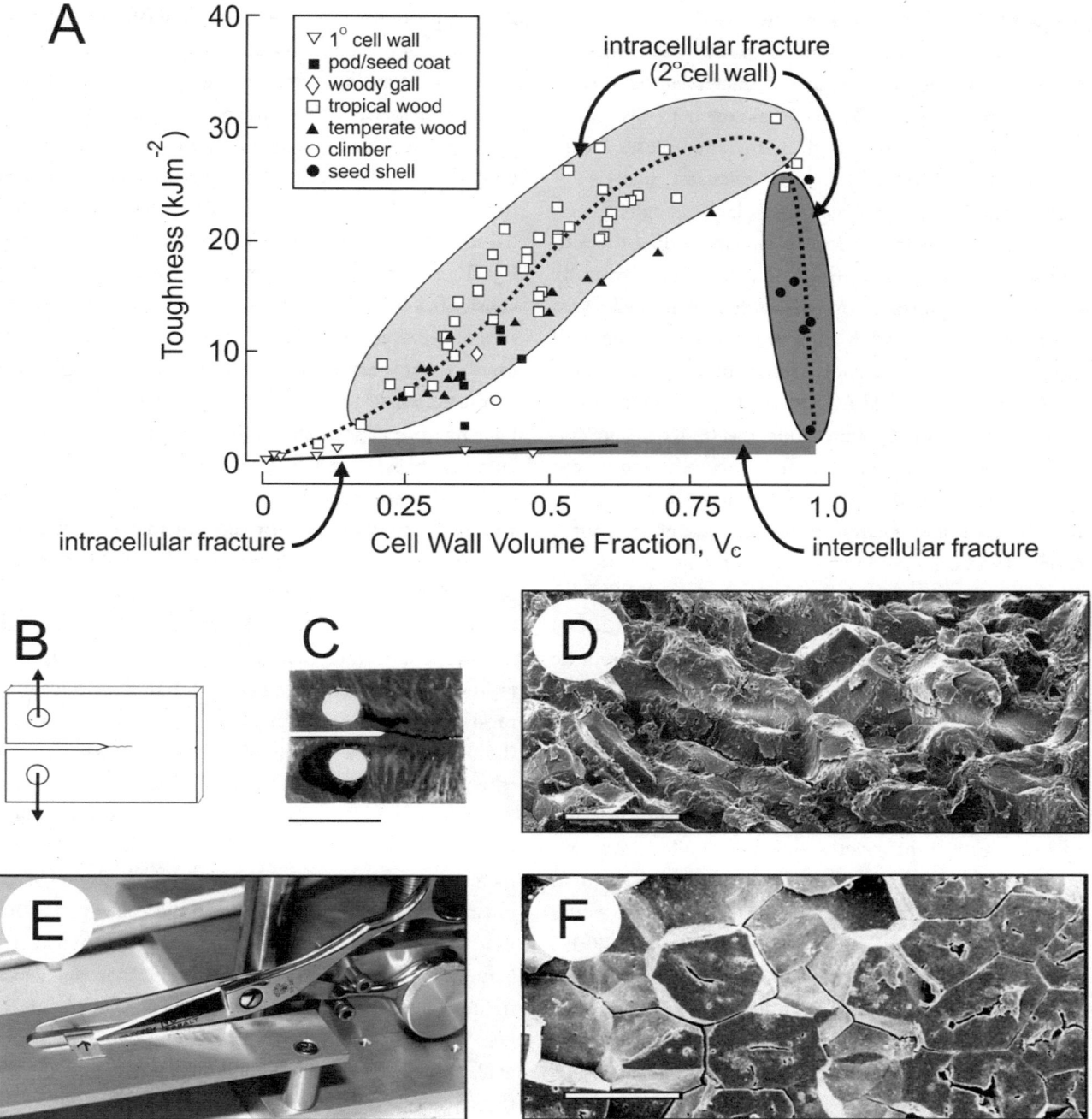

Figure 24.3. The crucial effect of the crack path on toughness in cellular materials. *A*, An overview of the toughness of plant tissues (partly schematic). Toughness depends critically on the amount of cell wall by volume in a tissue (called V_c in the figure). Free-running cracks in tissues with only primary cell walls nearly always break through cells, this being a cheaper fracture than intercellular failure. Such tissues have a low V_c. Cells with secondary cell walls tend to fail intercellularly when possible. *B*, A popular type of specimen in materials science called a "double cantilever beam" in which a notched specimen is pulled apart, so growing a crack to the right as shown. *C*, A real specimen derived from a seed shell produced in a laboratory, which has gradually failed to completion from a notch. Scale bar, 5 mm. *D*, The resulting crack surface from the specimen in C shows a "rubbly" fracture surface in which the crack has tracked between cells. Scale bar, 50 μm. *E*, An alternative test in which blades (on scissors) direct the crack on the same type of specimen. *F*, A fracture in which the crack has passed through cells rather than between them. Scale bar, 50 μm.

sue particularly disposed around leaf veins. In all but the lightest woods, cracks pass preferentially between cells (intercellular fracture) because intracellular fracture is much more costly (fig. 24.3a). That cost depends on the expression of a fracture mechanism called plastic buckling, which is exhibited by all but the densest (highest V_c) tissues. The latter, largely seed shells, lack enough of a cellular lumen (a central cavity) to buckle, leading them to be much less tough than woods that are often identically structurally organized. For cells with primary cell walls, it matters little whether fracture tests are conducted with or without blades (i.e., involve free-running or controlled cracks). So, for example, tests on the flesh of common apple varieties or carrot obtain similar results. Khan and Vincent (1993) found very little difference between wedge and notched tensile tests of apple pulp. Although there were marked differences in toughness, R, between apple varieties and orientation of a crack in the flesh, values recorded for specific varieties and orientations did not differ greatly between testing methods. Alvarez et al. (2000) used single-sided notched bending tests to measure the fracture toughness of several plant tissues, finding that apple pulp had a toughness of 39.6 (SD 10.5) J m^{-2} and raw carrot, 397.9 (SD 75.1) J m^{-2}. S. Williams et al. (2005) investigated apple pulp toughness with a scissors test and recorded very similar values of 57.0 (SD 17.8) J m^{-2}, despite using different varieties of apple. Similarly, Agrawal et al. (1997) recorded fracture toughness for raw carrot as 440.0 (SD 47.5) J m^{-2} using a wedge test, similar to values recorded using the scissors test by S. Williams et al. (2005) at 343.9 (SD 48.5) J m^{-2}, both studies essentially agreeing with tests that allow a free-running crack (Alvarez et al. 2000). There are other pertinent studies that could be cited including microtoming (Atkins and Vincent 1984). Variation in toughness in primary-cell-walled tissues, due to varying moisture content, turgidity (McGarry 1995), or the woodiness of older versus younger carrots (including cortex vs. medulla), must be anticipated, but we see nothing here that could be attributed to variation in the "official" mode of fracture in these tests. This is why simple cutting tests have been advocated for foods by both engineers and biologists.

For cells with secondary cell walls, the choice of cutting versus free-running tests makes a large difference, as described in figure 24.3. A free-running crack in a tissue of V_c = 0.95, produced by the technique shown in figure 24.3b,c, produces intercellular fracture (fig. 24.3d) and is a far cheaper option than a cutting test (fig. 24.3e), which produces intracellular fracture (fig. 24.3f). Why then do a cutting test on such tissue? All depends on what the primate does during feeding. Leaf-eating primates with crested molars cut through leaf veins in a manner that free-running crack tests cannot emulate. Very often, woody tissue is composed of fibers (long, thin cells of very high V_c). There is no other way to fracture such tissue without substantial intracellular fracture. The sharp-bladed incisors of aye-ayes and rodents represent tooth designs that force their way across cells in tissues of very high V_c. Many seed-eaters have blunt teeth in order to produce intercellular fracture in seed shells. Tests need to be tailored to circumstances.

In addressing point 2, it is necessary to observe that when a crack evolves in virtually any material other than mica or graphite, energy is expended not just in the cleavage of surface area but also on a zone around the crack tip, which becomes distorted (plastically deformed). The volume of deformed material around the crack tip often represents a greater sink of energy than does the surface production itself. So, if the production of cracks involves the distortion of a volume of surrounding material, then when tissues are tested at a thickness below that of the radius of this zone (i.e., below the radius of the volume of distorted material), the cost of producing a crack in that material will be correspondingly reduced. The effect is a linear dependence of toughness on thickness, up to a limiting thickness when the toughness will plateau off. This is known for many materials. For plant materials (plant tissues), the dependence of toughness on specimen thickness is usually shown below 0.5–1.0 mm (Lucas, Turner, et al. 2000). It seems to depend on cell size, though, such that tissues with very large cells, such as watermelon, show a dependence on thickness up to about 4 mm (Lucas et al. 1995). This cell-size phenomenon is well known in materials science for foamed materials. Berthaume (2016) dwells on the difference between results with scissors and those with a wedge. We attribute this to the thickness effect with scissors on thin specimens, particularly leaves with their laminae <1 mm. With thick-enough specimens, scissors values plateau to approximate those with wedges (Lucas et al. 1995). What you get with scissors is the toughness of that material in an object of that size.

Importantly, in structured materials (virtually any biological tissue), there is also a dependence of toughness on the length of a crack. This is seen because cracks encounter little resistance as they start off, but once they grow to a certain length, they start to encounter higher structural resistance in the tissue. So yet again there is a curve with short cracks that plateaus as cracks elongate. These are referred to in materials science as "r-curves." The mechanisms that produce this phenomenon are understood in many biological structures and are often replicated in bio-inspired, tough man-made materials (Munch et al. 2008). Both the thickness effect and the effect of crack length on the toughness of biological materials are well known, and their existence does not detract from the fundamental form of toughness as a concept.

What follows is some practical advice for the budding field experimenter to maintain accuracy in results and sanity during testing. It is preferable to perform cutting tests with a sharp blade free of major defects; keep the blade clean between tests, as this reduces the effects of friction and adhesion, allowing for more accurate estimations of fracture toughness (alcohol wipes are excellent for this purpose). It is also important to take care while producing samples for testing: ensuring that edges are straight and parallel enables the accurate measurements of fracture areas. If possible, use cutting aids, such as mandolin cutters or set squares, to ensure true geometric shapes for cutting. Most of all, do not rush, as a few high-quality results are far better than many low-grade ones.

Why measure toughness? Quite simply, together with the elastic modulus, it is the essential property to measure in order to gauge the resistance of a food to breakdown.

DIRTY FOODS

Foods are often viewed as a standalone entity, but they are produced and foraged as part of a wider environment, and this can affect the state of the food when ingested by a primate. One such variable that may be of chief concern to primates while feeding is the extent of external abrasive pollutants, as these contribute substantially to tooth wear. It is unlikely that food is clean; the silica and silicates that are prevalent all over the surface of the earth (Lutgens and Tarbuck 2000) will probably be present on foods that are derived terrestrially, and even foods that are elevated may collect windborne dust and grit typically made up of particles <100μm.

The reason such external pollutants could be a threat to primate teeth is down to their mechanics and how they interact with enamel. Evidence indicates that if teeth wear substantially, their functionality decreases, and this is likely to decrease the fitness of an individual (Cuozzo and Sauther 2006; S. King et al. 2005). While tooth wear can happen at a variety of scales (Lucas and van Casteren 2015; Ungar, this volume), by far the most prevalent is that of microwear. This process is the result of extremely small particles and multiple contacts with enamel either moving or removing material via rubbing or abrasion, respectively (Lucas, Omar, et al. 2013).

The mechanics of microwear is an area of research that is both fast paced and somewhat fractious, with competing views and theories on the efficacy of various external particulates to wear down teeth (Constantino, Borrero-Lopez, et al. 2015; Lucas, Omar, et al. 2013; Rodriguez-Rojas et al. 2020; van Casteren et al. 2020; Xia, Tian, et al. 2017; Xia, Zheng, et al. 2015). Using controlled experimentation, where naturally occurring particles were placed into sliding contact with enamel surfaces, Lucas, Omar, et al. (2013) were the first to progress a mechanical theory of microwear. They concluded that the effectiveness of a particle in abrading enamel is governed partly by the inherent mechanical properties that the particle possesses and partly by the critical angle of attack, which is dictated by the shear yield stress and toughness of the material (Atkins and Liu 2007; Lucas, Omar, et al. 2013; fig. 24.4a). According to this model of microwear formation, if a wear agent placed into sliding contact with a material possesses a higher hardness than the material being removed and also has a sufficient angle of attack at contact, then instantaneous removal of material by abrasion can result (fig. 24.4c). While plant-derived particulates such as phytoliths will be numerous in the environment, their mechanical properties do not lend themselves to such instantaneous wear (Kaiser et al. 2018; Lucas, Omar, et al. 2013; Sanson, Kerr, and Gross 2007; van Casteren et al. 2020). This is because of their lower hardness than mammalian dental enamel (Kaiser et al. 2018). Phytoliths may mark enamel by moving material around plastically (fig. 24.4b) but are unlikely to instantly abrade (Lucas, Omar, et al. 2013). It is indeed possible that this movement induces instability in a material, but it will not directly remove material, and therefore the rate of wear is likely much lower than that of other harder particulates that di-

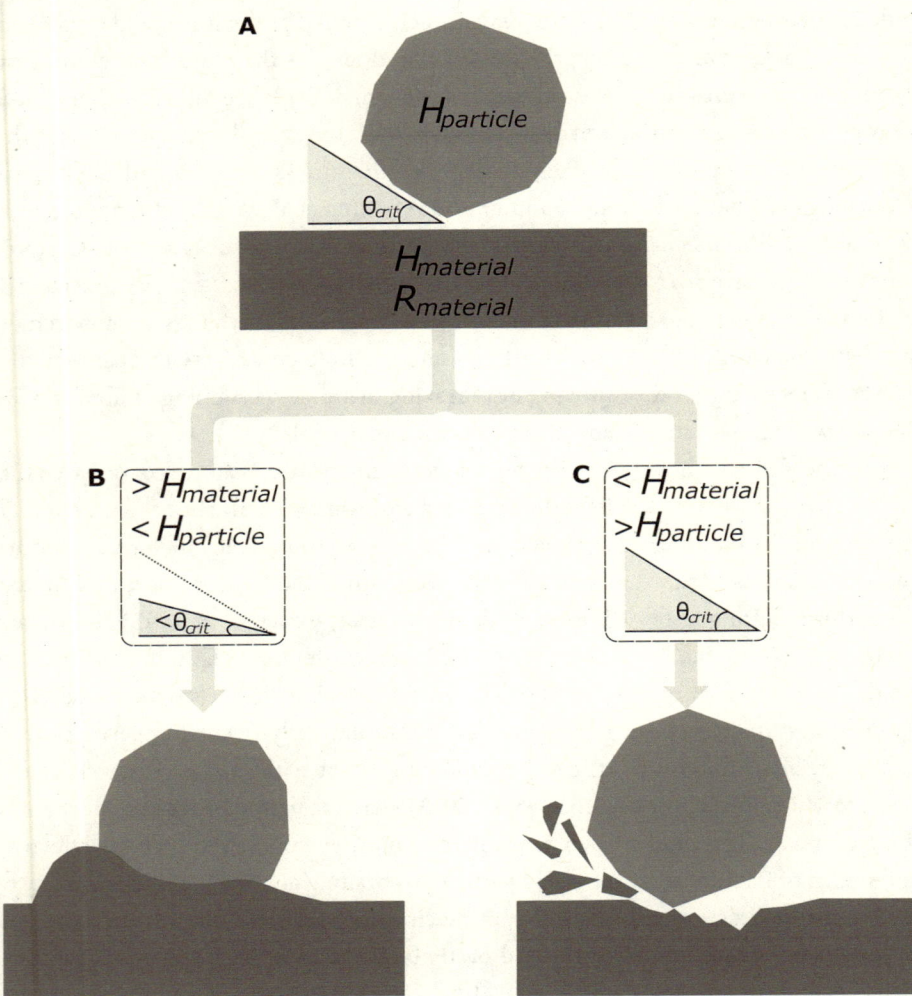

Figure 24.4. The main conditions under which a wear particulate can produce instantaneous wear. *A*, How a particle will interface with a material is dictated by two main factors: particle hardness ($H_{particle}$) and the critical angle of attack of particle asperities (θ_{crit}). θ_{crit} is dictated by the toughness ($R_{material}$) and shear yield stress ($H_{material}$) of the material being worn. *B*, If the abrading particle has a hardness lower than the material being worn or the angle of attack is lower than θ_{crit}, the material is deformed plastically rather than instantly removed. *C*, Conversely, if the abrading particle is of a higher hardness than the material being worn and the θ_{crit} is reached, then chips will break off and instantaneous wear will occur.

rectly abrade. External pollutants in the form of dust and grit are, however, many times harder than enamel (Lucas, Omar, et al. 2013; Lucas et al. 2014) and once fractured in the mouth are likely to contain the angularity needed to instantaneously remove enamel (Lucas et al. 2014), making these much more damaging to teeth than softer biologically derived particles.

However, Xia and colleagues (Xia, Tian, et al. 2017; Xia, Zheng, et al. 2015) claimed that abrasion of enamel can occur regardless of particle mechanical properties, so long as the contact pressure be sufficient to break the protein bonds holding enamel crystallites together. This model of microwear mechanics was formulated from a series of sliding experiments using metallic ball bearings and spheres of silicon dioxide instead of naturally occurring particulates like grit or phytoliths. Xia and colleagues (Xia, Tian, et al. 2017; Xia, Zheng, et al. 2015) ostensibly demonstrated that softer materials, lacking any obvious angularity, can produce extremely small enamel chips, seemingly produced by abrasion. The use of metallic spheres instead of naturally occurring wear particulates has been criticized by van Casteren, Lucas, et al. (2018). They demonstrated that oxide layers and work hardening of metals increase the hardness of the ball bearings used in the wear experiments of Xia, Zheng, et al. (2015) to levels above that of enamel. This suggests that chip formation followed a similar action as predicted by Lucas, Omar, et al. (2013). In addition, further sliding experiments using particles of woody seed shells failed to achieve instant abrasion, further corroborating predictions of the Lucas, Omar, et al. (2013) model of wear formation (van Casteren et al. 2020). Despite such lively debate, a universally accepted theory of microwear formation has yet to be proposed, and clearly more experimental work is needed to fully elucidate the underlying mechanics of dental microwear.

Research into primate feeding does now, in general, accept the role of grit and dust as a main protagonist in the wear of dental enamel (Fannin, Singels, et al. 2021), and recent work is highlighting the influence that these microscopic pollutants can have on primate feeding ecology. High concentrations of quartz in the soil are thought to produce high dental wear rates in mandrills (Galbany, Romero, et al. 2014). An increase in windborne dust has been linked to reduced chewing efficiency and increased tooth wear in western chimpanzees (Schulz-Kornas et al. 2019). Populations of howler monkeys in Nicaragua that live downwind of active volcanoes and must endure volcanic ash events have significantly more premolar wear than populations that do not (Spradley et al. 2016). It has even been suggested that the washing, brushing, and rubbing of food seen in terrestrial monkeys such as macaques and baboons is a behavioral device to avert grit ingestion (Fannin, Singels, et al. 2021). Yet interest in the understanding of these geologically borne particulates compared to those derived from plants is somewhat lacking.

While some studies have begun to measure the levels of external wear agents in primate diets and environments (Geissler et al. 2018; Ungar et al. 1995), there are still large gaps in our knowledge on the amount, composition, and morphology of these abrasives. If we are to learn more about the role of dust and grit in the feeding ecology of primates, field measurements are key, as these will give researchers firsthand evidence relating to the degree of contamination and its prevalence in individual niches.

In the field, there are three main types of measurement strategy. The first is passive dust collection, generally in the form of dust traps designed to collect airborne particles that may be widespread in the atmosphere. These traps can be set in varying locations to assess gradients in dust and grit concentration. Traps of this type were pioneered in a primatological context and used to great effect previously by Ungar et al. (1995), demonstrating the amount and sizes of atmospheric particulates found in both dry and wet forested areas. However, one weakness is that these traps are completely passive and take no heed of the plants—foods that may be extracted from gritty substrates or employ strategies that concentrate grit on their surfaces as an act of defense from herbivory (Lucas et al. 2014). Therefore, more recent studies tend to clean surfaces of foods eaten by primates to get a handle on the level of contamination these items present (Bender and Irwin 2014; Geissler et al. 2018). Care must be taken during cleaning as contaminants are likely to be very small, and it is essential that the cleaning method have the ability to collect particles of varying size ranges that may span several orders of magnitude. It is also possible to look at the feces of primates, which can be cored (to avoid contamination) and then ashed to remove the biological material. The remnant will consist just of the mineral component of the feces. This method allows one to see the actual ingested wear particulates. However, care must be taken when partitioning the mineral remnant of feces, as deciphering what mineral particle is plant derived and what has come from environmental sources is not always such an easy task. Luckily, novel methodologies have now been developed that employ heavy liquid floatation to separate and quantify crystalline and amorphous siliceous particulates, and these techniques have shown promising results in studies of wild and domesticated North American herbivores (Fannin, Laugier, et al. 2022). All three methods have advantages and weaknesses, so often a multifaceted approach will probably lead to a greater understanding of external wear particulates and how the mechanics of these could influence primate feeding and foraging.

Mammals such as primates appear to invest a great deal in protecting themselves from contamination by dust. As examples, hairs can be useful in intercepting particulates, and reflex actions such as sneezing can expel particles that have entered the upper respiratory tract and oral cavity. Histamines in the body can react to dust that actually breaches the integument, and an ever-present fluid coating in these chambers operates to trap particulates from membranes and surfaces, moving them in the respiratory tract via cilia. These form a suite of protective physiological adaptations (Madden 2014). Therefore, it is not unreasonable to expect that some other behavioral adaptations may also reduce the ingestion rate of potentially dangerous wear particulates during feeding. Broadly speaking, a primate must both detect and avoid abrasive pollutants. Detection will likely be a physiological response. Inside the mouth, it has been suggested that mechanoreceptors in the periodontal ligament are tuned specifically to detect forces in the very low Newton range, similar to those that are likely at contacts between teeth and small particles (Lucas et al. 2014; Trulsson and Essick 2010). This, coupled with an audible "crunching" fracture sound, could provide an early warning system for the potential of enamel loss through wear. Having said this, the study of both the perception and avoidance of such threats is in its infancy and requires more research (Lucas et al. 2014). There are also limited data that pertain to how primates may remove or avoid mineral particles in their daily forage, although this limitation is more telling about a lack of investigation than a lack of congruence between mineral contamination and extraoral feeding behavior. Videos of such behaviors not only will be useful for understanding how primates ingest and process foods orally but also may reveal behaviors linked to the cleaning of foods such as brushing or washing. While the wholescale mechanics of food is no doubt an important influence in primate feeding ecology, the contribution of particulate micromechanics and their effect on teeth is still to be fully understood and presents an exciting frontier for field researchers to cross.

THE INFLUENCE OF IN-FIELD MECHANICAL TESTING ON PRIMATOLOGY

In-field mechanical testing using similar methods to those showcased in this chapter has had considerable influence on the field of primatology. While it is beyond the scope of this chapter to review the primate food mechanics literature exhaustively, we have selected three areas of research where we believe dietary mechanics and in-field mechanical testing have been pivotal to furthering the field.

Food Mechanics and Tooth Morphology

Teeth are essentially agents of fracture, and the link between dietary mechanics and tooth morphology has been demonstrated theoretically and practically (Lucas 2004; Ungar, this volume). However, more data on specific food mechanical properties are needed to elucidate the complexities between dental form and food physics, especially given recent findings that traditional broad food categories such as folivory and frugivory are not necessarily associated with specific mechanical attributes of foods that primates eat (Coiner-Collier et al. 2016).

Several studies have incorporated estimates of food mechanical properties collected directly from the field in analyses of the musculoskeletal adaptations of primates. Wright (2005) showed that the craniofacial robusticity of capuchins, particularly *Sapajus* (previously *Cebus*) *apella*, as compared to other cebids and atelids, could sustain higher anterior tooth bite forces. This matches dietary data that B. Wright (2005) collected, showing that *S. apella* could sometimes target harder, tougher foods and breach them using the anterior dentition. E. Vogel, van Woerden, et al. (2008) compared the mechanical attributes of diets of single populations of chimpanzees and Bornean orangutans with respect to their molar morphology. Orangutans have thicker enamel with steeper molar-cusp slopes than chimpanzees, which have a much flatter occlusal surface and thinner enamel. Mechanical measurements on ingested foods demonstrated that orangutans, in general, consumed foods of a higher stiffness and toughness than those of chimpanzees. The researchers attributed this ability to the above dental traits. The reduced enamel thickness of chimpanzees is likely adaptive in that wear produces sharp-angled crests that can cut foliage, while the whole tooth retains a degree of flatness for compressing fluidal mesocarp. Vogel et al. (2014) have gone further and found significant differences between the diets of Bornean and Sumatran orangutans that match differences in their mandibular morphology. An intraspecific comparison in chimpanzees demonstrated some differences in mechanical demands of foods in the diet of forest (Ngogo, Uganda) versus savanna (Issa Valley, Tanzania) populations (van Casteren, Oeleze, et al. 2018). A recent study demonstrated that western lowland gorillas, whose diets are traditionally characterized by large amounts of fruit and tough, fibrous plant material, are surprisingly consuming the seeds of *Coula edulis* for three months a year. These large, hard, and mechanically challenging seeds are cracked open orally by gorillas, and mechanical measurements of the seeds suggest that in doing so they are regularly taxing their dentition to its upper limit (van Casteren, Wright, et al. 2019). In all these studies, the association of anatomical form with diet is made more concrete via quantification of the relevant aspects of diet.

Ingestive Biomechanics and Food Selection

Understanding how food mechanical properties influence the feeding behavior of primates is key to understanding the impact that the diet exerts not only on the feeding system but also on the wider suite of daily actions performed by primates. A series of studies on robust capuchins (*Sapajus* spp.) elegantly demonstrate how combining measurements of food mechanical properties and observed feeding behavior can help researchers decipher the level of influence that diet can exert on the daily life of a primate. In robust capuchins, ingestive behaviors seem to be food specific and driven, at least in part, by food mechanical properties and geometry (Laird, Wright, et al. 2020). Feeding postures were less varied when capuchins accessed foods of a higher mechanical challenge (Laird, Punjani, et al. 2022). Such results help morphologists pinpoint where and why functional adaptations may occur.

Food mechanical properties may also contribute insights about the wider societal makeup of primate species. Chalk-Wilayto et al. (2022) demonstrated that juvenile robust capuchins had reduced processing efficiency when accessing embedded foods that had high stress-limited indices. This implies that skill development needed to exploit such mechanically challenging dietary items may be an important constraint. Falótico, Valença, et al. (2022) researched stone tool use in three robust capuchin species, finding that tool selection was not correlated with food mechanical properties. Instead, the authors suggested that a mixture of raw material availability and cultural factors influenced the size and shape of stone tools used. In both of these examples, well-characterized food mechanical properties have permitted a more nuanced understanding of primate societies.

Primates may only identify fruits and leaves as food when these have specific characteristics, which in lay terms could be called "ripe" or "young," respectively. Since these items are not available year-round, primates need to have the sensory capacity to identify these edible stages of development (Dominy, Lucas, et al. 2001; Melin and Veilleux, this volume). In the context of fruit consumption, Dominy, Yeakel, et al. (2016) demonstrated that the critical sensory cue for the consumption of figs (of one species important in the diet of primates in Kibale) was likely not their color but rather mechanical information, as garnered from manual manipulation (squeezing related to the elastic modulus) and incisal biting (related to toughness). This represented a reliable proxy for the fructose content of a fruit. Glowacka et al. (2017) studied fracture toughness in the diet of mountain gorillas, suggesting that these primates can assess fracture toughness to select against it. This mechanical attribute was shown to be a better predictor of consumption than plant frequency, density, or biomass within an environment, indicating that this property imparts a hefty influence on mountain gorilla food selection. Furthermore, they suggested that gorillas are morphologically adapted for repetitive mastication of low-toughness plant material (Glowacka et al. 2017). Avoidance of high fracture toughness in leaves has also been observed in other primates in that they eat only the terminal parts of leaves, which are the least tough (Teaford, Lucas, et al. 2006).

Such investigations have now been extended to the deformability of leaves (Talebi et al. 2016). Young leaves are of a low stiffness and behave like membranes, exhibiting a great deal of floppiness compared to stiffer older leaves, which act like plates of material and maintain constant angulation of the leaf lamina to the sun. Such floppiness is a salient mechanical signal indicating the fracture toughness of leaf material, allowing primates in a forested environment to easily identify leaves presenting a low masticatory challenge and high protein content. This finding provides an indication of how a primate may judge the age of a leaf without having to continually monitor cohorts of leaves. Added to this, a membranous leaf may be "balled up" in the mouth, allowing more effective breakdown by nonbladed teeth like those possessed by frugivores who fall back on leaves.

Extinct Primate Feeding

While it is impossible to measure the mechanical properties of foods consumed by long-extinct primate species, it is necessary to understand the mechanical relationship between tooth and food in order to reconstruct diets from fossil remains. Primates provide excellent extant models for understanding the feeding systems of fossil hominins, yet these are just models and unlikely to provide an exemplary fit for extinct hominins (see further discussion in Ungar, this volume). This has meant that field primatologists have on occasion turned to modern-day dietary analogues, not necessarily consumed by modern-day primates, to gather mechanical data on predicted hominin diets.

One such food resource is underground storage organs (USOs), which have often been presented as a possible fallback food for fossil hominins. Dominy et al. (2008) sought to address the plausibility of such a prediction by gathering mechanical property data on USOs from sub-Saharan Africa. The results were consistent with dietary predictions based on the dental morphology, microwear analysis, and isotopic signatures of hominins, indicating the suitability of USOs as an important food. Alternative interpretations of hominin microwear signals and isotopic signatures have promoted the habitual consumption of grass stems and leaves, although such a diet would appear to be at odds with hominin tooth morphology. However, Paine et al. (2018) combined data on the nutritional content and fracture toughness of grasses and forbs from several African microhabitats. They found that some grasses showed a high protein content at a moderate fracture toughness, comparable to the leaves consumed by extant apes. While there are still unanswered questions about the efficacy of grass leaves as a main energy resource for large-bodied hominins (Yeakel et al. 2013), the authors claim these results indicate that grasses are a more plausible hominin food source than previously thought, widening the debate on what foods were driving the selection of megadontia in robust australopiths.

The incorporation of food mechanical property data in primatology has been widely used, and its influence on informing the studies of primate form, function, and behavior cannot be overlooked. Yet there is still much to learn from dietary mechanics, and with the introduction of more advanced testing equipment, the mechanical nuances of primate diets are likely to be increasingly elucidated, allowing for a clearer understanding of extinct and extant primate feeding and foraging.

SUMMARY AND CONCLUSIONS

Field mechanical measurements represent a growing and interesting area of research that allows researchers to peer further into the environmental influences driving primate morphology and adaptation in a manner that food chemistry alone cannot. In recent years, more accessible and easier-to-use methods have expanded data on the mechanics of primate diets, but there are still large descriptive holes in our knowledge that can be plugged only by further field experimentation. The approaches and methods we have advocated here allow the accurate quantification of key mechanical properties measured directly from primate foods. These data facilitate broad investigations both across and within primate species more intricately at the interface between teeth and the foods they process. We hope that field methods become even more simplified in the future, but whatever is developed is likely to require both force and displacement sensing and be displacement controlled (i.e., be like a universal testing machine).

Habitat, season, environmental disturbance (human or natural), and even social variables can influence the dietary mechanical landscape of primates. To understand functionally how these mechanical challenges influence ingestive behaviors and shape morphological characteristics, the mechanical makeup of foods is key. Food physics is a vital link between feeding behaviors observed in wild primates and the anatomical nuances and life-history clues conveyed in their paleontological remains. These data help us understand why certain foods are targeted by particular species and how these tissues are broken down. We can probe the dietary plasticity afforded by dental characteristics and explore the scenarios under which dental damage may occur. Whether one is trying to decipher feeding in extant primates or trying to piece together the clues extinct ones have left, without data on food mechanical properties, our understanding will remain vague and descriptive.

- There are still large gaps in our knowledge of the dietary mechanics of primates.
- The methods and approaches we outline here allow for the measurement of fundamental material properties comparable across biological tissues.
- Such properties can be used to investigate diverse dietary questions ranging from population-level studies to those concerning the direct interactions between teeth and food.
- Understanding food physics can lead to a deeper understanding of feeding behavior and morphology in both extinct and extant primate species.

25 Modeling Primate Nutrition

David Raubenheimer

As the many chapters in this book demonstrate, nutrition is closely associated with just about every aspect of biology, from biochemistry and physiology to behavior, morphology, life history, and the populations and ecological communities in which organisms interact. Nutrition therefore offers powerful opportunities for understanding the patterns and origins of biological diversity.

The immediate opportunity is that nutrition can contribute significantly to understanding the traits of animals, including mechanistic, functional, and ecological aspects. It can, for example, help explain sexual size dimorphism (Noonan et al. 2016), group size (Markham and Gesquiere 2017), the distribution of visual pigments (Melin, Hiramatsu, et al. 2014), patterns of movement (C. A. Johnson, Raubenheimer, Chapman, et al. 2017) including migration (Nie et al. 2015), key life history traits such as growth rates (J. Jones 2011) and variance in natural life span (Mattison and Vaughan 2017), and much else besides. More ambitious, but no less important, is that nutrition provides an integrative thread that can link aspects of an animal's biology that are normally considered the realms of different subdisciplines or disciplines (Zeisel et al. 2001). For example, it can help us understand the links between physiological homeostasis, life histories, foraging behavior, evolution, and population and community ecology (Raubenheimer, Simpson, and Tait 2012; Sperfeld et al. 2016). This integrative capacity of nutrition, I believe, might even turn out to be to mechanistic biology what natural selection is to functional biology: a framework that encompasses all of biology (S. Simpson and Raubenheimer 2012). In combination, the mechanistic reach of nutrition and the adaptive logic of natural selection make a formidable pairing.

There are, however, substantial challenges to overcome if nutrition can reach its potential and fulfill these roles. Nutrition is immensely complex, involving many food components that combine and interact in diverse ways in their links with animal physiology, development, behavior, and ecology. Furthermore, just as chemical components combine to form foods, so too do foods (and their chemical components) combine into meals, meals into diets, and ultimately diets into population-level dietary patterns. For many research purposes, these different levels in the "mixture hierarchy" that comprises nutrition are not substitutable, because they play distinct roles in the biology and ecology of the animal (Raubenheimer and Simpson 2016). For example, nutrients interact most closely with physiology, including taste receptors, appetites, and metabolism. Yet many proximal components of foraging, such as sensory ecology (Dominy, Lucas, et al. 2001; Melin and Veilleux, this volume), link directly to nonnutrient properties of foods such as color, while meals, which are structured combinations of foods, play an important role in integrating feeding both with the daily time budget and with digestive physiology (Raubenheimer and Simpson 1998; Sibly 1981). Ultimately, however, neither foods nor meals determine fitness-related outcomes such as growth, health, and reproduction, but diets do, and the mechanism is largely via their chemical constituents, particularly the nutrients. The interactions of animals with food environments thus involve all of these levels, and unraveling the details poses formidable challenges for nutritional ecology.

A powerful, if not indispensable, aid for navigating this complexity are models (Dunbar 2002). Among the many benefits of models is that they provide "proof-of-concept" tests of the logic of verbal explanations of how a phenomenon works (Servedio et al. 2014), force us to explicitly distinguish assumption from observation (Dunbar 2002), and enable us to make empirically testable predictions (Houlahan et al. 2017) and to extrapolate beyond the data to explore "what-if scenarios" (Dutreuil 2014). Models also provide important repositories for understanding (Houlahan et al. 2017), aids to communicating complex ideas with precision, and tools for integrating information and structuring interdisciplinary research (K. Schneider and Hoffmann 2011).

Compared with most other taxa, the primates offer good opportunities for realizing the potential of nutrition as an integrating factor in biology. This is partly because primates in the wild habituate to the presence of humans, enabling detailed and prolonged observations in an ecological context that would be difficult or impossible for most taxa. Consequently, the field has progressed from early qualitative and subsequently quantitative studies of diet to studies investigating relationships between diet and many aspects of primate biology, from ecology, evolution, and sociobiology through morphology, cognition, and behavior (Chivers and McConkey, this volume). More recently, primate foraging studies have shifted from food-level descriptions of diet or energy-based models such as optimal foraging theory (S. Altmann 2006; Sayers et al. 2010) to consideration of the detailed nutritional content of primate diets (Chivers and McConkey, this volume; Felton and Lambert, this volume). This shift to a nutritional ecology perspective offers powerful opportunities for consolidating the status of primates as a model system for integrative foraging research, especially if combined with nutritionally explicit modeling frameworks (Raubenheimer, Simpson, and Mayntz 2009).

In this chapter, I review an integrative approach for modeling nutrition, called nutritional geometry, and compare it to a second popular framework for foraging studies, optimal foraging theory. Originally developed in laboratory studies of insects, nutritional geometry has since been applied in several studies of primates in the wild and humans in industrial food environments. I begin by considering some theoretical aspects, including the place of nutritional geometry among the related concepts of "frameworks," "theories," and "models." In the same section, I examine the relationship between the terms "nutritional geometry" and "the geometric framework for nutrition" and propose a term that encompasses both: the "nutritional geometry framework" (NGF). This section might seem overly theoretical for a chapter on primate nutritional ecology, but it lays an important foundation for discussions that follow. I next introduce the basic concepts of nutritional geometry, using examples from laboratory studies of insects. The reason I use insects is that the concepts are best illustrated with experimental studies that establish causation, and there are very few studies that have applied the NGF in experimental studies of captive primates. Next, I consider other frameworks that have been applied to modeling diet and foraging in primates, with particular emphasis on how those models intersect with the NGF, and then review applications of the NGF to nonhuman primates. In the final section, I demonstrate the integrative capacity of the NGF using the primate species for which by far the largest amount of data exists: humans.

THE NUTRITIONAL GEOMETRY FRAMEWORK: THEORETICAL ASPECTS

Frameworks, Theories, and Models

Seemingly straightforward, the question "What is a model?" is in fact one of the most contentious in the philosophy of science (Godfrey-Smith 2006). I will not enter that discussion here, except to position nutritional geometry in relation to the concepts of "frameworks," "theories," and "models." Somewhat confusingly, all three of these are kinds of models, in the sense that they are idealized structures that help us understand how the world works (Godfrey-Smith 2006). It is nonetheless useful to distinguish the kinds of models that these terms represent, because this can provide scope and lucidity in analyzing complex phenomena such as nutrition. For clarity, in what follows I refer to "models" in the third sense mentioned above as "formal models."

For the present purposes, I follow Elinor Ostrom (2010) in her Nobel Prize lecture and distinguish "frameworks," "theories," and "formal models" according to their levels of generality. To paraphrase Ostrom, I consider a framework to be a structured conceptual schema containing the most general set of variables that a nutritional ecologist might want to use to examine foraging in a variety of contexts, such as different habitats, species, and developmental stages. Frameworks encompass theories that specify which components of the schema are useful to explain diverse outcomes and how these outcomes relate to each other. For example, as discussed below (section "Optimal Foraging Theory through the Nutritional Geometry Lens"), diet-breadth theory posits that important components for understanding the dietary generalist-specialist spectrum include not only the range of foods eaten but also the patterns of nutrient intake and the flexibility of physiological systems to handle variable nutrient intakes (see also Machovsky-Capuska, Senior, et al. 2016). Formal models make precise assumptions about a limited number of variables in the theory and generate predictions that can test these assumptions. An example is given below, where a geometric model is described testing the specific prediction that when feeding on nutritionally imbalanced foods, dietary generalists should have the physiological capacity to overeat the excessive nutrients to a greater extent than should specialists. Nesting these concepts from most general (framework) to most precise (formal model) in this way provides the equivalent of a "zoom lens" through which to view nutritional geometry at different levels of granularity.

From Wide-Angle to Macro: The Domains of Nutritional Geometry

Nutritional geometry is a framework developed for integrating theory from several disciplines to understand the role of nutrition in biology. The centerpiece, nutrition, is traditionally somewhat atheoretical (Doering and Stroehle 2015; Raubenheimer and Simpson 2016). Nutritional geometry has contributed structured theory for nutrition principally in two ways. First, it explicitly recognizes the complexity of nutrition by distinguishing its multiple dimensions (nutrients and nonnutrient components, such as plant secondary compounds) and multiple levels (foods, meals, diets, and dietary patterns), and it models the conjunction of dimensions and levels as a mixture hierarchy (as illustrated in "Proportions-Based Models of Human Nutrition").

The second way in which nutritional geometry contributes theory in nutrition is to link it to theory from other disciplines, such as behavior, physiology, evolution, and ecology, and to provide a nutritional context in which to interrelate the theory from these different fields. As mentioned in the introduction, the baseline concept of the mixture hierarchy facilitates this integration, because different levels of the hierarchy connect in particular ways with the animal-environment interface that is the main focus in nutritional ecology studies (Raubenheimer and Simpson 2016). Examples of such integration include the use of homeostasis theory from physiology in proximal interpretations of foraging and feeding (Raubenheimer, Simpson, and Tait 2012) and the use of life-history theory to explain why animals live longer when restricted by the quality or quantity of available foods (M. Reichard 2017). The integration of homeostasis theory and life-history theory offers powerful opportunities to understand the natural distributions of life spans (Raubenheimer, Simpson, et al. 2016). Likewise, behavioral, physiological, and ecological theory combine to explain the pattern of meal taking (Raubenheimer and Simpson 1998), evolutionary theory explains sex differences in nutritional priorities of animals (Maklakov et al. 2008; Morehouse et al. 2010), and niche theory explains how foods and diets influence the ecological distributions of animals (Kearney et al. 2010; Machovsky-Capuska, Senior, et al. 2016).

Through providing a framework in which theory from different fields can be interrelated in this way, nutritional geometry enables formal models to be constructed that test specific hypotheses at the interface of evolution, ecology, behavior, and physiology. For example, precise quantitative predictions can be generated to discriminate between caloric restriction and specific-nutrient explanations for the life-extending effects of moderate dietary restriction (Raubenheimer, Simpson, et al. 2016), to test the theory of ecological generalism (S. Simpson, Raubenheimer, et al. 2002), or to distinguish between energy maximization, toxin avoidance, and nutrient balancing explanations for foraging behavior (Felton, Felton, Lindenmayer, et al. 2009).

The elements of this nutritional framework and its constituent theories and models can be expressed in any representational medium, including verbally, in a diagram, or algebraically. However, geometry has the advantage that it combines the precision of mathematics with the intuitive accessibility of diagrams. Geometry also naturally lends itself to the multidimensional nature of nutrition.

A Note on Terminology: Nutritional Geometry versus Geometric Framework for Nutrition

An issue that is worth clarifying is the relationship between the terms "nutritional geometry" and "the geometric framework for nutrition," which are often used interchangeably. The geometric framework for nutrition, which was introduced in the 1990s (Raubenheimer and Simpson 1993; S. Simpson and Raubenheimer 1993), is a true framework in the sense defined above—it comprises a "structured conceptual schema containing the most general set of variables that a nutritional ecologist might want to use to examine foraging in a variety of contexts." As demonstrated in the next section, the representational medium for structuring the schema through interrelating the relevant variables is a particular form of geometry in which the axes are scaled in units of absolute amounts (e.g., kJ or g of protein vs. kJ or g of fat or carbohydrate). Accordingly, this approach is sometimes referred to as "amounts-based nutritional geometry."

Raubenheimer (2011) recommended a different but related approach for representing and interrelating the general variables of the geometric framework for nutrition, which is proportions-based (proportions-based nutritional geometry). As the term implies, in proportions-based geometry the representational medium is the mixture simplex, in which each axis represents the proportional contribution of a particular component to the sum of all components in the mixture (most commonly three). The simplex can take one of two forms (Raubenheimer 2011). More conventionally, it is an equilateral triangle, where each of the apices represents 100% of one of the three mixture components, and the value decreases with distance from the respective apex; the logic of the equilateral simplex is explained below, in "Foraging Frameworks in Primatology." While there are contexts in which this representation has advantages, Raubenheimer (2011) argued that in many circumstances a different representation, called the right-angled mixture triangle, is preferable. In this approach, two primary, or explicit, axes (the x- and y-axes) subtend an angle of 90°, and a third variable (called the implicit variable) decreases in value with distance from the origin of the x-y plot (as illustrated below).

Proportions-based nutritional geometry is not a different framework from the geometric framework but a different representational medium within it. The two approaches are complementary, in the sense that there are different circumstances in which each is preferable, and ultimately the most powerful insights can be obtained by applying both to the same system (as demonstrated below for humans). To encompass both amounts-based and proportions-based approaches (as well as any future representations that might be developed) while respecting their differences, I therefore propose that the term "nutritional geometry framework" (NGF) be used, within which proportions-based nutritional geometry is distinguished from amounts-based nutritional geometry.

Amounts-Based Nutritional Geometry

In this section, I demonstrate how the elements of the nutritional geometry framework—broadly, the general set of variables that Ostrom (2010) recognized as diagnostic of a framework (above)—are represented in amounts-based nutritional geometry.

The platform around which geometric models are constructed is a Cartesian space called a "nutrient space," which, as mentioned in the previous section, is constructed from axes that are scaled as absolute amounts. Figure 25.1a illustrates this in a hypothetical model involving two nutrients, protein (P) and carbohydrate (C). Within the nutrient space are plotted quantitative representations of key factors that are important for understanding how nutrition mediates the interaction between the animal and its environment.

The "intake target" represents the amounts and balance of the two nutrients that the animal's appetite systems will target. Foods are represented as lines, called "rails," which radiate from the origin into the nutrient space at an angle determined by the ratio of the nutrients the food contains. As an animal eats, it ingests the nutrients in the same proportion as they occur in the food, and the resulting change

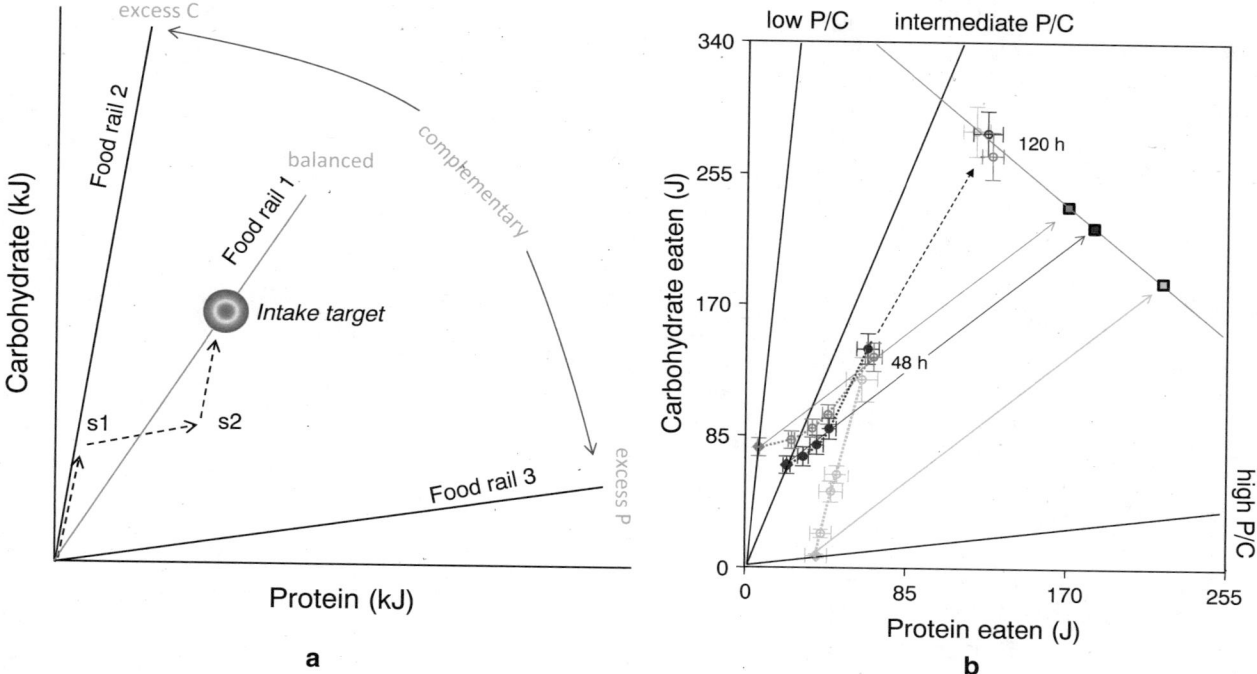

Figure 25.1. *a*, Schematic illustrating nutrient regulation in relation to dietary balance in the NGF, using a model involving a two-dimensional protein (P) and carbohydrate (C) nutrient space. The intake target (IT) represents the balance and amounts of macronutrients targeted by the regulatory mechanisms. Foods are represented by radials, called nutritional rails, projecting into the nutrient space at angles determined by the ratio of the nutrients they contain. As the animal eats, its nutritional state changes along a trajectory with the same slope as the nutritional rail for the food it is eating, with the distance moved along this trajectory being determined by the amount eaten. For example, by eating Food 2, the animal's state changes from the origin to state S1. The animal can achieve its target state by selecting Food 1, which is nutritionally balanced with respect to its target (the rail passes through the target), or else by mixing its intake from nutritionally complementary Foods 2 and 3 (illustrated by the dashed arrows). Thus, when in state S1 the animal is off course in relation to its intake target, but by switching to Food 2 it can move to S2, and by switching back to Food 1 it could reach the target. *b*, Geometric model distinguishing macronutrient balancing from energy prioritization in cockroaches (*Blatella germanica*). Solid diamonds represent mean ± SE intakes following 48 h of restriction to low, intermediate, or high P/C foods, and hollow circles represent macronutrient intakes when subsequently allowed to self-compose a diet from all three foods. The negative diagonal represents constant energy intake (energy isoline, calculated as P energy eaten + C energy eaten = constant). Under energy prioritization, all three groups are predicted to take the shortest trajectory to the energy isoline (head in parallel directions, as shown by the solid lines) to achieve the same energy intakes (reach the energy isoline) but different macronutrient ratios (spread across the isoline, indicated by squares). The data show that the cockroaches from the three experimental groups took different trajectories through the nutrient space to converge on the same nutrient intakes by 48 h and thereafter maintained the same trajectory (macronutrient balance) to the point where the experiment ended at 120 h (the black dashed arrow). *c*, Geometric model distinguishing macronutrient balancing from three alternative hypotheses in locusts (*Locusta migratoria*): feeding to constant volume (black squares), energy prioritization (medium gray squares), and protein maximization (light gray squares). Animals were fed for six days on one of four food pairings composed of (%P:%C) 7:14 + 28:14, 7:14 + 14:7, 14:28 + 28:14, or 14:28 + 14:7. Despite having access to different food pairings, all four groups of locusts arrived at a similar mean selected intake (black dots with error bars), demonstrating regulation of macronutrient balance. Data from P. Chambers, Simpson, and Raubenheimer (1995), as modified by Raubenheimer, Simpson, and Mayntz (2009).

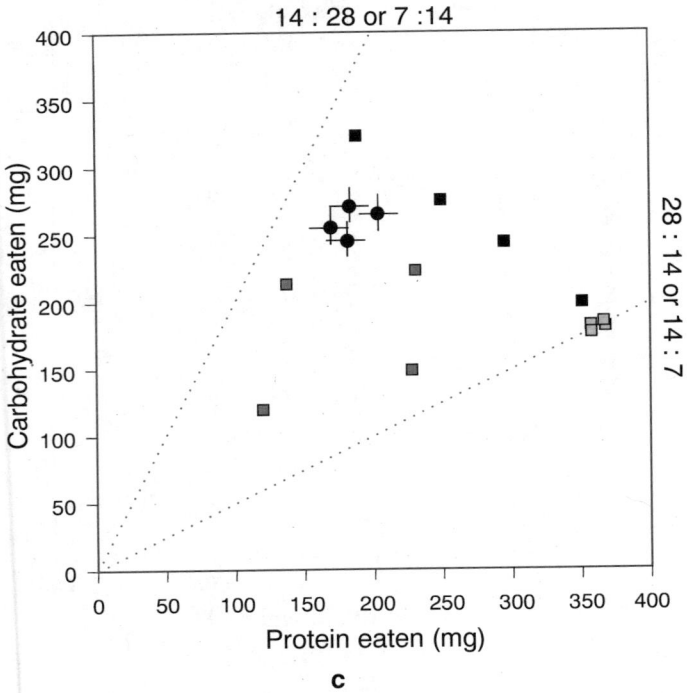

in its nutritional state is modeled as movement outward from the origin along the rail representing the food being eaten. The rate of movement and distance moved are determined by the rate of ingestion of the nutrients and the amounts eaten, respectively. If the rail intersects the target (Food 1 in fig. 25.1a), then by eating this food the animal can achieve its nutritional goal—such a food is said to be "nutritionally balanced" with respect to the nutrients in the model. In contrast, "nutritionally imbalanced" foods (Foods 2 and 3 in fig. 25.1a) do not intersect the target. While on its own an imbalanced food prevents the animal from reaching its regulatory target, it can nonetheless contribute toward a balanced diet if combined in appropriate proportions with other foods, provided those foods fall on the opposite side of the target to the first (fig. 25.1a). Such imbalanced foods that can be combined to compose a balanced diet are "nutritionally complementary" foods. Empirical examples of target selection through complementary feeding are given in figures 25.1b and 25.1c. In both examples, the models also explicitly stipulate the expected outcome under alternative hypotheses (e.g., energy maximization, protein maximization, and eating to fixed volume), a topic to which I return in the following section.

The intake target is an important empirical datum because it provides a homeostatic target on which to anchor predictions about an animal's nutritional responses (Raubenheimer, Simpson, and Tait 2012). It also represents an indispensable reference point against which to assess the effects of dietary imbalance on animals. When an animal is confined to an imbalanced diet that prevents it from reaching its intake target, it cannot satisfy its dietary goals in relation to all nutrients in the model and is forced to overeat some and undereat other nutrients (fig. 25.2a). The trade-off the animal adopts in this situation, termed a "rule of compromise," is best established by measuring the responses of the animal to a range of diets varying in the ratios of the nutrients in the model (figs. 25.2b and 25.3). In lab studies (e.g., fig. 25.3), the variation is created by manipulating the composition of experimental diets, whereas noninvasive field studies (discussed below) rely on natural variation in food composition. Like intake targets, measuring rules of compromise for animals in various circumstances provides important insight into their nutritional priorities and can help explain and predict their foraging and feeding activities.

Intake targets and rules of compromise are primarily measures of how appetite systems engage with foods in the process of eating. Appetites are important, because they represent a biological mechanism that links foraging and feeding choices to the specific nutritional needs of animals. The distinctive approach that the NGF brings

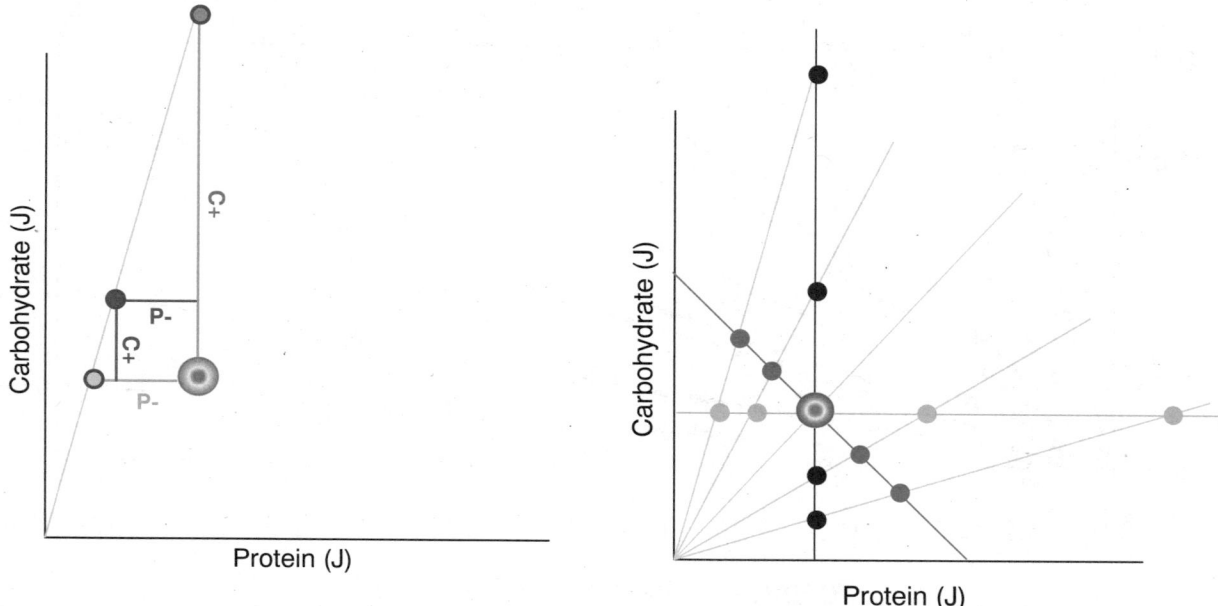

Figure 25.2. Schematic illustrating nutrient regulation in relation to dietary imbalance in nutritional geometry, using a model involving protein (P) and carbohydrate (C). *a*, When confined to a single nutritionally imbalanced food (i.e., with a rail that does not intersect the intake target), the animal needs to resolve a trade-off between overingesting one nutrient and underingesting the other. By feeding to the light gray point, it would meet its target for carbohydrate but suffer a shortage of protein (P−); at the medium gray point, it would meet its target for protein but overingest carbohydrate (C+); and at the dark gray point, it would spread the cost between a protein shortage and carbohydrate excess. *b*, Testing different experimental groups each on one of a range of foods varying in nutrient balance provides a description of how the animal resolves the trade-off between over- and underingesting nutrients when confined to imbalanced foods, termed a "rule of compromise." Three possibilities are illustrated: absolute prioritization of protein (i.e., feeding to the target coordinate for protein regardless of whether this involves over- or undereating carbohydrate; black symbols), absolute prioritization of carbohydrate (light gray symbols), and an intermediate response (medium gray symbols). An intermediate response with a slope of −1, as in this illustration, could mean one of two things. Either the animal is regulating to the point where it weights one unit of carbohydrate overconsumption (C+ in *a*) equal to one unit of protein shortage (P−), or it is feeding to a constant energy intake regardless of macronutrient balance (i.e., energy prioritization). These alternatives can be distinguished using an experiment equivalent to that in figure 25.1b, where the animals are provided free choice to compose the preferred diet. If in that circumstance they select an intake target (as did the cockroaches in fig. 25.1b), this would indicate equal weighting of nutrient surpluses and deficits, whereas intakes that are indifferent to macronutrient balance (the squares in fig. 25.1b) indicate energy prioritization. Many other configurations are possible. Adapted from Raubenheimer, Simpson, et al. (2016).

is that it explicitly recognizes that animals have separate appetite systems for different nutrients and models the ways that these appetite systems interact with each other to regulate feeding (Raubenheimer and Simpson 2020). In the case where the animal is unconstrained and thus able to select a balanced diet (as in figs. 25.1b and 25.1c), the appetite systems work together to achieve the target intake. When constrained to an imbalanced diet, however, appetite systems can be thought of as coming into conflict, and the rule of compromise describes the trade-off that has evolved to mediate this conflict (fig. 25.2).

Figure 25.3 provides an empirical example from a laboratory study of adult female beetles, which combines both target selection and the rule of compromise in a single model (K. Jensen et al. 2012).

The example in figure 25.3 also illustrates how relationships are modeled between nutrient intakes and other important variables that cannot be expressed in nutritional terms—for example, the consequences associated with specific nutrient intakes (e.g., survival and reproduction), the mechanistic causes (e.g., metabolic pathways), and behavioral correlates (e.g., foraging pat-

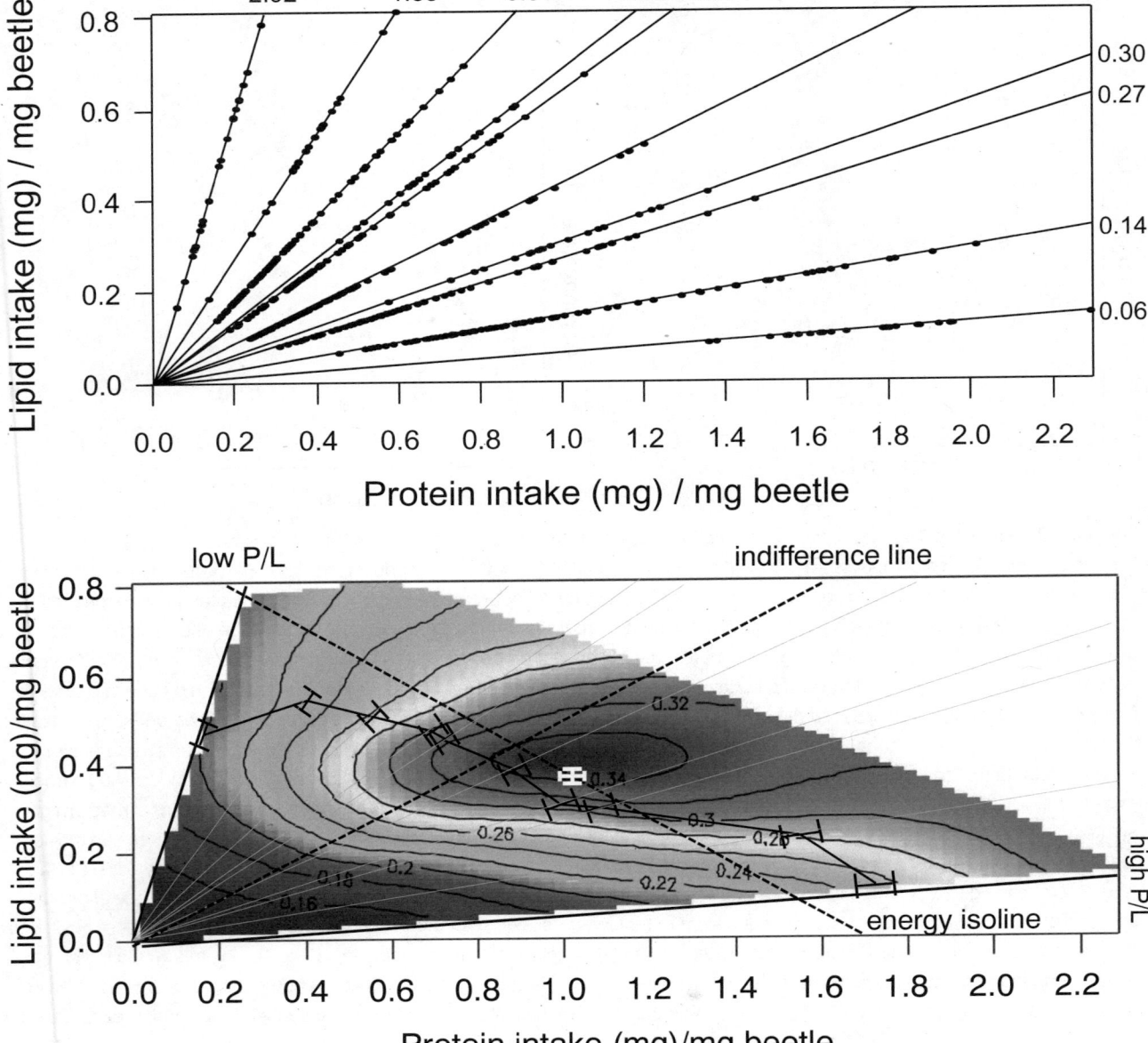

Figure 25.3. Testing adaptive hypotheses using the nutritional geometry framework. Diets of female *Anchomenus dorsalis* beetles were experimentally manipulated to span a wide range of lipid and protein intakes (*top*), and a response surface was superimposed relating egg production to diet (dark gray on right side of figure indicates high and light and dark gray closer to origin indicates low egg production (*bottom*). The solid radials projecting from the origin are nutritional rails representing the experimental diets (as in fig. 25.1a). Egg production showed a distinct peak, varying with both the balance (across nutritional rails) and amounts (along nutritional rails) of protein and lipid eaten. This demonstrates that energy is not a good proxy for fitness in this species, but macronutrition is. As predicted if fecundity was a selective factor in the evolution of macronutrient regulation, beetles allowed to compose a diet by combining the two extreme foods (labeled low P/L and high P/L) selected an intake target that corresponded with maximum egg production (the white cross, representing mean ± SE intakes of the self-selecting beetles). The dashed radial (indifference line) represents the dietary balance that would be associated with random or indifferent feeding from the high- and low-P/L foods. When confined to one of a range of foods varying in P/L balance and allowed to eat ad libitum (as schematically shown in fig. 25.2b), the beetles fed to the point on their respective rail that intersected the contour for maximum egg production that could be achieved given the macronutrient ratio of the experimental food. Modified from K. Jensen et al. (2012).

terns, as discussed further in "The NGF in Nonhuman Primate Studies," below). These are represented in NGF models using response surfaces, which relate the value of the nonnutritional variable to the nutritional composition of the ingested diet. Overall, this example shows that fecundity was strongly related to macronutrient balance, and when allowed to mix their intake from nutritionally complementary foods, the beetles selected the diet that maximized fecundity. Furthermore, when constrained from reaching the target, they showed a rule of compromise that provided the maximum fecundity that could be achieved given the macronutrient balance of the food to which they were confined.

Proportions-Based Nutritional Geometry

There are several reasons I recommend proportions-based nutritional geometry as a tool for extending the NGF (Raubenheimer 2011). First, data on proportional compositions often are easier to collect empirically than are reliable data on amounts, and they are more widely available in the literature for secondary analyses. This is especially true in the context of field studies. For example, in contrast with primates, for many other animals it is difficult or impossible to observe feeding over periods long enough to obtain the detailed measures needed for amounts-based geometric analyses (Remonti et al. 2016). However, in such cases the proportions of different foods in the diet can be estimated using indirect techniques, such as scat analysis, regurgitations, or gut content analysis of dead animals or regurgitates (Grainger et al. 2020; Machovsky-Capuska, Coogan, et al. 2016). Second, for many questions, amounts are irrelevant—for example, analyses of food compositions (Rothman, Chapman, and Van Soest 2012) or interspecific comparisons of diet compositions (Raubenheimer, Machovsky-Capuska, Chapman, et al. 2015). Third, proportional analysis enables the representation of three nutritional components in a two-dimensional plot. This is particularly important in macronutrition, because it enables protein, fat, and digestible carbohydrate to be represented separately in a model rather than via compound variables such as the energy content of fat and available carbohydrate combined (often called "nonprotein energy"). Fourth, as discussed further below, proportions-based nutritional geometry is an important tool for defining and modeling the nutritional mixture hierarchy discussed above. Finally, for all the above reasons, proportions-based measures provide greater integrative reach than measures of amounts, in the sense that they enable a wider range of factors to be combined and integrated within a single model than do amounts-based measures, as illustrated in the example of human nutrition below.

The logic of proportions-based geometric plots is explained in figure 25.4. As shown, a key feature of this approach is that it enables mixtures to be represented as points within the nutrient space and provides a geometry for combining mixtures into higher-order mixtures ("meta-mixtures")—for example, foods into meals and meals into diets. In so doing, it provides a powerful tool for understanding the food choices that animals make, as well as their causes and consequences, and for constructing predictive models for understanding and managing these choices (Coogan and Raubenheimer 2016).

FORAGING FRAMEWORKS IN PRIMATOLOGY

Primatology has traditionally had a reputation among behavioral ecologists of being rather atheoretical, although this is not altogether true (Dunbar 2002). With respect to foraging, primatologists have applied various models from the optimal foraging framework. Primatologists have also used mixture triangles, albeit, as I show below, in a fundamentally different application from their use in the NGF. In this section, I briefly introduce the optimal foraging theory (OFT) framework and various implementations that are relevant to primate foraging studies. Since this topic has been reviewed elsewhere from several angles (e.g., J. Lambert and Rothman 2015; Sayers et al. 2010), my coverage is brief and tailored specifically as a backdrop to the following section, which contrasts the OFT framework and its implementation in primate studies with the NGF. I close with a discussion of the use of mixture triangles in primatology.

Optimal Foraging Theory

Optimal foraging theory is a framework that aims to understand and predict the behavior of animals when foraging. The framework is grounded on the assumption that behavior evolves through natural selection and is

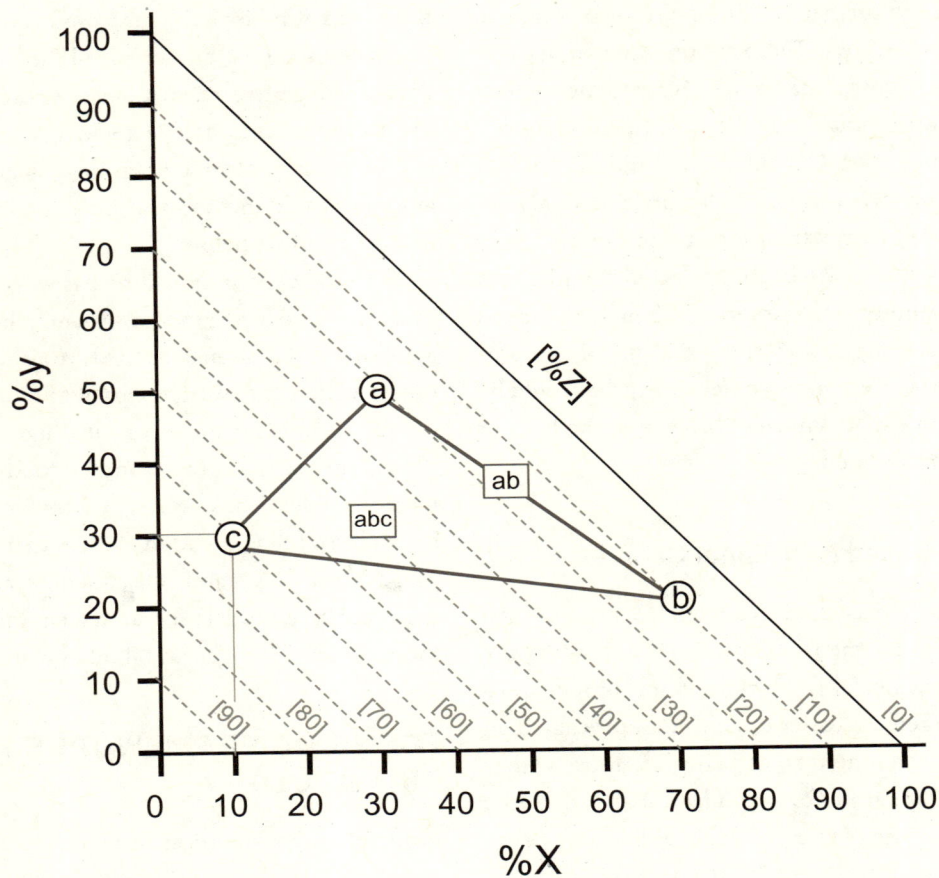

Figure 25.4. Logic of the right-angled mixture triangle. The three axes represent proportional contributions of three components (X, Y, and Z) to mixtures consisting of X, Y, and Z—for example, energy contributions of fat, carbohydrate, and protein to food (circles) or diets (rectangles). Percentages X and Y increase as usual along the x- and y-axes, respectively. Percentage Z increases across the negative diagonals, from 0% (the diagonal labeled [0]) to 100% (at the origin). Mixture "c," for example, comprises 10% X, 30% Y, and 60% Z. If two mixtures are combined into a metamixture—for example, foods "a" and "b" into a diet—then the resultant mixture is constrained to fall on the line connecting the parent mixtures (e.g., diet "ab"). If three foods are combined, then the accessible diet composition expands to a space (e.g., combining foods "a," "b," and "c" into diet "abc").

therefore optimized, in the same sense that organs such as eyes (Goldsmith 1990) and guts (Hume 2002) are optimized, to maximize the benefit-cost ratio in relation to their biological function. Mathematical models are used to construct scenarios and generate predictions to test these scenarios. Typically, an optimality model involves a "currency," or a measurable variable that is assumed to correlate with fitness, and an "objective function," which stipulates the hypothesized foraging goal of the animal in terms of the currency. The foraging currency is usually a single dietary component (most commonly energy or occasionally protein), the time required to obtain the dietary component, or some combination of the two (e.g.,

the rate of energy gain). The goal is typically to maximize the value of the currency (e.g., absolute or rate of energy gain) or minimize it (e.g., if the currency is time spent foraging). Clearly, however, an animal cannot ingest a particular nutritional currency in infinite amounts or at an infinite rate, and therefore a "constraints set" is stipulated to generate a prediction of what realistic level or rate of intake would represent the maximum possible (i.e., the optimal behavior).

One classical version of OFT models, the optimal diet model (also known as the prey choice model or contingency model), predicts how the choice of different foods should change as a function of their energy content, relative densities, handling times, and so on. Foods are ranked in terms of their profitability, evaluated as the ratio of nutritional benefit (usually energy content) and costs (the time it takes to capture and eat the food). Sets of foods that are more profitable than the average of all foods are considered the optimal diet. Another application of OFT models, the marginal value theorem, uses similar logic to predict the point at which an animal should leave a foraging patch—for example, to seek a more profitable patch or return to its nest.

Classical OFT models have been applied to human hunter-gatherers (Dusseldorp 2012; K. Hill 1988; Sayers et al. 2010), and their logic has been extended to explain human obesity in industrialized food systems (Lieberman 2006; Pyke 2017). They have also been applied to nonhuman primates, although surprisingly less frequently than might be expected (K. Potts, Baken, Levang, et al. 2016; Sayers et al. 2010).

Linear Programming

In the terminology developed above ("Frameworks, Theories, and Models"), linear programming (LP) is not a different framework from OFT but a different representational medium for implementing the concepts of OFT. In common with other OFT models, the goal is to identify the strategy that maximizes or minimizes some foraging currency, often energy intake or foraging time, subject to stipulated constraints. However, unlike many other OFT models, and in common with the NGF, LP models usually are expressed as a Cartesian space that integrates foods and food components such as energy, nutrients, and toxins.

Typically, the axes in an LP model represent the amounts of two different foods or food categories (e.g., leaves vs. fruits) in the diet of an animal (fig. 25.5), rather than amounts or proportions of nutrients as in the NGF. Based on the compositions of the foods, lines are drawn within the food space that represent constraints on the amounts of the two foods that can be eaten. For example, an animal might have a minimum requirement for sodium, which is met to one side of the relevant constraint line (i.e., for some combinations of the two foods eaten) but not on the other. It might likewise have a maximum tolerance of a toxin that is exceeded in diets that fall to one side but not the other of that constraint line. Constraints could also involve factors other than food components, such as maximum gut capacity or maximum time available for foraging. Collectively, the constraint lines delineate a region of diets that are adequate in the sense that they satisfy all constraints. An additional line is drawn in the same way as are constraint lines, relating amounts that would be gained of the foraging currency (e.g., energy) in various intake scenarios (i.e., combinations of the two foods eaten). The optimal diet is identified as the diet that maximizes (e.g., energy) or minimizes (e.g., foraging time) the currency. Although the constraints are usually linear, nonlinear constraints can also be modeled (S. Altmann 2006).

LP models were first applied to nonhuman primates by Stuart Altmann and colleagues (S. Altmann 1998, 2006; S. Altmann and Wagner 1978). They have also been used to model hunter-gatherer foraging (Belovsky 1987) and the diets of humans in industrial environments (e.g., Briend and Darmon 2000; Levesque et al. 2015).

Mixture Triangles

Chivers (1998) and Chivers and McConkey (chap. 1, this volume) use an equilateral mixture triangle to represent the proportional contributions of various food groups to the diets of 80 species spanning the main primate groups. The structure of the triangle is the same as the equilateral mixture triangle used in the NGF (Raubenheimer 2011), with one important difference. In the NGF application, foods are plotted as points within a space defined by axes representing nutrients, whereas in Chivers's application the axes represent food categories, and the points represent diets (combinations of foods; fig. 25.6). A limitation of scaling the axes as foods is that the utility is best suited for specialized feeders, where the number of foods in the diet is three, or else cases where the dietary components can usefully be categorized into three groups; for example, in Chivers's analysis, the categories were animal matter, fruits + seeds, and leaves + flowers + gum + sap. Three is a convenient number of dimensions for nutrition-based analyses because it encompasses the conventional classification of macronutrients into proteins, fats, and carbohydrates, the latter usually including both endogenously digested and microbe-digested components (Rothman, Chapman, and Van Soest 2012). There are various approaches for building higher-dimensional nutrient-based models, some of which are illustrated below ("The NGF in Nonhuman Primate Studies" and "Humans"; see also Blumfield et al. 2012; Raubenheimer 2011).

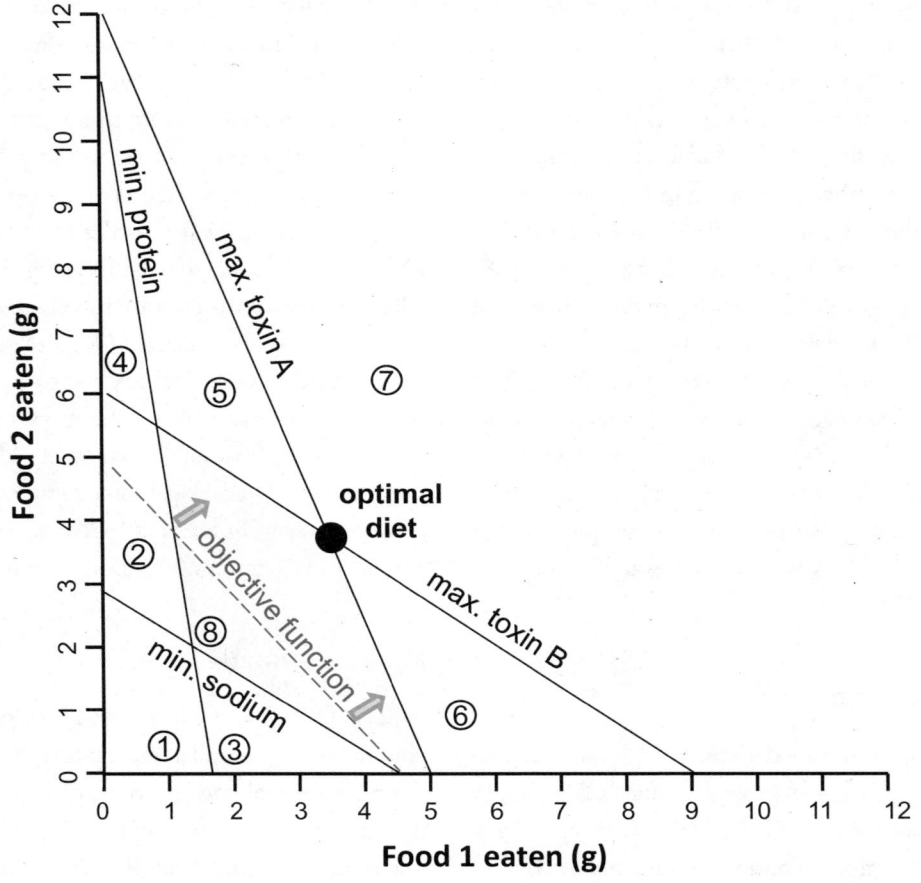

Figure 25.5. Hypothetical linear programming model. The axes model the amounts of two foods in the diet of an animal. Solid lines are constraint lines, which divide the space into regions delineating combinations of the two foods that are tolerable and are not tolerable, in this case set by minimum requirements for sodium and protein and maximal tolerance thresholds for two toxins. Constraints are derived using the equation $X + Y = $ constant, where the constant is the minimal (if a nutrient) or maximal (if a toxin) tolerable threshold, and X and Y are the amounts of the constraining constituent contributed from Foods 1 and 2, respectively. The shaded area represents the set of feasible diets—that is, those that contain both nutrients in sufficient quantities and do not exceed the tolerable threshold for either toxin. The line labeled "objective function" is derived in the same way as the constraint line, except this represents the currency to be maximized (often energy intake). The optimal diet is the diet that yields a maximum return of the currency while satisfying all constraints (i.e., while staying within the shaded region). Diets 1–7 are not feasible: 1 is deficient in protein and sodium, 2 is protein deficient, 3 is sodium deficient, 4 is protein deficient with excess toxin B, 5 has excess toxin B, 6 has excess toxin A, and 7 has excess of toxins A and B. Diet 8 is feasible, but it could be improved by increasing intake to increase energy gain, up to the point labeled "optimal diet." Modified from S. Altmann and Wagner (1978) and S. Altmann (1998).

OPTIMAL FORAGING THEORY THROUGH THE NUTRITIONAL GEOMETRY LENS

As discussed in the previous section, alongside the NGF, OFT is the main framework that has been applied to understand foraging decisions of primates. I now address the question of how the methods and concepts of these frameworks differ and where they overlap (see Raubenheimer and Simpson [2018] for further discussion on this).

Description versus Prediction

As noted by Sayers et al. (2010), OFT has been influential because it has transformed largely descriptive work into studies that use measurable variables to generate quantitative, testable predictions. I agree with this but

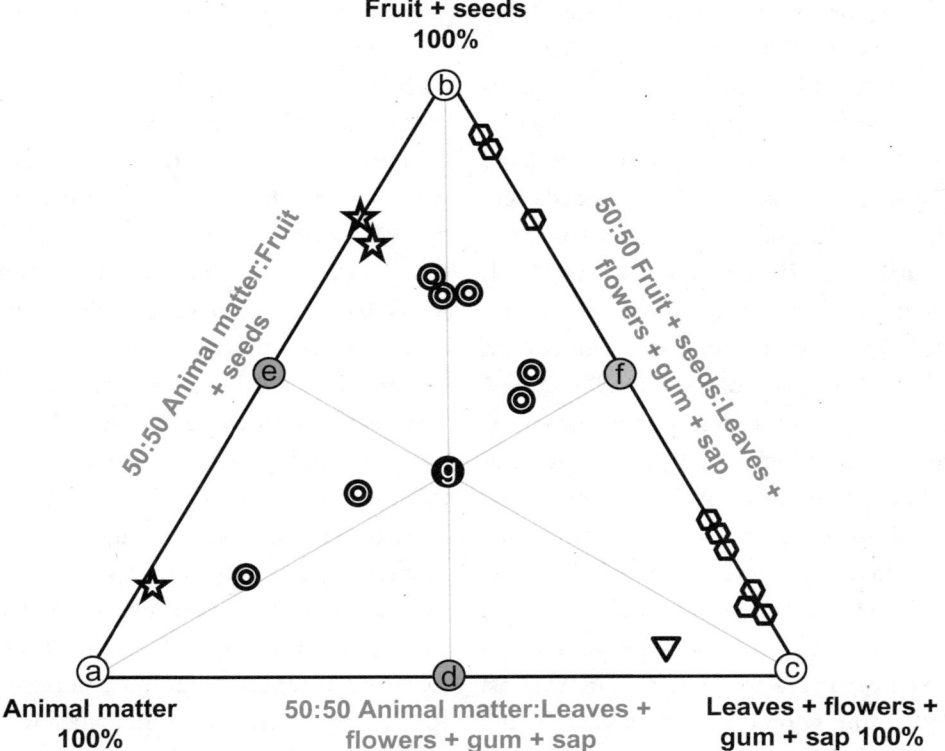

Figure 25.6. Equilateral mixture triangle representing the proportions of three food categories in the diets of primates (animal matter, fruit + seeds, and leaves + flowers + gum + sap). Hollow circles (*a*, *b*, and *c*) represent diets containing foods from only one of these categories. The gray radials projecting from the apices to midway along the opposite side of the triangle represent single-category axes, such that each category has a value of 100% at the apex and 0% where the axis intersects the opposite perpendicular; for example, animal matter is 100% at *a* and 0% at *f*. Gray circles (*d*, *e*, and *f*) therefore represent diets containing a 50%:50% mix of two categories. The intersection of the three single-category axes represents a diet containing a ⅓ mix of all three categories (labeled *g*). Also shown are actual diets of 20 primate species selected from 80 shown in figure 1 of Chivers (1998; reproduced in Chivers and McConkey, this volume). Hexagons show foli-frugivores, stars show fauni-frugivores, and the inverted triangle is a single fauni-folivore. Donuts represent species with diets that are composed of all three food categories.

caution against the widespread tendency to interpret "descriptive" in a pejorative sense while holding up prediction as the benchmark for behavioral studies. To summarize my conclusion before I present the argument, prediction is only as good as the description on which it is based, and in many cases, it is more valuable to invest in sound descriptive data rather than poorly grounded prediction.

Indeed, most field-based NGF studies have to date been largely descriptive. This is, however, not because the NGF is atheoretical or because it is structured in a way that does not lend itself to prediction. Instead, the reason is that too little is known at present about the patterns of diversity in primate nutritional regulatory responses, or about nutritional regulatory responses of animals in the wild more generally, to generate worthwhile predictions. As the number of studied cases increases, it will become possible to make targeted predictions inductively, of the sort "Species A, B, C, and D, which all show regulatory response I, are primarily frugivorous; species E is also frugivorous; if I is characteristic of frugivory, then E too will show this response." Such inductively derived predictions are important, because they help direct attempts to elucidate the patterns of association between regulatory responses and primate nutritional ecology. They are also an important prerequisite for the next step, which is to understand these patterns of diversity. For this, deductive predictions can be made, of the sort "If regulatory pattern I is causally associated with a low dietary protein energy ratio, rather than eating fruits per se, then species F, which also has a low dietary protein energy ratio but achieves this using acorns, will likewise show I."

Lab studies have shown how the NGF is used to progress in this way from observation to prediction. In the first study ever to use the NGF, Raubenheimer and Simpson (1993) asked the simple question "How do locusts regulate food intake in the face of macronutrient (protein/carbohydrate) imbalance?" There was no basis on which to make a prediction because this question had not previously been investigated for any animal of which I am aware. Results showed that the study species, *Locusta migratoria*, had an arc-shaped rule of compromise, which for geometric reasons we called "closest distance" optimization (Raubenheimer and Simpson 1997). An interesting biological aspect of this discovery is that it showed that, when feeding on macronutrient-imbalanced diets, the locusts settled on a large deficit of the deficient nutrient and avoided a relatively small excess of the surplus nutrient. Such avoidance of excesses stood in stark contrast with the energy maximization assumption of OFT, and we were interested to learn how generally this applies. We subsequently compared, in a single experiment, the rule of compromise of *L. migratoria* with that of another species of locust, *Schistocerca gregaria* (Raubenheimer and Simpson 2003). The reason we made this comparison is that *L. migratoria* is a dietary specialist (feeding only on grasses), whereas *S. gregaria* is a generalist folivore, and we wondered whether a generalist feeder might be more opportunistic than the specialist in consuming surplus nutrients when available (Raubenheimer and Simpson 1997). This was indeed the case: the generalist species fed to the point where the intake of the surplus nutrient equaled the deficit of the deficient nutrient, rather than the deficit being larger than the surplus. Accordingly, this rule of compromise, which is linear with slope negative 45° (the red configuration in fig. 25.2b) rather than arc-shaped, is called the "equal distance" pattern.

Several subsequent experiments that compared matched specialist- and generalist-feeding insects showed the same difference (reviewed in S. Simpson and Raubenheimer 2012), but insufficient cases have been studied to establish whether equal distance regulation is statistically associated with dietary generalism while controlling for the effect of phylogeny (G. Stone et al. 2011). S. Simpson, Raubenheimer, et al. (2002) capitalized on a powerful opportunity to test this in a context that is not confounded by phylogeny, using a system where the same genotype has two alternative developmental pathways. Locusts reared in high-density populations develop into adults that are morphologically, behaviorally, physiologically, and ecologically very different from those reared at low densities. The key difference for the present purposes is that in the wild, high-density-reared locusts have a broader diet than those reared at low density. Experiments showed, as predicted, that the generalist form had a pattern of macronutrient regulation more similar to the equal-distance pattern than did the more specialized form. The same has been demonstrated for caterpillars that have dietary generalist and specialist developmental pathways (K. Lee, Simpson, and Raubenheimer 2004).

More recently, Cui, Wang, Shao, et al. (2018) have drawn on this insect work to test the prediction that the high degree of ecological flexibility and dietary generalism in rhesus macaques (*Macaca mulatta*) will be associated with the pattern of macronutrient regulation observed in dietary generalist insects. Remarkably, this was found to be the case, providing the only known example of a primate with that pattern of macronutrient regulation. This finding has provided an opportunity to examine the ecological context of the generalist pattern of macronutrient regulation in this species and test the reasoning developed in insect studies to explain generalist macronutrient regulation (Cui, Zhang, et al. 2022).

Other experiments using the NGF have explicitly tested optimality predictions equivalent to those developed in OFT. S. Simpson, Sibly, et al. (2004) provide an empirical example of this, together with detailed modeling and discussion on theoretical aspects of constructing optimality models in the NGF. That model is equivalent to the model presented in figure 25.3, in which animals were confined to one of a range of foods to establish the consequences for growth and reproduction of eating different diets. The diets were formulated to independently achieve variation in macronutrient balance (restricting animals to different food rails) and absolute intakes (varying the amounts of each that the animal could eat). The dietary macronutrient composition that supported best performance was considered the intake target that an optimally foraging animal should select if fecundity was an important selective factor in the evolution of foraging. In both the experiments, a self-selecting group, which was experimentally restricted in neither macronutrient balance nor the amount they could eat, selected the diet that maximized reproductive output (illustrated in fig. 25.3).

Simplicity versus Complexity

OFT is sometimes considered a simpler alternative to the NGF. For example, Sayers et al. (2010) state, "While students of primate diet argue for approaches of greater and greater complexity (Felton et al. 2009a), it is possible, and perhaps even likely, that quantifying only several key variables from foraging theory (e.g., energy or protein gain, handling time, travel time) would be sufficient to explain much of the variance in primate feeding behavior." The "complex model" that Felton, Felton, Wood, Foley, et al. (2009) reviewed is the nutrient-balancing framework of the NGF, contrasting it with four single-currency models (energy maximization, protein maximization, avoidance of plant secondary metabolites, and limitations on the intake of dietary fiber). The alternative to which Sayers et al. (2010) refer, which involves key variables from foraging theory such as energy or protein gain, or handling time or travel time, references the OFT framework.

Contrasting OFT with the NGF based on complexity is wrong. The defining feature of the OFT framework, on which its influential contribution is based, is not its simplicity but the introduction of evolutionarily inspired mathematical modeling into foraging studies (Milinski 2014). This aspect of OFT is foundational to the NGF, which is explicitly based on the evolutionary approach that was pioneered in OFT. In the NGF, however, evolutionary principles are integrated with theory from other fields that are relevant to animal adaptation, such as homeostasis and epigenetic inheritance (Raubenheimer, Simpson, and Tait 2012).

Furthermore, OFT models need not be simple, and NGF models usually are not complex. For example, S. Altmann (1998) applied a model derived from OFT to baboon foraging that involved 72 constraint functions. In contrast, many NGF models involve only two variables, often protein versus nonprotein energy. Others involve three variables—for example, if different sources of nonprotein energy are distinguished, if components such as fiber or allelochemicals are included, or if a response variable such as reproductive output is included. The question is thus not how many variables are included in a model but which variables are included and how they are interrelated within the model to generate and test predictions.

Energy versus Nutrients

Perhaps the most obvious respect in which OFT appears simpler than the NGF is that the former models a single nutritional currency subject to constraints whereas the latter models two or more nutrients and their interactions. It is important to stress, however, that the NGF does not exclude the possibility of single-nutrient or energy maximization but can explicitly recognize and distinguish these outcomes from nutrient balancing. To demonstrate, I will use a hypothetical example that tests whether the foraging goal is energy maximization. For this, it is necessary first to distinguish two meanings of the word "energy" in primate foraging, since these have very different implications for predictions derived within the NGF. The first is total available energy, which is the sum of all dietary components that yield metabolic energy, including nonstructural carbohydrates, fats, protein, and the digestible component of fiber. The second sense in which "energy" is sometimes used refers to energy derived from sources other than protein, which is referred to as "nonprotein energy." I first discuss NGF representations of the prediction that total available energy is maximized, which is the usual sense in which "energy" is used in OFT models, and thereafter comment on the prediction that nonprotein energy is maximized.

If an animal were maximizing the intake of total available energy subject to a fixed constraint, such as finite gut capacity, as is commonly postulated in OFT models, then the plot of protein versus nonprotein energy intakes generated from standard NGF studies would resemble a negatively sloped linear array as demonstrated in the hypothetical example in figure 25.7. The reason is that this configuration of intakes represents feeding to constant energy regardless of whether that energy is derived from fat, carbohydrate, protein, or digestible fiber. Put another way, it suggests that the animal is treating these different dietary components as interchangeable sources of energy rather than as different nutrients. Furthermore, if energy intake was a proxy for fitness, as is commonly assumed in OFT models, then the relevant response surface (e.g., reproductive output, as in fig. 25.3) would rise

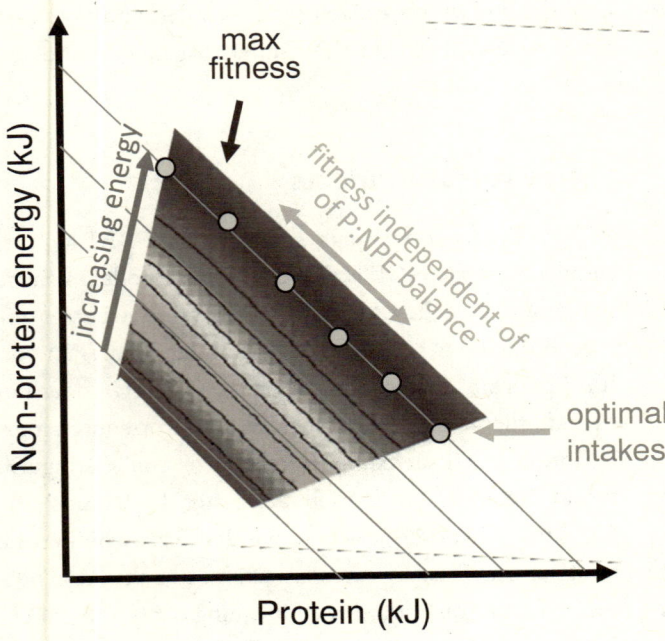

Figure 25.7. Hypothetical NGF model concomitantly testing the hypothesis of energy maximization in a forager and the assumption that energy intake is a proxy for fitness. The gray circles show that expected intakes under energy maximization would align on the highest possible (subject to constraints) energy isoline (negative diagonal, as in fig. 25.1b) but be indifferent as to the balance of macronutrients that contributed the energy (i.e., the position on the line). If energy intake were a proxy for fitness, then the fitness response surface (e.g., for reproduction) would have diagonal contours that increase in value (from bottom left to top right) with increasing energy intake, with no effect of macronutrient balance.

across contours parallel to the constant energy line, where it would reach a maximum (as shown in fig. 25.7). This model therefore not only tests the prediction of energy maximization but also directly tests whether energy is a reliable proxy for fitness. In the experiment reported in figure 25.3, this was found not to be the case, because the macromolecular source of energy (the ratio of protein to nonprotein energy) strongly influenced reproduction in the study species.

In field studies, however, things might be more complex than in tightly controlled laboratory studies that use chemically defined experimental foods. For example, in free-ranging primates it is possible that different constraints on energy intake apply for different foods, and this could distort the shape of the negative diagonal. High-protein leaves, for example, might contain higher levels of nondigestible fiber than low-protein fruits, and this could impose greater volumetric constraints on energy intakes in the high-protein (lower right) than low-protein (upper left) part of the plot. This is directly testable using a model that includes macronutrients as well as nondigestible fiber. Of course, the NGF does not create these complications but rather exposes and potentially resolves them.

Another potential complicating factor is that animals might be indifferent to the macronutritional sources of energy (i.e., show the negative diagonal pattern of intake) for reasons other than energy maximization. As discussed above, several studies have shown that the negative diagonal pattern is characteristic of generalist-feeding but not specialist-feeding insect herbivores (reviewed in S. Simpson and Raubenheimer 2012), and yet we know that the specific balance of macronutrients eaten is important for them. Two sources of evidence prove this. First, when provided with complementary foods, the insects select a particular protein:carbohydrate intake target (as in fig. 25.1), demonstrating that if given the opportunity they do discriminate between different energetic substrates. Second, when experimentally constrained from achieving the selected target—which is the circumstance in which they show the negative diagonal pattern—fitness varies with the balance of macronutrients that the animals eat, even if total energy intake is constant (see S. Simpson, Raubenheimer, et al. [2002] and Raubenheimer and Simpson [2003] for examples). This demonstrates that for these insects, the negative diagonal line does not represent energy maximization but rather is a rule of compromise (fig. 25.1) with a specific interpretation relating to the respective costs of ingesting surpluses and deficits (relative to the target intake) of protein and nonprotein energy. The relevant interpretation is that animals that show this pattern of regulation treat one unit of excess or deficit of protein calories as equivalent to one unit of excess or deficit of nonprotein calories, which is a very different thing from regarding all calorie intake as equivalent (S. Simpson, Sibly, et al. 2004).

The foregoing discussion applies to the case where the energy being modeled is total energy intake, comprising all utilizable sources of calories (fat, nonstructural carbohydrates, digestible fiber, and protein). As noted above, in primatology as well as some other fields, the term "energy" is also used to refer to nonprotein calories. Such cases, too, are readily detectable using an NGF plot of protein versus nonprotein energy intake, as a horizontal line equivalent to the light gray configuration in figure 25.2b. However, the same caveat applies as discussed above in relation to the diagonal shape indicating total energy maximization: this pattern could be interpreted as nonprotein energy maximization only if the animal did not select a specific protein/nonprotein energy target. If it did, as have all animals tested to date (S. Simpson and Raubenheimer 2012), then this pattern would represent strategic regulatory responses to constrained variation in dietary macronutrient balance—i.e., rules of compromise—rather than single-currency maximization.

FORAGING ACTIVITIES

To this point, the discussion around the NGF has involved not foraging behavior per se but nutrient intakes, which are the outcome of that behavior, whereas OFT often makes predictions about behavior (travel time, handling time, patch residency, etc.). It would be wrong, however, to conclude that behavior could not be explicitly integrated into NGF models. There are several examples where this has been done in lab studies using insects (reviewed by Raubenheimer and Simpson 2018; S. Simpson, Ribeiro, and González-Tokman 2018). These include tests of the influence on nutrient balancing of travel distance (Behmer, Cox, et al. 2003) and the relative frequencies of different foods (Behmer, Raubenheimer, and Simpson 2001), as well as tests of the optimal frequency (P. Chambers, Raubenheimer, and Simpson 1998) and sequence (Houston et al. 2011) of switching between complementary foods. As discussed in "The NGF in Nonhuman Primate Studies," there have also been nutritional-geometry-inspired field studies of primates that were explicitly behavioral.

GRAPHICAL REPRESENTATION

Beyond the fact that both LP and NGF models are graphical, these two approaches are fundamentally different. One difference is that in LP models the axes are scaled as foods with nutrients being represented as lines within the plot (fig. 25.5), whereas in the NGF the axes are nutrients and foods are represented as lines (nutritional rails in amounts-based plots—e.g., fig. 25.1b) or points in proportions-based models (e.g., fig. 25.4). As discussed in relation to mixture triangles above, scaling the axes as foods rather than nutrients has important implications for the two forms of models. First, the applicability of standard LP models is restricted to cases where few foods or food categories are dominant in the foraging strategies of the animals. In the NGF, by contrast, any number of foods can be examined, with the restriction being on the number of nutrients represented in a particular model. However, given the overriding importance of the macronutrients, which NGF models can accommodate, and the fact that most animals eat a wide variety of foods (Pineda-Munoz and Alroy 2014), I suspect that in most cases restricting the number of nutrients will be less of a limiting factor than restricting the number of foods in a model. Indeed, theory predicts that, in general, the number of nutritional dimensions that will substantially affect foraging is low (Raubenheimer and Simpson 2016). Furthermore, when warranted, NGF models can integrate any number of other food components, such as micronutrients (e.g., Blumfield et al. 2012; Nie et al. 2015), plant toxins (Provenza et al. 2003; S. Simpson and Raubenheimer 2001), and fiber (K. Lee, Raubenheimer, and Simpson 2004).

A more fundamental problem with LP models, and with OFT models more generally, is that they partition diet components into "currency" (the primary component that is being targeted) and "constraint" (something that limits the gain of the currency). The hypothetical model in figure 25.5, for example, assumes that maximal energy is the foraging target, and the need to meet minimum intakes of protein and sodium are constraints. There are many problems with this approach, which I will not discuss in detail; instead, I refer the interested reader to the cogent account of Illius et al. (2002). Suffice to say, these are sufficiently fundamental to include problems of definition—it is no trivial task, and often arbitrary, to decide that a given trait such as a minimum requirement for a particular nutrient is a constraint, whereas another, such as the targeted energy intake, is an adaptive strategy. Rather, as demonstrated in the balance concept of the NGF, the levels eaten of both protein and energy should be considered adaptive decisions, with an important goal being to balance the intake of these (the intake target, as in fig. 25.1). Where the animal cannot balance its macronutrient intake—for example, when confined to noncomplementary imbalanced foods—the adaptive response is to settle on a trade-off that represents the least-cost solution for resolving the conflicting requirements for the two nutrients (the rule of compromise; figs. 25.2 and 25.3). Furthermore, even if in a particular case a constraint could be satisfactorily defined, it is difficult to stipulate the level at which the variable becomes a constraint. A common approach is to do this through observing what the animal does in a particular circumstance (e.g., the minimal level of protein that it will eat). However, as pointed out previously (Illius et al. 2002; Owen-Smith 1996), this is circular; it explains what the animal does (eats a minimum of x units of protein) by observing what it does (eats a minimum of x units of protein). Finally, in LP models, such as that shown in figure 25.5, nutrient constraints are often represented as discrete thresholds below which the animal cannot cope and above which it derives no further benefit of additional intake (S. Altmann and Wagner 1978). However, for some nutrients, including protein, this is unrealistic. Figure 25.3 shows an example in which reproduction is related to protein intake via a complex, continuous function, which is at first positive and then becomes negative.

THE "PACKAGING PROBLEM" AND THE "DUAL"

In a series of publications, Stuart Altmann has articulated within the context of LP models an interesting set of nutritional concepts that relates closely to some core concepts in the NGF. One concept, which Altmann called the "packaging problem," references the fact all foods are inextricably linked packages of benefits (generally required nutrients) and costs (S. Altmann 1998, 2009). Costs are separated into intrinsic costs (nutritional surpluses and deficits, toxins, etc.) and extrinsic costs (such as time costs and predation risk involved in acquiring the food).

The packaging problem, as applied to intrinsic costs, is represented within the NGF in the fact that when the animal eats a particular food, it is constrained to move along the trajectory defined by the nutritional rail describing that food (fig. 25.1a). If one axis of the model represents a toxin and the other a nutrient (e.g., Provenza et al. 2003), then any amount of the food that is eaten involves a trade-off between cost (toxin ingestion) and benefit (nutrient gain), up to the point where the target intake for the nutrient is reached. However, if both axes represent nutrients, the situation is more complex. In this case, there is no packaging problem if the animal is eating a diet that is balanced with respect to the nutrients in the model (e.g., Food 1 in fig. 25.1a), because the animal satisfies its needs for the various nutrients simultaneously (reaches the intake target). If available foods are imbalanced, however, then the situation is more complex and dependent on the strategy of the animal. For animals that compose a balanced diet by switching between nutritionally imbalanced but complementary foods (e.g., Foods 2 and 3 in fig. 25.1a), there is no intrinsic cost, or there might even be benefits of the packaging "problem." For example, complementary feeding can enable the exploitation of a wider range of foods than a more specialized dietary strategy (Raubenheimer and Jones 2006). If, however, limited availability of suitable complementary food combinations prevents the animal from reaching its intake target, then the packaging problem forces the animal to overingest some nutrients or underingest others (or both) relative to the target. The rule of compromise is a measure of the strategy animals adopt in response to this problem. While much of the experimental research measuring this has examined animals' responses to "macronutrient packaging," studies have also considered nonnutrient food components, such as indigestible fiber and plant toxins, and their interaction with nutrients (reviewed in S. Simpson and Raubenheimer 2012). Where there are benefits to eating low levels of a toxin (i.e., "hormesis"), a model testing the interaction of toxins and nutrients is much the same as one testing two or more nutrients (Raubenheimer and Simpson 2009).

Another concept introduced to nutrition by Altmann in the context of linear programming is referred to in the mathematics of linear optimization as "the dual" (S. Altmann 1984). Altmann pointed out that intrinsic to the mathematics of linear programming is both a "primal" problem and a "dual" problem. The primal problem is addressed using the procedure illustrated in figure 25.5, in which the optimal diet is identified by solving the equations to provide maximum energy gain subject to stipulated constraints, such as meeting minimum nutrient requirements and avoiding lethal intakes of toxins. I will not go into the mathematical details (which are presented by S. Altmann 1984), except to say that solving the equations for the dual problem has the interesting biological implication that it enables the cost for the animal imposed by the constraint variables to be expressed in the same units as the optimized variable (i.e., energy). It thus provides an "exchange rate" whereby the cost to the animal of, for example, avoiding one additional unit of toxin intake can be expressed in terms of the reduced energy intake that this will entail, courtesy of the packaging problem.

The dual is closely related to the rule of compromise in the NGF, which measures the deficit of one nutrient the animal is prepared to tolerate to avoid ingesting one unit of excess of another (fig. 25.2). As discussed above, however, a key difference is that in the NGF the two nutrients are not stratified in the model as "currency" versus "constraint." Rather, they are treated as dynamically interacting variables, which the animal needs to balance in such a way that the net benefits are maximized given its current circumstances. When these circumstances enable it to reach the intake target, then what is balanced is the ratio of the nutrients in the diet. When it is ecologically constrained to eat an imbalanced diet, then what is balanced is the ratio of the amount overeaten of the excessive nutrient and the amount undereaten of the deficient nutrients in the model.

THE NGF IN NONHUMAN PRIMATE STUDIES

Several studies have applied the NGF to free-ranging nonhuman primates, demonstrating the importance of macronutrient balance in understanding the nutritional ecology of these animals. Examples include studies of intake target selection, rules of compromise, and spatial aspects of foraging behavior, such as patch selection. Some have applied only amounts-based models, others proportions-based models, and a few have combined the two categories of model in the same study.

Intake Target Selection

C. A. Johnson, Raubenheimer, Rothman, et al. (2013) provide a spectacular example of target selection in a free-ranging primate (fig. 25.8a). This study recorded daily macronutrient intakes over 30 consecutive days of a single adult female chacma baboon living on the outskirts of Cape Town. The baboon ate 69 different naturally occurring food types and 12 human-derived foods during the study. Her daily energy intakes varied over a fivefold range, from 400 to 2,000 kcal. However, in the plot of protein (P) versus nonprotein energy (NPE) intakes, daily diets scattered tightly ($r^2 = 0.7$) around a line of 1:5 P:NPE, suggesting that macronutrient balance was being conserved across foraging days. A valuable aspect of this study is the long-term nature of the observations. This first provides a sensitive test of homeostatic regulation, because it captures cases of complementary diet switches across successive days—for example, where a low P:NPE intake on one day is compensated by a high P:NPE on the following day. Second, it provides a record of diet over longer timescales, which are linked more tightly to fitness outcomes than are daily diets. A cumulative plot of the diet over 30 days showed a remarkably tight adherence to a P:NPE ratio of 1:5 (fig. 25.8b). C. A. Johnson, Raubenheimer, Rothman, et al. (2013) combined these amounts-based analyses with proportions-based analyses examining the contributions of natural and human-associated foods to the daily macronutrient intakes of the baboon.

Felton, Felton, Lindenmayer, et al. (2009) provided further evidence for intake target regulation in free-ranging primates. This study showed that spider monkeys (*Ateles chamek*) specifically targeted a small number of foods with similar macronutrient ratios. When these foods were not available in sufficient quantities, the monkeys combined nutritionally complementary foods to compose a diet that converged closely in macronutrient balance with the targeted foods. Studies of howler monkeys (*Alouatta pigra*; Righini 2014) and sifaka lemurs (*Propithecus diadema*; Irwin, Raharison, Raubenheimer, et al. 2015) also suggest that wild primates target a particular balance of macronutrients across a single day or across multiple days. More recently, Guo, Hou, et al. (2018) showed that the intake target selected by golden snub-nosed monkeys (*Rhinopithecus rexellana*) dynamically tracks changing requirements for specific nutrients.

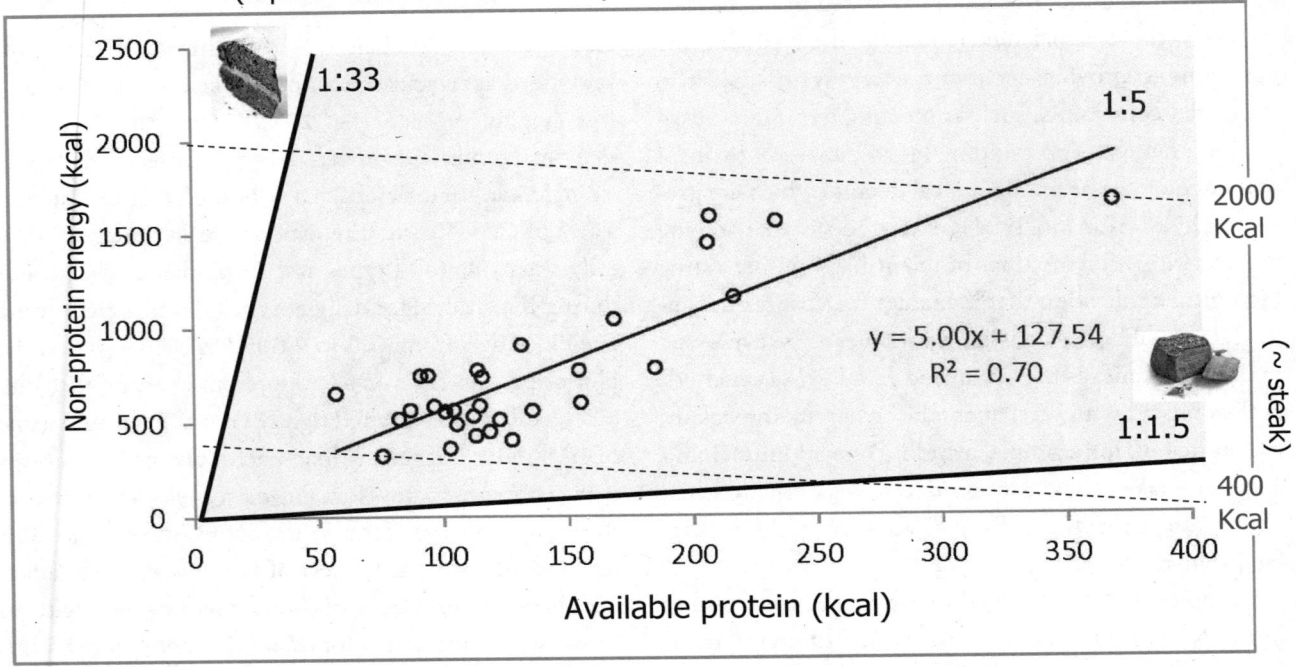

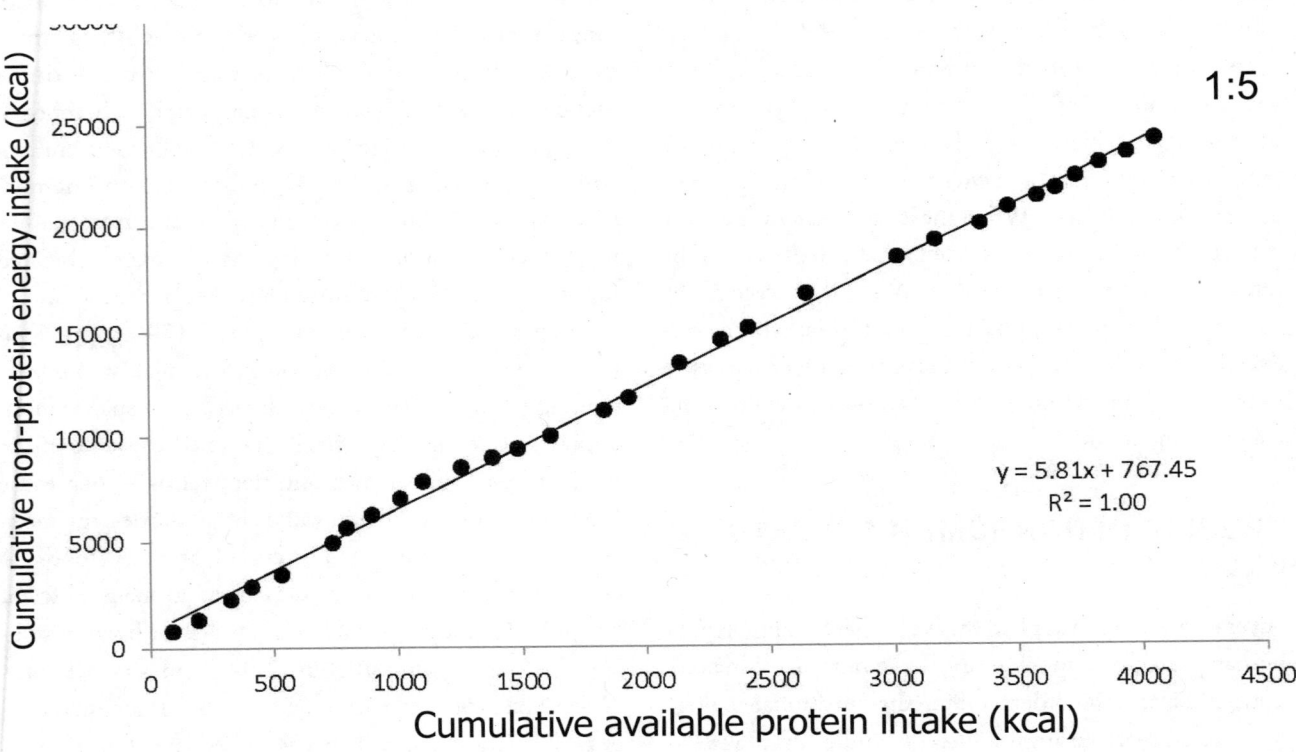

Figure 25.8. *a*, Daily intakes of an adult female chacma baboon observed over 30 days. Food macronutrient ratios (protein:nonprotein energy) varied widely, from 1:33 (equivalent to chocolate cake) to 1:1.5 (equivalent to lean steak), as did daily energy intakes (400–2,000 kcal/day). The daily diets, however, clustered around a ratio of protein:nonprotein energy of 1:5. *b*, The cumulative intake shows a 30-day diet with a remarkably tight macronutrient balance. From C. A. Johnson, Raubenheimer, Rothman, et al. (2013).

The study found that the target selected by the monkeys during the cold winter was higher specifically in carbohydrate and fat by an amount that closely matched the additional energetic requirements for thermoregulation, whereas protein intake did not differ seasonally.

While the above studies used amounts-based nutritional geometry to detect nutrient balancing, Raubenheimer, Machovsky-Capuska, Chapman, et al. (2015) showed how proportions-based nutritional geometry can be used for the same purpose. As an example, data on the diets of allopatric populations of mountain gorillas (*Gorilla beringei*) living in ecologically different habitats (Rothman, Plumptre, et al. 2007) were plotted using right-angled mixture triangles (fig. 25.9a). The results showed that the two gorilla populations composed nutritionally similar diets using different food combinations, as predicted under a nutrient balancing model of food selection. Furthermore, the dominant mode of nutrient balancing differed between the populations, with one predominately eating foods with composition that resembled the intake target and the other combining divergent complementary foods to balance the diet (fig. 25.9b).

An interesting case of macronutrient balancing concerns mammalian milk. Raubenheimer (2011) applied right-angled mixture triangles to literature data to examine the macronutrient and water composition of milk samples from common marmosets (*Callithrix jacchus*) and tufted capuchins (*Cebus apella*). In both cases, the percentage of energy contributed by protein remained remarkably constant compared with the fat:sugar ratio (fig. 25.10). Similar results were found for mountain gorillas (Raubenheimer, Gosby, and Simpson 2015). This likely reflects physiological regulation by the mother to provide the infant a diet balanced with respect to the protein energy ratio, a parameter achieved by nonsuckling animals through food selection and complementary feeding (Raubenheimer 2011). The high variability in the fat:carbohydrate ratio of the milk samples likely reflects the substitutability of these macronutrients as sources of nonprotein energy for stabilizing the protein energy ratio. Interchangeable use of fats and carbohydrates for stabilizing the protein energy ratio has been observed in the diet selection of adults of several species, including fish (Ruohonen et al. 2007), domesticated dogs (Hewson-Hughes et al. 2013), grizzly bears (Erlenbach et al. 2014), baboons (C. A. Johnson, Raubenheimer, Rothman, et al. 2013) and humans (S. Simpson and Raubenheimer 2005).

Responses to Dietary Constraint

The first study to measure the regulatory responses by free-ranging primates to constrained variation in dietary macronutrient balance was Felton, Felton, Raubenheimer, Simpson, Foley, et al. (2009), which showed that Peruvian spider monkeys (*Ateles chamek*) most strongly regulate protein intake while allowing nonprotein energy to vary with dietary macronutrient balance (fig. 25.11a). Rothman, Raubenheimer, and Chapman (2011) subsequently found that mountain gorillas (*Gorilla beringei*) show the opposite response, overeating protein to maintain nonprotein energy intake in periods when restricted to high-protein leaves (fig. 25.11b). Irwin, Raharison, Raubenheimer, et al. (2015) found that sifakas (*Propithecus diadema*) show neither the protein prioritization pattern of spider monkeys nor the prioritization of nonprotein energy found in mountain gorillas, but rather reduced their intake to maintain the target macronutrient balance (fig. 25.11c). In addition to using amounts-based nutritional geometry to examine the patterns of macronutrient intake, the latter study also used proportions-based nutritional geometry to investigate the distributions of macronutrients in the foods eaten by sifakas. As discussed above ("Description versus Prediction"), Cui, Wang, Shao, et al. (2018) found that rhesus macaques maintain a constant energy intake across wide range of dietary protein:nonprotein energy ratios.

Foraging Behavior

C. A. Johnson, Raubenheimer, Chapman, et al. (2017) combined amounts-based and proportions-based nutritional geometry to show that patch residency time in guerezas (*Colobus guereza*) is strongly influenced by macronutrient balancing, with macronutrient concentration also playing a role. The monkeys maintained a dietary protein:nonprotein energy balance of approximately 1:1.55 when feeding across patches and remained longest in patches that offered this ratio (fig. 25.12). DiGiorgio and Knott (2017) similarly

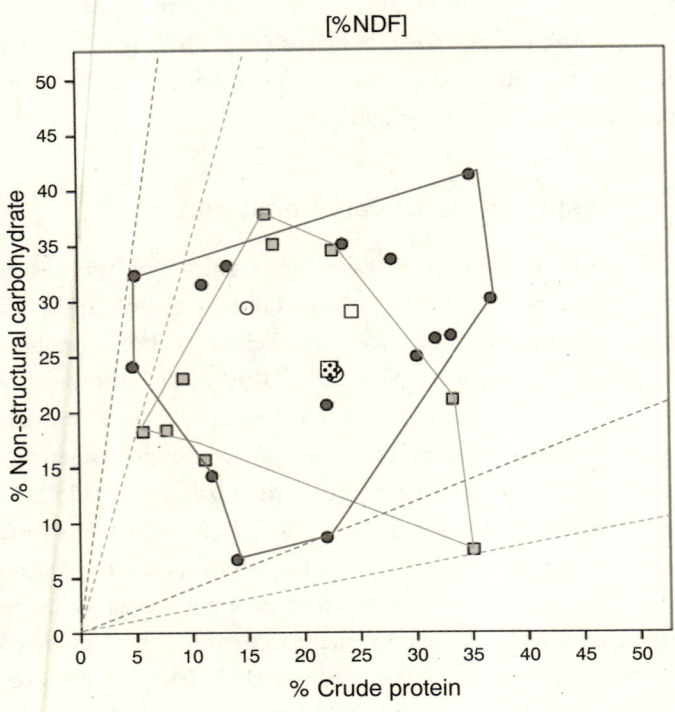
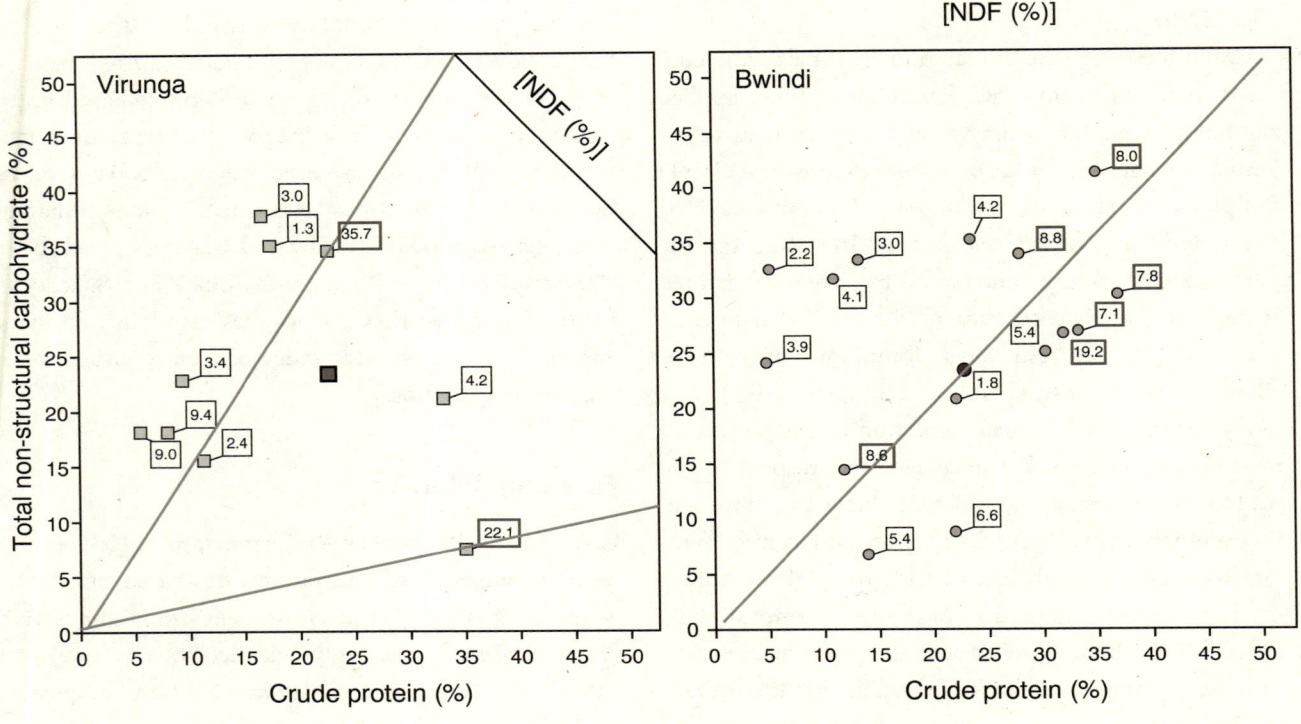

Figure 25.9 *(facing page)*. Use of the right-angled mixture triangle to detect nutrient balancing. *a*, Balance of protein, nonstructural carbohydrate, and neutral detergent fiber in the principal foods (those contributing >1% to the diet) and diet composition of two allopatric populations of mountain gorillas, in Virunga and Bwindi National Parks, Uganda. Circles represent the foods (dark gray) and diet composition (white with lines) of Bwindi gorillas, and squares represent the foods (light gray) and diet (polka-dot patterned) of Virunga gorillas. The hollow circle and square represent the expected diet composition of Bwindi and Virunga gorillas, respectively, if foods were eaten in proportion to their availability. The line joining the outermost foods from each site delineates the accessible space available to each gorilla population given its choice of foods (referred to by Machovsky-Capuska, Senior, et al. [2016] as the "food compositions niche"). Despite the foods differing between the sites, the composition of the diet ingested by the two populations of gorillas was closely similar but different from the expected diet if feeding was proportional to availability. *b*, Food and diet composition of Virunga (*left*) and Bwindi (*right*) gorillas plotted separately. Also shown is the percent contribution of each food to the diet, with the top-ranking foods that collectively contributed approximately 60% of the diet highlighted in bold boxes. The comparison shows that 58% of the Virunga diet comprised two foods, one with a protein:carbohydrate ratio that was considerably greater than that of the diet composition and the other with a smaller ratio. This demonstrates that the diet of Virunga gorillas was assembled to a large extent through complementary feeding (as illustrated in fig. 25.1a). By contrast, 60% of the diet of Bwindi gorillas was composed from six foods, all of which had a protein:carbohydrate ratio that closely resembled the diet composition. This suggests a stronger role in the selection by gorillas in Bwindi of foods that are balanced with respect to protein:carbohydrate, with a minimal role for complementary feeding. From Raubenheimer, Machovsky-Capuska, Chapman, et al. (2015).

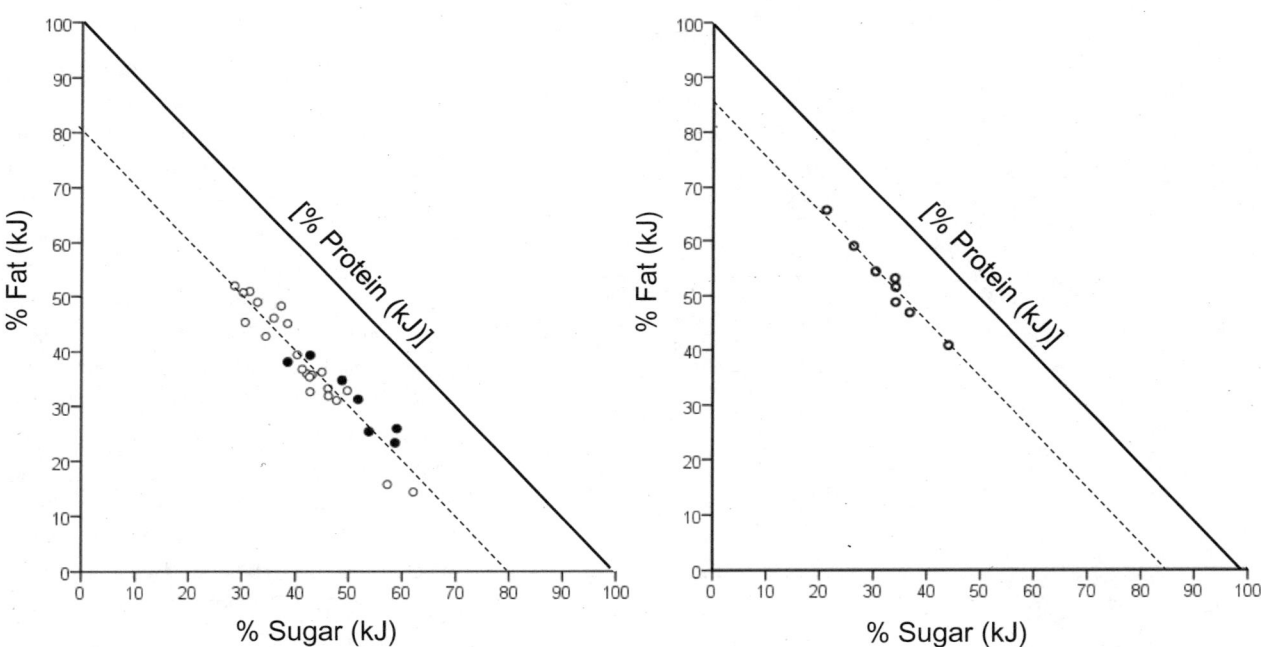

Figure 25.10. *a*, The macronutrient composition of milk samples from common marmosets (*Callithrix jacchus*), showing that the proportional protein content is maintained while the fat:carbohydrate ratio varies more widely. Black and hollow symbols represent samples collected from wild and captive animals, respectively. *b*, The same pattern is observed in milk samples from the tufted capuchin (*Cebus apella*). Modified from Raubenheimer (2011); data from M. L. Power, Verona, et al. (2008) (*a*) and L. Milligan (2010) (*b*).

found evidence that nutrient balancing influences foraging behavior in orangutans. They demonstrated that orangutans leave available fruit crops before they are depleted, that they eat nonfruit even when fruit is available nearby, and that they pass by fruit en route to eat other food types. This supports the hypothesis that nonfruit foods included in the diets of orangutans are not fallback options occasioned by fruit depletion but are complementary dietary components that are combined with fruits to compose a balanced diet (J. Lambert and Rothman 2015).

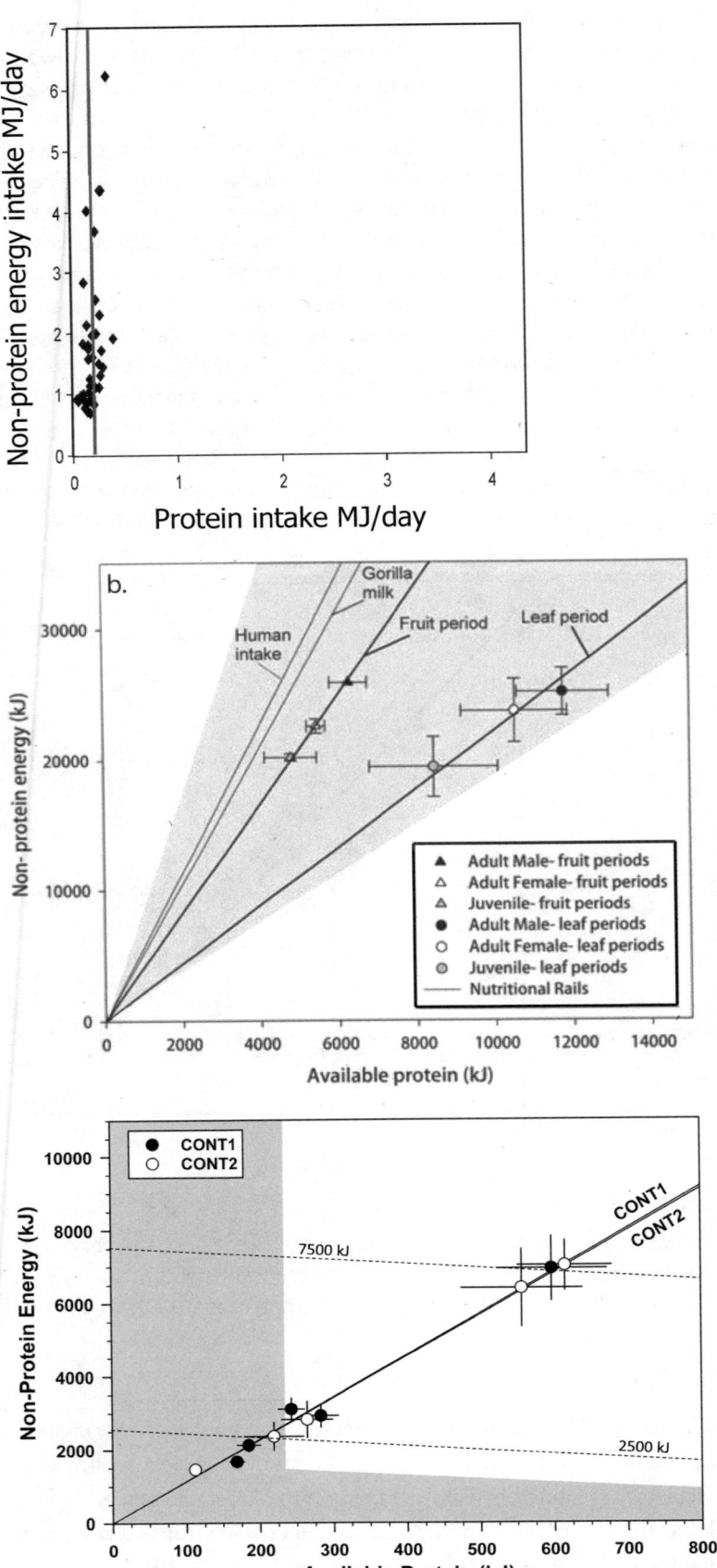

Figure 25.11. Regulatory responses by free-ranging primates to constrained variation in dietary macronutrient ratios. a, Peruvian spider monkeys (*Ateles chamek*) prioritize maintaining the target intake of protein, with fat and carbohydrate varying (modified from Felton, Felton, Raubenheimer, Simpson, Foley, et al. 2009). b, In periods when fruits and leaves are abundant (labeled "fruit period"), mountain gorilla (*Gorilla beringei*) juveniles, adult females, and adult males select an intake with approximately 19% of energy contributed by protein. When fruit is scarce ("leaf period"), and they are confined to a diet of 31% protein, all three age-sex groups maintain the intake of nonprotein energy constant while overeating protein (from Rothman, Raubenheimer, and Chapman 2011). c, Diademed sifakas (*Propithecus diadema*) have lower total energy intake in lean seasons (roughly 2,500 kJ) than in abundant seasons, but the dietary protein:nonprotein energy ratio remains constant. The data represent intakes averaged within five "seasons" separately for two different groups (CONT1 and CONT2). The solid radial lines are nutritional rails representing average protein:nonprotein energy ratios for the two groups, and the dashed diagonals are energy isolines representing 2,500 kJ and 7,500 kJ. The shaded area represents estimated minimal dietary requirements for protein (vertical) and energy (diagonal). Adapted from Irwin, Raharison, Raubenheimer, et al. (2015).

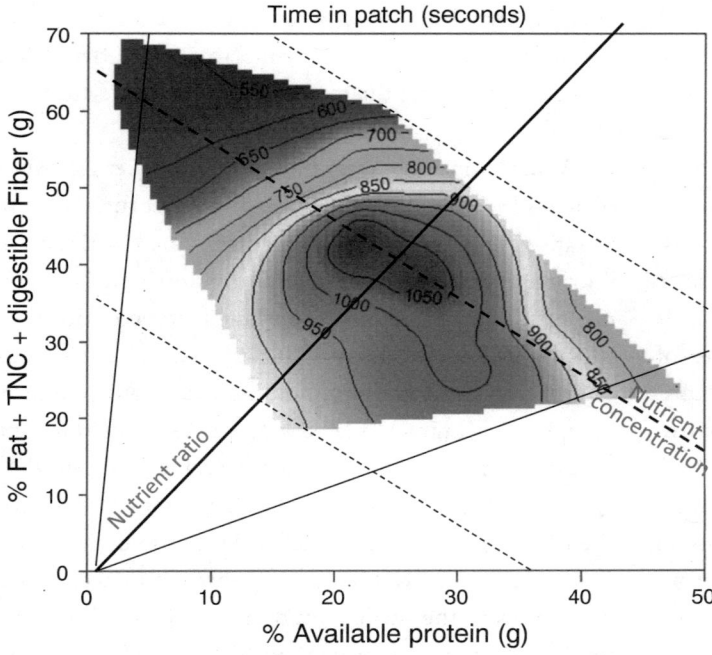

Figure 25.12. Response surface showing time spent feeding in patches by individual guereza as a function of the average dry weight composition of the foods in those patches. The way that the axes are scaled (g/100 g dry weight) means that the plot is effectively a right-angled mixture triangle in which the third axis is not an energetic nutrient but the sum of all nonenergetic nutrient components (principally nondigestible fiber). The response surface therefore shows the effect of nutrient balance (which changes across the solid radial lines) and nutrient concentration (which increases across the dashed diagonal lines). The monkeys spent the most time in patches with nutrient ratios similar to the target diet and intermediate nutrient concentrations (the intersection of the bold radial and diagonal). Modified from C. A. Johnson, Raubenheimer, Chapman, et al. (2017).

PRIORITIES FOR FIELD PRIMATE RESEARCH

Compared with many other taxa (e.g., insects), primates offer both challenges and opportunities in nutritional ecology studies. On the one hand, their cognitive and social complexity, combined with ethical considerations, weigh against their suitability for controlled laboratory studies. On the other hand, as demonstrated in the many studies discussed throughout this book, their propensity to habituate to the presence of humans, combined with other characteristics, makes the primates a model system for collecting data within the less well-controlled but more realistic context of field studies. An important challenge is to develop field-based approaches for moving beyond description and correlation and increasing the strength of causal inference in primate field studies. I now briefly discuss some approaches for doing this.

One approach, which I have discussed above ("Optimal Foraging Theory through the Nutritional Geometry Lens"), is to use predictive models to identify research questions and design studies for addressing them. Prediction strengthens the conclusions arising from studies because it sets them against independent precedents (inductive predictions) and links them to bodies of explanatory theory (deductive prediction). In turn, predictive studies feed back to the theory and in this way can hasten progress in understanding the ecology and evolution of foraging strategies, especially if they are quantitative (Platt 1964). An example of this is given above, in the study of macronutrient balancing by rhesus macaques ("Optimal Foraging Theory through a Nutritional Geometry Lens").

A second approach for strengthening causal inferences derived from field studies is to control in the study design for potential confounds where possible. This is, of course, common sense, but I raise it nonetheless because there are some specific examples from NGF-based studies that are worth emphasizing. One example is the importance of establishing that the recorded patterns of intake reflect the forager's preferences rather than constraints on those preferences set by ecological circumstances. This applies differently to studies of intake target selection and rules of compromise. For intake target selection, there should be constraints on neither the amount of food available ("quantitative constraints") nor the quality of foods available ("qualitative constraints"), because the intake target represents the nutritional regulatory goal of the animal and any factor that prevents it from achieving this goal could lead to incorrect conclusions. One way around this is to establish from ecological measures that foods are not constraining. For example, DiGiorgio and Knott (2017) established that Bornean orangutans frequently switch from fruit to nonfruit dietary items when fruits are still available, suggesting that inclusion of the nonfruit dietary items is by preference rather than constraint due to fruit depletion. Another approach is to estimate the target in more than one group of animals that are subject to different ecological circumstances. An example is provided in the comparison of Bwindi and Virunga mountain gorillas, which used different food combinations to achieve similar nutrient intakes in two separate habitats (fig. 25.9).

In contrast with intake targets, measuring rules of compromise requires that there are not quantitative constraints on food availability but that there are qualitative constraints. This is because, as illustrated in figure 25.2, rules of compromise represent a measure of ad libitum intake when an animal is restricted to imbalanced diets—in the terminology of the NGF, how far it chooses to move along an imbalanced nutritional rail. Clearly, this cannot be measured in circumstances where insufficient food is available for the animal to reach its preferred point on the imbalanced rail or where available foods contain other substances, such as nutrients not in the model, fiber, or toxins, that prevent the animal from achieving its macronutrient goals. As with intake targets, food shortage could be established using direct measures of food availability. Alternatively, indirect evidence could be used—for example, data showing that a second group of animals with higher nutrient requirements (e.g., larger animals or lactating animals) ate more within the same environment. Constraints due to other nutrients, toxins, or fiber can be tested by expanding the model to include these as model variables.

A third important prerequisite for interpreting results of field-based NGF studies is to ensure that rules of compromise are distinguished from cases where variation in nutrient intakes reflects differences in nutritional requirements. Unlike laboratory experiments, where foods are specifically selected or manipulated to achieve the variation in nutrient balance required for measuring rules of compromise, unmanipulated field studies rely on natural variation for this purpose. If the variation is achieved by spreading observations over different times of the year, then where possible additional evidence should be sought to ensure that the observed differences in intakes reflect differences in the quality (nutrient balance) of available foods rather than temporally changing intake targets. Temperate primates, for example, might have intake targets with higher nonprotein/protein energy ratios in winter to fuel the additional costs of thermoregulation (e.g., Guo, Hou, et al. 2018), and it could be an error to conclude from a geometric plot of seasonal intakes that the results reflect the protein prioritization rule of compromise (fig. 25.13). Wherever possible, therefore, comparisons should be sought in which there is good reason to suspect that variation in intakes is driven by variation in nutrient

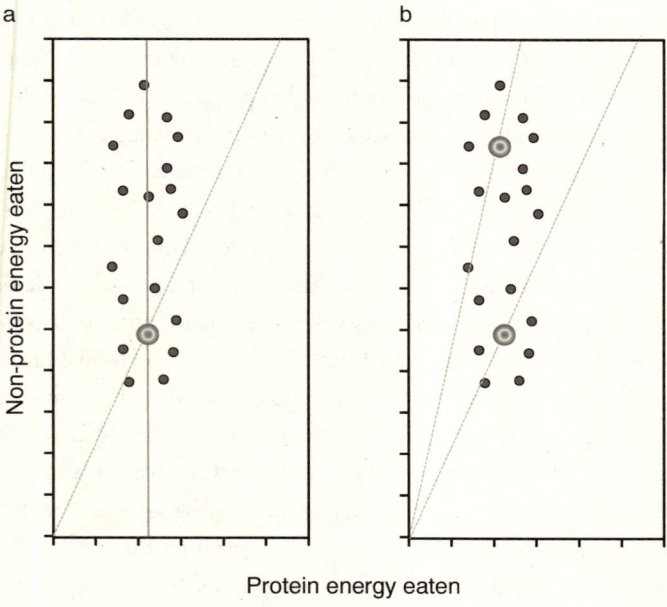

Figure 25.13. Scheme distinguishing rules of compromise from varying intake targets in the nutritional geometry framework. The solid dots are intakes from hypothetical field studies of primates, and the target symbols represent intake targets. *a*, All animals share a common intake target, and the spread of intakes therefore represents a rule of compromise, in this case protein prioritization (as in fig. 25.2). *b*, The spread of intakes is identical to that in *a*, but in this case it is due to subgroups of animals selecting different intake targets rather than representing a response to constrained dietary imbalance. This might come about if, for example, the scatter in intakes was derived from studying animals year-round, and there were seasonal differences in macronutrient requirements (e.g., increased fat and carbohydrate requirements for thermoregulation in winter).

supply rather than nutrient demand—for example, by comparing intakes at the same times of year where there is interannual variation in food supply. The study by Hou, Chapman, Rothman, et al. (2021) of golden snub-nosed monkeys provides the most comprehensive example of which I am aware in which relationships between rules of compromise, resource scarcity, and changing intake targets have been empirically disentangled.

Finally, a powerful way to unravel the net of causal factors underlying primate foraging is to expand the integrative scope and use the NGF to examine links between multiple factors. To date, the species of primate that best illustrates this point is *Homo sapiens*.

HUMANS

Humans provide a fascinating model species for nutritional ecology studies using the NGF. First, the problems of human nutrition are immense, with malnutrition (including undernutrition, overnutrition, and imbalanced nutrition) affecting at least a third of the world's population and being by far the greatest contributor to the global burden of disease (IFPRI 2016). This means that new insights can be evaluated against their real-world impact, in addition to other yardsticks, such as their ability to predict further research outcomes. Second, the practical importance of human nutrition has ensured that large amounts of resources and research have been invested in the field. Despite this effort, however, the problem has continued to worsen (Fitzmaurice et al. 2015; Mattei et al. 2015; Ng et al. 2014; Popkin 2015; Roth et al. 2015), suggesting that new approaches are needed in the field of human nutrition. Third, the modern epidemic of nutrition-related diseases has arisen through an interaction between recent changes in human food environments and the responses of human behavior and physiology to those changes. Yet, the discipline that specializes in understanding such relationships between biology and food environments, nutritional ecology, has until recently scarcely been engaged to help interpret the huge repository of data and knowledge that exists for human nutrition. Finally, such a rich body of data for a single species provides unprecedented opportunities for testing and elaborating nutritional ecology theory.

In this section, I provide a brief overview of recent research that has applied the NGF to human nutrition. My aim is specifically to capitalize on the abundant data available on humans to illustrate how the various components of the NGF can be combined to integrate diverse sources of information into a single model, in a way that is currently not possible for any other species. Several recent reviews provide more-detailed accounts of research that has applied the NGF to human nutrition (Raubenheimer, Machovsky-Capuska, Gosby, et al. 2014; Raubenheimer and Simpson 2016, 2019; S. Simpson, Le Couteur, James, et al. 2017).

Human Appetites

As discussed above ("Amounts-Based Nutritional Geometry"), appetite is a central component of the NGF. Several studies have applied nutritional geometry to address the question of how human appetites prioritize macronutrients. Evidence suggests that when given the opportunity in a controlled experimental setting, humans regulate to an intake target of approximately 15% protein energy (C. Campbell et al. 2016), a value remarkably similar to those recorded from observational population studies (S. Simpson and Raubenheimer 2020; fig. 25.14).

Several studies have addressed the question of how humans respond when experimentally restricted to diets manipulated to diverge from the target ratio (C. Campbell et al. 2016; Gosby, Conigrave, Lau, et al. 2011; Gosby, Conigrave, Raubenheimer, et al. 2013; Martens et al. 2013; Raubenheimer, Machovsky-Capuska, Gosby, et al. 2014). Overwhelmingly, the evidence suggests that the human response to macronutrient imbalance is to regulate protein intake most strongly, regardless of whether this involves over- or undereating nonprotein energy (on low and high protein:nonprotein energy diets, respectively) relative to the target (fig. 25.15). Population studies suggest that the same is true of free-ranging human populations (Bekelman, Santamaria-Ulloa, Dufour, Marin, and Dengo 2015; Bekelman, Santamaria-Ulloa, Dufour, Marin-Arias, and Dengo 2017; Grech et al. 2022; Martinez-Cordero et al. 2012; Martinez-Steele et al. 2018).

A corollary of the protein prioritization pattern of macronutrient regulation is that the strong tendency of human appetites to maintain constant absolute protein intake will result in increased fat and carbohydrate intake with decreasing dietary protein:nonprotein energy ratios. Conversely, on diets where this ratio is high, early satia-

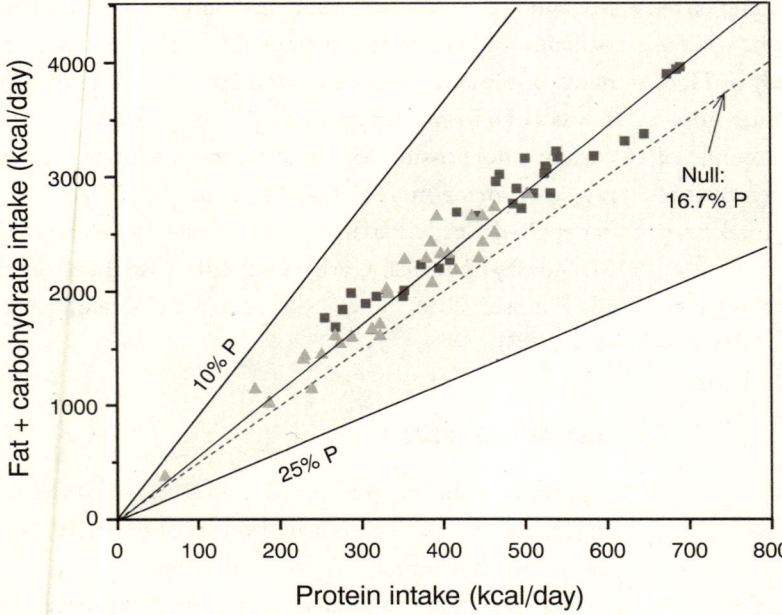

Figure 25.14. Macronutrient selection by humans. Sixty-three adults maintained in a residential experimental facility were allowed to self-select a diet over three days from menus containing foods manipulated to have 10%, 15%, or 25% of energy contributed by protein. This design allows any diet to be selected with a composition that falls within the shaded region. Both males (squares) and females (triangles) selected diets that clustered tightly around a mean protein energy content of 14.7%, which differed significantly ($P < 0.0001$) from the null expectation of 16.7%.

tion of the protein appetite and reluctance to overeat protein to gain additional fats and carbohydrates will result in low energy intake. This phenomenon has been called "protein leverage," to signify the fact that the strong human appetite for protein "leverages" the intake of fat and carbohydrate.

The protein leverage pattern of macronutrient regulation has given rise to an ecological hypothesis to explain the human obesity epidemic, the "protein leverage hypothesis" (PLH; Raubenheimer and Simpson 2019; S. Simpson and Raubenheimer 2005). The PLH postulates that variance in the protein energy ratio of the human diet can help explain variance in energy overconsumption and obesity. This provides a new focus in obesity research by directing attention to the factors in modern human environments that might influence dietary protein energy ratios. To investigate these, it is useful to expand the model using proportions-based nutritional geometry, as I now illustrate.

Proportions-Based Models of Human Nutrition

An advantage of proportions-based nutritional geometry is that it provides a framework for modeling the mixture hierarchy, referred to above ("Introduction" and "From Wide-Angle to Macro: The Domains of Nutritional Geometry"). This is illustrated in figure 25.16, which shows how dietary components, in this example macronutrients, combine into foods, meals, diets, and potentially population-level dietary patterns (as explained below in relation to fig. 25.17). It is important, however, to move beyond descriptions of mixtures and examine how these relate to broader aspects of human biology and the food environment with which it interacts. One example of this is to relate diet composition with human macronutrient requirements. To illustrate, I have superimposed on the nutrient space the recommended proportional macronutrient intakes for Australia and New Zealand, the Acceptable Macronutrient Distribution Ranges (AMDR; protein = 15%–25% of energy, fat = 20%–35%, and available carbohydrate = 45%–65%). Rather than represent these as separate recommendations, the mixture triangle enables them to be integrated into a single recommendation represented as the area defined by the conjunction of the three separate recommendations (the shaded polygon in fig. 25.16). Any food, meal, or diet that falls within this area is balanced with respect to the recommendation (meal 1 and diet 1), and any that falls outside (meals 2–7 and diet 2) is imbalanced.

While the axes of mixture triangles represent proportions, models generated using these can easily in-

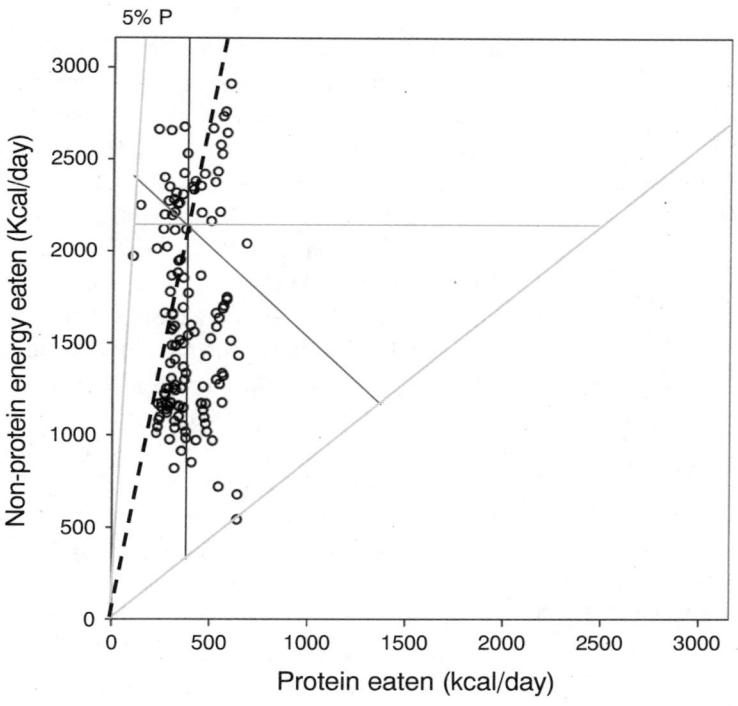

Figure 25.15. Regulatory responses of humans to constrained variation in dietary macronutrient ratios. The data are a literature compilation of ad libitum daily intakes by subjects restricted to one of 138 experimental diets, ranging between 5% and 54% energy from protein. The lines correspond with the regulatory strategies presented in figure 25.2b. The results show that in the face of dietary macronutrient imbalance, humans maintain protein intake relatively constantly while allowing nonprotein energy to vary more widely. Adapted from Raubenheimer and Simpson (2016).

tegrate variables expressed as amounts (e.g., calories or grams eaten). To illustrate, figure 25.17a shows the experimental data from figure 25.15 replotted as dietary macronutrient ratios. The figure shows that these data collectively span a wide range of macronutrient ratios, some of which are balanced with respect to the Australian/New Zealand AMDR (fall within the shaded polygon) and others are imbalanced in various dimensions. For comparison, I have superimposed the US AMDR on the plot, which spans the same ranges of carbohydrate and fat as the Australian/New Zealand AMDR but a wider range of permissible proportional protein intakes (10%–35% vs. 15%–25%) and therefore encompasses a wider range of the diets. Since in all the studies each subject received a single diet of fixed macronutrient composition (the experimental manipulation) and the response variable was the amount eaten, these data enable us to build a comprehensive picture of how three-dimensional dietary macronutrient compositions influenced absolute energy intake in the study population. To examine this, a response surface representing daily energy intakes is superimposed onto the proportional data (fig. 25.17b). The plot shows that energy intake increased with decreasing dietary percent protein (from right to left on the x-axis), whereas there was no effect of the ratio of fat:carbohydrate (increases up the y-axis) on energy intake.

In the same way that regions of the macronutrient space can be delineated based on proportional compositions, as illustrated for the AMDR, so too can the space be partitioned into regions based on variables expressed in the model as absolute amounts. For example, in figure 25.17b I superimposed on the surface plot a contour representing the value for the estimated average equilibrium energy intake (EEI) for the experimental subjects (the dashed line). This parameter represents the energy consumption that matches the average estimated energy expenditure, taking into account sex, body size, and activity levels (Raubenheimer, Gosby, and Simpson 2015). EEI is thus a threshold that distinguishes the regions in the space where subjects are in positive energy balance (eat more energy than they expend, to the left of the threshold) and negative energy balance (eat less energy than they expend, to the right). Including EEI in this way enables us to link energy balance to dietary macronutrient ratios and evaluate the implications for energy balance of complying with recommendations for dietary macronutrient ratios (i.e., the AMDR).

Overall, the model shows that in this population, ad libitum energy intake increased as the dietary protein

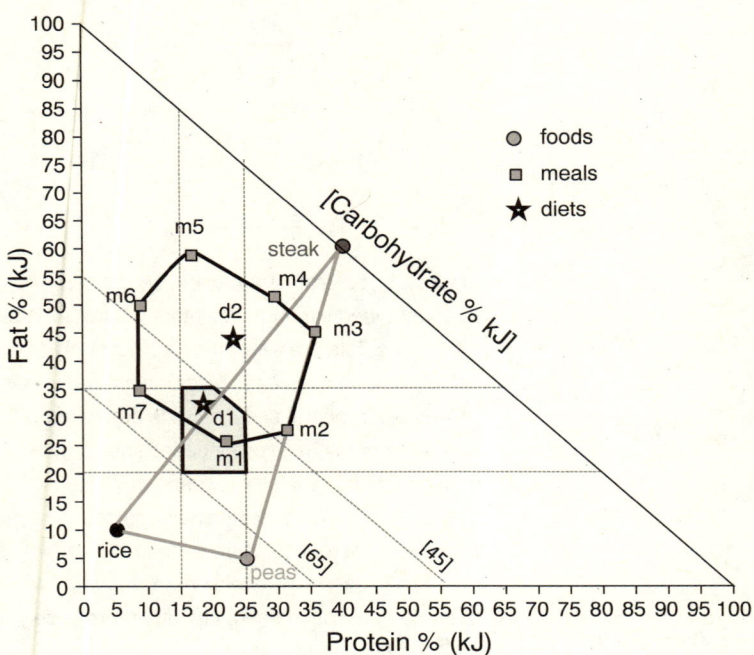

Figure 25.16. Nutritional mixture hierarchy modeled using the right-angled mixture triangle. Nutrients combine into foods (circles)—in this case peas, rice, and steak. Foods, in turn, combine into meals (squares, m1–m7) and meals into diets (stars, d1 and d2), following the logic illustrated in figure 25.4. Other mixtures, such as meals and population-level dietary patterns, could equally be represented in this way. The dashed lines represent the recommended ranges for the proportional contributions of protein (15%–25%), fat (20%–35%), and digestible carbohydrate (45%–65%; Acceptable Macronutrient Distribution Ranges [AMDR]) for the diets of Australians and New Zealanders. Any food, meal, or diet that falls within the conjunction of these ranges, shown by the shaded polygon, is thus balanced with respect to the recommendations—for example, meal m1 and diet d1—and all others are imbalanced. Modified from Raubenheimer and Simpson (2016).

energy ratio of the experimental foods decreased, as expected from protein leverage. The fact that the EEI line bisected the AMDR region is encouraging, because it suggests that the recommendation of 15%–25% protein energy in the diet is likely to lead to equilibrium energy intakes, whereas diets lower and higher in protein are associated with ingested energy excesses and deficits, respectively. This explains why high-protein diets, such as the Atkins, Protein Powder, Paleo, and Sugar Busters diets (shown in the figure), are recommended for weight loss: their macronutrient ratios are associated with spontaneously reduced energy intakes.

On the other hand, extending the model further suggests that benefits of high-protein diets for reducing energy consumption must be weighed against negative health impacts associated with these diets. Figure 25.17c shows a response surface for absolute protein intake associated with the experimental diets. Even though protein is regulated more strongly than nonprotein energy, absolute protein intake nonetheless increased with increasing dietary percent protein (left to right on the x-axis), albeit to a lesser extent than energy increased with decreasing percent protein (fig. 25.17b). There now exists strong evidence that excessive protein intake by humans and many other animals is associated with accelerated aging and premature onset of several diseases, including cancers (Fontana and Partridge 2015; Le Couteur, Solon-Biet, Cogger, et al. 2016; S. Simpson, Le Couteur, and Raubenheimer 2015). Consistent with this is the fact that dietary patterns associated with exceptionally long life spans and healthy metabolic profiles, such as the traditional Japanese Okinawan, Mediterranean, and Kitivan Islander diets, have notably low protein energy ratios (fig. 25.17c). The likely reason that these diets are not associated with energy overconsumption, as predicted by the model in figure 25.17b, is their high fiber content, which in humans is associated with satiety (E. Chambers et al. 2015; Raubenheimer, Gosby, and Simpson 2015).

This model demonstrates how human macronutrient appetites interact with food composition to influence energy intake, nutrient intake, and ultimately health. Most importantly, it situates human nutrition in an ecological context by highlighting questions about the environmental factors that influence dietary macronutrient compositions, which interact with appetites to drive energy and nutrient intakes. One such factor is national dietary guidelines. As shown in figure 25.17, the USDA AMDR spans a wider range of proportional protein intakes than the Australian/New Zealand AMDR,

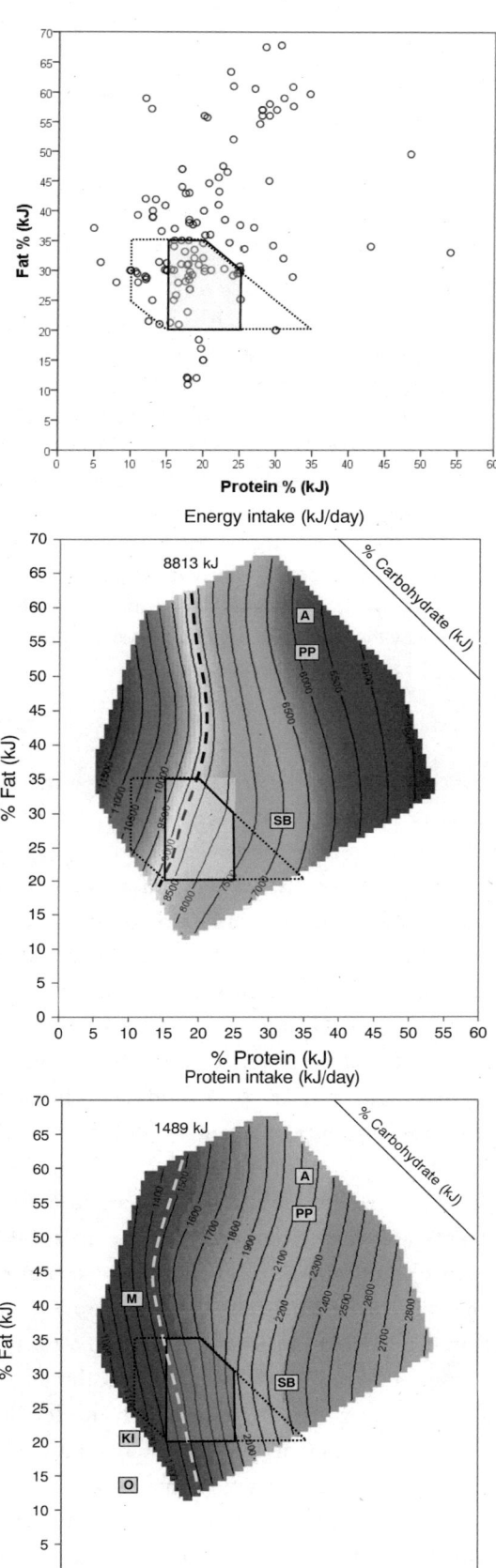

Figure 25.17. Integrating proportions and amounts in right-angled mixture triangles to examine relationships between dietary macronutrient balance, national recommendations for dietary macronutrient ratios, and absolute energy and nutrient intakes. *a*, Three-dimensional macronutrient compositions of experimental diets from the literature data plotted in figure 25.15. Some diets are balanced with respect to the Australian and New Zealand AMDR (fall within the shaded polygon), and others are imbalanced in various dimensions. For comparison, the US AMDR is also plotted (the dotted polygon), which encompasses a wider range of protein densities than the Australian and New Zealand recommendation. *b*, Response surface for measured absolute energy intakes associated with the dietary compositions plotted in *a*. Also plotted is the average equilibrium energy intakes for the population of experimental subjects, or the energy intake that matches the estimated energy expenditure (the dark dashed contour). Diets with compositions that fall to the left of this are predicted to lead to positive energy balance (intake > expenditure) and to the right negative energy balance (intake < expenditure). Consistent with this is the fact that high-protein weight-loss diets, such as the Atkins (A), Protein Power (PP), and Sugar Busters (SB) diets, fall within the area to the right associated with low ad libitum energy intakes (compositions from J. Anderson et al. 2000). *c*, Response surface showing absolute protein intakes associated with the same experimental diets, together with calculated approximate average protein requirements for the study population (the dashed contour). The model shows that the USDA AMDR spans areas predicted to be associated with both excessive energy intake (to the left of the protein axis) and excessive protein intakes (to the right). Also plotted are the approximate macronutrient profiles of three dietary patterns that are associated with long lifespans and healthy metabolic profiles, the Mediterranean diet (M), Kitivan Islanders diet (KI), and traditional Okinawan diet (O), all of which have low proportional protein content. Adapted from Raubenheimer, Gosby, and Simpson (2015).

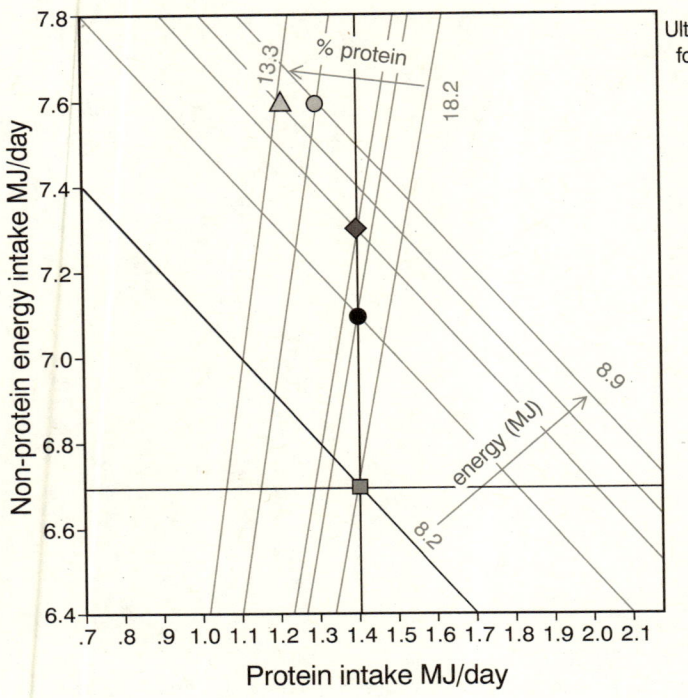

Figure 25.18. Diet survey data showing the relationship between the contribution of ultraprocessed foods to the diets of Americans, dietary macronutrient ratios, and absolute energy intake. The plot shows that increased intake of these foods (from medium gray box to light gray triangle) is associated with decreased proportional dietary protein content (the gray radials) and increasing absolute energy intake (the gray diagonals), as predicted by the protein leverage hypothesis. From Martinez-Steele et al. (2018).

encompassing regions associated with both excess energy intake (to the left of the protein axis) and excess protein intake (to the right). If this is substantiated using population data, it raises a case for reconsidering the protein values encompassed by the USDA AMDR. Another candidate ecological factor, which already is widely implicated in the obesity epidemic, is the incursion into human food systems of highly processed industrial foods (ultraprocessed foods; Monteiro et al. 2013). These are typically energy dense, low in nutrients, and engineered to be hyperpalatable (Moss 2013; S. Simpson and Raubenheimer 2014). Martinez-Steele et al. (2018) used the US National Health and Nutrition Survey data to examine the hypothesis that protein leverage is the mechanism through which ultraprocessed foods influence energy overconsumption. The results were highly consistent with the protein leverage hypothesis: dietary percent protein decreased and energy intake increased as the proportional contribution of ultraprocessed foods to the diet increased (fig. 25.18). Similar results are reported by Grech et al. (2022) for Australia. Another example concerns economics, which is arguably the most salient aspect of contemporary human food environments. To examine a role for this in excess energy consumption, Brooks et al. (2010) analyzed the relationship between the dollar cost of supermarket foods and their macronutrient composition. As shown in figure 25.19, the cost of foods was negatively associated with their protein content, suggesting that budgetary considerations might be another factor that entices humans to eat obesogenic diets. Other candidate environmental influences are discussed in Raubenheimer, Machovsky-Capuska, Gosby, et al. (2014), including an impact of rising atmospheric carbon dioxide on the macronutrient ratios of agricultural foods and a possible influence of feeding human infants high-protein milk formula.

Finally, an interesting question is how modern food environments and diets compare with the Paleolithic environments where most of our biological evolution took place. Figure 25.20a shows a reconstruction of the macronutrient composition of multiple Paleolithic dietary patterns (from Raubenheimer, Rothman, et al. 2014). These estimates were derived from detailed models that varied plant/animal subsistence ratios, included paleoecological information spanning a range of hunter-gatherer habitats, and were parameterized using data on the compositions of relevant East African plant and animal foods as well as realistic pathophysiological constraints (Kuipers et al. 2010). For comparison with

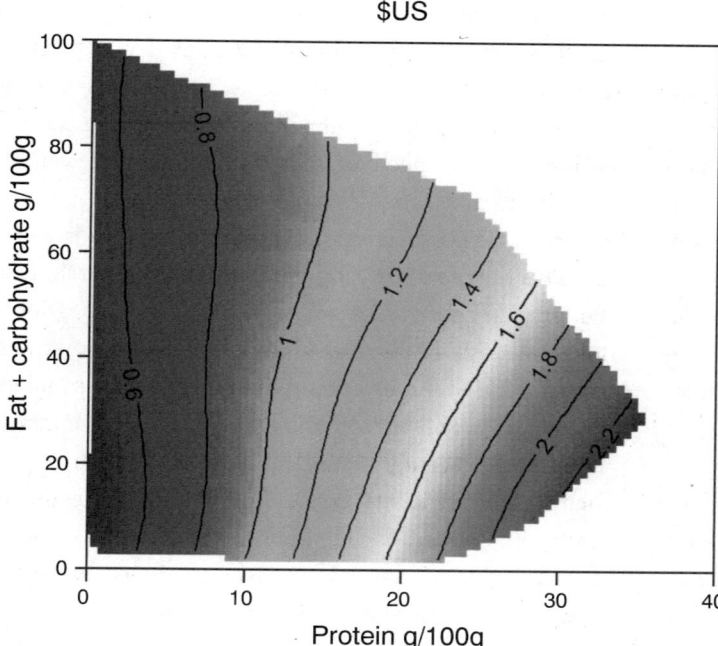

Figure 25.19. Relationship between macronutrient composition and the cost ($US) of 106 supermarket foods. The graph shows that the dollar cost increases with food protein density (left to right), but there is no effect of fat and carbohydrates (along the y-axis). Modified from Raubenheimer, Machovsky-Capuska, Gosby, et al. (2014).

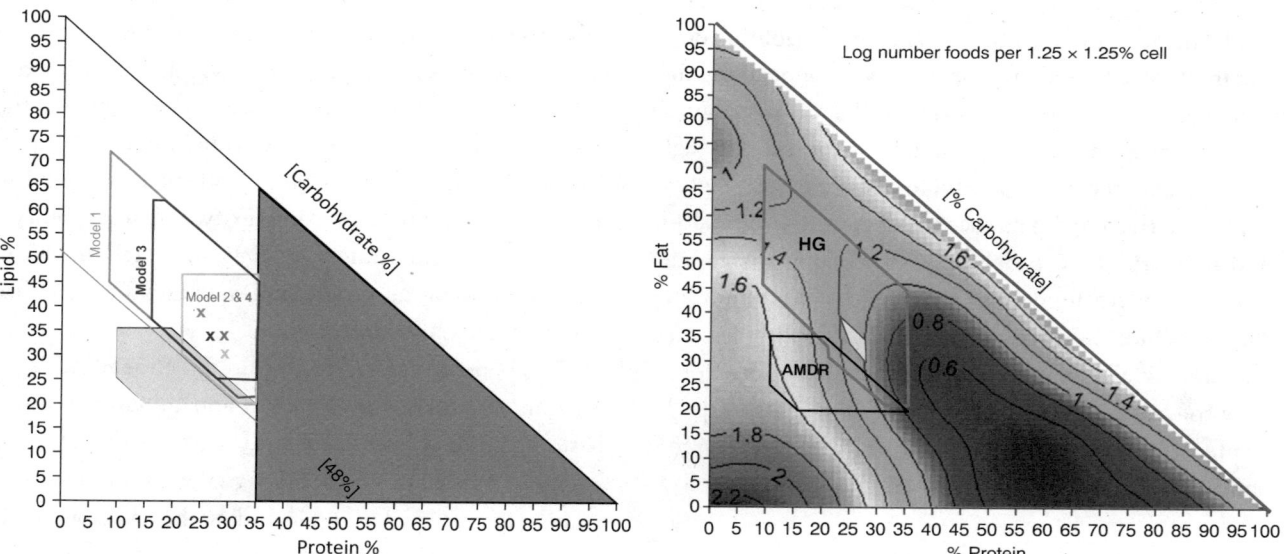

Figure 25.20. Comparison of macronutrient distributions in Paleolithic and modern food environments. a, Estimated macronutrient compositions of the diets of African Paleolithic hunter-gatherers compared with recommended intakes for the US (the US AMDR, light gray solid polygon). The light gray, medium gray, and dark gray polygons represent estimated maximal ranges of proportional macronutrient intakes under four different foraging models presented by Kuipers et al. (2010). Together, these models encompass a wide range of possible ecological and behavioral scenarios (e.g., nonselective vs. selective feeding). They also integrate physiological constraints—for example, the maximum proportion of protein energy tolerable over the long term by humans (the dark gray triangle shaded area). The X's represent the estimated median diets under the four models. Modified from Raubenheimer, Rothman, et al. (2014). b, As a representation of the modern food environment, a response surface is superimposed showing the logarithm of the counts of foods from the USDA food composition database that fall within 1.25% × 1.25% cells across the proportional macronutrient space (modified from Coogan and Raubenheimer 2016). The polygon with the light gray border represents the most inclusive model from a, and the small gray-filled polygon indicates the range covered by the most likely actual hunter-gatherer diets (X's in a).

macronutrient compositions of Paleolithic diets, the US AMDR region (as introduced in fig. 25.16b) is also plotted. A clear conclusion is that relative to the AMDR, the models of Kuipers et al. suggest that carbohydrate energy was low in the diets of Paleolithic foragers. Even the highest estimate from the most inclusive model places the carbohydrate intake of Paleolithic diets at 48%, which is very close to the lower limit recommended for carbohydrate intake in the US AMDR (45%). The median estimates for all four models (the crosses in the figure) had lower carbohydrate than the recommendation for contemporary American diets.

Figure 25.20b presents information on the macronutrient compositions of the contemporary American food environment (modified from Coogan and Raubenheimer 2016). The response surface represents the macronutrient compositions of 7,613 foods and dishes in the USDA food compositions database. This macronutrient distribution was quantified by partitioning the macronutrient space into 1.25% × 1.25% cells and counting the number of foods that fall within each. One striking conclusion is that unlike the macronutrient constraints on Paleolithic diets, the modern US food environment will support any diet composition. A second conclusion is that the distribution of modern foods is heavily weighted toward the high-carbohydrate region (red) and, to a lesser extent, high fat (orange). This highlights the role of culture—agriculture and industrial food processing—in making hyperabundant in modern food environments the macronutrients that were most limiting in the diets of Paleolithic hunter-gatherers. It also provides perspective on the question of how the protein energy ratio of modern diets might be diluted, leading to protein leverage and energy overconsumption, as shown in figure 25.18.

SUMMARY AND CONCLUSIONS

The fundamental importance of nutrition to living systems across scales from molecular and cellular biology to ecosystems science holds tremendous potential for deepening integration across the life sciences. Primates offer a model system for progressing this agenda. The amenability of many primates to detailed noninvasive field studies provides unparalleled opportunities for examining ecological and evolutionary aspects of foraging and nutrition in species with varied and complex social structures, and they have among their ranks the species that attracts by far the most research effort, humans. The benefits of combining the rich data for humans with the ecological and evolutionary perspectives offered by primatology cannot be overstated.

From the perspective of human health, there is widespread agreement that human nutrition research falls significantly short of its potential for solving the global nutrition crisis (Allison et al. 2015; Raubenheimer and Simpson 2016). An important reason is that the problem has been approached from the wrong level. Human nutrition science is centered on the medical paradigm, where chemistry (nutrients) meets physiology to generate health and disease. The crisis, however, is largely ecological and evolutionary—it has arisen out of changes to our food environment that are too recent, too rapid, and too extreme for a long-lived species to adapt. As I have tried to demonstrate in this chapter, ecological and evolutionary perspectives from the biological sciences can make fundamentally important contributions to conceptualizing and modeling human nutrition in this context by providing a systems perspective for integrating existing data and directing future research guided by fundamental theory from biology (see also Raubenheimer and Simpson 2016). Nonhuman primates are indispensable for this. Their diversity, biological complexity, and amenability to field studies offers a unique bridge over which theoretical and methodological progress in nutritional ecology can benefit human nutrition science.

The benefit is by no means unidirectional. Added to the immense influence of human nutrition in developing basic nutritional knowledge and methods, benefits that are more subtle but potentially equally important are beginning to emerge. My aim in ending this chapter with an overview of the application of the NGF to humans was to demonstrate how, given the right data, the framework provides a quantitative template for examining in detail the interactions between biological traits such as appetites and food environments. Its application to humans has played an important role in nuancing and extending the NGF, and in my own research it has been a powerful aid for asking focused questions in primate nutritional ecology. Human nutrition also provides a detailed experiment for examining the causes and consequences of mismatches between evolved biological systems and rapid environmental change.

In a world where human activities are substantially impacting not only our own environments and diets but all environments and the diets of many other species, it is more important than ever to understand, predict, and manage the effects of such changes on nonhuman primates, for their contributions not only to biodiversity and to ecological functioning but also to human nutrition research. With 60% of primate species listed as vulnerable, endangered, or critically endangered, many of which are predicted to go extinct by the end of the century in the absence of conservation efforts to protect them (Estrada, Garber, et al. 2017), there is no time to waste.

26 Reconstructing Fossil Primate Diets
Dental-Dietary Adaptations and Foodprints for Thought

Peter S. Ungar

Paleontologists use relationships between the diet and teeth of living vertebrates to infer diet from teeth of fossil species. In this chapter, I review the basic methods and theories developed to unravel and understand these relationships, with a focus on primates. I first consider dental-dietary adaptations and relationships between the material properties of the foods a species evolved to eat and tooth size, shape, and structure. I then consider primate feeding and the perspective that brings to relationships between tooth function and form. Such studies make clear that dental-dietary adaptations reflect what past species were capable of consuming, but not necessarily their food preferences. A discussion of foodprints, actual traces of past feeding activity, follows.

Foodprints give us clues as to what a specific individual alive in the past actually ate over a period of time. The ultimate goal is to, as George Gaylord Simpson (1926, 228) once wrote, "consider a very ancient and long extinct group of mammals not as bits of broken bone but as flesh and blood beings." Studies of dental ecology (sensu Cuozzo and Sauther 2012), or how primates in the wild use their teeth, suggest that both adaptive evidence and foodprints are needed for a comprehensive paleoecological approach to understanding food choice and foraging strategies in the past.

DENTAL-DIETARY ADAPTATIONS

Most studies of dental-dietary adaptation in extant primates focus on tooth size, crown shape, or enamel structure (see Ungar and Lucas 2010). Each of these lines of evidence is built on a separate set of theories, has the potential to teach us something about feeding adaptations of extinct primate species, and has its inherent limitations.

Tooth Size

The basic idea behind the relationship between tooth size and diet is that bigger teeth provide greater capacity at the front of the mouth to prepare food for ingestion and larger platforms in the back for processing it during mastication (Groves and Napier 1968). Incisor row length and molar (or summed cheek tooth) occlusal area are historically the most frequently considered, as these are easy to measure in a consistent manner; and data abound in the literature (see Ungar 2014 for review).

Incisor Size

There is indeed a tendency among higher primates for those with larger incisors to consume foods requiring front tooth use during ingestion. Anthropoids reported to feed on large, husked fruits requiring extensive incisal preparation, such as orangutans and chimpanzees, tend to have larger front teeth relative to their body masses than do primates that more often eat leaves or small berries, such as gorillas and gibbons (Hylander 1975). There is, however, an offset in incisor size among higher-level taxa; catarrhines have broader front teeth than platyrrhines, and hominids have wider ones than hylobatids, all else being equal (Eaglen 1984; Ungar 1996). Furthermore, among strepsirrhines, there is no consistent relationship between diet and incisor size, evidently because the lower anterior teeth function not just in food processing but also as a dental comb independent of diet (Eaglen 1986).

Molar Size

Relationships between molar size and diet are less clear. The focus to date has been on allometry—specifically, how occlusal area varies with body mass. Kleiber (1947) observed that while larger mammals need absolutely more energy to fuel bigger bodies, that need does not increase one-to-one with body size. He found instead that metabolic rate for mammals scales at the ¾ power to body mass ($M_b^{0.75}$)—an elephant that weighs 500,000 times as much as a mouse needs only about 19,000 times as much energy. Because larger mammals have less surface area per unit volume, they must burn fewer calories per unit mass to maintain their body temperatures. Pilbeam and Gould (1974) therefore argued that chewing surface area should also scale at $M_b^{0.75}$ to match the amount of food required to meet metabolic needs. Because a one-to-one relationship between area (two dimensions) and volume (three dimensions) would result in a scaling rate of $M_b^{0.67}$, that would mean occlusal area should be positively allometric. And there is some support for this among mammals, at least for rodents, pigs, and deer (S. Gould 1975).

In comparisons of primate species with similar diets, however, occlusal area actually does scale at $M_b^{0.67}$, 1:1 with body mass (Kay 1975). Their molar platforms might be slightly positively allometric for occlusal area across diet categories in some cases, but if so, it is likely only indirectly a scaling phenomenon; larger species tend to eat more leaves and other low-quality foods, which require more processing and select for larger chewing platforms.

What about Kleiber's rule, then? If larger primates have lower mass-specific metabolic rates, do they actually need smaller cheek teeth to process less food of a given type for a given body size? Fortelius (1988) proposed an elegant solution to reconcile theory and fact. Chewing rate scales at about $M_b^{-0.25}$, so larger animals chew more slowly than smaller ones. If food volume in the mouth scales 1:1 with body mass, the amount of food entering the gut per unit time should be $M_b^{1.0} \times M_b^{-0.25} = M_b^{0.75}$. In other words, because larger species chew more slowly, food entering the gut at any given time should match Kleiber's rule. In the end, reality has proved to be somewhat more complicated, but the general idea has stood the test of time reasonably well (see Ungar 2014 for review).

Allometry issues aside, there is a tendency for folivorous apes (*Gorilla* spp. and *Hylobates syndactylus*) and New World monkeys (e.g., *Alouatta* spp.) to have relatively more occlusal surface area than do closely related frugivores (Kay 1985). But the pattern does not hold for all higher-level primate taxa. Frugivorous Old World monkeys as a group actually have larger molars than do folivorous species (Ungar 1998). This might be explained by the fact that colobines have shorter faces than cercopithecines, perhaps an adaptation to improve leverage for chewing muscles (J. E. Scott 2011). Shorter faces mean less room for large teeth. But other questions remain. Why don't folivorous strepsirrhines have larger molars than closely related insectivores or frugivores (Strait 1993; Vinyard and Hanna 2005)? And why do females of sexually dimorphic species tend to have relatively larger cheek teeth than males (Harvey, Clutton-Brock, and Kavanagh 1978)? It can be difficult to infer diet for fossil species when the form-function relationship lacks consistency in living taxa (Kay and Cartmill 1977).

Crown Shape

Relationships between crown form and function are somewhat more consistent across the primate order (fig. 26.1). Cheek teeth operate as both guides for chewing motions and tools for fracturing food items. These correspond roughly to P. Butler's (1983) "internal" and "external" environments or A. Evans and Sanson's (2006) "geometry of occlusion" and "geometry of function."

G. Simpson (1933) developed the basic model for teeth as guides. He argued that chewing was largely about the way opposing teeth fit together, which depends on the shapes of tooth crowns. Crompton and Hiiemae (1969, 1970) worked out the details of how tooth form relates to mastication with in vivo studies of opossums using cineradiography, or X-ray movies. They found that vertical movement combined with steep crests running parallel to the plane of motion results in shearing. Vertical movement combined with broad cusps and deep basins, on the other hand, results in crushing. And the combination of crushing and shearing with small crests on a horizontal surface results in grinding. This work was later extended to

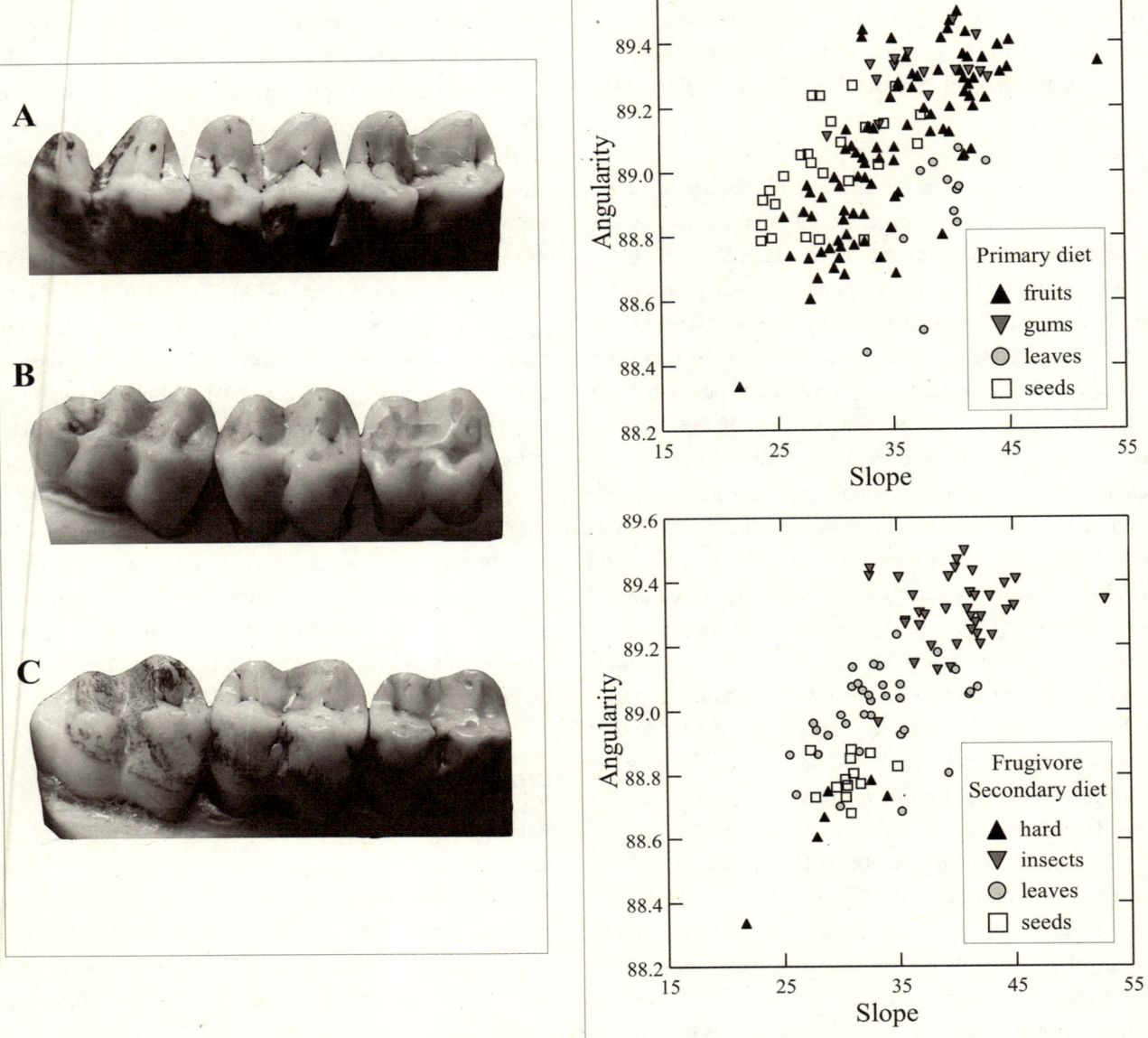

Figure 26.1. Dental topography. *Left*, compare crest lengths and crown relief on the molars of a hard-object-adapted mangabey (*A*), a frugivorous macaque (*B*), and a folivorous langur (*C*). *Right*, average molar slope versus angularity for primarily fruit-, gum-, leaf-, and seed-eating platyrrhine primates (*above*) and for frugivorous platyrrhines that supplement their diets with hard objects, insects, leaves, and seeds (*below*). All data points represent moderately worn M2s (data from Ungar, Healy, et al. 2018).

primates, whose chewing actions are similarly described by the combination of tooth movement and crown shape (Hiiemae and Kay 1972; Kay and Hiiemae 1974).

Lucas (2004), in contrast, has considered teeth as complex tools evolved for breaking food. One can think of chewing in terms of an evolutionary "arms race" (sensu Dawkins and Krebs 1979), with foods evolving ways to strengthen their tissues to resist fracture and teeth de-

veloping structures to overcome those defenses without themselves being broken in the process (Ungar 2008). Some foods are hard, with defenses to prevent a crack from starting. These are called "stress-limited" defenses. Others are tough, with defenses to resist the spread of a crack. These are called "displacement-limited" defenses (Lucas, Turner, et al. 2000). A tooth with rounded, hemispherical cusps is well suited to fracturing hard foods; it concentrates force but at the same time is blunt enough to protect the crown from being broken. Teeth with long, narrow blades work better on tough foods; cracks spread as they are wedged apart at the advancing tip. And pliable foods tend to spread across a blade for a compressive stress field and increased contact area that minimize risk to the tooth (Lucas 2004; Ungar and Lucas 2010).

MEASURING TOOTH FORM IN PRIMATES

Researchers use several tools for characterizing and comparing functional aspects of occlusal form. This section reviews some of the more popular approaches. Kay's (1984) shearing quotient (SQ) has long been the gold standard. The summed length of mesiodistal crests on a molar are plotted against tooth length for related species with similar diets, and a regression line is used to describe the relationship. The SQ is a measure of the deviation from that line and indicates whether a species in that group has longer or shorter crests than expected for that diet. Primates described as folivores, such as *Presbytis* and *Colobus*, tend to have higher SQs and longer crests than soft fruit eaters, such as *Macaca* and *Cercopithecus*; and among frugivores, those that eat hard foods, such as *Lophocebus*, tend to have lower SQs and blunter cusps (see fig. 26.1). Likewise, among insectivores, hard-object feeders, such as the beetle-eating *Galagoides*, have shorter shearing crests than the soft-caterpillar-eating *Arctocebus*. And the trend relating crest length and food properties holds true for strepsirrhines, platyrrhines, cercopithecoids, and hominoids (Anthony and Kay 1993; Kirk and Simons 2001; Meldrum and Kay 1997; Strait 1993). Other measurements of occlusal functional morphology, such as the radius of curvature of cusps and relative areas of basins to cusps, have yielded similarly promising results (e.g., Yamashita 1998).

One approach that has become increasingly common over the past few years is dental topographic analysis. This involves 3D modeling of occlusal surfaces and calculation of surface topography attributes. Dental topographic analysis has the advantage of allowing comparisons of whole crowns and variably worn teeth. Initial efforts included measurements of slope, relief, and angularity using geographic information systems (GIS) models (Ungar and Williamson 2000; Zuccotti et al. 1998). Results from such studies confirmed that folivorous primates tend to have more occlusal relief and more sloping, angular surfaces than do closely related frugivores, and that among frugivores, hard-object feeders have the least slope and relief (J. Bunn and Ungar 2009; M'Kirera and Ungar 2003; Ungar and Bunn 2008; Ungar and M'Kirera 2003; see fig. 26.1). Moreover, while teeth tend to get flatter with use for given species, differences among taxa seem to hold at comparable stages of wear. Researchers have even documented variation between individuals within species with different diets, particularly when considering those in anthropogenically disturbed habitats (Cuozzo, Head, et al. 2014; Yamashita et al. 2015). Ring-tailed lemurs (*Lemur catta*) in disturbed habitats, for example, show more rapid wear-related tooth shape change, presumably due to the consumption of mechanically challenging foods. Moreover, this approach has been useful for determining how wear sculpts teeth to maintain functional efficiency for breaking down foods (J. Dennis et al. 2004; Pampush, Spradley, et al. 2018; see below).

Efforts have continued to develop other functionally relevant characterizations of dental occlusal topography. Evans, Jernvall, and their colleagues, for example, have measured crown complexity by counting the number of contiguous areas of similar aspect (they call these "patches"; A. Evans 2013; A. Evans and Jernvall 2009; A. Evans et al. 2007). Herbivorous mammals with tough, fibrous diets tend to have higher patch counts, or more complex surfaces, than carnivores. This approach allows researchers to directly compare distantly related species and has been applied to both living and fossil primates (Boyer et al. 2010, 2012; J. Bunn et al. 2011; Godfrey et al. 2012; Ledogar et al. 2013; Winchester et al. 2014). Another technique gaining popularity is Dirichlet normal energy, which measures tooth surface curvature using tools from differential geometry (J. Bunn et al. 2011; Pampush, Morse, et al. 2022). This also separates primate species by diet, as folivore and insectivore teeth with taller, sharper cusps have higher values than those of frugivores (Winchester et al. 2014).

Tooth Structure

Researchers also look to tooth structure for dental adaptations related to diet. Teeth must concentrate and pass on the forces needed to break foods without themselves being broken in the process. Different diets mean different stress environments for teeth (see van Casteren and Lucas, this volume). This suggests that natural selection should favor internal architecture of teeth appropriate to foods eaten to buttress dental tissues against fracture. In theory, then, we should be able to look to internal structure to determine the stress environment a tooth type evolved in and, by extension, to understand something about the food a species is adapted to eat.

Enamel Thickness

Most functional studies of enamel structure of primates have focused on the thickness of the enamel cap that covers the crown. It has been suggested that primates with thick enamel evolved it to prolong the life of a tooth in an abrasive feeding environment, to strengthen that tooth against fracture given a hard-food diet, or some combination of the two (see Ungar 2015 for review). The abrasive-versus-hard foods debate has been central to paleoanthropology because many hominins had relatively thicker dental enamel than do living African apes (J. T. Robinson 1956). The idea that thick enamel evolved to prolong the use life of teeth given the grit-laden, abrasive foods that hominins ate on the ground has had many supporters (Simons 1976; Simons and Pilbeam 1972), with R. Smith and Pilbeam (1980) going so far as to propose that orangutans must have had terrestrial ancestors because they have thick enamel today. A new twist to the abrasive-diet hypothesis is the observation that primates with thick enamel also tend to consume plant foods high in biogenic silica, or phytoliths (Rabenold and Pearson 2011, 2014). Perhaps it is not a coincidence, then, that the type specimen of the most thickly enameled hominin, *Paranthropus boisei*, has much of its first molar occlusal enamel worn away despite its third molar not being fully erupted.

That said, most researchers who study enamel thickness in primates today consider it an adaptation to resist tooth failure given a hard-food diet. The association between life on the savanna and mechanically protected foods, such as hard, dry seeds, has deep roots too (C. Jolly 1970b). Besides, there is no clear association between terrestriality and enamel thickness in living primates (Kay 1981). Moreover, there is no need to invoke a hypothetical ground-dwelling ancestor for orangutans because they more often consume exceptionally hard foods than do other great apes (E. Vogel, van Woerden, et al. 2008). Thick enamel is evidently not compensation to match rapid tooth wear for orangutans today, as they typically wear their crowns less than other great apes (Dean et al. 1992). Indeed, the evidence for thick enamel as an adaptation for hard-object feeding among at least some primates seems beyond doubt (Constantino, Lee, et al. 2012; Constantino, Lucas, et al. 2009; Dumont 1995; J. Lambert et al. 2004; Pampush, Duque, et al. 2013; E. Vogel, van Woerden, et al. 2008).

Primate teeth are strong because they are composites (fig. 26.2). The outer enamel cap is about 97% hydroxyapatite mineral by weight—extremely hard but also brittle. Once a crack is initiated, it takes little work to propagate it through the crown. The dentin that underlies it, in contrast, is tougher and more elastic. Dentin is only about 70% mineral; most of the rest consists of protein/collagen fibers. A tooth's response to chewing loads can be changed by setting the proportion of these tissue types to best match a species' needs. Because enamel is so stiff, for example, increasing its relative contribution should mean less deformation and less risk of fracture with compression (Popowics et al. 2001). This does not mean, of course, that thick enamel is not adaptive for extending the life of a tooth too. As Constantino, Lee, and coauthors (2012) have suggested, large, hard objects, such as fruit pits, can select for thick enamel to stop cracks, whereas small, hard abrasives can select for thick enamel to mitigate the effects of tissue loss.

Part of the challenge in interpreting enamel thickness has been inconsistency among dental researchers in its characterization (Teaford 2007). The molar enamel of gorillas, for example, is thin when considered relative to body mass (Shellis et al. 1998). However, in absolute terms, which may be more relevant for functional studies since size is considered in the equation for the maximum load a tooth can bear without failing, gorillas actually have moderately thick enamel—close to that of orangutans (Constantino, Lucas, et al. 2009). In fact, gorillas and orangutans have similar absolute crown strength, greater than that of chimpanzees (G. Schwartz et al. 2020). In addition, the distribution of enamel within the crown differs between species (G. Schwartz 2000; Thiery et al. 2019). Chewing stresses vary across a tooth depending in large part on diet, and enamel should be laid out in a manner best suited to reinforce that tooth against whatever

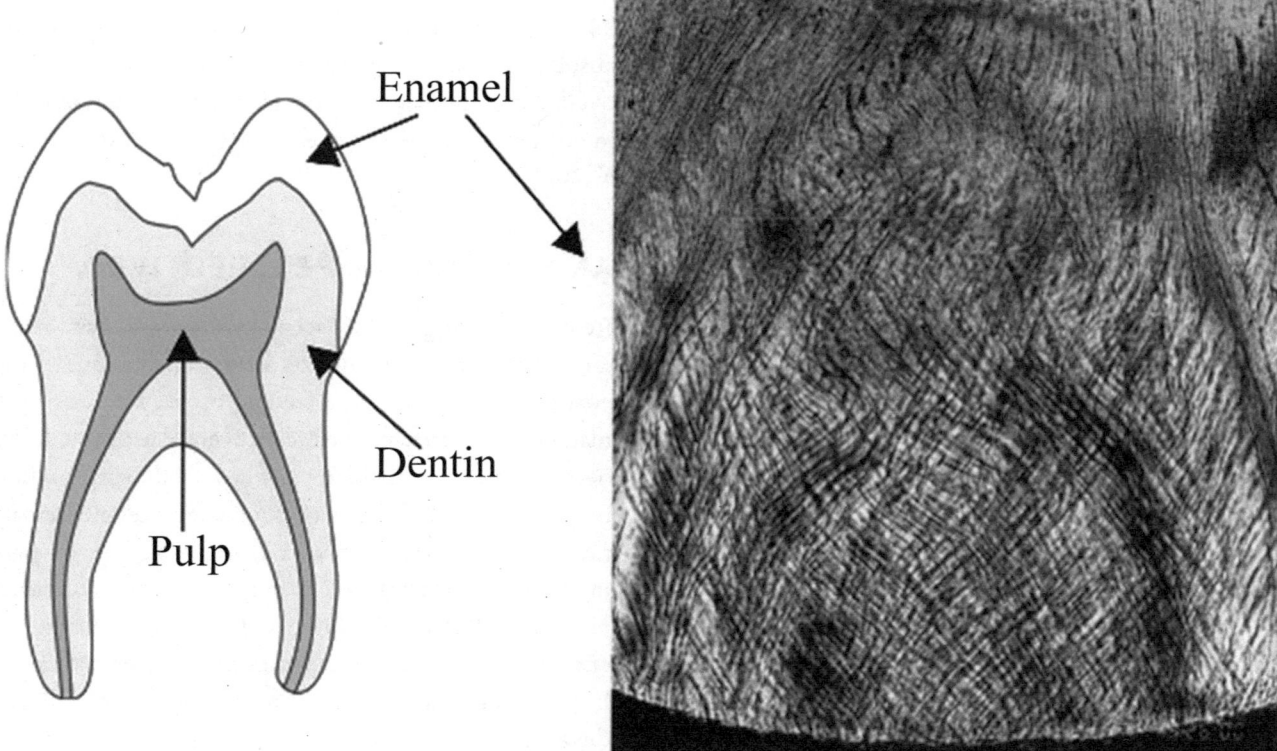

Figure 26.2. Enamel structure. Note the wriggling and interweaving of enamel prisms from dentin edge to occlusal surface on this *Archaeolemur* tooth (micrograph courtesy of Gary Schwartz).

stresses it encounters during chewing. A useful analogy is the position of the metal hoop surrounding a barrel, there to reinforce it and keep the weight of wine from cracking the wood (Lucas, Constantino, et al. 2008). Just as hoop placement can teach us something about the forces acting on a barrel, enamel buttressing can help us understand the forces acting on a tooth. So, while aggregate measures of enamel volume may hint at dental-dietary adaptation, we really need to understand its distribution across the crown to see the whole picture.

There are also challenges associated with the interpretation of enamel thickness. Besides the abrasive-versus-hard food debate, phylogeny can play an important role in enamel thickness, all else being equal (Dumont 1995). Moreover, competing needs might actually select for thinner enamel in some cases and in some parts of the crown (Ungar 2008). Many mammalian herbivore species have "secondary morphology" (Fortelius 1985), wherein teeth require wear to become functionally efficient. Guinea pigs (*Cavia porcellus*), for example, grind their teeth in utero so that they are worn and ready to go shortly after birth (Teaford and Walker 1983). Sharp edges can appear as abrupt interfaces formed between harder enamel and softer dentin exposed with wear (see Ungar 2005). Thus, there is selection for thinner enamel in specific parts of the crown to literally sculpt a tooth by wear to make or maintain the best tool for the job of fracturing whatever food a species is adapted to eat (Ungar 2011). It is no wonder, then, that while occlusal slope and relief drop with wear, angularity (or jaggedness) tends to be maintained throughout most of molar wear in strepsirrhines, platyrrhines, cercopithecoids, and hominoids (see J. Bunn and Ungar 2009; Cuozzo, Head, et al. 2014; J. Dennis et al. 2004; Ungar and M'Kirera 2003; and below).

Enamel Microstructure

Another way to strengthen enamel is to modify its microscopic structure. Enamel forms from long, thin hydroxyapatite nanofibers, each about 60 nm in diameter.

Primates bundle these by the thousands into rods, or prisms, each measuring 2–10 μm across. Those prisms are, in turn, packed together in rows, with long axes running from the dentin out toward the tooth surface. Their course tends to be straight and parallel as they approach the outer edge, but the angle at which they strike the surface varies, and this affects both wear resistance and stiffness (Macho and Shimizu 2009; Shimizu et al. 2005). Researchers are only now beginning to determine how this relates to the forces of mastication, food fracture properties, and dietary abrasiveness.

Equally important to considerations of enamel microstructure is what happens deep beneath the surface of the crown. Because dentin flexes more than the enamel that overlies it, cracks tend to start at or near the boundary between the two tissue types; and they are prone to spread along planes of weakness between adjacent rows of prisms (Lucas, Constantino, et al. 2008). As a result, primates and many other mammals tend to have layers of prisms that wriggle in waves along their paths near the enamel-dentin junction. That forces a crack to change directions, which increases the work it takes to fragment the enamel, thereby strengthening the tooth. Adjacent layers of prisms can also interweave and be stacked horizontally, vertically, or in a zig-zag manner (fig. 26.2). Stacks of layers can themselves be arranged into layers of layers to further strengthen enamel (Maas and Dumont 1999; Rensberger 2000). This complex microstructure can stop a crack in its path by absorbing the energy needed to propagate it.

Work on the relationships between primate dental microstructure and diet remains underdeveloped, but it is encouraging that orangutans, which are evidently adapted for hard-food consumption, have more decussated enamel than do other great apes (Macho et al. 2003). Another interesting finding is that there seems to be a body size threshold of sorts, around 1.5–2 kg, for decussation, presumably because smaller primates generate less force during chewing (Teaford, Maas, and Simons 1996). Other studies also show promise. For example, a recent analysis showed that *Sapajus* has more histological complexity of its canine enamel than does *Cebus*, consistent with the robust capuchins using their anterior teeth more for hard-object feeding (Hogg and Elokda 2021).

Hardness and stiffness also vary with depth and side of a tooth, given differences in local chemistry (levels of mineralization, organic matter, and water content; Braly et al. 2007; Cuy et al. 2002; Darnell et al. 2010; J. Lee et al. 2010). Nevertheless, it seems that enamel thickness and absolute size are key to maximum "safe" bite force given overlapping mechanical properties for species that differ in enamel structure and chemistry (Constantino, Lee, et al. 2012).

AN ECOLOGICAL PERSPECTIVE

Remarkable progress has been made toward characterizing and understanding dental-dietary adaptations. If our goal is to reconstruct the feeding ecology of fossil primates rather than what their teeth evolved for, though, it is necessary to understand not just how teeth work but how they are used. While this may sound like a matter of semantics, the difference is profound. There is a movement gaining momentum to put ecology front and center in studies of primate teeth (Cuozzo, Ungar, and Sauther 2012). The focus is squarely on how teeth respond to the environment, both through natural selection over deep time (dental-dietary adaptations) and in life (foodprints). As Cuozzo and Sauther (2012, 168) describe it, "A dental ecology approach, using the comparative method and information from living forms, can transcend temporal boundaries, and use extant forms to understand even complex ecological and behavioral aspects of fossil species through teeth." This represents a clear attempt to meet G. Simpson's (1926) challenge to consider fossil primates not as broken bits of bone and teeth but as "flesh and blood beings" alive in the past.

Most researchers who study fossil primates and dental-dietary adaptations accept that relationships between teeth and diet are complex; however, this complexity is difficult to appreciate without the perspective of a field primatologist or at least the experience of watching lemurs or monkeys or apes in the wild eat through an entire phenological cycle. Some examples follow.

Species-Specific Dietary Patterns

The species-specific dietary adaptation model (Sussman 1978) is the default model for relationships between diet and teeth. Sussman (1987, 152) wrote, "Each primate species is relatively fixed in certain features of its dietary preferences; although plant or animal species eaten may differ in different localities, the types of items eaten and the degree of diversity of the diet remain quite constant." This is because, he continued, "dietary patterns are ultimately dependent on the morphological and physiological adaptations of the species." Many paleontologists assume this model holds when they look to teeth to reconstruct diets of fossil species.

Sussman's work was initially intended to determine how closely related, sympatric species partition their niches (Sussman 2014). He considered brown lemurs (*Eulemur fulvus*) and ring-tailed lemurs in three forest patches: one with the first species but not the second, another with the second species but not the first, and a third with both. The species overlapped in their diets (both consumed fruits, leaves, flowers, and bark), but the brown lemurs ate more leaves, and the ring-tailed lemurs consumed more fruit. The ring-tailed lemurs also had larger day ranges and used a greater number of plant species. These differences were consistent between sites and seasons, whether the same plant taxa were available at a given time and place or not. Sussman (1977) interpreted these results to imply that lemurs sought out and ate the variety and types of foods to which they were adapted regardless of where, when, or with whom they were eating.

While at first glance, one might think this bodes well for the inference of diet from dental adaptations, the two lemur species have similarly long molar shearing crests (Kay, Sussman, and Tattersall 1978). Kay, Sussman, and Tattersall suggested that the apparent inconsistency between form and function might have something to do with the way diets were reported or categorized. They also proposed that leaves might put stronger selective pressure on shearing crest length than fruit and asked, "How much folivory is needed to select for folivorous dental morphology?" (Kay, Sussman, and Tattersall 1978, 126). Perhaps shearing crest length reflects not how many leaves you eat but whether you need to eat them at all. Kay, Sussman, and Tattersall had hit upon a profound and important new idea, though it would remain undeveloped for decades, until researchers began considering the dental ecology of gorillas.

Fallback Adaptations and Liem's Paradox

Gorillas have highly specialized, sharp molars with long crests and thin tooth enamel. Early studies of mountain gorillas (*Gorilla beringei beringei*) on the peaks of the Virunga Massif suggested a strong dental form-function relationship. Gorillas there live in montane forest and woodland habitats that average two miles in altitude, and they regularly climb to higher altitudes to eat stems, leaves, and pith of nonwoody plants on or near the ground (Fossey and Harcourt 1977; D. Watts 1984). This terrestrial herbaceous vegetation (THV) is plentiful and available year-round. It makes sense, then, that gorillas have massive guts to break down fiber-rich plant parts; strong, deep jaws to withstand heavy chewing of tough, fibrous plant parts; and sharp molar teeth, with long crests for shearing and slicing leaves and stems (e.g., Chivers and Hladik 1980; A. Taylor 2003; Uchida 1998; Ungar and M'Kirera 2003).

However, researchers have come to consider the Virunga mountain gorillas as a peripheral, marginal population living in the most extreme habitat and eating the most extreme diet of any of the wild apes. In fact, gorillas (*Gorilla gorilla gorilla*) are more abundant in the lowland rainforests of the Congo Basin, a thousand miles to the west (Mehlman 2008). Remis (1997) found that gorillas at Bai Hokou in the Dzanga-Ndoki National Park in the Central African Republic have diets that vary over the course of a year; they eat mostly fleshy fruits during much of the rainy season. Other foods, such as tree leaves, stems, and bark, as well as THV, are eaten year-round, but they dominate the diet only during the dry season. The gorillas at Bai Hokou actually prefer fruit flesh and eat it when they can get it. Remis (2002) followed up her work in the Central African Republic with feeding experiments at the San Francisco Zoo that confirmed gorilla preferences for fleshy fruits (i.e., mango, cantaloupe, and corn) over more fibrous vegetation (i.e., broccoli, cabbage, kale, and celery).

There remains little doubt that western lowland gorillas prefer foods high in sugar and low in fiber when given a choice. They are selective and consume mostly ripe fruit during seasonal fruiting peaks (e.g., Doran-Sheehy, Mongo, Lodwick, and Conklin-Brittain 2009; M. Robbins, Nkurunungi, and McNeilage 2006). Mountain gorillas in fact also consume fruit flesh when it is available. Those in Uganda's Bwindi Impenetrable National Park that live at lower altitudes where fruit is available eat more fruit when they can get it (M. Robbins and McNeilage 2003; Stanford and Nkurunungi 2003). The same is true for Grauer's gorillas (*Gorilla beringei graueri*) found in the lowlands and moderate elevations in the Democratic Republic of the Congo (Yamagiwa and Basabose 2009). In fact, there is a strong correlation between altitude, fruit availability, and fruit consumption across gorilla ranges (M. Robbins, Nkurunungi, and McNeilage 2006).

It is evident, then, that gorillas have evolved specialized adaptations to consume tough, fibrous foods, and they do so when necessary—but they prefer soft, sugary fruit flesh. They are, in this sense, adapted to fallback foods, defined here following A. Marshall and Wrangham (2007) as those of poor nutritional quality and high abundance that are consumed when preferred resources are scarce (see Knott and DiGiorgio, this volume, and Marshall, this volume, for additional discussion of defining fallback foods). As Warren Kinzey (1978, 378) noted many years ago, "When a food item is critical for survival, even though not part of the primary specialization, it will influence the selection of dental features." He was referring at the time to South American titi monkeys (*Cheracebus torquatus* and *Plecturocebus moloch*), but the same holds for gorillas, particularly those at lower altitudes.

This brings up an apparent paradox in dental-dietary adaptations, one commonly identified with ichthyologist Karel Liem. Liem noted, with reference to Minkley's cichlid fish (*Herichthys minckleyi*), that some have flat, pebble-like teeth adapted to crack hard snail shells, but the fish pass right by the snails when softer foods are available (Liem and Kaufman 1984). The cichlids avoid the very foods to which they are adapted given the choice. The paradox in this case is, as Liem (1980, 295) wrote, that "the most specialized taxa are not only remarkable specialists in a narrow sense, but also jacks-of-all-trades." More specialized teeth can actually lead to broader, more generalized diets, at least when dental morphology does not limit the ability to consume preferred foods (B. Robinson and Wilson 1998). This seems to be the very same phenomenon identified in lowland gorillas.

But what about higher-altitude mountain gorillas? Their diet is dominated year-round by THV evidently not because they want to eat low-quality, fibrous foods but because they have to. The populations in the Virunga Massif are surrounded by some of the highest human densities on the African continent (Mehlman 2008). Hunting and forest clearing for firewood and cultivation have kept gorillas there to altitudes where there is little food other than THV available. In this case, their anatomy allows them to eat nonpreferred foods year-round. These gorillas might be referred to as "perpetual fallback feeders" (Ungar 2004), regularly consuming less desired foods because those favored elsewhere are simply unavailable to them. There may be very little difference in dental functional morphology between a Virunga and a Bwindi mountain gorilla, or, for that matter, between the cheek teeth of *Gorilla beringei beringei* and *G. gorilla*. However, their diets differ greatly, driven by variation in resource availability, which should also be considered when inferring function from teeth of fossil primates.

Opposite Ends of the Spectrum

The purposeful avoidance of fallback foods when possible is indeed common among primates. Gray-cheeked mangabeys (*Lophocebus albigena*) and sympatric red-tailed guenons (*Cercopithecus ascanius*) in Uganda's Kibale National Park give us an especially extreme example (J. Lambert 2007; J. Lambert et al. 2004). The mangabeys at Kibale (fig. 26.3) have thick tooth enamel compared with most other primates—much thicker than that of the guenons that live alongside them (see Kay 1981). Nevertheless, both monkeys favor soft fruit flesh and have similar diets most of the time. That changes, however, at times of extreme resource stress, such as during strong El Niño events when guenons fall back on leaves but mangabeys eat more hard seeds and bark. J. Lambert and her colleagues (2004, 367) conclude, "It is not so much what is consumed most commonly (i.e., soft, fleshy fruit) that selects for enamel thickness, but the hardness of foods that are consumed infrequently when other, more preferred foods are not available." For the mangabeys at Kibale, thick enamel may confer an advantage taken as infrequently as once in a generation.

But there are also primate species for which preferred foods are mechanically challenging, and specialized teeth do reflect a dietary "primary specialization" (sensu Kinzey 1978). The sooty mangabeys (*Cercocebus atys*) in the Ivory Coast's Taï National Park present a case in point (fig. 26.3). Many of the primates at Taï feed on the sweet flesh of *Sacoglottis gabonensis* fruit during the months that fruiting bursts make them available (see McGraw, Vick, and Daegling 2014a and references therein). The pits of these fruits are rot resistant and can be found on the forest floor year-round. Of the 11 primate species at Taï,

Figure 26.3. Mangabeys. Compare the sooty mangabey, *Cercocebus atys*, from Taï (*left*) with the gray-cheeked mangabey, *Lophocebus albigena*, from Kibale (*right*). Sooty mangabey image courtesy of Scott McGraw, and gray-cheeked mangabey image courtesy of Alaine Houle.

only the mangabeys eat *Sacoglottis* nuts on a regular basis (fig. 26.3). They crack open the hard endocarp with their cheek teeth and eat the softer seeds within; *Sacoglottis* nut shells are harder than cherry pits and popcorn kernels (Daegling et al. 2011). Despite the mechanical challenges of consuming *Sacoglottis* nuts, these nuts are still the most abundant item in the mangabey diet, and the monkeys consume them every month of the year. While niche partitioning with the other primates at Taï could play a role in limiting their diets to items foraged mostly from leaf litter on the forest floor (McGraw 2007), *Sacoglottis* seeds are clearly not fallback foods (McGraw, Vick, and Daegling 2014a).

The mangabeys at Taï, like those at Kibale, have very thick cheek-tooth enamel (McGraw, Pampush, and Daegling 2012). This dental specialization is well suited to resisting the stresses associated with cracking *Sacoglottis* nuts (Daegling et al. 2011; McGraw, Pampush, and Daegling 2012) and gives the mangabeys access to a year-round resource unavailable to the other primates in the park, except for the occasional tool-wielding chimpanzee (Boesch-Achermann and Boesch 1993). Indeed, Gilbert (2013) has argued that the clade including mangabeys may have evolved a shift to a hard-object-feeding niche to avoid competition with other forest-living cercopithecines. In this case, dental specialization reflects dietary specialization—the primary specialization (sensu Kinzey 1978).

The mangabeys at Taï and Kibale have among the thickest cheek-tooth enamel in the primate order (Pampush, Duque, et al. 2013). And while this seems to be an adaptation to hard-object feeding in both cases, the selective pressures involved, and their ecological implications, represent opposite ends of the spectrum. Mangabeys are not monophyletic (E. Harris and Disotell 1998), and *Cercocebus* and *Lophocebus* may have evolved their extra-thick enamel separately but in parallel (McGraw, Vick, and Daegling 2014a). Perhaps ancestors of the sooty mangabey evolved thick enamel for a specialized terrestrial hard-object-feeding niche, whereas those of the gray-cheeked mangabey evolved it to fall back on hard foods at the rare times when preferred softer ones were unavailable. The differences in their foraging strategies are ecologically important but not evident from dental functional morphology. In other words, one solution can follow from two different problems. This needs to be considered by paleoprimatologists when interpreting dental-dietary adaptations.

Subtle Advantages

Orangutans present another example of feeding advantage conferred by thick cheek-tooth enamel. It has long been assumed that these apes have thick enamel as an adaptation for hard-object feeding (Kay 1981). E. Vogel and colleagues' (2008) dental ecology work bears this out. While orangutans at Tuanan (*Pongo pygmaeus wurmbii*) more often eat soft, fleshy fruits, they still regularly consume hard-shelled seeds and husked fruits when and where they can get them. They crush hard *Mezzettia parviflora* seeds, for example, between their strong, thickly enameled cheek teeth (Lucas, Peters, and Arrandale 1994). These are not fallback foods by the usual definition, both because they are not available all the time and because orangutans seem to prefer them. Therefore, the relationship between dental form and diet for the Tuanan orangutans is different from that of either the mangabeys at Taï, which eat hard foods year-round, or those at Kibale, for whom hard foods are a fallback resource. Such case studies illustrate the complex nature of the relationship between feeding strategy and dental-dietary adaptation.

My own work on Sumatran orangutans (*Pongo abelii*) and sympatric primates at Ketambe provides another example. Orangutans on Sumatra, like their congeners on Borneo, are not hard-object specialists. In fact, the Sumatran species may actually have mandibular adaptations less specialized for mechanically challenging foods than do the Bornean species (A. Taylor 2006), and most of the fruits they eat are soft and fleshy. Nevertheless, Sumatran orangutans still have thick tooth enamel and powerful jaws compared with most other primates. Orangutans at Ketambe do eat hard-shelled fruits more often than do sympatric macaques (*Macaca fascicularis*), gibbons (*Hylobates lar*), and leaf monkeys (*Presbytis thomasi*; Ungar 1995), but the advantage conferred to orangutans by thick enamel can be quite subtle. The fruit of *Gnetum* cf. *latifolium* has a hard shell and pulpy innards surrounding several large, soft seeds. Gibbons, macaques, and leaf monkeys all break through the husk with their incisor teeth and pry pieces off to expose the edible parts, a slow and tedious process—it takes more than a minute for them to eat a single fruit. Orangutans, on the other hand, crush *Gnetum* fruits between opposing teeth to split them open, enabling the consumption of several fruits per minute (Ungar 1994b).

Additionally, *Gnetum* fruits harden as they ripen, and gibbons, macaques, and leaf monkeys eat only the softer, unripe ones. I saw several of these primates try, unsuccessfully, to pierce the husk of a ripening fruit, then drop it uneaten. On the other hand, orangutans continued to visit *Gnetum* plants and eat their fruits up to weeks after the other primates in the forest were forced to abandon them. Strong, thickly enameled teeth and powerful jaws gave the orangutans an advantage over other primates. Such studies of dental ecology demonstrate that the relationship between teeth and diet can be a complicated one, with the devil in the details. These studies also demonstrate that it is especially challenging to interpret foraging and feeding strategies from dental functional morphology alone.

FOODPRINTS

There can be little doubt that primates with thick enamel are better suited to breaking hard objects, such as *Sacoglottis* pits or *Gnetum* shells, whereas those with long, sharp molar crests are better suited to shearing tough foods, such as wild celery stalks or leaves. However, again, adaptations like these can be for preferred foods or fallback items. Such foods can be eaten on very rare occasions or all the time. Tooth size, shape, and structure reflect the sorts of foods a species is capable of eating, or even the sorts of foods a species has evolved to eat, but not necessarily the proportion in which they are eaten. For some functional morphologists, this is enough, depending on the questions to be addressed. For paleoecologists, though, stopping here can be limiting. The proportions of different foods eaten, both daily and throughout the year, are important to know if the goal is to understand not just dental adaptations but food preferences and foraging strategies in the past. Dental ecologists aim to understand not just how teeth work but also the relationship between dental-dietary adaptation and how teeth are actually used on a regular basis.

Fortunately, there are tools available to work out how teeth are used. These fall within the domain of what can be termed "foodprints" (Ungar 2017), traces of foods eaten by individuals in life preserved on or in fossil teeth. These, like dental-dietary adaptations, come in three types: food debris, tooth chemistry, and dental wear, especially microwear.

Food Debris

Plant microfossils are very rarely associated with fossil primate teeth. The probability drops with transport prior to deposition, exposure prior to recovery, and specimen handling and cleaning after. Furthermore, it can be difficult to have confidence that plant microfossils are actually food debris in even the best of circumstances. That said, plant microfossils have, on occasion, been found embedded in enamel at the ends of microscopic scratches that evidently formed in life during a chewing bout (Ciochon et al. 1990), or preserved between a tooth surface and dental calculus covering it (Henry, Brooks, and Piperno 2014; Henry, Ungar, et al. 2012). Researchers have found starch grains, but the most common plant microfossils on teeth are phytoliths. These are small, amorphous silica particles, formed as monosilicic acid is absorbed from groundwater and taken up in a plant's cell walls (Piperno 2006). The shape and surface ornamentation of a phytolith depend on those cells and can often be identified to plant taxon. It is difficult to know whether a candidate phytolith was endogenous to a plant food or came from dust or grit on it, but a comparison of types and proportions with those in the depositional matrix surrounding a fossil can hint at probability.

Ancient DNA of long-extinct animals can also be preserved within or behind dental calculus on teeth, albeit even more rarely and typically in fossils from more recently extinct species (e.g., Weyrich et al. 2017). These can provide a list of actual species eaten by an individual in the past. It is even possible in some such cases to get a glimpse at the oral microbiome (e.g., Yates et al. 2021). Some of the best evidence, though, which is even rarer, is food remains preserved as stomach contents, as for Eocene specimens from the Messel site in Germany (Franzen and Wilde 2003). Such finds are extremely unusual in the primate fossil record.

Stable Isotope Analyses

A more common approach to paleodiet studies, at least for early hominins and other similarly ancient fossil primates, involves stable isotope analyses of elements found in fossil teeth. If foods vary predictably in proportions of isotopes of a given element, that should be reflected in teeth built using raw materials from those foods. The most commonly considered chemical signature is the ratio of ^{13}C to ^{12}C (denoted as $\delta^{13}C$ when considered relative to an international standard). Low-latitude tropical grasses and sedges use the C_4 photosynthetic pathway, which discriminates less against ^{13}C than the C_3 pathway used by trees, bushes, shrubs, and forbs found in the same places—and the teeth of animals that eat different proportions of C_3 and C_4 plants differ in $\delta^{13}C$ values accordingly. Paleoanthropologists have used this, for example, to trace the shift from forest/woodland resources to open-savanna foods during hominin evolution (Sponheimer, Alemseged, et al. 2013; J. G. Wynn et al. 2020).

While most nonhuman primates live in forested settings and consume exclusively C_3 plants, there are still subtle yet predictable differences in $\delta^{13}C$ values for their foods (Schoeninger 2009). Plants in evergreen forests, and the primates that eat them, tend to have lower $\delta^{13}C$ values than those in more open woodland settings because lower light levels and higher proportions of CO_2 respired from organic detritus in the soil combine to decrease the ratio of ^{13}C to ^{12}C (Schoeninger, Iwaniec, and Glander 1997; Schoeninger, Iwaniec, and Nash 1998; N. van der Merwe and Medina 1989, 1991). Primates in forests and woodlands likewise have lower $\delta^{13}C$ values than do those in savannas (Schoeninger, Most, et al. 2016). A canopy effect within evergreen forests has also been reported; $\delta^{13}C$ values increase with height (Farquhar et al. 1989; N. van der Merwe and Medina 1989, 1991). That said, the canopy height signal can be obscured by other factors, such as plant parts eaten and the ages of those parts (Blumenthal, Rothman, et al. 2015; Crowley, Carter, et al. 2010; Krigbaum et al. 2013). Still, there are predictable differences in some cases. For example, western lowland gorillas have lower average but more variable $\delta^{13}C$ values than sympatric chimpanzees, which makes sense given their consumption of terrestrial herbaceous vegetation and a seasonal shift to more fruit when available (Macho and

Lee-Thorp 2014; Oelze, Head, et al. 2014). This probably relates to the canopy effect, but it is also true that fruits are generally enriched in ^{13}C compared with leaves (Cernusak et al. 2009).

Primate diet researchers have considered other elements too, especially oxygen. Water molecules with lighter ^{16}O evaporate more quickly than those with heavier ^{18}O. Therefore, δ^{18}O values increase higher in the canopy given lower humidity. This also explains why leaves have relatively more ^{18}O (Yakir 1992). Indeed, oxygen isotopes can sometimes separate primates by canopy height and distinguish folivores from frugivores (Carter and Bradbury 2015; Krigbaum et al. 2013; Sternberg et al. 1989). Nitrogen has also been examined. The ratio of ^{15}N to ^{14}N increases through the food chain from plant to herbivore to carnivore (Schoeninger and DeNiro 1984). Among primates, insectivores do tend to have higher δ^{15}N values than herbivores. Furthermore, among herbivores, the more legumes in the diet, the less ^{15}N (Macho and Lee-Thorp 2014; Schoeninger, Iwaniec, and Glander 1997; Schoeninger, Iwaniec, and Nash 1998). And δ^{15}N values also tend to be lower in the lower canopy (Lowry et al. 2021). There are few studies of nitrogen isotopes of fossil species, though, except for the recent fossil record (e.g., M. Richards and Trinkaus 2009), because the preservation of nitrogen tends to be poor for most fossils, at least compared with that of oxygen and carbon (Kohn and Cerling 2002).

Primate isotope researchers still have much work ahead of them (see also Sponheimer and Crowley, this volume). Studies of living animals often rely on hair, feces, or bone collagen. Isotope ratio values need to be calibrated to tooth enamel for fossil studies because different tissue types are enriched in one or another isotope to different degrees, all else being equal (Crowley, Carter, et al. 2010). The details also need to be better worked out, even for the commonly considered elements. More work is also needed on less studied elements, such as sulfur (M. Richards, Fuller, et al. 2003), strontium (Copeland et al. 2011), and calcium (J. Martin, Tacail, et al. 2020), which may hold additional potential for primate paleodiet reconstruction.

Dental Wear

The final major category of foodprint to consider is tooth wear. Recall occlusal surface sculpting for secondary morphology (sensu Fortelius 1985). Because dentin is softer than enamel, sharp edges form at the interface between tissue types, where thinner parts of the enamel cap wear through to the underlying horn. This can lead to a jagged, complex surface well suited to shearing or grinding tough vegetation. It likely explains why gorillas, for example, tend to have relatively, albeit not necessarily absolutely, thin enamel (Ungar 2008; Ungar and M'Kirera 2003; Ungar and Williamson 2000) and why lengths of compensatory shearing blades on wild sifaka molars remain nearly constant despite a drop in overall crown relief (S. King et al. 2005). Most studies of tooth wear related to diet, though, have focused on patterns of dental macrowear and microwear.

Dental Macrowear and Diet

Ruminants, rather than primates, have been the taxa of choice for most dental macrowear analyses. Some ruminant studies have considered wear gradients to determine the abrasiveness of food and habitat, or diet "quality" given that tough, fibrous foods require more chew cycles per mouthful (Kaiser and Schulz 2006; Kubo and Yamada 2014). Others have looked to mesowear (Fortelius and Solounias 2000), with the idea that attrition (tooth-tooth contact) tends to form sharp-edged facets as teeth slide past one another, but abrasion (tooth-food/abrasive contact) tends to blunt them (Popowics and Fortelius 1997). Grass-eating herbivores of many shapes and sizes indeed have more rounded occlusal surfaces than do closely related browsers (K. Butler, Louys, and Travouillon 2014; Fortelius and Solounias 2000; Kaiser and Solounias 2003).

Diet-related studies of gross tooth wear are less well developed for primates. P. Morse et al. (2013) looked to variation among monkey species at Taï and found diet-related differences in macrowear between teeth best explained by differential use in food processing. Furthermore, Galbany and colleagues (Galbany, Altmann, et al. 2011; Galbany, Imanizabayo, et al. 2016) have found associations between wear gradients and consumption of gritty corms by baboons and roots by gorillas. Quartz-rich soils seem to exacerbate dental macrowear (Galbany, Romero, et al. 2014), though food properties can also play an important role in rates of wear (Cuozzo, Head, et al. 2014; Yamashita et al. 2015). Furthermore, Galbany, Twahirwa, et al. (2020) recently found that frugivores tend to have lower tooth wear rates than hard-object feeders and folivores, though enamel thickness and phylogeny affect rates of wear too.

Dental Microwear Analysis

Most studies of primate tooth wear have involved microwear, the micron-scale scratches and pits that form on a tooth's surface as the result of its use (fig. 26.4). While there have been studies of incisor teeth (e.g., J. Kelley 1990; A. Ryan 1981; Ungar 1994a) and the buccal edges of molars (Galbany, Estebaranz, et al. 2009; Galbany, Moya-Sola, and Perez-Perez 2005), most analyses have focused on molar occlusal surfaces. The pattern in principle reflects the material properties of foods eaten and therefore gives us the best chance of direct comparison of results with those from studies of occlusal form. The goal is to determine whether the individual animals whose teeth we find actually ate the sorts of foods we infer they evolved to eat. This can lead into the realm of dental paleoecology and even food choice and foraging strategy in the past.

The basic idea behind occlusal surface microwear was from the outset that when opposing teeth slide past one another (during shearing or grinding), chewing should cause scratches as abrasives get dragged along the surface, whereas when lower teeth are pressed against uppers (crushing), pits should result (Grine 1977). These scratches and pits, in turn, can be used to reconstruct the fracture properties of foods, assuming tough items are sheared or ground (causing scratches) and hard items are crushed (causing pitting). Microwear results on living primates with known diets have been more or less as expected. Teaford and Walker (1984) pioneered the effort, measuring pits and scratches on photomicrographs generated using scanning electron microscopes (SEMs). They found that folivores have a high proportion of scratches relative to pits on their molars, hard-object feeders evince more pits, and soft-fruit eaters tend to be intermediate in scratch-to-pit ratio. Other studies followed to work out the details (see Calandra and Merceron 2016; DeSantis 2016; Teaford, Ungar, and Grine 2013; Ungar 2015, 2019).

That said, there has been debate in recent years about the etiology of molar microwear. Some have suggested that exogenous quartz grit and dust are more likely to cause tooth wear than are foods themselves (Lucas, Omar, et al. 2013; Sanson, Kerr, and Gross 2007), which has led to speculation that microwear patterning may reflect habitat rather than diet. Others have claimed, though, that endogenous abrasives in foods (e.g., phytoliths) cause microwear, too (G. Baker, Jones, and Wardrop 1959; C. Fox et al. 1994; Gügel et al. 2001; Rabenold and Pearson 2011; Teaford and Byrd 1989; Walker et al. 1978; Xia, Zheng, et al. 2015). Clearly, the etiology of microwear is complex, with abrasive quantity, abrasive type, and food properties all playing a role in pattern formation (Ackermans et al. 2020; Hua, Chen, and Ungar 2020; Schulz-Kornas et al. 2020; Teaford, Ross, et al. 2021). Nevertheless, the diet signal tends not to be overwhelmed by other factors (N. F. Adams et al. 2020; Burgman et al. 2016; Merceron et al. 2016; H. Rodrigues et al. 2009). For example, an analysis of communities of past mammals shows no consistent regional effect across taxa (Ungar, Scott, and Steininger 2016). The reason for this is evidently that, regardless of whether grit and dust or phytoliths are the agents of wear, microwear patterns still reflect masticatory movements associated with specific food properties (Hua, Brandt, et al. 2015).

Recent efforts in microwear research have focused on finding new approaches to characterizing and comparing microwear textures between species. Because most SEMs represent a 3D surface in two dimensions, information is lost in translation. Shadowing effects give the illusion of depth to an SEM photomicrograph, and the image obtained depends on the slope and aspect chosen for the surface relative to the electron beam. Just as casting light on an object from different angles gives different perspectives, scratches and pits appear and fade as a specimen is rotated on an SEM stage, which can make consistency of imaging a challenge. Also, feature-based surface characterizations are prone to measurement error, which often runs between about 5% and 10% (Grine et al. 2002). The surface represented by a photomicrograph can have hundreds of features, the boundaries of which are irregular and commonly overlap the borders of others and the edges of the field of view. Identifying all the features on a surface and determining their endpoints can be difficult. Taken together, the 3D-to-2D problem and measurement error can introduce noise that obscures the diet signal and limits the potential of microwear.

Dental microwear texture analysis has emerged as an alternative approach to reduce that noise (Percher et al. 2018; R. Scott, Ungar, Bergstrom, Brown, Childs, et al.

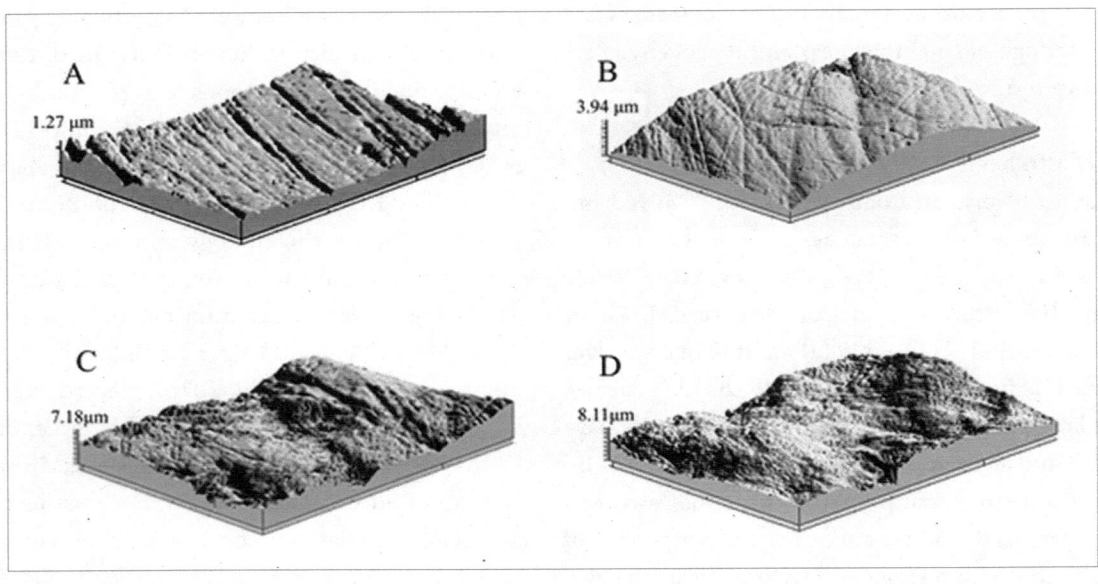

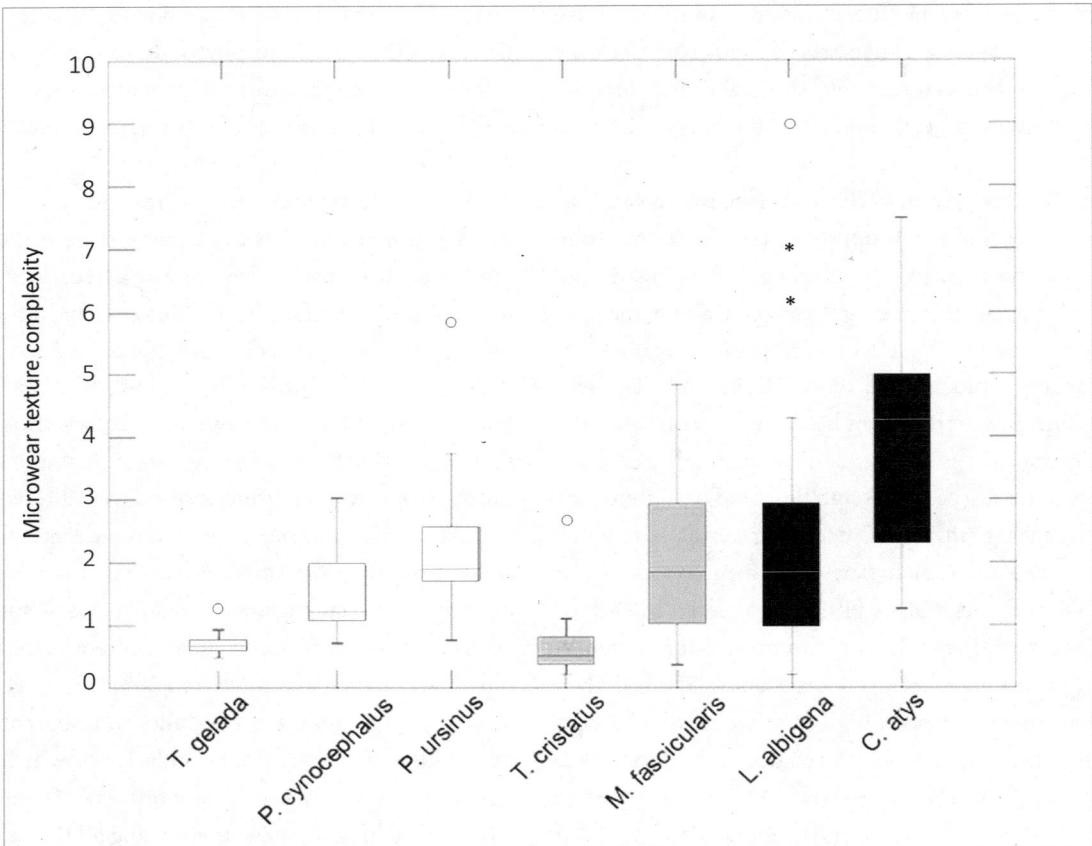

Figure 26.4. Molar occlusal microwear. *Top*: digital elevation models of the teeth of a leaf-eating howler monkey, *Alouatta palliata* (*A*); a grass-eating gelada, *Theropithecus gelada* (*B*); a hard-object-feeding capuchin, *Cebus apella* (*C*); and a gray-cheeked mangabey, *Lophocebus albigena* (*D*). All images represent an area 138 μm × 104 μm. *Bottom*: area-scale fractal complexity plots for various primates. The *C. atys* data are for Taï monkeys (Daegling et al. 2011), and the others are from R. Scott, Teaford, and Ungar (2012). The boxes represent the central 50% of values, with the edges indicating the first and third quartiles and the horizontal lines representing medians. The whiskers are the range of values within 1.5 times the interquartile range from the box edges, the asterisks are values between 1.5 and 3 times the interquartile range from the box edges, and open circles are values beyond that.

2006; Ungar, Brown, et al. 2003). Confocal profilometry is combined with surface texture measurements for a whole-surface characterization of microwear patterning. The details can be found in the literature (Ungar, Scott, et al. 2007), but suffice it to say that measures are taken from point clouds without the observer error inherent in identifying and measuring hundreds of tiny scratches and pits on a photomicrograph. One approach uses scale-sensitive fractal analysis, with texture complexity (change in surface roughness with scale of observation) and anisotropy (directionality of wear) as especially useful variables. Cratered surfaces dominated by pits of many sizes are complex, and those with aligned scratches are aniosotropic. Studies of primate microwear textures confirm differences by diet (fig. 26.4). Hard-object feeders tend to have complex surfaces, whereas tough-food eaters more often have anisotropic ones (J. R. Scott et al. 2009; R. Scott, Teaford, and Ungar 2012). Similar differences are found for comparisons of species within many other mammalian orders (see Calandra and Merceron 2016; DeSantis 2016; Ungar 2015 for review). The other common approach involves a whole battery of standard texture measurements using the ISO 25178 protocol (see Purnell et al. 2013; Schulz et al. 2010). These too have proved useful for distinguishing diets from microwear texture patterns.

The reduction of observer error in measurement has been key to working out not just central tendencies in diet but also dispersion (R. Scott, Ungar, Bergstrom, Brown, Grine, et al. 2005). We can, in principle, understand from the distribution of microwear texture patterns for a sample something about the foraging strategies of extinct species—like whether individuals preferred mechanically challenging foods or ate them only on occasion. The basic idea is that microwear features tend to be, at most, a few microns deep and are typically erased and replaced by others as a tooth wears (Teaford and Oyen 1988; Teaford, Ungar, et al. 2017, 2020). This turnover leads to what has been called the "last supper" phenomenon (Grine 1986), wherein microwear surfaces are not time-averaged over life but rather reflect a snapshot of diet over days or weeks. Enough snapshots can lead to a sense of variation in diet, assuming our sample is representative of the population or species of interest.

If hard-object consumption yields a heavily pitted surface, for example, a randomly selected sample of fallback feeders should show a broad distribution of complexity values, most in line with soft-food eaters but a few out toward the upper extreme. Hard-object specialists, in contrast, should have a higher median complexity value and fewer individuals at the low end of the spectrum. Consider the mangabey species discussed above. *Cercocebus atys* specimens from Taï have heavily pitted, complex molar microwear surface textures. Their median fractal complexity value is substantially higher than that of *Lophocebus albigena* specimens examined (Daegling et al. 2011; R. Scott, Teaford, and Ungar 2012), an expected result for a hard-object specialist (fig. 26.4). *Lophocebus albigena*, on the other hand, has microwear values clustered at the low end of the complexity spectrum, with a few outliers near the upper extreme of the *C. atys* distribution, as predicted for a hard-object fallback feeder. While their teeth may be similar in shape and enamel thickness, microwear texture clearly separates the two species. In other words, microwear has the potential to distinguish a primary from a secondary adaptation (sensu Kinzey 1978) when dental form cannot. Therefore, microwear can help us understand not just how tooth form relates to function but how adaptation relates to ecology (see Ungar 2009 for discussion).

FINAL THOUGHTS

Dental functional morphologists can surely benefit from the primatological perspective—but dental ecologists likewise can benefit from the paleontological one. As L. P. Hartley (1953) wrote in *The Go-Between*, "The past is a foreign country; they do things differently there." If our sights are limited to the living primates, we are fated to what historians call presentism, a narrow, distorted view of the past—in the worst case, a replica of the present (J. Kelley 1993; Ungar 2007b; Walker 2007). Of course, without the present, paleobiological interpretation is left in undecipherable chaos (L. Martin 1991). Fortunately, the combination of foodprints and dental-dietary adaptations can help us recognize evolutionary novelties, unique conditions in a past that was different from today.

Consider *Paranthropus boisei*, long considered a hard-object feeder because of its big, flat, thickly enameled teeth and massive jaws driven by powerful chewing muscles. Its foodprints tell a different story. Wispy microwear scratches and a steep molar wear gradient are consistent with grinding rather than crushing, and a C_4 isotope signature indicates tropical grasses and sedges rather than forest nuts and seeds. These data suggest dental-dietary adaptations used for grinding tough, abrasive foods on the open savanna (Peterson et al. 2018; Ungar and Hlusko 2016; Ungar and Sponheimer 2011). However, there is no precedent for this form-function relationship among the living primates. With the limited number of extant primate species today, perhaps that is unsurprising. Pandas, which specialize on hard but tough bamboo, have large molars with blunt cusps; and those teeth have simple, anisotropic microwear surface textures dominated by scratches rather than pits (Donohue et al. 2013). Perhaps we need to look beyond today's primates for appropriate analogues.

The combination of foodprints and dental-dietary adaptations may also help us understand evolutionary mismatches, such as primates exceeding the limits of their dental-dietary adaptations. Consider the ring-tailed lemurs at Beza Mahafaly, with their long molar crests and thin enamel. Cuozzo and Sauther (2012) showed that some fall back on hard but tough, abrasive tamarind fruits, which leave their teeth extremely worn if not broken or lost. The structure and shape of their teeth are clearly not adapted to such foods, but they nevertheless consume them. This seems to be related to limited food choice given rapid habitat change due to anthropogenic disturbance. As Cuozzo and Sauther suggest, this could give us a model for understanding some apparent mismatches between dental-dietary adaptations and foodprints in the fossil record, such as thin enamel but extreme tooth wear for the Plio-Pleistocene monkey, *Cercopithecoides kimeui*. Perhaps that too relates to environmental change—in this case, increased aridity resulting in a more abrasive diet (Jablonski and Leakey 2008).

SUMMARY AND CONCLUSIONS

We can learn a lot about the diets of fossil primates from their teeth. Dental-dietary adaptations, including tooth size, shape, and structure, lead to an understanding of something about the foods a species in the past was capable of eating, perhaps even what it was adapted to eat. Foodprints, including adherent plant microfossils and ancient DNA, isotope ratios, and microwear, help us resolve something about how those teeth were actually used by individuals in life. Combining dental-dietary adaptations and foodprints with insights from ecological studies of living primates and other mammals may even offer insights into foraging strategies and relationships between an extinct species and its environment.

- Dental-dietary adaptations reflect the sorts of food a species evolved to eat. Lines of evidence include tooth size, shape, and structure.
- Adaptive evidence reflects the types of foods a species is capable of eating but not necessarily the frequencies with which they are eaten. Dental-dietary adaptations can reflect selection for preferred foods or fallback items.
- Foodprints are traces left by foods eaten by individuals in life. Common foodprints include plant microfossils and ancient DNA, stable isotope ratios, and dental microwear textures.
- The combination of dental-dietary adaptations and foodprints with a dental ecology perspective offers our best hope for understanding fossils as "flesh-and-blood beings" alive in the past.

27 Food and Primate Carrying Capacity

Andrew J. Marshall

Carrying capacity is a widely invoked ecological concept. Although frequently misunderstood and often criticized (Caughley 1979; Del Monte-Luna et al. 2004; Dhondt 1988; Hui 2006), it is still commonly referenced in discussions of population ecology, conservation biology, agricultural production, fisheries, livestock management, and human population growth (e.g., Catton 1987; J. E. Cohen 1995; Fagen 1988; McLeod 1997; D. Price 1999). In this chapter, I present a review of the carrying capacity concept and its importance for primates, focusing on the link between food and primate carrying capacity. I begin with a review of the concept of carrying capacity, including definitions of the concept, discussion of its significance, and consideration of the effects of variation in carrying capacity on primate individuals, groups, and populations. I then critique how we assess carrying capacity for wild primate populations and consider factors that complicate attempts to do so. I follow with a brief discussion of the regulation and limitation of primate populations and the relative importance of top-down and bottom-up effects. I then consider the relationship between food and primate carrying capacity, including a discussion of conceptual issues related to the determinants of primate carrying capacity and an overview of empirical evidence. I conclude with a summary of current understanding of the relationship between food and primate carrying capacity and identification of some important outstanding questions.

WHAT IS CARRYING CAPACITY?

The term "carrying capacity" is used widely but rarely defined explicitly. This might lead one to assume that definition is unnecessary and that the concept is clear and unambiguous. Indeed, most who have taken an introductory ecology course have a general notion of what the term means: loosely defined, carrying capacity is the number of individuals that an environment can support in the long term (C. Krebs 2001; Ricklefs 2008). This is the way the term is used in most ecology textbooks and in the popular media (D. Price 1999). But closer examination reveals the concept to be far less clear and easy to define. Most who have written about carrying capacity have noted that the term is used in different ways by different people (Catton 1987; Caughley 1979; Dasmann 1964; Dhondt 1988; R. Edwards and Fowle 1955; D. Price 1999). This imprecision has created much confusion and has led to assessments that the concept is "seriously flawed" (D. Price 1999, 17) and "so complex it should be used only in the most superficial way with the public" (Giles 1978, 195, in Dhondt 1988). Many have proposed rigorous definitions of the term "carrying capacity" or have advocated its replacement with more precise alternatives (e.g., Caughley 1979; Dasmann 1964; Del Monte-Luna et al. 2004; R. Edwards and Fowle 1955; Milne 1962; Hui 2006), but the use of the term, and confusion surrounding it, continues.

Much of the modern confusion over the carrying capacity concept can be traced to its use to describe two distinct things. One use of the term refers to the maximum number of individuals that an environment could theoretically support based on the availability of resources (Leopold 1933; Milne 1962; Verhulst 1838). According to this use of the term, carrying capacity is a property of the environment, or more accurately, the interaction between an environment and the organisms that inhabit it. Thus, all other things being equal, doubling the food availability in the habitat of a food-limited species would double its carrying capacity. This definition is of great utility for wildlife managers and conservationists, as carrying capacity may then be a useful measure of habitat quality. A second use of the term defines carrying capacity as the observed equilibrium size of a population—in other words, it is a description of a specific population in a particular place over a certain period (Errington 1934; May 1981). This sense of the term "carrying capacity" is rooted in classic logistic population growth models (Lotka 1956; Verhulst 1838). In these models, populations are maintained at carrying capacity by density dependence in births or deaths (or both) such that populations above carrying capacity decline and those below it increase in size. This use of the term "carrying capacity" is most useful to theoretical and population ecologists, although it is also used in more applied contexts to calculate sustainable offtake rates (Caughley 1976; J. G. Robinson and Redford 1991) and estimate extinction risk (P. Foley 1994; Punt 2000).

Although conceptually related, these two distinct uses of the term can create confusion (Hui 2006). Under the first definition, it is perfectly reasonable to describe a stable population as existing below the environment's carrying capacity due, for example, to human hunting or predation (e.g., A. Marshall, Engström, et al. 2006; V. Wynne-Edwards 1970). Under the second definition, this would be nonsensical as the stable population size would be defined as the population's carrying capacity. This confusion is probably due to Eugene Odum's unifying these two distinct meanings under the same term in his seminal book *Fundamentals of Ecology* (Odum 1953), a unification that has persisted in most ecology texts ever since (Dhondt 1988).

The term "carrying capacity" is not commonly used in the primatological literature. A systematic Google Scholar search of all uses of the term in the full text of articles in the top two primatology journals shows that the term has been used in 39 papers in the *American Journal of Primatology* and 63 papers in the *International Journal of Primatology* (representing 1.3% and 2.6% of the papers published in these journals, respectively). (An additional 7 papers in *AJP* and 9 papers in *IJP* include the phrase "carrying capacity," but only in the titles of cited sources and not in their text. Note: this analysis was conducted on March 3, 2016.) As with its more general use in scientific literature, carrying capacity is rarely defined explicitly. Full-text searches revealed that only one paper in the *American Journal of Primatology* and two papers in the *International Journal of Primatology* explicitly defined the term when used, and in the latter case one of the two papers used the term to refer to the number of individuals that could feed in a tree ("patch carrying capacity"; Melin, Hiramatsu, et al. 2014, 264) rather than in the standard sense of population carrying capacity. The two other explicit definitions of carrying capacity provided highlight the lack of consistency in usage. One paper defined the term by quoting directly from Caughley: "The natural limit of a population set by resources in a particular environment" (Caughley 1976 in Grueter, Ndamiyabo, et al. 2013, 268)—in other words, carrying capacity is defined as a habitat attribute. The second paper stated, "Carrying capacity . . . is the average value around which the size of a population fluctuates" (E. Zucker and Clarke 2003, 88)—thus, a description of a population. When the term "carrying capacity" is used in the primatological literature, it is usually in the context of assertions that populations are above, at, or below carrying capacity, suggesting that the first of the two conceptions of carrying capacity (as a habitat attribute) is the more common. Of course, when populations are stable and food limited, the two uses of the carrying capacity concept—as a habitat attribute and as a population description—will be equivalent.

Carrying capacity is usually invoked by primatologists as a synonym for habitat quality, typically in the context of resource availability (Dobson and Lyles 1989; A. Marshall, Salas, et al. 2007; Struhsaker 2008). Accordingly, high-quality habitats have high carrying capacities because they have sufficient food to support high population densities, and poor-quality habitats have low carrying capacities as they have relatively little available food and therefore support low population densities (Chapman and Chapman 1999; Hanya, Yoshihiro, et al. 2004; A. Marshall 2010). Primatologists (as well as field ecologists and wildlife managers) routinely equate carrying capacity, population density, and habitat quality (Bock and Jones 2004; Chivers 2001; Pérot and Villard 2009; K. Potts, Chapman, and

Lwanga 2009; Villard and Part 2004; Wasserman and Chapman 2003). While this is often quite reasonable, care should be taken as there are certain circumstances in which density is a misleading indicator of habitat quality (N. Hobbs and Hanley 1990; Lengyel 2006; Van Horne 1981, 1983; Weldon and Haddad 2005). In a widely cited paper, Van Horne (1983) described several key environmental characteristics and species attributes that can lead to a disconnection between observed population density and the importance of a habitat for the maintenance of a population. The major environmental types that she listed were those in which there are extreme temporal and spatial fluctuations in habitat quality over a scale that permits or necessitates the mass migration of individuals between habitat types. Van Horne (1983) also predicted a decoupling of habitat quality and density in generalist species with strong social dominance hierarchies and high reproductive capacities whose populations can move to track fluctuations in resource availability. Although a range of vertebrates may inhabit these types of environments or exhibit mass migrations or large fluctuations in population size (N. Hobbs and Hanley 1990), primates seem among the least likely groups to do so. Compared to many vertebrates (especially temperate-zone birds and mammals generally studied by authors voicing concerns about disconnects between population density and carrying capacity), primates occupy relatively stable environments, experience relatively modest temporal fluctuations in population size, and very rarely exhibit large-scale spatial migrations. Thus, there seems to be little reason to fear equating carrying capacity, population density, and habitat quality for most primate populations under natural conditions.

There are, however, instances in which observed population densities may well not reflect carrying capacity, even for primates. Specifically, human disturbance—notably logging or fragmentation—can lead to compression effects in which populations are crowded into densities far exceeding those that the habitat could support under equilibrium conditions (Mathewson et al. 2008; Schmiegelow et al. 1997; Siex 2005). Similarly, human hunting can reduce population densities to levels substantially lower than a habitat could support in the absence of hunting (A. Marshall, Engström, et al. 2006; Muchaal and Ngandjui 1999; Peres 2000a). In both instances, given adequate time, we expect that population densities would return to habitat-specific baselines if these sources of disturbance were removed (provided that the temporarily elevated or depressed population densities did not subject populations to new ecological pressures, such as density-dependent disease or Allee effects).

Given the complexity of the carrying capacity concept, and the fact that it is invoked to refer to different things by different people, the most prudent course would seem to be to explicitly define the term any time it is used. Thus, in the context of this chapter, carrying capacity is defined in the habitat-attribute sense: a habitat's carrying capacity is the stable population density that the habitat could support in the long term under natural conditions (i.e., those ecological conditions of food availability, disease, and predation that characterized the population prior to the introduction of chainsaws, modern weapons, industrial agriculture, and other technological advancements that have changed human population densities). As noted above, for stable, food-limited populations, this will be same as the population-description sense of the term (i.e., K in logistic population models), but the key distinction is that the definition of the term used here permits discussion of stable populations that may be held well below carrying capacity for extended periods. This seems to be the most useful definition of the term in the context of discussing the relationship between food and primate carrying capacity.

WHY IS UNDERSTANDING CARRYING CAPACITY IMPORTANT?

Compared to ecologists studying other vertebrates, primatologists rarely explicitly discuss carrying capacity (see above). The concept is nevertheless important in the field, as it fundamentally underpins topics of great interest to many primatologists. For example, a great deal of field research on wild primates is focused, either directly or indirectly, on understanding the ecological forces promoting and constraining gregariousness (e.g., Chapman and Chapman 2000a; Crook and Gartlan 1966; Isbell and Van Vuren 1996; Janson and Goldsmith 1995; Schülke and Ostner 2012; Janson, this volume). The most commonly invoked ecological mechanisms that influence primate group size are predation, infanticide, and inter- and intragroup competition for resources (Boinski, Sughrue, et al. 2002; Sterck et al. 1997; van Schaik 1983; Wrangham 1980). Debates over the relative importance of these mechanisms often explicitly invoke carrying capacity. For instance, some have hypothesized that "overcrowding" (i.e., populations being above carrying capacity) can explain some less appealing aspects of primate behavior such as infanticide and intergroup aggression (Bartlett et al. 1993; R. Ferguson 2011). Others have based explorations of alternative constraints on grouping on the assumption that folivorous primates routinely live at densities far below those that the environment would seem to be able to support (Crockett and Janson 2000; Yeager and Kirkpatrick 1998).

Primatological discussions of the effects of variation in habitat quality are based (often tacitly) on the carrying capacity concept. For example, the low primate species richness and diversity of Asian forests has been hypothesized to result from the relatively low carrying capacity of these forests for primates (likely due to the low general productivity of forests dominated by mast fruiting of trees in the Dipterocarpaceae; Janzen 1974; Reed and Bidner 2004; Terborgh and van Schaik 1987). Differences in carrying capacity between Borneo and Sumatra are thought to explain variation in sociality and culture within and between the two *Pongo* species (Delgado and van Schaik 2000; van Schaik, Marshall, and Wich 2009). Variation in carrying capacity across smaller spatial scales has also been shown to exert strong local effects on primate diets, group size, demographic composition, ranging, and social behavior (Chapman and Chapman 1999; Hanya, Yoshihiro, et al. 2004; Iwamoto and Dunbar 1983; A. Marshall 2010).

Much discussion of primate conservation is also intimately related to the carrying capacity concept. The extensive literature on the effects of logging on primates is largely concerned with determining the magnitude and direction of the effects of timber removal on the carrying capacity of tropical forests (A. D. Johns and Skorupa 1987; Meijaard et al. 2005; Chapman et al., this volume). J. G. Robinson and Redford's (1991) widely used index to assess whether harvest rates of primates and other Neotropical forest mammals are sustainable is based on a population's density at carrying capacity and its intrinsic rate of increase. Finally, primate populations living in habitats with high carrying capacities are frequently targeted as priorities for conservation (A. Marshall, Lacy, et al. 2009; Nishida, Wrangham, Jones, et al. 2000). These brief examples demonstrate that the concept of carrying capacity is widely applicable to theoretical and applied primatology.

WHAT ARE THE EFFECTS OF CARRYING CAPACITY ON PRIMATES?

The carrying capacity of primate habitats can vary dramatically, and this variation is likely to have important effects. The consequences of populations being above or below carrying capacity are easy to predict in most instances. If a population is below carrying capacity, intraspecific feeding competition is expected to be relatively weak, and consequently individual primates should be in relatively good condition and population growth rates should be positive. Conversely, when populations are above carrying capacity, individuals should be in poorer condition (e.g., due to intensified competition for resources) and population growth rates will be negative. (Density relative to carrying capacity will likely also affect dispersal and have other important behavioral effects; e.g., Butynski 1990.) Less easy to predict, however, are what the effects of variation in carrying capacity might be for populations at carrying capacity. Here, the influence of carrying capacity on primate individuals, groups, and populations is considered.

How Does Carrying Capacity Affect Individuals?

It can be tempting to assume that individuals living in high-quality habitats (i.e., those with higher carrying capacities) are in some sense "better off." Indeed, there is some empirical evidence to indicate that this is the case (see also Knott and DiGiorgio, this volume). For example, female chimpanzees (*Pan troglodytes* spp.) in high-quality habitats have shorter interbirth intervals (Knott 2001), white-bearded gibbons (*Hylobates albibarbis*) in high-quality habitats have higher reproductive success (A. Marshall 2010), and white-headed langurs (*Trachypithecus leucocephalus*) living in high-quality habitats spend more time playing than they do in low-quality habitats (Z. Li and Rogers 2004). Compared to Bornean orangutans (*Pongo pygmaeus*), Sumatran orangutans (*Pongo abelii*) live in forests with higher carrying capacities (Husson et al. 2009) and eat diets of higher quality (Russon, Wich, et al. 2009), experience less frequent and less extreme periods of food shortage (A. Marshall, Lacy, et al. 2009;

Wich, Vogel, et al. 2011), and may exhibit life histories that indicate that they have evolved under conditions of lower extrinsic mortality (van Schaik, Marshall, and Wich 2009; Wich, de Vries, et al. 2009). But there is no theoretical reason to necessarily predict a positive relationship between carrying capacity and individual fitness. If carrying capacity is determined by food availability, then the potential benefits of being in a habitat with more food may be offset by the increased number of conspecific competitors present. This is the logic underlying the "ideal free distribution" (Fretwell and Lucas 1969): because animals distribute themselves in proportion to the availability of food, carrying capacity does not affect per capita food availability or fitness. The ideal free distribution has rarely been examined in primates (Chapman, Rothman, and Lambert 2012), particularly outside the context of mating, but several tests suggest that some primate species do not neatly conform to its predictions (e.g., baboons, *Papio cynocephalus ursinus*: Cowlishaw 1997; red colobus monkeys, *Procolobus rufomitratus*: Snaith and Chapman 2008; Hamlyn's monkeys, *Cercopithecus hamlyni*: Willoughby 1996 in R. Young 1998; moustached tamarins, *Saguinus mystax*: A. Smith, Buchanan-Smith, et al. 2005). Indeed, it may seem implausible that gregarious and highly intelligent species such as primates would ever follow so simple a pattern as the ideal free distribution. Nevertheless, at least two species appear to do so (Japanese macaques, *Macaca fuscata*: Nakagawa 1990; red leaf monkeys, *Presbytis rubicunda*: A. Marshall 2010). These complexities suggest that broad conclusions about how variation in carrying capacity might affect individuals are not yet warranted; rather, they must be empirically examined on a species-by-species basis.

How Does Carrying Capacity Affect Groups?

Primatologists have generally demonstrated much greater interest in understanding factors that influence group size than those that affect carrying capacity or population density. For example, there have been more than 18 times as many articles published containing the words "group size" as those containing "carrying capacity" in the *American Journal of Primatology* and over 11 times as many in the *International Journal of Primatology*; "group size" is mentioned roughly twice as often as "population density" in these journals. Group size is thought to be related to (i.e., to be shaped by or to influence) a wide range of ecological variables in primates (e.g., feeding competition and the risks of infanticide, predation, and disease), although it is plausible that some of these variables may be at least as importantly influenced by population density as they are by group size. Moreover, the relationship between group size and population density is not necessarily positive or linear, because population densities can be achieved in multiple ways. For example, group size may remain constant and density variation may be due to differences in home range size or the degree of home range overlap (fig. 27.1a). This implies that in low-quality habitats (i.e., those with low carrying capacities), animals require a larger area than in high-quality habitats in order to gain access to sufficient resources to support a group (Dunbar 1987). Alternatively, home range size may remain relatively constant (e.g., due to locomotor constraints; Chapman and Chapman 2000a), and density differences may be due to variation in group size between habitats (fig. 27.1c). A third possibility is that density is due to both home range size and group size; in such instances, both group size and home range size would be correlated with density when the effect of the other variable is held constant statistically (fig. 27.1b). These three possibilities imply distinct, testable predictions (fig. 27.1d–f).

Distinguishing between group-size-mediated and home-range-mediated variation in population density requires that we have estimates of both home range size and group size for the same species inhabiting areas with distinct carrying capacities. Ideally, carrying capacity should be estimated through systematic surveys of population density (e.g., via line-transect distance sampling) rather than estimated based on group size, home range size, and the degree of home range overlap so the variables are independent. There are relatively few species for which such data are available. Work on populations of gibbons (*Hylobates albibarbis*) and red leaf monkeys (*Presbytis rubicunda*) at Gunung Palung National Park, West Kalimantan, Indonesia, suggests that, in both species, density variation is primarily mediated by group size rather than home range size (A. Marshall 2004, 2010; fig. 27.2a–d). Data from other taxa suggest that group-size-mediated variation in population density is common. For instance, in black-and-white colobus monkeys (*Colobus guereza*) at Bole Valley, Ethiopia, home range sizes are relatively invariant while group

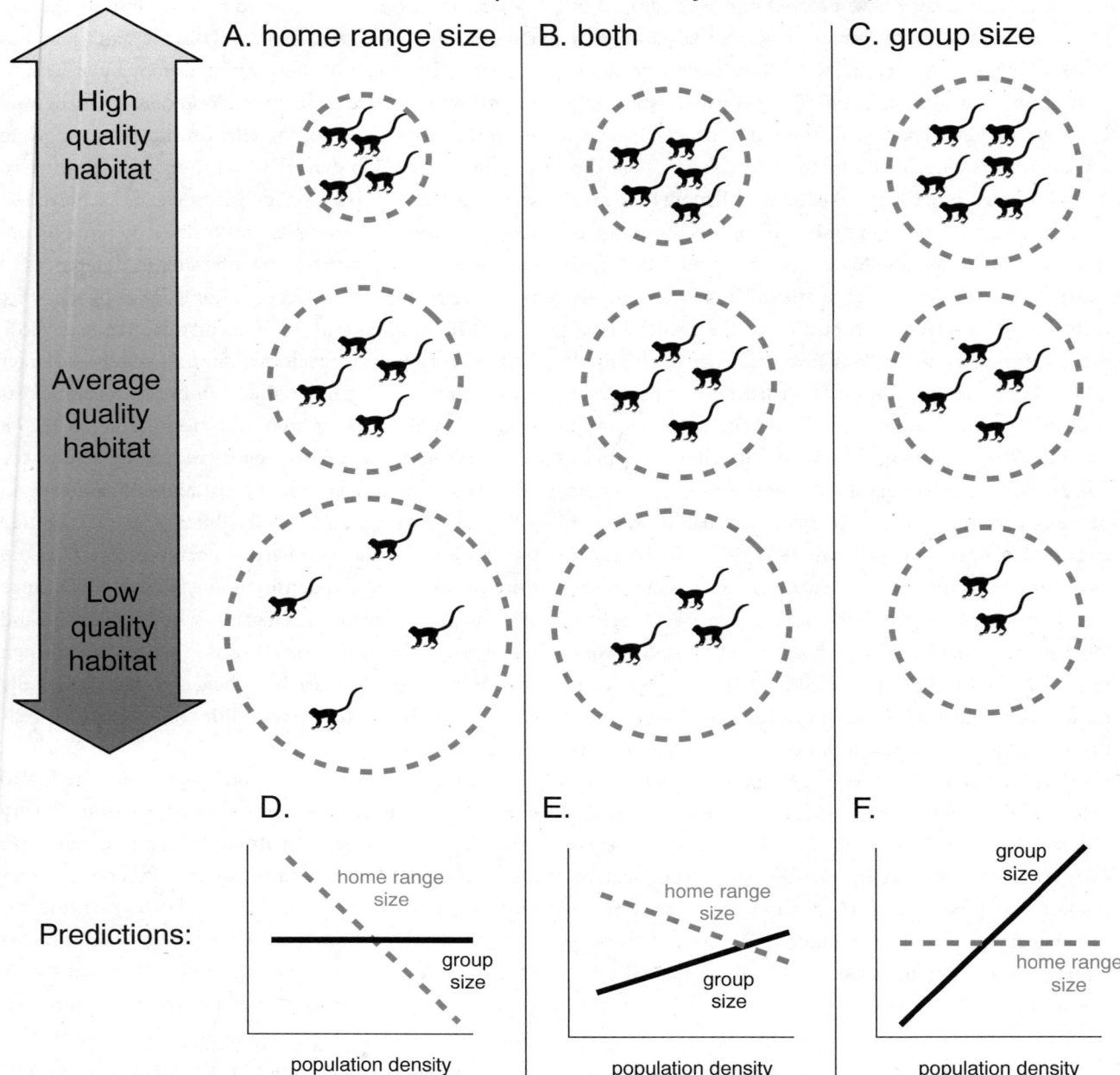

Figure 27.1. Schematic depiction of mechanisms influencing carrying capacity. Population density may vary because habitat quality influences home range size (A), group size (C), or both home range size and group size (B). These three alternatives imply distinct, testable predictions about the relationships between population density, home range size, and group size. The home-range-mediated model of density variation predicts that home range size is inversely correlated with population density and there is no correlation between group size and population density (D). The group-size-mediated model predicts positive correlations between group size and population density and no relationship between home range size and density (F). Both factors may work in tandem, leading to intermediate predictions (E). Bogdan Bocianowski made the primate icon, which was downloaded from http://phylopic.org and is used under a Creative Commons license (CC BY-SA 3.0).

sizes are higher in high-quality habitats (Dunbar 1987). Black howler (*Alouatta pigra*) populations exhibited larger group sizes (and larger, not smaller, home ranges) in high-density populations (Ostro, Silver, Koontz, Horwich, et al. 2001; Ostro, Silver, Koontz, Young, et al. 1999). The large differences in chimpanzee (*Pan troglodytes*) population densities between Kanyawara and Ngogo in Kibale National Park, Uganda, are largely due to differences in group size, as estimates of home range are comparable for the two sites (D. Watts 2012b). Many other studies report higher group sizes in higher-density populations, which is consistent with a group-size-mediated model of variation in population density—although the lack of data on home range sizes often precludes assessment of the relative importance of group size and home range in determining density (e.g., red colobus, *Procolobus gordonorum*: Struhsaker, Marshall, et al. 2004; ring-tailed lemurs, *Lemur catta*: A. Jolly, Dobson, et al. 2002; Hanuman langurs, *Presbytis entellus*: A. Koenig, Beise, et al. 1998).

Population density is not, however, always primarily mediated by group size; there is evidence of home-range-mediated density variation in some taxa. For example, Takasaki (1981) found that home ranges of Japanese macaques (*Macaca fuscata*) were smaller in high-quality evergreen habitats than in deciduous forests; reanalysis of these data confirms that density variation is predicted by home range size but not group size (fig. 27.2e, f). Variation in black-and-white colobus (*Colobus guereza*) density across sites in East Africa suggests a similar pattern (fig. 27.2g, h; data from Dunbar 1987), as does variation in proboscis monkey (*Nasalis larvatus*) population density across sites in Borneo (Boonratana 2000). Variation in density across habitat types of varying quality within sites also suggests that for some species, high densities are achieved primarily by differences in home range size rather than group size (e.g., vervet monkeys, *Cercopithecus aethiops*, at the Masai-Amboseli Game Reserve in Kenya: Struhsaker 1967; mangabeys, *Cercocebus albigena*, at Kibale Forest, Uganda: Freeland 1979a). Variation in indri (*Indri indri*) populations among sites also suggests that home ranges are smaller in high-density populations in this species, although data on group sizes are needed to distinguish the relative importance of group-size-mediated and home-range-mediated variation (Glessner and Britt 2005).

In some cases, variation in population density may be mediated by both group size and home range size (fig. 27.1b, e). For example, a model of density variation in Japanese macaques that includes both home range and group size vastly outperforms the home-range-only model ($\Delta AIC_c = 883$, $\omega > 0.999$; again using data from Takasaki 1981), with home range exhibiting a negative effect and group size a positive one, as expected under a hybrid model. It is rare to have a sample with as many independent data points as Takasaki's data set does, making it difficult to conduct robust multivariate analyses in other instances.

In addition to sample size limitations, ecological factors often complicate interpretation of recorded patterns. For example, the data presented by Pinto et al. (1993) suggest no clear relationship between group size and population density in five species of Brazilian primates in Atlantic forests, suggesting the possibility that density variation is primarily mediated by differences in home ranges (which were not measured)—although it is difficult to assess the extent to which these patterns are influenced by habitat fragmentation and hunting. Population density in Mentawi snub-nosed langurs appears to be mediated by both group size and home range size, with group sizes being larger and home ranges much smaller in the high-density Grunka site than in the low-density Sarabua site (Watanabe 1981), although this relationship is likely confounded by the intensity of hunting pressure. Compression in habitat fragments likely also plays an important role in mediating the relationships between group size, home range size, and population density (e.g., Dunbar 1987). Finally, the spatial scale of comparisons can influence results. For instance, comparisons of guereza groups with the Bole Valley suggest group-size-mediated density variation in this taxon, while comparisons across sites in East Africa suggest that home-range-mediated density variation is more important (Dunbar 1987). Similarly, while broad comparisons across Japanese macaque populations indicate primarily home-range-mediated variation in population density (Takasaki 1981, 1984), comparisons within habitat types, those at smaller spatial scales, and those that consider temporal variation reveal more complex patterns (Hanya, Kiyono, Yamada, et al. 2006; Yamagiwa and Hill 1998).

The relationship between population density, group size, and home (or day) range has been considered in

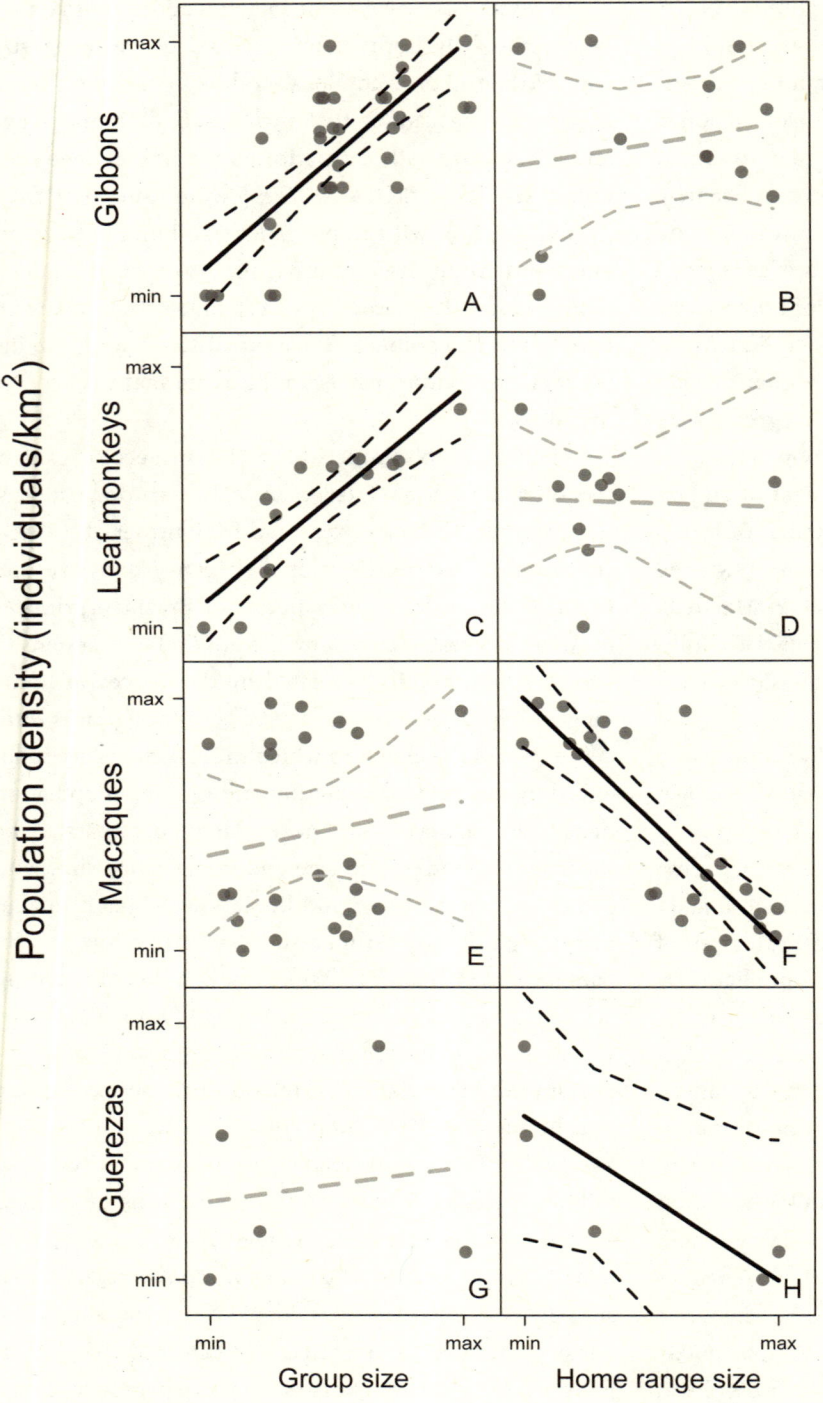

Figure 27.2. Mechanisms influencing population density (i.e., carrying capacity) differ among taxa. Gibbons (A, B) and leaf monkeys (C, D) at Gunung Palung National Park exhibit group-size-mediated density variation (data from A. Marshall 2004, 2010). Macaques across Japan (E, F; data from Takasaki 1981) and guerezas at Bole Valley, Ethiopia (G, H; data from Dunbar 1987), exhibit home-range-mediated density variation. Raw population density data in all plots have been rescaled to facilitate comparisons. The lines in each panel plot ordinary least squares regression lines and 95% confidence intervals (95% CIs are outside the plotting range in G). For each species, models of density predicted by group size (A, C, E, G) and home range size (B, D, F, H) were compared; black lines indicate the better model as determined by formal model comparison using Akaike's information criteria corrected for small sample sizes (AIC_c), and gray lines indicate the worse model. Black and gray regression lines are also significant and nonsignificant, respectively, using a standard null hypothesis significance testing framework. Gibbons: model A (mA, R^2_{adj} = 0.57, n = 33 groups), mB (R^2_{adj} = −0.06, n = 12 groups), mA > mB (ΔAIC_c = 2, model weight for A [ω_{mA}] = 0.73); leaf monkeys: mC (R^2_{adj} = 0.75, n = 13 groups), mD (R^2_{adj} = −0.12, n = 10 groups), mC > mD (ΔAIC_c = 11.4, ω_{mC} > 0.99). In B and D, home range size estimates control for differences in sampling intensity among groups; sample sizes in B and D are lower than in A and C because accurate home range estimates were available for only a subset of groups. Macaques: mE (R^2_{adj} = −0.03, n = 22 sites), mF (R^2_{adj} = 0.67, n = 22 sites), mF > mE (ΔAIC_c = 25, ω_{mF} > 0.999). Following Takasaki (1981), the data in E and F have been $\log_{10}$ transformed, and only data from undisturbed and lightly disturbed sites are plotted. Guerezas: mG (R^2_{adj} = −0.30, n = 5 sites), mH (R^2_{adj} = 0.60, n = 5 sites), mH > mG (ΔAIC_c = 5.9, ω_{mH} > 0.95).

several interspecific analyses (with each taxon contributing a single datum, e.g., Chapman and Chapman 2000a; Clutton-Brock and Harvey 1979; Wrangham, Gittleman, and Chapman 1993; Yeager and Kirkpatrick 1998), but I am unaware of any systematic analyses of interspecific variation in the relationship between these variables within species. In other words, it is not known whether there are systematic differences between taxa in which population density is group-size-mediated and those in which it is home-range-size-mediated. A systematic analysis of this question would entail compiling multiple data points on many species, characterizing the relationship within each species, and comparing the intraspecific patterns between species (akin to Janson and Goldsmith's [1995] analysis of relative ranging costs). Such an effort is beyond the scope of this chapter but may reveal interesting variation due to diet, body size, degree of home range overlap, ranging costs, social system, or the extent to which a species is food limited.

Carrying capacity is likely to affect primate groups in ways other than simply the number of individuals they contain or how large an area they occupy. It may well affect group cohesiveness, social relationships, or the rates and nature of intergroup interactions. Unfortunately, the effects of carrying capacity on the simplest attributes of primate groups (e.g., their size and the extent of their home ranges) are not well known, suggesting that it may be some time before the more subtle group-level effects of variation in carrying capacity are characterized.

How Does Carrying Capacity Affect Populations?

Assessing the effects of variation in carrying capacity on populations requires the collection of data from multiple populations of a species living at different carrying capacities. There is a rich history of comparative research in primatology, and much of this work has focused on populations living in habitats of different quality (Ganzhorn 1992; McKey 1978; Oates, Whitesides, et al. 1990; Wasserman and Chapman 2003). It is therefore perhaps surprising that understanding of the effects of variation in carrying capacity on primate populations is remarkably limited. There are probably at least two reasons for this. First, much of this work is aimed at identifying the causes of variation in carrying capacity rather than the consequences of this variation (i.e., carrying capacity is the dependent rather than the independent variable). Second, much research that addresses the consequences of variation in population density is focused on the role of population size relative to local carrying capacity rather than on the role of variation in carrying capacity per se (Bartlett et al. 1993; Butynski 1990; Pope 1998).

Even when good data are available on multiple populations living at different population densities and it is reasonable to conclude that each population is at carrying capacity, it can be difficult to determine whether observed differences are causally related to differences in carrying capacity rather than being due to differences in some alternative ecological factor. Observations of single primate populations occupying a range of habitats of different qualities can provide some useful insights here, because variation in predation, disease, and biogeographic factors that confound comparisons among distant sites are largely controlled (Chapman and Chapman 1999; Hanya, Yoshihiro, et al. 2004; Iwamoto and Dunbar 1983; A. Marshall 2010). Despite the potential benefits of such work, it is not completely analogous to studying distinct populations at different carrying capacities, because the dispersal of individuals among habitats can alter population dynamics and confound our ability to extrapolate results to entire isolated populations living at a given carrying capacity. For instance, in heterogeneous environments primate populations can persist at very low, stable population densities in sink habitats due to immigration from nearby sources (A. Marshall 2009; Ohsawa and Dunbar 1984; Siex and Struhsaker 1999). We can study the effects of living at low population densities on individuals and groups in sink subpopulations, but this is not analogous to studying a self-contained population at a comparable carrying capacity.

Despite these complications, it is reasonable to hypothesize from first principles that carrying capacity will have important population-level effects. For instance, given equal habitat area, population sizes in areas with high carrying capacity will be greater, and many predictions can be made based on variation in population size (e.g., susceptibility to decline due to demographic stochasticity or rates of cultural innovation). In addition, the potential importance of several density-dependent ecological forces might be expected to vary as a function of carrying capacity (e.g., populations in environments with high carrying capacities may be more at risk from emerging diseases, pathogens, or proximity-induced social stress).

POPULATION REGULATION AND LIMITATION

It is a truism that animal populations cannot increase indefinitely and that their numbers must be somehow held "in check"—but precisely how this occurs has been one of the most contentious questions in ecology over the last century. Key components of the debate are the topics of population limitation and population regulation (Caughley and Sinclair 1994; T. C. White 2001). Limiting factors are ecological forces that set the equilibrium density of a population, whereas regulatory factors are ecological forces that keep a population near its equilibrium density (Caughley and Sinclair 1994; C. Krebs 2002). Regulatory factors must, by definition, be both biotic and density dependent, because they act to increase or decrease birth or death rates based on a population's size relative to carrying capacity (Nicholson 1933; Turchin 1999). Limiting factors can be either density dependent or density independent and either biotic or abiotic (Caughley and Sinclair 1994; Enright 1976). Population ecologists have hotly debated the relative importance of population limitation and regulation, whether density-dependent or density-independent factors are more important, and the merits of models of weak versus strong density dependence (C. Krebs 1995; B. Murray 1999; Turchin 1999; T. C. White 2001; Wolda 1995). Related issues, such as whether populations are generally at equilibrium (C. Krebs 1992; Wolda 1992) and how population dynamics are affected by spatial heterogeneity (Pastor et al. 1997) and environmental stochasticity (McLeod 1997), have also received substantial attention.

C. Krebs (2002) suggested that most studies in population ecology can be characterized as seeking to identify factors that (1) stabilize density at carrying capacity, (2) prevent population growth above carrying capacity, or (3) set carrying capacity itself. The first two questions are primarily about population regulation, while the third relates to population limitation. He argued the last of these is the least well studied and understood, despite being of great importance in applied ecology (C. Krebs 2002). Krebs's view was that population limitation has been largely neglected "because it is theoretically very dull" (C. Krebs 2002, 5), because anything that is either a source of mortality or alters birth rates can limit populations (Caughley and Sinclair 1994; Enright 1976). Despite this, many more studies in primate population ecology have focused on the question of what limits populations (factor 3) than on questions of population regulation (factors 1 and 2; but see, e.g., Dittus 1979). This may be because population limitation is more tractable to study—as it does not require demonstration of density dependence—or of more direct relevance to conservation (C. Krebs 2002). Following this convention, here the focus is on whether and how primate populations are limited by food (i.e., what sets carrying capacity), and the question of whether food constrains primate populations (i.e., whether food is a density-dependent regulatory factor) is largely ignored. There is little reason to assume that populations are both limited and regulated by the same ecological factor. Thus, evidence bearing on the importance of food as a limiting resource is of limited relevance to the question of whether food regulates primate populations.

TOP-DOWN VERSUS BOTTOM-UP LIMITS ON POPULATIONS

Any source of mortality—predation, food limitation, disease, parasites, and so on—may differ among habitats and therefore contribute to differences in density among them. Thus, there are many candidate ecological factors to consider when assessing what limits populations. A great deal of attention has been paid to debating whether populations are limited primarily by top-down or bottom-up mechanisms (Matson and Hunter 1992). Proponents of top-down models argue that populations are primarily limited by predators (Fretwell 1977; Hairston et al. 1960; Terborgh 1988), while adherents to bottom-up models place a primacy on resources (e.g., food and nutrients) as the determinants of population density (M. Hunter and Price 1992; E. Wilson 1987). As with other dichotomous debates about ecological mechanisms (e.g., whether neutral or niche mechanisms primarily determine community structure), over time researchers have come to recognize the importance of both forces, and focus has shifted from debates over which mechanism occurs in nature to studies that seek to determine the relative importance of each under various conditions and at various levels of analysis

(M. Hunter and Price 1992; M. E. Power 1992; Schmitz 2008; G. Wang et al. 2006). Nevertheless, ecologists still often use the terms "top-down" and "bottom-up" forces as useful shorthand when discussing mechanisms affecting population density (e.g., Hopcraft et al. 2010).

Individual primates can die of hunger or be killed by predators, and selection to avoid either fate has likely had huge effects on primate behavior and socioecology (for a discussion of the "landscape of fear," see Hill, this volume). But which ecological force, bottom-up food limitation or top-down predator limitation, primarily limits primate populations? It is reasonable to expect variation among taxa, especially due to dietary adaptations or differences in susceptibility to predation, but it also seems reasonable to hypothesize a priori that most primate species will be limited by food and not by predators. First, there is considerable theoretical support for the hypothesis that, for large-bodied, long-lived consumers with slow rates of population increase, like primates, populations are primarily limited by the availability of resources (Floyd et al. 1996; Schoener 1973, 1974). Second, primate species' relatively slow life histories imply low rates of extrinsic mortality (including that due to predation). Finally, primate species in adjacent habitats may live at stable, long-term carrying capacities that differ by an order of magnitude or more (e.g., gibbons, macaques, and leaf monkeys at Cabang Panti in Gunung Palung National Park; A. Marshall, Beaudrot, and Wittmer 2014); it seems unlikely that these large density differences are due to different predation pressures—particularly when the home ranges of individual predators can overlap multiple adjacent habitats (e.g., clouded leopards at the same site; M. L. Allen et al. 2016). Thus, the hypothesis that primates are chiefly limited by bottom-up mechanisms and not top-down pressures is plausible. It is also consistent with a substantial body of empirical work (see below); I am unaware of empirical data suggesting that primate populations are limited by predation pressure.

While the focus in this chapter is primarily on food as a limiting factor, it is worth noting that primate populations are likely regulated by bottom-up forces as well. As noted above, regulatory mechanisms must have density-dependent effects around a population's carrying capacity. Predation can cause substantial mortality across a range of primate species (Boinski and Chapman 1995; Hart 2007; Isbell 1994a), but little evidence exists that primate mortality due to predation is density dependent. In addition, most predation targets infant or juvenile primates (Cheney 1987), whose loss likely has relatively little demographic impact. This assertion is consistent with the results of stage-based models of primate population viability, which conclude that the size of primate populations is most sensitive to adult, not juvenile, survivorship (e.g., Dobson and Lyles 1989; A. Marshall, Boyko, et al. 2009). An alternative possibility is that primate density is regulated by disease or parasite load. While these factors are normally density dependent (Freeland 1979b; May and Anderson 1979), the fitness costs associated with parasites and disease are not always easy to demonstrate for primates (e.g., C. Davies et al. 1991; Freeland 1979b; Hausfater and Watson 1976; Milton 1996). Furthermore, demonstration that a disease or parasite influences individual mortality or fecundity does not necessarily imply that it regulates a population (C. Krebs 1995). Outbreaks of epidemic disease can clearly have massive demographic effects. For example, an outbreak of yellow fever led to a dramatic short-term reduction in howler monkey population size on Barro Colorado Island (Collias and Southwick 1952), and more recently, an Ebola epidemic decimated great ape populations across central Africa (Bermejo, Rodríguez-Teijeiro, et al. 2006; P. Walsh et al. 2003). As large as the effects of such outbreaks are, there is little reason to assume that they regulate primates around an equilibrium carrying capacity. At present, there appears to be little theoretical or empirical support for the hypotheses that primate density is typically regulated by mortality or fertility changes due to predation, disease, or parasites. In contrast, competition between primate consumers for limited food resources suggests that food availability will be strongly density dependent around K, at least during periods of food shortage (Cant 1980).

FOOD AND PRIMATE CARRYING CAPACITY

The Lack of a General Theory

Primatologists have long assumed that food availability is the primary determinant of primate abundance (e.g., Cant 1980; Hanya and Chapman 2013; Oates, Whitesides, et al. 1990; Terborgh and van Schaik 1987), a view shared by many researchers who study other vertebrates (Boutin 1990; Schoener 1973; T. C. White 2001). Indeed, it is rare that alternative mechanisms are even considered. The widespread acceptance of the hypothesis that food sets carrying capacity belies the rudimentary state of knowledge on this topic. While understanding of the determinants of abundance is improving for some specific taxa (e.g., Japanese macaques: Hanya, Kiyono, Yamada, et al. 2006; Hanya, Yoshihiro, et al. 2004; red colobus monkeys: Chapman and Chapman 1999; Chapman, Schoof, et al. 2015), we are far from having a general theory of primate abundance that would allow us to predict population density from first principles. A simple thought experiment highlights our lack of a general theory in this domain. We start with the plausible assumption that food sets the carrying capacity of a primate taxon. Next, we make the unrealistic assumption that all important aspects of food quality and quantity for this taxon can be captured in a single variable, "food availability," ignoring known complexities related to issues such as nutrient balancing, the packaging problem, and the likely multivariate nature of ecological drivers of primate abundance (S. Altmann 2009; Hanya, Stevenson, et al. 2011; J. Lambert and Rothman 2015; Raubenheimer, this volume) and other factors that must be important, such as the frequency of food fluctuations in relation to the taxon's life span and their spatial extent relative to the species' home ranges. We further assume that the variable "food availability" is assessed in two distinct habitats (x and y) in an accurate, unbiased way and that our measure is equally valid in each—neither of which is a trivial undertaking (A. Marshall, Ancrenaz, et al. 2009; A. Marshall and Wich 2013). Finally, we assume that while our two habitats differ in food availability, they are otherwise ecologically identical.

If food availability in habitat x is stable and always greater than the stable amount available in habitat y, then we would easily predict that x is higher-quality habitat and should support a higher population density than habitat y (fig. 27.3a). Adding cyclical variation in food abundance in the two habitats would not change this prediction, whether the cyclical variation is comparable among habitats (fig. 27.3b) or differs in frequency (fig. 27.3c) or amplitude (fig. 27.3d). But it is unclear what we might predict in alternative scenarios that are perfectly ecologically plausible. For instance, if habitat x had higher mean food availability than habitat y but experienced a more extreme period of food shortage (fig. 27.3e), which of these two habitats would be higher quality for primates? Alternatively, we might imagine a scenario in which habitat x had a higher mean food availability than habitat y and that the periods of food scarcity were equally extreme in magnitude in the two habitats, but that the period of food shortage was of longer duration in habitat x (fig. 27.3f). Again, which habitat should support a higher density of our taxon? There is no obvious answer to either of these questions. Furthermore, even if we could make a plausible a priori prediction based on ecological theory, it is likely that this prediction might differ based on the body size or life-history strategy of our taxon. Our knowledge of (and theory about) this most fundamental issue in primate population ecology is surprisingly limited.

Common Analytical Approaches

The assumption that primates are food limited is rather more common than empirical demonstration that they are. Ecologists have typically employed two analytical approaches to identify how food availability limits population density: assessing the correlation between measures of food abundance and density across space or over time (C. Krebs 2002; Krohne 2001); Hanya and Chapman (2013) call these "static" and "dynamic" approaches, respectively. Both approaches require the assumption that the population under consideration is at its equilibrium carrying capacity.

The first approach entails documenting variation in carrying capacity of the same taxon across habitats (the dependent variable), measuring the aspect or aspects of food availability thought to limit populations (the independent, predictor variables), and examining the strength of correlations between them. Indices of food availability that are statistically significant positive predictors of population density—and ideally predictors that explain a large amount of variation in abundance—are assumed to be the primary determinants of carrying capacity (e.g., A. Marshall 2010; A. Marshall and Leighton 2006). This is the most common method employed by primatologists, and there are numerous examples of primate population density (or biomass) being correlated with indices of food availability across sites. Several studies conducted at large spatial scales report positive relationships between the total biomass of frugivorous primates and measures of fruit fall (e.g., Hanya, Stevenson, et al. 2011; Kinnaird and

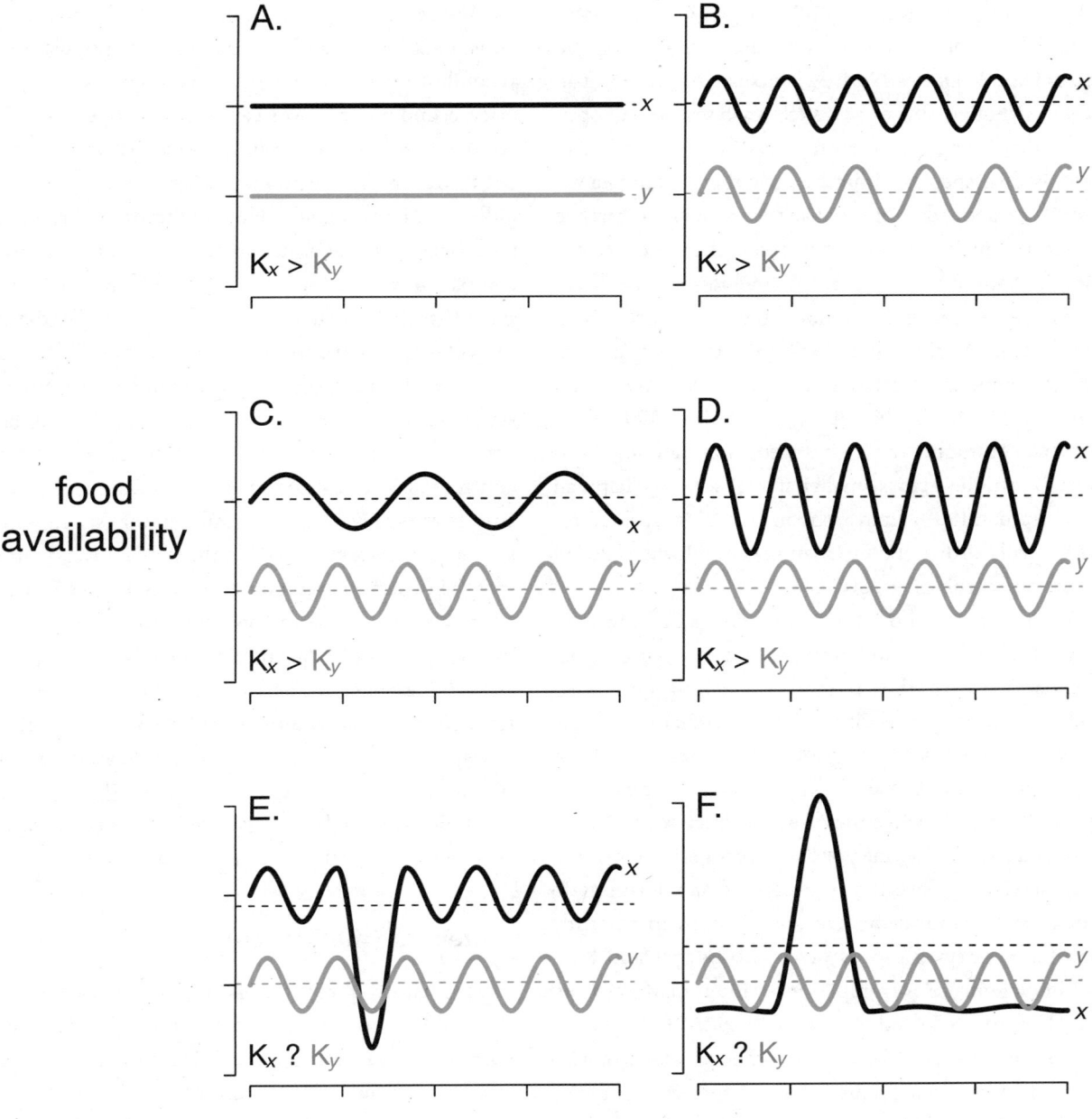

Figure 27.3. Schematic depictions of food availability over time in two habitats (x and y) in six ecological scenarios. In each panel, the solid lines depict food availability in habitats x (black) and y (gray), and the horizontal dashed lines indicate mean food availability in each habitat following the same color scheme. In A–D, food availability in habitat x is consistently higher than in habitat y, despite differences in patterns of temporal variation. Thus, in these four scenarios, it is easy to predict that habitat x will be of higher quality than habitat y and support higher population densities of a given primate taxon (i.e., the carrying capacity [K] of habitat x will be greater than that of habitat y). It is less easy to predict which of the two habitats would sustain a higher population density in the bottom two panels. In E and F, habitat x exhibits higher mean food availability than habitat y, but the period of food scarcity is more extreme (E) or of longer duration (F) in habitat x than in habitat y. Current theory does not furnish a strong a priori prediction about which habitat will have a higher carrying capacity or how fundamental biological attributes (e.g., body size and life history) might influence a species' carrying capacity under these scenarios.

O'Brien 2005; P. Stevenson 2001) or between the stem density of food plants and the population density of single species (e.g., A. Davies 1994; A. Marshall, Ancrenaz, et al. 2009). Although suggestive, such analyses are suboptimal in identifying the determinants of carrying capacity for individual species, either because they lump data for multiple taxa together or because the substantial distance among sites makes it difficult to assess the effects of potentially confounding variables, including differences among sites in research methods, biogeography, human disturbance, and the presence or abundance of competitors (Chapman, Gautier-Hion, et al. 1999; Gupta and Chivers 1999; A. Marshall 2004; Struhsaker 1999). Such potential confounds can be reduced, although not eliminated, by studies across smaller spatial scales (Chapman and Chapman 1999; Hanya, Kiyono, Yamada, et al. 2006; A. Marshall 2010; K. Potts, Chapman, and Lwanga 2009; Rovero and Struhsaker 2007).

The second method entails examining the effects of temporal changes in food availability on carrying capacity. In some cases, the amount of a putatively limiting food resource is altered directly by researchers (typically via supplementation), and in others changes result from natural processes. Primate food supplies are rarely deliberately altered, and when they are it is typically not for the expressed goal of studying population ecology (Hanya and Chapman 2013). Nevertheless, results of manipulations do strongly support the general notion that food importantly determines carrying capacity (see also Borries and Koenig, this volume). For example, populations of *Macaca mulatta* (Koford 1965) and *M. fuscata* (Itani 1975) have been shown to grow 9% to 16% annually when provisioned with additional food. In a detailed study of the effects of food provisioning in a troop of *M. fuscata*, Mori (1979) documented a drastic increase in the total number of females and the per capita birth rate during a seven-year period of intensive provisioning. When the food supplementation was substantially reduced, group size and birth rate quickly dropped (Mori 1979). Observations of naturally occurring variation in food supply also indicate that food sets carrying capacity (see also Emery Thompson, this volume). For example, many populations have been shown to decline, sometimes dramatically, during periods of food scarcity (e.g., *Macaca sinica*: Dittus 1979; *Cercopithecus aethiops*: P. Lee and Hauser 1998; Struhsaker 1973; *Papio cynocephalus*: J. Altmann et al. 1985; Hausfater 1975).

Although valuable, both methods have several limitations. First, the assumption that primate populations are at equilibrium is often incorrect, as compression, hunting, disease, and other factors can result in populations living above or below their carrying capacity (Mathewson et al. 2008; Rovero and Struhsaker 2007). Second, it is often difficult to convincingly rule out alternative explanations for observed spatial or temporal variation. For instance, in comparisons across spatial scales, it is often unsafe to assume that sites are fundamentally similar, other than with respect to the candidate predictor variables. Third, it is not easy to infer causation from observed correlations across space or over time. Even when putative limiting factors are experimentally altered, we rarely have appropriate control populations to truly infer causation. Finally, these studies typically employ a null-hypothesis significance testing framework and seek only to reject the hypothesis that the observed predictor has precisely no effect on the outcome variable. More powerful is to compare evidence for alternative candidate models using formal information criteria. Demonstration that a single predictor has a significant effect (or that multiple predictors are significant in a multiple regression) does not allow us to assess the effects of alternative competing models (e.g., that populations are limited by total food availability versus fallback food availability versus preferred food availability).

Protein-to-Fiber Ratios

The protein-to-fiber model is commonly raised in discussions of limitations on primate abundance. It is another example of the static approach (sensu Hanya and Chapman 2013) mentioned above, but it warrants specific discussion due to its widespread use and critique (Chapman et al., this volume; Felton and Lambert, this volume). The logic is simple: mammalian herbivores seek to optimize the trade-off between ingesting valuable protein and avoiding the nonnutritious ballast of plant structural compounds (Milton 1979). Therefore, they should actively seek and selectively feed on leaves with a high ratio of protein to fiber. The protein-to-fiber ratio of leaves is commonly cited as a primary factor influencing leaf choice among primate folivores (Chapman, Chapman, Naughton-Treves, et al. 2004; Hanya and Bernard 2012; McKey, Gartlan, et al. 1981; Milton 1981b; Wasserman and Chapman 2003), and habitat-wide indices of the protein-to-fiber ratio of leaves have frequently been shown to predict the biomass of folivorous monkeys (Chapman, Chapman, Naughton-Treves, et al. 2004; Ganzhorn 1992; McKey, Gartlan, et al. 1981; Waterman, Ross, et al. 1988). There are methodological reasons to quibble with the ways in which the protein-to-fiber model is commonly employed in the context of primate feeding selectivity (e.g., "preference" analyses are based not on preference but instead on importance, and competing hypotheses are rarely entertained), but here the focus is on concerns with its application to understanding primate carrying capacity.

Protein-to-fiber ratios may well have utility as a relatively quick assay of habitat suitability (although there are reasons for caution in this realm, too, especially if the results are used in conservation management; Gogarten, Guzman, et al. 2012), but we should be careful not to assume that we have conclusively identified the determinants of carrying capacity for several reasons. First, protein-to-fiber ratios are often calculated for mature leaves (which are rarely consumed by most folivorous primates; e.g., Ganzhorn 1992; Waterman, Ross, et al. 1988) or for plants that are common but not necessarily known to be consumed by primates (Chapman, Chapman, Bjorndal, et al. 2002; Oates, Whitesides, et al. 1990; Waterman, Ross, et al. 1988). Second, the sample sizes of leaves analyzed are typically quite small, masking the substantial variation among individual trees within a site (Chapman, Chapman, Rode, et al. 2003). Third, assays have typically measured total nitrogen, which is not always well correlated with the ecologically relevant measure, available nitrogen (I. Wallis et al. 2012). Fourth, tannin complexes can lead to an overestimation of fiber content (Makkar and Singh 1995; I. Wallis et al. 2012), meaning that data that we interpret as evidence for avoiding fiber might instead be due to avoidance of tannins (I. Wallis et al. 2012). Fifth, new evidence that folivores do not necessarily seek protein (Felton, Felton, Lindenmayer, et al. 2009; Rothman, Raubenheimer, and Chapman 2011) coupled with the fact that levels of protein in leaves far exceed primate nutritional requirements (Chapman, Rothman, and Lambert 2012; Oftedal 1991) suggests that protein might not be as important for folivorous primates as is generally assumed. Sixth, there are well-established statistical and inferential limitations of using ratios in analyses (Kronmal 1993; K. Pearson 1896; Smith, Aboitiz, et al. 2005b). Seventh, the correlation between protein-to-fiber ratios and primate abundance does not imply that the two are causally linked (Chapman, Chapman, Bjorndal, et al. 2002; Hanya and Chapman 2013)—this correlation-causation problem plagues most work in primate ecology (including mine). Finally, the protein-to-fiber model is conceptually fuzzy. Initial discussions of protein-to-fiber ratios were in the context of feeding selectivity; plants with high ratios were identified as preferred foods (McKey, Gartlan, et al. 1981; Milton 1979), but hypotheses about the mechanism by which protein-to-fiber ratios influence folivore biomass relate to their importance as fallback foods (A. Davies 1994). (Fallback foods are a distinct class of resource defined as foods eaten when preferred foods are unavailable; see below and the discussion in Knott and DiGiorgio, this volume.) The matter is further complicated by the fact that, as noted above, protein-to-fiber ratios are typically measured for mature leaves, which are generally avoided (very low preference) and rarely serve as fallback foods (Chapman, Chapman, Naughton-Treves, et al. 2004; Hanya and Chapman 2013). Thus, while protein-to-fiber ratios are often good predictors of folivorous primate biomass, methodological limitations and the lack of a clear mechanism linking the two suggest that we should be circumspect when assessing whether the ratio is the key ecological factor limiting carrying capacity.

Other Nutritional Approaches

The protein-to-fiber ratio model uses coarse-grained nutritional analysis of macronutrients as an index of the quality of available foods. Variants on this theme were widely employed in the wildlife literature in the 1970s and 1980s (e.g., N. Hobbs and Swift 1985; Mentis and Duke 1976; Rowe-Rowe and Scotcher 1986). Although it is intuitively appealing to use an index of the availability of a putatively limiting macro- or micronutrient to predict animal abundance, these models have fallen out of favor among wildlife ecologists due to unrealistic assumptions, failure to incorporate stochasticity, and often large discrepancies between the predicted and observed population densities (McLeod 1997). Although increasing attention is being paid to studies of the nutritional components of primate diets (Felton, Felton, Lindenmayer, et al. 2009; J. Lambert and Rothman 2015; Righini 2017; Rothman, Raubenheimer, and Chapman 2011), this work has rarely been directly tied to studies of primate carrying capacity. Rode, Chapman, McDowell, et al. (2006) showed that redtail monkey densities in five groups across three sites were closely related to their intake of copper but not to their overall food intake or to the stem density of food trees, suggesting that the availability of this micronutrient might limit population density. E. Vogel, Harrison, et al. (2015) compared two nearby orangutan populations and showed that orangutans at the site with higher population density had diets that provided more metabolizable energy, again consistent with the hypothesis that nutritional differences might underpin differences in abundance. Although suggestive despite their small sample sizes, neither study measured the availability of nutrients in the environment, which would be a crucial link in the chain of evidence necessary to establish that these populations were limited by the nutritional quality of foods available in their habitats. Additional fieldwork and laboratory analyses—accompanied by more detailed data on the nutritional requirements of wild primates—will be required to determine whether this will be a fruitful line of inquiry to elucidate the determinants of primate carrying capacity.

Identifying Critical Periods and Crucial Food Resources

The observation that primate population sizes tend to be relatively stable despite dramatic temporal variation in the quantity and quality of foods suggests that carrying capacity is set by relatively short, critical periods or by specific foods. This implies that while feeding competition might be crucially important in determining population size, it occurs only sporadically (Cant 1980; Hanya and Chapman 2013; A. Marshall, Cannon, and Leighton 2009; Schoener 1974). Of considerable interest has been discussion of whether primate populations are primarily limited by the availability of high-quality foods during good times or by the amount of food during periods of scarcity—in other words, whether preferred or fallback foods set carrying capacity (e.g., Hanya and Chapman 2013; Hanya, Kiyono, Yamada, et al. 2006; A. Marshall and Leighton 2006; A. Marshall, Boyko, et al. 2009). Preferred foods are those that are used more than would be expected based on their availability (i.e., they are positively selected; Clink et al. 2017; Dillis et al. 2015; M. Leighton 1993); they are generally considered high-quality food items. Fallback foods are resources used when preferred resources are scarce or absent (J. Lambert 2007; A. Marshall and Wrangham 2007) and are generally considered to be of lower nutritional quality than preferred foods (see further discussion in Knott and DiGiorgio, this volume). In theory, either class of food might set carrying capacity (A. Marshall and Leighton 2006; A. Marshall, Boyko, et al. 2009; A. Marshall 2010). Preferred foods provide substantial energy to primates and are often crucial for reproduction. Thus, habitats with abundant preferred foods might support higher population densities (i.e., have higher carrying capacities) than those in which they are scarce (J. Altmann et al. 1985; Balcomb et al. 2000). Alternatively, because fallback foods can sustain individuals during periods of resource scarcity (i.e., help them survive "ecological crunches"; Cant 1980; Peres 2000b; Terborgh 1986), they may be the primary determinants of carrying capacity.

There is empirical evidence in support of each of these hypotheses. The preferred food hypothesis is supported by correlations across sites in Southeast Asia between preferred food trees and the biomass of gibbons (*Hylobates lar*; Caldecott 1980; Mather 1992 in Chivers 2001) and colobine monkeys (*Presbytis* spp.; A. Davies 1994). Similarly, *Presbytis* densities are higher in habitats where the availability of preferred leaves and seeds is high (A. Marshall 2010; Oates, Whitesides, et al. 1990; Waterman, Ross, et al. 1988). The fallback food hypothesis also has empirical support. The density of several primate species is highly correlated with the stem density of known fallback food plants (e.g., gibbons, *Hylobates albibarbis*: A. Marshall and Leighton 2006; orangutans, *Pongo abelii*: Wich, Buij, and van Schaik 2004; gorillas, *Gorilla gorilla gorilla*: Doran et al. 2002) or food availability during periods of resource scarcity (orangutans, *Pongo* spp.; A. Marshall, Ancrenaz, et al. 2009). Annual habitat use by Yunnan snub-nosed monkeys, *Rhinopithecus bieti*, is primarily driven by the availability of fallback foods, not preferred foods (Grueter, Li, et al., 2012), and fluctuations in availability of fallback foods during periods of resource scarcity appear to dictate habitat use by apes in Lopé Reserve, Gabon (L. White et al. 1995).

There are several possible explanations for the observed variation in support of the alternative preferred and fallback food hypotheses. First, observed variation may be due to real taxonomic differences among primate species based on biological characteristics. For example, fallback foods might be crucially important for populations of species exhibiting relatively risk-averse life-history strategies that place a premium on survivorship, while preferred foods determine carrying capacity for species with faster life histories in which reproductive rates are more important than survivorship in determining individual fitness (A. Marshall, Boyko, et al. 2009; A. Marshall 2010). Different results might also be related to interspecific differences in metabolic rates or the ability to store fat during periods of resource abundance to enhance survival during resource-poor periods (Hanya and Chapman 2013; Hanya, Yoshihiro, et al. 2004; Schmid and Ganzhorn 1996). Second, differences among habitats in the quality of fallback foods might also dictate whether fallback or preferred foods set carrying capacity; when fallback foods are insufficient to sustain populations for extended periods, their carrying capacities will be set by the abundance of other resources (Hanya and Chapman 2013; A. Marshall, Boyko, et al. 2009). The proposed distinction between staple and filler fallback foods might have utility in explaining some of this variation (M. E. Harrison and Marshall 2011; A. Marshall and Wrangham 2007). Third, if ecological forces other than food influence carrying capacity, then the explanatory power of single measures of food abundance is unlikely to be very strong (Chapman, Schoof, et al. 2015; A. Marshall, Boyko, et al. 2009). Finally, differences might be due, at least in part, to other, less interesting sources of variation—such as measurement error, differences in study duration, or the fact that only one possible explanation was explored in an analysis. Data on the role of food in setting carrying capacity from a wider range of wild primate species and populations are necessary to assess the relative importance of these sources of variation. In addition, development of models that consider food quantity and quality in more nuanced ways (Hanya and Chapman 2013; J. Lambert 2007; J. Lambert and Rothman 2015; A. Marshall and Wrangham 2007) and case studies of populations with long-term data on multiple ecological factors (e.g., Chapman, Schoof, et al. 2015) will be valuable.

SUMMARY AND CONCLUSIONS

Carrying capacity is a key ecological concept, albeit a misunderstood and often ill-defined one. It is central to both theoretical and applied primatology, but despite its fundamental importance it is relatively understudied. There is a clear need for both more theory and more empirical work in this realm, but at present several broad generalizations appear to be warranted:

- The carrying capacity of primate habitats can vary substantially under natural conditions, sometimes by orders of magnitude.
- Carrying capacity can affect primates at the levels of individuals, groups, and populations, although the precise nature of these effects varies substantially.
- Primate population ecologists have generally focused on population limitation, not population regulation. These two distinct concepts are often conceptually blurred in the primate literature.
- Most primate populations are likely limited by bottom-up rather than top-down mechanisms; they may well be regulated by them as well, although evidence is scant at present.
- While multiple strands of evidence strongly suggest that food is important in setting primate carrying capacity, we lack a general theory of primate population ecology that would allow us to predict population density from first principles.
- Several important questions about primate carrying capacity remain unanswered, including what specific aspects of food quantity and quality set carrying capacity, by what mechanisms these act, how and why their effects vary among species, and whether there is a systematic difference among taxa as a function of body size, life history, dietary guild, or phylogeny.
- Understanding of primate carrying capacity would likely be substantially improved if future work were to move beyond verbal arguments to build explicit models (M. Johnson 2007), test alternative competing models of carrying capacity, and identify causal links between the nutritional quality of available foods and primate abundance. Current and anticipated future ecological changes in primate habitats—for example, due to human disturbance (hunting, logging, etc.), climate change, or invasive species—will provide a multitude of natural experiments that could also advance knowledge of how and why food influences primate carrying capacities.

28 Climate Change and Primate Nutritional Ecology

Jessica M. Rothman, John B. Makombo, and Mitchell T. Irwin

It is now the Anthropocene, and human-caused global changes are altering the abundance and distribution of biodiversity across the world's landscapes and seascapes (Crutzen 2006; S. Lewis and Maslin 2015). There is a pressing need to understand how different species, including primates, will adapt to changing conditions caused by humans (Behie et al. 2019). Global climate changes along with other human-caused pressures affect primate diet and nutrition in fundamentally important ways. The primary way that primate feeding is expected to vary in response to global change is through changes in plant community structure, composition, and nutrient availability, which in turn affect primates in both direct (e.g., through feeding) and indirect ways (e.g., through landscape changes that force shifts in home ranges). In addition, altered temperature and rainfall could affect primate physiology, including in ways that alter food intake, particularly for those primates in highly seasonal environments.

Although such climate change could benefit, harm, or have no consequence to primates, there is good reason to suspect it will be harmful, at least for many species. First, the potential geographic shifts in primate habitat due to climate change are not inconsequential (Korstjens and Hillyer 2016), and these shifts can interact with habitat loss to strongly reduce available habitat for primates (Estrada, Garber, et al. 2017; Gouveia et al. 2016). Second, shifts in temperature and rainfall often have adverse effects on primate feeding patterns and nutritional intake, which affect primate demography (A. Bernard and Marshall 2020; Rothman, Chapman, Struhsaker, et al. 2015). Since adequate dietary intake is essential for growth and reproduction, adult primates could survive but forego reproduction if they lack sufficient resources. Conservation managers need to proactively prepare for changes in primate ranges, particularly for vulnerable and endangered species that are already in decline. In addition, managers need to anticipate the potential for altered human-wildlife interactions as primates move from their routinely used habitats into other human-dominated landscapes.

Here, we discuss the aspects of climate change that affect primate nutritional ecology, with a focus on changes in environmental food resource quality and availability due to water and temperature stress, and the potential physiological mismatches in the ingestion and digestion of foods that result from these changes. We also discuss how conservation managers can proactively mitigate the effects of global change on primate nutrition through resistance, resilience, and transformation.

PRIMATES AND CLIMATE CHANGE

Despite many long-term studies of primates, primatologists still have little knowledge of how primates are responding to climate change, as well as the intrinsic and extrinsic factors that determine their vulnerability to such changes (Korstjens and Hillyer 2016). Since climate change is not unfolding evenly across continents and ecoregions, it is difficult to predict the susceptibility of particular primate habitats, which include not only rainforests but also savannas, temperate forests, flooded areas, karst caves, and semiarid areas. Primates living in extreme habitats that are already under stress might be even more susceptible, since climate change disproportionately affects these areas. In addition to changes in baseline temperature and rainfall, climate change is increasing the frequency of high-impact events and extreme weather, such as heat waves, droughts, hurricanes, fires, and floods (Seneviratne et al. 2021). While primates are affected by these events in different ways, primate persistence will depend on their behavioral and ecological flexibility—in other words, their ability to obtain adequate energy and nutrients in acute situations to avoid starvation and their capacity to adapt to habitat changes resulting from these events. Predicting the vulnerability of primates to global change is difficult, though Korstjens and Hillyer (2016) note that diet and dietary specialization are important factors.

Climate change also increases the frequency of extreme weather events. It is thought that 16% of primates are vulnerable to cyclones and 22% are susceptible to droughts (L. Zhang et al. 2019), and some evidence suggests that frugivores are the most vulnerable. The direct physical effects of high-energy storms (hurricanes or cyclones) can destroy growing flowers or fruit, which erases the entire year's food crop for plants that set fruits once annually, whereas leaves are often still available (Behie and Pavelka 2005; Pavelka and Behie 2005). In general, frugivores are typically more vulnerable to the effects of climate change than folivores because fruiting is more sensitive to weather patterns than leafing (Behie and Pavelka 2012a; L. Zhang et al. 2019).

As a starting point, it is important to consider that global climate change is not a single process acting on a single variable. The multifaceted physical changes considered under the umbrella term "climate change" affect primates through different pathways, ranging from direct pressures such as thermal and water stress to indirect pathways acting through other species in the ecosystem (Rosenblatt and Schmitz 2016).

Thermal Stress

Elevated temperatures can have direct physiological effects on primates that affect their feeding patterns, which are likely to be the most serious in already-hot habitats and those with limited canopy cover. Some primates use behavioral flexibility to react to thermal conditions. For example, when vervet monkeys (*Chlorocebus pygerythrus*) are hot, they swim to cool off (McFarland, Bartlett, et al. 2020), and many primates seek shade under trees or in caves during hotter weather (R. Hill 2006; Pruetz 2007). Some usually diurnal primates, such as chimpanzees (*Pan troglodytes verus*), show increased nocturnal activity and feeding in the hot savanna (Pruetz 2018), and cathemeral primates such as *Eulemur* species can react to hotter conditions by shifting the balance toward more nocturnal activity (Razanaparany and Sato 2020). Redtail monkeys (*Cercopithecus ascanius*) living in both rainforests and savannas reduced their daily travel distances during hotter conditions (McLester et al. 2019). Though this behavioral flexibility is important to understand, future conditions may exceed the ability to adapt for some primate populations—and the effects of these shifts on foraging opportunities are not yet understood but are potentially important.

Thermal stress can also have indirect effects, through temperature's effects on food resources. There are well-documented effects of heat stress on many plant species that form the basis of primate diets, though few studies have directly investigated the plants in primate habitats. Forests, which cover more than 30% of the earth's land surface (FAO 2010), are the predominant habitat for most primates. Because trees have variable life history strategies that are phenotypically plastic, changes in temperatures have both positive and negative effects on tree growth and reproduction. However, there is clear evidence from long-term studies that phenological cycles are disrupted due to heat stress and that tree mortality can increase as a result of climate change (C. Allen et al. 2010). In terms of food production, altered temperatures may cause (1) increases or decreases in food availability (i.e., the amplitude of the peaks) and (2) altered timing of phenological events (including changes in synchrony and congruence at the level of the plant community). There is already mounting evidence that climate change has affected plant and animal communities (J. M. Cohen et al. 2018; Ovaskainen et al. 2013).

The most common plant family in primate diets is Moraceae, which comprises predominantly figs, and at least 114 fig species are consumed by primates (Lim et al. 2021). It has long been known that *Ficus* fruits are particularly critical for primates, as they are widespread, calcium-rich, and energy-dense and provide an important fallback food (Felton, Felton, Wood, and Lindenmayer 2008; O'Brien, Kinnaird, and Dierenfeld 1998; Wrangham, Conklin, Etot, et al. 1993). This pantropical genus is speciose, with more than 700 species (Jevanandam et al. 2013). While little is known about the effects of heat on the majority of *Ficus* species, limited experimental studies on a handful of species suggest that they are one of the more robust genera, capable of tolerating a wide range of temperatures (Krause et al. 2013). However, their obligate pollinator is affected by increased temperatures: an experimental test of four species of fig wasps in Singapore suggested that an increase of 3°C or more will decrease the life span of adult females (Jevanandam et al. 2013). It is not known whether the fig wasps can eventually adapt, but reduced pollination due to declining fig wasps could lead to *Ficus* declines. These declines would be consequential to primate (and other vertebrate) communities as figs are arguably a keystone resource (F. Lambert and Marshall 1991; Lim et al. 2021). Regardless, it is important to remember that primate diets are diverse, and even *Ficus* is only one small component. Many primates are likely to see some food sources decline while others increase, and feeding flexibility will help mitigate nutritional shortfalls.

Some climate-change-driven effects on primate food intake emerge from a different mechanism: physiological responses to elevated heat that affect the propensity to feed (Beale et al. 2018). Heat stress in relation to animal nutrient intake has been extensively studied in livestock for management purposes, because of the need to efficiently produce human food (Beede and Collier 1986). Decades of experimental research has shown that heat stress increases energy requirements, depresses food intake, and reduces growth rates, milk yields, egg production, and reproductive performance in various domestic species, such as cattle, swine, and poultry (Beede and Collier 1986; J. Johnson et al. 2015). Heat stress is also teratogenic (J. Johnson et al. 2015). While agricultural managers can change the conditions for their managed livestock, wild primates need to adapt to thermal conditions by changing feeding behaviors or reducing thermal stress where possible. Nevertheless, these studies on domestic animals can provide insight and frameworks for study in wild systems if detailed data can be collected.

One reason that higher temperatures cause animals to feed less is the phenomenon known as "diet-induced thermogenesis": feeding causes heat production due to the body's work in digesting, transporting, and storing nutrients (Youngentob et al. 2021) as well as detoxifying or otherwise mitigating the effects of plant secondary metabolites (PSMs; Beale et al. 2018). In other words, animals face a difficult trade-off between the need to acquire nutrients and the physiologically harmful effects of overheating. Korstjens and Hillyer (2016) suggest that primates mitigate this pressure by increasing resting or digesting time, reducing mobility, shifting their diets, and perhaps including nonnutritive food sources that may adsorb excess PSMs. However, in warmer climates (and especially for diurnal species), where this dilemma is already somewhat acute, increased temperatures could push vulnerable primate populations outside physiological limits and lead to extirpation. Youngentob et al. (2021) note several risk factors that make a species more sensitive to reductions in food intake: obtaining most of their water from food, having a diet high in fiber and low in more readily digested nutrients, having a restricted time for feeding (perhaps due to inter- or intraspecific competition), having low energy reserves, having narrow diets, and consuming PSMs (Youngentob et al. 2021).

A related problem lies in the effects of PSMs on internal thermoregulation mechanisms themselves, because some PSMs are thermogenic and thermodisruptive—that is, they disrupt temperature homeostasis (Beale et al. 2018). For example, some PSMs activate the heat and cold thermoreceptors in different body regions. Thus, the ingestion of PSMs could signal to the brain that heat is detected, causing herbivores to feel hotter and therefore eat less. An example that Beale et al. (2018) provide that is relevant to human nutrition is that of chili peppers, whereby eating the capsaicin causes a feeling of heat. While most PSMs in many primate foods have not yet been identified, thermoregulatory processes could be affected by these compounds if present. Wild ginger, or *Aframomum* spp., are eaten by many African primates, including lowland and mountain gorillas, olive baboons, and chimpanzees (Furuichi, Hashimoto, and Tashiro 2001; C. A. Johnson, Swedell, and Rothman 2012; Rogers, Abernethy, et al. 2004; fig. 28.1), and are known to have PSMs that bind to hot and cold receptors, which could affect their foraging behavior (Beale et al. 2018; Morera et al. 2012).

Figure 28.1. Olive baboon eating *Aframomum* fruit.

To date, there has been little research on how diet selection is influenced by temperature specifically, in large part because the varying factors affecting food availability are difficult to disentangle in natural ecosystems. Longitudinal research documenting dietary change is logistically difficult, and cross-sectional studies comparing geographically separated populations will often show dietary contrasts, yet these exist for many reasons and cannot be simply related to local temperature. Yakushima macaques (*Macaca fuscata yakui*) changed their diet to insects from fruits and mature leaves during hotter days (Agetsuma 1995); however, increasing temperatures change insect abundance, so the dietary change could simply reflect an opportunistic strategy. Primates' behavioral flexibility and capacity for adaptive learning may allow them to "outmaneuver" climate change in some habitats through microhabitat selection, dietary shifts, and other behavioral changes. However, we know little about the limits to that adaptive flexibility in the wild.

Water Stress

While many primates are found in tropical rainforests, where water is abundant, others are limited by the amount of water in their habitats. In general, it is thought that increasing rainfall causes higher plant productivity and promotes higher plant and animal diversity, yet at very high rainfall the opposite trends can occur (Kay, Madden, et al. 1997). Thus, the effects really depend on the starting point. More arid sites could benefit from increased rainfall but suffer from further restriction, while rain-heavy sites would see the opposite. In general, changes in the amount and timing of water delivery could act in two ways to limit primate populations: directly through dehydration and indirectly through changes in food availability. Both have important consequences for primate nutrition, as water intake is essential, either directly through drinking or through their diets in the form of preformed water in plant material (Jequier and Constant 2009; NRC 2003; Pontzer, Brown, Wood, et al. 2021).

In habitats where plant growth is limited by water availability, a further decrease in rainfall leads to reduced plant growth and reproduction, thus altering the availability of leaves, fruit, flowers, and other plant parts (D. P. Anderson et al. 2005; Freitas et al. 2016; van Schaik, Terborgh, and Wright 1993). However, even in habitats where water is not strictly limiting but acts as a cue for initiating flowering and fruiting, changes in water availability could reduce the production of flowers and fruit (McLaren and McDonald 2005). As many plants may prioritize leaf production (which is essential) over flowering and fruiting (which are not essential to survival and often require more water), primate folivores may be more resilient to these changes than frugivores. In extreme conditions, water scarcity can lead to die-offs of trees (C. Allen et al. 2010) and eventually species turnover in the plant community (replacement with more drought-tolerant species). Finally, it is possible that reduced rainfall in already-wet sites could bring benefits (i.e., increased productivity) through increased sun penetration and reduced soil leaching. As is true for studies of temperature's effects, studies in tropical ecosystems are few, and we are largely guided by agricultural research.

During extreme droughts, the consequences for primate populations can be dire. Death occurs when water and the food supply are severely limited, causing starvation and dehydration (Campos, Kalbitzer, et al. 2020; L. Gould, Sussman, and Sauther 1999; Hamilton 1985). Primates faced with water shortage often change their feeding and ranging behavior to cope with reduced water or food supply. Northern muriquis (*Brachyteles hypoxanthus*) fissioned temporarily into smaller parties and used the ground more often during a drought, demonstrating their resilience to the effects of changing food availability (Strier 2021a). Common brown lemurs (*Eulemur fulvus*) switched to moisture-rich foods during seasonal drought conditions (Sato, Ichino, and Hanya 2014), and capuchins (*Cebus capucinus*) centered their home range use on waterholes in response to water scarcity (Campos and Fedigan 2008). Ring-tailed lemurs switched their main diet to mature leaves, which have more water than their usual diet (Mertl-Millhollen et al. 2003). Black howler monkeys (*Alouatta pigra*) spend more time drinking in places where the climate is hotter (Dias et al. 2014). Specialist primates that are inflexible in their diets, such as the greater bamboo lemur (*Prolemur simus*), could be more vulnerable to dry spells because they are confined to just one food source, bamboo, which is already nutritionally poor; they are additionally thought to be extinction-prone compared to other primates because climate change will contract their range (Eronen et al. 2017).

In some cases, behavioral modifications are not sufficient, and die-offs occur, with infants and juveniles being particularly vulnerable. During a severe drought in Beza-Mahafaly Special Reserve, Madagascar, 20% of female ring-tailed lemurs (*Lemur catta*) and 80% of infants in the reserve died, probably due to malnourishment; in general, immature mortality is much higher in drought years than in nondrought years (L. Gould, Sussman, and Sauther 1999). Similarly, in the Parc National du W du Niger, Niger, olive baboons (*Papio anubis*) faced severe consequences with reduced water supply; over 60 animals died during a single drought (Poche 1976). In a seasonally dry tropical forest in Costa Rica, infant white-faced capuchins (*Cebus capucinus imitator*) had a threefold increase in the risk of death during drought conditions (Campos, Kalbitzer, et al. 2020). This was attributed to the lack of drinkable water sources; the authors indicate that if standing water was present, it was frequently visited, but if no standing water was available, the capuchins experienced extreme water stress that could lead to infant mortality (Campos, Kalbitzer, et al. 2020). Campos, Kalbitzer, and colleagues (2020) also analyzed a long-term data set on Geoffrey's spider monkeys (*Ateles geoffroyi*) to assess the effects of drought on these (nonsympatric) species. Spider monkeys coped with water stress in different ways than capuchins; for example, spider monkeys were reluctant to use standing water as a drinking source. During extreme drought, they stopped reproducing, which could be related to declines in fruit availability caused by the drought (Campos, Kalbitzer, et al. 2020).

If infants and juveniles do make it through extreme

water events or other catastrophic environmental events, there could be lifetime consequences. Savanna baboons (*Papio cynocephalus*) with low-ranking mothers experience larger declines in fertility during adulthood if they are born under drought conditions (Beehner et al. 2006; Lea et al. 2015). Social factors appear to influence the relationships between health and stress (Alberts 2019; Snyder-Mackler, Burger, et al. 2020). In the case of the savanna baboons, dominance rank seems to buffer the effects of early life adversity due to drought, perhaps because the higher-ranking individuals had access to the highest-quality foods available during these periods (Lea et al. 2015).

Interestingly, primates living near human agriculture or in urban areas are buffered by the effects of droughts. Hanuman langurs (*Semnopithecus entellus*) in a protected area in India were affected by a drought that wiped out 50% of the population from 1999 to 2001 (Waite et al. 2007). The urban langur populations were not affected by the same drought because they were provisioned by humans, likely due to their sacred status in human societies (Waite et al. 2007), indicating that this supplemental food source was lifesaving.

Conversely, unusually high rainfall or flooding can also be detrimental. High mortality of capuchins (72%–77% of the population) was associated with nutritional stress that resulted from flooding on Barro Colorado Island in Panama (Milton and Giacalone 2013). The authors speculated that an acute shortage of dietary protein due to the lack of arthropods was responsible for starvation of the capuchins; this did not affect the sympatric howler monkeys that do not eat insects (Milton and Giacalone 2013).

Finally, it is important to note that precipitation patterns do not necessarily vary gradually. The interconnectedness of weather patterns means that global weather systems can oscillate between discrete states. The best-known example of this is the El Niño–Southern Oscillation (ENSO), which manifests as discrete warm or cool periods across the tropical Pacific Ocean and causes large-scale changes in temperature and precipitation. A. Dunham, Erhart, and Wright (2011) used a long-term data set from Ranomafana, Madagascar, to demonstrate that ENSO phase affected the rainfall experienced and the fecundity of Milne-Edwards's sifakas (*Propithecus edwardsi*), almost certainly acting through variation in food availability or quality (or both). Given the observed impacts on reproduction, it is important to understand the consequences of ENSO shifts more precisely as well as realize that future environmental change can be abrupt, as interconnected systems flip to a new state instead of changing continuously and gradually.

Increasing Atmospheric CO_2

Decades of research in plant physiology has demonstrated that, along with solar radiation and water availability, the gas composition of the growing environment affects plant growth and the chemical composition of plant tissues (Jump and Penuelas 2005; Parmesan and Hanley 2015; E. Robinson et al. 2012). Based on our current understanding of these influences, it is thought that increases in atmospheric CO_2 will cause reduced nutritional quality of plants. This drop is a result of decreased leaf protein concentrations and increased levels of nonstructural and structural carbohydrates (E. Robinson et al. 2012). Overall, the ratio of macronutrients to fiber in both mature and young leaves is predicted to decline in response to global climate change, which could have physiological consequences for leaf-eating primates. As noted for previous stressors, in natural ecosystems it is hard to tease apart the effects of increased rainfall, temperatures, and CO_2 on plant chemistry. In Kibale National Park, Uganda, a forest that hosts high densities of primates, mature and young leaves sampled in an undisturbed area over 15–30 years from the same trees and the same tree community showed a 10%–15% increase in fiber concentrations. Young leaves that were sampled 15 years apart (from the same species in the same tree community) declined in protein concentrations (Rothman, Chapman, Struhsaker, et al. 2015). The colobus monkeys that mainly feed on these leaves can tolerate high concentrations of fiber as a result of foregut fermentation (Chivers 1994), so this change may not affect them; however, there are many other primates that regularly include young leaves in their diets when preferred foods are scarce, meaning that their lean-season nutritional bottlenecks become even more constrained. These include redtail monkeys, gray-,cheeked mangabeys, chimpanzees, and blue monkeys (Chapman, Struhsaker, et al. 2010; Conklin-Brittain, Wrangham, and Hunt 1998; Struhsaker 1969; Uwimbabazi et al. 2021).

PSMs, on the other hand, seem to be more idiosyncratic in their responses to atmospheric composition (Lindroth 2012), with examples of increases, decreases, and no change across species and specific PSMs (Bidart-Bouzat and Imeh-Nathaniel 2008; E. Robinson et al. 2012). As primates tend to be generalists with diverse diets, the specific effects of climate change on PSM ingestion are likely to be mixed. When species do experience a net increase in PSM intake, this will be problematic for primates that do not have detoxification mechanisms or are unable to mix their diets to dilute their effects (Stalenberg et al., this volume). Thus, the additive effect of more PSMs in primate diets means that the diets are less palatable, less digestible, and potentially toxic. A related problem can also occur if those PSMs that increase in the diet are compounds that upset thermoregulatory balance (described previously), as increased activation of heat receptors could cause animals to feel hotter and eat less (Beale et al. 2017).

Interactions with Life-History Schedules and Hibernation

The previous sections have dealt mainly with how climate change can affect the abundance and chemistry of the foods on which primates rely. However, an additional threat lies in altered timing of the availability of those food resources. This has been shown to be particularly important for birds and mammals that migrate (Portner and Farrell 2008), as migration is often triggered proximately by physical cues such as day length, yet the timing of food availability at the destination may be altered. Primates generally do not migrate, so they could be immune to many of these effects, but some species have evolved to have life-history events aligned with annual variation in food availability. The most obvious example of this is the Malagasy lemurs (Janson and Verdolin 2005). For species that time the initiation of reproduction to unchanging physical cues such as day length but for whom the timing of peaks in food availability is altered, the match between nutrient need and availability may be disrupted. For example, sifakas (genus *Propithecus*) strictly conceive in the season of high resource abundance (November–January), birth when both leaves and fruit are at lowest availability (May–July), and wean during the season of highest fruit and leaf availability (January–March). Across lemur species, most are just as strict in their timing, but they appear to share in common the timing of weaning when resources are abundant (P. Wright 1999). A change in the timing of food availability could lead to additional stresses during gestation, lactation, and weaning that could lead to starvation or reproductive failure.

It also deserves special mention that increasing temperatures will affect the schedules of the obligate hibernating dwarf lemurs (*Cheirogaleus* spp.) as well as some mouse lemurs (*Microcebus* spp.; Dausmann, Glos, et al. 2004; Dausmann, Glos, and Heldmaier 2009; Kobbe and Dausmann 2009). The timing of hibernation seems to depend on the interactive effects of food availability and temperature, though the triggers are largely unknown (Dausmann, Glos, and Heldmaier 2009; Fietz and Ganzhorn 1999). Hibernators rely on fat accumulation to survive through this period, which lasts three to seven months depending on the species (Blanco, Dausmann, et al. 2018). Interestingly, in captivity, the increase in food availability, photoperiod, and temperature causes a normally hibernating primate to express torpor in shorter bouts than those in the wild; when husbandry changed, these lemurs resembled their wild counterparts more closely, suggesting that these are the cues that determine hibernation (Blanco, Greene, et al. 2021). Disrupting the match between the timing of cues and actual resource availability could cause starvation, reproductive disruptions, and population decline.

NUTRITIONAL ECOLOGY, CONSERVATION, AND MANAGEMENT IN PRACTICE

It has long been realized that knowledge of species' nutritional requirements is important for in situ habitat management and conservation. Understanding the available quantity and quality of potential foods is critical for determining the habitat requirements of primates and whether they thrive or fail within them. This includes an understanding of not only what is needed but also how primates choose foods, what their digestive constraints are, and how they learn about and adjust to changes in their food and water supply. However, one aspect of primate populations that remains poorly understood is how close they are to nutritional thresholds—tipping points beyond which populations cannot survive (Barnosky et al. 2012). In some richer habitats, primates could probably withstand substantial change, showing reduced population density or behavioral shifts. In already-marginal habitats, or habitats already suffering the combined effects of habitat destruction, hunting, and resource loss, climate change could prompt the extinction of primates, especially those that already occur in small ranges. For example, droughts that occur in areas where there is concomitant habitat destruction force already-weakened dehydrated southern patas monkeys (*Erythrocebus baumstarki*) outside of protected areas and into agroecological matrices to find water holes (de Jong and Butynski 2021). At these water resource areas, they are especially vulnerable to predation and harassment by domestic dogs and humans. If they are not killed, they are often exposed to waterborne zoonotic diseases (de Jong and Butynski 2021). Managers must therefore proactively plan for these types of shifts to prevent catastrophic and irreversible losses.

There are several ways that wildlife managers can act to mitigate the effects of climate change on primate habitats, and no single management strategy will fit all situations (A. Bernard and Marshall 2020). Recognizing that management consumes financial resources, there is a need to prioritize actions based on the susceptibility of habitats to climate change and the likelihood that the actions will be effective. These management actions should be flexible, so that they are adaptive over time, and should be evidence based, wherever possible relying on both the social and biological sciences (J. Setchell, Fairet, et al. 2017).

A goal of conservation managers is to ensure that suitable habitat exists for primates if their original habitat is degraded or used for another area (J. Harris et al. 2006). In addition to understanding species requirements in terms of space, locomotion, and safety, successful restoration requires knowledge of primate species' dietary preferences, nutritional requirements, and foraging constraints (N. Hobbs and Swift 1985). While the roots of traditional restoration ecology focused on either plants or animals, an integrative approach is needed to understand how animals meet nutritional needs within their environment for conservation and to manage both the habitat and the animals accordingly (McAlpine et al. 2016; Raubenheimer, Simpson, and Tait 2012).

To address the effects of climate change on the food supplies of primates, following Millar et al. (2007), strategies include three main options: resistance options, whereby the protection of the most valuable resources is prioritized; resilience options, whereby restoration can occur after impacts are already noted; and response/transition options, whereby a shift of the managed area from one state to a more desirable state is facilitated. In addition, larger programs to offset greenhouse emissions are needed (Millar et al. 2007). Employing these options can be a large task depending on the scale of the intervention and involves institutional coordination and in-depth research (Heller and Zavaleta 2009; LeDee et al. 2021) as well as integrating the support of human communities both locally and nationally (Chazdon et al. 2020; E. Riley and Fuentes 2011; Waters et al. 2018).

Resistance

Despite shortfalls in some areas (Laurance, Carolina Useche, et al. 2012; Le Saout et al. 2013), protected areas that have both law enforcement and community support remain the most effective way to conserve the habitat of primates and perhaps buffer the effects of climate change (Gaston et al. 2008; Geldmann et al. 2013; Struhsaker, Struhsaker, and Siex 2005). In some cases, flagship critically endangered species such as mountain gorillas (*Gorilla beringei*; fig. 28.2) can be targeted for "extreme" conservation, whereby increased protection through law enforcement, community conservation, and intensified veterinary care for individual animals leads to massive population recoveries even when faced with climate change (M. Robbins, Gray, Fawcett, et al. 2011). In this case, food sources can be monitored and buffer zones created to mitigate human-wildlife interaction (N. Seiler and Robbins 2016). However, gorillas present an unusual case; it is well recognized that this approach is usually not practical due to financial constraints. Resistance options are focused not necessarily on monitoring individual primates but on trying to protect the habitat broadly and thereby avoid the changes that might befall primates. Here, the importance of long-term studies is clear (Kappeler et al. 2012; Strier 2010); understanding the variability in food resources will better enable managers to target specific areas for increased protection (Mainka and Howard 2010).

Wildlife managers frequently consider the prevention or reversal of biological invasions of exotic species to ensure that habitats return to their native state. Where invasions of nonindigenous species have already occurred, eradication is often the only management tool

Figure 28.2. Mountain gorilla female and her infant in Bwindi Impenetrable National Park, Uganda.

available to enable natural states (Prior et al. 2018). Invasions of exotic species threaten native biodiversity, and as such, this eradication has many benefits where it is biologically feasible. For example, plant removal is viewed as positive because it allows native plants and primate foods to persist (Pebsworth, MacIntosh, et al. 2012). However, in some instances, primates depend heavily on exotic species as foods, so a good understanding of their nutritional implications in their diets is needed (Gerard et al. 2015). For example, endangered samango monkeys (*Cercopithecus albogularis labiatus*) in the Eastern Cape province of South Africa rely heavily on the seeds of the invasive black wattle (*Acacia mearnsii*; Wimberger et al. 2017), and the invasive tree *Solanum mauritianum*, native to South America, contributes up to 10% of diademed sifaka (*Propithecus diadema*) diet in the lean season at Tsinjoarivo, Madagascar, and supplies appreciable protein (Irwin 2008; Thurau et al. 2021). Blue monkeys (*Cercopithecus mitis*) in Kakamega, Kenya, have a diverse diet; they eat over 440 different food items (Takahashi et al. 2019) but are still considered nutritional specialists because they eat a consistent balance of protein to energy during most of the year (Takahashi et al. 2021). For these monkeys that vary in the types and amounts of nonnatural foods they eat, exotic foods provide more protein and nonstructural carbohydrates than their natural diets, suggesting that exotic foods could be options for them and perhaps other primates in degraded forests (Takahashi et al. 2023).

Resilience

Resilience of primate habitats in the context of primate nutrition means that their food supply is able to resist, absorb, or transform after stress; thus, plant performance is an important measure (Ibanez et al. 2019). For example, to what extent can particular tree species recover or persist after a drought, a flood, or continued exposure to altered rainfall and temperature conditions? Many of these are unknowns in primatology and should be the focus of future research in adaptive management. One well-documented case study that provides insight into this question is a hurricane that made landfall in Belize in October 2001 and had significant impacts on the local black howler monkeys (*Alouatta pigra*; Pavelka and Behie 2005). After the hurricane, the mortality of trees in the area reached 35% for the major foods eaten by the monkeys, and the monkeys switched their diets to mature leaves because their fruit trees were not available (Behie and Pavelka 2012a). However, the population of fruit trees slowly recovered. The authors documented all the specific trees that monkeys ate, as well as the incurred damage from the hurricane on each tree type and species. An assessment over six years posthurricane demonstrated that fruit consumption had the strongest effect on monkey population recovery (Behie and Pavelka 2013); it was the recovery of the fruiting trees that allowed monkey populations to persist. Because the authors have data from before and after the hurricane, they can determine whether the forest community returns to its previous state and how this affects primates. There are additional case studies from around the world: Ratsimbazafy (2002) documented how black-and-white ruffed lemurs (*Varecia variegata*) barely survived a destructive cyclone that caused extensive tree damage at the coastal forest of Manombo in southeastern Madagascar. These frugivorous lemurs ceased reproduction for multiple years but eventually recovered as their fruit resources recovered. These examples are important to demonstrate the extent of resilience for some species, but they remain anecdotal. Longitudinal studies of primates through environmental stress remain few—in part because some stresses cannot be predicted (such as cyclones) and because such studies require long-term dedication and funding.

In addition to understanding the differential vulnerability of ecoregions and habitats, it is important to understand that primate species themselves vary in the degree to which they are resilient, flexible, and able to cope with changing conditions (Strier 2021a). For example, it is expected that generalist primates such as *Papio* spp. will perform better when areas become degraded or transformed as a result of global change (S. Hill and Winder 2019). However, the ability of individual animals to survive in a particular area does not necessarily imply long-term success of a population, as was the case with howler monkeys (*Alouatta caraya*) in Argentina (Bicca-Marques et al. 2020); a snapshot population strategy is insufficient to judge viability.

Response/Transition

In this final category, managers admit that ecosystem change is inevitable and consciously accommodate this change rather than resisting it. Within a habitat stand or protected area, change in the plant composition may be unavoidable; this is where knowledge of the dietary tolerances of primate inhabitants is critical for managers wanting to identify species at risk of extinction. Using these baseline data along with population monitoring, it should be possible to identify which species will be unable to persist in the habitat post-transition. Some of those species exist in sufficient protected habitat elsewhere, and the local extirpation can be tolerated. If that is not the case, the options for those species are likely to be few, intensive, and expensive.

Range shifts are one way that species facing this challenge can be preserved in the wild, and it is possible that management interventions can help speed this process by introducing those species to habitats outside their historic range but where conditions will allow their survival. In some cases, suitable habitat does not exist, but instead the area needs to be restored to an acceptable state. Habitat restoration could be considered where there is available land near the original primate habitat, whether it was abandoned, purchased, or converted for this purpose (Chapman, Bicca-Marques, et al. 2020). If it is near the original habitat, then primates could extend their range naturally or through managed translocations. Habitat restoration can also be used in buffer zones that can be constructed as a conservation tool to reduce human-wildlife conflict. In all of these scenarios, aboveground and belowground ecosystem processes need to be considered (Kardol and Wardle 2010). Soil communities are important because they will supply the plants with needed nutrients, and the plant communities will in turn feed primates that will hopefully return to the area. In addition, specific trees can be planted to restore areas that have been previously logged to help the area return to its natural state to best promote adequate primate nutrition. Here, it is preferable to have knowledge of the foods that primates in a particular area eat, so that they can be considered for restoration plantings. For example, Steffens (2020) compiled a list of all known foods eaten by 56 wild lemurs, which included more than 1,000 species. Though lemurs generally find these foods acceptable, they may not be appropriately nutritionally balanced, which should be considered. The life-history traits of these plants are important, particularly if the plant is invasive (Steffens 2020). In some cases, but not all (Albert-Daviaud et al. 2020), primates act as important seed dispersers to facilitate this process (McConkey, Prasad, et al. 2012; Moses and Semple 2011).

Only a few studies have the long-term data to assess primate recoveries after habitat restoration. In Kibale, Uganda, portions of the forest experienced low- and high-intensity logging between 1968 and 1969 (Struhsaker 1997). This same area has been subsequently protected since then and allowed to regenerate, and it has been monitored for about 45 years to assess primate population recovery in the area. Though the forest is still regenerating, populations of five of six monkey species studied recovered over this time (Chapman, Omeja, et al. 2018). This research relied on detailed knowledge about the quality and quantity of foods eaten, as well as the nutritional factors that predict primate biomass, to understand the processes that predict primate recovery. Estimating and documenting phenological and nutritional parameters during the restoration process (e.g., L. Garcia et al. 2014) is important because these measures often serve as predictors of primate abundance and biomass (A. Marshall, Beaudrot, and Wittmer 2014; E. Vogel, Harrison, et al. 2015). For example, orangutan (*Pongo pygmaeus wurmbii*) density differed depending on the nutritional quality of foods within the habitat (E. Vogel, Harrison, et al. 2015). Intraspecific variability in nutrients can also play a role in driving population densities, as seen in Berenty Reserve, Madagascar, where ring-tailed lemurs adapt their ranges according to the nutritional quality of tamarind trees (Mertl-Millhollen et al. 2003). While few studies have investigated this problem in primates, intraspecific variation in nutrients and toxins has the potential to influence primate densities, as it does for other mammals (DeGabriel, Moore, Felton, et al. 2014; B. Moore, Lawler, et al. 2010; B. Moore, Wallis, Pala-Paul, et al. 2004)—and this understanding must be increased to better execute management interventions.

SUMMARY AND CONCLUSIONS

Climate change has multifaceted and long-term consequences for primates. Thermal stress has direct physiological effects on primate feeding. Droughts and changes in water availability can affect primates in two main ways: directly through dehydration and indirectly through changes in resource availability. Increased carbon dioxide in primate habitats can also lead to changes in food supply. Long-term studies are an important means to quantify plant-primate interactions, because comparisons with past baselines will be important in detecting ecological and behavioral change; data archiving and sharing will be of critical importance to these efforts.

- Nutritional ecology studies should endeavor to be long-term and include measures of temperature and rainfall at different scales. Combining data on food availability, plant life histories, primate activity patterns, and primate nutrient intake on a global scale will be important to proactively mitigate these climate change effects on primate habitats wherever possible.
- The physiological tolerance of herbivores to PSMs as well as the nutritional compositions of plants are likely to change with climate change. Plants in primate habitats should be monitored for their density, diversity, and nutritional and PSM composition in addition to more commonly used measures such as food availability, vegetation structure, and tree growth.
- Further research is needed to clarify the magnitude and consequences of the complex interactions involving

plant nutrients and primate feeding habits in relation to climate change. Documenting and understanding changes in vegetation, land use, and climatic variables will be important to proactively manage primate habitats and ensure primate health.
- Primatologists should work with conservation managers to determine the best actions to protect the health and survival of primates in the face of global change. Global climate change is underway, and planners must make careful choices—sometimes resisting that change, sometimes promoting resilience to it, and sometimes embracing the inevitable and mitigating the consequences.

29 Primate Foraging Strategies Modulate Responses to Anthropogenic Change and Thus Primate Conservation

Colin A. Chapman, Kim Valenta, Fabiola Espinosa-Gómez, Amélie Corriveau, and Sarah Bortolamiol

On October 17, 2022, the World Wildlife Fund announced that the world had lost 69% of its biodiversity over the past 50 years, and it is estimated that we are losing approximately 11,000–58,000 species annually (Dirzo et al. 2014; WWF 2022). Primate species may also have been lost (McGraw 2005; Oates, Abedi-Lartey, et al. 2000), and many are on the brink of extinction (Estrada, Garber, et al. 2017; Mittermeier, Reuter, et al. 2022). In fact, it is estimated that close to 50% of the world's primates are at risk of extinction (Estrada 2013; Estrada, Garber, et al. 2017; Mittermeier, Reuter, et al. 2022), with the IUCN Red List database considering 42% critically endangered. Primates face threats on multiple fronts, but likely the most significant is habitat loss (Chapman and Peres 2021). Between 2000 and 2012, it is estimated that 2.3 million km^2 of forest were lost globally (approximately the size of the Democratic Republic of the Congo), and, in the tropics, forest loss increased by 2,101 km^2 per year (M. Hansen et al. 2013). The loss is greatest in South America and Africa (fig. 29.1).

Primate species loss corresponds with an increase in agriculture and cropland in tropical countries, which expanded by 48,000 km^2 per year between 1999 and 2008, largely at the expense of forest (Phalan et al. 2013). One estimate suggests that approximately 1 billion ha of additional land (an area larger than Canada), primarily in developing countries, will need to be converted to agriculture by 2050 to meet the demands of the growing human population (Laurance, Sayer, and Cassman 2014). Ultimately, these changes are driven by increased human population size and consumption rates. The UN Population Division estimates that the world's population is expected to rise from 7 billion in 2011 to 9 billion in 2050. Making the situation more dire for primates is the fact that, in the three primate range regions, the human population growth rate between 1980 and 2005 was 2.7% per year, well above that of European countries (0.2% per year; Estrada 2013). Thus, the human population will double in regions with primates in less than 30 years. In a global analysis using a 1 km^2 resolution of the changes that occurred between 1993 and 2009, the human population size increased by 23%, the world economy grew by 153%, and the human footprint increased by 9% (Venter et al. 2016). This has resulted in 75% of the earth's land surface being impacted, with most of the unexploited areas being in the far north and in desert ecosystems. All these measures suggest that the situation for primates is grave and conservation actions must be initiated—but which ones?

In the past, conservation biologists have typically responded to change and attempted to take corrective action after negative situations have occurred (Caughley 1994; Chapman and Peres 2001), but it would be much more effective if researchers were able to predict negative changes prior to their occurrence and proactively prevent animal population declines. It can be easier to prevent declines than to rebuild populations. Furthermore, population decline can lead to further detrimental effects such as increased risk of disease in small populations or loss of

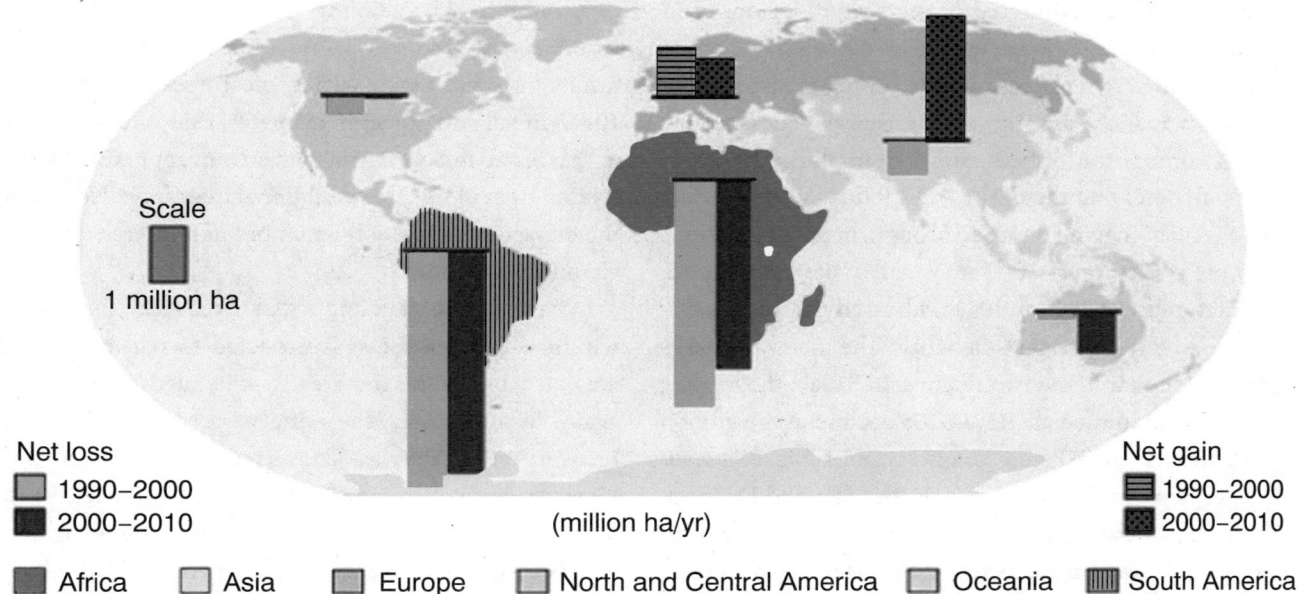

Figure 29.1. Annual change in forest area by region, 1990–2010 (from FAO 2010; reprinted with permission). These net gain values include plantation forest and forests lost to fire, but positive gains cannot represent regeneration of old-growth forest, as this would take hundreds of years.

genetic potential (i.e., inbreeding causes loss of genetic diversity) and must be considered in conservation and management plans. To predict declines and not simply respond to changes, conservation biologists must find general patterns across taxa and locations. However, predicting change has proved difficult.

We are ethically obligated to take conservation action, and conservation practitioners need to predict negative changes prior to their occurrence. It is common sense that understanding foraging strategies and nutritional needs is a tool that would allow conservation biologists to predict which species are most vulnerable to negative human activities and that this understanding would facilitate effective action. It would be desirable if broad generalizations could be made; for example, frugivores are more vulnerable than folivores (Oates 1996). With such generalizations in hand, setting conservation priorities and constructing management plans could be easily done. However, such generalizations have proved elusive, and, in our opinion, they are impractical to make at this time. This is because any particular population is faced with a myriad of anthropogenic and natural changes, and thus identifying the effect of one factor is impractical. For example, consider the following comparison between two populations. One folivore population (Population 1) is in an area where the local community does not hunt, is only slightly adversely impacted by climate change (Rothman, Chapman, Struhsaker, et al. 2015), and benefits from the secondary growth associated with logging (Coley 1983; Ganzhorn 1992; Oates 1996). This population would be expected to increase in abundance after the disturbance caused by timber extraction. A second population (Population 2) of the same species that is hunted during the logging operation (Brodie, Giordano, et al. 2015), is in an area with the same degree of climate change, and similarly benefits from logging would be expected to decrease in abundance after logging because of hunting. As it is so difficult to quantify historical hunting patterns, the effect of the hunting may be largely unknown and unquantified. Therefore, a comparison of the two populations examining the effects of logging would produce little meaningful understanding. Similarly, comparisons of Population 1 where the timber extraction removed mostly nonfood trees with Population 2 where mostly food trees were removed would not be meaningful without those details of what was removed. If Populations 1 and 2 experienced similar hunting, climate change, and logging, but Population 1 was monitored 1 year after the logging and

Population 2 was monitored 10 years after logging, again comparisons would be meaningless. We suggest that the details needed to make meaningful comparisons are often unknown, making large literature reviews of questionable use. A comparison of two studies exemplifies the divergent outcomes that can occur. A. D. Johns (1992) studied the effects of logging on frugivorous primate populations in dipterocarp forests in Peninsular Malaysia, while E. Bennett and Dahaban (1995) addressed the same question in dipterocarp forests in Sabah. The intensity of logging was similar in the two regions. In Sabah, the logging produced an immediate 35%–70% decline in the gibbon (*Hylobates muelleri*) and langur populations (*Presbytis* spp., only partially frugivorous; E. Bennett and Dahaban 1995). In contrast, survival of the same genera in Peninsular Malaysia was much greater (10% decline to an increase of 74%; A. D. Johns 1983).

We do not attempt to state generalities, such as asserting that folivores have a lower risk of vulnerability than frugivores (Oates 1996). Rather, our objective is to outline trends in global threats to primates and predict how primates with different dietary needs and foraging strategies will respond to each threat. We consider three global threats: habitat destruction through logging and fragmentation, climate change, and bushmeat hunting. We conclude by pointing out gaps in our knowledge and by considering future research directions that would be helpful if foraging studies are to contribute to the conservation of primates.

HABITAT DESTRUCTION THROUGH LOGGING AND FOREST FRAGMENTATION

Trends

Deforestation rates are staggering (M. Hansen et al. 2013). Much of this loss can be either directly or indirectly attributed to logging and activities that timber extraction facilitated. Globally, logging shows no signs of slowing down, and thus more land that sustains primates is being destroyed each year (Estrada 2013; Estrada, Garber, et al. 2017; D. Kim et al. 2015). Between 1960 and 2010, industrial round-wood production increased from 28 million m^3 to 155 million m^3 in Central and South America, from 23 million m^3 to 71 million m^3 in sub-Saharan Africa, and from 15 million m^3 to 30 million m^3 in Southeast Asia (Estrada 2013). Approximately one cubic meter results from harvesting two tropical trees that are about 10–15 m tall and 60 cm in diameter; thus, the 30 million m^3 harvested in South Asia comes from approximately 60 million trees of this size. Additional trees are killed during the extraction process; the number of total tree deaths is alarming.

When logging is conducted, a recovery time is set within which the forest is expected to regain its basal area, at which time the area is scheduled to be logged again. In most cases, these logged areas are not left to recover; rather, they are converted to agricultural land. However, if they are not further disturbed, recent studies demonstrate that tree regeneration following logging is slower than expected or even arrested entirely. In Kibale National Park, Uganda, many abandoned logging gaps showed little forest recovery 40 years after selective logging was concluded (Chapman and Chapman 2004; Lawes and Chapman 2006; J. Paul et al. 2004). In Budongo Forest Reserve, Uganda, Plumptre (1996) found that 50 years of regeneration was insufficient for forest structure to recover to unlogged levels. Similarly, data from Bolivia indicate that the growth rate of timber trees was insufficient for similar wood volumes to be cut in the next planned harvest, and estimated recoverable volumes in the second harvest ranged from 4% to 28% of the potentially harvestable volume in the first cycle (Dauber et al. 2005). Slow rates of recovery reduce the value of logged forests. Furthermore, when logging operations revisit a concession that they previously logged after the rotation period has ended, they harvest a broader range of species than they had previously, including species, for example, from the important *Ficus* genus. Large-scale commercial harvesting of *Ficus* has been documented in Bolivia and to a lesser extent in Peru and Brazil. In Bolivia, extraction volumes of fig trees increased steadily since records began in 1998 but plateaued in the economic downturn of 2008–2009 (Felton, Felton, Rumiz, et al. 2013). Small-scale removal of *Ficus* has also been documented in Uganda (Felton, Felton, Rumiz, et al. 2013), where *Ficus* are considered important food sources, including as fallback foods, for many frugivorous primates (J. Lambert 2007, 2009) and are also important foods for folivores (Fashing 2001; Pavelka and Behie 2005).

Predicted Responses of Primates with Different Foraging Strategies to Habitat Destruction

Logging results in the removal of the targeted timber trees and substantial damage to other trees because of the damage done when getting equipment to the felled tree, when the targeted tree falls, and when the timber tree is taken out. The consequences of logging for primates are many, they are often interacting, and the processes that are altered change on different timescales. Initially, logging (1) causes a reduction of food resources for all primates; (2) increases travel costs as animals attempt to find new food sources (Rode, Chapman, McDowell, et al. 2006) and navigate around canopy gaps (Gebo and Chapman 1995)—some species may reduce group size, permitting reduced travel costs, but others may not be able to because of the increased predation risk associated with smaller groups (Norscia, Carrai, and Borgogini-Tarli 2006); (3) causes increased intergroup conflict for territorial species, possibly involving lethal intergroup contact (J. Mitani and Watts 2005a); and (4) causes changes in stress and health status (T. Gillespie et al. 2005). Within a few years of logging, the forest begins to show significant signs of regeneration, and primates with different dietary needs are differentially able to take advantage of the regrowth. Folivorous primates often benefit first following disturbance by feeding on this regeneration, as the leaves of secondary-growth tree species are often high in protein, low in fiber, and only poorly chemically protected (Coley 1983; but see Gogarten, Guzman, et al. 2012). At this time in the forest regeneration process, folivore abundance may be higher than in neighboring old-growth areas (A. Davies 1994; Fashing 2011; Oates 1996; Plumptre and Reynolds 1994).

The patterns of change in folivore populations following logging provide a useful illustration of how the evaluation of responses to logging depends on the temporal scale evaluated. Immediately following logging, folivore populations likely decline as a result of the disturbance that the logging causes, possible bushmeat harvest associated with the presence of the loggers, and loss of feeding trees. After a few years of secondary growth, trees that are trying to grow rapidly and thus do not invest in toxic leaf defenses (Coley 1983) are available at significant densities, and the folivore populations will likely rise (Oates 1977; Struhsaker 1997). Over time, these secondary-growth trees will senesce and die out (Chapman, Jacob, et al. 2010), which may in turn lead to a decline in folivore populations. The extent and speed at which secondary-growth trees arrive and grow depend on the evolutionary history of the trees in the region (P. Richards 1996). For example, hurricane forests that are disturbed every decade or so will be rapidly colonized by many secondary-growth species that grow rapidly, while other forests will respond more slowly (Boucher 1990; Chapman, Chapman, Kaufman, et al. 1999; Ganzhorn 1995b; P. Richards 1996). The appropriate temporal scale to use will depend on the question being asked. It should also be kept in mind that any pattern of animal abundance that these responses to logging generate can be significantly modified or even totally reversed by the effects of bushmeat hunting or climate change. In addition to these direct effects (i.e., the removal of feeding species), industrial logging leads to the construction of roads that promote total deforestation through subsequent agricultural development and cattle ranching (R. Butler and Laurance 2008; Laurance, Clements, et al. 2014).

The nutritional requirements of specific primates should be used by managers to guide pre- and postlogging practices. For example, important feeding trees should be left standing in selective logging operations, and loggers should use directional felling to reduce impacts on important food resources (Putz et al. 2000). Adopting such management practices may result in lowering the population declines of primates that are negatively impacted by logging or speeding population recovery. For folivores, there is a robust model predicting their abundance or biomass that can be applied to their management. Milton (1979) suggested that the protein and fiber content of leaves was an important criterion for leaf choice for small-bodied arboreal mammals such as colobines. Fiber is fermented by symbiotic microbes in the gastrointestinal tract, and some fiber components are only partially digestible; thus, animals do not select a high-fiber diet (McNab 2002). In contrast, nitrogen is a limiting nutrient in most environments and is predominantly found in protein; thus, herbivores should choose foods high in protein and low in fiber (T. C. White 1993). This influential idea has been successfully applied, and the index of protein to fiber of dominant trees in a forest predicts the abundance and biomass of folivorous monkeys at both local (Chapman and Chapman 2002; Ganzhorn 2002) and regional scales (Chapman, Chapman, Naughton-Treves, et al. 2004; A. Davies 1994; Fashing et al. 2007; Oates, Whitesides, et al. 1990; Waterman, Ross, et al. 1988). Combining all existing data, this ratio accounts for 87% of the variance in colobine biomass (Chapman, Chapman, Naughton-Treves, et al. 2004). Similarly, an assessment across nine locations in Southeast Asia revealed a correlation between the abundance of leguminous trees and *Presbytis* biomass (A. Davies 1994), presumably because of their high leaf protein content. However, the protein-to-fiber model is controversial due to four issues (reviewed by I. Wallis et al. 2012). First, the studies used to test this protein/fiber model are all correlative (Chapman, Chapman, Jacob, et al. 2010; Chapman, Struhsaker, et al. 2010; I. Wallis et al. 2012). Second, the protein levels of leaves that colobines feed on in the wild are higher than their believed requirements (Oftedal 1991; Rothman, Raubenheimer, and Chapman 2011). Third, available nitrogen should be used to evaluate these ideas, and this is typically not done (Rothman, Chapman, and Pell 2008; Rothman, Chapman, and Van Soest 2012; I. Wallis et al. 2012). Finally, new empirical data call into question the generality of this model (Chapman, Chapman, Jacob, et al. 2010; Chapman, Struhsaker, et al. 2010; Gogarten, Guzman, et al. 2012).

With respect to predicting frugivore abundance, there are fewer comparative papers and no general model; however, frugivore abundance is related to the density of fruit resources. For example, a correlation was found across six sites between chimpanzee (*Pan troglodytes*) nest density and large, fleshy-fruit tree density (Balcomb et al. 2000; see also Chapman, Chapman, and Gillespie, et al. 2002 for a similar result with a frugivorous/insectivorous primate), and chimpanzee density is higher where fig basal area is higher (Bortolamiol et al. 2014; but see Lacroux et al. 2022 for insights into other factors affecting chimpanzee habitat use). In general, it is well known that figs are important for many species of primates, other mammals, and birds (Bleher et al. 2003; Diaz-Martin et al. 2014; Janzen 1979a), while their commercial harvesting is on the rise (Felton, Felton, Rumiz, et al. 2013). Figs contain many essential nutrients and minerals (O'Brien, Kinnaird, and Dierenfeld 1998) and can provide a nutritionally balanced staple food in some areas (Felton, Felton, Wood, Foley, et al. 2009). While there is no evidence that figs act as keystone species (Chapman, Chapman, Struhsaker, et al. 2005; Gautier-Hion and Michaloud 1989), ripe fruits are eaten by many frugivorous primates (O'Brien, Kinnaird, and Dierenfeld 1998), and folivores often eat unripe fruits or fig leaves (Dasilva 1994; Milton, Morrison, et al. 1982).

Not only is there loss in the total surface area of old-growth forest, but it is also being fragmented at an accelerating rate (L. Marsh and Chapman 2013; UNEP 2001). Forest fragmentation often leaves primates with a very low diversity of food species (exacerbated by climate change; see below). This is illustrated by two examples from Western Uganda, where deforestation outside of national parks has been extensive. A group of over 20 red colobus (*Piliocolobus tephrosceles*) living along a riverine strip of forest with extremely low plant diversity spent 91.9% of their foraging time eating from one species of tree (Chapman, Chapman and Gillespie 2002). This group survived under these conditions for at least four years, until the riverine forest was largely cleared for timber. Similarly, in a very degraded fragment, a group of four to six red colobus survived in an area containing only seven trees with a diameter at breast height of 10 cm or greater, and they survived for over 15 years (Chapman, Ghai, et al. 2013). In some cases, remaining forest fragments may be too small to meet the nutritional requirements of primates, forcing them to forage in the matrix. In Mexico, black howlers (*Alouatta pigra*) live in fragments as small as 0.4 ha, and the paucity of essential foods and specific nutrients probably forces them to travel to scattered trees in pastures. This matrix foraging puts them at risk of predation, including from domesticated animals such as dogs (Pozo-Montuy et al. 2013). The construction of corridors of trees is one means of removing this danger, and corridors have been a major aspect of a number of conservation projects (T. Jones et al. 2012; Noss 1995). In restoration projects, these corridors could be planted with important food trees, and the choice of these trees can be directly guided by the knowledge gained by research focused on the feeding ecology conducted on primate species of special concern. A complication in many areas of the world is that primates in fragments are hunted, further decreasing the conservation value of forest fragments (Benchimol and Peres 2013a, 2013b).

CLIMATE CHANGE

Trends

The earth's climate has warmed significantly as the result of human actions. The Intergovernmental Panel on Climate Change estimates that the earth's climate has warmed by 1.2°C since industrialization, and by the end of the 21st century the earth's mean surface temperature is set to increase by at least 1.5°C (IPCC 2021). Rising temperature alters global patterns of circulation, which affects rainfall patterns, but changes do not occur uniformly around the earth (Graham et al. 2016). Given where primates occur, they will experience 10% more warming than the global mean (Graham et al. 2016). Precipitation changes will also likely be quite varied across primate ranges (from >7.5% increases per °C of global warming to >7.5% decreases; Graham et al. 2016). To date, there have been numerous documented shifts in plant and animal distributions, population abundance, life history, and survival of species in response to climate change (Hannah et al. 2002; Newman et al. 2011; Pounds et al. 1999). Short-term extremes such as droughts and floods will affect plants, animals, and human populations, and such events will come more frequently and be more intense with climate change. In fact, droughts have already increased in frequency and intensity since the 1970s (N. Watts et al. 2021).

While changes that will have an obvious effect on primate populations are of great concern (e.g., loss of habitat because of drying and heat stress), it is also important to proactively seek to understand the subtle, unexpected, or cascading effects of climate change (see also Rothman et al., this volume). One such unanticipated effect of climate change may be alterations in the nutritional quality of plant parts, in particular the leafy food resources of folivorous primates (Coley et al. 2002). Current knowledge of how plants respond to changes in climate (CO_2 levels, temperature, and rainfall) is based on greenhouse experiments, and results vary depending on plant species and soil nutrients; however, in general, experiments demonstrate that increased temperature and elevated CO_2 levels result in a reduction in leaf protein and an increase in fiber (Buse et al. 1998; Curtis and Wang 1998; Dury et al. 1998; Kanowski 2001; E. Robinson et al. 2012; Zvereva and Kozlov 2006). In addition, analysis of tropical trees along a rainfall gradient revealed that nitrogen content (mostly protein) decreased and nitrogen-to-fiber ratios decreased with increasing precipitation (Santiago and Mulkey 2005). Similar effects on leaf chemistry have been found for increasing temperature (Craine, Elmore, Olson, et al. 2010; Weih and Karlsson 2001). A significant proportion of the nitrogen in a leaf is protein, so the protein-to-fiber model described earlier predicts that this nitrogen decline will have negative consequences for the nutrition of folivorous primates.

For primates that rely on insects, studies have found that insect populations respond to climate change with changes in range, abundance, and phenology (reviewed in Andrew et al. 2013). In fact, one study of insects in southern Africa forecasts reduced insect diversity as a consequence of abiotic changes resulting from climate change (Pio et al. 2014).

Predicted Responses of Primates with Different Foraging Strategies to Climate Change

Understanding the consequences of climate change for primates may be the most important question that primate conservation biologists must address in the next decade; however, our ability to predict primate responses to climate change is very poor (P. Wright 2007). It is relatively easy to imagine that in areas becoming hotter and drier, primates will become physiologically stressed (McFarland, Barrett, Boner, et al. 2014), food trees will die, and primates will die along with them or be forced to move if fragmentation is not severe enough to make that impossible. This is supported by data from Amboseli National Park, Kenya, where the average daily maximum temperature increased by 0.275°C per year (an order of magnitude greater than that predicted by climate change models; J. Altmann et al. 2002) and there was a dramatic loss of tree cover (J. Altmann et al. 2002), which may have driven the concomitant decline in local vervet populations and movement of baboon populations (Struhsaker 1973).

The hotter temperatures associated with climate change may trigger a short-term regulatory response in primates that may change what they need to eat (Raubenheimer, Simpson, and Tait 2012). For example, primates living under hotter temperatures may select foods with a higher water or salt content. Many primates have been shown to have low salt intake and to eat unusual foods, such as decaying wood, soil, aquatic plants, or eucalyptus bark, to obtain salt (Oates 1978; Rode, Chapman, Chapman, et al. 2003; Rothman, Van Soest, and Pell 2006). Added salt demand under high-temperature conditions may be stressful to animals.

In contrast to the situation where the climate gets hotter and drier, what happens when the climate gets wetter is not clear. Kibale National Park, Uganda, has experienced climate change well above the global average. The area receives 300 mm more rainfall per year than at the start of the century, and the average maximum monthly temperature has increased by 4.4°C in the last 40 years (Rothman, Chapman, Struhsaker, et al. 2015). Corresponding with this change in climate, several tree species stopped fruiting (Chapman, Chapman, Struhsaker, et al. 2005), meaning that there was less fruit available for frugivores. For example, *Trilepisium madagascariense* (formerly *Bosqueia phoberos*) stopped fruiting at a site to the north of the park but continued to fruit at a drier site to the south; this has corresponded with a decline over time in blue monkey populations (*Cercopithecus mitis*) but not redtail monkey (*C. ascanius*) or mangabey populations (*Lophocebus albigena*; Chapman, Balcomb, et al. 2000; Chapman, Struhsaker, et al. 2010).

Few studies have quantified changes in the quality of the leaves to test the generality of the greenhouse experiments because of the needed duration of monitoring. However, over the last 30 years, the quality of leaves eaten by some folivores has been described. In agreement with greenhouse experiments, Rothman, Chapman, Struhsaker, et al. (2015) show a general increase in fiber and a decline in protein in nonexperimental trees compared to data collected 15 and 30 years previously. Because many folivores select leaves with high protein-to-fiber ratios, declining leaf quality could have a major impact on folivore abundance. Based on the predictive model between colobine biomass and the protein-to-fiber ratio of mature leaves from common tree species in an area (Chapman, Chapman, Naughton-Treves, et al. 2004), a 31% decline in colobus monkey abundance would be predicted. However, this decline has not been seen, possibly because of a change in the composition of the old-growth forest at this site (Chapman, Struhsaker, et al. 2010; Gogarten, Jacob, et al. 2014), because the population can be flexible in what it eats, or because the decline may be yet to come.

Not only may climate change affect fruiting periodicity, the composition of leaves, and what primates require in their foods; it also may change what is available for them to eat. Experiments suggest that the warming associated with climate change reduces plant diversity (Gedan and Bertness 2009), which could affect the foraging strategies and persistence of all primates. Climate change models have also predicted reduced diversity of fruiting angiosperms, which will reduce fruit availability for frugivorous primates (Vamosi and Wilson 2008). In addition to directly affecting species important in primate diets, climate change is expected to contribute to the increasing warming and aridification of the tropics and to interact synergistically with human-altered landscapes to reduce important wildlife habitat (Brodie, Post, and Laurance 2012).

BUSHMEAT HUNTING

Trends

To most primatologists, the bushmeat trade is undesirable and objectionable and should be stopped, but since it will not be stopped in the foreseeable future, it is something that nutritional ecologists must study to understand its consequences (R. D. Harrison 2011). Estimates of the extent of the trade are poor. However, E. Bennett et al. (2000) estimated that six million animals were hunted annually in Malaysian Borneo, which is approximately 36 animals per km^2, while in Africa, four million metric tons of bushmeat were extracted each year from the Congo basin alone (Fa and Brown 2009). This trade means that even many national parks do not function as safe havens, and species in parks can still be driven to extinction through hunting (Oates, Abedi-Lartey, et al. 2000). In a global analysis of 60 parks, Laurance, Carolina Useche, et al. (2012) documented that researchers considered only approximately half of all reserves to have been effective over the last 20–30 years, while the remainder of the reserves had experienced an alarming erosion of biodiversity, which included a loss of primate species because of hunting (Oates, Abedi-Lartey, et al. 2000). This phenomenon is poignantly illustrated by a study in Taï National Park, Cote d'Ivoire, where a park-wide survey illustrated that, regardless of primate species, density was 100 times higher near the protected research station and tourism site than in the remainder of the park (N'Goran et al. 2012).

Ironically, reduced primate densities through hunting can, in the short term, lead to less competition for food, enabling animals to select the most nutritious foods. However, in the long term, the decrease in primate seed dispersers can lead to declines in primate food trees (Chapman and Onderdonk 1998; Peres and Dolman 2000). Hunters tend to target the larger primates, and evidence suggests that, with regard to forest-wide seed dispersal, removal of large-bodied primates is only partially offset by increases in smaller-bodied primates (Peres and Dolman 2000).

Predicted Responses of Primates with Different Foraging Strategies to Bushmeat Hunting

Managing wildlife populations for harvest has been a topic of inquiry for hundreds of years, dating back to the hunting reserves of kings (R. D. Harrison 2011) As a result, there is a wealth of information to predict the impact of hunting, which varies as a function of body size, life history, and diet (J. Lambert 1998; Leigh 1994a). This information has been refined to provide guidelines for sustainable hunting (Bodmer et al. 1997; Mayor et al. 2016), although we are unaware of any community of primates that are hunted sustainably. Given this wealth of data, we are not going to address this issue, except to state that hunters tend to target large-bodied animals (more value for the cost of the bullet and easier to transport per kg to market). This means that the hunted animals are often folivores, as large body size is a requirement of fiber digestion (McNab 2002), and these animals are often vulnerable because of their relatively large group size and sedentary lifestyle. Large-bodied folivores are particularly vulnerable because of their relatively slow life history. Some regions of South America are exceptions to this as folivore diversity is low, and hunters in some regions do not target howler monkeys (*Alouatta* spp.); however, hunting in other regions has caused local extinctions (Crockett 1998).

FUTURE DIRECTIONS

To guide future research, it is important not only to understand how the current situation is negatively affecting primates but also to identify positive trends that indicate opportunities for researchers and conservation biologists. Globally, there are two positive trends that offer opportunities for primate conservation: forest change trends and the formation of new protected areas. In most primate host countries, degraded forests now exceed the area covered by old-growth forests (FAO 2005). In fact, it is estimated that, in the 1990s, secondary forest replaced at least 1 of every 6 hectares of deforested old-growth forest (S. Wright and Muller-Landau 2006). Secondary forests now represent approximately 35% of all remaining tropical forests (Emrich et al. 2000). The year 2008 marked the first time more people lived in cities than in rural settings, and the rate of urbanization is increasing. The UN Population Division estimates that 90% of the world's population growth between 2000 and 2030 will occur in cities in the developing world. Urbanization leads to declines in rural populations in many countries, particularly in those countries where the population growth rate has declined (A. Jacob et al. 2008; S. Wright and Muller-Landau 2006). Abandoned rural land offers great primate conservation opportunities because many of the plants that regenerate earliest are palatable and nutritious and bear plentiful fruit. There are a variety of trajectories for future land use: abandoned land can be converted into huge agricultural monocultures, such as palm oil plantations (Linder 2013), or to agroecosystems where primate conservation is possible to varying degrees (Estrada, Raboy, and Oliveira 2012), or to agricultural land with fragments and corridors (Pozo-Montuy et al. 2013). Alternatively, the land could be allowed to regenerate to natural forest, which offers greater potential for the persistence of primates. Baya and Storch (2010) surveyed a village site in Korup National Park, Cameroon, that was abandoned seven to eight years previously and found populations of all eight species of primates that occur in the region; in addition, sighting frequency was not significantly different from that in other sectors of the park surveyed in 2004–2005 (Linder 2008). In Kibale National Park, Uganda, seven years after an area of grassland was replanted with trees as part of a carbon offset program, all species of diurnal primates were present in high numbers, including the endangered red colobus and chimpanzee (Omeja, Lawes, et al. 2016; Omeja, Obua, et al. 2012). Regeneration in this area of Kibale National Park has been extensive (fig. 29.2), which offers hope for the future. In carbon offset proj-

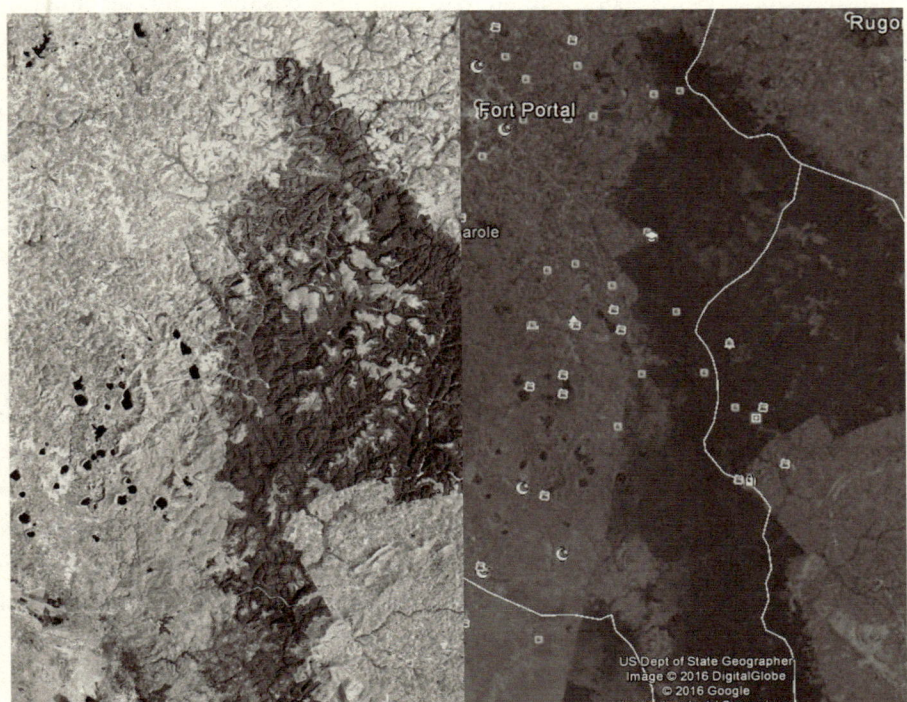

Figure 29.2. A satellite image of Kibale National Park, Uganda, in 1993 and a similar image taken from Google Earth in 2016 showing the extent of forest regeneration occurring in 23 years. The forested area is dark gray, and the agricultural and grazing land is light gray (the dark areas to the west of the park are crater lakes).

ects, if primatologists know the nutritional requirements of endangered primates, they can guide what species are planted. Urbanization and abandoned rural land represent a significant opportunity, and we must proactively influence land use policy and manage these lands to make the largest conservation gains.

A second positive trend is the continued formation of new protected areas. Since 1992, protected areas have grown steadily, increasing by an average of 2.5% in total area and 1.4% in number of sites annually (Butchart et al. 2010; Rands et al. 2010). By 2006, protected areas covered 24 million km^2, in 133,000 designated areas (Butchart et al. 2010; Rands et al. 2010). However, these positive developments need to be viewed from a balanced perspective (Andam et al. 2008; Joppa and Pfaff 2009, 2010; Joppa et al. 2008). For example, it is estimated that 20% of vertebrate taxa recognized as threatened by the IUCN do not live in protected areas (A. Rodrigues et al. 2004). Also, although protected areas are normally effective at protecting land from being cleared, they are less effective at eliminating logging, human-created fire, and bushmeat hunting (Bruner et al. 2001; Chapman and Peres 2001; Hartter et al. 2011; Oates 1996). Protected areas are also not necessarily effective at maintaining the processes that generate and maintain biodiversity (T. B. Smith et al. 1993). About one-half of all protected areas are experiencing an erosion of biodiversity (Laurance, Carolina Useche, et al. 2012). Also, it is estimated that more than two-thirds of critical sites for biodiversity have incomplete or no protection (Butchart et al. 2010). This means that local communities need to be meaningfully integrated into conservation efforts, so that some of their needs can be met and positive park-people relations can be achieved. Researching how to make the largest conservation gains for primates from existing and new conservation areas is a clear research priority and will involve working closely with the local communities.

SUMMARY

Many primate populations are threatened as the result of human actions. We are now ethically obligated to take conservation action. To do this, conservation practitioners need to predict negative changes prior to their occurrence. In this chapter, we have used our knowledge of primate nutritional ecology and foraging strategies to make general predictions as to how primates could respond to three global threats: habitat destruction through logging and fragmentation, climate change, and bushmeat hunting.

Logging and fragmentation have been extensive over the last few decades and are showing few signs of slowing down, except in areas where there is little or no forest left. In general, folivores typically do better in logged and fragmented areas than frugivores, but this is not always the case, and some frugivores can do well in these disturbed habitats. Climate change is altering the fruiting patterns of trees, creating periods of fruit scarcity, and decreasing the quality of leaf resources; these changes correspond to declines in insect abundance and lead to some forested areas currently supporting primates being unable to support closed-canopy forest. The bushmeat trade is a large industry that is decimating many primate populations. Hunters tend to target large-bodied animals, which means the hunted animals are often folivores, as large body size is a requirement of fiber digestion, and these animals are often vulnerable because of their relatively large group size and sedentary lifestyle. They can also unsustainably target large-bodied frugivores.

Despite all the research on primate foraging and nutritional ecology published to date, we conclude that we have insufficient knowledge to predict how a particular species in a particular location will respond to disturbance. Thus, there are many ways that new academic studies can contribute to the future conservation of primates.

CONCLUSIONS

- No matter what the nutritional needs of a species or how they meet those needs, from a conservation perspective, it is critical not to remove the food resources that meet those needs.
- It is a paradox that bushmeat hunting lowers primate population density, reduces competition, and thus increases the amount of food available per individual; thus, processes that have negative effects at a population level can be beneficial at an individual level. However, in no way can this be considered beneficial for a primate population.
- Primate species that have a flexible foraging strategy will be impacted less by forest degradation than those with a rigid foraging strategy or specific dietary needs; however, in general, at the present time, we have insufficient knowledge of nutritional ecology to state a priori which species are flexible and which are not.
- It is our hope that we have illustrated the importance of maintaining forests and the food resources they contain and that this will provide researchers, conservation biologists, and managers motivation to maintain more forests, work on connecting fragments with corridors, provide guidance to restoration projects, and, in general, consider foraging needs whenever a forest is going to be altered.

Afterword

Alison Richard

The pleasure of writing an afterword, I have just discovered, is the interest of learning many things I didn't know, with none of the work involved in discovering and assembling them. This is an amazing book. I had not realized until now that I have been waiting for it for many years.

Two developments since I began my own field research long ago on lemur feeding (not nutritional) ecology jump off its pages. The first is simply how much more is known today. This is not necessarily inevitable. Most primates live at low densities, and many are hard to see or follow—zebra fish in a lab tank are a picnic by comparison, and it would not be altogether surprising if interest in studying wild primates had waned. Moreover, the study of nonhuman primates sits uncomfortably at the intersection of biology and anthropology. Do we "really" study them for what they can tell us about ourselves, or because they are interesting mammals? For some, the inclusion of primate research in anthropology is a stretch; for others, its inclusion in biology is problematic, with a whiff of anthropomorphism in the air. This book decisively lays to rest my fears that primate research would be sidelined for these reasons.

The second, extremely exciting development illuminated by this book is the kind of research now undertaken. Scientific and technological advances coupled with new questions and considerable ingenuity bring the lab to the field and vice versa in ways unimaginable in the past. Where once the study of primate feeding (and, with rare exceptions, it was mainly just that) meant collecting observations of behavior, today it (routinely, I'm tempted to say) encompasses energetics and nutrition, hormones and microbiomes, phytochemistry and physiology as well. This has produced a quantum leap in our understanding of the what, how, and why. The journey is far from over, as the editors signal in their preface, but this book is surely a major milestone along the way.

The volume's length (and weight, no doubt) already qualifies it as a doorstop, and I have no criticism of the editors' decision to frame and end it as they did: you can't cover everything. The role of primates in forest function and threats to their survival run as threads through many chapters, and, indeed, the preface ends with a clarion call to connect academic research to actionable conservation solutions. Neither issue is embedded in the book's overall organization, however, and my hope is that the editors will take up these key issues head-on in another volume.

One final observation: even in the quite recent past, an editorial preface would not have concluded with reflections on access and equity and a forthright self-assessment of limited success in this regard. These reflections, clearly heartfelt, throw down a gauntlet to us all. The fact that these issues are being thought about and cared about represents progress of a different kind but of equal importance to the scientific advances so brilliantly described in the body of the book.

Alison Richard
October 2022

Acknowledgments

We are humbled by the number of individuals who have shaped and guided the content of this volume. As we described in the preface, our contribution stands on the shoulders of scholars and practitioners who have been working in the field and laboratory for decades. The list of names of those who have contributed to our current understanding of how and what animals eat is vastly longer than we could ever possibly assemble. We view what follows as a partial compilation.

For reviews, professional and personal support throughout this process, mentorship and wisdom over the years, and assistance in the field and in the laboratory, the editors and authors would like to thank Shauhin Alavi, Andy Allan, Camilo Flores Amasifuén, Stanly Ambrose, Carol Augsburger, Alec Ayers, Andrea Baden, Elizabeth Ballare, Carly Batist, Joel Berger, Paco Bertolani, Nazaire Bonang, Kelly Boyer-Ontl, Timothy Bransford, Rebecca Brittain, Aaron Brownell, Jackson Bryer, Richard Byrne, Elhadj Camara, Mboule Camara, Waly Camara, Fernando Campus, Santiago Cassalet, Clayton Clement, Ben Coleman, Paul Constantino, Mark Cook, Kat Coops, Marina Cords, Andy Cunningham, Frank Cuozzo, Grace Davis, Lucas Delezene, Eric Delson, John Dennis, Ellen Dierenfeld, Nate Dominy, Diane Doran-Sheehy, Wendy Erb, Agaba Erimosi, Damien Farine, Adam Felton, John Fleagle, Dax Garber, Jenni Garber, Sara Garber, Eva Garrett, Maja Gašperšič, Ian Gilby, Ken Glander, Paul Griffiths, Fred Grine, Mary Hartig, Carrie Healey, Claire Hemingway, James Higham, Alain Houle, Kevin Hunt, Karin Isler, Jerry Jacka, Karline Janmaat, Susannah Johnson-Fulton, Dondo Kante, Richard Kaseregenyu, Francis Katurama, Richard Kay, Jacques Keita, Kathleen Koops, Kristin Krueger, Laura LaBarge, Phyllis Lee, Steve Leigh, Mark Leighton, Amy Lu, James Magaro, David Marks, Ikki Matsuda, Scott McGraw, Bill McGrew, Chrissy Mckenney, Antoine Mende, Katie Milton, Francis M'Kirera, Alysse Moldawer, Carson Murray, Hillary Musinguzi, Daniel Naumenko, Maria van Noordwijk, Marilyn Norconk, Kate Nowak, Caitlin O'Connell, Olav Oftedal, Kerry Ossi-Lupo, Jenny Paltan, Ed Parker, Alejandro Pérez-Pérez, Alex Piel, Mike Power, Didik Prasetyo, Ilya Raskin, Matt Ravosa, Melissa Remis, Sabiiti Richard, Gabrielle Rosenthal, Michel Sadiakho, Ethan Schreiber, Jessica Scott, Robert Scott, Tatang Mitra Setia, Oliver Shulke, Jeroen B. Smaers, Abba Sonko, Fiona Stewart, Peter Stirling, Tom Struhsaker, Larissa Swedell, Mark Teaford, Ney Shahuano Tello, Lucia Torrez, Alexa Ugarte, Kim Valenta, Allan Walker, Chris Wall, Malcolm Watford, Julie Wieczkowski, Erik Willems, Malcolm Williamson, Heiko Wittmer, Stephen Wooten, Richard Wrangham, Barth Wright, Ny Yamushita, Sireen El Zaatari, the people of Fongoli and Djendji, the citizens of Kanyawara, and all researchers who have contributed to the current knowledge of primate diets worldwide.

For financial and infrastructural support, the editors and authors gratefully acknowledge the Alexander von Humboldt-Stiftung Program; ARCUS foundation; Deutscher Akademischer Austauschdienst; Deutsche Forschungsgemeinschaft; American Society of Primatologists; Anglia Ruskin University; Canada Research

Chairs program 889; CIFAR Fellowship of the Northwestern University Humans and the Microbiome Program; Claude Leon Foundation; Conservation, Food and Health Foundation; Department of Human Biological Evolution (Harvard University); Disney Conservation Fund; Durham University; Earthwatch Institute; Eijkman Institute for Molecular Biology; EU COFUND; Fonds Québécois de la Recherche sur la Nature et les Technologies; Foundation for Anthropological Research; Graduate Center CUNY; Hunter College CUNY; Gunung Palung National Park office (BTNGP); International Development Research Center; International Primate Action Fund; Iowa State University; Indonesian Institute of Sciences (LIPI); Lajuma Research Centre; Leakey Foundation; Leverhulme Trust; Leibniz-Gemeinschaft; Makerere University Biological Field Station (Uganda); Nacey Maggioncalda Foundation; National Geographic Society; National Science and Engineering Research Council Discovery 890 Grant and Accelerator Supplement; National Science Foundation (USA); Natural Environment Research Council (Canada); Natural Science and Engineering Research Council of Canada; New York Consortium in Evolutionary Primatology; Norwegian University of Life Sciences; Packard Foundation; Packard Foundation Fellowship Center for the Advanced Study of Collective Behavior (University of Konstanz); Primate Action Fund; Primate Conservation Inc.; Republic of Senegal Department of Eaux et Forets; Rutgers—the State University of New Jersey and Center for Human Evolutionary Studies; Smithsonian National Zoological Park (Nutrition Laboratory); Texas State University; Uganda Wildlife Authority; University of Calgary; Universitas Nasional (Indonesia); Universitas Tanjungpura; University of Colorado Boulder; University of East Anglia; US Fish and Wildlife Service; Vilas Research Professorship from the Vilas Trust (University of Wisconsin–Madison); Wenner-Gren Foundation; Wildlife Conservation Society; and several anonymous donors.

Finally, and most importantly, we thank the citizens of the beautiful countries in which we all work, the animals themselves, and the families who have supported us. Our families include not just our immediate human kin but also the kin of planet Earth: all creatures, great and small.

Literature Cited

Aagaard, K. M., Ma, J., Antony, K. M., Ganu, R., Petrosino, J. F., and Versalovic, J. (2014). The placenta harbors a unique microbiome. *Science and Translational Medicine*, 6, 237ra265.

Abu Baker, M. A. (2015). Titrating the cost of plant toxins against predators: A case study with common duikers, *Sylvicapra grimmia*. *Journal of Chemical Ecology*, 41, 924–28.

Abu Baker, M. A., and Brown, J. S. (2010). Islands of fear: Effects of wooded patches on habitat suitability of the striped mouse in a South African grassland. *Functional Ecology*, 24, 1313–22.

Abu Baker, M. A., and Brown, J. S. (2014). Foraging and habitat use of common duikers, *Sylvicapra grimmia*, in a heterogeneous environment within the Soutpansberg, South Africa. *African Journal of Ecology*, 52, 318–27.

Abu Baker, M. A., Emerson, S. E., and Brown, J. S. (2015). Foraging and habitat use of eastern cottontails (*Sylvilagus floridanus*) in an urban landscape. *Urban Ecosystems*, 18, 977–87.

Abrahams, M. V., & Dill, L. M. (1989). A determination of the energetic eqivalence of the risk of predation. *Ecology*, 70, 999–1007.

Abwe, E. E., Morgan, B. J., Doudja, R., Kentatchime, F., Mba, F., Dadjo, A., et al. (2020). Dietary ecology of the Nigeria–Cameroon chimpanzee (*Pan troglodytes ellioti*). *International Journal of Primatology*, 41, 81–104.

Achenbach, G. G., and Snowdon, C. T. (2002). Costs of caregiving: Weight loss in captive adult male cotton-top tamarins (*Saguinus oedipus*) following the birth of infants. *International Journal of Primatology*, 23, 179–89.

Ackermans, N. L., Winkler, D. E., Martin, L. F., Kaiser, T. M., Clauss, M., and Hatt, J. M. (2020). Dust and grit matter: Abrasives of different size lead to opposing dental microwear textures in experimentally fed sheep (*Ovis aries*). *Journal of Experimental Biology* 223, jeb220442.

Adams, A. L., Dickinson, K. J. M., Robertson, B. C., and van Heezik, Y. (2013). An evaluation of the accuracy and performance of lightweight GPS collars in a suburban environment. *PLoS ONE*, 8, e68496.

Adams, L. G., Farley, S. D., Stricker, C. A., Demma, D. J., Roffler, G. H., Miller, D. C., et al. (2010). Are inland wolf–ungulate systems influenced by marine subsidies of Pacific salmon? *Ecological Applications*, 20, 251–62.

Adams, N. F., Gray, T., and Purnell, M. A. 2020. Dietary signals in dental microwear of predatory small mammals appear unaffected by extremes in environmental abrasive load. *Palaeogeography, Palaeoclimatology, Palaeoecology*, 558, 109929.

Adams, N. R. (1990). Permanent infertility in ewes exposed to plant estrogens. *Australian Veterinary Journal*, 67, 197–201.

Adams, N. R. (1995). Organizational and activational effects of phytoestrogens on the reproductive-tract of the ewe. *Proceedings of the Society for Experimental Biology and Medicine*, 208, 87–91.

Adelman, J. S., Moyers, S. C., and Hawley, D. M. (2014). Using remote biomonitoring to understand heterogeneity in immune-responses and disease-dynamics in small, free-living animals. *American Zoologist*, 54, 377–386.

Adeyemi, S. B., Ogundele, K. O., and Animasaun, M. A. (2014). Influence of drying methods on the proximate and phytochemical composition of *Moringa oleifera*. *Advances in Natural and Applied Science*, 2, 1–5.

Adipietro, K. A., Mainland, J. D., and Matsunami, H. (2012). Functional evolution of mammalian odorant receptors. *PLoS Genetics*, 8, e1002821.

Adlercreutz, H. (1995). Phytoestrogens—epidemiology and a possible role in cancer protection. *Environmental Health Perspectives*, 103, 103–12.

Afacan, N. J., Fjell, C. D., and Hancock, R. E. (2012). A systems biology approach to nutritional immunology—focus on innate immunity. *Molecular Aspects of Medicine*, 33(1), 14–25.

Afshar, M., and Giraledeau, L. A. (2014). A unified modelling approach for producer-scrounger games in complex ecological conditions. *Animal Behaviour*, 96, 167–76.

Agetsuma, N. (1995). Dietary selection by Yakushima macaques (*Macaca fuscata yakui*): The influence of food availability and temperature. *International Journal of Primatology*, 16, 611–28.

Agostini, C., Marangoni, F., Lammardo, A. M., Galli, C., Giovannini, M., and Riva, E. (2001). Long-chain polyunsaturated fatty acid concentrations in human hindmilk are constant throughout twelve months of lactation. In D. S. Newburg (Ed.), *Bioactive Components of Human Milk* (pp. 157–61). Kluwer Academic/Plenum.

Agostini, I. (2017). Experimental testing of reciprocal effects of nutrition and parasitism in wild black capuchin monkeys. *Scientific Reports*, 7, 12778.

Agostini, I., Holzmann, I., and Di Bitetti, M. S. (2010). Are howler monkey species ecologically equivalent? Trophic niche overlap in syntopic *Alouatta guariba clamitans* and *Alouatta caraya*. *American Journal of Primatology*, 72, 173–86.

Agostini, I., Holzmann, I., and Di Bitetti, M. S. (2012). Influence of seasonality, group size, and presence of a congener on activity patterns of howler monkeys. *Journal of Mammalogy*, 93, 645–57.

Agrawal, K. R., Lucas, P. W., Prinz, J. F., and Bruce, I. C. (1997). Mechanical properties of foods responsible for resisting food breakdown in the human mouth. *Archives of Oral Biology*, 42, 19.

Aguiar, L. M., and Moro-Rios, R. F. (2009). The direct observational method and possibilities for Neotropical carnivores: An invitation for the rescue of a classical method spread over the primatology. *Zoologia*, 26, 587–93.

Aguiar, L. M., dos Reis, N. R., Ludwig, G., and Rocha, V. J. (2003). Dieta, área de vida, vocalizações e estimativas populacionais de *Alouatta guariba* em um remanescente florestal no norte do estado do Paraná. *Neotropical Primates*, 11, 78.

Aguilar-Melo, A. R., Calme, S., Pinacho-Guendulain, B., Smith-Aguilar, S. E., & Ramos-Fernández, G. (2020). Ecological and social determinants of association and proximity patterns in the fission–fusion society of spider monkeys (*Ateles geoffroyi*). *American Journal of Primatology*, 82(1), e23077.

Aiello, L. C., and Wheeler, P. (1995). The expensive-tissue hypothesis: The brain and the digestive system in human and primate evolution. *Current Anthropology*, 36, 199–221.

Akers, R. M. (2002). *Lactation and the Mammary Gland*. Iowa State Press.

Akhtar, M. S., Ashino, R., Oota, H., Ishida, H., Niimura, Y., Touhara, K., et al. (2022). Genetic variation of olfactory receptor gene family in a Japanese population. *Anthropological Science*, 130(2), 93–106.

Alatalo, R. V., Gustafsson, L., and Lundberg, A. (1986). Interspecific competition and niche changes in tits (*Parus* spp.): Evaluation of nonexperimental data. *American Naturalist*, 127, 819–34.

Alba-Mejia, L., Caillaud, D., Montenegro, O. L., Sanchez-Palomino, P., and Crofoot, M. C. (2013). Spatiotemporal interactions among three neighboring groups of free-ranging white-footed tamarins (*Saguinus leucopus*) in Colombia. *International Journal of Primatology*, 34, 1281–97.

Albert, A., Hambuckers, A., Culot, L., Savini, T., and Huynen, M. C. (2013). Frugivory and seed dispersal by northern pig-tailed macaques (*Macaca leonina*) in Thailand. *International Journal of Primatology*, 34, 170–93.

Albert, A., McConkey, K. R., Savini, T., and Huynen, M. C. (2013). The value of disturbance-tolerant cercopithecine monkeys as seed dispersers in degraded habitats. *Biological Conservation*, 170, 300–310.

Albert-Daviaud, A., Buerki, S., Onjalalaina, G. E., Perillo, S., Rabarijaona, R., Razafindratsima, O. H., et al. (2020). The ghost fruits of Madagascar: Identifying dysfunctional seed dispersal in Madagascar's endemic flora. *Biological Conservation*, 242, 108438.

Alberts, S. C. (2019). Social influences on survival and reproduction: Long-term study of baboons. *Journal of Animal Ecology*, 88, 47–66.

Alberts, S. C., and Altmann, J. (2006). The evolutionary past and research future: Environmental variation and life history flexibility in a primate lineage. In L. Swedell and S. R. Leigh (Eds.), *Reproduction and Fitness in Baboons: Behavioral, Ecological, and Life History Perspectives* (pp. 277–303). Springer.

Alberts, S. C., Altmann, J., and Wilson, M. L. (1996). Mate guarding constrains foraging activity of male baboons. *Animal Behaviour*, 51, 1269–77.

Alberts, S. C., Hollister-Smith, J. A., Mututua, R. S., Sayialel, S. N., Muruthi, P. M., Warutere, J. K., et al. (2005). Seasonality and long-term change in a savanna environment. In D. K. Brockman and C. P. van Schaik (Eds.), *Seasonality in Primates: Studies of Living and Extinct Human and Non-human Primates* (pp. 157–96). Cambridge University Press.

Albiach-Serrano, A. (2015). Cooperation in primates: A critical methodological review. *Interaction Studies*, 163, 361–82.

Albon, S. D., and Langvatn, R. (1992). Plant phenology and the benefits of migration in a temperate ungulate. *Oikos*, 65, 502–13.

Alexander, R. D. (1974). The evolution of social behaviour. *Annual Review of Ecology and Systematics*, 5, 325–83.

Ali, Y. B., Verger, R., and Abousalhan, A. (2012). Lipases or esterases: Does it really matter? Toward a new bio-physico-chemical classification. *Methods in Molecular Biology*, 861, 31–51.

Aliyu, M., Zohora, F. T., Anka, A. U., Ali, K., Maleknia, S., Saffarioun, M., et al. (2022). Interleukin-6 cytokine: An overview of the immune regulation, immune dysregulation, and therapeutic approach. *International Immunopharmacology*, 111, 109130.

Allan, A. T. L., and Hill, R. A. (2018). What have we been looking at? A call for consistency in studies of primate vigilance. *American Journal of Physical Anthropology*, 165, 4–22.

Allan, A. T. L. and Hill, R. A. (2021). Definition and interpretation effects: How different vigilance definitions can produce varied results. *Animal Behaviour*, 180, 197–208.16.

Allan, A. T. L., Bailey, A., and Hill, R. A. (2020). Habituation is not neutral or equal: Individual differences in tolerance suggest an overlooked personality trait. *Science Advances*, 6, eaaz0870.

Allan, A. T. L., Bailey, A., and Hill, R. A. (2021). Consistency in the flight and visual orientation distances of habituated chacma baboons after an observed leopard predation: Do flight initiation distance methods always measure perceived predation risk? *Ecology and Evolution*, 11, 15404–154.

Allan, A. T. L., White, A., and Hill, R. A. (2022). Intolerant baboons avoid observer proximity, creating biased inter-individual association patterns. *Scientific Reports*, 12, 8077.

Allen, C. D., Macalady, A. K., Chenchouni, H., Bachelot, D., McDowell, N., Vennetier, M., et al. (2010). A global overview of drought and heat-induced tree mortality reveals emerging climate change risks for forests. *Forest Ecology and Management*, 259, 660–84.

Allen, M. E. (1989). *Nutritional Aspects of Insectivory*. Michigan State University.

Allen, M. L., Wittmer, H. U., Setiawan, E., Jaffe, S., and Marshall, A. J. (2016). Scent marking in Sunda clouded leopards (*Neofelis diardi*): Novel observations close a key gap in understanding felid communication behaviours. *Scientific Reports*, 6, 35433.

Allen, W. L., Stevens, M., and Higham, J. P. (2014). Character displacement of Cercopithecini primate visual signals. *Nature Communications*, 5, 4266.

Allen-Blevins, C. R., Sela, D. A., and Hinde, K. (2015). Milk bioactives may manipulate microbes to mediate parent–offspring conflict. *Evolution, Medicine and Public Health*, 2015, 106–21.

Allison, D. B., Bassaganya-Riera, J., Burlingame, B., Brown, A. W., le Coutre, J., Dickson, S. L., et al. (2015). Goals in nutrition science 2015–2020. *Frontiers in Nutrition*, 2, 26.

Allritz, M., Call, J., Schweller, K., McEwen, E. S., de Guinea, M., Janmaat, K. R., et al. (2022). Chimpanzees (*Pan troglodytes*) navigate to find hidden fruit in a virtual environment. *Science Advances*, 8(25), eabm4754.

Alofs, K. M., and Polivka, K. M. (2004). Microhabitat-scale influences of resources and refuge on habitat selection by an estuarine opportunist fish. *Marine Ecology Progress Series*, 271, 297–306.

Alp, R. (1993). Meat eating and ant dipping by wild chimpanzees in Sierra Leone. *Primates*, 34, 463–68.

Alp, R. (1997). "Stepping-sticks" and "seat-sticks": New types of tools used wild chimpanzees (*Pan troglodytes*) in Sierra Leone. *American Journal of Primatology*, 41, 45–52.

Altemus, M., Rao, B., Dhabhar, F. S., Ding, W. H., and Granstein, R. (2001). Stress-induced changes in skin barrier function in healthy women. *Journal of Investigative Dermatology*, 117(2), 309–17.

Altendorf, K. B., Laundré, J. W., Gonzalez, C. A. L., and Brown, J. S. (2001). Assessing effects of predation risk on foraging behavior of mule deer. *Journal of Mammalogy*, 82, 430–39.

Altmann, J. (1974). Observational study of behavior: Sampling methods. *Behaviour*, 49, 227–65.

Altmann, J. (1980). *Baboon Mothers and Infants*. Harvard University Press.

Altmann, J. (1983). Costs of reproduction in baboons (*Papio cynocephalus*). In W. P. Aspey and S. I. Lustick (Eds.), *Behavioral Energetics: The Cost of Survival in Vertebrates* (pp. 67–88). Ohio State University Press.

Altmann, J., and Alberts, S. C. (2003). Variability in reproductive success viewed from a life-history perspective in baboons. *American Journal of Human Biology*, 15, 401–9.

Altmann, J., and Alberts, S. C. (2005). Growth rates in a wild primate population: Ecological influences and maternal effects. *Behavioral Ecology and Sociobiology*, 57, 490–501.

Altmann, J., and Samuels, A. (1992). Costs of maternal care: Infant-carrying in baboons. *Behavioral Ecology and Sociobiology*, 29, 391–98.

Altmann, J., Alberts, S. C., and Roy, S. B. (2002). Dramatic change in local climate patterns in Amboseli basin, Kenya. *African Journal of Ecology*, 40, 248–51.

Altmann, J., Altmann, S. A., and Hausfater, G. (1978). Primate infant's effects on mother's future reproduction. *Science*, 201, 1028–30.

Altmann, J., Hausfater, G., and Altmann, S. A. (1985). Demography of Amboseli baboons, 1963–1983. *American Journal of Primatology*, 8, 113–25.

Altmann, J., Hausfater, G., and Altmann, S. A. (1988). Determinants of reproductive success in savannah baboons, *Papio cynocephalus*. In T. H. Clutton-Brock (Ed.), *Reproductive Success: Studies of Individual Variation in Contrasting Breeding Systems* (pp. 403–18). University of Chicago Press.

Altmann, J., Schoeller, D., Altmann, S. A., Muruthi, P., and Sapolsky, R. M. (1993). Body size and fatness of free-living baboons reflect food availability and activity levels. *American Journal of Primatology*, 30, 149–61.

Altmann, S. A. (1967). Social communication among Primates. In *Social Communication among Primates*. Chicago University Press.

Altmann, S. A. (1974). Baboons, space, time, and energy. *American Zoologist*, 14, 221–48.

Altmann, S. A. (1984). What is the dual of the energy-maximization problem? *American Naturalist*, 123, 433–41.

Altmann, S. A. (1991). Diets of yearling female primates (*Papio cynocephalus*) predict lifetime fitness. *Proceedings of the National Academy of Sciences of the United States of America*, 88, 420–23.

Altmann, S. A. (1998). *Foraging for Survival: Yearling Baboons in Africa*. University of Chicago Press.

Altmann, S. A. (2006). Primate foraging adaptations: Two research strategies. In G. Hohmann, M. M. Robbins, and C.

Boesch (Eds.), *Feeding Ecology in Apes and Other Primates* (pp. 243–62). Cambridge University Press.

Altmann, S. A. (2009). Fallback foods, eclectic omnivores, and the packaging problem. *American Journal of Physical Anthropology*, 140, 615–29.

Altmann, S. A., and Altmann, J. (1970). *Baboon Ecology*. University of Chicago Press.

Altmann, S. A., and Wagner, S. S. (1978). A general model of optimal diet. *Recent Advances in Primatology*, 4, 407–14.

Alvarez, M. D., Saunders, D. E. J., Vincent J. F V., and Jeronimidis, G. (2000). An engineering method to evaluate the crisp texture of fruit and vegetables. *Journal of Texture Studies*, 348, 363–372.

Amato, K. R. (2016). Incorporating the gut microbiota into models of human and non-human primate ecology and evolution. *American Journal of Physical Anthropology*, 159, 196–215.

Amato, K. R., Clayton, J. B., and Hale, V. L. (2022) The Colobine gut microbiota: New perspectives on the nutrition and health of a specialized subfamily of primates. In I. Matsuda, C. G. Grueter, and J. A. Teichroeb (Eds), *The Colobines: Natural History, Behaviour and Ecological Diversity*. Cambridge University Press (pp. 78–93).

Amato, K. R., Leigh, S. R., Kent, A., Mackie, R. I., Yeoman, C. J., Stumpf, R. M., et al. (2014). The role of gut microbes in satisfying the demands of adult female and juvenile wild, black howler monkeys (*Alouatta pigra*). *American Journal of Physical Anthropology*, 155, 652–64.

Amato, K. R., Leigh, S. R., Kent, A., Mackie, R. I., Yeoman, C. J., Stumpf, R. M., et al. (2015). The gut microbiota appears to compensate for seasonal diet variation in the wild black howler monkey (*Alouatta pigra*). *Microbial Ecology*, 69, 434–43.

Amato, K. R., Mallot, E. K., Lambert, J. E., McDonald, D., Gomez, A., Metcalf, J. L., et al. (2019). Convergence of human and Old World monkey gut microbiomes demonstrates the importance of human ecology over phylogeny. *Genome Biology*, 20, 201.

Amato, K. R., Martinez-Mota, R., Righini, N., Raguet-Schofield, M. L., Corcione, F. P., Marini, E., et al. (2016). Phylogenetic and ecological factors impact the gut microbiota of two Neotropical primate species. *Oecologia*, 180(3), 717–33.

Amato, K. R., Sanders, J. G., Song, S. J., Nute, M., Metcalf, J. L., Thompson, L. R., et al. (2018). Evolutionary trends in host physiology outweigh dietary niche in structuring primate gut microbiomes. *ISME*, 13, 576–87.

Amato, K. R., Ulanov, A., Ju, K. S., and Garber, P. A. (2017). Metabolomic data suggest regulation of black howler monkey (*Alouatta pigra*) diet composition at the molecular level. *American Journal of Primatology*, 79, e22616.

Amato, K. R., Yeoman, C. J., Kent, A., Carbonero, F., Righini, N., Estrada, A. E., et al. (2013). Habitat degradation impacts primate gastrointestinal microbiomes. *ISME Journal*, 7, 1344–53.

Ambrose, S. H. (2001). Paleolithic technology and human evolution. *Science*, 291, 1748–53.

Ambrose, S. H., and DeNiro, M. J. (1986). The isotopic ecology of East African mammals. *Oecologia*, 69, 395–406.

Ambrose, S. H., and Norr, L. (1993). Experimental evidence for the relationship of the carbon isotope ratios of whole diet and dietary protein to those of bone collagen and carbonate. In P. D. J. B. Lambert and P. D. G. Grupe (Eds.), *Prehistoric Human Bone* (pp. 1–37). Springer, Berlin Heidelberg.

Ambrose, S. H., Butler, B. M., Hanson, D. B., Hunter-Anderson, R. L., and Krueger, H. W. (1997). Stable isotopic analysis of human diet in the Marianas Archipelago, Western Pacific. *American Journal of Physical Anthropology*, 104, 343–61.

Ament, S. A., Yang, Y., and Robinson, G. E. (2010). Nutritional regulation of division of labor in honey bees: Towards a systems biology perspective. *Wiley Interdisciplinary Reviews: Systems Biology and Medicine*, 2, 566–76.

Amundson, R., Austin, A. T., Schuur, E. A. G., Yoo, K., Matzek, V., Kendall, C., et al. (2003). Global patterns of the isotopic composition of soil and plant nitrogen. *Global Biogeochemical Cycles*, 17, 1031.

Andam, K. S., Ferraro, P. J., Pfaff, A., Sanchez-Azofeifa, G. A., and Robalino, J. A. (2008). Measuring the effectiveness of protected area networks in reducing deforestation. *Proceedings of the National Academy of Sciences of the United States of America*, 105, 16089–94.

Andelman, S. J. (1986). Ecological and social determinants of cercopithecine mating patterns. In D. R. Rubenstein and R. W. Wrangham (Eds.), *Ecological Aspects of Social Evolution: Birds and Mammals* (pp. 201–16). Princeton University Press.

Andelt, W. F., Kie, J. G., Knowlton, F. F., and Cardwell, K. (1987). Variation in coyote diets associated with season and successional changes in vegetation. *Journal of Wildlife Management*, 51, 273–77.

Anderson, C. R., and Lindzey, F. G. (2003). Estimating cougar predation rates from GPS location clusters. *Journal of Wildlife Management*, 67, 307–16.

Anderson, D. M. W., Weiping, W., and Lewis, G. P. (1990). The composition and properties of eight gum exudates (Leguminosae) of American origin. *Biochemical Systematics and Ecology*, 18, 39–42.

Anderson, D. P., Nordheim, E. V., Boesch, C., and Moermond, T. C. (2002). Factors influencing fission-fusion grouping in chimpanzees in the Tai National Park, Cote d'Ivoire. In C. Boesch, G. Hohmann, and L. Marchant (Eds.), *Behavioral Diversity in Chimpanzees and Bonobos* (pp. 90–101). Cambridge University Press.

Anderson, D. P., Nordheim, E. V., Moermond, T. C., Gone Bi, Z. B., and Boesch, C. (2005). Factors influencing tree phenology in Tai National Park, Cote d'Ivoire. *Biotropica*, 37, 631–40.

Anderson, H. B., Emery Thompson, M., Knott, C. D., and Perkins, L. (2008). Fertility and mortality patterns of captive

Bornean and Sumatran orangutans: Is there a species difference in life history? *Journal of Human Evolution*, 54, 34–42.

Anderson, J., Konz, E., and Jenkins, D. (2000). Health advantages and disadvantages of weight reducing diets: A computer analysis and critical review. *Journal of the American College of Nutrition*, 19, 578–90.

Anderson, S. M., Rudolph, M. C., McManaman, J. L., and Neville, M. C. (2007). Key stages in mammary gland development. Secretory activation in the mammary gland: It's not just about milk protein synthesis! *Breast Cancer Research*, 9, 204.

Andlauer, W., Martena, M. J., and Furst, P. (1999). Determination of selected phytochemicals by reversed-phase high-performance liquid chromatography combined with ultraviolet and mass spectrometric detection. *Journal of Chromatography A*, 849, 341–48.

Andreas, N. J., Kampmann, B., and Le-Doare, K. M. (2015). Human breast milk: A review on its composition and bioactivity. *Early Human Development*, 91, 629–35.

Andrew, N. R., Hill, S. J., Binns, M., Habibullah-Bahar, M. D., Ridley, E. V., Jung, M. P., et al. (2013). Assessing insect responses to climate change: What are we testing for? Where should we be heading? *Peer Journal*, 1, e11.

Ang, K. Y., Lucas, P. W., and Tan, H. T. W. (2008). Novel way of measuring the fracture toughness of leaves and other thin films using a single inclined razor blade. *New Phytologist*, 177, 830–37.

Ann, D. K., and Lin, H. H. (1993). Macaque salivary proline-rich protein: Structure, evolution, and expression. *Critical Reviews in Oral Biology and Medicine*, 4, 545–55.

Anson, J. R., Dickman, C. R., Boonstra, R., and Jessop, T. S. (2013). Stress triangle: Do introduced predators exert indirect costs on native predators and prey? *PLoS ONE*, 8, e60919.

Anthony, M. R. L., and Kay, R. F. (1993). Tooth form and diet in ateline and alouattine primates: Reflections on the comparative method. *American Journal of Science*, 293A, 356–82.

Antonow-Schlorke, I., Schwab, M., Cox, L. A., Li, C., Stuchlik, K., Witte, O. W., et al. (2011). Vulnerability of the fetal primate brain to moderate reduction in maternal global nutrient availability. *Proceedings of the National Academy of Sciences of the United States of America*, 108, 3011–16.

Apelt, U. (1995). *Zum Einfluss von Alter, Rang und Verwandtschaft auf die Sozialbeziehungen weiblicher Hanuman Languren (Presbytis entellus) in Ramnagar, Nepal*. Georg-August-Universitaet.

Apostolopoulou, A., Haidich, A. B., Kofina, K., Manzanares, W., Bouras, E., Tsaousi, G., et al. (2020). Effects of glutamine supplementation on critically ill patients: Focus on efficacy and safety—an overview of systematic reviews. *Nutrition*, 78, 110960.

Arago, L. E. O. C., Malhi, Y., Metcalfe, D. B., Silva-Espejo, J. E., Jimenez, E., Navarrete, D., et al. (2009). Above- and belowground net primary productivity across ten Amazonian forests on contrasting soils. *Biogeosciences*, 6, 2759–78.

Archie, E. A., Altmann, J., and Alberts, S. C. (2014). Costs of reproduction in a long-lived female primate: Injury risk and wound healing. *Behavioral Ecology and Sociobiology*, 68, 1183–93.

Arekar, K., Sambandam, S., and Karanth, P. K. (2021). Integrative taxonomy confirms the species status of the Himalayan langurs, *Semnopithecus schistaceus* Hodgson, 1840. *Journal of Zoological Systematics and Evolutionary Research*, 59, 543–56.

Arias-Del Razo, I., Hernandez, L., Laundré, J. W., and Velasco-Vazquez, L. (2012). The landscape of fear: Habitat use by a predator (*Canis latrans*) and its main prey (*Lepus californicus* and *Sylvilagus audubonii*). *Canadian Journal of Zoology*, 90, 683–93.

Aristizabal, J. F., Rothman, J. M., García-Fería, L. M., and Serio-Silva, J. C. (2016). Contrasting time-based and weight-based estimates of protein and energy intake of black howler monkeys (*Alouatta pigra*). *American Journal of Primatology*, 79, e22611.

Arlet, M. E., and Isbell, L. A. (2009). Variation in behavioral and hormonal responses of adult male gray-cheeked mangabeys (*Lophocebus albigena*) to crowned eagles (*Stephanoaetus coronatus*) in Kibale National Park, Uganda. *Behavioral Ecology and Sociology*, 63, 491.

Arlet, M. E., Isbell, L. A., Kaasik, A., Molleman, F., Chancellor, R. L., Chapman, C. A., et al. (2015). Determinants of reproductive performance among female gray-cheeked mangabeys (*Lophocebus albigena*) in Kibale National Park, Uganda. *International Journal of Primatology*, 36, 55–73.

Arnold, C., Matthews, L. J., and Nunn, C. L. (2010). The 10kTrees website: A new online resource for primate phylogeny. *Evolutionary Anthropology*, 19, 114–18.

Arnold, K., and Zuberbühler, K. (2006). The alarm-calling system of adult male putty-nosed monkeys, *Cercopithecus nictitans martini*. *Animal Behaviour*, 72, 643–53.

Arnold, K., Pohlner, Y., and Zuberbühler, K. (2008). A forest monkey's alarm call series to predator models. *Behavioral Ecology and Sociobiology*, 62, 549–59.

Arnold, S. J. (1994). Multivariate inheritance and evolution: A review of concepts. In C. R. B. Boake (Ed.), *Quantitative Genetic Studies of Behavioral Evolution* (pp. 17–48). University of Chicago Press.

Aronsen, G. P., Beuerlein, M. M., Watts, D. P., and Bribiescas, R. G. (2015). Redtail and red colobus monkeys show intersite urinary cortisol concentration variation in Kibale National Park, Uganda. *Conservation Physiology*, 3, cov006.

Arrieta, M. C., Arévalo, A., Stiemsma, L., Dimitriu, P., Chico, M. E., Loor, S., et al. (2017). Associations between infant fungal and bacterial dysbiosis and childhood atopic wheeze in a nonindustrialized setting. *Journal of Allergy Clinical Immunology*, 142, 424–34.

Arseneau-Robar, T. J. M., Anderson, K. A., Vasey, E. N., Sicotte, P., and Teichroeb, J. A. (2022). Think fast! Vervet monkeys assess the risk of being displaced by a dominant competitor when making foraging decisions. *Frontiers in Ecology and Evolution*, 10, 1–14.

Artis, D. (2008). Epithelial-cell recognition of commensal bacteria and maintenance of immune homeostasis in the gut. *Nature Reviews in Immunology*, 8, 411–20.

Asensio, N., Brockelman, W. Y., Malaivijitnond, S., and Reichard, U. H. (2014). White-handed gibbon (*Hylobates lar*) core area use over a short-time scale. *Biotropica*, 46, 461–69.

Asensio, N., Brockelman, W., Malaivijitnond, S., and Reichard, U. (2011). Gibbon travel paths are goal oriented. *Animal Cognition*, 14, 395–405.

Asensio, N., Korstjens, A. H., and Aureli, F. (2009). Fissioning minimizing ranging costs in spider monkeys: A multiple-level approach. *Behavioral Ecology and Sociology*, 63, 649–59.

Ashby, M. F., and Jones, D. R. H. (2006). *Engineering Materials 2*. 3rd ed. Butterworth-Heinemann.

Ashby, M. F., and Jones, D. R. H. (2011). *Engineering Materials 1*. 4th ed. Butterworth-Heinemann.

Ashton, P. S. (1988). Dipterocarp biology as a window to the understanding of tropical forest structure. *Annual Review of Ecology and Systematics*, 19, 347–70.

Ashton, P. S., Givnish, T. J., and Appanah, S. (1988). Staggered flowering in the Dipterocarpaceae: New insights into floral induction and the evolution of mast fruiting. *American Naturalist*, 132, 44–66.

Asner, G. P., and Martin, R. E. (2009). Airborne spectranomics: Mapping canopy chemical and taxonomic diversity in tropical forests. *Frontiers in Ecology and the Environment*, 7, 269–76.

Asner, G. P., Martin, R. E., Knapp, D. E., Tupayachi, R., Anderson, C., Carranza, L., et al. (2011). Spectroscopy of canopy chemicals in humid tropical forests. *Remote Sensing of Environment*, 115, 3587–98.

Asquith, P. J. (1989). Provisioning and the study of free-ranging primates: History, effects, and prospects. *Yearbook of Physical Anthropology*, 32, 129–58.

Association of Official Analytical Chemists. (1990). *Official Methods of Analysis of the Association of Official Analytical Chemists* (15th ed.). Association of Analytical Chemists.

Astaras, C., Mühlenberg, M., and Waltert, M. (2008). Note on drill (*Mandrillus leucophaeus*) ecology and conservation status in Korup National Park, Southwest Cameroon. *American Journal of Primatology*, 70, 306–10.

Atarashi, K., Nishimura, A., Shima, T., Umesaki, Y., Yamamoto, M., Onoue, M., et al. (2008). ATP drives lamina propria T(H)17 cell differentiation. *Nature*, 455(7214), 808–12.

Atickem, A., Loe, L. E., and Stenseth, N. C. (2014). Individual heterogeneity in use of human shields by mountain nyala. *Ethology*, 120, 715–25.

Atkins, A. G. (1982). Topics in indentation hardness. *American Metal Science*, 16, 127–37.

Atkins, A. G. (2009). *The Science and Engineering of Cutting*. Elsevier.

Atkins, A. G., and Liu, J. H. (2007). Toughness and the transition between cutting and rubbing in abrasive contacts. *Wear*, 262, 146–59.

Atkins, A. G., and Vincent, J. F. V. (1984). An instrumented microtome for improved histological sections and the measurement of fracture toughness. *Journal of Materials Science Letters*, 3, 310–12.

Atkinson, C., Frankenfeld, C. L., and Lampe, J. W. (2005). Gut bacterial metabolism of the soy isoflavone daidzein: Exploring the relevance to human health. *Experimental Biology and Medicine*, 230, 155–70.

Atkinson, S., Alston-Mills, B., Lönnerdal, B.O., and Neville, M. C. (1995). Major minerals and ionic constituents of human and bovine milks. In Handbook of milk composition (pp. 593–622). Academic Press.

Atsalis, S. (1999). Seasonal fluctuations in body fat and activity levels in a rain-forest species of mouse lemur, *Microcebus rufus*. *International Journal of Primatology*, 20, 883–910.

Au, D. W., and Pitman, R. L. (1986). Seabird interactions with dolphins and tuna in the eastern tropical Pacific. *Condor*, 88, 304–17.

Aureli, F., Schaffer, C. A., Boesch, C., Bearder, S. K., Call, J., Chapman, C. A., et al. (2008). Fission-fusion dynamics: New research frameworks. *Current Anthropology*, 49, 627–54.

Ausman, L. M., and Hayes, K. (1974). Vitamin E deficiency anemia in Old and New World monkeys. *American Journal of Clinical Nutrition*, 27(10), 1141–51.

Austin, C., Smith, T. M., Bradman, A., Hinde, K., Joannes-Boyau, R., Bishop, D., et al. (2013). Barium distributions in teeth reveal early-life dietary transitions in primates. *Nature*, 498, 216–19.

Austin, P. J., Suchar, L. A., Robbins, C. T., and Hagerman, A. E. (1989). Tannin-binding proteins in saliva of deer and their absence in saliva of sheep and cattle. *Journal of Chemical Ecology*, 15, 1335–47.

Avitsur, R., Padgett, D. A., and Sheridan, J. F. (2006). Social interactions, stress, and immunity. *Neurologic Clinics*, 24(3), 483–91.

Ayers, A. M., Allan, A. T. L., Howlett, C., Tordiffe, A. S. W., Williams, K. S., Williams, S. T., et al. (2020). Illuminating movement: Nocturnal activity patterns in chacma baboons (*Papio ursinus*). *Journal of Zoology*, 310, 287–97.

Ayliffe, L. K., Cerling, T. E., Robinson, T., West, A. G., Sponheimer, M., Passey, B. H., et al. (2004). Turnover of carbon isotopes in tail hair and breath CO_2 of horses fed an isotopically varied diet. *Oecologia*, 139, 11–22.

Ayres, J. S., and Schneider, D. S. (2009). The role of anorexia in resistance and tolerance to infections in drosophila. *PLoS Biology*, 7(7) e1000150.

Ayres, M. P., Clausen, T. P., MacLean, S. F., Redman, A. M., and Reichardt, P. B. (1997). Diversity of structure and antiherbivore activity in condensed tannins. *Ecology*, 78, 1696–712.

Azad, M. B., Nickel, N. C., Bode, L., Brockway, M., Brown, A., Chambers, C., et al. (2021). Breastfeeding and the origins of health: Interdisciplinary perspectives and priorities. *Maternal and Child Nutrition*, 17(2), e13109.

Baden, A. L., Webster, T. H., and Kamilar, J. M. (2016). Resource seasonality and reproduction predict fission–fusion dynamics in black-and-white ruffed lemurs (*Varecia variegata*). *American Journal of Primatology*, 78, 256–79.

Bădescu, I., Katzenberg, M. A., Watts, D. P., and Sellen, D. W. (2017). A novel fecal stable isotope approach to determine the timing of age-related feeding transitions in wild infant chimpanzees. *American Journal of Physical Anthropology*, 162, 285–99.

Bădescu, I., Watts, D. P., Katzenberg, M. A., and Sellen, D. W. (2022). Maternal lactational investment is higher for sons in chimpanzees. *Behavioral Ecology and Sociobiology*, 76(3), 1–19.

Badrian, N., and Malenky, R. K. (1984). Feeding ecology of *Pan paniscus* in the Lomako Forest, Zaire. In R. L. Susman (Ed.), *The Pygmy Chimpanzee: Evolutionary Biology and Behavior* (pp. 275–99). Plenum.

Badrian, N., Badrian, A., and Susman, R. L. (1981). Preliminary observations on the feeding behavior of *Pan paniscus* in the Lomako forest of central Zaïre. *Primates*, 22, 173–81.

Badyaev, A. V. (2002). Growing apart: An ontogenetic perspective on the evolution of sexual size dimorphism. *Trends in Ecology and Evolution*, 17(8), 369–78.

Bailey, I., Myatt, J. P., Wilson, A.M.2013. Group hunting within the Carnivora: Physiological, cognitive and environmental influences on strategy and cooperation. *Behavioral Ecology and Sociobiology*, 67, 1–17.

Bailey, M., and Coe, C. L. (1999). Maternal separation disrupts the integrity of the intestinal microflora in infant rhesus monkeys. *Developmental Psychobiology*, 35(2), 146–55.

Bailey, M., West, K. P., and Black, R. A. (2015). The epidemiology of global micronutrient deficiencies. *Annals of Nutrition and Metabolism*, 66, 22–33.

Bailey, R. W. (1973). Water in herbage. In G. W. Butler and R. W. Bailey (Eds.), *Chemistry and Biochemistry of Herbage* (Vol. 2). (pp. 13–24) Academic Press.

Bailey, R. W., and Ulyatt, M. J. (1970). Pasture quality and ruminant nutrition, II. Carbohydrate and lignin composition of detergent extracted residues from pasture grasses and legumes. *Journal of Agricultural Research*, 13, 591–604.

Baker, A. J., and Woods, F. (1992). Reproduction of the emperor tamarin (*Saguinus imperator*) in captivity, with comparisons to cotton-top and golden lion tamarins. *American Journal of Primatology*, 26, 1–10.

Baker, G., Jones, L. H. P., and Wardrop, I. D. (1959). Cause of wear in sheep's teeth. *Nature*, 184(4698), 1583–84.

Balasse, M. (2002). Reconstructing dietary and environmental history from enamel isotopic analysis: Time resolution of intra-tooth sequential sampling. *International Journal of Osteoarchaeology*, 12, 155–65.

Balcomb, S. R., Chapman, C. A., and Wrangham, R. W. (2000). Relationship between chimpanzee (*Pan troglodytes*) density and large, fleshy-fruit tree density: Conservation implications. *American Journal of Primatology*, 51, 197–203.

Bales, K., O'Herron, M., Baker, A. J., and Dietz, J. M. (2001). Sources of variability in numbers of live births in wild golden lion tamarins (*Leontopithecus rosalia*). *American Journal of Primatology*, 54, 211–21.

Ballard, O., and Morrow, A. L. (2013). Human milk composition: Nutrients and bioactive factors. *Pediatric Clinics of North America*, 60, 49–74.

Ballare, E. F. (2021). *Health Effects of Rehabilitation and Reintroduction in Bornean Orangutans (Pongo pygmaeus wurmbii)*. Rutgers, the State University of New Jersey.

Ballhorn, D. J., Kautz, S., and Rakotoarivelo, F. P. (2009). Quantitative variability of cyanogenesis in *Cathariostachys madagascariensis*—the main food plant of bamboo lemurs in southeastern Madagascar. *American Journal of Primatology*, 71, 305–15.

Balter, V. (2004). Allometric constraints on Sr/Ca and Ba/Ca partitioning in terrestrial mammalian trophic chains. *Oecologia*, 139, 83–88.

Ban, S. D., Boesch, C., and Janmaat, K. R. L. (2014). Taï chimpanzees anticipate revisiting high-valued fruit trees from further distances. *Animal Cognition*, 17, 1353–64.

Baniel, A., Petrullo, L., Mercer, A., Reitsema, L., Sams, S., Beehner, J. C., et al. (2022). Maternal effects on early-life gut microbiota maturation in a wild nonhuman primate. *Current Biology*, 32, 4508–20.

Bankoff, R. J., Jerjos, M., Hohman, B., Lauterbur, M. E., Kistler, L., and Perry, G. H. (2017). Testing convergent evolution in auditory processing genes between echolocating mammals and the aye-aye, a percussive-foraging primate. *Genome Biology and Evolution*, 9, 1978–89.

Banuls, C., de Maranon, A. M., Veses, S., Castro-Vega, I., Lopez-Domenech, S., Salom-Vendrell, C., et al. (2019). Malnutrition impairs mitochondrial function and leukocyte activation. *Nutrition Journal*, 18(1), 89.

Barbour, M. G. (2007). Stable oxygen isotope composition of plant tissue: A review. *Functional Plant Biology*, 34, 83–94.

Barboza, M., Pinzon, J., Wickramasinghe, S., Froehlich, J. W., Moeller, I., Smilowitz, J. T., et al. (2012). Glycosylation of human milk lactoferrin exhibits dynamic changes during early lactation enhancing its role in pathogenic bacteria-host interactions. *Molecular and Cellular Proteomics*, 11, M111-015248.

Barboza, P. S., Parker, K. L., and Hume, I. D. (2009). *Integrative Wildlife Nutrition*. Springer.

Bark, S., Holm, I., Hakansson, I., and Wretlind, A. (1976). Nitrogen-sparing effect of fat emulsion compared with glucose in the postoperative period. *Acta Chirurgica Scandinavica*, 142(6), 423–27.

Barks, S. K., Calhoun, M. E., Hopkins, W. D., Cranfield, M. R., Mudakikwa, A., Stoinski, T. et al. (2015). Brain organization of gorillas reflects species differences in ecology. *American Journal of Physical Anthropology*, 156, 252–62.

Barnard, C. J., and Sibly, R. M. (1981). Producers and scroungers: a general model and its application to captive flocks of house sparrows. *Animal Behaviour*, 29, 543–50.

Barnard, C. J., Thompson, D. B. A., and Stephens, H. (1982). Time budgets, feeding efficiency and flock dynamics in mixed species flocks of lapwings, golden plovers and gulls. *Behaviour*, 80, 44–69.

Barnett, A. A., Almeida, T., Spironello, W. R., Silva, W. S., MacLarnon, A., and Ross, C. (2012). Terrestrial foraging by *Cacajao melanocephalus ouakary* (Primates) in Amazonian Brazil: Is choice of seed patch size and position related to predation risk? *Folia Primatologica*, 83, 126–39.

Barnett, A., Boyle, S., Pinto, L., Lourenco, W., Almeida, T., Silva, W., et al. (2012). Primary seed dispersal by three Neotropical seed-predating primates (*Cacajao melanocephalus ouakary*, *Chiropotes chiropotes* and *Chiropotes albinasus*). *Journal of Tropical Ecology*, 28, 543–55.

Barnett, A. A., Ronchi-Teles, B., Almeida, T., Deveny, A., Schiel-Baracuhy, V., Sousa Silva, W., et al. (2013). Arthropod predation by a specialist seed predator, the golden-backed uacari (*Cacajao melanocephalus ouakary*, Pitheciidae) in Brazilian Amazonia. *International Journal of Primatology*, 34, 470–85.

Barnett, A., Ronchi-Teles, B., Silva, W., Andrade, R., Almeida, T., Bezerra, B., et al. (2017). Covert carnivory? A seed-predating primate, the golden-backed uacari, shows preferences for insect-infested fruits. *Journal of Zoological Research*, 1, 16–31.

Barnosky, A. D., Hadly, E. A., Bascompte, J., Berlow, E. L., Brown, J. H., Fortelius, M., et al. (2012). Approaching a state shift in Earth's biosphere. *Nature*, 486, 52–58.

Barrickman, N. L., Bastian, M. L., Isler, K., and van Schaik, C. P. (2008). Life history costs and benefits of encephalization: a comparative test using data from long-term studies of primates in the wild. *Journal of Human Evolution*, 54, 568–590.

Barraclough, D. J., Conroy, M. L., and Lee, D. (2004). Prefrontal cortex and decision making in a mixed strategy game. *Nature Neuroscience*, 7, 404–10.

Barrett, B. J., Monteza-Moreno, C. M., Dogandžić, T., Zwyns, N., Ibáñez, A., and Crofoot, M. C. (2018). Habitual stone-tool-aided extractive foraging in white-faced capuchins, *Cebus capucinus*. *Royal Society Open Science*, 5, 181002.

Barrett, L., Dunbar, R. I. M., and Dunbar, P. (1995). Sources of variability in numbers of live births in wild golden lion tamarins (*Leontopithecus rosalia*). *Animal Behaviour*, 49, 805–10.

Barrett, L., Halliday, J., and Henzi, S. P. (2006). The ecology of motherhood: The structuring of lactation costs by chacma baboons. *Journal of Animal Ecology*, 75, 875–86.

Barrett, L., Henzi, S. P., and Lycett, J. E. (2006). Whose life is it anyway? Maternal investment, developmental trajectories, and life history strategies in baboons. In L. Swedell and S. R. Leigh (Eds.), *Reproduction and Fitness in Baboons: Behavioral, Ecological, and Life History Perspectives* (pp. 199–224). Springer.

Barri, F. R., Roldan, N., Navarro, J. L., and Martella, M. B. (2012). Effects of group size, habitat and hunting risk on vigilance and foraging behaviour in the Lesser Rhea (*Rhea pennata pennata*). *Emu*, 112, 67–70.

Bartlett, T. Q., Sussman, R. W., and Cheverud, J. M. (1993). Infant killing in primates: A review of observed cases with specific reference to the sexual selection hypothesis. *American Anthropologist*, 95, 958–90.

Barton, R. A. (2006). Olfactory evolution and behavioral ecology in primates. *American Journal of Primatology*, 6, 545–58.

Barton, R. A., Byrne, R. W., and Whiten, A. (1996). Ecology, feeding competition and social structure in baboons. *Behavioral Ecology and Sociobiology*, 38, 321–29.

Barton, R. A., and Capellini, I. (2011). Maternal investment, life histories, and the costs of brain growth in mammals. *Proceedings of the National Academy of Sciences of the United States of America*, 108, 6169–74.

Barton, R. A., and Whiten, A. (1993). Feeding competition among female olive baboons *Papio anubis*. *Animal Behavior*, 46, 777–89.

Barton, R. A., and Whiten, A. (1994). Reducing complex diets to simple rules—food selection by olive baboons. *Behavioral Ecology and Sociobiology*, 35, 283–93.

Barton, R. A., Whiten, A., Strum, S. C., Byrne, R. W., and Simpson, R. J. (1992). Habitat use and resource availability in baboons. *Animal Behaviour*, 42, 831–44.

Basabose, K., and Yamagiwa, J. (1997). Predation on mammals by the chimpanzees in the montane forest of Kahuzi. *Primates*, 38, 45–55.

Basille, M., Fortin, D., Dussault, C., Bastille-Rousseau, G., Ouellet, J. P., and Courtois, R. (2015). Plastic response of fearful prey to the spatiotemporal dynamics of predator distribution. *Ecology*, 96, 2622–31.

Basset, Y., Aberlenc, H. P., and Delvare, G. (1992). Abundance and stratification of foliage arthropods in a lowland rain forest of Cameroon. *Ecological Entomology*, 17, 310–18.

Bassili, H. R., and Deitel, M. (1981). Nutritional support in long term intensive care with special reference to ventilator patients: A review. *Canadian Anesthesia Society Journal*, 28(1), 17–21.

Bastard, J. P., Maachi, M., Lagathu, C., Kim, M. J., Caron, M., Vidal, H., et al. (2006). Recent advances in the relationship between obesity, inflammation, and insulin resistance. *European Cytokine Network*, 17(1), 4–12.

Bastos, M., Madeiros, K., Jones, G., and Bezerra, B. (2018). Small but wise: Common marmosets (*Callithrix jacchus*) use acoustic signals as cues to avoid interactions with blonde capuchin monkeys (*Sapajus flavius*). *American Journal of Primatology*, 80, e22744.

Bates, L. A., and Byrne, R. W. (2009). Sex differences in the movement patterns of free-ranging chimpanzees (*Pan troglodytes schweinfurthii*): Foraging and border checking. *Behavioral Ecology and Sociobiology*, 64, 247–55.

Batool, R., Butt, M. S., Sultan, M. T., Saeed, F., and Naz, R. (2015). Protein-energy malnutrition: A risk factor for various ailments. *Critical Reviews in Food Science and Nutrition*, 55(2), 242–53.

Bau, A. M., Ernert, A., Schenk, L., Wiegand, S., Martus, P., Grueters, A., et al. (2009). Is there a further acceleration in the age at onset of menarche? A cross-sectional study of 1840 school children focusing on age and bodyweight at the onset of menarche. *European Journal of Endocrinology*, 160, 107–13.

Bauer, E., Williams, B. A., Smidt, H., Verstegen, M. W., and Mosenthin, R. (2006). Influence of the gastrointestinal microbiota on development of the immune system in young animals. *Current Issues in Intestinal Microbiology*, 7, 35–51.

Bauer, M. E., and Fuente, M. (2016). The role of oxidative and inflammatory stress and persistent viral infections in immunosenescence. *Mechanisms of Aging and Development*, 158, 27–37.

Baya, L., and Storch, I. (2010). Status of diurnal primate populations at the former settlement of a displaced village in Cameroon. *American Journal of Primatology*, 72, 645–52.

Bazazi, S., Romanczuk, P., Thomas, S., Schimansky-Geier, L., Hale, J. J., Miller, G. A., et al. (2011). Nutritional state and collective motion: From individuals to mass migration. *Proceedings of the Royal Society B: Biological Sciences*, 278, 356–63.

Beale, C. M., and Monaghan, P. (2004). Human disturbance: people as predation-free predators? *Journal of Applied Ecology*, 41(2), 335–43.

Beale, P. K., Marsh, K. J., Foley, W. J., and Moore, B. D. (2018). A hot lunch for herbivores: Physiological effects of elevated temperatures on mammalian feeding ecology. *Biological Reviews*, 93, 674–92.

Beard, J. (2007). Recent evidence from human and animal studies regarding iron status and infant development. *Journal of Nutrition*, 137, 524S–30S.

Bearder, S. K., and Martin, R. D. (1980). Acacia gum and its use by bushbabies, *Galago senegalensis* (Primates: Lorisidae). *International Journal of Primatology*, 1, 103–28.

Bearder, S. K., Nekaris, K. A. I., and Buzzell, C. (2002). Dangers in the night: Are some nocturnal primates afraid of the dark? In L. Miller (Ed.), *Eat or Be Eaten: Predator Sensitive Foraging among Primates* (pp. 21–43). Cambridge University Press.

Bearder, S. K., Nekaris, K. A. I., and Curtis, D. J. (2006). A re-evaluation of the role of vision in the activity and communication of nocturnal primates. *Folia Primatologica*, 77, 50–71.

Beaudrot, L., Rejmánek, M., and Marshall, A. J. (2013). Dispersal modes affect tropical forest assembly across trophic levels. *Ecography*, 36, 984–93.

Beck, K. L., Weber, D., Phinney, B. S., Smilowitz, J. T., Hinde, K., Lönnerdal, B., et al. (2015). Comparative proteomics of human and macaque milk reveals species-specific nutrition during postnatal development. *Journal of Proteome Research*, 14, 2143–57.

Beck, V., Unterrieder, E., Krenn, L., Kubelka, W., and Jungbauer, A. (2003). Comparison of hormonal activity (estrogen, androgen and progestin) of standardized plant extracts for large scale use in hormone replacement therapy. *Journal of Steroid Biochemistry*, 84, 259–68.

Bedoya-Perez, M. A., Carthey, A. J. R., and Mella, V. S. A. (2013). A practical guide to avoid giving up on giving-up densities. *Behavioral Ecology and Sociobiology*, 67, 1541–1553.

Bedoya-Perez, M. A., Issa, D. D., Banks, P. B., and McArthur, C. (2014). Quantifying the response of free-ranging mammalian herbivores to the interplay between plant defense and nutrient concentrations. *Oecologia*, 175, 1167–77.

Beecher, G. R. (2003). Overview of dietary flavonoids: Nomenclature, occurrence and intake. *Journal of Nutrition*, 133, 3248s–54s.

Beede, D. K., and Collier, R. J. (1986). Potential nutritional strategies for intensively managed cattle during thermal stress. *Journal of Animal Science*, 62, 543–54.

Beehner, J. C., Onderdonk, D. A., Alberts, S. C., and Altmann, J. (2006). The ecology of conception and pregnancy failure in wild baboons. *Behavioral Ecology*, 17, 741–50.

Beghini, F., Pasolli, E., Truong, T. D., Putignani, L., Caccio, S. M., and Segata, N. (2017). Large scale comparative metagenomics of *Blastocystis*, a common member of the human gut microbiome. *ISME Journal*, 11, 2848–63.

Behie, A. M., and Pavelka, M. S. M. (2005). The short-term effects of a hurricane on the diet and activity of black howlers (*Alouatta pigra*) in Monkey River, Belize. *Folia Primatologica*, 76, 1–9.

Behie, A. M., and Pavelka, M. S. M. (2012a). Food selection in the black howler monkey following habitat disturbance: Implications for the importance of mature leaves. *Journal of Tropical Ecology*, 28, 153–60.

Behie, A. M., and Pavelka, M. S. M. (2012b). The role of minerals in food selection in a black howler monkey (*Alouatta pigra*) population in Belize following a major hurricane. *American Journal of Primatology*, 74, 1054–63.

Behie, A. M., and Pavelka, M. S. M. (2013). Interacting roles of diet, cortisol levels, and parasites in determining population density of Belizean howler monkeys in a hurricane damaged forest fragment. In L. K. Marsh and C. A. Chapman (Eds.), *Primates in Fragments: Complexity and Resilience* (pp. 447–56). Springer.

Behie, A. M., Teichroeb, J. A., and Malone, N. (2019). Changing priorities for primate conservation and research in the Anthropocene. In A. M. Behie and N. Malone (Eds.), *Primate Research and Conservation in the Anthropocene (pp.1–14)*. Cambridge University Press.

Behmer, S. T., Raubenheimer, D., and Simpson, S. J. (2001). Frequency-dependent food selection in locusts: A geometric analysis of the role of nutrient balancing. *Animal Behaviour*, 61, 995–1005.

Behmer, S., Cox, E., Raubenheimer, D., and Simpson, S. J. (2003). Food distance and its effect on nutrient balancing in a mobile insect herbivore. *Animal Behaviour*, 66, 665–75.

Behrens, M., and Meyerhof, W. (2013). Bitter taste sensitivity in humans and chimpanzees. In *Encyclopedia of Life Sciences*. John Wiley and Sons.

Behringer, V., Borchers, C., Deschner, T., Mostl, E., Selzer, D., and Hohmann, G. (2013). Measurements of salivary alpha amylase and salivary cortisol in hominoid primates reveal within-species consistency and between-species differences. *PLoS ONE*, 8, e60773.

Behringer, V., Deimel, C., Stevens, J. M. G., Kreyer, M., Lee, S. M., Hohmann, G., et al. (2021). Cell-mediated immune ontogeny is affected by sex but not environmental context in a long-lived primate species. *Frontiers in Ecology and Evolution*, 9, 12.

Behringer, V., Muller-Klein, N., Strube, C., Schulke, O., Heistermann, M., and Ostner, J. (2021). Responsiveness of fecal immunoglobulin A to HPA-axis activation limits its use for mucosal immunity assessment. *American Journal of Primatology*, 83(12), 13.

Behringer, V., Preis, A., Wu, D. F., Crockford, C., Leendertz, F. H., Wittig, R. M., et al. (2020). Urinary cortisol increases during a respiratory outbreak in wild chimpanzees. *Frontiers in Veterinary Science*, 7, 9.

Behringer, V., Stevens, J. M., Leendertz, F. H., Hohmann, G., and Deschner, T. (2017). Validation of a method for the assessment of urinary neopterin levels to monitor health status in non-human-primate species. *Frontiers in Physiology*, 8, 51.

Behringer, V., Stevens, J. M. G., Wittig, R. M., Crockford, C., Zuberbühler, K., Leendertz, F. H., et al. (2019). Elevated neopterin levels in wild, healthy chimpanzees indicate constant investment in unspecific immune system. *BMC Zoology*, 4, 7.

Beisel, W. R. (1975). Metabolic response to infection. *Annual Review of Medicine*, 26, 9–20.

Bekelman, T. A., Santamaria-Ulloa, C., Dufour, D. L., Marin-Arias, L., and Dengo, A. L. (2017). Using the protein leverage hypothesis to understand socioeconomic variation in obesity. *American Journal of Human Biology*, 29, e22953.

Bekelman, T. A., Santamaria-Ulloa, C., Dufour, D. L., Marin, L. I., and Dengo, A. L. (2015). Using the protein leverage hypothesis to understand obesity among urban Costa Rican women. *American Journal of Human Biology*, 27, 261–61.

Bekoff, M. (1972). The development of social interaction, play, and metacommunication in mammals: An ethological perspective. *Quarterly Review of Biology*, 47, 412–34.

Bell, A. V., Hinde, K., and Newson, L. (2013). Who was helping? The scope for female cooperative breeding in early *Homo*. *PLoS ONE*, 8, e83667.

Belovsky, G. E. (1978). Diet optimization in a generalist herbivore—the moose. *Theoretical Population Biology*, 14, 105–34.

Belovsky, G. E. (1986). Optimal foraging and community structure—implications for a guild of generalist grassland herbivores. *Oecologia*, 70, 35–52.

Belovsky, G. E. (1987). Hunter-gatherer foraging: A linear programming approach. *Journal of Anthropological Archaeology*, 6, 29–76.

Belovsky, G. E., and Jordan, P. A. (1978). Time-energy budget of a moose. *Theoretical Population Biology*, 14, 76–104.

Belwood, J. J. (1990). Anti-predator defences and ecology of Neotropical forest katydids, especially the Pseudophyllinae. In W. J. Bailey and D. C. F. Rentz (Eds.), *The Tettigoniidae: Biology, Systematics and Evolution* (pp. 8–26). Springer.

Bender, C. R., and Irwin, M. (2014). What's grit got to do with it: An analysis of grit accumulation in the canopy of a fragmented forest of Madagascar? *American Journal of Physical Anthropolgy* (Suppl. 58), 153, 76.

Ben-David, M., and Flaherty, E. A. (2012). Stable isotopes in mammalian research: A beginner's guide. *Journal of Mammalogy*, 93, 312–28.

Ben Shaul, D. M. (1963). The composition of the milk of wild animals. *International Zoo Yearbook*, 4, 333–42.

Benavidez, K. M., Harris, T. R., Wasserman, M. D., Chapman, C. A., Leitman, D. C., and Rothman, J. M. (2015). Intergroup variation in estrogenic plant consumption for the black-and-white colobus monkey of Kibale National Park, Uganda. *American Journal of Physical Anthropology*, 156, 83.

Benchimol, M., and Peres, C. A. (2013a). Anthropogenic modulators of species-area relationships in Neotropical primates: A continental-scale analysis of fragmented forest landscapes. *Diversity and Distributions*, 19, 1339–52.

Benchimol, M., and Peres, C. A. (2013b). Predicting primate local extinctions with "real-world" forest fragments: A Pan-Neotropical analysis. *American Journal of Primatology*, 76, 1–14.

Bendlin, B. B., Canu, E., Willette, A., Kastman, E. K., McLaren, D. G., Kosmatka, K. J., et al. (2011). Effects of aging and calorie restriction on white matter in rhesus macaques. *Neurobiology of Aging*, 32(12), 2319.e1–2319.e11.

Benhaiem, S., Delon, M., Lourtet, B., Cargnelutti, B., Aulagnier, S., Hewison, A. J. M., et al. (2008). Hunting increases vigilance levels in roe deer and modifies feeding site selection. *Animal Behaviour*, 76, 611–18.

Benie, T., and Thieulant, M. L. (2003). Interaction of some traditional plant extracts with uterine oestrogen or progestin receptors. *Phytotherapy Research*, 17, 756–60.

Benitez-Lopez, A., Alkemade, R., and Verweij, P. A. (2010). The impacts of roads and other infrastructure on mammal and bird populations: A meta-analysis. *Biological Conservation*, 143, 1307–16.

Bennett, A. F., and Gorman, G. C. (1979). Population density and energetics of lizards on a tropical island. *Oecologia*, 42, 339–58.

Bennett, E. L., and Dahaban, Z. (1995). Wildlife responses to disturbances in Sarawak and their implications for forest management. In R. B. Primack and T. E. Lovejoy (Eds.), *Ecology, Conservation, and Management of Southeast Asian Rainforests* (pp. 66–86). Yale University Press.

Bennett, E. L., Nyaoi, A., and Sompud, J. (2000). Saving Borneo's bacon: The sustainability of hunting in Sarawak and Sabah. In J. Robinson and E. Bennett (Eds.), *Hunting for Sustainability in Tropical Forests* (pp. 305–24). Columbia University Press.

Bennett, R. N., and Wallsgrove, R. M. (1994). Secondary metabolites in plant defense-mechanisms. *New Phytologist*, 127, 617–33.

Bennick, A. (2002). Interaction of plant polyphenols with salivary proteins. *Critical Reviews in Oral Biology and Medicine*, 13, 184–96.

Benoit, J., Fernandez, V., Manger, P. R., and Rubidge, B. S. (2017). Endocranial casts of pre-mammalian therapsids reveal an unexpected neurological diversity at the deep evolutionary root of mammals. *Brain, Behavior and Evolution*, 90(4), 311–33.

Benoit, J., Manger, P. R., and Rubidge, B. S. (2016). Palaeoneurological clues to the evolution of defining mammalian soft tissue traits. *Scientific Reports*, 6, 25604.

Bentley, G. R., Harrigan, A. M., and Ellison, P. T. (1998). Dietary composition and ovarian function among Lese horticulturalist women of the Ituri Forest, Democratic Republic of Congo. *European Journal of Clinical Nutrition*, 52, 261–70.

Bentley-Condit, V. K. (2009). Food choices and habitat use by the Tana River yellow baboons (*Papio cynocephalus*): A preliminary report on five years of data. *American Journal of Primatology*, 71, 732–36.

Beran, M. J. (2011). Chimpanzees (*Pan troglodytes*) show the isolation effect during serial list recognition memory tests. *Animal Cognition*, 14, 637–45.

Bercovitch, F. B. (1983). Time budgets and consortships in olive baboons (*Papio anubis*). *Folia Primatologica*, 41, 180–90.

Bercovitch, F. B., and Nürnberg, P. (1996). Socioendocrine and morphological correlates of paternity in rhesus macaques (*Macaca mulatta*). *Journal of Reproduction and Fertility*, 107, 59–68.

Berger, J. (2007). Fear, human shields and the redistribution of prey and predators in protected areas. *Biology Letters*, 3, 620–23.

Berger, M. M. (2014). The 2013 Arvid Wretlind lecture: Evolving concepts in parenteral nutrition. *Clinical Nutrition*, 33(4), 563–70.

Bergeson, D. J. (1998). Patterns of suspensory feeding in *Alouatta palliata*, *Ateles geoffroyi*, and *Cebus capucinus*. In E. Strasser, J. G. Fleagle, A. L. Rosenberger, and H. M. McHenry (Eds.), *Primate Locomotion* (pp. 45–60). Springer.

Bergstrom, M. L. (2015). *Seasonal Effects on the Nutrition and Energetic Condition of Female White-Faced Capuchin Monkeys*. University of Calgary.

Bergstrom, M. L., Hogan, J. D., Melin, A. D., and Fedigan, L. M. (2019). The nutritional importance of invertebrates to female *Cebus capucinus imitator* in a highly seasonal tropical dry forest. *American Journal of Physical Anthropology*, 170, 207–16.

Bergstrom, M. L., Kalbitzer, U., Campos, F. A., Melin, A. D., Emery Thompson, M., and Fedigan, L. M. (2020). Non-invasive estimation of the costs of feeding competition in a Neotropical primate. *Hormones and Behavior*, 118, 104632.

Bergstrom, M. L., Emery Thompson, M., Melin, A. D., and Fedigan, L. M. (2017). Using urinary parameters to estimate seasonal variation in the physical condition of female white-faced capuchin monkeys (*Cebus capucinus imitator*). *American Journal of Physical Anthropology*, 163(4), 707–15.

Berkey, C. S., Gardner, J. D., Frazier, A. L., and Colditz, G. A. (2000). Relation of childhood diet and body size to menarche and adolescent growth in girls. *American Journal of Epidemiology*, 152, 446–52.

Berman, C. M. (1989). Trapping activities and mother-infant relationships on Cayo Santiago: A cautionary tale. *Puerto Rico Health Sciences Journal*, 8, 73–78.

Bermejo, M., Illera, G., and Pi, J. S. (1994). Animals and mushrooms consumed by bonobos (*Pan paniscus*): New records from Liungu (Ikela), Zaire. *International Journal of Primatology*, 27, 879–98.

Bermejo, M., Rodríguez-Teijeiro, J. D., Illera, G., Barroso, A., Vilá, C., and Walsh, P. D. (2006). Ebola outbreak killed 5000 gorillas. *Science*, 314, 1564.

Bernard, A. B., and Marshall, A. J. (2020). Assessing the state of knowledge of contemporary climate change and primates. *Evolutionary Anthropology*, 29, 317–31.

Bernard, D. A. I., Kwabena, O. D., Osei, G. A., Daniel, S. A. E., and Sandra, A. (2014). The effect of different drying methods on the phytochemicals and radical scavenging activity of Ceylon cinnamon (*Cinnamomum zeylanicum*) plant parts. *European Journal of Medicinal Plants*, 4, 1324–35.

Berthaume, M. A. (2016). Food mechanical properties and dietary ecology. *American Journal of Physical Anthropology*, 159(suppl.), 79–104.

Bertram, J. E. (2004). New perspectives on brachiation mechanics. *American Journal of Physical Anthropology*, 39, 100–117.

Besold, A. N., Culbertson, E. M., and Culotta, V. C. (2016). The Yin and Yang of copper during infection. *Journal of Biological Inorganic Chemistry*, 21(2), 137–44.

Bessa, J., Sousa, C., and Hockings, K. J. (2015). Feeding ecology of chimpanzees (*Pan troglodytes verus*) inhabiting a forest-mangrove-savanna-agricultural matrix at Caiquene-Cadique, Cantanhez National Park, Guinea-Bissau. *American Journal of Primatology*, 77, 651–65.

Bhaskaram, P. (2001). Immunobiology of mild micronutrient deficiencies. *British Journal of Nutrition*, 85(S2), S75–S80.

Bhat, H. R. (1990). Additional information on tool use by lion-tailed macaques. *Lion-Tailed Macaque Newsletter*, 7, 6.

Bi, J., Knyazikhin, Y., Choi, S., Park, T., Barichivich, J., Ciais, P., et al. (2015). Sunlight mediated seasonality in canopy structure and photosynthetic activity of Amazonian rainforests. *Environmental Research Letters*, 10, 064014.

Bicca-Marques, J.C. (2003). How do howler monkeys cope with habitat fragmentation?. In: Marsh, L.K. (eds) Primates in Fragments. Springer, Boston, MA. (pp. 283-303).

Bicca-Marques, J. C., and Garber, P. A. (2003). Experimental field study of the relative costs and benefits to wild tamarins (*Saguinus imperator* and *S. fuscicollis*) of exploiting contestable food patches as single- and mixed-species troops. *American Journal of Primatology*, 60, 139–53.

Bicca-Marques, J. C., and Garber, P. A. (2004). Use of spatial, visual, and olfactory information during foraging in wild nocturnal and diurnal anthropoids: A field experiment comparing *Aotus, Callicebus,* and *Saguinus. American Journal of Primatology,* 62, 171–87.

Bicca-Marques, J. C., and Garber, P. A. (2005). Use of social and ecological information in tamarin foraging decisions. *International Journal of Primatology,* 26, 1321–44.

Bicca-Marques, J. C., Chaves, O. M., and Hass, G. P. (2020). Howler monkey tolerance to habitat shrinking: Lifetime warranty or death sentence? *American Journal of Primatology,* 82, e23089.

Bidart-Bouzat, M. G., and Imeh-Nathaniel, A. (2008). Global change effects on plant chemical defenses against insect herbivores. *Journal of Integrative Plant Biology,* 50, 1339–54.

Biebouw, K. (2009). *Revealing the Behavioural Ecology of the Elusive Hairy-Eared Dwarf Lemur* (Allocebus trichotis). Oxford Brookes University.

Bielby, J., Mace, G. M., Bininda-Emonds, O. R. P., Cardillo, M., Gittleman, J. L., Jones, K. E., et al. (2007). The fast-slow continuum in mammalian life history: An empirical reevaluation. *American Naturalist,* 169, 748–57.

Bila-Isa, I. (2003). Bonobos dig termite mounds: A field example of tool use by wild bonobos of the Etate, northern sector of the Salonga National Park. Bonobo Workshop: Behaviour, Ecology and Conservation of Wild Bonobos, Inuyama, Japan.

Bilenger, M. (1995). Chemical factors influencing food choice of howler monkey (*Alouatta palliata*). *Turkish Journal of Zoology,* 19, 291–303.

Binford, L. R. (1981). *Bones: Ancient Men and Modern Myths* (1st ed.). Academic Press.

Bingham, H. C. (1932). *Gorillas in a Native Habitat.* Carnegie Institution of Washington, 66 pp.

Biro, D., Haslam, M., and Rutz, D. (2013). Tool use as adaptation. *Philosophical Transactions of the Royal Society of London Series B—Biological Sciences,* 368, 20120408.

Biro, D., Inoue-Nakamura, N., Tonooka, R., Yamakoshi, G., Sousa, C., and Matsuzawa, T. (2003). Cultural innovation and transmission of tool use in wild chimpanzees: Evidence from field experiments. *Animal Cognition,* 6, 213–23.

Bishop, A. (1962). Control of the hand in lower primates. *Annals of the New York Academy of Sciences,* 102, 316–37.

Bissell, H. (2014). Nutritional implications of the high-elevation lifestyle of *Rhinopithecus bieti.* In N. B. Grow, S. Gursky-Doyen, and A. Krzton (Eds.), *High Altitude Primates* (pp. 199–210). Springer.

Bistrian, B. R., Blackburn, G. L., Scrimshaw, N. S., and Flatt, J. P. (1975). Cellular immunity in semistarved states in hospitalized adults. *American Journal of Clinical Nutrition,* 28(10), 1148–55.

Bistrian, B. R., George, D. T., Blackburn, G. L., and Wannemacher, R. W. (1981). The metabolic response to yellow fever immunization: Protein-sparing modified fast. *American Journal of Clinical Nutrition,* 34(2), 229–37.

Bistrian, B. R., Winterer, J. C., Blackburn, G. L., and Scrimshaw, N. S. (1977). Failure of yellow fever immunization to produce a catabolic response in individuals fully adapted to a protein-sparing modified fast. *American Journal of Clinical Nutrition,* 30(9), 1518–22.

Bjorkholm, B., Bok, C. M., Lundin, A., Rafter, J., Hibberd, M. L., and Pettersson, S. (2009). Intestinal microbiota regulate xenobiotic metabolism in the liver. *PLoS ONE,* 4, e6958.

Bjorndal, K. A. (1991). Diet mixing—nonadditive interactions of diet items in an omnivorous fresh-water turtle. *Ecology,* 72, 1234–41.

Black, R. E., Allen, L. H., Bhutta, Z. A., Caulfield, L. E., de Onis, M., Ezzati, M., et al. (2008). Maternal and child undernutrition: Global and regional exposures and health consequences. *Lancet,* 371, 243–60.

Blackburn, D. G., Hayssen, V., and Murphy, C. J. (1989). The origins of lactation and the evolution of milk: A review with new hypotheses. *Mammal Review,* 19, 1–26.

Blackburn, G. L., and Bistrian, B. R. (1976). Nutritional care of the injured and/or septic patient. *Surgical Clinics of North America,* 56(5), 1195–224.

Blackmore, H. L., and Ozanne, S. E. (2015). Programming of cardiovascular disease across the life-course. *Journal of Molecular and Cellular Cardiology,* 83, 122–30.

Blanco, M. B., Dausmann, K. H., Faherty, S., and Yoder, A. D. (2018). Tropical heterothermy is "cool": The expression of daily torpor and hibernation in primates. *Evolutionary Anthropology,* 27, 147–61.

Blanco, M. B., Greene, L. K., Schopler, R., Williams, C. V., Lynch, D., Browning, J., et al. (2021). On the modulation and maintenance of hibernation in captive dwarf lemurs. *Scientific Reports,* 11, 5740.

Bleher, B., Potgieter, C. J., Johnson, D. N., and Bohning-Gaese, K. (2003). The importance of figs for frugivores in South African coastal forest. *Journal of Tropical Ecology,* 19, 375–86.

Bleicher, S. S. (2017). The landscape of fear conceptual framework: Definition and review of current applications and misuses. *PeerJ,* 5, e3772.

Blok, W. L., Katan, M. B., and van der Meer, J. W. (1996). Modulation of inflammation and cytokine production by dietary (n-3) fatty acids. *Journal of Nutrition,* 126(6), 1515–33.

Blomquist, C. H., Lima, P. H., and Hotchkiss, J. R. (2005). Inhibition of 3 alpha-hydroxysteroid dehydrogenase (3 alpha-HSD) activity of human lung microsomes by genistein, daidzein, coumestrol and C-18-, C-19- and C-21-hydroxysteroids and ketosteroids. *Steroids,* 70, 507–14.

Blomquist, G. E., Hinde, K., and Newmark, L. A. M. (2017). Diet predicts milk composition in primates. *BioRxiv,* 2017, 197004.

Blumenthal, S. A., Chritz, K. L., Rothman, J. M., and Cerling, T. E. (2012). Detecting intraannual dietary variability in wild mountain gorillas by stable isotope analysis of feces. *Proceedings of the National Academy of Sciences of the United States of America,* 109, 21277–82.

Blumenthal, S. A., Rothman, J. M., Chritz, K. L., and Cerling, T. E. (2016). Stable isotopic variation in tropical forest plants for applications in primatology. *American Journal of Primatology*, 78, 1041–54.

Blumfield, M., Hure, A., MacDonald-Wicks, L., Smith, R., Simpson, S. J., Raubenheimer, D., et al. (2012). The association between the macronutrient content of maternal diet and the adequacy of micronutrients during pregnancy in the Women and Their Children's Health (WATCH) Study. *Nutrients*, 4, 1958–76.

Bo, S., Durazzo, M., Gambino, R., Berutti, C., Milanesio, N., Caropreso, A., et al. (2008). Associations of dietary and serum copper with inflammation, oxidative stress, and metabolic variables in adults. *Journal of Nutrition*, 138(2), 305–10.

Bocherens, H. (2009). Neandertal dietary habits: Review of the isotopic evidence. In J. J. Hublin and M. P. Richards (Eds.), *The Evolution of Hominin Diets: Integrating Approaches to the Study of Paleolithic Subsistence* (pp. 241–50). Springer.

Bocherens, H., Drucker, D. G., Billiou, D., Patou-Mathis, M., and Vandermeersch, B. (2005). Isotopic evidence for diet and subsistence pattern of the Saint-Césaire I Neanderthal: Review and use of a multi-source mixing model. *Journal of Human Evolution*, 49, 71–87.

Bock, C. E., and Jones, Z. F. (2004). Avian habitat evaluation: Should counting birds count? *Frontiers in Ecology and the Environment*, 2, 403–10.

Bode, L. (2012). Human milk oligosaccharides: Every baby needs a sugar mama. *Glycobiology*, 22, 1147–62.

Bodmer, R. E., Eisenberg, J. F., and Redford, K. H. (1997). Hunting and the likelihood of extinction of Amazonian mammals. *Conservation Biology*, 11, 460–66.

Boesch, C. (1978). New observations on chimpanzees of Tai Forest, Ivory Coast. *Terre et la Vie*, 32, 195–201.

Boesch, C. (1991a). The effects of leopard predation on grouping patterns in forest chimpanzees. *Behaviour*, 117, 221–42.

Boesch, C. (1991b). Teaching among wild chimpanzees. *Animal Behaviour*, 41, 530–32.

Boesch, C. (1994). Cooperative hunting in wild chimpanzees. *Animal Behaviour*, 48, 653–68.

Boesch, C. (2003). Is culture a golden barrier between human and chimpanzee? *Evolutionary Anthropology*, 12, 82–91.

Boesch, C., and Boesch, H. (1982). Optimisation of nut-cracking with natural hammers by wild chimpanzees. *Behaviour*, 83, 265–86.

Boesch, C., and Boesch, H. (1984). Mental maps in wild chimpanzees: An analysis of hammer transports for nut-cracking. *Primates*, 25, 160–70.

Boesch, C., and Boesch, H. (1989). Hunting behaviour of wild chimpanzees in the Taï National Park. *American Journal of Physical Anthropology*, 78, 547–73.

Boesch, C., and Boesch, H. (1990). Tool use and tool making in wild chimpanzees. *Folia Primatologica*, 54, 86–99.

Boesch, C., and Boesch-Achermann, H. (2000). *The Chimpanzees of the Tai Forest: Behavioural Ecology and Evolution*. Oxford University Press.

Boesch, C., Head, J., and Robbins, M. M. (2009). Complex tool sets for honey extraction among chimpanzees in Loango National Park, Gabon. *Journal of Human Evolution*, 56, 560–69.

Boesch, C., Kalan, A. K., Agbor, A., Arandjelovic, M., Dieguez, P., Lapeyre, V., et al. (2017). Chimpanzees routinely fish for algae with tools during the dry season in Bakoun, Guinea. *American Journal of Primatology*, 79, e22613.

Boesch, C., Marchesi, P., Marchesi, N., Fruth, B., and Joulian, F. (1994). Is nut cracking in wild chimpanzees a cultural behaviour? *Journal of Human Evolution*, 26, 325–38.

Boesch-Achermann, H., and Boesch, C. (1993). Tool use in wild chimpanzees: New light from dark forests. *Current Directions in Psychological Research*, 2(1), 18–21.

Bogart, S. L., and Pruetz, J. D. (2008). Ecological context of savanna chimpanzee (*Pan troglodytes verus*) termite fishing at Fongoli, Senegal. *American Journal of Primatology*, 70, 605–12.

Bogart, S. L., and Pruetz, J. D. (2011). Insectivory of savanna chimpanzees (*Pan troglodytes verus*) at Fongoli Senegal. *American Journal of Physical Anthropology*, 145, 11–20.

Bogin, B. (1997). Evolutionary hypotheses for human childhood. *Yearbook of Physical Anthropology*, 40, 63–89.

Boinski, S. (1987). Birth synchrony in squirrel monkeys (*Saimiri oerstedi*): A strategy to reduce neonatal predation. *Behavioral Ecology and Sociobiology*, 21, 393–400.

Boinski, S. (1988). Sex differences in the foraging behavior of squirrel monkeys in a seasonal habitat. *Behavioral Ecology and Sociobiology*, 32, 177–86.

Boinski, S. (1989). The positional behavior and substrate use of squirrel monkeys: Ecological implications. *Journal of Human Evolution*, 18, 659–77.

Boinski, S., and Chapman, C. A. (1995). Predation on primates: Where are we and what's next? *Evolutionary Anthropology*, 4, 1–3.

Boinski, S., and Fowler, N. (1989). Seasonal patterns in a tropical lowland forest. *Biotropica*, 21, 223–33.

Boinski, S., Kauffman, L., Westoll, A., Stickler, C. M., Cropp, S., and Ehmke, E. (2003). Are vigilance, risk from avian predators and group size consequences of habitat structure? A comparison of three species of squirrel monkey (*Saimiri oerstedii*, *S. boliviensis*, and *S. sciureus*). *Behaviour*, 140, 1421–67.

Boinski, S., and Timm, R. M. (1985). Predation by squirrel monkeys and double-toothed kites on tent-making bats. *American Journal of Primatology*, 9, 121–27.

Boinski, S., Sughrue, K., Selvaggi, L., Quatrone, R., Henry, M., and Cropp, S. (2002). An expanded test of the ecological model of primate social evolution: Competitive regimes and female bonding in three species of squirrel monkeys (*Saimiri oerstedii*, *S. boliviensis*, and *S. sciureus*). *Behaviour*, 139, 227–61.

Bolen, R. H., and Green, S. M. (1997). Use of olfactory cues in foraging by owl monkeys (*Aotus nancymai*) and capuchin monkeys (*Cebus apella*). *Journal of Comparative Physiology*, 111, 152–58.

Bolter, D. R. (2004). *Anatomical Growth Patterns in Colobine*

Monkeys and Implications for Primate Evolution. University of California.

Bompas, A., Kendall, G., and Sumner, P. (2013). Spotting fruit versus picking fruit as the selective advantage of human colour vision. *Perception,* 4, 84–94.

Bonneaud, C., Mazuc, J., Gonzalez, G., Haussy, C., Chastel, O., Faivre, B., et al. (2003). Assessing the cost of mounting an immune response. *American Naturalist,* 161(3), 367–79.

Bonnell, T. R., Campenni, M., Chapman, C. A., Gogarten, J. F., Reyna-Hurtado, R. A., Teichroeb, J. A., et al. (2013). Emergent group level navigation: An agent-based evaluation of movement patterns in a folivorous primate. *PLoS ONE,* 8, e78264.

Bonnell, T. R., Chapman, C. A., and Sengupta, R. (2016). Interaction between scale and scheduling choices in simulations of spatial agents. *International Journal of Geographical Information Science,* 30, 2075–88.

Bonnell, T. R., Henzi, S. P., and Barrett, L. (2019). Functional social structure in baboons: Modeling interactions between social and environmental structure in group-level foraging. *Journal of Human Evolution,* 126, 14–23.

Bonnell, T. R., Sengupta, R., Chapman, C. A., and Goldberg, T. L. (2010). An agent-based model of red colobus resources and disease dynamics implicates key resource sites as hot spots of disease transmission *Ecological Modeling,* 221, 2491–500.

Boonratana, R. (2000). Ranging behavior of proboscis monkeys (*Nasalis larvatus*) in the Lower Kinabatangan, Northern Borneo. *International Journal of Primatology,* 21, 497–518.

Boonstra, R., and Krebs, C. J. (2006). Population limitation of the northern red-backed vole in the boreal forests of northern Canada. *Journal of Animal Ecology,* 75, 1269–84.

Boose, K. J., White, F. J., and Meinelt, A. (2013). Sex differences in tool use acquisition in bonobos (*Pan paniscus*). *American Journal of Primatology,* 75, 917–26.

Booth, A., Magnuson, A., Fouts, J., and Foster, M. T. (2016). Adipose tissue: An endocrine organ playing a role in metabolic regulation. *Hormone Molecular Biology and Clinical Investigation.*26, 25–42.

Bora, P. S., Rocha, R. V. M., Narain, N., Moreira-Monteiro, A. C., and Moreira, R. (2003). Characterization of principal nutritional components of Brazilian oil palm (*Eliaes guineensis*) fruits. *Bioresource Technology,* 87, 1–5.

Borg, C., Majolo, B., Qarro, M., and Semple, S. (2014). A comparison of body size, coat condition and endoparasite diversity of wild Barbary macaques exposed to different levels of tourism. *Anthrozoos,* 27, 49–63.

Bornman, L., Baladi, S., Richard, M. J., Tyrrell, R. M., and Polla, B. S. (1999). Differential regulation and expression of stress proteins and ferritin in human monocytes. *Journal of Cellular Physiology,* 178(1), 1–8.

Borries, C., and Koenig, A. (2000). Infanticide in Hanuman langurs: Social organization, male migration, and weaning age. In C. P. van Schaik and C. H. Janson (Eds.), *Infanticide by Males and Its Implications* (pp. 99–122). Cambridge University Press.

Borries, C., Gordon, A. D., and Koenig, A. (2013). Beware of primate life history data: A plea for data standards and a repository. *PLoS ONE,* 8, e67200.

Borries, C., Koenig, A., and Winkler, P. (2001). Variation of life history traits and mating patterns in female langur monkeys (*Semnopithecus entellus*). *Behavioral Ecology and Sociobiology,* 50, 391–402.

Borries, C., Larney, E., Lu, A., Ossi, K., and Koenig, A. (2008). Costs of group size: Lower developmental and reproductive rates in larger groups of leaf monkeys. *Behavioral Ecology,* 19, 1186–91.

Borries, C., Lodwick, J. L., Salmi, R., Koenig, A. (2022). Phenotypic plasticity rather than ecological risk aversion or folivory can explain variation in gorilla life history. *Frontiers in Ecology and Evolution,* 10, 873557.

Borries, C., Lu, A., Ossi-Lupo, K., Larney, E., and Koenig, A. (2011). Primate life histories and dietary adaptations: A comparison of Asian colobines and macaques. *American Journal of Physical Anthropology,* 144, 286–99.

Borries, C., Lu, A., Ossi-Lupo, K., Larney, E., and Koenig, A. (2014). The meaning of weaning in wild Phayre's leaf monkeys: Last nipple contact, survival, and independence. *American Journal of Physical Anthropology,* 154, 291–301.

Borries, C., Sandel, A. A., Koenig, A., Fernandez-Duque, E., Kamilar, J. M., Amoroso, C. R., et al. (2016). Transparency, usability, and reproducibility: Guiding principles for improving comparative databases using primates as examples. *Evolutionary Anthropology,* 25, 232–38.

Borries, C., Sommer, V., and Srivastava, A. (1991). Dominance, age, and reproductive success in free-ranging female Hanuman langurs. *International Journal of Primatology,* 12, 230–57.

Bortolamiol, S., Cohen, M., Potts, K. B., Rwaburindore, P., Kasenene, J., Seguya, A., et al. (2014). Suitable habitats for endangered frugivorous mammals: Small-scale comparison, regeneration forest and chimpanzee density in Kibale National Park, Uganda. *PLoS ONE,* 9, e102177.

Bosch, G., Hagen-Plantinga, E. A., and Hendricks, W. H. (2015). Dietary nutrient profiles of wild wolves: Insights for optimal dog nutrition? *British Journal of Nutrition,* 113, S40–S54.

Bostian, K. A., Blackburn, B. S., Wannemacher, R. W. Jr., McGann, V. G., Beisel, W. R., and Dupont, H. L. (1976). Sequential changes in the concentration of specific serum proteins during typhoid fever infection in man. *Journal of Laboratory and Clinical Medicine,* 87(4), 577–85.

Boucher, D. H. (1990). Growing back after hurricanes: Catastrophes may be critical to rain forest dynamics. *Bioscience,* 40, 163–66.

Bourlat, S. J., Juliusdottir, T., Lowe, C. J., Freeman, R., Aronowicz, J., Kirschner, M., et al. (2006). Deuterostome phylogeny reveals monophyletic chordates and the new phylum Xenoturbellida. *Nature,* 444, 85–88.

Boutin, S. (1990). Food supplementation experiments with terrestrial vertebrates: Patterns, problems, and the future. *Canadian Journal of Zoology*, 68, 203–20.

Bowen, G. J., James, R., Ehleringer, J. R., Chesson, L. A., Thompson, A. H., Podlesak, D. W., et al. (2009). Dietary and physiological controls on the hydrogen and oxygen isotope ratios of hair from mid-20th century indigenous populations. *American Journal of Physical Anthropology*, 139, 494–504.

Bowen, G. J., Wassenaar, L. I., and Hobson, K. A. (2005). Global application of stable hydrogen and oxygen isotopes to wildlife forensics. *Oecologia*, 143, 337–48.

Bowen, W. D., Oftedal, O. T., and Boness, D. J. (1985). Birth to weaning in 4 days: Remarkable growth in the hooded seal, *Cystophora cristata*. *Canadian Journal of Zoology*, 63, 2841–46.

Bowler, M., and Bodmer, R. E. (2011). Diet and food choice in Peruvian red uakaris (*Cacajao calvus ucayalii*): Selective or opportunistic seed predation? *International Journal of Primatology*, 32, 1109–22.

Boyer, D. M., Evans, A. R., and Jernvall, J. (2010). Evidence of dietary differentiation among late Paleocene–early Eocene plesiadapids (Mammalia, Primates). *American Journal of Physical Anthropology*, 142, 194–210.

Boyer, D., Ramos-Fernandez, G., Miramontes, O., Mateos, J. L., Cocho, G., Larralde, H., et al. (2006). Scale-free foraging by primates emerges from their interaction with a complex environment. *Proceedings of the Royal Society B: Biological Sciences*, 282, 1743–1750.

Boyer, D. M., Scott, C. S., and Fox, R. C. (2012). New craniodental material of *Pronothodectes gaoi* Fox (Mammalia,"Plesiadapiformes") and relationships among members of Plesiadapidae. *American Journal of Physical Anthropology*, 147, 511–50.

Boyle, S. A., Thompson, C. L., Deluycker, A., Alvarez, S. J., Alvim, T. H. G., Aquino, R., et al. (2016). Geographic comparison of plant genera used in frugivory among the pitheciids *Cacajao*, *Callicebus*, *Chiropotes*, and *Pithecia*. *American Journal of Primatology*, 78, 493–506.

Bradford, P. G., and Awad, A. B. (2007). Phytosterols as anticancer compounds. *Molecular Nutrition and Food Research*, 51, 161–70.

Bradley, B. J., and Lawler, R. R. (2011). Linking genotypes, phenotypes, and fitness in wild primate populations. *Evolutionary Anthropology*, 20, 104–19.

Bradley, B. J., Stiller, M., Doran-Sheehy, D., Vigilant, L., and Poinar, H. (2007). Plant DNA sequences from feces: Potential means for assessing diets of wild primates. *American Journal of Primatology*, 69, 699–705.

Braly, A., Darnell, L. A., Mann, A. B., Teaford, M. F., and Weihs, T. P. (2007). The effect of prism orientation on the indentation testing of human molar enamel. *Archives of Oral Biology*, 52(9), 856–60.

Branch-Mays, G. L., Dawson, D. R., Gunsolley, J. C., Reynolds, M. A., Ebersole, J. L., Novak, K. F., et al. (2008). The effects of a calorie-reduced diet on periodontal inflammation and disease in a non-human primate model. *Journal of Periodontology*, 79(7), 1184–91.

Bravo, J. A., Forsythe, P., Chew, M. V., Escaravage, E., Savignac, H. M., Dinan, T. G., et al. (2011). Ingestion of *Lactobacillus* strain regulates emotional behavior and central GABA receptor expression in a mouse via the vagus nerve. *Proceedings of the National Academy of Sciences of the United States of America*, 108, 16050–55.

Brawand, D., Wahli, W., and Kaessmann, H. (2008). Loss of egg yolk genes in mammals and the origin of lactation and placentation. *PLoS Biology*, 6, e63.

Brenna, J. T. (2002). Efficiency of conversion of α-linolenic acid to long chain n-3 fatty acids in man. *Current Opinion in Clinical Nutrition and Metabolic Care*, 5, 127–32.

Brenowitz, E. A. (1986). Environmental influences on acoustic and electric animal communication. *Brain, Behavior and Evolution*, 28, 32–42.

Brent, L. J. N., Semple, S., MacLarnon, A., Ruiz-Lambides, A., Gonzalez-Martinez, J., and Platt, M. L. (2014). Personality traits in rhesus macaques (*Macaca mulatta*) are heritable but do not predict reproductive output. *International Journal of Primatology*, 35, 188–209.

Breslin, P. A. S. (2013). An evolutionary perspective on food and human taste. *Current Biology*, 23, R409–R418.

Brett, F. L., Turner, T. R., Jolly, C. J., and Cauble, R. G. (1982). Trapping baboons and vervet monkeys from wild, free-ranging populations. *Journal of Wildlife Management*, 46, 164–74.

Breuer, T., Ndoundou-Hockemba, M., and Fishlock, V. (2005). First observation of tool use in wild gorillas. *PLoS Biology*, 3, 2041–43.

Bridge, E. S., Thorup, K., Bowlin, M. S., Chilson, P. B., Diehl, R. H., Fleron, R. W., et al. (2011). Technology on the move: Recent and forthcoming innovations for tracking migratory birds. *BioScience*, 61, 689–98.

Briend, A., and Darmon, N. (2000). Determining limiting nutrients by linear programming: A new approach to predict insufficient intakes from complementary foods. *Pediatrics*, 106, 1288–90.

Briskin, D. P. (2000). Medicinal plants and phytomedicines: Linking plant biochemistry and physiology to human health. *Plant Physiology*, 124, 507–14.

Britt, A., Randriamandratonirina, N. J., Glasscock, K. D., and Iambana, B. R. (2002). Diet and feeding behaviour of *Indri indri* in a low-altitude rain forest. *Folia Primatologica*, 73, 225–39.

Broad, K. D., Curley, J. P., and Keverne, E. B. (2006). Mother–infant bonding and the evolution of mammalian social relationships. *Philosophical Transactions of the Royal Society of London Series B—Biological Sciences*, 361, 2199–214.

Brockelman, W. Y. (2011). Rainfall patterns and unpredictable fruit production in seasonally dry evergreen forest and their effects on gibbons. In W. J. McShea, S. J. Davies, and N. Bhumpakphan (Eds.), *The Ecology and Conservation of Seasonally Dry Forests in Asia* (pp. 195–216). Smithsonian Institution Scholarly Press.

Brockman, D. K., and van Schaik, C. P. (2005). Seasonality and reproductive function. In D. K. Brockman and C. P. van Schaik (Eds.), *Seasonality in Primates: Studies of Living and Extinct Human and Non-human Primates* (pp. 269–305). Cambridge University Press.

Brockman, D. K., Harrison, R. O., and Nadler, T. (2009). Conservation of douc langurs in Vietnam: An assessment of Agent Orange exposure in douc langurs (*Pygathrix*) at the Endangered Primate Rescue Center, Cuc Phuong National Park, Vietnam. *Vietnamese Journal of Primatology*, 1, 45–64.

Brodie, J. F., Giordano, A. J., Zipkin, E. F., Bernard, H., Mohd-Azlan, J., and Ambu, L. (2015). Correlation and persistence of hunting and logging impacts on tropical rainforest mammals. *Conservation Biology*, 29, 110–21.

Brodie, J. F., Post, E., and Laurance, W. F. (2012). Climate change and tropical biodiversity: A new focus. *Trends in Ecology and Evolution*, 27, 145–50.

Bromage, T. G., Hogg, R. T., Lacruz, R. S., and Hou, C. (2012). Primate enamel evinces long period biological timing and regulation of life history. *Journal of Theoretical Biology*, 305, 131–44.

Bronikowski, A. M., Altmann, J., Brockman, D. K., Cords, M., Fedigan, L. M., Pusey, A. E., et al. (2011). Aging in the natural world: Comparative data reveal similar mortality patterns across primates. *Science*, 331, 1325–28.

Bronikowski, A. M., Cords, M., Alberts, S. C., Altmann, J., Brockman, D. K., Fedigan, L. M., et al. (2016). Female and male life tables for seven wild primate species. *Scientific Data*, 3, 160006.

Bronson, F. H. (1989). *Mammalian Reproductive Biology*. University of Chicago Press.

Brooks, R. C., Simpson, S. J., and Raubenheimer, D. (2010). The price of protein: Combining evolutionary and economic analysis to understand excessive energy consumption. *Obesity Reviews*, 11, 887–94.

Brown, D., Kays, R., Wikelski, M., Wilson, R., and Kimley, A. (2013). Observing the unwatchable through acceleration logging of animal behavior. *Animal Biotelemetry*, 1, 1–16.

Brown, J. (1982). Optimal group size in territorial animals. *Journal of Theoretical Biology*, 95, 793–810.

Brown, J. S. (1988). Patch use as an indicator of habitat preference, predation risk and competition. *Behavioral Ecology and Sociobiology*, 22, 37–47.

Brown, J. S. (1992). Patch use under predation risk: I. Models and predictions. *Annales Zoologici Fennici*, 29, 301–9.

Brown, J. S. (2000). Foraging ecology of animals in response to heterogeneous environments. In M. J. Hutchings, E. A. John, and A. J. A. Stewart (Eds.), *The Ecological Consequences of Environmental Heterogeneity* (pp. 181–214). Cambridge University Press.

Brown, J. S., and Kotler, B. P. (2004). Hazardous duty pay and the foraging cost of predation. *Ecology Letters*, 7, 999–1014.

Brown, J. S., and Kotler, B. P. (2007). Foraging and the ecology of fear. In D. Stephens, J. S. Brown, and R. C. Ydenberg (Eds.), *Foraging: Behavior and Ecology* (pp. 437–82). University of Chicago Press.

Brown, J. S., Kotler, B. P., and Mitchell, W. A. (1994). Foraging theory, patch use, and the structure of a Negev Desert granivore community. *Ecology*, 75, 2286–300.

Brown, J. S., Laundré, J. W., and Gurung, M. (1999). The ecology of fear: Optimal foraging, game theory, and trophic interactions. *Journal of Mammalogy*, 80, 385–99.

Brown, K. S., Marean, C. W., Jacobs, Z., Schoville, B. J., Oestmo, S., Fisher, E. C., et al. (2012). An early and enduring advanced technology originating 71,000 years ago in South Africa. *Nature*, 491, 590–93.

Brown, N. M., Galandi, S. L., Summer, S. S., Zhao, X. H., Heubi, J. E., King, E. C., et al. (2014). S-(-)equol production is developmentally regulated and related to early diet composition. *Nutrition Research*, 34, 401–9.

Brucker, R. M., and Bordenstein, S. R. (2012). Speciation by symbiosis. *Trends in Ecology and Evolution*, 27(8), 443–51.

Brumm, H., Voss, K., Köllmer, I., and Todt, D. (2004). Acoustic communication in noise: Regulation of call characteristics in a New World monkey. *Journal of Experimental Biology*, 207, 443–48.

Bruhn, J. M. (1934). The respiratory metabolism of infrahuman primates. *American Journal of Physiology*, 110, 477–84.

Bruhn, J. M., Benedict, F. G. (1936). The respiratory metabolism of the chimpanzee. *Proceedings of the American Academy of the Arts and Sciences*, 71, 259–326.

Brun, L., Schneider, J., Carrió, E. M., Dongre, P., Taberlet, P., van de Waal, E., et al. (2022). Focal vs. fecal: Seasonal variation in the diet of wild vervet monkeys from observational and DNA metabarcoding data. *Ecology and Evolution*, 12, e9358.

Bruner, A. G., Gullison, R. E., Rice, R. E., and da Fonseca, G. A. B. (2001). Effectiveness of parks in protecting tropical biodiversity. *Science*, 291, 125–28.

Bryant, J. P., Wieland, G. D., Reichardt, P. B., Lewis, V. E., and McCarthy, M. C. (1983). Pinosylvin methyl ether deters snowshoe hare feeding on green alder. *Science*, 222, 1023–25.

Bryer, M. A. H., Chapman, C. A., and Rothman, J. M. (2013). Diet and polyspecific associations affect spatial patterns among redtail monkeys (*Cercopithecus ascanius*). *Behavior*, 150, 277–93.

Bryer, M. A. H., Chapman, C. A., Raubenheimer, D., Lambert, J. E., and Rothman, J. M. (2015). Macronutrient and energy contributions of insects to the diet of a frugivorous monkey (*Cercopithecus ascanius*). *International Journal of Primatology*, 36, 839–54.

Bshary, R. (2001). Diana monkeys, *Cercopithecus diana*, adjust their anti-predator response behaviour to human hunting strategies. *Behavioral Ecology and Sociobiology*, 50, 251–56.

Bucar, F. (2013). Phytoestrogens in plants: With special reference to isoflavones. In V. R. Preedy (Ed.), *Isoflavones: Chemistry,*

Analysis, Function and Effects (pp. 14–27). Royal Society of Chemistry.

Buchanan-Smith, H. M. (1991). A field study on the red-bellied tamarin, *Saguinus l. labiatus*, in Bolivia. *International Journal of Primatology*, 12, 259–76.

Buchin, K., Sijben, S., van Loon, E., Sapir, N., Mercier, S., Marie Arseneau, T., et al. (2015). Deriving movement properties and the effect of the environment from the Brownian bridge movement model in monkeys and birds. *Movement Ecology*, 3, 1–11.

Buckley, J., Maunder, R. J., Foey, A., Pearce, J., Val, A. L., and Sloman, K. A. (2010). Biparental mucus feeding: A unique example of parental care in an Amazonian cichlid. *Journal of Experimental Biology*, 213, 3787–95.

Bufalo, F. S., Galetti, M., and Culot, L. (2016). Seed dispersal by primates and implications for the conservation of a biodiversity hotspot, the Atlantic forest of South America. *International Journal of Primatology*, 37, 333–49.

Buffington, S. A., Di Prisco, G. V., Auchtung, T. A., Ajami, N. J., Petrosino, J. F., and Costa-Mattioli, M. (2016). Microbial reconstitution reverses maternal diet induced social and synaptic deficits in offspring. *Cell*, 165, 1762–75.

Bugir, C. K., Butynski, T. M., and Hayward, M.W. (2021). Prey preferences of the chimpanzee (*Pan troglodytes*). *Nature Ecology and Evolution*, 2021, 7138–46.

Bump, J. K., Fox-Dobbs, K., Bada, J. L., Koch, P. L., Peterson, R. O., and Vucetich, J. A. (2007). Stable isotopes, ecological integration and environmental change: Wolves record atmospheric carbon isotope trend better than tree rings. *Proceedings of the Royal Society B: Biological Sciences*, 274, 2471–80.

Bunce, J. A., Isbell, L. A., Grote, M. N., and Jacobs, G. H. (2011). Color vision variation and foraging behavior in wild Neotropical titi monkeys (*Callicebus brunneus*): Possible mediating roles for spatial memory and reproductive status. *International Journal of Primatology*, 32, 1058–75.

Bunn, H. T. (2007). Meat made us human. In P. Ungar (Ed.), *Evolution of the Human Diet: The Known, the Unknown, and the Unknowable* (pp. 191–211). Oxford University Press.

Bunn, H. T., and Ezzo, J. A. (1993). Hunting and scavenging by Plio-Pleistocene hominids: Nutritional constraints, archaeological patterns, and behavioural implications. *Journal of Archaeological Sciences*, 20, 365–98.

Bunn, H. T., and Pickering, T. (2010). Bovid mortality profiles in paleoecological context falsify hypotheses of endurance running and passive scavenging by early Pleistocene hominins. *Quaternary Research*, 74, 395–404.

Bunn, J. M., and Ungar, P. S. (2009). Dental topography and diets of four Old World monkey species. *American Journal of Primatology*, 71(6), 466–77.

Bunn, J. M., Boyer, D. M., Lipman, Y., St Clair, E. M., Jernvall, J., and Daubechies, I. (2011). Comparing Dirichlet normal surface energy of tooth crowns, a new technique of molar shape quantification for dietary inference, with previous methods in isolation and in combination. *American Journal of Physical Anthropology*, 145(2), 247–61.

Bunning, H., Rapkin, J., Belcher, L., Archer, C. R., Jensen, K., and Hunt, J. (2015). Protein and carbohydrate intake influence sperm number and fertility in male cockroaches, but not sperm viability. *Proceedings of the Royal Society B: Biological Sciences*, 282, 20142144.

Burdge, G. C. (2006). Metabolism of alpha-linolenic acid in humans. *Prostaglandins, Leukotrienes and Essential Fatty Acids*, 75, 161–68.

Burgess, M. L., McFarlin, S. C., Mudakikwa, A., Cranfield, M. R., and Ruff, C. B. (2018). Body mass estimation in hominoids: Age and locomotor effects. *Journal of Human Evolution*, 115, 36–46.

Burgess, M. A., and Chapman, C. A. (2005). Tree leaf chemical characters: Selective pressures by folivorous primates and invertebrates. *African Journal of Ecology*, 43, 242–50.

Burgman, J. H. E., Leichliter, J., Avenant, N. L., and Ungar, P. S. (2016). Dental microwear of sympatric rodent species sampled across habitats in southern Africa: Implications for environmental influence. *Integrative Zoology*, 11(2), 111–27.

Burkhardt, R. W. (2014). Tribute to Tinbergen: Putting Niko Tinbergen's "four questions" in historical context. *Ethology*, 120, 215–23.

Burnham, P., Dadhania, D., Heyang, M., Chen, F., Westblade, L. F., Suthanthiran, M., et al. (2018). Urinary cell-free DNA is a versatile analyte for monitoring infections of the urinary tract. *Nature Communications*, 9(1), 2412.

Burrows, A. M., and Nash, L. T. (2010). *The Evolution of Exudativory in Primates*. Springer.

Burrows, A. M., Hartstone-Rose, A., and Nash, L. T. (2015). Exudativory in the Asian loris, *Nycticebus*: Evolutionary divergence in the toothcomb and M3. *American Journal of Physical Anthropology*, 158, 663–72.

Buse, A., Good, J. E. G., Dury, S., and Perrins, C. M. (1998). Effects of elevated temperature and carbon dioxide on the nutritional quality of leaves of oak (*Quercus robur* L.) as food for the winter moth (*Operophtera brumata* L.). *Functional Ecology*, 12, 742–49.

Bushdid, C., Magnasco, M. O., Vosshall, L. B., and Keller, A. (2014). Humans can discriminate more than 1 trillion olfactory stimuli. *Science*, 343, 1370–72.

Busia, L., Schaffner, C. M., and Aureli, F. (2017). Relationship quality affects fission decisions in wild spider monkeys (*Ateles geoffroyi*). *Ethology*, 123, 405–11.

Busia, L., Schaffner, C. M., Rothman, J. M., and Aureli, F. (2016). Do fruit nutrients affect subgrouping patterns in wild spider monkeys (*Ateles geoffroyi*)? *International Journal of Primatology*, 37, 738–51.

Butchart, S. H. M., Walpole, M., Collen, B., van Strien, A., Scharlemann, J. P. W., Almond, R. E. A., et al. (2010).

Global biodiversity: Indicators of recent declines. *Science*, 328, 1164–68.

Butler, K., Louys, J., and Travouillon, K. (2014). Extending dental mesowear analyses to Australian marsupials, with applications to six Plio-Pleistocene kangaroos from southeast Queensland. *Palaeogeography, Palaeoclimatology, Palaeoecology*, 408, 11–25.

Butler, P. M. (1983). Evolution and mammalian dental morphology. *Journal de Biologie Buccale*, 11(4), 285–302.

Butler, R. A., and Laurance, W. F. (2008). New strategies for conserving tropical forests. *Trends in Ecology and Evolution*, 23, 469–72.

Butt, N., Seabrook, L., Maron, M., Law, B. S., Dawson, T. P., Syktus, J., et al. (2015). Cascading effects of climate extremes on vertebrate fauna through changes to low-latitude tree flowering and fruiting phenology. *Global Change Biology*, 21, 3267–77.

Butte, N. F., and Hopkinson, J. M. (1998). Body composition changes during lactation are highly variable among women. *Journal of Nutrition*, 128, 381S–385S.

Butynski, T. (1982). Vertebrate predation by primates: A review of hunting patterns and prey. *Journal of Human Evolution*, 11, 421–30.

Butynski, T. M. (1988). Guenon birth seasons and correlates with rainfall and food. In A. Gautier-Hion, F. Bourliére, J. P. Gautier, and J. Kingdon (Eds.), *A Primate Radiation: Evolutionary Biology of the African Guenons* (pp. 284–322). Cambridge University Press.

Butynski, T. M. (1990). Comparative ecology of blue monkeys (*Cercopithecus mitis*) in high- and low-density subpopulations. *Ecological Monographs*, 60, 1–26.

Byrne, R. W. (1979). Memory for urban geography. *Quarterly Journal of Experimental Psychology*, 31, 147–54.

Byrne, R. W. (1997). The technical intelligence hypothesis: An additional evolutionary stimulus to intelligence? In A. Whiten and R. W. Byrne (Eds.), *Machiavellian Intelligence II: Extensions and Evaluations* (pp. 289–311). Cambridge University Press.

Byrne, R. W. (2000). How monkeys find their way: Leadership, coordination, and cognitive maps of African baboons. In S. Boinski and P. A. Garber (Eds.), *On the Move: How and Why Animals Travel in Groups* (pp. 491–518). University of Chicago Press.

Byrne, R. W., and Whiten, A. (1988). *Machiavellian Intelligence: Social Expertise and the Evolution of Intellect in Monkeys, Apes and Humans*. Oxford Science.

Byrne, R. W., Hobaiter, C., and Klailova, M. (2011). Local traditions in gorilla manual skill: Evidence for observational learning of behavioral organization. *Animal Cognition*, 14, 683–93.

Byrne, R. W., Noser, R., Bates, L. A., and Jupp, P. E. (2009). How did they get here from there? Detecting changes of direction in terrestrial ranging. *Animal Behaviour*, 77, 619–31.

Byrne, R. W., Sanz, C. M., and Morgan, D. B. (2013). Chimpanzees plan their tool use. In C. M. Sanz, J. Call, and C. Boesch (Eds.), *Tool Use in Animals: Cognition and Ecology* (pp. 48–64). Cambridge University Press.

Cabana, F., Dierenfeld, E. S., Wirdateti, W., Donati, G., and Nekaris, K. A. I. (2017a). Exploiting a readily available but hard to digest resource: A review of exudativorous mammals identified thus far and how they cope in captivity. *Integrative Zoology*, 13, 94–111.

Cabana, F., Dierenfeld, E. S., Wirdateti, W., Donati, G., and Nekaris, K. A. I. (2017b). The seasonal feeding ecology of the Javan slow loris (*Nycticebus javanicus*). *American Journal of Physical Anthropology*, 162, 768–81.

Cabana, F., Dierenfeld, E. S., Wirdateti, W., Donati, G., and Nekaris, K. A. I. (2017c). Slow lorises (*Nycticebus* spp.) really are slow: A study of food passage rates. *International Journal of Primatology*, 38, 900–913.

Cabrera-Rubio, R., Collado, M. C., Laitinen, K., Salminen, S., Isolauri, E., and Mira, A. (2012). The human milk microbiome changes over lactation and is shaped by maternal weight and mode of delivery. *American Journal of Clinical Nutrition*, 96(3), 544–51.

Cagnacci, F., Boitani, L., Powell, R. A., and Boyce, M. S. (2010). Animal ecology meets GPS-based radiotelemetry: A perfect storm of opportunities and challenges. *Philosophical Transactions of the Royal Society of London Series B—Biological Sciences*, 365, 2157–62.

Caillaud, D., Crofoot, M. C., Scarpino, S. V., Jansen, P. A., Garzon-Lopez, C. X., Winklelhagen, A. J. S., et al. (2010). Modeling the spatial distribution and fruiting pattern of a key tree species in a Neotropical forest: Methodology and potential applications. *PLoS ONE*, 5, e15002.

Caine, N. G., Addington, R. L., and Windfelder, T. L. (1995). Factors affecting the rates of food calls given by red-bellied tamarins. *Animal Behaviour*, 50, 53–60.

Caine, N. G., and Mundy, N. I. (2000). Demonstration of a foraging advantage for trichromatic marmosets (Callithrix geoffroyi) dependent on food colour. *Proceedings of the Royal Society of London. Series B: Biological Sciences*, 267(1442), 439–44.

Caine, N. G., Osorio, D., and Mundy, N. I. (2010). A foraging advantage for dichromatic marmosets (*Callithrix geoffroyi*) at low light intensity. *Biology Letters*, 6(1), 36–38.

Calafat, A. M., Ye, X. Y., Wong, L. Y., Reidy, J. A., and Needham, L. L. (2008). Exposure of the US population to bisphenol A and 4-tertiary-octylphenol: 2003–2004. *Environmental Health Perspectives*, 116, 39–44.

Calandra, I., and Merceron, G. 2016. Dental microwear texture analysis in mammalian ecology. *Mammal Review*, 46(3), 215–28.

Caldecott, J. O. (1980). Habitat quality and populations of two sympatric gibbons (Hylobatidae) on a mountain in Malaya. *Folia Primatologica*, 33, 291–309.

Caldecott, J. O. (1986). Mating patterns, societies and the ecogeography of macaques. *Animal Behaviour*, 34, 208–20.

Calder, P. C. (1997). n-3 polyunsaturated fatty acids and cytokine production in health and disease. *Annals of Nutrition and Metabolism*, 41(4), 203–34.

Calder, P. C. (1998a). Dietary fatty acids and lymphocyte functions. *Proceedings of the Nutrition Society*, 57(4), 487–502.

Calder, P. C. (1998b). Dietary fatty acids and the immune system. *10th International Congress on Immunology*, vols. 1 and 2, 813–19.

Calder, P. C. (2013). n-3 fatty acids, inflammation and immunity: New mechanisms to explain old actions. *Proceedings of the Nutrition Society*, 72(3), 326–36.

Calder, P. C., and Field, C. J. (2002). Fatty acids, inflammation and immunity. In P. C. Calder, C. J. Field, and H. S. Gill (Eds.), *Nutrition and Immune Function* (pp. 57–92). CABI.

Calder, P. C., and Grimble, R. F. (2002). Polyunsaturated fatty acids, inflammation and immunity. *European Journal of Clinical Nutrition*, 56(suppl. 3), S14–19.

Calder, W. A. (1996). *Size, Function, and Life History*. Courier Dover.

Calvert, J. J. (1985). Food selection by western gorillas (*G. g. gorilla*) in relation to food chemistry. *Oecologia*, 65, 236–46.

Cameron, E. Z. (1998). Is suckling behaviour a useful predictor of milk intake? A review. *Animal Behaviour*, 56(3), 521–32.

Cameron, E. Z., Stafford, K. J., Linklater, W. L., and Veltman, C. J. (1999). Suckling behaviour does not measure milk intake in horses, *Equus caballus*. *Animal Behaviour*, 57(3), 673–78.

Campbell, A. F., and Sussman, R. W. (1994). The value of radio tracking in the study of Neotropical rainforest monkeys. *American Journal of Primatology*, 32, 291–301.

Campbell, C. P., Raubenheimer, D., Badaloo, A. V., Gluckman, P. D., Martinez, C., Gosby, A., et al. (2016). Developmental contributions to macronutrient selection: A randomized controlled trial in adult survivors of malnutrition. *Evolution, Medicine and Public Health*, 2016, 158–69.

Campbell, J. L., Eisemann, J. H., Williams, C. V., and Glenn, K. M. (2000). Description of the gastrointestinal tract of five lemur species: *Propithecus tattersalli*, *Propithecus verreauxi coquereli*, *Varecia variegata*, *Hapalemur griseus*, and *Lemur catta*. *American Journal of Primatology*, 52, 133–42.

Campbell, M. C., Ranciara, A., Froment, A., Hirbo, J., Omar, S., Bodo, J. M., et al. (2012). Evolution of functionally diverse alleles associated with PTC bitter taste sensitivity in Africa. *Molecular Biology and Evolution*, 29, 1141–53.

Campbell-Smith, G., Campbell-Smith, M., Singleton, I., and Linkie, M. (2011). Raiders of the lost bark: Orangutan foraging strategies in a degraded landscape. *PLoS ONE*, 6, e20962.

Campera, M., Balestri, M., Besnard, F., Phelps, M., Rakotoarimanana, F., Nijman, V., et al. (2021). The influence of seasonal availability of young leaves on dietary niche separation in two ecologically similar folivorous lemurs. *Folia Primatologica*, 92, 139–50.

Campos, F. A., and Fedigan, L. M. (2008). Behavioral adaptations to heat stress and water scarcity in white-faced capuchins (*Cebus capucinus*) in Santa Rosa National Park, Uganda. *American Journal of Physical Anthropology*, 138, 101–11.

Campos, F. A., and Fedigan, L. M. (2014). Spatial ecology of perceived predation risk and vigilance behavior in white-faced capuchins. *Behavioural Ecology*, 25, 477–86.

Campos, F. A., Bergstrom, M. L., Childers, A., Hogan, J. D., Jack, K. M., Melin, A. D., et al. (2014). Drivers of home range characteristics across spatiotemporal scales in a Neotropical primate, *Cebus capucinus*. *Animal Behaviour*, 91, 93–109.

Campos, F. A., Jack, K. M., and Fedigan, L. M. (2015). Climate oscillations and conservation measures regulate white-faced capuchin population growth and demography in a regenerating tropical dry forest in Costa Rica. *Biological Conservation*, 186, 204–13.

Campos, F. A., Kalbitzer, U., Melin, A. D., Hogan, J. D., Cheves, S. E., Murillo-Chacon, E., et al. (2020). Differential impact of severe drought on infant mortality in two sympatric Neotropical primates. *Royal Society Open Science*, 7, 200302.

Campos, F. A., Morris, W. F., Alberts, S. C., Altmann, J., Brockman, D. B., Cords, M., et al. (2017). Does climate variability influence the demography of wild primates? Evidence from long-term life-history data in seven species. *Global Change Biology*, 23, 4907–21.

Canale, C. I., Ozgula, A., Allaine, D., and Cohas, A. (2016). Differential plasticity of size and mass to environmental change in a hibernating mammal. *Global Change Biology*, 22, 3286–303.

Canale, C. I., Perret, M., and Henry, P. Y. (2012). Torpor use during gestation and lactation in a primate. *Naturwissenschaften*, 99, 159–63.

Canale, G. R., Guidorizzi, C. E., Kierulff, M. C. M., and Gatto, C. (2009). First record of tool use by wild populations of the yellow-breasted capuchin monkey (*Cebus xanthosternos*) and new records for the bearded capuchin (*Cebus libidinosus*). *American Journal of Primatology*, 71, 366–72.

Cancelliere, E., DeAngelis, N., Nkurunungi, J. B., Raubenheimer, D., and Rothman, J. M. (2014). Minerals in the foods eaten by mountain gorillas (*Gorilla beringei*). *PLoS ONE*, 9, 1–11.

Cannon, C. H., and Leighton, M. (1994). Comparative locomotory ecology of gibbons and macaques: Selection of canopy elements for crossing gaps. *American Journal of Physical Anthropology*, 93, 505–24.

Cannon, C. H., and Leighton, M. (1996). Comparative locomotor ecology of gibbons and macaques: Does brachiation minimize travel costs? *Tropical Biodiversity*, 3, 261–67.

Cannon, C. H., Peart, D. R., Leighton, M., and Kartawinata, K. (1994). The structure of lowland rainforest after selective logging in West Kalimantan, Indonesia. *Forest Ecology and Management*, 67, 49–68.

Cant, J. G. H. (1980). What limits primates? *Primates*, 21, 538–44.

Cant, J. G. H. (1987). Positional behavior of female Bornean orangutans (*Pongo pygmaeus*). *American Journal of Primatology*, 12, 71–90.

Cantu-Salazar, L., Hidalgo-Mihart, M. G., Lopez-Gonzalez, C. A., and Gonzalez-Romero, A. (2005). Diet and food resource use by the pygmy skunk (*Spilogale pygmaea*) in the tropical dry forest of Chamela, Mexico. *Journal of Zoology*, 267(3), 283–89.

Caperos, J. M., Morcillo, A., Peláez, F., Fidalgo, A., and Sánchez, S. (2012). The effect of infant body mass on carrier travel speed in cotton-top tamarins (*Saguinus oedipus*). *International Journal of Primatology*, 33, 447–59.

Caporaso, J. G., Lauber, C. L., Costello, E. K., Berg-Lyons, D., Gonzalez, A., Stombaugh, J., et al. (2011). Moving pictures of the human microbiome. *Genome Biology and Evolution*, 12, R50.

Caraco, T., and Giraldea, L. A. (1991). Social foraging: Producing and scrounging in a stochastic environment. *Journal of Theoretical Biology*, 153, 559–83.

Carboni, S., Dezeure, J., Cowlishaw, G., Huchard, E., and Marshall, H. H. (2022). Stable isotopes reveal the effects of maternal rank and infant age on weaning dynamics in wild chacma baboons. *Animal Behaviour*, 193, 21–32.

Cardenas-Gonzalez, M., Srivastava, A., Pavkovic, M., Bijol, V., Rennke, H. G., Stillman, I. E., et al. (2017). Identification, confirmation, and replication of novel urinary microRNA biomarkers in lupus nephritis and diabetic nephropathy. *Clinical Chemistry*, 63(9), 1515–26.

Carlo, T. A., and Yang, S. (2011). Network models of frugivory and seed dispersal: Challenges and opportunities. *Acta Oecologica*, 37, 619–24.

Carlson, B. A., Rothman, J. M., and Mitani, J. C. (2013). Diurnal variation in nutrients and chimpanzee foraging behavior. *American Journal of Primatology*, 75, 342–49.

Carlson, S. E. (1999). Long-chain polyunsaturated fatty acids and development of human infants. *Acta Paediatrica*, 88, 72–77.

Carlson, S. E. (2001). Docosahexaenoic acid and arachidonic acid in infant development. *Seminars in Fetal and Neonatal Medicine*, 6, 437–49.

Carmody, R. N., Cone, E., Wrangham, R. W., and Secor, S. M., (2009). Cooking and the net energy value of meat: Implications for human evolution. *Integrative and Comparative Biology*. 49, E27.

Carne, C., Semple, S., Morrogh-Bernard, H., Zuberbühler, K., and Lehmann, J. (2013). Predicting the vulnerability of great apes to disease: The role of superspreaders and their potential vaccination. *PLoS ONE*, 8(12) e84642.

Carne, C., Semple, S., Morrogh-Bernard, H., Zuberbühler, K., and Lehmann, J. (2014). The risk of disease to great apes: Simulating disease spread in orang-utan (*Pongo pygmaeus wurmbii*) and chimpanzee (*Pan troglodytes schweinfurthii*) association networks. *PLoS ONE*, 9(4) e95039.

Carpenter, C. R. (1934). A field study of the behavior and social relations of howling monkeys. *Comparative Psychological Monographs*, 10, 1–168.

Carpenter, C. R. (1940). A field study in Siam of the behavior and social relations of the gibbon, *Hylobates lar*. *Comparative Psychological Monographs*, 16, 1–212.

Carpenter, K. J. (1994). *Protein and Energy: A Study of Changing Ideas in Nutrition*. Cambridge University Press.

Carrai, V., Borgogini-Tarli, S. M., Huffman, M. A., and Bardi, M. (2003). Increase in tannin consumption by sifaka (*Propithecus verreauxi verreauxi*) females during the birth season: A case for self-medication in prosimians? *Primates*, 44, 61–66.

Carrascal, L. M., and Alonso, C. L. (2006). Habitat use under latent predation risk: A case study with wintering forest birds. *Oikos*, 112, 51–62.

Carreau, C., Flouriot, G., Bennetau-Pelissero, C., and Potier, M. (2008). Enterodiol and enterolactone, two major diet-derived polyphenol metabolites have different impact on ER alpha transcriptional activation in human breast cancer cells. *Journal of Steroid Biochemistry*, 110, 176–85.

Carretero-Pinzon, X., Defler, T. R., and Ferrari, S. F. (2003). Observations of black-capped capuchins (*Cebus apella*) feeding on an owl monkey (*Aotus brumbacki*) in the Colombian llanos. *Neotropical Primates*, 16, 53.

Carson, R. (1962). *Silent Spring*. Houghton Mifflin Riverside.

Carter, M. L., and Bradbury, M. W. (2016). Oxygen isotope ratios in primate bone carbonate reflect amount of leaves and vertical stratification in the diet. *American Journal of Primatology*, 78, 1086–97.

Carthey, A. J. R., and Banks, P. B. (2015). Foraging in groups affects giving-up densities: Solo foragers quit sooner. *Oecologia*, 178, 707–13.

Cartmill, M. (1974). Rethinking primate origins. *Science*, 184, 436–43.

Cartmill, M. (1992). New views on primate origins. *Evolutionary Anthropology*, 1, 105–11.

Carvalho, S. (2011). Extensive surveys of chimpanzee stone tools: From the telescope to the magnifying glass. In T. Matsuzawa, T. Humle, and Y. Sugiyama (Eds.), *The Chimpanzees of Bossou and Nimba* (pp. 145–55). Springer.

Cary, M. E., Valentine, B., and White, G. L. (2003). The effects of confinement environment on reproductive efficiency in the baboon. *Contemporary Topics in Laboratory Animal Science*, 42, 35–39.

Caselli, C. B., and Freire Setz, E. Z. (2011). Feeding ecology and activity pattern of black-fronted titi monkeys (*Callicebus nigrifrons*) in a semideciduous tropical forest of southern Brazil. *Primates*, 52, 351–59.

Casey, C. E., Smith, A. C., and Zhang, P. (1995). Microminerals in human and animal milks. In R. G. Jensen (Ed.), *Handbook of Milk Composition* (pp. 622–74). Academic Press.

Cash, J. F. (2013). *Feeding and Ranging Ecology of Grey-Cheeked*

Mangabeys (*Lophocebus albigena*) *at Nyungwe Forest Reserve, Rwanda*. California State University at Fullerton.

Casper, R. M. (2009). Guidelines for the instrumentation of wild birds and mammals. *Animal Behaviour*, 78, 1477–83.

Castellanos, H. G. (1995). *Feeding Behaviour of Ateles belzebuth (E. Geoffroy 1806) (Cebidae: Atelinae) in Tawadu Forest Southern Venezuela*. University of Exeter.

Castro, S. C. d. N., Souto, A. d. S., Schiel, N., Biondi, L. M., and Caselli, C. B. (2017). Techniques used by bearded capuchin monkeys (*Sapajus libidinosus*) to access water in a semi-arid environment of North-Eastern Brazil. *Folia Primatologica*, 88, 267–73.

Catton, W. R. (1987). The world's most polymorphic species. *BioScience*, 37, 413–19.

Caughley, G. (1976). Wildlife management and the dynamics of ungulate populations. *Applied Biology*, 1, 183–246.

Caughley, G. (1979). What is this thing called carrying capacity. In M. S. Boyce and L. D. Hayden-Wing (Eds.), *North American Elk: Ecology, Behavior and Management* (pp. 2–8). University of Wyoming Press.

Caughley, G. (1994). Directions in conservation biology. *Journal of Animal Ecology*, 63, 215–44.

Caughley, G., and Sinclair, A. R. E. (1994). *Wildlife Ecology and Management*. Blackwell Science.

Cavalcanti, S. M. C., and Gese, E. M. (2010). Kill rates and predation patterns of jaguars (*Panthera onca*) in the southern Pantanal, Brazil. *Journal of Mammalogy*, 91, 722–36.

Caves, E. M., Brandley, N. C., and Johnsen, S. (2018). Visual acuity and the evolution of signals. *Trends in Ecolgy and Evolution*, 33(5), 358–72.

Cederroth, C. R., Auger, J., Zimmermann, C., Eustache, F., and Nef, S. (2010). Soy, phyto-oestrogens and male reproductive function: A review. *International Journal of Andrology*, 33, 304–16.

Cederroth, C. R., Zimmerman, C., Beny, J. L., Schaad, O., Combepine, C., Descombes, P., et al. (2010). Potential detrimental effects of a phytoestrogen-rich diet on male fertility in mice. *Molecular and Cellular Endocrinology*, 321, 152–60.

Ceesay, S. M., Prentice, A. M., Cole, T. J., Foord, F., Poskitt, E.M.E., Weaver, L.T., and Whitehead, R. G. (1997). Effects on birth weight and perinatal mortality of maternal dietary supplements in rural Gambia: 5 year randomised controlled trial. *BMJ* 315(7111):786–90.

Cerling, T. E., Bernasconi, S. M., Hofstetter, L. S., Jaggi, M., Wyss, F., Rudolf von Rohr, C., et al. (2021). CH_4/CO_2 ratios and carbon isotope enrichment between diet and breath in herbivorous mammals. *Frontiers in Ecology and Evolution*, 9, 638568.

Cerling, T. E., and Harris, J. M. (1999). Carbon isotope fractionation between diet and bioapatite in ungulate mammals and implications for ecological and paleoecological studies. *Oecologia*, 120, 347–63.

Cerling, T. E., Hart, J. A., and Hart, T. B. (2004). Stable isotope ecology in the Ituri Forest. *Oecologia*, 138, 5–12.

Cerling, T. E., Manthi, F. K., Mbua, E. N., Leakey, L. N., Leakey, M. G., Leakey, R. E., et al. (2013). Stable isotope-based diet reconstructions of Turkana Basin hominins. *Proceedings of the National Academy of Sciences of the United States of America*, 110, 10501–6.

Cerling, T. E., Mbua, E., Kirera, F. M., Manthi, F. K., Grine, F. E., Leakey, M. G., et al. (2011). Diet of *Paranthropus boisei* in the early Pleistocene of East Africa. *Proceedings of the National Academy of Sciences of the United States of America*, 108, 9337–41.

Cerling, T. E., and Viehl, K. (2004). Seasonal diet changes of the forest hog (*Hylochoerus meinertzhageni* Thomas) based on the carbon isotopic composition of hair. *African Journal of Ecology*, 42(2), 88–92.

Cernusak, L. A., Tcherkez, G., Keitel, C., Cornwell, W. K., Santiago, L. S., Knohl, A., et al. (2009). Why are non-photosynthetic tissues generally ^{13}C enriched compared with leaves in C_3 plants? Review and synthesis of current hypotheses. *Functional Plant Biology*, 36, 199–213.

Chadderdon, S. M., Belcik, J. T., Bader, L., Kirigiti, M. A., Peters, D. M., Kievit, P., et al. (2014). Proinflammatory endothelial activation detected by molecular imaging in obese nonhuman primates coincides with onset of insulin resistance and progressively increases with duration of insulin resistance. *Circulation*, 129(4), 471–78.

Chalk-Wilayto, J., Fogaça, M. D., Wright, B. W., van Casteren, A., Fragaszy, D. M., Izar, P., et al. (2022). Effects of food material properties and embedded status on food processing efficiency in bearded capuchins. *American Journal of Biological Anthropology*, 178, 617–35.

Chai, H., Lee, J. J.W., Kwon, J. Y., Lucas, P. W., and Lawn, B. R. (2009). A simple model for enamel fracture from margin cracks. *Acta Biomaterialia*, 5, 1663–67.

Chamberlain, C. P., Waldbauer, J. R., Fox-Dobbs, K., Newsome, S. D., Koch, P. L., Smith, D. R., et al. (2005). Pleistocene to recent dietary shifts in California condors. *Proceedings of the National Academy of Sciences of the United States of America*, 102, 16707–11.

Chamberlain, J., Nelson, G., and Milton, K. (1993). Fatty-acid profiles of major food sources of howler monkeys (*Alouatta palliata*) in the Neotropics. *Experentia*, 49, 820–24.

Chambers, E. S., Morrison, D. J., and Frost, G. (2015). Control of appetite and energy intake by SCFA: What are the potential underlying mechanisms? *Proceedings of the Nutrition Society*, 74, 328–36.

Chambers, P. G., Raubenheimer, D., and Simpson, S. J. (1998). The functional significance of switching interval in food mixing by *Locusta migratoria*. *Journal of Insect Physiology*, 44, 77–85.

Chambers, P., Simpson, S. J., and Raubenheimer, D. (1995). Behavioural mechanisms of nutrient balancing in *Locusta migratoria*. *Animal Behaviour*, 50, 1513–23.

Chamoun, E., Mutch, D. M., Allen-Vercoe, E., Buchholz, A. C., Duncan, A. M., Spriet, L. L., et al. (2018). A review of the associations between single nucleotide polymorphisms in taste receptors, eating behaviors, and health. *Critical Reviews in Food Science and Nutrition*, 58, 194–207.

Chancellor, R. L., and Isbell, L. A. (2009). Food site residence time and female competitive relationships in wild gray-cheeked mangabeys (*Lophocebus albigena*). *Behavioral Ecology and Sociobiology*, 63, 1447–58.

Chandra, R. K. (1983). Numerical and functional deficiency in T-helper cells in protein-energy malnutrition. *Clinical and Experimental Immunology*, 51(1), 126–32.

Chang, N. C., Su, H. H., and Lee, L. L. (2016). Effects of dietary fiber on gut retention time in captive *Macaca cyclopis*, *Macaca fascicularis*, *Hylobates lar*, and *Pongo pygmaeus* and the germination of ingested seeds. *International Journal of Primatology*, 37, 671–87.

Changizi, M. A., and Shimojo, S. (2008). "X-ray vision" and the evolution of forward-facing eyes. *Journal of Theoretical Biology*, 254, 756–67.

Changizi, M. A., Zhang, Q., and Shimojo, S. (2006). Bare skin, blood and the evolution of primate colour vision. *Biology Letters*, 2, 217–21.

Chapman, C. A. (1985). The influence of habitat on behaviour in a group of St-Kitts green monkeys. *Journal of Zoology*, 206, 311–20.

Chapman, C. A. (1995). Primate seed dispersal: Coevolution and conservation implications. *Evolutionary Anthropology*, 4, 74–82.

Chapman, C. A., Balcomb, S. R., Gillespie, T., Skorupa, J., and Struhsaker, T. T. (2000). Long-term effects of logging on African primate communities: A 28 year comparison from Kibale National Park, Uganda. *Conservation Biology*, 14, 207–17.

Chapman, C. A., Bicca-Marques, J. C., Dunham, A. E., Fan, P., Fashing, P. J., Gogarten, J. F., et al. (2020). Primates can be a rallying symbol to promote tropical forest restoration. *Folia Primatologica*, 91, 669–87.

Chapman, C. A., Bonnell, T. R., Gogarten, J. F., Lambert, J. E., Omeja, P. A., Twinomugisha, D., et al. (2013). Are primates ecosystem engineers? *International Journal of Primatology*, 34, 1–14.

Chapman, C. A., Bonnell, T. R., Sengupta, R., Goldberg, T. L., and Rothman, J. M. (2013). Is *Markhamia lutea*'s abundance determined by animal foraging? *Forest Ecology and Management*, 308, 62–66.

Chapman, C. A., and Chapman, L. J. (1990). Dietary variability in primate populations. *Primates*, 31, 121–28.

Chapman, C. A., and Chapman, L. J. (1999). Implications of small scale variation in ecological conditions for the diet and density of red colobus monkeys. *Primates*, 40, 215–31.

Chapman, C. A., and Chapman, L. J. (2000a). Determinants of group size in primates: The importance of travel costs. In S. Boinski and P. Garber (Eds.), *On the move: How and why animals travel in groups* (pp. 24–42). University of Chicago Press.

Chapman, C. A., and Chapman, L. J. (2000b). Interdemic variation in mixed-species association patterns: Common diurnal primates of Kibale National Park, Uganda. *Behavioral Ecology and Sociobiology*, 47, 129–39.

Chapman, C. A., and Chapman, L. J. (2002). Foraging challenges of red colobus monkeys: Influence of nutrients and secondary compounds. *Comparative Biochemistry and Physiology A—Molecular and Integrative Physiology*, 133, 861–75.

Chapman, C. A., and Chapman, L. J. (2004). Unfavorable successional pathways and the conservation value of logged tropical forest. *Biodiversity and Conservation*, 13, 2089–105.

Chapman, C. A., Chapman, L. J., Bjorndal, K. A., and Onderdonk, D. A. (2002). Application of protein-to-fiber ratios to predict colobine abundance on different spatial scales. *International Journal of Primatology*, 23, 283–310.

Chapman, C. A., Chapman, L. J., and Gillespie, T. R. (2002). Scale issues in the study of primate foraging: Red colobus of Kibale National Park. *American Journal of Physical Anthropology*, 117, 349–63.

Chapman, C. A., Chapman, L. J., Jacob, A. L., Rothman, J. M., Omeja, P., Reyna-Hurtado, R., et al. (2010). Tropical tree community shifts: Implications for wildlife conservation. *Biological Conservation*, 143, 366–74.

Chapman, C. A., Chapman, L. J., Kaufman, L., and Zanne, A. E. (1999). Potential causes of arrested succession in Kibale National Park, Uganda: Growth and mortality of seedlings. *African Journal of Ecology*, 37, 81–92.

Chapman, C. A., Chapman, L. J., Naughton-Treves, L., Lawes, M. J., and McDowell, L. R. (2004). Predicting folivorous primate abundance: Validation of a nutritional model. *American Journal of Primatology*, 62, 55–69.

Chapman, C. A., Chapman, L. J., Rode, K. D., Hauck, E. M., and McDowell, L. R. (2003). Variation in the nutritional value of primate foods: Among trees, time periods, and areas. *International Journal of Primatology*, 24, 317–33.

Chapman, C. A., Chapman, L. J., Struhsaker, T. T., Zanne, A. E., Clark, C. J., and Poulsen, J. R. (2005). A long-term evaluation of fruiting phenology: Importance of climate change. *Journal of Tropical Ecology*, 21, 31–45.

Chapman, C. A., Chapman, L. J., Wrangham, R., Hunt, K., Gebo, D., and Gardner, L. (1992). Estimators of fruit abundance of tropical trees. *Biotropica*, 24, 527–31.

Chapman, C. A., Chapman, L. J., Zanne, A. E., Poulsen, J. R., and Clark, C. J. (2005). A 12-year phenological record of fruiting: Implications for frugivore populations and indicators of climate change. In J. L. Dew and J. P. Boubli (Eds.), *Tropical Fruits and Frugivores: The Search for Strong Interactors* (pp. 75–92). Springer.

Chapman, C. A., and Fedigan, L. M. (1990). Dietary differences between neighboring *Cebus capucinus* groups: Local traditions, food availability or responses to food profitability? *Folia Primatologica*, 54, 177–86.

Chapman, C., Gautier-Hion, A., Oates, J. F., and Onderdonk, D. A. (1999). African primate communities: Determinants of structure and threats to survival. In J. G. Fleagle, C. Janson, and K. Reed (Eds.), *Primate Communities* (pp. 1–37). Cambridge University Press.

Chapman, C. A., Ghai, R. R., Jacob, A. L., Koojo, S. M., Reyna-Hurtado, R., Rothman, J. M., et al. (2013). Going, going, gone: A 15-year history of the decline of primates in forest fragments near Kibale National Park, Uganda. In L. K. Marsh and C. A. Chapman (Eds.), *Primates in Fragments: Complexity and Resilience*. Springer

Chapman, C. A., and Lefebvre, L. (1990). Manipulating foraging group size: Spider monkey food calls at fruiting trees. *Animal Behavior*, 39, 891–96.

Chapman, C. A., Omeja, P. A., Kalbitzer, U., Fan, P., and Lawes, M. J. (2018). Restoration provides hope for faunal recovery: Changes in primate abundance over 45 years in Kibale National Park, Uganda. *Tropical Conservation Science*, 11, 1940082918787376.

Chapman, C. A., and Onderdonk, D. A. (1998). Forests without primates: Primate/plant codependency. *American Journal of Primatology*, 45, 127–41.

Chapman, C. A., and Peres, C. A. (2001). Primate conservation in the new millennium: The role of scientists. *Evolutionary Anthropology*, 10, 16–33.

Chapman, C. A., and Peres, C. A. (2021). Primate conservation: Lessons learned in the last 20 years can guide future efforts. *Evolutionary Anthropology*, 30, 345–61.

Chapman, C. A., and Rothman, J. M. (2009). Within-species differences in primate social structure: Evolution of plasticity and phylogenetic constraints. *Primates*, 50, 12–22.

Chapman, C. A., Rothman, J. M., and Lambert, J. E. (2012). Food as a selective force in primates. In J. C. Mitani, J. Call, P. M. Kappeler, R. A. Palombit, and J. B. Silk (Eds.), *The Evolution of Primate Societies* (pp. 149–68). University of Chicago Press.

Chapman, C. A., and Russo, S. E. (2007). Primate seed dispersal: Linking behavioral ecology with forest community structure. In C. J. Campbell, A. Fuentes, K. C. MacKinnon, M. Panger, and S. K. Bearder (Eds.), *Primates in Perspective* (pp. 510–25). Oxford University Press.

Chapman, C. A., Schoof, V. A. M., Bonnell, T. R., Gogarten, J. F., and Calmé, S. (2015). Competing pressures on populations: Long-term dynamics of food availability, food quality, disease, stress and animal abundance. *Philosophical Transactions of the Royal Society of London Series B—Biological Sciences*, 370, 20140112.

Chapman, C. A., Snaith, T. V., and Gogarten, J. F. (2014). How ecological conditions affect the abundance and social organization of folivorous monkeys. In J. Yamagiwa and L. Karczmarski (Eds.), *Primates and Cetaceans: Field Research and Conservation of Complex Mammalian Societies* (pp. 3–23). Springer.

Chapman, C. A., Steiniche, T., Benavidez, K. M., Sarkar, D., Amato, K., Serio Silva, J. C., et al. (2022). The chemical landscape of tropical mammals in the Anthropocene. *Biological Conservation*, 269, 109522.

Chapman, C. A., Struhsaker, T. T., Skorupa, J. P., Snaith, T. V., and Rothman, J. M. (2010). Understanding long-term primate community dynamics: Implications of forest change. *Ecological Applications*, 20, 179–91.

Chapman, C. A., Wasserman, M. D., Gillespie, T. R., Speirs, M. L., Lawes, M. J., Saj, T. L., et al. (2006). Do nutrition, parasitism, and stress have synergistic effects on red colobus populations living in forest fragments? *American Journal of Physical Anthropology*, 131, 525–34.

Chapman, C. A., Wrangham, R., and Chapman, L. J. (1994). Indices of habitat-wide fruit abundance in tropical forests. *Biotropica*, 26, 160–71.

Chapman, C. A., Wrangham, R. W., and Chapman, L. J. (1995a). A complex social structure with fission-fusion properties can emerge from a simple foraging model. *Behavioral Ecology and Sociology*, 36, 59–70.

Chapman, C. A., Wrangham, R. W., and Chapman, L. J. (1995b). Ecological constraints on group size: An analysis of spider monkey and chimpanzee subgroups. *Behavioral Ecology and Sociobiology*, 36, 59–70.

Chapman, C. A., Wrangham, R. W., Chapman, L., Kennard, D., and Zanne, A. (1999). Fruit and flower phenology and two sites in Kibale National Park, Uganda. *Journal of Tropical Ecology*, 15, 189–211.

Charbonneau, M. R., Blanton, L. V., DiGiulio, D. B., Relman, D. A., Lebrilla, C. B., Mills, D. A., et al. (2016). A microbial perspective of human developmental biology. *Nature*, 535, 48–55.

Charles-Dominique, P. (1977). *Ecology and Behaviour of Nocturnal Primates: Prosimians of Equatorial West Africa*. Columbia University Press.

Charnov, E. L. (1976). Optimal foraging, the marginal value theorem. *Theoretical Population Biology*, 9, 129–36.

Charnov, E. L. (2001). Evolution of mammal life histories. *Evolutionary Ecology Research*, 3, 521–35.

Charnov, E. L. (2009). Optimal (plastic) life histories in growing versus stable populations. *Evolutionary Ecology Research*, 11, 983–87.

Charnov, E. L., and Berrigan, D. (1993). Why do female primates have such long lifespans and so few babies? Or life in the slow lane. *Evolutionary Anthropology*, 1, 191–94.

Chaves, Ó. M., and Bicca-Marques, J. C. (2016). Feeding strategies of brown howler monkeys in response to variations in food availability. *PLoS ONE*, 11, e0145819.

Chaves, Ó. M., and Bicca-Marques, J. C. (2017). Crop feeding by brown howlers (*Alouatta guariba clamitans*) in forest fragments: The conservation value of cultivated species. *International Journal of Primatology*, 38, 263–81.

Chaves, Ó. M., Stoner, K. E., Ángeles-Campos, S., and Arroyo-Rodrígues, V. (2011). Wood consumption by Geoffroyi's spider monkeys and its role in mineral supplementation. *PLoS ONE*, 6, e25070.

Chaves, Ó. M., Stoner, K. E., and Arroyo-Rodrígues, V. (2012). Differences in diet between spider monkey groups living in forest fragments and continuous forest in Mexico. *Biotropica*, 44, 105–13.

Chaves, Ó. M., Stoner, K. E., Arroyo-Rodrígues, V., and Estrada, A. (2011). Effectiveness of spider monkeys (*Ateles geoffroyi vellerosus*) as seed dispersers in continuous and fragmented rain forests in southern Mexico. *International Journal of Primatology*, 32, 177–92.

Chazdon, R. L., Cullen, L., Padua, S. M., and Valladares Padua, C. (2020). People, primates and predators in the Pontal: From endangered species conservation to forest and landscape restoration in Brazil's Atlantic Forest. *Royal Society Open Science*, 7, 200939.

Cheeke, P. R., and Dierenfeld, E. S. (2010). *Comparative Animal Nutrition and Metabolism*. CABI.

Chen, H., Yao, H., Yang, W., Xiang, Z., Ostner, J., and Cristóbal-Azkarate, J. (2021). Validation of a fecal T3 metabolite assay for measuring energetics in wild golden snub-nosed monkeys (*Rhinopithecus roxellana*). *International Journal of Primatology*, 42, 759–63.

Cheney, D. L. (1981). Intergroup encounters among free-ranging vervet monkeys. *Folia Primatologica*, 35, 124–46.

Cheney, D. (1987). Predation. In B. B. Smuts, D. L. Cheney, R. M. Seyfarth, R. W. Wrangham, and T. T. Struhsaker (Eds.), *Primate Societies* (pp. 227–39). University of Chicago Press.

Cheney, D. L., and Seyfarth, R. M. (1987). The influence of intergroup competition on the survival and reproduction of female vervet monkeys. *Behavioral Ecology and Sociobiology*, 21, 375–86.

Cheney, D. L., Seyfarth, R. M., Andelman, S. J., and Lee, P. C. (1986). Factors affecting reproductive success in vervet monkeys. In T. Clutton-Brock (Ed.), *Reproductive Success* (pp. 384–402). Chicago University Press.

Chester, E. M., Fender, E., and Wasserman, M. D. (2020). Screening for phytoestrogens using a cell-based estrogen receptor β reporter assay. *Journal of Visualized Experiments*, 160, e61005.

Chevalier-Skolnikoff, S. (1990). Tool use by wild cebus monkeys at Santa Rosa National Park, Costa Rica. *Primates*, 31, 375–83.

Cheverud, J. M., Wilson, P., and Dittus, W. P. J. (1992). Primate population studies at Polonnaruwa III: Somatometric growth in a natural population of toque macaques (*Macaca sinica*). *Journal of Human Evolution*, 23, 51–77.

Chiarello, A. G. (1993). Activity pattern of the brown howler monkey *Alouatta fusca*, Geoffroy 1812, in a forest fragment of southeastern Brazil. *Primates*, 34, 289–93.

Chiarello, A. G. (1994). Diet of the brown howler monkey *Alouatta fusca* in a semi-deciduous forest fragment of southeastern Brazil. *Primates*, 35, 25–34.

Chiarello, A. G. (1995). Grooming in brown howler monkeys, *Alouatta fusca*. *American Journal of Primatology*, 35, 73–81.

Childress, M. J., and Lung, M. A. (2003). Predation risk, gender and the group size effect: Does elk vigilance depend upon the behaviour of conspecifics? *Animal Behaviour*, 66, 389–98.

Chirico, G., Marzollo, R., Cortinovis, S., Fonte, C., and Gasparoni, A. (2008). Antiinfective properties of human milk. *Journal of Nutrition*, 138, 1801S–1806S.

Chivers, D. J. (1969). On the daily behavior and spacing of howling monkey groups. *Folia Primatologica*, 10, 48–102.

Chivers, D. J. (1974). The siamang in Malaya: A field study of a primate in tropical rainforest. *Contributions to Primatology*, 4, 1–335.

Chivers, D. J. (1980). *Malayan Forest Primates*. Plenum.

Chivers, D. J. (1991). Species differences in tolerance to environmental change. In H. O. Box (Ed.), *Primate Responses to Environmental Change* (pp. 5–37). Chapman and Hall.

Chivers, D. J. (1992). Diets and guts. In S. Jones, R. Martin, and D. Pilbeam (Eds.), *The Cambridge Encyclopedia of Human Evolution* (pp. 60–62). Cambridge University Press.

Chivers, D. J. (1994). Functional anatomy of the gastrointestinal tract. In A. G. Davies and J. F. Oates (Eds.), *Colobine Monkeys: Their Ecology, Behaviour, and Evolution* (pp. 205–28). Cambridge University Press.

Chivers, D. J. (1998). Measuring food intake in wild animals: Primates. *Proceedings of the Nutrition Society*, 57, 321–32.

Chivers, D. J. (2001). The swinging singing apes: Fighting for food and family in far-east forests. In *The Apes: Challenges for the 21st Century* (pp. 1–28). Chicago Zoological Society.

Chivers, D. J., and Hladik, C. M. (1980). Morphology of the gastrointestinal tract in primates: Comparisons with other mammals in relation to diet. *Journal of Morphology*, 166, 337–86.

Chivers, D. J., and Hladik, C. M. (1984). Diet and gut morphology in primates. In Chivers, D. J., Wood, B. A., and Bilsborough, A. (eds.), *Food Acquisition and Processing in Primates* (pp. 213-30). Plenum Press.

Chivers, D. J., Wood, B. A., and Bilsborough, A. (1984). *Food Acquisition and Food Processing in Primates*. Springer Science.

Chiyo, P. I., Lee, P. C., Moss, C. J., Archie, E. A., Hollister-Smith, J. A., and Alberts, S. C. (2011). No risk, no gain: Effects of crop raiding and genetic diversity on body size in male elephants. *Behavioral Ecology*, 22, 552–58.

Choo, G. M., Waterman, P. G., McKey, D. B., and Gartlan, J. S. (1981). A simple enzyme assay for dry-matter digestibility and its value in studying food selection by generalist herbivores. *Oecologia*, 49, 170–78.

Christian, L. M., Graham, J. E., Padgett, D. A., Glaser, R., and Kiecolt-Glaser, J. K. (2006). Stress and wound healing. *Neuroimmunomodulation*, 13(5–6), 337–46.

Chua, W. K., and Oyen, M. L. (2009). Viscoelastic properties of membranes measured by spherical indentation. *Cellular and Molecular Bioengineering*, 2, 49–56.

Cimprich, D. A., and Grubb, T. C. Jr. (1994). Consequences for Carolina chickadees of foraging with tufted titmice in winter. *Ecology*, 75, 1615–25.

Ciochon, R. L., Piperno, D. R., and Thompson, R. G. (1990). Opal phytoliths found on the teeth of the extinct ape *Gigantopithecus blacki*: Implications for paleodietary studies. *Proceedings of the National Academy of Sciences of the United States of America, 87(20), 8120–24.

Cipolletta, C., Spagnoletti, N., Todd, A., Robbins, M. M., Cohen, H., and Pacyna, S. (2007). Termite feeding by *Gorilla gorilla gorilla* at Bai Hokou, Central African Republic. *International Journal of Primatology*, 28, 457–76.

Cipollini, M. L., and Levey, D. J. (1997). Secondary metabolites of fleshy vertebrate-dispersed fruits: Adaptive hypotheses and implications for seed dispersal. *American Naturalist*, 150, 346–72.

Ciuti, S., Northrup, J. M., Muhly, T. B., Simi, S., Musiani, M., Pitt, J. A., et al. (2012). Effects of humans on behaviour of wildlife exceed those of natural predators in a landscape of fear. *PLoS ONE*, 7, e50611.

Clark, C. W., and Mangel, M. (1986). The evolutionary advantages of group foraging. *Theoretical Population Biology*, 30, 45–75.

Clausen, T. P., Reichardt, P. B., and Bryant, J. P. (1986). Pinosylvin and pinosylvin methyl ether as feeding deterrents in green alder. *Journal of Chemical Ecology*, 12, 2117–31.

Clavel, T., Fallani, M., Lepage, P., Levenez, F., Mathey, J., Rochet, V., et al. (2005). Isoflavones and functional foods alter the dominant intestinal microbiota in postmenopausal women. *Journal of Nutrition*, 135, 2786–92.

Clavel, T., Henderson, G., Alpert, C. A., Philippe, C., Rigottier-Gois, L., Dore, J., et al. (2005). Intestinal bacterial communities that produce active estrogen-like compounds enterodiol and enterolactone in humans. *Applied and Environmental Microbiology*, 71, 6077–85.

Clauss, M., Schwarm, A., Ortmann, S., Streich, W. J., and Hummel, J. (2007). A case of non-scaling in mammalian physiology? Body size, digestive capacity, food intake, and ingesta passage in mammalian herbivores. *Comparative Biochemistry and Physiology Part A: Molecular & Integrative Physiology*, 148(2), 249–65.

Cleland, T. P., and Power, M. L. (2022). Variation in milk proteins across lactation in *Pongo pygmaeus* and *Gorilla gorilla*. *Journal of Proteome Research*, 21, 2647–54.

Clemens, E. T., and Stevens, C. E. (1980). A comparison of gastrointestinal transit time in ten species of mammal. *The Journal of Agricultural Science*, 94(3), 735–37.

Clementz, M. T., Holden, P., and Koch, P. L. (2003). Are calcium isotopes a reliable monitor of trophic level in marine settings? *International Journal of Osteoarchaeology*, 13, 29–36.

Clinchy, M., Sheriff, M. J., and Zanette, L. Y. (2013). Predator-induced stress and the ecology of fear. *Functional Ecology*, 27, 56–65.

Cline, J. M., and Wood, C. E. (2009). Estrogen/isoflavone interactions in cynomolgus macaques (*Macaca fascicularis*). *American Journal of Primatology*, 71, 722–31.

Clink, D. J., Dillis, C., Feilen, K. L., Beaudrot, L., and Marshall, A. J. (2017). Dietary diversity, feeding selectivity, and responses to food scarcity of two sympatric Bornean primates (*Hylobates albibarbis* and *Presbytis rubicunda rubida*). *PLoS ONE*, 12, e0173369.

Clotfelter, E. D., Bell, A. M., and Levering, K. R. (2004). The role of animal behaviour in the study of endocrine-disrupting chemicals. *Animal Behaviour*, 68, 665–76.

Clowes, G. H. Jr., O'Donnell, T. F., Blackburn, G. L., and Maki, T. N. (1976). Energy metabolism and proteolysis in traumatized and septic man. *Surgical Clinics of North America*, 56(5), 1169–84.

Clutton-Brock, T. H. (1977a). *Primate Ecology: Studies of Feeding and Ranging in Lemurs, Monkeys and Apes*. Academic Press.

Clutton-Brock, T. H. (1977b). Some aspects of intraspecific variation in feeding and ranging behaviour in primates. In T. H. Clutton-Brock (Ed.), *Primate Ecology* (pp. 539–56). Academic Press.

Clutton-Brock, T. H. (1988). *Reproductive Success: Studies of Individual Variation in Contrasting Breeding Systems*. University of Chicago Press.

Clutton-Brock, T. H. (1991). *The Evolution of Parental Care*. Princeton University Press.

Clutton-Brock, T. H. (2016). *Mammal Societies*. Wiley.

Clutton-Brock, T. H., and Harvey, P. H. (1976). Evolutionary rules and primate societies. In P. P. G. Bateson and R. A. Hinde (Eds.), *Growing Points in Ethology* (pp. 195–237). Cambridge University Press.

Clutton-Brock, T. H., and Harvey, P. H. (1977). Primate ecology and social organization. *Journal of Zoology*, 183, 1–39.

Clutton-Brock, T., and Harvey, P. (1979). Home range size, population density and phylogeny in primates. In I. S. Bernstein and E. O. Smith (Eds.), *Primate Ecology and Human Origins: Ecological Influences on Social Organisation* (pp. 201–14). Garland.

Clutton-Brock, T. H., and Harvey, P. H. (1980). Primates, brains and ecology. *Journal of Zoology*, 190, 309–23.

Clutton-Brock, T., and Janson, C. H. (2012). Primate socioecology at the crossroads: Past, present, and future. *Evolutionary Anthropology*, 21, 136–50.

Clutton-Brock, T. H., Price, O. F., and MacColl, A. D. C. (1992). Mate retention, harassment, and the evolution of ungulate leks. *Behavioural Ecology*, 3, 234–42.

Cobo-Sánchez, L., Pizarro-Monzo, M., Cifuentes-Alcobendas, G., Jiménez García, B., Abellán Beltrán, N., Courtenay, L. A., et al. (2022). Computer vision supports primary access to meat by early *Homo* 1.84 million years ago. *PeerJ*, 10, e14148.

Cockburn, A., Legge, S., and Double, M. C. (2002). Sex ratios in birds and mammals: Can the hypotheses be disentangled? In I. C. W. Hardy (Ed.), *Sex Ratios: Concepts and Research Methods* (pp. 266–86). Cambridge University Press.

Codron, D., Lee-Thorp, J. A., Sponheimer, M., de Ruiter, D., and Codron, J. (2006). Inter- and intrahabitat dietary variability of chacma baboons (*Papio ursinus*) in South African savannas based on fecal delta C-13, delta N-15, and %N. *American Journal of Physical Anthropology*, 129, 204–14.

Codron, J., Codron, D., Lee-Thorp, J. A., Sponheimer, M., Bond, W. J., de Ruiter, D., et al. (2005). Taxonomic, anatomical, and spatio-temporal variations in the stable carbon and nitrogen isotopic compositions of plants from an African savanna. *Journal of Archaeological Sciences*, 32, 1757–72.

Codron, J., Codron, D., Lee-Thorp, J. A., Sponheimer, M., Kirkman, K., Duffy, K. J., et al. (2011). Landscape-scale feeding patterns of African elephant inferred from carbon isotope analysis of feces. *Oecologia*, 165, 89–99.

Codron, J., Codron, D., Sponheimer, M., Kirkman, K., Duffy, K. J., Raubenheimer, E. J., et al. (2012). Stable isotope series from elephant ivory reveal lifetime histories of a true dietary generalist. *Proceedings of the Royal Society B: Biological Sciences*, 279, 2433–41.

Codron, J., Lee-Thorp, J. A., Sponheimer, M., and Codron, D. (2013). Plant stable isotope composition across habitat gradients in a semi-arid savanna: Implications for environmental reconstruction. *Journal of Quaternary Science*, 28, 301–10.

Coelho Jr, A. M., 1974. Socio-bioenergetics and sexual dimorphism in primates. *Primates*. 15, 263-269.

Coelho, A. M., Bramblett, C. A., and Quick, L. B. (1979). Activity patterns in howler and spider monkeys: An application of socio-bioenergetic methods. In: Berntein, I. S., Smith, E. O. (Eds.), *Primate Ecology and Human Origins* (pp. 175–99). Gartland.

Coe, C. L., Lubach, G. R., and Shirtcliff, E. A. (2007). Maternal stress during pregnancy predisposes for iron deficiency in infant monkeys impacting innate immunity. *Pediatric Research*, 61, 520–24.

Coffey, D. S. (2001). Similarities of prostate and breast cancer: Evolution, diet, and estrogens. *Urology*, 57, 31–38.

Cofran, Z. (2018). Brain size growth in wild and captive chimpanzees (*Pan troglodytes*). *American Journal of Primatology*, 80, e22876.

Cohen, J. E. (1995). How many people can the earth support? *Sciences*, 35, 18–23.

Cohen, J. M., Lajeunesse, M. J., and Rohr, J. R. (2018). A global synthesis of animal phenological responses to climate change. *Nature Climate Change*, 8, 224–28.

Coiner-Collier, S., Scott, R. S., Chalk-Wilayto, J., Cheyne, S. M., Constantino, P., Dominy, N. J., et al. (2016). Primate dietary ecology in the context of food mechanical properties. *Journal of Human Evolution*, 98, 103–18.

Colas, R. A., Souza, P. R., Walker, M. E., Burton, M., Zaslona, Z., Curtis, A. M., et al. (2018). Impaired production and diurnal regulation of vascular RvDn-3 DPA increase systemic inflammation and cardiovascular disease. *Circulation Research*, 122(6), 855–63.

Colborn, T., Saal, F. S. V., and Soto, A. M. (1993). Developmental effects of endocrine-disrupting chemicals in wildlife and humans. *Environmental Health Perspectives*, 101, 378–84.

Coleman, A., Richardson, D., Schechter, R., and Blumstein, D. T. (2008). Does habituation to humans influence predator discrimination in Gunther's dik-diks (*Madoqua guentheri*)? *Biology Letters*, 4, 250–52.

Coleman, B. T., and Hill, R. A. (2014). Living in a landscape of fear: The impact of predation, resource availability and habitat structure on primate range use. *Animal Behaviour*, 88, 165–73.

Coleman, B. T., Setchell, J. M. and Hill, R. A. (2021) Seasonal variation in the behavioural ecology of samango monkeys (*Cercopithecus albogularis schwarzi*) in a southern latitude montane environment. *Primates*, 62, 1005–18.

Coleman, M. N. (2009). What do primates hear? A meta-analysis of all known nonhuman primate behavioral audiograms. *International Journal of Primatology*, 30, 55–91.

Coley, P. (1983). Herbivory and defensive characteristics of tree species in a lowland tropical forest. *Ecological Monographs*, 53, 209–33.

Coley, P., Massa, M., Lovelock, C. E., and Winter, K. (2002). Effects of elevated CO_2 on foliar chemistry of saplings of nine species of tropical tree. *Oecologia*, 133, 62–69.

Collias, N., and Southwick, C. (1952). A field study of population density and social organization in howling monkeys. *Proceedings of the American Philosophical Society*, 96, 143–56.

Colquhoun, I. C. (1993). The socioecology of *Eulemur macaco*: A preliminary report. In C. P. van Schaik and P. M. Kappeler (Eds.), *Lemur Social Systems and Their Ecological Basis* (pp. 11–23). Springer.

Conard, N. J., Serengeli, J., Bohner, U., Starkovich, B. M., Miller, C. E., Urban, B., et al. (2015). Excavations at Schoningen and paradigm shifts in human evolution. *Journal of Human Evolution*, 89, 1–17.

Conklin, N. L., and Wrangham, R. W. (1994). The value of figs to a hind-gut fermenting frugivore—a nutritional analysis. *Biochemical Systematics and Ecology*, 22, 137–51.

Conklin-Brittain, N. L., Dierenfeld, E. S., Wrangham, R. W., Norconk, M., and Silver, S. C. (1999). Chemical protein analysis: A comparison of Kjeldahl crude protein and total ninhydrin protein from wild, tropical vegetation. *Journal of Chemical Ecology*, 25, 2601–22.

Conklin-Brittain, N. L., Knott, C. D., and Wrangham, R. W. (2006). Energy intake by wild chimpanzees and orangutans: Methodological considerations and a preliminary comparison. In G. Hohmann, M. M. Robbins, and C. Boesch (Eds.), *Feeding Ecology in Apes and Other Primates: Ecological, Physical and Behavioral Aspects* (pp. 445–71). Cambridge University Press.

Conklin-Brittain, N. L., Wrangham, R. W., and Hunt, K. D. (1998). Dietary response of chimpanzees and cercopithecines to seasonal variation in fruit abundance. II. Macronutrients. *International Journal of Primatology*, 19, 971–98.

Conradt, L., Krause, J., Couzin, I. D., and Roper, T. J. (2009). "Leading according to need" in self-organizing groups. *American Naturalist*, 173, 304–12.

Conradt, L., and List, C. (2009). Group decision making in humans and animals. *Royal Society Philosophical Transactions Biological Sciences*, 364, 719–852.

Conradt, L., and Roper, T. J. (2003). Group decision-making in animals. *Nature*, 421, 155–58.

Conradt, L., and Roper, T. J. (2005). Consensus decision making in animals. *Trends in Ecology and Evolution*, 20, 449–56.

Conradt, L., and Roper, T. J. (2007). Democracy in animals: The evolution of shared group decisions. *Proceedings of the Royal Society B: Biological Sciences*, 274, 2317–26.

Conradt, L., and Roper, T. J. (2009). Conflicts of interest and the evolution of decision sharing. *Royal Society Philosophical Transactions Biological Sciences*, 364, 807–19.

Conroy, M. E., and Walker, W. A. (2008). Intestinal immune health. *Nestle Nutrition Workshop Series: Pediatric Program*, 62, 111–21.

Constantino, P. J., Borrero-Lopez, O., Pajares, A., and Lawn, B. R. (2015). Simulation of enamel wear for reconstruction of diet and feeding behavior in fossil animals: A micromechanics approach. *BioEssays*, 38, 89–99.

Constantino, P. J., Lee, J. J. W., Gerbig, Y., Hartstone-Rose, A., Talebi, M., Lawn, B. R., et al. (2012). The role of tooth enamel mechanical properties in primate dietary adaptation. *American Journal of Physical Anthropology*, 148(2), 171–77.

Constantino, P. J., Lucas, P. W., Lee, J. J. W., and Lawn, B. R. (2009). The influence of fallback foods on great ape tooth enamel. *American Journal of Physical Anthropology*, 140(4), 653–60.

Coogan, S. C., Machovsky-Capuska, G. E., Senior, A. M., Martin, J. M., Major, R. E., and Raubenheimer, D. (2017). Macronutrient selection of free-ranging urban Australian white ibis (*Threskiornis moluccus*). *Behavioral Ecology*, 28, 1021–29.

Coogan, S., and Raubenheimer, D. (2016). Might macronutrient requirements influence grizzly bear-human conflict? Insights from nutritional geometry. *Ecosphere*, 7, e01204.

Coogan, S. C., Raubenheimer, D., Stenhouse, G. B., Coops, N. C., and Nielsen, S. E. (2018). Functional macronutritional generalism in a large omnivore, the brown bear. *Ecology and Evolution*, 28, 1021–29.

Cook, N. (1939). Notes on captive *Tarsius carbonarious*. *Journal of Mammalogy*, 20, 173–178.

Cook, S. C., and Behmer, S. T. (2010). Macronutrient regulation in the tropical terrestrial ant *Ectatomma ruidum* (Formicidae): A field study in Costa Rica. *Biotropica*, 42, 135–39.

Cook, S. C., Eubanks, M. D., Gold, R. E., and Behmer, S. T. (2010). Colony-level macronutrient regulation in ants: Mechanisms, hoarding and associated costs. *Animal Behaviour*, 79, 135–39.

Cooper, W. C., Good, R. A., and Mariani, T. (1974). Effects of protein insufficiency on immune responsiveness. *American Journal of Clinical Nutrition*, 27(6), 647–64.

Copeland, S. R., Sponheimer, M., de Ruiter, D. J., Lee-Thorp, J. A., Codron, D., le Roux, P. J., et al. (2011). Strontium isotope evidence for landscape use by early hominins. *Nature*, 474(7349), 76–U100.

Corbin, G. D., and Schmid, J. (1995). Insect secretions determine habitat use patterns by a female lesser mouse lemur (*Microcebus murinus*). *American Journal of Primatology*, 37, 317–24.

Cordain, L., Brand-Miller, J., Eaton, S. B., Mann, N., Holt, S. H. A., and Speth, J. D. (2000). Plant to animal subsistence ratios and macronutrient energy estimations in worldwide samples of hunter-gatherer diets. *American Journal of Clinical Nutrition*, 71, 682–92.

Cords, M. (1986). Interspecific and intraspecific variation in diet of two forest guenons, *Cercopithecus ascanius* and *C. mitis*. *Journal of Animal Ecology*, 55, 811–27.

Cords, M. (1987). Mixed species association of *Cercopithecus* monkeys in the Kakamega Forest, Kenya. *University of California Publications in Zoology*, 117, 1–109.

Cords, M. (1990). Mixed-species association of East African guenons: General patterns or specific examples? *American Journal of Primatology*, 21, 101–14.

Cords, M., and Chowdhury, S. (2010). Life history of *Cercopithecus mitis stuhlmanni* in the Kakamega Forest, Kenya. *International Journal of Primatology*, 31, 433–55.

Cords, M., and Wursig, B. (2014). A mix of species: Associations of heterospecifics among primates and dolphins. In J. Yamagiwa and L. Karczmarksi (Eds.), *Primates and Cetaceans: Field Research and Conservation of Complex Mammalian Societies* (pp. 409–32). Springer Science and Business Media.

Cork, S. J., and Foley, W. J. (1991). Digestive and metabolic strategies of arboreal folivores in relation to chemical defenses in temperate and tropical forests. In R. T. Palo and C. T. Robbins (Eds.), *Plant Defenses against Mammalian Herbivory* (pp. 133–66). CRC.

Cork, S. J., and Kenagy, G. J. (1989). Nutritional value of hypogeous fungus for a forest-dwelling ground squirrel. *Ecology*, 70, 577–86.

Corlett, R. T. (2011). How to be a frugivore (in a changing world). *Acta Oecologica*, 37, 674–81.

Corlett, R. T., and Lucas, P. W. (1990). Alternative seed-handling strategies in primates: Seed spitting by long-tailed macaques (*Macaca fascicularis*). *Oecologia*, 82, 166–71.

Corlett, R. T., and Primack, R. B. (2011). *Tropical Rain Forests: An Ecological and Biogeographical Comparison* (2nd ed.). Wiley-Blackwell.

Corso, J., Bowler, M., Heymann, E. W., Roos, C., and Mundy, N. I. (2016). Highly polymorphic colour vision in a New World monkey with red facial skin, the bald uakari (*Cacajao calvus*). *Proceedings of the Royal Society B: Biological Sciences*, 283, 20160067.

Costello, E. K., Lauber, C. L., Hamady, M., Fierer, N., Gordon, J. I., and Knight, R. (2009). Bacterial community variation in human body habitats across space and time. *Science*, 326, 1694–97.

Côté, I. M., and Poulin, R. (1995). Parasitism and group size in social animals: A meta-analysis. *Behavioural Ecology*, 6, 159–65.

Cotter, S. C., Reavey, C. E., Tummala, Y., Randall, J. L., Holdbrook, R., Ponton, F., et al. (2019). Diet modulates the relationship between immune gene expression and functional immune responses. *Insect Biochemistry and Molecular Biology*, 109, 128–41.

Cotter, S. C., Simpson, S. J., Raubenheimer, D., and Wilson, K. (2011). Macronutrient balance mediates trade-offs between immune function and life history traits. *Functional Ecology*, 25(1), 186–98.

Courser, W. D., and Dinsmore, J. J. (1975). Foraging associates of white ibis. *Auk*, 92, 599–601.

Coutsoudis, A., Kiepiela, P., Coovadia, H. M., and Broughton, M. (1992). Vitamin-A supplementation enhances specific IgG antibody-levels and total lymphocyte numbers while improving morbidity in measles. *Pediatric Infectious Disease Journal*, 11(3), 203–9.

Couzin, I. D., and Krause, J. (2003). Self-organization and collective behavior in vertebrates. In P. J. B. Slater, J. S. Rosenblatt, C. T. Snowdon, and T. J. Roper (Eds.), *Advances in the Study of Behavior* (vol. 32, pp. 1–75). Academic Press.

Couzin, I. D., Ioannou, C. C., Demirel, G., Gross, T., Torney, C. J., Hartnett, A., et al. (2011). Uninformed individuals promote Democratic consensus in animal groups. *Science*, 334, 1578–80.

Couzin, I. D., Krause, J., Franks, N. R., and Levin, S. A. (2005). Effective leadership and decision-making in animal groups on the move. *Nature*, 433, 513–16.

Cowling, R. A., Potts, A. J., Franklin, J., Midgley, J. F., Englebrecht, F., and Marean, C.W. (2020). Describing a drowned Pleistocene ecosystem: Last glacial maximum vegetation reconstruction of the Palaeo-Agulhas Plain. *Quaternary Science Reviews*, 235, 105866.

Cowlishaw, G. (1997). Trade-offs between foraging and predation risk determine habitat use in a desert baboon population. *Animal Behaviour*, 53, 667–86.

Cowlishaw, G. (1998). The role of vigilance in the survival and reproductive strategies of desert baboons. *Behaviour*, 135, 431–52.

Cowlishaw, G., and Dunbar, R. I. M. (2000). *Primate Conservation Biology*. University of Chicago Press.

Craine, J. M., Elmore, A. J., Aidar, M. P. M., Bustamante, M., Dawson, T. E., Hobbie, E. A., et al. (2009). Global patterns of foliar nitrogen isotopes and their relationships with climate, mycorrhizal fungi, foliar nutrient concentrations, and nitrogen availability. *New Phytologist*, 183, 980–92.

Craine, J. M., Elmore, A. J., Olson, K. C., and Tolleson, D. R. (2010). Climate change and cattle nutritional stress. *Global Change Biology*, 16, 2901–11.

Creel, S. (2011). Toward a predictive theory of risk effects: Hypotheses for prey attributes and compensatory mortality. *Ecology*, 92, 2190–95.

Creel, S., and Christianson, D. (2008). Relationships between direct predation and risk effects. *Trends in Ecology and Evolution*, 23, 194–201.

Creel, S., and Creel, N. M. (1995). Communal hunting and pack size in African wild dogs (*Lycaon pictus*). *Animal Behaviour*, 50, 1325–39.

Creel, S., Christianson, D., Liley, S., and Winnie, J. A. (2007). Predation risk affects reproductive physiology and demography of elk. *Science*, 315, 960.

Creel, S., Fox, J. E., Hardy, A., Sands, J., Garrott, B., and Peterson, R. O. (2002). Snowmobile activity and glucocorticoid stress responses in wolves and elk. *Conservation Biology*, 16, 809–14.

Creel, S., Schuette, P., and Christianson, D. (2014). Effects of predation risk on group size, vigilance, and foraging behavior in an African ungulate community. *Behavioral Ecology*, 25, 773–84.

Creel, S., Winnie, J. A., and Christianson, D. (2013). Underestimating the frequency, strength and cost of antipredator responses with data from GPS collars: An example with wolves and elk. *Ecology and Evolution*, 3, 5189–200.

Creel, S., Winnie, J., Maxwell, B., Hamlin, K., and Creel, M. (2005). Elk alter habitat selection as an antipredator response to wolves. *Ecology*, 86, 3387–97.

Crespi, B., and Summers, K. (2005). Evolutionary biology of cancer. *Trends in Ecology and Evolution*, 20, 545–52.

Cresswell, W., and Quinn, J. L. (2013). Contrasting risks from different predators change the overall nonlethal effects of predation risk. *Behavioral Ecology*, 24, 871–76.

Cristescu, B., Stenhouse, G. B., and Boyce, M. S. (2013). Perception of human-derived risk influences choice at top of the food chain. *PLoS ONE*, 8, e82738.

Crockett, C. M. (1998). Conservation biology of the genus *Alouatta*. *International Journal of Primatology*, 19, 549–78.

Crockett, C. M., and Janson, C. H. (2000). Infanticide in red howlers: Female group size, male membership, and a possible link to folivory. In C. P. van Schaik and C. H. Janson (Eds.), *Infanticide by Males and Its Implications*. Cambridge University Press.

Crockett, C. M., and Rudran, R. (1987). Red howler monkey birth data II: Interannual, habitat, and sex comparisons. *American Journal of Primatology*, 13, 369–84.

Croes, B. M., Laurance, W. F., Lahm, S. A., Tchignoumba, L., Alonso, A., Lee, M. E., et al. (2007). The influence of hunting on antipredator behavior in central African monkeys and duikers. *Biotropica*, 39, 257–63.

Crofoot, M. C. (2013). The cost of defeat: Capuchin groups travel further, faster and later after losing conflicts with neighbors. *American Journal of Physical Anthropology*, 152, 79–85.

Crofoot, M. C. (2021). "Next-gen" tracking in primatology: Opportunities and challenges. In C. A. Shaffer, F. L. Dolins, J. R. Hickey, N. Nibbelink, and L. M. Porter (Eds.), *GPS and GIS for Primatologists: A Practical Guide to Spatial Analysis.* (pp. 42–63) Cambridge University Press.

Crofoot, M. C., Lambert, T. D., Kays, R., and Wikelski, M. C. (2010). Does watching a monkey change its behaviour? Quantifying observer effects in habituated wild primates using automated radiotelemetry. *Animal Behaviour,* 80, 475–80.

Crompton, A. W., and Hiiemae, K. (1969). Functional occlusion in tribosphenic molars. *Nature,* 222(5194), 678–79.

Crompton, A. W., and Hiiemae, K. (1970). Molar occlusion and mandibular movements during occlusion in the American opossum, *Didelphis marsupialis. Zoological Journal of the Linnaean Society,* 49, 21–47.

Crook, J. H. (1964). The evolution of social organisation and visual communication in the weaver birds (Ploceinae). *Behaviour,* 10, 1–178.

Crook, J. H., and Gartlan, J. C. (1966). Evolution of primate societies. *Nature,* 210, 1200–1203.

Crowley, B. (2012). Stable isotope techniques and applications for primatologists. *International Journal of Primatology,* 33, 673–701.

Crowley, B. E. (2014). Oxygen isotope values in bone carbonate and collagen are consistently offset for New World monkeys. *Biology Letters,* 10, 20140759.

Crowley, B. E., Blanco, M. B., Arrigo-Nelson, S. J., and Irwin, M. T. (2013). Stable isotopes document resource partitioning and effects of forest disturbance on sympatric cheirogaleid lemurs. *Naturwissenschaften,* 100, 943–56.

Crowley, B. E., Carter, M. L., Karpanty, S. M., Zihlman, A. L., Koch, P. L., and Dominy, N. J. (2010). Stable carbon and nitrogen isotope enrichment in primate tissues. *Oecologia,* 164(3), 611–26.

Crowley, B. E., and Godfrey, L. R. (2013). Why all those spines? Anachronistic defences in the Didiereoideae against now extinct lemurs. *South African Journal of Science,* 109, Art. 1346.

Crowley, B. E., Godfrey, L. R., Bankoff, R. J., Perry, G. H., Culleton, B. J., Kennett, D. J., et al. (2017). Island-wide aridity did not trigger recent megafaunal extinctions in Madagascar. *Ecography,* 40, 901–12.

Crowley, B. E., Godfrey, L. R., Guilderson, T. P., Zermeño, P., Koch, P. L., and Dominy, N. J. (2012). Extinction and ecological retreat in a community of primates. *Proceedings of the Royal Society B: Biological Sciences,* 279, 3597–605.

Crowley, B. E., McGoogan, K. C., and Lehman, S. M. (2012). Edge effects on foliar stable isotope values in a Madagascan tropical dry forest. *PLoS ONE,* 7, e44538.

Crowley, B. E., Melin, A. D., Yeakel, J. D., and Dominy, N. J. (2015). Do oxygen isotope values in collagen reflect the ecology and physiology of Neotropical mammals? *Frontiers in Ecology and Evolution,* 3, 127.

Crowley, B. E., Rasoazanabary, E., and Godfrey, L. R. (2014). Stable isotopes complement focal individual observations and confirm dietary variability in reddish-gray mouse lemurs (*Microcebus griseorufus*) from southwestern Madagascar. *American Journal of Physical Anthropology,* 155, 77–90.

Crowley, B. E., Reitsema, L. J., Oelze, V. M., and Sponheimer, M. (2016). Advances in primate stable isotope ecology—achievements and future prospects. *American Journal of Primatology,* 78, 995–1003.

Crowley, B. E., Thorén, S., Rasoazanabary, E., Vogel, E. R., Barrett, M. A., Zohdy, S., et al. (2011). Explaining geographical variation in the isotope composition of mouse lemurs (*Microcebus*). *Journal of Biogeography,* 28, 2106–21.

Crumière, A. J., Stephenson, C. J., Nagel, M., and Shik, J. Z. (2020). Using nutritional geometry to explore how social insects navigate nutritional landscapes. *Insects,* 11, 53.

Crutzen, P. J. (2006). The "anthropocene." In *Earth System Science in the Anthropocene* (pp. 13–18). Springer.

Cruzat, V. F., Krause, M., and Newsholme, P. (2014). Amino acid supplementation and impact on immune function in the context of exercise. *Journal of the International Society of Sports Nutrition,* 11(1), 61.

Cryan, J. F., O'Riordan, K. J., Cowan, C. S. M., Sandhu, K. V., Bastiaanssen, T. F. S., Boehme, M., et al. (2019). The microbiota-gut-brain axis. *Physiological Review,* 99, 1877–2013.

Cui, Z., Shao, Q., Grueter, C. C., Wang, Z., Lu, J., and Raubenheimer, D. (2019). Dietary diversity of an ecological and macronutritional generalist primate in a harsh high-latitude habitat, the Taihangshan macaque (*Macaca mulatta tcheliensis*). *American Journal of Primatology,* 81, e22965.

Cui, Z. W., Wang, Z. L., Shao, Q., Raubenheimer, D., and Lu, J. Q. (2018). Macronutrient signature of dietary generalism in an ecologically diverse primate in the wild. *Behavioral Ecology,* 29(4), 804–13.

Cui, Z., Wang, Z., Zhang, S., Wang, B., Lu, J., and Raubenheimer, D. (2020). Living near the limits: Effects of interannual variation in food availability on diet and reproduction in a temperate primate, the Taihangshan macaque (*Macaca mulatta tcheliensis*). *American Journal of Primatology,* 82, e23080.

Cui, Z. W., Zhang, Y., Yan, J. B., Zhang, Y. F., Dong, Y., Ren, C., et al. (2022). What does it mean to be a macronutrient generalist? A five-year case study in wild rhesus macaques (*Macaca mulatta*). *Zoological Research,* 43(6), 935–39.

Cunningham, E., and Janson, C. (2007). Integrating information about location and value of resources by white-faced saki monkeys (*Pithecia pithecia*). *Animal Cognition,* 10, 293–304.

Cunningham, E., and Janson, C. H. (2013). Effect of fruit scarcity on use of spatial memory in a seed predator, white-faced saki (*Pithecia pithecia*). *International Journal of Primatology,* 34, 808–22.

Cunningham, E. P., Edmonds, D., Stalter, L., and Janal, M. N. (2021). Ring-tailed lemurs (*Lemur catta*) use olfaction to locate distant fruit. *American Journal of Physical Anthropology*, 175(1), 300–307.

Cunningham, J. J. (1995). Body composition and nutrition support in pediatrics: What to defend and how soon to begin. *Nutrition in Clinical Practice*, 10(5), 177–82.

Cunningham-Rundles, S. (2002). Evaluation of the effects of nutrients on immune function. In P. Calder, C. Field, and H. Gill (Eds.), *Nutrition and Immune Function* (pp. 21–39). CAB International.

Cunningham-Rundles, S., McNeeley, D. F., and Moon, A. (2005). Mechanisms of nutrient modulation of the immune response. *Journal of Allergy and Clinical Immunology*, 115(6), 1119–28.

Cuozzo, F. P., and Sauther, M. L. (2006). Severe wear and tooth loss in wild ring-tailed lemurs (*Lemur catta*): a function of feeding ecology, dental structure, and individual life history. *Journal of Human Evolution*, 51, 490-505.

Cuozzo, F. P., and Sauther, M. L. (2012). What is dental ecology? *American Journal of Physical Anthropology*, 148(2), 163–70.

Cuozzo, F. P., Head, B. R., Sauther, M. L., Ungar, P. S., and O'Mara, M. T. (2014). Sources of tooth wear variation early in life among known aged wild ring-tailed lemurs (*Lemur catta*) at the Bezà Mahafaly Special Reserve, Madagascar. *American Journal of Primatology*, 76, 1037–48.

Cuozzo, F. P., Ungar, P. S., and Sauther, M. L. (2012). Primate dental ecology: How teeth respond to the environment. *American Journal of Physical Anthropology*, 148(2), 159–62.

Curran, L. M. (2000). Impact of logging and El Niño on the forests of Indonesia: Reproductive failure of rain forest trees. *Environmental Science and Policy*, 7, 1–7.

Curtis, P., and Wang, X. W. (1998). A meta-analysis of elevated CO_2 effects on woody plant mass, form, and physiology. *Oecologia*, 113, 299–313.

Cuy, J. L., Mann, A. B., Livi, K. J., Teaford, M. F., and Weihs, T. P. (2002). Nanoindentation mapping of the mechanical properties of human molar tooth enamel. *Archives of Oral Biology*, 47(4), 281–91.

Daan, S., Masman, D., and Groenewold, A. (1990). Avian basal metabolic rates: Their association with body composition and energy expenditure in nature. *American Journal of Physiology—Regulatory, Integrative, and Comparative Physiology*. 259, R333-R340.

da Costa, R. R., Vreeburg, S. M., Shik, J. Z., Aanen, D. K., and Poulsen, M. (2019). Can interaction specificity in the fungus-farming termite symbiosis be explained by nutritional requirements of the fungal crop? *Fungal Ecology*, 38, 54–61.

Daegling, D. J., McGraw, W. S., Ungar, P. S., Pampush, J. D., Vick, A. E., and Bitty, E. A. (2011). Hard-object feeding in sooty mangabeys (*Cercocebus atys*) and interpretation of early hominin feeding ecology. *PLoS ONE*, 6(8), e23095.

Dalerum, F., and Angerbjorn, A. (2005). Resolving temporal variation in vertebrate diets using naturally occurring stable isotopes. *Oecologia*, 144, 647–58.

Dallas, D. C., and German, J. B. (2017). Enzymes in human milk. In Isolauri, E., Sherman, P. M. and Walker, W, A. (eds.), *Intestinal Microbiome: Functional Aspects in Health and Disease* (pp. 129–36). Karger.

Dallas, D. C., Guerrero, A., Khaldi, N., Borghese, R., Bhandari, A., Underwood, M. A., et al. (2014). A peptidomic analysis of human milk digestion in the infant stomach reveals protein-specific degradation patterns. *Journal of Nutrition*, 144(6), 815–20.

Dammhahn, M., and Kappeler, P. M. (2008). Comparative feeding ecology of sympatric *Microcebus berthae* and *M. murinus*. *International Journal of Primatology*, 29, 1567–89.

Dammhahn, M., and Kappeler, P. M. (2014). Stable isotope analyses reveal dense trophic species packing and clear niche differentiation in a Malagasy primate community. *American Journal of Physical Anthropology*, 153, 249–59.

Dangles, O., Irschick, D., Chittka, L., and Casas, J. (2009). Variability in sensory ecology: Expanding the bridge between physiology and evolutionary biology. *Quarterly Review of Biology*, 84, 51–74.

Danish, L. M., Heistermann, M., Agil, M., and Engelhardt, A. (2015). Validation of a novel collection device for non-invasive urine sampling from free-ranging animals. *PLoS ONE*, 10(11), e0142051.

Dansgaard, W. (1964). Stable isotopes in precipitation. *Tellus*, 16, 436–68.

Darden, D. B., Kelly, L. S., Fenner, B. P., Moldawer, L. L., Mohr, A. M., and Efron, P. A. (2021). Dysregulated immunity and immunotherapy after sepsis. *Journal of Clinical Medicine*, 10(8), 1742.

Darimont, C. T., Paquet, P. C., and Reimchen, T. E. (2007). Stable isotopic niche predicts fitness of prey in a wolf–deer system. *Biological Journal of the Linnean Society*, 90, 125–37.

Darnell, L. A., Teaford, M. F., Livi, K. J., and Weihs, T. P. (2010). Variations in the mechanical properties of *Alouatta palliata* molar enamel. *American Journal of Physical Anthropology*, 141(1), 7–15.

Darrah, C. H., Van Soest, P. J., and Fick, G. W. (1977). Microwave treatment and heat damage artifacts in forages. *Agronomy Journal*, 69, 120–21.

Dart, R. A. (1957). *The Osteodontokeratic Culture of Australopithecus prometheus.* Transvaal Museum.

Darvell, B. W. (2009). *Materials Science for Dentistry,9th ed.* Woodhead, Cambridge.

Darvell, B. W., Lee, P. K. D., Yuen, T. D. B., & Lucas, P. W. (1996). A portable fracture toughness tester for biological materials. *Measurement Science and Technology*, 7, 954–62.

Darwin C. 1861. *On the Origin of Species*. London: John Murray.

Das, U. N. (2011). Lipoxins as biomarkers of lupus and other inflammatory conditions. *Lipids in Health and Disease*, 10, 76.

Dasilva, G. L. (1992). The western black-and-white colobus as a

Dasilva, G. L. (1994). Diet of *Colobus polykomos* on Tiwai Island—selection of food in relation to its seasonal abundance and nutritional quality. *International Journal of Primatology*, 15, 655–80.

Dasmann, R. F. (1964). *Wildlife Biology*. Wiley and Sons.

Dattilo, W., Serio-Silva, J. C., Chapman, C. A., and Rico-Gray, V. (2014). Highly nested diets in intrapopulation monkey-resource food webs. *American Journal of Primatology*, 76, 670–78.

Dauber, E., Fredericksen, T. S., and Pena, M. (2005). Sustainability of timber harvesting in Bolivian tropical forests. *Forest Ecology and Management*, 214, 294–304.

Dausch Ibañez, D., Hernandez Salazar, L. T., and Laska, M. (2019). Taste responsiveness of spider monkeys to dietary ethanol. *Chemical Senses*, 44(8), 631–38.

Dausmann, K. H., Glos, J., and Heldmaier, G. (2009). Energetics of tropical hibernation. *Journal of Comparative Physiology B*, 179, 345–57.

Dausmann, K. H., Glos, J., Ganzhorn, J. U., and Heldmaier, G. (2004). Hibernation in a tropical primate. *Nature*, 429, 825–26.

Davenport, T. R. B., de Luca, D. W., Bracebridge, C. E., Machaga, S. J., Mpunga, N. E., Kibure, O., et al. (2010). Diet and feeding patterns in the kipunji (*Rungwecebus kipunji*) in Tanzania's Southern Highlands: A first analysis. *Primates*, 51, 213–20.

Davies, A. C., Radford, A. N., and Nicol, C. J. (2014). Behavioural and physiological expression of arousal during decision-making in laying hens. *Physiology & Behavior*, 123, 93–99.

Davies, A. G. (1991). Seed-eating by red leaf monkeys (*Presbytis rubicunda*) in dipterocarp forest of northern Borneo. *International Journal of Primatology*, 12, 119–44.

Davies, A. G. (1994). Colobine populations. In A. G. Davies and J. F. Oates (Eds.), *Colobine Monkeys: Their Ecology, Behaviour, and Evolution* (pp. 285–310). Cambridge University Press.

Davies, A. G., and Baillie, I. C. (1988). Seed-eating by red leaf monkeys (*Presbytis rubicunda*) in Sabah, Northern Borneo. *Biotropica*, 20, 252–58.

Davies, A. G., Bennett, E. L., and Waterman, P. G. (1988). Food selection by two Southeast Asian colobine monkeys (*Presbytis rubicunda* and *Presbytis melalophos*) in relation to plant chemistry. *Biological Journal of the Linnean Society*, 34, 33–56.

Davies, A. G., Oates, J. F., and Dasilva, G. L. (1999). Patterns of frugivory in three West African colobine monkeys. *International Journal of Primatology*, 20, 327–57.

Davies, C. R., Ayres, J. M., Dye, C., and Deane, L. M. (1991). Malaria infection rate of Amazonian primates increases with body weight and group size. *Functional Ecology*, 5, 655–62.

Davis, G. H., Crofoot, M. C., and Farine, D. R. (2022). Using optimal foraging theory to infer how groups make collective decisions. *Trends in Ecology and Evolution*, 37, 942–52.

Davis, M. J., Thokala, S., Xing, X., Hobbs, N. T., Miller, M. W., Han, R., et al. (2013). Testing the functionality and contact error of a GPS-based wildlife tracking network. *Wildlife Society Bulletin*, 37, 855–61.

Dawkins, R., and Krebs, J. R. (1979). Arms races between and within species. *Proceedings of the Royal Society of London B: Biological Sciences*, 205(1161), 489–511.

de Abreu, T., Tavares, M. C. H., Bretas, R., Rodrigues, R. C., Pissinati, A., and Aversi-Ferreira, T. A. (2021). Comparative anatomy of the encephalon of new world primates with emphasis for the Sapajus sp. *PLoS One*. 16, e0256309.

de Andrade Carneiro, L., Moreno, T. B., Fernandes, B. D., Souza, C. M. M., Bastos, T. S., Félix, A. P., et al. (2021). Effects of two dietary fiber levels on nutrient digestibility and intestinal fermentation products in captive brown howler monkeys (*Alouatta guariba*). *American Journal of Primatology*, 83, e23238.

De Block, M., and Stoks, R. (2008). Short-term larval food stress and associated compensatory growth reduce adult immune function in a damselfly. *Ecological Entomology*, 33(6), 796–801.

DeCasien, A. R., and Higham, J. P. (2019). Primate mosaic brain evolution reflects selection on sensory and cognitive specialization. *Nature Ecology & Evolution*, 3(10), 1483–93.

DeCasien, A. R., Williams, S. A., and Higham, J. P. (2017). Primate brain size is predicted by diet but not sociality. *Nature ecology & evolution*, 1(5), 0112.

de Heinzelin, J., Clark, J. D., White, T. W., Hart, W., Renne, P., Woldegabriel, G., et al. (1999). Environment and behavior of 2.5-million-year-old Bouri hominids. *Science*, 284, 625–29.

de Jong, Y. A., and Butynski, T. M. (2021). Is the southern patas monkey *Erythrocebus baumstarki* Africa's next primate extinction? Reassessing taxonomy, distribution, abundance and conservation. *American Journal of Primatology*, 83, e23316.

De la Fuente, M. F., Schiel, N., Bicca-Marques, J. C., Caselli, C. B., Souto, A., and Garber, P. A. (2019). Balancing contest competition, scramble competition, and social tolerance at feeding sites in wild common marmosets (*Callithrix jacchus*). *American Journal of Primatology*, 81(4), e22964.

De la Fuente, M. F., Sueur, C., Garber, P. A., Bicca-Marques, J. C., Souto, A., and Schiel, N. (2022). Foraging networks and social tolerance in a cooperatively breeding primate (*Callithrix jacchus*). *Journal of Animal Ecology*, 91(1), 138–53.

de Luna, A. G., Link, A., Montes, A., Alfonso, F., Mendieta, L., and Di Fiore, A. (2017). Increased folivory in brown spider monkeys *Ateles hybridus* living in a fragmented forest in Colombia. *Endangered Species Research*, 32, 123–34.

DePasquale, A. N., Webb, S. E., Williamson, R. E., Fedigan, L. M., and Melin, A. D. (2021). Testing the niche differentiation hypothesis in wild capuchin monkeys with polymorphic color vision. *Behavioral Ecology*, 32(4), 599–608.

DePasquale, A., Hogan, J. D., Guadamuz Araya, C., Dominy, N. J., and Melin, A. D. (2022). Aeroscapes and the sensory ecology of olfaction in a tropical dry forest. *Frontiers in Ecology and Evolution*, 10, 347.

de Ruiter, J. (1986). The influence of group size on predator scanning and foraging behaviour of wedgecapped capuchin monkeys (*Cebus olivaceus*). *Behaviour*, 98, 240–58.

de Weerth, C., Aatsinki, A. K., Azad, M. B., Bartol, F. F., Bode, L., Collado, M. C., et al. (2022). Human milk: From complex tailored nutrition to bioactive impact on child cognition and behavior. *Critical Reviews in Food Science and Nutrition*, 63, 7945-7982.

Dean, M. C. (2010). Retrieving chronological age from dental remains of early fossil hominins to reconstruct human growth in the past. *Philosophical Transactions of the Royal Society of London Series B—Biological Sciences*, 365, 3397–410.

Dean, M. C., Jones, M. E., and Pilley, J. R. (1992, Jan). The natural history of tooth wear, continuous eruption and periodontal disease in wild shot great apes. *Journal of Human Evolution*, 22(1), 23–39.

Dearing, D. M. (2013). Temperature-dependent toxicity in mammals with implications for herbivores: A review. *Journal of Comparative Physiology B*, 183, 43–50.

Dearing, D. M., Foley, W. J., and McLean, S. (2005). The influence of plant secondary metabolites on the nutritional ecology of herbivorous terrestrial vertebrates. *Annual Review of Ecology, Evolution and Systematics*, 36, 169–89.

Deblauwe, I., and Janssens, G. P. J. (2008). New insights in insect prey choice by chimpanzees and gorillas in southeast Cameroon: The role of nutritional value. *American Journal of Physical Anthropology*, 135, 42–55.

DeBose, J. L., and Nevitt, G. A. (2008). The use of odors at different spatial scales: Comparing birds with fish. *Journal of Chemical Ecology*, 34, 867–81.

Deecke, V. (2012). Tool-use in the brown bear (*Ursus arctos*). *Animal Cognition*, 15, 725–30.

Defler, T. R. (1985). Contiguous distribution of two species of *Cebus* monkeys in El Tuparro National Park, Colombia. *American Journal of Primatology*, 8, 101–12.

DeGabriel, J. L., Moore, B. D., Felton, A. M., Ganzhorn, J. U., Stolter, C., Wallis, I. R., et al. (2014). Translating nutritional ecology from the laboratory to the field: Milestones in linking plant chemistry to population regulation in mammalian browsers. *Oikos*, 123, 298–308.

DeGabriel, J. L., Moore, B. D., Foley, W. J., and Johnson, C. N. (2009). The effects of plant defensive chemistry on nutrient availability predict reproductive success in a mammal. *Ecology*, 90, 711–19.

DeGabriel, J. L., Moore, B. D., Shipley, L. A., Krockenberger, A. K., Wallis, I. R., Johnson, C. N., et al. (2009). Inter-population differences in the tolerance of a marsupial folivore to plant secondary metabolites. *Oecologia*, 161, 539–48.

DeGabriel, J. L., Wallis, I. R., Moore, B. D., and Foley, W. J. (2008). A simple, integrative assay to quantify nutritional quality of browses for herbivores. *Oecologia*, 156, 107–16.

Del Monte-Luna, P., Brook, B. W., Zetina-Rejón, M. J., and Cruz-Escalona, V. H. (2004). The carrying capacity of ecosystems. *Global Ecology and Biogeography*, 13, 485–95.

Del Prado, M., Villalponda, S., Elizondo, A., Rodríguez, M., Demmelmair, H., and Koletzko, B. (2001). Contribution of dietary and newly formed arachidonic acid to human milk lipids in women eating a low-fat diet. *American Journal of Clinical Nutrition*, 74, 242–47.

Dela, J. D. S. (2007). Seasonal food use strategies of *Semnopithecus vetulus nestor*, at Panadura and Piliyandala, Sri Lanka. *International Journal of Primatology*, 28, 3.

Delano, M. J., and Moldawer, L. L. (2006). The origins of cachexia in acute and chronic inflammatory diseases. *Nutrition in Clinical Practice*, 21(1), 68–81.

Delehanty, B., and Boonstra, R. (2009). Impact of live trapping on stress profiles of Richardson's ground squirrel (*Spermophilus richardsonii*). *General and Comparative Endocrinology*, 160, 176–82.

Delgado, R. A., and van Schaik, C. P. (2000). The behavioral ecology and conservation of the orangutan (*Pongo pygmaeus*): A tale of two islands. *Evolutionary Anthropology*, 9, 201–18.

Delsuc, F., Metcalf, J. L., Parfrey, L. W., Song, S. J., Gonzalez, A., and Knight, R. (2014). Convergence of gut microbiomes in myrmecophagous mammals. *Molecular Ecology*, 23(6), 1301–17.

Dennis, J. C., Ungar, P. S., Teaford, M. F., and Glander, K. E. (2004). Dental topography and molar wear in *Alouatta palliata* from Costa Rica. *American Journal of Physical Anthropology*, 125(2), 152–61.

Dennis, T. E., Chen, W. C., Koefood, I. M., Lacoursiere, C. J., Walker, M. M., Laube, P., et al. (2010). Performance characteristics of small Global-Positioning-System tracking collars for terrestrial animals. *Wildlife Biology in Practice*, 6, 14–31.

DePaolo, D. J. (2004). Calcium isotopic variations produced by biological, kinetic, radiogenic and nucleosynthetic processes. *Review in Mineralogy and Geochemistry*, 55, 255–88.

Derrickson, E. M. (1992). Comparative reproductive strategies of altricial and precocial eutherian mammals. *Functional Ecology*, 6, 57–65.

DeSantis, L. R. G. 2016. Dental microwear textures: Reconstructing diets of fossil mammals. *Surface Topography: Metrology and Properties*, 4, 023002.

Deschner, T., Fuller, B. T., Oelze, V. M., Boesch, C., Hublin, J. J., Mundry, R., et al. (2012). Identification of energy consumption and nutritional stress by isotopic and elemental analysis of urine in bonobos (*Pan paniscus*). *Rapid Communications in Mass Spectrometry*, 26, 69–77.

Deschner, T., Fuller, B. T., Oelze, V., Ortmann, S., Richard, M. P., and G., H. (2010). Monitoring nutritional stress with urinary delta[15]N and C/N ratios in captive bonobos. *American Journal of Physical Anthropology*, suppl. 50, 93.

DeSilva, J. M. (2011). A shift toward birthing relatively large infants early in human evolution. *Proceedings of the National*

Academy of Sciences of the United States of America, 108, 1022–27.

Dettmer, A. M., Allen, J. M., Jaggers, R. M., and Bailey, M. T. (2019). A descriptive analysis of gut microbiota composition in differentially reared infant rhesus monkeys (*Macaca mulatta*) across the first 6 months of life. *American Journal of Primatology*, 81(10–11), e22969.

DeVore, I. (1965). *Primate Behavior: Field Studies of Monkeys and Apes*. Holt, Rinehart and Winston.

DeVore, I., and Washburn, S. (1963). Baboon ecology and human evolution. In F. Bourliére and C. F. Howell (Eds.), *African Ecology and Human Evolution* (pp. 335–67). Routledge.

Dew, J. L. (2005). Foraging, food choice, and food processing by sympatric ripe-fruit specialists: *Lagothrix lagotricha poeppigii* and *Ateles belzebuth belzebuth*. *International Journal of Primatology*, 26, 1107–35.

Dew, J. L., and Wright, P. (1998). Frugivory and seed dispersal by four species of primates in Madagascar's eastern rain forest. *Biotropica*, 30, 425–37.

Dewan, P., Kaur, I. R., Faridi, M. M., and Agarwal, K. N. (2009). Cytokine response to dietary rehabilitation with curd (Indian dahi) and leaf protein concentrate in malnourished children. *Indian Journal of Medical Research*, 130(1), 31–36.

Dewey, K. G. (1997). Energy and protein requirements during lactation. *Annual Review of Nutrition*, 17, 19–36.

Dewi, T., Imron, M. A., Lukmandaru, G., Hedger, K., Campera, M., and Nekaris, K. A. I. (2022). The sticky tasty: The nutritional content of the exudativorous diet of the Javan slow loris in a lowland forest. *Primates*, 63, 93–102.

Dezeure, J., Baniel, A., Carter, A., Cowlishaw, G., Godelle, B., and Huchard, E. (2021). Birth timing generates reproductive trade-offs in a non-seasonal breeding primate. *Proceedings of the Royal Society B: Biological Sciences*, 288, 20210286.

Dhondt, A. A. (1988). Carrying capacity: A confusing concept. *Acta Oecologica*, 9, 337–46.

Di Bitetti, M. S., and Janson, C. H. (2000). When will the stork arrive? Patterns of birth seasonality in Neotropical primates. *American Journal of Primatology*, 50, 109–30.

Di Bitetti, M. S., and Janson, C. H. (2001). Social foraging and the finder's share in capuchin monkeys, *Cebus apella*. *Animal Behaviour*, 62, 47–56.

Di Fiore, A. (2003). Ranging behavior and foraging ecology of lowland woolly monkeys (*Lagothrix lagotricha poeppigii*) in Yasuni National Park, Ecuador. *American Journal of Primatology*, 59, 47–66.

Di Fiore, A. (2004). Diet and feeding ecology of woolly monkeys in a western Amazonian rain forest. *International Journal of Primatology*, 25, 767–801.

Di Fiore, A., and Rendall, D. (1994). Evolution of social organization: A reappraisal for primates by using phylogenetic methods. *Proceedings of the National Academy of Sciences of the United States of America*, 91, 9941–45.

Di Fiore, A., and Rodman, P. S. (2001). Time allocation patterns of lowland woolly monkeys (*Lagothrix lagotricha poeppigii*) in a Neotropical terra firma forest. *International Journal of Primatology*, 22, 449–80.

Di Fiore, A., and Suarez, S. A. (2007). Route-based travel and shared routes in sympatric spider and woolly monkeys: Cognitive and evolutionary implications. *Animal Cognition*, 10, 317–29.

Di Stefano, A., Scata, M., Vijayakumar, S., Angione, C., La Corte, A., and Lio, P. (2019). Social dynamics modeling of chrono-nutrition. *PLoS Computational Biology*, 15, e1006714.

Dias, P. A. D., and Rangel-Negrín, A. (2015). Diets of howler monkeys. In M. M. Kowalewski, P. A. Garber, L. Cortés-Ortiz, B. Urbani, and D. Youlatos (Eds.), *Howler Monkeys: Behavior, Ecology, and Conservation* (pp. 21–56). Springer.

Dias, P. A. D., Rangel-Negrín, A., and Canales-Espinosa, D. (2011). Effects of lactation on the time-budgets and foraging patterns of female black howlers (*Alouatta pigra*). *American Journal of Physical Anthropology*, 145(1), 137–46.

Dias, P. A. D., Rangel-Negrin, A., Coyohua-Fuentes, A., and Canales-Espinosa, D. (2014). Factors affecting the drinking behavior of black howler monkeys (*Alouatta pigra*). *Primates*, 55, 1–4.

Diaz-Martin, Z., Swamy, V., Terborgh, J., Alvarex-Loayza, P., and Cornejo, F. (2014). Identifying keystone plant resources in an Amazonian forest using a long-term fruit-fall record. *Journal of Tropical Ecology*, 30, 291–301.

Dibakou, S. E., Basset, D., Souza, A., Charpentier, M., and Huchard, E. (2019). Determinants of variations in fecal neopterin in free-ranging mandrills. *Frontiers in Ecology and Evolution*, 7

Dierenfeld, E. S., du Toit, R., and Braselton, W. E. (1995). Nutrient composition of selected browses consumed by black rhinoceros (*Diceros bicornis*) in the Zambezi Valley, Zimbabwe. *Journal of Zoo and Wildlife Medicine*, 26, 220–30.

Dietz, J. M., Baker, A. J., and Miglioretti, D. (1994). Seasonal variation in reproduction, juvenile growth, and adult body mass in golden lion tamarins (*Leontopithecus rosalia*). *American Journal of Primatology*, 34, 115–32.

Dietz, J. M., Peres, C. A., and Pinder, L. (1997). Foraging ecology and use of space in wild golden lion tamarins (*Leontopithecus rosalia*). *American Journal of Primatology*, 41, 289–305.

Digby, L. J. (1995). Infant care, infanticide, and female reproductive strategies in polygynous groups of common marmosets (*Callithrix jacchus*). *Behavioral Ecology and Sociobiology*, 37, 51–61.

DiGiorgio, A. L., and Knott, C. D. (2017). Orangutans, fruit, and the geometric framework—fruit and non-fruit choice in wild *Pongo pygmaeus wurmbi*. 86th Annual Meeting of the American Association of Physical Anthropologists, New Orleans, LA.

DiGiorgio, A. L., Ma, Y., Upton, E., Gopal, S., Robinson, N., Susanto, T. W., et al. (2023). Famished frugivores or choosy consumers: A generalist frugivore (wild Bornean orangutans, *Pongo pygmaeus wurmbii*) leaves available fruit for non-fruit foods. *International Journal of Primatology*. 44, 377–98.

DiGiorgio, A. L., Susanto, T. W., and Knott, C. D. (2021). Behavioral evidence of wild Bornean orangutans navigating to non-fruit foods: implications for fallback foods. *American Journal of Physical Anthropology*, 174(S71).

Dill, L. M., and Fraser, A. H. G. (1984). Risk of predation and the feeding behavior of juvenile coho salmon (*Oncorhynchus kisutsch*). *Behavioral Ecology and Sociobiology*, 16, 65–71.

Dillis, C., Beaudrot, L., Clink, D. J., Feilen, K. L., Wittmer, H. U., and Marshall, A. J. (2015). Modeling the ecological and phenological predictors of fruit consumption by gibbons (*Hylobates albibarbis*). *Biotropica*, 47, 85–93.

Dils, R. R. (1986). Comparative aspects of milk fat synthesis. *Journal of Dairy Science*, 69, 904–10.

Dinets, V., Brueggen, J. C., and Brueggen, J. D. (2013). Crocodilians use tools for hunting. *Ethology, Ecology and Evolution*, 27, 74–78.

Dinsmore, M. P., Strier, K. B., and Louis, E. E. Jr. (2021). The influence of seasonality, anthropogenic disturbances, and cyclonic activity on the behavior of northern sportive lemurs (*Lepilemur septentrionalis*) at Montagne des Français, Madagascar. *American Journal of Primatology*, 83(12), e23333.

Dirzo, R., Young, H. S., Galetti, M., Ceballos, G., Isaac, N. J. B., and Collen, B. (2014). Defaunation in the Anthropocene. *Science*, 345, 401–6.

Dittus, W. P. (1979). The evolution of behaviors regulating density and age-specific sex ratios in a primate population. *Behaviour*, 69, 265–301.

Dittus, W. P. J. (1984). Toque macaque food calls: Semantic communication concerning food distribution in the environment. *Animal Behaviour*, 32, 470–77.

Dittus, W. P. J. (2013). Arboreal adaptations of body fat in wild toque macaques (*Macaca sinica*) and the evolution of adiposity in primates. *American Journal of Physical Anthropology*, 152, 333–44.

Dixon, R. A. (2004). Phytoestrogens. *Annual Review of Plant Biology*, 55, 225–61.

Dixson, A., and Fleming, D. (1981). Parental behaviour and infant development in owl monkeys (*Aotus trivirgatus griseimembra*). *Journal of Zoology*, 194, 25–39.

D'mello, J. P. F. (1992). Chemical constraints to the use of tropical legumes in animal nutrition. *Animal Feed Science and Technology*, 38, 237–61.

Dobson, A. P., and Lyles, A. (1989). The population dynamics and conservation of primate populations. *Conservation Biology*, 3, 362–80.

Doering, G., and Stroehle, A. (2015). Nutritional biology: A neglected basic discipline of nutritional science. *Genes and Nutrition*, 10, 55.

Dolado, R., Cooke, C., and Beltran, F. S. (2016). How many for lunch today? Seasonal fission-fusion dynamics as a feeding strategy in wild red-capped mangabeys (*Cercocebus torquatus*). *Folia Primatologica*, 87, 197–212.

Dolby, A. S., and Grubb, T. C. Jr. (1998). Benefits to satellite members in mixed-species foraging groups: An experimental analysis. *Animal Behaviour*, 56, 501–9.

Dolins, F. L. (2009). Captive cotton-top tamarins' (*Saguinus oedipus oedipus*) use of landmarks to localize hidden food items. *American Journal of Primatology*, 71, 316–23.

Dolins, F. L., Klimowicz, C., Kelley, J., and Menzel, C. R. (2014). Using virtual reality to investigate comparative spatial cognitive abilities in chimpanzees and humans. *American Journal of Primatology*, 76, 496–513.

Dolins, F. L., and Mitchell, R. W. (2010). Linking spatial cognition and spatial perception. In F. L. Dolins and R. W. Mitchell (Eds.), *Spatial Cognition, Spatial Perception: Mapping the Self and Space* (pp. 1–31). Cambridge University Press.

Dolotovskaya, S., and Heymann, E. W. (2020). Do less or eat more: Strategies to cope with costs of parental care in a pair-living monkey. *Animal Behaviour*, 163, 163–73.

Dominguez-Rodrigo, M. (2012). Conceptual premises in experimental design and their bearing on the use of analogy: A critical example from experiments on cut marks. In M. Dominguez-Rodrigo (Ed.), *Stone Tools and Fossil Bones: Debates in the Archaeology of Human Origins* (pp. 191–211). Cambridge University Press.

Domínguez-Rodrigo, M., Baquedano, E., Organista, E., Cobo-Sánchez, L., Mabulla, A., Maskara, V., et al. (2021). Early Pleistocene faunivorous hominins were not kleptoparasitic, and this impacted the evolution of human anatomy and socio-ecology. *Scientific Reports*, 11, 16135.

Dominguez-Rodrigo, M., and Barba, R. (2006). New estimates of tooth mark and percussion mark frequencies at the FLK Zinj site: the carnivore-hominid-carnivore hypothesis falsified. *Journal of Human Evolution*, 50, 170–94.

Dominguez-Rodrigo, M., Bunn, H., Mabulla, A. Z. P., Basquedono, E., Uribellarena, E., Pérez-Gonzaléz, A., et al. (2014). On meat eating and human evolution: A taphonomic analysis of BK 46 (Upper Bed, I. I., Olduvai Gorge, Tanzania) and its bearing on hominid megafaunal consumption. *Quaternary International*, 322, 129–52.

Dominguez-Rodrigo, M., Pickering, T. R., Semaw, S., and Rogers, M. J. (2005). Cutmarked bones from Pliocene archaeological sites at Gona, Afar, Ethiopia: Implications for the functions of the world's oldest stone tools. *Journal of Human Evolution*, 42, 109–21.

Dominy, N. J. (2004). Fruits, fingers, and fermentation: The sensory cues available to foraging primates. *Integrative and Comparative Biology*, 44, 295–303.

Dominy, N. J., and Lucas, P. W. (2001). Ecological importance of trichromatic vision to primates. *Nature*, 410, 363–66.

Dominy, N.J., Lucas, P.W. and N.S. Noor. 2006. Primate sensory systems and foraging behaviour. In Hohmann, G., Robbins, M.M. & C. Boesch (eds.). *Feeding ecology in apes and other primates* (pp. 489–509). Cambridge University Press.

Dominy, N. J., Lucas, P. W., Osorio, D., and Yamashita, N. (2001). The sensory ecology of primate food perception. *Evolutionary Anthropology*, 10, 171–86.

Dominy, N. J., and Melin, A. D. (2020). Liminal light and primate evolution. *Annual Review of Anthropology*, 49, 257–76.

Dominy, N. J., Vogel, E. R., Yeakel, J. D., Constantino, P., and Lucas, P. W. (2008). Mechanical properties of plant underground storage organs and implications for dietary models of early hominins. *Evolutionary Biology*, 35, 159–75.

Dominy, N. J., Yeakel, J. D., Bhat, U., Ramsden, L., Wrangham, R. W., and Lucas, P. W. (2016). How chimpanzees integrate sensory information to select figs. *Interface Focus*, 6, 20160001.

Donati, G., Kesch, K., Ndremifidy, K., Schmidt, S. L., Ramanamanjato, J. B., Borgogini-Tarli, S. M., et al. (2011). Better few than hungry: Flexible feeding ecology of collared lemurs *Eulemur collaris* in littoral forest fragments. *PLoS ONE*, 6, e19807.

Donatti, C. I., Guimarães, P. R., Galetti, M., Pizo, M. A., Marquitti, F. M. D., and Dirzo, R. (2011). Analysis of a hyperdiverse seed dispersal network: Modularity and underlying mechanisms. *Ecology Letters*, 14, 773–81.

Donnelly, J. E., Hill, J. O., Jacobsen, D. J., Potteiger, J., Sullivan, D. K., Johnson, S. L., et al. (2003). Effects of a 16-month randomized controlled exercise trial on body weight and composition in young, overweight men and women: the Midwest Exercise Trial. *Archives of Internal Medicine*, 163, 1343–50.

Donnet-Hughes, A., Perez, P. F., Dore, J., Leclerc, M., Levenez, F., Benyacoub, J., et al. (2010). Potential role of the intestinal microbiota of the mother in neonatal immune education. *Proceedings of the Nutrition Society*, 69, 407–15.

Donohue, S. L., DeSantis, L. R. G., Schubert, B. W., and Ungar, P. S. (2013). Was the giant short-faced bear a hyper-scavenger? A new approach to the dietary study of ursids using dental microwear textures. *PLoS ONE*, 8(10), e77531.

Donovan, S. M. (2006). Role of human milk components in gastrointestinal development: Current knowledge and future needs. *Journal of Pediatrics*, 149, S49–S61.

Donovan, S. M., and Odle, J. (1994). Growth factors in milk as mediators of infant development. *Annual Review of Nutrition*, 14, 147–67.

Dooley, H. M., and Judge, D. S. (2015). Kloss gibbon (*Hylobates klossii*) behavior facilitates the avoidance of human predation in the Peleonan Forest, Siberut Island, Indonesia. *American Journal of Primatology*, 77, 296–308.

Doran, D. M. (1996). Comparative positional behavior of the African apes. In W. C. McGrew, L. F. Marchant, and T. Nishida (Eds.), *Great Ape Societies* (pp. 213–24). Cambridge University Press.

Doran, D. (1997). Influences of seasonality on activity patterns, feeding behavior, ranging, and grouping patterns in Taï chimpanzees. *International Journal of Primatology*, 18, 183–206.

Doran, D. M., McNeilage, A., Greer, D., Bocian, C., Mehlman, P., and Shah, N. (2002). Western lowland gorilla diet and resource availability: New evidence, cross-site comparisons, and reflections on indirect sampling methods. *American Journal of Primatology*, 58, 91–116.

Doran-Sheehy, D. M., Mongo, P., Lodwick, J. L., Salmi, R., and Borries, C. (2012). Life history of wild western gorillas (*Gorilla gorilla*): New data and cross-site comparisons indicate gorillas do not grow and reproduce as fast as you think. *American Journal of Physical Anthropology*, 147, 133.

Doran-Sheehy, D., Mongo, P., Lodwick, J., and Conklin-Brittain, N. L. (2009). Male and female western gorilla diet: Preferred foods, use of fallback resources, and implications for ape versus Old World monkey foraging strategies. *American Journal of Physical Anthropology*, 140(4), 727–38.

Dore, K. M., Hansen, M. F., Klegarth, A. R., Fitchel, C., Koch, F., Springer, A., et al. (2020). Review of GPS collar deployments and performance on nonhuman primates. *Primates*, 61, 373–87.

Dos Santos, G. G., Batool, S., Hastreiter, A., Sartori, T., Nogueira-Pedro, A., Borelli, P., et al. (2017). The influence of protein malnutrition on biological and immunomodulatory aspects of bone marrow mesenchymal stem cells. *Clinical Nutrition*, 36(4), 1149–57.

dos Santos-Barnett, T. C., Cavalcante, T., Boyle, S. A., Matte, A. L., Bezerra, B. M., de Oliveira, T. G., et al. (2022). Pulp Fiction: Why some populations of ripe-fruit specialists Ateles chamek and A. marginatus prefer insect-infested foods. *International Journal of Primatology*, 43(3), 384-408.

Doyle, G. A., Andersson, A., and Bearder, S. K. (1971). Reproduction in the lesser bushbaby (*Galago senegalensis moholi*) under semi-natural conditions. *Folia Primatologica*, 14, 15–22.

Doyle, G. A., Martin, R. D., and Niemitz, C. (1979). Outline of the behavior of *Tarsius bancanus*. In G. A. Doyle and R. D. Martin (Eds.), *The Study of Prosimian Behaviour* (pp. 631–60). Academic Press.

Doyle, J. J., and Luckow, M. A. (2003). The rest of the iceberg: Legume diversity and evolution in a phylogenetic context. *Plant Physiology*, 131, 900–910.

Dröscher, I., and Kappeler, P. M. (2014). Competition for food in a solitarily foraging folivorous primate (*Lepilemur leucopus*)? *American Journal of Primatology*, 76, 842–54.

Dröscher, I., Rothman, J. M., Ganzhorn, J. U., and Kappeler, P. M. (2016). Nutritional consequences of folivory in a small-bodied lemur (*Lepilemur leucopus*): Effects of season and reproduction on nutrient balancing. *American Journal of Physical Anthropology*, 160, 197–207.

Druce, D. J., Brown, J. S., Castley, J. G., Kerley, G. I. H., Kotler, B. P., Slotow, R., et al. (2006). Scale-dependent foraging costs: Habitat use by rock hyraxes (*Procavia capensis*) determined using giving-up densities. *Oikos*, 115, 513–25.

Duc, H. M., Baxter, G. S., and Page, M. J. (2009). Diet of *Pygathrix nigripes* in southern Vietnam. *International Journal of Primatology*, 30, 15–28.

Dudley, R. (2002). Fermenting fruit and the historical ecology of ethanol ingestion: Is alcoholism in modern humans an evolutionary hangover? *Addiction*, 97, 381–88.

Dugas, L. R., Harders, R., Merrill, S., Ebersole, K., Shoham, D. A., Rush, E. C., et al. (2011). Energy expenditure in adults living in developing compared with industrialized countries: a meta-analysis of doubly labeled water studies. *American Journal of Clinical Nutrition*, 93, 427–41.

Dufour, D. L., and Sauther, M. L. (2002). Comparative and evolutionary dimensions of the energetics of human pregnancy and lactation. *American Journal of Human Biology*, 14, 584–602.

Dumont, E. R. (1995). Enamel thickness and dietary adaptation among extant primates and chiropterans. *Journal of Mammalogy*, 76(4), 1127–36.

Dumont, E. R., Ryan, T. M., and Godfrey, L. R. (2011). The *Hadropithecus* conundrum reconsidered, with implications for interpreting diet in fossil hominins. *Proceedings of the Royal Society B: Biological Sciences*, 278, 3654–61.

Dunbar, R. I. M. (1984). *Reproductive Decisions—an Economic Analysis of Gelada Baboon Social Strategies*. Princeton University Press.

Dunbar, R. (1987). Habitat quality, population dynamics, and group composition in colobus monkeys (*Colobus guereza*). *International Journal of Primatology*, 8, 299–329.

Dunbar, R. (1988). *Primate Social Systems: Studies in Behavioral Adaptation*. Springer.

Dunbar, R. I. M. (1992). Neocortex size as a constraint on group size in primates. *Journal of Human Evolution*, 20, 469–93.

Dunbar, R. I. M. (1995). Neocortex size and group size in primates: A test of the hypothesis. *Journal of Human Evolution*, 28, 287–96.

Dunbar, R. I. M. (1998). The social brain hypothesis. *Evolutionary Anthropology*, 6, 178–90.

Dunbar, R. I. M. (2002). Modelling primate behavioral ecology. *International Journal of Primatology*, 23, 785–819.

Dunbar, R. I. M. (2003). The social brain: Mind, language, and society in evolutionary perspective. *Annual Review of Anthropology*, 32, 163–81.

Dunbar, R. I. M., and Bose, U. (1991). Adaptation to grass-eating in gelada baboons. *Primates*, 32, 1–7.

Dunbar, R. I. M., and Dunbar, P. (1988). Maternal time budgets of gelada baboons. *Animal Behaviour*, 36, 970–80.

Dunbar, R. I. M., and Shultz, S. (2007). Understanding primate brain evolution. *Philosophical Transactions of the Royal Society B: Biological Sciences*, 362, 649-658.

Duncan, S. H., Scott, K. P., Ramsay, A. G., Harmsen, H. J. M., Welling, G. W., Stewart, C. S., et al. (2003). Effects of alternative dietary substrates on competition between human colonic bacteria in an anaerobic fermentor system. *Applied and Environmental Microbiology*, 69, 1136–42.

Dunham, A. E., Erhart, E. M., and Wright, P. C. (2011). Global climate cycles and cyclones: Consequences for rainfall patterns and lemur reproduction in southeastern Madagascar. *Global Change Biology*, 17, 219–27.

Dunham, A. E., Erhart, E. M., Overdorff, D. J., and Wright, P. C. (2008). Evaluating effects of deforestation, hunting, and El Niño events on a threatened lemur. *Biological Conservation*, 141, 287–97.

Dunham, N. T. (2017). Feeding ecology and dietary flexibility of *Colobus angolensis palliatus* in relation to habitat disturbance. *International Journal of Primatology*, 38, 553–71.

Dunham, N., and Rodriguez-Saona, L. (2018). Nutrient intake and balancing among female *Colobus angolensis palliatus* inhabiting structurally distinct forest areas: Effects of group, season, and reproductive state. *American Journal of Primatology*, 80, e22878.

Dunsworth, H. M., and Eccleston, L. (2015). The evolution of difficult childbirth and helpless hominin infants. *Annual Review of Anthropology*, 44, 55–69.

Dunsworth, H. M., Warrener, A. G., Deacon, T., Ellison, P. T., and Pontzer, H. (2012). Metabolic hypothesis for human altriciality. *Proceedings of the National Academy of Sciences of the United States of America*, 109, 15212–16.

Dury, S. J., Good, J. E. G., Perrins, C. M., Buse, A., and Kaye, T. (1998). The effects of increasing CO_2 on oak leaf palatability and the implications for herbivorous insects. *Global Change Biology*, 4, 55–61.

Dusseldorp, G. L. (2012). Studying prehistoric hunting proficiency: Applying optimal foraging theory to the middle Palaeolithic and middle Stone Age. *Quaternary International*, 252, 3–15.

Dussutour, A., Latty, T., Beekman, M., and Simpson, S. J. (2010). Amoeboid organism solves complex nutritional challenges. *Proceedings of the National Academy of Sciences of the United States of America*, 107, 4607–11.

Dussutour, A., and Simpson, S. J. (2009). Communal nutrition in ants. *Current Biology*, 19, 740–44.

Dutreuil, S. (2014). What good are abstract and what-if models? Lessons from the Gaia hypothesis. *History and Philosophy of the Life Sciences*, 36, 16–41.

Dykes, L., and Rooney, L. W. (2006). Sorghum and millet phenols and antioxidants. *Journal of Cereal Science*, 44, 236–51.

Dzowela, B. H., Hove, L., and Mafongoya, P. L. (1995). Effect of drying method on chemical composition and in vitro digestibility of multi-purpose tree and shrub fodders. *Tropical Grasslands*, 29, 263–69.

Eaglen, R. H. (1984). Incisor size and diet revisited: The view from a platyrrhine perspective. *American Journal of Physical Anthropology*, 64(3), 263–75.

Eaglen, R. H. (1986). Morphometrics of the anterior dentition in strepsirhine primates. *American Journal of Physical Anthropology*, 71(2), 185–201.

Eberle, M., and Kappeler, P. M. (2004). Sex in the dark: Determinants and consequences of mixed male mating tactics in *Microcebus murinus*, a small solitary nocturnal primate. *Behavioral Ecology and Sociobiology*, 57, 77–90.

Eberle, M., and Kappeler, P. M. (2006). Family insurance: Kin selection and cooperative breeding in a solitary primate (*Microcebus murinus*). *Behavioral Ecology and Sociobiology*, 60, 582–88.

Ebersole, K. E., Dugas, L. R., Durazo-Arvizut, R. A., Adeyemo, A. A., Tayo, B. O., Omotade, O. O., et al. (2008). Energy expenditure and adiposity in Nigerian and African-American women. *Obesity*, 16, 2148–54.

Edet, E. E., Eka, O. U., and Ifon, E. T. (1985). Chemical evaluation of the nutritive value of seeds of African breadfruit (*Treculia africana*). *Food Chemistry*, 17, 41–47.

Edwards, M. S., and Ullrey, D. E. (1999a). Effect of dietary fiber concentration on apparent digestibility and digesta passage in non-human primates. I. Ruffed lemurs (*Varecia variegata variegata* and *V. v. rubra*). *Zoo Biology*, 18, 529–36.

Edwards, M. S., and Ullrey, D. E. (1999b). Effect of dietary fiber concentration on apparent digestibility and digesta passage in non-human primates. II. Hindgut- and foregut-fermenting folivores. *Zoo Biology*, 18, 537–49.

Edwards, R., and Fowle, C. D. (1955). The concept of carrying capacity. In J. B. Trefethen (Ed.), *Transactions of the Twentieth North American Wildlife Conference* (pp. 589–602). Wildlife Management Institute.

Edwards, W., Lonsdorf, E. V., and Pontzer, H. (2017). Total energy expenditure in captive capuchins (*Sapajus apella*). *American Journal of Primatology*, 79, e22638.

Eerkens, J. W., and Bartelink, E. J. (2013). Sex-biased weaning and early childhood diet among middle holocene hunter–gatherers in Central California. *American Journal of Physical Anthropology*, 152, 471–83.

Eggert, A. K., Otte, T., and Muller, J. K. (2008). Starving the competition: A proximate cause of reproductive skew in burying beetles (*Nicrophorus vespilloides*). *Proceedings of the Royal Society B: Biological Sciences*, 275, 2521–28.

Ehleringer, J., Field, C., Lin, Z., and Kuo, C. (1986). Leaf carbon isotope and mineral composition in subtropical plants along an irradiance cline. *Oecologia*, 70, 520–26.

Eisenberg, J., Muckenheim, N. A., and Rudran, R. (1972). The relation between ecology and social structure in primates. *Science*, 176, 863–74.

Eisner, T., and Davis, J. A. (1967). Mongoose throwing and smashing millipedes. *Science*, 155, 577–79.

Ejidike, B. N., and Okosodo, F. E. (2007). Food and feeding habits of the thick-tailed Galago (*Otelemur* [*Otolemur*] *crassicaudatus*) in Okomu National Park, Edo State. *Journal of Fisheries International*, 2, 231–33.

Elder, A. A. (2013). *Competition among Three Primate Species at Way Canguk, Sumatra, Indonesia*. Stony Brook University.

Elenkov, I. J., and Chrousos, G. P. (2002). Stress hormones, proinflammatory and antiinflammatory cytokines, and autoimmunity. *Neuroendocrine Immune Basis of the Rheumatic Diseases II, Proceedings*, 966, 290–303.

Elenkov, I. J., Iezzoni, D. G., Daly, A., Harris, A. G., and Chrousos, G. P. (2005). Cytokine dysregulation, inflammation and well-being. *Neuroimmunomodulation*, 12(5), 255–69.

Elinav, E., Strowig, T., Kau, A. L., Henao-Mejia, J., Thaiss, C. A., Booth, C. J., et al. (2011). NLRP6 inflammasome regulates colonic microbial ecology and risk for colitis. *Cell*, 145(5), 745–57.

Ellis, W., FitzGibbon, S., Pye, G., Whipple, B., Barth, B., Johnston, S., et al. (2015). The role of bioacoustic signals in koala sexual selection: Insights from seasonal patterns of associations revealed with GPS-proximity units. *PLoS ONE*, 10, e0130657.

Ellison, P. T. (1990). Human ovarian function and reproductive ecology: New hypotheses. *American Anthropologist*, 92, 933–52.

Ellison, P. T. (2001). *On Fertile Ground—a Natural History of Human Reproduction*. Harvard University Press.

Ellison, P. T. (2003). Energetics and reproductive effort. *American Journal of Human Biology*, 15, 342–51.

Elser, J. J., Fagan, W. F., Denno, R. F., Dobberfuhl, D. R., Folarin, A., Huberty, A., et al. (2000). Nutritional constraints in terrestrial and freshwater food webs. *Nature*, 408, 578–80.

Ely, J. J., Dye, B., Frels, W. I., Fritz, J., Gagneux, P., Khun, H. H., et al. (2005). Subspecies composition and founder contribution of the captive US chimpanzee (*Pan troglodytes*) population. *American Journal of Primatology*, 67, 223–41.

Embar, K., Kotler, B. P., and Mukherjee, S. (2011). Risk management in optimal foragers: The effect of sightlines and predator type on patch use, time allocation, and vigilance in gerbils. *Oikos*, 120, 1657–66.

Emerson, S. E., and Brown, J. S. (2012). Using giving-up densities to test for dietary preferences in primates: An example with Samango monkeys (*Cercopithecus* (*nictitans*) *mitis erythrarchus*). *International Journal of Primatology*, 33, 1420–38.

Emerson, S. E., and Brown, J. S. (2013). Identifying preferred habitats of samango monkeys (*Cercopithecus* (*nictitans*) *mitis erythrarchus*) through patch use. *Behavioural Processes*, 100, 214–21.

Emerson, S. E., and Brown, J. S. (2015). The influence of food chemistry on food-safety tradeoffs in samango monkeys. *Journal of Mammalogy*, 96, 237–44.

Emerson, S. E., Brown, J. S., and Linden, J. D. (2011). Identifying Sykes' monkeys', *Cercopithecus albogularis erythrarchus*, axes of fear through patch use. *Animal Behaviour*, 81, 455–62.

Emery Thompson, M. (2005). Reproductive endocrinology of wild female chimpanzees (*Pan troglodytes schweinfurthii*):

Methodological considerations and the role of hormones in sex and conception. *American Journal of Primatology*, 67, 137–58.

Emery Thompson, M. (2013a). Comparative reproductive energetics of human and nonhuman primates. *Annual Review of Anthropology*, 42, 287–304.

Emery Thompson, M. (2013b). Reproductive ecology of female chimpanzees. *American Journal of Primatology*, 75, 222–37.

Emery Thompson, M. (2017). Energetics of feeding, social behavior, and life history in non-human primates. *Hormones and Behavior*, 91, 84–96.

Emery Thompson, M., and Knott, C. D. (2008). Urinary C-peptide of insulin as a non-invasive marker of energy balance in wild orangutans. *Hormones and Behavior*, 53(4), 526–35.

Emery Thompson, M., Fox, S. A., Berghänel, A., Sabbi, K. H., Phillips-Garcia, S., Enigk, D. K., et al. (2020). Wild chimpanzees exhibit humanlike aging of glucocorticoid regulation. *Proceedings of the National Academy of Sciences of the United States of America*, 117(15), 8424–30.

Emery Thompson, M., Kahlenberg, S. M., Gilby, I. C., and Wrangham, R. W. (2007). Core area quality is associated with variance in reproductive success among female chimpanzees at Kibale National Park. *Animal Behaviour*, 73, 501–12.

Emery Thompson, M., Machanda, Z. P., Fox, S. A., Sabbi, K. H., Otali, E., Thompson Gonzalez, N., et al. (2020). Evaluating the impact of physical frailty during ageing in wild chimpanzees (*Pan troglodytes schweinfurthii*). *Philosophical Transactions of the Royal Society of London B—Biological Sciences*, 375(1811), 20190607.

Emery Thompson, M., Muller, M. N., Sabbi, K., Machanda, Z. P., Otali, E., and Wrangham, R. W. (2016). Faster reproductive rates trade off against offspring growth in wild chimpanzees. *Proceedings of the National Academy of Sciences of the United States of America*, 113, 7780–85.

Emery Thompson, M., Muller, M. N., and Wrangham, R. W. (2012a). The energetics of lactation and the return to fecundity in wild chimpanzees. *Behavioral Ecology*, 23, 1234–41.

Emery Thompson, M., Muller, M. N., and Wrangham, R. W. (2012b). Technical note: Variation in muscle mass in wild chimpanzees—application of a modified urinary creatinine method. *American Journal of Physical Anthropology*, 149(4), 622–27.

Emery Thompson, M., Muller, M. N., and Wrangham, R. W. (2014). Male chimpanzees compromise the foraging success of their mates in Kibale National Park, Uganda. *Behavioral Ecology and Sociobiology*, 68, 1973–83.

Emery Thompson, M., Muller, M. N., Wrangham, R. W., Lwanga, J. S., and Potts, K. B. (2009). Urinary C-peptide tracks seasonal and individual variation in energy balance in wild chimpanzees. *Hormones and Behavior*, 55, 299–305.

Emery Thompson, M., Wilson, M. L., Gobbo, G., Muller, M. N., and Pusey, A. E. (2008). Hyperprogesteronemia in response to *Vitex fischeri* consumption in wild chimpanzees (*Pan troglodytes schweinfurthii*). *American Journal of Primatology*, 70, 1064–71.

Emery Thompson, M., and Wrangham, R. W. (2006). Comparison of sex differences in gregariousness in fission-fusion species: Reducing bias by standardizing for party size. In N. E. Newton-Fisher, H. Notman, V. Reynolds, and J. D. Paterson (Eds.), *Primates of Western Uganda* (pp. 209–26). Springer.

Emery Thompson, M., and Wrangham, R. W. (2008). Diet and reproductive function in wild female chimpanzees (*Pan troglodytes schweinfurthii*) at Kibale National Park, Uganda. *American Journal of Physical Anthropology*, 135, 171–81.

Emidio, R. A., and Ferreira, R. G. (2012). Energetic payoff of tool use for capuchin monkeys in the Caatinga: Variation by season and habitat type. *American Journal of Primatology*, 74, 332–43.

Emlen, J. M. (1966). Role of time and energy in food preference. *American Naturalist*, 100, 611-617.

Emrich, A., Pokorny, B., and Sepp, C. (2000). *The Significance of Secondary Forest Management for Development Policy*. GTZ.

Endler, J. A., Westcott, D. A., Madden, J. R., and Robson, T. (2005). Animal visual systems and the evolution of color patterns: Sensory processing illuminates signal evolution. *Evolution (International Journal of Organic Biology)*, 59, 1795–818.

Engelen, L., and de Wijk, R. (2012). Oral processing and texture perception. In J. Chen and L. Engelen (Eds.), *Food Oral Processing: Fundamentals of Eating and Sensory Perception* (pp. 159–77). John Wiley and Sons.

Engelhardt, A., Heistermann, M., Hodges, J. K., Nuernberg, P., and Niemitz, C. (2006). Determinants of male reproductive success in wild long-tailed macaques (*Macaca fascicularis*)—male monopolisation, female mate choice or post-copulatory mechanisms? *Behavioral Ecology and Sociobiology*, 59, 740–52.

Engels, F. M. (1996). Developments in application of light and scanning electron microscopy techniques for cell wall degradation studies. *Netherlands Journal of Agricultural Science*, 44, 357–73.

Engström, M. T., Arvola, J., Nenonen, S., Virtanen, V. T. J., Leppa, M. M., Tahtinen, P., et al. (2019). Structural features of hydrolyzable tannins determine their ability to form insoluble complexes with bovine serum albumin. *Journal of Agricultural and Food Chemistry*, 67(24), 6798–808.

Enjapoori, A. K., Grant, T. R., Nicol, S. C., Lefebvre, C. M., Nicholas, K. R., and Sharp, J. A. (2014). Monotreme lactation protein is highly expressed in monotreme milk and provides antimicrobial protection. *Genome Biology and Evolution*, 6, 2754–73.

Enright, J. (1976). Climate and population regulation. *Oecologia*, 24, 295–310.

Eppley, T. M., Balestri, M., Campera, M., Rabenantoandro, J., Ramanamanjato, J. B., Randriatafika, F., et al. (2017). Ecological flexibility as measured by the use of pioneer and exotic plants by two lemurids: *Eulemur collaris* and *Hapalemur meridionalis*. *International Journal of Primatology*, 38, 338–57.

Eppley, T. M., Hall, K., Donati, G., and Ganzhorn, J. U. (2015). An unusual case of affiliative association of a female *Lemur catta* in a *Hapalemur meridionalis* social group. *Behaviour*, 152, 1041–61.

Eppley, T. M., Hoeks, S., Chapman, C. A., Ganzhorn, J. U., Hall, K., Owen, M. A., and Adams, D. B. (2022). Factors influencing terrestriality in primates of the Americas and Madagascar. *Proceedings of the National Academy of Sciences*, 119, e2121105119.

Eppley, T. M., Tan, C. L., Arrigo-Nelson, S. J., Donati, G., Ballhorn, D. J., and Ganzhorn, J. U. (2017). High energy or protein concentrations in food as possible offsets for cyanide consumption by specialized bamboo lemurs in Madagascar. *International Journal of Primatology*, 38, 881–99.

Eppley, T. M., Verjans, E., and Donati, G. (2011). Coping with low-quality diets: A first account of the feeding ecology of the southern gentle lemur, *Hapalemur meridionalis*, in the Mandena littoral forest, southeast Madagascar. *Primates*, 52, 673–703.

Erb, W. M., Borries, C., Lestari, N. S., and Hodges, J. K. (2012). Annual variation in ecology and reproduction of wild simakobu (*Simias concolor*). *International Journal of Primatology*, 33, 1406–19.

Erhart, E. M., and Overdorff, D. J. (2008). Spatial memory during foraging in prosimian primates: *Propithecus edwardsi* and *Eulemur fulvus rufus*. *Folia Primatologica*, 79, 185–96.

Erhart, E. M., Tecot, S. R., and Grassi, C. (2018). Interannual variation in diet, dietary diversity, and dietary overlap in three sympatric strepsirrhine species in southeastern Madagascar. *International Journal of Primatology*, 39, 289–311.

Erickson, C. J. (1991). Percussive foraging in the aye-aye, *Daubentonia madagascariensis*. *Animal Behavior*, 41, 793–801.

Erickson, C. J., Nowicki, S., Dollar, L., and Goehring, N. (1998). Percussive foraging: Stimuli for prey location by aye-ayes (*Daubentonia madagascariensis*). *International Journal of Primatology*, 19, 111–22.

Eriksson, O. (2016). Evolution of angiosperm seed disperser mutualisms: The timing of origins and their consequences for coevolutionary interactions between angiosperms and frugivores. *Biological Reviews*, 91, 168–86.

Erlenbach, J. A., Rode, K. D., Raubenheimer, D., and Robbins, C. T. (2014). Macronutrient optimization and energy maximization determine diets of brown bears. *Journal of Mammalogy*, 95, 160–68.

Eronen, J. T., Zohdy, S., Evans, A. R., Tecot, S. R., Wright, P. C., and Jernvall, J. (2017). Feeding ecology and morphology make a bamboo specialist vulnerable to climate change. *Current Biology*, 27, 3384–89.

Errington, P. L. (1934). Vulnerability of bob-white populations to predation. *Ecology*, 15, 110–27.

Erwin, D. H., and Davidson, E. H. (2002) The last common bilateral ancestor. *Development* 129, 3021–3032.

Espinosa-Gómez, F., García, J. S., Rosales, S. G., Wallis, I. R., Chapman, C. A., Mávil, J. M., et al. (2015). Howler monkeys (*Alouatta palliata mexicana*) produce tannin-binding salivary proteins. *International Journal of Primatology*, 36, 1086–100.

Espinosa-Gómez, F., Gómez-Rosales, S., Wallis, I. R., Canales-Espinosa, D., and Hernández-Salazar, L. (2013). Digestive strategies and food choice in mantled howler monkeys *Alouatta palliata mexicana*: Bases of their dietary flexibility. *Journal of Comparative Physiology B*, 183, 1089–100.

Espinosa-Gómez, F. C., Serio-Silva, J. C., Santiago-Garcia, J. D., Sandoval-Castro, C. A., Hernández-Salazar, L. T., Mejia-Varas, F., et al. (2018). Salivary tannin-binding proteins are a pervasive strategy used by the folivorous/frugivorous black howler monkey. *American Journal of Primatology*, 80, e22737.

Estrada, A. (2013). Socioeconomic context of primate conservation: Population, poverty, global economic demands, and sustainable land use. *American Journal of Primatology*, 75, 30–45.

Estrada, A., Garber, P. A., Rylands, A. B., Roos, C., Fernandez-Duque, E., Di Fiore, A., et al. (2017). Impending extinction crisis of the world's primates: Why primates matter. *Science Advances*, 3, e1600946.

Estrada, A., Juan-Solano, S., Martinez, T. O., and Coates-Estrada, R. (1999). Feeding and general activity patterns of a howler monkey (*Alouatta palliata*) troop living in a forest fragment at Los Tuxtlas, Mexico. *American Journal of Primatology*, 48, 167–83.

Estrada, A., Raboy, B. E., and Oliveira, L. C. (2012). Agroecosystems and primate conservation in the tropics: A review. *American Journal of Primatology*, 74, 696–711.

Evans, A. R. (2013). Shape descriptors as ecometrics in dental ecology. *Hystrix, the Italian Journal of Mammalogy*, 24(1), 133–40.

Evans, A. R., and Jernvall, J. (2009). Patterns and constraints in carnivoran and rodent dental complexity and tooth size. *Journal of Vertebrate Paleontology*, 29, 24A.

Evans, A. R., and Sanson, G. D. (2006). Spatial and functional modeling of carnivore and insectivore molariform teeth. *Journal of Morphology*, 267, 649–62.

Evans, A. R., Wilson, G. P., Fortelius, M., and Jernvall, J. (2007). High-level similarity of dentitions in carnivorans and rodents. *Nature*, 445(7123), 78–81.

Evans, K. D., Foley, W. J., Chapman, C. A., Rothman J. M. (2021). Deconstructing protein in the diet and biomass of colobine primates. *International Journal of Primatology*, 42, 283–300.

Eveleth, P. B., and Tanner, J. M. (1990). *Worldwide Variation in Human Growth*. Cambridge University Press.

Fa, J. E., and Brown, D. (2009). Impacts of hunting on mammals in African tropical moist forests: A review and synthesis. *Mammal Review*, 39, 231–64.

Fabricant, S. A., and Herberstein, M. E. (2015). Hidden in plain orange: Aposematic coloration is cryptic to a colorblind insect predator. *Behavioural Ecology*, 26, 38–44.

Fagan, W. F., Lewis, M. A., Auger-Méthé, M., Avgar, T., Benhamou, S., Breed, G., et al. (2013). Spatial memory and animal movement. *Ecology Letters*, 16, 1316–29.

Fagen, R. (1988). Population effects of habitat change: A quantitative assessment. *Journal of Wildlife Management*, 52, 41–46.

Fahy, G. E., Richards, M., Riedel, J., Hublin, J. J., and Boesch, C. (2013). Stable isotope evidence of meat eating and hunting specialization in adult male chimpanzees. *Proceedings of the National Academy of Sciences of the United States of America*, 110, 5829–33.

Fairbanks, L., and McGuire, M. (1995). Maternal condition and the quality of maternal care in vervet monkeys. *Behaviour*, 132, 733–54.

Falcy, M. R., and Danielson, B. J. (2013). A complex relationship between moonlight and temperature on the foraging behavior of the Alabama beach mouse. *Ecology*, 94, 2632–37.

Falótico, T., Siqueira, J. O., and Ottoni, B. (2017). Digging up food: Excavation stone tool use by wild capuchin monkeys. *Scientific Reports*, 7, 6278.

Falótico, T., Valença, T., Verderane, M. P. and Fogaça, M. D. (2022). Stone tools differences across three capuchin monkey populations: Food's physical properties, ecology, and culture. *Scientific Reports*, 12, 14365.

Fan, L., Lindsley, S. R., Comstock, S. M., Takahashi, D. L., Evans, A. E., He, G. W., et al. (2013). Maternal high-fat diet impacts endothelial function in nonhuman primate offspring. *International Journal of Obesity*, 37(2), 254–62.

Fan, P. F., and Jiang, X. L. (2008). Effects of food and topography on ranging behavior of black crested gibbon (*Nomascus concolor jingdongensis*) in Wuliang Mountain, Yunnan, China. *American Journal of Primatology*, 70, 871–78.

Fan, P. F., Ai, H. S., Fei, H. L., Zhang, D., and Yuan, S. D. (2013). Seasonal variation of diet and time budget of eastern hoolock gibbons (*Hoolock leuconedys*) living in a northern montane forest. *Primates*, 54, 137–46.

Fan, P. F., Fei, H. L., Scott, M. B., Zhang, W., and Ma, C. Y. (2011). Habitat and food choice of the critically endangered cao vit gibbon (*Nomascus nasutus*) in China: Implications for conservation. *Biological Conservation*, 144, 2247–54.

Fan, P. F., Ni, Q. Y., Sun, G. Z., Huang, B., and Jiang, X. L. (2009). Gibbons under seasonal stress: The diet of the black crested gibbon (*Nomascus concolor*) on Mt. Wuliang, Central Yunnan, China. *Primates*, 50, 37–44.

Fan, Y., Wei, C., Xiao, W., Zhang, W., Wang, N., Chuang, P. Y., et al. (2014). Temporal profile of the renal transcriptome of HIV-1 transgenic mice during disease progression. *PLoS ONE*, 9(3), e93019.

Fannin, L. D., Laugier, E. J., van Casteren, A., Greenwood, S. L., and Dominy, N. J. (2022). Differentiating siliceous particulate matter in the diets of mammalian herbivores. *Methods in Ecology and Evolution*, 13, 2198–208.

Fannin, L. D., Singels, E., Esler, K. J., and Dominy, N. J. (2021). Grit and consequence. *Evolutionary Anthropology*, 30, 375–84.

FAO. (2004). *Human energy requirements*. Report of a Joint FAO/WHO/UNU Expert Consultation. FAO Food and Nutrition Technical Report Series No.1 Food and Agriculture Organization of the United Nations: Rome.

FAO. (2005). *Global Forest Resources Assessment 2005: Progress towards Sustainable Forest Management*. FAO Forestry Paper 147.

FAO. (2010). *Global Forest Resources Assessment 2010*.

Fardi, S., Sauther, M., Cuozzo, F. P., Jacky, I. A., and Bernstein, R. M. (2017). The effect of extreme weather events on hair cortisol and body weight in a wild ring-tailed lemur population (*Lemur catta*) in southwestern Madagascar. *American Journal of Primatology*, 80, e22731.

Farine, D. R., Aplin, L. M., Sheldon, B. C., and Hoppit, W. (2015). Interspecific social networks promote information transmission in wild songbirds. *Proceedings of the Royal Society B: Biological Sciences*, 282, 20142804.

Farine, D. R., Garroway, C. J., and Sheldon, B. C. (2012). Social network analysis of mixed-species flocks: Exploring the structure and evolution of interspecific social behaviour. *Animal Behaviour*, 84, 1271–77.

Farine, D. R., and Milburn, P. J. (2013). Social organisation of thornbill-dominated mixed-species flocks using social network analysis. *Behavioral Ecology and Sociology*, 67, 321–30.

Farine, D. R., Strandburg-Peshkin, A., Berger-Wolf, T., Ziebart, B., Brugere, I., Li, J., et al. (2016). Both nearest neighbours and long-term affiliates predict individual locations during collective movement in wild baboons. *Scientific Reports*, 6, 27704.

Farquhar, G. D., Ehleringer, J. R., and Hubick, K. T. (1989). Carbon isotope discrimination and photosynthesis. *Annual Review of Plant Physiology and Plant Molecular Biology*, 40, 503–37.

Fashing, P. J. (2001). Feeding ecology of guerezas in the Kakamega Forest, Kenya: The importance of Moraceae fruit in their diet. *International Journal of Primatology*, 22, 579–609.

Fashing, P. J. (2011). African colobines: Their behavior, ecology, and conservation. In C. Campbell, A. Fuentes, K. MacKinnon, S. Bearder, and R. Stumpf (Eds.), *Primates in Perspective* (2nd ed., pp. 203–29). Oxford University Press.

Fashing, P. J., Dierenfeld, E. S., and Mowry, C. B. (2007). Influence of plant and soil chemistry on food selection, ranging patterns, and biomass of *Colobus guereza* in Kakamega Forest, Kenya. *International Journal of Primatology*, 28, 673–703.

Fay, J. M., and Carroll, R. W. (1994). Chimpanzee tool use for honey and termite extraction in Central Africa. *American Journal of Primatology*, 34, 309–17.

Fedigan, L. (1990). Vertebrate predation in *Cebus capucinus*: Meat eating in a Neotropical monkey. *Folia Primatologica*, 54, 196–205.

Fedigan, L. M., Carnegie, S. D., and Jack, K. M. (2008). Predictors of reproductive success in female white-faced capuchins (*Cebus capucinus*). *American Journal of Physical Anthropology*, 137, 82–90.

Fedigan, L. M., Fedigan, L., Chapman, C. A., and Glander, K. E. (1988). Spider monkey home ranges: A comparison of radio telemetry and direct observation. *American Journal of Primatology*, 16, 19–29.

Fedigan, L. M., Melin, A. D., Addicott, J. F., and Kawamura, S. (2014). The heterozygote superiority hypothesis for polymorphic color vision is not supported by long-term fitness data from wild Neotropical monkeys. *PLoS ONE*, 9, e84872.

Fedriani, J. M., and Boulay, R. (2006). Foraging by fearful frugivores: Combined effect of fruit ripening and predation risk. *Functional Ecology*, 20, 1070–79.

Feeny, P. (1970). Seasonal changes in oak leaf tannins and nutrients as a cause of spring feeding by winter moth caterpillars. *Ecology*, 51, 565–81.

Fehlmann, G., O'Riain, M. J., Hopkins, P. W., O'Sullivan, J., Holton, M. D., Shepard, E. L. C., et al. (2017). Identification of behaviours from accelerometer data in a wild social primate. *Animal Biotelemetry*, 5, 6.

Fehlmann, G., O'Riain, M. J., Kerr-Smith, C., Hailes, S., Luckman, A., and Shepard, E. L. C. (2017). Extreme behavioural shifts by baboons exploiting risky, resource-rich, human-modified environments. *Scientific Reports*, 7, 15057.

Felton, A. M., Felton, A., Lindenmayer, D. B., and Foley, W. J. (2009). Nutritional goals of wild primates. *Functional Ecology*, 23, 70–78.

Felton, A. M., Felton, A., Raubenheimer, D., Simpson, S. J., Foley, W. J., Wood, J. T., et al. (2009). Protein content of diets dictates the daily energy intake of a free-ranging primate. *Behavioral Ecology*, 20(4), 685–90.

Felton, A. M., Felton, A., Raubenheimer, D., Simpson, S. J., Krizsan, S. J., Hedwall, P. O., et al. (2016). The nutritional balancing act of a large herbivore: An experiment with captive moose (*Alces alces*). *PLoS ONE*, 11, e0150870.

Felton, A. M., Felton, A., Rumiz, D. I., Pena-Claros, M., Villaroel, N., Chapman, C. A., et al. (2013). Commercial harvesting of *Ficus* timber—an emerging threat to frugivorous wildlife and sustainable forestry. *Biological Conservation*, 159, 96–100.

Felton, A. M., Felton, A., Wood, J. T., and Lindenmayer, D. B. (2008). Diet and feeding ecology of the Peruvian spider monkey (*Ateles chamek*) in a Bolivian semi-humid forest: The importance of *Ficus* as a staple food resource. *International Journal of Primatology*, 29, 379–403.

Felton, A. M., Felton, A., Wood, J. T., Foley, W. J., Raubenheimer, D., Wallis, I. R., et al. (2009). Nutritional ecology of spider monkeys (*Ateles chamek*) in lowland Bolivia: How macronutrient balancing influences food choices. *International Journal of Primatology*, 30, 675–96.

Felton, A. M., Wam, H. K., Felton, A., Simpson, S. J., Stolter, C., Hedwall, P.-O., et al. (2021). Macronutrient balancing in free-ranging populations of moose. *Ecology and Evolution*, 11, 11223–40.

Feret, J. B., and Asner, G. P. (2011). Spectroscopic classification of tropical forest species using radiative transfer modeling. *Remote Sensing of Environment*, 115, 2415–22.

Ferguson, R. B. (2011). Born to live: Challenging killer myths. In R. W. Sussman and C. R. Cloninger (Eds.), *Origins of Altruism and Cooperation* (pp. 249–70). Springer.

Ferguson, S. H., and McLoughlin, P. D. (2000). Effect of energy availability, seasonality, and geographic range on brown bear life history. *Ecography*, 23, 193–200.

Fernandes, G. (2008). Progress in nutritional immunology. *Immunologic Research*, 40(3), 244–61.

Fernandes, M. E. B. (1991). Tool use and predation of oysters (*Crassostrea rhizophorae*) by the tufted capuchin, *Cebus apella apella*, in brackish water mangrove swamp. *Primates*, 32, 529–31.

Fernandez, A. A., and Morris, M. R. (2007). Sexual selection and trichromatic color vision in primates: Statistical support for the preexisting-bias hypothesis. *American Naturalist*, 170, 10–20.

Fernandez, E. V., Li, Z., Zheng, W., Ding, Y., Sun, D., and Che, Y. (2014). Intraspecific host selection of Pere David's deer by cattle egret in Dafent, China. *Behavioral Processes*, 105, 36–39.

Fernandez-Duque, E., Juárez, C. P., and Di Fiore, A. (2008). Adult male replacement and subsequent infant care by male and siblings in socially monogamous owl monkeys (*Aotus azarai*). *Primates*, 49, 81–84.

Fernandez-Duque, E., Valeggia, C. R., and Mendoza, S. P. (2009). The biology of paternal care in human and nonhuman primates. *Annual Review of Anthropology*, 38, 115–30.

Ferrari, S. F. (1988). *The Behaviour and Ecology of the Buffy-Headed Marmoset*, Callithrix flaviceps (*O. Thomas 1903*). University College London.

Ferrari, S. F., and Lopes, M. A. (2002). Fruit rejection by tufted capuchins (*Cebus apella*, Primates, Cebinae) during the predation of *Cariniana micrantha* seeds: Suboptimal or just "wasteful" foraging behavior. *Revista de Etologia*, 4, 3–9.

Ferraro, J. V., Plummer, T. W., Pobiner, B. L., Oliver, J. S., Bishop, L. C., Braun, D. R., et al. (2013). Earliest archaeological evidence of persistent hominin carnivory. *PLoS ONE*, 8, e62174.

Ferretti, P., Pasolli, E., Tett, A., Asnicar, F., Gorfer, V., Fedi, S., et al. (2018). Mother-to-infant microbial transmission from different body sites shapes the developing infant gut microbiome. *Cell Host and Microbe*, 24, 133–45.

Fichtel, C., Perry, S., and Gros-Louis, J. (2005). Alarm calls of white-faced capuchin monkeys: An acoustic analysis. *Animal Behaviour*, 70, 165–76.

Fierer, N., and Jackson, R. B. (2006). The diversity and biogeography of soil bacterial communities. *Proceedings of the National Academy of Sciences of the United States of America*, 103, 626–31.

Fietz, J. (2003). Pair living and mating strategies in the fat-tailed dwarf lemur (*Cheirogaleus medius*). In U. H. Reichard and C. Boesch (Eds.), *Monogamy: Mating Strategies and Partnerships* in *Birds, Humans and Other Mammals* (pp. 214–31). Cambridge University Press.

Fietz, J., and Ganzhorn, J. U. (1999). Feeding ecology of the hibernating primate *Cheirogaleus medius*: How does it get so fat? *Oecologia*, 121, 157–64.

Filho, R. F., Viega, S., and Bezerra, B. (2020). Bearded capuchin (*Sapajus libidinosus*) predation on a rock cavy (*Kerodon rupestris*). *Primates*, 62, 463–66.

Finarelli, J. A., 2009. Does encephalization correlate with life history or metabolic rate in Carnivora? *Biology Letters*, 6, 350–53.

Fiorenza, L., Bennazi, S., Henry, A. G., Salazar-Garcia, D. C., Blasco, R., Picin, A., et al. (2015). To meat or not to meat? New perspectives on Neandertal ecology. *Yearbook of Physical Anthropology*, 59, 43–71.

Fischer, K. E., and Chapman, C. A. (1993). Frugivores and fruit syndromes: Differences in patterns at the genus and species level. *Oikos*, 66, 472–82.

FitzGibbon, C. D. (1990). Mixed-species grouping in Thomson's and Grant's gazelles: The antipredator benefits. *Animal Behaviour*, 39, 1116–26.

Flatt, J. P., and Blackburn, G. L. (1974). The metabolic fuel regulatory system: Implications for protein-sparing therapies during caloric deprivation and disease. *American Journal of Clinical Nutrition*, 27(2), 175–87.

Fleagle, J. G. (1980). Locomotion and posture. In D. J. Chivers (Ed.), *Malayan Forest Primates: Ten Years' Study in Tropical Rain Forest* (pp. 191–208). Plenum.

Fleagle, J. G. (1984). Primate locomotion and diet. In D. J. Chivers, B. A. Wood, and A. Bilsborough (Eds.), *Food Acquisition and Processing in Primates* (pp. 105–17). Plenum.

Fleagle, J. G. (1988). *Primate Adaptation and Evolution* (1st ed.). Academic Press.

Fleagle, J. G. (1998). *Primate Adaptation and Evolution* (2nd ed.). Academic Press.

Fleagle, J. G., Janson, C. H., and Reed, K. E. (1999). *Primate Communities*. Cambridge University Press.

Fleming, T. H., Breitwisch, R., and Whitesides, G. H. (1987). Patterns of tropical vertebrate frugivore diversity. *Annual Review of Ecology and Systematics*, 18, 91–109.

Fleming, T. H., and Kress, W. J. (2011). A brief history of fruits and frugivores. *Acta Oecologica*, 37, 521–30.

Fletcher, Q. E., and Boonstra, R. (2006). Impact of live trapping on the stress response of the meadow vole (*Microtus pennsylvanicus*). *Journal of Zoology*, 270, 473–78.

Flint, H. J., and Bayer, E. A. (2008). Plant cell wall breakdown by anaerobic microorganisms from the mammalian digestive tract. *Annals of the New York Academy of Science*, 1125, 280–88.

Flint, H. J., Bayer, E. A., Rincon, M. T., Lamed, R., and White, B. A. (2008). Polysaccharide utilization by gut bacteria: Potential for new insights from genomic analysis. *Nature*, 6, 121–31.

Flint, H. J., Duncan, S. H., Scott, K. P., and Louis, P. (2007). Interactions and competition within the microbial community of the human colon: Links between diet and health. *Environmental Microbiology*, 9(5), 1101–11.

Flörchinger, M., Braun, J., Böhning-Gaese, K., and Schaefer, H. M. (2010). Fruit size, crop mass, and plant height explain differential fruit choice of primates and birds. *Oecologia*, 164, 151–61.

Floyd, R., Sheppard, A. W., and De Barro, P. J. (1996). *Frontiers of Population Ecology*. CSIRO Collingwood.

Foerster, S., Cords, M., and Monfort, S. L. (2011). Social behavior, foraging strategies, and fecal glucocorticoids in female blue monkeys (*Cercopithecus mitis*): Potential fitness benefits of high rank in a forest guenon. *American Journal of Primatology*, 73, 870–82.

Foerster, S., Kithome, K., Cords, M., and Monfort, S. L. (2015). Social status and helminth infections in female forest guenons (*Cercopithecus mitis*). *American Journal of Physical Anthropology*, 158, 55–66.

Foley, P. (1994). Predicting extinction times from environmental stochasticity and carrying capacity. *Conservation Biology*, 8, 124–37.

Foley, W. J., and McArthur, C. (1994). The effects and costs of allelochemicals for mammalian herbivores: An ecological perspective. In D. J. Chivers and P. Langer (Eds.), *The Digestive System in Mammals: Food, Form and Function* (pp. 370–91). Cambridge University Press.

Foley, W. J., and Moore, B. D. (2005). Plant secondary metabolites and vertebrate herbivores—from physiological regulation to ecosystem function. *Current Opinion in Plant Biology*, 8, 430–35.

Foley, W. J., McIlwee, A., Lawler, I., Aragones, L., Woolnough, A. P., and Berding, N. (1998). Ecological applications of near-infrared reflectance spectroscopy—a tool for rapid, cost-effective prediction of the composition of plant and animal tissues and aspects of animal performance. *Oecologia*, 116, 293–305.

Fonne-Pfister, R., and Meyer, U. A. (1988). Xenobiotic and endobiotic inhibitors of cytochrome P-450dbl function, the target of the debrisoquine/sparteine type polymorphism. *Biochemical Pharmacology*, 37, 3829–35.

Fonseca Aldana, M. L. (2019). Arthropod and fruit availability related to the diet of the critically endangered Colombian highland woolly monkey (*Lagothrix lagotricha lugens*). Master's thesis, Universidad de los Andes.

Fontana, L., and Partridge, L. (2015). Promoting health and longevity through diet: From model organisms to humans. *Cell*, 161, 106–18.

Forbey, J. S., Dearing, M. D., Gross, E. M., Orians, C. M., Sotka, E. E., and Foley, W. J. (2013). A pharm-ecological perspective of terrestrial and aquatic plant-herbivore interactions. *Journal of Chemical Ecology*, 39, 465–80.

Forbey, J. S., and Foley, W. J. (2009). PharmEcology: A pharmacological approach to understanding plant-herbivore interactions—an introduction to the symposium. *Integrative and Comparative Biology*, 49, 267–73.

Forbey, J. S., Harvey, A. L., Huffman, M. A., Provenza, F. D., Sullivan, R., and Tasdemir, D. (2009). Exploitation of secondary metabolites by animals: A response to homeostatic challenges. *Integrative and Comparative Biology*, 49, 314–28.

Ford, S. M., and Davis, L. C. (1992). Systematics and body size: Implications for feeding adaptations in New World monkeys. *American Journal of Physical Anthropology*, 88, 415–68.

Forin-Wiart, M. A., Hubert, P., Sirguey, P., and Poulle, M. L. (2015). Performance and accuracy of lightweight and low-cost GPS data loggers according to antenna positions, fix intervals, habitats and animal movements. *PLoS ONE*, 10, e0129271.

Formenty, P., Boesch, C., Wyers, M., Steiner, C., Donati, F., Dind, F., et al. (1999). Ebola virus outbreak among wild chimpanzees living in a rain forest of Cote d'Ivoire. *Journal of Infectious Diseases*, 179, S120–S126.

Forsythe, P., Sudo, N., Dinan, T., Taylor, V. H., and Bienenstock, J. (2010). Mood and gut feelings. *Brain, Behavior, and Immunity*, 24, 9–16.

Fortelius, M. (1985). Ungulate cheek teeth: Developmental, functional and evolutionary interrelations. *Acta Zoologica Fennica*, 180, 1–76.

Fortelius, M. (1988). Isometric scaling of mammalian cheek teeth is also true metabolic scaling. In D. E. Russell, J.-P. Santoro, and D. Sigogneau-Russell (Eds.), *Teeth Revisited: Proceedings of the VIIth International Symposium on Dental Morphology, Paris, 1986* (pp. 458–62). Muséum National d'Histoire Naturelle.

Fortelius, M., and Solounias, N. (2000, 2000). Functional characterization of ungulate molars using the abrasion-attrition wear gradient: A new method for reconstructing paleodiets. *American Museum Novitates*, 3301, 1–36.

Fortin, D., Beyer, H. L., Boyce, M. S., Smith, D. W., Duchesne, T., and Mao, J. S. (2005). Wolves influence elk movements: Behavior shapes a trophic cascade in Yellowstone National Park. *Ecology*, 86, 1320–30.

Fossey, D., and Harcourt, A. H. (1977). Feeding ecology of free ranging mountain gorillas. In T. H. Clutton-Brock (Ed.), *Primate Ecology: Studies of Feeding and Ranging Behaviour in Lemurs, Monkeys and Apes* (pp. 415–47). Academic Press.

Foster, J. A., and McVey Neufeld, K. A. (2013). Gut-brain axis: How the microbiome influences anxiety and depression. *Cell*, 36(5), 305–12.

Fox, C. L., Perezperez, A., and Juan, J. (1994). Dietary information through the examination of plant phytoliths on the enamel surface of human dentition. *Journal of Archaeological Science*, 21(1), 29–34.

Fox, E. A., and bin'Muhammad, I. (2002). New tool use by wild Sumatran orangutans (*Pongo pygmaeus abelii*). *American Journal of Physical Anthropology*, 119, 186–88.

Fox, E. A., Sitompul, A. F., and van Schaik, C. P. (1999). Intelligent tool use in wild Sumatran orangutans. In S. T. Parker, R. W. Mitchell, and H. L. Miles (Eds.), *The Mentalities of Gorillas and Orangutans: Comparative Perspectives* (pp. 99–116). Cambridge University Press.

Fox, E. A., van Schaik, C. P., Sitampul, A., and Wright, D. N. (2004). Inter- and intra-population differences in orangutan (*Pongo pygmaeus*) activity and diet: Implications for the invention of tool use. *American Journal of Physical Anthropology*, 125, 162–74.

Fox, J. E., Starcevic, M., Jones, P. E., Burow, M. E., and McLachlan, J. A. (2004). Phytoestrogen signaling and symbiotic gene activation are disrupted by endocrine-disrupting chemicals. *Environmental Health Perspectives*, 112, 672–77.

Fragaszy, D. M. (1998). How non-human primates use their hands. In K. J. Connolly (Ed.), *The Psychobiology of the Hand* (pp. 77–96). Mac Keith.

Fragaszy, D. M., and Bard, K. (1997). Comparison of development and life history in *Pan* and *Cebus*. *International Journal of Primatology*, 18, 683–701.

Fragaszy, D. M., Izar, P., Liu, Q., Eshcar, Y., Young, L. A., and Visalberghi, E. (2016). Body mass in wild bearded capuchins (*Sapajus libidinosus*): Ontogeny and sexual dimorphism. *American Journal of Primatology*, 78, 473–84.

Fragaszy, D. M., Schwarz, S., and Shimosaka, D. (1982). Longitudinal observations of care and development of infant titi monkeys (*Callicebus moloch*). *American Journal of Primatology*, 2, 191–200.

Fragaszy, D. M., Visalberghi, E., and Fedigan, L. M. (2004). *The Complete Capuchin: The Biology of the Genus Cebus*. Cambridge University Press.

Franca, E. L., Bitencourt, R. V., Fujimori, M., de Morais, T. C., Calderon, I. D., and Honorio-França, A. C. (2011). Human colostral phagocytes eliminate enterotoxigenic *Escherichia coli* opsonized by colostrum supernatant. *Journal of Microbiology, Immunology, and Infection*, 44, 1–7.

Francisco, T. M., Couto, D. R., Zanuncio, J. C., Serrão, J. E., Silva, I. d. O., and Boere, V. (2014). Vegetable exudates as food for *Callithrix* spp. (Callitrichidae): Exploratory patterns. *PLoS ONE*, 9, e112321.

Frantzis, A., & Herzing, D. L. (2002). Mixed-species associations of striped dolphins (*Stenella coeruleoalba*), short-beaked common dolphins (*Delphinus delphis*), and Risso's dolphins (*Grampus griseus*) in the Gulf of Corinth (Greece, Mediterranean Sea). *Aquatic Mammals*, 28(2), 188–97.

Franzen, J. L., and Wilde, V. (2003). First gut content of a fossil primate. *Journal of Human Evolution*, 44(3), 373–78.

Fraser, W. D., Taggart, D. P., Fell, G. S., Lyon, T. D., Wheatley, D., Garden, O. J., et al. (1989). Changes in iron, zinc, and copper concentrations in serum and in their binding to transport proteins after cholecystectomy and cardiac surgery. *Clinical Chemistry*, 35(11), 2243–47.

Freckleton, R. P. (2002). On the misuse of residuals in ecology:

Regression of residuals vs. multiple regression. *Journal of Animal Ecology*, 71, 542–45.

Fredricksson, G. M., and Wich, S. A. (2006). Frugivory in sun bears (*Helarctos malayanus*) is linked to El Niño-related fluctuations in fruiting phenology, East Kalimantan, Indonesia. *Biological Journal of the Linnean Society*, 89, 489–508.

Freed, B. Z. (2006). Polyspecific associations of crowned lemurs and Sanford's lemurs in Madagascar. In L. Gould and M. L. Sauther (Eds.), *Lemurs*. Springer.

Freeland, W. (1979a). Mangabey (*Cercocebus albigena*) social organization and population density in relation to food use and availability. *Folia Primatologica*, 32, 108–24.

Freeland, W. J. (1979b). Primate social groups as biological islands. *Ecology*, 60, 719–728.

Freeland, W. J., and Janzen, D. H. (1974). Strategies in herbivory by mammals: The role of plant secondary compounds. *American Naturalist*, 961, 269–89.

Freeland-Graves, J. H., Sanjeevi, N., and Lee, J. J. (2015). Global perspectives on trace element requirements. *Journal of Trace Elements in Medicine and Biology*, 31, 135–41.

Freese, C. H. (1977). Food habits of white-faced capuchins *Cebus capucinus* (Primates: Cebidae) in Santa Rosa National Park, Costa Rica. *Brenesia*, 10, 43–56.

Freestone, P. P., Haigh, R. D., and Lyte, M. (2007). Blockade of catecholamine-induced growth by adrenergic and dopaminergic receptor antagonists in *Escherichia coli* O157:H7, *Salmonella enterica* and *Yersinia enterocolitica*. *BMC Microbiology*, 7, 8.

Freitas, C., Costa, F. R., Barbosa, C. E., and Cintra, R. (2016). Restriction limits and main drivers of fruit production in palm in central Amazonia. *Acta Oecologia*, 77, 75–84.

French, A. R., and Smith, T. B. (2005). Importance of body size in determining dominance hierarchies among diverse tropical frugivores 1. *Biotropica*, 37(1), 96–101.

Frenzel, H., Bohlender, J., Pinsker, K., Wohlleben, B., Tank, J., Lechner, S. G., et al. (2012). A genetic basis for mechanosensory traits in humans. *PLoS Biology*, 10, e1001318.

Fretwell, S. D. (1972). *Populations in a Seasonal Environment*. Princeton University Press.

Fretwell, S. D. (1977). The regulation of plant communities by the food chains exploiting them. *Perspectives in Biology and Medicine*, 20, 169–85.

Fretwell, S. D., and Lucas, H. (1969). On territorial behavior and other factors influencing habitat distribution in birds. I. Theoretical development. *Acta BioTheoretica*, 19, 16–36.

Fretwell, S. D., and Lucas, H. L. (1970). On territorial behavior and other factors influencing habitat distribution in birds. *Acta BioTheoretica*, 19, 16–36.

Freund, R.J., Wilson, W.J. (1998). *Regression analysis*. Academic Press.

Frey, J. C., Rothman, J. M., Pell, A. N., Nizeyi, J. B., Cranfield, M. R., and Angert, E. R. (2006). Fecal bacterial diversity in a wild gorilla. *Applied and Environmental Microbiology*, 72, 3788–92.

Frias, A. E., Morgan, T. K., Evans, A. E., Rasanen, J., Oh, K. Y., Thornburg, K. L., et al. (2011). Maternal high-fat diet disturbs uteroplacental hemodynamics and increases the frequency of stillbirth in a nonhuman primate model of excess nutrition. *Endocrinology*, 152(6), 2456–64.

Frid, A., and Dill, L. (2002). Human-caused disturbance stimuli as a form of predation risk. *Conservation Ecology*, 6, 11.

Friis, H., Mwaniki, D., Omondi, B., Muniu, E., Magnussen, P., Geissler, W., et al. (1997). Serum retinol concentrations and *Schistosoma mansoni*, intestinal helminths, and malarial parasitemia: A cross-sectional study in Kenyan preschool and primary school children. *American Journal of Clinical Nutrition*, 66(3), 665–71.

Frisch, R. E., and Revelle, R. (1971). Height and weight at menarche and a hypothesis of menarche. *Archives of Disease in Childhood*, 46, 695–701.

Frisch, R. E., Revelle, R., and Cook, S. (1973). Components of weight at menarche and the initiation of the adolescent growth spurt in girls: Estimated total water, lean body weight and fat. *Human Biology*, 45, 469–83.

Fritz, J., Hummel, J., Kienzle, E., Arnold, C., Nunn, C., and Clauss, M. (2009). Comparative chewing efficiency in mammalian herbivores. *Oikos*, 118, 1623–32.

Fruth, B., and Hohmann, G. (2018). Food sharing across borders: first observation of intercommunity meat sharing by bonobos at LuiKotale, DRC. *Human Nature*, 29(2), 91–103.

Fruth, B., and Hohmann, G. (2002). How bonobos handle hunts and harvests: Why share food? In Boesch, C., Hohmann, G. and Marchant, L. F. (Eds.). *Behavioural Diversity in Chimpanzees and Bonobos* (pp 231–43). Cambridge University Press.

Frydrych, L. M., Bian, G., O'Lone, D. E., Ward, P. A., and Delano, M. J. (2018). Obesity and type 2 diabetes mellitus drive immune dysfunction, infection development, and sepsis mortality. *Journal of Leukocyte Biology*, 104(3), 525–34.

Fuentes, A. (2012). Ethnoprimatology and the anthropology of the human-primate interface. *Annual Review of Anthropology*, 41, 101–17.

Fujii, J. A., Ralls, K., and Tinker, M. T. (2015). Ecological drivers of variation in tool-use frequency across sea otter populations. *Behavioral Ecology*, 26, 519–26.

Fujita, S., Sugiura, H., Mitsunaga, F., and Shimizu, K. (2004). Hormone profiles and reproductive characteristics in wild female Japanese macaques (*Macaca fuscata*). *American Journal of Primatology*, 64, 367–75.

Fukatsu, K., and Kudsk, K. A. (2011). Nutrition and gut immunity. *Surgical Clinics of North America*, 91(4), 755–70, vii.

Fukuda, S., Toh, H., Hase, K., Oshima, K., Nakanishi, Y., Yoshimura, K., et al. (2011). Bifidobacteria can protect from enteropathogenic infection through production of acetate. *Nature*, 469, 543–49.

Fülöp, A., Németh, Z., Kocsis, B., Deák-Molnár, B., Bozsoky, T., and Barta, Z. (2019). Personality and social foraging tactic use in free-living Eurasian tree sparrows (*Passer montanus*). *Behavioral Ecology*, 30, 894–903.

Furnell, S., Blanchard, M. L., Crompton, R. H., and Sellers, W. I. (2015). Locomotor ecology of *Propithecus verreauxi* in Kirindy Mitea National Park. *Folia Primatologica*, 86, 223–30.

Furuichi, T., Hashimoto, C., and Tashiro, Y. (2001). Fruit availability and habitat use by chimpanzees in the Kalinzu Forest, Uganda: Examination of fallback foods. *International Journal of Primatology*, 22, 929–45.

Furuichi, T., Sanz, C., Koops, K., Sakamaki, T., Ryu, H., Tokuyama, N., et al. (2015). Why do wild bonobos not use tools like chimpanzees do? *Behaviour*, 152, 425–60.

Fuzessy, L., Balbuena, J. A., Nevo, O., Tonos, J., Papinot, B., Park, D., et al. (2023). Friends or foes? Plant-animal coevolutionary history is driven by both mutualistic and antagonistic interactions. (unpublished manuscript)

Fuzessy, L. F., Cornelissen, T. G., Janson, C., and Silveira, F. A. O. (2016). How do primates affect seed germination? A meta-analysis of gut passage effects on Neotropical plants. *Oikos*, 125, 1069–80.

Gaboriau-Routhiau, V., Rakotobe, S., Lecuyer, E., Mulder, I. E., Lan, A., Bridonneau, C., et al. (2009). The key role of segmented filamentous bacteria in the coordinated maturation of gut helper T cell responses. *Immunity*, 31(4), 677–89.

Gabunia, L., Vekua, A., Lordkipanidze, D., Swisher, C., Ferring, R., Justus, A., et al. (2000). Earliest Pleistocene hominid cranial remains from Dmanisi, Republic of Georgia: Taxonomy, geological setting, and age. *Science*, 288, 1019–25.

Galbany, J., Altmann, J., Perez-Perez, A., and Alberts, S. C. (2011). Age and individual foraging behavior predict tooth wear in Amboseli baboons. *American Journal of Physical Anthropology*, 144(1), 51–59.

Galbany, J., Estebaranz, F., Martinez, L. M., and Perez-Perez, A. (2009). Buccal dental microwear variability in extant African Hominoidea: Taxonomy versus ecology. *Primates*, 50(3), 221–30.

Galbany, J., Imanizabayo, O., Romero, A., Vecellio, V., Glowacka, H., Cranfield, M. R., et al. (2016). Tooth wear and feeding ecology in mountain gorillas from Volcanoes National Park, Rwanda. *American Journal of Physical Anthropology*, 159(3), 457–65.

Galbany, J., Moya-Sola, S., and Perez-Perez, A. (2005). Dental microwear variability on buccal tooth enamel surfaces of extant Catarrhini and the Miocene fossil *Dryopithecus laietanus* (Hominoidea). *Folia Primatologica (Basel)*, 76(6), 325–41.

Galbany, J., Romero, A., Mayo-Aleson, M., Itsoma, F., Gamarra, B., Perez-Perez, A., et al. (2014). Age-related tooth wear differs between forest and savanna primates. *PLoS ONE*, 9, e94938.

Galbany, J., Tung, J., Altmann, J., and Alberts, S. C. (2015). Canine length in wild male baboons: Maturation, aging and social dominance rank. *PLoS ONE*, 10, e0126415.

Galbany, J., Twahirwa, J. C., Baiges-Sotos, L., Kane, E. E., Tuyisingize, D., Kaleme, P., et al. (2020). Dental macrowear in catarrhine primates: Variability across species. In C. W. Schmidt and J. T. Watson (Eds.), *Dental Wear in Evolutionary and Biocultural Contexts* (pp. 11–37). Academic Press.

Galdikas, B. M. F. (1988). Orangutan diet, range, and activity at Tanjung Puting, Central Borneo. *International Journal of Primatology*, 9, 1–35.

Galdikas, B. M. F., and Ashbury, A. (2013). Reproductive parameters of female orangutans (*Pongo pygmaeus wurmbii*), 1971–2011, a 40-year study at Tanjung Puting National Park, Central Kalimantan, Indonesia. *Primates*, 54, 61–72.

Galetti, M., and Pedroni, F. (1994). Seasonal diet of capuchin monkeys (*Cebus apella*) in a semideciduous forest in southeast Brazil. *Journal of Tropical Ecology*, 10, 27–39.

Galland, R. B., and Polk, H. C. (1982). Non-specific stimulation of host defenses against a bacterial challenge in malnourished hosts. *British Journal of Surgery*, 69(11), 665–68.

Galli, M., and Oyen, M. L. (2008). Spherical indentation of a finite poroelastic coating. *Applied Physics Letters*, 93, 031911.

Ganas, J., and Robbins, M. M. (2004). Intrapopulation differences in ant eating in the mountain gorillas of Bwindi Impenetrable National Park, Uganda. *Primates*, 45, 275–78.

Ganas, J., Ortmann, S., and Robbins, M. M. (2008). Food preferences of wild mountain gorillas. *American Journal of Primatology*, 70, 927–38.

Ganas, J., Ortmann, S., and Robbins, M. M. (2009). Food choices of the mountain gorilla in Bwindi Impenetrable National Park, Uganda: The influence of nutrients, phenolics and availability. *Journal of Tropical Ecology*, 25, 123–34.

Ganzhorn, J. U. (1988). Food partitioning among Malagasy primates. *Oecologia*, 75, 436–50.

Ganzhorn, J. U. (1989a). Niche separation of 7 lemur species in the eastern rainforest of Madagascar. *Oecologia*, 79, 279–86.

Ganzhorn, J. U. (1989b). Primate species separation in relation to secondary plant chemicals. *Human Evolution*, 4, 125–32.

Ganzhorn, J. U. (1992). Leaf chemistry and the biomass of folivorous primates in tropical forests—test of a hypothesis. *Oecologia*, 91, 540–47.

Ganzhorn, J. U. (1995a). Cyclones over Madagascar: Fate or fortune? *Ambio*, 24, 124–25.

Ganzhorn, J. U. (1995b). Low-level forest disturbance effects on primary production, leaf chemistry, and lemur populations. *Ecology*, 76, 2084–96.

Ganzhorn, J. U. (2002). Distribution of a folivorous lemur in relation to seasonally varying food resources: Integrating quantitative and qualitative aspects of food characteristics. *Oecologia*, 131, 427–35.

Ganzhorn, J. U., and Abraham, J. P. (1991). Possible role of plantations for lemur conservation in Madagascar: Food for folivorous species. *Folia Primatologica*, 56, 171–76.

Ganzhorn, J. U., Abraham, J. P., and Razanahoera-Rakotomalala, M. (1985). Some aspects of the natural history and food selection of *Avahi laniger*. *Primates*, 26, 452–63.

Ganzhorn, J. U., Arrigo-Nelson, S. J., Boinski, S., Bollen, A., Carrai, V., Derby, A., et al. (2009). Possible fruit protein effects on primate communities in Madagascar and the Neotropics. *PLoS ONE*, 4, e8523.

Ganzhorn, J. U., Arrigo-Nelson, S. J., Carrari, V., Chalise, M. K., Donati, G., Droescher, I., et al. (2017). The importance of protein in leaf selection of folivorous primates. *American Journal of Primatology*, 79, 1–13.

Ganzhorn, J. U., Pietsch, T., Fietz, J., Gross, S., Schmid, J., and Steiner, N. (2004). Selection of food and ranging behaviour in a sexually monomorphic folivorous lemur: *Lepilemur ruficaudatus*. *Journal of Zoology*, 263, 393–99.

Ganzhorn, J. U., Rakotondranary, S. J., and Ratovomanana, Y. R. (2011). Habitat description and phenology. In J. M. Setchell and D. J. Curtis (Eds.), *Field and Laboratory Methods in Primatology: A Practical Guide* (pp. 51–68). Cambridge University Press.

Ganzhorn, J., and Wright, P. (1994). Temporal patterns in primate leaf eating: The possible role of leaf chemistry. *Folia Primatologica*, 63, 203–8.

Garber, P. A. (1980). Locomotor behavior and feeding ecology of the Panamanian tamarin (*Saguinus oedipus geoffroyi*, Callitrichidae, Primates). *International Journal of Primatology*, 1, 185–201.

Garber, P. A. (1984). Proposed nutritional importance of plant exudates in the diet of the Panamanian tamarin, *Saguinus oedipus geoffroyi*. *International Journal of Primatology*, 5, 1–15.

Garber, P. A. (1986). The ecology of seed dispersal in two species of callitrichid primates (*Saguinus mystax* and *Saguinus fuscicollis*). *American Journal of Primatology*, 10, 155–70.

Garber, P. A. (1987). Foraging strategies among living primates. *Annual Review of Anthropology*, 16, 339–64.

Garber, P. A. (1988a). Diet, foraging patterns, and resource defense in a mixed species troop of *Saguinus mystax* and *Saguinus fuscicollis* in Amazonian Peru. *Behaviour*, 105, 18–34.

Garber, P. A. (1988b). Foraging decisions during nectar feeding by tamarin monkeys (*Saguinus mystax* and *Saguinus fuscicollis*, Callitrichidae, Primates) in Amazonian Peru. *Biotropica*, 20, 100–106.

Garber, P. A. (1989). Role of spatial memory in primate foraging patterns: *Saguinus mystax* and *Saguinus fuscicollis*. *American Journal of Primatology*, 12, 203–16.

Garber, P. A. (1992). Vertical clinging, small body size, and the evolution of feeding adaptations in the Callitrichinae. *American Journal of Physical Anthropology*, 88, 469–82.

Garber, P. A. (1997). One for all and breeding for one: Cooperation and competition as a tamarin reproductive strategy. *Evolutionary Anthropology*, 5, 187–99.

Garber, P. A. (2000). Evidence for the use of spatial, temporal, and social information by some primate foragers. In S. Boinski and P. A. Garber (Eds.), *On the Move: How and Why Animals Travel in Groups* (pp. 261–98). University of Chicago Press.

Garber, P. A., Bicca-Marques, J. C., and Azevedo-Lopes, M. A. O. (2009). Primate cognition: Integrating social and ecological information in decision-making. In P. A. Garber, A. Estrada, J. C. Bicca-Marques, E. Heymann, and K. B. Strier (Eds.), *South American Primates: Comparative Perspectives in the Study of Behavior, Ecology, and Conservation* (pp. 365–85). Springer.

Garber, P. A., and Brown, E. (2006). Use of landmark cues to locate feeding sites in wild capuchin monkeys (*Cebus capucinus*): An experimental field study. In A. Estrada, P. A. Garber, M. Pavelka, and L. Luecke (Eds.), *New Perspectives in the Study of Mesoamerican Primates: Distribution, Ecology, Behavior and Conservation* (pp. 311–32). Kluwer.

Garber, P. A., and Dolins, F. L. (1996). Evidence for use of spatial and perceptual information and rule-based foraging in wild moustached tamarins. In A. L. Rosenberger and P. A. Garber (Eds.), *Adaptive Radiation of Neotropical Primates* (pp. 201–16). Plenum.

Garber, P. A., and Dolins, F. L. (2014). Primate spatial strategies and cognition: Introduction to this special issue. *American Journal of Primatology*, 76, 393–98.

Garber, P. A., Estrada, A., Bicca-Marques, J. C., Heymann, E. W., and Strier, K. B. (2009). *South American Primates: Comparative Perspectives in the Study of Behavior, Ecology, and Conservation*. Springer.

Garber, P. A., Gomes, D. F., and Bicca-Marques, J. C. (2012). Experimental field study of problem-solving using tools in free ranging capuchins (*Sapajus nigritus*, formerly *Cebus nigritus*). *American Journal of Primatology*, 74, 344–58.

Garber, P. A., and Hannon, B. (1993). Modeling monkeys: A comparison of computer generated and naturally occurring foraging patterns in two species of Neotropical primates. *International Journal of Primatology*, 14, 827–52.

Garber, P. A., and Jelinek, P. E. (2006). Travel patterns and spatial mapping in Nicaraguan mantled howler monkeys (*Alouatta palliata*). In A. Estrada, P. A. Garber, M. Pavelka, and L. Luecke (Eds.), *New Perspectives in the Study of Mesoamerican Primates: Distribution, Ecology, Behavior and Conservation* (pp. 287–309). Kluwer.

Garber, P. A., Mallott, E. K., Porter, L. M., & Gomez, A. (2019). The gut microbiome and metabolome of saddleback tamarins (*Leontocebus weddelli*): Insights into the foraging ecology of a small-bodied primate. *American Journal of Primatology*, 81(10-11), e23003.

Garber, P. A., and Paciulli, L. (1997). Experimental field study of spatial memory and learning in wild capuchin monkeys (*Cebus capucinus*). *Folia Primatologica*, 68, 236–53.

Garber, P. A., and Porter, L. M. (2010). The ecology of exudate production and exudate feeding in *Saguinus* and *Callimico*. In A. M. Burrows and L. T. Nash (Eds.), *The Evolution of Exudativory in Primates* (pp. 89–108). Springer.

Garber, P. A., and Porter, L. M. (2014). Navigating in small-scale space: The role of landmarks and resource monitoring in understanding saddleback tamarin travel. *American Journal of Primatology*, 76, 447–59.

Garber, P. A., Righini, N., and Kowalewski, M. M. (2015). Evidence of alternative dietary syndromes and nutritional goals in the genus *Alouatta*. In M. M. Kowalewski, P. A. Garber, L. Cortés-Ortiz, B. Urbani, and D. Youlatos (Eds.), *Howler Monkeys: Behavior, Ecology, and Conservation* (pp. 85–109). Springer.

Garbino, G. S. T., da Silva, L. H., Amaral, R. G., Rezenda, G. C., Pereira, V. J. A., and Culot, L. (2020). Predation of treefrogs (Anura: Hylidae) with toxic skin by the black lion tamarin (*Leontopithecus chrysopygus*, Callitrichinae). *Primates*, 61, 567–72.

Garcia, C., Lee, P. C., and Rosetta, L. (2006). Dominance and reproductive rates in captive female olive baboons, *Papio anubis*. *American Journal of Physical Anthropology*, 131, 64–72.

Garcia, L. C., Hobbs, R. J., Maees dos Santos, F. A., and Rodrigues, R. R. (2014). Flower and fruit availability along a forest restoration gradient. *Biotropica*, 46(1), 114–23.

Garcia, M., Power, M. L., and Moyes, K. M. (2017). Immunoglobulin A and nutrients in milk from great apes throughout lactation. *American Journal of Primatology*, 79(3), e22614.

Gareau, M. G., Jury, J., MacQueen, G., Sherman, P. M., and Perdue, M. H. (2007). Probiotic treatment of rat pups normalises corticosterone release and ameliorates colonic dysfunction induced by maternal separation. *Gut*, 56, 1522–28.

Garey, J. D. (1984). A possible role for secondary plant-compounds in the regulation of primate breeding cycles. *American Journal of Physical Anthropology*, 63, 160.

Garrett, E. C., and Steiper, M. E. (2014). Strong links between genomic and anatomical diversity in both mammalian olfactory chemosensory systems. *Proceedings of the Royal Society B: Biological Sciences*, 281, 20132828.

Garrett, W. S., Lord, G. M., Punit, S., Lugo-Villarino, G., Mazmanian, S. K., Ito, S., et al. (2007). Communicable ulcerative colitis induced by T-bet deficiency in the innate immune system. *Cell*, 131(1), 33–45.

Garrido, D., Ruiz-Moyano, S., Kirmiz, N., Davis, J. C., Totten, S. M., Lemay, D. G., et al. (2016). A novel gene cluster allows preferential utilization of fucosylated milk oligosaccharides in *Bifidobacterium longum longum* SC596. *Scientific Reports*, 6, 35045.

Gaston, K. J., Jackson, S. F., Cantu-Salazer, L., and Cruz-Pinon, G. (2008). The ecological performance of protected areas. *Annual Review of Ecology, Evolution, and Systematics*, 39, 93–113.

Gathua, M. (2000). The effects of primates and squirrels on seed survival of a canopy tree, *Afzelia quanzensis*, in Arabuko-Sokoke Forest, Kenya 1. *Biotropica*, 32(1), 127–32.

Gaulin, S. J. C. (1979). A Jarman/Bell model of primate feeding niches. *Human Evolution*, 7, 1–20.

Gaulin, S. J. C., and Gaulin, C. K. (1982). Behavioural ecology of *Alouatta seniculus* in Andean cloud forest. *International Journal of Primatology*, 3, 1–32.

Gaulin, S. J. C., and Konner, M. J. (1977). On the natural diets of primates, including humans. In R. J. Wurtman and J. J. Wurtman (Eds.), *Nutrition and the Brain* (pp. 1–86). Raven.

Gautier-Hion, A. (1980). Seasonal variations of diet related to species and sex in a community of *Cercopithecus* monkeys. *Journal of Animal Ecology*, 49, 237–69.

Gautier-Hion, A. (1988). Polyspecific associations among forest guenons: Ecological, behavioral and evolutionary aspects. In A. Gautier-Hion, F. Bourliere, J. P. Gautier, and J. Kingdon (Eds.), *A Primate Radiation: Evolution of the African Guenons* (pp. 452–76). Cambridge University Press.

Gautier-Hion, A., Duplantier, J. M., Quris, R., Feer, F., Sourd, C., Decoux, J. P., et al. (1985). Fruit characters as a basis of fruit choice and seed dispersal in a tropical forest vertebrate community. *Oecologia*, 65, 324–37.

Gautier-Hion, A., Gautier, J. P., and Maisels, F. (1993). Seed dispersal versus seed predation: An inter-site comparison of two related African monkeys. *Vegetatio*, 107, 237–44.

Gautier-Hion, A., and Maisels, F. (1994). Mutualism between a leguminous tree and large African monkeys as pollinators. *Behavioral Ecology and Sociobiology*, 34, 203–10.

Gautier-Hion, A., and Michaloud, G. (1989). Are figs always keystone resources for tropical frugivorous vertebrates? A test in Gabon. *Ecology*, 70, 1826–33.

Gautier-Hion, A., Ouris, R., and Gautier-Hion, J. P. (1983). Monospecific vs polyspecific life: A comparative study of foraging and antipredatory tactics in a community of *Cercopithecus* monkeys *Behavioral Ecology and Sociobiology*, 12, 325–35.

Gavrilets, S., Auerbach, J., and Van Vugt, M. (2016). Convergence to consensus in heterogeneous groups and the emergence of informal leadership. *Scientific Reports*, 6, 29704.

Gaynor, K. M., Brown, J. S., Middleton, A. D., Power M. E., and Brashares J. S. (2019). Landscapes of fear: Spatial patterns of risk perception and response. *Trends in Ecology and Evolution*, 34(4), 355–68.

Gaynor, K. M., and Cords, M. (2012). Antipredator and social monitoring functions of vigilance behaviour in blue monkeys. *Animal Behaviour*, 84, 531–37.

Gaynor, K. M., McInturff A. and Brashares J. S. (2022). Contrasting patterns of risk from human and non-human predators shape temporal activity of prey. *Journal of Animal Ecology*, 91(1), 46–60.

Gebo, D. L., and Chapman, C. A. (1995). Habitat, annual, and seasonal effects on positional behavior in red colobus monkeys. *American Journal of Physical Anthropology*, 96, 73–82.

Gedan, K. B., and Bertness, M. D. (2009). Experimental warming causes rapid loss of plant diversity in New England salt marshes. *Ecology Letters*, 12, 842–48.

Geddes, D. T., and Prescott, S. L. (2013). Developmental origins of health and disease: The role of human milk in preventing disease in the 21st century. *Journal of Human Lactation*, 29, 123–27.

Geerah, D. R., O'Hagan, R. P., Wirdateti, W., and Nekaris, K. A.

I. (2019). The use of ultrasonic communication to maintain social cohesion in the Javan slow loris (*Nycticebus javanicus*). *Folia Primatologica*, 90(5), 392–403.

Geffroy, B., Samia, D. S. M., Bessa, E., and Blumstein, D. T. (2015). How nature-based tourism might increase prey vulnerability to predators. *Trends in Ecology and Evolution*, 30, 755–65.

Gehrig, J. L., Venkatesh, S., Chang, H. W., Hibberd, M., Kung, V. L., Chen, J., et al. (2019). Effects of microbiota-directed foods in gnotobiotic animals and undernourished children. *Science*, 365, 139.

Geissler, E., Daegling, D. J., and McGraw, W. S. (2018). Forest floor leaf cover as a barrier for dust accumulation in Tai National Park: Implications for primate dental wear studies. *International Journal of Primatology*, 39, 633–45.

Geissmann, T. (1991). Reassessment of age of sexual maturity in gibbons (*Hylobates* spp.). *American Journal of Primatology*, 23, 11–22.

Geissmann, T., and Orgeldinger, M. (1995). Neonatal weight in gibbons (*Hylobates* spp.). *American Journal of Primatology*, 37(3), 179–89.

Geldmann, J., Barnes, M., Coad, L., Craigie, I. D., Hockings, M., and Burgess, N. D. (2013). Effectiveness of terrestrial protected areas in reducing habitat loss and population declines. *Biological Conservation*, 161, 230–38.

Gelman, A., and Rubin, D. B. (1992). Inference from iterative simulation using multiple sequences. *Statistical Science*, 457–72.

Georgiev, A. V. (2012). *Energetic costs of reproductive effort in male chimpanzees*. Harvard University Press.

Génard, M., and Bruchou, C. (1992). Multivariate analysis of within-tree factors accounting for the variation of peach fruit quality. *Scientia Horticulturae*, 52, 37–51.

Génin, F. G. S., Masters, J. C., and Ganzhorn, J. U. (2010). Gummivory in cheirogaleids: Primitive retention or adaptation to hypervariable environments? In A. M. Burrows and L. T. Nash (Eds.), *The Evolution of Exudativory in Primates* (pp. 123–40). Springer.

Genton, B., Al-Yaman, F., Ginny, M., Taraika, J., and Alpers, M. P. (1998). Relation of anthropometry to malaria morbidity and immunity in Papua New Guinean children. *American Journal of Clinical Nutrition*, 68(3), 734–41.

Gentry, A. H., and Emmons, L. H. (1987). Geographical variation in fertility, phenology, and composition of the understory of Neotropical forests. *Biotropica*, 19, 216–27.

Georgiadis, N. J., and McNaughton, S. J. (1990). Elemental and fibre contents of savanna grasses: Variation with grazing, soil type, season and species. *Journal of Applied Ecology*, 27, 623–34.

Georgiev, A. V., Muehlenbein, M. P., Prall, S. P., Thompson, M. E., and Maestripieri, D. (2015a). Innate immune function and oxidative stress as measures of male quality in Cayo Santiago rhesus macaques. *American Journal of Physical Anthropology*, 156, 143–43.

Georgiev, A. V., Muehlenbein, M. P., Prall, S. P., Thompson, M. E., and Maestripieri, D. (2015b). Male quality, dominance rank, and mating success in free-ranging rhesus macaques. *Behavioral Ecology*, 26(3), 763–72.

Georgiev, A. V., Russell, A., Emery Thompson, M., Otali, E., Muller, M. N., and Wrangham, R. W. (2014). The foraging costs of mating effort in male chimpanzees (*Pan troglodytes schweinfurthii*). *International Journal of Primatology*, 35, 725–45.

Georgiev, A. V., Thompson, M. E., Lokasola, A. L., and Wrangham, R. W. (2011). Seed predation by bonobos (*Pan paniscus*) at Kokolopori, Democratic Republic of the Congo. *Primates*, 52, 309–14.

Gerard, A., Ganzhorn, J. U., Kull, C. A., and Carriere, S. M. (2015). Possible roles of introduced plants for native vertebrate conservation: The case of Madagascar. *Restoration Ecology*, 23, 768–75.

Gerhold, K. A., Pellegrino, M., Tsunozaki, M., Morita, T., Leitch, D. B., Tsuruda, P. R., et al. (2013). The star-nosed mole reveals clues to the molecular basis of mammalian touch. *PLoS ONE*, 8, e55001.

German, J. B., Dillard, C. J., and Ward, R. E. (2002). Bioactive components in milk. *Current Opinion in Clinical Nutrition and Metabolic Care*, 5, 653–58.

Gesquiere, L. R., Altmann, J., Archie, E. A., and Alberts, S. C. (2018). Interbirth intervals in wild baboons: Environmental predictors and hormonal correlates. *American Journal of Physical Anthropology*, 166, 107–26.

Ghazoul, J., and Sheil, D. (2010). *Tropical Rain Forest Ecology, Diversity, and Conservation*. Oxford University Press.

Gibson, K. R. (1986). Cognition, brain size, and the extraction of embedded food resources. In J. G. Else and P. C. Lee (Eds.), *Primate Ontogeny, Cognition, and Social Behavior* (pp. 93–103). Cambridge University Press.

Gibson, L. J., and Ashby, M. F. (1999). *Cellular Solids Structure and Properties* (2nd ed.) Cambridge University Press.

Gibson, L. (2011). Possible shift in macaque trophic level following a century of biodiversity loss in Singapore. *Primates*, 52, 217–20.

Gibson, M., and Mace, R. (2006). An energy-saving development initiative increases birth rate and childhood malnutrition in rural Ethiopia. *PLoS Medicine*, 3, 1–10.

Gibson, R. A., and Makrides, M. (1999). Long-chain polyunsaturated fatty acids in breast milk: Are they essential? In D. S. Newburg (Ed.), *Bioactive Components of Human Milk* (pp. 375–83). Plenum.

Gil, M. A., Emberts, Z., Jones, H., and St. Mary, C. M. (2017). Social information on fear and food drives animal grouping and fitness. *American Naturalist*, 189, 227–41.

Gilbert, C. C. (2013). Cladistic analysis of extant and fossil African papionins using craniodental data. *Journal of Human Evolution*, 64(5), 399–433.

Gilby, I. (2006). Meat sharing among the Gombe chimpanzees:

Harassment and reciprocal exchange. *Animal Behaviour*, 71, 953–63.

Gilby, I. C., and Connor, R. R. (2010). The role of intelligence in group hunting: Are chimpanzees different from other social predators? In E. V. Lonsdorf, S. R. Ross, and T. Matsuzawa (Eds.), *The Mind of the Chimpanzee: Ecological and Experimental Perspectives* (pp. 220–33). University of Chicago Press.

Gilby, I. C., and Wawrzyniak, D. (2018). Meat eating by wild chimpanzees (*Pan troglodytes schweinfurthii*): Effects of prey age on carcass consumption sequence. *International Journal of Primatology*, 39, 127–40.

Gilby, I. C., and Wrangham, R. W. (2007). Risk-prone hunting by chimpanzees (*Pan troglodytes*) increases during periods of high diet quality. *Behavioral Ecology and Sociobiology*, 61, 1771–79.

Gilby, I. C., Eberle, L. E., and Wrangham, R. W. (2008). Economic profitability of social predation among wild chimpanzees: Individual variation promotes cooperation. *Animal Behaviour*, 75, 351–60.

Gilby, I. C., Eberle, L. E., Pintea, L., and Pusey, A. E. (2006). Ecological and social influences on the hunting behavior of wild chimpanzees (*Pan troglodytes schweinfurthii*). *Animal Behaviour*, 72, 169–80.

Gilby, I. C., Emery Thompson, M., Ruane, J. D., and Wrangham, R. W. (2010). No evidence of short-term exchange of meat for sex among chimpanzees. *Journal of Human Evolution*, 59, 44–53.

Gilby, I. C., Machanda, Z. P., Mjungu, D., Rosen, J., Muller, M. N., Pusey, A. E., et al. (2015). "Impact hunters" catalyze cooperative hunting in two wild chimpanzee communities. *Philosophical Transactions of the Royal Society of London Series B—Biological Sciences*, 370, 2015005.

Gilby, I. C., Machanda, Z. P., O'Malley, R. C., Murray, C. M., Lonsdorf, E. V., Walker, K., et al. (2017). Predation by female chimpanzees: Toward an understanding of sex differences in meat acquisition in the last common ancestor of *Pan* and *Homo*. *Journal of Human Evolution*, 110, 82–94.

Gill, J. A., Sutherland, W. J., and Watkinson, A. R. (1996). A method to quantify the effects of human disturbance on animal populations. *Journal of Applied Ecology*, 33, 786–92.

Gillespie, M. J., Stanley, D., Chen, H., Donald, J. A., Nicholas, K. R., Moore, R. J., et al. (2012). Functional similarities between pigeon "milk" and mammalian milk: Induction of immune gene expression and modification of the microbiota. *PLoS ONE*, 7, e48363.

Gillespie, T. R. (2006). Noninvasive assessment of gastrointestinal parasite infections in free-ranging primates. *International Journal of Primatology*, 27, 1129–43.

Gillespie, T. R., and Chapman, C. A. (2001). Determinants of group size in the red colobus monkey (*Procolobus badius*): An evaluation of the generality of the ecological-constraints model. *Behavioral Ecology and Sociobiology*, 50, 329–38.

Gillespie, T. R., Chapman, C. A., and Greiner, E. C. (2005). Effects of logging on gastrointestinal parasite infections and infection risk in African primates. *Journal of Applied Ecology*, 42, 699–707.

Gilmore, D., and Cook, B. (1981). *Environmental Factors in Mammal Reproduction*. Macmillan.

Giraldeau, L. A., and Caraco, T. (2000). *Social Foraging Theory*. Princeton University Press.

Giraldeau, L. A., and Dubois, F. (2008). Social foraging and the study of exploitative behavior. *Advances in the Study of Behavior*, 38, 59–104.

Girard-Buttoz, C., Heistermann, M., Rahmi, E., Marzec, A., Agil, M., Fauzan, P. A., et al. (2014). Mate-guarding constrains feeding activity but not energetic status of wild male long-tailed macaques (*Macaca fascicularis*). *Behavioral Ecology and Sociobiology*, 68, 583–95.

Girard-Buttoz, C., Higham, J. P., Heistermann, M., Wedegärtner, S., Maestripieri, D., and Engelhardt, A. (2011). Urinary C-peptide measurement as a marker of nutritional status in macaques. *PLoS ONE*, 6, e18042.

Girvetz, E. H., and Greco, S. E. (2007). How to define a patch: A spatial model for hierarchically delineating organism-specific habitat patches. *Landscape Ecology*, 22, 1131–42.

Gittleman, J. L., and Thompson, S. D. (1988). Energy allocation in mammalian reproduction. *American Zoologist*, 28, 863–75.

Glander, K. E. (1978). Howling monkey feeding behavior and plant secondary compounds: A study of strategies. In C. G. Montgomery (Ed.), *The Ecology of Arboreal Folivores* (pp. 561–74). Smithsonian Institution Press.

Glander, K. E. (1980). Reproduction and population-growth in free-ranging mantled howling monkeys. *American Journal of Physical Anthropology*, 53, 25–36.

Glander, K. E. (1981). Feeding patterns in mantled howling monkeys. In A. C. Kamil and T. D. Sargent (Eds.), *Foraging Behavior: Ecological, Ethological, and Psychological Approaches* (pp. 231–59). Garland STPM.

Glander, K. E. (1982). The impact of plant secondary compounds on primate feeding behavior. *American Journal of Physical Anthropology*, 25, 1–18.

Glander, K. E. (1994). Nonhuman primate self-medication with wild plant foods. In N. L. Etkin (Ed.), *Eating on the Wild Side: The Pharmacologic, Ecologic and Social Implications of Using Noncultigens* (pp. 227–39). University of Arizona Press.

Glander, K. E., Fedigan, L. M., Fedigan, L., and Chapman, C. (1991). Field methods for capture and measurement of three monkey species in Costa Rica. *Folia Primatologica*, 57, 70–82.

Glander, K. E., and Rabin, D. P. (1983). Food choice from endemic North Carolina tree species by captive prosimians (*Lemur fulvus*). *American Journal of Primatology*, 5, 221–29.

Glander, K. E., Wright, P. C., Daniels, P. S., and Merenlender, A. M. (1992). Morphometrics and testicle size of rain forest lemur species from southeastern Madagascar. *Journal of Human Evolution*, 22, 1–17.

Glander, K. E., Wright, P. C., Siegler, D. S., Randrianasolo, V., and Randrianasolo, B. (1989). Consumption of cyanogenic bamboo by a newly discovered species of bamboo lemur. *American Journal of Primatology*, 19, 119–24.

Glenndinning, J. I. (1994). Is the bitter rejection response always adaptive? *Physiology and Behavior*, 56, 1217–27.

Glessner, K. D. G., and Britt, A. (2005). Population density and home range size of *Indri indri* in a protected low altitude rain forest. *International Journal of Primatology*, 26, 855–72.

Glick, B. R. (1995). The enhancement of plant-growth by free-living bacteria. *Canadian Journal of Microbiology*, 41, 109–17.

Glowacka, H., McFarlin, S. C., Vogel, E. R., Stoinski, T. S., Ndagijimana, F., Tuyisingize, D., Mudakikwa, A., and Schwartz G. T. (2017). Toughness of the Virunga mountain gorilla (*Gorilla beringei beringei*) diet across an altitudinal gradient. *American Journal of Primatology*, 79, e22661.

Gluckman, P. D., Hanson, M. A., Cooper, C., and Thornburg, K. L. (2008). Effect of in utero and early-life conditions on adult health and disease. *New England Journal of Medicine*, 359, 61–73.

Gochman, S. R., Brown, M. B., and Dominy, N. J. (2016). Alcohol discrimination and preferences in two species of nectar-feeding primate. *Royal Society Open Science*, 3, 160217.

Godfrey, L. R., Crowley, B. E., and Dumont, E. R. (2011). Thinking outside the box: A lemur's take on hominin craniodental evolution. *Proceedings of the National Academy of Sciences of the United States of America*, 108, E742.

Godfrey, L. R., Winchester, J. M., King, S. J., Boyer, D. M., and Jernvall, J. (2012). Dental topography indicates ecological contraction of lemur communities. *American Journal of Physical Anthropology*, 148(2), 215–27.

Godfrey, L. R., Crowley, B. E., Muldoon, K. M., Kelley, E. A., King, S. J., Best, A. W., et al. (2016). What did *Hadropithecus* eat, and why should paleoanthropologists care? *American Journal of Primatology*, 78, 1098–112.

Godfrey, L. R., Samonds, K. E., Jungers, W. L., and Sutherland, M. R. (2001). Teeth, brains, and primate life histories. *American Journal of Physical Anthropology*, 114, 192–214.

Godfrey-Smith, P. (2006). The strategy of model-based science. *Biology and Philosophy*, 21, 725–40.

Goering, H. K., and Van Soest, P. J. (1970). Forage fiber analysis (apparatus, reagents, procedures, and some applications). In Agricultural Research Service (Ed.), *Agriculture Handbook No. 379*. USDA.

Goering, H. K., Van Soest, P. J., and Hemken, R. W. (1973). Relative susceptibility of forages to heat damage as affected by moisture, temperature and pH. *Journal of Dairy Science*, 56, 137–43.

Gogarten, J. F., Guzman, M., Chapman, C. A., Jacob, A. L., Omeja, P. A., and Rothman, J. M. (2012). What is the predictive power of the colobine protein-to-fiber model and its conservation value? *Tropical Conservation Science*, 5, 381–93.

Gogarten, J. F., Jacob, A. L., Ghai, R. R., Rothman, J. M., Twinomugisha, D., Wasserman, M. D., et al. (2014). Group size dynamics over 15+ years in an African forest primate community. *Biotropica*, 47, 101–12.

Goldberg, G. R., Prentice, A. M., Coward, W. A., Davies, H. L., Murgatroyd, P. R., Sawyer, M. B., et al. (1991). Longitudinal assessment of the components of energy balance in well-nourished lactating women. *American Journal of Clinical Nutrition*, 54, 788–98.

Goldberg, T. L., Gillepsie, T. R., Rwego, I. B., Estoff, E. L., and Chapman, C. A. (2008). Forest fragmentation as cause of bacterial transmission among nonhuman primates, humans and livestock, Uganda. *Emerging Infectious Diseases*, 14, 1375–82.

Goldizen, A. W. (1987a). Facultative polyandry and the role of infant-carrying in wild saddle-back tamarins (*Saguinus fuscicollis*). *Behavioral Ecology and Sociobiology*, 20, 99–109.

Goldizen, A. W. (1987b). Tamarins and marmosets: Communal care of offspring. In B. B. Smuts, D. L. Cheney, R. M. Seyfarth, R. W. Wrangham, and T. T. Struhsaker (Eds.), *Primate Societies* (pp. 34–43). University of Chicago Press.

Goldizen, A. W. (1990). A comparative perspective on the evolution of tamarin and marmoset social systems. *International Journal of Primatology*, 11, 68–83.

Goldizen, A. W., Terborgh, J., Cornejo, F., Porras, D., and Evans, R. (1988). Seasonal food shortage, weight loss, and the timing of births in saddle-back tamarins (*Saguinus fuscicollis*). *Journal of Animal Ecology*, 57, 893–901.

Goldsmith, T. H. (1990). Optimization, constraint, and history in the evolution of eyes. *Quarterly Review of Biology*, 65, 281–322.

Golub, M., Gershwin, M., Hurley, L., Baly, D., and Hendrickx, A. (1984). Studies of marginal zinc deprivation in rhesus monkeys: II. Pregnancy outcome. *American Journal of Clinical Nutrition*, 39, 879–87.

Golub, M. S., Hogrefe, C. E., Tarantal, A. F., Germann, S. L., Beard, J. L., Georgieff, M. K., et al. (2006). Diet-induced iron deficiency anemia and pregnancy outcome in rhesus monkeys. *American Journal of Clinical Nutrition*, 83, 647–56.

Gonçalves, B. d. A., Pires Lima, L. C., and Aguiar, L. M. (2022). Diet diversity and seasonality of robust capuchins (*Sapajus* sp.) in a tiny urban forest. *American Journal of Primatology*, 84, e23396.

Gonzalez, N. T., Otali, E., Machanda, Z., Muller, M. N., Wrangham, R., and Thompson, M. E. (2020). Urinary markers of oxidative stress respond to infection and late-life in wild chimpanzees. *PLoS ONE*, 15(9), e0238066.

González-Zamora, A., Arroyo-Rodrígues, V., Chaves, Ó. M., Sánchez-López, S., Stoner, K. E., and Riba-Hernández, P. (2009). Diet of spider monkeys (*Ateles geoffroyi*) in Mesoamerica: Current knowledge and future directions. *American Journal of Primatology*, 71, 8–20.

Goodale, E., Beauchamp, G., and Ruxton, G. D. (2017). *Mixed Species Groups of Animals: Behavior, Community Structure, and Conservation*. Academic Press.

Goodale, E., and Kotagama, S. W. (2005). Testing the roles of species in mixed-species bird flocks of a Sri Lankan rain forest. *Journal of Tropical Ecology*, 21(6), 669–76.

Goodale, E., Sridhar, H., Sieving, K. E., Bangal, P., Colorado, Z., G. J., Farine, D. R., et al. (2020). Mixed company: A framework for understanding the composition and organization of mixed-species animal groups. *Biological Reviews*, 95, 889–910.

Goodall, J. (1963). Feeding behaviour of wild chimpanzees: A preliminary report. *Symposia of the Zoological Society of London*, 10, 39–47.

Goodall, J. (1964). Tool-using and aimed throwing in a community of free-living chimpanzees. *Nature*, 201, 1264–66.

Goodall, J. (1968). The behaviour of free-living chimpanzees in the Gombe Stream Reserve. *Animal Behaviour Monographs*, 1, 161–311.

Goodall, J. (1986). *The Chimpanzees of Gombe: Patterns of Behavior*. Harvard University Press.

Goodrich, J. K., Waters, J. L., Poole, A. C., Sutter, J. L., Koren, O., Blekhman, R., et al. (2014). Human genetics shape the gut microbiome. *Cell*, 159, 789–99.

Goran, M. I., and Poehlman, E. T. (1992). Endurance training does not enhance total energy expenditure in healthy elderly persons. *American Journal of Physiology-Endocrinology and Metabolism*, 263(5), E950-E957.

Gordon, A. D. (2006). Scaling of size and dimorphism in primates II: Macroevolution. *International Journal of Primatology*, 27, 63–105.

Gore, A. C., Chappell, V. A., Fenton, S. E., Flaws, J. A., Nadal, A., Prins, G. S., et al. (2015). EDC-2: The Endocrine Society's second scientific statement on endocrine-disrupting chemicals. *Endocrine Reviews*, 36, E1–E150.

Gore, M. A., Brandes, F., Kaup, F. J., Lenzner, R., Mothes, T., and Osman, A. A. (2001). Callitrichid nutrition and food sensitivity. *Journal of Medical Primatology*, 30(3), 179–84.

Gosby, A. K., Conigrave, A., Lau, N. S., Iglesias, M. A., Hall, R. M., Jebb, S. A., et al. (2011). Testing protein leverage in lean humans: A randomised controlled experimental study. *PLoS ONE*, 6, e25929.

Gosby, A. K., Conigrave, A., Raubenheimer, D., and Simpson, S. J. (2013). Protein leverage and energy intake. *Obesity Reviews*, 15, 183–99.

Goto, K., Fukuda, K., Senda, A., Saito, T., Kimura, K., Glander, K. E., et al. (2010). Chemical characterization of oligosaccharides in the milk of six species of New and Old World monkeys. *Glycoconjugate Journal*, 27, 703–15.

Goulart, V. D., Boubli, J. P., and Young, R. J. (2017). Medium/long wavelength sensitive opsin diversity in Pitheciidae. *Scientific Reports*, 7(1), 7737.

Gould, L., Constabel, P., Mellway, R., and Rambeloarivony, H. (2009). Condensed tannin intake in spiny-forest-dwelling *Lemur catta* at Berenty reserve, Madagascar, during reproductive periods. *Folia Primatologica*, 80, 249–63.

Gould, L., Power, M. L., Ellwanger, N., and Rambeloarivony, H. (2011). Feeding behavior and nutrient intake in spiny forest-dwelling ring-tailed lemurs (*Lemur catta*) during early gestation and early to mid-lactation periods: Compensating in a harsh environment. *American Journal of Physical Anthropology*, 145, 469–79.

Gould, L., Sussman, R. W., and Sauther, M. L. (1999). Natural disasters and primate populations: The effects of a 2-year drought on a naturally occurring population of ring-tailed lemurs (*Lemur catta*) in Southwestern Madagascar. *International Journal of Primatology*, 20, 69–84.

Gould, S. J. (1975). On scaling of tooth size in mammals. *American Zoologist*, 15(2), 351–62.

Gould, S. J. (2002). *The Structure of Evolutionary Theory*. Belknap Press of Harvard University Press.

Goulden, M. L., Miller, S. D., Da Rocha, H. R., Menton, M. C., de Freitas, H. C., and de Sousa, C. A. D. (2004). Diel and seasonal patterns of tropical forest CO_2 exchange. *Ecological Applications*, 14, 42–54.

Gouveia, S. F., Souza-Alves, J. P., Rattis, L., Dobrovolski, R., Jerusalinsky, L., Beltrão-Mendes, R., et al. (2016). Climate and land use changes will degrade the configuration of the landscape for titi monkeys in eastern Brazil. *Global Change Biology*, 22(6), 2003–12.

Grafen, A. (1989). The phylogenetic regression. *Philosophical Transactions of the Royal Society of London Series B—Biological Sciences*, 326, 119–57.

Graham, T. L., Matthews, H. D., and Turner, S. E. (2016). A global-scale evaluation of primate exposure and vulnerability to climate change. *International Journal of Primatology*, 37, 158–74.

Grainger, R., Peddemors, V. M., Raubenheimer, D., and Machovsky-Capuska, G. E. (2020). Diet composition and nutritional niche breadth variability in juvenile white sharks (*Carcharodon carcharias*). *Frontiers in Marine Science*, 7, 1–20.

Grand, T. I. (1972). A mechanical interpretation of terminal branch feeding. *Journal of Mammalogy*, 53, 198–201.

Grant, P., and Ramasamy, S. (2012). An update on plant derived anti-androgens. *International Journal of Endocrinology and Metabolism*, 10(2), 497.

Grassi, C. (2006). Variability in habitat, diet, and social structure of *Hapalemur griseus* in Ranomafana National Park, Madagascar. *American Journal of Physical Anthropology*, 131, 50–63.

Grayson, B. E., Levasseur, P. R., Williams, S. M., Smith, M. S., Marks, D. L., and Grove, K. L. (2010). Changes in melanocortin expression and inflammatory pathways in fetal offspring of nonhuman primates fed a high-fat diet. *Endocrinology*, 151(4), 1622–32.

Grech, A., Sui, Z., Rangan, A., Simpson, S. J., Coogan, S. C. P., and Raubenheimer, D. (2022). Macronutrient (im)balance drives energy intake in an obesogenic food environment: An ecological analysis. *Obesity*, 30, 2156–66.

Green, S. J., Boruff, B. J., Niyigaba, P., Ndikubwimana, I., and Grueter, C. C. (2020). Chimpanzee ranging responses to fruit availability in a high-elevation environment. *American Journal of Primatology*, 82, e23119.

Greenham, J. R., Grayer, R. J., Harborne, J. B., and Reynolds, V. (2007). Intra- and interspecific variations in vacuolar flavonoids among *Ficus* species from the Budongo Forest, Uganda. *Biochemical Systematics and Ecology*, 35, 81–90.

Greenwald, A. M. (2017). *Isotopic Reconstruction of Weaning Age and Childhood Diet among Ancient California Foragers: Life History Strategies and Implications for Demographics, Resource Intensification, and Social Organization.* University of California, Davis.

Gregory, T., Mullett, A., and Norconk, M. A. (2014). Strategies for navigating large areas: A GIS spatial ecology analysis of the bearded saki monkey, *Chiropotes sagulatus*, in Suriname. *American Journal of Primatology*, 76, 586–95.

Gregory, W. K. (1910). *The Orders of Mammals* (vol. 27). American Museum of Natural History.

Gresl, T. A., Baum, S. T., and Kemnitz, J. W. (2000). Glucose regulation in captive *Pongo pygmaeus abelii*, *P.-p. pygmaeus*, and *P.-p. abelii* x *P-p. pygmaeus* orangutans. *Zoo Biology*, 19(3), 193–208.

Grether, G. F., Palombit, R. A., and Rodman, P. S. (1992). Gibbon foraging decisions and the marginal value model. *International Journal of Primatology*, 13, 1-17.

Grieneisen, L., Dasari, M., Gould, T. J., Bjork, J. R., Grenier J-C, Yotova, V., et al. (2021). Gut microbiome heritability is nearly universal but environmentally contingent. *Science*, 373 (6551), 181–86.

Griffiths, B. M., Bowler, M., Kolowski, J., Stabach, J., Benson, E. L., and Gilmore, M. P. (2022). Revisiting optimal foraging theory (OFT) in a changing Amazon: Implications for conservation and management. *Human Ecology*, 50, 545–58.

Griffiths, M., Green, B., Leckie, R. M., Messer, M., and Newgrain, K. W. (1984). Constituents of platypus and echidna milk, with particular reference to the fatty acid complement of the triglycerides. *Australian Journal of Biological Sciences*, 37, 323–30.

Grine, F. E. (1977). Postcanine tooth function and jaw movement in the gomphodont cynodont *Diademodon* (Reptilia; Therapsida). *Palaeontologica Africana*, 20(123), 135.

Grine, F. E. (1986). Dental evidence for dietary differences in *Australopithecus* and *Paranthropus*: A quantitative analysis of permanent molar microwear. *Journal of Human Evolution*, 15(8), 783–822.

Grine, F. E., and Kay, R. F. (1988). Early hominid diets from quantitative image analysis of dental microwear. *Nature*, 333, 765–68.

Grine, F. E., Ungar, P. S., and Teaford, M. F. (2002). Error rates in dental microwear quantification using scanning electron microscopy. *Scanning*, 24(3), 144–53.

Grine, F. E., Ungar, P. S., and Teaford, M. F. (2006). Was the Early Pliocene hominin "*Australopithecus*" *anamensis* a hard object feeder? *South African Journal of Science*, 102, 301–10.

Grodwohl, J. B. (2019). Animal behavior, population biology and the modern synthesis (1955–1985). *Journal of the History of Biology*, 52, 597–633.

Groer, M. W., Morgan, K. H., Louis-Jacques, A., and Miller, E. M. (2020). A scoping review of research on the human milk microbiome. *Journal of Human Lactation*, 36(4), 628–43.

Grover, C. D., Kay, A. D., Monson, J. A., Marsh, T. C., and Howay, D. A. (2007). Linking nutrition and behavioural dominance: Carbohydrate scarcity limits aggression and activity in Argentine ants. *Proceedings of the Royal Society B: Biological Sciences*, 274, 2951–57.

Groves, C. P., and Napier, J. R. (1968,). Dental dimensions and diet in australopithecines. *Proc. VIII International Congress of the Anthropological and Ethnological Sciences*, 3, 273–76.

Grow, N., Gursky, S., and Duma, Y. (2013). Altitude and forest edges influence the density and distribution of pygmy tarsiers (*Tarsius pumilus*). *American Journal of Primatology*, 75, 464–77.

Grow, N. B., Gursky-Doyen, S., and Krzton, A. (2014). *High Altitude Primates*. Springer Science and Business Media.

Gruber, T. (2013). Historical hypotheses of chimpanzee tool use behaviour in relation to natural and human-induced changes in an East African rain forest. *Revue de Primatologie*, 5, 66.

Gruber, T., Clay, Z., and Zuberbühler, K. (2010). A comparison of bonobo and chimpanzee tool use: Evidence for a female bias in the *Pan* lineage. *Animal Behaviour*, 80, 1023–33.

Gruber, T., Potts, K. B., Krupenye, C., Byrne, M., Mackworth-Young, C., McGrew, W. C., et al. (2012). The influence of ecology on chimpanzee (*Pan troglodytes*) cultural behavior: A case study of five Ugandan chimpanzee communities. *Journal of Comparative Physiology*, 126, 446–57.

Grueter, C. C., Deschner, T., Behringer, V., Fawcett, K., and Robbins, M. M. (2014). Socioecological correlates of energy balance using urinary C-peptide measurements in wild female mountain gorillas. *Physiology and Behavior*, 127, 13–19.

Grueter, C. C., Li, D. Y., Ren, B. P., Wei, F. W., and van Schaik, C. P. (2009). Dietary profile of *Rhinopithecus bieti* and its socioecological implications. *International Journal of Primatology*, 30, 601–24.

Grueter, C. C., Li, D., Ren, B., Wei, F., Xiang, Z., and van Schaik, C. P. (2009). Fallback foods of temperate-living primates: A case study on snub-nosed monkeys. *American Journal of Physical Anthropology*, 140, 700–715.

Grueter, C. C., Li, D., Ren, B., Xiang, Z., and Li, M. (2012). Food abundance is the main determinant of high-altitude range use in snub-nosed monkeys. *International Journal of Zoology*, 739419.

Grueter, C. C., Ndamiyabo, F., Plumptre, A. J., Abavandimwe, D., Mundry, R., Fawcett, K. A., et al. (2013). Long-term temporal and spatial dynamics of food availability for endangered mountain gorillas in Volcanoes National Park, Rwanda: Dynamics of mountain gorilla food availability. *American Journal of Primatology*, 75, 267–80.

Grueter, C. C., Robbins, A. M., Abavandimwe, D., Vecellio, V., Ndagijimana, F., Ortmann, S., et al. (2016). Causes, mechanisms, and consequences of contest competition among female mountain gorillas in Rwanda. *Behavioral Ecology*, 27(3), 766–76.

Grueter, C. C., Robbins, A. M., Abavandimwe, D., Vecellio, V., Ndagijimana, F., Stoinski, T. S., et al. (2018). Quadratic relationships between group size and foraging efficiency in a herbivorous primate. *Scientific Reports*, 8, 16718.

Grund, C., Neumann, C., Zuberbühler, K., Gruber, T. (2019). Necessity creates opportunities for chimpanzee tool use. *Behavioral Ecology* 30(4), 1136–44.

Grutzmacher, K., Keil, V., Leinert, V., Leguillon, F., Henlin, A., Couacy-Hymann, E., et al. (2018). Human quarantine: Toward reducing infectious pressure on chimpanzees at the Tai Chimpanzee Project, Cote d'Ivoire. *American Journal of Primatology*, 80(1) e22619.

Guedes, D., Young, R. J., and Strier, K. B. (2008). Energetic costs of reproduction in female northern muriquis, *Brachyteles hypoxanthus* (Primates: Platyrrinhi: Atelidae). *Revista Brasiliera de Zoológica*, 25, 587–93.

Guerra, B., and Vickery, W. L. (1998). How do red squirrels, *Tamiasciurus hudsonicus*, and eastern chipmunks, *Tamias striatus*, coexist? *Oikos*, 83, 139–44.

Guerrero-Bosagna, C. M., and Skinner, M. K. (2014). Environmental epigenetics and phytoestrogen/phytochemical exposures. *Journal of Steroid Biochemistry and Molecular Biology*, 139, 270–76.

Guevara, E. E., Veilleux, C. C., Saltonstall, K., Caccone, A., Mundy, N. I., and Bradley, B. J. (2016). Potential arms race in the coevolution of primates and angiosperms: Brazzein sweet proteins and gorilla taste receptors. *American Journal of Physical Anthropology*, 161, 181–85.

Gügel, I. L., Grupe, G., and Kunzelmann, K. H. (2001). Simulation of dental microwear: Characteristic traces by opal phytoliths give clues to ancient human dietary behavior. *American Journal of Physical Anthropology*, 114(2), 124–38.

Guillot, D. (2011). Forelimb suspensory gait characteristics of wild *Lagothrix poeppigii* and *Ateles belzebuth*: Developing video-based methodologies in free-ranging primates. In K. D'Aout and E. E. Vereecke (Eds.), *Primate Locomotion: Linking Field and Laboratory Research* (pp. 247–69). Springer.

Guillotin, M., Dubost, G., and Sabatier, D. (2009). Food choice and food competition among the three major primate species of French Guiana. *Journal of Zoology*, 233, 551–79.

Gumert, M., and Malaivijitnond, S. (2012). Marine prey processed with stone tools by Burmese long-tailed macaques (*Macaca fascicularis aurea*) in intertidal habitats. *American Journal of Physical Anthropology*, 149, 447–57.

Gunst, N., Boinski, S., and Fragaszy, D. M. (2008). Acquisition of foraging competence in wild brown capuchins (*Cebus apella*), with special reference to conspecifics' foraging artefacts as an indirect social influence. *Behaviour*, 145, 195–229.

Guo, S., Hou, R., Garber, P., Raubenheimer, D., Righini, N., Ji, W., et al. 2018. Nutrient-specific compensation for seasonal cold stress in a free-ranging temperate colobine monkey. *Functional Ecology*, 32(9), 2170–80.

Guo, S., Li, B., and Watanabe, K. (2007). Diet and activity budget of *Rhinopithecus roxellana* in the Qinling Mountains, China. *Primates*, 48, 268–76.

Gupta, A., and Chivers, D. J. (1999). Biomass and use of resources in south and south-east Asian primate communities. In J. G. Fleagle, C. Janson, and K. Reed (Eds.), *Primate Communities* (pp. 38–54). Cambridge University Press.

Gupta, A. K., and Kumar, A. (1994). Feeding ecology and conservation of the Phayre's leaf monkey *Presbytis phayrei* in northeast India. *Biological Conservation*, 69, 301–6.

Gursky, S. (2000). Effect of seasonality on the behavior of an insectivorous primate, *Tarsius spectrum*. *International Journal of Primatology*, 21, 477–95.

Gursky, S. (2003). Lunar philia in a nocturnal primate. *International Journal of Primatology*, 24, 351–67.

Gursky, S. (2007). Tarsiiformes. In C. J. Campbell, A. Fuentes, K. C. MacKinnon, M. Panger, and S. K. Bearder (Eds.), *Primates in Perspective* (pp. 73–84). Oxford University Press.

Gurven, M., Kaplan, H., and Gutierrez, M. (2006). How long does it take to become a proficient hunter? Implications for the evolution of extended development and long life span. *Journal of Human Evolution*, 51, 451–70.

Gurven, M. D., Trumble, B. C., Stieglitz, J., Yetish, G., Cummings, D., Blackwell, A. D., et al. (2016). High resting metabolic rate among Amazonian forager-horticulturalists experiencing high pathogen burden. *American Journal of Physical Anthropology*, 161, 414–425.

Gurven, M., and Walker, R. (2006). Energetic demand of multiple dependents and the evolution of slow human growth. *Proceedings of the Royal Society B: Biological Sciences*, 273, 835–41.

Habig, B., Doellman, M. M., Woods, K., Olansen, J., and Archie, E. A. (2018). Social status and parasitism in male and female vertebrates: A meta-analysis. *Scientific Reports*, 8, 3629.

Hadi, S., Ziegler, T., Waltert, M., Syamsuri, F., Mühlenberg, M., and Hodges, J. K. (2012). Habitat use and trophic niche overlap of two sympatric colobines, *Presbytis potenziani* and *Simias concolor*, on Siberut Island, Indonesia. *International Journal of Primatology*, 33, 218–32.

Hadidian, J., and Bernstein, I. S. (1979). Female reproductive cycles and birth data from an Old World monkey colony. *Primates*, 20, 429–42.

Hadley, M. E. (2000). *Endocrinology*. Prentice Hall.

Hadfield, J. D. (2010). MCMC methods for multi-response generalized linear mixed models: the MCMCglmm R package. *Journal of Statistical Software*, 33, 1–22.

Haemaelaeinen, A., Heistermann, M., Fenosoa, Z. S. E., and

Kraus, C. (2014). Evaluating capture stress in wild gray mouse lemurs via repeated fecal sampling: Method validation and the influence of prior experience and handling protocols on stress responses. *General and Comparative Endocrinology*, 195, 68–79.

Haeusler, M., Grunstra, N. D. S., Martin, R. D., Krenn, V. A., Fornai, C., and Webb, N. M. (2021). The obstetrical dilemma hypothesis: There's life in the old dog yet. *Biological Reviews*, 96, 2031–57.

Hagerman, A. E., and Robbins, C. T. (1993). Specificity of tannin-binding salivary proteins relative to diet selection by mammals. *Canadian Journal of Zoology*, 71, 628–33.

Hairston, N. G., Smith, F. E., and Slobodkin, L. B. (1960). Community structure, population control, and competition. *American Naturalist*, 94, 421–25.

Hall, J. A., Bouladoux, N., Sun, C. M., Wohlfert, E. A., Blank, R. B., Zhu, Q., et al. (2008). Commensal DNA limits regulatory T cell conversion and is a natural adjuvant of intestinal immune responses. *Immunity*, 29(4), 637–49.

Hamada, Y., Hayakawa, S., Suzuki, J., and Okhura, S. (1999). Adolescent growth and development in Japanese macaques (*Macaca fuscata*): Punctuated adolescent growth spurt by season. *Primates*, 40, 439–52.

Hamilton, W. J. (1985). Demographic consequences of a food and water shortage to desert Chacma baboons (*Papio ursinus*). *International Journal of Primatology*, 6, 451–62.

Hamilton, W. J., Buskirk, R. E., and Buskirk, W. H. (1978). Omnivory and utilization of food sources by chacma baboons, *Papio ursinus*. *American Naturalist*, 112, 911–24.

Hamilton, W. J., and Busse, C. (1978). Primate carnivory and its significance to human diets. *BioScience*, 28, 761–66.

Hamilton, W. J., and Busse, C. (1982). Social dominance and predatory behavior of chacma baboons. *Journal of Human Evolution*, 11, 568–73.

Hammond, P., Lewis-Bevan, L., Biro, D., and Carvalho, S. (2022) Risk perception and terrestriality in primates: A quasi-experiment through habituation of chacma baboons (*Papio ursinus*) in Gorongosa National Park, Mozambique. *American Journal of Biological Anthropology*, 179(1), 48–59.

Handley, L. L., Austin, A. T., Robinson, D., Scrimgeour, C. M., Raven, J. A., Heaton, T. H. E., et al. (1999). The ^{15}N natural abundance (δ^{15}N) of ecosystem samples reflects measures of water availability. *Australian Journal of Plant Physiology*, 26, 185–99.

Hannah, L., Midgley, G. F., Lovejoy, T., Bond, W. J., Bush, M., Lovett, J. C., et al. (2002). Conservation of biodiversity in a changing climate. *Conservation Biology*, 16, 264–68.

Hansen, C. H. F., Nielsen, D. S., Kverka, M., Zakostelska, Z., Klimesova, K., Hudcovic, T., et al. (2012). Patterns of early gut colonization shape future immune responses of the host. *PLoS ONE*, 7, e34043.

Hansen, M. C., Potapov, P. V., Moore, R. J., Hancher, M., Turubanova, S. A., Tyukavina, A., et al. (2013). High-resolution global maps of 21st-century forest cover change. *Science*, 342, 850–53.

Hanson, A. M., Hall, M. B., Porter, L. M., and Lintzenich, B. (2006). Composition and nutritional characteristics of fungi consumed by *Callimico goeldii* in Pando, Bolivia. *International Journal of Primatology*, 27, 323–46.

Hanson, A. M., Hodge, K. T., and Porter, L. M. (2003). Mycophagy among primates. *Mycologist*, 17, 6–10.

Hanson, M., and Gluckman, P. (2014). Early developmental conditioning of later health and disease: Physiology or pathophysiology? *Physiology Review*, 94, 1027–76.

Hanya, G. (2004). Diet of a Japanese macaque troop in the coniferous forest of Yakushima. *International Journal of Primatology*, 25, 55–72.

Hanya, G., and Bernard, H. (2012). Fallback foods of red leaf monkeys (*Presbytis rubicunda*) in Danum Valley, Borneo. *International Journal of Primatology*, 33, 322–37.

Hanya, G., and Bernard, H. (2015). Different roles of seeds and young leaves in the diet of red leaf monkeys (*Presbytis rubicunda*): Comparisons of availability, nutritional properties, and associated feeding behavior. *International Journal of Primatology*, 36, 177–93.

Hanya, G., and Chapman, C. A. (2013). Linking feeding ecology and population abundance: A review of food resource limitation on primates. *Ecological Research*, 28, 183–90.

Hanya, G., Fuse, M., Aiba, S. I., Takafumi, H., Tsujino, R., Agetsuma, N., et al. (2014). Ecosystem impacts of folivory and frugivory by Japanese macaques in two temperate forests in Yakushima. *American Journal of Primatology*, 76, 596–607.

Hanya, G., Kiyono, M., Takafumi, H., Tsujino, R., and Agetsuma, N. (2007). Mature leaf selection of Japanese macaques: Effects of availability and chemical content. *Journal of Zoology*, 273, 140–47.

Hanya, G., Kiyono, M., Yamada, A., Suzuki, K., Furukawa, M., Yoshida, Y., et al. (2006). Not only annual food abundance but also fallback food quality determines the Japanese macaque density: Evidence from seasonal variations in home range size. *Primates*, 47, 275–78.

Hanya, G., Ménard, N., Qarro, M., Tattou, M. I., Fuse, M., Vallet, D., et al. (2011). Dietary adaptations of temperate primates: Comparisons of Japanese and Barbary macaques. *Primates*, 52, 187–98.

Hanya, G., Stevenson, P., van Noordwijk, M., Te Wong, S., Kanamori, T., Kuze, N., et al. (2011). Seasonality in fruit availability affects frugivorous primate biomass and species richness. *Ecography*, 34, 1009–17.

Hanya, G., Yoshihiro, S., Zamma, K., Matsubara, H., Ohtake, M., Kubo, R., et al. (2004). Environmental determinants of the altitudinal variations in relative group densities of Japanese macaques on Yakushima. *Ecological Research*, 19, 485–93.

Harborne, J. B. (1984). *Phytochemical Methods: A Guide to Modern Techniques of Plant Analysis* (2nd ed.). Chapman and Hall.

Harborne, J. B. (1988). *Introduction to Ecological Biochemistry*. Academic Press.

Harborne, J. B. (2000). Arsenal for survival: Secondary plant products. *Taxon*, 49, 435–49.

Harborne, J. B. (2001). Twenty-five years of chemical ecology. *Natural Product Reports*, 18, 361–79.

Harcourt, A. H., and Harcourt, S. A. (1984). Insectivory by gorillas. *Folia Primatologica*, 43, 229–33.

Harcourt, A., and Stewart, K. (2007). *Gorilla Society: Conflict, Compromise, and Cooperation between the Sexes*. University of Chicago Press.

Harding, R. A. (1975). Meat eating and hunting in baboons. In R. H. Tuttle (Ed.), *Socioecology and Psychology of Primates* (pp. 246–57). Mouton.

Harding, R. A. (1981). An order of omnivores: Nonhuman primate diets in the wild. In R. A. Harding and G. Teleki (Eds.), *Omnivorous Primates* (pp. 191–214). Columbia University Press.

Harding, R. S. O. (1976). Ranging patterns of a troop of baboons (*Papio anubis*) in Kenya. *Folia Primatologica*, 25, 143–85.

Hardus, M. E., Lameira, A. R., Zulfa, A., Atmoko, S. S. U., de Vries, H., and Wich, S. A. (2012). Behavioral, ecological, and evolutionary aspects of meat-eating by Sumatran orangutans (*Pongo abelii*). *International Journal Primatology*, 33, 287–304.

Harel, R., Alavi, S., Ashbury, A. M., Aurisano, J., Berger-Wolf, T., Davis, G. H., et al. (2022). Life in 2.5 D: Animal movement in the trees. *Frontiers in Ecology and Evolution*, 10, 801850.

Harel, R., Carter Loftus, J., and Crofoot, M. C. (2021). Locomotor compromises maintain group cohesion in baboon troops on the move. *Proceedings of the Royal Society B: Biological Sciences*, 288, 20210839.

Harper, J. L. (1977). *Population Biology of Plants*. Academic Press.

Harris, E. E., and Disotell, T. R. (1998. Nuclear gene trees and the phylogenetic relationships of the mangabeys (Primates: Papionini). *Molecular Biology and Evolution*, 15(7), 892–900.

Harris, J. A., Hobbs, R. J., Higgs, E., and Aronson, J. (2006). Ecological restoration and global climate change. *Restoration Ecology*, 14(2), 170–76.

Harris, T. R. (2006). Between-group contest competition for food in a highly folivorous population of black and white colobus monkeys (*Colobus guereza*). *Behavioral Ecology and Sociobiology*, 61, 317–29.

Harris, T. R., and Chapman, C. A. (2007). Variation in diet and ranging of black and white colobus monkeys in Kibale National Park, Uganda. *Primates*, 48, 208–21.

Harris, T., Chapman, C., and Monfort, S. (2009). Small folivorous primate groups exhibit behavioral and physiological effects of food scarcity. *Behavioural Ecology*, 21, 46–56.

Harrison, M. E. (2009). *Orang-utan Feeding Behaviour in Sabangua, Central Kalimantan* [PhD Dissertation]. University of Cambridge, Cambridge, UK.

Harrison, M. E., and Marshall, A. J. (2011). Strategies for the use of fallback foods in apes. *International Journal of Primatology*, 32, 531–65.

Harrison, M. E., Morrogh-Bernard, H. C., and Chivers, D. J. (2010). Orangutan energetics and the influence of fruit availability in the nonmasting peat-swamp forest of Sabangau, Indonesian Borneo. *International Journal of Primatology*, 31(4), 585–607.

Harrison, M. E., Vogel, E. R., Morrough-Bernard, H. C., and van Noordwijk, M. A. (2009). Methods for calculating activity budgets compared: A case study using orangutans. *American Journal of Primatology*, 71, 353–58.

Harrison, M. J. S. (1983). Age and sex differences in the diet and feeding strategies of the green monkey, *Cercopithecus sabaeus*. *Animal Behavior*, 31, 969–77.

Harrison, M. J. S. (1984). Optimal foraging strategies in the diet of the green monkey, *Cercopithecus sabaeus*, at Mount-Assirik, Senegal. *International Journal of Primatology*, 5, 435–71.

Harrison, R. D. (2011). Emptying the forest: Hunting and the extirpation of wildlife from tropical nature reserves. *BioScience*, 61, 919–24.

Harrison, R. M., Phillippi, P. P., Swan, K. F., and Henson, M. C. (1999). Effect of genistein on steroid hormone production in the pregnant rhesus monkey. *Proceedings of the Society for Experimental Biology and Medicine*, 222, 78–84.

Hart, D. (2007). Predation on primates: A biogeographical analysis. In S. Gursky-Doyen and K. A. I. Nekaris (Eds.), *Primate Anti-predator Strategies* (pp. 27–59). Springer Science and Business Media.

Harten, L., Matalon, Y., Galli, N., Navon, H., Dor, R., and Yovel, Y. (2018). Persistent producer-scrounger relationships in bats. *Science Advances*, 4, e1603293.

Hartenstein, V., and Martinez, P. (2019). Structure, development, and evolution of the digestive system. *Cell and Tissue Research*, 377, 289–92.

Hartter, J., Ryan, S. J., Southworth, J., and Chapman, C. A. (2011). Landscapes as continuous entities: Forest disturbance and recovery in the Albertine Rift landscape. *Landscape Ecology*, 26, 877–90.

Hartwell, K. S., Notman, H., Kalbitzer, U., Chapman, C. A., and Pavelka, M. M. S. M. (2021). Fruit availability has a complex relationship with fission–fusion dynamics in spider monkeys. *Primates*, 62, 165–75.

Harvey, P. H., and Clutton-Brock, T. H. (1985). Life history variation in primates. *Evolution*, 39, 559–81.

Harvey, P. H., Clutton-Brock, T. H., and Kavanagh, M. (1978). Sexual dimorphism in primate teeth. *Journal of Zoology*, 186, 475–85.

Harvey, P. H., Pagel, M. D., and Rees, J. A. (1991). Mammalian metabolism and life histories. *American Naturalist*, 137, 556–66.

Harvey, P. H., Promislow, D. E. L., and Read, A. F. (1989). Causes and correlates of life history differences among mammals. In

V. Standen and R. A. Foley (Eds.), *Comparative Socioecology: The Behavioural Ecology of Humans and Other Mammals* (pp. 305–18). Blackwell Scientific.

Hashimoto, C., Furuichi, T., and Tashiro, Y. (2001). What factors affect the size of chimpanzee parties in the Kalinzu Forest, Uganda? Examination of fruit abundance and number of estrous females. *International Journal of Primatology*, 22, 947–59.

Hashimoto, C., Isaji, M., Koops, K., and Furuichi, T. (2015). First records of tool-set use for ant-dipping by eastern chimpanzees (*Pan troglodytes schweinfurthii*) in the Kalinzu Forest Reserve, Uganda. *Primates*, 56, 301–5.

Haslam, M., Gumert, M. D., Biro, D., Carvalho, S., and Malaivijitnond, S. (2013). Use-wear patterns on wild macaque stone tools reveal their behavioural history. *PLoS ONE*, 8, 1–8.

Hauser, M. D. (1988). Invention and social transmission: New data from wild vervet monkeys. In R. W. Byrne and A. Whiten (Eds.), *Machiavellian Intelligence: Social Expertise and the Evolution of Intellect in Monkeys, Apes, and Humans* (pp. 327–43). Oxford University Press.

Hauser, M. D., and Fairbanks, L. A. (1988). Mother-offspring conflict in vervet monkeys: Variation in response to ecological conditions. *Animal Behaviour*, 36, 802–13.

Hausfater, G. (1975). Dominance and reproduction in baboons (*Papio cynocephalus*). *Contributions to Primatology*, 7, 1–150.

Hausfater, G. (1976). Predatory behavior of yellow baboons. *Behaviour*, 56, 44–68.

Hausfater, G., and Bearce, W. H. (1976). Acacia tree exudates: Their composition and use as a food source by baboons. *African Journal of Ecology*, 14, 241–43.

Hausfater, G., and Watson, D. F. (1976). Social and reproductive correlates of parasite ova emissions by baboons. *Nature*, 262, 688–89.

Havmøller, L. W., Loftus, J. C., Havmøller, R. W., Alavi, S. E., Caillaud, D., Grote, M. N., et al. (2021) Arboreal monkeys facilitate foraging of terrestrial frugivores. *Biotropica*, 53, 1685–97.

Hawes, J. E., Calouro, A. M., and Peres, C. A. (2013). Sampling effort in Neotropical primate diet studies: Collective gains and underlying geographic and taxonomic biases. *International Journal of Primatology*, 34, 1081–104.

Hawes, J. E., and Peres, C. A. (2014). Ecological correlates of trophic status and frugivory in Neotropical primates. *Oikos*, 123, 365–77.

Hawkes, K. (2001). Is meat the hunter's property? Big game, ownership, and explanations of hunting and sharing. In C. S. Stanford and H. T. Bunn (Eds.), *Meat Eating and Human Evolution* (pp. 219–36). Oxford University Press.

Hawkes, K. (2003). Grandmothers and the evolution of human longevity. *American Journal of Human Biology*, 15, 380–400.

Hawkes, K., and Bliege Bird, R. (2003). Showing off, handicap signaling, and the evolution of men's work. *Evolutionary Anthropology*, 11, 58–68.

Hawkes, K., O'Connell, J. F., and Blurton-Jones, N. E. (1997). Hadza women's time allocation, offspring production, and the evolution of long postmenopausal lifespans. *Current Anthropology*, 8, 551–77.

Hawkes, K., O'Connell, J. F., Blurton-Jones, N. E., Charnov, E. L., and Alvarez, H. (1998). Grandmothering, menopause, and the evolution of human life histories. *Proceedings of the National Academy of Sciences of the United States of America*, 95, 1336–39.

Hawkins, E., and Papworth, S. (2022). Little evidence to support the risk–disturbance hypothesis as an explanation for responses to anthropogenic noise by pygmy marmosets (*Cebuella niveiventris*) at a tourism site in the Peruvian Amazon. *International Journal of Primatology*, 43, 1110–32.

Hawkins, R. B., Raymond, S. L., Stortz, J. A., Horiguchi, H., Brakenridge, S. C., Gardner, A., et al. (2018). Chronic critical illness and the persistent inflammation, immunosuppression, and catabolism syndrome. *Frontiers of Immunology*, 9, 1511.

Hayakawa, T., Sugawara, T., Go, Y., Udono, T., Hirai, H., and Imai, H. (2012). Eco-geographical diversification of bitter taste receptor genes (*TAS2Rs*) among subspecies of chimpanzees (*Pan troglodytes*). *PLoS ONE*, 7, e43277.

Hayakawa, T., Suzuki-Hashido, N., Matsui, A., and Go, Y. (2014). Frequent expansions of the bitter taste receptor gene repertoire during evolution of mammals in the Euarchontoglires clade. *Molecular Biology and Evolution*, 31, 2018–31.

Hayden, S., Bekaert, M., Crider, T. A., Mariani, S., Murphy, W. J., and Teeling, E. C. (2010). Ecological adaptation determines functional mammalian olfactory subgenomes. *Genome Research*, 20, 1–9.

Hayes, K. (1974). Hemolytic anemia of premature infants associated with vitamin E deficiency. Animal model: Hemolytic anemia in monkeys deficient in vitamin E. *American Journal of Pathology*, 77(1), 123–26.

Hayes, T. B., Anderson, L. L., Beasley, V. R., de Solla, S. R., Iguchi, T., Ingraham, H., et al. (2011). Demasculinization and feminization of male gonads by atrazine: Consistent effects across vertebrate classes. *Journal of Steroid Biochemistry*, 127, 64–73.

Hayssen, V., and Blackburn, D. G. (1985). α-Lactalbumin and the origins of lactation. *Evolution*, 39, 1147–49.

Hazeldine, J., Lord, J. M., and Hampson, P. (2015). Immunesenescence and inflammaging: A contributory factor in the poor outcome of the geriatric trauma patient. *Ageing Research Reviews*, 24(pt. B), 349–57.

Heck, L., Crowley, B., Thorén, S., and Radespiel, U. (2016). Determinants of isotopic variation in two sympatric mouse lemur species from northwestern Madagascar. In S. M. Lehman, U. Radespiel, and E. Zimmerman (Eds.), *The Dwarf and Mouse Lemurs of Madagascar: Biology, Behavior and Conservation Biogeography of the Cheirogaleidae* (pp. 281–304). Cambridge University Press.

Hedges, K. J., and Abrahams, M. V. (2015). Hypoxic refuges, predator-prey interactions and habitat selection by fishes. *Journal of Fish Biology*, 86, 288–303.

Hedges, R. E. M., Clement, J. G., Thomas, C. D. L., and O'Connell, T. C. (2007). Collagen turnover in the adult femoral mid-shaft: Modeled from anthropogenic radiocarbon tracer measurements. *American Journal of Physical Anthropology*, 133, 808–16.

Heesen, M., Rogahn, S., Ostner, J., and Schülke, O. (2013). Food abundance affects energy intake and reproduction in frugivorous female Assamese macaques. *Behavioral Ecology and Sociobiology*, 67, 1053–66.

Heesen, M., Macdonald, S., Ostner, J., and Schülke, O. (2015). Ecological and social determinants of group cohesiveness and within-group spatial position in wild Assamese macaques. *Ethology*, 121, 270–83.

Heesy, C. P. (2008). Ecomorphology of orbit orientation and the adaptive significance of binocular vision in primates and other mammals. *Brain, Behavior and Evolution*, 71, 54–67.

Heffner, R. S. (2004). Primate hearing from a mammalian perspective. *Anatomical Record A: Discoveries in Molecular, Cellular, and Evolutionary Biology*, 281A, 1111–22.

Hegab, I. M., Pan, H., Dong, J., Wang, A., Yin, B., Yang, S., et al. (2014). Effects of physical attributes and chemical composition of novel foods on food selection by Norway rats (*Rattus norvegicus*). *Journal of Pest Science*, 87, 99–106.

Hegnauer, R. (1994). *Chemotaxonomie der Pflanzen: Band 11a—Leguminosae*. Birkhäuser.

Heiduck, S. (1997). Food choice in masked titi monkeys (*Callicebus personatus melanochir*): Selectivity or opportunism? *International Journal of Primatology*, 18, 487–502.

Heistermann, M., and Higham, J. P. (2015). Urinary neopterin, a non-invasive marker of mammalian cellular immune activation, is highly stable under field conditions. *Scientific Reports*, 5, 16308.

Heldstab, S. A., van Schaik, C. P., and Isler, K. (2017). Getting fat or getting help? How female mammals cope with energetic constraints on reproduction. *Frontiers in Zoology*, 14, 29.

Heller, N. E., and Zavaleta, E. S. (2009). Biodiversity management in the face of climate change: A review of 22 years of recommendations. *Biological Conservation*, 142, 14–32.

Hemingway, C. A. (1996). Morphology and phenology of seeds and whole fruit eaten by Milne-Edwards' sifaka, *Propithecus diadema edwardsi*, in Ranomafana National Park, Madagascar. *International Journal of Primatology*, 17, 637–59.

Hemingway, C. A. (1998). Selectivity and variability in the diet of Milne-Edwards' sifakas (*Propithecus diadema edwardsi*): Implications for folivory and seed-eating. *International Journal of Primatology*, 19, 355–77.

Hemingway, C. A. (1999). Time budgets and foraging in a Malagasy primate: Do sex differences reflect reproductive condition and female dominance? *Behavioral Ecology and Sociobiology*, 45, 311–22.

Hemingway, C. A., and Bynum, N. (2005). The influence of seasonality on primate diet and ranging. In D. Brockman and C. P. van Schaik (Eds.), *Seasonality in Primates: Studies of Living and Extinct Human and Non-human Primates* (pp. 57–104). Cambridge University Press.

Hendershott, R., Behiel, A., and Rawson, B. (2016). Seasonal variation in the activity and dietary budgets of Cat Ba langurs (*Trachypithecus poliocephalus*). *International Journal of Primatology*, 37, 586–604.

Hendrickx, A. G., and Dukelow, W. R. (1995). Reproductive biology. In B. T. Bennett, C. R. Abee, and R. Henrickson (Eds.), *Nonhuman Primates in Biomedical Research: Biology and Management* (pp. 147–91). Academic Press.

Hendriksma, H. P., and Shafir, S. (2016). Honey bee foragers balance colony nutritional deficiencies. *Behavioral Ecology and Sociobiology*, 70, 509–17.

Henery, M. L., Moran, G. F., Wallis, I. R., and Foley, W. J. (2007). Identification of quantitative trait loci influencing foliar concentrations of terpenes and formylated phloroglucinol compounds in *Eucalyptus nitens*. *New Phytologist*, 176, 82–95.

Henry, A. G., Brooks, A. S., and Piperno, A. R. (2014). Plant foods and the dietary ecology of Neandertals and early modern humans. *Journal of Human Evolution*, 69, 44–54.

Henry, A. G., Ungar, P. S., Passey, B. H., Sponheimer, M., Rossouw, L., Bamford, M., et al. (2012). The diet of *Australopithecus sediba*. *Nature*, 487(7405), 90–93.

Henshilwood, C. S., D'Errico, F., Marean, C. W., Milo, R. G., and Yates, R. (2001). An early bone tool industry from the Middle Stone Age at Blombos Cave, South Africa: Implications for the origins of modern human behavior, symbolism, and language. *Journal of Human Evolution*, 41, 631–88.

Hernandez-Aguilar, R. A., Moore, J., and Pickerings, T. R. (2007). Savanna chimpanzees use tools to harvest the underground storage organs of plants. *Proceedings of the National Academy of Sciences of the United States of America*, 104, 19210–13.

Hernell, O., Timby, N., Domellöf, M., and Lönnerdal, B. (2016). Clinical benefits of milk fat globule membranes for infants and children. *Journal of Pediatrics*, 173, S60–S65.

Herrera, C. M. (1998). Long-term dynamics of Mediterranean frugivorous birds and fleshy fruits: A 12-year study. *Ecological Monographs*, 68, 511–38.

Hewlett, B. S., and Winn, S. (2014). Allomaternal nursing in humans. *Current Anthropology*, 55, 200–229.

Hewson-Hughes, A. K., Hewson-Hughes, V. L., Colyer, A., Miller, A. T., McGrane, S. J., Hall, S. R., et al. (2013). Geometric analysis of macronutrient selection in breeds of the domestic dog, *Canis lupus familiaris*. *Behavioral Ecology*, 24, 293–304.

Heymann, E. W. (2006). The neglected sense—olfaction in primate behavior, ecology, and evolution. *American Journal of Primatology*, 68, 519–24.

Heymann, E. W. (2011). Florivory, nectarivory, and pollination—a review of primate-flower interactions. *Ecotropica*, 17, 41–52.

Heymann, E. W. (2013). Can seeds help to expel parasites? A comment on the Garber-Kitron (1997) hypothesis. *International Journal of Primatology*, 34, 445–49.

Heymann, E. W., and Buchanan-Smith, H. M. (2000). The behavioural ecology of mixed-species troops of callitrichine primates. *Biological Reviews*, 75, 169–90.

Heymann, E. W., and Hsia, S. S. (2015). Unlike fellows: A review of primate-non-primate associations. *Biological Reviews*, 90, 142–56.

Heymann, E. W., and Soini, P. (1999). Offspring number in pygmy marmosets, *Cebuella pygmaea*, in relation to group size and the number of adult males. *Behavioral Ecology and Sociobiology*, 400–404.

Heymann, E. W., Knogge, C., and Tirado Herrera, E. R. (2000). Vertebrate predation by sympatric tamarins, *Saguinus mystax* and *Saguinus fuscicollis*. *American Journal of Primatology*, 51, 153–58.

Hicks, T. C. (2010). *A Chimpanzee Mega-culture? Exploring Behavioral Continuity in* Pan troglodytes schweinfurthii *across Northern DR Congo*. University of Amsterdam.

Hicks, T. C., Kühl, H. S., Boesch, C., Dieguez, P., Ayimisin, A. E., Martin Fernandez, R., et al. (2019). Bili-Uéré: A chimpanzee behavioural realm in northern DR Congo. *Folia Primatologica*, 90, 3–64.

Hiddleston, W. A. (1978). The production of the common marmoset, *Callithrix jacchus*, as a laboratory animal. In D. J. Chivers and W. Lane-Petters (Eds.), *Recent Advances in Primatology* (pp. 174–81). Academic Press.

Higham, J. P., Heistermann, M., and Maestripieri, D. (2011). The energetics of male–male endurance rivalry in free-ranging rhesus macaques, *Macaca mulatta*. *Animal Behaviour*, 81, 1001–7.

Higham, J. P., Kraus, C., Stahl-Hennig, C., Engelhardt, A., Fuchs, D., and Heistermann, M. (2015). Evaluating noninvasive markers of nonhuman primate immune activation and inflammation. *American Journal of Physical Anthropology*, 158(4), 673–84.

Higham, J. P., and Maestripieri, D. (2014). The costs of reproductive success in male rhesus macaques (*Macaca mulatta*) on Cayo Santiago. *International Journal of Primatology*, 35, 661–76.

Higham, J. P., Ross, C., Warren, Y., Heistermann, M., and MacLarnon, A. M. (2007). Reduced reproductive function in wild baboons (*Papio hamadryas anubis*) related to natural consumption of the African black plum (*Vitex doniana*). *Hormones and Behavior*, 52, 384–90.

Higham, J. P., Stahl-Hennig, C., and Heistermann, M. (2020). Urinary suPAR: A non-invasive biomarker of infection and tissue inflammation for use in studies of large free-ranging mammals. *Royal Society Open Science*, 7(2), 191825.

Higham, J. P., Vitale, A. B., Rivera, A. M., Ayala, J. E., and Maestripieri, D. (2010). Measuring salivary analytes from free-ranging monkeys. *Physiology and Behavior*, 101(5), 601–7.

Higham, J. P., Warren, Y., Adanu, J., Umaru, B. N., MacLarnon, A. M., Sommer, V., et al. (2009). Living on the edge: Life history of olive baboons at Gashaka-Gumti National Park, Nigeria. *American Journal of Primatology*, 71, 293–304.

Hiiemae, K. (1978). Mammalian mastication: A review of the activity of jaw muscles and the movements they produce in chewing. In P. M. Butler and K. A. Joysey (Eds.), *Development, Function and Evolution of Teeth* (pp. 359–98). Academic Press.

Hiiemae, K., and Kay, R. F. (1972). Trends in evolution of primate mastication. *Nature*, 240(5382), 486–87.

Hilário, R. R., and Ferrari, S. F. (2010). Feeding ecology of a group of buffy-headed marmosets (*Callithrix flaviceps*): Fungi as a preferred resource. *American Journal of Primatology*, 72, 515–21.

Hilgartner, R., Zinner, D., and Kappeler, P. M. (2008). Life history traits and parental care in *Lepilemur ruficaudatus*. *American Journal of Primatolology*, 70, 2–11.

Hill, C. M. (2000). Conflict of interest between people and baboons: Crop raiding in Uganda. *International Journal of Primatology*, 21, 299–315.

Hill, C. M. (2017). Primate crop feeding behavior, crop protection, and conservation. *International Journal of Primatology*, 38, 385–400.

Hill, D. A. (1997). Seasonal variation in the feeding behavior and diet of Japanese macaques (*Macaca fuscata yakui*) in lowland forest of Yakushima. *American Journal of Primatology*, 43, 305–22.

Hill, K. (1982). Hunting and human evolution. *Journal of Human Evolution*, 11, 521–44.

Hill, K. (1988). Macronutrient modifications of optimal foraging theory—an approach using indifference curves applied to some modern foragers. *Human Ecology*, 16, 157–97.

Hill, K., and Hurtado, A. M. (1996). *Ache Life History: The Ecology and Demography of a Foraging People*. Aldine de Gruyter.

Hill, K., Barton, M., and Hurtado, A. M. (2009). The emergence of human uniqueness: Characters underlying behavioral modernity. *Evolutionary Anthropology*, 18, 187–200.

Hill, M. J. (1997). Intestinal flora and endogenous vitamin synthesis. *European Journal of Cancer Prevention*, 6(suppl. 1), S43–S45.

Hill, R. A. (2005). Thermal constraints on activity scheduling and habitat choice in baboons. *American Journal of Physical Anthropology*, 129, 242–49.

Hill, R. A., and Cowlishaw, G. (2002). Foraging female baboons exhibit similar patterns of antipredator vigilance across two populations. In L. E. Miller (Ed.), *Eat or Be Eaten: Predator Sensitive Foraging among Primates* (pp. 187–204). Cambridge University Press.

Hill, R. A., and Dunbar, R. I. M. (1998). An evaluation of the roles of predation rate and predation risk as selective pressures on primate grouping behaviour. *Behaviour*, 135, 411–30.

Hill, R. A., and Lee, P. C. (1998). Predation risk as an influence on group size in cercopithecoid primates: Implications for social structure. *Journal of Zoology*, 245, 447–56.

Hill, R. A., and Weingrill, T. (2007). Predation risk and habitat use in chacma baboons (*Papio hamadryas ursinus*). In

S. Gursky-Doyen and K. A. I. Nekaris (Eds.), *Primate Antipredator Strategies* (pp. 339–54). Springer.

Hill, R. A., Barrett, L., Gaynor, D., Weingrill, T., Dixon, P., Payne, H., et al. (2003). Day length, latitude and behavioural (in)flexibility in baboons (*Papio cynocephalus ursinus*). *Behavioral Ecology and Sociobiology*, 53, 278–86.

Hill, R. A., Lycett, J. E., and Dunbar, R. I. M. (2000). Ecological and social determinants of birth intervals in baboons. *Behavioral Ecology*, 11(5), 560–64.

Hill, S. E., and Winder, I. C. (2019). Predicting the impacts of climate change on *Papio* baboon biogeography: Are widespread, generalist primates "safe"? *Journal of Biogeography*, 46, 1380–405.

Hill, W. C. O. (1958). Pharynx, oesophagus, stomach, small and large intestine: Form and position. *Primatologia*, 3, 139–207.

Hinde, K. (2009). Richer milk for sons but more milk for daughters: Sex-biased investment during lactation varies with maternal life history in rhesus macaques. *American Journal of Human Biology*, 21, 512–19.

Hinde, K. (2014). The potential wonders of other's milk: Commentary on "Allomaternal nursing in humans." *Current Anthropology*, 55, 216–17.

Hinde, K. (2015). Motherhood. *Emerging Trends in the Social and Behavioral Sciences*, 2015, 1–16.

Hinde, K. (2017). Colostrum through a cultural lens. International Milk Genomics Consortium. https://www.milkgenomics.org/?splash=colostrum-cultural-lens.

Hinde, K., and Capitanio, J. P. (2010). Lactational programming? Mother's milk energy predicts infant behavior and temperament in rhesus macaques (*Macaca mulatta*). *American Journal of Primatology*, 72, 522–29.

Hinde, K., Carpenter, A. J., Clay, J. S., and Bradford, B. J. (2014). Holsteins favor heifers, not bulls: biased milk production programmed during pregnancy as a function of fetal sex. *PloS One*, 9(2), e86169.

Hinde, K., Foster, A. B., Landis, L. M., Rendina, D., Oftedal, O. T., and Power, M. L. (2013). Daughter dearest: Sex-biased calcium in mother's milk among rhesus macaques. *American Journal of Physical Anthropology*, 151, 144–50.

Hinde, K., and German, J. B. (2012). Food in an evolutionary context: Insights from mother's milk. *Journal of the Science of Food and Agriculture*, 92, 2219–23.

Hinde, K., and Lewis, Z. T. (2015). Mother's littlest helpers. *Science*, 348, 1427–28.

Hinde, K., and Milligan, L. (2011). Primate milk: Proximate mechanisms and ultimate perspectives. *Evolutionary Anthropology*, 20, 9–23.

Hinde, K., Power, M. L., and Oftedal, O. T. (2009). Rhesus macaque milk: Magnitude, sources, and consequences of individual variation over lactation. *American Journal of Physical Anthropology*, 128, 148–57.

Hiramatsu, C., Goda, N., and Komatsu, H. (2011). Transformation from image-based to perceptual representation of materials along the human ventral visual pathway. *NeuroImage*, 57, 482–94.

Hiramatsu, C., Melin, A. D., Allen, W. L., Dubuc, C., and Higham, J. P. (2017). Experimental evidence that primate trichromacy is well suited for detecting primate social colour signals. *Proceedings of the Royal Society B: Biological Sciences*, 284, 20162458.

Hiramatsu, C., Melin, A. D., Aureli, F., Schaffner, C. M., Vorobyev, M., and Kawamura, S. (2009). Interplay of olfaction and vision in fruit foraging of spider monkeys. *Animal Behaviour*, 77, 1421–26.

Hiramatsu, C., Melin, A. D., Aureli, F., Schaffner, C. M., Vorobyev, M., Matsumoto, Y., et al. (2008). Importance of achromatic contrast in short-range fruit foraging of primates. *PLoS ONE*, 3, e3356.

Hirata, S., Yamamoto, S., Takemoto, H., and Matsuzawa, T. (2010). A case report of meat and fruit sharing in a pair of wild bonobos. *Pan Africa News*, 17, 21–23.

Hirsch, B. T. (2010). Tradeoff between travel speed and olfactory food detection in ring-tailed coatis (*Nasua nasua*). *Ethology*, 116, 671–79.

Hirsch, B. T., Tujague, M. P., Di Blanco, Y. E., Di Bitetti, M. S., and Janson, C. H. (2013). Comparing capuchins and coatis: Causes and consequences of differing movement ecology in two sympatric mammals. *Animal Behaviour*, 86, 331–38.

Hiwatashi, T., Okabe, Y., Tsutsui, T., Hiramatsu, C., Melin, A. D., Oota, H., et al. (2010). An explicit signature of balancing selection for color-vision variation in New World monkeys. *Molecular Biology and Evolution*, 27, 453–64.

Hladik, C. M. (1973). Alimentation et activité d'un groupe de chimpanzés réintroduits en forêt gabonaise. *La Terre et la vie*, 27, 343-413.

Hladik, A., and Hladik, C. M. (1969). Rapports trophiques entre végétation et primates dans la forêt de Barro Colorado (Panama). *La Terre et la Vie*, 23, 25–117.

Hladik, A., and Hladik, C. M. (1977). Occurrence of alkaloids in rain-forest plants and its ecological significance—preliminary results of a screening survey in Gabon Terre et la Vie. *Revue D'Ecologie Appliquee*, 31, 515–55.

Hladik, C. M. (1973). Feeding and activity of a group of chimpanzees reintroduced in the Gabon Forest. *La Terre et la Vie*, 27, 343–413.

Hladik, C. M. (1975). Ecology, diet, and social patterning in Old and New World primates. In R. A. Tuttle (Ed.), *Socioecology and Psychology of Primates* (pp. 3–35). Mouton.

Hladik, C. M. (1977a). Chimpanzees of Gabon and chimpanzees of Gombe: Some comparative data on the diet. In T. H. Clutton-Brock (Ed.), *Primate Ecology: Studies of Feeding and Ranging Behavior in Lemurs, Monkey and Apes* (pp. 481–501). Academic Press.

Hladik, C. M. (1977b). A comparative study of the feeding strategies of two sympatric species of leaf monkeys: *Presbytis senex* and *Presbytis entellus*. In T. H. Clutton-Brock (Ed.), *Primate*

Ecology: Studies of Feeding and Ranging Behaviour in Lemurs, Monkeys and Apes (pp. 323–53). Academic Press.

Hladik, C. M. (1978). Adaptive strategies of primates in relation to leaf eating. In G. G. Montgomery (Ed.), *The Ecology of Arboreal Folivores* (pp. 373–95). Smithsonian Institution Press.

Hladik, C. M. (1979). Diet and ecology of prosimians. In G. A. Doyle and R. D. Martin (Eds.), *The Study of Prosimian Behavior* (pp. 307–57). Academic Press.

Hladik, C. M., Charles-Dominique, P., and Petter, J. J. (1980). Feeding strategies of five nocturnal prosimians in the dry forest of the west coast of Madagascar. In P. Charles-Dominique, H. M. Cooper, A. Hladik, C. M. Hladik, E. Pages, G. F. Pariente, et al. (Eds.), *Nocturnal Malagasy Primates: Ecology, Physiology, and Behavior* (pp. 41–73). Academic Press.

Hladik, C. M., and Simmen, B. (1998). Taste perception and feeding behavior in nonhuman primates and human populations. *Evolutionary Anthropology*, 5, 25–71.

Hobaiter, C., Poisot, T., Zuberbühler, K., Hoppitt, W., and Gruber, T. (2014). Social network analysis shows direct evidence for social transmission of tool use in wild chimpanzees. *PLoS Biology*, 12, 1–12.

Hobaiter, C., Samuni, L., Mullins, C., Akankwasa, W. J., and Zuberbühler, K. (2017). Variation in hunting behavior in neighboring chimpanzee communities in the Budongo Forest, Uganda. *PLoS ONE*, 12, e0178065.

Hobbs, C. J., Jones, R. E., and Plymate, S. R. (1992). The effects of sex-hormone binding globulin (Shbg) on testosterone transport into the cerebrospinal-fluid. *Journal of Steroid Biochemistry*, 42, 629–35.

Hobbs, N. T., and Hanley, T. A. (1990). Habitat evaluation: Do use/availability data reflect carrying capacity? *Journal of Wildlife Management*, 54, 515–22.

Hobbs, N. T., and Swift, D. M. (1985). Estimates of habitat carrying capacity incorporating explicit nutritional constraints. *Journal of Wildlife Management*, 49, 814–22.

Hochman, V., and Kotler, B. P. (2007). Patch use, apprehension, and vigilance behavior of Nubian ibex under perceived risk of predation. *Behavioural Ecology*, 18, 368–74.

Hockings, K. J., and MacLennan, M. R. (2012). From forest to farm: Systematic review of cultivar feeding by chimpanzees—management implications for wildlife in anthropogenic landscapes. *PLoS ONE*, 7, e33391.

Hodges, J. K., and Heistermann, M. (2011). Field endocrinology: Monitoring hormonal changes in free-ranging primates. In J. M. Setchell and D. J. Curtis (Eds.), *Field and Laboratory Methods in Primatology* (pp. 353–70). Cambridge University Press.

Hodgkison, R., Ayasse, M., Kalko, E. K. V., Häberlein, C., Schulz, S., Mustapha, W. A. W., et al. (2007). Chemical ecology of fruit bat foraging behavior in relation to the fruit odors of two species of Paleotropical bat-dispersed figs (*Ficus hispida* and *Ficus scortechinii*). *Journal of Chemical Ecology*, 33, 2097–110.

Hodgkiss, S., Thetford, E., Waitt, C. D., and Nijman, V. (2010). Female reproductive parameters in the Javan gibbon (*Hylobates moloch*). *Zoo Biology*, 29, 449–56.

Hoellein Less, E. (2012). *Adiposity in Zoo Gorillas (Gorilla gorilla gorilla): The Effects of Diet and Behavior*. Case Western Reserve University.

Hoffman, C. L., Higham, J. P., Heistermann, M., Coe, C. L., Prendergast, B. J., and Maestripieri, D. (2011). Immune function and HPA axis activity in free-ranging rhesus macaques. *Physiology and Behavior*, 104(3), 507–14.

Hoffman, C. L., Ruiz-Lambides, A. V., Maldonado, E., Gerald, M. S., and Maestripieri, D. (2008). Sex differences in survival costs of reproduction in a promiscuous primate. *Behavioral Ecology and Sociobiology*, 62, 1711–18.

Hoffman, L. C. (2008). The yield and nutritional value of meat from African ungulates, Camelidae, rodents, ratites, and reptiles. *Meat Science*, 80, 94–100.

Hoffman, L. C., and Ferreira, A. V. (2004). Chemical composition of two muscles of the common duiker (*Sylvicapra grimmia*). *Journal of the Science of Food and Agriculture*, 84(12), 1541–44.

Hoffman, M. L., Beran, M. J., and Washburn, D. A. (2009). Memory for "what," "where," and "when" information in rhesus monkeys (*Macaca mulatta*). *Journal of Experimental Psychology: Animal Behavior Processes*, 35, 143–52.

Hoffman, T. S., and O'Riain, M. J. (2012). Monkey management: Using spatial ecology to understand the extent and severity of human-baboon conflict in the Cape peninsula, South Africa. *Ecology and Society*, 17, 13.

Hoffmann, J. N., Montag, A. G., and Dominy, N. J. (2004). Meissner corpuscles and somatosensory acuity: The prehensile appendages of primates and elephants. *Anatomical Record A: Discoveries in Molecular, Cellular, and Evolutionary Biology*, 281A, 1138–47.

Hofman, C. A., Rick, T. C., Maldonado, J. E., Collins, P. W., Erlandson, J. M., Fleischer, R. C., et al. (2016). Tracking the origins and diet of an endemic island canid (*Urocyon littoralis*) across 7300 years of human cultural and environmental change. *Quaternary Science Review*, 146, 147–60.

Hofreiter, M., Kreuz, E., Eriksson, J., Schubert, G., and Hohmann, G. (2010). Vertebrate DNA in fecal samples from bonobos and gorillas: Evidence for meat consumption or artefact? *PLoS ONE*, 5, e9419.

Hogan, J. D., Fedigan, L. M., Hiramatsu, C., Kawamura, S., and Melin, A. D. (2018). Trichromatic perception of flower colour improves resource detection among New World monkeys. *Scientific Reports*, 8(1), 10883.

Hogan, J. D., Melin, A. D., Mosdossy, K. N., and Fedigan, L. M. (2016). Seasonal importance of flowers to Costa Rican capuchins (*Cebus capucinus imitator*): Implications for plant and primate. *American Journal of Physical Anthropology*, 161, 591–602.

Hogg, R. T., and Elokda, A. 2021. Quantification of enamel decussation in gracile and robust capuchins (*Cebus, Sapa-*

jus, Cebidae, Platyrrhini). *American Journal of Primatology*, 83(5), e23246.

Hohmann, G. (1988). A case of simple tool use in wild liontailed macaques (*Macaca silenus*). *Primates*, 29, 565–67.

Hohmann, G. (2009). The diets of non-human primates: Frugivory, food processing, and food sharing. In J. J. Hublin and M. P. Richards (Eds.), *The Evolution of Hominin Diets: Integrating Approaches to the Study of Paleolithic Subsistence* (pp. 1–14). Springer.

Hohmann, G., Fowler, A., Sommer, V., and Ortmann, S. (2006). Frugivory and gregariousness of Salonga bonobos and Gashaka chimpanzees: The abundance and nutritional quality of fruit. In G. Hohmann (Ed.), *Feeding Ecology in Apes and Other Primates* (pp. 123–59). Cambridge University Press.

Hohmann, G., and Fruth, B. (2003). Culture in bonobos? Between-species and within-species variation in behavior. *Current Anthropology*, 44, 563–71.

Hohmann, G., and Fruth, B. (2008). New records on prey capture and meat eating by bonobos at Lui Kotale, Salonga National Park, Democratic Republic of Congo. *Folia Primatologica*, 79, 103–10.

Hohmann, G., Potts, K. B., N'Guesson, A., Fowler, A., Mundry, R., Ganzhorn, J. G., et al. (2010). Plant foods consumed by *Pan*: Exploring the variation of nutritional ecology across Africa. *American Journal of Physical Anthropology*, 141, 476–85.

Holbrook, S. J., and Schmitt, R. J. (1988). The combined effects of predation risk and food reward on patch selection. *Ecology*, 69, 125-134.

Holekamp, K. E., Smale, L., and Szykman, M. (1996). Rank and reproduction in the female spotted hyaena. *Journal of Reproduction and Fertility*, 108, 229–37.

Holmes, S. M., Gordon, A. D., Louis, E. E., and Johnson, S. E. (2016). Fission-fusion dynamics in black-and-white ruffed lemurs may facilitate both feeding strategies and communal care of infants in a spatially and temporally variable environment. *Behavioral Ecology and Sociobiology*, 70, 1949–60.

Hongo, S., Nakashima, Y., Akomo-Okoue, E. F., and Mindonga-Nguelet, F. L. (2018). Seasonal change in diet and habitat use in wild mandrills (*Mandrillus sphinx*). *International Journal of Primatology*, 39, 27–48.

Honorio Coronado, E. N., Baker, T. R., Phillips, T. R., Phillips, O. L., Pitman, N. C. A., Pennington, R. T., et al. (2009). Multiscale comparisons of tree composition in Amazonian terra firme forests. *Biogeosciences*, 6, 2719–31.

Hooper, L. V., Littman, D. R., and Macpherson, A. J. (2012). Interactions between the microbiota and the immune system. *Science*, 336, 1268–73.

Hooper, P. L., Gurven, M., Winking, J., and Kaplan, H. S. (2015). Inclusive fitness and differential productivity across the life course determine intergenerational transfers in a small-scale human society. *Proceedings of the Royal Society B: Biological Sciences*, 282, 20142808.

Hoover, K. C. (2010). Smell with inspiration: The evolutionary significance of olfaction. *American Journal of Physical Anthropology*, 143 (suppl.), 63–74.

Hoover, K. C., Gokcumen, O., Qureshy, Z., Bruguera, E., Savangsuksa, A., Cobb, M., et al. (2015). Global survey of variation in a human olfactory receptor gene reveals signatures of non-neutral evolution. *Chemical Senses*, 40, 481–88.

Hopcraft, J. G. C., Olff, H., and Sinclair, A. R. E. (2010). Herbivores, resources and risks: Alternating regulation along primary environmental gradients in savannas. *Trends in Ecology and Evolution*, 25, 119–28.

Hopkins, H. C. (1983). The taxonomy, reproductive biology and economic potential of *Parkia* (Leguminosae: Mimosoideae) in Africa and Madagascar. *Botanical Journal of the Linnean Society*, 87, 135–67.

Hopkins, W. D., Reamer, L., Mareno, M. C., and Schapiro, S. J. (2015). Genetic basis in motor skill and hand preference for tool use in chimpanzees (*Pan troglodytes*). *Proceedings of the Royal Society B: Biological Sciences*, 282, 20141223.

Hosaka, K., Nakamura, M., Takahata Y. 2020. Longitudinal changes in the targets of chimpanzee (*Pan troglodytes*) hunts at Mahale Mountains National Park: How and why did they begin to intensively hunt red colobus (*Piliocolobus rufomitratus*) in the 1980s? *Primates*, 61, 391–401.

Hosaka, K., Nishida, T., Hamai, M., Matsumoto-Oda, A., Uehara, S. (2002). Predation of mammals by the chimpanzees of the Mahale Mountains, Tanzania. In Galdikas, B. M. F., Briggs, N. E., Sheeran, L. K., Shapiro, G. L., and Goodall, J. (eds.). *All Apes Great and Small. Developments in Primatology: Progress and Prospects* (pp. 107–130). Springer.

Hoshino, J. (1985). Feeding ecology of mandrills (*Mandrillus sphinx*) in Campo Animal Reserve, Cameroon. *Primates*, 26, 248–73.

Hosking, C. J., Raubenheimer, D., Charleston, M. A., Simpson, S. J., and Senior, A. M. (2019). Macronutrient intakes and the lifespan-fecundity trade-off: A geometric framework agent-based model. *Journal of the Royal Society Interface*, 16, 20180733.

Hosokawa, T., Nikoh, N., Koga, R., Satio, M., Tanahashi, M., Meng, X. Y., et al. (2012). Reductive genome evolution, host–symbiont co-speciation and uterine transmission of endosymbiotic bacteria in bat flies. *ISME Journal*, 6, 577.

Hostinar, C. E., Lachman, M. E., Mroczek, D. K., Seeman, T. E., and Miller, G. E. (2015). Additive contributions of childhood adversity and recent stressors to inflammation at midlife: Findings from the MIDUS study. *Developmental Psychology*, 51(11), 1630–44.

Hou, R., Chapman, C. A., Jay, O., Guo, S., Li, B., and Raubenheimer, D. (2020). Cold and hungry: Combined effects of low temperature and resource scarcity on an edge-of-range temperate primate, the golden snub-nose monkey. *Ecography*, 43(11), 1672–82.

Hou, R., Chapman, C. A., Rothman, J. M., Zhang, H., Huang, K., Guo, S., et al. (2020). The geometry of resource constraint:

An empirical study of the golden snub-nosed monkey. *Journal of Animal Ecology*, 90, 751–65.

Hou, R., He, S., Wu, F., Chapman, C. A., Pan, R., Garber, P. A., et al. (2018). Seasonal variation in diet and nutrition of the northern-most population of *Rhinopithecus roxellana*. *American Journal of Primatology*, 80(4), e22755.

Houlahan, J. E., McKinney, S. T., Anderson, T. M., and McGill, B. J. (2017). The priority of prediction in ecological understanding. *Oikos*, 126, 1–7.

Houle, A., Chapman, C. A., and Vickery, W. L. (2007). Intratree variation in fruit production and implications for primate foraging. *International Journal of Primatology*, 28, 1197–217.

Houle, A., Chapman, C. A., and Vickery, W. L. (2010). Intratree vertical variation of fruit density and the nature of contest competition in frugivores. *Behavioral Ecology and Sociobiology*, 64, 429–41.

Houle, A., Conklin-Brittain, N. L., and Wrangham, R. W. (2014). Vertical stratification of the nutritional value of fruit: Macronutrients and condensed tannins. *American Journal of Primatology*, 76, 1207–32.

Houle, A., Vickery, W. L., and Chapman, C. A. (2006). Testing mechanisms of coexistence among two species of frugivorous primates. *Journal of Animal Ecology*, 75, 1034–44.

Houston, A. I., Higginson, A. D., and McNamara, J. M. (2011a). Optimal foraging for multiple nutrients in an unpredictable environment. *Ecology Letters*, 14, 1101–7.

Hoverstad, T., and Midtvedt, T. (1986). Short-chain fatty acids in germ free mice and rats. *Journal of Nutrition*, 116, 1772–76.

Hovey, R. C., Trott, J. F., and Vonderhaar, B. K. (2002). Establishing a framework for the functional mammary gland: From endocrinology to morphology. *Journal of Mammary Gland Biology and Neoplasia*, 7, 17–38.

Howard, A. M., Nibbelink, N. P., Madden, M., Young, L. A., Bernardes, S., and Fragaszy, D. M. (2015). Landscape influences on the natural and artificially manipulated movements of bearded capuchin monkeys. *Animal Behaviour*, 106, 59–70.

Hrdy, S. B. (1977). *The Langurs of Abu: Female and Male Strategies of Reproduction*. Harvard University Press.

Hrdy, S. B. (1999). *Mother Nature—a History of Mothers, Infants, and Natural Selection*. Pantheon Books.

Hrdy, S. B. (2007). Evolutionary context of human development: The cooperative breeding model. In C. S. Carter, L. Ahnert, K. E. Grossmann, S. B. Hrdy, M. E. Lamb, S. W. Porges, et al. (Eds.), *Attachment and Bonding: A New Synthesis* (pp. 39–68). MIT Press.

Hrdy, S. B. (2011). *Mothers and Others: The Evolutionary Origins of Mutual Understanding*. Harvard University Press.

Hrdy, S. B. (2016). Development plus social selection in the emergence of "emotionally modern" humans. In C. L. Meehan and A. N. Crittenden (Eds.), *Childhood—Origins, Evolution, and Implications* (pp. 11–44). School for Advanced Research Press.

Hrdy, S. B., and Hrdy, D. B. (1976). Hierarchical relations among female Hanuman langurs (Primates: Colobinae, *Presbytis entellus*). *Science*, 193, 913–15.

Hua, L.-C., Brandt, E. T., Meullenet, J.-F., Zhou, Z. R., and Ungar, P. S. (2015). Technical note: An in vitro study of dental microwear formation using the BITE Master II chewing machine. *American Journal of Physical Anthropology*, 158(4), 769–75.

Hua, L. C., Chen, J., and Ungar, P. S. (2020). Diet reduces the effect of exogenous grit on tooth microwear. *Biosurface and Biotribology*, 6, 48–52.

Huang, J., Guerrero, A., Parker, E., Strum, J. S., Smilowitz, J. T., German, J. B., et al. (2015). Site-specific glycosylation of secretory immunoglobulin A from human colostrum. *Journal of Proteome Research*, 14, 1335–49.

Huang, M. C., and Brenna, J. T. (2001). On the relative efficacy of α-linolenic acid and preformed docosahexaenoic acid as substrates for tissue docosahexaenoate during perinatal development. In D. Mostofsky, S. Yehuda, and N. Salem (Eds.), *Fatty Acids: Physiological and Behavioral Functions* (pp. 99–113). Humana Press.

Huang, X., Hancock, D. P., Gosby, A. K., McMahon, A. C., Solon, S. M. C., Le Couteur, D. G., et al. (2013). Effects of dietary protein to carbohydrate balance on energy intake, fat storage, and heat production in mice. *Obesity*, 21(1), 85–92.

Huang, Z. P., Huo, S., Yang, S., Cui, L., and Xiao, W. (2010). Leaf choice in black-and-white snub-nosed monkeys *Rhinopithecus bieti* is related to the physical and chemical properties of leaves. *Current Zoology*, 56, 643–49.

Huang, Z. P., Scott, M. B., Li, Y. P., Ren, G. P., Xiang, Z. F., Cui, L. W., et al. (2017). Black-and-white snub-nosed monkey (*Rhinopithecus bieti*) feeding behavior in a degraded forest fragment: Clues to a stressed population. *Primates*, 58, 517–24.

Huang, F. Y., Sutcliffe, M. P., and Grabenhorst, F. (2021). Preferences for nutrients and sensory food qualities identify biological sources of economic values in monkeys. *Proceedings of the National Academy of Sciences*, 118(26), e2101954118.

Hübener, F., and Laska, M. (1998). Assessing olfactory performance in an Old World primate, *Macaca nemestrina*. *Physiology and Behavior*, 64, 521–27.

Huffman, M. A. (1997). Current evidence for self-medication in primates: A multidisciplinary perspective. *Yearbook of Physical Anthropology*, 40, 171–200.

Huffman, M. A. (2003). Animal self-medication and ethnomedicine: Exploration and exploitation of the medicinal properties of plants. *Proceedings of the Nutrition Society*, 62, 371–81.

Huffman, M. A. (2015). Chimpanzee self-medication: a historical perspective of the key findings. In Nakamura, K. Hosaka, N. Itoh, and K. Zamma. *Mahale chimpanzees: 50 Years of Research* (pp. 340–53). Cambridge University Press.

Huffman, M. A., and Hirata, S. (2004). An experimental study of leaf swallowing in captive chimpanzees: Insights into the origin of a self-medicative behavior and the role of social learning. *Primates*, 45, 113–18.

Huffman, M. A., and Kalunde, M. S. (1993). Tool-assisted predation on a squirrel by a female chimpanzee in the Mahale Mountains, Tanzania. *Primates*, 34, 93–98.

Hughes, D. A. (2002). Antioxidant vitamins and immune function. In P. C. Calder (Ed.), *Nutrition and Immune Function* (pp. 171–91). CABI, in association with the Nutrition Society.

Hughes, G. M., Teeling, E. C., and Higgins, D. G. (2014). Loss of olfactory receptor function in hominin evolution. *PLoS ONE*, 9, e84714.

Huguet, R., Sabadic, P., Caceres, I., Diez, C., Rosell, J., Bennasar, M., et al. (2013). Successful subsistence strategies of the first humans in southwestern France. *Quaternary International*, 295, 168–82.

Hui, C. (2006). Carrying capacity, population equilibrium, and environment's maximal load. *Ecological Modelling*, 192, 317–20.

Hume, I. D. (2002). Digestive strategies of mammals. *Acta Zoologica Sinica*, 48, 1–19.

Humle, T. (1999). New record of fishing for termites (*Macrotermes*) by the chimpanzees of Bossou (*Pan troglodytes verus*), Guinea. *Pan Africa News*, 6, 3–4.

Humle, T. (2003). *Culture and Variation in Wild Chimpanzee Behaviour: A Study of Three Communities in West Africa*. University of Stirling.

Humle, T. (2011). The tool repertoire of Bossou chimpanzees. In T. Matsuzawa, T. Humle, and Y. Sugiyama (Eds.), *Chimpanzees of Bossou and Nimba* (pp. 61–71). Springer.

Humle, T., and Matsuzawa, T. (2002). Ant-dipping among the chimpanzees of Bossou, Guinea, and some comparisons with other sites. *American Journal of Primatology*, 58, 133–48.

Humphrey, N. (1976). The social function of intellect. In P. P. G. Bateson and R. A. Hinde (Eds.), *Growing Points in Ethology* (pp. 303–17). Cambridge University Press.

Hunt, G. R., and Gray, R. D. (2004). The crafting of hook tools by wild New Caledonian crows. *Proceedings of the Royal Society B: Biological Sciences*, 271, S88–S90.

Hunt, K. D., and McGrew, W. C. (2002). Chimpanzees in the dry habitats of Assirik, Senegal and Semliki Wildlife Reserve, Uganda. In C. Boesch, G. Hohmann, and L. Marchant (Eds.), *Behavioural Diversity in Chimpanzees and Bonobos* (pp. 35–51). Cambridge University Press.

Hunter, L. T. B., and Skinner, J. D. (1998). Vigilance behaviour in African ungulates: The role of predation pressure. *Behaviour*, 135, 195–211.

Hunter, M. D., and Price, P. W. (1992). Playing chutes and ladders: Heterogeneity and the relative roles of bottom-up and top-down forces in natural communities. *Ecology*, 73, 724–32.

Husson, S. J., Wich, S. A., Marshall, A. J., Dennis, R. D., Ancrenaz, M., Brassey, R., et al. (2009). Orangutan distribution, density, abundance and impacts of disturbance. In S. A. Wich, S. S. Utami Atmoko, T. Mitra Setia, and C. P. van Schaik (Eds.), *Orangutans: Geographic Variation in Ecology and Conservation* (pp. 77–96). Oxford University Press.

Huxley, J. (1942). *Evolution: The Modern Synthesis*. George Allen and Unwin.

Hylander, W. L. (1975). Incisor size and diet in anthropoids with special reference to Cercopithecidae. *Science*, 189(4208), 1095–98.

Ibanez, I., Acharya, K., Juno, E., Karounos, C., Lee, B. R., McCollum, C., et al. (2019). Forest resilience under global environmental change: Do we have the information we need? A systematic review. *PLoS ONE*, 14, e0222207.

Ibrahim, M. K., Zambruni, M., Melby, C. L., and Melby, P. C. (2017). Impact of childhood malnutrition on host defense and infection. *Clinical Microbiology Reviews*, 30(4), 919–71.

Ichinohe, T., Pang, I. K., Kumamoto, Y., Peaper, D. R., Ho, J. H., Murray, T. S., et al. (2011). Microbiota regulates immune defense against respiratory tract influenza A virus infection. *Proceedings of the National Academy of Sciences of the United States of America*, 108(13), 5354–59.

IFPRI. (2016). *Global Nutrition Report 2016: From Promise to Impact—Ending Malnutrition by 2030*.

Ihobe, H. (1992). Observations on the meat-eating behavior of wild bonobos (*Pan paniscus*) at Wamba, Republic of Zaire. *Primates*, 33, 247–50.

Illius, A. W., Tolkamp, B. J., and Yearsley, J. (2002). The evolution of the control of food intake. *Proceedings of the Nutrition Society*, 61, 465–72.

Imafidon, G. I., and Sosulski, E. W. (1990). Nonprotein nitrogen contents of animal and plant foods. *Journal of Agricultural and Food Chemistry*, 38, 114–18.

Imai, H., Suzuki, N., Ishimaru, Y., Sakurai, T., Yin, L., Pan, W., et al. (2012). Functional diversity of bitter taste receptor TAS2R16 in primates. *Biology Letters*, 8, 652–56.

Ingmanson, E. J. (1996). Tool-using behavior in wild *Pan paniscus*: Social and ecological consideration. In A. R. Russon, K. A. Bard, and S. T. Parker (Eds.), *Reaching into Thought: The Minds of the Great Apes* (pp. 190–210). Cambridge University Press.

Innis, S. M. (2003). Perinatal biochemistry and physiology of long-chain polyunsaturated fatty acids. *Journal of Pediatrics*, 143, 1–8.

IPCC. (2021). *Climate Change 2021: The Physical Science Basis*. Contribution of Working Group I to the Sixth Assessment Report of the Intergovernmental Panel on Climate Change. V. Masson-Delmotte, P. Zhai, A. Pirani, S. L. Connors, C. Péan, S. Berger, et al. (Eds.). Cambridge University Press.

Irwin, M. T. (2008). Diademed sifaka (*Propithecus diadema*) ranging and habitat use in continuous and fragmented forest: Higher density but lower viability in fragments? *Biotropica*, 40, 231–40.

Irwin, M. T. (2008). Feeding ecology of *Propithecus diadema* in forest fragments and continuous forest. *International Journal of Primatology*, 29, 95–115.

Irwin, M. T., Raharison, F., Rakotoarimanana, H., Razandrakoto, E., Ranaivoson, E., Rakotofanala, J., et al. (2007). Diademed sifakas (*Propithecus diadema*) use olfaction to forage for the inflorescences of subterranean parasitic plants (Balanophoraceae: *Langsdorffia* sp., and Cytinaceae: *Cytinus* sp.). *American Journal of Primatology*, 69, 471–76.

Irwin, M. T., Raharison, J. L., Chapman, C. A., Junge, R. E., and Rothman, J. M. (2017). Minerals in the foods and diet of diademed sifakas: Are they nutritional challenges? *American Journal of Primatology*, 79, 1-14.

Irwin, M. T., Raharison, J. L., Raubenheimer, D. R., Chapman, C. A., and Rothman, J. M. (2015). The nutritional geometry of resource scarcity: Effects of lean seasons and habitat disturbance on nutrient intakes and balancing in wild sifakas. *PLoS ONE*, 10, e0128046.

Irwin, M. T., Raharison, J. L., Raubenheimer, D., Chapman, C. A., and Rothman, J. M. (2014). Nutritional correlates of the "lean season": Effects of seasonality and frugivory on the nutritional ecology of diademed sifakas. *American Journal of Physical Anthropology*, 153, 78–91.

Isbell, L. A. (1991). Contest and scramble competition: Patterns of female aggression and ranging behavior among primates. *Behavioral Ecology*, 2, 143–55.

Isbell, L. A. (1994a). Predation on primates: Ecological patterns and evolutionary consequences. *Evolutionary Anthropology*, 3, 61–71.

Isbell, L. A. (1994b). Vervets, leopards and researchers—Reply. *Natural History*, 103, 8.

Isbell, L. A. (1994c). The vervets' year of doom. *Natural History*, 103, 48–55.

Isbell, L. A. (1998). Diet for a small primate: Insectivory and gummivory in the (large) patas monkey (*Erythrocebus patas pyrrhonotus*). *American Journal of Primatology*, 45, 381–98.

Isbell, L. (2004). Is there no place like home? Ecological bases of female dispersal and philopatry and their consequences for the formation of kin groups. In B. Chapais and C. M. Berman (Eds.), *Kinship and Behavior in Primates* (pp. 71–108). Oxford University Press.

Isbell, L. A. (2012). Re-evaluating the ecological constraints model with red colobus monkeys (*Procolobus rufomitratus tephrosceles*). *Behaviour*, 149, 493–529.

Isbell, L. A., and Bidner, L. R. (2016). Vervet monkey (*Chlorocebus pygerythrus*) alarm calls to leopards (*Panthera pardus*) function as a predator deterrent. *Behaviour*, 153, 591–606.

Isbell, L. A., Bidner, L. R., Van Cleave, E. K., Matsumoto-Oda, A., and Crofoot, M. C. (2018). GPS-identified vulnerabilities of savannah-woodland primates to leopard predation and their implications for early hominins. *Journal of Human Evolution*, 118, 1–13.

Isbell, L. A., Pruetz, J. D., and Young, T. P. (1998). Movements of vervets (*Cercopithecus aethiops*) and patas monkeys (*Erythrocebus patas*) as estimators of food resource size, density and distribution. *Behavioral Ecology and Sociobiology*, 42, 123–33.

Isbell, L. A., and Van Vuren, D. (1996). Differential costs of locational and social dispersal and their consequences for female group-living primates. *Behaviour*, 133, 1–36.

Isbell, L. A., and Young, T. P. (1993). Human presence reduces predation in a free-ranging vervet monkey population in Kenya. *Animal Behaviour*, 45, 1233–35.

Isbell, L. A., and Young, T. P. (2002). Ecological models of female social relationships in primates: Similarities, disparities, and some directions for future clarity. *Behaviour*, 139, 177–202.

Isbell, L. A., and Young, T. P. (2007). Interspecific and temporal variation of ant species within *Acacia drepanolobium* ant domatia, a staple food of patas monkeys (*Erythrocebus patas*) in Laikipia, Kenya. *American Journal of Primatology*, 69, 1387–1398.

Isbell, L. A., Young, T. P., Enstam Jaffe, K., Carlson, A. A., and Chancellor, R. L. (2009). Demography and life histories of sympatric patas monkeys, *Erythrocebus patas*, and vervets, *Cercopithecus aethiops*, in Laikipia, Kenya. *International Journal of Primatology*, 30, 103–24.

Islam, M. A., and Feeroz, M. M. (1992). Ecology of hoolock gibbon of Bangladesh. *Primates*, 33, 451–64.

Isler, K., Kirk, E. C., Miller, J. M. A., Albrecht, G. A., Gelvin, B. R., and Martin, R. D. (2008). Endocranial volumes of primate species: Scaling analyses using a comprehensive and reliable data set. *Journal of Human Evolution*, 55, 967–78.

Isler, K., and van Schaik, C. P. (2006). Metabolic costs of brain size evolution. *Biology Letters*, 2, 557–60.

Isler, K., and van Schaik, C. P. (2009). The expensive brain: a framework for explaining evolutionary changes in brain size. *Journal of Human Evolution*, 57, 392–400.

Isler, K., and van Schaik, C. P. (2012). Allomaternal care, life history and brain size evolution in mammals. *Journal of Human Evolution*, 63, 52–63.

Itani, J. (1975). Twenty years with Mount Takasaki monkeys. In G. Bermant and D. G. Lindburg (Eds.), *Primate Utilization and Conservation* (pp. 101–25). Wiley Interscience.

Itoigawa, A., Fierro, F., Chaney, M. E., Lauterbur, M. E., Hayakawa, T., Tosi, A. J., et al. (2021). Lowered sensitivity of bitter taste receptors to β-glucosides in bamboo lemurs: an instance of parallel and adaptive functional decline in TAS2R16? *Proceedings of the Royal Society B*, 288(1948), 20210346.

Ivanov, I. I., Atarashi, K., Manel, N., Brodie, E. L., Shima, T., Karaoz, U., et al. (2009). Induction of intestinal Th17 cells by segmented filamentous bacteria. *Cell*, 139(3), 485–98.

Ivosevic, B., Han, Y. G., Cho, Y., and Kwon, O. (2015). The use of conservation drones in ecology and wildlife research. *Journal of Ecology and Environment*, 38, 113–18.

Iwamoto, T., and Dunbar, R. I. M. (1983). Thermoregulation, habitat quality and the behavioural ecology of gelada baboons. *Journal of Animal Ecology*, 52, 357–66.

Iyer, S. S., Chatraw, J. H., Tan, W. G., Wherry, E. J., Becker, T. C., Ahmed, R., et al. (2012). Protein energy malnutrition impairs homeostatic proliferation of memory CD8 T cells. *Journal of Immunology*, 188(1), 77–84.

Izar, P., Peternelli-dos-Santos, L., Rothman, J. M., Raubenheimer, D., Presotto, A., Gort, G., et al. (2022). Stone tools improve diet quality in wild monkeys. *Current Biology*, 32, 4088–92.e3.

Izawa, K. (1993). Soil-eating by *Alouatta* and *Ateles*. *International Journal of Primatology*, 14, 229–42.

Izawa, K., and Mizuno, A. (1977). Palm-fruit cracking behavior of wild black-capped capuchin (*Cebus apella*). *Primates*, 18, 773–92.

Jablonski, N. G., and Leakey, M. G. (2008). Systematic paleontology of the small colobines. In N. G. Jablonski and M. G. Leakey (Eds.), *Koobi Fora Research Project, the Fossil Monkeys* (vol. 6, pp. 12–30). California Academy of Sciences.

Jack, K. M., Lenz, B. B., Healan, E., Rudman, S., Schoof, V. A. M., and Fedigan, L. (2008). The effects of observer presence on the behavior of *Cebus capucinus* in Costa Rica. *American Journal of Primatology*, 70, 490–94.

Jackson, C. J. C., Dini, J. P., Lavandier, C., Rupasinghe, H. P. V., Faulkner, H., Poysa, V., et al. (2002). Effects of processing on the content and composition of isoflavones during manufacturing of soy beverage and tofu. *Process Biochemistry*, 37, 1117–23.

Jackson, C. P., and Reichard, U. H. (2021). Pygmy marmoset exudate feeding stimulates exudate production. *Folia Primatologia*, 92, 175–82.

Jackson, R. L., Greiwe, J. S., and Schwen, R. J. (2011). Emerging evidence of the health benefits of S-equol, an estrogen receptor beta agonist. *Nutrition Reviews*, 69, 432–48.

Jacob, A. L., Vaccaro, I., Sengupta, R., Hartter, J., and Chapman, C. A. (2008). How can conservation biology best prepare for declining rural population and ecological homogenization? *Tropical Conservation Science*, 1, 307–20.

Jacob, J., and Brown, J. S. (2000). Microhabitat use, giving-up densities and temporal activity as short- and long-term antipredator behaviors in common voles. *Oikos*, 91, 131–38.

Jacob, K. D., Noren Hooten, N., Trzeciak, A. R., and Evans, M. K. (2013). Markers of oxidant stress that are clinically relevant in aging and age-related disease. *Mechanisms of Ageing and Development*, 134(3–4), 139–57.

Jacobs, G. H. (1996). Primate photopigments and primate color vision. *Proceedings of the National Academy of Sciences of the United States of America*, 93, 577–81.

Jacobs, G. H. (2009). Evolution of colour vision in mammals. *Philosophical Transactions of the Royal Society of London Series B—Biological Sciences*, 364, 2957–67.

Jacobs, G. H. (2013). Losses of functional opsin genes, short-wavelength cone photopigments, and color vision—a significant trend in the evolution of mammalian vision. *Visual Neuroscience*, 30, 39–53.

Jacobs, G. H., Neitz, M., Deegan, J. F., and Neitz, J. (1996). Trichromatic colour vision in New World monkeys. *Nature*, 382, 156–58.

Jacobs, R. L., MacFie, T. S., Spriggs, A. N., Baden, A. L., Morelli, T. L., Irwin, M. T., et al. (2017). Novel opsin gene variation in large-bodied, diurnal lemurs. *Biology Letters*, 13, 20170050.

Jacobs, R. L., Veilleux, C. C., Louis, E. E., Herrera, J. P., Hiramatsu, C., Frankel, D. C., et al. (2019). Less is more: lemurs (*Eulemur* spp.) may benefit from loss of trichromatic vision. *Behavioral Ecology and Sociobiology*, 73, 1–17.

Jacobsen, N. (1970). Salivary amylase II: alpha amylase in salivary glands of the *Macaca irus* monkey, the *Cercopithecus aethiops* monkey, and man. *Caries Research*, 4, 200–205.

Jaeger, S. R., McRae, J. F., Bava, C. M., Beresford, M. K., Hunter, D., Jia, Y., et al. (2013). A Mendelian trait for olfactory sensitivity affects odor experience and food selection. *Current Biology*, 23, 1601–5.

Jaeggi, A., and Gurven, M. (2013). Natural cooperators: Food sharing in humans and other primates. *Evolutionary Anthropology*, 22, 186–95.

Jahoor, F., Badaloo, A., Reid, M., and Forrester, T. (2008). Protein metabolism in severe childhood malnutrition. *Annals of Tropical Paediatrics*, 28(2), 87–101.

Jakubas, W. J., and Gullion, G. W. (1990). Coniferyl benzoate in quaking aspen: A ruffed grouse feeding deterrent. *Journal of Chemical Ecology*, 16, 1077–87.

Jaman, M. F., and Huffman, M. A. (2013). The effect of urban and rural habitats and resource type on activity budgets of commensal rhesus macaques (*Macaca mulatta*) in Bangladesh. *Primates*, 54, 49–59.

James, W. P. T., and Schofield, E. C. (1990). Human energy requirements: a manual for planners and nutritionists (No. JAM 641 (BV 907.3)). The Food and Agriculture Organization of the United Nations (FAO).

Janiak, M. C. (2016). Digestive enzymes of human and nonhuman primates. *Evolutionary Anthropology: Issues, News, and Reviews*, 25(5), 253–66.

Janiak, M. C., Chaney, M. E., and Tosi, A. J. (2017). Evolution of acidic mammalian chitinase genes (CHIA) is related to body mass and insectivory in primates. *Molecular Biology and Evolution*, 35, 607–22.

Janmaat, K. R. L., Ban, S. D., and Boesch, C. (2013a). Chimpanzees use long-term spatial memory to monitor large fruit trees and remember feeding experiences across seasons. *Animal Behaviour*, 86, 1183–205.

Janmaat, K. R. L., Ban, S. D., and Boesch, C. (2013b). Taï chimpanzees use botanical skills to discover fruit: What we can learn from their mistakes. *Animal Cognition*, 16, 851–60.

Janmaat, K., Boesch, C., Byrne, R., Chapman, C. A., Gone Bí, Z. B., Head, J. S., et al. (2016). Spatial-temporal complexity of chimpanzee food: How cognitive adaptations can counteract the ephemeral nature of ripe fruit. *American Journal of Primatology*, 78, 626–45.

Janmaat, K. R. L., Byrne, R. W., and Zuberbühler, K. (2006a). Evidence for a spatial memory of fruiting states of rainforest trees in wild mangabeys. *Animal Behaviour*, 72, 797–807.

Janmaat, K. R. L., Byrne, R. W., and Zuberbühler, K. (2006b). Primates take weather into account when searching for fruits. *Current Biology*, 16, 1232–37.

Janmaat, K., and Chancellor, R. (2010). Exploring new areas: How important is long-term spatial memory for mangabey (*Lophocebus albigena johnstonii*) foraging efficiency? *International Journal of Primatology*, 31, 863–86.

Janmaat, K. R. L., Chapman, C. A., Meijer, R., and Zuberbühler, K. (2012). The use of fruiting synchrony by foraging mangabey monkeys: A "simple tool" to find fruit. *Animal Cognition*, 15, 83–96.

Janmaat, K. R., de Guinea, M., Collet, J., Byrne, R. W., Robira, B., van Loon, E., et al. (2021). Using natural travel paths to infer and compare primate cognition in the wild. *IScience*, 24(4), 102343.

Janmaat, K. R. L., Polansky, L., Ban, S. D., and Boesch, C. (2014). Wild chimpanzees plan their breakfast time, type, and location. *Proceedings of the National Academy of Sciences of the United States of America*, 111, 16343–48.

Jansen, P. A., Bohlman, S. A., Garzon-Lopez, C. X., Oiff, H., Muller-Landau, H. C., and Wright, S. J. (2008). Large-scale spatial variation in palm fruit abundance across a tropical moist forest estimated from high-resolution aerial photographs. *Ecography*, 31, 33–42.

Janson, C. H. (1983). Adaptation of fruit morphology to dispersal agents in a Neotropical forest. *Science*, 219, 187–89.

Janson, C. (1984a). Female choice and mating system of the brown capuchin monkey *Cebus apella* (Primates: Cebidae). *Zeitschrift fur Tierpsychologie*, 65, 177–200.

Janson, C. (1984b). Reconciling rigor and range: observations, experiments, and quasi-experiments in field primatology. *International Journal of Primatology*, 33, 520–41.

Janson, C. H. (1985). Aggressive competition and individual food consumption in wild brown capuchin monkeys (*Cebus apella*). *Behavioral Ecology and Sociobiology*, 18, 125–38.

Janson, C. (1986a). Capuchin counterpoint. *Natural History*, 95, 44–53.

Janson, C. H. (1986b). The mating system as a determinant of social evolution in capuchin monkeys. In J. Else and P. C. Lee (Eds.), *Primate Ecology and Conservation* (pp. 169–79). Cambridge University Press.

Janson, C. H. (1988a). Food competition in brown capuchin monkeys (*Cebus apella*): Quantitative effects of group size and tree productivity. *Behaviour*, 105, 53–76.

Janson, C. H. (1988b). Intra-specific food competition and primate social structure: A synthesis. *Behaviour*, 105, 1–17.

Janson, C. H. (1990a). Ecological consequences of individual spatial choice in foraging groups of brown capuchin monkeys, *Cebus apella*. *Animal Behaviour*, 40, 922–34.

Janson, C. H. (1990b). Social correlates of individual spatial choice in foraging groups of brown capuchin monkeys, *Cebus apella*. *Animal Behaviour*, 40, 910–21.

Janson, C. H. (1996). Towards an experimental socioecology of primates: Examples from Argentine brown capuchin monkeys (*Cebus apella nigritus*). In M. A. Norconk, A. Rosenberger, and P. Garber (Eds.), *Adaptive Radiations of Neotropical Primates* (pp. 309–25). Plenum.

Janson, C. H. (1998a). Experimental evidence for spatial memory in foraging wild capuchin monkeys, *Cebus apella*. *Animal Behaviour*, 55, 1229–43.

Janson, C. H. (1998b). Testing the predation hypothesis for vertebrate sociality: Prospects and pitfalls. *Behaviour*, 135, 389–410.

Janson, C. H. (2000). Primate socio-ecology: The end of a golden age. *Evolutionary Anthropology*, 9, 73–86.

Janson, C. H. (2007). Experimental evidence for route integration and strategic planning in wild capuchin monkeys. *Animal Cognition*, 10, 341–56.

Janson, C. H. (2014). Death of the (traveling) salesman: Primates do not show clear evidence of multi-step route planning. *American Journal of Primatology*, 76, 410–20.

Janson, C. H. (2016). Capuchins, space, time and memory: An experimental test of what-where-when memory in wild monkeys. *Proceedings of the Royal Society B: Biological Sciences*, 283, 20161432.

Janson, C., Baldovino, M., and Di Bitetti, M. (2012). *The Group Life Cycle and Demography of Brown Capuchin Monkeys (Cebus [apella] nigritus) in Iguazu National Park, Argentina*. Springer.

Janson, C. H., and Boinski, S. (1992). Morphological and behavioral adaptations for foraging in generalist primates: The case of the cebines. *American Journal of Physical Anthropology*, 88, 483–98.

Janson, C. H., and Byrne, R. (2007). What wild primates know about resources: Opening up the black box. *Animal Cognition*, 10, 357–67.

Janson, C. H., and Chapman, C. A. (1999). Resources and primate community structure. In J. G. Fleagle, C. Janson and K. E. Reed (Eds.), *Primate Communities* (pp. 237–68). Cambridge University Press.

Janson, C. H., and Emmons, L. H. (1990). Ecological structure of the nonflying mammal community at Cocha Cashu Biological Station, Manu National Park, Peru. In A. H. Gentry (Ed.), *Four Neotropical Rainforests* (pp. 314–38). Yale University Press.

Janson, C. H., and Goldsmith, M. (1995). Predicting group size in primates: Foraging costs and predation risk. *Behavioural Ecology*, 6, 326–36.

Janson, C. H., Terborgh, J., and Emmons, L. H. (1981). Nonflying mammals as pollinating agents in the Amazonian forest. *Biotropica*, 13, 1–6.

Janson, C. H., and van Schaik, C. P. (1988). Recognizing the many faces of primate food competition: Methods. *Behaviour*, 105, 165–86.

Janson, C. H., and van Schaik, C. (1993). Ecological risk aversion in juvenile primates: Slow and steady wins the race. In M. E. Pereira and L. Fairbanks (Eds.), *Juvenile Primates: Life History, Development, and Behavior* (pp. 57–74). Oxford University Press.

Janson, C., and Verdolin, J. (2005). Seasonality of primate births in relation to climate. In D. K. Brockman and C. P. van Schaik (Eds.), *Seasonality in Primates: Studies of Living and Extinct Human and Non-human Primates* (pp. 307–50). Cambridge University Press.

Janzen, D. H. (1971). Seed predation by animals. *Annual Review of Ecology and Systematics*, 2, 465–92.

Janzen, D. H. (1974). Tropical blackwater rivers, animals, and mast fruiting by the Dipterocarpaceae. *Biotropica*, 6, 69–103.

Janzen, D. H. (1979a). How to be a fig. *Annual Review of Ecology and Systematics*, 10, 13–51.

Janzen, D. H. (1979b). New horizons in the biology of plant defenses. In G. A. Rosenthal and D. H. Janzen (Eds.), *Herbivores: Their Interactions with Secondary Plant Metabolites* (pp. 331–50). Academic Press.

Janzen, D. H. (1980). When is it coevolution? *Evolution*, 34, 611–12.

Jaouen, K., Balter, V., Herrscher, E., Lamboux, A., Telouk, P., and Albarède, F. (2012). Fe and Cu stable isotopes in archeological human bones and their relationship to sex. *American Journal of Physical Anthropology*, 148, 334–40.

Jaouen, K., Beasley, M., Schoeninger, M., Hublin, J. J., and Richards, M. P. (2016). Zinc isotope ratios of bones and teeth as new dietary indicators: Results from a modern food web (Koobi Fora, Kenya). *Scientific Reports*, 6, 26281.

Jarcho, M. R., Power, M. L., Layne-Colon, D. G., and Tardif, S.D. (2013). Digestive efficiency mediated by serum calcium predicts bone mineral density in the common marmoset (*Callithrix jacchus*). *American Journal of Primatology*, 75(2), 153–60.

Jarvey, J. C., Low, B. S., Pappano, D. J., Bergman, T. J., and Beehner, J. C. (2018). Graminivory and fallback foods: Annual diet profile of geladas (*Theropithecus gelada*) living in the Simien Mountains National Park, Ethiopia. *International Journal of Primatology*, 39, 105–26.

Jay, P. C. (1968). *Primates: Studies in Adaptation and Variability*. Holt, Rinehart, and Winston.

Jayakody, S., Sibbald, A. M., Gordon, I. J., and Lambin, X. (2008). Red deer *Cervus elephus* vigilance behaviour differs with habitat and type of human disturbance. *Wildlife Biology*, 14, 81–91.

Jeannoda, V., Rakotonirina, O., Randrianarivo, H., Rakoto, D., Wright, P. C., and Hladik, C. M. (2003). The toxic principle of the bamboo eaten by *Hapalemur aureus* is not neutralized by soil consumption. *Revue D'ecologie—la Terre et la Vie*, 58, 151–53.

Jefferson, W. N., Patisaul, H. B., and Williams, C. J. (2012). Reproductive consequences of developmental phytoestrogen exposure. *Reproduction*, 143, 247–60.

Jensen, K., Mayntz, D., Toft, S., Clissold, F. J., Hunt, J., Raubenheimer, D., et al. (2012). Optimal foraging for specific nutrients in predatory beetles. *Proceedings of the Royal Society B: Biological Sciences*, 279, 2212–18.

Jensen, L. M., Wallis, I. R., and Foley, W. J. (2015). The relative concentrations of nutrients and toxins dictate feeding by a vertebrate browser, the greater glider *Petauroides volans*. *PLoS ONE*, 10, e0121584.

Jensen, L. M., Wallis, I. R., Marsh, K. J., Moore, B. D., Wiggins, N. L., and Foley, W. J. (2014). Four species of arboreal folivore show differential tolerance to a secondary metabolite. *Oecologia*, 176, 251–58.

Jensen, R. G., Bitman, J., Carlson, S. E., Couch, S. C., Hamosh, M., and Newberg, D. S. (1995). Milk lipids: A. Human milk lipids. In G. Meurant (Ed.), *Handbook of Milk Composition* (pp. 295–542). Academic Press.

Jequier, E., and Constant, F. (2009). Water as an essential nutrient: The physiological basis of hydration. *European Journal of Clinical Nutrition*, 64, 115–23.

Jevanandam, N., Goh, A. G. R., and Corlett, R. T. (2013). Climate warming and the potential extinction of fig wasps, the obligate pollinators of figs. *Biology Letters*, 9, 20130041.

Jiang, P., Josue, J., Li, X., Glaser, D., Li, W., Brand, J. G., et al. (2012). Major taste loss in carnivorous mammals. *Proceedings of the National Academy of Sciences of the United States of America*, 109, 4956–61.

Jim, S., Ambrose, S. H., and Evershed, R. P. (2004). Stable carbon isotopic evidence for differences in the dietary origin of bone cholesterol, collagen and apatite: Implications for their use in palaeodietary reconstruction. *Geochimica et Cosmochimica Acta*, 68, 61–72.

Jin, L., Hinde, K., and Tao, L. (2011). Species diversity and relative abundance of lactic acid bacteria in the milk of rhesus monkeys (*Macaca mulatta*). *Journal of Medical Primatology*, 40(1), 52–58.

Jin, T., Wang, D. Z., Zhao, Q., Yin, L., Qin, D. G., Ran, W. Z., et al. (2009). Reproductive parameters of wild *Trachypithecus leucocephalus*: Seasonality, infant mortality and interbirth interval. *American Journal of Primatology*, 71, 558–66.

Joannes-Boyau, R., Adams, J. W., Austin, C., Arora, M., Moffat, I., Herries, A. I., et al. (2019). Elemental signatures of *Australopithecus africanus* teeth reveal seasonal dietary stress. *Nature*, 572(7767), 112–15.

Joannes-Boyau, R., Adams, J., Austin, C., Arora, M., Moffat, I., Herries, A., et al. (2020). Cyclic elemental signatures suggest nursing adaptations to food stress in *Australopithecus africanus*. OSF Preprints, unpublished manuscript.

Johansson, M. E., Holmen Larsson, J. M., and Hansson, G. C. (2011). The two mucus layers of colon are organized by the MUC2 mucin, whereas the outer layer is a legislator of host-microbial interactions. *Proceedings of the National Academy of Sciences of the United States of America*, 108, 4659–65.

Johns, A. D. (1983). Tropical forest primates and logging—can they co-exist? *Oryx*, 17, 114–18.

Johns, A. D. (1986). Effects of selective logging on the behavioral ecology of West Malaysian primates. *Ecology*, 67, 684–94.

Johns, A. D. (1992). Vertebrate responses to selective logging: Implications for the design of logging systems. *Philosophical Transactions of the Royal Society of London Series B—Biological Sciences*, 335, 437–42.

Johns, A. D., and Skorupa, J. P. (1987). Responses of rain-forest primates to habitat disturbance: A review. *International Journal of Primatology*, 8, 157–91.

Johns, T. (1999). The chemical ecology of human ingestive behaviors. *Annual Review of Anthropology*, 28, 27–50.

Johnson, A. J., Vangay, P., Al-Ghalith, G. A., Hillmann, B. M., Ward, T. L., Shields-Cutler, R. R., et al. (2019). Daily sampling reveals personalized diet-microbiome associations in humans. *Cell Host and Microbe*, 25, 789–802.

Johnson, C. A., Raubenheimer, D., Chapman, C. A., Tombak, K. J., Reid, A. J., and Rothman, J. M. (2017). Macronutrient balancing affects patch departure by guerezas (*Colobus guereza*). *American Journal of Primatology*, 79(4), 1–9.

Johnson, C. A., Raubenheimer, D., Rothman, J. M., Clarke, D., and Swedell, L. (2013). 30 days in the life: Daily nutrient balancing in a wild chacma baboon. *PLoS ONE*, 8, e70383.

Johnson, C. A., Swedell, L., and Rothman, J. M. (2012). Feeding ecology of olive baboons (*Papio anubis*) in Kibale National Park, Uganda: Preliminary results on diet and food selection. *African Journal of Ecology*, 50, 367–70.

Johnson, C. D., Kudsk, K. A., Fukatsu, K., Renegar, K. B., and Zarzaur, B. L. (2003). Route of nutrition influences generation of antibody-forming cells and initial defense to an active viral infection in the upper respiratory tract. *Annals of Surgery*, 237(4), 565–73.

Johnson, J. S., Abuajamieh, M., Sanz Fernandez, M. V., Seibert, J. T., Stoakes, S. K., Nteeba, J., et al. (2015). Thermal stress alters postabsorbtive metabolism during pre-and postnatal development. In V. Sejian, J. Gaughan, L. Baumgard, and C. Prasad (Eds.), *Climate Change Impact on Livestock: Adaptation and Mitigation* (pp. 61–79). Springer, New Delhi.

Johnson, M. D. (2007). Measuring habitat quality: A review. *Condor*, 109, 489–504.

Johnson, S. D., Burgoyne, P. M., Harder, L. D., and Dötterl, S. (2011). Mammal pollinators lured by the scent of a parasitic plant. *Proceedings of the Royal Society B: Biological Sciences*, 278, 2303–10.

Johnson, S. D., Pauw, A., and Midgley, J. (2001). Rodent pollination in the African lily *Massonia depressa* (Hyacinthaceae). *American Journal of Botany*, 88, 1768–73.

Johnson, S. E. (2003). Life history and the competitive environment: Trajectories of growth, maturation, and reproductive output among chacma baboons. *American Journal of Physical Anthropology*, 120, 83–98.

Joly-Radko, M., & Zimmermann, E. (2010). Seasonality in gum and honeydew feeding in gray mouse lemurs. In: Burrows AM, Nash LT (eds) The evolution of exudativory in primates. (pp. 141-153). Springer.

Jolly, A. (1984). The puzzle of female feeding priority. In M. Small (Ed.), *Female Primates: Studies by Women Primatologists* (pp. 197–215). A. R. Liss.

Jolly, A. (2003). *Lemur catta*, ring-tailed lemur, Maky. In S. Goodman and J. Benstead (Eds.), *The Natural History of Madagascar* (pp. 1329–31). University of Chicago Press.

Jolly, A. J. (1972). *The Evolution of Primate Behavior*. Macmillan.

Jolly, A., Dobson, A., Rasamimanana, H., Walker, J., O'Connor, S., Solberg, M., et al. (2002). Demography of *Lemur catta* at Berenty Reserve, Madagascar: Effects of troop size, habitat and rainfall. *International Journal of Primatology*, 23, 327–53.

Jolly, C. J. (1970a). *Hadropithecus*: A lemuroid small-object feeder. *Man*, 5, 619–26.

Jolly, C. J. (1970b). The seed-eaters: A new model of hominid differentiation based on a baboon analogy. *Man*, 5, 5–26.

Joly, M., and Zimmerman, E. (2011). Do solitary foraging nocturnal mammals plan their routes? *Biology Letters*, 7, 638–40.

Jonas, W., and Woodside, B. (2016). Physiological mechanisms, behavioral and psychological factors influencing the transfer of milk from mothers to their young. *Hormones and Behavior*, 77, 167–81.

Jones, B. C., Smith, A. D., Bebus, S. E., and Schoech, S. J. (2016). Two seconds is all it takes: European starlings (*Sturnus vulgaris*) increase levels of circulating glucocorticoids after witnessing a brief raptor attack. *Hormones and Behavior*, 78, 72–78.

Jones, J. H. (2011). Primates and the evolution of long, slow life histories. *Current Biology*, 21, R708–R717.

Jones, M., Mandelik, Y., and Dayan, T. (2001). Coexistence of temporally partitioned spiny mice: Roles of habitat structure and foraging behavior. *Ecology*, 82, 2164–76.

Jones, O. A. H., Maguire, M. L., Griffin, J. L., Dias, D. A., Spurgeon, D. J., and Svendsen, C. (2013). Metabolomics and its use in ecology. *Austral Ecology*, 38, 713–20.

Jones, R. J., Ludlow, M. M., Troughton, J. H., and Blunt, C. G. (1979). Estimation of the proportion of C_3 and C_4 plant species in the diet of animals from the ratio of natural ^{12}C and ^{13}C isotopes in the faeces. *Journal of Agricultural Science*, 92, 91–100.

Jones, R. L., and Hanson, H. C. (1985). *Mineral Licks, Geophagy, and Biogeochemistry of North American Ungulates*. Iowa State Press.

Jones, T., Bamford, A. J., Ferrol-Schulte, D., Hieronimo, P., McWilliam, N., and Rovero, F. (2012). Vanishing wildlife corridors and options for restoration: A case study from Tanzania. *Tropical Conservation Science*, 5, 463–74.

Jönsson, K. I. (1997). Capital and income breeding as alternative tactics of resource use in reproduction. *Oikos*, 78, 57–66.

Jooste, E., Pitman, R. T., van Hoven, W., and Swanepoel, L. H. (2012). Unusually high predation on chacma baboons (*Pa-

pio ursinus) by female leopards (*Panthera pardus*) in the Waterberg Mountains, South Africa. *Folia Primatologica*, 83, 353–60.

Joppa, L. N., Loarie, S. R., and Pimm, S. L. (2008). On the protection of "protected areas." *Proceedings of the National Academy of Sciences of the United States of America*, 105, 6673–78.

Joppa, L. N., and Pfaff, A. (2009). High and far: Biases in the location of protected areas. *PLoS ONE*, 4, e8273.

Joppa, L., and Pfaff, A. (2010). Reassessing the forest impacts of protection: The challenge of nonrandom location and a corrective method. *Annals of the New York Academy of Sciences*, 1185, 135–49.

Judd, T. M., and Sherman, P. W. (1996). Naked mole-rats recruit colony mates to food sources. *Animal Behaviour*, 52, 957–69.

Julliot, C. (1996a). Fruit choice by red howler monkeys (*Alouatta seniculus*) in a tropical rain forest. *American Journal of Primatology*, 40, 261–82.

Julliot, C. (1996b). Seed dispersal by red howling monkeys (*Alouatta seniculus*) in the tropical rain forest of French Guiana. *International Journal of Primatology*, 17, 239–58.

Jump, A. S., and Penuelas, J. (2005). Running to stand still: Adapation and the response of plants to climate change. *Ecology Letters*, 8, 1010–20.

Jung, L., Mourthe, I., Grelle, C. E. V., Strier, K. B., and Boubli, J. P. (2015). Effects of local habitat variation on the behavioral ecology of two sympatric groups of brown howler monkey (*Alouatta clamitans*). *PLoS ONE*, 10, e0129789.

Jung, S., Rickert, D. A., Deak, N. A., Aldin, E. D., Recknor, J., Johnson, L. A., et al. (2003). Comparison of Kjeldahl and Dumas methods for determining protein contents of soybean products. *Journal of the American Oil Chemists' Society*, 8, 1169–73.

Jungers, W. L., and Susman, R. L. (1984). Body size and skeletal allometry in African apes. In R. L. Susman (Ed.), *The Pygmy Chimpanzee—Evolutionary Biology and Behavior* (pp. 131–77). Plenum.

Kaas, J. H. (2013). The evolution of brains from early mammals to humans. *Wiley Interdisciplinary Review of Cognition Science*, 4, 33–45.

Kagoro-Rugunda, G. (2020). Fruits' nutrient composition and their influence on consumption by chimpanzees in Kalinzu Forest, South Western Uganda. *Open Journal of Ecology*, 10, 289–302.

Kahlenberg, S. M., Emery Thompson, M., and Wrangham, R. W. (2008). Female competition over core areas in *Pan troglodytes schweinfurthii*, Kibale National Park, Uganda. *International Journal of Primatology*, 29, 931–47.

Kaiser, T. M., and Schulz, E. (2006). Tooth wear gradients in zebras as an environmental proxy: A pilot study. *Mitteilungen aus dem Hamburgischen Zoologischen Museum und Institut*, 103, 187–210.

Kaiser, T. M., and Solounias, N. (2003). Extending the tooth mesowear method to extinct and extant equids. *Geodiversitas*, 25(2), 321–45.

Kaiser, T. M., Braune, C., Kalinka, G. and Schulz-Kornas, E. (2018). Nano-indentation of native phytoliths and dental tissues: Implications for herbivore-plant combat and dental wear proxies. *Evolutionary Systematics*, 2, 55–63.

Kalko, E. K. V., Herre, E. A., and Handley, C. O. (1996). Relation of fig fruit characteristics to fruit-eating bats in the New and Old World tropics. *Journal of Biogeography*, 23, 565–76.

Källander, H. (2005). Commensal association of waterfowl with feeding swans. *Waterbirds*, 326–30.

Kamakoti, V., Kinnamon, D., Choi, K. H., Jagannath, B., and Prasad, S. (2018). Fully electronic urine dipstick probe for combinatorial detection of inflammatory biomarkers. *Future Science OA*, 4(5), FSO301.

Kamilar, J. M., and Baden, A. L. (2014). What drives flexibility in primate social organization? *Behavioral Ecology and Sociobiology*, 68, 1677–92.

Kamilar, J. M., Beaudrot, L., and Reed, K. E. (2015). Climate and species richness predict the phylogenetic structure of African mammal communities. *PLoS ONE*, 10, e0121808.

Kamilar, J. M., and Paciulli, L. M. (2008). Examining the extinction risk of specialized folivores: a comparative study of colobine monkeys. *American Journal of Primatology*, 70(9), 816–27.

Kanamori, T., Kuze, N., Bernard, H., Malim, T. P., and Kohshima, S. (2010). Feeding ecology of Bornean orangutans (*Pongo pygmaeus morio*) in Danum Valley, Sabah, Malaysia: A 3-year record including two mast fruitings. *American Journal of Primatology*, 72, 820–40.

Kane, E. E., and McGraw, W. S. (2017). Dietary variation in Diana monkeys (*Cercopithecus diana*): The effects of polyspecific associations. *Folia Primatologica*, 88, 455–82.

Kang, W., and Kudsk, K. A. (2007). Is there evidence that the gut contributes to mucosal immunity in humans? *Journal of Parenteral and Enteral Nutrition*, 31(3), 246–58.

Kano, T. (1996). Male rank order and copulation rate in a unit-group of bonobos at Wamba, Zaire. In W. C. McGrew, L. F. Marchant, and T. Nishida (Eds.), *Great Ape Societies* (pp. 135–45). Cambridge University Press.

Kanowski, J. (2001). Effects of elevated CO_2 on foliar chemistry of seedlings of two rainforest trees from north-east Australia: Implications for folivorous marsupials. *Austral Ecology*, 26, 165–72.

Kao, Y. C., Zhou, C. B., Sherman, M., Laughton, C. A., and Chen, S. (1998). Molecular basis of the inhibition of human aromatase (estrogen synthetase) by flavone and isoflavone phytoestrogens: A site-directed mutagenesis study. *Environmental Health Perspectives*, 106, 85–92.

Kapahi, P., Chen, D., Rogers, A. N., Katewa, S. D., Li, P. W., Thomas, E. L., et al. (2010). With TOR, less is more: A key role for the conserved nutrient-sensing TOR pathway in aging. *Cell Metabolism*, 11(6), 453–65.

Kapetanovic, R., and Cavaillon, J. M. (2007). Early events in innate immunity in the recognition of microbial pathogens. *Expert Opinion on Biological Therapy*, 7(6), 907–18.

Kaplan, H., Hill, K., Lancaster, J., and Hurtado, A. M. (2000). A theory of human life history evolution: Diet, intelligence, and longevity. *Evolutionary Anthropology*, 9, 156–85.

Kaplin, B. A., Munyaligoga, V., and Moermond, T. C. (1998). The influence of temporal changes in fruit availability on diet composition and seed handling in blue monkeys (*Cercopithecus mitis doggetti*). *Biotropica*, 30, 56–71.

Kappeler, P. M. (1991). Patterns of sexual dimorphism in body weight among prosimian primates. *Folia Primatologica*, 57, 132–46.

Kappeler, P. M. (1996). Causes and consequences of life-history variation among strepsirhine primates. *American Naturalist*, 148, 868–91.

Kappeler, P. M. (1998). Nests, tree holes, and the evolution of primate life histories. *American Journal of Primatology*, 46, 7–33.

Kappeler, P. M. (1999). Lemur social structure and convergence in primate socioecology. In P. C. Lee (Ed.), *Comparative Primate Socioecology* (pp. 273–99). Cambridge University Press.

Kappeler, P. M., van Schaik, C. P., and Watts, D. P. (2012). The values and challenges of long-term field studies. In P. M. Kappeler and D. P. Watts (Eds.), *Long-Term Field Studies of Primates* (pp. 3–20). Springer.

Kappeler, P. M., and Watts, D. P. (2012). *Long-Term Field Studies of Primates*. Springer.

Karapanou, O., and Papadimitriou, A. (2010). Determinants of menarche. *Reproductive Biology and Endocrinology*, 8, 115.

Karasov, W. H., and Martinez del Rio, C. (2007). *Physiological Ecology: How Animals Process Energy, Nutrients, and Toxins*. Princeton University Press.

Kardol, P., and Wardle, D. A. (2010). How understanding aboveground–belowground linkages can assist restoration ecology. *Trends in Ecology and Evolution*, 25(11), 670–79.

Karpanty, S. (2006). Direct and indirect impacts of raptor predation on lemurs in Southeastern Madagascar. *International Journal of Primatology*, 27, 239–61.

Kaspari, M. (2020). The seventh macronutrient: How sodium shortfall ramifies through populations, food webs and ecosystems. *Ecology Letters*, 23, 1153-1168.

Katz, K., and Naug, D. (2016). Dancers and followers in a honeybee colony differently prioritize individual and colony nutritional needs. *Animal Behaviour*, 119, 69–74.

Kau, A. L., Ahern, P. P., Griffin, N. W., Goodman, A. L., and Gordon, J. I. (2011). Human nutrition, the gut microbiome and the immune system. *Nature*, 474, 327–36.

Kaumanns, W., Rohrhuber, B., and Zinner, D. P. (1989). Reproductive parameters in a newly established colony of hamadryas baboons (*Papio hamadryas*). *Primate Report*, 24, 25–33.

Kaur, T., Singh, J., Tong, S., Humphrey, C., Clevenger, D., and Tan, W. (2009). Descriptive epidemiology of fatal respiratory outbreaks and detection of a human-related metapneumovirus in wild chimpanzees (*Pan troglodytes*) at Mahale Mountains National Park, Western Tanzania. *American Journal of Primatology*, 71(4), 364–64.

Kawamura, S., Hiramatsu, C., Melin, A. D., Schaffner, C. M., Aureli, F., and Fedigan, L. M. (2012). Polymorphic color vision in primates: Evolutionary considerations. In H. Hirai, H. Imai, and Y. Go (Eds.), *Post-genome Biology of Primates* (pp. 93–120). Springer.

Kawamura, S., and Kubotera, N. (2004). Ancestral loss of short wave-sensitive cone visual pigment in lorisiform prosimians, contrasting with its strict conservation in other prosimians. *Journal of Molecular Evolution*, 58, 314–21.

Kawamura, S., and Melin, A. D. (2017). Evolution of genes for color vision and the chemical senses in primates. In N. Saitou (Ed.), *Evolution of the Human Genome I: The Genomes and Genes* (pp. 1–60). Springer.

Kay, R. F. (1975a). Allometry and early hominids (comment). *Science*, 189(4196), 63–63.

Kay, R. F. (1975b). The functional adaptations of primate molar teeth. *American Journal of Physical Anthropology*, 43, 195–216.

Kay, R. F., and Sheine, W. S. (1979). On the relationship between chitin particle size and digestibility in the primate *Galago senegalensis*. *American Journal of Physical Anthropology*, 50:301–308.

Kay, R. F. (1981). The nut-crackers: A new theory of the adaptations of the Ramapithecinae. *American Journal of Physical Anthropology*, 55(2), 141–51.

Kay, R. F. (1984). On the use of anatomical features to infer foraging behavior in extinct primates. In P. S. Rodman and J. G. H. Cant (Eds.), *Adaptations for Foraging Behavior in Nonhuman Primates: Contributions to an Organismal Biology of Prosimians, Monkeys, and Apes* (pp. 21–53). Columbia University Press.

Kay, R. F. (1985). Dental evidence for the diet of *Australopithecus*. *Annual Review of Anthropology*, 14, 315–41.

Kay, R. F., and Cartmill, M. (1977). Cranial morphology and adaptations of *Palaechthon nacimienti* and other Paromomyidae (Plesiadapoidea, Primates), with a description of a new genus and species. *Journal of Human Evolution*, 6(1), 19.

Kay, R. F., and Davies, A. G. (1994). Digestive physiology. In Davies, A. G., and Oates, J. F. *Colobine monkeys: Their ecology, behaviour and evolution* (pp. 229–49) Cambridge University Press.

Kay, R. F., and Hiiemae, K. M. (1974). Jaw movement and tooth use in recent and fossil primates. *American Journal of Physical Anthropology*, 40(2), 227–56.

Kay, R. F., Madden, R. H., van Schaik, C. P., and Higdon, D. (1997). Primate species richness is determined by plant productivity: Implications for conservation. *Proceedings of the National Academy of Sciences of the United States of America*, 94, 13023–27.

Kay, R. F., Sussman, R. W., and Tattersall, I. (1978). Dietary and dental variations in genus *Lemur*, with comments concerning dietary-dental correlations among Malagasy primates. *American Journal of Physical Anthropology*, 49(1), 119–27.

Kays, R., Crofoot, M. C., Jetz, W., and Wikelski, M. (2015). Terrestrial animal tracking as an eye on life and planet. *Science*, 348, aaa2478.

Kays, R., Tilak, S., Crofoot, M., Fountain, T., Obando, D., Ortega, A., et al. (2011). Tracking animal location and activity with an automated radio telemetry system in a tropical rainforest. *Computer Journal*, 54, 1931–48.

Kearney, M., Simpson, S. J., Raubenheimer, D., and Helmuth, B. (2010). Modelling the ecological niche from functional traits. *Philosophical Transactions of the Royal Society of London Series B—Biological Sciences*, 365, 3469–83.

Keenan, R. A., Moldawer, L. L., Yang, R. D., Kawamura, I., Blackburn, G. L., and Bistrian, B. R. (1982). An altered response by peripheral leukocytes to synthesize or release leukocyte endogenous mediator in critically ill, protein-malnourished patients. *Journal of Laboratory and Clinical Medicine*, 100(6), 844–57.

Keith, A., (1891). Anatomical notes on Malay apes. *Journal of the Straits Branch of the Royal Asiatic Society*, 23, 77-93.

Kelber, A., and Roth, L. S. V. (2006). Nocturnal colour vision—not as rare as we might think. *Journal of Experimental Biology*, 209, 781–88.

Kelley, D. S., and Daudu, P. A. (1993). Fat intake and immune response. *Progress in Food and Nutrition Science*, 17(1), 41–63.

Kelley, D. S., Dougherty, R. M., Branch, L. B., Taylor, P. C., and Iacono, J. M. (1992). Concentration of dietary N-6 polyunsaturated fatty acids and the human immune status. *Clinical Immunology and Immunopathology*, 62(2), 240–44.

Kelley, D. S., Hubbard, N. E., and Erickson, K. L. (2005). Regulation of human immune and inflammatory responses by dietary fatty acids. *Advances in Food and Nutrition Research*, 50, 101–38.

Kelley, J. (1990). Incisor microwear and diet in three species of *Colobus*. *Folia Primatologica*, 55, 73–84.

Kelley, J. (1993). Taxonomic implications of sexual dimorphism in *Lufengpithecus*. In W. H. Kimbel and L. B. Martin (Eds.), *Species, Species Concepts, and Primate Evolution* (pp. 429–58). Plenum.

Kelley, J., and Schwartz, G. T. (2012). Life-history inference in the early hominins *Australopithecus* and *Paranthropus*. *International Journal of Primatology*, 33, 1332–63.

Kelly, R. L. (2007). *The Foraging Spectrum: Diversity in Hunter-Gatherer Lifeways*. Percheron.

Kemnitz, J., Eisele, S., Lindsay, K., Engle, M., Perelman, R., and Farrell, P. (1984). Changes in food intake during menstrual cycles and pregnancy of normal and diabetic rhesus monkeys. *Diabetologia*, 26, 60–64.

Kemp, C., and Kaplan, G. (2012). Olfactory cues modify and enhance responses to visual cues in the common marmoset (*Callithrix jacchus*). *Journal of Primatology*, 1, 102–14.

Kempf, E. (2009). Patterns of water use in primates. *Folia Primatologica* 80, 275e294.

Kennedy, G. E. (2005). From the ape's dilemma to the weanling's dilemma: Early weaning and its evolutionary context. *Journal of Human Evolution*, 48, 123–45.

Kenward, B., Schloegl, C., Rutz, C., Weir, A. A. S., Bugnyar, T., and Kacelnik, A. (2011). On the evolutionary and ontogenetic origins of tool-oriented behaviour in New Caledonian crows (*Corvus moneduloides*). *Biological Journal of the Linnean Society*, 102, 870–77.

Kerhoas, D., Perwitasari-Farajallah, D., Agil, M., Widdig, A., and Engelhardt, A. (2014). Social and ecological factors influencing offspring survival in wild macaques. *Behavioral Ecology*, 25, 1164–72.

Kerr, B., and Feldman, W. C. (2003). Carving the cognitive niche: Optimal learning strategies in homogeneous and heterogeneous environments. *Journal of Theoretical Biology*, 220, 169–88.

Keusch, G. T. (2003). The history of nutrition: Malnutrition, infection and immunity. *Journal of Nutrition*, 133(1), 336S–340S.

Key, C., and Ross, C. (1999). Sex differences in energy expenditure in non-human primates. *Proceedings of the National Academy of Sciences of the United States of America*, 266, 2479–85.

Khan, A. A., and Vincent, J. F. V. (1993). Anisotropy in the fracture properties of apple flesh as investigated by crack-opening tests. *Journal of Material Science*, 28, 45–51.

Kielan-Jaworowska, Z., Cifelli, R. L., and Luo, Z. X. (2005). *Mammals from the Age of Dinosaurs: Origins, Evolution, and Structure*. Columbia University Press.

Kiffner, C., Kioko, J., Leweri, C., and Krause, S. (2014). Seasonal patterns of mixed species groups in large East African mammals. *PLoS ONE*, 9, e113446.

Kifle, Z., and Bekele, A. (2021). Feeding ecology and diet of the southern geladas (*Theropithecus gelada obscurus*) in human-modified landscape, Wollo, Ethiopia. *Ecology and Evolution*, 11(16), 11373–86.

Kim, D. H., Sexton, J. O., and Townshend, J. R. G. (2015). Accelerated deforestation in the humid tropics from the 1990s to the 2000s. *Geophysical Research Letters*, 42, 3495–501.

Kim, E. H.-J., Corrigan, V. K., Wilson, A. J., Hedderley, D. I., and Morgenstern, M. P. (2012). Fundamental fracture properties associated with sensory hardness of brittle solid foods. *Journal of Texture Studies*, 43, 49–62.

Kim, H. K., Choi, Y. H., and Verpoorte, R. (2010). NMR-based metabolomic analysis of plants. *Nature Protocols*, 5, 536–49.

Kim, M., Morrison, M., and Yu, K. (2011). Status of the phylogenetic diversity census of ruminal microbiomes. *FEMS Microbial Ecology*, 76(1), 49–63.

Kim, S., Lappan, S., and Choe, J. C. (2011). Diet and ranging behavior of the endangered Javan gibbon (*Hylobates moloch*) in a submontane tropical rainforest. *American Journal of Primatology*, 73, 270–80.

Kim, S., Lappan, S., and Choe, J. C. (2012). Responses of Javan gibbon (*Hylobates moloch*) groups in submontane forest to monthly variation in food availability: Evidence for variation on a fine spatial scale. *American Journal of Primatology*, 74, 1154–67.

Kimura, T., and Hamada, Y. (1995). Retardation of bone development in the Koshima troop of Japanese macaques. *Primates*, 36, 91–100.

Kimura, T., and Hamada, Y. (1996). Growth of wild and laboratory born chimpanzees. *Primates*, 37, 237–51.

Kinani, J. F., and Zimmerman, D. (2015). Tool use for food acquisition in a wild mountain gorilla (*Gorilla beringei beringei*). *American Journal of Primatology*, 77, 353–57.

King, A. J., and Cowlishaw, G. (2009). Leaders, followers and group decision-making. *Communicative and Integrative Biology*, 2, 147–50.

King, A. J., Douglas, C. M. S., Huchard, E., Isaac, N. J. B., and Cowlishaw, G. (2008). Dominance and affiliation mediate despotism in a social primate. *Current Biology*, 18, 1833–38.

King, A. J., Isaac, N. J., and Cowlishaw, G. (2009). Ecological, social and reproductive factors shape producer-scrounger dynamics in baboons. *Behavioral Ecology*, 20, 1039–49.

King, A. J., Sueur, C., Huchard, E., and Cowlishaw, G. (2011). A rule-of-thumb based on social affiliation explains collective movements in desert baboons. *Animal Behaviour*, 82, 1337–45.

King, S. J., Arrigo-Nelson, S. J., Pochron, S. T., Semprebon, G. M., Godfrey, L. R., Wright, P. C., et al. (2005). Dental senescence in a long-lived primate links infant survival to rainfall. *Proceedings of the National Academy of Sciences of the United States of America*, 102, 16579–83.

Kinnaird, M. F., and O'Brien, T. G. (2005). Fast foods of the forest: The influence of figs on primates and hornbills across Wallace's line. In J. L. Dew and J. P. Boubli (Eds.), *Tropical Fruits and Frugivores: The Search for Strong Interactors*. Springer.

Kinzey, W. G. (1978). Feeding behavior and molar features in two species of titi monkey. In D. J. Chivers and J. Herbert (Eds.), *Recent Advances in Primatology, Volume 1, Behavior* (pp. 373–85). Academic Press.

Kinzey, W. G. (1987). *Evolution of Human Behavior: Primate Models*. State University of New York Press.

Kinzey, W. G. (1992). Dietary and dental adaptations in the Pitheciinae. *American Journal of Physical Anthropology*, 88, 499–514.

Kinzey, W. G., and Norconk, M. A. (1990). Hardness as a basis of fruit choice in two sympatric primates. *American Journal of Physical Anthropology*, 81, 5–15.

Kinzey, W. G., and Norconk, M. A. (1993). Physical and chemical-properties of fruit and seeds eaten by *Pithecia* and *Chiropotes* in Surinam and Venezuela. *International Journal of Primatology*, 14, 207–27.

Kinzey, W. G., Rosenberger, A. L., Heisler, P. S., Prowse, D. L., and Trilling, J. S. (1977). A preliminary field investigation of the yellow handed titi monkey, *Callicebus torquatus torquatus*, in Northern Peru. *Primates*, 18, 159–81.

Kirchhof, J., and Hammerschmidt, K. (2006). Functionally referential alarm calls in tamarins (*Saguinus fuscicollis* and *Saguinus mystax*)—evidence from playback experiments. *Ethology*, 112, 346–54.

Kirk, E. C., and Simons, E. L. (2001). Diets of fossil primates from the Fayum Depression of Egypt: A quantitative analysis of molar shearing. *Journal of Human Evolution*, 40(3), 203–29.

Kirkpatrick, R. C., and Grueter, C. C. (2010). Snub-nosed monkeys: Multilevel societies across varied environments. *Evolutionary Anthropology*, 19, 98–113.

Kirkpatrick, R. C., Long, Y. C., Zhong, T., and Xiao, L. (1998). Social organization and range use in the Yunnan snub-nosed monkey *Rhinopithecus bieti*. *International Journal of Primatology*, 19, 13–51.

Kirkwood, J., and Underwood, S. (1984). Energy requirements of captive cotton-top tamarins (*Saguinus oedipus oedipus*). *Folia Primatologica*, 42, 180–87.

Kirmani, S. N., Banks, P. B., and McArthur, C. (2010). Integrating the costs of plant toxins and predation risk in foraging decisions of a mammalian herbivore. *Oecologia*, 164, 349–56.

Kleiber, M. (1932). Body size and metabolism. *Hilgardia*, 6, 315–49.

Kleiber, M. (1947). Body size and metabolic rate. *Physiological Review*, 27, 511–41.

Klein, H., Bocksberger, G., Baas, P., Bunel, S., Théleste, E., Pika, S., et al. (2021). Hunting of mammals by central chimpanzees (*Pan troglodytes troglodytes*) in the Loango National Park, Gabon. *Primates*, 62, 267–78.

Klein, L. D., Breakey, A. A., Scelza, B., Valeggia, C., Jasienska, G., and Hinde, K. (2017). Concentrations of trace elements in human milk: Comparisons among women in Argentina, Namibia, Poland, and the United States. *PLoS ONE*, 12, e0183367.

Klein, L. D., Huang, J., Quinn, E. A., Martin, M. A., Breakey, A. A., Gurven, M., et al. (2018). Variation among populations in the immune protein composition of mother's milk reflects subsistence pattern. *Evolution, Medicine, and Public Health*, 230–45.

Kluge, M., Razanoelisoa, B., and Brulfert, J. (2001). Implications of genotypic diversity and phenotypic plasticity in the ecophysiological success of CAM plants, examined by studies on the vegetation of Madagascar. *Plant Biology*, 3, 214–22.

Knapka, J. (2003). Nutrient Requirements of Nonhuman Primates: Second Revised Edition. *Lab Animal*, 32, 26.

Knight, R. S., and Siegfried, W. R. (1983). Inter-relationships between type, size and colour of fruits and dispersal in southern African trees. *Oecologia*, 56, 405–12.

Knott, C. D. (1997). Field collection and preservation urine in orangutans and chimpanzees. *Tropical Biodiversity*, 4(1), 95–102.

Knott, C. D. (1998). Changes in orangutan caloric intake, energy balance, and ketones in response to fluctuating fruit availability. *International Journal of Primatology*, 19, 1061–79.

Knott, C. D. (1999a). Orangutan behavior and ecology. In P. Dolhinow and A. Fuentes (Eds.), *The Nonhuman Primates* (pp. 50–57). Mountain View.

Knott, C. D. (1999b). *Reproductive, Physiological and Behavioral Responses of Orangutans in Borneo to Fluctuations in Food Availability*. Harvard University.

Knott, C. D. (2001). Female reproductive ecology of the apes: Implications for human evolution. In P. Ellison (Ed.), *Reproductive Ecology and Human Evolution* (pp. 429–63). Aldine de Gruyter.

Knott, C. D. (2005). Energetic responses to food availability in the great apes: Implications for hominin evolution. In D. K. Brockman and C. P. van Schaik (Eds.), *Primate Seasonality: Implications for Human Evolution* (pp. 351–78). Cambridge University Press.

Knott, C. D., and Harwell, F. S. (2020). Ecological risk aversion and the evolution of great ape life histories. In L. Hopper and S. Ross (Eds.), *Chimpanzees in Context* (pp. 1–35). University of Chicago Press.

Knott, C. D., Thompson, M. E., and Wich, S. A. (2009). The ecology of reproduction in wild orangutans. In S. A. Wich, S. S. Utami, T. Mitra Setia, and C. P. van Schaik (Eds.), *Orangutans: Geographic Variation in Behavioral Ecology and Conservation* (pp. 171–88). Oxford University Press.

Kobayashi, Y., Koike, S., Miyaji, M., Hata, H., and Tanaka, K. (2006). Hindgut microbes, fermentation and their seasonal variations in Hokkaido native horses compared to light horses. *Ecological Research*, 21, 285–91.

Kobbe, S., and Dausmann, K. H. (2009). Hibernation in Malagasy mouse lemurs as a strategy to counter environmental challenges. *Naturwissenschaften*, 96, 1221–27.

Koch, F., Ganzhorn, J. U., Rothman, J. M., Chapman, C. A., and Fichtel, C. (2017). Sex and seasonal differences in diet and nutrient intake in Verreaux's sifakas (*Propithecus verreauxi*). *American Journal of Primatology*, 79, 1–10.

Koenig, A. (2000). Competitive regimes in forest-dwelling Hanuman langur females (*Semnopithecus entellus*). *Behavioral Ecology and Sociobiology*, 48, 93–109.

Koenig, A., and Borries, C. (2001). Socioecology of Hanuman langurs: The story of their success. *Evolutionary Anthropology*, 10, 122–37.

Koenig, A. (2002). Competition for resources and its behavioral consequences among female primates. *International Journal of Primatology*, 23, 759–783.

Koenig, A., and Borries, C. (2006). The predictive power of socioecological models: A reconsideration of resource characteristics, agonism, and dominant hierarchies. In G. Hohmann, M. M. Robbins, and C. Boesch (Eds.), *Feeding Ecology in Apes and Other Primates* (pp. 263–84). Cambridge University Press.

Koenig, A., and Borries, C. (2009). The lost dream of ecological determinism: Time to say goodbye? Or a white queen's proposal? *Evolutionary Anthropology*, 185, 166–74.

Koenig, A., Beise, J., Chalise, M. K., and Ganzhorn, J. U. (1998). When females should contest for food—testing the hypotheses about resource density, distribution, size, and quality with Hanuman langurs (*Presbytis entellus*). *Behavioral Ecology and Sociobiology*, 42, 225–37.

Koenig, A., Borries, C., Chalise, M. K., and Winkler, P. (1997). Ecology, nutrition, and timing of reproductive events in an Asian primate, the Hanuman langur (*Presbytis entellus*). *Journal of Zoology*, 243, 215–35.

Koenig, A., Scarry, C. J., Wheeler, B. C., and Borries, C. (2013). Variation in grouping patterns, mating systems and social structure: What socio-ecological models attempt to explain. *Philosophical Transactions of the Royal Society of London Series B—Biological Sciences*, 368, 20120348.

Koenig, J. E., Spor, A., Scalfone, N., Fricker, A. D., Stombaugh, J., Knight, R., et al. (2011). Succession of microbial consortia in the developing infant gut microbiome. *Proceedings of the National Academy of Sciences of the United States of America*, 108(suppl. 1), 4578–85.

Koford, C. B. (1965). Population dynamics of rhesus monkeys on Cayo Santiago. In I. DeVore (Ed.), *Primate Behavior* (pp. 160–74). Holt, Reinhart and Winston.

Kohl, K. D., and Dearing, D. M. (2012). Experience matters: Prior exposure to plant toxins enhances diversity of gut microbes in herbivores. *Ecology Letters*, 15, 1008–15.

Kohl, K. D., and Dearing, D. M. (2016). The woodrat gut microbiota as an experimental system for understanding microbial metabolism of dietary toxins. *Frontiers in Microbiology*, 7, 1165.

Kohl, K. D., Weiss, R. B., Cox, J., Dale, C., and Dearing, D. M. (2014). Gut microbes of mammalian herbivores facilitate intake of plant toxins. *Ecology Letters*, 17, 1238–46.

Kohn, M. J. (1996). Predicting animal $\delta^{18}O$: Accounting for diet and physiological adaptation. *Geochimica et Cosmochimica Acta*, 60, 4811–29.

Kohn, M. J., and Cerling, T. E. (2002). Stable isotope compositions of biological apatite. *Phosphates: Geochemical, Geobiological, and Materials Importance*, 48, 455–88.

Kohn, M. J., Schoeninger, M. J., and Valley, J. W. (1996). Herbivore tooth oxygen isotope compositions: Effects of diet and physiology. *Geochimica et Cosmochimica Acta*, 60, 3889–96.

Kohrs, M., Harper, A., and Kerr, G. (1976). Effects of a low-protein diet during pregnancy of the rhesus monkey. I. Reproductive efficiency. *American Journal of Clinical Nutrition*, 29, 136–45.

Koj, A. (1970). Synthesis and turnover of acute-phase reactants. Paper presented at the Ciba Foundation Symposium—Energy Metabolism in Trauma.

Kokaly, R. F., Asner, G. P., Ollinger, S. V., Martin, M. E., and Wessman, C. A. (2009). Characterizing canopy biochemistry from imaging spectroscopy and its application to ecosystem studies. *Remote Sensing of Environment*, 113, S78–S91.

Koletzko, B., Thiel, I., and Abiodun, P. O. (1992). The fatty acid composition of human milk in Europe and Africa. *Journal of Pediatrics*, 120, S62–S70.

Konar, N. (2013). Non-isoflavone phytoestrogenic compound

contents of various legumes. *European Food Research and Technology*, 236, 523–30.

König, B., and Markl, H. (1987). Maternal care in house mice. *Behavioral Ecology and Sociobiology*, 20, 1–9.

Kool, K. M. (1992). Food selection by the silver leaf monkey, *Trachypithecus auratus sondaicus*, in relation to plant chemistry. *Oecologia*, 90, 527–33.

Kool, K. M. (1993). The diet and feeding-behavior of the silver leaf monkey (*Trachypithecus auratus sondaicus*) in Indonesia. *International Journal of Primatology*, 14, 667–700.

Koops, K., Furuichi, T., and Hashimoto, C. (2015). Chimpanzees and bonobos differ in intrinsic motivation for tool use. *Scientific Reports*, 5, 11356.

Koops, K., McGrew, W. C., and Matsuzawa, T. (2010). Do chimpanzees (*Pan troglodytes*) use cleavers and anvils to fracture *Treculia africana* fruits? Preliminary data on a new form of percussive technology. *Primates*, 51, 175–78.

Koops, K., McGrew, W. C., and Matsuzawa, T. (2013). Ecology of culture: Do environmental factors influence foraging tool use in wild chimpanzees, *Pan troglodytes verus*? *Animal Behaviour*, 85, 175–85.

Koops, K., Wrangham, R. W., Cumberlidge, N., Fitzgerald, M. A., van Leeuwen, K. A., Rothman, J. M., et al. (2019). Crab-fishing by chimpanzees in the Nimba Mountains, Guinea. *Journal of Human Evolution*, 133, 230–41.

Koops, K., Schöning, C., McGrew, W. C., and Matsuzawa, T. (2015). Chimpanzees prey on army ants at Seringbara, Nimba Mountains, Guinea: Predation patterns and tool use characteristics. *American Journal of Primatology*, 77, 319–29.

Koops, K., Visalberghi, E., and van Schaik, C. P. (2014). The ecology of primate material culture. *Biology Letters*, 10, 20140508.

Kooros, S. J., Goossens, B., Sterck, E. H. M., Kenderdine, R., Malim, P. T., Ramirez Saldivar, D. A., et al. (2022) External environmental conditions impact nocturnal activity levels in proboscis monkeys (*Nasalis larvatus*) living in Sabah, Malaysia. *American Journal of Primatology*, 84, e23423.

Korb, J., and Linsenmair, K. E. (2002). Evaluation of predation risk in the collectively foraging termite *Macrotermes bellicosus*. *Insectes Sociaux*, 49, 264–69.

Koricheva, J., and Barton, K. E. (2012). Temporal changes in plant secondary metabolite production: Patterns, causes and consequences. In G. R. Iason, M. Dicke, and S. E. Hartley (Eds.), *The Ecology of Plant Secondary Metabolites* (pp. 34–55). Cambridge University Press.

Korstjens, A. H., and Hillyer, A. P. (2016). Primates and climate change: A review of current knowledge. In S. A. Wich and A. J. Marshall (Eds.), *An introduction to primate conservation* (pp. 175–92). Oxford University Press.

Korstjens, A. M., Lehmann, J., and Dunbar, R. I. M. (2010). Resting time as an ecological constraint on primate biogeography. *Animal Behavior*, 79, 361–74.

Kortlandt, A., and Holzhaus, E. (1987). New data on the use of stone tools by chimpanzees in Guinea and Liberia. *Primates*, 28, 473–96.

Koskela, E., Mappes, T., Niskanen, T., and Rutkowska, J. (2009). Maternal investment in relation to sex ratio and offspring number in a small mammal—a case for Trivers and Willard theory? *Journal of Animal Ecology*, 78, 1007–14.

Kotler, B. P. (1985). Owl predation on desert rodents which differ in morphology and behaviour. *Journal of Mammalogy*, 66, 824–28.

Kotler, B. P., and Brown, J. S. (1999). Mechanisms of coexistence of optimal foragers as determinants of local abundances and distributions of desert granivores. *Journal of Mammalogy*, 80, 361–74.

Kotler, B. P., Brown, J. S., and Knight, M. H. (1999). Habitat and patch use by hyraxes: There's no place like home? *Ecology Letters*, 2, 82–88.

Kotler, B. P., Brown, J., Mukherjee, S., Berger-Tal, O., and Bouskila, A. (2010). Moonlight avoidance in gerbils reveals a sophisticated interplay among time allocation, vigilance and state-dependent foraging. *Proceedings of the Royal Society B: Biological Sciences*, 277, 1469–74.

Kotliar, N. B., and Wiens, J. A. (1990). Multiple scales of patchiness and patch structure: A hierarchical framework for the study of heterogeneity. *Oikos*, 59, 253–60.

Kowalewski, M., and Zunino, G. E. (2004). Birth seasonality in *Alouatta caraya* in northern Argentina. *International Journal of Primatology*, 25, 383–400.

Koyabu, D. B., and Endo, H. (2009). Craniofacial variation and dietary adaptations of African colobines. *Journal of Human Evolution*, 56, 525–36.

Koyabu, D. B., and Endo, H. (2010). Craniodental mechanics and diet in Asian colobines: Morphological evidence of mature seed predation and sclerocarpy. *American Journal of Physical Anthropology*, 142, 137–48.

Koyama, N., Takahata, Y., Huffman, M. A., Norikoshi, K., and Suzuki, H. (1992). Reproductive parameters of female Japanese macaques: Thirty years data from the Arashiyama troops, Japan. *Primates*, 33, 33–47.

Kozlowski, T. T., Kramer, P. J., and Pallardy, S. G. (1991). *The Physiological Ecology of Woody Plants*. Academic Press.

Kraft, T. S., Venkataraman, V. V., Wallace, I. J., Crittenden, A. N., Holowka, N. B., Stieglitz, J., et al. (2021). The energetics of uniquely human subsistence strategies. *Science*, 374(6575), eabf0130.

Krajmalnik-Brown, R., Lozupone, C., Kang, D. W., and Adams, J. B. (2015). Gut bacteria in children with autism spectrum disorders: Challenges and promise of studying how a complex community influences a complex disease. *Microbial Ecology*, 26, 26914.

Kramer, K. L. (2005). Children's help and the pace of reproduction: Cooperative breeding in humans. *Evolutionary Anthropology*, 14, 224–37.

Kramer, K. L. (2010). Cooperative breeding and its significance to the demographic success of humans. *Annual Review of Anthropology*, 39, 417–36.

Kraus, J. B., and Strier, K. B. (2022). Geographic, climatic, and phylogenetic drivers of variation in colobine activity budgets. *Primates*, 63, 647–58.

Krause, G. H., Cheesman, A. W., Winter, K., Krause, B., and Virgo, A. (2013). Thermal tolerance, net CO_2 exchange and growth of a tropical tree species, *Ficus insipida*, cultivated at elevated daytime and nighttime temperatures. *Journal of Plant Physiology*, 170, 822–27.

Krebs, C. J. (1992). Population regulation revisited. *Ecology*, 73, 714–15.

Krebs, C. J. (1995). Two paradigms of population regulation. *Wildlife Research*, 22, 1–10.

Krebs, C. J. (1999). *Ecological Methodology* (2nd ed.). Addison Wesley Longman.

Krebs, C. J. (2001). *Ecology*. Benjamin Cummings.

Krebs, C. J. (2002). Beyond population regulation and limitation. *Wildlife Research*, 29, 1–10.

Krebs, J. R. (1978). Optimal foraging: Decision rules for predators. In J. R. Krebs and N. B. Davies (Eds.), *Behavioral Ecology: An Evolutionary Approach* (pp. 23–63). Sinauer Associates.

Krebs, J. R., MacRoberts, M. H., and Cullen, J. M. (1972). Flocking and feeding in the great tit *Parus major*: An experimental study. *Ibis*, 114, 507–30.

Kricher, J. (2011). *Tropical Ecology*. Princeton University Press.

Krief, S., Cibot, M., Bortomiol, S., Seguya, A., Krief, J. M., and Masi, S. (2014). Wild chimpanzees on the edge: Nocturnal activities in croplands. *PLoS ONE*, 9, 109925.

Krief, S., Huffman, M. A., Sevenet, T., Hladik, C. M., Grellier, P., Loiseau, P. M., et al. (2006). Bioactive properties of plant species ingested by chimpanzees (*Pan troglodytes schweinfurthii*) in the Kibale National Park, Uganda. *American Journal of Primatology*, 68, 51–71.

Krief, S., Iglesias-González, A., Appenzeller, B. M., Rachid, L., Beltrame, M., Asalu, E., et al. (2022). Chimpanzee exposure to pollution revealed by human biomonitoring approaches. *Ecotoxicology and Environmental Safety*, 233, 113341.

Kriewalt, F. H., and Hendrickx, A. G. (1968). Reproductive parameters of the baboon. *Laboratory Animal Science*, 18, 361–70.

Krigbaum, J., Berger, M. H., Daegling, D. J., and McGraw, W. S. (2013). Stable isotope canopy effects for sympatric monkeys at Taï Forest, Côte d'Ivoire. *Biology Letters*, 9, 20130466.

Krishnamani, R., and Mahaney, W. C. (2000). Geophagy among primates: Adaptive significance and ecological consequences. *Animal Behaviour*, 59, 899–915.

Krishnan, A. V., Stathis, P., Permuth, S. F., Tokes, L., and Feldman, D. (1993). Bisphenol-A—an estrogenic substance is released from polycarbonate flasks during autoclaving. *Endocrinology*, 132, 2279–86.

Krofel, M., Skrbinsek, T., and Kos, I. (2013). Use of GPS location clusters analysis to study predation, feeding, and maternal behavior of the Eurasian lynx. *Ecological Research*, 28, 103–16.

Krohne, D. T. (2001). *General Ecology*. Brooks/Cole.

Kronmal, R. A. (1993). Spurious correlation and the fallacy of the ratio standard revisited. *Journal of the Royal Statistical Society A: Statistics in Society*, 156, 379–92.

Krul, E. S. (2019). Calculation of nitrogen-to-protein conversion factors: A review with a focus on soy protein. *Journal of the American Oil Chemists' Society*, 96, 339–64.

Kryiakou, K., Blackhurst, D. M., Parkington, J. E., and Marais, A. D. (2016). Marine and terrestrial foods as a source of brain-selective nutrients for early modern humans in the southwestern cape, South Africa. *Journal of Human Evolution*, 97, 86–96.

Krykbaev, R., Fitz, L. J., Padmalatha, S. R., Winkler, A., Xuan, D., Yang, X., et al. (2010). Evolutionary and biochemical differences between human and monkey acidic mammalian chitinases. *Gene*, 452(1), 63–71.

Kubena, K. S., and McMurray, D. N. (1996). Nutrition and the immune system: A review of nutrient-nutrient interactions. *Journal of the American Dietetic Association*, 96(11), 1156–64.

Kubo, M. O., and Yamada, E. (2014). The inter-relationship between dietary and environmental properties and tooth wear: Comparisons of mesowear, molar wear rate, and hypsodonty index of extant sika deer populations. *PLoS ONE*, 9(6), e90745.

Kudsk, K. A. (2006). Immunonutrition in surgery and critical care. *Annual Review of Nutrition*, 26, 463–79.

Kudsk, K. A., Wu, Y., Fukatsu, K., Zarzaur, B. L., Johnson, C. D., Wang, R., et al. (2000). Glutamine-enriched total parenteral nutrition maintains intestinal interleukin-4 and mucosal immunoglobulin A levels. *Journal of Parenteral and Enteral Nutrition*, 24(5), 270–75.

Kuehn, M., Welsch, H., Zahnert, T., and Hummel, T. (2008). Changes of pressure and humidity affect olfactory function. *European Archives of Oto-rhino-laryngology*, 265, 299–302.

Kuhlisch, C., and Pohnert, G. (2015). Metabolomics in chemical ecology. *Nature Product Reports*, 32, 937–55.

Kuijper, D. P. J., Bubnicki, J. W., Churski, M., Mols, B., and van Hooft, P. (2015). Context dependence of risk effects: Wolves and tree logs create patches of fear in an old-growth forest. *Behavioral Ecology*, 26, 1558–68.

Kuiper, G. G. J. M., Lemmen, J. G., Carlsson, B., Corton, J. C., Safe, S. H., van der Saag, P. T., et al. (1998). Interaction of estrogenic chemicals and phytoestrogens with estrogen receptor beta. *Endocrinology*, 139, 4252–63.

Kuipers, R. S., Luxwolda, M. F., Dijck-Brouwer, D. A. J., Eaton, S. B., Crawford, M. A., Cordain, L., et al. (2010). Estimated macronutrient and fatty acid intakes from an East African Paleolithic diet. *British Journal of Nutrition*, 104, 1666–87.

Kulahci, I. G., Dornhaus, A., and Papaj, D. R. (2008). Multimodal signals enhance decision making in foraging bumblebees. *Proceedings of the Royal Society B: Biological Sciences*, 275, 797–802.

Kumar, A. (1987). *The Ecology and Population Dynamics of the Lion-Tailed Macaque (*Macaca silenus*) in South India*. University of Cambridge.

Kummer, H. (1971). *Primate Societies: Group Techniques of Ecological Adaptation*. Aldine-Atherton.

Kummer, H., and Goodall, J. (1985). Conditions of innovative behaviour in primates. *Philosophical Transactions of the Royal Society of London Series B—Biological Sciences*, 308, 203–14.

Kunz, C., and Lönnerdal, B. (1993). Protein composition of rhesus monkey milk: Comparison to human milk. *Comparative Biochemistry and Physiology A—Molecular and Integrative Physiology*, 104, 793–97.

Kuo, A. H., Li, C., Huber, H. F., Schwab, M., Nathanielsz, P. W., and Clarke, G. D. (2017). Maternal nutrient restriction during pregnancy and lactation leads to impaired right ventricular function in young adult baboons. *Journal of Physiology*, 595(13), 4245–60.

Kupfer, A., Muller, H., Antoniazzi, M. M., Jared, C., Greven, H., Nussbaum, R. A., et al. (2006). Parental investment by skin feeding in a caecilian amphibian. *Nature*, 440, 926.

Kupsch, D., Waltert, M., and Heymann, E. W. (2014). Forest type affects prey foraging of saddleback tamarins, *Saguinus nigrifrons*. *Primates*, 44, 403–13.

Kuroda, S., Suzuki, S., and Nishihara, T. (1996). Preliminary report on predatory behavior and meat sharing in Tschego chimpanzees (*Pan troglodytes troglodytes*) in the Ndoki Forest, northern Congo. *Primates*, 37, 253–59.

Kuzawa, C. W., Chugani, H. T., Grossman, L. I., Lipovich, L., Muzik, O., Hof, P. R., et al. (2014). Metabolic costs and evolutionary implications of human brain development. *Proceedings of the National Academy of Sciences of the United States of America*, 111, 13010–15.

Kuze, N., Dellatore, D., Banes, G. L., Pratje, P., Tajima, T., and Russon, A. E. (2012). Factors affecting reproduction in rehabilitant female orangutans: Young age at first birth and short inter-birth interval. *Primates*, 53, 181–92.

Kwiatt, A. (2017). Food, feeding, and foraging: Using stable isotope analysis as a methodology in the study of urban primate dietary patterns. In Dore, K. M., Riley E. P., Fuentes, A. (Eds.) *Ethnoprimatology: A practical guide to research at the human–nonhuman primate interface* (pp. 56–69). Cambridge University Press.

Kyriacou, K., Blackhurst, D. M., Parkington, J. E., and Marais, A. D. (2016). Marine and terrestrial foods as a source of brain-selective nutrients for early modern humans in the southwestern Cape, South Africa. *Journal of Human Evolution*, 97, 86–96.

LaBare, K. M., Broyles, S. B., and Klotz, R. L. (2000). Exploring nectar biology to learn about pollinators. *American Biology Teacher*, 62, 292–96.

LaBarge, L. R., Allan, A. T. L., Berman, C. M., Hill, R. A., and Margulis, S. W. (2022). Cortisol metabolites vary with environmental conditions, predation risk, and human shields in a wild primate, *Cercopithecus albogularis*. *Hormones and Behavior*, 145, 105237.

LaBarge, L. R., Allan, A. T. L., Berman C. M., Margulis, S. W., and Hill, R. A. (2020). Reactive and pre-emptive spatial cohesion in a social primate. *Animal Behaviour*, 163, 115–26.

LaBarge, L. R., Hill, R. A., Berman C. M., Margulis, S. W., and Allan, A. T. L. (2020). Anthropogenic influences on primate antipredator behaviour and implications for research and conservation. *American Journal of Primatology*, 82, e23087.

Lackey, D. E., and Olefsky, J. M. (2016). Regulation of metabolism by the innate immune system. *Nature Reviews Endocrinology*, 12(1), 15–28.

Lacroux, C., Pouydebat, E., Rossignol, M., Durand, S., Aleeje, A., Asalu, E., et al. 2022. Repellent activity against *Anopheles gambiae* of the leaves of nesting trees in the Sebitoli chimpanzee community of Kibale National Park, Uganda. *Malaria Journal*, 21, 1–11.

Lahann, P. (2007). Feeding ecology and seed dispersal of sympatric cheirogaleid lemurs (*Microcebus murinus*, *Cheirogaleus medius*, *Cheirogaleus major*) in the littoral rainforest of southeast Madagascar. *Journal of Zoology*, 271, 88–98.

Laird, M. F., Punjani, Z., Oshay, R. R., Wright, B. W., Fogaça, M. D., Casteren, A., et al. (2022). Feeding postural behaviors and food geometric and material properties in bearded capuchin monkeys (*Sapajus libidinosus*). *American Journal of Biological Anthropology*, 178, 3–16.

Laird, M. F., Wright, B. W., Rivera, A. O., Fogaça, M. D., van Casteren, A., Fragaszy, D. M., et al. (2020). Ingestive behaviors in bearded capuchins (*Sapajus libidinosus*). *Scientific Reports*, 10, 1–15.

Laland, K. N., and Janik, V. M. (2006). The animal cultures debate. *Trends in Ecology and Evolution*, 21, 542–47.

Laland, K. N., Kendal, J. R., and Kendal, R. L. (2009). Animal culture: Problems and solutions. In K. N. Laland and B. G. Galef (Eds.), *The Question of Animal Culture* (pp. 174–97). Harvard University Press.

Lambers, H., Chapin, F. S., and Pons, T. L. (1998). *Plant Physiological Ecology*. Springer.

Lambert, F. R., and Marshall, A. G. (1991). Keystone characteristics of bird-dispersed *Ficus* in a Malaysian lowland rainforest. *Journal of Ecology*, 79, 793–809.

Lambert, J. E. (1998). Primate digestion: Interactions among anatomy, physiology, and feeding ecology. *Evolutionary Anthropology*, 7, 8–20.

Lambert, J. E. (1999). Seed handling in chimpanzees (*Pan troglodytes*) and redtail monkeys (*Cercopithecus ascanius*): Implications for understanding hominoid and cercopithecine fruit processing strategies and seed dispersal. *American Journal of Physical Anthropology*, 109, 365–86.

Lambert, J. E. (2005). Competition, predation and the evolution of the cercopithecine cheek pouch: The case of *Cercopithecus* and *Lophocebus*. *American Journal of Physical Anthropology*, 126, 183–92.

Lambert, J. E. (2007). Primate nutritional ecology: Feeding biology and diet at ecological and evolutionary scales. In C. J. Campbell, A. Fuentes, K. C. MacKinnon, M. Panger, and S. K. Bearder (Eds.), *Primates in Perspective* (pp. 482–95). Oxford University Press.

Lambert, J. E. (2007). Seasonality, fallback strategies, and natural selection: A chimpanzee and cercopithecoid model for interpreting the evolution of hominin diet. In P. S. Ungar (Ed.), *Evolution of the Human Diet: The Known, the Unknown, and the Unknowable* (pp. 324–43). Oxford University Press.

Lambert, J. E. (2009). Primate fallback strategies as adaptive phenotypic plasticity: Scale, process, and pattern. *American Journal of Physical Anthropology*, 140, 759–66.

Lambert, J. E. (2011). Primate nutritional ecology: Feeding biology and diet at ecological and evolutionary scales. In C. J. Campbell, A. Fuentes, and K. C. MacKinnon (Eds.), *Primates in Perspective* (pp. 512–22). Oxford University Press.

Lambert, J. E. (2014). Evolutionary biology of ape and monkey feeding and nutrition. In W. Henke and I. Tattersall (Eds.), *Handbook of Paleoanthropology* (pp. 1–27). Springer Berlin Heidelberg.

Lambert, J. E., and Fellner, V. (2012). In vitro fermentation of dietary carbohydrates consumed by African apes and monkeys: Preliminary results for interpreting microbial and digestive strategy. *International Journal of Primatology*, 33, 263–81.

Lambert, J. E., and Garber, P. A. (1998). Evolutionary and ecological implications of primate seed dispersal. *American Journal of Primatology*, 45, 9–28.

Lambert, J. E., and Rothman, J. M. (2015). Fallback foods, optimal diets, and nutritional targets: Primate responses to varying food availability and quality. *Annual Review of Anthropology*, 44, 493–512.

Lambert, J. E., Chapman, C. A., Wrangham, R. W., and Conklin-Brittain, N. L. (2004). Hardness of cercopithecine foods: Implications for the critical function of enamel thickness in exploiting fallback foods. *American Journal of Physical Anthropology*, 125, 363–68.

Lancaster, J. B., and Lancaster, C. S. (1983). Parental investment: The hominid adaptation. In D. Ortner (Ed.), *How Humans Adapt* (pp. 33–58). Smithsonian Press.

Landau, S., Glasser, T., and Dvash, L. (2006). Monitoring nutrition in small ruminants with the aid of near infrared reflectance spectroscopy (NIRS) technology: A review. *Small Ruminant Research*, 61, 1–11.

Lande, R. (1982). A quantitative genetic theory of life history evolution. *Ecology*, 63, 607–15.

Landeck, G., and Garriga, J. G. (2016). The oldest hominin butchery in European mid-latitudes at the Jaramillo site of Untermassfeld (Thuringia, Germany). *Journal of Human Evolution*, 94, 53–71.

Landete, J. M. (2012). Plant and mammalian lignans: A review of source, intake, metabolism, intestinal bacteria and health. *Food Research International*, 46, 410–24.

Landete-Castillejos, T., García, A., López-Serrano, F. R., and Gallego, L. (2005). Maternal quality and differences in milk production and composition for male and female Iberian red deer calves (*Cervus elaphus hispanicus*). *Behavioral Ecology and Sociobiology*, 57, 267–74.

Lane, A. A., McGuire, M. K., McGuire, M. A., Williams, J. E., Lackey, K. A., Hagen, E. H., et al. (2019). Household composition and the infant fecal microbiome: The INSPIRE study. *American Journal of Biological Anthropology*, 169, 526–39.

Langenheim, J. H. (1994). Higher-plant terpenoids—a phytocentric overview of their ecological roles. *Journal of Chemical Ecology*, 20, 1223–80.

Langenheim, J. H. (2003). *Plant Resins—Chemistry, Evolution, Ecology and Ethnobotany*. Timber.

Langer, P. (2003). Lactation, weaning period, food quality, and digestive tract differentiations in Eutheria. *Evolution*, 57, 1196–215.

Langer, P. (2008). The phases of maternal investment in eutherian mammals. *Zoology*, 111, 148–62.

Langlois, M. R., Delanghe, J. R., Steyaert, S. R., Everaert, K. C., and De Buyzere, M. L. (1999). Automated flow cytometry compared with an automated dipstick reader for urinalysis. *Clinical Chemistry*, 45(1), 118–22.

Lanjouw, A. (2002). Behavioural adaptations to water scarcity in Tongo chimpanzees. In C. Boesch, G. Hohmann, and L. Marchant (Eds.), *Behavioural Diversity in Chimpanzees and Bonobos* (pp. 52–60). Cambridge University Press.

Lapcik, O., Vitkova, M., Klejdus, B., Al-Maharik, N., and Adlercreutz, H. (2004). Immunoassay for biochanin A. *Journal of Immunological Methods*, 294, 155–63.

Lappan, S. (2008). Male care of infants in a siamang (*Symphalangus syndactylus*) population including socially monogamous and polyandrous groups. *Behavioral Ecology and Sociobiology*, 62, 1307–17.

Lappan, S. (2009). The effects of lactation and infant care on adult energy budgets in wild siamangs (*Symphalangus syndactylus*). *American Journal of Physical Anthropology*, 140, 290–301.

Larsen, N., Vogensen, F. K., van den Berg, F. W. J., Nielsen, D. S., Andreasen, A. S., Pedersen, B. K., et al. (2010). Gut microbiota in human adults with type 2 diabetes differs from non-diabetic adults. *PLoS ONE*, 5(2), e9085.

Laska, M., and Hudson, R. (1993). Assessing olfactory performance in a New World primate, *Saimiri sciureus*. *Physiology and Behavior*, 53, 89–95.

Laska, M., and Seibt, A. (2002). Olfactory sensitivity for aliphatic alcohols in squirrel monkeys and pigtail macaques. *Journal of Experimental Biology*, 205, 1633–43.

Laska, M., Freist, P., and Krause, S. (2007). Which senses play a role in nonhuman primate food selection? A comparison between squirrel monkeys and spider monkeys. *American Journal of Primatology*, 69, 282–94.

Laska, M., Salazar, L. T. H., and Luna, E. R. (2000). Food preferences and nutrient composition in captive spider monkeys, *Ateles geoffroyi*. *International Journal of Primatology*, 21, 671–83.

Laska, M., Seibt, A., and Weber, A. (1999). "Microsmatic" primates revisited: Olfactory sensitivity in the squirrel monkey. *Chemical Senses*, 25, 47–53.

Latifi, R. (2011). Nutritional therapy in critically ill and injured patients. *Surgical Clinics of North America*, 91(3), 579–93.

Latombe, G., Fortin, D., and Parrott, L. (2014). Spatio-temporal dynamics in the response of woodland caribou and moose to the passage of grey wolf. *Journal of Animal Ecology*, 83, 185–98.

Lau, J. A., and Lennon, J. T. (2011). Evolutionary ecology of plant-microbe interactions: Soil microbial structure alters selection on plant traits. *New Phytologist*, 192, 215–24.

Launchbaugh, K., Provenza, F. D., and Burritt, E. A. (1993). How herbivores track variable environments: Response to variability of phytotoxins. *Journal of Chemical Ecology*, 19, 1047–56.

Laundré, J. W., Hernandez, L., and Altendorf, K. B. (2001). Wolves, elk, and bison: Reestablishing the "landscape of fear" in Yellowstone National Park, USA. *Canadian Journal of Zoology*, 79, 1401–9.

Laundré, J. W., Hernández, L., and Ripple, W. J. (2010). The landscape of fear: Ecological implications of being afraid. *Open Ecology Journal*, 3, 1–7.

Laundré, J. W., Hernandez, L., Lopez Medina, P., Campanella, A., Lopez-Portillo, J., Gonzalez-Romero, A., et al. (2014). The landscape of fear: The missing link to understand top-down and bottom-up controls of prey abundance? *Ecology*, 95, 1141–52.

Laurance, W. F., Carolina Useche, D., Rendeiro, J., Kalka, M., Bradshaw, C. J. A., Sloan, S. P., et al. (2012). Averting biodiversity collapse in tropical forest protected areas. *Nature*, 489, 290–94.

Laurance, W. F., Clements, G. R., Sloan, S. P., O'Connell, C. S., Mueller, N. D., Goosem, M., et al. (2014). A global strategy for road building. *Nature*, 513, 229–32.

Laurance, W. F., Sayer, J., and Cassman, K. G. (2014). Agriculture expansion and its impacts on tropical nature. *Trends in Ecology and Evolution*, 29, 107–16.

Lawes, M. J. (1991). Diet of samango monkeys (*Cercopithecus mitis erythrarchus*) in the Cape Vidal dune forest, South Africa. *Journal of Zoology*, 224, 149–73.

Lawes, M. J., and Chapman, C. A. (2006). Does the herb *Acanthus pubescens* and / or elephants suppress tree regeneration in disturbed Afrotropical forests? *Forest Ecology and Management*, 221, 274–84.

Lawler, I. R., Eschler, M., Schliebs, D. M., and Foley, W. J. (1999). Relationship between chemical functional groups on *Eucalyptus* secondary metabolites and their effectiveness as marsupial antifeedants. *Journal of Chemical Ecology*, 25, 2561–73.

Lawler, I. R., Foley, W. J., Eschler, B. M., Pass, D. M., and Handasyde, K. (1998). Intraspecific variation in *Eucalyptus* secondary metabolites determines food intake by folivorous marsupials. *Oecologia*, 116, 160–69.

Lawler, I. R., Stapley, J., Foley, W. J., and Eschler, B. M. (1999). Ecological example of conditioned flavor aversion in plant-herbivore interactions: Effect of terpenes of *Eucalyptus* leaves on feeding by common ringtail and brushtail possums. *Journal of Chemical Ecology*, 25, 401–15.

Lawrence, J. M. (2003). Preliminary report on the natural history of brown titi monkeys (*Callicebus brunneus*) at the Los Amigos Research Station, Madre de Dios, Peru. *American Journal of Physical Anthropology*, suppl. 36, 136.

Lazzaro, B. P., and Little, T. J. (2009). Immunity in a variable world. *Philosophical Transactions of the Royal Society B—Biological Sciences*, 364(1513), 15–26.

Le, H. T., Hoang, D. M., and Covert, H. H. (2019). Diet of the Indochinese silvered langur (*Trachypithecus germaini*) in Kien Luong Karst area, Kien Giang Province. *American Journal of Primatology*, 81, e23041.

Le Couteur, D. G., Tay, S. S., Solon-Biet, S., Bertolino, P., McMahon, A. C., Cogger, V. C., et al. (2015). The influence of macronutrients on splanchnic and hepatic lymphocytes in aging mice. *Journals of Gerontology Series A—Biological Sciences and Medical Sciences*, 70(12), 1499–507.

Le Couteur, D., Solon-Biet, S., Cogger, V., Gokarn, R., McMahon, A., Raubenheimer, D., et al. (2016). Mechanisms linking low-protein, high-carbohydrate diets with age-related health. *Gerontologist*, 54, 497.

Le Couteur, D., Solon-Biet, S., McMahon, A., Raubenheimer, D., and Simpson, S. (2014). Dietary protein to carbohydrate ratio, not caloric intake, determines ageing and lifespan in ad libitum-fed mice. *Australasian Journal on Ageing*, 33, 30–30.

Le Gros Clark, W. E. (1959). *History of the Primates*. Phoenix Books.

Le Saout, S., Hoffmann, M., Shi, Y., Hughes, A., Bernard, C., Brooks, T. M., et al. (2013). Protected areas and effective biodiversity conservation. *Science*, 803–5.

Lea, A. J., Altmann, J., Alberts, S. C., and Tung, J. (2015). Developmental constraints in a wild primate. *American Naturalist*, 185.

Leather, S. R. (2008). *Insect Sampling in Forest Ecosystems*. John Wiley and Sons.

Lebeau, C. W., Nielson, R. M., Hallingstad, E. C., and Young, D. P. (2015). Daytime habitat selection by resident golden eagles (*Aquila chrysaetos*) in southern Idaho, USA. *Journal of Raptor Research*, 49, 29–42.

Leblond, M., Dussault, C., and Ouellet, J. P. (2013). Impacts of human disturbance on large prey species: Do behavioral reactions translate to fitness consequences? *PLoS ONE*, 8, e73695.

LeDee, O. E., Handler, S. D., Hoving, C. L., Swanston, C. W., and Zuckerberg, B. (2021). Preparing wildlife for climate change: How far have we come? *Journal of Wildlife Management*, 85, 7–16.

Ledogar, J. A., Winchester, J. M., St. Clair, E. M., and Boyer, D. M. (2013). Diet and dental topography in pitheciine seed predators. *American Journal of Physical Anthropology*, 150(1), 107–21.

Lee, A. E., and Cowlishaw, G. (2017). Switching spatial scale reveals dominance-dependent social foraging tactics in a wild primate. *PeerJ*, 5, e3452.

Lee, J. J., Morris, D., Constantino, P. J., Lucas, P. W., Smith, T. M., and Lawn, B. R. (2010). Properties of tooth enamel in great apes. *Acta Biomaterialia*, 6(12), 4560–65.

Lee, J. W., Constantino, P., Lucas P.W. and Lawn, B. R. (2011). Fracture in teeth: a diagnostic for inferring tooth function and diet. *Biological Reviews*, 86, 959–74.

Lee, K. P., Behmer, S. T., Simpson, S. J., and Raubenheimer, D. (2002). A geometric analysis of nutrient regulation in the generalist caterpillar *Spodoptera littoralis* (Boisduval). *Journal of Insect Physiology*, 48(6), 655–65.

Lee, K. P., Cory, J. S., Wilson, K., Raubenheimer, D., and Simpson, S. J. (2006). Flexible diet choice offsets protein costs of pathogen resistance in a caterpillar. *Proceedings of the Royal Society B: Biological Sciences*, 273(1588), 823–29.

Lee, K. P., Raubenheimer, D., and Simpson, S. J. (2004). The effects of nutritional imbalance on compensatory feeding for cellulose-mediated dietary dilution in a generalist caterpillar. *Physiological Entomology*, 29, 108–17.

Lee, K. P., Simpson, S. J., and Raubenheimer, D. (2004). A comparison of nutrient regulation between solitarious and gregarious phases of the specialist caterpillar, *Spodoptera exempta* (Walker). *Journal of Insect Phsyiology*, 50, 1171–80.

Lee, K. P., Simpson, S. J., and Wilson, K. (2008). Dietary protein-quality influences melanization and immune function in an insect. *Functional Ecology*, 22(6), 1052–61.

Lee, K. P., Simpson, S. J., Clissold, F. J., Brooks, R., Ballard, J. W. O., Taylor, P. W., et al. (2008). Lifespan and reproduction in *Drosophila*: New insights from nutritional geometry. *Proceedings of the National Academy of Sciences of the United States of America*, 105(7), 2498–503.

Lee, P. C. (1987). Nutrition, fertility and maternal investment in primates. *Journal of Zoology, London*, 213, 409–22.

Lee, P. C. (1996). The meanings of weaning: Growth, lactation, and life history. *Evolutionary Anthropology*, 5, 87–98.

Lee, P. C. (1999). Comparative ecology of postnatal growth and weaning among haplorhine primates. In P. C. Lee (Ed.), *Comparative Primate Socioecology* (pp. 111–36). Cambridge University Press.

Lee, P. C. (2003). Innovation as a behavioural response to environmental challenges: A cost and benefit approach. In S. M. Reader and K. N. Laland (Eds.), *Animal Innovation* (pp. 261–77). Oxford University Press.

Lee, P. C. (2012). Growth and investment in hominin life history evolution: Patterns, processes, and outcomes. *International Journal of Primatology*, 33, 1309–31.

Lee, P. C., and Bowman, J. E. (1995). Influence of ecology and energetics on primate mothers and infants. In C. R. Pryce, R. D. Martin, and D. E. Skuse (Eds.), *Motherhood in Human and Nonhuman Primates: Biosocial Determinants* (pp. 47–58). Karger.

Lee, P. C., and Hauser, M. D. (1998). Long-term consequences of changes in territory quality on feeding and reproductive strategies of vervet monkeys. *Journal of Animal Ecology*, 67, 347–58.

Lee, P. C., and Kappeler, P. M. (2003). Socioecological correlates of phenotypic plasticity of primate life histories. In P. M. Kappeler and M. E. Pereira (Eds.), *Primate Life Histories and Socioecology* (pp. 41–65). Cambridge University Press.

Lee, P. C., Bussiére, L. F., Webber, C. E., Poole, J. H., and Moss, C. J. (2013). Enduring consequences of early experiences: 40 year effects on survival and success among African elephants (*Loxodonta africana*). *Biology Letters*, 9, 20130011.

Lee, P. C., Majluf, P., and Gordon, I. J. (1991). Growth, weaning and maternal investment from a comparative perspective. *Journal of Zoology*, 225, 99–114.

Lee, Y. K., and Mazmanian, S. K. (2010). Has the microbiota played a critical role in the evolution of the adaptive immune system? *Science*, 330, 1768–73.

Lee-Thorp, J. A., Sealy, J. C., van der Merwe, N. J. (1989). Stable carbon isotope ratio differences between bone collagen and bone apatite, and their relationship to diet. *Journal of Archaeology Science*, 16, 585–99.

Lee-Thorp, J., and Sponheimer, M. (2006). Contributions of biogeochemistry to understanding hominin dietary ecology. *American Journal of Physical Anthropology*, 131, 131–48.

Lee-Thorp, J., Sponheimer, M., Passey, B. H., de Ruiter, D. J., and Cerling, T. E. (2010). Stable isotopes in fossil hominin tooth enamel suggest a fundamental dietary shift in the Pliocene. *Philosophical Transactions of the Royal Society of London Series B—Biological Sciences*, 365, 3389–96.

Lee-Thorp, J. A., van der Merwe, N. J., and Brain, C. K. (1994). Diet of *Australopithecus robustus* at Swartkrans from stable carbon isotopic analysis. *Journal of Human Evolution*, 27, 361–72.

Lefebvre, L., Reader, S. M., and Sol, D. (2004). Brains, innovations and evolution in birds and primates. *Brain, Behavior and Evolution*, 63, 233–46.

Lefévre, C. M., Sharp, J. A., and Nicholas, K. R. (2010). Evolution of lactation: Ancient origin and extreme adaptations of the lactation system. *Annual Review of Genomics and Human Genetics*, 11, 219–38.

Lehmann, A., and Scheffler, C. (2016). What does the mean menarcheal age mean? An analysis of temporal pattern in variability in a historical Swiss population from the 19th and 20th centuries. *American Journal of Human Biology*, 28, 705–13.

Leichliter, J. N., Lüdecke, T., Foreman, A. D., Duprey, N. N., Winkler, D. E., Kast, E. R., et al. (2021) Nitrogen isotopes in tooth enamel record diet and trophic level enrichment: Results from a controlled feeding experiment. *Chemical Geology*, 563, 120047.

Leigh, S. R. (1994a). Ontogenetic correlates of diet in anthropoid primates. *American Journal of Physical Anthropology*, 94, 499–522.

Leigh, S. R. (1994b). Relations between captive and noncaptive weights in anthropoid primates. *Zoo Biology*, 13, 21–43.

Leigh, S. R. (2001). Evolution of human growth. *Evolutionary Anthropology*, 10, 223–36.

Leigh, S. R. (2004). Brain growth, life history, and cognition in primate and human evolution. *American Journal of Primatology*, 62, 139–64.

Leigh, S. R., and Blomquist, G. E. (2011). Life history. In C. J. Campbell, A. Fuentes, K. C. MacKinnon, S. K. Bearder, and R. Stumpf (Eds.), *Primates in Perspective* (2nd ed., pp. 418–28). Oxford University Press.

Leighton, M. (1993). Modeling dietary selectivity by Bornean orangutans—evidence for integration of multiple criteria in fruit selection. *International Journal of Primatology*, 14, 257–313.

Leighton, M., and Leighton, D. (1983). Vertebrate responses to fruiting seasonality within a Bornean rain forest. In S. L. Sutton, T. C. Whitmore, and A. C. Chadwick (Eds.), *Tropical Rain Forest: Ecology and Management* (pp. 181–96). Blackwell Scientific.

Leighton, M., and Leighton, D. R. (1982). The relationship of size of feeding aggregate to size of food patch: Howler monkeys (*Alouatta palliata*) feeding in *Trichilia cipo* fruit trees on Barro Colorado Island. *Biotropica*, 14, 81–90.

Leighton, P. A., Horrocks, J. A., and Kramer, D. L. (2010). Conservation and the scarecrow effect: Can human activity benefit threatened species by displacing predators? *Biological Conservation*, 143, 2156–63.

Lemay, D. G., Ballard, O. A., Hughes, M. A., Morrow, A. L., Horseman, N. D., and Nommsen-Rivers, L. A. (2013). RNA sequencing of the human milk fat layer transcriptome reveals distinct gene expression profiles at three stages of lactation. *PLoS ONE*, 8, e67531.

Lemay, D. G., Dillard, C. J., and German, J. B. (2007). Food structure for nutrition. In E. Dickinson and M. Leser (Eds.), *Food Colloids: Self-Assembly and Material Science* (pp. 1–15). Royal Society of Chemistry.

Lemay, D. G., Lynn, D. J., Martin, W. F., Neville, M. C., Casey, T. M., Rincon, G., et al. (2009). The bovine lactation genome: Insights into the evolution of mammalian milk. *Genome Biology*, 10, R43.

Lemorini, C., Plummer, T. W., Braun, D. R., Krittenden, A. N., Lichtfield, P. W., Bishop, L. C., et al. (2014). Old stones' song: Use-wear experiments and analysis of the Oldowan quartz and quartzite assemblage from Kanjera South (Kenya). *Journal of Human Evolution*, 72, 10–25.

Lendvai, A. Z., Barta, Z., Liker, A., and Bokony, V. (2004). The effect of energy reserves on social foraging: Hungry sparrows scrounge more. *Proceedings of the Royal Society B: Biological Sciences*, 271, 2467–72.

Lengyel, S. (2006). Spatial differences in breeding success in the pied avocet *Recurvirostra avosetta*: Effects of habitat on hatching success and chick survival. *Journal of Avian Biology*, 37, 381–95.

Leopold, A. (1933). *Game Management*. Charles Scribner's Sons.

Leopold, A. S., Erwin, M., Oh, J., and Browning, B. (1976). Phytoestrogens—adverse effects on reproduction in California quail. *Science*, 191, 98–100.

Leshem, A., Liwinski, T., and Elinav, E. (2020). Immune-microbiota interplay and colonization resistance in infection. *Molecular Cell*, 78, 597–613.

Letunic, I., and Bork, P. (2021). Interactive Tree of Life v5: An online tool for phylogenetic tree display and annotation. *Nucleic Acids Research*, 39, W293–96.

Leutenegger, W. (1970). Beziehungen zwischen der Neugeborenengroesse und dem Sexualdimorphismus am Becken bei simischen Primaten. *Folia Primatologica*, 12, 224–35.

Leutenegger, W. (1979). Evolution of litter size in primates. *American Naturalist*, 114, 525–31.

Leutenegger, W. (1982). Encephalization and obstetrics in primates with particular reference to human evolution. In E. Armstrong and D. Falk (Eds.), *Primate Brain Evolution: Methods and Concepts* (pp. 85–95). Springer.

Levesque, S., Delisle, H., and Agueh, V. (2015). Contribution to the development of a food guide in Benin: Linear programming for the optimization of local diets. *Public Health Nutrition*, 18, 622–31.

Levey, D. J. (1988). Spatial and temporal variation in Costa Rican fruit and fruit-eating bird abundance. *Ecological Monographs*, 58, 251–69.

Levin, N. E., Simpson, S. W., Quade, J., Cerling, T. E., and Frost, S. R. (2008). Herbivore enamel carbon isotopic composition and the environmental context of *Ardipithecus* at Gona, Ethiopia. *Geological Society of America Special Papers*, 446, 215–34.

Lewis, M. C., West, A. G., and O'Riain, M. J. (2018). Isotopic assessment of marine food consumption by natural-foraging chacma baboons on the Cape Peninsula, South Africa. *American Journal of Physical Anthropology*, 165, 77–93.

Lewis, R. J., and Kappeler, P. W. (2005). Seasonality, body condition, and timing of reproduction in *Propithecus verreauxi verreauxi* in the Kirindy Forest. *American Journal of Primatology*, 67, 347–64.

Lewis, S. L., and Maslin, M. A. (2015). Defining the Anthropocene. *Nature*, 519, 171–80.

Ley, R. E., Backhed, F., Turnbaugh, P. J., Lozupone, C., Knight, R., and Gordon, J. I. (2005). Obesity alters gut microbial ecology. *Proceedings of the National Academy of Sciences of the United States of America*, 102(31), 11070–75.

Ley, R. E., Hamady, M., Lozupone, C., Turnbaugh, P. J., Ramey, R. R., Bircher, J. S., et al. (2008). Evolution of mammals and their gut microbes. *Science*, 320(5883), 1647–51.

Li, C., Levitz, M., Hubbard, G., Jenkins, S., Han, V., Ferry, R., et al. (2007). The IGF axis in baboon pregnancy: Placental and systemic responses to feeding 70% global ad libitum diet. *Placenta*, 28, 1200–1210.

Li, D., Ren, B., Hu, J., Zhang, Q., Yang, Y., Grueter, C. C., et al. (2014). Geophagy of Yunnan snub-nosed monkeys (*Rhinopithecus bieti*) at Xiangguqing in the Baimaxueshan Nature Reserve, China. *North-Western Journal of Zoology*, 10, 293–99.

Li, H., Gao, D., Cao, Y., and Xu, H. (2008). A high gamma-aminobutyric acid-producing *Lactobacillus brevis* isolated from Chinese traditional paocai. *Annals of Microbiology*, 58(4), 649–53.

Li, J., Kudsk, K. A., Janu, P., and Renegar, K. B. (1997). Effect of glutamine-enriched total parenteral nutrition on small intestinal gut-associated lymphoid tissue and upper respiratory tract immunity. *Surgery*, 121(5), 542–49.

Li, M., Wang, B., Zhang, M., Rantalainen, M., Wang, S., Zhou, H., et al. (2008). Symbiotic gut microbes modulate human metabolic phenotypes. *Proceedings of the National Academy of Sciences of the United States of America*, 105(6), 2117–22.

Li, Y., Li, X., Zhou, X., Yan, J., Zhu, X., Pan, J., et al. (2013). Impact of sepsis on the urinary level of interleukin-18 and cystatin C in critically ill neonates. *Pediatric Nephrology*, 28(1), 135–44.

Li, Y., Ma, G., Zhou, Q., Li, Y., and Huang, Z. (2020). Nutrient contents predict the bamboo-leaf-based diet of Assamese macaques living in limestone forests of southwest Guangxi, China. *Ecology and Evolution*, 10, 5570–81.

Li, Y. M. (2006). Seasonal variation of diet and food availability in a group of Sichuan snub-nosed monkeys in Shennongjia Nature Reserve, China. *American Journal of Primatology*, 68, 217–33.

Li, Z., and Rogers, E. (2004). Habitat quality and activity budgets of white-headed langurs in Fusui, China. *International Journal of Primatology*, 25, 41–54.

Licitra, G., Hernandez, T. M., and VanSoest, P. J. (1996). Standardization of procedures for nitrogen fractionation of ruminant feeds. *Animal Feed Science and Technology*, 57, 347–58.

Lieberman, L. S. (2006). Evolutionary and anthropological perspectives on optimal foraging in obesogenic environments. *Appetite*, 47, 3–9.

Liechti, R., and Farmer, E. E. (2002). The jasmonate pathway. *Science*, 296, 1649–50.

Liem, K. F. (1980). Adaptive significance of intraspecific and interspecific differences in the feeding repertoires of cichlid fishes. *American Zoologist*, 20(1), 295–314.

Liem, K. F., and Kaufman, L. S. (1984). Intraspecific macroevolution: Functional biology of the polymorphic cichlid species *Cichlasoma minckleyi*. In A. A. Echelle and I. Kornfield (Eds.), *Evolution of Species Flocks* (pp. 203–15). University of Maine Press.

Lihoreau, M., Buhl, J., Charleston, M. A., Sword, G. A., Raubenheimer, D., and Simpson, S. J. (2015). Nutritional ecology beyond the individual: A conceptual framework for integrating nutrition and social interactions. *Ecology Letters*, 18, 273–86.

Lihoreau, M., Charleston, M. A., Senior, A. M., Clissold, F. J., Raubenheimer, D., Simpson, S. J., et al. (2017). Collective foraging in spatially complex nutritional environments. *Philosophical Transactions of the Royal Society B*, 372, 20160238.

Lihoreau, M., Clarke, I. M., Buhl, J., Sumpter, D. J., and Simpson, S. J. (2016). Collective selection of food patches in *Drosophila*. *Journal of Experimental Biology*, 219, 668–75.

Lihoreau, M., Gómez-Moracho, T., Pasquaretta, C., Costa, J. T., and Buhl, J. (2018). Social nutrition: an emerging field in insect science. *Current Opinion in Insect Science*, 28, 73–80.

Lihoreau, M., and Rivault, C. (2011). Local enhancement promotes cockroach feeding aggregations. *PLoS ONE*, 6, e22048.

Lile, C. W., McLester, E., Stewart, F. A., and Piel, A. K. (2020). Red-tailed monkeys (*Cercopithecus ascanius*) prey on and mob birds in the Issa Valley, western Tanzania. *Primates*, 61, 563–66.

Lim, J. Y., Wasserman, M. D., Veen, J., Despres-Einspenner, M. L., and Kissling, W. D. (2021). Ecological and evolutionary significance of primates' most consumed plant families. *Proceedings of the Royal Society B: Biological Sciences*, 288, 20210737

Lima, E. M., and Ferrari, S. F. (2003). Diet of a free-ranging group of squirrel monkeys (*Saimiri sciureus*) in eastern Brazilian Amazonia. *Folia Primatologica*, 74, 150–58.

Lima, S. L., and Bednekoff, P. A. (1999). Temporal variation in danger drives antipredator behavior: The predation risk allocation hypothesis. *American Naturalist*, 153, 649–59.

Lima, S. L., and Dill, L. M. (1990). Behavioural decisions made under the risk of predation—a review and prospectus. *Canadian Journal of Zoology*, 68, 619–40.

Lin, E., Kotani, J. G., and Lowry, S. F. (1998). Nutritional modulation of immunity and the inflammatory response. *Nutrition*, 14(6), 545–50.

Lindburg, D. G. (1987). Seasonality of reproduction in primates. In G. Mitchell and J. Erwin (Eds.), *Behavior, Cognition, and Motivation* (pp. 167–218). Alan R. Liss.

Lindburg, D. G., Lyles, A. M., and Czekala, N. M. (1989). Status and reproductive potential of lion-tailed macaques in captivity. *Zoo Biology Supplements*, 1, 5–16.

Linder, J. M. (2008). *The Impact of Hunting on Primates in Korup National Park, Cameroon: Implications for Primate Conservation*. City University of New York.

Linder, J. M. (2013). African primate diversity threatened by "New Wave" of industrial oil palm expansion. *African Primates*, 8, 25–38.

Lindroth, R. L. (2012). Atmospheric change, plant secondary metabolites and ecological interactions. In G. R. Iason, M. Dicke, and S. E. Hartley (Eds.), *The Ecology of Plant Secondary Metabolites: From Genes to Global Processes* (pp. 120–53). Cambridge University Press.

Lindshield, S., Danielson, B. J., Rothman, J. M., and Pruetz, J. D. (2017). Feeding in fear? How adult male western chimpanzees (*Pan troglodytes verus*) adjust to predation and savanna habitat pressures. *American Journal of Physical Anthropology*, 163(3), 480–96.

Lindshield, S., Rothman, J. M., Ortmann, S., and Pruetz, J. D. (2021). Western chimpanzees (*Pan troglodytes verus*) access a nutritionally balanced, high energy, and abundant food, baobab (*Adansonia digitata*) fruit, with extractive foraging and reingestion. *American Journal of Primatology*, 83(9), e23307.

Lindstedt, S. L., and Boyce, M. S. (1985). Seasonality, fasting endurance and body size in mammals. *American Naturalist*, 125, 873–78.

Ling, W. H., and Jones, P. J. H. (1995). Dietary phytosterols—a review of metabolism, benefits and side-effects. *Life Science*, 57, 195–206.

Link, A., and Di Fiore, A. (2006). Seed dispersal by spider monkeys and its importance in the maintenance of Neotropical rain-forest diversity. *Journal of Tropical Ecology*, 22, 235–46.

Link, A., and Di Fiore, A. (2013). Effects of predation risk on the grouping patterns of white-bellied spider monkeys (*Ateles belzebuth belzebuth*) in western Amazonia. *American Journal of Physical Anthropology*, 150, 579–90.

Link, A., and Stevenson, P. (2004). Fruit dispersal syndromes in animal disseminated plants at Tinigua National Park, Colombia. *Revista Chilena de Historia Natural*, 77, 319–34.

Linn, C. Jr., Campbell, M., and Roelofs, W. (1986). Male moth sensitivity to multicomponent pheromones: Critical role of female-released blend in determining the functional role of components and active space of the pheromone. *Journal of Chemical Ecology*, 12, 659–68.

Liu, G., Walter, L., Tang, S., Tan, X., Shi, F., Pan, H., et al. (2014). Differentiated adaptive evolution, episodic relaxation of selective constraints, and pseudogenization of umami and sweet taste genes *TAS1Rs* in catarrhine primates. *Frontiers in Zoology*, 11, 79.

Liu, J., and Li, M. (2006). Phenotypic flexibility of metabolic rate and organ masses among tree sparrows *Passer montanus* in seasonal acclimatization. *Acta Zoologica Sinica*. 52, 469–77.

Liu, R., Amato, K., Hou, R., Gomez, A., Dunn, D. W., Zhang, J., et al. (2022). Specialised digestive adaptations within the hindgut of a colobine monkey. *Innovation*, 3, 100207.

Liu, X., Li, X., Xia, B., Jin, X., Zou, Q., Zeng, Z., et al. (2021). High-fiber diet mitigates maternal obesity-induced cognitive and social dysfunction in the offspring via gut-brain axis. *Cell Metabolism*, 33, 923–38.

Liu, X., Stanford, C. B., Yang, J., Yao, H., and Li, Y. (2013). Foods eaten by the Sichuan snub-nosed monkey (*Rhinopithecus roxellana*) in Shennongjia National Nature Reserve, China, in relation to nutritional chemistry. *American Journal of Primatology*, 75, 860–71.

Livoreil, B., and Giraldeau, L. A. (1997). Patch departure decisions by spice finches foraging singly or in groups. *Animal Behaviour*, 54, 967–77.

Lochmiller, R. L., and Deerenberg, C. (2000). Trade-offs in evolutionary immunology: Just what is the cost of immunity? *Oikos*, 88, 87–98.

Lodge, E., Ross, C., Ortmann, S., and MacLarnon, A. M. (2013). Influence of diet and stress on reproductive hormones in Nigerian olive baboons. *General and Comparative Endocrinology*, 191, 146–54.

Lodwick, J. L., and Salmi, R. (2019). Nutritional composition of the diet of the western gorilla (*Gorilla gorilla*): Interspecific variation in diet quality. *American Journal of Primatology*, 81, e23044.

Lohrich, T., Behringer, V., Wittig, R. M., Deschner, T., and Leendertz, F. H. (2018). The use of neopterin as a noninvasive marker in monitoring diseases in wild chimpanzees. *Ecohealth*, 15(4), 792–803.

Lomáscolo, S. B., and Schaefer, H. M. (2010). Signal convergence in fruits: A result of selection by frugivores? *Journal of Evolutionary Biology*, 23, 614–24.

Lomáscolo, S. B., Speranza, P., and Kimball, R. T. (2008). Correlated evolution of fig size and color supports the dispersal syndromes hypothesis. *Oecologia*, 156, 783–96.

Lone, K., Loe, L. E., Gobakken, T., Linnell, J. D. C., Odden, J., Remmen, J., et al. (2014). Living and dying in a multi-predator landscape of fear: Roe deer are squeezed by contrasting pattern of predation risk imposed by lynx and humans. *Oikos*, 123, 641–51.

Lone, K., Loe, L. E., Meisingset, E. L., Stamnes, I., and Mysterud, A. (2015). An adaptive behavioural response to hunting: Surviving male red deer shift habitat at the onset of the hunting season. *Animal Behaviour*, 102, 127–38.

Long, C. A. (1969). The origin and evolution of mammary glands. *BioScience*, 19, 519–23.

Lönnerdal, B. (2000). Regulation of mineral and trace elements in human milk: Exogenous and endogenous factors. *Nutrition Reviews*, 58, 223–29.

Lönnerdal, B. (2003). Nutritional and physiologic significance of human milk proteins. *American Journal of Clinical Nutrition*, 77, 1537S–43S.

Lönnerdal, B., and Atkinson, S. (1995). Nonprotein nitrogen factors in human milk. In R. G. Jensen (Ed.), *Handbook of Milk Composition* (pp. 351–87). Academic Press.

Lönnerdal, B., Erdmann, P., Thakkar, S. K., Sauser, J., and Destaillats, F. (2017). Longitudinal evolution of true protein, amino acids and bioactive proteins in breast milk: A developmental perspective. *Journal of Nutritional Biochemistry*, 41, 1–11.

Lönnerdal, B., Keen, C. L., Glazier, C. E., and Anderson, J. (1984). A longitudinal study of rhesus monkey (*Macaca mulatta*) milk composition: Trace elements, minerals, protein, carbohydrate, and fat. *Pediatric Research*, 18, 911–14.

Lopez-Gutierrez, N., Romero-Gonzalez, R., Frenich, A. G., and Vidal, J. L. M. (2014). Identification and quantification of the main isoflavones and other phytochemicals in soy based nutraceutical products by liquid chromatography-orbitrap high resolution mass spectrometry. *Journal of Chromatography A*, 1348, 125–36.

Lotka, A. J. (1956). Elements of physical biology. *Science Progress in the Twentieth Century (1919–1933)*, 21, 341–43.

Loudon, J. E., Grobler J. P., Sponheimer, M., Moyer, K., Lorenz, J. G., and Turner, T. R. (2014). Using stable carbon and nitrogen isotope compositions of vervet monkey (*Chlorocebus pygerythrus*) to examine questions in ethnoprimatology. *PLoS ONE*, 1–7, e100758.

Loudon, J. E., Grobler, J. P., Sponheimer, M., Moyer, K., Lorenz, J. G., and Turner, T. R. (2016). Using stable carbon and nitrogen isotope compositions of vervet monkey (*Chlorocebus pygerythrus*) to examine questions in ethnoprimatology. *PLoS ONE*, 9, e100758.

Loudon, J. E., Sandberg, P. A., Wrangham, R. W., Fahey, B., and Sponheimer, M. (2016). The stable isotope ecology of *Pan* in Uganda and beyond. *American Journal of Primatology*, 78, 1070–85.

Lowry, B. E., Wittig, R. M., Pittermann, J., and Oelze, V. M. (2021). Stratigraphy of stable isotope ratios and leaf structure within an African rainforest canopy with implications for primate isotope ecology. *Scientific Reports*, 11(1), 14222.

Loy, J. (1970). Behavioral responses of free-ranging Rhesus monkeys to food shortage. *American Journal of Physical Anthropology*, 33, 263–72.

Lu, A., Beehner, J. C., Czekala, N. M., Koenig, A., Larney, E., and Borries, C. (2011). Phytochemicals and reproductive function in wild female Phayre's leaf monkeys (*Trachypithecus phayrei crepusculus*). *Hormones and Behavior*, 59, 28–36.

Lu, A., Bergman, T. J., McCann, C., Stinespring-Harris, A., and Beehner, J. C. (2016). Growth trajectories in wild geladas (*Theropithecus gelada*). *American Journal of Primatology*, 78, 707–19.

Lu, A., Borries, C., Caselli, A., and Koenig, A. (2013). Effects of age, reproductive state, and the number of competitors on the dominance dynamics of wild female Hanuman langurs. *Behaviour*, 150, 485–523.

Lu, A., Petrullo, L., Carrera, S., Feder, J., Schneider-Crease, I., and Snyder-Mackler, N. (2019). Developmental responses to early-life adversity: Evolutionary and mechanistic perspectives. *Evolutionary Anthropology*, 28(5), 249–66.

Lucas, P. W. (1994). Categorisation of food items relevant to oral processing. In D. J. Chivers and P. Langer (Eds.), *The Digestive System in Mammals* (pp. 197–218). Cambridge University Press.

Lucas, P. W. (2004). *Dental Functional Morphology: How Teeth Work*. Cambridge University Press.

Lucas, P. W., Beta, T., Darvell, B. W., Dominy, N. J., Essackjee, H. C., Lee, P. K. D., et al. (2001). Field kit to characterize physical, chemical and spatial aspects of potential foods of primates. *Folia Primatologica*, 72, 11–15.

Lucas, P. W., Darvell, B. W., Lee, P. K. D., Yuen, T. D. B., and Choong, M. F. (1995). Toughness of plant cell walls. *Philosophical Transactions of the Royal Society B*, 348, 363–72.

Lucas, P., Constantino, P., Wood, B., and Lawn, B. (2008). Dental enamel as a dietary indicator in mammals. *Bioessays*, 30(4), 374–85.

Lucas, P. W., Copes, L., Constantino, P. J., Vogel, E. R., Chalk, J., Talebi, M., et al. (2011). Measuring the toughness of primate foods and its ecological value. *International Journal of Primatology*, 33, 598–610.

Lucas, P. W., Corlett, R. T., Dominy, N. J, Essackjee, H. C., Ramsden, L., Riba-Hernandez, P., et al. (2011). Dietary analysis II: Chemistry. In J. Setchell, & D. Curtis, (Eds.), *Field and laboratory methods in primatology*, 2nd ed. (pp. 255–70). Cambridge University Press.

Lucas, P. W., and Luke, D. A. (1984). Chewing it over: Basic principles of food breakdown. In D. J. Chivers, B. A. Wood, and A. Bilsborough (Eds.), *Food Acquisition and Processing in Primates* (pp. 283–301). Plenum.

Lucas, P. W., Omar, R., Al-Fadhalah, K., Almusallam, A. S., Henry, A. G., Michael, S., et al. (2013). Mechanisms and causes of wear in tooth enamel: Implications for hominin diets. *Journal of the Royal Society, Interface*, 10(80), 20120923.

Lucas, P. W., Osorio, D., Yamashita, N., Prinz, J. F., Dominy, N. J., and Darvell, B. W. (2011). Dietary analysis I: Physics. In J. Setchell and D. Curtis (Eds.), *Field and Laboratory Methods in Primatology*, 2nd ed. (pp. 255–70). Cambridge University Press.

Lucas, P. W., and Pereira, B. (1991). Thickness effect in cutting systems. *Journal of Materials Science Letters*, 10, 235–36.

Lucas, P. W., Peters, C. R., and Arrandale, S. R. (1994). Seed-breaking forces exerted by orangutans with their teeth in captivity and a new technique for estimating forces produced in the wild. *American Journal of Physical Anthropology*, 94(3), 365–78.

Lucas, P. W., Turner, I. M., Dominy, N. J., and Yamashita, N. (2000). Mechanical defences to herbivory. *Annals of Botany*, 86(5), 913–20.

Lucas, P. W., and van Casteren, A. (2015). The wear and tear of teeth. *Medical Principles and Practice*, 24, 3–13.

Lucas, P. W., van Casteren, A., Al-Fadhalah, K., Almusallam, A. S., Henry, A. G., Michael, S., et al. (2014). The role of dust, grit and phytoliths in tooth wear. *Annales Zoologici Fennici*, 51, 143–52.

Luccas, V., Izar, P. (2021) Black capuchin monkeys dynamically adjust group spread throughout the day. *Primates*, 62, 789–99 (2021).

Lucore, J. M., Marshall, A. J., Brosnan, S. F., and Benitez, M. E. (2022). Validating urinary neopterin as a biomarker of immune response in captive and wild capuchin monkeys. *Frontiers in Veterinary Science*, 9, 918036.

Lüffe, T. M., Tirado Herrera, E. R., Nadjafzadeh, M., Berles, P., Smith, A. C., Knogge, C., et al. (2018). Seasonal variation and an "outbreak" of frog predation by tamarins. *Primates*, 59, 549–52.

Luhrs, M. L., Dammhahn, M., Kappeler, P. M., and Fichtel, C. (2009). Spatial memory in the grey mouse lemur (*Microcebus murinus*). *Animal Cognition*, 12, 599–609.

Luncz, L. V., and Boesch, C. (2015). The extent of cultural variation between adjacent chimpanzee (*Pan troglodytes verus*) communities: A microecological approach. *American Journal of Physical Anthropology*, 156, 67–75.

Lundin, A., Bok, C. M., Aronsson, L., Bjorkholm, B., Gustafsson, J. A., Pott, S., et al. (2008). Gut flora, Toll-like receptors and nuclear receptors: A tripartite communication that tunes innate immunity in large intestine. *Cell Microbiology*, 10(5), 1093–103.

Lunn, P. G., Austin, S., Prentice, A. M., and Whitehead, R. G. (1984). The effect of improved nutrition on plasma prolactin concentrations and postpartum infertility in lactating Gambian women. *American Journal of Clinical Nutrition*, 39, 227–35.

Luo, Z., Zhou, S., Yu, W., Yu, H., Yang, J., Tian, Y., et al. (2015). Impacts of climate change on the distribution of Sichuan snub-nosed monkeys (*Rhinopithecus roxellana*) in Shennongjia area, China. *American Journal of Primatology*, 77, 135–51.

Lupo, K. D. (2012). On early hominin meat eating and carcass acquisition strategies: Still relevant after all these years. In: Dominguez-Rodrigo, M. (Ed.) *Stone tools and fossil bones: debates in the archaeology of human origins* (pp. 115–51). Cambridge University Press.

Lutgens, F. K. and Tarbuck, E. J. (2000). *Essentials of geology*, 7th ed. Prentice Hall.

Luz, B., Cormie, A. B., and Schwarcz, H. P. (1990). Oxygen isotope variations in phosphate of deer bones. *Geochimica et Cosmochimica Acta*, 54, 1723–28.

Lwanga, J. S., Struhsaker, T. T., Struhsaker, P. J., Butynski, T. M., and Mitani, J. C. (2011). Primate population dynamics over 32.9 years at Ngogo, Kibale National Park, Uganda. *American Journal of Primatology*, 73(10), 997–1011.

Lyke, M. M., Di Fiore, A., Fierer, N., Madden, A. A., and Lambert, J. E. (2019). Metagenomic analyses reveal previously unrecognized variation in the diets of sympatric Old World monkey species. *PLoS ONE*, 14(6), e0218245.

Lyte, M., Fitzgerald, P., and Roshchina, V. (2010). Evolutionary considerations of neurotransmitters in microbial, plant, and animal cells. In M. Lyte and P. P. E. Freestone (Eds.), *Microbial Endocrinology: Interkingdom Signaling in Infectious Disease and Health* (pp. 17–52). Springer.

Ma, C., Liao, J., and Fan, P. (2016). Food selection in relation to nutritional chemistry of Cao Vit gibbons in Jingxi, China. *Primates*, 58, 1–12.

Maas, M. C., and Dumont, E. R. (1999). Built to last: The structure, function, and evolution of primate dental enamel. *Evolutionary Anthropology*, 8(4), 133–52.

MacDonald, G. J. (1971). Reproductive patterns of three species of macaques. *Fertility and Sterility*, 22, 373–77.

Macel, M., van Dam, N. M., Keurent, J., and Joost, J. (2010). Metabolomics: The chemistry between ecology and genetics. *Molecular Ecology Resources*, 10, 583–93.

Macho, G. A., and Lee-Thorp, J. A. (2014). Niche partitioning in sympatric *Gorilla* and *Pan* from Cameroon: Implications for life history strategies and for reconstructing the evolution of hominin life history. *PLoS ONE*, 9(7), e102794.

Macho, G. A., and Shimizu, D. (2009). Dietary adaptations of South African australopiths: Inference from enamel prism attitude. *Journal of Human Evolution*, 57(3), 241–47.

Macho, G. A., Jiang, Y., and Spears, I. R. (2003, Jul). Enamel microstructure: A truly three-dimensional structure. *Journal of Human Evolution*, 45(1), 81–90.

Machovsky-Capuska, G. E., Coogan, S. C. P., Simpson, S. J., and Raubenheimer, D. (2016). Motive for killing: What drives prey choice in wild predators? *Ethology*, 122, 703–11.

Machovsky-Capuska, G. E., Miller, M. G., Silva, F. R., Amiot, C., Stockin, K. A., Senior, A. M., et al. (2018). The nutritional nexus: Linking niche, habitat variability and prey composition in a generalist marine predator. *Journal of Animal Ecology*, 87, 1286–98.

Machovsky-Capuska, G. E., Senior, A. M., Simpson, S. J., and Raubenheimer, D. (2016). The multidimensional nutritional niche. *Trends in Ecology and Evolution*, 31, 355–65.

MacIntosh, A. J. J., Huffman, M. A., Nishiwaki, K., and Miyabe-Nishiwaki, T. (2012). Urological screening of a wild group of Japanese macaques (*Macaca fuscata yakui*): Investigating trends in nutrition and health. *International Journal of Primatology*, 33(2), 460–78.

Mackay, D., and Fraser, A. (2000). Bioaccumulation of persistent organic chemicals: Mechanisms and models. *Environmental Pollution*, 110, 375–91.

Mackie, R. I. (2002). Mutualistic fermentative digestion in the gastrointestinal tract: Diversity and evolution. *Integrative and Comparative Biology*, 42, 319–26.

Mackie, R. I., Sghir, A., and Gaskins, H. R. (1999). Developmental microbial ecology of the neonatal gastrointestinal tract. *American Journal of Clinical Nutrition*, 69, 1035S–45S.

MacKinnon, J. (1974). The behaviour and ecology of wild orangutans (*Pongo pygmaeus*). *Animal Behaviour*, 22, 3–74.

MacKinnon, J. R., and MacKinnon, K. S. (1980). Niche differentiation in a primate community. In D. J. Chivers (Ed.), *Malayan Forest Primates* (pp. 167–90). Plenum.

MacLarnon, A. M., Chivers, D. J., and Martin, R. D. (1986). Gastrointestinal allometry in primates and other mammals including new species. In J. G. Else and P. C. Lee (Eds.), *Primate Ecology and Conservation* (pp. 75–85). Cambridge University Press.

Madden, R. H. (2014). *Hypsodonty in mammals: evolution, geomorphology, and the role of earth surface processes*. Cambridge University Press.

Maestripieri, D., and Georgiev, A. V. (2016). What cortisol can tell us about the costs of sociality and reproduction among free-ranging rhesus macaque females on Cayo Santiago. *American Journal of Primatology*, 78(1), 92–105.

Magee, B. (1973). *Popper*. Fontana-Collins.

Magness, J. R., and Taylor G. F. (1925). An improved type of pressure tester for the determination of fruit maturity. *USDA Circular*, 350, 1–8.

Magliocca, F., and Gautier-Hion, A. (2002). Mineral content as a basis for food selection by western lowland gorillas in a forest clearing. *American Journal of Primatology*, 57, 67–77.

Mahaney, W. C., Aufreiter, S., and BHancock, G. V. (1995). Mountain gorilla geophagy—a possible seasonal behaviour for dealing with the effects of dietary-changes. *International Journal of Primatology*, 16, 475–88.

Mahumud, R. A., Uprety, S., Wali, N., Renzaho, A. M., and Chitekwe, S. (2022). The effectiveness of interventions on nutrition social behaviour change communication in improving child nutritional status within the first 1000 days: Evidence from a systematic review and meta-analysis. *Maternal and Child Nutrition*, 18(1), e13286.

Mainka, S. A., and Howard, G. W. (2010). Climate change and invasive species: Double jeopardy. *Integrative Zoology*, 5, 102–11.

Mainland, J. D., Keller, A., Li, Y. R., Zhou, T., Trimmer, C., Snyder, L. L., et al. (2014). The missense of smell: Functional variability in the human odorant receptor repertoire. *Nature Neuroscience*, 17, 114–20.

Mair, W., Piper, M. D., and Partridge, L. (2005). Calories do not explain extension of life span by dietary restriction in *Drosophila*. *PLoS Biology*, 3(7), e223.

Maisels, F., Gautier-Hion, A., and Gautier, J. P. (1994). Diets of two sympatric colobines in Zaire: More evidence on seed-eating in forests on poor soils. *International Journal of Primatology*, 15, 681–701.

Majolo, B., Lehmann, J., de Bortoli Vizioli, A., and Schino, G. (2012). Fitness-related benefits of dominance in primates. *American Journal of Physical Anthropology*, 147, 652–60.

Majolo, B., Vizioli, A. D. B., and Schino, G. (2008). Costs and benefits of group living in primates: Group size effects on behaviour and demography. *Animal Behaviour*, 76, 1235–47.

Makin, D. F., Payne, H. F., Kerley, G. I. H., and Shrader, A. M. (2012). Foraging in a 3-D world: How does predation risk affect space use of vervet monkeys? *Journal of Mammalogy*, 93, 422–28.

Makkar, H. P. S., Borowy, N. K., Becker, K., and Degen, A. (1995). Some problems in fiber determination of a tannin-rich forage (*Acacia saligna* leaves) and their implications in in-vivo studies. *Animal Feed Science and Technology*, 55, 67–76.

Makkar, H. P. S., and Singh, B. (1995). Determination of condensed tannins in complexes with fibre and proteins. *Journal of the Science of Food and Agriculture*, 69, 129–32.

Maklakov, A. A., Simpson, S. J., Zajitschek, F., Hall, M. D., Dessmann, J., Clissold, F. J., et al. (2008). Sex-specific fitness effects of nutrient intake on reproduction and lifespan. *Current Biology*, 18, 1062–66.

Malaivijitnond, S., Lekprayoon, C., Randavanittj, N., Panha, S., Cheewatham, C., and Hamada, Y. (2007). Stone-tool usage by Thai long-tailed macaques (*Macaca fascicularis*). *American Journal of Primatology*, 69, 227–33.

Malenky, R., and Stiles, E. W. (1991). Distribution of terrestrial herbaceous vegetation and its consumption by *Pan paniscus* in the Lomako Forest, Zaire. *American Journal of Primatology*, 23, 153–69.

Malenky, R., and Wrangham, R. W. (1994). A quantitative comparison of terrestrial herbaceous food consumption by *Pan paniscus* in the Lomako Forest, Zaire and *Pan troglodytes* in the Kibale Forest, Uganda. *American Journal of Primatology*, 32, 1–12.

Malenky, R. K., Wrangham, R., Chapman, C. A., and Vineberg, E. O. (1993). Measuring chimpanzee food abundance. *Tropics*, 2, 231–44.

Mallott, E. K., and Amato, K. R. (2021). Host specificity of the gut microbiome. *Nature Reviews Microbiology*, 19, 639–53.

Mallott, E. K., Borries, C., Koenig, A., Amato, K. R., and Lu, A. (2020). Reproductive hormones mediate changes in the gut microbiome during pregnancy and lactation in Phayre's leaf monkeys. *Scientific Reports*, 10(1), 1–9.

Mallott, E., Garber, P., and Malhi, R. (2017). Integrating feeding behavior, ecological data, and DNA barcoding to identify developmental differences in invertebrate foraging strategies in white-faced capuchins (*Cebus capucinus*). *American Journal of Physical Anthropology*, 162, 241–54.

Mallott, E. K., Garber, P. A., and Malhi, R. S. (2018). trnL outperforms rbcL as a DNA metabarcoding marker when compared with the observed plant component of the diet of wild white-faced capuchins (*Cebus capucinus*, Primates). *PLoS ONE*, 13, e0199556.

Malone, M. A., MacLatchy, L. M., Mitani, J. C., Kityo, R., and Kingston, J. D. (2021). A chimpanzee enamel-diet $\delta^{13}C$ enrichment factor and a refined enamel sampling strategy: Implications for dietary reconstructions. *Journal of Human Evolution*, 159, 103062.

Mankowski, R. T., Anton, S. D., Ghita, G. L., Brumback, B., Darden, D. B., Bihorac, A., et al. (2022). Older adults demonstrate biomarker evidence of the persistent inflammation, immunosuppression, and catabolism syndrome (PICS) after sepsis. *Journal of Gerontology A, Biological Sciences and Medical Sciences*, 77(1), 188–96.

Manly, B. F. L., McDonald, L., Thomas, D., McDonald, T. L., and Erickson, W. P. (2007). *Resource Selection by Animals: Statistical Design and Analysis for Field Studies*. Springer Science and Business Media.

Mann, J., and Patterson, E. M. (2013). Tool use by aquatic animals. *Philosophical Transactions of the Royal Society B-Biological Sciences*, 368(1630), 20120424.

Mann, J., Sargeant, B. L., Watson-Capps, J. J., Gibson, Q. A., Heithaus, M. R., Connor, R. C., et al. (2008). Why do dolphins carry sponges? *PLoS ONE*, 3, 3868.

Mannowetz, N., Miller, M. R., and Lishko, P. V. (2017). Regulation of the sperm calcium channel CatSper by endogenous steroids and plant triterpenoids. *Proceedings of the National Academy of Sciences of the United States of America*, 114, 5743–48.

Mannu, M., and Ottoni, E. B. (2009). The enhanced tool-kit of two groups of wild bearded capuchin monkeys in the Caatinga: Tool making, associative use, and secondary tools. *American Journal of Primatology*, 71, 242–51.

Manor, R., and Saltz, D. (2003). Impact of human nuisance disturbance on vigilance and group size of a social ungulate. *Ecological Applications*, 13, 1830–34.

Manor, R., and Saltz, D. (2005). Effects of human disturbance on use of space and flight distance of mountain gazelles. *Journal of Wildlife Management*, 69, 1683–90.

Manzel, A., Muller, D. N., Hafler, D. A., Erdman, S. E., Linker, R. A., and Kleinewietfeld, M. (2014). Role of "Western diet" in inflammatory autoimmune diseases. *Current Allergy and Asthma Reports*, 14(1), 404.

Marais, E. (1969). *The Soul of the Ape*. Atheneum.

Marchant, L., and McGrew, W. (2005). Percussive technology: Chimpanzee baobab smashing and the evolutionary modelling of hominin knapping. In V. Roux and B. Bril (Eds.), *Stone Knapping: The Necessary Conditions for a Uniquely Hominin Behavior* (pp. 341–50). Cambridge University Press.

Marcobal, A., and Sonnenburg, J. L. (2012). Human milk oligosaccharide consumption by intestinal microbiota. *Clinical Microbiology and Infection*, 18(suppl. s4), 12–15.

Marean, C. W. (2011). Coastal South Africa and the co-evolution of the modern human lineage and coastal adaptations. In N. Bicho, J. A. Haws, and L. G. Davis (Eds.), *Trekking the Shore: Changing Coastlines and the Antiquity of Coastal Settlement* (pp. 421–40). Springer.

Marean, C. W. (2014). The origins and significance of coastal resource use in Africa and Western Eurasia. *Journal of Human Evolution*, 77, 17–40.

Marean, C. W. (2015). An evolutionary anthropological perspective on modern human origins. *Annual Review of Anthropology*, 44, 533–56.

Marean, C. W. (2016). The transition to foraging for dense and predictable resources and its impact on the evolution of modern humans. *Philosophical Transactions of the Royal Society of London Series B—Biological Sciences*, 371, 20150229.

Marín-Gómez, O. H. (2008). Consumo de néctar por *Aotus lemurinus* y su rol como posible polinizador de las flores de *Inga edulis* (Fabales: Mimosoideae). *Neotropical Primates*, 15, 30–32.

Markham, A. C., and Altmann, J. (2008). Remote monitoring of primates using automated GPS technology in open habitats. *American Journal of Primatology*, 70, 495–99.

Markham, A. C., and Gesquiere, L. R. (2017). Costs and benefits of group living in primates: An energetic perspective. *Philosophical Transactions of the Royal Society of London Series B—Biological Sciences*, 372, 20160239.

Markham, A. C., Alberts, S. C., and Altmann, J. (2012). Intergroup conflict: Ecological predictors of winning and consequences of defeat in a wild primate population. *Animal Behaviour*, 84, 399–403.

Markham, A. C., Gesquiere, L. R., Bellenger, J. P., Alberts, S. C., and Altmann, J. (2011). White monkey syndrome and presumptive copper deficiency in wild savannah baboons. *American Journal of Primatology*, 73(11), 1160–68.

Markham, A. C., Guttal, V., Alberts, S. C., and Altmann, J. (2013). When good neighbors don't need fences: Temporal landscape partitioning among baboon social groups. *Behavioral Ecology and Sociobiology*, 67, 875–84.

Marks, D. L., Swain, T., Goldstein, S., Richard, A., and Leighton, M. (1988). Chemical correlates of rhesus monkey food choice—the influence of hydrolyzable tannins. *Journal of Chemical Ecology*, 14, 213–35.

Marler, P. (1973). A comparison of vocalizations of red-tailed monkeys and blue monkeys *Cercopithecus ascanius* and *C. mitis*, in Uganda. *Zeitschrift fur Tierpsychologie*, 33, 223–47.

Marlowe, F. (2000). Paternal investment and the human mating system. *Behavioural Processes*, 51, 45–61.

Marlowe, F. (2001). Male contribution to diet and female reproductive success among foragers. *Current Anthropology*, 42, 755–59.

Marlowe, F. W. (2005). Hunter-gatherers and human evolution. *Evolutionary Anthropology*, 14, 54–67.

Marsh, C. W. (1981). Ranging behavior and its relation to diet selection in Tana River red colobus (*Colobus badius rufomitratus*). *Journal of Zoology, London*, 195, 473–92.

Marsh, K. (2006). *Feeding by Marsupial Folivores in Response to Plant Chemical Defences*. Australian National University.

Marsh, K., Foley, W. J., Cowling, A., and Wallis, I. R. (2003). Differential susceptibility to *Eucalyptus* secondary compounds explains feeding by the common ringtail (*Pseudocheirus peregrinus*) and common brushtail possum (*Trichosurus vulpecula*). *Journal of Comparative Physiology B*, 173, 69–78.

Marsh, K. J., Moore, B. D., Wallis, I. R., and Foley, W. J. (2014). Feeding rates of a mammalian browser confirm the predictions of a "foodscape" model of its habitat. *Oecologia*, 174, 873–82.

Marsh, K. J., Wallis, I. R., Andrew, R. L., and Foley, W. J. (2006). The detoxification limitation hypothesis: Where did it come from and where is it going? *Journal of Chemical Ecology*, 32, 1247–66.

Marsh, K. J., Wallis, I. R., McLean, S., Sorensen, J. S., and Foley, W. J. (2006). Conflicting demands on detoxification pathways influence how common brushtail possums choose their diets. *Ecology*, 87, 2103–12.

Marsh, K. J., Yin, B., Singh, I. P., Saraf, I., Choudhary, A., Au, J., et al. (2015). From leaf metabolome to in vivo testing: Identifying antifeedant compounds for ecological studies of marsupial diets. *Journal of Chemical Ecology*, 41, 1–7.

Marsh, L. K., and Chapman, C. A. (2013). *Primates in Fragments—Complexity and Resilience*. Springer.

Marshall, A. J. (2004). *Population Ecology of Gibbons and Leaf Monkeys across a Gradient of Bornean Forest Types*. Harvard University Press.

Marshall, A. J. (2009). Are montane forests demographic sinks for Bornean white-bearded gibbons *Hylobates albibarbis*? *Biotropica*, 41, 257–67.

Marshall, A. J. (2010). Effect of habitat quality on primate populations in Kalimantan: Gibbons and leaf monkeys as case studies. In S. Gursky and J. Supriatna (Eds.), *Indonesian Primates* (pp. 157–77). Springer.

Marshall, A. J., and Leighton, M. (2006). How does food availability limit the population density of white-bearded gibbons? In G. Hohmann, M. M. Robbins, and C. Boesch (Eds.), *Feeding Ecology in Apes and Other Primates* (pp. 311–33). Cambridge University Press.

Marshall, A. J., and Wich, S. A. (2013). Characterization of primate environments through assessment of plant phenology. In E. Sterling, N. Bynum, and M. Blair (Eds.), *Primate Ecology and Conservation: A Handbook of Techniques* (pp. 103–27). Oxford University Press.

Marshall, A. J., and Wrangham, R. W. (2007). Evolutionary consequences of fallback foods. *International Journal of Primatology*, 28, 1219–35.

Marshall, A. J., Ancrenaz, M., Brearley, F. Q., Fredericksson, G. M., Ghaffar, N., Heydon, M., et al. (2009). The effects of forest phenology and floristics on populations of Bornean and Sumatran orangutans. In S. A. Wich, S. S. Utami, T. Mitra Setia, and C. P. van Schaik (Eds.), *Orangutans: Geographic Variation in Behavioral Ecology and Conservation* (pp. 97–118). Oxford University Press.

Marshall, A. J., Beaudrot, L., and Wittmer, H. U. (2014). Responses of primates and other frugivorous vertebrates to plant resource variability over space and time at Gunung Palung National Park. *International Journal of Primatology*, 35, 1178–201.

Marshall, A. J., Boyko, C. M., Feilen, K. L., Boyko, R. H., and Leighton, M. (2009). Defining fallback foods and assessing their importance in primate ecology and evolution. *American Journal of Physical Anthropology*, 140, 603–14.

Marshall, A. J., Cannon, C. H., and Leighton, M. (2009). Competition and niche overlap between gibbons (*Hylobates albibarbis*) and other frugivorous vertebrates in Gunung Palung National Park, West Kalimantan, Indonesia. In D. Whittaker and S. Lappan (Eds.), *The Gibbons* (pp. 161–88). Springer.

Marshall, A. J., Engström, L. M., Pamungkas, B., Meijaard, E., and Stanley, S. A. (2006). The blowgun is mightier than the chainsaw in determining population density of Bornean orangutans (*Pongo pygmaeus morio*) in the forests of East Kalimantan. *Biological Conservation*, 129, 566–78.

Marshall, A. J., Lacy, R., Ancrenaz, M., Byers, O., Husson, S. J., Leighton, M., et al. (2009). Orangutan population biology, life history, and conservation. In S. A. Wich, S. S. Utami Atmoko, T. Mitra Setia, and C. P. van Schaik (Eds.), *Orangutans: Geographic Variation in Ecology and Conservation* (pp. 311–26). Oxford University Press.

Marshall, A. J., Salas, L. A., Stephens, S., Engstrom, L., Meijaard, E., and Stanley, S. A. (2007). Use of limestone karst forests by Bornean orangutans (*Pongo pygmaeus morio*) in the Sangkulirang Peninsula, East Kalimantan, Indonesia. *American Journal of Primatology*, 69, 212–19.

Marshall, H. H., Carter, A. J., Ashford, A., Rowcliffe, J. M., and Cowlishaw, G. (2015). Social effects on foraging behavior and success depend on local environmental conditions. *Ecology and Evolution*, 5, 475–92.

Marston, C. G., Wilkinson, D. M., Sponheimer, M., Codron, D., Codron, J., and O'Regan, H. J. (2020). "Remote" behavioural ecology: Do megaherbivores consume vegetation in proportion to its presence in the landscape? *PeerJ*, 8, e8622.

Martens, E. A., Lemmens, S. G., and Westerterp-Plantenga, M. S. (2013). Protein leverage affects energy intake of high-protein diets in humans. *American Journal of Clinical Nutrition*, 97, 86–93.

Martin, C. K., Johnson, W. D., Myers, C. A., Apolzan, J. W., Earnest, C. P., Thomas, D. M., et al. (2019). Effect of different doses of supervised exercise on food intake, metabolism, and non-exercise physical activity: The E-MECHANIC randomized controlled trial. *American Journal of Clinical Nutrition*, 110(3), 583–92.

Martin, J. E., Vance, D., and Balter, V. (2014). Natural variation of magnesium isotopes in mammal bones and teeth from two South African trophic chains. *Geochimica et Cosmochimica Acta*, 130, 12–20.

Martin, J. E., Vance, D., and Balter, V. (2015). Magnesium stable isotope ecology using mammal tooth enamel. *Proceedings of the National Academy of Sciences of the United States of America*, 112, 430–35.

Martin, J. E., Tacail, T., Braga, J., Cerling, T. E., and Balter, V. 2020. Calcium isotopic ecology of Turkana Basin hominins. *Nature Communications*, 11(1), 3587.

Martin, L. (1991). Paleoanthropology: Teeth, sex and species. *Nature*, 352(6331), 111–12.

Martin, M. (2017). Mixed-feeding in humans: Evolution and current implications. In *Breastfeeding* (pp. 140–54). Routledge.

Martin, M. A., and Sela, D. A. (2013). Infant gut microbiota: Developmental influences and health outcomes. In K. B. H. Clancy, K. Hinde, and J. N. Rutherford (Eds.), *Building Babies: Primate Development in Proximate and Ultimate Perspective* (pp. 233–56). Springer.

Martin, M., Keith, M., Olmedo, S., Edwards, D., Barrientes, A., Pan, A., et al. (2022). Cesarean section and breastfeeding outcomes in an Indigenous Qom community with high breastfeeding support. *Evolution, Medicine, and Public Health*, 10(1), 36–46.

Martin, P., and Bateson, P. (2021). *Measuring Behaviour: An Introductory Guide* (4th ed.). Cambridge University Press.

Martin, R. D. (1982). Allometric approaches to the evolution of the primate nervous system. In E. Armstrong and D. Falk (Eds.), *Primate Brain Evolution: Methods and Concepts* (pp. 39–56). Springer.

Martin, R. D. (1984a). Body size, brain size and feeding strategies. In D. J. Chivers, B. A. Wood, and A. Bilsborough (Eds.), *Food Acquisition and Processing in Primates* (pp. 73–103). Plenum.

Martin, R. D. (1984b). Scaling effects and adaptive strategies in mammalian lactation. *Symposia of the Zoological Society of London*, 51, 87–117.

Martin, R. D. (1996). Scaling of the mammalian brain: The maternal energy hypothesis. *News in Physiological Sciences*, 11, 149–56.

Martin, R. D. (2007). The evolution of human reproduction: A primatological perspective. *Yearbook of Physical Anthropology*, 134, 59–84.

Martin, R. D., and Bearder, S. K. (1979). Radio bush baby. *Natural History*, 88, 77–80.

Martin, R. D., and MacLarnon, A. M. (1985). Gestation period, neonatal size and maternal investment in placental mammals. *Nature*, 313, 220–23.

Martin, R. D., Chivers, D. J., MacLarnon, A. M., and Hladik, C. M. (1985). Gastrointestinal allometry in primates and other mammals. In W. L. Jungers (Ed.), *Size and Scaling in Primate Biology* (pp. 61–89). Springer.

Martin-Ordas, G., Bernstein, D., and Call, J. (2013). Memory for distant past events in chimpanzees and orangutans. *Current Biology*, 23, 1438–41.

Martinez-Cordero, C., Kuzawa, C. W., Sloboda, D. M., Stewart, J., Simpson, S. J., and Raubenheimer, D. (2012). Testing the protein leverage hypothesis in a free-living human population. *Appetite*, 59, 312–15.

Martinez-Steele, E., Raubenheimer, D., Simpson, S. J., Baraldi, L. G., and Monteiro, C. A. (2018). Ultra-processed foods, protein leverage and energy intake in the USA. *Public Health Nutrition*, 21, 114–24.

Martins, E. P., and Hansen, T. F. (1997). Phylogenies and the comparative method: A general approach to incorporating phylogenetic information into the analysis of interspecific data. *American Naturalist*, 149, 646–67.

Martins, Q., Horsnell, W. G. C., Titus, W., Rautenbach, T., and Harris, S. (2011). Diet determination of the Cape Mountain leopards using global positioning system location clusters and scat analysis. *Journal of Zoology*, 283, 81–87.

Masette, M., Isabirye-Basuta, G., Baranga, D., and Chemurot, M. (2015). Levels of tannins in fruit diet of grey-cheeked mangabeys (*Lophocebus ugandae*, Groves) in Lake Victoria Basin forest reserves. *Journal of Ecology and the Natural Environment*, 7, 146–57.

Masette, M., Isabirye-Basuta, G., Baranga, D., Chapman, C. A., and Rothman, J. M. (2015). The challenge of interpreting primate diets: Mangabey foraging on *Blighia unijugata* fruit in relation to changing nutrient content. *African Journal of Ecology*, 53, 259–67.

Masi, S., Cipolletta, C., and Robbins, M. M. (2009). Western lowland gorillas (*Gorilla gorilla gorilla*) change their activity patterns in response to frugivory. *American Journal of Primatology*, 71, 91–100.

Masi, S., Gustafsson, E., Saint Jalme, M., Narat, V., Todd, A., Bromsel, M. C., et al. (2012). Unusual feeding behavior in wild great apes, a window to understand origins of self-medication in humans: Role of sociality and physiology on learning process. *Physiology and Behavior*, 105, 337–49.

Masi, S., Mundry, R., Ortmann, S., Cipolletta, C., Boitani, L., and Robbins, M. M. (2015). The influence of seasonal frugivory on nutrient and energy intake in wild western gorillas. *PLoS ONE*, 10, e0129254.

Masette, M., Isabiyre-Bastua, G., Baranga, D., Chapman, C. A., and Rothman, J. M. (2014). The challenge of interpreting primate diets: mangabey foraging on *Blighia unijugata* fruit in relation to changing nutrient content. *African Journal of Ecology*, 53:259–267.

Massad, T. J., Richards, L. A., Philbin, C., Yamaguchi, L. F., Kato, M. J., Jeffrey, C. S., et al. (2022). The chemical ecology of tropical forest diversity: Environmental variation, chemical similarity, herbivory, and richness. *Ecology*, 103(9), e3762.

Massey, F. P., Ennos, A. R., and Hartley, S. E. (2006). Silica in grasses as a defence against insect herbivores: Contrasting effects on folivores and a phloem feeder. *Journal of Animal Ecology*, 75, 595–603.

Matassa, C. M., and Trussell, G. C. (2011). Landscape of fear influences the relative importance of consumptive and non-consumptive predator effects. *Ecology*, 92, 2258–66.

Mathewson, P. D., Spehar, S. N., Meijaard, E., Sasmirul, A., and Marshall, A. J. (2008). Evaluating orangutan census techniques using nest decay rates: Implications for population estimates. *Ecological Applications*, 18, 208–21.

Mathy, J. W., and Isbell, L. A. (2001). The relative importance of size of food and interfood distance in eliciting aggression in captive rhesus macaques (*Macaca mulatta*). *Folia Primatologica*, 72, 268–77.

Matson, P. A., and Hunter, M. D. (1992). Special feature: The relative contributions to top-down and bottom-up forces in population and community ecology. *Ecology*, 73, 723.

Matsubara, M. (2003). Costs of mate guarding and opportunistic mating among wild male Japanese macaques. *International Journal of Primatology*, 24, 1057–75.

Matsuda, I., Chapman, C. A., and Clauss, M. (2019). Colobine forestomach anatomy and diet. *Journal of Morphology*, 280(11), 1608–16.

Matsuda, I., Tuuga, A., Bernard, H., Sugau, J., and Hanya, G. (2013). Leaf selection by two Bornean colobine monkeys in relation to plant chemistry and abundance. *Scientific Reports*, 3, 1873.

Matsumoto, S., Setoyama, H., and Umesaki, Y. (1992). Differential induction of major histocompatability complex molecules on mouse intestine by bacterial colonization. *Gastroenterology*, 103, 1777–82.

Matsumoto-Oda, A., and Oda, R. (1998). Changes in the activity budget of cycling female chimpanzees. *American Journal of Primatology*, 46, 157–66.

Matsusaka, T., Nishie, H., Shimada, M., Kutsukake, N., Zamma, K., Nakamura, M., et al. (2006). Tool-use for drinking water by immature chimpanzees of Mahale: Prevalence of an unessential behavior. *Primates*, 47, 113–22.

Matsuzawa, T. (1991). Grammar of action in stone-tool use by wild chimpanzees. *Reichorui Kenkyu*, 7, 160.

Matsuzawa, T. (1994). Field experiments on use of stone tools by chimpanzees in the wild. In R. W. Wrangham, W. C. McGrew, F. B. M. de Waal, and P. G. Heltne (Eds.), *Chimpanzee Cultures* (pp. 351–70). Harvard University Press.

Matsuzawa, T. (1999). Tool use and culture in wild chimpanzees. *International Journal of Psychology*, 35(3), 105.

Matthews, J. K., Ridley, A., Niyigaba, P., Kaplin, B. A., and Grueter, C.C. (2019). Chimpanzee feeding ecology and fallback food use in the montane forest of Nyungwe National Park, Rwanda *American Journal of Primatology*, 81, e22971.

Mattison, J. A., Roth, G. S., Beasley, T. M., Tilmont, E. M., Handy, A. M., Herbert, R. L., et al. (2012). Impact of caloric restriction on health and survival in rhesus monkeys from the NIA study. *Nature*, 489(7415), 318–21.

Mattison, J. A., and Vaughan, K. L. (2017). An overview of nonhuman primates in aging research. *Experimental Gerontology*, 94, 41–45.

Mattson, W. J. (1980). Herbivory in relation to plant nitrogen content. *Annual Review of Ecology and Systematics*, 11, 119–61.

Mau, M., de Almeida, A. M., Coelho, A. V., and Suedekum, K. H. (2011). First Identification of tannin-binding proteins in saliva of *Papio hamadryas* using MS/MS mass spectrometry. *American Journal of Primatology*, 73, 896–902.

Mau, M., Südekum, K. H., Johann, A., Silwa, A., and Kaiser, T. M. (2009). Saliva of the graminivorous *Theropithecus gelada* lacks proline-rich proteins and tannin-binding capacity. *American Journal of Primatology*, 71, 663–69.

Mau, M., Südekum, K. H., Johann, A., Sliwa, A., and Kaiser, T. M. (2010). Indication of higher salivary α-amylase expression in hamadryas baboons and geladas compared to chimpanzees and humans. *Journal of Medical Primatology*, 39(3), 187–90.

May, R. M. (1981). Models for single populations. In R. M. May (Ed.), *Theoretical Ecology* (pp. 5–29). Blackwell Scientific.

May, R. M., and Anderson, R. M. (1979). Population biology of infectious diseases: Part II. *Nature*, 280, 455–61.

Mayack, C., and Naug, D. (2013). Individual energetic state can prevail over social regulation of foraging in honeybees. *Behavioral Ecology and Sociobiology*, 67, 929–36.

Mayer, P., Geissler, K., Valent, P., Ceska, M., Bettelheim, P., and Liehl, E. (1991). Recombinant human interleukin 6 is a potent inducer of the acute phase response and elevates the blood platelets in nonhuman primates. *Experimental Hematology*, 19(7), 688–96.

Mayor, P., El Bizri, H., Bodmer, R. E., and Bowler, M. (2016). Reproductive biology for the assessment of hunting sustainability of rainforest mammal populations through the participation of local communities. *Conservation Biology*, 31, 912-923.

Mayr, E. (1961). Cause and effect in biology. *Science*, 134, 1501–6.

Mazmanian, S. K., Liu, C. H., Tzianabos, A. O., and Kasper, D. L. (2005). An immunomodulatory molecule of symbiotic bacteria directs maturation of the host immune system. *Cell*, 122, 107–18.

Mazoir, N., Benharref, A., Bailen, M., Reina, M., and Gonzalez-Coloma, A. (2008). Bioactive triterpene derivatives from latex of two *Euphorbia* species. *Phytochemistry*, 69, 1328–38.

McAlpine, C., Catterall, C. P., Nally, R. M., Lindenmayer, D., Reid, J. L., Holl, K. D., et al. (2016). Integrating plant- and animal-based perspectives for more effective restoration of biodiversity. *Frontiers in Ecology and the Environment*, 14(1), 37–45.

McArt, S. H., Spalinger, D. E., Collins, W. B., Schoen, E. R., Stevenson, T., and Bucho, M. (2009). Summer dietary nitrogen availability as a potential bottom-up constraint on moose in south-central Alaska. *Ecology*, 90, 1400–1411.

McArt, S. H., Spalinger, D. E., Kennish, J. M., and Collins, W. B. (2006). A modified method for determining tannin–protein precipitation capacity using accelerated solvent extraction (ASE) and microplate gel filtration. *Journal of Chemical Ecology*, 32, 1367–77.

McArthur, C., Banks, P. B., Boonstra, R., and Forbey, J. S. (2014). The dilemma of foraging herbivores: Dealing with food and fear. *Oecologia*, 176, 677–89.

McArthur, C., Orlando, P., Banks, P. B., and Brown, J. S. (2012). The foraging tightrope between predation risk and plant toxins: A matter of concentration. *Functional Ecology*, 26, 74–83.

McArthur, C., Sanson, G. D., and Beal, A. M. (1995). Salivary proline-rich proteins in mammals: Roles in oral homeostasis and counteracting dietary tannin. *Journal of Chemical Ecology*, 21, 663–91.

McCabe, G. M., and Emery Thompson, M. (2013). Reproductive seasonality in wild Sanje mangabeys (*Cercocebus sanjei*), Tanzania: Relationship between the capital breeding strategy and infant survival. *Behaviour*, 150, 1399–429.

McCabe, G. M., and Fedigan, L. M. (2007). Effects of reproductive status on energy intake, ingestion rates, and dietary composition of female *Cebus capucinus* at Santa Rosa, Costa Rica. *International Journal of Primatology*, 28, 837–51.

McCabe, G. M., Fernández, D., and Ehardt, C. L. (2013). Ecology of reproduction in Sanje mangabeys (*Cercocebus sanjei*): Dietary strategies and energetic condition during a high fruit period. *American Journal of Primatology*, 75, 1196–208.

McCall, A. C., and Irwin, R. E. (2006). Florivory: The intersection of pollination and herbivory. *Ecological Letters*, 9, 1351–65.

McCann, C. M. (1995). *Social Factors Affecting Reproductive Success in Female Gelada Baboons (Theropithecus gelada)*. City University of New York.

McCarthy, M. S., Lester, J. D., and Stanford, C. B. (2017). Chimpanzees (*Pan troglodytes*) flexibly use introduced species for nesting and bark feeding in a human-dominated habitat. *International Journal of Primatology*, 38, 321–37.

McClellan, H. L., Miller, S. J., and Hartmann, P. E. (2008). Evolution of lactation: Nutrition v. protection with special reference to five mammalian species. *Nutrition Research*, 21, 97–116.

McConkey, K. R. (2000). Primary seed shadow generated by gibbons in the rain forests of Barito Ulu, central Borneo. *American Journal of Primatology*, 52, 13–29.

McConkey, K. R., Aldy, F., Ario, A., and Chivers, D. J. (2002). Selection of fruit by gibbons (*Hylobates mulleri* × *agilis*) in the rain forests of central Borneo. *International Journal of Primatology*, 23, 123–45.

McConkey, K. R., Ario, A., Aldy, F., and Chivers, D. J. (2003). Influence of forest seasonality on gibbon food choice in the rain forests of Barito Ulu, central Borneo. *International Journal of Primatology*, 24, 19–32.

McConkey, K. R., and Brockelman, W. Y. (2011). Nonredundancy in the dispersal network of a generalist tropical forest tree. *Ecology*, 92, 1492–502.

McConkey, K. R., Brockelman, W. Y., and Saralamba, C. (2014). Mammalian frugivores with different foraging behavior can show similar seed dispersal effectiveness. *Biotropica*, 46, 647–51.

McConkey, K. R., Prasad, S., Corlett, R. T., Campos-Arceiz, A., Brodie, J. F., Rogers, H., and Santamaria, L. (2012). Seed dispersal in changing landscapes. *Biological Conservation*, 146, 1–13.

McCurdy, C. E., Bishop, J. M., Williams, S. M., Grayson, B. E., Smith, M. S., Friedman, J. E., et al. (2009). Maternal high-fat diet triggers lipotoxicity in the fetal livers of nonhuman primates. *Journal of Clinical Investigation*, 119(2), 323–35.

McDougall, P. (2012). Is passive observation of habituated animals truly passive? *Journal of Ethology*, 30, 219–23.

McDowell, L. R. (2003). *Minerals in Animal and Human Nutrition*. Elsevier Science B. V.

McDowell, N. G., Bond, B. J., Dickman, L. T., Ryan, M. G., and Whitehead, D. (2011). Relationships between tree height and carbon isotope discrimination. In F. C. Meinzer, B. Lachenbruch, and T. E. Dawson (Eds.), *Size- and Age-Related Changes in Tree Structure and Function: Tree Physiology* (pp. 255–86). Springer.

McEwen, B. S. (2000). Allostasis and allostatic load: Implications for neuropsychopharmacology. *Neuropsychopharmacology*, 22(2), 22.22.

McEwen, B. S., and Wingfield, J. C. (2003). The concept of allostasis in biology and biomedicine. *Hormones and Behavior*, 43(1), 2–15.

McFarland, R., Barrett, L., Boner, R., Freeman, N. J., and Henzi, S. P. (2014). Behavioral flexibility of vervet monkeys in response to climatic and social variability. *American Journal of Physical Anthropology*, 154, 357–64.

McFarland, R., Barrett, L., Fuller, A., Henzi, P., Maloney, S. K., Mitchell, D., et al. (2017). Feverish monkeys get kicked when they're down. *American Journal of Physical Anthropology*, 162, 126–26.

McFarland, R., Bartlett, L., Costello, M., Fuller, A., Hetem, R. S., Maloney, S. K., et al. (2020). Keeping cool in the heat: Behavioral thermoregulation and body temperature patterns in wild vervet monkeys. *American Journal of Physical Anthropology*, 171, 407–18.

McFarland, R., Henzi, S. P., Barrett, L., Bonnell, T., Fuller, A., Young, C., et al. (2021). Fevers and the social costs of acute infection in wild vervet monkeys. *Proceedings of the National Academy of Sciences of the United States of America*, 118(44), e2107881118.

McFarland Symington, M.M. (1988). Demography, ranging patterns, and activity budgets of black spider monkeys (*Ateles paniscus chamek*) in the Manu National Park, Peru. *American Journal of Primatology*, 15, 45–67.

McGarry, A. (1995). Cellular basis of tissue toughness in carrot (*Daucus carota* L.) storage roots. *Annals of Botany*, 75, 157–63.

McGeachin, R. L., and Akin, J. R. (1982). Amylase levels in the tissues and body fluids of several primate species. *Comparative Biochemistry and Physiology Part A*, 72(1), 267–69.

McGraw, W. S. (2005). Update on the search for Miss Waldron's red colobus monkey. *International Journal of Primatology*, 26, 605–19.

McGraw, W. S. (2007). Positional behavior and habitat use of Taï monkeys. In W. S. McGraw, K. Zuberhuhler, and R. Noe (Eds.), *The Monkeys of the Tai Forest: An African Primate Community* (pp. 223–55). Cambridge University Press.

McGraw, W. S., and Daegling, D. J. (2012). Primate feeding and foraging: Integrating studies of behavior and morphology. *Annual Review of Anthropology*, 41, 203–19.

McGraw, W. S., and Daegling, D. J. (2020). Diet, feeding behavior, and jaw architecture of Taï monkeys: Congruence and chaos in the realm of functional morphology. *Evolutionary Anthropology: Issues, News, and Reviews*, 29(1), 14–28. https://doi.org/10.1002/evan.21799.

McGraw, W. S., Pampush, J. D., and Daegling, D. J. (2012). Brief communication: Enamel thickness and durophagy in mangabeys revisited. *American Journal of Physical Anthropology*, 147(2), 326–33.

McGraw, W. S., Vick, A. E., and Daegling, D. J. (2011). Sex and age differences in the diet and ingestive behaviors of sooty mangabeys (*Cercocebus atys*) in the Tai forest, Ivory coast. *American Journal of Physical Anthropology*, 144, 140–53.

McGraw, W. S., Vick, A. E., and Daegling, D. J. (2014a). Dietary variation and food hardness in sooty mangabeys (*Cercocebus atys*): Implications for fallback foods and dental adaptation. *American Journal of Physical Anthropology*, 154, 413–23.

McGrew, W. C. (1974). Tool use by wild chimpanzees in feeding upon driver ants. *Journal of Human Evolution*, 3, 501–8.

McGrew, W. C. (1979). Evolutionary implications of sex differences in chimpanzee predation and tool use. In D. A. Hamburg and E. R. McCown (Eds.), *The Great Apes: Perspectives on Human Evolution* (vol. 5, pp. 441–63). Benjamin Cummings.

McGrew, W. C. (1992). *Chimpanzee Material Culture*. Cambridge University Press.

McGrew, W. C. (2013). Is primate tool use special? Chimpanzee and New Caledonian crow compared. *Philosophical Transactions of the Royal Society of London Series B—Biological Sciences*, 368, 1–8.

McGrew, W. C. (2014). The "other faunivory" revisited: Insectivory in human and nonhuman primates and the evolution of the human diet. *Journal of Human Evolution*, 71, 4–11.

McGrew, W. C., Baldwin, P. J., and Tutin, C. E. G. (1988). Diet of wild chimpanzees (*Pan troglodytes verus*) at Mt. Assirik, Senegal: 1. Composition. *American Journal of Primatology*, 16, 213–26.

McGrew, W. C., and Collins, D. A. (1985). Tool use by wild chimpanzees (*Pan troglodytes*) to obtain termites (*Macrotermes herus*) in the Mahale Mountains, Tanzania. *American Journal of Primatology*, 9, 47–62.

McGrosky, A., Meloro, C., Navarrete, A., Heldstab, S. A., Kitchener, A. C., Isler, K., et al. (2019). Gross intestinal morphometry and allometry in primates. *American Journal of Primatology*, 81(8), e23035. https://doi.org/10.1002/ajp.23035.

McIlwee, A. M., Lawler, I. R., Cork, S. J., and Foley, W. J. (2001). Coping with chemical complexity in mammal-plant interactions: Near-infrared spectroscopy as a predictor of *Eucalyptus* foliar nutrients and of the feeding rates of folivorous marsupials. *Oecologia*, 128, 539–48.

McIntyre, N. E., and Wiens, J. A. (1999). Interactions between landscape structure and animal behavior: The roles of heterogeneously distributed resources and food deprivation on movement patterns. *Landscape Ecology*, 14, 437–47.

McKay, L. I., and Cidlowski, J. A. (2003). Corticosteroids: Physiologic and pharmacologic effects of corticosteroids. In D. W. Kufe, R. E. Pollock, R. R. Weichselbaum, R. C. Bast, T. S. Gansler, J. F. Holland, et al. (Eds.), *Holland-Frei Cancer Medicine* (6th ed., p. 2900). BC Decker.

McKey, D. B. (1978). Soils, vegetation, and seed-eating by black colobus monkeys. In G. G. Montgomery (Ed.), *The Ecology of Arboreal Folivores* (pp. 423–37). Smithsonian Institution Press.

McKey, D. B., Gartlan, J. S., Waterman, P. G., and Choo, G. M. (1981). Food selection by black colobus monkeys (*Colobus satanas*) in relation to plant chemistry. *Biological Journal of the Linnean Society*, 16, 115–46.

McKey, D. B., Waterman, P. G., Gartlan, J. S., and Struhsaker, T. T. (1978). Phenolic content of vegetation in two African rain forests: Ecological implications. *Science*, 202, 61–64.

McKinney, T. (2011). The effects of provisioning and crop-raiding on the diet and foraging activities of human-commensal white-faced capuchins (*Cebus capucinus*). *American Journal of Primatology*, 73, 439–48.

McLachlan, J. A. (2001). Environmental signaling: What embryos and evolution teach us about endocrine disrupting chemicals. *Endocrine Reviews*, 22, 319–41.

McLaren, K. P., and McDonald, M. A. (2005). Seasonal patterns of flowering and fruiting in dry tropical forest in Jamaica. *Biotropica*, 37, 584–90.

McLean, K. A., Trainor, A. M., Asner, G. P., Crofoot, M. C., Hopkins, M. E., Campbell, C. J., et al. (2016). Movement patterns of three arboreal primates in a Neotropical moist forest explained by LiDAR-estimated canopy structure. *Landscape Ecology*, 31, 1–14.

McLennan, M. R., and Ganzhorn, J. U. (2017). Nutritional characteristics of wild and cultivated foods for chimpanzees (*Pan troglodytes*) in agricultural landscapes. *International Journal of Primatology*, 38, 122–50.

McLennan, M. R., Spagnoletti, N., and Hockings, K. J. (2017). The implications of primate behavioral flexibility for sustainable human–primate coexistence in anthropogenic habitats. *International Journal of Primatology*, 38, 105–21.

McLeod, S. R. (1997). Is the concept of carrying capacity useful in variable environments? *Oikos*, 79, 529–42.

McLester, E. (2022). Golden-bellied mangabeys (*Cercocebus chrysogaster*) consume and share mammalian prey at Lui Kotale, DRC. *Journal of Tropical Ecology*, 38, 254–58.

McLester, E., Brown, M., Stewart, F. A., and Piel, A. K. (2019). Food abundance and weather influence habitat-specific ranging patterns in forest- and savanna mosaic-dwelling red-tailed monkeys (*Cercopithecus ascanius*). *American Journal of Physical Anthropology*, 170, 217–31.

McManaman, J. L. (2009). Formation of milk lipids: A molecular perspective. *Journal of Clinical Lipidology*, 4, 391–401.

McManaman, J. L., and Neville, M. C. (2003). Mammary physiology and milk secretion. *Advanced Drug Delivery Reviews*, 55, 629–41.

McNab, B. K. (1980). Food habits, energetics, and the population biology of mammals. *American Naturalist*, 116, 106–24.

McNab, B. K. (1986). The influence of food habits on the energetics of eutherian mammals. *Ecological Monographs*, 56, 1–19.

McNab, B. K. (2002). *The Physiological Ecology of Vertebrates: A View from Energetics*. Cornell University Press.

McNab, B. K. (2010). Geographic and temporal correlations of mammalian size reconsidered: A resource rule. *Oecologia*, 164, 13–23.

McNaughton, S. J., Tarrants, J. L., McNaughton, M. M., and Davis, R. H. (1985). Silica as a defense against herbivory and a growth promoter in African grasses. *Ecology*, 66, 528–35.

McRae, J. F., Jaeger, S. R., Bava, C. M., Beresford, M. K., Hunter, D., Jia, Y., et al. (2013). Identification of regions associated with variation in sensitivity to food-related odors in the human genome. *Current Biology*, 23, 1596–600.

Meagher, L. P., and Beecher, G. R. (2000). Assessment of data on the lignan content of foods. *Journal of Food Composition and Analysis*, 13, 935–47.

Meder, A. (1986). Physical and activity changes associated with pregnancy in captive lowland gorillas (*Gorilla gorilla gorilla*). *American Journal of Primatology*, 11, 111–16.

Medina, E., and Minchin, P. (1980). Stratification of δ¹³C values of leaves in Amazonian rain forests. *Oecologia*, 45, 377–78.

Meehan, C. L., Lackey, K. A., Hagen, E. H., Williams, J. E., Roulette, J., Helfrecht, C., et al. (2018). Social networks, cooperative breeding, and the human milk microbiome. *American Journal of Human Biology*, 30(4), e23131.

Mehlman, P. T. (2008). Current status of wild gorilla populations and strategies for their conservation. In T. S. Stoinski, H. D. Steklis, and P. T. Mehlman (Eds.), *Conservation in the 21st Century: Gorillas as a Case Study* (pp. 3–54). Springer Science.

Meier, B., and Albignac, R. (1991). Rediscovery of *Allocebus trichotis* Günther 1875 (Primates) in northeast Madagascar. *Folia Primatologica*, 56, 57–63.

Meijaard, E., Sheil, D., Nasi, R., Augeri, D., Rosenbaum, B., Iskander, D., et al. (2005). *Life after Logging: Reconciling Wildlife Conservation and Production Forestry in Indonesian Borneo*. CIFOR.

Meise, K., Franks, D. W., and Bro-Jørgensen, J. (2019). Using social network analysis of mixed-species groups in African savannah herbivores to assess how community structure responds to environmental change. *Philosophical Transactions of the Royal Society B*, 347, 20190009.

Mekota, A. M., Grupe, G., Ufer, S., and Cuntz, U. (2006). Serial analysis of stable nitrogen and carbon isotopes in hair: Monitoring starvation and recovery phases of patients suffering from anorexia nervosa. *Rapid Communications in Mass Spectrometry*, 20, 1604–10.

Meldrum, D. J., and Kay, R. F. (1997). *Nuciruptor rubricae*, a new pitheciin seed predator from the Miocene of Colombia. *American Journal of Physical Anthropology*, 102(3), 407–27.

Melin, A. D., Chiou, K. L., Walco, E. R., Bergstrom, M. L., Kawamura, S., and Fedigan, L. M. (2017). Trichromacy increases fruit intake rates of wild capuchins (*Cebus capucinus imitator*). *Proceedings of the National Academy of Sciences of the United States of America*, 114, 10402–7.

Melin, A. D., Crowley, B. E., Brown, S. T., Wheatley, P. V., Moritz, G. L., Yu, F. T. Y., et al. (2014). Technical note: Calcium and carbon stable isotope ratios as paleodietary indicators. *American Journal of Physical Anthropology*, 154, 633–43.

Melin, A. D., Fedigan, L. M., Hiramatsu, C., Hiwatashi, T., Parr, N., and Kawamura, S. (2009). Fig foraging by dichromatic and trichromatic *Cebus capucinus* in a tropical dry forest. *International Journal of Primatology*, 30, 753.

Melin, A. D., Fedigan, L. M., Hiramatsu, C., Sendall, C. L., and Kawamura, S. (2007). Effects of colour vision phenotype on insect capture by a free-ranging population of white-faced capuchins, *Cebus capucinus*. *Animal Behaviour*, 73, 205–14.

Melin, A. D., Hiramatsu, C., Parr, N. A., Matsushita, Y., Kawamura, S., and Fedigan, L. M. (2014). The behavioral ecology of color vision: Considering fruit conspicuity, detection distance and dietary importance. *International Journal of Primatology*, 35, 258–87.

Melin, A. D., Khetpal, V., Matsushita, Y., Zhou, K., Campos, F. A., Welker, B., et al. (2017). Howler monkey foraging ecology suggests convergent evolution of routine trichromacy as an adaptation for folivory. *Ecology and Evolution*, 7, 1421–34.

Melin, A. D., Kline, D. W., Hickey, C. M., and Fedigan, L. M. (2013). Food search through the eyes of a monkey: A functional substitution approach for assessing the ecology of primate color vision. *Vision Research*, 86, 87–96.

Melin, A. D., Kline, D. W., Hiramatsu, C., and Caro, T. (2016). Zebra stripes through the eyes of their predators, zebras, and humans. *PLoS ONE*, 11, e0145679.

Melin, A. D., Matsushita, Y., Moritz, G. L., Dominy, N. J., and Kawamura, S. (2013). Inferred L/M cone opsin polymorphism of ancestral tarsiers sheds dim light on the origin of anthropoid primates. *Proceedings of the Royal Society B: Biological Sciences*, 280, 20130189.

Melin, A. D., Moritz, G. L., Fosbury, R. A. E., Kawamura, S., and Dominy, N. J. (2012). Why aye-ayes see blue. *American Journal of Primatology*, 74, 185–92.

Melin, A. D., Nevo, O., Shirasu, M., Williamson, R. E., Garrett, E. C., Endo, M., et al. (2019). Fruit scent and observer colour vision shape food-selection strategies in wild capuchin monkeys. *Nature Communications*, 10, 2407.

Melin, A. D., Shirasu, M., Matsushita, Y., Myers, M. S., Bergstrom, M. L., Venkataraman, V., et al. (2015). Examining the links among fruit signals, nutritional value, and the sensory behaviors of wild capuchin monkeys (*Cebus capucinus*) *American Journal of Physical Anthropology*, 156, 223–24.

Melin, A. D., Veilleux, C. C., Janiak, M. C., Hiramatsu, C., Sánchez-Solano, K. G., Lundeen, I. K., et al. (2022). Anatomy and dietary specialization influence sensory behaviour among sympatric primates. *Proceedings of the Royal Society B*, 289(1981), 20220847.

Melin, A. D., Webb, S. E., Williamson, R. E., and Chiou, K. L. (2018). Data collection in field primatology: a renewed look at measuring foraging behaviour. In Kalbitzer, U., and Jack, K. M. (Eds.) *Primate life histories, sex roles, and adaptability: essays in honour of Linda M. Fedigan* (pp. 161–92). Springer.

Melin, A. D., Wells, K., Moritz, G. L., Kistler, L., Orkin, J. D., Timm, R. M., et al. (2016). Euarchontan opsin variation brings new focus to primate origins. *Molecular Biology and Evolution*, 33, 1029–41.

Melin, A. D., Young, H. C., Mosdossy, K. N., and Fedigan, L. M. (2014). Seasonality, extractive foraging and the evolution of primate sensorimotor intelligence. *Journal of Human Evolution*, 71, 77–86.

Mella, V. S. A., Banks, P. B., and McArthur, C. (2014). Negotiating multiple cues of predation risk in a landscape of fear: What scares free-ranging brushtail possums? *Journal of Zoology*, 294, 22–30.

Mella, V. S. A., Ward, A. J. W., Banks, P. B., and McArthur, C. (2015). Personality affects the foraging response of a mammalian herbivore to the dual costs of food and fear. *Oecologia*, 177, 293–303.

Mellars, P. A. (1996). *The Neanderthal Legacy: An Archaeological Perspective from Western Europe.* Princeton University Press.

Ménard, N., Motsch, P., Delahaye, A., Saintvanne, A., Le Flohic, G., Dupé, S., et al. (2013). Effect of habitat quality on the ecological behaviour of a temperate-living primate: Time-budget adjustments. *Primates*, 54, 217–28.

Ménard, N., Motsch, P., Delahaye, A., Saintvanne, A., Le Flohic, G., Dupé, S., et al. (2014). Effect of habitat quality on diet flexibility in Barbary macaques. *American Journal of Primatology*, 76, 679–93.

Mendes, F. D. C., Cardoso, R. M., Ottoni, E. B., Izar, P., Villar, D. N. A., and Marquezan, R. F. (2015). Diversity of nutcracking tool sites used by *Sapajus libidinosus* in Brazilian Cerrado. *American Journal of Primatology*, 77, 535–46.

Mendes, S. L. (1989). Estudo ecológico de *Alouatta fusca* (Primates: Cebidae) na Estação Biológica de Caratinga, MG. *Revista Nordestina de Biologia*, 6, 71–104.

Menezes, J. C. T., and Martini, M. A. (2017). Predators of bird nests in the Neotropics: A review. *Journal of Field Ornithology*, 88, 99–114.

Mentis, M. T., and Duke, R. R. (1976). Carrying capacities of natural veld in Natal for large wild herbivores. *South African Journal of Wildlife Research*, 6, 65–74.

Menzel, C. R. (1991). Cognitive aspects of foraging in Japanese monkeys. *Animal Behaviour*, 41, 397–402.

Menzel, C. R., Savage-Rumbaugh, E. S., and Menzel, E. W. (2002). Bonobo (*Pan paniscus*) spatial memory and communication in a 20-hectare forest. *International Journal of Primatology*, 23, 601–19.

Menzel, E. W. (1973). Chimpanzee spatial memory organization. *Science*, 182, 943–45.

Mercader, J., Barton, H., Gillespie, J., Harris, J., Kuhn, S., Tyler, R., et al. (2007). 4,300-year-old chimpanzee sites and the origins of percussive stone technology. *Proceedings of the National Academy of Sciences of the United States of America*, 104, 3043–48.

Mercader, J., Panger, M., and Boesch, C. (2002). Excavation of a chimpanzee stone tool site in the African rainforest. *Science*, 296, 1452–55.

Merceron, G., Ramdarshan, A., Blondel, C., Boisserie, J. R., Brunetiere, N., Francisco, A., et al. (2016). Untangling the environmental from the dietary: Dust does not matter. *Proceedings of the Royal Society B: Biological Sciences*, 283(1838), 20161032.

Merrill, A. L., and Watt, B. K. (1955). Energy value of foods: Basis and derivation. In Agricultural Research Service (Ed.), *Agricultural Handbook No. 74.* USDA.

Mertl-Millhollen, A. S., Moret, E. S., Felantsoa, D., Rasamimanana, H., Blumenfeld-Jones, K. C., and Jolly, A. (2003). Ring-tailed lemur home ranges correlate with food abundance and nutritional content at a time of environmental stress. *International Journal of Primatology*, 24(5), 969–85.

Mesa, F., Magan-Fernandez, A., Munoz, R., Papay-Ramirez, L., Poyatos, R., Sanchez-Fernandez, E., et al. (2014). Catecholamine metabolites in urine, as chronic stress biomarkers, are associated with higher risk of chronic periodontitis in adults. *Journal of Periodontology*, 85(12), 1755–62.

Messaoudi, M., Lalonde, R., Violle, N., Javelot, H., Desor, D., Nejdi, A., et al. (2010). Assessment of psychotropic-like properties of a probiotic formulation (*Lactobacillus helveticus* R0052 and *Bifidobacterium longum* R0175) in rats and human subjects. *British Journal of Nutrition*, 105, 755–64.

Messeri, P. (1978). Some observations on a littoral troop of yellow baboons. *Monitore Zoologico Italiano*, 12, 69.

Metzler, M., Pfeiffer, E., and Hildebrand, A. A. (2010). Zearalenone and its metabolites as endocrine disrupting chemicals. *World Mycotoxin Journal*, 3, 385–401.

Meunier, H., Leca, J. B., Deneubourg, J. L., and Petit, O. (2006). Group movement decisions in capuchin monkeys: The utility of an experimental study and a mathematical model to explore the relationship between individual and collective behaviours. *Behaviour*, 143, 1511–27.

Meyer, A. L. S., and Pie, M. R. (2022). Climate change estimates surpass rates of climatic niche evolution in primates. *International Journal of Primatology*, 43, 40–56.

Meyer, D., Zeileis, A., and Hornik, K. (2016). vcd: Visualising categorical data. R package version 1.4-3.

Meyerhof, W., Batram, C., Kuhn, C., Brockhoff, A., Chudoba, E., Bufe, B., et al. (2010). The molecular receptive ranges of human TAS2R bitter taste receptors. *Chemical Senses*, 35, 157–70.

Meyrier, E., Jenni, L., Boetsch, Y., Strebel, S., Erne, B., and Tablado, Z. (2017). Happy to breed in the city? Urban food resources limit reproductive output in western jackdaws. *Ecology and Evolution*, 7, 1363–74.

Michael, H., Amimo, J. O., Rajashekara, G., Saif, L. J., and Vlasova, A. N. (2022). Mechanisms of kwashiorkor-associated immune suppression: Insights from human, mouse, and pig studies. *Frontiers in Immunology*, 13, 826268.

Michel, E. S., Flinn, E. B., Demarais, S., Strickland, B. K., Wang, G., and Dacus, C. M. (2016). Improved nutrition cues switch from efficiency to luxury phenotypes for a long-lived ungulate. *Ecology and Evolution*, 6, 7276–85.

Michod, R. E. (1979). Evolution of life histories in response to age-specific mortality factors. *American Naturalist*, 113, 531–50.

Milinski, M. (2014). The past and the future of behavioral ecology. *Behavioral Ecology*, 25, 680–84.

Millar, C. I., Stephenson, N. L., and Stephens, S. L. (2007). Climate change and forests of the future: Managing in the face of uncertainty. *Ecological Applications*, 17(8), 2145–51.

Miller, K. E., Bales, K. L., Ramos, J. H., and Dietz, J. M. (2006). Energy intake, energy expenditure, and reproductive costs of female wild golden lion tamarins (*Leontopithecus rosalia*). *American Journal of Primatology*, 68, 1037–53.

Miller, R. C., Brindle, E., Holman, D. J., Shofer, J., Klein, N. A., Soules, M. R., et al. (2004). Comparison of specific gravity and creatinine for normalizing urinary reproductive hormone concentrations. *Clinical Chemistry*, 50(5), 924–32.

Milligan, L. (2010). Milk composition of captive tufted capuchins (*Cebus apella*). *American Journal of Primatology*, 72, 81–86.

Milligan, L. A., and Bazinet, R. P. (2008). Evolutionary modifications of human milk composition: Evidence from long-chain polyunsaturated fatty acid composition of anthropoid milks. *Journal of Human Evolution*, 55, 1086–95.

Milligan, L. A., Rapoport, S. I., Cranfield, M. R., Dittus, W., Glander, K. E., Oftedal, O. T., et al. (2008). Fatty acid composition of anthropoid milks. *Comparative Biochemistry and Physiology A—Molecular and Integrative Physiology*, 149, 74–82.

Milligan, S. R., Kalita, J. C., Heyerick, A., Rong, H., De Cooman, L., and De Keukeleire, D. (1999). Identification of a potent phytoestrogen in hops (*Humulus lupulus l.*) and beer. *Journal of Clinical Endocrinology and Metabolism*, 84, 2249–52.

Milne, A. (1962). On a theory of natural control of insect population. *Journal of Theoretical Biology*, 3, 19–50.

Milton, K. (1978). Behavioral adaptations to leaf-eating by the mantled howler monkey (*Alouatta palliata*). In G. G. Montgomery (Ed.), *The Ecology of Arboreal Folivores* (pp. 535–49). Smithsonian Institution Press.

Milton, K. (1979). Factors influencing leaf choice by howler monkeys: A test of some hypotheses of food selection by generalist herbivores. *American Naturalist*, 114, 362–78.

Milton, K. (1980). *The Foraging Strategy of Howler Monkeys: A Study in Primate Economics*. Columbia University Press.

Milton, K. (1981a). Distribution patterns of tropical plant foods as an evolutionary stimulus to primate mental development. *American Anthropologist*, 83, 534–48.

Milton, K. (1981b). Food choice and digestive strategies of two sympatric primate species. *American Naturalist*, 117, 496–505.

Milton, K. (1984). Habitat, diet, and activity patterns of free-ranging woolly spider monkeys (*Brachyteles arachnoides* E. Geoffroy 1806). *International Journal of Primatology*, 5, 491–514.

Milton, K. (1986). Digestive physiology in primates. *Physiology*, 1, 76–79.

Milton, K. (1987). Primate diets and gut morphology: Implications for hominid evolution. In M. Harris and E. B. Ross (Eds.), *Food and Evolution: Toward a Theory of Human Food Habits* (pp. 93–115). Temple University Press.

Milton, K. (1991). Leaf change and fruit production in 6 Neotropical Moraceae species. *Journal of Ecology*, 79, 1–26.

Milton, K. (1993). Diet and primate evolution. *Scientific American*, 269, 86–93.

Milton, K. (1996). Effects of bot fly (*Alouattamyia baeri*) parasitism on a free-ranging howler monkey (*Alouatta palliata*) population in Panama. *Journal of Zoology*, 239, 39–63.

Milton, K. (1998). Physiological ecology of howlers (*Alouatta*): Energetic and digestive considerations and comparison with the Colobinae. *International Journal of Primatology*, 19, 513–48.

Milton, K. (1999). A hypothesis to explain the role of meat-eating in human evolution. *Evolutionary Anthropology*, 8, 11–21.

Milton, K. (2000a). Back to basics: Why foods of wild primates have relevance for modern human health. *Nutrition*, 16(7), 480–83.

Milton, K. (2000b). Quo vadis? Tactics of food search and group movements in primates and other animals. In S. Boinski and P. A. Garber (Eds.), *On the Move: How and Why Animals Travel in Groups* (pp. 375–417). University of Chicago Press.

Milton, K. (2003a). The critical role played by animal source foods in human (*Homo*) evolution. *Journal of Nutrition*, 133, 3886S–92S.

Milton, K. (2003b). Micronutrient intakes of wild primates: Are humans different? *Comparative Biochemistry and Physiology A—Molecular and Integrative Physiology*, 136, 47–59.

Milton, K. (2008). Macronutrient patterns of 19 species of Panamanian fruits from Barro Colorado Island. *Neotropical Primates*, 15, 1–7.

Milton, K., and Demment, M. W. (1988). Digestion and passage kinetics of chimpanzees fed high and low fiber diets and comparison with human diet. *Journal of Nutrition*, 118, 1082–88.

Milton, K., and Dintzis, F. R. (1981). Nitrogen-to-protein conversion factors for tropical plant-samples. *Biotropica*, 13, 177–81.

Milton, K., and Giacalone, J. (2013). Differential effects of unusual climatic stress on capuchin (*Cebus capucinus*) and howler monkey (*Alouatta palliata*) populations on Barro Colorado Island, Panama. *American Journal of Primatology*, 76, 249–61.

Milton, K., and Jenness, R. (1987). Ascorbic acid content of neotropical plant parts available to wild monkeys and bats. *Experientia*, 43, 339–42.

Milton, K., and May, M. L. (1976). Body weight, diet and home range area in primates. *Nature*, 259, 459–62.

Milton, K., Morrison, D. W., Estribi, M. A., and Windsor, D. M. (1982). Fruiting phenologies of two Neotropical *Ficus* species. *Ecology*, 63, 752–62.

Milton, K., Van Soest, P. J., and Robertson, J. B. (1980). Digestive efficiencies of wild howler monkeys. *Physiological Zoology*, 53, 402–9.

Min, B. R., Barry, T. N., Attwood, G. T., and McNabb, W. C. (2003). The effect of condensed tannins on the nutrition and health of ruminants fed fresh temperate forages: A review. *Animal Feed Science and Technology*, 106, 3–19.

Minot, S., Bryson, A., Chehoud, C., Wu, G. D., Lewis, J. D., and Bushman, F. D. (2013). Rapid evolution of the human gut virome. *Proceedings of the National Academy of Sciences of the United States of America*, 110(30), 12450–55.

Minot, S., Sinha, R., Chen, J., Li, H., Keilbaugh, S. A., Wu, G. D., et al. (2011). The human gut virome: Inter-individual vari-

ation and dynamic response to diet. *Genome Research*, 21, 1616–25.
Minta, S. C., Minta, K. A., and Lott, D. F. (1992). Hunting associations between badgers (*Taxidea taxus*) and coyotes (*Canis latrans*). *Journal of Mammalogy*, 73, 814–20.
Miranda, J. M. D., and Passos, F. C. (2004). Hábito alimentar de *Alouatta guariba* (Humboldt) (Primates, Atelidae) em Floresta de Araucária, Paraná, Brasil. *Revista Brasiliera de Zoologica*, 21, 821–26.
Mirón, L. L., Herrera, L. G., Ramirez, N., and Hobson, K. A. (2006). Effect of diet quality on carbon and nitrogen turnover and isotopic discrimination in blood of a New World nectarivorous bat. *Journal of Experimental Biology*, 209, 541–48.
Mirzaei, M. K., Khan, M. A. A., Ghosh, P., Taranu, Z. E., Taguer, M., Ru, J., et al. (2020). Bacteriophages isolated from stunted children can regulate gut bacterial communities in an age-specific manner. *Cell Host and Microbe*, 27, 199–212.
Mitani, J. C., and Rodman, P. S. (1979). Territoriality: The relation of ranging pattern and home range size to defendability, with an analysis of territoriality among primate species. *Behavioral Ecology and Sociobiology*, 5, 241–51.
Mitani, J. C., and Watts, D. (1997). The evolution of non-maternal caretaking among anthropoid primates: Do helpers help? *Behavioral Ecology and Sociobiology*, 40, 213–20.
Mitani, J. C., and Watts, D. P. (1999). Demographic influences on the hunting behavior of chimpanzees. *American Journal of Physical Anthropology*, 109, 439–54.
Mitani, J. C., and Watts, D. P. (2001). Why do chimpanzees hunt and share meat? *Animal Behaviour*, 61, 915–24.
Mitani, J. C., and Watts, D. P. (2005a). Correlates of territorial boundary patrol behaviour in wild chimpanzees. *Animal Behaviour*, 70, 1079–86.
Mitani, J. C., and Watts, D. P. (2005b). Primate hunting seasonality. In D. Brockman and C. P. van Schaik (Eds.), *Seasonality in Primates: Studies of Living and Extinct Human and Non-human Primates* (pp. 215–42). Cambridge University Press.
Mitani, J. C., and Watts, D. P. (2005c). Seasonality in hunting by non-human primates. In D. K. Brockman and C. P. van Schaik (Eds.), *Seasonality in Primates: Studies of Living and Extinct Human and Non-human Primates* (pp. 215–41). Cambridge University Press.
Mitani, J. C., Watts, D., and Muller, M. (2002). Recent developments in the study of wild chimpanzee behavior. *Evolutionary Anthropology*, 11, 9–25.
Mitani, M. (1991). Niche overlap and polyspecific associations among sympatric cercopithecids in the Campo Animal Reserve, southwestern Cameroon. *Primates*, 32, 137–51.
Mitchell, C. L., Boinski, S., and van Schaik, C. P. (1991). Competitive regimes and female bonding in two species of squirrel monkeys (*Saimiri oerstedi* and *S. sciureus*). *Behavioral Ecology and Sociobiology*, 28, 55–60.
Mitoulas, L. R., Kent, J. C., Cox, D. B., Owens, R. A., Sherriff, J. L., and Hartmann, P. E. (2002). Variation in fat, lactose and protein in human milk over 24 h and throughout the first year of lactation. *British Journal of Nutrition*, 88, 29–37.
Mittermeier, R. A., Reuter, K. E., Rylands, A. B., Jerusalinsky, L., Schwitzer, C., Strier, K. B., et al. (2022). *Primates in Peril: The World's 25 Most Endangered Primates, 2022–2023*. IUCN Primate Specialist Group, International Primatological Society, Rewild.
Mittermeier, R. A., Rylands, A. B., and Wilson, D. E. (2013). *Primates* (vol. 3). Lynx.
Mittermeier, R. A., and van Roosmalen, M. G. M. (1981). Preliminary observations on habitat utilization and diet in eight Surinam monkeys. *Folia Primatologica*, 36, 1–39.
M'Kirera, F., and Ungar, P. S. (2003). Occlusal relief changes with molar wear in *Pan troglodytes troglodytes* and *Gorilla gorilla gorilla*. *American Journal of Primatology*, 60(2), 31–41.
Moehrenschlager, A., Macdonald, D. W., and Moehrenschlager, C. (2003). Reducing capture-related injuries and radio-collaring effects on swift foxes. In L. Carbyn and M. A. Sovada (Eds.), *The Swift Fox: Ecology and Conservation of Swift Foxes in a Changing World* (pp. 107–13). University of Regina Press.
Moen, R., Cohen, Y., and Pastor, J. (1998). Linking moose population and plant growth models with a moose energetics model. *Ecosystems*, 1, 52–63.
Moen, R., Pastor, J., and Cohen, Y. (1997). A spatially explicit model of moose foraging and energetics. *Ecology*, 78, 505–21.
Moldawer, A. M., Ungaro, R., Setia, T. M., and Vogel, E. R. (2014). A non-invasive methodology for monitoring physiological and endocrinological responses to stressors in correlation to skeletal protein breakdown in wild orangutans. *American Journal of Physical Anthropology*, 153, 188.
Moldawer, L. L., and Copeland, E. M. (1997). Proinflammatory cytokines, nutritional support, and the cachexia syndrome. *Cancer*, 79(9), 1828–39.
Mole, S., Butler, L., and Iason, G. (1990). Defense against dietary tannin in herbivores: A survey for proline rich salivary proteins in mammals. *Biochemical Systematics and Ecology*, 18, 287–93.
Moll, R. J., Redilla, K. M., Mudumba, T., Muneza, A. B., Gray, S. M., Abade, L., et al. (2017). The many faces of fear: A synthesis of the methodological variation in characterizing predation risk. *Journal of Animal Ecology*, 86, 749–65.
Monteiro, C. A., Levy, R. B., Claro, R., Martins, A. P. B., Louzada, M. L. C., Baraldi, L. G., et al. (2013). Ultra-processed food and drink products and obesity: A new hypothesis, and evidence. *Annals of Nutrition and Metabolism*, 63, 1007.
Montgomery, G. G. (1978). *The Ecology of Arboreal Folivores*. Smithsonian Institution Press.
Moore, B. D., and Foley, W. J. (2005). Tree use by koalas in a chemically complex landscape. *Nature*, 435, 488–90.
Moore, B. D., Lawler, I. R., Wallis, I. R., Beale, C. M., and Foley, W. J. (2010). Palatability mapping: A koala's eye view of spatial variation in habitat quality. *Ecology*, 91, 3165–76.

Moore, B. D., Wallis, I. R., Marsh, K. J., and Foley, W. J. (2004). The role of nutrition in the conservation of the marsupial folivores of eucalypt forests. In D. Lunney (Ed.), *Conservation of Australia's Forest Fauna* (pp. 549–75). Royal Zoological Society of NSW.

Moore, B. D., Wallis, I. R., Pala-Paul, J., Brophy, J. J., Willis, R. H., and Foley, W. J. (2004). Antiherbivore chemistry of *Eucalyptus*—cues and deterrents for marsupial folivores. *Journal of Chemical Ecology*, 30, 1743–69.

Moore, J., Black, J., Hernandez-Aguilar, R. A., Idani, G. I., Piel, A., and Stewart, F. (2017). Chimpanzee vertebrate consumption: savanna and forest chimpanzees compared. *Journal of Human Evolution*, 112, 30-40.

Moore, S. E., Cole, T. J., Poskitt, E. M. E., Sonko, B. J., Whitehead, R. G., McGregor, I. A., et al. (1997). Season of birth predicts mortality in rural Gambia. *Nature*, 388, 434–34.

Mora, J. R., Iwata, M., and von Andrian, U. H. (2008). Vitamin effects on the immune system: Vitamins A and D take centre stage. *Nature Reviews Immunology*, 8(9), 685–98.

Morabia, A., Costanza, M. C., and World Health Organization. (1998). International variability in ages at menarche, first livebirth and menopause. *American Journal of Epidemiology*, 148, 1195–205.

Moraes, P. L. R., de Carvalho, O., and Strier, K. B. (1998). Population variation in patch and party size in muriquis (*Brachyteles arachnoides*). *International Journal of Primatology*, 19, 325–37.

Morand, S., and Harvey, P. H. (2000). Mammalian metabolism, longevity and parasite species richness. *Proceedings of the Royal Society of London B: Biological Sciences*, 267, 1999–2003.

Morand-Ferron, J., Wu, G. M., and Giraldeau, L. A. (2011). Persistent individual differences in tactic use in producer-scrounger game are group dependent *Animal Behaviour*, 82, 811–16.

Morehouse, N. I., Nakazawa, T., Booher, C. M., Jeyasingh, P. D., and Hall, M. D. (2010). Sex in a material world: Why the study of sexual reproduction and sex-specific traits should become more nutritionally-explicit. *Oikos*, 119, 766–78.

Moreira, E., Burghi, G., and Manzanares, W. (2018). Update on metabolism and nutrition therapy in critically ill burn patients. *Medicina Intensiva (English Edition)*, 42(5), 306–16.

Moreira, L. A., Duytschaever, G., Higham, J. P., and Melin, A. D. (2019). Platyrrhine color signals: New horizons to pursue. *Evolutionary Anthropology*, 28(5), 236–48.

Morera, E., De Petrocellis, L., Morera, L., Schiano-Moriello, A., Nalli, M., DiMarzo, V. et al. (2012). Synthesis and biological evaluation of [6]-gingerol analogues as transient receptor potential channel TRPV1 and TRPA1 modulators. *Bioorganic and Medicinal Chemistry Letters*, 22, 1674–77.

Moret, Y., and Schmid-Hempel, P. (2000). Survival for immunity: The price of immune system activation for bumblebee workers. *Science*, 290(5494), 1166–68.

Mori, A. (1979). Analysis of population changes by measurement of body weight in the Koshima troop of Japanese monkeys. *Primates*, 20, 371–97.

Morimoto, J. (2022). Nutrigonometry II: Experimental strategies to maximize nutritional information in multidimensional performance landscapes. *Ecology and Evolution*, 12(8), e9174.

Morimoto, J., and Lihoreau, M. (2019). Quantifying nutritional trade-offs across multidimensional performance landscapes. *American Naturalist*, 193, E168–E181.

Moritz, G. L. (2015). *Primate Origins through the Lens of Functional and Degenerate Cone Opsins*. Dartmouth College.

Moritz, G. L., Melin, A. D., Tuh Yit Tu, F., Bernard, H., Ong, P. S., and Dominy, N. J. (2014). Niche convergence suggests functionality of the nocturnal fovea. *Frontiers in Integrated Neuroscience*, 8, 61.

Moritz, G. L., Ong, P. S., Perry, G. H., and Dominy, N. J. (2017). Functional preservation and variation in the cone opsin genes of nocturnal tarsiers. *Philosophical Transactions of the Royal Society of London Series B—Biological Sciences*, 372, 20160075.

Morris, C. R., Hamilton-Reeves, J., Martindale, R. G., Sarav, M., and Ochoa Gautier, J. B. (2017). Acquired amino acid deficiencies: A focus on arginine and glutamine. *Nutrition in Clinical Practice*, 32(suppl. 1), 30S–47S.

Morris, D. W. (2005). Habitat-dependent foraging in a classic predator-prey system: A fable from snowshoe hares. *Oikos*, 109, 239–54.

Morrogh-Bernard, H. C., Foitová, I., Yeen, Z., Wilkin, P., De Martin, R., Rárová, L., et al. (2017). Self-medication by orangutans (*Pongo pygmaeus*) using bioactive properties of *Dracaena cantleyi*. *Scientific Reports*, 7(1), 16653.

Morse, D. H. (1970). Ecological aspects of some mixed-species foraging flocks of birds. *Ecological Monographs*, 40, 119–68.

Morse, P. E., Daegling, D. J., McGraw, W. S., and Pampush, J. D. (2013). Dental wear among cercopithecid monkeys of the Tai Forest, Cote d'Ivoire. *American Journal of Physical Anthropology*, 150(4), 655–65.

Mosdossy, K. N., Melin, A. D., and Fedigan, L. M. (2015). Quantifying seasonal fallback on invertebrates, pith, and bromeliad leaves by white-faced capuchin monkeys (*Cebus capucinus*) in a tropical dry forest. *American Journal of Physical Anthropology*, 158, 67–77.

Moser, E. I., Kropff, E., and Moser, M. B. (2008). Place cells, grid cells, and the brain's spatial representation system. *Annual Review of Neuroscience*, 31, 69–89.

Moses, K., and Semple, S. (2011). Primary seed dispersal by the black-and-white ruffed lemur (*Varecia variegata*) in the Manombo forest, South-East Madagascar. *Journal of Tropical Ecology*, 27, 529–38.

Moss, M. (2013). *Salt Sugar Fat: How the Food Giants Hooked Us*. Random House.

Moura, A. C., and Lee, P. C. (2004). Capuchin stone tool use in Caatinga dry forest. *Science*, 206, 1909.

Mowlana, F., Heath, M. R., van der Bilt, A., and van der Glas, H. W. (1994). Assessment of chewing efficiency: A comparison of particle size distribution determined using optical scanning and sieving of almonds. *Journal of Oral Rehabilitation*, 21, 545–51.

Mowry, C. B., Decker, B. S., and Shure, D. J. (1996). The role of phytochemistry in dietary choices of Tana River red colobus monkeys (*Procolobus badius rufomitratus*). *International Journal of Primatology*, 17, 63–84.

Muchaal, P. K., and Ngandjui, G. (1999). Impact of village hunting on wildlife populations in the Western Dja Reserve, Cameroon. *Conservation Biology*, 13, 385–96.

Muchlinski, M. N. (2010). Ecological correlates of infraorbital foramen area in primates. *American Journal of Physical Anthropology*, 141, 131–41.

Mudau, F. N., and Ngezimana, W. (2014). Effect of different drying methods on chemical composition and antimicrobial activity of bush tea (*Athrixia phylicoides*). *International Journal of Agricultural Biology*, 16, 1011–14.

Muegge, B. D., Kuczynski, J., Knights, D., Clemente, J. C., Gonzalez, A., Fontana, L., et al. (2011). Diet drives convergence in gut microbiome functions across mammalian phylogeny and within humans. *Science*, 332(6032), 970–74.

Muehlenbein, M. P. (2006). Intestinal parasite infections and fecal steroid levels in wild chimpanzees. *American Journal of Physical Anthropology*, 130(4), 546–50.

Muehlenbein, M. (2008). Human immune functions are energetically costly. *American Journal of Physical Anthropology*, 137, S47, 158–59.

Muehlenbein, M. P. (2015). Metabolic and endocrine changes during immune activation. *American Journal of Physical Anthropology*, 156, 231–31.

Muehlenbein, M. P., Campbell, B. C., Richards, R. J., Svec, F., Falkenstein, K. P., Murchison, M. A., et al. (2003). Leptin, body composition, adrenal and gonadal hormones among captive male baboons. *Journal of Medical Primatology*, 32(6), 320–24.

Muehlenbein, M. P., and Lewis, C. M. (2013). Health assessment and epidemiology. In Sterling, E. J. and Blair, M. E. (Eds.) *Primate Ecology and Conservation: A Handbook of Techniques* (pp. 40–57). Oxford University Press.

Muehlenbein, M. P., and Watts, D. P. (2010). The costs of dominance: Testosterone, cortisol and intestinal parasites in wild male chimpanzees. *BioPsychoSocial Medicine*, 4, 21.

Mueller-Harvey, I. (2006). Unravelling the conundrum of tannins in animal nutrition and health. *Journal of the Science of Food and Agriculture*, 86, 2010–37.

Muhly, T. B., Semeniuk, C., Massolo, A., Hickman, L., and Musiani, M. (2011). Human activity helps prey win the predator-prey space race. *PLoS ONE*, 6, e17050.

Muirhead-Thompson, R. (2012). *Trap Responses of Flying Insects: The Influence of Trap Design on Capture Efficiency*. Academic Press.

Muletz-Wolz, C. R., Kurata, N. P., Himschoot, E. A., Wenker, E. S., Quinn, E. A., Hinde, K., et al. (2019). Diversity and temporal dynamics of primate milk microbiomes. *American Journal of Primatology*, 81(10–11), e22994.

Müller, E. (1985). Basal metabolic rates in primates—the possible role of phylogenetic and ecological factors. *Comparative Biochemistry and Physiology A—Molecular and Integrative Physiology*, 81, 707–11.

Muller, M., and Emery Thompson, M. (2012). Mating, parenting and male reproductive strategies. In J. C. Mitani, J. Call, P. M. Kappeler, R. Palombit, and J. Silk (Eds.), *The Evolution of Primate Societies* (pp. 387–411). University of Chicago Press.

Muller, N., Heistermann, M., Strube, C., Schulke, O., and Ostner, J. (2017). Age, but not anthelmintic treatment, is associated with urinary neopterin levels in semi-free ranging Barbary macaques. *Scientific Reports*, 7, 41973.

Muller-Klein, N., Heistermann, M., Strube, C., Franz, M., Schulke, O., and Ostner, J. (2019). Exposure and susceptibility drive reinfection with gastrointestinal parasites in a social primate. *Functional Ecology*, 33(6), 1088–98.

Müller-Schwarze, D. (2006). *Chemical Ecology of Vertebrates*. Cambridge University Press.

Munch, E., Launey, M. E., Alsem, D. H., Saiz, E., Tomsia, A. P., and Ritchie, R. O. (2008). Tough, bio-inspired hybrid materials. *Science*, 322, 1516–20.

Munds, R. A., Cooper, E. B., Janiak, M. C., Lam, L. G., DeCasien, A. R., Bauman Surratt, S., et al. (2022). Variation and heritability of retinal cone ratios in a free-ranging population of rhesus macaques. *Evolution*, 76(8), 1776–89.

Mundy, N. I., Morningstar, N. C., Baden, A. L., Fernandez-Duque, E., Dávalos, V. M., and Bradley, B. J. (2016). Can colour vision re-evolve? Variation in the X-linked opsin locus of cathemeral Azara's owl monkeys (*Aotus azarae azarae*). *Frontiers in Zoology*, 13, 9.

Munn, A. J., Dawson, T. J., and McLeod, S. R. (2011). Feeding biology of two functionally different foregut-fermenting mammals, the marsupial red kangaroo (*Macropus rufus*) and the ruminant sheep (*Ovis aries*): How physiological ecology can inform land management. *Journal of Zoology*, 283, 298–298.

Munn, C. A. (1986). Birds that "cry wolf." *Nature*, 319, 143.

Munoz, D., Estrada, A., Naranjo, E., and Ochoa, S. (2006). Foraging ecology of howler monkeys in a cacao (*Theobroma cacao*) plantation in Comalcalco, Mexico. *American Journal of Primatology*, 68, 127–42.

Murphy, D., Lea, S. E. G., and Zuberbühler, K. (2013). Male blue monkey alarm calls encode predator type and distance. *Animal Behaviour*, 85, 119–25.

Murr, C., Widner, B., Wirleitner, B., and Fuchs, D. (2002). Neopterin as a marker for immune system activation. *Current Drug Metabolism*, 3, 175–87.

Murray, B. G. (1999). Can the population regulation controversy be buried and forgotten? *Oikos*, 84, 148–52.

Murray, C. J., Barber, R. M., Foreman, K. J., Abbasoglu Ozgoren, A., Abd-Allah, F., Abera, S. F., et al. (2015). Global, regional, and national disability-adjusted life years (DALYs) for 306 diseases and injuries and healthy life expectancy (HALE) for 188 countries, 1990–2013: Quantifying the epidemiological transition. *Lancet*, 386(10009), 2145–91.

Murray, C. M., Eberly, L. E., and Pusey, A. E. (2006). Foraging strategies as a function of season and rank among wild female chimpanzees (*Pan troglodytes*). *Behavioral Ecology*, 17, 1020–28.

Murray, C. M., Lonsdorf, E. V., Eberle, L. E., and Pusey, A. E. (2009). Reproductive energetics in free-living female chimpanzees (*Pan troglodytes schweinfurthii*). *Behavioral Ecology*, 20, 1211–16.

Murray, P. (1975). The role of cheek pouches in cercopithecine monkey adaptive strategy. In R. A. Tuttle (Ed.), *Primate Functional Morphology and Evolution* (pp. 151–94). Mouton.

Murton, R. K., Isaacson, A. J., and Westwood, N. J. (1971). Significance of gregarious feeding behaviour and adrenal stress in a population of wood-pigeons *Columba palumbus*. *Journal of Zoology*, 165, 53–57.

Musey, P. I., Adlercreutz, H., Gould, K. G., Collins, D. C., Fotsis, T., Bannwart, C., et al. (1995). Effect of diet on lignans and isoflavonoid phytoestrogens in chimpanzees. *Life Science*, 57, 655–64.

Mutschler, T. (2002). Alaotran gentle lemur: Some aspects of its behavioral ecology. *Evolutionary Anthropology*, 11, 101–4.

Nabhani, Z. A., and Eberl, G. (2020). Imprinting of the immune system by the microbiota early in life. *Mucosal Immunology*, 13, 183–89.

Nacionales, D. C., Gentile, L. F., Vanzant, E., Lopez, M. C., Cuenca, A., Cuenca, A. G., et al. (2014). Aged mice are unable to mount an effective myeloid response to sepsis. *Journal of Immunology*, 192(2), 612–22.

Nacionales, D. C., Szpila, B., Ungaro, R., Lopez, M. C., Zhang, J., Gentile, L. F., et al. (2015). A detailed characterization of the dysfunctional immunity and abnormal myelopoiesis induced by severe shock and trauma in the aged. *Journal of Immunology*, 195(5), 2396–407.

Nadjafzadeh, M. N., and Heymann, E. W. (2008). Prey foraging of red titi monkeys, *Callicebus cupreus*, in comparison to sympatric tamarins, *Saguinus mystax* and *Saguinus fuscicollis*. *American Journal of Physical Anthropology*, 135, 56–63.

Nadjafzadeh, M., Hofer, H., and Krone, O. (2016). Sit-and-wait for large prey: Foraging strategy and prey choice of white-tailed eagles. *Journal of Ornithology*, 157, 165–78.

Nadler, T. (2003). Wiederentdeckung des oestlichen schwarzen gibbons (*Nomascus nasutus*) *Vietnam Zool Garten (NF)*, 73, 65–73.

Nagy, J. D., Victor, E. M., and Cropper, J. H. (2007). Why don't all whales have cancer? A novel hypothesis resolving Peto's paradox. *Integrative and Comparative Biology*, 47(2), 317–28.

Nagy, K. A., Girard, I. A., and Brown, T. K. (1999). Energetics of free-ranging mammals, reptiles, and birds. *Annual Review of Nutrition*, 19, 247–77.

Nagy, K. A., and Milton, K. (1979a). Aspects of dietary quality, nutrient assimilation and water-balance in wild howler monkeys (*Alouatta palliata*). *Oecologia*, 39, 249–58.

Nagy, K. A., and Milton, K. (1979b). Energy metabolism and food consumption by wild howler monkeys (*Alouatta palliata*). *Ecology*, 60, 475–80.

Nakagawa, N. (1990). Choice of food patches by Japanese monkeys (*Macaca fuscata*). *American Journal of Primatology*, 21, 17–29.

Nakamura, M., and Itoh, N. (2008). Hunting with tools by Mahale chimpanzees. *Pan Africa News*, 15, 3–6.

Nakashita, R., Hamada, Y., Hirasaki, E., Suzuki, J., and Oi, T. (2013). Characteristics of stable isotope signature of diet in tissues of captive Japanese macaques as revealed by controlled feeding. *Primates*, 54, 271–81.

Nam, C. M., Oh, K. W., Lee, K. H., Jee, S. H., Cho, S. Y., Shim, W. H., et al. (2003). Vitamin C intake and risk of ischemic heart disease in a population with a high prevalence of smoking. *Journal of the American College of Nutrition*, 22(5), 372–78.

Napal, G. N. D., Defagó, M. T., Valladares, G. R., and Palacios, S. M. (2010). Response of *Epilachna paenulata* to two flavonoids, pinocembrin and quercetin, in a comparative study. *Journal of Chemical Ecology*, 36, 898–904.

Nash, L. T. (1986). Dietary, behavioral, and morphological aspects of gummivory in primates. *Yearbook of Physical Anthropology*, 29, 113–37.

Nash, L. T. (2007). Moonlight and behavior in nocturnal and cathemeral primates, especially *Lepilemur leucopus*: Illuminating possible anti-predator efforts. In S. Gursky-Doyen and K. A. I. Nekaris (Eds.), *Primate Anti-predator Strategies* (pp. 173–205). Springer.

Nash, L. T., and Burrows, A. (2010). Introduction: Advances and remaining sticky issues in the understanding of exudativory in primates. In A. M. Burrows and L. T. Nash (Eds.), *The Evolution of Exudativory in Primates* (pp. 1–23). Springer.

Nash, L. T., and Whitten, P. L. (1989). Preliminary observations on the role of acacia gum chemistry in acacia utilization by *Galago senegalensis* in Kenya. *American Journal of Primatology*, 17, 27–39.

Nash, T. H. (2008). *Lichen Biology* (2nd ed.). Cambridge University Press.

Nater, A., Mattle-Greminger, M. P., Nurcahyo, A., Nowak, M. G., de Manuel, M., Desai, T., et al. (2017). Morphometric, behavioral, and genomic evidence for a new orangutan species. *Current Biology*, 27, 3487–98.

Nathan, R., Getz, W. M., Revilla, E., Holyoak, M., Kadmon, R., Saltz, D., et al. (2008). A movement ecology paradigm for unifying organismal movement research. *Proceedings of the National Academy of Sciences of the United States of America*, 105, 19052–59.

Naughton-Treves, L. (1998). Predicting patterns of crop damage by wildlife around Kibale National Park, Uganda. *Conservation Biology*, 12, 156–68.

Naughton-Treves, L., Treves, A., Chapman, C., and Wrangham, R. (1998). Temporal patterns of crop-raiding by primates: Linking food availability in croplands and adjacent forest. *Journal of Applied Ecology*, 35, 596–606.

Navarrete, A., Van Schaik, C. P., and Isler, K. (2011). Energetics and the evolution of human brain size. *Nature*, 480(7375), 91–93.

Negrey, J. D., Behringer, V., Langergraber, K. E., and Deschner, T. (2021). Urinary neopterin of wild chimpanzees indicates that cell-mediated immune activity varies by age, sex, and female reproductive status. *Scientific Reports*, 11(1), 9298.

Negrey, J. D., Reddy, R. B., Scully, E. J., Phillips-Garcia, S., Owens, L. A., Langergraber, K. E., et al. (2019). Simultaneous outbreaks of respiratory disease in wild chimpanzees caused by distinct viruses of human origin. *Emerging Microbes and Infections*, 8(1), 139–49.

Negro, M., Giardina, S., Marzani, B., and Marzatico, F. (2008). Branched-chain amino acid supplementation does not enhance athletic performance but affects muscle recovery and the immune system. *Journal of Sports Medicine and Physical Fitness*, 48(3), 347–51.

Neha, S. A., Khatun, U. H. and Ul Hasan, M. A. (2021) Resource partitioning and niche overlap between hoolock gibbon (*Hoolock hoolock*) and other frugivorous vertebrates in a tropical semi-evergreen forest. *Primates* 62, 331–42.

Neish, A. S. (2009). Microbes in gastrointestinal health and disease. *Gastroenterology*, 136(1), 65–80.

Neiss, J. H., Leithauser, F., Adler, G., and Reimann, J. (2008). Commensal gut flora drives the expansion of proinflammatory CD4 T cells in the colonic lamina propria under normal and inflammatory conditions. *Journal of Immunology*, 180, 559–68.

Nekaris, K. A. I. (2005). Foraging behaviour of the slender loris (*Loris lydekkerianus lydekkerianus*): Implications for theories of primate origins. *Journal of Human Evolution*, 49, 289–300.

Nekaris, K. A. I., and Bearder, S. K. (2007). The lorisiform primates of Asia and mainland Africa: Diversity shrouded in darkness. In C. J. Campbell, A. Fuentes, K. C. MacKinnon, M. Panger, and S. K. Bearder (Eds.), *Primates in Perspective*. Oxford University Press.

Nekaris KA-I, Jayewardene J. 2003. Pilot study and conservation status of the slender loris (*Loris tardigradus* and *L. lydekkeriannus*) in Sri Lanka. *Primate Conservation*, 19, 83–90.

Nekaris, K. A. I., and Rasmussan, D. T. (2003). Diet and feeding behavior of Mysore slender lorises. *International Journal of Primatology*, 24, 33–46.

Nekaris, K. A. I., Moore, R. S., Rode, E. J., and Fry, B. G. (2013). Mad, bad and dangerous to know: The biochemistry, ecology and evolution of slow loris venom. *Journal of Venomous Animals and Toxins including Tropical Diseases*, 19, 21.

Nekaris, K. A., Campera, M., Chimienti, M., Murray, C., Balestri, M., and Showell, Z. (2022). Training in the dark: Using target training for non-invasive application and validation of accelerometer devices for an endangered primate (*Nycticebus bengalensis*). *Animals*, 12(4), 411.

Nelson, E. H., Matthews, C. E., and Rosenheim, J. A. (2004). Predators reduce prey population growth by inducing changes in prey behavior. *Ecology*, 85, 1853–58.

Nelson, R. J. (2011). *An Introduction to Behavioral Endocrinology*. Sinauer Associates.

Nelson, R. S., Lonsdorf, E. V., Terio, K. A., Wellens, K. R., Lee, S. M., and Murray, C. M. (2022). Drinking frequency in wild lactating chimpanzees (*Pan troglodytes schweinfurthii*) and their offspring. *American Journal of Primatology*, 84(6), e23371.

Nelson, S. V. (2007). Isotopic reconstructions of habitat change surrounding the extinction of *Sivapithecus*, a Miocene hominoid, in the Siwalik Group of Pakistan. *Palaeogeography, Palaeoclimatology, Palaeoecology*, 243, 204–22.

Neri-Arboleda, I., Scott, P., and Arboleda, N. P. (2002). Home ranges, spatial movements and habitat associations of the Philippine tarsier (*Tarsius syrichta*) in Corella, Bohol. *Journal of Zoology*, 257, 387–402.

Nersesian, C. L., Banks, P. B., and McArthur, C. (2011). Titrating the cost of plant toxins against predators: Determining the tipping point for foraging herbivores. *Journal of Animal Ecology*, 80, 753–60.

Nettleton, J. A., Villalpando, S., Cassani, R. S., and Elmadfa, I. (2013). Health significance of fat quality in the diet. *Annals of Nutrition and Metabolism*, 63(1–2), 96–102.

Neuman, H., Debelius, J. W., Knight, R., and Koren, O. (2015). Microbial endocrinology: The interplay between the microbiota and the endocrine system. *FEMS Microbiology Review*, 39, 509–21.

Neville, M. C., Anderson, S. M., McManaman, J. L., Badger, T. M., Bunik, M., Crume, T., et al. (2012). Lactation and neonatal nutrition: Defining and refining the critical questions. *Journal of Mammary Gland Biology and Neoplasia*, 17, 167–88.

Neville, M. C., McFadden, T. B., and Forsyth, I. (2002). Hormonal regulation of mammary differentiation and milk secretion. *Journal of Mammary Gland Biology and Neoplasia*, 7, 49–66.

Neville, M. C., and Morton, J. (2001). Physiology and endocrine changes underlying human lactogenesis II. *Journal of Nutrition*, 131, 3005S–3008S.

Nevo, O., and Heymann, E. W. (2015). Led by the nose: Olfaction in primate feeding ecology. *Evolutionary Anthropology*, 24, 137–48.

Nevo, O., Heymann, E. W., Schulz, S., and Ayasse, M. (2016). Fruit odor as a ripeness signal for seed-dispersing primates? A case study on four Neotropical plant species. *Journal of Chemical Ecology*, 42, 323–28.

Nevo, O., Orts Garri, R., Hernández-Salazar, L. T., Schulz, S.,

Heymann, E. W., Ayasse, M., et al. (2015). Chemical recognition of fruit ripeness in spider monkeys (*Ateles geoffroyi*). *Scientific Reports*, 5, 14895.

Nevo, O., Razafimandimby, D., Jeffrey, J. A. J., Schulz, S., and Ayasse, M. (2018). Fruit scent as an evolved signal to primate seed dispersal. *Science Advances*, 4(10), eaat4871.

Nevo, O., and Valenta, K. (2018). The ecology and evolution of fruit odor: implications for primate seed dispersal. *International Journal of Primatology*, 39, 338–55.

Nevo, O., Valenta, K., Razafimandimby, D., Melin, A. D., Ayasse, M., and Chapman, C. A. (2018). Frugivores and the evolution of fruit colour. *Biology Letters*, 14(9), 20180377.

Newburg, D. S., and Neubauer, S. H. (1995). Carbohydrates in milks: Analysis, quantities, and significance. In R. G. Jensen (Ed.), *Handbook of Milk Composition* (pp. 273–349). Academic Press.

Newburg, D. S., and Walker, W. A. (2007). Protection of the neonate by the innate immune system of developing gut and of human milk. *Pediatric Research*, 61, 2–8.

Newby, J. R., Mills, L. S., Ruth, T. K., Pletscher, D. H., Mitchell, M. S., Quigley, H. B., et al. (2013). Human-caused mortality influences spatial population dynamics: Pumas in landscapes with varying mortality risks. *Biological Conservation*, 159, 230–39.

Newman, J. A., Anand, A., Henry, H. A. L., Hunt, S., and Gedalof, Z. (2011). *Climate Change Biology*. CABI.

Newsholme, E. A., and Carrie, A. L. (1994). Quantitative aspects of glucose and glutamine metabolism by intestinal cells. *Gut*, 35(suppl. 1), S13–17.

Newton, P. (1992). Feeding and ranging patterns of forest Hanuman langurs (*Presbytis entellus*). *International Journal of Primatology*, 13, 245–85.

Newton-Fisher, N. E., Notman, H., and Reynolds, V. (2002). Hunting of mammalian prey by Budongo Forest chimpanzees. *Folia Primatologica*, 73(5), 281–83.

Newton-Fisher, N. E., Reynolds, V., and Plumptre, A. J. (2000). Food supply and chimpanzee (*Pan troglodytes schweinfurthii*) party size in Budongo Forest Reserve. *International Journal of Primatology*, 21, 613–28.

Neyrolles, O., Wolschendorf, F., Mitra, A., and Niederweis, M. (2015). Mycobacteria, metals, and the macrophage. *Immunological Reviews*, 264(1), 249–63.

Ng, F. S. P. (1977). Gregarious flowering of dipterocarps in Kepong, 1976. *Malaysian Forester*, 40, 126–37.

Ng, S., Lasekan, O., Muhammad, K. S., Hussain, N., and Sulaiman, R. (2015). Physiochemical properties of Malaysian-grown tropical almond nuts (*Terminalia catappa*). *Journal of Food Science and Technology*, 52, 6623–30.

Ngo, V., Gorman, J. C., De la Fuente, M. F., Souto, A., Schiel, N., and Miller, C. T. (2022). Active vision during prey capture in wild marmoset monkeys. *Current Biology*, 32(15), 3423–28.

N'Goran, P. K., Boesch, C., Mundry, R., N'Goran, E. L., Herbinger, I., Yapi, F. E., et al. (2012). Hunting, law enforcement, and African primate conservation. *Conservation Biology*, 26, 565–71.

N'Guesson, A. K., Ortmann, S., and Boesch, C. (2009). Daily energy balance and protein gain among *Pan troglodytes verus* in the Tai National Park, Cote d'Ivoire. *International Journal of Primatology*, 30, 481–96.

Nguyen, N. (2013). Primate behavioral endocrinology. In E. J. Sterling, N. Bynum and M. E. Blair (Eds.), *Primate Ecology and Conservation: A Handbook of Techniques* (pp. 224–37). Oxford University Press.

Ni, Q. Y., Huang, B., Liang, Z. L., Wang, X. W., and Jiang, X. L. (2014). Dietary variability in the western black crested gibbon (*Nomascus concolor*) inhabiting an isolated and disturbed forest fragment in Southern Yunnan, China. *American Journal of Primatology*, 76, 217–29.

Nica, V., Popp, R. A., Crisan, T. O., and Joosten, L. A. B. (2022). The future clinical implications of trained immunity. *Expert Review of Clinical Immunology*, 18, 1125-1134.

Nicholson, A. J. (1933). The balance of animal populations. *Journal of Animal Ecology*, 2, 131–78.

Nickle, D. A., and Castner, J. L. (1995). Strategies utilized by katydids (Orthoptera: Tettigoniidae) against diurnal predators in rainforests of northeastern Peru. *Journal of Orthoptera Research*, 4, 75–88.

Nickle, D. A., and Heymann, E. W. (1996). Predation on Orthoptera and other orders of insects by tamarin monkeys, *Saguinus mystax mystax* and *Saguinus fuscicollis nigrifrons* (Primates: Callitrichidae), in northeastern Peru. *Journal of Zoology, London*, 239, 799–819.

Nicol, L. E., Grant, W. R., Comstock, S. M., Nguyen, M. L., Smith, M. S., Grove, K. L., et al. (2013). Pancreatic inflammation and increased islet macrophages in insulin-resistant juvenile primates. *Journal of Endocrinology*, 217(2), 207–13.

Nie, Y., Zhang, Z., Raubenheimer, D., Elser, J. J., Wei, W., and Wei, F. (2015). Obligate herbivory in an ancestrally carnivorous lineage: The giant panda and bamboo from the perspective of nutritional geometry. *Functional Ecology*, 29, 26–34.

Niemitz, C. (1984). *Biology of Tarsiers*. Gustac Fischer.

Nievergelt, C. M., and Martin, R. D. (1999). Energy intake during reproduction in captive common marmosets (*Callithrix jacchus*). *Physiology and Behavior*, 65, 849–54.

Nigi, H. (1976). Some aspects related to conception of the Japanese monkey (*Macaca fuscata*). *Primates*, 17, 81–87.

Niimura, Y., Matsui, A., and Touhara, K. (2018). Acceleration of olfactory receptor gene loss in primate evolution: possible link to anatomical change in sensory systems and dietary transition. *Molecular Biology and Evolution*, 35(6), 1437–50.

Nikolich-Žugich, J., and Messaoudi, I. (2005). Mice and flies and monkeys too: Caloric restriction rejuvenates the aging immune system of non-human primates. *Experimental Gerontology*, 40(11), 884–93.

Nimeitz, C. (1979). Outline of the behavior of *Tarsius bancanus*.

In G. Doyle and R. Martin (Eds.), *The Study of Prosimian Behavior* (pp. 631–60). Academic Press.

Ning, W. H., Guan, Z. H., Huang, B., Fan, P. F., and Jiang, X. L. (2019). Influence of food availability and climate on behavior patterns of western black crested gibbons (*Nomascus concolor*) at Mt. Wuliang, Yunnan, China. *American Journal of Primatology*, 81, e23068.

Nishida, T. (1968). The social group of wild chimpanzees in the Mahali Mountains. *Primates*, 9, 167–224.

Nishida, T., and Hiraiwa-Hasegawa, M. (1987). Chimpanzees and bonobos: Cooperative relationships among males. In: B. B. Smuts, D. L. Cheney, R. M. Seyfarth, R. W. Wrangham, and T. T. Struhsaker, (Eds.), *Primate Societies* (pp. 87–104). University of Chicago Press.

Nishida, T., Matsusaka, T., and McGrew, W. C. (2009). Emergence, propagation or disappearance of novel behavioral patterns in the habituated chimpanzees of Mahale: A review. *Primates*, 50, 23–36.

Nishida, T., and Uehara, S. (1983). Natural diet of chimpanzees (*Pan troglodytes schweinfurthii*): Long-term record from the Mahale Mountains, Tanzania. *African Studies Monographs*, 3, 109–30.

Nishida, T., Wrangham, R. W., Goodall, J., and Uehara, S. (1983). Local differences in plant-feeding habits of chimpanzees between the Mahale Mountains and Gombe National Park, Tanzania. *Journal of Human Evolution*, 12, 457–80.

Nishida, T., Wrangham, R., Jones, J., Marshall, A., and Wakibara, J. (2000). Do chimpanzees survive the 21st century? *The Apes: Challenges for the 21st Century*, Chicago, IL.

Nishijima, S., Nagata, N., Kiguchi, Y., Kojima, Y., Miyoshi-Akiyama, Kimura, M., et al. (2022). Extensive gut virome variation and its associations with host and environmental factors in a population-level cohort. *Nature Communications*, 13, 5252.

Nissen, H. W. (1956). Individuality in the behavior of chimpanzees. *American Anthropologist*, 58, 407–13.

Niven, L. (2013). A diachronic evaluation of Neandertal cervid exploitation and site use at Peche de L'Azè, I. V., France. In J. L. Clark and J. D. Speth (Eds.), *Zooarchaeology and Modern Human Origins: Human Hunting Behavior during the Later Pleistocene* (pp. 151–62). Springer.

Noë, R., and Bshary, R. (1997). The formation of red colobus-diana monkey associations under predation pressure from chimpanzees. *Proceedings of the Royal Society B: Biological Sciences*, 264, 253–59.

Noonan, M. J., Johnson, P. J., Kitchener, A. C., Harrington, L. A., Newman, C., and Macdonald, D. W. (2016). Sexual size dimorphism in musteloids: An anomalous allometric pattern is explained by feeding ecology. *Ecology and Evolution*, 6, 8495–501.

Norconk, M. A. (1986). *Interactions between Primate Species in a Neotropical Forest: Mixed-Species Troops of Saguinus mystax and S. fuscicollis (Callitrichidae)*. University of California, Los Angeles.

Norconk, M. A. (1990). Mechanisms promoting stability in mixed *Saguinus mystax* and *S. fuscicollis* troops. *American Journal of Primatology*, 21, 159–70.

Norconk, M. (2011). Sakis, uakaris, and titi monkeys: Behavioral diversity in a radiation of primate seed predators. In C. J. Campbell, A. Fuentes, K. C. MacKinnon, S. K. Bearder, and R. M. Stumpf (Eds.), *Primates in Perspective* (2nd ed., pp. 122–39). Oxford University Press.

Norconk, M., and Conklin-Brittain, N. L. (2004). Variation on frugivory: The diet of Venezuelan white-faced sakis. *International Journal of Primatology*, 25, 1–26.

Norconk, M. A., and Conklin-Brittain, N. L. (2016). Bearded saki feeding strategies on an island in Lago Guri, Venezuela. *American Journal of Primatology*, 78, 507–22.

Norconk, M. A., and Veres, M. (2011). Physical properties of fruit and seeds ingested by primate seed predators with emphasis on sakis and bearded sakis. *Anatomical Record*, 294, 2092–111.

Norconk, M. A., Grafton, B. W., and Conklin-Brittain, N. L. (1998). Seed dispersal by Neotropical seed predators. *American Journal of Primatology*, 45, 103–26.

Norconk, M. A., Grafton, B. W., and McGraw, W. S. (2013). Morphological and ecological adaptations to seed predation—a primate-wide perspective. In L. M. Veiga, A. A. Barnett, S. F. Ferrari, and M. A. Norconk (Eds.), *Evolutionary Biology and Conservation of Titis, Sakis and Uacaris* (pp. 55–70). Cambridge University Press.

Norconk, M. A., Wright, B. W., Conklin-Brittain, N. L., and Vinyard, C. J. (2009). Mechanical and nutritional properties of food as factors in platyrrhine dietary adaptations. In P. A. Garber, A. Estrada, J. C. Bicca-Marques, E. W. Heymann, and K. B. Strier (Eds.), *South American Primates: Comparative Perspectives in the Study of Behavior, Ecology, and Conservation* (pp. 279–319). Springer.

Norden, N., Chave, J., Belbenoit, P., Caubére, A., Châtelet, P., Forget, P. M., et al. (2007). Mast fruiting is a frequent strategy in woody species of eastern South America. *PLoS ONE*, 2, e1709.

Normand, E., and Boesch, C. (2009). Sophisticated Euclidean maps in forest chimpanzees. *Animal Behaviour*, 77, 1195–201.

Normand, E., Ban, S. D., and Boesch, C. (2009). Forest chimpanzees (*Pan troglodytes verus*) remember the location of numerous fruit trees. *Animal Cognition*, 12, 797–807.

Norscia, I., Carrai, V., and Borgognini-Tarli, S. M. (2006). Influence of dry season and food quality and quantity on behavior and feeding strategy of *Propithecus verreauxi* in Kirindy, Madagascar. *International Journal of Primatology*, 27, 1001–22.

Norscia, I., Ramanamanjato, J. B., and Ganzhorn, J. U. (2012). Feeding patterns and dietary profile of nocturnal southern woolly lemurs (*Avahi meridionalis*) in southeast Madagascar. *International Journal of Primatology*, 33, 150–67.

Norton, G. W., Rhine, R. J., Wynn, G. W., and Wynn, R. D. (1987). Baboon diet: A five-year study of stability and variability in the plant feeding and habitat of the yellow baboons (*Papio cynocephalus*) of Mikumi National Park, Tanzania. *Folia Primatologica*, 48, 78–120.

Norum, J. K., Lone, K., Linnell, J. D. C., Odden, J., Loe, L. E., and Mysterud, A. (2015). Landscape of risk to roe deer imposed by lynx and different human hunting tactics. *European Journal of Wildlife Research*, 61, 831–40.

Noser, R., and Byrne, R. W. (2007). Mental maps in chacma baboons (*Papio ursinus*): Using inter-group encounters as a natural experiment. *Animal Cognition*, 10, 331–40.

Noser, R., and Byrne, R. W. (2010). How do wild baboons (*Papio ursinus*) plan their routes? Travel among multiple high-quality food sources with inter-group competition. *Animal Cognition*, 13, 145–55.

Noser, R., and Byrne, R. W. (2014). Change point analysis of travel routes reveals novel insights into foraging strategies and cognitive maps of wild baboons. *American Journal of Primatology*, 76, 399–409.

Noss, R. F. (1995). *Maintaining Ecological Integrity in Representative Reserve Networks*. World Wildlife Fund Canada.

Noverr, M. C., and Huffnagle, G. B. (2004). Does the microbiota regulate immune responses outside the gut? *Trends in Microbiology*, 12(12), 562–68.

Nowack, J., Mzilikazi, N., and Dausmann, K. H. (2010). Torpor on demand: Heterothermy in the non-lemur primate *Galago moholi*. *PLoS ONE*, 5, e10797.

Nowack, J., Wippich, M., Mzilikazi, N., and Dausmann, K. H. (2013). Surviving the cold, dry period in Africa: Behavioral adjustments as an alternative to heterothermy in the African lesser bushbaby (*Galago moholi*). *International Journal of Primatology*, 34, 49–64.

Nowak, K., Hill, R. A., Wimberger, K., and le Roux, A. (2016). Risk-taking in Samango monkeys in relation to humans at two sites in South Africa. In T. M. Waller (Ed.), *Ethnoprimatology: Primate Conservation in the 21st Century* (pp. 301–14). Springer International.

Nowak, K., le Roux, A., Richards, S. A., Scheijen, C. P. J., and Hill, R. A. (2014). Human observers impact habituated samango monkeys' perceived landscape of fear. *Behavioral Ecology*, 25, 1199–204.

Nowak, K., Richards, S. A., le Roux, A., and Hill, R. A. (2016). The influence of live-capture on the risk perceptions of habituated samango monkeys. *Journal of Mammalogy*, 97, 1461–68.

Nowak, K., Wimberger, K., Richards, S. A., Hill, R. A., and le Roux, A. (2017). Samango monkeys (*Cercopithecus albogularis labiatus*) manage risk in a highly seasonal, human-modified landscape in Amathole Mountains, South Africa. *International Journal of Primatology*, 38, 194–206.

NRC (National Research Council). (2003) *Nutrient Requirements of Nonhuman Primates* (2nd ed.). National Academies Press.

Ntiamoa-Baidu, Y., 1997. Wildlife and Food Security in Africa: FAO Conservation

O'Neal, T. J., Friend, D. M., Guo, J., Hall, K. D., and Kravitz, A.V. (2017). Increases in physical activity result in diminishing increments in daily energy expenditure in mice. *Current Biology*, 27(3), 423-430.

Nunn, C. L. (2011). *The Comparative Approach in Evolutionary Anthropology and Biology*. University of Chicago Press.

Nunn, C. L., and Altizer, S. (2006). *Infectious Diseases in Primates: Behavior, Ecology and Evolution*. Oxford University Press.

Nunn, C. L., Thrall, P. H., Stewart, K., and Harcourt, A. H. (2008). Emerging infectious diseases and animal social systems. *Evolutionary Ecology*, 22(4), 519–43.

Oates, J. F. (1977). The guereza and its food. In T. H. Clutton-Brock (Ed.), *Primate Ecology* (pp. 276–321). Academic Press.

Oates, J. F. (1978). Water-plant and soil consumption by guereza monkeys (*Colobus guereza*)—relationship with minerals and toxins in the diet. *Biotropica*, 10, 241–53.

Oates, J. F. (1988). The diet of the olive colobus monkey, *Procolobus verus*, in Sierra Leone. *International Journal of Primatology*, 9, 457–78.

Oates, J. F. (1996). Habitat alteration, hunting, and the conservation of folivorous primates in African forests. *Australian Journal of Ecology*, 21, 1–9.

Oates, J. F., Abedi-Lartey, M., McGraw, W. S., and Struhsaker, T. T. (2000). Extinction of a West African red colobus monkey. *Conservation Biology*, 14, 1526–32.

Oates, J. F., Swain, T., and Zantovska, J. (1977). Secondary compounds and food selection by colobus monkeys. *Biochemical Systematics and Ecology*, 5, 317–21.

Oates, J. F., Waterman, P. G., and Choo, G. M. (1980). Food selection by the South Indian leaf-monkey, *Presbytis johnii*, in relation to leaf chemistry. *Oecologia*, 45, 45–56.

Oates, J. F., and Whitesides, G. H. (1990). Association between olive colobus (*Procolobus verus*), Diana guenons (*Cercopithecus diana*), and other forest monkeys in Sierra Leone. *American Journal of Primatology*, 21, 129–46.

Oates, J. F., Whitesides, G. H., Davies, A. G., Waterman, P. G., Green, S. M., Dasilva, G. L., et al. (1990). Determinants of variation in tropical forest primate biomass—new evidence from West-Africa. *Ecology*, 71, 328–43.

Obasi, N. A., Ukadilonu, J., Eze, E., Akubugwo, E. I., and Okorie, U. C. (2012). Proximate composition, extraction, characterization and comparative assessment of coconut (*Cocos nucifera*) and melon (*Colocynthis citrullus*) seeds and seed oils. *Pakistan Journal of Biological Sciences*, 15, 1–9.

Oberdorster, E., Clay, M. A., Cottam, D. M., Wilmot, F. A., McLachlan, J. A., and Milner, M. J. (2001). Common phytochemicals are ecdysteroid agonists and antagonists: A possible evolutionary link between vertebrate and invertebrate steroid hormones. *Journal of Steroid Biochemistry*, 77, 229–38.

O'Brien, T. G., Kinnaird, M., and Dierenfeld, E. S. (1998). What's so special about figs? *Nature*, 392, 668.

O'Brien, T. G., Kinnaird, M. F., Nurcahyo, A., Prasetyaningrum, M., and Iqbal, M. (2003). Fire, demography and the persistence of siamang (*Symphalangus syndactylus*: Hylobatidae) in a Sumatran rainforest. *Animal Conservation*, 6, 115–21.

O'Connell, C. A., DiGiorgio, A. L., Ugarte, A. D., Brittain, R. S. A., Naumenko, D. J., Utami Atmoko, S. S., et al. (2021). Wild Bornean orangutans experience muscle catabolism during episodes of fruit scarcity. *Scientific Reports*, 11, 10185.

O'Connell, T. C., and Hedges, R. E. M. (1999). Investigations into the effect of diet on modern human hair isotopic values. *American Journal of Physical Anthropology*, 108, 409–25.

Ode, K. L., Gray, H. L., Ramel, S. E., Georgieff, M. K., and Demerath, E. W. (2012). Decelerated early growth in infants of overweight and obese mothers. *Journal of Pediatrics*, 161, 1028–34.

Odum, E. (1953). *Fundamentals of Ecology*. Saunders.

Oelze, V. M. (2016). Reconstructing temporal variation in great ape and other primate diets: A methodological framework for isotope analyses in hair. *American Journal of Primatology*, 78, 1004–16.

Oelze, V. M., Douglas, P. H., Stephens, C. R., Surbeck, M., Behringer, V., Richards, M. P., et al. (2016). The steady state great ape? Long term isotopic records reveal the effects of season, social rank and reproductive status on bonobo feeding behavior. *PLoS ONE*, 11, e0162091.

Oelze, V. M., Fahy, G., Hohmann, G., Robbins, M. M., Leinert, V., Lee, K., et al. (2016). Comparative isotope ecology of African great apes. *Journal of Human Evolution*, 101, 1–16.

Oelze, V. M., Fuller, B. T., Richards, M. P., Fruth, B., Surbeck, M., Hublin, J. J., et al. (2011). Exploring the contribution and significance of animal protein in the diet of bonobos by stable isotope ratio analysis of hair. *Proceedings of the National Academy of Sciences of the United States of America*, 108, 9792–97.

Oelze, V. M., Head, J. S., Robbins, M. M., Richards, M., and Boesch, C. (2014). Niche differentiation and dietary seasonality among sympatric gorillas and chimpanzees in Loango National Park (Gabon) revealed by stable isotope analysis. *Journal of Human Evolution*, 66, 95–106.

Oelze, V. M., Percher, A. M., Nsi Akoué, G., El Ksabi, N., Willaume, E., and Charpentier, M. J. (2020). Seasonality and interindividual variation in mandrill feeding ecology revealed by stable isotope analyses of hair and blood. *American Journal of Primatology*, 82, e23206.

Oftedal, O. T. (1984). Milk composition, milk yield and energy output at peak lactation: A comparative review. *Symposia of the Zoological Society of London*, 51, 33–85.

Oftedal, O. T. (1985). Pregnancy and lactation. In R. J. Hudson and R. G. White (Eds.), *Bioenergetics of Wild Herbivores* (pp. 215–38). CRC.

Oftedal, O. T. (1991). The nutritional consequences of foraging in primates—the relationship of nutrient intakes to nutrient-requirements. *Philosophical Transactions of the Royal Society of London Series B—Biological Sciences*, 334, 161–70.

Oftedal, O. T. (2000). Use of maternal reserves as a lactation strategy in large mammals. *Proceedings of the Nutrition Society*, 59(1), 99–106.

Oftedal, O. T. (2002a). The mammary gland and its origin during synapsid evolution. *Journal of Mammary Gland Biology and Neoplasia*, 7, 225–52.

Oftedal, O. T. (2002b). The origin of lactation as a water source for parchment-shelled eggs. *Journal of Mammary Gland Biology and Neoplasia*, 7, 253–66.

Oftedal, O. T. (2012). The evolution of milk secretion and its ancient origins. *Animal*, 6, 355–68.

Oftedal, O. T. (2013). Origin and evolution of the major constituents of milk. In P. L. H. McSweeney and P. E. Fox (Eds.), *Advanced Dairy Chemistry* (4th ed., vol. 1A, pp. 1–42). Springer.

Oftedal, O. T., and Iverson, S. J. (1995). Comparative analysis of nonhuman milks: Phylogenetic variation in the gross composition of milks. In R. G. Jensen (Ed.), *Handbook of Milk Composition* (pp. 749–89). Academic Press.

Ogaro, F. O., Orinda, V. A., Onyango, F. E., and Black, R. E. (1993). Effect of vitamin A on diarrhoeal and respiratory complications of measles. *Tropical and Geographical Medicine*, 45(6), 283–86.

Ogden, C. L., Yanovski, S. Z., Carroll, M. D., and Flegal, K. M. (2007). The epidemiology of obesity. *Gastroenterology*, 132(6), 2087–102.

Ohashi, G. (2006). Behavioral repertoire of tool use in the wild chimpanzees at Bossou. In T. Matsuzawa, M. Tomonaga, and M. Tanaka (Eds.), *Cognitive Development in Chimpanzees* (pp. 439–51). Springer.

Ohashi, G. (2015). Pestle-pounding and nut-cracking by wild chimpanzees at Kpala, Liberia. *Primates*, 56, 113–17.

Ohsawa, H., and Dunbar, R. I. M. (1984). Variations in the demographic structure and dynamics of gelada baboon populations. *Behavioral Ecology and Sociobiology*, 15, 231–40.

Okecha, A. A., and Newton-Fisher, N. E. (2006). The diet of olive baboons (*Papio anubis*) in the Budongo Forest. In N. E. Newton-Fisher, H. Notman, J. D. Paterson, and V. Reynolds (Eds.), *Primates of Western Uganda* (pp. 61–73). Springer.

O'Leary, M. H. (1981). Carbon isotope fractionation in plants. *Phytochemistry*, 20, 553–67.

Olender, T., Waszak, S. M., Viavant, M., Khen, M., Ben-Asher, E., Reyes, A., et al. (2012). Personal receptor repertoires: Olfaction as a model. *BMC Genomics*, 13, 414.

Oliveira, A. C. M., Himelbloom, B., Crapo, C. A., Vorholt, C., Fong, Q., and RaLonde, R. (2006). Quality of Alaskan maricultured oysters (*Crassostrea gigas*): A one-year survey. *Journal of Food Science and Technology*, 71, C532–C542.

Olsson, O., Brown, J. S., and Smith, H. (2001). Gain curves in depletable food patches: A test of five models with European starlings. *Evolutionary Ecology Research*, 3, 285–310.

O'Mahony, C., Scully, P., O'Mahony, D., Murphy, S., O'Brien, F., Lyons, A., et al. (2008). Commensal-induced regulatory T cells mediate protection against pathogen-stimulated NF-KB activation. *PLoS Pathogens*, 4, e1000112.

O'Malley, R. C., and Power, M. L. (2012). Nutritional composition of actual and potential insect prey for the Kasekela chimpanzees of Gombe National Park, Tanzania. *American Journal of Physical Anthropology*, 149, 493–503.

O'Malley, R. C., and Power, M. L. (2014). The energetic and nutritional yield from insectivory for Kasekela chimpanzees. *Journal of Human Evolution*, 71, 46–58.

O'Malley, R. C., Wallauer, W., Murray, C. M., and Goodall, J. (2012). The appearance and spread of ant fishing among the Kasekela chimpanzees of Gombe: A possible case of inter-community cultural transmission. *Current Anthropology*, 53, 650–63.

Omeja, P. A., Lawes, M. J., Corriveau, A., Valenta, K., Paim, F. P., and Chapman, C. A. (2016). Recovery of the animal and plant communities across large scales in Kibale National Park, Uganda. *Biotropica*, 48, 770–79.

Omeja, P. A., Obua, J., Rwetsiba, A., and Chapman, C. A. (2012). Biomass accumulation in tropical lands with different disturbance histories: Contrasts within one landscape and across regions. *Forest Ecology and Management*, 269, 293–300.

Ometto, J. P. H., Flanagan, L. B., Martinelli, L. A., and Ehleringer, J. R. (2005). Oxygen isotope ratios of waters and respired CO_2 in Amazonian forest and pasture ecosystems. *Ecological Applications*, 15, 58–70.

Onderdonk, D. A., and Chapman, C. A. (2000). Coping with forest fragmentation: The primates of Kibale National Park, Uganda. *International Journal of Primatology*, 21, 587–611.

Ong, L., McConkey, K. R., and Campos-Arceiz, A. (2022). The ability to disperse large seeds, rather than body mass alone, defines the importance of animals in a hyper-diverse seed dispersal network. *Journal of Ecology*, 110(2), 313–26.

O'Neill, M. C. (2012). Gait-specific metabolic costs and preferred speeds in ring-tailed lemurs (*Lemur catta*), with implications for the scaling of locomotor costs. *American Journal of Physical Anthropology*, 149, 356–64.

Onoda, Y., Westoby, M., Adler, P. B., Choong, A. M. F., Clissold, F. J., Cornelissen, J. H. C., et al. (2011). Global patterns of leaf mechanical properties. *Ecology Letters*, 14, 301–12.

O'Regan, H., Chenery, C., Lamb, A., Stevens, R. E., Rook, L., and Elton, S. (2008). Modern macaque dietary heterogeneity assessed using stable isotope analysis of hair and bone. *Journal of Human Evolution*, 55, 617–26.

Oriol-Cotterill, A., Valeix, M., Frank, L. G., Riginos, C., and Macdonald, D. W. (2015). Landscapes of coexistence for terrestrial carnivores: The ecological consequences of being downgraded from ultimate to penultimate predator by humans. *Oikos*, 124, 1263–73.

Orkin, J. D., Montague, M. J., Tejada-Martinez, D., De Manuel, M., Del Campo, J., Cheves Hernandez, S., et al. (2021). The genomics of ecological flexibility, large brains, and long lives in capuchin monkeys revealed with fecalFACS. *Proceedings of the National Academy of Sciences*, 118(7), e2010632118.

Ortmann, S., Bradley, B. J., Stolter, C., and Ganzhorn, J. U. (2006). Estimating the quality and composition of wild animal diets—a critical survey of methods. In G. Hohmann, M. M. Robbins, and C. Boesch (Eds.), *Feeding Ecology in Apes and Other Primates: Ecological, Physical, and Behavioral Aspects* (pp. 397–420). Cambridge University Press.

Osman, N. A., Abdul-Latiff, M. A. B., Mohd-Ridwan, A. R., Yaakop, S., Nor, S. M., and Md-Zain, B. M. (2020). Diet composition of the wild stump-tailed macaque (*Macaca arctoides*) in Perlis State Park, Peninsular Malaysia, using a chloroplast tRNL DNA metabarcoding approach: A preliminary study. *Animals*, 10, 2215.

Osorio, D., Smith, A. C., Vorobyev, M., and Buchanan-Smith, H. M. (2004). Detection of fruit and the selection of primate visual pigments for color vision. *American Naturalist*, 164, 696–708.

Osorio, D., and Vorobyev, M. (1996). Colour vision as an adaptation to frugivory in primates. *Proceedings of the Royal Society B: Biological Sciences*, 263, 596–99.

Ostlund, R. E. (2002). Phytosterols in human nutrition. *Annual Review of Nutrition*, 22, 533–49.

Ostro, L. E. T., Silver, S. C., Koontz, F. W., and Young, T. P. (2000). Habitat selection by translocated black howler monkeys in Belize. *Animal Conservation*, 3, 175–81.

Ostro, L. E., Silver, S. C., Koontz, F. W., Horwich, R. H., and Brockett, R. (2001). Shifts in social structure of black howler (*Alouatta pigra*) groups associated with natural and experimental variation in population density. *International Journal of Primatology*, 22, 733–48.

Ostro, L. E., Silver, S. C., Koontz, F. W., Young, T. P., and Horwich, R. H. (1999). Ranging behavior of translocated and established groups of black howler monkeys *Alouatta pigra* in Belize, Central America. *Biological Conservation*, 87, 181–90.

Ostrom, E. (2010). Beyond markets and states: Polycentric governance of complex economic systems. *American Economic Review*, 100, 641–72.

Osvath, M., and Osvath, H. (2008). Chimpanzee (*Pan troglodytes*) and orangutan (*Pongo abelii*) forethought: Self-control and pre-experience in the face of future tool use. *Animal Cognition*, 11, 661–74.

Ottoni, E. B. (2015). Tool use traditions in nonhuman primates: The case of tufted capuchin monkeys. *Human Ethology Bulletin*, 30, 22–40.

Ottoni, E. B., and Izar, P. (2008). Capuchin monkey tool use: Overview and implications. *Evolutionary Anthropology*, 17, 171–78.

Ovaskainen, O., Skorokhodova, S., Yakovleva, M., Sukhov, A., Kutenkov, A., Kutenkova, N., et al. (2013). Community-level phenology response to climate change. *Proceedings of the National Academy of Sciences of the United States of America*, 110, 13434–39.

Overdorff, D. J. (1993a). Ecological and reproductive correlates to range use in red-bellied lemurs (*Eulemur rubriventer*) and rufous lemurs (*Eulemur fulvus rufus*). In P. M. Kappeler and J.

U. Ganzhorn (Eds.), *Lemur Social Systems and Their Ecological Basis* (pp. 167–78). Springer.

Overdorff, D. J. (1993b). Similarities, differences, and seasonal patterns in the diets of *Eulemur rubriventer* and *Eulemur fulvus rufus* in the Ranomafana National Park, Madagascar. *International Journal of Primatology*, 14, 721–53.

Overdorff, D. J., Strait, S. G., and Telo, A. (1997). Seasonal variation in activity and diet in a small-bodied folivorous primate, *Hapalemur griseus*, in southeastern Madagascar. *American Journal of Primatology*, 43, 211–23.

Owen-Smith, N. (1996). Circularity in linear programming models of optimal diet. *Oecologia*, 108, 259–61.

Oyen, M. L., Ferguson, V. L., Bembey, A. K., Bushby, A. J., and Boyde, A. (2008). Composite bounds on the elastic modulus of bone. *Journal of Biomechanics*, 41(11), 2585–88.

Oyen, O. J. (1979). Tool-use in free-ranging baboons of Nairobi National Park. *Primates*, 20, 595–97.

Ozanne, C. M. P., Bell, J. R., and Weaver, D. G. (2011). Collecting arthropods and arthropod remains for primate studies. In J. M. Setchell and D. J. Curtis (Eds.), *Field and Laboratory Methods in Primatology: A Practical Guide* (pp. 271–85). Cambridge University Press.

Pablo-Rodríguez, M., Hernández-Salazar, L., Aureli, F., and Schaffner, C. M. (2015). The role of sucrose and sensory systems in fruit selection and consumption of *Ateles geoffroyi* in Yucatan, Mexico. *Journal of Tropical Ecology*, 31, 213–19.

Packer, C., Collins, D. A., and Eberle, L. E. (2000). Problems with primate sex ratios. *Philosophical Transactions of the Royal Society of London Series B—Biological Sciences*, 355, 1627–35.

Paim, F. P., Chapman, C. A., de Queiroz, H. L., and Paglia, A. P. (2017). Does resource availability affect the diet and behavior of the vulnerable squirrel monkey, *Saimiri vanzolinii*? *International Journal of Primatology*, 38, 572–87.

Paine, O. C. C., Koppa, A., Henry, A. G., Leichliter, J. N., Codron, D., Codron, J., et al. (2018). Grass leaves as potential hominin dietary resources. *Journal of Human Evolution*, 117, 44–52.

Pajic, P., Pavlidis, P., Dean, K., Neznanova, L., Romano, R. A., Garneau, D., et al. (2019). Independent amylase gene copy number bursts correlate with dietary preferences in mammals. *Elife*, 8, e44628.

Pal, A., Kumara, H. N., Sarathi Mishra, P., Velankar, A. D., and Singh, M. (2017). Extractive foraging and tool-aided behaviors in the wild Nicobar long-tailed macaque (*Macaca fascicularis umbrosus*). *Primates*, 59, 173–83.

Palacios, E., and Rodriguez, A. (2001). Ranging pattern and use of space in a group of red howler monkeys (*Alouatta seniculus*) in a southeastern Colombian rainforest. *American Journal of Primatology*, 55, 233–51.

Palma, C., Cassone, A., Serbousek, D., Pearson, C. A., and Djeu, J. Y. (1992). Lactoferrin release and interleukin-1, interleukin-6, and tumor necrosis factor production by human polymorphonuclear cells stimulated by various lipopolysaccharides: relationship to growth inhibition of *Candida albicans*. *Infection and Immunity*, 60(11), 4604–11.

Palmer, B., Jones, R. J., Wina, E., and Tangendjala, B. (2000). The effect of sample drying conditions on estimates of condensed tannin and fibre content, dry matter digestibility, nitrogen digestibility and PEG binding of *Calliandra calothyrsus*. *Animal Feed Science and Technology*, 87, 29–40.

Palminteri, S., Powell, G. V. N., Asner, G. P., and Peres, C. A. (2012). LiDAR measurements of canopy structure predict spatial distribution of a tropical mature forest primate. *Remote Sensing of Environment*, 127, 98–105.

Palminteri, S., Powell, G. V., and Peres, C. A. (2012). Advantages of granivory in seasonal environments: Feeding ecology of an arboreal seed predator in Amazonian forests. *Oikos*, 121, 1896–904.

Palmquist, A. (2020). Demedicalizing breastmilk: The discourses, practices, and identities of informal milk sharing. In *Ethnographies of Breastfeeding* (pp. 23–44). Routledge.

Palmquist, A. E., and Doehler, K. (2016). Human milk sharing practices in the US. *Maternal and Child Nutrition*, 12, 278–90.

Palombit, R. A. (1997). Inter- and intraspecific variation in the diets of sympatric siamang (*Hylobates syndactylus*) and lar gibbons (*Hylobates lar*). *Folia Primatologica*, 68, 321–27.

Pampush, J. D., Duque, A. C., Burrows, B. R., Daegling, D. J., Kenney, W. F., and McGraw, W. S. (2013). Homoplasy and thick enamel in primates. *Journal of Human Evolution*, 64(3), 216–24.

Pampush, J. D., Daegling, D. J., Vick, A. E., McGraw, W. S., Covey, R. M., and Rapoff, A. J. (2011). Technical note: converting durometer data into elastic modulus in biological materials. *American Journal of Physical Anthropology*, 146, 650–53.

Pampush, J. D., Morse, P. E., Fuselier, E. J., Skinner, M. M., and Kay, R. F. 2022. Sign-oriented Dirichlet normal energy: Aligning dental topography and dental function in the R-package molaR. *Journal of Mammalian Evolution*, 29, 713–32.

Pampush, J. D., Spradley, J. P., Morse, P. E., Griffith, D., Gladman, J. T., Gonzales, L. A., et al. 2018. Adaptive wear-based changes in dental topography associated with atelid (Mammalia: Primates) diets. *Biological Journal of the Linnean Society*, 124(4), 584–606.

Pang, W. W., and Hartmann, P. E. (2007). Initiation of human lactation: Secretory differentiation and secretory activation. *Journal of Mammary Gland Biology and Neoplasia*, 12, 211–21.

Panger, M. (1997). *Hand Preference and Object-Use in Free-Ranging White Faced Capuchin Monkeys* (Cebus capucinus) *in Costa Rica*. University of California, Berkeley.

Panger, M. (2007). Tool use and cognition in primates. In C. J. Campbell, A. Fuentes, K. C. MacKinnon, M. Panger, and S. K. Bearder (Eds.), *Primates in Perspective* (pp. 665–77). Oxford University Press.

Panger, M. A., Perry, S., Rose, L., Gros-Louis, J., Vogel, E., MacKinnon, K. C., et al. (2002). Cross-site differences in foraging behavior of white-faced capuchins (*Cebus capucinus*). *American Journal of Physical Anthropology*, 119, 52–66.

Panter-Brick, C., Lotstein, D. S., and Ellison, P. T. (1993). Seasonality of reproductive function and weight loss in rural Nepali women. *Human Reproduction*, 8, 684–490.

Paola Juarez, C., Alejandro Rotundo, M., Berg, W., and Fernandez-Duque, E. (2011). Costs and benefits of radio-collaring on the behavior, demography, and conservation of owl monkeys (*Aotus azarai*) in Formosa, Argentina. *International Journal of Primatology*, 32, 69–82.

Papathakis, P. C., Singh, L. N., and Manary, M. J. (2016). How maternal malnutrition affects linear growth and development in the offspring. *Molecular and Cellular Endocrinology*, 435, 40–47.

Papworth, S., Bose, A. S., Barker, J., Schel, A. M., and Zuberbühler, K. (2008). Male blue monkeys alarm call in response to danger experienced by others. *Biology Letters*, 4, 472–75.

Papworth, S., Milner-Gulland, E. J., and Slocombe, K. (2013). Hunted woolly monkeys (*Lagothrix poeppigii*) show threat-sensitive responses to human presence. *PLoS ONE*, 8, e62000.

Parfrey, L. W., Walters, W. A., and Knight, R. (2011). Microbial eukaryotes in the human microbiome: Ecology, evolution, and future directions. *Frontiers in Microbiology*, 2, 153.

Parfrey, L. W., Walters, W. A., Lauber, C., Clemente, J. C., Berg-Lyons, D., Teiling, C., et al. (2014). Communities of microbial eukaryotes in the mammalian gut within the context of environmental eukaryotic diversity. *Frontiers in Microbiology*, 5, 298.

Park, A. J., Collins, J., Blennerhassett, P., Ghia, J. E., Verdu, E. F., Bercik, P., et al. (2013). Altered colonic function and microbiota profile in a mouse model of chronic depression. *Neurogastroenterology and Motility*, 25(9), 733-e575.

Parker, E. J., Hill, R. A. and Koyama, N. F. (2022) Behavioural responses to spatial variation in predation risk and resource availability in an arboreal primate. *Ecosphere*, 13, e3945.

Parker, K. L., Barboza, P. S., and Gillingham, M. P. (2009). Nutrition integrates environmental responses of ungulates. *Functional Ecology*, 23, 57–69.

Parker, S. T., and Gibson, K. R. (1977). Object manipulation, tool use and sensorimotor intelligence as feeding adaptations in cebus monkeys and great apes. *Journal of Human Evolution*, 6, 623–41.

Parker, T. H., Nakagawa, S., Gurevitch, J., and IIEE. (2016). Promoting transparency in evolutionary biology and ecology. *Ecology Letters*, 19, 726–28.

Parmesan, C., and Hanley, M. E. (2015). Plants and climate change: Complexities and surprises. *Annals of Botany*, 116, 849–64.

Parnell, A. C., Inger, R., Bearhop, S., and Jackson, A. L. (2010). Source partitioning using stable isotopes: Coping with too much variation. *PLoS ONE*, 5, e9672.

Parra, R. (1978). Comparison of foregut and hindgut fermentation in herbivores. In G. G. Montgomery (Ed.), *The Ecology of Arboreal Folivores* (pp. 205–30). Smithsonian Institution Press.

Parrado-Rosselli, A., Machado, J. L., and Prieto-López, T. (2006). Comparison between two methods for measuring fruit production in a tropical forest. *Biotropica*, 38, 267–71.

Parsons, P. E., and Taylor, C. R. (1977). Energetics of brachiation versus walking: A comparison of a suspended and an inverted pendulum mechanism. *Physiological Zoology*, 50, 182–88.

Pasquaretta, C., Gomez-Moracho, T., Heeb, P., and Lihoreau, M. (2018). Exploring interactions between the gut microbiota and social behavior through nutrition. *Genes*, 9, 534.

Pass, D. M., Foley, W. J., and Bowden, B. (1998). Vertebrate herbivory on *Eucalyptus*—identification of specific feeding deterrents for common ringtail possums (*Pseudocheirus peregrinus*) by bioassay-guided fractionation of *Eucalyptus ovata* foliage. *Journal of Chemical Ecology*, 24, 1513–27.

Passey, B. H., Robinson, T. F., Ayliffe, L. K., Cerling, T. E., Sponheimer, M., Dearing, D. M., et al. (2005). Carbon isotope fractionation between diet, breath CO_2, and bioapatite in different mammals. *Journal of Archaeological Sciences*, 32, 1459–70.

Pastor, J., Moen, R., and Cohen, Y. (1997). Spatial heterogeneities, carrying capacity, and feedbacks in animal-landscape interactions. *Journal of Mammalogy*, 78, 1040–52.

Patisaul, H. B., and Adewale, H. B. (2009). Long-term effects of environmental endocrine disruptors on reproductive physiology and behavior. *Frontiers in Behavioral Neuroscience*, 3, 10.

Patisaul, H. B., and Bateman, H. L. (2008). Neonatal exposure to endocrine active compounds or an ER agonist increases adult anxiety and aggression in gonadally intact male rats. *Hormones and Behavior*, 53, 580–88.

Patisaul, H. B., Luskin, J. R., and Wilson, M. E. (2004). A soy supplement and tamoxifen inhibit sexual behavior in female rats. *Hormones and Behavior*, 45, 270–77.

Patrono, L. V., Samuni, L., Corman, V. M., Nourifar, L., Rothemeier, C., Wittig, R. M., et al. (2018). Human coronavirus OC43 outbreak in wild chimpanzees, Cote d'Ivoire, 2016. *Emerging Microbes and Infections*, 7, 118.

Patterson, E. M., Krzyszczyk, E., and Mann, J. (2016). Age-specific foraging performance and reproduction in tool-using wild bottlenose dolphins. *Behavioral Ecology*, 27, 401–10.

Patterson, S. K., Hinde, K., Bond, A. B., Trumble, B. C., Strum, S. C., and Silk, J. B. (2021). Effects of early life adversity on maternal effort and glucocorticoids in wild olive baboons. *Behavioral Ecology and Sociobiology*, 75(8), 114.

Paudel, P. K., and Kindlmann, P. (2012). Human disturbance is a major determinant of wildlife distribution in Himalayan midhill landscapes of Nepal. *Animal Conservation*, 15, 283–93.

Paul, A., and Kuester, J. (1987). Sex ratio adjustment in a seasonally breeding primate species: Evidence from the Barbary macaque population at Affenberg Salem. *Ethology*, 74, 117–32.

Paul, J. R., Randle, A. M., Chapman, C. A., and Chapman, L. J. (2004). Arrested succession in logging gaps: Is tree seedling growth and survival limiting? *African Journal of Ecology*, 42, 245–51.

Pavelka, M. S. M., and Behie, A. M. (2005). The effect of Hurricane Iris on the food supply of black howlers (*Alouatta pigra*) in southern Belize. *Biotropica*, 37, 102–8.

Pays, O., Ekori, A., and Fritz, H. (2014). On the advantages of mixed-species groups: Impalas adjust their vigilance when associated with larger prey herbivores. *Ethology*, 120, 1207–16.

Pazol, K., and Cords, M. (2005). Seasonal variation in feeding behavior, competition and female social relationships in a forest dwelling guenon, the blue monkey (*Cercopithecus mitis stuhlmanni*) in Kakamega Forest, Kenya. *Behavioral Ecology and Sociobiology*, 58, 220–46.

Peacor, S. D., Barton, B. T., Kimbro, D. L., Sih, A., and Sheriff, M. J. (2020) A framework and standardized terminology to facilitate the study of predation-risk effects. *Ecology*, 101(12), e03152.

Pearson, D. L., and Derr, J. A. (1986). Seasonal patterns of lowland forest flood arthropod abundance in southeastern Peru. *Biotropica*, 18, 244–56.

Pearson, K. (1896). Mathematical contributions to the theory of evolution—on a form of spurious correlation which may arise when indices are used in the measurement of organs. *Proceedings of the Royal Society London*, 60, 489–98.

Pebsworth, P. A., Bardi, M., and Huffman, M. A. (2012). Geophagy in chacma baboons: Patterns of soil consumption by age class, sex, and reproductive state. *American Journal of Primatology*, 74, 48–57.

Pebsworth, P. A., Hillier, S., Wendler, R., Glahn, R., Ta, C. A. K., Arnason, J. T., and Young, S. L. (2019). Geophagy among East African chimpanzees: consumed soils provide protection from plant secondary compounds and bioavailable iron. *Environmental Geochemistry and Health*, 41, 2911–27.

Pebsworth, P. A., Huffman, M. A., Lambert, J. E., and Young, S. L. (2019). Geophagy among nonhuman primates: A systematic review of current knowledge and suggestions for future directions. *American Journal of Physical Anthropology*, 168, 164–94.

Pebsworth, P. A., MacIntosh, A. J. J., Morgan, H. R., and Huffman, M. A. (2012). Factors influencing the ranging behavior of chacma baboons (*Papio hamadryas ursinus*) living in a human-modified habitat. *International Journal of Primatology*, 33, 872–87.

Pebsworth, P. A., Morgan, H. R., and Huffman, M. A. (2012). Evaluating home range techniques: Use of Global Positioning System (GPS) collar data from chacma baboons. *Primates*, 53, 345–55.

Peck, M. D., Babcock, G. F., and Alexander, J. W. (1992). The role of protein and calorie restriction in outcome from *Salmonella* infection in mice. *Journal of Parenteral and Enteral Nutrition*, 16(6), 561–65.

Peck, M., Thorn, J., Mariscal, A., Baird, A., Tirira, D., and Kniveton, D. (2011). Focusing conservation efforts for the critically endangered brown-headed spider monkey (*Ateles fusciceps*) using remote sensing, modeling, and playback survey methods. *International Journal of Primatology*, 32, 134–48.

Peckre, L. R., Defolie, C., Kappeler, P. M., and Fichtel, C. (2018). Potential self-medication using millipede secretions in red-fronted lemurs: combining anointment and ingestion for a joint action against gastrointestinal parasites? *Primates*, 59, 483–94.

Peleg, M. (1980). A note on the sensitivity of fingers, tongue and jaws as mechanical sensing instruments. *Journal of Texture Studies*, 10, 245–51.

Pels, R. J., Bor, D. H., Woolhandler, S., Himmelstein, D. U., and Lawrence, R. S. (1989). Dipstick urinalysis screening of asymptomatic adults for urinary tract disorders. II. Bacteriuria. *Journal of the American Medical Association*, 262(9), 1221–24.

Penry, D. L. (1993). Digestive constraints on diet selection. In R. Hughes (Ed.), *Diet Selection: An Interdisciplinary Approach to Foraging Behaviour* (pp. 32–55). Blackwell Scientific.

Percher, A. M., Merceron, G., Nsi Akoue, G., Galbany, J., Romero, A., and Charpentier, M. J. (2018). Dental microwear textural analysis as an analytical tool to depict individual traits and reconstruct the diet of a primate. *American Journal of Physical Anthropology*, 165(1), 123–38.

Pereira, M. E. (1993). Seasonal adjustment of growth rate and adult body weight in ringtailed lemurs. In J. U. Ganzhorn and P. M. Kappeler (Eds.), *Lemur Social Systems and Their Ecological Basis* (pp. 205–21). Springer.

Pereira, S., Henderson, D., Hjelm, M., Hård, T., Salazar, L. T. H., and Laska, M. (2021). Taste responsiveness of chimpanzees (*Pan troglodytes*) and black-handed spider monkeys (*Ateles geoffroyi*) to eight substances tasting sweet to humans. *Physiology & Behavior*, 238, 113470.

Pereira, M. E., and Leigh, S. R. (2003). Modes of primate development. In P. M. Kappeler and M. E. Pereira (Eds.), *Primate Life History and Socioecology* (pp. 149–76). University of Chicago Press.

Peres, C. A. (1991). Seed predation of *Cariniana micrantha* (Lecythidaeceae) by brown capuchin monkeys in Central Amazonia. *Biotropica*, 23, 262–70.

Peres, C. A. (1992a). Consequences of joint-territoriality in a mixed-species group of tamarin monkeys. *Behaviour*, 123, 220–46.

Peres, C. A. (1992b). Prey-capture benefits in a mixed-species group of Amazonian tamarins, *Saguinus fuscicollis* and *Saguinus mystax*. *Behavioral Ecology and Sociobiology*, 31, 339–47.

Peres, C. A. (1993). Diet and feeding ecology of saddle-back (*Saguinus fuscicollis*) and moustached (*S. mystax*) tamarins in Amazonian terra firme forest. *Journal of Zoology, London*, 230, 567–92.

Peres, C. A. (1994a). Diet and feeding ecology of gray woolly monkeys (*Lagothrix lagotricha cana*) in central Amazonia:

Comparisons with other atelines. *International Journal of Primatology*, 15, 333–72.

Peres, C. A. (1994b). Primate responses to phenological changes in an Amazonian terra firme forest. *Biotropica*, 26, 98–112.

Peres, C. A. (1996). Food patch structure and plant resource partitioning in interspecific associations of Amazonian tamarins. *International Journal of Primatology*, 17, 695.

Peres, C. A. (2000a). Effects of subsistence hunting on vertebrate community structure in Amazonian forests. *Conservation Biology*, 14, 240–53.

Peres, C. A. (2000b). Identifying keystone plant resources in tropical forests: The case of gums from *Parkia* pods. *Journal of Tropical Ecology*, 16, 287–317.

Peres, C. A., and Dolman, P. M. (2000). Density compensation in Neotropical primate communities: Evidence from 56 hunted and nonhunted Amazonian forest of varying productivity. *Oecologia*, 122, 175–89.

Periquet, S., Todd-Jones, L., Valeix, M., Stapelkamp, B., Elliot, N., Wijers, M., et al. (2012). Influence of immediate predation risk by lions on the vigilance of prey of different body size. *Behavioral Ecology*, 23, 970–76.

Pérot, A., and Villard, M. A. (2009). Putting density back into the habitat-quality equation: Case study of an open-nesting forest bird. *Conservation Biology*, 23, 1550–57.

Perry, G. H., Martin, R. D., and Verrelli, B. C. (2007). Signatures of functional constraint at aye-aye opsin genes: The potential of adaptive color vision in a nocturnal primate. *Molecular Biology and Evolution*, 24, 1963–70.

Perry, S. (2008). *Manipulative Monkeys: The Capuchins of Lomas Barbudal*. Harvard University Press.

Perry, S. (2009). Conformism in the food processing techniques of white-faced capuchin monkeys (*Cebus capucinus*). *Animal Cognition*, 12, 705–16.

Perry, S., and Manson, J. H. (2003). Traditions in monkeys. *Evolutionary Anthropology*, 12, 71–81.

Persson, A., and Stenberg, M. (2006). Linking patch-use behavior, resource density, and growth expectations in fish. *Ecology*, 87, 1953–59.

Peters, R. H. (1986). *The Ecological Implications of Body Size*. Cambridge University Press.

Peterson, A., Abella, E. F., Grine, F. E., Teaford, M. F., and Ungar, P. S. (2018). Microwear textures of *Australopithecus africanus* and *Paranthropus robustus* molars in relation to paleoenvironment and diet. *Journal of Human Evolution*, 119, 42–63.

Peterson, J., Dwyer, J., Adlercreutz, H., Scalbert, A., Jacques, P., and McCullough, M. L. (2010). Dietary lignans: Physiology and potential for cardiovascular disease risk reduction. *Nutrition Reviews*, 68, 571–603.

Petit, O., Gautrais, J., Leca, J. B., Theraulaz, G., and Deneubourg, J. L. (2009). Collective decision-making in white-faced capuchin monkeys. *Proceedings of the Royal Society B: Biological Sciences*, 276, 3495–503.

Petrullo, L., Baniel, A., Jorgensen, M. J., Sams, S., Snyder-Mackler, N., and Lu, A. (2022). The early life microbiota mediates maternal effects on offspring growth in a nonhuman primate. *iScience*, 25(3), 103948.

Petrullo, L., Jorgensen, M. J., Snyder-Mackler, N., and Lu, A. (2019). Composition and stability of the vervet monkey milk microbiome. *American Journal of Primatology*, 81(10–11), e22982.

Pettang, C. (Ed.), (2016). *Decision support for construction cost control in developing countries*. IGI Global.

Petter, J. J., Schilling, A., and Pariente, G. F. (1971). Observation écoéthologiques sur deux lémurien malagaches nocturnes: *Phaner furcifer* et *Microcebus coquereli*. *Terre La Vie*, 25, 287–327.

Petto, A. J., LaReau-Alves, M. N., Ellison, P. T., and Abbruzzese, M. C. (1995). Reproduction in captive Taiwan macaques (*Macaca cyclopis*) in comparison to other common macaque species. *Zoo Biology*, 14, 331–46.

Phalan, B., Bertzky, M., Butchart, S. H. M., Donald, P. F., Scharlemann, J. P. W., Stattersfield, A., et al. (2013). Crop expansion and conservation priorities in tropical countries. *PLoS ONE*, 8, e51759.

Phiapalath, P., Borries, C., and Suwanwaree, P. (2011). Seasonality of group size, feeding, and breeding in wild red-shanked douc langurs (Lao PDR). *American Journal of Primatology*, 73, 1134–44.

Philip, A. M., Kim, S. D., and Vijayan, M. M. (2012). Cortisol modulates the expression of cytokines and suppressors of cytokine signaling (SOCS) in rainbow trout hepatocytes. *Developmental and Comparative Immunology*, 38(2), 360–67.

Phillips, C. A., and McGrew, W. C. (2014). Macroscopic inspection of ape feces: What's in a quantification method? *American Journal of Primatology*, 76, 539–50.

Phillips, K. (1998). Tool use in wild capuchin monkeys (*Cebus albifrons trinitatis*). *American Journal of Primatology*, 46, 259–61.

Phillips, K. A., Grafton, B., and Haas, M. E. (2003). Tap-scanning for invertebrates by capuchins (*Cebus apella*). *Folia Primatologica*, 74, 162–64.

Phillips, K. A., and Hopkins, W. D. (2007). Exploring the relationship between cerebellar asymmetry and handedness in chimpanzees (*Pan troglodytes*) and capuchins (*Cebus apella*). *Neuropsychologia*, 45, 2333–39.

Phillips-Conroy, J. E., and Jolly, C. J. (1988). Dental eruption schedules of wild and captive baboons. *American Journal of Primatology*, 15, 17–29.

Piaget, J. (1952). *The Origins of Intelligence in Children*. International Universities Press.

Pichard, G., and Van Soest, P. J. (1977, November). Protein solubility of ruminant feeds. In *Proceedings of the Cornell Nutrition Conference* (vol. 91). Cornell University.

Pickering, T. R., and Bunn, H. T. (2012). Meat foraging by Pleistocene African hominins: Tracking behavioral evolution beyond baseline inferences of early access to carcasses. In M. Dominguez-Rodrigo (Ed.), *Stone Tools and Fossil Bones: Debates in the Archaeology of Human Origins* (pp. 152–73). Cambridge University Press.

Pickett, S. B., Bergey, C. M., and Di Fiore, A. (2012). A metagenomic study of primate insect diet diversity. *American Journal of Primatology*, 74, 622–31.

Piedrahita, P., Meise, K., Werner, C., Krüger, O., and Trillmich, F. (2014). Lazy sons, self-sufficient daughters: Are sons more demanding? *Animal Behaviour*, 98, 69–78.

Piep, M., Radespiel, U., Zimmerman, E., Schmidt, S., and Siemers, B. M. (2008). The sensory basis of prey detection in captive-born grey mouse lemurs, *Microcebus murinus*. *Animal Behaviour*, 75, 871–78.

Pieracci, F. M., and Barie, P. S. (2005). Iron and the risk of infection. *Surgical Infections*, 6 (suppl. 1), S41–46.

Pierre, J. F., Heneghan, A. F., Lawson, C. M., Wischmeyer, P. E., Kozar, R. A., and Kudsk, K. A. (2013). Pharmaconutrition review: Physiological mechanisms. *Journal of Parenteral and Enteral Nutrition*, 37(suppl. 5), 51S–65S.

Pigman, W. W. (1943). Classification of carbohydrates. *Journal of Research of the National Institute of Standards and Technology*, 30, 257–65.

Pika, S., Klein, H., Bunel, S., Baas, P., Théleste, E., and Deschner, T. (2019). Wild chimpanzees (*Pan troglodytes troglodytes*) exploit tortoises (*Kinixys erosa*) via percussive technology. *Scientific Reports*, 9, 7661.

Pilbeam, D., and Gould, S. J. (1974). Size and scaling in human evolution. *Science*, 186(4167), 892–901.

Pimley, E. R., Bearder, S. K., and Dixson, A. F. (2005). Home range analysis of *Perodicticus potto edwardsi* and *Sciurocheirus cameronensis*. *International Journal of Primatology*, 26, 191–206.

Pinacho-Guendulain, B., and Ramos-Fernández, G. (2017). Influence of fruit availability on the fission-fusion dynamics of spider monkeys (*Ateles geoffroyi*). *International Journal of Primatology*, 38, 466–84.

Pineda-Munoz, S., and Alroy, J. (2014). Dietary characterization of terrestrial mammals. *Proceedings of the Royal Society B: Biological Sciences*, 281, 20141173.

Pineda-Munoz, S., Evans, A. R., and Alroy, J. (2016). The relationship between diet and body mass in terrestrial mammals. *Paleobiology*, 42, 659–69.

Pinto, L., Costa, C., Strier, K., and Da Fonseca, G. (1993). Habitat, density and group size of primates in a Brazilian tropical forest. *Folia Primatologica*, 61, 135–43.

Pio, D. V., Engler, R., Linder, H. P., Monadjem, A., Cotterill, F. P., Taylor, P. J., et al. (2014). Climate change effects on animal and plant phylogenetic diversity in southern Africa. *Global Change Biology*, 20, 1538–49.

Piperno, D. R. (2006). *Phytoliths: A Comprehensive Guide for Archaeologists and Paleoecologists*. Alta Mira.

Pirarat, N., Kobayashi, T., Katagiri, T., Maita, M., and Endo, M. (2006). Protective effects and mechanisms of a probiotic bacterium *Lactobacillus rhamnosus* against experimental *Edwardsiella tarda* infection in tilapia (*Oreochromis niloticus*). *Veterinary Immunology and Immunopathology*, 113, 339–47.

Pittet, F., Johnson, C., and Hinde, K. (2017). Age at reproductive debut: Developmental predictors and consequences for lactation, infant mass, and subsequent reproduction in rhesus macaques (*Macaca mulatta*). *American Journal of Physical Anthropology*, 164, 457–76.

Plante, S., Colchero, F., and Calmé, S. (2014). Foraging strategy of a Neotropical primate: How intrinsic and extrinsic factors influence destination and residence time. *Journal of Animal Ecology*, 83, 116–25.

Platt, J. R. (1964). Strong inference. *Science*, 146, 347–53.

Plavcan, J. M., and van Schaik, C. P. (1997). Interpreting hominid behavior on the basis of sexual dimorphism. *Journal of Human Evolution*, 32, 345–74.

Plummer, M., Best, N., Cowles, K., and Vines, K. (2006). CODA: convergence diagnosis and output analysis for MCMC. *R news*, 6(1), 7–11.

Plumptre, A. J. (1996). Changes following 60 years of selective harvesting in the Budongo Forest Reserve, Uganda. *Forest Ecology and Management*, 89, 101–13.

Plumptre, A. J., and Reynolds, V. (1994). The effect of selective logging on the primate populations in the Budongo Forest Reserve, Uganda. *Journal of Applied Ecology*, 31, 631–41.

Poché, R. M. (1976). Notes on primates in Parc National du W du Niger, West Africa. *Mammalia*, 40, 2, 187–98.

Poirier, A. C., Waterhouse, J. S., Watsa, M., Erkenswick, G. A., Moreira, L. A., Tang, J., et al. (2021). On the trail of primate scent signals: A field analysis of callitrichid scent-gland secretions by portable gas chromatography-mass spectrometry. *American Journal of Primatology*, 83(3), e23236.

Polansky, L., and Boesch, C. (2013). Long-term changes in fruit phenology in a lowland tropical rainforest are not explained by rainfall. *Biotropica*, 45, 434–40.

Polansky, L., and Robbins, M. M. (2013). Generalized additive mixed models for disentangling long-term trends, local anomalies, and seasonality in fruit tree phenology. *Ecology and Evolution*, 3, 3141–51.

Pollick, A. S., Gouzoules, H., and de Waal, F. B. M. (2005). Audience effects on food calls in captive brown capuchin monkeys, *Cebus apella*. *Animal Behaviour*, 70, 1273–81.

Pond, C. M. (1977). The significance of lactation in the evolution of mammals. *Evolution*, 31, 177–99.

Pond, C. M. (1984). Physiological and ecological importance of energy storage in the evolution of lactation: Evidence for a common pattern of anatomical organization of adipose tissue in mammals. *Symposia of the Zoological Society of London*, 51, 1–32.

Pontes, A. R. M., Monteiro da Cruz, M. A. O. (1995). Home range, intergroup transfers, and reproductive status of common marmosets *Callithrix jacchus* in a forest fragment in North-Eastern Brazil. *Primates*, 36, 335–47.

Ponton, F., Morimoto, J., Robinson, K., Kumar, S., Cotter, S., Wilson, K., et al. (2020). Macronutrients modulate survival to infection and immunity in *Drosophila*. *Journal of Animal Ecology*, 89, 460–70.

Ponton, F., Wilson, K., Cotter, S. C., Raubenheimer, D., and Simpson, S. J. (2011). Nutritional immunology: A multidimensional approach. *PLoS Pathogens*, 7(12), e1002223.

Ponton, F., Wilson, K., Holmes, A. J., Cotter, S. C., Raubenheimer, D., and Simpson, S. J. (2013). Integrating nutrition and immunology: A new frontier. *Journal of Insect Physiology*, 59(2), 130–37.

Pontzer, H. (2015). Energy expenditure in humans and other primates: a new synthesis. *Annual Review of Anthropology*, 44, 169–87.

Pontzer, H. (2017). The crown joules: energetics, ecology, and evolution in humans and other primates. *Evolutionary Anthropology*, 26, 12–24.

Pontzer, H. (2018). Energy constraint as a novel mechanism linking exercise and health. *Physiology*, 33, 384–93.

Pontzer, H., Brown, M. H., Raichlen, D. A., Dunswoth, H. M., Hare, B., Walker, K., et al. (2016). Metabolic acceleration and the evolution of human brain size and life history. *Nature*, 533, 390–92.

Pontzer, H., Brown, M. H., Wood, B. M., Raichlen, D. A., Mabulla, A. Z. P., Harris, J. A., et al. (2021). Evolution of water conservation in humans. *Current Biology*, 31, 1804–10.

Pontzer, H., Durazo-Arvizu, R., Dugas, L. R., Plange-Rhule, J., Bovet, P., Forrester, T. E., et al. (2016). Constrained total energy expenditure and metabolic adaptation to physical activity in adult humans. *Current Biology*, 26, 410–17.

Pontzer, H., and Kamilar, J. M. (2009). Great ranging associated with greater reproductive investment in mammals. *Proceedings of the National Academy of Sciences*, 106(1), 192–96.

Pontzer, H., and McGrosky, A. (2022). Balancing growth, reproduction, maintenance, and activity in evolved energy economies. *Current Biology*, 32(12), R709–R719.

Pontzer, H., Raichlen, D. A., Wood, B. M., Emery Thompson, M., Racette, S. B., Mabulla, A. Z., et al. (2015). Energy expenditure and activity among Hadza hunter-gatherers. *American Journal of Human Biology*, 27(5), 628–37.

Pontzer, H., Raichlen, D. A., Wood, B. M., Mabulla, A. Z., Racette, S. B., and Marlowe, F. W. (2012). Hunter-gatherer energetics and human obesity. *PLoS ONE*, 7, p.e40503.

Pontzer, H., Raichlen, D. A., Shumaker, R. W., Ocobock, C., and Wich, S. A. (2010). Metabolic adaptation for low energy throughput in orangutans. *Proceedings of the National Academy of Sciences of the United States of America*, 107(32), 14048–52.

Pontzer, H., and Wrangham, R. W. (2006). Ontogeny of ranging in wild chimpanzees. *International Journal of Primatology*, 27, 295–309.

Pontzer, H., Yamada, Y., Sagayama, H., Ainslie, P. N., Andersen, L. F., Anderson, L. J., et al. (2021). Daily energy expenditure through the human life course. *Science*, 373(6556), 808–12.

Poole, K., Herget, R., Lapatsina, L., Ngo, H. D., and Lewin, G. R. (2014). Tuning Piezo ion channels to detect molecular-scale movements relevant for fine touch. *Nature Communications*, 5, 3520.

Pope, T. R. (1998). Effects of demographic change on group kin structure and gene dynamics of populations of red howling monkeys. *Journal of Mammalogy*, 79, 692–712.

Popkin, B. M. (2015). Nutrition transition and the global diabetes epidemic. *Current Diabetes Reports*, 15, 64–64.

Popowics, T. E., and Fortelius, M. (1997). On the cutting edge: Tooth blade sharpness in herbivorous and faunivorous mammals. *Annales Zoologici Fennici*, 34(2), 73–88.

Popowics, T. E., Rensberger, J. M., and Herring, S. W. (2001). The fracture behaviour of human and pig molar cusps. *Archives of Oral Biology*, 46(1), 1–12.

Poppitt, S. D., Prentice, A. M., Jéquier, E., Schutz, Y., and Whitehead, R. G. (1993). Evidence of energy sparing in Gambian women during pregnancy: A longitudinal study using whole-body calorimetry. *American Journal of Clinical Nutrition*, 57, 353–64.

Port, M., Hildenbrandt, H., Pen, I., Schülke, O., Ostner, J., and Weissing, F. J. (2020). The evolution of social philopatry in female primates. *American Journal of Physical Anthropology*, e24123.

Port, M., Schülke, O., and Ostner, J. (2017). From individual to group territoriality: Competitive environments promote the evolution of sociality *American Naturalist*, 189, E46–E57.

Porter, J., Craven, B., Khan, R. M., Chang, S. J., Kang, I., Judkewitz, B., et al. (2007). Mechanisms of scent-tracking in humans. *Nature Neuroscience*, 10, 27–29.

Porter, L. M. (2001). Dietary differences among sympatric Callitrichinae in northern Bolivia: *Callimico goeldii*, *Saguinus fuscicollis* and *S. labiatus*. *International Journal of Primatology*, 22, 961–92.

Porter, L. M., and Garber, P. A. (2010). Mycophagy and its influence on habitat use and ranging patterns in *Callimico goeldii*. *American Journal of Physical Anthropology*, 142, 468–75.

Porter, L. M., and Garber, P. A. (2013). Foraging and spatial memory in Weddell's saddleback tamarins (*Saguinus fuscicollis weddelli*) when moving between distant and out-of-sight goals. *International Journal of Primatology*, 34, 30–48.

Porter, L. M., Garber, P. A., Boesch, C., and Janmaat, K. R. L. (2021). Using GIS to examine long-term foraging strategies in tamarins and chimpanzees. In C. A. Shaffer, F. L. Dolins, J. R. Hickey, N. Nibbelink, and L. M. Porter (Eds.), *Spatial Analysis in Field Primatology: Applying GIS at Varying Scales*. Cambridge University Press.

Porter, L. M., Gilbert, C. C., and Fleagle, J. G. (2014). Diet and phylogeny in primate communities. *International Journal of Primatology*, 35, 1144–63.

Portman, O. W. (1970). Nutritional requirements (NRC) of nonhuman primates. In R. Harris (Ed.), *Feeding and Nutrition of Nonhuman Primates* (pp. 87–115). Academic Press.

Portner, H. O., and Farrell, A. P. (2008). Physiology and climate change. *Science*, 332, 690–92.

Post, D. G. (1984). Is optimization the optimal approach to primate foraging? In P. S. Rodman and J. G. H. Cant (Eds.), *Adaptations for Foraging in Nonhuman Primates: Contributions to an Organismal Biology of Prosimians, Monkeys and Apes* (pp. 280–303). Columbia University Press.

Post, D. G., Hausfater, G., and McCuskey, S. A. (1980). Feeding behavior of yellow baboons (*Papio cynocephalus*): Relationship to age, gender and dominance rank. *Folia Primatologica*, 34, 170–95.

Potts, K. B., Baken, E., Levang, A., and Watts, D. P. (2016). Ecological factors influencing habitat use by chimpanzees at Ngogo, Kibale National Park, Uganda. *American Journal of Primatology*, 78, 432–40.

Potts, K. B., Baken, E., Ortmann, S., Watts, D. P., and Wrangham, R. W. (2015). Variability in population density is paralleled by large differences in foraging efficiency in chimpanzees (*Pan troglodytes*). *International Journal of Primatology*, 36, 1101–19.

Potts, K. B., Chapman, C. A., and Lwanga, J. S. (2009). Floristic heterogeneity between forested sites in Kibale National Park, Uganda: Insights into the fine-scale determinants of density in a large-bodied frugivorous primate. *Journal of Animal Ecology*, 78, 1269–77.

Potts, K. B., Watts, D. P., and Wrangham, R. W. (2011). Comparative feeding ecology of two communities of chimpanzees (*Pan troglodytes*) in Kibale National Park, Uganda. *International Journal of Primatology*, 32, 669–90.

Potts, R. (2004). Paleo environmental basis of cognitive evolution in great apes. *American Journal of Primatology*, 62, 209–28.

Poucet, B. (1993). Spatial cognitive maps in animals: New hypotheses on their structure and neural mechanisms. *Psychological Review*, 100, 163–82.

Poulsen, J. R., Clark, C. J., Connor, E. F., and Smith, T. B. (2002). Differential resource use by primates and hornbills: Implications for seed dispersal. *Ecology*, 83, 228–40.

Poulsen, J., Clark, C. J., and Smith, T. B. (2001). Seasonal variation in the feeding ecology of the grey-cheeked mangabey (*Lophocebus albigena*) in Cameroon. *American Journal of Primatology*, 54, 91–105.

Pounds, J. A., Fogden, M. P. L., and Campbell, J. H. (1999). Biological response to climate change on a tropical mountain. *Nature*, 398, 611–15.

Povey, S., Cotter, S. C., Simpson, S. J., Lee, K. P., and Wilson, K. (2009). Can the protein costs of bacterial resistance be offset by altered feeding behaviour? *Journal of Animal Ecology*, 78(2), 437–46.

Powe, C. E., Knott, C. D., and Conklin-Brittain, N. (2010). Infant sex predicts breast milk energy content. *American Journal of Human Biology*, 22, 50–54.

Powell, G. V. (1985). Sociobiology and adaptive significance of interspecific foraging flocks in the Neotropics. *Ornithological Monographs*, 713–32.

Powell, G. V. (1989). On the possible contribution of mixed species flocks to species richness in neotropical avifaunas. *Behavioral Ecology and Sociobiology*, 24(6), 387–93.

Power, M. E. (1992). Top-down and bottom-up forces in food webs: Do plants have primacy. *Ecology*, 73, 733–46.

Power, M. L. (2010) Nutritional and digestive challenges to being a gum-feeding primate. In: Burrows, A. M., and Nash, L. T. (Eds) *The evolution of exudativory in primates* (pp. 25–44). Springer.

Power, M. L., and Schulkin, J. (2009). *The Evolution of Obesity*. Johns Hopkins University Press.

Power, M. L., and Schulkin, J. (2016). *Milk: The Biology of Lactation*. JHU Press.

Power, M. L., Oftedal, O. T., and Tardif, S. D. (2002). Does the milk of callitrichid monkeys differ from that of larger anthropoids? *American Journal of Primatology*, 56, 117–27.

Power, M. L., Tardif, S. D., Layne, D. G., and Schulkin, J. (1999). Ingestion of calcium solutions by common marmosets (*Callithrix jacchus*). *American Journal of Primatology*, 47, 255–61.

Power, M. L., Verona, C., Ce, R. M., and Oftedal, O. T. (2008). The composition of milk from free-living common marmosets (*Callithrix jacchus*) in Brazil. *American Journal of Primatology*, 70, 78–83.

Powzyk, J. A., and Mowry, C. B. (2003). Dietary and feeding differences between sympatric *Propithecus diadema diadema* and *Indri indri*. *International Journal of Primatology*, 24, 1143–62.

Powzyk, J. A., and Mowry, C. B. (2006). The feeding ecology and related adaptations of *Indri indri*. In L. Gould and M. L. Sauther (Eds.), *Lemurs* (pp. 353–68). Springer.

Pozo-Montuy, G., Serio-Silva, J. C., Chapman, C. A., and Bonilla-Sánchez, Y. M. (2013). Resource use in a landscape matrix by an arboreal primate: Evidence of supplementation in *Alouatta pigra*. *International Journal of Primatology*, 34, 714–31.

Prange, S., Jordan, T., Hunter, C., and Gehrt, S. D. (2006). New radiocollars for the detection of proximity among individuals. *Wildlife Society Bulletin*, 34(5), 1333–44.

Preisser, E. L., Bolnick, D. I., and Benard, M. F. (2005). Scared to death? The effects of intimidation and consumption in predator-prey interactions. *Ecology*, 86, 501–9.

Preisser, E. L., Orrock, J. L., and Schmitz, O. J. (2007). Predator hunting mode and habitat domain alter nonconsumptive effects in predator-prey interactions. *Ecology*, 88, 2744–51.

Prentice, A. (1994). Maternal calcium requirements during pregnancy and lactation. *American Journal of Clinical Nutrition*, 59, 477S–482S.

Prentice, A., Jarjou, L., Cole, T. J., Stirling, D. M., Dibba, B., and Fairweather-Trait, S. (1995). Calcium requirements of lactating Gambian mothers: Effects of a calcium supplement on breast-milk calcium concentration, maternal bone mineral content, and urinary calcium excretion. *American Journal of Clinical Nutrition*, 62, 58–67.

Prentice, A., Lunn, P., Watkinson, M., and Whitehead, R. (1983). Dietary supplementation of lactating Gambian women. II. Effect on maternal health, nutritional status and biochemistry. *Human Nutrition and Clinical Nutrition*, 37, 65–74.

Prentice, A. M., and Goldberg, G. R. (2000). Energy adaptations in human pregnancy: Limits and long-term consequences. *American Journal of Clinical Nutrition*, 71, 1226S–1232S.

Prentice, A. M., and Prentice, A. (1988). Energy costs of lactation. *Annual Review of Nutrition*, 8, 63–79.

Prentice, A. M., and Prentice, A. (1995). Evolutionary and environmental influences on human lactation. *Proceedings of the Nutrition Society*, 54, 391–400.

Prentice, A. M., Roberts, S., Watkinson, M., Whitehead, R. G., Paul, A., Prentice, A., et al. (1980). Dietary supplementation of Gambian nursing mothers and lactational performance. *Lancet*, 316, 886–88.

Prentice, A. M., Whitehead, R. G., Roberts, S. B., and Paul, A. A. (1981). Long-term energy balance in child-bearing Gambian women. *American Journal of Clinical Nutrition*, 34, 2790–99.

Presotto, A., and Izar, P. (2010). Spatial reference of black capuchin monkeys in Brazilian Atlantic Forest: Egocentric or allocentric? *Animal Behaviour*, 80, 125–32.

Price, D. (1999). Carrying capacity reconsidered. *Population and Environment*, 21, 5–26.

Price, E. C. (1992a). Changes in the activity of captive cotton-top tamarins (*Saguinus oedipus*) over the breeding cycle. *Primates*, 33, 99–106.

Price, E. C. (1992b). The costs of infant carrying in captive cotton-top tamarins. *American Journal of Primatology*, 26, 23–33.

Price, M. V., Waser, N. M., and Bass, T. A. (1984). Effects of moonlight on microhabitat use by desert rodents. *Journal of Mammalogy*, 65, 353–56.

Price, S. A., and Hopkins, S. S. B. (2015). The macroevolutionary relationship between diet and body mass across mammals. *Biological Journal of the Linnean Society*, 115, 173–84.

Price, T., Wadewitz, P., Cheney, D., Seyfarth, R., Hammerschmidt, K., and Fischer, J. (2015). Vervets revisited: A quantitative analysis of alarm call structure and context specificity. *Scientific Reports*, 5, 13220.

Prichard, G., and Van Soest, P. J. (1977). Protein solubility of ruminant feeds. Cornell Nutrition Conference for Feed Manufactures, Ithaca, NY.

Pride, E. (2005). Optimal group size and seasonal stress in ring-tailed lemurs (*Lemur catta*). *Behavioral Ecology*, 16, 550–60.

Prior, K. M., Adams, D. C., Klepzig, K. D., and Hulcr, J. (2018). When does invasive species removal lead to ecological recovery: Implications for management success. *Biological Invasions*, 20, 267–83.

Priston, N. E. C., Wyper, R. M., and Lee, P. C. (2012). Buton macaques (*Macaca ochreata brunnescens*): Crops, conflict, and behavior on farms. *American Journal of Primatology*, 74, 29–36.

Proffitt, K. M., Grigg, J. L., Hamlin, K. L., and Garrott, R. A. (2009). Contrasting effects of wolves and human hunters on elk behavioral responses to predation risk. *Journal of Wildlife Management*, 73, 345–56.

Proffitt, K. M., Gude, J. A., Hamlin, K. L., and Messer, M. A. (2013). Effects of hunter access and habitat security on elk habitat selection in landscapes with a public and private land matrix. *Journal of Wildlife Management*, 77, 514–24.

Pronovost, G. N., and Hsiao, E. Y. (2019). Perinatal interactions between the microbiome, immunity, and neurodevelopment. *Immunity*, 50, 18–36.

Provenza, F. D. (1995). Postingestive feedback as an elementary determinant of food preference and intake in ruminants. *Journal of Range Management*, 48, 2–17.

Provenza, F. D., Villalba, J. J., Dziba, L. E., Atwood, S. B., and Banner, R. E. (2003). Linking herbivore experience, varied diets, and plant biochemical diversity. *Small Ruminant Research*, 49, 257–74.

Pruetz, J. D. (2006). Feeding ecology of savanna chimpanzees (*Pan troglodytes verus*) at Fongoli, Senegal. In G. Hohmann, M. M. Robbins, and C. Boesch (Eds.), *Feeding Ecology of Apes and Other Primates: Ecological, Physical and Behavioral Aspects* (pp. 161–82). Cambridge University Press.

Pruetz, J. D. (2007). Evidence of cave use by savanna chimpanzees (*Pan troglodytes verus*) at Fongoli, Senegal: Implications for thermoregulatory behavior. *Primates*, 48, 316–19.

Pruetz, J. D. (2018). Nocturnal behavior by a diurnal ape, the West African chimpanzee (*Pan troglodytes verus*), in a savanna environment at Fongoli, Senegal. *American Journal of Physical Anthropology*, 166, 541–48.

Pruetz, J. D., and Bertolani, P. (2007). Savanna chimpanzees, *Pan troglodytes*, hunt with tools. *Current Biology*, 17, 1–6.

Pruetz, J. D., Bogart, S. L., and Lindshield, S. M. (2020). Extractive foraging in an extreme environment: tool and proto-tool use by chimpanzees at Fongoli, Senegal. In Hopper, L. M., and Ross, S. R. (Eds.) *Chimpanzees in Context: A Comparative Perspective on Chimpanzee Behavior, Cognition, Conservation, and Welfare* (pp. 391–409). University of Chicago Press.

Pruetz, J. D., and Isbell, L. A. (1999). What makes a food contestable? Food properties and contest competition in vervets and patas monkeys in Laikipia, Kenya. *American Journal of Physical Anthropology (Supplement S29)*, 110, 225–26.

Pruetz, J. D., Bertolani, P., Boyer Ontil, K., Lindshield, S., Shelley, M., and Wessling, E. G. (2015). New evidence on the tool-assisted hunting exhibited by chimpanzees (*Pan troglodytes verus*) in a savanna habitat at Fongoli, Senegal. *Royal Society Open Science*, 2, 140507.

Pulliam, H., and Caraco, T. (1984). Living in groups: Is there an optimal group size? In J. Krebs and N. Davies (Eds.), *Behavioural Ecology: An Evolutionary Approach* (pp. 122–47). Sinauer Assoc.

Pullinger, G. D., van Diemen, P. M., Carnell, S. C., Davies, S. H., Lyte, M., and Stevens, M. P. (2010). 6-hydroxydopamine-mediated release of norepinephrine increases faecal excretion of *Salmonella enterica* serovar typhimurium in pigs. *Veterinary Research*, 41, 68.

Punt, A. E. (2000). Extinction of marine renewable resources: A demographic analysis. *Population Ecology*, 42, 19–27.

Purba, L. H. P. S., Widayati, K. A., Tsutsui, K., Suzuki-Hashido, N., Hayakawa, T., Nila, S., et al. (2017). Functional characterization of the TAS2R38 bitter taste receptor for phenylthiocarbamide in colobine monkeys. *Biology Letters*, 13(1), 20160834.

Purnell, M. A., Crumpton, N., Gill, P. G., Jones, G., and Rayfield, E. J. 2013. Within-guild dietary discrimination from 3-D textural analysis of tooth microwear in insectivorous mammals. *Journal of Zoology*, 291(4), 249–57.

Pusey, A. E. (2012). Magnitude and sources of variation in female reproductive performance. In J. C. Mitani, J. Call, P. M. Kappeler, R. A. Palombit, and J. B. Silk (Eds.), *The Evolution of Primate Societies* (pp. 343–66). University of Chicago Press.

Pusey, A. E., Oehlert, G. W., Williams, J. M., and Goodall, J. (2005). Influence of ecology and social factors on body mass of wild chimpanzees. *International Journal of Primatology*, 26, 3–31.

Putz, F. E., Dykstra, D. P., and Heinrich, R. (2000). Why poor logging practices persist in the tropics. *Conservation Biology*, 14, 505–8.

Pyke, G. H. (1984). Optimal foraging theory—a critical review. *Annual Review of Ecology and Systematics*, 15, 523–75.

Pyke, G. H. (2017). Do humans forage optimally and what does this mean for zoology at the table? *Australian Zoologist*, 39, 17–25.

Pyke, G. H., Pulliam, H. R., and Charnov, E. L. (1977). Optimal foraging—selective review of theory and tests. *Quarterly Review of Biology*, 52, 137–54.

Qazzaz, S., Mamattah, J., Ashcroft, T., and McFarlane, H. (1981). The development and nature of immune deficit in primates in response to malnutrition. *British Journal of Experimental Pathology*, 62(5), 452.

Qi, X. G., Garber, P. A., Ji, W., Huang, Z. P., Huang, K., Zhang, P., et al. (2014). Satellite telemetry and social modeling offer new insights into the origin of primate multilevel societies. *Nature Communications*, 5, 5296.

Quam, R. M., de Ruiter, D. J., Masali, M., Arsuaga, J. L., Martínez, I., and Moggi-Cecchi, J. (2013). Early hominin auditory ossicles from South Africa. *Proceedings of the National Academy of Sciences of the United States of America*, 110, 8847–51.

Quéméré, E., Hilbert, F., Miquel, C., Lhuillier, E., Rasolondraibe, E., Champeau, J., et al. (2013). A DNA metabarcoding study of a primate dietary diversity and plasticity across its entire fragmented range. *PLoS ONE*, 8, e58971.

Quenette, P. Y. (1990). Functions of vigilance behaviour in mammals—a review. *Acta Oecologica*, 11, 801–18.

Quesnel, L., MacKay, A., Forsyth, D. M., Nicholas, K. R., and Festa-Bianchet, M. (2017). Size, season and offspring sex affect milk composition and juvenile survival in wild kangaroos. *Journal of Zoology*, 302, 252–62.

Quinlan, K. P., and Hayani, K. C. (1996). Vitamin A and respiratory syncytial virus infection: Serum levels and supplementation trial. *Archives of Pediatrics and Adolescent Medicine*, 150(1), 25–30.

Quinn, E. A. (2013). No evidence for sex biases in milk macronutrients, energy, or breastfeeding frequency in a sample of Filipino mothers. *American Journal of Physical Anthropology*, 152, 209–16.

Quinn, E. A. (2014). Too much of a good thing: Evolutionary perspectives on infant formula fortification in the United States and its effects on infant health. *American Journal of Human Biology*, 26, 10–17.

Quinn, E. A. (2017). Milk, medium chain fatty acids and human evolution. In *Breastfeeding* (pp. 112–26). Routledge.

Quinn, E. A. (2021). Centering human milk composition as normal human biological variation. *American Journal of Human Biology*, 33(1), e23564.

Quinn, E. A., Diki Bista, K., and Childs, G. (2016). Milk at altitude: Human milk macronutrient composition in a high-altitude adapted population of Tibetans. *American Journal of Physical Anthropology*, 159(2), 233–43.

R Development Core Team. (2014). *R: A Language and Environment for Statistical Computing*.

Rabenold, D., and Pearson, O. M. (2011). Abrasive, silica phytoliths and the evolution of thick molar enamel in primates, with implications for the diet of *Paranthropus boisei*. *PLoS ONE*, 6(12), e28379.

Rabenold, D., and Pearson, O. M. (2014). Scratching the surface: A critique of Lucas et al. (2013)'s conclusion that phytoliths do not abrade enamel. *Journal of Human Evolution*, 74, 130–33.

Raboin, D. L., Baden, A. L., and Rothman, J. M. (2021). Maternal feeding benefits of allomaternal care in black-and-white colobus (*Colobus guereza*). *American Journal of Primatology*, 83(10), e23327.

Radespiel, U., Ehresmann, P., and Zimmermann, E. (2003). Species-specific usage of sleeping sites in two sympatric mouse lemur species (*Microcebus murinus* and *M. ravelobensis*) in northwestern Madagascar. *American Journal of Primatology*, 59, 139–51.

Rafacz, M. L., Margulis, S. W., Santymire, R. M. (2013). Hormonal and behavioral patterns of reproduction in female hylobatids. *Animal Reproduction Science*, 137, 103–112.

Rafferty, K. L., Teaford, M. F., and Jungers, W. L. (2002). Molar microwear of subfossil lemurs: Improving the resolution of dietary inferences. *Journal of Human Evolution*, 43, 645–57.

Ragir, S., Rosenberg, M., and Tierno, P. (2000). Gut morphology and the avoidance of carrion among chimpanzees, baboons, and early hominins. *Journal of Anthropological Research*, 56, 477–512.

Raguso, R. A., and Willis, M. A. (2005). Synergy between visual and olfactory cues in nectar feeding by wild hawkmoths, *Manduca sexta*. *Animal Behaviour*, 69, 407–18.

Rahaman, M. M., Jaman, M. F., Khatun, M. T., Mahmood, I. A., Alam, M. M., Hossain, M. S., et al. (2015). Feeding ecology of the northern plains sacred langur *Semnopithecus entellus* (Dufresne) in Jessore, Bangladesh: Dietary composition, seasonal and age-sex differences. *Asian Primates Journal*, 5, 24–39.

Rahaman, H., Srihari, K., and Krishnamoorthy, R. V. (1975). Polysaccharide digestion in cheek pouches of the bonnet macaque. *Primates*, 16,175–80.

Raichlen, D. A., Pontzer, H., Harris, J. A., Mabulla, A. Z., Marlowe, F. W., Josh Snodgrass, J., et al. (2017). Physical activity patterns and biomarkers of cardiovascular disease risk in hunter-gatherers. *American Journal of Human Biology*, 29, e22919.

Raichlen, D. A., Wood, B. M., Gordon, A. D., Mabulla, A. Z. P., Marlowe, F. W., and Pontzer, H. (2014). Evidence of Lévy walk foraging patterns in human hunter-gatherers. *Proceedings of the National Academy of Sciences of the United States of America*, 111, 728–33.

Rajpurohit, L. S., and Mohnot, S. M. (1991). The process of weaning in Hanuman langurs *Presbytis entellus entellus*. *Primates*, 32, 213–18.

Rakotondranary, S. J., Struck, U., Knoblauch, C., and Ganzhorn, J. U. (2011). Regional, seasonal and interspecific variation in ^{15}N and ^{13}C in sympatric mouse lemurs. *Naturwissenschaften*, 98, 909–17.

Ralls, K., Sanchez, J. N., Savage, J., Coonan, T. J., Hudgens, B. R., and Cypher, B. L. (2013). Social relationships and reproductive behavior of island foxes inferred from proximity logger data. *Journal of Mammalogy*, 94, 1185–96.

Ramírez, M. (1985). Feeding ecology of the moustached tamarin *Saguinus mystax*. In M. T. Mello (Ed.), *Primatologia no Brasil* (vol. 2, pp. 211–12). Fundação Biodiversitas.

Ramos-Fernandez, G., Boyer, D., and Gomez, V. P. (2006). A complex social structure with fission-fusion properties can emerge from a simple foraging model *Behavioral Ecology and Sociobiology*, 60, 536–49.

Ramsier, M. A., and Dominy, N. J. (2010). A comparison of auditory brainstem responses and behavioral estimates of hearing sensitivity in *Lemur catta* and *Nycticebus coucang*. *American Journal of Primatology*, 72, 217–33.

Ramsier, M. A., and Dominy, N. J. (2012). Receiver bias and the acoustic ecology of aye-ayes (*Daubentonia madagascariensis*). *Communicative and Integrative Biology*, 5, 637–40.

Ramsier, M. A., Cunningham, A. J., Moritz, G. L., Finneran, J. J., Williams, C. V., Ong, P. S., et al. (2012). Primate communication in the pure ultrasound. *Biology Letters*, 8, 508–11.

Ramsier, M. A., and Rauschecker, J. P. (2017). Primate audition: reception, perception, and ecology. In: Quam, R., and Popper A. (Eds.) *Primate hearing and communication* (pp. 47–77). Springer Handbook of Auditory Research, Springer International Publishing.

Randrianambinina, B., Rakotondravony, D., Radespiel, U., and Zimmerman, E. (2003). Seasonal changes in general activity, body mass and reproduction of two small nocturnal primates: A comparison of the golden brown mouse lemur (*Microcebus ravelobensis*) in northwestern Madagascar and the brown mouse lemur (*Microcebus rufus*) in eastern Madagascar. *Primates*, 44, 321–31.

Rands, M. R. W., Adams, W. M., Bennun, L., Butchart, S. H. M., Clements, A., Coomes, D., et al. (2010). Biodiversity conservation: Challenges beyond 2010. *Science*, 329, 1298–303.

Rangel-Negrín, A., Alfaro, J. L., Valdez, R. A., Romano, M. C., and Serio-Silva, J. C. (2009). Stress in Yucatan spider monkeys: Effects of environmental conditions on fecal cortisol levels in wild and captive populations. *Animal Conservation*, 12, 496–502.

Raño, M., Kowalewski, M.M., Cerezo, A.M., and Garber, P.A. (2016). Determinants of daily path length in black and gold howler monkeys (*Alouatta caraya*) in northeastern Argentina. *American Journal of Primatology*, 78, 825–37.

Rasa, O. (1973). Prey capture, feeding techniques, and their ontogeny in the African dwarf mongoose, *Helogale undulata rufula*. *Zeitschrift fur Tierpsychologie*, 32, 449–88.

Rasoazanabary, E. (2004). A preliminary study of mouse lemurs in the Beza Mahafaly Special Reserve, southwest Madagascar. *Lemur News*, 9, 4–7.

Ratsimbazafy, J. (2002). *On the Brink of Extinction and the Process of Recovery: Responses of Black-and-White Ruffed Lemurs (Varecia variegata variegata) to Disturbance in Manombo Forest, Madagascar*. Stony Brook University.

Raubenheimer, D. (2011). Toward a quantitative nutritional ecology: The right-angled mixture triangle. *Ecological Monographs*, 81, 407–27.

Raubenheimer, D., Gosby, A. K., and Simpson, S. J. (2015). Integrating nutrients, foods, diets, and appetites with obesity and cardiometabolic health. *Obesity*, 23, 1741–42.

Raubenheimer, D., and Jones, S. A. (2006). Nutritional imbalance in an extreme generalist omnivore: Tolerance and recovery through complementary food selection. *Animal Behaviour*, 71, 1253–62.

Raubenheimer, D., Machovsky-Capuska, G. E., Chapman, C. A., and Rothman, J. M. (2015). Geometry of nutrition in field studies: An illustration using wild primates. *Oecologia*, 177(1), 223–34.

Raubenheimer, D., Machovsky-Capuska, G. E., Gosby, A. K., and Simpson, S. J. (2014). Nutritional ecology of obesity: From humans to companion animals. *British Journal of Nutrition*, 113, S26–S39.

Raubenheimer, D., and Rothman, J. M. (2013). Nutritional ecology of entomophagy in humans and other primates. *Annual Review of Entomology*, 58, 141–60.

Raubenheimer, D., Rothman, J. M., Pontzer, H., and Simpson, S. J. (2014). Macronutrient contributions of insects to the diets of hunter-gatherers: A geometric analysis. *Journal of Human Evolution*, 71, 70–76.

Raubenheimer, D., and Simpson, S. J. (1993). The geometry of compensatory feeding in the locust. *Animal Behaviour*, 45, 953–64.

Raubenheimer, D., and Simpson, S. J. (1997). Integrative models of nutrient balancing: Application to insects and vertebrates. *Nutrition Research Reviews*, 10, 151–79.

Raubenheimer, D., and Simpson, S. J. (1998). Nutrient transfer functions: The site of integration between feeding behaviour and nutritional physiology. *Chemoecology*, 8, 61–68.

Raubenheimer, D., and Simpson, S. J. (1999). Integrating nutrition: A geometrical approach. *Entomologia Experimentalis et Applicata*, 91, 67–82.

Raubenheimer, D., and Simpson, S. J. (2003). Nutrient balancing in grasshoppers: Behavioural and physiological correlates of dietary breadth. *Journal of Experimental Biology*, 206(10), 1669–81.

Raubenheimer, D., and Simpson, S. J. (2006). The challenge of supplementary feeding: Can geometric analysis help save the kakapo? *Notornis*, 53, 100–111.

Raubenheimer, D., and Simpson, S. J. (2009). Nutritional PharmEcology: Doses, nutrients, toxins, and medicines. *Integrative and Comparative Biology*, 49, 329–37.

Raubenheimer, D., and Simpson, S. J. (2016). Nutritional ecology and human health. *Annual Review of Nutrition*, 36, 603–26.

Raubenheimer, D., and Simpson, S. J. (2018). Nutritional ecology and foraging theory. *Current Opinion in Insect Science*, 27, 38–45.

Raubenheimer, D., and Simpson, S. J. (2019). Protein leverage: Theoretical foundations and ten points of clarification. *Obesity*, 27(8), 1225–38.

Raubenheimer, D., and Simpson, S. J. (2020). *Eat Like the Animals*. Houghton Mifflin.

Raubenheimer, D., Simpson, S. J., and Mayntz, D. (2009). Nutrition, ecology and nutritional ecology: Toward an integrated framework. *Functional Ecology*, 23, 4–16.

Raubenheimer, D., Simpson, S. J., and Tait, A. H. (2012). Match and mismatch: Conservation physiology, nutritional ecology and the timescales of biological adaptation. *Philosophical Transactions of the Royal Society of London Series B—Biological Sciences*, 367, 1628–46.

Raubenheimer, D., Simpson, S. J., Le Couteur, D. G., Solon-Biet, S. M., and Coogan, S. C. P. (2016). Nutritional ecology and the evolution of aging. *Experimental Gerontology*, 86, 50–61.

Raubenheimer, D., Zemke-White, W. L., Phillips, R. J., and Clements, K. D. (2005). Algal macronutrients and food selection by the omnivorous marine fish *Girella tricuspidata*. *Ecology*, 86, 2601–10.

Raupp, M. (1985). Effects of leaf toughness on mandibular wear of the leaf beetle, *Plagiodera versicolora*. *Ecological Entomology*, 10, 73–79.

Razafindratsima, O. H., Sato, H., Tsuji, Y., and Culot, L. (2018). Advances and frontiers in primate seed dispersal. *International Journal of Primatology*, 39, 315–20.

Razanaparany, P. T., and Sato, H. (2020). Abiotic factors affecting the cathemeral activity of *Eulemur fulvus* in the dry deciduous forest of north-western Madagascar. *Folia Primatologica*, 91, 463–80.

Reader, S. M., and Laland, K. N. (2002). Social intelligence, innovation and enhanced brain size in primates. *Proceedings of the National Academy of Sciences of the United States of America*, 99, 4436–41.

Reader, S. M., and MacDonald, K. (2003). Environmental variability and primate behavioural flexibility. In S. M. Reader and K. N. Laland (Eds.), *Animal Innovation* (pp. 83–116). Oxford University Press.

Reader, S. M., Hager, Y., and Laland, K. N. (2011). The evolution of primate general and cultural intelligence. *Philosophical Transactions of the Royal Society of London Series B—Biological Sciences*, 366, 1017–27.

Rech, M., To, L., Tovbin, A., Smoot, T., and Mlynarek, M. (2014). Heavy metal in the intensive care unit: A review of current literature on trace element supplementation in critically ill patients. *Nutrition in Clinical Practice*, 29(1), 78–89.

Recio, M. R., Mathieu, R., Denys, P., Sirguey, P., and Seddon, P. J. (2011). Lightweight GPS-tags, one giant leap for wildlife tracking? An assessment approach. *PLoS ONE*, 6, e28225.

Reddy, B. S., Pleasants, J. R., and Wostmann, B. S. (1972). Effect of intestinal microflora on iron and zinc metabolism, and on activities of metalloenzymes in rats. *Journal of Nutrition*, 102, 101–7.

Redford, K. H., Bouchardet da Fonseca, G. A., and Lacher, T. E. (1984). The relationship between frugivory and insectivory in primates. *Primates*, 25, 433–440.

Redondo, R. A. F., and Santos, F. R. (2006). Evolutionary studies on an α-amylase gene segment in bats and other mammals. *Genetica*, 126, 199–213.

Reed, K. E., and Bidner, L. R. (2004). Primate communities: Past, present, and possible future. *American Journal of Physical Anthropology*, 125, 2–39.

Rees, W. D. W., and Turnberg, L. A. (1981). Biochemical aspects of gastric secretion. *Clinics in Gastroenterology*, 10(3), 521–54.

Regan, B. C., Julliot, C., Simmen, B., Viénot, F., Charles-Dominique, P., and Mollon, J. D. (2001). Fruits, foliage and the evolution of primate colour vision. *Philosophical Transactions of the Royal Society of London Series B—Biological Sciences*, 356, 229–83.

Reichard, M. (2017). Evolutionary perspectives on ageing. *Seminars in Cell and Developmental Biology*, 70, 99–107.

Reichard, U. H., Ganpanakngan, M., and Barelli, C. (2012).

White-handed gibbons of Khao Yai: Social flexibility, complex reproductive strategies, and a slow life history. In P. M. Kappeler and D. P. Watts (Eds.), *Long-Term Field Studies of Primates* (pp. 237–58). Springer.

Reichardt, P. B., Bryant, J. P., Clausen, T. P., and Wieland, G. D. (1984). Defense of winter-dormant Alaska paper birch against snowshoe hares. *Oecologia*, 65, 58–69.

Reiches, M. W., Ellison, P. T., Lipson, S. F., Sharrock, K. C., Gardiner, E., and Duncan, L. G. (2009). Pooled energy budget and human life history. *American Journal of Human Biology*, 21, 421–29.

Reid, R. B., Crowley, B. E., and Haupt, R. (2023). The prospects of poop: A review of past achievements and future possibilities in fecal isotope analysis. *Biological Reviews*, 98, 2091–113.

Reimann, W., and Zimmerman, E. (2002). Feeding ecology of females of two sympatric lemurs in northwestern Madagascar: *Microcebus murinus* and *Microcebus ravelobensis*. *Zoology*, 105, 19.

Reiner, W. B., Petzinger, C., Power, M. L., Hyeroba, D., and Rothman, J. M. (2014). Fatty acids in mountain gorilla diets: Implications for primate nutrition and health. *American Journal of Primatology*, 76, 281–88.

Reisland, M. A., and Lambert, J. E. (2016). Sympatric apes in sacred forests: Shared space and habitat use by humans and endangered Javan gibbons (*Hylobates moloch*). *PLoS ONE*, 11, e28225.

Reiter, K., Balling, G., Bonelli, V., Pabst von Ohain, J., Braun, S. L., Ewert, P., et al. (2018). Neutrophil gelatinase-associated lipocalin reflects inflammation and is not a reliable renal biomarker in neonates and infants after cardiopulmonary bypass: A prospective case-control study. *Cardiology in the Young*, 28(2), 243–51.

Reitsema, L. J. (2012). Introducing fecal stable isotope analysis in primate weaning studies. *American Journal of Primatology*, 74(10), 926–39.

Reitsema, L. J. (2015). Laboratory and field methods for stable isotope analysis in human biology. *American Journal of Human Biology*, 27, 593–604.

Reitsema, L. J., and Muir, A. B. (2015). Growth velocity and weaning delta^{15}N "dips" during ontogeny of *Macaca mulatta*. *American Journal of Physical Anthropology*, 157, 347–57.

Reitsma, L. J., Patrick, K. A., and Muir, A. B. (2016). Interindividual variation in weaning among rhesus macaques (*Macaca mulatta*): Serum stable isotope indicators of suckling duration and lactation. *American Journal of Primatology*, 78, 1113–34.

Remis, M. J. (1997). Western lowland gorillas (*Gorilla gorilla gorilla*) as seasonal frugivores: Use of variable resources. *American Journal of Primatology*, 43(2), 87–109.

Remis, M. J. (2002). Food preferences among captive western gorillas (*Gorilla gorilla gorilla*) and chimpanzees (*Pan troglodytes*). *International Journal of Primatology*, 23, 231–49.

Remis, M. J. (2006). The role of taste in food selection by African apes: Implications for niche separation and overlap in tropical forests. *Primates*, 47, 56–64.

Remis, M. J., and Dierenfeld, E. S. (2004). Digestive passage, digestibility, and behavior in captive gorillas under two dietary regimes. *International Journal of Primatology*, 25, 825–45.

Remis, M. J., and Kerr, M. E. (2002). Taste responses to fructose and tannic acid among gorillas (*Gorilla gorilla gorilla*). *International Journal of Primatology*, 23, 251–61.

Remis, M. J., Dierenfeld, E. S., Mowry, C. B., and Carroll, R. W. (2001). Nutritional aspects of western lowland gorilla (*Gorilla gorilla gorilla*) diet during seasons of fruit scarcity at Bai Hokou, Central African Republic. *International Journal of Primatology*, 22, 807–36.

Remonti, L., Balestrieri, A., Raubenheimer, D., and Saino, N. (2016). Functional implications of omnivory for dietary nutrient balance. *Oikos*, 125, 1233–40.

Ren, B. P., Li, D. Y., Garber, P. A., and Li, M. (2012). Fission-fusion behavior in Yunnan snub-nosed monkeys (*Rhinopithecus bieti*) in Yunnan, China. *International Journal of Primatology*, 33, 1096–109.

Rensberger, J. M. (2000). Pathways to functional differentiation in mammalian enamel. In M. F. Teaford, M. M. Smith, and M. W. J. Ferguson (Eds.), *Development, Function, and Evolution of Teeth* (pp. 252–68). Cambridge University Press.

Reyes, T. M., and Coe, C. L. (1998). The proinflammatory cytokine network: Interactions in the CNS and blood of rhesus monkeys. *American Journal of Physiology—Regulatory Integrative and Comparative Physiology*, 274(1), R139–R144.

Reynard, L. M., and Hedges, R. E. M. (2008). Stable hydrogen isotopes of bone collagen in palaeodietary and palaeoenvironmental reconstruction. *Journal of Archaeological Sciences*, 35, 1934–42.

Reynard, L. M., Henderson, G. M., and Hedges, R. E. M. (2010). Calcium isotope ratios in animal and human bone. *Geochimica et Cosmochimica Acta*, 74, 3735–50.

Reynaud, J., Guilet, D., Terreaux, R., Lussignol, M., and Walchshofer, N. (2005). Isoflavonoids in non-leguminous families: An update. *Natural Product Reports*, 22, 504–15.

Reynolds, V. (2005). *The Chimpanzees of the Budongo Forest: Ecology, Behaviour, and Conservation*. Oxford University Press.

Reynolds, V., Lloyd, A. W., Babweteera, F., and English, C. J. (2009). Decaying *Raphia farinifera* palm trees provide a source of sodium for wild chimpanzees in the Budongo forest, Uganda. *PLoS ONE*, 4, e6194.

Reynolds, V., Plumptre, A. J., Greenham, J., and Harborne, J. (1998). Condensed tannins and sugars in the diet of chimpanzees (*Pan troglodytes schweinfurthii*) in the Budongo Forest, Uganda. *Oecologia*, 115, 331–36.

Rezende, E. L., Gomes, F. R., Chappell, M. A., and Garland Jr, T. (2009). Running behavior and its energy cost in mice selectively bred for high voluntary locomotor activity. *Physiological and Biochemical Zoology*, 82, 662–79.

Reznick, D. N., Butler, M. J. I., and Rodd, H. (2001). Life-history evolution in guppies. VII. The comparative ecology of high- and low-predation environments. *American Naturalist*, 157, 126–40.

Rhine, R. J., Norton, G. W., Wynn, G. M., Wynn, R. D., and Rhine, H. B. (1986). Insect and meat eating among infant and adult baboons (*Papio cynocephalus*) of Mikumi National Park, Tanzania. *American Journal of Physical Anthropology*, 70, 105–18.

Riba-Hernández, P., and Stoner, K. E. (2005). Massive destruction of *Symphonia globulifera* (Clusiaceae) flowers by Central American spider monkeys (*Ateles geoffroyi*). *Biotropica*, 37, 274–78.

Richard, A. F. (1985). *Primates in Nature*. W. H. Freeman.

Richard, A. F., and Nicoll, M. E. (1987). Female social dominance and basal metabolism in a Malagasy primate, *Propithecus verreauxi*. *American Journal of Primatology*, 12, 309–14.

Richard, A. F., Dewar, R. E., Schwartz, M., and Ratsirarson, J. (2000). Mass change, environmental variability and female fertility in wild *Propithecus verreauxi*. *Journal of Human Evolution*, 39, 381–91.

Richards, L. A., Dyer, L. A., Forister, M. L., Smilanich, A. M., Dodson, C. D., Leonard, M. D., et al. (2015). Phytochemical diversity drives plant–insect community diversity. *Proceedings of the National Academy of Sciences*, 112(35), 10973–78.

Richards, L. A., and Windsor, D. M. (2007). Seasonal variation of arthropod abundance in gaps and the understorey of a lowland moist forest in Panama. *Journal of Tropical Ecology*, 23, 169–76.

Richards, M. P., and Trinkaus, E. (2009). Isotopic evidence for the diets of European Neanderthals and early modern humans. *Proceedings of the National Academy of Sciences of the United States of America*, 106(38), 16034–39.

Richards, M. P., Fuller, B. T., Sponheimer, M., Robinson, T., and Ayliffe, L. (2003). Sulphur isotopes in palaeodietary studies: A review and results from a controlled feeding experiment. *International Journal of Osteoarchaeology*, 13(1–2), 37–45.

Richards, M. P., Pettitt, P. B., Trinkaus, E., Smith, F. H., Paunovic, M., and Karavanic, I. (2000). Neanderthal diet at Vindija and Neanderthal predation: The evidence from stable isotopes. *Proceedings of the National Academy of Sciences of the United States of America*, 97, 7663–66.

Richards, P. W. (1996). *The Tropical Rain Forest* (2nd ed.). Cambridge University Press.

Richter, C., Gras, P., Hodges, K., Ostner, J., and Schülke, O. (2015). Feeding behavior and aggression in wild Siberut macaques (*Macaca siberu*) living under low predation risk. *American Journal of Primatology*, 77, 741–52.

Richter, D., Grün, R., Renaud, J. B., Steele, T. E., Amani, F., Rué Fernades, P., et al. (2017). The age of the hominin fossils from Djebel Irhoud, Morocco, and the origins of the Middle Stone Age. *Nature*, 546, 293–96.

Ricklefs, R. E. (2008). *The Economy of Nature*. Macmillan.

Ricklefs, R. E., and Wikelski, M. C. (2002). The physiology/life-history nexus. *Trends in Ecology and Evolution*, 17, 462–68.

Ridley, H. (1930). *The Dispersal of Plants throughout the World*. L. Reeve.

Riedel, C. U., Schwiertz, A., and Egert, M. (2014). The stomach and small and large intestinal microbiomes. In J. Marchesi (Ed.), *The Human Microbiota and Microbiome* (pp. 1–19). CABI.

Riehl, C., and Jara, L. (2009). Natural history and reproductive biology of the communally breeding greater ani (*Crotophaga major*) at Gatún Lake, Panama. *Wilson Journal of Ornithology*, 121, 679e687.

Riek, A. (2008). Relationship between milk energy intake and growth rate in suckling mammalian young at peak lactation: An updated meta-analysis. *Journal of Zoology*, 274, 160–70.

Riek, A. (2011). Allometry of milk intake at peak lactation. *Mammal Biology*, 76, 3–11.

Riek, A. (2021). Comparative phylogenetic analysis of milk output at peak lactation. *Comparative Biochemistry and Physiology Part A: Molecular and Integrative Physiology*, 257, 110976.

Rieucau, G., Vickery, W. L., and Doucet, G. J. (2009). A patch use model to separate effects of foraging costs on giving-up densities: An experiment with white-tailed deer (*Odocoileus virginianus*). *Behavioral Ecology and Sociobiology*, 63, 891–97.

Righini, N. (2014). *Primate Nutritional Ecology: The Role of Food Selection, Energy Intake, and Nutrient Balancing in Mexican Black Howler Monkey* (Alouatta pigra) *Foraging Strategies*. University of Illinois at Urbana-Champaign.

Righini, N. (2017). Recent advances in primate nutritional ecology. *American Journal of Primatology*, 79, 1–5.

Righini, N., Garber, P. A., and Rothman, J. M. (2017). The effects of plant nutritional chemistry on food selection of Mexican black howler monkeys (*Alouatta pigra*): The role of lipids. *American Journal of Primatology*, 79, e22524.

Rijksen, H. D. (1978). *A Field Study on Sumatran Orangutans* (Pongo pygmaeus abelii *Lesson 1827*): *Ecology, Behaviour and Conservation*. Wageningen University.

Riley, A. P. (1994). Determinants of adolescent fertility and its consequences for maternal health, with special reference to rural Bangladesh. *Annals of the New York Academy of Sciences*, 709, 86–100.

Riley, E. P. (2007). Flexibility in diet and activity patterns of *Macaca tonkeana* in response to anthropogenic habitat alteration. *International Journal of Primatology*, 28, 107–33.

Riley, E. P., and Fuentes, A. (2011). Conserving social-ecological systems in Indonesia: Human and nonhuman primate interconnections in Bali and Sulawesi. *American Journal of Primatology*, 73, 62–74.

Riley, E. P., Tolbert, B., and Farida, W. R. (2013). Nutritional content explains the attractiveness of cacao to crop raiding Tonkean macaques. *Current Zoology*, 59, 160–69.

Rimm, E. B., Stampfer, M. J., Ascherio, A., Giovannucci, E., Colditz, G. A., and Willett, W. C. (1993). Vitamin E consumption and the risk of coronary heart disease in men. *New England Journal of Medicine*, 328(20), 1450–56.

Rincon, J. C., Efron, P. A., and Moldawer, L. L. (2022). Immunopathology of chronic critical illness in sepsis survivors: Role of abnormal myelopoiesis. *Journal of Leukocyte Biology*, 112, 1525-1534.

Riopelle, A. J., and Hale, P. A. (1975). Nutritional and environmental factors affecting gestation length in rhesus monkeys. *American Journal of Clinical Nutrition*, 28, 1170–76.

Riopelle, A. J., Hill, C. W., and Li, S. C. (1975). Protein deprivation in primates. V. Fetal mortality and neonatal status of infant monkeys born of deprived mothers. *American Journal of Clinical Nutrition*, 28, 989–93.

Riopelle, A. J., Hill, C. W., Li, S. C., Wolf, R. H., Seibold, H. R., and Smith, J. (1975). Protein deficiency in primates. IV. Pregnant rhesus monkey. *American Journal of Clinical Nutrition*, 28, 20–28.

Ripple, W. J., and Beschta, R. L. (2003). Wolf reintroduction, predation risk, and cottonwood recovery in Yellowstone National Park. *Forest Ecology and Management*, 184, 299–313.

Ripple, W. J., and Beschta, R. L. (2004). Wolves and the ecology of fear: Can predation risk structure ecosystems? *BioScience*, 54, 755–66.

Rismiller, P. D., and McKelvey, M. W. (2009). Activity and behaviour of lactating echidnas (*Tachyglossus aculeatus multiaculeatus*) from hatching of egg to weaning of young. *Australian Journal of Zoology*, 57, 265–73.

Rivera, S., Sattler, F. R., Boyd, H., Auffenberg, T., Nakao, S., and Moldawer, L. L. (2001). Urinary cytokines for assessing inflammation in HIV-associated wasting. *Cytokine*, 13(5), 305–13.

Robbins, C. T. (1993). *Wildlife Feeding and Nutrition* (2nd ed.). Academic Press.

Robbins, C. T., Felicetti, L. A., and Sponheimer, M. (2005). The effect of dietary protein quality on nitrogen isotope discrimination in mammals and birds. *Oecologia*, 144, 534–40.

Robbins, C. T., Fortin, J. K., Rode, K. D., Farley, S. D., Shipley, L. A., and Felicetti, L. A. (2007). Optimizing protein intake as a foraging strategy to maximize mass gain in an omnivore. *Oikos*, 116, 1675–82.

Robbins, C. T., Hanley, T. A., Hagerman, A. E., Hjeljord, O., Baker, D. L., Schwartz, C. C., et al. (1987). Role of tannins in defending plants against ruminants—reduction in protein availability. *Ecology*, 68, 98–107.

Robbins, M. M. (1999). Male mating patterns in wild multimale mountain gorilla groups. *Animal Behaviour*, 57(5), 1013–20.

Robbins, M. M. (2008). Feeding competition and agonistic relationships among Bwindi *Gorilla beringei*. *International Journal of Primatology*, 29, 999–1018.

Robbins, M. M., and McNeilage, A. (2003). Home range and frugivory patterns of mountain gorillas in Bwindi Impenetrable National Park, Uganda. *International Journal of Primatology*, 24(3), 467–91.

Robbins, M. M., Gray, M., Fawcett, K. A., Nutter, F. B., Uwingeli, P., Mburanumwe, I., et al. (2011). Extreme conservation leads to recovery of the Virunga mountain gorillas. *PLoS ONE*, 6, e19788.

Robbins, M. M., Gray, M., Kagoda, E., and Robbins, A. M. (2009). Population dynamics of the Bwindi mountain gorillas. *Biological Conservation*, 142, 2886–95.

Robbins, M. M., Nkurunungi, J. B., and McNeilage, A. (2006). Variability of the feeding ecology of eastern gorillas. In G. Hohmann, M. M. Robbins, and C. Boesch (Eds.), *Feeding Ecology in Apes and Other Primates* (pp. 25–48). Cambridge University Press.

Robbins, M. M., Robbins, A. M., Gerald-Steklis, N., and Steklis, H. D. (2005). Long-term dominance relationships in female mountain gorillas: Strength, stability and determinants of rank. *Behaviour*, 142, 779–809.

Robert, K. A., and Braun, S. (2012). Milk composition during lactation suggests a mechanism for male biased allocation of maternal resources in the tammar wallaby (*Macropus eugenii*). *PLoS ONE*, 7, e51099.

Roberts, E. K., Lu, A., Bergman, T. J., and Beehner, J. C. (2017). Female reproductive parameters in wild geladas (*Theropithecus gelada*). *International Journal of Primatology*, 38, 1–20.

Roberts, S. B., Cole, T. J., and Coward, W. A. (1985). Lactational performance in relation to energy intake in the baboon. *American Journal of Clinical Nutrition*, 41, 1270–76.

Robinette, S. L., Brüschweiler, R., Schroeder, F. C., and Edison, A. S. (2011). NMR in metabolomics and natural products research: Two sides of the same coin. *Accounts of Chemical Research*, 45, 288–97.

Robinson, B. W., and Wilson, D. S. (1998). Optimal foraging, specialization, and a solution to Liem's paradox. *American Naturalist*, 151(3), 223–35.

Robinson, E. A., Ryan, G. D., and Newman, J. A. (2012). A meta-analytical review of the effects of elevated CO_2 on plant-arthropod interactions highlights the importance of interacting environmental and biological variables. *New Phytologist*, 194, 321–36.

Robinson, J. G. (1986). Seasonal variation in use of time and space by the wedge-capped capuchin monkey, *Cebus olivaceus*: Implications for foraging theory. *Smithsonian Contributions to Zoology*, 431, 1–60.

Robinson, J. G., and Redford, K. H. (1986). Body size, diet, and population density of Neotropical forest mammals. *American Naturalist*, 128, 665–80.

Robinson, J. G., and Redford, K. H. (1991). Sustainable harvest of Neotropical forest mammals. In J. G. Robinson and K. H. Redford (Eds.), *Neotropical Wildlife Use and Conservation* (pp. 415–29). University of Chicago Press.

Robinson, J. T. (1956). *The Dentition of the Australopithecinae*. Transvaal Museum Memoir 9. Transvaal Museum.

Robson, S. L., van Schaik, C. P., and Hawkes, K. (2006). The derived features of human life history. In K. Hawkes and R. R. Paine (Eds.), *The Evolution of Human Life History* (pp. 17–44). School of American Research Press.

Rocha, V. J., Aguiar, L. M., Ludwig, G., Hilst, C. L. S., Teixeira, G. M., Svoboda, W. K., et al. (2007). Techniques and trap models for capturing wild tufted capuchins. *International Journal of Primatology*, 28, 231–43.

Rode, K. D., Chapman, C. A., Chapman, L. J., and McDowell, L. R. (2003). Mineral resource availability and consumption by *Colobus* in Kibale National Park, Uganda. *International Journal of Primatology*, 24, 541–73.

Rode, K. D., Chapman, C. A., McDowell, L. R., and Stickler, C. (2006). Nutritional correlates of population density across habitats and logging intensities in redtail monkeys (*Cercopithecus ascanius*). *Biotropica*, 38, 625–34.

Rödel, M. O., and Ernst, R. (2004). Measuring and monitoring amphibian diversity in tropical forests: I. An evaluation of methods with recommendations for standardization. *Ecotropica*, 10, 1–14.

Rödel, M. O., Range, F., Seppanen, J., and Noe, R. (2002). Caviar in the rain forest: Monkeys as frogspawn predators in Tai National Park, Ivory Coast. *Journal of Tropical Ecology*, 18, 289–94.

Rodman, P. S. (1978). Diets, densities, and distributions of Bornean primates. In G. G. Montgomery (Ed.), *The Ecology of Arboreal Folivores* (pp. 253–62). Smithsonian Institution Press.

Rodman, P. S. (1988). Resources and group sizes of primates. In C. N. Slobodchikoff (Ed.), *The Ecology of Social Behavior* (pp. 83–108). Academic Press.

Rodman, P. S., and Cant, J. G. H. (1984). *Adaptations for Foraging in Nonhuman Primates*. Columbia University Press.

Rodrigues, A. S. L., Akcakaya, H. R., Andelman, S. J., Bakarr, M. I., Boitani, L., Brooks, T. M., et al. (2004). Global gap analysis: Priority regions for expanding the global protected-area network. *Bioscience*, 54, 1092–100.

Rodrigues, H. G., Merceron, G., and Viriot, L. (2009). Dental microwear patterns of extant and extinct Muridae (Rodentia, Mammalia): Ecological implications. *Naturwissenschaften*, 96(4), 537–42.

Rodrigues, M. A. (2017). Female spider monkeys (*Ateles geoffroyi*) cope with anthropogenic disturbance through fission-fusion dynamics. *International Journal of Primatology*, 38, 835–55.

Rodriguez Cervilla, J., Fraga, J. M., Garcia Riestra, C., Fernandez Lorenzo, J. R., and Martinez Soto, I. (1998). Neonatal sepsis: Epidemiologic indicators and relation to birth weight and length of hospitalization time. *Anales Espanoles de Pediatria*, 48(4), 401–8.

Rodríguez-Hidalgo, A. R., Saladie, P., Olle, A., Arsuaga, J. L., Bermúdez de Castro, J. M., and Carbonell, E. (2017). Human predatory behavior and the social implications of communal hunting based on evidence from the TD10.2 bison bone bed at Gran Dolina (Atapuerca, Spain). *Journal of Human Evolution*, 105, 89–122.

Rodriguez-Rojas, F., Borrero-Lopez, O., Constantino, P. J., Henry, A. G., and Lawn, B. R. (2020). Phytoliths can cause tooth wear. *Journal of the Royal Society, Interface*, 17, 20200613.

Rodseth, L., Wrangham, R. W., Harrigan, A. M., and Smuts, B. B. (1991). The human community as a primate society. *Current Anthropology*, 32, 221–54.

Roebroeks, W., and Verpoorte, A. (2009). A "language-free" explanation for differences between the European Middle and Upper Paleolithic record. In C. Knight (Ed.), *Studies in the Evolution of Language: Cradle of Language* (pp. 150–66). Oxford University Press.

Roff, D. A. (1992). *Life History Evolution*. Sinauer Associates.

Rogala, J. K., Hebblewhite, M., Whittington, J., White, C. A., Coleshill, J., and Musiani, M. (2011). Human activity differentially redistributes large mammals in the Canadian Rockies National Parks. *Ecology and Society*, 16, 16.

Rogers, M. E., Abernethy, K., Bermejo, M., Cipolletta, C., Doran, D., McFarland, K., et al. (2004). Western gorilla diet: A synthesis from six sites. *American Journal of Primatology*, 64, 173–92.

Rogers, M. E., Maisels, F., Williamson, E. A., Fernandez, M., and Tutin, C. E. G. (1990). Gorilla diet in the Lope reserve, Gabon—a nutritional analysis. *Oecologia*, 84, 326–39.

Roh, M. S., Moldawer, L. L., Ekman, L. G., Dinarello, C. A., Bistrian, B. R., Jeevanandam, M., et al. (1986). Stimulatory effect of interleukin-1 upon hepatic metabolism. *Metabolism*, 35(5), 419–24.

Rolls, E. T., and Wirth, S. (2018). Spatial representations in the primate hippocampus, and their functions in memory and navigation. *Progress in Neurobiology*, 171, 90–113.

Romero, L. M., Dickens, M. J., and Cyr, N. E. (2009). The reactive scope model—a new model integrating homeostasis, allostasis, and stress. *Hormones and Behavior*, 55(3), 375–89.

Ronsted, N., Weiblen, G. D., Clement, W. L., Zerega, N. J. C., and Savolainen, V. (2008). Reconstructing the phylogeny of figs (*Ficus*, Moraceae) to reveal the history of the fig pollination mutualism. *Symbiosis*, 45, 45–55.

Rose, L. M. (1994). Sex differences in diet and foraging behavior in white-faced capuchins (*Cebus capucinus*). *International Journal of Primatology*, 15, 95–114.

Rose, L. M. (1997). Vertebrate predation and food sharing in *Cebus* and *Pan*. *International Journal of Primatology*, 18, 727–65.

Rose, L. M. (2001). Meat and the early human diet: Insights from Neotropical primate studies. In C. S. Stanford and H. T. Bunn (Eds.), *Meat Eating and Human Evolution* (pp. 141–59). Oxford University Press.

Rose, L. M., and Fedigan, L. M. (1995). Vigilance in white-face capuchins, *Cebus cupucinus*, in Costa Rica. *Animal Behaviour*, 49, 63–70.

Rose, L. M., Perry, S., Panger, M. A., Jack, K., Manson, J. H., Gros-Lewis, J., et al. (2003). Interspecific interactions between *Cebus capucinus* and other species: Data from three Costa Rican sites. *International Journal of Primatology*, 24, 759–96.

Roseboom, T. J., de Rooij, S., and Painter, R. (2006). The Dutch famine and its long-term consequences for adult health. *Early Human Development*, 82, 485–91.

Roseboom, T. J., Painter, R. C., van Abeelen, A. F. M., Veenendaal, M. V. E., and de Rooij, S. R. (2011). Hungry in the womb: What are the consequences? Lessons from the Dutch famine. *Maturitas*, 70, 141–45.

Roseboom, T. J., van der Meulen, J. H., Ravelli, A. C., Osmond, C., Barker, D. J., and Bleker, O. P. (2001). Effects of prenatal exposure to the Dutch famine on adult disease in later life: An overview. *Molecular and Cellular Endocrinology*, 185, 93–98.

Rosenbaum, S., Zeng, S., Campos, F. A., Gesquiere, L. R., Altmann, J., Alberts, S. C., et al. (2020). Social bonds do not mediate the relationship between early adversity and adult glucocorticoids in wild baboons. *Proceedings of the National Academy of Sciences of the United States of America*, 117(33), 20052–62.

Rosenberger, A. L. (1992). Evolution of feeding niches in New World monkeys. *American Journal of Physical Anthropology*, 88, 525–62.

Rosenberger, A. L., Cooke, S. B., Halenar, L. B., Tejedor, M. F., Hartwig, W. C., Novo, N. M., et al. (2015). Fossil alouattines and the origins of *Alouatta*: Craniodental diversity and interrelationships. In M. M. Kowalewski, P. A. Garber, L. Cortes-Ortiz, B. Urbani, and D. Youlatos (Eds.), *Howler Monkeys* (pp. 21–54). Springer.

Rosenberger, A. L., Halenar, L., and Cooke, S. B. (2011). The making of platyrrhine semifolivores: Models for the evolution of folivory in primates. *Anatomical Record*, 294, 2112–30.

Rosenberger, A. L., and Kinzey, W. G. (1976). Functional patterns of molar occlusion in platyrrhine primates. *American Journal of Physical Anthropology*, 45, 281–97.

Rosenberger, A. L., and Strier, K. B. (1989). Adaptive radiation of the ateline primates. *Journal of Human Evolution*, 18, 717–50.

Rosenblatt, A. E., and Schmitz, O. J. (2016). Climate change, nutrition, and bottom-up and top-down food web processes. *Trends in Ecology and Evolution*, 31, 965–75.

Rosetta, L., Lee, P. C., and Garcia, C. (2011). Energetics during reproduction: A doubly labeled water study of lactating baboons. *American Journal of Physical Anthropology*, 144, 661–68.

Rosqvist, F., Smedman, A., Lindmark-Mansson, H., Paulsson, M., Petrus, P., Straniero, S., et al. (2015). Potential role of milk fat globule membrane in modulating plasma lipoproteins, gene expression, and cholesterol metabolism in humans: A randomized study. *American Journal of Clinical Nutrition*, 102, 20–30.

Ross, C. (1992). Basal metabolic rate, body weight and diet in primates: an evaluation of the evidence. *Folia Primatologica*, 58, 7-23.

Ross, C. (1998). Primate life histories. *Evolutionary Anthropology*, 5, 54–63.

Ross, C. (2001). Park or ride? Evolution of infant carrying in primates. *International Journal of Primatology*, 22, 749–71.

Ross, C. F., Iriarte-Diaz, J., and Nunn, C. L. (2012). Innovative approaches to the relationship between diet and mandibular morphology in primates. *International Journal of Primatology*, 33, 632–60.

Ross, C. F., and Kirk, E. C. (2007). Evolution of eye size and shape in primates. *Journal of Human Evolution*, 52, 294–313.

Ross, C., and MacLarnon, A. (2000). The evolution of non-maternal care in anthropoid primates: A test of the hypotheses. *Folia Primatologica*, 71, 93–113.

Ross, C. F., Washington, R. L., Eckhardt, A., Reed, D. A., Vogel, E. R., Dominy, N. J., et al. (2009). Ecological consequences of scaling of chew cycle duration and daily feeding time in Primates. *Journal of Human Evolution*, 56, 570–85.

Ross, R., and Janssen, I. A. N. (2001). Physical activity, total and regional obesity: dose-response considerations. *Medicine & Science in Sports & Exercise*, 33(6), S521-S527.

Roth, G. A., Nguyen, G., Forouzanfar, M. H., Mokdad, A. H., Naghavi, M., and Murray, C. J. L. (2015). Estimates of global and regional premature cardiovascular mortality in 2025. *Circulation*, 132, 1270–82.

Rothe, H., Koenig, A., Darms, K., and Siess, M. (1987). Analysis of litter size in a colony of the common marmoset (*Callithrix jacchus*). *Zeitschrift für Saeugetierkunde*, 52, 227–35.

Rothman, J. M. (2015). Nutritional geometry provides new insights into the interaction between food quality and demography in endangered wildlife. *Functional Ecology*, 29, 3–4.

Rothman, J. M., Chapman, C. A., Hansen, J. L., Cherney, D. J., and Pell, A. N. (2009). Rapid assessment of the nutritional value of foods eaten by mountain gorillas: Applying near-infrared reflectance spectroscopy to primatology. *International Journal of Primatology*, 30, 729–42.

Rothman, J. M., Chapman, C. A., and Pell, A. N. (2008). Fiber-bound nitrogen in gorilla diets: Implications for estimating dietary protein intake of primates. *American Journal of Primatology*, 70, 690–94.

Rothman, J. M., Chapman, C. A., Struhsaker, T. T., Raubenheimer, D., Twinomugisha, D., and Waterman, P. G. (2015). Long-term declines in nutritional quality of tropical leaves. *Ecology*, 96, 873–78.

Rothman, J. M., Chapman, C. A., and Van Soest, P. J. (2012). Methods in primate nutritional ecology: A user's guide. *International Journal of Primatology*, 33(3), 542–66.

Rothman, J. M., DePasquale, A. N., Evans, K. D., and Raboin, D. L. (2022). Colobine nutritional ecology. In I. Matsuda, C. C. Grueter, and J. A. Teichroeb (Eds.), *The Colobines: Natural History, Behaviour and Ecological Diversity* (pp. 94–107). Cambridge University Press.

Rothman, J. M., Dierenfeld, E. S., Hintz, H. F., and Pell, A. N. (2008). Nutritional quality of gorilla diets: Consequences of age, sex, and season. *Oecologia*, 155, 111–22.

Rothman, J. M., Dierenfeld, E. S., Molina, D. O., Shaw, A. V., Hintz, H. F., and Pell, A. N. (2006). Nutritional chemistry of foods eaten by gorillas in Bwindi Impenetrable National Park, Uganda. *American Journal of Primatology*, 68, 675–91.

Rothman, J. M., Dusinberre, K., and Pell, A. N. (2009). Condensed tannins in the diets of primates: A matter of methods? *American Journal of Primatology*, 71, 70–76.

Rothman, J. M., Pell, A. N., and Bowman, D. D. (2009). How does diet quality affect the parasite ecology of mountain gorillas? In M. A. Huffman and C. A. Chapman (Eds.), *Primate Parasite Ecology: The Dynamics and Study of Host-Parasite Relationships* (pp. 441–62). Cambridge University Press.

Rothman, J. M., Plumptre, A. J., Dierenfeld, E. S., and Pell, A. N. (2007). Nutritional composition of the diet of the gorilla (*Gorilla beringei*): A comparison between two montane habitats. *Journal of Tropical Ecology*, 23, 673–82.

Rothman, J. M., Raubenheimer, D., Bryer, M. A. H., Takahashi, M., and Gilbert, C. C. (2014). Nutritional contributions of insects to primate diets: Implications for primate evolution. *Journal of Human Evolution*, 71, 59–69.

Rothman, J. M., Raubenheimer, D., and Chapman, C. A. (2011). Nutritional geometry: Gorillas prioritize non-protein energy while consuming surplus protein. *Biology Letters*, 7, 847–49.

Rothman, J. M., Van Soest, P. J., and Pell, A. N. (2006). Decaying wood is a sodium source for mountain gorillas. *Biology Letters*, 2, 321–24.

Rothman, J. M., Vogel, E. R., and Blumenthal, S. A. (2013). Diet and nutrition. In E. J. Sterling, N. Bynum and M. E. Blair (Eds.), *Primate Ecology and Conservation: A Handbook of Techniques* (pp. 195–212). Oxford University Press.

Round, J. L., and Mazmanian, S. K. (2010). Inducible Fox p3+ regulatory T-cell development by a commensal bacterium of the intestinal microbiota. *Proceedings of the National Academy of Sciences of the United States of America*, 107, 12204–9.

Rovero, F., and Struhsaker, T. T. (2007). Vegetative predictors of primate abundance: Utility and limitations of a fine-scale analysis. *American Journal of Primatology*, 69, 1242–56.

Rowe, N. (1996). *The pictorial guide to the living primates*. Pogonias Press.

Rowell, T. E., and Chalmers, N. R. (1970). Reproductive cycles of the mangabey *Cercocebus albigena*. *Folia Primatologica*, 12, 264–72.

Rowe-Rowe, D. T., and Scotcher, J. S. B. (1986). Ecological carrying capacity of the Natal Drakensberg for wild ungulates. *South African Journal of Wildlife Research*, 16, 12–16.

Rowland, I. R., Wiseman, H., Sanders, T. A. B., Adlercreutz, H., and Bowey, E. A. (2000). Interindividual variation in metabolism of soy isoflavones and lignans: Influence of habitual diet on equol production by the gut microflora. *Nutrition and Cancer*, 36, 27–32.

Rozlog, L. A., Kiecolt-Glaser, J. K., Marucha, P. T., Sheridan, J. F., and Glaser, R. (1999). Stress and immunity: Implications for viral disease and wound healing. *Journal of Periodontology*, 70(7), 786–92.

Rubin, D. C., and Umanath, S. (2015). Event memory: A theory of memory for laboratory, autobiographical, and fictional events. *Psychology Review*, 122, 1–23.

Rudel, L. L., Parks, J. S., and Sawyer, J. K. (1995). Compared with dietary monounsaturated and saturated fat, polyunsaturated fat protects African green monkeys from coronary artery atherosclerosis. *Arteriosclerosis, Thrombosis, and Vascular Biology*, 15(12), 2101–10.

Rudolph, M. C., Neville, M. C., and Anderson, S. M. (2007). Lipid synthesis in lactation: Diet and the fatty acid switch. *Journal of Mammary Gland Biology and Neoplasia*, 12, 269–81.

Rui, L. (2014). Energy metabolism in the liver. *Comprehensive Physiology*, 4, 177–97.

Ruiz-Navarro, A., Barberá, G. G., Albaladejo, J., and Querejeta, J. I. (2016). Plant $\delta^{15}N$ reflects the high landscape-scale heterogeneity of soil fertility and vegetation productivity in a Mediterranean semiarid ecosystem. *New Phytologist*, 212, 1030–43.

Ruohonen, K., Simpson, S. J., and Raubenheimer, D. (2007). A new approach to diet optimisation: A re-analysis using European whitefish (*Coregonus lavaretus*). *Aquaculture*, 267, 147–56.

Ruppert, E.E., Fox, R.S., and Barnes, R.D. (2004). Introduction to Eumetazoa. *Invertebrate Zoology* (7 ed.) pp. 99–103. Brooks / Cole.

Rushmore, J., Leonhardt, S. D., and Drea, C. M. (2012). Sight or scent: Lemur sensory reliance in detecting food quality varies with feeding ecology. *PLoS ONE*, 7, e41558.

Ruslin, F., Matsuda, I. and Md-Zain, B. M. (2019). The feeding ecology and dietary overlap in two sympatric primate species, the long-tailed macaque (*Macaca fascicularis*) and dusky langur (*Trachypithecus obscurus obscurus*), in Malaysia. *Primates*, 60, 41–500.

Russak, S. M. (2014). Using patch focals to study uninhabited dry-habitat chimpanzees (*Pan troglodytes schweinfurthii*) and sympatric fauna at Issa, Ugalla, Tanzania. *International Journal of Primatology*, 35, 1202–21.

Russell, G. A., and Chappell, M. A., (2007). Is BMR repeatable in deer mice? Organ mass correlates and the effects of cold acclimation and natal altitude. *Journal of Comparative Physiology B*, 177, 75–87.

Russo, S. E., and Chapman, C. A. (2010). Primate seed dispersal: Linking behavioral ecology with forest community structure. In C. Campbell, A. Fuentes, K. Mackinnon, S. Bearder, and R. Stumpf (Eds.), *Primates in Perspective* (pp. 523–34). Oxford University Press.

Russo, S., Campbell, C., Dew, J., Stevenson, P., and Suarez, S. (2005). A multi-forest comparison of dietary preferences and seed dispersal by *Ateles* spp. *International Journal of Primatology*, 26, 1017–37.

Russon, A. E., Compost, A., Kuncoro, P., and Feria, A. (2014). Orangutan fish eating, primate aquatic fauna eating, and their implications for the origins of ancestral hominin fish eating. *Journal of Human Evolution*, 77, 50–63.

Russon, A. E., Wich, S. A., Ancrenaz, M., Kanamori, T., Knott, C. D., Kuze, N., et al. (2009). Geographic variation in orangutan diets. In S. A. Wich, S. S. Utami Atmoko, T. Mitra Setia, and C. P. van Schaik (Eds.), *Orangutans: Geographic Variation in Ecology and Conservation* (pp. 135–56). Oxford University Press.

Rutz, C., Bluff, L. A., Reed, N., Troscianko, J., Newton, J., Inger, R., et al. (2010). The ecological significance of tool use in New Caledonian crows. *Science*, 329, 1523–26.

Rutz, C., Klump, B. C., Komarczyk, L., Leighton, R., Kramer, J., Wischnewski, S., et al. (2016). Discovery of species-wide tool use in the Hawaiian crow. *Nature*, 537, 403–7.

Rutz, C., and St. Clair, J. J. H. (2012). The evolutionary origins and ecological context of tool use in New Caledonian crows. *Behavioural Processes*, 89, 153–65.

Ruxton, G. D., and Lima, S. L. (1997). Predator-induced breeding suppression and its consequences for predator-prey population dynamics. *Proceedings of the Royal Society B: Biological Sciences*, 264, 409–15.

Ryan, A. S. (1981). Anterior dental microwear and its relationship to diet and feeding behavior in three African primates (*Pan troglodytes troglodytes, Gorilla gorilla gorilla,* and *Papio hamadryas*). *Primates*, 22, 533–50.

Ryan, N. T. (1976). Metabolic adaptations for energy production during trauma and sepsis. *Surgical Clinics of North America*, 56(5), 1073–90.

Rylands, A. B., Heymann, E. W., Lynch Alfaro, J., Buckner, J. C., Roos, C., Matauschek, C., et al. (2016). Taxonomic review of the New World tamarins (Primates: Callitrichidae). *Zoological Journal of the Linnean Society*, 177, 1003–28.

Sabatier, D. (1985). Saisonnalité et determinisme du pic de fructification en forêt guyanaise. *Revue d'Ecologie (Terre et Vie)*, 40, 289–320.

Sabatini, L. M., Warner, T. F., Saitho, E., and Azen, E. A. (1989). Tissue distribution of RNAs for cystatins, histatins, statherin, and proline-rich salivary proteins in human and macaques. *Journal of Dental Research*, 68, 1138–45.

Sacco, A. J., Granatosky, M. C., Laird, M. F., and Milich, K. M. (2021). Validation of a method for quantifying urinary C-peptide in platyrrhine monkeys. *General and Comparative Endocrinology*, 300, 113644.

Sacks, G. S. (1999). Glutamine supplementation in catabolic patients. *Annals of Pharmacotherapy*, 33(3), 348–54.

Sacramento, T. S., and Bicca-Marques, J. C. (2022). Scrounging marmosets eat more when the finder's share is low without changing their searching effort. *Animal Behaviour*, 183, 117–25.

Sadleir, R. M. S. (1969). *The Ecology of Reproduction in Wild and Domestic Mammals*. Methuen.

Sáez-Plaza, P., Michałowski, T., Navas, M. J., Asuero, A. G., and Wybraniec, S. (2013). An overview of the Kjeldahl method of nitrogen determination, part I: Early history, chemistry of the procedure, and titrimetric finish. *Critical Reviews in Analytical Chemistry*, 43, 4, 178–223.

Sáez-Plaza, P., Navas, M. J., Wybraniec, S., Michałowski, T., and Asuero, A. G. (2013). An overview of the Kjeldahl method of nitrogen determination part II: Sample preparation, working scale, instrumental finish, and quality control. *Critical Reviews in Analytical Chemistry*, 43, 224–72.

Saito, C. (1988). Cost of lactation in the Malagasy primate *Propithecus verreauxi*: Estimates of energy intake in the wild. *Folia Primatologica*, 69, 414.

Sakai, S. (2002). General flowering in lowland mixed dipterocarp forests of South-east Asia. *Biological Journal of the Linnean Society*, 75, 233–47.

Sakamaki, T., Maloueki, U., Bakaa, B., Bongoli, L., Kasalevo, P., Terada, S., et al. (2016). Mammals consumed by bonobos (*Pan paniscus*): New data from the Iyondji Forest, Tshuapa, Democratic Republic of Congo. *Primates*, 5, 295–301.

Salmi, R., Presotto, A., Scarry, C. J., Hawman, P., and Doran-Sheehy, D. M. (2020). Spatial cognition in western gorillas (*Gorilla gorilla*): an analysis of distance, linearity, and speed of travel routes. *Animal Cognition*, 23, 545–57.

Salomon, M., Mayntz, D., and Lubin, Y. (2008). Colony nutrition skews reproduction in a social spider. *Behavioral Ecology*, 19, 605–11.

Salminen, J. P., and Karonen, M. (2011). Chemical ecology of tannins and other phenolics: we need a change in approach. *Functional Ecology*, 25(2), 325–38.

Salzman, N. H., Ghosh, D., Huttner, K. M., Paterson, Y., and Bevins, C. L. (2003). Protection against enteric salmonellosis in transgenic mice expressing a human intestinal defensin. *Nature*, 422, 522–26.

Samuel, B. S., and Gordon, J. I. (2006). A humanized gnotobiotic mouse model of host-Archaeal-bacterial mutualism. *Proceedings of the National Academy of Sciences of the United States of America*, 103, 10011–16.

Samuni, L., Preis, A., Deschner, T., Crockford, C., and Wittig, R. M. (2018). Reward of labor coordination and hunting success in wild chimpanzees. *Communications Biology*, 1, 138.

Sánchez-Giraldo, C., and Daza, J. M. (2019). Getting better temporal and spatial ecology data for threatened species: Using lightweight GPS devices for small primate monitoring in the northern Andes of Colombia. *Primates*, 60(1), 93–102.

Sand, H., Zimmerman, B., Wabakken, P., Andren, H., and Pedersen, H. C. (2005). Using GPS technology and GIS cluster analyses to estimate kill rates in wolf-ungulate ecosystems. *Wildlife Society Bulletin*, 33, 914–25.

Sandberg, P. A., Loudon, J. E., and Sponheimer, M. (2012). Stable isotope analysis in primatology: A critical review. *American Journal of Primatology*, 74, 969–89.

Sanders, J. G., Beichman, A. C., Roman, J., Scott, J. J., Emerson, D., McCarthy, J. J., et al. (2015). Baleen whales host a unique gut microbiome with similarities to both carnivores and herbivores. *Nature Communications*, 6, 8285.

Sanders, J. G., Powell, S., Kronauer, D. J. C., Vasconcelos, H. L., Fredrickson, M. E., and Pierce, N. E. (2014). Stability and phylogenetic correlation in gut microbiota: Lessons from ants and apes. *Molecular Ecology*, 23(6), 1268–83.

Sanson, G. D., Kerr, S. A., and Gross, K. A. (2007). So silica phytoliths really wear mammalian teeth? *Journal of Archaeological Science*, 34, 526–31.

Sanson, G., Read, J., Aranwela, N., Clissold, F., and Peeters, P. (2001). Measurement of leaf biomechanical properties in studies of herbivory: opportunities, problems and procedures. *Austral Ecology*, 26, 535–46.

Santacruz, A., Collado, M. C., Garcia-Valdes, L., Segura, M. T., Martin-Lagos, J. A., Anjos, T., et al. (2010). Gut microbiota composition is associated with body weight, weight gain and biochemical parameters in pregnant women. *British Journal of Nutrition*, 104, 83–92.

Santhosh, K., Kumura, H. N., Velankar, A. D., and Sinha, A. (2015). Ranging behavior and resource use by lion-tailed macaques (*Macaca silenus*) in selectively logged forests. *International Journal of Primatology*, 36, 288–310.

Santiago, L. S., and Mulkey, S. S. (2005). Leaf productivity along a precipitation gradient in lowland Panama: Patterns from leaf to ecosystem. *Trees*, 19, 349–56.

Santos, G. A. d. S. D., Bianchini, E., and Reis, N. R. d. (2013). Seasonal variation of consumption of the species used as fruit source by brown howler monkeys (*Alouatta clamitans*) in southern Brazil. *Biota Neotropica*, 13, 148–53.

Santos-Buelga, C., and Scalbert, A. (2000). Proanthocyanidins and tannin-like compounds—nature, occurrence, dietary intake and effects on nutrition and health. *Journal of the Science of Food and Agriculture*, 80, 1094–117.

Sanz, C. M., and Morgan, D. B. (2007). Chimpanzee tool technology in the Goualougo Triangle, Republic of Congo. *Journal of Human Evolution*, 52, 420–33.

Sanz, C., and Morgan, D. B. (2009). Flexible and persistent tool-using strategies in honey-gathering by wild chimpanzees. *International Journal of Primatology*, 30, 411–27.

Sanz, C. M., and Morgan, D. B. (2013). Ecological and social correlates of chimpanzee tool use. *Philosophical Transactions of the Royal Society of London Series B—Biological Sciences*, 368, 1–14.

Sanz, C. M., Call, J., and Morgan, D. (2009). Design complexity in termite-fishing tools of chimpanzees (*Pan troglodytes*). *Biology Letters*, 5, 293–96.

Sanz, C. M., Deblauwe, I., Tagg, N., and Morgan, D. B. (2014). Insect prey characteristics affecting regional variation in chimpanzee tool use. *Journal of Human Evolution*, 71, 28–37.

Sanz, C. M., Morgan, D., and Gulick, S. (2004). New insights into chimpanzees, tools, and termites from the Congo basin. *American Naturalist*, 164, 567–81.

Saraswat, R., Sinha, A., and Radhakrishna, S. (2015). A god becomes a pest? Human-rhesus macaque interactions in Himachal Pradesh, northern India. *European Journal of Wildlife Research*, 61, 435–43.

Sardans, J., Penuelas, J., and Rivas-Ubach, A. (2011). Ecological metabolomics: Overview of current developments and future challenges. *Chemoecology*, 21, 191–225.

Sasaki, A., Fukuda, H., Shiida, N., Tanaka, N., Furugen, A., Ogura, J., et al. (2015). Determination of omega-6 and omega-3 PUFA metabolites in human urine samples using UPLC/MS/MS. *Analytical and Bioanalytical Chemistry*, 407(6), 1625–39.

Sato, H., Ichino, S., and Hanya, G. (2014). Dietary modification by common brown lemurs (*Eulemur fulvus*) during seasonal drought conditions in western Madagascar. *Primates*, 55, 219–30.

Sato, H., Santini, L., Patel, E. R., Campera, M., Yamashita, N., Colquhoun, I. C., et al. (2016). Dietary flexibility and feeding strategies of *Eulemur*: A comparison with *Propithecus*. *International Journal of Primatology*, 37, 109–29.

Sauerwald, T. U., Demmelmair, H., and Koletzko, B. (2001). Polyunsaturated fatty acid supply with human milk. *Lipids*, 36, 991–96.

Sauther, M. L. (1994). Wild plant use by pregnant and lactating ringtail lemurs, with implications for early hominid foraging. In N. L. Etkin (Ed.), *Eating on the Wild Side: The Pharmacologic, Ecologic, and Social Implications of Using Noncultigens* (pp. 240–46). University of Arizona Press.

Savage, A., Snowdon, C. T., Giraldo, L. H., and Soto, L. H. (1996). Parental care patterns and vigilance in wild cotton-top tamarins (*Saguinus oedipus*). In M. A. Norconk, A. L. Rosenberger, and P. A. Garber (Eds.), *Adaptive Radiations of Neotropical Primates* (pp. 187–99). Springer.

Sävendahl, L., and Underwood, L. E. (1997). Decreased interleukin-2 production from cultured peripheral blood mononuclear cells in human acute starvation 1. *Journal of Clinical Endocrinology and Metabolism*, 82(4), 1177–80.

Sayers, K. (2013). On folivory, competition, and intelligence: Generalisms, overgeneralizations, and models of primate evolution. *Primates*, 54, 111–24.

Sayers, K., Norconk, M. A., and Conklin-Brittain, N. L. (2010). Optimal foraging on the roof of the world: Himalayan langurs and the classical prey mode. *American Journal of Physical Anthropology*, 141, 337–57.

Sazima, C., Krajewski, J. P., Bonaldo, R. M., and Sazima, I. (2007). Nuclear-follower foraging associations of reef fishes and other animals at an oceanic archipelago. *Environmental Biology of Fishes*, 80, 351.

Scanlan, P. D., Stensvold, C. R., Rajilic-Stoianovic, M., Heilig, H. G. H. J., De Vos, W. M., O'Toole, P. W., et al. (2014). The microbial eukaryote *Blastocystis* is a prevalent and diverse member of the healthy human gut microbiota. *FEMS Microbiology Ecology*, 60(1), 326–30.

Scarry, C. J. (2013). Between-group contest competition among tufted capuchin monkeys, *Sapajus nigritus*, and the role of male resource defence. *Animal Behaviour*, 85, 931–39.

Scelza, B. A., and Hinde, K. (2019). Crucial contributions. *Human Nature*, 30(4), 371–97.

Schafrank, L. A., Washabaugh, J. R., and Hoke, M. K. (2020). An examination of breastmilk composition among high altitude Peruvian women. *American Journal of Human Biology*, 32(6), e23412.

Schetz, M., Casaer, M. P., and Van den Berghe, G. (2013). Does artificial nutrition improve outcome of critical illness? *Critical Care*, 17(1), 302.

Schillaci, M. A., Castellini, J. M., Stricker, C. A., Jones-Engel, L., Lee, B. P. Y. H., and O'Hara, T. M. (2014). Variation in hair $\delta^{13}C$ and $\delta^{15}N$ values in long-tailed macaques (*Macaca fascicularis*) from Singapore. *Primates*, 55, 25–34.

Schlabritz-Loutsevich, N. E., Dudley, C. J., Gomez, J. J., Nevill, H., Smith, B. K., Jenkins, S. L., et al. (2007). Metabolic adjustments to moderate maternal nutrient restriction. *British Journal of Nutrition*, 98, 276–84.

Schmid, J., and Ganzhorn, J. U. (1996). Resting metabolic rates of *Lepilemur ruficaudatus*. *American Journal of Primatology*, 38, 169–74.

Schmid, J., and Kappeler, P. M. (1998). Fluctuating sexual dimorphism and differential hibernation by sex in a primate, the gray mouse lemur (*Microcebus murinus*). *Behavioral Ecology and Sociobiology*, 43, 125–32.

Schmid, J., and Speakman, J. R. (2009). Torpor and energetic consequences in free-ranging grey mouse lemurs (*Microcebus murinus*): A comparison of dry and wet forests. *Naturwissenschaften*, 96, 609–20.

Schmidt, D. A., Kerley, M. S., Dempsey, J. L., Porton, I. J., Porter, J. H., Griffin, M. E., et al. (2005). Fiber digestibility by the orangutan (*Pongo abelii*): In vitro and in vivo. *Journal of Zoo and Wildlife Medicine*, 36, 571–80.

Schmidt, K. A. (2000). Interactions between food chemistry and predation risk in fox squirrels. *Ecology*, 81, 2077–85.

Schmiegelow, K. F. A., Machtans, C. S., and Hannon, S. J. (1997). Are boreal birds resilient to forest fragmentation? An experimental study of short-term community responses. *Ecology*, 78, 1914–32.

Schmitz, O. J. (2008). Herbivory from individuals to ecosystems. *Annual Review of Ecology and Systematics*, 39, 133–52.

Schmitz, O. J., Hamback, P. A., and Beckerman, A. P. (2000). Trophic cascades in terrestrial systems: A review of the effects of carnivore removals on plants. *American Naturalist*, 155, 141–53.

Schmitz, O. J., Krivan, V., and Ovadia, O. (2004). Trophic cascades: The primacy of trait-mediated indirect interactions. *Ecology Letters*, 7, 153–63.

Schneider, E. R., Mastrotto, M., Laursen, W. J., Schulz, V. P., Goodman, J. B., Funk, O. H., et al. (2014). Neuronal mechanism for acute mechanosensitivity in tactile-foraging waterfowl. *Proceedings of the National Academy of Sciences of the United States of America*, 111, 14941–46.

Schneider, K., and Hoffman, I. (2011). Nutrition ecology: A concept for systemic nutrition research and integrative problem solving. *Ecology of Food and Nutrition*, 50, 1–17.

Schoener, T. W. (1971). Theory of feeding strategies. *Annual Review of Ecology and Systematics*, 2, 369–404.

Schoener, T. W. (1973). Population growth regulated by intraspecific competition for energy or time: Some simple representations. *Theoretical Population Biology*, 4, 56–84.

Schoener, T. W. (1974). Resource partitioning in ecological communities. *Science*, 185, 27–39.

Schoeninger, M. J. (2009). $\delta^{13}C$ values reflect aspects of primate ecology in addition to diet. In J. J. Hublin and M. Richards (Eds.), *The Evolution of Hominin Diets* (pp. 121–27). Springer Science.

Schoeninger, M. J. (2010). Toward a $\delta^{13}C$ isoscape for primates. In J. B. West, G. J. Bowen, T. E. Dawson, and K. P. Tu (Eds.), *Isoscapes: Understanding Movement, Pattern, and Process on Earth through Isotope Mapping* (pp. 319–33). Springer.

Schoeninger, M. J., and DeNiro, M. J. (1984). Nitrogen and carbon isotopic composition of bone collagen from marine and terrestrial animals. *Geochimica et Cosmochimica Acta*, 48, 625–39.

Schoeninger, M. J., DeNiro, M. J., and Tauber, H. (1983). Stable nitrogen isotope ratios of bone collagen reflect marine and terrestrial components of prehistoric human diet. *Science*, 220, 1381–83.

Schoeninger, M. J., Iwaniec, U. T., and Glander, K. E. (1997). Stable isotope ratios indicate diet and habitat use in New World monkeys. *American Journal of Physical Anthropology*, 103, 69–83.

Schoeninger, M. J., Iwaniec, U. T., and Nash, L. T. (1998). Ecological attributes recorded in stable isotope ratios of arboreal prosimian hair. *Oecologia*, 113, 222–30.

Schoeninger, M. J., Moore, J., and Sept, J. M. (1999). Subsistence strategies of two "savanna" chimpanzee populations: The stable isotope evidence. *American Journal of Primatology*, 49, 297–314.

Schoeninger, M. J., Most, C. A., Moore, J. J., and Somerville, A. D. (2016). Environmental variables across *Pan troglodytes* study sites correspond with the carbon, but not the nitrogen, stable isotope ratios of chimpanzee hair. *American Journal of Primatology*, 78, 1055–69.

Schöning, C., Humle, T., Möbius, Y., and McGrew, W. C. (2008). The nature of culture: Technological variation in chimpanzee predation on army ants revisited. *Journal of Human Evolution*, 55, 48–59.

Schoonaert, K., D'Aout, K., Samuel, D., Talloen, W., Nauwelaerts, S., Kivell, T. L., et al. (2016). Gait characteristics and spatiotemporal variables of climbing in bonobos (*Pan paniscus*). *American Journal of Primatology*, 78, 1165–77.

Schreier, A. L., and Grove, M. (2010). Ranging patterns of hamadryas baboons: Random walk analyses. *Animal Behavior*, 80, 75–87.

Schreier, A. L., and Grove, M. (2014). Recurrent patterning in the daily foraging routes of hamadryas baboons (*Papio hamadryas*): Spatial memory in large-scale versus small-scale space. *American Journal of Primatology*, 76, 421–35.

Schreier, B. M., Harcourt, A. H., Coppeto, S. A., and Somi, M. F. (2009). Interspecific competition and niche separation in primates: A global analysis. *Biotropica*, 41, 283–91.

Schuelke, O. (1997). *Haremhalter und Jungesellen—Nahrungsoekologie, Lokomotion und differenzielles Energiebudget adulter Langurenmaennchen* (Presbytis entellus) *in Jodhpur (Rajasthan), Indien*. Georg-August-Universitaet.

Schuelke, O. (2001). Differential energy budget and monopolization potential of harem holders and bachelors in Hanuman langurs (*Semnopithecus entellus*): Preliminary results. *American Journal of Primatology*, 55, 57–63.

Schülke, O., and Ostner, J. (2012). Ecological and social influences on sociality. In J. C. Mitani, J. Call, P. M. Kappeler, R. A. Palombit, and J. B. Silk (Eds.), *The Evolution of Primate Societies* (pp. 195–219). University of Chicago Press.

Schulz, E., Calandra, I., and Kaiser, T. M. (2010). Applying tribology to teeth of hoofed mammals. *Scanning*, 32(4), 162–82.

Schulz-Kornas, E., Stuhlträger, J., Clauss, M., Wittig, R. M., and Kupczik, K. (2019). Dust affects chewing efficiency and tooth wear in forest dwelling Western chimpanzees (*Pan troglodytes verus*). *American Journal of Physical Anthropology*, 169(1), 66–77.

Schulz-Kornas, E., Winkler, D. E., Clauss, M., Carlsson, J., Ackermans, N. L., Martin, L. F., et al. (2020). Everything matters: Molar microwear texture in goats (*Capra aegagrus hircus*) fed diets of different abrasiveness. *Palaeogeography, Palaeoclimatology, Palaeoecology*, 552, 109783.

Schupp, E. W., Jordano, P., and Gómez, J. M. (2010). Seed dispersal effectiveness revisited: A conceptual review. *New Phytologist*, 188, 333–53.

Schurr, M. R., Fuentes, A., Luecke, E., Cortes, J., and Shaw, E. (2012). Intergroup variation in stable isotope ratios reflects anthropogenic impact on the Barbary macaques (*Macaca sylvanus*) of Gibraltar. *Primates*, 53, 31–40.

Schwartz, G. T. (2000). Taxonomic and functional aspects of the patterning of enamel thickness distribution in extant large-bodied hominoids. *American Journal of Physical Anthropology*, 111(2), 221–44.

Schwartz, G. T., McGrosky, A., and Strait, D. S. 2020. Fracture mechanics, enamel thickness and the evolution of molar form in hominins. *Biology Letters*, 16(1), 20190671.

Schwartz, S. M., and Kemnitz, J. W. (1992). Age- and gender-related changes in body size, adiposity, and endocrine and metabolic parameters in free-ranging rhesus macaques. *American Journal of Physical Anthropology*, 89, 109–21.

Scott, J. E. (2011). Folivory, frugivory, and postcanine size in the Cercopithecoidea revisited. *American Journal of Physical Anthropology*, 146(1), 20–27.

Scott, J. R., Godfrey, L. R., Jungers, W. L., Scott, R. S., Simons, E. L., Teaford, M. F., et al. (2009). Dental microwear texture analysis of two families of subfossil lemurs from Madagascar. *Journal of Human Evolution*, 54, 405–16.

Scott, R. S., Teaford, M. F., and Ungar, P. S. (2012). Dental microwear texture and anthropoid diets. *American Journal of Physical Anthropology*, 147(4), 551–79.

Scott, R. S., Ungar, P. S., Bergstrom, T. S., Brown, C. A., Childs, B. E., Teaford, M. F., et al. (2006). Dental microwear texture analysis: Technical considerations. *Journal of Human Evolution*, 51, 339–49.

Scott, R. S., Ungar, P. S., Bergstrom, T. S., Brown, C. A., Grine, F. E., Teaford, M. F., et al. (2005). Dental microwear texture analysis reflects diets of living primates and fossil hominins. *Nature*, 436, 693–95.

Scrimshaw, N. S., Taylor, C. E., and Gordon, J. E. (1959). Interactions of nutrition and infection. *American Journal of the Medical Sciences*, 237(3), 367–403.

Sealy, J., Armstrong, R., and Schrire, C. (1995). Beyond lifetime averages: Tracing life histories through isotopic analysis of different calcified tissues from archaeological human skeletons. *Antiquity*, 69, 290–300.

Searle, K. R., Hobbs, N. T., and Gordon, I. J. (2007). It's the "foodscape", not the landscape: Using foraging behavior to make functional assessments of landscape condition. *Israel Journal of Ecology and Evolution*, 53, 297–316.

Searle, K. R., Thompson Hobbs, N., and Shipley, L. A. (2005). Should I stay or should I go? Patch departure decisions by herbivores at multiple scales. *Oikos*, 111, 417–24.

Sears, R., and Mace, R. (2008). Who keeps children alive? A review of the effects of kin on child survival. *Evolution of Human Behavior*, 29, 1–18.

Sedio, B. E. (2017). Recent breakthroughs in metabolomics promise to reveal the cryptic chemical traits that mediate plant community composition, character evolution and lineage diversification. *New Phytologist*, 214, 952–58.

Sedio, B. E., Archibold, A. D., Echeverri, J. C. R., Debyser, C., and Wright, S. J. (2019). A comparison of inducible, ontogenetic, and interspecific sources of variation in the foliar metabolome in tropical trees. *PeerJ*, 7, e7536.

Sehgal, P. B. (1990). Interleukin-6: A regulator of plasma protein gene expression in hepatic and non-hepatic tissues. *Molecular Biology and Medicine*, 7(2), 117–30.

Seiler, M., Holderied, M., and Schwitzer, C. (2014). Habitat selection and use in the critically endangered Sahamalaza sportive lemur *Lepilemur sahamalazensis* in altered habitat. *Endangered Species Research*, 24, 273–86.

Seiler, M., Schwitzer, C., Gamba, M., and Holderied, M. W. (2013). Interspecific semantic alarm call recognition in the solitary Sahamalaza sportive lemur, *Lepilemur sahamalazensis*. *PLoS ONE*, 8, e67397.

Seiler, N., and Robbins, M. M. (2016). Factors influencing ranging on community land and crop raiding by mountain gorillas. *Animal Conservation*, 19, 176–88.

Sejrsen, K., and Purup, S. (1997). Influence of prepubertal feeding level on milk yield potential of dairy heifers: A review. *Journal of Animal Science*, 75, 828–35.

Sekirov, I., Russel, S. I., Antunes, C. M., and Finlay, B. B. (2010).

Gut microbiota in health and disease. *Physiological Reviews*, 90, 859–904.

Sellen, D. W. (2007). Evolution of infant and young child feeding: Implications for contemporary public health. *Annual Review of Nutrition*, 27, 123–48.

Sellen, D. W. (2009). Evolution of human lactation and complementary feeding: Implications for understanding contemporary cross-cultural variation. In G. R. Goldberg, A. Prentice, A. Prentice, S. Filteau, and K. Simondon (Eds.), *Breast-Feeding: Early Influences on Later Health*. Springer Netherlands.

Sellen, D. W., and Smay, D. B. (2001). Relationship between subsistence and age at weaning in "preindustrial" societies. *Human Nature*, 12, 47–87.

Sellers, W. I., and Crompton, R. H. (2004). Automatic monitoring of primate locomotor behaviour using accelerometers. *Folia Primatologica*, 75, 279–93.

Semaw, S., Rogers, M. J., Quade, J., Renne, P. R., Butler, R. F., Dominguez-Rodrigo, M., et al. (2003). 2.6-million-year-old stone tools and associated bones from OGS-6 and OGS-7, Gona, Afar, Ethiopia. *Journal of Human Evolution*, 45, 169–77.

Sempaio, M., and Ferrari, S. F. (2005). Predation of an infant titi monkey (*Callicebus molloch*) by a tufted capuchin (*Cebus apella*). *Folia Primatologica*, 76, 113–15.

Sender, R., Fuchs, S., and Milo, R. (2016). Are we really vastly outnumbered? Revisiting the ratio of bacterial to host cells in humans. *Cell*, 164(3), 337–40.

Seneviratne, S. I., X. Zhang, M. Adnan, W. Badi, C. Dereczynski, A. Di Luca, S., et al. (2021). Weather and climate extreme events in a changing climate. In Masson-Delmotte, V., Zhai, P. Pirani, A., Connors, S. L., Péan, C., Berger, S., et al. (Eds.) *Climate Change 2021: The physical science basis. Contribution of working group I to the Sixth Assessment Report of the Intergovernmental Panel on Climate Change* (pp. 1513-1766). Cambridge University Press.

Sengupta, A., McConkey, K. R., and Radhakrishna, S. (2015). Primates, provisioning and plants: Impacts of human cultural behaviors on primate ecological functions. *PLoS ONE*, 10, e0140961.

Senior, A. M., Charleston, M. A., Lihoreau, M., Buhl, J., Raubenheimer, D., and Simpson, S. J. (2015). Evolving nutritional strategies in the presence of competition: A geometric agent-based model. *PLoS Computational Biology*, 11, e1004111.

Senior, A. M., Grueber, C. E., Machovsky-Capuska, G., Simpson, S. J., and Raubenheimer, D. (2016). Macronutritional consequences of food generalism in an invasive mammal, the wild boar. *Mammalian Biology*, 81, 523–26.

Senior, A. M., Lihoreau, M., Buhl, J., Raubenheimer, D., and Simpson, S. J. (2016). Social network analysis and nutritional behavior: An integrated modeling approach. *Frontiers in Psychology*, 7, 18.

Senior, A. M., Lihoreau, M., Charleston, M. A., Buhl, J., Raubenheimer, D., and Simpson, S. J. (2016). Adaptive collective foraging in groups with conflicting nutritional needs. *Royal Society Open Science*, 3, 150638.

Serckx, A., Kuhl, H. S., Beudels-Jamar, R. C., Poncin, P., Bastin, J. F., and Huynen, M. C. (2015). Feeding ecology of bonobos living in forest-savannah mosaics: Diet seasonal variation and importance of fallback foods. *American Journal of Primatology*, 77, 948–62.

Serio-Silva, J. C., Olguin, E. J., Garcia-Feria, L., Tapia-Fierro, K., and Chapman, C. A. (2015). Cascading impacts of anthropogenically driven habitat loss: Deforestation, flooding, and possible lead poisoning in howler monkeys (*Alouatta pigra*). *Primates*, 56, 29–35.

Serio-Silva, J. C., Rico-Gray, V., Hernández-Salazar, L. T., and Espinosa-Gómez, R. (2002). The role of *Ficus* (Moraceae) in the diet and nutrition of a troop of Mexican howler monkeys, *Alouatta palliata mexicana*, released on an island in southern Veracruz, Mexico. *Journal of Tropical Ecology*, 18, 913–28.

Servedio, M. R., Brandvain, Y., Dhole, S., Fitzpatrick, C. L., Goldberg, E. E., Stern, C. A., et al. (2014). Not just a theory—the utility of mathematical models in evolutionary biology. *PLoS Biology*, 12, e1002017.

Setchell, J. M., Fairet, E., Shutt, K., Waters, S., and Bell, S. (2017). Biosocial conservation: Integrating biological and ethnographic methods to study human-primate interactions *International Journal of Primatology*, 38, 401–26.

Setchell, J. M., Wickings, E. J., and Knapp, L. A. (2006). Signal content of red facial coloration in female mandrills (*Mandrillus sphinx*). *Proceedings of the Royal Society B: Biological Sciences*, 273, 2395–400.

Setchell, K. D. R., and Clerici, C. (2010a). Equol: History, chemistry, and formation. *Journal of Nutrition*, 140, 1355s–62s.

Setchell, K. D. R., and Clerici, C. (2010b). Equol: Pharmacokinetics and biological actions. *Journal of Nutrition*, 140, 1363s–68s.

Severance, E. G., Yolken, R. H., and Eaton, W. W. (2014). Autoimmune diseases, gastrointestinal disorders and the microbiome in schizophrenia: More than a gut feeling. *Schizophrenia Research*, 176, 23–25.

Seyfarth, R. M., and Cheney, D. L. (2012). The evolutionary origins of friendship. *Annual Review of Psychology*, 63, 153–77.

Seyfarth, R. M., and Cheney, D. L. (2015). Social cognition. *Animal Behaviour*, 103, 191–202.

Seyfarth, R. M., Cheney, D. L., and Marler, P. (1980a). Monkey responses to 3 different alarm calls—evidence of predator classification and semantic communication. *Science*, 210, 801–3.

Seyfarth, R. M., Cheney, D. L., and Marler, P. (1980b). Vervet monkey alarm calls—semantic communication in a free-ranging primate. *Animal Behaviour*, 28, 1070–94.

Sganga, G., Siegel, J. H., Brown, G., Coleman, B., Wiles, C. E., Belzberg, H., et al. (1985). Reprioritization of hepatic plasma protein release in trauma and sepsis. *Archives of Surgery*, 120(2), 187–99.

Sha, J. C. M., Chua, S. C., Chew, P. T., Ibrahim, H., Lua, H. K., Fung, T. K., et al. (2017). Small-scale variability in a mosaic tropical rainforest influences habitat use of long-tailed macaques. *Primates*, 59, 163–71.

Shaffer, C. A. (2013a). Ecological correlates of ranging behavior in bearded sakis (*Chiropotes sagulatus*) in a continuous forest in Guyana. *International Journal of Primatology*, 34, 515–32.

Shaffer, C. A. (2013b). GIS analysis of patch use and group cohesiveness of bearded sakis (*Chiropotes sagulatus*) in the upper Essequibo Conservation Concession, Guyana. *American Journal of Physical Anthropology*, 150, 235–46.

Shaffer, C. A. (2014). Spatial foraging in free ranging bearded sakis: Traveling salesmen or Lévy walkers? *American Journal of Primatology*, 76, 472–84.

Shahidi, F., and Ambigaipalan, P. (2015). Phenolics and polyphenolics in foods, beverages and spices: Antioxidant activity and health effects—a review. *Journal of Functional Foods*, 18, 820–97.

Shanahan, M., So, S., Compton, S. G., and Corlett, R. (2001). Fig-eating by vertebrate frugivores: A global review. *Biological Reviews*, 76, 529–72.

Shannon, G., Cordes, L. S., Hardy, A. R., Angeloni, L. M., and Crooks, K. R. (2014). Behavioral responses associated with a human-mediated predator shelter. *PLoS ONE*, 9, e94630.

Shapiro, L. J., Kemp, A. D., and Young, J. W. (2016). Effects of substrate size and orientation on quadrupedal gait kinematics in mouse lemurs (*Microcebus murinus*). *Journal of Experimental Zoology Part A: Ecological Genetics and Physiology*, 325, 329–43.

Sharp, Z. (2006). *Principles of Stable Isotope Geochemistry* (1st ed.). Prentice Hall.

Sharpe, R. M., Martin, B., Morris, K., Greig, I., McKinnell, C., McNeilly, A. S., et al. (2002). Infant feeding with soy formula milk: Effects on the testis and on blood testosterone levels in marmoset monkeys during the period of neonatal testicular activity. *Human Reproduction*, 17, 1692–703.

Sheehan, R. L. and Papworth, S. (2019) Human speech reduces pygmy marmoset (*Cebuella pygmaea*) feeding and resting at a Peruvian tourist site, with louder volumes decreasing visibility. *American Journal of Primatology*, 81, e22967.

Sheine, W. S., and Kay, R. F. (1977). An analysis of chewed food particle size and its relationship to molar structure in the primates *Cheirogaleus medius* and *Galago senegalensis* and the insectivoran *Tupaia glis*. *American Journal of Physical Anthropology*, 47, 15–20.

Shellis, R. P., Beynon, A. D., Reid, D. J., and Hiiemae, K. M. (1998). Variations in molar enamel thickness among primates. *Journal of Human Evolution*, 35(4–5), 507–22.

Shelmidine, N., Borries, C., and McCann, C. (2009). Patterns of reproduction in Malayan silvered leaf monkeys at the Bronx Zoo. *American Journal of Primatology*, 71, 852–59.

Shen, B., Avila-Flores, R., Liu, Y., Rossiter, S. J., and Zhang, S. (2011). Prestin shows divergent evolution between constant frequency echolocating bats. *Journal of Molecular Evolution*, 73, 109–15.

Shepherd, A. A. (2008). Nutrition through the life-span. Part 2: Children, adolescents and adults. *British Journal of Nursing*, 17(21), 1332–38.

Shepherd, A. (2009). Nutrition through the life span. Part 3: Adults aged 65 years and over. *British Journal of Nursing*, 18(5), 301–2, 304–7.

Sheriff, M. J., Krebs, C. J., and Boonstra, R. (2011). From process to pattern: How fluctuating predation risk impacts the stress axis of snowshoe hares during the 10-year cycle. *Oecologia*, 166, 593–605.

Sherrow, H. M. (2005). Tool use in insect foraging by the chimpanzees of Ngogo, Kibale National Park, Uganda. *American Journal of Primatology*, 65, 377–83.

Sherry, D. S., and Ellison, P. T. (2007). Potential applications of urinary C-peptide of insulin for comparative energetics research. *American Journal of Physical Anthropology*, 133(1), 771–78.

Sherwen, S. L., Magrath, M. J. L., Butler, K. L., and Hemsworth, P. H. (2015). Little penguins, *Eudyptula minor*, show increased avoidance, aggression and vigilance in response to zoo visitors. *Applied Animal Behaviour Science*, 168, 71–76.

Shik, J. Z., Concilio, A., Kaae, T., and Adams, R. M. (2018). The farming ant *Sericomyrmex amabilis* nutritionally manages its fungal symbiont and its social parasite. *Ecological Entomology*, 43, 440–46.

Shik, J. Z., and Silverman, J. (2013). Towards a nutritional ecology of invasive establishment: Aphid mutualists provide better fuel for incipient Argentine ant colonies than insect prey. *Biological Invasions*, 15, 829–36.

Shimada, T. (2006). Salivary proteins as a defense against dietary tannins. *Journal of Chemical Ecology*, 32, 1149–63.

Shimada, T., Nishii, E., and Saitoh, T. (2011). Interspecific differences in tannin intakes of forest-dwelling rodents in the wild revealed by a new method using fecal proline content. *Journal of Chemical Ecology*, 37, 1277–84.

Shimada, T., Saitoh, T., Sasaki, E., Nishitani, Y., and Osawa, R. (2006). Role of tannin-binding salivary proteins and tannase-producing bacteria in the acclimation of the Japanese wood mouse to acorn tannins. *Journal of Chemical Ecology*, 32, 1165–80.

Shimizu, D., Macho, G. A., and Spears, I. R. (2005). Effect of prism orientation and loading direction on contact stresses in prismatic enamel of primates: Implications for interpreting wear patterns. *American Journal of Physical Anthropology*, 126(4), 427–34.

Shipley, L. A. (2010). Fifty years of food and foraging in moose: Lessons in ecology from a model herbivore. *Alces*, 46, 1–13.

Shipley, L. A., Blomquist, S., and Danell, K. (1998). Diet choices made by free-ranging moose in northern Sweden in relation to plant distribution, chemistry, and morphology. *Canadian Journal of Zoology*, 76, 1722–33.

Shipley, L. A., Illius, A. W., Danell, K., Hobbs, N. T., and Spalinger, D. E. (1999). Predicting bite size selection of mammalian herbivores: A test of a general model of diet optimization. *Oikos*, 84, 55–68.

Shopland, J. M. (1987). Food quality, spatial deployment, and the intensity of feeding interference in yellow baboons (*Papio cyanocephalus*). *Behavioral Ecology and Sociobiology*, 21, 149–56.

Shrader, A. M., Brown, J. S., Kerley, G. I. H., and Kotler, B. P. (2008). Do free-ranging domestic goats show "landscapes of fear"? Patch use in response to habitat features and predator cues. *Journal of Arid Environments*, 72, 1811–19.

Shrader, A. M., Kotler, B. P., Brown, J. S., and Kerley, G. I. H. (2008). Providing water for goats in arid landscapes: Effects on feeding effort with regard to time period, herd size and secondary compounds. *Oikos*, 117, 466–72.

Shulaev, V., Cortes, D., Miller, G., and Mittler, R. (2008). Metabolomics for plant stress response. *Physiologia Plantarum*, 132, 199–208.

Shultz, S., Noe, R., McGraw, W. S., and Dunbar, R. I. M. (2004). A community-level evaluation of the impact of prey behavioural and ecological characteristics on predator diet composition. *Proceedings of the Royal Society B: Biological Sciences*, 271, 725–32.

Shumaker, R. W., Walkup, K. R., and Beck, B. B. (2011). *Animal Tool Behavior: The Use and Manufacture of Tools by Animals*. Johns Hopkins University Press.

Shumaker, R. W., Wich, S. A., and Perkins, L. (2008). Reproductive life history traits of female orangutan (*Pongo* spp.). In S. Atsalis, S. W. Margulis, and P. R. Hof (Eds.), *Primate Reproductive Aging* (pp. 147–61). Karger.

Sibly, R. M. (1981). Strategies of digestion and defecation. In C. R. Townsend and P. Calow (Eds.), *Physiological Ecology* (pp. 109–39). Blackwell Scientific.

Sibly, R. M., and Brown, J. H. (2007). Effects of body size and lifestyle on evolution of mammal life histories. *Proceedings of the National Academy of Sciences*, 104(45), 17707–12.

Siddiqui, J. A., Pothuraju, R., Jain, M., Batra, S. K., and Nasser, M. W. (2020). Advances in cancer cachexia: Intersection between affected organs, mediators, and pharmacological interventions. *Biochimica et Biophysica Acta—Reviews on Cancer*, 1873(2), 188359.

Siemers, B. M. (2012). The sensory ecology of foraging for animal prey. In R. A. Tuttle (Ed.), *Leaping Ahead: Advances in Prosimian Biology* (pp. 257–63). Springer.

Siemers, B. M., Goerlitz, H. R., Robsomanitrandrasana, E., Piep, M., Ramanamanjato, J. B., Takotondravony, D., et al. (2007). Sensory basis of food detection in wild *Microcebus murinus*. *International Journal of Primatology*, 28, 291–304.

Siew, E. D., Ikizler, T. A., Gebretsadik, T., Shintani, A., Wickersham, N., Bossert, F., et al. (2010). Elevated urinary IL-18 levels at the time of ICU admission predict adverse clinical outcomes. *Clinical Journal of the American Society of Nephrology*, 5(8), 1497–505.

Siex, K. S. (2005). Habitat destruction, population compression and overbrowsing by the Zanzibar red colobus monkeys (*Procolobus kirkii*). In J. D. Paterson and J. Wallis (Eds.), *Commensalism and Conflict: The Human-Primate Interface* (pp. 294–337). American Society of Primatologists.

Siex, K. S., and Struhsaker, T. T. (1999). Ecology of the Zanzibar red colobus monkey: Demographic variability and habitat stability. *International Journal of Primatology*, 20, 163–92.

Sigg, H., and Stolba, A. (1981). Home range and daily march in a hamadryas baboon troop. *Folia Primatologica*, 36, 40–75.

Sih, A. (1980). Optimal behavior: can foragers balance two conflicting demands? *Science*, 2010, 1041-1043.

Silk, J. B. (1986). Eating for two: Behavioral and environmental correlates of gestation length among free-ranging baboons (*Papio cynocephalus*). *International Journal of Primatology*, 7, 583–602.

Silk, J. (2002). Practice random acts of aggression and senseless acts of intimidation: The logic of status contests in social groups. *Evolutionary Anthropology*, 11, 221–25.

Silk, J. B., Brosnan, S. F., Henrichs, J., Lambeth, S. P., and Shapiro, S. (2013). Chimpanzees share food for many reasons: The role of kinship, reciprocity, social bonds, and harassment of food transfers. *Animal Behaviour*, 85, 941–47.

Silva, L. P., Santana, L. M., and de Melo, F. R. (2021). Effect of seasonality on the feeding behavior of Martins' bare-faced tamarin *Saguinus martinsi martinsi* (Primates: Callitrichidae) in the Brazilian Amazon. *Primate Conservation*, 35, 37–45.

Silva, Y. P., Bernardi, A., and Frozza, R. L. (2020). The role of short-chain fatty acids from gut microbiota in gut-brain communication. *Frontiers in Endocrinology*, 11, 25.

Silver, S. C., Ostro, L. E. T., Yeager, C. P., and Horwich, R. (1998). Feeding ecology of the black howler monkey (*Alouatta pigra*) in northern Belize. *American Journal of Primatology*, 45, 263–79.

Simmen, B., and Rasamimanana, H. (2018). Energy (im-)balance in frugivorous lemurs in southern Madagascar: A preliminary study in *Lemur catta* and *Eulemur rufifrons × collaris*. *Folia Primatologica*, 89(6), 382–96.

Simmen, B., Bayart, F., Marez, A., and Hladik, A. (2007). Diet, nutritional ecology and birth season of *Eulemur macao* in an anthropogenic forest in Madagascar. *International Journal of Primatology*, 28, 1253–66.

Simmen, B., Hladik, A., Ramasiarisoa, P. L., Iaconelli, S., and Hladik, C. M. (1999). Taste discrimination in lemurs and other primates, and the relationships to distribution of plant allelochemicals in different habitats of Madagascar. In B. Rakotosamimanana, H. Rasamimanana, J. U. Ganzhorn, and S.

M. Goodman (Eds.), *New Directions in Lemur Studies* (pp. 201–19).

Simmen, B., Peronny, S., Hladik, A., and Marez, A. (2006). Diet quality and taste perception of plant secondary metabolites by *Lemur catta*. In A. Jolly, R. W. Sussman, N. Koyama, and H. Rasamimanana (Eds.), *Ringtailed Lemur Biology*: Lemur catta *in Madagascar* (pp. 160–83). Springer US.

Simmen, B., Sauther, M. L., Soma, T., Rasamimanana, H., Sussman, R. W., Jolly, A., et al. (2006). Plant species fed on by *Lemur catta* in gallery forests of the southern domain of Madagascar. In A. Jolly, R. W. Sussman, N. Koyama, and H. Rasamimanana (Eds.), *Ringtailed Lemur Biology*: Lemur catta *in Madagascar* (pp. 55–68). Springer US.

Simmen, B., Tarnaud, L., and Hladik, A. (2012). Leaf nutritional quality as a predictor of primate biomass: Further evidence of an ecological anomaly within prosimian communities in Madagascar. *Journal of Tropical Ecology*, 28, 141–51.

Simmen, B., Tarnaud, L., Bayart, F., Hladik, A., Thiberge, A. L., Jaspart, S., et al. (2005). Secondary metabolite contents in the forests of Mayotte and Madagascar, and their incidence on two leaf-eating lemur-species (*Eulemur* spp.). *Revue d'Ecologie—La Terre et la Vie*, 60, 297–324.

Simmen, B., Tarnaud, L., Marez, A., and Hladik, A. (2014). Leaf chemistry as a predictor of primate biomass and the mediating role of food selection: A case study in a folivorous lemur (*Propithecus verreauxi*). *American Journal of Primatology*, 76, 563–75.

Simon, N. G., Kaplan, J. R., Hu, S., Register, T. C., and Adams, M. R. (2004). Increased aggressive behavior and decreased affiliative behavior in adult male monkeys after long-term consumption of diets rich in soy protein and isoflavones. *Hormones and Behavior*, 45, 278–84.

Simondon, K. B., Simon, I., and Simondon, F. (1997). Nutritional status and age at menarche of Senegalese adolescents. *Annals of Human Biology*, 24, 521–32.

Simons, E. L. (1976). Nature of transition in dental mechanism from pongids to hominids. *Journal of Human Evolution*, 5(5), 511–28.

Simons, E. L., and Pilbeam, D. R. (1972). Hominoid paleoprimatology. In R. Tuttle (Ed.), *The Functional and Evolutionary Biology of the Primates* (pp. 36–62). Aldine.

Simpson, G. G. (1926). Mesozoic Mammalia, IV: The multituberculates as living animals. *American Journal of Science*, 11, 228–50.

Simpson, G. G. (1933). Paleobiology of Jurassic mammals. *Paleobiologica*, 5, 127–58.

Simpson, S. J., and Raubenheimer, D. (1993). A multi-level analysis of feeding behaviour: The geometry of nutritional decisions. *Philosophical Transactions of the Royal Society of London Series B—Biological Sciences*, 342, 381–402.

Simpson, S. J., and Raubenheimer, D. (1995). The geometric analysis of feeding and nutrition—a user's guide. *Journal of Insect Physiology*, 41, 545–53.

Simpson, S. J., and Raubenheimer, D. (2001). The geometric analysis of nutrient-allelochemical interactions: A case study using locusts. *Ecology*, 82, 422–39.

Simpson, S. J., and Raubenheimer, D. (2005). Obesity: The protein leverage hypothesis. *Obesity Reviews*, 6, 133–42.

Simpson, S. J., and Raubenheimer, D. (2009). Macronutrient balance and lifespan. *Aging*, 1(10), 875–80.

Simpson, S. J., and Raubenheimer, D. (2012). *The Nature of Nutrition: A Unifying Framework from Animal Adaptation to Human Obesity*. Princeton University Press.

Simpson, S. J., and Raubenheimer, D. (2014). Perspective: Tricks of the trade. *Nature*, 508, 566–66.

Simpson, S. J., and Raubenheimer, D. (2020). The power of protein. *American Journal of Clinical Nutrition*, 112, 6–7.

Simpson, S. J., Clissold, F. J., Lihoreau, M., Ponton, F., Wilder, S. M., and Raubenheimer, D. (2015). Recent advances in the integrative nutrition of arthropods. *Annual Review of Entomology*, 60, 293–311.

Simpson, S. J., Le Couteur, D. G., and Raubenheimer, D. (2015). Putting the balance back in diet. *Cell*, 161, 18–23.

Simpson, S. J., Le Couteur, D. G., James, D. E., George, J., Gunton, J. E., Solon-Biet, S. M., et al. (2017). The geometric framework for nutrition as a tool in precision medicine. *Nutrition and Healthy Aging*, 4, 217–26.

Simpson, S. J., Raubenheimer, D., Behmer, S. T., Whitworth, A., and Wright, G. A. (2002). A comparison of nutritional regulation in solitarious- and gregarious-phase nymphs of the desert locust, *Schistocerca gregaria*. *Journal of Experimental Biology*, 205, 121–29.

Simpson, S. J., Raubenheimer, D., Charleston, M. A., and Clissold, F. J. (2010). Modelling nutritional interactions: from individuals to communities. *Trends in Ecology & Evolution*, 25(1), 53–60.

Simpson, S. J., Ribeiro, C., and González-Tokman, D. M. (2018). Insect feeding behavior. In A. Córdoba-Aguilar, D. González-Tokman, and I. González-Santoyo (Eds.), *Insect Behaviour: From Mechanisms to Ecological and Evolutionary Consequences*. Oxford University Press.

Simpson, S. J., Sibly, R. M., Lee, K. P., Behmer, S. T., and Raubenheimer, D. (2004). Optimal foraging when regulating intake of multiple nutrients. *Animal Behaviour*, 68, 1299–311.

Simpson, S. J., Sword, G. A., Lorch, P. D., and Couzin, I. D. (2006). Cannibal crickets on a forced march for protein and salt. *Proceedings of the National Academy of Sciences of the United States of America*, 103, 4152–56.

Singer, E., Marko, L., Paragas, N., Barasch, J., Dragun, D., Muller, D. N., et al. (2013). Neutrophil gelatinase-associated lipocalin: Pathophysiology and clinical applications. *Acta Physiologica*, 207(4), 663–72.

Singh, H., and Gallier, S. (2017). Nature's complex emulsion: The fat globules of milk. *Food Hydrocolloids*, 68, 81–89.

Singh, M., Roy, K., and Singh, M. (2011). Resource partitioning in sympatric langurs and macaques in tropical rainforests of the central Western Ghats, south India. *American Journal of Primatology*, 73, 335–46.

Singh, U., Devaraj, S., and Jialal, I. (2005). Vitamin, E., oxidative stress, and inflammation. *Annual Review of Nutrition*, 25, 151–74.

Siva-Jothy, M. T., and Thompson, J. J. W. (2002). Short-term nutrient deprivation affects immune function. *Physiological Entomology*, 27(3), 206–12.

Skibiel, A. L., Downing, L. M., Orr, T. J., and Hood, W. R. (2013). The evolution of the nutrient composition of mammalian milks. *Journal of Animal Ecology*, 82, 1254–64.

Skinner, M. M., and Wood, B. (2006). The evolution of modern human life history. In K. Hawkes and R. R. Paine (Eds.), *The Evolution of Human Life History* (pp. 331–64). School of American Research Press.

Skopec, M. M., Hagerman, A. E., and Karasov, W. H. (2004). Do salivary proline-rich proteins counteract dietary hydrolyzable tannin in laboratory rats? *Journal of Chemical Ecology*, 30, 1679–92.

Skvarla, M. J. (2015). *Sampling Terrestrial Arthropod Biodiversity: A Case Study in Arkansas*. University of Arkansas.

Sleeman, J. M., Cameron, K., Mudakikwa, A. B., Nizeyi, J. B., Anderson, S., Cooper, J. E., et al. (2000). Field anesthesia of free-living mountain gorillas (*Gorilla gorilla beringei*) from the Virunga Volcano Region, central Africa. *Journal of Zoo and Wildlife Medicine*, 31, 9–14.

Slot, M., Garcia, M. N., and Winter, K. (2016). Temperature response of CO_2 exchange in three tropical species. *Functional Plant Biology*, 43, 468–78.

Smith, A. C. (2000). Interspecific differences in prey captured by associating saddleback (*Saguinus fuscicollis*) and moustached (*Saguinus mystax*) tamarins. *Journal of Zoology*, 251, 315–24.

Smith, A. C. (2010a). Exudativory in primates: Interspecific patterns. In A. M. Burrows and L. T. Nash (Eds.), *The Evolution of Exudativory in Primates* (pp. 45–87). Springer.

Smith, A. C. (2010b). Influences on gum feeding in primates. In A. M. Burrows and L. T. Nash (Eds.), *The Evolution of Exudativory in Primates* (pp. 109–21). Springer.

Smith, A. C., Buchanan-Smith, H. M., Surridge, A. K., Osorio, D., and Mundy, N. I. (2003). The effect of colour vision status on the detection and selection of fruits by tamarins (*Saguinus* spp.). *Journal of Experimental Biology*, 206(18), 3159–65.

Smith, A. C., Buchanan-Smith, H. M., Surridge, A. K., and Mundy, N. I. (2005). Factors affecting group spread within wild mixed-species troops of saddleback and moustached tamarins. *International Journal of Primatology*, 26, 337–55.

Smith, A. C., Kelez, S., and Buchanan-Smith, H. M. (2004). Factors affecting vigilance within wild mixed-species troops of saddleback (*Saguinus fuscicollis*) and moustached tamarins (*S. mystax*). *Behavioral Ecology and Sociobiology*, 56, 18–25.

Smith, A. C., Surridge, A. K., Prescott, M. J., Osorio, D., Mundy, N. I., and Buchanan-Smith, H. M. (2012). Effect of colour vision status on insect prey capture efficiency of captive and wild tamarins (*Saguinus* spp.). *Animal Behaviour*, 83(2), 479–86.

Smith, B. N., and Epstein, S. (1971). Two categories of $^{13}C/^{12}C$ ratios for higher plants. *Plant Physiology*, 47, 380–84.

Smith, C. C., Morgan, M. E., and Pilbeam, D. (2010). Isotopic ecology and dietary profiles of Liberian chimpanzees. *Journal of Human Evolution*, 58, 43–55.

Smith, F., and Montgomery, R. (1959). *The Chemistry of Plant Gums and Mucilages and Some Related Polysaccharides*. Reinhold.

Smith, J. M., and Smith, A. C. (2013). An investigation of ecological correlates with hand and foot morphology in callitrichid primates. *American Journal of Physical Anthropology*, 152, 447–58.

Smith, L. W., Link, A., and Cords, M. (2008). Cheek pouch use, predation risk, and feeding competition in blue monkeys (*Cercopithecus mitis stuhlmanni*). *American Journal of Physical Anthropology*, 137, 334–41.

Smith, R., Aboitiz, F., Schrter, C., Barton, R., Denenberg, V., Fitch, R. H., et al. (2005). Relative size versus controlling for size: interpretation of ratios in research on sexual dimorphism in the human corpus callosum. *Current Anthropology*, 46, 249–273.

Smith, R. J., and Jungers, W. L. (1997). Body mass in comparative primatology. *Journal of Human Evolution*, 32, 523–59.

Smith, R. J., and Leigh, S. R. (1998). Sexual dimorphism in primate neonatal body mass. *Journal of Human Evolution*, 34, 173–201.

Smith, T. B., Bruford, M. W., and Wayne, R. K. (1993). Leaf productivity along a precipitation gradient in lowland Panama: Patterns from leaf to ecosystem. *Biodiversity Letters*, 1, 164–67.

Smith, T. M., Austin, C., Green, D. R., Joannes-Boyau, R., Bailey, S., Dumitriu, D., et al. (2018). Wintertime stress, nursing, and lead exposure in Neanderthal children. *Science Advances*, 4(10), eaau9483.

Smith, T. M., Austin, C., Hinde, K., Vogel, E. R., and Arora, M. (2017). Cyclical nursing patterns in wild orangutans. *Science Advances*, 3, e1601517.

Smith, T. M., Machanda, Z., Bernard, A. B., Donovan, R. M., Papakyrikos, A. M., Muller, M. N., et al. (2013). First molar eruption, weaning, and life history in living wild chimpanzees. *Proceedings of the National Academy of Sciences of the United States of America*, 110(8), 2787–91.

Smith, T. M., Tafforeau, P., Reid, D. J., Grün, R., and Eggins, S. (2007). Earliest evidence of modern human life history in North African early *Homo sapiens*. *Proceedings of the National Academy of Sciences of the United States of America*, 104, 6128–33.

Smolensky, N. L., and Fitzgerald, L. A. (2010). Distance sampling underestimates population densities of dune-dwelling lizards. *Journal of Herpetology*, 44, 372–81.

Smuts, B. B., Cheney, D. L., Seyfarth, R. M., Wrangham, R. W., and Struhsaker, T. T. (Eds.). (1987). *Primate Societies*. University of Chicago Press.

Smythe, N. (1982). The seasonal abundance of night-flying insects in a Neotropical forest. In E. G. Leigh, A. S. Rand, and D. M. Windsor (Eds.), *The Ecology of a Tropical Forest: Seasonal Rhythms and Long-Term Changes* (pp. 309–18). Smithsonian Institution Press.

Snaith, T. V., and Chapman, C. A. (2005). Towards an ecological solution to the folivore paradox: Patch depletion as an indicator of within-group scramble competition in red colobus monkeys (*Piliocolobus tephrosceles*). *Behavioral Ecology and Sociobiology*, 59, 185–90.

Snaith, T. V., and Chapman, C. A. (2008). Red colobus monkeys display alternative behavioral responses to the costs of scramble competition. *Behavioral Ecology*, 19, 1289–96.

Snaith, T. V., Chapman, C. A., Rothman, J. M., and Wasserman, M. D. (2008). Bigger groups have fewer parasites and similar cortisol levels: A multi-group analysis in red colobus monkeys. *American Journal of Primatology*, 70, 1072–80.

Snodgrass, J. J., Leonard, W. R., and Robertson, M. L. (2007). Primate bioenergetics: an evolutionary perspective. In Ravosa, M. J., Dagosto, M. (Eds.), *Primate origins: adaptations and evolution* (pp. 703–37). Springer USA.

Snyder-Mackler, N., Burger, J. R., Gaydosh, L., Belsky, D. W., Noppert, G. A., Campos, F. A., et al. (2020). Social determinants of health and survival in humans and other animals. *Science*, 368, 6493.

Snyder-Mackler, N., Majoros, W. H., Yuan, M. L., Shaver, A. O., Gordon, J. B., Kopp, G. H., et al. (2016). Efficient genome-wide sequencing and low-coverage pedigree analysis from noninvasively collected samples. *Genetics*, 203, 699–714.

Sohal, R. S., and Forster, M. J. (2014). Caloric restriction and the aging process: A critique. *Free Radical Biology and Medicine*, 73, 366–82.

Sol, D., Bacher, S., Reader, S. M., and Lefebvre, L. (2008). Brain size predicts the success of mammal species introduced into novel environments. *American Naturalist*, 172, S63–S71.

Sol, D., Duncan, R. P., Blackburry, T. M., Cassey, P., and Lefebvre, L. (2005). Big brains, enhanced cognition, and response of birds to novel environments. *Proceedings of the National Academy of Sciences of the United States of America*, 102, 5460–65.

Solon-Biet, S. M., McMahon, A. C., Ballard, J. W. O., Ruohonen, K., Wu, L. E., Cogger, V. C., et al. (2014). The ratio of macronutrients, not caloric intake, dictates cardiometabolic health, aging, and longevity in ad libitum-fed mice. *Cell Metabolism*, 19(3), 418–30.

Soltis, J., King, L. E., Douglas-Hamilton, I., Vollrath, F., and Savage, A. (2014). African elephant alarm calls distinguish between threats from humans and bees. *PLoS ONE*, 9, e89403.

Sommer, A., Katz, J., and Tarwotjo, I. (1984). Increased risk of respiratory disease and diarrhea in children with preexisting mild vitamin A deficiency. *American Journal of Clinical Nutrition*, 40(5), 1090–95.

Sommer, V., Srivastava, A., and Borries, C. (1992). Cycles, sexuality, and conception in free-ranging langurs (*Presbytis entellus*). *American Journal of Primatology*, 28, 1–27.

Son, V. D. (2003). Diet of *Macaca fascicularis* in a mangrove forest, Vietnam. *Laboratory Primate Newsletter*, 42, 1–5.

Sorensen, J. S., McLister, J. D., and Dearing, M. D. (2005). Novel plant secondary metabolites impact dietary specialists more than generalists (*Neotoma* spp.). *Ecology*, 86, 140–54.

Sorokowska, A., Sorokowski, P., Hummel, T., and Huanca, T. (2013). Olfaction and environment: Tsimane' of Bolivian rainforest have lower threshold of odor detection than industrialized German people. *PLoS ONE*, 8, e69203.

Soultoukis, G. A., and Partridge, L. (2016). Dietary protein, metabolism, and aging. *Annual Review of Biochemistry*, 85, 5–34.

Souto, A., Bione, C. B. C., Bastos, M., Bezerra, B. M., Fragaszy, D., and Schiel, N. (2011). Critically endangered blonde capuchins fish for termites and use new techniques to accomplish the task. *Biology Letters*, 7, 532–35.

Souza-Alves, J. P., Barbosa, G. V., and Hilário, R. R. (2020). Tree-gouging by marmosets (Primates: *Callitrichidae*) enhances tree turnover. *Biotropica*, 52, 808–12.

Souza-Alves, J. P., Chagas, R. R. D., Santana, M. M., Boyle, S. A., and Bezerra, B. M. (2021). Food availability, plant diversity, and vegetation structure drive behavioral and ecological variation in endangered Coimbra-Filho's titi monkeys. *American Journal of Primatology*, 83, e23237.

Souza-Alves, J. P., Fontes, I. P., Chagas, R. R. D., and Ferrari, S. F. (2011). Seasonal versatility in the feeding ecology of a group of titis (*Callicebus coimbrai*) in the northern Brazilian Atlantic forest. *American Journal of Primatology*, 73, 1199–209.

Spagnoletti, N., Visalberghi, E., Ottoni, E., Izar, P., and Fragaszy, D. (2011). Stone tool use by adult wild bearded capuchin monkeys (*Cebus libidinosus*): Frequency, efficiency and tool selectivity. *Journal of Human Evolution*, 61, 97–107.

Spagnoletti, N., Visalberghi, E., Verderane, M. P., Ottoni, E., Izar, P., and Fragaszy, D. (2012). Stone tool use in wild bearded capuchin monkeys, *Cebus libidinosus*: Is it a strategy to overcome food scarcity? *Animal Behaviour*, 83, 1285–94.

Sparvoli, L. G., Cortez, R. V., Daher, S., Padilha, M., Sun, S. Y., Nakamura, M. U., et al. (2020). Women's multisite microbial modulation during pregnancy. *Microbial Pathogenesis*, 147, 104230.

Sperfeld, E., Raubenheimer, D., and Wacker, A. (2016). Bridging factorial and gradient concepts of resource co-limitation: Towards a general framework applied to consumers. *Ecology Letters*, 19, 201–15.

Speth, J. D. (2013). Middle Paleolithic large mammal hunting in the southern Levant. In J. L. Clark and J. D. Speth (Eds.), *Zooarchaeology and Modern Human Origins: Human Hunting Behavior during the Later Pleistocene* (pp. 19–43). Springer.

Speth, J., and Clark, J. (2006). Hunting and overhunting in the Levantine Late Middle Palaeolithic. *Before Farming*, 2006, 1–42.

Spiegel, O., and Crofoot, M. C. (2016). The feedback between where we go and what we know—information shapes movement, but movement also impacts information acquisition. *Current Opinion in Behavioral Sciences*, 12, 90–96.

Spiegel, O., Harel, R., Getz, W. M., and Nathan, R. (2013). Mixed strategies of griffon vultures' (*Gyps fulvus*) response to food deprivation lead to a hump-shaped movement pattern. *Movement Ecology*, 1, 1–12.

Sponheimer, M., Alemseged, Z., Cerling, T. E., Grine, F. E., Kimbel, W. H., Keakey, M. G., et al. (2013). Isotopic evidence of early hominin diets. *Proceedings of the National Academy of Sciences of the United States of America*, 110, 10513–18.

Sponheimer, M., and Lee-Thorp, J. A. (1999). Oxygen isotopes in enamel carbonate and their ecological significance. *Journal of Archaeological Sciences*, 26, 723–28.

Sponheimer, M., and Lee-Thorp, J. A. (2001). The oxygen isotope composition of mammalian enamel carbonate from Morea Estate, South Africa. *Oecologia*, 126, 153–57.

Sponheimer, M., Loudon, J. E., Codron, D., Howells, M. E., Pruetz, J., Codron, J., et al. (2006). Do "savanna" chimpanzees consume C_4 resources? *Journal of Human Evolution*, 51, 128–33.

Sponheimer, M., Robinson, T., Ayliffe, L., Roeder, B., Hammer, J., Passey, B., et al. (2003). Nitrogen isotopes in mammalian herbivores: Hair δ15N values from a controlled feeding study. *International Journal of Osteoarchaeology*, 13, 80–87.

Spor, A., Koren, O., and Ley, R. E. (2011). Unravelling the effects of the environment and host genotype on the gut microbiome. *Nature Reviews*, 9, 279–90.

Spradley, J. P., Glander, K. E., and Kay, R. F. (2016). Dust in the wind: How climate variables and volcanic dust affect rates of tooth wear in central American howling monkeys. *American Journal of Physical Anthropology*, 159, 210–22.

Sprague, D. S., Kabaya, M., and Hagihara, K. (2004). Field testing a global positioning system (GPS) collar on a Japanese monkey: Reliability of automatic GPS positioning in a Japanese forest. *Primates*, 45, 151–54.

Sreelakshmi, K. R., Manjusha, L., Vartak, V. R., and Venkateshwarlu, G. (2016). Variation in proximate composition and fatty acid profiles of mud crab meat with regard to sex and body parts. *Indian Journal of Fisheries*, 63, 147–50.

Sridhar, H., Beauchamp, G., and Shanker, K. (2009). Why do birds participate in mixed-species foraging flocks? A large-scale synthesis. *Animal Behaviour*, 78, 337–47.

Sridhara, S., McConkey, K., Prasad, S., and Corlett, R. T. (2016). Frugivory and seed dispersal by large herbivores of Asia. In F. S. Ahrestani and M. Sankaran (Eds.), *The Ecology of Large Herbivores in South and Southeast Asia* (vol. 225) (pp. 121–50). Springer Science+Business Media.

Stacey, P. B. (1986). Group size and foraging efficiency in yellow baboons. *Behavioral Ecology and Sociobiology*, 18, 175–87.

Stammers, A. L., Lowe, N., Medina, M. W., Patel, S., Dykes, F., et al. (2015). The relationship between zinc intake and growth in children aged 1–8 years: a systematic review and meta-analysis. *European Journal of Clinical Nutrition*, 69, 2, 147–53.

Stander, P. E. (1992). Cooperative hunting in lions: The role of the individual. *Behavioral Ecology and Sociobiology*, 29, 445–54.

Stanford, C. B. (1991). The diet of the capped langur (*Presbytis pileata*) in a moist deciduous forest in Bangladesh. *International Journal of Primatology*, 12, 199–216.

Stanford, C. B. (1995). The influence of chimpanzee predation on group size and anti-predator behaviour in red colobus monkeys. *Animal Behaviour*, 49, 577–87.

Stanford, C. B. (1996). The hunting ecology of wild chimpanzees: Implications for the evolutionary ecology of Pliocene hominids. *American Anthropologist*, 98, 96–113.

Stanford, C. B. (1998). *Chimpanzee and Red Colobus: The Ecology of Predator and Prey*. Harvard University Press.

Stanford, C. B., and Nkurunungi, J. B. (2003). Behavioral ecology of sympatric chimpanzees and gorillas in Bwindi Impenetrable National Park, Uganda: Diet. *International Journal of Primatology*, 24(4), 901–18.

Stanford, C. B., Wallis, J., Matema, H. E., and Goodall, J. (1994). Patterns of predation by chimpanzees on red colobus monkeys in Gombe National Park, 1982–1991. *American Journal of Physical Anthropology*, 94, 213–28.

Stanford, C. B., Wallis, J., Mpongo, E., and Goodall, J. (1994). Hunting decisions in wild chimpanzees. *Behaviour*, 131, 1–18.

Stankowich, T. (2008). Ungulate flight responses to human disturbance: A review and meta-analysis. *Biological Conservation*, 141, 2159–73.

Stapp, P. (2002). Stable isotopes reveal evidence of predation by ship rats on seabirds on the Shiant Islands, Scotland. *Journal of Applied Ecology*, 39, 831–40.

Starr, C., and Nekaris, K. A. I. (2013). Obligate exudativory characterises the diet of the pygmy slow loris *Nycticebus pygmaeus*. *American Journal of Primatology*, 75, 1054–61.

Stearns, S. C. (1992). *The Evolution of Life Histories*. Oxford University Press.

Stears, K., and Shrader, A. M. (2015). Increases in food availability can tempt oribi antelope into taking greater risks at both large and small spatial scales. *Animal Behaviour*, 108, 155–64.

Steenbeek, R., and van Schaik, C. P. (2001). Competition and group size in Thomas's langurs (*Presbytis thomasi*): The folivore paradox revisited. *Behavioral Ecology and Sociobiology*, 49, 100–110.

Steffee, W. P., Goldsmith, R. S., Pencharz, P. B., Scrimshaw, N. S., and Young, V. R. (1976). Dietary protein intake and dynamic aspects of whole body nitrogen metabolism in adult humans. *Metabolism*, 25(3), 281–97.

Steffens, K. J. E. (2020). Lemur food plants as options for forest restoration in Madagascar. *Restoration Ecology*, 28, 1517–27.

Stein, Z., and Susser, M. (1975). The Dutch famine, 1944–1945, and the reproductive process. I. Effects on six indices at birth. *Pediatric Research*, 9, 70–76.

Steingraber, S. (2007). *The Falling Age of Puberty in US Girls: What We Know, What We Need to Know*. Breast Cancer Fund.

Stelmasiak, M., Balan, B. J., Mikaszewska-Sokolewicz, M., Niewinski, G., Kosalka, K., Szczepanowska, E., et al. (2021). The relationship between the degree of malnutrition and changes in selected parameters of the immune response in critically ill patients. *Central European Journal of Immunology, 46*(1), 82–91.

Stephens, D. W., and Krebs, J. R. (1986). *Foraging Theory*. Princeton University Press.

Stephens, P. A., Boyd, I. L., McNamara, J. M., and Houston, A. I. (2009). Capital breeding and income breeding: Their meaning, measurement, and worth. *Ecology, 90*, 2057–67.

Stephenson, P. J., Speakman, J. R., and Racey, P. A., (1994). Field metabolic rate in two species of shrew-tenrec, *Microgale dobsoni* and *M. talazaci*. *Comparative Biochemistry and Physiology Part A: Physiology, 107*, 283–7.

Sterck, E. H. M. (1997). Determinants of female dispersal in Thomas langurs. *American Journal of Primatology, 42*, 179–98.

Sterck, E. H. M. (1998). Female dispersal, social organization, and infanticide in langurs: Are they linked to human disturbance? *American Journal of Primatology, 44*, 235–54.

Sterck, E. H. M. (2002). Predator sensitive foraging in Thomas langurs. In L. E. Miller (Ed.), *Eat or Be Eaten: Predator Sensitive Foraging among Primates* (pp. 74–91). Cambridge University Press.

Sterck, E. H. M. (2012). The behavioral ecology of colobine monkeys. In J. C. Mitani, J. Call, P. M. Kappeler, R. A. Palombit, and J. B. Silk (Eds.), *The Evolution of Primate Societies* (pp. 65–90). University of Chicago Press.

Sterck, E. H. M., Watts, D. P., and van Schaik, C. P. (1997). The evolution of female social relationships in nonhuman primates. *Behavioral Ecology and Sociobiology, 41*, 291–309.

Sterling, E. J. (1993). Patterns of range use and social organization in aye-ayes (*Daubentonia madagascarensis*) on Nosy Magabe. In Kappler, P., and Ganzhorn, J. U. (Eds.). *Lemur social systems and their ecological basis* (pp 1–10). Springer USA.

Sterling, E. J., and Povinelli, D. J. (1999). Tool use, aye-ayes, and sensorimotor intelligence. *Folia Primatologica, 70*, 8–16.

Sterling, E. J., Dierenfeld, E. S., Ashbourne, C. J., and Feistner, A. T. C. (1994). Dietary intake, food composition and nutrient intake in wild and captive populations of *Daubentonia madagascariensis*. *Folia Primatologica, 62*, 115–24.

Stern, M., and Goldstone, R. (2005). Red colobus as prey: The leaping habits of five sympatric Old World monkeys. *Folia Primatologica, 76*, 100–112.

Sternberg, L. S. L., Mulkey, S. S., and Wright, S. J. (1989). Oxygen isotope ratio stratification in a tropical moist forest. *Oecologia, 81*, 51–56.

Steudel, K. (2000). The physiology and energetics of movement: Effects on individuals and groups. In S. Boinski and P. A. Garber (Eds.), *On the Move: How and Why Animals Travel in Groups* (pp. 9–23). University of Chicago Press.

Stevens, C. E., and Hume, I. D. (1995). *Comparative physiology of the vertebrate digestive system*. Cambridge University Press.

Stevens, J. F., and Page, J. E. (2004). Xanthohumol and related prenylflavonoids from hops and beer: To your good health! *Phytochemistry, 65*, 1317–30.

Stevenson, P. R. (2004). Phenological patterns of woody vegetation at Tinigua Park, Colombia: Methodological comparisons with emphasis on fruit production. *Caldasia, 26*, 125–50.

Stevenson, P. R., Quiñones, M. J., and Ahumada, J. A. (1998). Effects of fruit patch availability on feeding subgroup size and spacing patterns in four primate species at Tinigua National Park, Columbia. *International Journal of Primatology, 19*, 313–24.

Stevenson, P., Quiñones, M. J., and Ahumada, J. A. (2000). Influence of fruit availability on ecological overlap among four Neotropical primates at Tinigua National Park, Colombia. *Biotropica, 32*, 533–44.

Stewart, C. M., Kothari, P. D., Mouliere, F., Mair, R., Somnay, S., Benayed, R., et al. (2018). The value of cell-free DNA for molecular pathology. *Journal of Pathology, 244*(5), 616–27.

Steyaert, S. M. J. G., Leclerc, M., Pelletier, F., Kindberg, J., Brunberg, S., Swenson, J. E., et al. (2016). Human shields mediate sexual conflict in a top predator. *Proceedings of the Royal Society B: Biological Sciences, 283*, 20160906.

Stiling, P., and Cornelissen, T. (2007). How does elevated carbon dioxide (CO_2) affect plant-herbivore interactions? A field experiment and meta-analysis of CO_2-mediated changes on plant chemistry and herbivore performance. *Global Change Biology, 13*, 1823–42.

Stiner, M. (2013). An unshakable Middle Paleolithic? Trends versus conservatism in the predatory niche and their social ramifications. *Current Anthropology, 54*, S288–S304.

Stini, W. A., Weber, C. W., Kemberling, S. R., and Vaughn, L. A. (1980). Bioavailability of nutrients in human breast milk as compared to formula. *Studies in Physical Anthropology, 6*, 32–35.

Stoinski, T. S., Perdue, B. M., Breuer, T., and Hoff, M. P. (2013). Variability in the developmental life history of the genus *Gorilla*. *American Journal of Physical Anthropology, 152*, 165–72.

Stone, A. I. (2007). Responses of squirrel monkeys to seasonal changes in food availability in an Eastern Amazonian forest. *American Journal of Primatology, 69*, 142–57.

Stone, A. I. (2014). Is fatter sexier? Reproductive strategies of male squirrel monkeys (*Saimiri sciureus*). *International Journal of Primatology, 35*, 628–42.

Stone, A. I., and Ruivo, L. V. P. (2020). Synchronization of weaning time with peak fruit availability in squirrel monkeys (*Saimiri collinsi*) living in Amazonian Brazil. *American Journal of Primatology, 82*, e23139.

Stone, A. I., Castro, P. H. G., Monteiro, F. O. B., Ruivo, L. P., and de Sousa e Silva, J. (2015). A novel method for capturing and monitoring a small Neotropical primate, the squirrel monkey (*Saimiri collinsi*). *American Journal of Primatology*, 77, 239–45.

Stone, G. N., Nee, S., and Felsenstein, J. (2011). Controlling for non-independence in comparative analysis of patterns across populations within species. *Philosophical Transactions of the Royal Society of London Series B—Biological Sciences*, 366, 1410–24.

Stoner, K. E. (1996). Habitat selection and seasonal patterns of activity and foraging of mantled howling monkeys (*Alouatta palliata*) in northeastern Costa Rica. *International Journal of Primatology*, 17, 1–30.

Stoner, K. E., Riba-Hernández, P., and Lucas, P. W. (2005). Comparative use of color vision for frugivory by sympatric species of platyrrhines. *American Journal of Primatology*, 67, 399–409.

Strait, S. G. (1993). Differences in occlusal morphology and molar size in frugivores and faunivores. *Journal of Human Evolution*, 25(6), 471–84.

Strait, S. G., and Overdorff, D. J. (1996). Physical properties of fruits eaten by Malagasy primates. *American Journal of Physical Anthropology*, 22 (supplement), 224.

Strandburg-Peshkin, A., Farine, D. R., Couzin, I. D., and Crofoot, M. C. (2015a). Shared decision-making drives collective movement in wild baboons. *Science*, 348, 1358–61.

Strandburg-Peshkin, A., Farine, D. R., Couzin, I. D., and Crofoot, M. C. (2015b). The wisdom of baboon decisions—response. *Science*, 349, 935–36.

Strandburg-Peshkin, A., Farine, D. R., Crofoot, M. C., and Couzin, I. D. (2017). Habitat and social factors shape individual decisions and emergent group structure during baboon collective movement. *eLife*, 6, e19505.

Strandin, T., Babayan, S. A., and Forbes, K. M. (2018). Reviewing the effects of food provisioning on wildlife immunity. *Philosophical Transactions of the Royal Society of London Series B—Biological Sciences*, 373, 20170088.

Streleckiene, G., Reid, H. M., Arnold, N., Bauerschlag, D., and Forster, M. (2018). Quantifying cell free DNA in urine: Comparison between commercial kits, impact of gender and inter-individual variation. *Biotechniques*, 64(5), 225–30. https://doi.org/10.2144/btn-2018-0003.

Strickland, J. D. H., and Parsons, T. R. (1972). *A Practical Handbook of Seawater Analysis*. Fisheries Board of Canada.

Strier, K. B. (1989). Effects of patch size on feeding associations in muriquis (*Brachyteles arachnoides*). *Folia Primatologica*, 52, 70–77.

Strier, K. B. (1991). Diet in one group of woolly spider monkeys, or muriquis (*Brachyteles arachnoides*). *American Journal of Primatology*, 23, 113–26.

Strier, K. B. (1992). Atelinae adaptations—behavioral strategies and ecological constraints. *American Journal of Physical Anthropology*, 88, 515–24.

Strier, K. B. (1993). Menu for a monkey. *Natural History*, 102, 34–43.

Strier, K. B. (2007). *Primate Behavioral Ecology* (3rd ed.). Allyn and Bacon.

Strier, K. B. (2009). Seeing the forest through the seeds: Mechanisms of primate behavioral diversity from individuals to populations and beyond. *Current Anthropology*, 50, 213–28.

Strier, K. B. (2010). Long-term field studies: Positive impacts and unintended consequences. *American Journal of Primatology*, 72, 772–78.

Strier, K. B. (2017). What does variation in primate behavior mean? *American Journal of Physical Anthropology*, 162, 4–14.

Strier, K. B. (2021a). The limits of resilience. *Primates*, 62, 861–68.

Strier, K. B. (2021b) *Primate Behavioral Ecology*, 6th ed. Routledge. Companion website: https://routledgetextbooks.com/textbooks/9780367222888/.

Strier, K. B., and Boubli, J. P. (2006). A history of long-term research and conservation of northern muriquis (*Brachyteles hypoxanthus*) at the Estação Biológica de Caratinga/RPPN-FMA. *Primate Conservation*, 20, 53–63.

Strier, K. B., Lee, P. C., and Ives, A. R. (2014). Behavioral flexibility and the evolution of primate social states. *PLoS ONE*, 9(12), e114099.

Strier, K. B., Possamai, C. B., and Mendes, S. L. (2015). Dispersal patterns of northern muriquis: Implications for social dynamics, life history, and conservation. In T. Furuichi, F. Aureli, and J. Yamagiwa (Eds.), *Dispersing Primate Females* (pp. 3–22). Springer.

Struhsaker, T. T. (1967). Ecology of vervet monkeys (*Cercopithecus aethiops*) in the Masai-Amboseli Game Reserve, Kenya. *Ecology*, 48, 891–904.

Struhsaker, T. T. (1969). Correlates of ecology and social organization among African cercopithecines. *Folia Primatologica*, 11, 80–118.

Struhsaker, T. T. (1973). A recensus of vervet monkeys in the Masai-Amboseli Game Reserve, Kenya. *Ecology*, 54, 930–32.

Struhsaker, T. T. (1981). Polyspecific associations among tropical rain-forest primates. *Zeitschrift fur Tierpsychologie*, 57, 268–304.

Struhsaker, T. T. (1997). *Ecology of an African Rain Forest: Logging in Kibale and the Conflict between Conservation and Exploitation*. University of Florida Press.

Struhsaker, T. T. (1999). Primate communities in Africa: The consequence of long-term evolution or the artifact of recent hunting. In J. G. Fleagle, C. Janson, and K. Reed (Eds.), *Primate Communities* (pp. 289–94). Cambridge University Press.

Struhsaker, T. T. (2008). Demographic variability in monkeys: Implications for theory and conservation. *International Journal of Primatology*, 29, 19–34.

Struhsaker, T., and Leland, L. (1977). Palm-nut smashing by *Cebus a. apella* in Columbia. *Biotropica*, 9, 124–26.

Struhsaker, T. T., Marshall, A. R., Detwiler, K., Siex, K., Ehardt,

C., Lisbjerg, D. D., et al. (2004). Demographic variation among Udzungwa red colobus in relation to gross ecological and sociological parameters. *International Journal of Primatology*, 25, 615–58.

Struhsaker, T. T., Struhsaker, P. J., and Siex, K. S. (2005). Conserving Africa's rain forests: Problems in protected areas and solutions. *Biological Conservation*, 123, 45–54.

Strum, S. C. (1981). Processes and products of change: Baboon predatory behavior at Gilgil, Kenya. In R. S. O. Harding and G. Teleki (Eds.), *Omnivorous Primates: Gathering and Hunting in Human Evolution* (pp. 255–302). Columbia University Press.

Strum, S. C. (1991). Weight and age in wild olive baboons. *American Journal of Primatology*, 25, 219–37.

Strum, S. C. (2010). The development of primate raiding: Implications for management and conservation. *International Journal of Primatology*, 31, 133–56.

Strum, S. C., and Western, D. (1982). Variations in fecundity with age and environment in olive baboons (*Papio anubis*). *American Journal of Primatology*, 3, 61–76.

Stueckle, S., and Zinner, D. (2008). To follow or not to follow: Decision making and leadership during the morning departure in chacma baboons. *Animal Behaviour*, 75, 1995–2004.

Stulp, G., Barrett, L., Tropf, F. C., and Mills, M. (2015). Does natural selection favour taller stature among the tallest people on earth? *Proceedings of the Royal Society B: Biological Sciences*, 282, 20150211.

Stumpf, W. E. (2006). The dose makes the medicine. *Drug Discovery Today*, 11, 550–55.

Stutz, R. S., Croak, B. M., Proschogo, N., Banks, P. B., and McArthur, C. (2017). Olfactory and visual plant cues as drivers of selective herbivory. *Oikos*, 126, 259–68.

Suarez, S. A. (2014). Ecological factors predictive of wild spider monkey (*Ateles belzebuth*) foraging decisions in Yasuní, Ecuador. *American Journal of Primatology*, 76, 1185–95.

Suarez, S. A., Karro, J., Kipper, J., Farler, D., McElroy, B., Rogers, B. C., et al. (2014). A comparison of computer-generated and naturally-occurring foraging patterns in route-network-constrained spider monkeys. *American Journal of Primatology*, 76, 460–71.

Sudo, N. (2006). Stress and gut microbiota: Does postnatal microbial colonization programs the hypothalamic-pituitary-adrenal system for stress response? *International Congress Series*, 1287, 350–54.

Sueur, C. (2011). Group decision-making in chacma baboons: Leadership, order and communication during movement. *BMC Ecology*, 11, 26.

Sueur, C., Deneubourg, J. L., and Petit, O. (2012). From social network (centralized vs. decentralized) to collective decision-making (unshared vs. shared consensus). *PLoS ONE*, 7, e32566.

Sueur, C., Deneubourg, J. L., Petit, O., and Couzin, I. D. (2010). Differences in nutrient requirements imply a non-linear emergence of leaders in animal groups. *PLoS Computational Biology*, 7, e32566.

Sueur, C., MacIntosh, A. J., Jacobs, A. T., Watanabe, K., and Petit, O. (2013). Predicting leadership using nutrient requirements and dominance rank of group members. *Behavioral Ecology and Sociobiology*, 67, 457–70.

Sueur, C., Petit, O., and Deneubourg, J. L. (2009). Selective mimetism at departure in collective movements of *Macaca tonkeana*: An experimental and theoretical approach. *Animal Behaviour*, 78, 1087–95.

Sugawara, T., Go, Y., Udono, T., Morimura, N., Tomonaga, M., Hirai, H., et al. (2011). Diversification of bitter taste receptor gene family in western chimpanzees. *Molecular Biology and Evolution*, 28, 921–31.

Sugawara, T., and Imai, H. (2012). Post-genome biology of primates focusing on taste perception. In H. Hirai, H. Imai, and Y. Go (Eds.), *Post-genome Biology of Primates* (pp. 79–91). Springer.

Sugiyama, Y. (1989). Local variation of tool and tool behavior among wild chimpanzee populations. In Y. Sugiyama (Ed.), *Behavioral Studies of Wild Chimpanzees at Bossou, Guinea* (pp. 1–15). Kyoto University Primate Research Institute.

Sugiyama, Y. (1994). Tool use by wild chimpanzees. *Nature*, 367, 327.

Sugiyama, Y. (1997). Social tradition and the use of tool-composites by wild chimpanzees. *Evolutionary Anthropology*, 6, 23–27.

Sugiyama, Y. (2015). Influence of provisioning on primate behavior and primate studies. *Mammalia*, 79, 255–65.

Sugiyama, Y., and Koman, J. (1979). Tool-using and -making behavior in wild chimpanzees at Bossou, Guinea. *Primates*, 20, 513–24.

Sugiyama, Y., and Koman, J. (1987). A preliminary list of chimpanzees' alimentation at Boussou, Guinea. *Primates*, 28, 133–47.

Sugiyama, Y., and Ohsawa, H. (1982). Population dynamics of Japanese monkeys with special reference to the effect of artificial feeding. *Folia Primatologica*, 39, 238–63.

Sullivan, E. L., Nousen, E. K., and Chamlou, K. A. (2014). Maternal high fat diet consumption during the perinatal period programs offspring behavior. *Physiology and Behavior*, 123, 236–42.

Sullivan, E. L., Smith, M. S., and Grove, K. L. (2011). Perinatal exposure to high-fat diet programs energy balance, metabolism and behavior in adulthood. *Neuroendocrinology*, 93(1), 1–8.

Sullivan, J. T. (1973). Drying and storing herbage as hay. In G. W. Butler and R. W. Bailey (Eds.), *Chemistry and Biochemistry of Herbage*. Academic Press.

Sun, B., Xu, X., Xia, Y., Cheng, Y., Mao, S., Xiang, X., et al. (2021). Variation of gut microbiome in free-ranging female Tibetan macaques (*Macaca thibetana*) across different reproductive states. *Animals*, 11(1), 39.

Sun, H. T., Zhang, J., Hou, N., Zhang, X. Q., Wang, J., and Bai, Y. X. (2014). Spontaneous periodontitis is associated with metabolic syndrome in rhesus monkeys. *Archives of Oral Biology*, 59(4), 386–92.

Surbeck, M., Fowler, A., Deimel, C., and Hohmann, G. (2009). Evidence for the consumption of diurnal, arboreal primates by bonobos, *Pan paniscus*. *American Journal of Primatology*, 71, 171–74.

Surbeck, M., and Hohmann, G. (2008). Primate hunting by bonobos at Lui Kotale, Salongo National Park. *Current Biology*, 18, R906–7.

Sussman, R. W. (1977). Feeding behavior in *Lemur catta* and *Lemur fulvus*. In T. H. Clutton-Brock (Ed.), *Primate Ecology: Studies in Feeding and Ranging Behavior in Lemurs, Monkeys and Apes* (pp. 1–36). Academic Press.

Sussman, R. W. (1978). Foraging patterns of nonhuman primates and nature of food preferences in Man. *Federation Proceedings*, 37(1), 55–60.

Sussman, R. W. (1987). Morpho-physiological analysis of diets: Species-specific dietary patterns in primates and human dietary adaptations. In W. G. Kinzey (Ed.), *The Evolution of Human Behaviour: Primate Models* (pp. 151–79). State University of New York Press.

Sussman, R. W. (1991). Primate origins and the evolution of angiosperms. *American Journal of Primatology*, 23, 209–23.

Sussman, R. W. (2014). The lure of lemurs to an anthropologist. In K. B. Strier (Ed.), *Primate Ethnographies* (pp. 34–45). Pearson.

Sussman, R. W., and Garber, P. A. (2011). Cooperation, collective action, and competition in primate social interactions. In C. J. Campbell, A. Fuentes, K. C. MacKinnon, S. K. Bearder, and R. S. M. Stumpf (Eds.), *Primates in Perspective* (2nd ed., pp. 587–99). Oxford University Press.

Sussman, R. W., and Tattersall, I. (1976). Cycles of activity, group composition, and diet of *Lemur mongoz mongoz* Linnaeus 1766 in Madagascar. *Folia Primatologica*, 26, 270–83.

Sussman, R. W., Rasmussen, D. T., and Raven, P. H. (2013). Rethinking primate origins again. *American Journal of Primatology*, 75, 95–106.

Suwanvecho, U., Brockelman, W. Y., Nathalang, A., Santon, J., Matmoon, U., Somnuk, R., et al. (2018). High interannual variation in the diet of a tropical forest frugivore (*Hylobates lar*). *Biotropica*, 50(2), 346–56.

Suzuki, A. (1969). An ecological study of chimpanzees in a savanna woodland. *Primates*, 10, 103–48.

Suzuki, A. (1975). The origin of hominid hunting: A primatological perspective. In R. A. Tuttle (Ed.), *Socioecology and Psychology of Primates* (pp. 259–78). Mouton.

Suzuki, K., Harasawa, R., Yoshitake, Y., and Mitsuoka, T. (1983). Effect of crowding and heat stress on intestinal flora, body weight gain, and feed efficiency of growing rats and chicks. *Nippon Juigaku Zasshi*, 45, 331–38.

Suzuki, S., Kuroda, S., and Nishihara, T. (1995). Tool-set for termite-fishing by chimpanzees in the Ndoki forest, Congo. *Behaviour*, 132, 219–35.

Suzuki, S., Noma, N., and Izawa, K. (1998). Inter-annual variation of reproductive parameters and fruit availability in two populations of Japanese macaques. *Primates*, 39, 313–24.

Suzuki-Hashido, N., Hayakawa, T., Matsui, A., Go, Y., Ishimaru, Y., Misaka, T., et al. (2015). Rapid expansion of phenylthiocarbamide non-tasters among Japanese macaques. *PLoS ONE*, 10, e0132016.

Svihus, B., and Holand, Ø. (2000). Lichen polysaccharides and their relation to reindeer/caribou nutrition. *Journal of Range Management*, 53, 642–48.

Swap, R., Aranibar, J., Bowty, P., Gilhooly, I. W., and Macho, S. A. (2004). Natural abundance of ^{13}C and ^{15}N in C_3 and C_4 vegetation of southern Africa: Patterns and implications. *Global Change Biology*, 10, 350–58.

Swapna, N., Radhakrishna, S., Gupta, A. K., and Kumar, A. (2010). Exudativory in the Bengal slow loris (*Nycticebus bengalensis*) in Trishna Wildlife Sanctuary, Tripura, northeast India. *American Journal of Primatology*, 72, 113–21.

Swedell, L., Hailemeskel, G., and Schreier, A. (2008). Composition and seasonality of diet in wild hamadryas baboons: Preliminary findings from Filoha. *Folia Primatologica*, 79, 476–90.

Swedell, L., Leedom, L., Saunders, J., and Pines, M. (2014). Sexual conflict in a polygynous primate: Costs and benefits of a male-imposed mating system. *Behavioral Ecology and Sociobiology*, 68, 263–73.

Swinburn, B. A., Sacks, G., Hall, K. D., McPherson, K., Finegood, D. T., Moodie, M. L., et al. (2011). The global obesity pandemic: Shaped by global drivers and local environments. *Lancet*, 378(9793), 804–14.

Symington, M. (1986). Ecological determinants of fission-fusion sociality in *Ateles* and *Pan*. In J. G. Else and P. C. Lee (Eds.), *Primate Ecology and Conservation* (pp. 181–90). Cambridge University Press.

Symington, M. M. (1988). Food competition and foraging party size in the black spider monkey (*Ateles paniscus chamek*). *Behaviour*, 105, 117–32.

Symington, M. M. (1990). Fission-fusion social organization in *Ateles* and *Pan*. *International Journal of Primatology*, 11, 47–61.

Ta, C. A. K., Pebsworth, P. A., Liu, R., Hillier, S., Gray, N., Arnason, J. T., et al. (2018). Soil eaten by chacma baboons adsorbs polar plant secondary metabolites representative of those found in their diet. *Environmental Geochemistry and Health*, 40, 803–13.

Tabor, D. (1950). *The Hardness of Metals*. Clarendon.

Tadesse, S. A., and Kotler, B. P. (2012). Impact of tourism on Nubian ibex (*Capra nubiana*) revealed through assessment of behavioral indicators. *Behavioral Ecology*, 23, 1257–62.

Takahashi, M. (2018). *The Nutritional Ecology of Adult Female Blue Monkeys,* Cercopithecus mitis *in the Kakamega Forest, Kenya* [PhD dissertation]. Columbia University, New York, USA.

Takahashi, M. Q., Rothman, J. M., and Cords, M. (2023). The role of non-natural foods in the nutritional strategies of monkeys in a human-modified mosaic landscape. *Biotropica,* 55(1), 106–18.

Takahashi, M. Q., Rothman, J. M., Raubenheimer, D., and Cords, M. (2019). Dietary generalists and nutritional specialists: Feeding strategies of adult female blue monkeys (*Cercopithecus mitis*) in the Kakamega Forest, Kenya. *American Journal of Primatology,* 81, e23016.

Takahashi, M. Q., Rothman, J. M., Raubenheimer, D., and Cords, M. (2021). Daily protein prioritization and long-term nutrient balancing in a dietary generalist, the blue monkey. *Behavioral Ecology,* 32(2), 223–35.

Takahata, Y., Hasegawa, T., and Nishida, T. (1984). Chimpanzee predation in the Mahale Mountains from August 1979 to May 1982. *International Journal of Primatology,* 5, 213–33.

Takahata, Y., Suzuki, S., Agetsuma, N., Okayasu, N., Sugiura, H., Takahashi, H., et al. (1998). Reproduction of wild Japanese macaque females at Yakushima and Kinkazan Islands: A preliminary report. *Primates,* 39, 339–49.

Takasaki, H. (1981). Troop size, habitat quality, and home range area in Japanese macaques. *Behavioral Ecology and Sociobiology,* 9, 277–81.

Takasaki, H. (1984). A model for relating troop size and home range area in a primate species. *Primates,* 25, 22–27.

Takemoto, H. (2003). Phytochemical determination for leaf food choice by wild chimpanzees in Guinea, Bossou. *Journal of Chemical Ecology,* 29, 2551–573.

Takenaka, M., Hayashi, K., Yamada, G., Ogura, T., Ito, M., Milner, A. M., et al. (2022). Behavior of snow monkeys catching fish to survive the winter. *Scientific Reports,* 12, 20324.

Takenoshita, Y., Sprague, D., and Iwasaki, N. (2005). Factors affecting success rate and accuracy of GPS collar positioning for free-ranging Japanese macaque. *Primate Research,* 21, 107–19.

Talebi, M., Bastos, A., and Lee, P. C. (2005). Diet of southern muriquis in continuous Brazilian Atlantic Forest. *International Journal of Primatology,* 26, 1175–87.

Talebi, M. G., Sala, E. A., Carvalho, B., Villani, G., Canela, G., Lucas, P. W., et al. (2016). Membrane-plate transition in leaves as an influence on dietary selectivity and tooth form. *Journal of Human Evolution,* 98, 18–26.

Talham, G. L., Jiang, H. Q., Bos, N. A., and Cebra, J. J. (1999). Segmented filamentous bacteria are potent stimuli of a physiologically normal state of the murine gut mucosal immune system. *Infectious Immunology,* 67, 1992–2000.

Tambling, C. J., Cameron, E. Z., Du Toit, J. T., and Getz, W. M. (2010). Methods for locating African lion kills using Global Positioning System movement data. *Journal of Wildlife Management,* 74, 549–56.

Tan, A. W. Y., Luncz, L. V., Haslam, M., Malaivijitnond, S., and Gumert, M. D. (2016). Complex processing of prickly pear cactus (*Opuntia* sp.) by free-ranging long-tailed macaques: Preliminary analysis for hierarchical organisation. *Primates,* 57, 141–47.

Tan, A., Tan, S. H., Vyas, D., Malaivijitnond, S., and Gumert, M. D. (2015). There is more than one way to crack an oyster: Identifying variation in Burmese long-tailed macaque (*Macaca fascicularis aurea*) stone-tool use. *PLoS ONE,* 10, e0124733.

Tan, C. L. (1999). Group composition, home range size, and diet of three sympatric bamboo lemur species (genus *Hapalemur*) in Ranomafana National Park, Madagascar. *International Journal of Primatology,* 20, 547–66.

Tan, Y., and Li, W. H. (1999). Vision: Trichromatic vision in prosimians. *Nature,* 402, 36.

Tan, Y., Yoder, A. D., Yamashita, N., and Li, W. H. (2005). Evidence from opsin genes rejects nocturnality in ancestral primates. *Proceedings of the National Academy of Sciences of the United States of America,* 102, 14712–16.

Tang, L., and Schwarzkopf, L. (2013). Foraging behaviour of the peaceful dove (*Geopelia striata*) in relation to predation risk: Group size and predator cues in a natural environment. *Emu,* 113, 1–7.

Tang, W. J., Fernandez, J. G., Sohn, J. J., and Amemiya, C. T. (2015). Chitin is endogenously produced in vertebrates. *Current Biology,* 25(7), 897–900.

Tanner, J. M. (1990). *Fetus into Man: Physical Growth from Conception to Maturity, Revised and Enlarged.* Harvard University Press.

Tao, N., Wu, S., Kim, J., An, H. J., Hinde, K., Power, M. L., et al. (2011). Evolutionary glycomics: Characterization of milk oligosaccharides in primates. *Journal of Proteome Research,* 10, 1548–57.

Tardif, S. D. (1994). Relative energetic cost of infant care in small-bodied Neotropical primates and its relation to infant-care patterns. *American Journal of Primatology,* 34, 133–43.

Tardif, S. D., and Jaquish, C. E. (1997). Number of ovulations in the marmoset monkey (*Callithrix jacchus*): Relation to body weight, age and repeatability. *American Journal of Primatology,* 42, 323–29.

Tardif, S., Power, M., Layne, D., Smucny, D., and Ziegler, T. (2004). Energy restriction initiated at different gestational ages has varying effects on maternal weight gain and pregnancy outcome in common marmoset monkeys (*Callithrix jacchus*). *British Journal of Nutrition,* 92, 841–49.

Tardif, S. D., Power, M., Oftedal, O. T., Power, R. A., and Layne, D. G. (2001). Lactation, maternal behavior and infant growth in common marmoset monkeys (*Callithrix jacchus*): Effects of maternal size and litter size. *Behavioral Ecology and Sociobiology,* 51, 17–25.

Tardif, S. D., Smucny, D. A., Abbott, D. H., Mansfield, K. G., Schultz-Darken, N. J., and Yamamoto, M. E. (2003). Reproduction in captive common marmosets (*Callithrix jacchus*). *Comparative Medicine*, 53, 364–68.

Tayek, J. A., and Blackburn, G. L. (1984). Goals of nutritional support in acute infections. *American Journal of Medicine*, 76(5A), 81–90.

Taylor, A. B. (2003). Ontogeny and function of the masticatory complex in *Gorilla*: Functional, evolutionary, and taxonomic implications. In A. B. Taylor and M. L. Goldsmith (Eds.), *Gorilla Biology: A Multidisciplinary Perspective* (pp. 132–93). Cambridge University Press.

Taylor, A. B. (2006). Feeding behavior, diet, and the functional consequences of jaw form in orangutans, with implications for the evolution of *Pongo*. *Journal of Human Evolution*, 50, 377–93.

Taylor, R. A., Ryan, S. J., Brashares, J. M., and Johnson, L. R. (2016). Hunting, food subsidies, and mesopredator release: The dynamics of crop-raiding baboons in a managed landscape. *Ecology*, 97, 951–60.

Teaford, M. F. (2007). What do we know and not know about diet and enamel structure? In P. S. Ungar (Ed.), *Evolution of the Human Diet: The Known, the Unknown, and the Unknowable.* (pp. 56–76). Oxford University Press.

Teaford, M. F., and Byrd, K. E. (1989). Differences in tooth wear as an indicator of changes in jaw movement in the guinea pig *Cavia porcellus*. *Archive of Oral Biology*, 34(12), 929–36.

Teaford, M. F., Lucas, P. W., Ungar, P. S., and Glander, K. E. (2006). Mechanical defenses in leaves eaten by Costa Rican howling monkeys (*Alouatta palliata*). *American Journal of Physical Anthropology*, 129, 99–104.

Teaford, M. F., Maas, M. C., and Simons, E. L. (1996). Dental microwear and microstructure in early Oligocene primates from the Fayum, Egypt: Implications for diet. *American Journal of Physical Anthropology*, 101(4), 527–43.

Teaford, M. F., and Oyen, O. J. (1988). In vivo and in vitro turnover in dental microwear. *American Journal of Physical Anthropology*, 75(2), 279–79.

Teaford, M. F., Ross, C. F., Ungar, P. S., Vinyard, C. J., and Laird, M. F. (2021). Grit your teeth and chew your food: Implications of food material properties and abrasives for rates of dental microwear formation in laboratory *Sapajus apella* (Primates). *Palaeogeography, Palaeoclimatology, Palaeoecology*, 583.110644.

Teaford, M. F., Ungar, P. S., and Grine, F. E. (2013). Dental microwear and paleoecology. In M. Sponheimer, J. A. Lee-Thorp, K. E. Reed, and P. S. Ungar (Eds.), *Early Hominin Paleoecology* (pp. 251–80). University of Colorado Press.

Teaford, M. F., Ungar, P. S., Taylor, A. B., Ross, C. F., and Vinyard, C. J. (2017). In vivo rates of dental microwear formation in laboratory primates fed different food items. *Biosurface and Biotribology*, 3(4), 166–73.

Teaford, M. F., Ungar, P. S., Taylor, A. B., Ross, C. F., and Vinyard, C. J. (2020). The dental microwear of hard-object feeding in laboratory *Sapajus apella* and its implications for dental microwear formation. *America Journal of Physical Anthropology*, 171(3), 439–55.

Teaford, M. F., and Walker, A. (1983). Dental microwear in adult and still-born guinea pigs (*Cavia porcellus*). *Archive of Oral Biology*, 28(11), 1077–81.

Teaford, M. F., and Walker, A. (1984). Quantitative differences in dental microwear between primate species with different diets and a comment on the presumed diet of *Sivapithecus*. *American Journal of Physical Anthropology*, 64(2), 191–200.

Tebbich, S., Taborsky, M., Fessl, B., and Dvorak, M. (2002). The ecology of tool-use in the woodpecker finch (*Cactospiza pallida*). *Ecology Letters*, 5, 656–64.

Teelen, S. (2007). Primate abundance along five transect lines at Ngogo, Kibale National Park, Uganda. *American Journal of Primatology*, 69, 1030–44.

Teelen, S. (2008). Influence of chimpanzee predation on the red colobus population at Ngogo, Kibale National Park, Uganda. *Primates*, 49, 41–49.

Teichroeb, J. A., and Sicotte, P. (2012). Cost-free vigilance during feeding in folivorous primates? Examining the effect of predation risk, scramble competition, and infanticide threat on vigilance in ursine colobus monkeys (*Colobus vellerosus*). *Behavioral Ecology and Sociobiology*, 66, 453–66.

Teichroeb, J. A., and Vining, A. Q. (2019). Navigation strategies in three nocturnal lemur species: Diet predicts heuristic use and degree of exploratory behavior. *Animal Cognition*, 22, 343–54.

Teichroeb, J. A., White, M. M. J., and Chapman, C. A. (2015). Vervet (*Chlorocebus pygerythrus*) intragroup spatial positioning: Dominants trade-off predation risk for increased food acquisition. *International Journal of Primatology*, 36, 154–76.

Teleki G. (1975). Primate subsistence patterns: Collector-predators and hunter-gatherers. *Journal of Human Evolution*, 4, 125–84.

Tello, J. G. (2003). Frugivores at a fruiting *Ficus* in south-eastern Peru. *Journal of Tropical Ecology*, 19, 717–21.

Tennie, C., Gilby, I. C., and Mundry, R. (2009). The meat-scrap hypothesis: Small quantities of meat may promote cooperative hunting in wild chimpanzees (*Pan troglodytes*). *Behavioral Ecology and Sociobiology*, 63, 421–31.

Tennie, C., O'Malley, R. C., and Gilby, I. C. (2014). Why do chimpanzees hunt? Considering the benefits and costs of acquiring and consuming vertebrate and invertebrate prey. *Journal of Human Evolution*, 71, 38–45.

Ter Hofstede, H. M., and Ratcliffe, J. M. (2016). Evolutionary escalation: The bat-moth arms race. *Journal of Experimental Biology*, 219, 1589–602.

Ter Steege, H., Pitman, N. C. A., Phillips, O. L., Chave, J., Sabatier, D., Duque, A., et al. (2006). Continental-scale patterns of canopy tree composition and function across Amazonia. *Nature*, 443, 444–47.

Terborgh, J. (1983). *Five new world primates: a study in comparative ecology*. Princeton University Press.

Terborgh, J. (1986). Keystone plant resources in the tropical forest. In M. Soulé (Ed.), *Conservation Biology: The Science of Scarcity and Diversity* (pp. 330–44). Sinauer and Associates.

Terborgh, J. (1986). The social systems of New World primates: An adaptionist view. In J. G. Else and P. C. Lee (Eds.), *Primate Ecology and Conservation* (pp. 199–211). Cambridge University Press.

Terborgh, J. (1987). Mixed flocks and troops—costs and benefits of polyspecific groups of birds and monkeys. *International Journal of Primatology*, 8, 453.

Terborgh, J. (1988). The big things that run the world—a sequel to E. O. Wilson. *Conservation Biology*, 2, 402–3.

Terborgh, J. (1990). Mixed flocks and polyspecific associations: costs and benefits of mixed groups to birds and monkeys. *American Journal of Primatology*, 21(2), 87–100.

Terborgh, J., and Goldizen, A. W. (1985). On the mating system of the cooperatively breeding saddle-backed tamarin (*Saguinus fuscicollis*). *Behavioral Ecology and Sociobiology*, 16, 293–99.

Terborgh, J., and Janson, C. H. (1986). The socioecology of primate groups. *Annual Review of Ecology and Systematics*, 17, 111–36.

Terborgh, J., and van Schaik, C. P. (1987). Convergence vs. nonconvergence in primate communities. In L. H. R. Gee and P. S. Giller (Eds.), *Organization of Communities, Past and Present* (pp. 205–26). Blackwell Scientific.

Terranova, C. J., and Coffman, B. S. (1997). Body weight of wild and captive lemurs. *Zoo Biology*, 16, 17–30.

Thackeray, J. F., Henzi, S. P., and Brain, C. (1996). Stable carbon and nitrogen isotope analysis of bone collagen in *Papio cynocephalus ursinus*: Comparison with ungulates and *Homo sapiens* from southern and East African environments. *South African Journal of Science*, 92, 209–12.

Thakkar, S. K., Giuffrida, F., Cristina, C. H., Castro, C. A., Mukherjee, R., Tran, L. A., et al. (2013). Dynamics of human milk nutrient composition of women from Singapore with a special focus on lipids. *American Journal of Human Biology*, 25, 770–79.

Thalmann, U. (2001). Food resource characteristics in two nocturnal lemurs with different social behavior: *Avahi occidentalis* and *Lepilemur edwardsi*. *International Journal of Primatology*, 22, 287–324.

Tham, D. M., Gardner, C. D., and Haskell, W. L. (1998). Clinical review 97—potential health benefits of dietary phytoestrogens: A review of the clinical, epidemiological, and mechanistic evidence. *Journal of Clinical Endocrinology and Metabolism*, 83, 2223–35.

Thatcher, H. R., Downs, C. T., and Koyama, N. F. (2020). Understanding foraging flexibility in urban vervet monkeys, *Chlorocebus pygerythrus*, for the benefit of human-wildlife coexistence. *Urban Ecosystems*, 23(6), 1349–57.

Theuerkauf, J., and Rouys, S. (2008). Habitat selection by ungulates in relation to predation risk by wolves and humans in the Bialowieza Forest, Poland. *Forest Ecology and Management*, 256, 1325–32.

Thierry, B., Iwaniuk, A. N., and Pellis, S. M. (2000). The influence of phylogeny on the social behaviour of macaques (Primates: Cercopithecidea, genus *Macaca*). *Ethology*, 106, 713–28.

Thiery, G., Gillet, G., Lazzari, V., Merceron, G., and Guy, F. (2017). Was *Mesopithecus* a seed eating colobine? Assessment of cracking, grinding and shearing ability using dental topography. *Journal of Human Evolution*, 112, 79–92.

Thiery, G., Guy, F., and Lazzari, V. (2019). La variabilité morphologique en paléoanthropologie: De nouvelles approches, de nouveaux enjeux? *Bulletins et Mémoires de la Société d'Anthropologie de Paris*, 31, 52–59.

Thomas, D. M., Bouchard, C., Church, T., Slentz, C., Kraus, W. E., Redman, L. M., et al. (2012). Why do individuals not lose more weight from an exercise intervention at a defined dose? An energy balance analysis. *Obesity Reviews*, 13, 835–47

Thomas, F., Renaud, F., Benefice, E., de Meeues, T., and Guegan, J. F. (2001). International variability of ages at menarche and menopause: Patterns and main determinants. *Human Biology*, 73, 271–90.

Thompson, C. L., Robl, N. J., Melo, L. C. d. O., Valença-Montenegro, M. M., Valle, Y. B. M., de Oliveira, M. A. B., et al. (2013). Spatial distribution and exploitation of trees gouged by common marmosets (*Callithrix jacchus*). *International Journal of Primatology*, 34, 65–85.

Thompson, D. B. A., and Barnard, C. J. (1984). Prey selection by plovers: Optimal foraging in mixed-species groups. *Animal Behaviour*, 32, 554–63.

Thompson, J. C., Carvalho, S., Marean, C. W., and Alemseged, Z. (2019). Origins of the human predatory pattern: The transition to large-animal exploitation by early hominins. *Current Anthropology*, 60, 1–23.

Thorén, S., Quietzsch, F., Schwochow, D., Sehen, L., Meusel, C., Meares, K., et al. (2011). Seasonal changes in feeding ecology and activity patterns of two sympatric mouse lemur species, the gray mouse lemur (*Microcebus murinus*) and the golden-brown mouse lemur (*M. ravelobensis*), in northwestern Madagascar. *International Journal of Primatology*, 32, 566–86.

Thorpe, S. K. S., Crompton, R. H., and Alexander, R. M. (2007). Orangutans use compliant branches to lower the energetic cost of locomotion. *Biology Letters*, 3(3), 253–56.

Thurau, E. G., Rahajanirina, A. N., and Irwin, M. T. (2021). Condensed tannins in the diet of folivorous diademed sifakas and the gap between crude and available protein. *American Journal of Primatology*, 83, e23239.

Tieszen, L. L., and Fagre, T. (1993). Effect of diet quality and composition on the isotopic composition of respiratory CO_2, bone collagen, bioapatite, and soft tissues. In J. B. Lambert and G. Grupe (Eds.), *Prehistoric Human Bone: Archaeology at the Molecular Level* (pp. 121–55). Springer.

Tilden, C. D., and Oftedal, O. T. (1997). Milk composition reflects pattern of material care in prosimian primates. *American Journal of Primatology*, 41, 195–211.

Tinkel, J., Hassanain, H., and Khouri, S. J. (2012). Cardiovascular antioxidant therapy: A review of supplements, pharmacotherapies, and mechanisms. *Cardiology in Review*, 20(2), 77–83.

Tipnee, S., Ramaraj, R., and Unpaprom, Y. (2015). Nutritional evaluation of edible freshwater green macroalga *Spirogyra varians*. *Emergent Life Sciences Research*, 1, 1–7.

Tkaczynski, P., MacLarnon, A., and Ross, C. (2014). Associations between spatial position, stress and anxiety in forest baboons *Papio anubis*. *Behavioural Processes*, 108, 1–6.

Tomasello, M., and Call, J. (1997). *Primate Cognition*. Oxford University Press.

Tombak, K. J., Wikberg, E. C., Rubenstein, D. I., and Chapman, C. A. (2019). Reciprocity and rotating social advantage among females in egalitarian primate societies. *Animal Behaviour*, 157, 189–200.

Tomkiewicz, S. M., Fuller, M. R., Kie, J. G., and Bates, K. K. (2010). Global positioning system and associated technologies in animal behaviour and ecological research. *Philosophical Transactions of the Royal Society of London Series B—Biological Sciences*, 365, 2163–76.

Tomlinson, S., Arnall, S. G., Munn, A., Bradshaw, S. D., Maloney, S. K., Dixon, K. W., et al. (2014). Applications and implications of ecological energetics. *Trends in Ecology & Evolution*, 29(5), 280–90.

Tomoko, K., Kuze, N., Bernard, H., Malim, T. P., and Kohshima, S. (2010). Feeding ecology of Bornean orangutans (*Pongo pygmaeus morio*) in Danum Valley, Sabah, Malaysia: A 3-year record including two mast fruitings. *American Journal of Primatology*, 72, 820–40.

Tonos, J., Razafindratsima, O. H., Fenosoa, Z. S. E., and Dunham, A. E. (2022). Individual-based networks reveal the highly skewed interactions of a frugivore mutualist with individual plants in a diverse community. *Oikos*, 2022(2).

Torigoe, T. (1985). Comparison of object manipulation among 74 species of non-human primates. *Primates*, 26, 182–94.

Torralvo, K., Rabelo, R. M., Andrade, A., and Botero-Arias, R. (2017). Tool use by Amazonian capuchin monkeys during predation on caiman nests in a high-productivity forest. *Primates*, 58, 279–83.

Torregrossa, A. M., and Dearing, M. D. (2009). Nutritional toxicology of mammals: Regulated intake of plant secondary compounds. *Functional Ecology*, 23, 48–56.

Torres de Assumpção, C. (1981). *Cebus apella* and *Brachyteles arachnoides* (Cebidae) as potential pollinators of *Mabea fistulifera* (Euphorbiaceae). *Journal of Mammalogy*, 62, 386–88.

Toth, A. L., and Robinson, G. E. (2005). Worker nutrition and division of labour in honeybees. *Animal Behaviour*, 69, 427–35.

Touitou, S., Heistermann, M., Schülke, O., and Ostner, J. (2021). The effect of reproductive state on activity budget, feeding behavior, and urinary C-peptide levels in wild female Assamese macaques. *Behavioral Ecology and Sociobiology*, 75(9), 1–22.

Tracy, B. F., and McNaughton, S. J. (1995). Elemental analysis of mineral lick soils from the Serengeti National Park, the Konza Prairie and Yellowstone National Park. *Ecography*, 18, 91–94.

Trapanese, C. (2018). *Spatial foraging in primates: strategies and mechanisms of decision-making* (Doctoral dissertation, Sorbonne Paris Cité).

Trapanese, C., Meunier, H., and Masi, S. (2019). What, where and when: Spatial foraging decisions in primates. *Biological Reviews*, 94, 483–502.

Trayford, H. R., and Farmer, K. H. (2012). An assessment of the use of telemetry for primate reintroductions. *Journal for Nature Conservation*, 20, 311–25.

Treves, A. (1999). Has predation shaped the social systems of arboreal primates? *Journal of Primatology*, 20, 35–67.

Treves, A., Drescher, A., and Snowdon, C. T. (2003). Maternal watchfulness in black howler monkeys (*Alouatta pigra*). *Ethology*, 109, 135–46.

Trimmer, C., Keller, A., Murphy, N. R., Snyder, L. L., Willer, J. R., Nagai, M. H., et al. (2019). Genetic variation across the human olfactory receptor repertoire alters odor perception. *Proceedings of the National Academy of Sciences*, 116(19), 9475–80.

Trisomboon, H., Malaivijitnond, S., Watanabe, G., and Taya, K. (2005). Ovulation block by *Pueraria mirifica*—a study of its endocrinological effect in female monkeys. *Endocrine*, 26, 33–39.

Trivers, R. L. (1972). Parental investment and sexual selection. In B. Campbell (Ed.), *Sexual Selection and the Descent of Man* (pp. 136–79). Aldine.

Trock, B. J., Hilakivi-Clarke, L., and Clarke, R. (2006). Meta-analysis of soy intake and breast cancer risk. *Journal of the National Cancer Institute*, 98, 459–71.

Troscianko, J., von Bayern, A. M. P., Chappell, J., Rutz, C., and Martin, G. R. (2012). Extreme binocular vision and a straight bill facilitate tool use in New Caledonian crows. *Nature Communications*, 3, 1110.

Truchet, S., and Honvo-Nouéto, E. (2017). Physiology of milk secretion. *Best Practice and Research Clinical Endocrinology and Metabolism*, 31, 367–84.

Trulsson, M., and Essick, G. K. (2010). Sensations evoked by microstimulation of single mechanoreceptive afferents innervating the human face and mouth. *Journal of Neurophysiology*, 72, 1734–44.

Tsao, R., and Deng, Z. Y. (2004). Separation procedures for naturally occurring antioxidant phytochemicals. *Journal of Chromatography B*, 812, 85–99.

Tsuji, Y., Hanya, G., and Grueter, C. C. (2013). Feeding strategies of primates in temperate and alpine forests: Comparison of Asian macaques and colobines. *Primates*, 54, 201–15.

Tsujino, R., and Yumoto, T. (2008). Topography-specific seed dispersal by Japanese macaques in a lowland forest on Yakushima Island, Japan. *Journal of Animal Ecology*, 78, 119–25.

Tsurim, I., Kotler, B. P., Gilad, A., Elazary, S., and Abramsky, Z. (2010). Foraging behavior of an urban bird species: Molt gaps, distance to shelter, and predation risk. *Ecology*, 91, 233–41.

Tsutsui, K., Otoh, M., Sakurai, K., Suzuki-Hashido, N., Hayakawa, T., Misaka, T., et al. (2016). Variation in ligand responses of the bitter taste receptors TAS2R1 and TAS2R4 among New World monkeys. *BMC Evolutionary Biology*, 16, 208.

Tubbs, C., Hartig, P., Cardon, M., Varga, N., and Milnes, M. (2012). Activation of southern white rhinoceros (*Ceratotherium simum simum*) estrogen receptors by phytoestrogens: Potential role in the reproductive failure of captive-born females? *Endocrinology*, 153, 1444–52.

Tubbs, C. W., Moley, L. A., Ivy, J. A., Metrione, L. C., LaClaire, S., Felton, R. G., et al. (2016). Estrogenicity of captive southern white rhinoceros diets and their association with fertility. *General and Comparative Endocrinology*, 238, 32–38.

Tucker, D. J., Wallis, I. R., Bolton, J. M., Marsh, K. J., Rosser, A. A., Brereton, I. M., et al. (2010). A metabolomic approach to identifying chemical mediators of mammal–plant interactions. *Journal of Chemical Ecology*, 36, 727–35.

Tuen, A. A., and Brown, J. S. (1996). Evaluating habitat suitability for tree squirrels in a suburban environment. *Malaysian Applied Biology*, 25, 1–8.

Tung, J., Archie, E. A., Altmann, J., and Alberts, S. C. (2016). Cumulative early life adversity predicts longevity in wild baboons. *Nature Communications*, 7, 11181.

Turchin, P. (1999). Population regulation: A synthetic view. *Oikos*, 84, 153–59.

Turnbaugh, P. J., Ridaura, V. K., Faith, J. J., Rey, F. E., Knight, R., and Gordon, H. A. (2009). The effect of diet on the human gut microbiome: A metagenomic analysis in humanized gnotobiotic mice. *Science Translational Medicine*, 1, 6ra14.

Turner, T. R., Danzy, C., J., Nisbett, A., and Gray, J. P. (2016). A comparison of adult body size between captive and wild vervet monkeys (*Chlorocebus aethiops sabaeus*) on the island of St. Kitts. *Primates*, 57, 211–20.

Tutin, C. E. G., and Fernandez, M. (1992). Insect-eating by sympatric lowland gorillas (*Gorilla g. gorilla*) and chimpanzees (*Pan t. troglodytes*) in the Lopé Reserve, Gabon. *American Journal of Primatology*, 28, 29–40.

Tutin, C. E. G., Fernandez, M., Rogers, M. E., Williamson, E. A., and McGrew, W. C. (1991). Foraging profiles of sympatric lowland gorillas and chimpanzees in the Lope Reserve, Gabon. *Philosophical Transactions of the Royal Society of London Series B—Biological Sciences*, 334, 179–86.

Tutin, C. E. G., Ham, R. M., White, L. J. T., and Harrison, M. J. S. (1997). The primate community of the Lope Reserve, Gabon: Diets, responses to fruit scarcity, and effects on biomass. *American Journal of Primatology*, 42, 1–24.

Tweheyo, M., Lye, K. A., and Weladji, R. B. (2004). Chimpanzee diet and habitat selection in the Budongo Forest Reserve, Uganda. *Forest Ecology and Management*, 188, 267–78.

Twigger, A. J., Hepworth, A. R., Lai, C. T., Chetwynd, E., Stuebe, A. M., Blancafort, P., et al. (2015). Gene expression in breast-milk cells is associated with maternal and infant characteristics. *Scientific Reports*, 10, 12933.

Uchida, A. (1998). Variation in tooth morphology of *Gorilla gorilla*. *Journal of Human Evolution*, 34(1), 55–70.

Uden, P., and Van Soest, P. J. (1982). The determination of digesta particle size in some herbivores. *Animal Feed Science and Technology*, 7, 35–44.

Uehara, M., Lapcik, O., Hampl, R., Al-Maharik, N., Makela, T., Wahala, K., et al. (2000). Rapid analysis of phytoestrogens in human urine by time-resolved fluoroimmunoassay. *Journal of Steroid Biochemistry*, 72, 273–82.

Uehara, S. (1982). Seasonal changes in the techniques employed by wild chimpanzees in the Mahale Mountains, Tanzania, to feed on termites (*Pseudacanthotermes spiniger*). *Folia Primatologica*, 37, 44–76.

Uehara, S. (1997). Predation on mammals by the chimpanzee (*Pan troglodytes*). *Primates*, 38, 193–214.

Uehara, S., Nishida, T., Hamai, M., Hasegawa, T., Hayaki, H., Huffman, M. A., et al. (1992). Characteristics of predation by the chimpanzees in the Mahale Mountains National Park, Tanzania. In T. Nishida, W. C. McGrew, P. Marler, M. Pickford, and F. B. M. de Waal (Eds.), *Topics in Primatology*, vol. 1, *Human Origins* (pp. 143–58). University of Tokyo Press.

Ulijaszek, S. J. (2002). Comparative energetics of primate fetal growth. *American Journal of Human Biology*, 14, 603–8.

UNEP. (2001). *An Assessment of the Status of the World's Remaining Closed Forests*.

Ungar, P. S. (1994a). Incisor microwear of Sumatran anthropoid primates. *American Journal of Physical Anthropology*, 94(3), 339–63.

Ungar, P. S. (1994b). Patterns of ingestive behavior and anterior tooth use differences in sympatric anthropoid primates. *American Journal of Physical Anthropology*, 95(2), 197–219.

Ungar, P. S. (1995). Fruit preferences of four sympatric primate species at Ketambe, northern Sumatra, Indonesia. *International Journal of Primatology*, 16, 221–45.

Ungar, P. S. (1996). Relationship of incisor size to diet and anterior tooth use in sympatric Sumatran anthropoids. *American Journal of Primatology*, 38(2), 145–56.

Ungar, P. S. (1998). Dental allometry, morphology, and wear as evidence for diet in fossil primates. *Evolutionary Anthropology*, 6(6), 205–17.

Ungar, P. S. (2004). Dental topography and diets of *Australopithecus afarensis* and early *Homo*. *Journal of Human Evolution*, 46(5), 605–22.

Ungar, P. S. (2005). Reproductive fitness and tooth wear: Milking as much as possible out of dental topographic analysis.

Proceedings of the National Academy of Sciences of the United States of America, 102(46), 16533–34.

Ungar, P. S. (2007a). *Dental Functional Morphology: The Known, the Unknown, and the Unknowable.* Oxford University Press.

Ungar, P. S. (2007b). Limits to knowledge on the evolution of hominin diet. In P. S. Ungar (Ed.), *Early Hominin Diets: The Known, the Unknown and the Unknowable* (pp. 395–408). Oxford University Press.

Ungar, P. S. (2008). Materials science: Strong teeth, strong seeds. *Nature*, 452, 703–5.

Ungar, P. S. (2009). Tooth form and function: Insights into adaptation through the analysis of dental microwear. In T. Koppe, G. Meyer, and K. W. Alt (Eds.), *Comparative Dental Morphology* (vol. 13, pp. 38–43). Karger.

Ungar, P. S. (2011). Dental evidence for the diets of Plio-Pleistocene hominins. *Yearbook of Physical Anthropology*, 54, 47–62.

Ungar, P. S. (2014). Dental allometry in mammals: A retrospective. *Acta Zoologica Fennici*, 51, 177–87.

Ungar, P. S. (2015). Mammalian dental function and wear: A review. *Biosurface and Biotribology*, 1, 25–41.

Ungar, P. S. (2017). *Evolution's Bite: A Story of Teeth, Diet, and Human Origins.* Princeton University Press.

Ungar, P. S. (2019). Inference of diets of early hominins from primate molar form and microwear. *Journal of Dental Research*, 98(4), 398–405.

Ungar, P. S., Brown, C. A., Bergstrom, T. S., and Walkers, A. (2003). Quantification of dental microwear by tandem scanning confocal microscopy and scale-sensitive fractal analyses. *Scanning*, 25(4), 185–93.

Ungar, P. S., and Bunn, J. M. (2008). Primate dental topographic analysis and functional morphology. In J. D. Irish and G. C. Nelson (Eds.), *Technique and Application in Dental Anthropology* (pp. 253–65). Cambridge University Press.

Ungar, P. S., Healy, C., Karme, A., Teaford, M. F., and Fortelius, M. (2018). Dental topography and diets of platyrrhine primates. *Historical Biology*, 30, 64-75.

Ungar, P. S., and Hlusko, L. J. (2016). The evolutionary path of least resistance. *Science*.6294, 29–30.

Ungar, P. S., and Lucas, P. W. (2010). Tooth form and function in biological anthropology. In C. S. Larsen (Ed.), *A Companion to Biological Anthropology.* (pp. 516–29). Wiley-Blackwell.

Ungar, P. S., and M'Kirera, F. (2003). A solution to the worn tooth conundrum in primate functional anatomy. *Proceedings of the National Academy of Sciences of the United States of America*, 100(7), 3874–77.

Ungar, P. S., Scott, J. R., and Steininger, C. M. (2016). Dental microwear and environments of Plio-Pleistocene bovids from southern and eastern Africa. *South African Journal of Science*, 3/4, 134–38.

Ungar, P. S., Scott, R. S., Scott, J. R., and Teaford, M. F. (2007). Dental microwear analysis: Historical perspectives and new approaches. In J. D. Irish and G. C. Nelson (Eds.), *Dental Anthropology* (pp. 389–425). Cambridge University Press.

Ungar, P. S., and Sponheimer, M. (2011). The diets of early hominins. *Science,* 334(6053), 190–93.

Ungar, P. S., Teaford, M. F., Glander, K. E., and Pastor, R. F. (1995). Dust accumulation in the canopy: A potential cause of dental microwear in primates. *American Journal of Physical Anthropology*, 97, 93–99.

Ungar, P. S., and Williamson, M. (2000). Exploring the effects of tooth wear on functional morphology: A preliminary study using dental topographic analysis. *Paleontologica Electronica*, 3, 1–18.

Urashima, T., Katayama, T., Sakanaka, M., Fukuda, K., and Messer, M. (2022). Evolution of milk oligosaccharides: Origin and selectivity of the ratio of milk oligosaccharides to lactose among mammals. *Biochimica et Biophysica Acta—General Subjects*, 1866(1), 130012.

Urbani, B. (1998). An early report on tool use by Neotropical primates. *Neotropical Primates*, 6, 123–24.

Urbani, B. (2009). *Spatial Mapping in Wild White-Faced Capuchin Monkeys (Cebus capucinus).* University of Illinois.

Urbano, F., Cagnacci, F., Calenge, C., Dettki, H., Cameron, A., and Neteler, M. (2010). Wildlife tracking data management: A new vision. *Philosophical Transactions of the Royal Society of London Series B—Biological Sciences*, 365, 2177–85.

Urlacher, S. S., Snodgrass, J. J., Dugas, L. R., Madimenos, F. C., Sugiyama, L. S., Liebert, M. A., et al. (2021). Childhood daily energy expenditure does not decrease with market integration and is not related to adiposity in Amazonia. *Journal of Nutrition*, 151(3), 695–704.

Urlacher, S. S., Snodgrass, J. J., Dugas, L. R., Sugiyama, L. S., Liebert, M. A., Joyce, C. J., et al. (2019). Constraint and trade-offs regulate energy expenditure during childhood. *Science Advances*, 5(12), eaax1065.

Uruakpa, F. O., Ismond, M. A., and Akobondu, E. N. (2002). Colostrum and its benefits: A review. *Nutrition Research*, 22, 755–67.

Uwimbabazi, M., Raubenheimer, D., Tweheyo, M., Basuta, G. I., Conklin-Brittain, N. L., Wrangham, R. W., et al. (2021). Nutritional geometry of female chimpanzees (*Pan troglodytes*). *American Journal of Primatology*, 83, e23269.

Uwimbabazi, M., Rothman, J. M., Basuta, G. I., Machanda, Z. P., Conklin-Brittain, N. L., and Wrangham, R. W. (2019). Influence of fruit availability on macronutrient and energy intake by female chimpanzees. *African Journal of Ecology*, 57, e12636.

Valeix, M., Loveridge, A. J., Chamaille-Jammes, S., Davidson, Z., Murindagomo, F., Fritz, H., et al. (2009). Behavioral adjustments of African herbivores to predation risk by lions: Spatiotemporal variations influence habitat use. *Ecology*, 90, 23–30.

Valenta, K., Brown, K. A., Rafaliarison, R. R., Styler, S. A., Jackson, D., Lehman, S. M., et al. (2015). Sensory integration during foraging: The importance of fruit hardness, colour, and odour to brown lemurs. *Behavioral Ecology and Sociobiology*, 69, 1855–65.

Valenta, K., Burke, R. J., Styler, S. A., Jackson, D. A., Melin, A. D., and Lehman, S. M. (2013). Colour and odour drive fruit selection and seed dispersal by mouse lemurs. *Scientific Reports*, 3, 2424.

Valenta, K., Nevo, O., Martel, C., and Chapman, C. A. (2017). Plant attractants: Integrating insights from pollination and seed dispersal ecology. *Evolution and Ecology*, 31, 249–67.

Valero, A., and Byrne, R. W. (2007). Spider monkey ranging patterns in Mexican subtropical forest: Do travel routes reflect planning? *Animal Cognition*, 10, 305–15.

Vamosi, J. C., and Wilson, J. R. U. (2008). Nonrandom extinction leads to elevated loss of angiosperm evolutionary history. *Ecology Letters*, 11, 1047–53.

Van Calsteren, K., de Catte, L., Devlieger, R., Chai, D. C., and Amant, F. (2009). Sonographic biometrical normograms and estimation of fetal weight in the baboon (*Papio anubis*). *Journal of Medical Primatology*, 38, 321–27.

van Casteren, A., Lucas, P. W., Strait, D. S., Michael, S., Bierwisch, N., Schwarzer, N., et al. (2018). Evidence that metallic proxies are unsuitable for assessing the mechanics of microwear formation and a new theory of the meaning of microwear. *Royal Society Open Science*, 5, 171699.

van Casteren, A., Oelze, V. M., Angedakin, S., Kalan, A. K., Kambi, M., Boesch, C., et al. (2018). Food mechanical properties and isotopic signatures in forest versus savannah dwelling eastern chimpanzees. *Communications Biology*, 1(1), 109.

van Casteren, A., Strait, D. S., Swain, M. V., Michael, S., Thai, L. A., Philip, S. M., et al., (2020). Hard plant tissues do not contribute meaningfully to dental microwear: Evolutionary implications. *Scientific Reports*, 10(1), 582.

van Casteren, A., Venkataraman, V., Ennos, A. R., and Lucas, P.W. (2016). Novel developments in field mechanics. *Journal of Human Evolution*, 98, 5-17

van Casteren, A., Wright, E., Kupczik, K., and Robbins, M. M. (2019). Unexpected hard-object feeding in western lowland *gorillas*. *American Journal of Physical Anthropology*, 170, 433–38.

Van der Bilt, A., Engelen, L., Pereira, L. J., van der Glas, H. W., and Abbink, J. H. (2006). Oral physiology and mastication. *Physiology and Behavior*, 89, 22–27.

Van der Merwe, M., and Brown, J. S. (2008). Mapping the landscape of fear of the cape ground squirrel (*Xerus inauris*). *Journal of Mammalogy*, 89, 1162–69.

Van der Merwe, N. J., and Medina, E. (1989). Photosynthesis and $^{13}C/^{12}C$ ratios in Amazonian rainforests. *Geochimica et Cosmochimica Acta*, 53(5), 1091–94.

Van der Merwe, N. J., and Medina, E. (1991). The canopy effect, carbon isotope ratios and foodwebs in Amazonia. *Journal of Archaeological Science*, 18(3), 249–59.

Van der Wel, H., Larson, G., Hladik, A., Hladik, C. M., Hellekant, G., and Glaser, D. (1989). Isolation and characterization of pentadin, the sweet principle of *Pentadiplandra brazzeana* Baillon. *Chemical Senses*, 14, 75–79.

Van Doorn, A. C., O'Riain, M. J., and Swedell, L. (2010). The effects of extreme seasonality of climate and day length on the activity budget and diet of semi-commensal chacma baboons (*Papio ursinus*) in the Cape Peninsula of South Africa. *American Journal of Primatology*, 72, 104–12.

Van Gils, J. A., and Tijsen, W. (2007). Short-term foraging costs and long-term fueling rates in central-place foraging swans revealed by giving-up exploitation times. *American Naturalist*, 169, 609–20.

Van Hooff, J. A. R. A. M., and van Schaik, C. P. (1994). Male bonds: Affiliative relationships among nonhuman primate males. *Behaviour*, 130, 309–37.

Van Horne, B. (1981). Demography of *Peromyscus maniculatus* populations in seral stages of coastal coniferous forest in southeast Alaska. *Canadian Journal of Zoology*, 59, 1045–61.

Van Horne, B. (1983). Density as a misleading indicator of habitat quality. *Journal of Wildlife Management*, 47, 893–901.

Van Lawick-Goodall, J. (1968). The behaviour of free-living chimpanzees in the Gombe Stream Reserve. *Animal Behaviour Monographs*, 1, 165–311.

Van Noordwijk, A. J., and de Jong, G. (1986). Acquisition and allocation of resources: Their influence on variation in life history tactics. *American Naturalist*, 128, 137–42.

Van Noordwijk, M. A., Utami Atmoko, S. S., Knott, C. D., Kuze, N., Morrogh-Bernard, H. C., Oram, F., et al. (2018). The slow ape: High infant survival and long interbirth intervals in wild orangutans. *Journal of Human Evolution*, 125, 38–49.

Van Noordwijk, M. A., and van Schaik, C. P. (1987). Competition among female long-tailed macaques, *Macaca fascicularis*. *Animal Behaviour*, 35, 577–89.

Van Noordwijk, M. A., and van Schaik, C. P. (1999). The effects of dominance rank and group size on female lifetime reproductive success in wild long-tailed macaques, *Macaca fascicularis*. *Primates*, 40, 105–30.

Van Noordwijk, M. A., Willems, E. P., Utami Atmoko, S. S., Kuzawa, C., and van Schaik, C. P. (2013). Multi-year lactation and its consequences in Bornean orangutans (*Pongo pygmaeus wurmbii*). *Behavioral Ecology and Sociobiology*, 67, 805–14.

Van Roosmalen, M. G. M., Mittermeier, R. A., and Fleagle, J. G. (1988). Diet of the northern bearded saki (*Chiropotes satanas chiropotes*): A Neotropical seed predator. *American Journal of Primatology*, 14, 11–35.

Van Schaik, C. P. (1983). Why are diurnal primates living in groups? *Behaviour*, 87, 120–44.

Van Schaik, C. P. (1986). Phenological changes in a Sumatran rainforest. *Journal of Tropical Ecology*, 2, 327–47.

Van Schaik, C. P. (1989). The ecology of social relationships amongst female primates. In V. Standen and R. A. Foley (Eds.), *Comparative Socioecology: The Behavioral Ecology of Humans and Other Mammals* (pp. 195–218). Blackwell.

Van Schaik, C. P. (1999). The socioecology of fission-fusion sociality in orangutans. *Primates*, 40, 69–86.

Van Schaik, C. P. (2000). Vulnerability to infanticide by males: Patterns among mammals. In C. P. van Schaik and C. H. Janson (Eds.), *Infanticide by Males and Its Implications* (pp. 61–71). Cambridge University Press.

Van Schaik, C. P., Ancrenaz, M., Borgen, G., Galdikas, B., Knott, C. D., Singleton, I., et al. (2003). Orangutan cultures and the evolution of material culture. *Science*, 299, 102–5.

Van Schaik, C. P., Ancrenaz, M., Djojoasmoro, R., Knott, C. D., Morrough-Bernard, H. C., Odom, N., et al. (2009). Orangutan cultures revisited. In S. A. Wich, S. S. Utami Atmoko, T. M. Setia, and C. P. van Schaik (Eds.), *Orangutans: Geographic Variation in Behavioral Ecology and Conservation* (pp. 299–309). Oxford University Press.

Van Schaik, C. P., Barrickman, N. L., Bastian, M. L., Krakauer, E. B., and van Noordwijk, M. A. (2006). Primate life histories and the role of brains. In K. Hawkes and R. R. Paine (Eds.), *The Evolution of Human Life History* (pp. 127–54). School of American Research Press.

Van Schaik, C. P., and Brockman, D. K. (2005). Seasonality in primate ecology, reproduction, and life history: An overview. In D. K. Brockman and C. P. van Schaik (Eds.), *Seasonality in Primates* (pp. 3–20). Cambridge University Press.

Van Schaik, C. P., Damerius, L., and Isler, K. (2013). Wild orangutan males plan and communicate their travel direction one day in advance. *PLoS ONE*, 8, e74896.

Van Schaik, C. P., Deaner, R. O., and Merrill, M. Y. (1999). The conditions for tool use in primates: Implications for the evolution of material culture. *Journal of Human Evolution*, 36, 719–41.

Van Schaik, C. P., Fox, E. A., and Sitompul, A. F. (1996). Manufacture and use of tools in wild Sumatran orangutans: Implications for human evolution. *Naturwissenschaften*, 83, 186–88.

Van Schaik, C. P., Isler, K., and Burkart, J. M. (2012). Explaining brain size variation: From social to cultural brain. *Trends in Cognitive Sciences*, 16, 278–84.

Van Schaik, C. P., and Knott, C. D. (2001). Geographic variation in tool use on *Neesia* fruits in orangutans. *American Journal of Physical Anthropology*, 114, 331–42.

Van Schaik, C. P., Laland, K. N., and Galef, B. G. (2009). Geographic variation in the behavior of wild great apes: Is it really cultural? In K. N. Laland and B. G. Galef (Eds.), *The Question of Animal Culture* (pp. 70–98). Harvard University Press.

Van Schaik, C. P., Marshall, A. J., and Wich, S. A. (2009). Geographic variation in orangutan behavior and biology. In S. A. Wich, S. S. Utami Atmoko, T. Mitra Setia, and C. P. van Schaik (Eds.), *Orangutans: Geographic Variation in Ecology and Conservation* (pp. 351–61). Oxford University Press.

Van Schaik, C. P., and Pfannes, K. R. (2005). Tropical climates and phenology: A primate perspective. In D. K. Brockman and C. P. van Schaik (Eds.), *Seasonality in Primates: Studies of Living and Extinct Human and Non-human Primates* (pp. 23–54). Cambridge University Press.

Van Schaik, C. P., and Pradhan, G. R. (2003). A model for tool-use traditions in primates: Implications for the coevolution of culture and cognition. *Journal of Human Evolution*, 44, 645–64.

Van Schaik, C. P., Priatna, A., and Priatna, D. (1995). Population estimates and habitat preferences of orangutans based on line transects of nests. In R. D. Nadler, B. F. Galdikas, L. K. Sheeran, and N. Rosen (Eds.), *Neglected Ape* (pp. 129–47). Plenum.

Van Schaik, C. P., Terborgh, J. W., and Wright, S. J. (1993). The phenology of tropical forests: Adaptive significance and consequences for primary consumers. *Annual Review of Ecology and Systematics*, 24, 353–77.

Van Schaik, C. P., and van Hooff, J. (1983). On the ultimate causes of primate social systems. *Behaviour*, 85, 1–2.

Van Schaik, C. P., and van Noordwijk, M. A. (1985). Interannual variability in fruit abundance and the reproductive seasonality in Sumatran long-tailed macaques (*Macaca fascicularis*). *Journal of Zoology, London*, 206, 533–49.

Van Schaik, C. P., and van Noordwijk, M. A. (1988). Scramble and contest in feeding competition among female long-tailed macaques (*Macaca fascicularis*). *Behaviour*, 105, 77–98.

Van Schaik, C. P., van Noordwijk, M. A., and Wich, S. (2006). Innovation in wild Bornean orangutans (*Pongo pygmaeus wurmbii*). *Behaviour*, 143, 837–76.

Van Soest, P. J. (1978). Dietary fibers: Their definition and nutritional properties. *American Journal of Clinical Nutrition*, 31, S12–S20.

Van Soest, P. J. (1994). *Nutritional Ecology of the Ruminant* (2nd ed.). Cornell University Press.

Van Soest, P. J. (2015). *The Detergent System for Analysis of Foods and Feeds*. Cornell University Press.

Van Soest, P. J., and Jones, L. H. P. (1968). Effect of silica in forages upon digestibility. *Journal of Dairy Science*, 51, 1644–48.

Van Soest, P. J., and Mason, V. C. (1991). The influence of the Maillard reaction upon the nutritive value of fibrous feeds. *Animal Feed Science and Technology*, 32, 45–53.

Vance, C. K., Tolleson, D. R., Kinoshita, K., Rodriguez, J., and Foley, W. J. (2016). Near infrared spectroscopy in wildlife and biodiversity. *Journal of Near Infrared Spectroscopy*, 24, 1–25.

Vandeleest, J. J., Beisner, B. A., Hannibal, D. L., Nathman, A. C., Capitanio, J. P., Hsieh, F., et al. (2016). Decoupling social status and status certainty effects on health in macaques: A network approach. *PeerJ*, 4, e2394.

Vandercone, R. P., Dinadh, C., Wijethunga, G., Ranawana, K., and Rasmussan, D. T. (2012). Dietary diversity and food selection in Hanuman langurs (*Semnopithecus entellus*) and purple-faced langurs (*Trachypithecus vetulus*) in the Kaludiyapokuna Forest Reserve in the dry zone of Sri Lanka. *International Journal of Primatology*, 33, 1382–405.

vanWieren, S. E. (1996). Do large herbivores select a diet that maximizes short-term energy intake rate? *Forest Ecology and Management*, 88, 149–56.

Varpe, O. (2017). Life history adaptations to seasonality. *Integrative and Comparative Biology*, 57, 943–60.

Vasey, N. (2000). Niche separation in *Varecia variegata rubra* and *Eulemur fulvus albifrons*: I. Interspecific patterns. *American Journal of Physical Anthropology*, 112, 411–31.

Vasey, N. (2005). Activity budgets and activity rhythms in red ruffed lemurs (*Varecia rubra*) on the Masoala Peninsula, Madagascar: Seasonality and reproductive energetics. *American Journal of Primatology*, 66, 23–44.

Vasey, N. (2006). Impact of seasonality and reproduction on social structure, ranging patterns, and fission-fusion social organization in red ruffed lemurs. In L. Gould and M. L. Sauther (Eds.), *Lemurs—Developments in Primatology: Progress and Prospect* (pp. 275–304). Springer.

Vehrencamp, S. L. (1983). A model for the evolution of despotic versus egalitarian societies. *Animal Behaviour*, 31, 667–82.

Veilleux, C. C. (2020). Seeing in the dark: visual function and ecology of lorises and pottos. In: Nekaris, K. A. I., and Burrows, A. M. (Eds.). (2020). *Evolution, Ecology and Conservation of Lorises and Pottos* (pp. 174–86). Cambridge University Press.

Veilleux, C. C., Dominy, N. J., and Melin, A. D. (2022). The sensory ecology of primate food perception, revisited. *Evolutionary Anthropology*, 31, 281–301.

Veilleux, C. C., and Bolnick, D. A. (2009). Opsin gene polymorphism predicts trichromacy in a cathemeral lemur. *American Journal of Primatology*, 71, 86–90.

Veilleux, C. C., Garrett, C. E., Pajic, P., Saitou, M., Ochieng, J., Dagsaan, L. D., et al. (2023). Human subsistence and signatures of selection on chemosensory genes. *Communications Biology*, 6, 1–12.

Veilleux, C. C., Jacobs, R. L., Cummings, M. E., Louis, E. E., and Bolnick, D. A. (2014). Opsin genes and visual ecology in a nocturnal folivorous lemur. *International Journal of Primatology*, 35, 88–107.

Veilleux, C. C., Kawamura, S., Montague, M. J., Hiwatashi, T., Matsushita, Y., Fernandez-Duque, E., et al. (2021). Color vision and niche partitioning in a diverse neotropical primate community in lowland Amazonian Ecuador. *Ecology and Evolution*, 11(10), 5742–58.

Veilleux, C. C., and Kirk, E. C. (2009). Visual acuity in the cathemeral strepsirrhine *Eulemur macaco flavifrons*. *American Journal of Primatology*, 71, 343–52.

Veilleux, C. C., and Kirk, E. C. (2014). Visual acuity in mammals: Effects of eye size and ecology. *Brain, Behavior and Evolution*, 83, 43–53.

Veilleux, C. C., Louis, E. E., and Bolnick, D. A. (2013). Nocturnal light environments influence color vision and signatures of selection on the *OPN1SW* opsin gene in nocturnal lemurs. *Molecular Biology and Evolution*, 30, 1420–37.

Veilleux, C. C., Scarry, C. J., Di Fiore, A., Kirk, E. C., Bolnick, D. A., and Lewis, R. J. (2016). Group benefit associated with polymorphic trichromacy in a Malagasy primate (*Propithecus verreauxi*). *Scientific Reports*, 6, srep38418.

Venable, E. M., Machanda, Z., Hagberg, L., Lucore, J., Otali, E., Rothman, J. M., et al. (2020). Wood and meat as complementary sources of sodium for Kanyawara chimpanzees *American Journal of Physical Anthropology*, 172, 41–47.

Venkataraman, V. V., Kraft, T. S., Dominy, N. J., and Endicott, K. M. (2017). Hunter-gatherer residential mobility and the marginal value of rainforest patches. *Proceedings of the National Academy of Sciences of the United States of America*, 114, 3097–102.

Venter, O., Sanderson, E. W., Magrach, A., Allan, J. R., Beher, J., Jones, K. R., et al. (2016). Sixteen years of change in the global terrestrial human footprint and implications for biodiversity conservation. *Nature Communications*, 7, 12558.

Verdolin, J. L. (2006). Meta-analysis of foraging and predation risk trade-offs in terrestrial systems. *Behavioral Ecology and Sociobiology*, 60, 457–64.

Verendeev, A., Thomas, C., McFarlin, S. C., Hopkins, W. D., Phillips, K. A., and Sherwood, C. C. (2015). Comparative analysis of Meissner's corpuscles in the fingertips of primates. *Journal of Anatomy*, 227, 72–80.

Verhulst, P. E. (1838). Notice sur la loi que la population suit dans accroissement. *Correspondance Mathématique et Physique*, 10, 113–21.

Vickery, W. L., Giraledeau, L. A., Templeton, J. J., Kramer, D. L., and Chapman, C. A. (1991). Producers, scrounger, and group foraging. *American Naturalist*, 137, 847–63.

Vicsek, T., Czirok, A., Ben-Jacob, E., Cohen, I., and Shochet, O. (1995). Novel type of phase transition in a system of self-driven particles. *Physical Review Letters*, 75, 1226.

Victora, C. G., Bahl, R., Barros, A. J., França, G. V., Horton, S., Krasevec, J., et al. (2016). Breastfeeding in the 21st century: Epidemiology, mechanisms, and lifelong effect. *Lancet*, 387, 475–90.

Videan, E. N., Fritz, J., Schwandt, M., and Howell, S. (2005). Neighbor effect: Evidence of affiliative and agonistic social contagion in captive chimpanzees (*Pan troglodytes*). *American Journal of Primatology*, 68, 131–44.

Vidueiros, S. M., Fernandez, I., Slobodianik, N., Roux, M. E., and Pallaro, A. (2008). Nutrition disorder and immunologic parameters: Study of the intestinal villi in growing rats. *Nutrition*, 24(6), 575–81.

Vijay-Kumar, M., Aitken, J. D., Carvalho, F. A., Cullender, T. C., Mwangi, S., Srinivasan, S., et al. (2010). Metabolic syndrome and altered gut microbiota in mice lacking Toll-like Receptor 5. *Science*, 328(5975), 228–31.

Villalba, J. J., and Provenza, F. D. (2005). Foraging in chemically diverse environments: Energy, protein, and alternative foods influence ingestion of plant secondary metabolites by lambs. *Journal of Chemical Ecology*, 31, 123–38.

Villalba, J. J., and Provenza, F. D. (2007). Self-medication and homeostatic behaviour in herbivores: Learning about the benefits of nature's pharmacy. *Animal*, 1, 1360–70.

Villard, M. A., and Part, T. (2004). Don't put all your eggs in real nests: A sequel to Faaborg. *Conservation Biology*, 18, 371–72.

Vincent, J. F. V., Saunders, D. E. J., and Beyts, P. (2002). The use of critical stress intensity factor to quantify "hardness" and "crunchiness" objectively. *Journal of Texture Studies*, 33, 149–59.

Vinjamuri, A., Davis, J. C., Totten, S. M., Wu, L. D., Klein, L. D., Martin, M., et al. (2022). Human milk oligosaccharide compositions illustrate global variations in early nutrition. *The Journal of Nutrition*, 152(5), 1239–53.

Vinyard, C. J., and Hanna, J. (2005). Molar scaling in strepsirrhine primates. *Journal of Human Evolution*, 49(2), 241–69.

Virginia, R. A., and Delwiche, C. C. (1982). Natural ^{15}N abundance of presumed N_2-fixing and non-N_2-fixing plants from selected ecosystems. *Oecologia*, 54, 317–25.

Visalberghi, E., Fragaszy, D., Izar, P., and Ottoni, E. B. (2006). Tool use in wild bearded capuchin monkeys (*Cebus libidinosus*): New findings and hypotheses. *Folia Primatologica*, 77, 276–77.

Visalberghi, E., Fragaszy, D., Ottoni, E. B., Izar, P., de Oliveira, M. G., and Andrade, F. R. D. (2007). Characteristics of hammer stones and anvils used by wild bearded capuchin monkeys (*Cebus libidinosus*) to crack open palm nuts. *American Journal of Physical Anthropology*, 132, 426–44.

Visalberghi, E., Sabbatini, G., Spagnoletti, N., Andrade, F. R. D., Ottoni, E., Izar, P., et al. (2008). Physical properties of palm fruits processed with tools by wild bearded capuchins (*Cebus libidinosus*). *American Journal of Primatology*, 70, 884–91.

Visconti, A., LeRoy, C. I., Rosa, F., Rossi, N., Martin, T. C., Mohney, R. P., et al. (2019). Interplay between the human gut microbiome and host metabolism. *Nature Communications*, 10, 4505.

Visser, M. E., and Both, C. (2005). Shifts in phenology due to global climate change: The need for a yardstick. *Proceedings of the Royal Society B: Biological Sciences*, 272, 2561–69.

Vivar, O. I., Saunier, E. F., Leitman, D. C., Firestone, G. L., and Bjeldanes, L. F. (2010). Selective activation of estrogen receptor-beta target genes by 3,3′-diindolylmethane. *Endocrinology*, 151, 1662–67.

Voet, D., Voet, J. G., and Pratt, C. W. (2016). *Fundamentals of biochemistry: life at the molecular level*. John Wiley & Sons.

Vogel, E. R. (2005). Rank differences in energy intake rates in white-faced capuchin monkeys, *Cebus capucinus*: The effects of contest competition. *Behavioral Ecology and Sociobiology*, 58, 333–44.

Vogel, E. R., Alavi, S. E., Utami Atmoko, S. S., van Noordwijk, M. A., Bransford, T. D., Erb, W., et al. (2016). Nutritional ecology of wild Bornean orangutans (*Pongo pygmaeus wurmbii*) in a peat swamp habitat: Effects of age, sex, and season. *American Journal of Primatology*, 79, 1–20.

Vogel, E. R., Crowley, B. E., Knott, C. D., Blakely, M. D., Larsen, M. D., and Dominy, N. J. (2012). A noninvasive method for estimating nitrogen balance in free-ranging primates. *International Journal of Primatology*, 33, 567–87.

Vogel, E., and Dominy, N. J. (2011). Ecological methods. In C. J. Campbell, A. Fuentes, K. C. MacKinnon, S. K. Bearder, and R. M. Stumpf (Eds.), *Primates in Perspective* (pp. 367–76). Oxford University Press.

Vogel, E. R., Haag, L., Mitri-Setia, T., van Schaik, C. P., and Dominy, N. J. (2009). Foraging and ranging behavior during a fallback episode: *Hylobates albibarbis* and *Pongo pygmaeus wurmbii* compared. *American Journal of Physical Anthropology*, 140, 716–26.

Vogel, E. R., Harrison, M. E., Zulfa, A., Bransford, T. D., Alavi, S. E., Husson, S., et al. (2015). Nutritional differences between two orangutan habitats: Implications for population density. *PLoS ONE*, 10(10), e0138612.

Vogel, E. R., and Janson, C. H. (2007). Predicting the frequency of food-related agonism in white-faced capuchin monkeys (*Cebus capucinus*), using a novel focal-tree method. *American Journal of Primatology*, 69, 533–50.

Vogel, E. R., and Janson, C. H. (2011). Quantifying primate food distribution and abundance for socioecological studies: An objective consumer-centered method. *International Journal of Primatology*, 32, 737–54.

Vogel, E. R., Knott, C. D., Crowley, B. E., Blakely, M. D., Larsen, M. D., and Dominy, N. J. (2012). Bornean orangutans on the brink of protein bankruptcy. *Biology Letters*, 8, 333–36.

Vogel, E. R., Munch, S. B., and Janson, C. H. (2007). Understanding escalated aggression over food resources in white-faced capuchin monkeys. *Animal Behaviour*, 74, 71–80.

Vogel, E. R., Naumenko, D. J., Bransford, T. D., and Atmoko, S. S. U. (2019). The benefits of negative energy balance? Oxidative stress and inflammation in wild Bornean orangutans (*Pongo pygmaeus wurmbii*). *American Journal of Physical Anthropology*, 168, 259–59.

Vogel, E. R., Neitz, M., and Dominy, N. J. (2007). Effect of color vision phenotype on the foraging of wild white-faced capuchins, *Cebus capucinus*. *Behavioural Ecology*, 18, 292–97.

Vogel, E. R., Rothman, J. M., Moldawer, A. M., Bransford, T. D., Emery Thompson, M. E., van Noordwijk, M. A., et al. (2015). Coping with a challenging environment: Nutritional balancing, health, and energetics in wild Bornean orangutans. *American Journal of Physical Anthropology*, 156, 314–15.

Vogel, E. R., van Woerden, J. T., Lucas, P. W., Atmoko, S. S. U., van Schaik, C. P., and Dominy, N. J. (2008). Functional ecology and evolution of hominoid molar enamel thickness: *Pan troglodytes schweinfurthii* and *Pongo pygmaeus wurmbii*. *Journal of Human Evolution*, 55(1), 60–74.

Vogel, E. R., Zulfa, A., Hardus, M., Wich, S. A., Dominy, N. J., and Taylor, A. B. (2014). Food mechanical properties, feeding ecology, and the mandibular morphology of wild orangutans. *Journal of Human Evolution*, 75, 110–24.

Vogel, J. C., and van der Merwe, N. J. (1977). Isotopic evidence for early maize cultivation in New York State. *American Antiquity*, 42, 238–42.

Vorbach, C., Capecchi, M. R., and Penninger, J. M. (2006). Evolution of the mammary gland from the innate immune system? *Bioessays*, 28, 606–16.

Wachter, B., Schabel, M., and Noë, R. (1997). Diet overlap and poly-specific associations of red colobus and Diana monkeys in the Taï National Park, Ivory Coast. *Ethology*, 103, 514–26.

Waga, I. C., Dacier, A. K., Pinha, P. S., and Tavares, M. C. H. (2006). Spontaneous tool use by wild capuchin monkeys (*Cebus libidinosus*) in the Cerrado. *Folia Primatologica*, 77, 337–44.

Wagner, I., Ganzhorn, J. U., Kalko, E. K. V., and Tschapka, M. (2015). Cheating on the mutualistic contract: Nutritional gain through seed predation in the frugivorous bat *Chiroderma villosum* (Phyllostomidae). *Journal of Experimental Biology*, 218, 1016–21.

Waite, T. A., Chhangani, A. K., Campbell, L. G., Rajpurohit, L. S., and Mohnot, S. M. (2007). Sanctuary in the city: Urban monkeys buffered against catastrophic die-off during ENSO-related drought. *EcoHealth*, 4, 278–86.

Wakefield, M. L., Hickmott, A. J., Brand, C. M., Takaoka, I. Y., Meador, L. M., Waller, M. T., et al. (2019). New observations of meat eating and sharing in wild bonobos (*Pan paniscus*) at Iyema, Lomako Forest Reserve, Democratic Republic of the Congo. *Folia Primatologica*, 90(3), 179–89.

Walczyk, T., and Von Blanckenburg, F. (2002). Natural iron isotope variations in human blood. *Science*, 295, 2065–66.

Walker, A. (2007). Early hominin diets: Overview and historical perspective. In P. S. Ungar (Ed.), *Evolution of the Human Diet: The Known, the Unknown, and the Unknowable* (pp. 3–10). Oxford University Press.

Walker, A., Hoeck, H. N., and Perez, L. (1978). Microwear of mammalian teeth as an indicator of diet. *Science*, 201(4359), 908–10.

Wallace, F. A., Miles, E. A., Evans, C., Stock, T. E., Yaqoob, P., and Calder, P. C. (2001). Dietary fatty acids influence the production of Th1-but not Th2-type cytokines. *Journal of Leukocyte Biology*, 69(3), 449–57.

Wallace, G. E., and Hill, C. M. (2012). Crop damage by primates: Quantifying the key parameters of crop-raiding events. *PLoS ONE*, 7, e46636.

Wallace, R. B. (2005). Seasonal variations in diet and foraging behavior of *Ateles chamek* in a southern Amazonian tropical forest. *International Journal of Primatology*, 26, 1053–75.

Wallis, I. R., and Goldingay, R. L. (2014). Does a sap feeding marsupial choose trees with specific chemical characteristics? *Austral Ecology*, 39, 973–83.

Wallis, I. R., Edwards, M. J., Windley, H., Krockenberger, A. K., Felton, A. M., Quenzer, M., et al. (2012). Food for folivores: Nutritional explanations linking diets to population density. *Oecologia*, 169, 281–91.

Wallis, J. (1997). A survey of reproductive parameters in the free-ranging chimpanzees of Gombe National Park. *Journal of Reproduction and Fertility*, 109, 297–307.

Wallis, S. J. (1983). Sexual behavior and reproduction of *Cercocebus albigena johnstonii* in Kibale Forest, West Uganda. *International Journal of Primatology*, 4, 153–66.

Walrand, S., Moreau, K., Caldefie, F., Tridon, A., Chassagne, J., Portefaix, G., et al. (2001). Specific and nonspecific immune responses to fasting and refeeding differ in healthy young adult and elderly persons. *American Journal of Clinical Nutrition*, 74(5), 670–78.

Walsh, C. M., Bautista, D. M., and Lumpkin, E. A. (2015). Mammalian touch catches up. *Current Opinion in Neurobiology*, 34, 133–39.

Walsh, P. D., Abernethy, K. A., Bermejo, M., Beyers, R., De Wachter, P., Akou, M. E., et al. (2003). Catastrophic ape decline in western equatorial Africa. *Nature*, 422, 611–14.

Walter, J., and Ley, R. (2011). The human gut microbiome: ecology and recent evolutionary changes. *Annual Review of Microbiology*, 65, 411–29.

Wan, J. M.-F., Haw, M. P., and Blackburn, G. L. (1989). Nutrition, immune function, and inflammation: An overview. *Proceedings of the Nutrition Society*, 48(03), 315–35.

Wang, G., Hobbs, N. T., Boone, R. B., Illius, A. W., Gordon, I. J., Gross, J. E., et al. (2006). Spatial and temporal variability modify density dependence in populations of large herbivores. *Ecology*, 87, 95–102.

Wang, S., Steiniche, T., Romanak, K. A., Johnson, E., Quirós, R., Mutegeki, R., et al. (2019). Atmospheric occurrence of legacy pesticides, current use pesticides, and flame retardants in and around protected areas in Costa Rica and Uganda. *Environmental Science and Technology*, 53(11), 6171–81.

Wang, S., Steiniche, T., Rothman, J. M., Wrangham, R. W., Chapman, C. A., Mutegeki, R., et al. (2020). Feces are effective biological samples for measuring pesticides and flame retardants in primates. *Environmental Science and Technology*, 54(19), 12013–23.

Wang, S., Tu, H., Wan, J., Chen, W., Liu, X., Luo, J., et al. (2016). Spatio-temporal distribution and natural variation of metabolites in citrus fruits. *Food Chemistry*, 199, 8–17.

Wang, X., Bowyer, K. P., Porter, R. R., Breneman, C. B., and Custer, S. S. (2017). Energy expenditure responses to exercise training in older women. *Physiological Reports*, 5(15), e13360.

Wang, X. L., Rainwater, D. L., Mahaney, M. C., and Stocker, R. (2004). Cosupplementation with vitamin E and coenzyme Q10 reduces circulating markers of inflammation in baboons. *American Journal of Clinical Nutrition*, 80(3), 649–55.

Wang, Z., Heshka, S., Zhang, K., Boozer, C. N., and Heymsfield, S. B. (2001). Resting energy expenditure: systematic organization and critique of prediction methods. *Obesity Research*, 9(5), 331–36.

Ward, R. E., Ninonuevo, M., Mills, D. A., Lebrilla, C. B., and German, J. B. (2006). In vitro fermentation of breast milk oligosaccharides by *Bifidobacterium infantis* and *Lactobacillus gasseri*. *Applied and Environmental Microbiology*, 72, 4497–99.

Wardhaugh, C. W. (2014). The spatial and temporal distributions of arthropods in forest canopies: Uniting disparate patterns with hypotheses for specialisation. *Biological Reviews*, 89, 1021–41.

Warren, R. D., and Crompton, R. H. (1997). A comparative study of the ranging behaviour, activity rhythms and sociality of *Lepilemur edwardsi* (Primates, Lepilemuridae) and *Avahi occidentalis* (Primates, Indriidae) at Ampijoroa, Madagascar. *Journal of Zoology*, 243, 397–415.

Warren, Y., Higham, J. P., Maclarnon, A. M., and Ross, C. (2011). Crop-raiding and commensalism in olive baboons: The costs and benefits of living with humans. In V. Sommer and C. Ross (Eds.), *Primates of Gashaka* (pp. 359–84). Springer.

Waser, P. M. (1977). Feeding, ranging and group size in the mangabey *Cercocebus albigena*. In T. H. Clutton-Brock (Ed.), *Primate Ecology: Studies of Feeding And Ranging Behavior in Lemurs, Monkeys and Apes* (pp. 183–222). Academic Press.

Waser, P. M. (1987). Interactions among primate species. In B. B. Smuts, D. L. Cheney, R. M. Seyfarth, R. W. Wrangham, and T. T. Struhsaker (Eds.), *Primate Societies* (pp. 210–26). University of Chicago Press.

Wasser, S. K., and Barash, D. P. (1983). Reproductive suppression among female mammals: Implications for biomedicine and sexual selection theory. *Quarterly Review of Biology*, 58, 513–38.

Wasser, S. K., and Starling, A. K. (1988). Proximate and ultimate causes of reproductive suppression among female yellow baboons at Mikumi National Park, Tanzania. *American Journal of Primatology*, 16, 97–121.

Wasserman, M. D., and Chapman, C. A. (2003). Determinants of colobine monkey abundance: The importance of food energy, protein and fibre content. *Journal of Animal Ecology*, 72, 650–59.

Wasserman, M. D., Chapman, C. A., Milton, K., Gogarten, J. F., Wittwer, D. J., and Ziegler, T. E. (2012). Estrogenic plant consumption predicts red colobus monkey (*Procolobus rufomitratus*) hormonal state and behavior. *Hormones and Behavior*, 62, 553–62.

Wasserman, M. D., Chapman, C. A., Milton, K., Goldberg, T. L., and Ziegler, T. E. (2013). Physiological and behavioral effects of capture darting on red colobus monkeys (*Procolobus rufomitratus*) with a comparison to chimpanzee (*Pan troglodytes*) predation. *International Journal of Primatology*, 34, 1020–31.

Wasserman, M. D., Milton, K., and Chapman, C. A. (2013). The roles of phytoestrogens in primate ecology and evolution. *International Journal of Primatology*, 34, 861–78.

Wasserman, M. D., Taylor-Gutt, A., Rothman, J. M., Chapman, C. A., Milton, K., and Leitman, D. C. (2012). Estrogenic plant foods of red colobus monkeys and mountain gorillas in Uganda. *American Journal of Physical Anthropology*, 148, 88–97.

Watanabe, K. (1981). Variations in group composition and population density of the two sympatric Mentawaian leaf-monkeys. *Primates*, 22, 145–60.

Waterman, P. G. (1984). Food acquisition and processing as a function of plant chemistry. In D. J. Chivers, B. A. Wood, and A. Bilsborough (Eds.), *Food Acquisition and Processing in Primates* (pp. 177–211). Springer US.

Waterman, P. G., and McKey, D. (1989). Herbivory and secondary compounds in rain-forest plants. In H. Lieth and M. J. A. Werger (Eds.), *Ecosystems of the World 14B: Tropical Rain Forest Ecosystems—Biogeographical and Ecological Studies* (pp. 513–36). Elsevier.

Waterman, P. G., Ross, J. A. M., and McKey, D. B. (1984). Factors affecting the levels of some phenolic compounds, digestibility and nitrogen content of the mature leaves of *Barteria fistulosa* (Passifloraceae). *Journal of Chemical Ecology*, 10, 387–401.

Waterman, P. G., Ross, J. A. M., Bennett, E. L., and Davies, A. G. (1988). A comparison of the floristics and leaf chemistry of the tree flora in two Malaysian rain forests and the influence of leaf chemistry on populations of colobine monkeys in the Old World. *Biological Journal of the Linnean Society*, 34, 1–32.

Waters, S., Watson, T., Bell, S., and Setchell, J. M. (2018). Communicating for conservation: Circumventing conflict with communities over domestic dog ownership in north Morocco. *European Journal of Wildlife Research*, 64, (69), 2018.

Watts, D. P. (1984). Composition and variability of mountain gorilla diets in the central Virungas. *American Journal of Primatology*, 7, 323–56.

Watts, D. P. (1998). Long-term habitat use by mountain gorillas (*Gorilla gorilla beringei*). 2. Reuse of foraging areas in relation to resource abundance, quality and depletion. *International Journal of Primatology*, 19, 681–702.

Watts, D. P. (2008). Tool use by chimpanzees at Ngogo, Kibale National Park, Uganda. *International Journal of Primatology*, 29, 83–94.

Watts, D. P. (2012a). The apes: Taxonomy, biogeography, life history, and behavioral ecology. In J. C. Mitani, J. Call, P. M. Kappeler, R. A. Palombit, and J. B. Silk (Eds.), *The Evolution of Primate Societies* (pp. 113–42). University of Chicago Press.

Watts, D. P. (2012b). Long-term research on chimpanzee behavioral ecology in Kibale National Park, Uganda. In P. M. Kappeler and D. P. Watts (Eds.), *Long-Term Field Studies of Primates* (pp. 313–38). Springer Berlin Heidelberg.

Watts, D. P. (2020). Meat-eating by nonhuman primates: A review and synthesis. *Journal of Human Evolution*, 149, 102882.

Watts, D. P., and Amsler, S. A. (2013). Chimpanzee-red colobus encounter rates show a red colobus population decline associated with predation by chimpanzees at Ngogo. *American Journal of Primatology*, 75, 927–37.

Watts, D. P., and Mitani, J. C. (2002a). Hunting and meat sharing by chimpanzees at Ngogo, Kibale National Park, Uganda. In G. Hohmann and C. Boesch (Eds.), *Behavioral Diversity in Chimpanzees and Bonobos* (pp. 244–58). Cambridge University Press.

Watts, D. P., and Mitani, J. C. (2002b). Hunting by chimpanzees at Ngogo, Kibale National Park, Uganda. *International Journal of Primatology*, 23, 1–29.

Watts, D. P., and Mitani, J. C. (2015). Hunting and prey switching by chimpanzees (*Pan troglodytes schweinfurthii*) at Ngogo. *International Journal of Primatology*, 36, 728–48.

Watts, N., Amann, M., Arnell, N., Ayeb-Karlsson, S., Beagley, J., Belesova, K., et al. (2021). The 2020 report of the *Lancet* Countdown on health and climate change: Responding to converging crises. *Lancet*, 397(10269), 129–70.

Weckle, A., Hou, Z. C., Chen, C. Y., Xing, J., Sterner, K. N., Baker, J. L., et al. (2012). Adaptive evolution and ancestral resurrection of anthropoid estrogen receptor beta. *American Journal of Physical Anthropology*, 147, 299.

Wehncke, E. V., Hubbell, S. P., Fosters, R. B., and Dalling, J. W. (2003). Seed dispersal patterns produced by white-faced monkeys: Implications for the dispersal limitation of Neotropical tree species. *Journal of Ecology*, 91, 677–85.

Weih, M., and Karlsson, P. S. (2001). Growth response of mountain bird to air and soil temperature: Is increase leaf-nitrogen content acclimation to lower air temperature? *New Phytologist*, 150, 147–55.

Weldon, A. J., and Haddad, N. M. (2005). The effects of patch shape on indigo buntings: Evidence for an ecological trap. *Ecology*, 86, 1422–31.

Welker, B. J., Koenig, W., Pietsch, M., and Adams, R. P. (2007). Feeding selectivity by mantled howler monkeys (*Alouatta palliata*) in relation to leaf secondary chemistry in *Hymenaea courbaril*. *Journal of Chemical Ecology*, 33, 1186–96.

Wells, J. C. K., and Stock, J. T. (2011). Re-examining heritability: Genetics, life history and plasticity. *Trends in Endocrinology and Metabolism*, 22, 421–28.

Wells, J. C. K., DeSilva, J. M., and Stock, J. T. (2012). The obstetric dilemma: An ancient game of Russian roulette, or a variable dilemma sensitive to ecology? *Yearbook of Physical Anthropology*, 149, 40–71.

Wells, J. C., Jonsdottir, O. H., Hibberd, P. L., Fewtrell, M. S., Thorsdottir, I., Eaton, S., et al. (2012). Randomized controlled trial of 4 compared with 6 mo of exclusive breastfeeding in Iceland: Differences in breast-milk intake by stable-isotope probe. *American Journal of Clinical Nutrition*, 96, 73–79.

Welp, T., Rushen, J., Kramer, D. L., Festa-Bianchet, M., and de Passille, A. M. (2004). Vigilance as a measure of fear in dairy cattle. *Applied Animal Behaviour Science*, 87, 1–13.

Wen, L., Ley, R. E., Volchkov, P. Y., Stranges, P. B., Avanesyan, L., Stonebraker, A. C., et al. (2008). Innate immunity and intestinal microbiota in the development of type 1 diabetes. *Nature*, 455(7216), 1109–13.

West, H. E., and Capellini, I. (2016). Male care and life history traits in mammals. *Nature Communications*, 7, 11854.

Westoby, M. (1974). Analysis of diet selection by large generalist herbivores. *American Naturalist*, 108, 290–304.

Weyrich, L. S., Duchene, S., Soubrier, J., Arriola, L., Llamas, B., Breen, J., et al. (2017). Neanderthal behaviour, diet, and disease inferred from ancient DNA in dental calculus. *Nature*, 544, 357–61.

Wheatley, B. P. (1982). Energetics of foraging in *Macaca fascicularis* and *Pongo pygmaeus* and a selective advantage of large body size in the orangutan. *Primates*, 23, 348–63.

Wheatley, B. P. (1987). The evolution of large body size in orangutans: A model for hominoid divergence. *American Journal of Primatology*, 13, 313–24.

Wheatley, B. P. (1988). Cultural behaviour and extractive foraging in *Macaca fascicularis*. *Current Anthropology*, 29, 516–19.

Wheeler, B. C., and Fischer, J. (2012). Functionally referential signals: A promising paradigm whose time has passed. *Evolutionary Anthropology*, 21, 195–205.

Wheeler, B. C., Scarry, C. J., and Koenig, A. (2013). Rates of agonism among female primates: A cross-taxon perspective. *Behavioral Ecology*, 24, 1369–80.

Whelan, J. (2008). The health implications of changing linoleic acid intakes. *Prostaglandins, Leukotrienes and Essential Fatty Acids*, 79(3–5), 165–67.

White, C. A., Feller, M. C., and Bayley, S. (2003). Predation risk and the functional response of elk-aspen herbivory. *Forest Ecology and Management*, 181, 77–97.

White, L. J. T., Rogers, M. E., Tutin, C. E. G., Williamson, E. A., and Fernandez, M. (1995). Herbaceous vegetation in different forest types in the Lopé Reserve, Gabon: Implications for keystone food availability. *African Journal of Ecology*, 33, 124–41.

White, T. C. R. (1993). *The Inadequate Environment: Nitrogen and the Abundance of Animals*. Springer.

White, T. C. R. (2001). Opposing paradigms: Regulation or limitation of populations? *Oikos*, 93, 148–52.

White, T. D., Ambrose, S. H., Suwa, G., Su, D. F., DeGusta, D., Bernor, R. L., et al. (2009). Macrovertebrate paleontology and the Pliocene habitat of *Ardipithecus ramidus*. *Science*, 326, 67–93.

Whiten, A., Byrne, R. W., Barton, R. A., Waterman, P. G., and Henzi, S. P. (1991). Dietary and foraging strategies of baboons. *Philosophical Transactions of the Royal Society of London Series B—Biological Sciences*, 334, 187–97.

Whiten, A., Goodall, J., McGrew, W. C., Nishida, T., Reynolds, V., Sugiyama, Y., et al. (1999). Cultures in chimpanzees. *Nature*, 399, 682–85.

Whiten, A., Goodall, J., McGrew, W. C., Nishida, T., Reynolds, V., Sugiyama, Y., et al. (2001). Charting cultural variation in chimpanzees. *Behaviour*, 138, 1481–516.

Whiten, A., Hinde, R. A., Laland, K. N., and Stringer, C. B. (2011). Culture evolves. *Philosophical Transactions of the Royal Society B: Biological Sciences*, 366(1567), 938–48.

Whiten, A., and van Schaik, C. P. (2007). The evolution of animal "cultures" and social intelligence. *Philosophical Transactions of the Royal Society of London Series B—Biological Sciences*, 362, 603–20.

Whitmore, T. C. (1986). *Tropical Rain Forests of the Far East*. Oxford University Press.

Whitten, A. J. (1982). Diet and feeding of Kloss gibbons on Siberut Island, Indonesia. *Folia Primatologica*, 37, 177–208.

Whitten, P. L. (1982). *Effects of Patch Quality and Feeding Subgroup Size on Feeding Success in Vervet Monkeys (Cercopithecus aethiops)*. Harvard University.

Whitten, P. L. (1983a). Diet and dominance among female vervet monkeys (*Cercopithecus aethiops*). *American Journal of Primatology*, 5, 139–59.

Whitten, P. L. (1983b). Flowers, fertility and females. *American Journal of Physical Anthropology*, 60, 269–70.

Whitten, P. L., and Patisaul, H. B. (2001). Cross-species and interassay comparisons of phytoestrogen action. *Environmental Health Perspectives*, 109, 5–20.

Whitten, P. L., and Turner, T. R. (2009). Endocrine mechanisms of primate life history trade-offs: Growth and reproductive maturation in vervet monkeys. *American Journal of Human Biology*, 21, 754–61.

Wich, S. A., Buij, R., and van Schaik, C. P. (2004). Determinants of orangutan density in the dryland forests of the Leuser ecosystem. *Primates*, 45, 177–82.

Wich, S. A., De Vries, H., Ancrenaz, M., Perkins, L., Shumaker, R., Suzuki, A., et al. (2009). Orangutan life history variation. In S. A. Wich, S. S. Utami Atmoko, T. Mitra Setia, and C. P. van Schaik (Eds.), *Orangutans: Geographic Variation in Ecology and Conservation* (pp. 65–75). Oxford University Press.

Wich, S. A., Shumaker, R. W., Perkins, L., and de Vries, H. (2009). Captive and wild orangutan (*Pongo* sp.) survivorship: A comparison and the influence of management. *American Journal of Primatology*, 71, 680–86.

Wich, S. A., Utami-Atmoko, S. S., Mitra Setia, T., Djoyosudharmo, S., and Geurts, M. L. (2006). Dietary and energetic responses of *Pongo abelii* to fruit availability fluctuations. *International Journal of Primatology*, 27, 1535–50.

Wich, S. A., Utami Atmoko, S. S., Mitra Setia, T., Rijksen, H. D., Schuermann, C., van Hooff, J. A. R. A. M., et al. (2004). Life history of wild Sumatran orangutans (*Pongo abelii*). *Journal of Human Evolution*, 47, 385–98.

Wich, S. A., and van Schaik, C. P. (2000). The impact of El Niño on mast fruiting in Sumatra and elsewhere in Malesia. *Journal of Tropical Ecology*, 16, 563–77.

Wich, S. A., Vogel, E. R., Larsen, M. D., Fredriksson, G., Leighton, M., Yeager, C. P., et al. (2011). Forest fruit production is higher on Sumatra than on Borneo. *PLoS ONE*, 6, e21278.

Wickham, H. (2009). *ggplot2: Elegant Graphics for Data Analysis*. R Project for Statistical Computing.

Widayati, K. A., Yan, X., Suzuki-Hashido, N., Itoigawa, A., Purba, L. H. P. S., Fahri, et al. (2019). Functional divergence of the bitter receptor TAS2R38 in Sulawesi macaques. *Ecology and Evolution*, 9(18), 10387–403.

Wieczkowski, J. (2009). Brief communication: Puncture and crushing resistance scores of Tana River mangabey (*Cercocebus galeritus*) diet items. *American Journal of Physical Anthropology*, 140, 572–77.

Wieczkowski, J. (2013). The value of measuring food availability on the ground for a semiterrestrial frugivore, the Tana River mangabey (*Cercocebus galeritus*) of Kenya. *International Journal of Primatology*, 34, 973–85.

Wiens, F., Zitzmann, A., and Hussein, N. A. (2006). Fast food for slow lorises: Is low metabolism related to secondary compounds in high-energy plant diet? *Journal of Mammalogy*, 87, 790–98.

Wiens, J. A. (1976). Population responses to patchy environments. *Annual Review of Ecology and Systematics*, 7, 81–120.

Wiggins, N. L., McArthur, C., and Davies, N. W. (2006). Diet switching in a generalist mammalian folivore: Fundamental to maximising intake. *Oecologia*, 147, 650–57.

Wiggins, N. L., McArthur, C., McLean, S., and Boyle, R. (2003). Effects of two plant secondary metabolites, cineole and gallic acid, on nightly feeding patterns of the common brushtail possum. *Journal of Chemical Ecology*, 29, 1447–64.

Wikoff, W. R., Anfora, A. T., Liu, J., Schultz, P. G., Lesley, S. A., Peters, E. C., et al. (2009). Metabolomics analysis reveals large effects of gut microflora on mammalian blood metabolites. *Proceedings of the National Academy of Sciences of the United States of America*, 106(10), 3698–703.

Wilde, C. J., Addey, C. V., Boddy, L. M., and Peaker, M. (1995). Autocrine regulation of milk secretion by a protein in milk. *Biochemical Journal*, 305, 51–58.

Wilde, C. J., and Peaker, M. (1990). Autocrine control in milk secretion. *Journal of Agricultural Science*, 114, 235–38.

Wiles, P. G., Gray, I. K., and Kissling, R. C. (1998). Routine analysis of proteins by Kjeldahl and Dumas methods: Review and interlaboratory study using dairy products. *Journal of the Association of Official Analytical Chemists*, 81, 620–32.

Willems, E. P., and Hill, R. A. (2009). Predator-specific landscapes of fear and resource distribution: Effects on spatial range use. *Ecology*, 90, 546–55.

Willems, E. P., Barton, R. A., and Hill, R. A. (2009). Remotely sensed productivity, regional home range selection, and local range use by an omnivorous primate. *Behavioral Ecology*, 20, 985–92.

Willette, A. A., Bendlin, B. B., McLaren, D. G., Canu, E., Kastman, E. K., Kosmatka, K. J., et al. (2010). Age-related changes in neural volume and microstructure associated with interleukin-6 are ameliorated by a calorie-restricted diet in old rhesus monkeys. *Neuroimage*, 51(3), 987–94.

Williams, H. J., and Safi, K. (2021). Certainty and integration of options in animal movement. *Trends in Ecology and Evolution*, 36(11), 990–99.

Williams, L. H. (1954). The feeding habits and food preferences of Acrididae and the factors which determine them. *Transactions of the Royal Entomological Society of London*, 105, 423–54.

Williams, N. I., Helmreich, D. L., Parfitt, D. B., Caston-Balderrama,

A., and Cameron, J. L. (2001). Evidence for a causal role of low energy availability in the induction of menstrual cycle disturbances during strenuous exercise training. *Journal of Clinical Endocrinology and Metabolism*, 86, 5184–93.

Williams, S. H., Wright, B. W., Van Den Truong, Daubert, C. R., & Vinyard, C. J. (2005). Mechanical properties of foods used in experimental studies of primate masticatory function. *American Journal of Primatology*, 67, 329–46.

Williamson, E. A., and Feistner, A. T. C. (2011). Habituating primates: Processes, techniques, variables and ethics. In J. M. Setchell and D. J. Curtis (Eds.), *Field and Laboratory Methods in Primatology: A Practical Guide* (pp. 33–50). Cambridge University Press.

Williamson, E. A., Tutin, C. E. G., Rogers, M. E., and Fernandez, M. (1990). Composition of the diet of lowland gorillas at Lopé in Gabon. *American Journal of Primatology*, 21, 265–77.

Willis, E. A., Herrmann, S. D., Honas, J. J., Lee, J., Donnelly, J. E., and Washburn, R. A. (2014). Nonexercise energy expenditure and physical activity in the Midwest Exercise Trial 2. *Sports Medicine and Science in and Exercise*, 46(12), 2286.

Willis, E. O., and Oniki, Y. (1978). Birds and army ants. *Annual Review of Ecology and Systematics*, 9, 243–63.

Willis, M. A. (2008). Chemical plume tracking behavior in animals and mobile robots. *Navigation*, 55, 127–35.

Willson, M. F. (1983). *Plant reproductive ecology*. John Wiley & Sons.

Willson, M. F., Irvine, A. K., and Walsh, N. G. (1989). Vertebrate dispersal syndromes in some Australian and New Zealand plant communities, with geographic comparisons. *Biotropica*, 21, 133–47.

Wilsea, M., Johnson, K. L., and Ashby, M. F. (1975). Indentation of foamed plastics. *International Journal of Mechanical Sciences*, 17, 457–60.

Wilson, E. O. (1975). *Sociobiology*. Belknap.

Wilson, E. O. (1987). The little things that run the world (the importance and conservation of invertebrates). *Conservation Biology*, 1, 344–46.

Wilson, K., and Cotter, S. C. (2013). Host parasite interactions and the evolution of immune defense. In H. J. Brockmann, T. J. Roper, M. Naguib and L. Barrett (Eds.), *Advances in the Study of Behavior* (pp. 81–174). Academic Press.

Wilson, R. P., Shepard, E. L. C., and Liebsch, N. (2008). Prying into the intimate details of animal lives: use of a daily diary on animals. *Endangered Species Research*, 4(1-2), 123–37.

Wimberger, K., Nowak, K., and Hill, R. A. (2017). Reliance on exotic plants by two groups of threatened Samango monkeys, *Cercopithecus albogularis labiatus*, at their southern range limit. *International Journal of Primatology*, 38, 151–71.

Winchester, J. M., Boyer, D. M., St Clair, E. M., Gosselin-Ildari, A. D., Cooke, S. B., and Ledogar, J. A. (2014). Dental topography of platyrrhines and prosimians: Convergence and contrasts. *American Journal of Physical Anthropology*, 153(1), 29–44.

Windle, C. P., Baker, H. F., Ridley, R. M., Oerke, A. K., and Martin, R. D. (1999). Unrearable litters and prenatal reduction of litter size in the common marmoset (*Callithrix jacchus*). *Journal of Medical Primatology*, 28, 73–83.

Windley, H. R., Starrs, D., Stalenberg, E., Rothman, J. M., Ganzhorn, J. U., and Foley, W. J. (2022). Plant secondary metabolites and primate food choices: A meta-analysis and future directions. *American Journal of Primatology*, 84, e23397.

Winfield, R. D. (2014). Caring for the critically ill obese patient: Challenges and opportunities. *Nutrition in Clinical Practice*, 29(6), 747–50.

Winfield, R. D., Delano, M. J., Cuenca, A. G., Cendan, J. C., Lottenberg, L., Efron, P. A., et al. (2012). Obese patients show a depressed cytokine profile following severe blunt injury. *Shock*, 37(3), 253–56.

Wink, M. (2008). Plant secondary metabolism: Diversity, function and its evolution. *Natural Product Communications*, 3, 1205–16.

Winkel-Shirley, B. (2002). Biosynthesis of flavonoids and effects of stress. *Current Opinion in Plant Biology*, 5, 218–23.

Winkelmann, R. D. (1963). Nerve endings in the skin of primates. In J. Buetnner-Janusch (Ed.), *Evolutionary and Genetic Biology of Primates* (vol. 1, pp. 229–59). Academic Press.

Winter, K. (1979). $\Delta^{13}C$ values of some succulent plants from Madagascar. *Oecologia*, 40, 103–12.

Wintergerst, E. S., Maggini, S., and Hornig, D. H. (2007). Contribution of selected vitamins and trace elements to immune function. *Annals of Nutrition and Metabolism*, 51(4), 301–23.

Wirz, A., and Riviello, M. C. (2008). Reproductive parameters of a captive colony of capuchin monkeys (*Cebus apella*) from 1984 to 2006. *Primates*, 49, 265–70.

Wittig, R. M., and Boesch, C. (2003). Food competition and linear dominance hierarchy among female chimpanzees of the Tai National Park. *International Journal of Primatology*, 24, 847–67.

Wolda, H. (1980). Seasonality of tropical insects. *Journal of Animal Ecology*, 49, 277–90.

Wolda, H. (1992). Trends in abundance of tropical forest insects. *Oecologia*, 89, 47–52.

Wolda, H. (1995). The demise of the population regulation controversy? *Researches on Population Ecology*, 37, 91–93.

Wolf, J., Rose-John, S., and Garbers, C. (2014). Interleukin-6 and its receptors: A highly regulated and dynamic system. *Cytokine*, 70(1), 11–20.

Wolfe, B. M., Culebras, J. M., Sim, A. J., Ball, M. R., and Moore, F. D. (1977). Substrate interaction in intravenous feeding: Comparative effects of carbohydrate and fat on amino acid utilization in fasting man. *Annals of Surgery*, 186(4), 518–40.

Wolfe, R. R., Shaw, J. H., and Durkot, M. J. (1983). Energy metabolism in trauma and sepsis: The role of fat. *Progress in Clinical and Biological Research*, 111, 89–109.

Wolff, J. O., and van Horn, T. (2003). Vigilance and foraging patterns of American elk during the rut in habitats with and without predators. *Canadian Journal of Zoology*, 81, 266–71.

Wolovich, C. K., Evans, S., and French, J. A. (2008). Dads do not pay for sex but do buy the milk: Food sharing and reproduction in owl monkeys (*Aotus* spp.). *Animal Behaviour*, 75, 1155–63.

Wolovich, C. K., Perea-Rodriguez, J. P., and Fernandez-Duque, E. (2008). Food transfers to young and mates in wild owl monkeys (*Aotus azarai*). *American Journal of Primatology*, 70, 211.

Wong, A. C. N., Holmes, A., Ponton, F., Lihoreau, M., Wilson, K., Raubenheimer, D., et al. (2015). Behavioral microbiomics: A multi-dimensional approach to microbial influence on behavior. *Frontiers in Microbiology*, 6, 1359.

Wood, B., and Gilby, I. (2017). From *Pan* to man the hunter: Hunting and meat sharing by chimpanzees, humans, and our common ancestor. In Muller, M. N., Wrangham, R. W. and Pilbeam, D. (Eds.). *Chimpanzees and Human Evolution* (pp. 339–82). Harvard University Press.

Wood, B. M., and Marlowe, F. W. (2013). Household and kin provisioning by Hadza men. *Human Nature*, 24, 280–317.

Wood, B. M., Watts, D. P., Mitani, J. C., and Langergraber, K. E. (2017). Favorable ecological circumstances promote life expectancy in chimpanzees similar to that of human hunter-gatherers. *Journal of Human Evolution*, 105, 41–56.

Woodhams, D. C., Bletz, M. C., Becker, C. G., Bender, H. A., Buitrago-Rosas, D., Diebboll, H., et al. (2020). Host-associated microbiomes are predicted by immune system complexity and climate. *Genome Biology*, 23, 1–20.

Woods, D. L. (1989). The constraint of maternal nutrition on the trajectory of fetal growth in humans. In M. N. Bruton (Ed.), *Alternative Life-History Styles of Animals* (pp. 459–64). Kluwer.

Workman, C. (2010). Diet of the Delacour's langur (*Trachypithecus delacouri*) in Van Long Nature Reserve, Vietnam. *American Journal of Primatology*, 72, 317–24.

Worman, C. O. D., and Chapman, C. A. (2005). Seasonal variation in the quality of a tropical ripe fruit and the response of three frugivores. *Journal of Tropical Ecology*, 21, 689–97.

Wostmann, B. S., Larkin, C., Moriarty, A., and Bruckner-Kardoss, E. (1983). Dietary intake, energy metabolism, and excretory losses of adult male germfree Wistar rats. *Laboratory Animal Science*, 33, 46–50.

Wrangham, R. W. (1977). Feeding behavior of chimpanzees in Gombe National Park, Tanzania. In T. H. Clutton-Brock (Ed.), *Primate Ecology: Studies of Feeding and Ranging Behaviour in Lemurs, Monkeys, and Apes* (pp. 503–38). Academic Press.

Wrangham, R. W. (1979). On the evolution of ape social systems. *Social Science Information*, 18, 334–68.

Wrangham, R. W. (1980). An ecological model of female-bonded primate groups. *Behaviour*, 75, 262–300.

Wrangham, R. W. (1986). Ecology and social relationships in two species of chimpanzee. In D. I. Rubenstein and R. W. Wrangham (Eds.), *Ecological Aspects of Social Evolution: Birds and Mammals* (pp. 352–78). Princeton University Press.

Wrangham, R. W., Conklin-Brittain, N. L., and Hunt, K. D. (1998). Dietary response of chimpanzees and cercopithecines to seasonal variation in fruit abundance: I. Antifeedants. *International Journal of Primatology*, 19, 949–70.

Wrangham, R. W., Conklin, N. L., Chapman, C. A., and Hunt, K. D. (1991). The value of fibrous foods to chimpanzees. *Philosophical Transactions of the Royal Society of London Series B—Biological Sciences*, 334, 171–78.

Wrangham, R. W., Conklin, N. L., Etot, G., Obua, J., Hunt, K. D., Hauser, M. D., et al. (1993). The value of figs to chimpanzees. *International Journal of Primatology*, 14, 243–56.

Wrangham, R. W., Gittleman, J. L., and Chapman, C. A. (1993). Constraints on group size in primates and carnivores: Population density and day-range as assays of exploitation competition. *Behavioral Ecology and Sociobiology*, 32, 199–209.

Wrangham, R. W., Machanda, Z. P., Weaver, C., and Muller, M. N. (2014). The toxin-reduction hypothesis for geophagy: Evidence from pregnant chimpanzees. *American Journal of Physical Anthropology*, 153, 277.

Wrangham, R., Rogers, M. E., and Basuta, G. I. (1993). Ape food density in the ground layer in Kibale Forest, Uganda. *African Journal of Ecology*, 31, 37–49.

Wrangham, R. W., and Smuts, B. (1980). Sex differences in the behavioral ecology of chimpanzees in the Gombe National Park, Tanzania. *Journal of Reproduction and Fertility*, 28, 13–31.

Wrangham, R. W., and Van Zinnicq Bergmann Riss, E. (1990). Rates of predation on mammals by Gombe chimpanzees, 1972–1975. *Primates*, 31(2), 157–70.

Wrangham, R. W., and Waterman, P. G. (1981). Feeding behavior of vervet monkeys on *Acacia tortilis* and *Acacia xanthophloea*—with special reference to reproductive strategies and tannin production. *Journal of Animal Ecology*, 50, 715–31.

Wrangham, R. W., and Waterman, P. G. (1983). Condensed tannins in fruits eaten by chimpanzees. *Biotropica*, 15, 217–22.

Wright, B. W. (2005) Craniodental biomechanics and dietary toughness in the genus *Cebus*. *Journal of Human Evolution*, 48, 473–92.

Wright, E., and Robbins, M. M. (2014). Proximate mechanisms of contest competition among female Bwindi mountain gorillas (*Gorilla beringei beringei*). *Behavioral Ecology and Sociobiology*, 68, 1785–97.

Wright, E., Robbins, A. M., and Robbins, M. M. (2014). Dominance rank differences in the energy intake and expenditure of female Bwindi mountain gorillas. *Behavioral Ecology and Sociobiology*, 68, 957–70.

Wright, P. (1999). Lemur traits and Madagascar ecology: Coping with an island environment. *Yearbook of Physical Anthropology*, 42, 31–72.

Wright, P. C. (1990). Patterns of paternal care in primates. *International Journal of Primatology*, 11, 89–102.

Wright, P. C. (2007). Considering climate change in lemur conservation. In L. Gould and M. L. Sauther (Eds.), *Lemurs: Ecology and Adaptation* (pp. 385–401). Springer.

Wright, P., King, S., Baden, A., and Jernvall, J. (2008). Aging in wild female lemurs: Sustained fertility with increased infant mortality. In S. Atsalis, S. W. Margulis, and P. R. Hof (Eds.), *Primate Reproductive Aging: Cross-Taxon Perspectives* (pp. 17–28). Karger.

Wright, S. J., and Muller-Landau, H. C. (2006). The future of tropical forest species. *Biotropica*, 38, 287–301.

Wrogemann, D., Radespiel, U., and Zimmerman, E. (2001). Comparison of reproductive characteristics and changes in body weight between captive populations of rufous and gray mouse lemurs. *International Journal of Primatology*, 22, 91–108.

Wronka, I., and Pawlinska-Chmara, R. (2005). Menarcheal age and socio-economic factors in Poland. *Annals of Human Biology*, 32, 630–38.

Wu, A. H., Yu, M. C., Tseng, C. C., and Pike, M. C. (2008). Epidemiology of soy exposures and breast cancer risk. *British Journal of Cancer*, 98, 9–14.

Wu, D. F., Behringer, V., Wittig, R. M., Leendertz, F. H., and Deschner, T. (2018). Urinary neopterin levels increase and predict survival during a respiratory outbreak in wild chimpanzees (Tai National Park, Cote d'Ivoire). *Scientific Reports*, 8, 9.

Wu, G. D., Chen, J., Hoffmann, C., Bittinger, K., Chen, Y. Y., Keilbaugh, S. A., et al. (2011). Linking long-term dietary patterns with gut microbial enterotypes. *Science*, 334, 105–8.

Wunderlich, R. E., Lawler, R. R., and Williams, A. E. (2011). Field and experimental approaches to the study of locomotor ontogeny in *Propithecus verreauxi*. In K. D'Aout and E. E. Vereecke (Eds.), *Primate Locomotion: Linking Field and Laboratory Research* (pp. 135–54). Springer.

WWF. (2022). *Living Planet Report 2022—Building a Nature-Positive Society*. R. E. A. Almond, M. Grooten, D. Juffe Bignoli, and T. Petersen (Eds.). WWF, Gland, Switzerland.

Wyers, M., Formenty, P., Cherel, Y., Guigand, L., Fernandez, B., Boesch, C., et al. (1999). Histopathological and immunohistochemical studies of lesions associated with Ebola virus in a naturally infected chimpanzee. *Journal of Infectious Diseases*, 179, S54–S59.

Wynn, J. G., Alemseged, Z., Bobe, R., Grine, F. E., Negash, E. W., and Sponheimer, M. 2020. Isotopic evidence for the timing of the dietary shift toward C_4 foods in eastern African *Paranthropus*. *Proceedings of the National Academy of Sciences of the United States of America*, 117(36), 21978–84.

Wynn, J. L., Guthrie, S. O., Wong, H. R., Lahni, P., Ungaro, R., Lopez, M. C., et al. (2015). Postnatal age is a critical determinant of the neonatal host response to sepsis. *Molecular Medicine*, 21, 496–504.

Wynne-Edwards, K. E. (2001). Evolutionary biology of plant defenses against herbivory and their predictive implications for endocrine disruptor susceptibility in vertebrates. *Environmental Health Perspectives*, 109, 443–48.

Wynne-Edwards, V. C. (1970). Feedback from food resources to population regulation. In A. Watson (Ed.), *Animal Populations in Relation to their Food Resources* (pp. 413–27). Blackwell.

Xia, J., Tian, Z. R., Hua, L., Chen, L., Zhou, Z., Qian, L., et al. (2017). Enamel crystallite strength and wear: Nanoscale responses of teeth to chewing loads. *Journal of the Royal Society Interface*, 14, 20170456.

Xia, J., Zheng, J., Huang, D. D., Tian, Z. R., Chen, L., Zhou, Z. R., et al. (2015). New model to explain tooth wear with implications for microwear formation and diet reconstruction. *Proceedings of the National Academy of Sciences of the United States of America*, 112(34), 10669–72.

Xiang, Z., Huo, S., Xiao, W., Quan, R., and Grueter, C. C. (2007). Diet and feeding behavior of *Rhinopithecus bieti* at Xiaochangdu, Tibet: Adaptations to a marginal environment. *American Journal of Primatology*, 69, 1141–58.

Yakir, D. (1992). Variations in the natural abundance of oxygen-18 and deuterium in plant carbohydrates. *Plant, Cell, and Environment*, 15, 1005–20.

Yamagiwa, J., and Basabose, A. K. (2009). Fallback foods and dietary partitioning among *Pan* and *Gorilla*. *American Journal of Physical Anthropology*, 140(4), 739–50.

Yamagiwa, J., and Hill, D. A. (1998). Intraspecific variation in the social organization of Japanese macaques: Past and present scope of field studies in natural habitats. *Primates*, 39, 257–63.

Yamakoshi, G. (1998). Dietary responses to fruit scarcity of wild chimpanzees at Bossou, Guinea: Possible implications for ecological importance of tool use. *American Journal of Physical Anthropology*, 106, 283–95.

Yamakoshi, G., and Sugiyama, Y. (1995). Pestle-pounding behavior of wild chimpanzees at Bossou, Guinea: A newly observed tool-using behavior. *Primates*, 36, 489–500.

Yamamoto, S., Yamakoshi, G., Humle, T., and Matsuzawa, T. (2008). Invention and modification of a new tool use behavior: Ant-fishing in trees by a wild chimpanzee (*Pan troglodytes verus*) at Bossou, Guinea. *American Journal of Primatology*, 70, 699–702.

Yamashita, N. (1998). Functional dental correlates of food properties in five Malagasy lemur species. *American Journal of Physical Anthropology*, 106(2), 169–88.

Yamashita, N. (2003). Food procurement and tooth use in two sympatric lemur species. *American Journal of Physical Anthropology*, 121, 125–33.

Yamashita, N. (2008). Chemical properties of the diets of two lemur species in southwestern Madagascar. *International Journal of Pharmacology*, 29, 339–64.

Yamashita, N., Cuozzo, F. P., Sauther, M. L., Fitzgerald, E., Riemenschneider, A., and Ungar, P. S. (2015). Mechanical food properties and dental topography differentiate three populations of *Lemur catta* in southwest Madagascar. *Journal of Human Evolution*, 98, 66-75.

Yasukochi, Y., and Satta, Y. (2015). Molecular evolution of the CYP2D subfamily in primates: Purifying selection on substrate recognition sites without the frequent or long-tract gene conversion. *Genome Biology and Evolution*, 7, 1053–67.

Yates, J. A. F., Velsko, I. M., Aron, F., Posth, C., Hofman, C. A., Austin, R. M., et al. (2021). The evolution and changing ecology of the African hominid oral microbiome. *Proceedings of the National Academy of Sciences of the United States of America*, 118(20), e2021655118.

Yatsunenko, T., Rey, F. E., Manary, M. J., Trehan, I., Dominguez-Bello, M. G., Contreras, M., et al. (2012). Human gut microbiome viewed across age and geography. *Nature*, 486(7402), 222–27.

Ydenberg, R. C., Brown, J. S., and Stephens, D. W. (2007). Foraging: An overview. In D. W. Stephens, J. S. Brown, and R. C. Ydenberg (Eds.), *Foraging: Behavior and Ecology* (pp. 1–28). University of Chicago Press.

Yeager, C. P. (1989). Feeding ecology of the proboscis monkey (*Nasalis larvatus*). *International Journal of Primatology*, 10, 497–530.

Yeager, C. P. (1996). Feeding ecology of the long-tailed macaque (*Macaca fascicularis*) in Kalimantan Tengah, Indonesia. *International Journal of Primatology*, 17, 51–62.

Yeager, C. P., and Kirkpatrick, R. C. (1998). Asian colobine social structure: Ecological and evolutionary constraints. *Primates*, 39, 147–55.

Yeager, C. P., Silver, S. C., and Dierenfeld, E. S. (1997). Mineral and phytochemical influences on foliage selection by the proboscis monkey (*Nasalis larvatus*). *American Journal of Primatology*, 41, 117–28.

Yeakel, J. D., Dominy, N. J., Koch, P. J., and Mangel M. (2013). Functional morphology, stable isotopes, and human evolution: A model of consilience. *Evolution*, 68, 190–203.

Yépez, P., De La Torre, S., and Snowdon, C. T. (2005). Interpopulation differences in exudate feeding of pygmy marmosets in Ecuadorian Amazonia. *American Journal of Primatology*, 66, 145–58.

Yiming, L. (2006). Seasonal variation of diet and food availability in a group of Sichuan snub-nosed monkeys in Shennongjia Nature Reserve, China. *American Journal of Primatology*, 68(3), 217–33.

Yom-Tov, Y., and Geffen, E. (2011). Recent spatial and temporal changes in body size of terrestrial vertebrates: Probable causes and pitfalls. *Biological Reviews*, 86, 531–41.

Yoneda, M. (1984). Comparative studies on vertical separation, foraging behavior and traveling mode of saddle-backed tamarins (*Saguinus fuscicollis*) and red-chested moustached tamarins (*Saguinus labiatus*) in Northern Bolivia. *Primates*, 25, 414–22.

Yoshikawa, M., and Ogawa, H. (2015). Diet of savanna chimpanzees in the Ugalla area, Tanzania. *African Studies Monographs*, 36, 189–209.

Youlatos, D. (2002). Positional behavior of black spider monkeys (*Ateles paniscus*) in French Guiana. *International Journal of Primatology*, 23, 1071–93.

Young, A. L., Richard, A. F., and Aiello, L. C. (1990). Female dominance and maternal investment in strepsirhine primates. *American Naturalist*, 135, 473–88.

Young, C., Majolo, B., Heistermann, M., Schuelke, O., and Ostner, J. (2013). Male mating behaviour in relation to female sexual swellings, socio-sexual behaviour and hormonal changes in wild Barbary macaques. *Hormones and Behavior*, 63, 32–39.

Young, H., Fedigan, L. M., and Addicott, J. F. (2008). Look before leaping: Foraging selectivity of capuchin monkeys on acacia trees in Costa Rica. *Oecologia*, 155, 85–92.

Young, R. J. (1998). Behavioural studies of guenons *Cercopithecus* spp at Edinburgh Zoo. *International Zoo Yearbook*, 36, 49–56.

Young, S. L., Sherman, P. W., Lcuks, J. B., Pelto, G. H., and Rowe, L. (2011). Why on Earth? Evaluating hypotheses about the physiological functions of human geophagy. *Quarterly Review of Biology*, 86, 97–120.

Youngblut, N. D., Reischer, G. H., Walters, W., Schuster, N., Walzer, C., Stalder, G., et al. (2019). Host diet and evolutionary history explain different aspects of gut microbiome diversity among vertebrate clades. *Nature Communications*, 10, 2200.

Youngentob, K. N., Lindenmayer, D., Marsh, K. J., Krockenberger, A. K., and Foley, W. J. (2021). Food intake: An overlooked driver of climate change casualties? *Trends in Ecology and Evolution*, 36, 676–78.

Youssef, K. M., and Mokhtar, S. M. (2014). Effect of drying methods on the antioxidant capacity, color and phytochemicals of *Portulaca oleracea* L. leaves. *Journal of Nutrition and Food Sciences*, 4, 322.

Yunger, J. A., Meserve, P. L., and Gutierrez, J. R. (2002). Small-mammal foraging behavior: Mechanisms for coexistence and implication for population dynamics. *Ecological Monographs*, 72, 561–77.

Zausa, D., Koné, I., and Ouattara, K. (2018). The optimal foraging strategy used by Campbell's monkeys, *Cercopithecus campbelli*, in the dry season in the Taï National Park (Côte d'Ivoire). *International Journal of Research in BioSciences*, 7, 8–20.

Zeder, M. A. (2012). The Broad Spectrum Revolution at 40: Resource diversity, intensification, and an alternative to optimal foraging explanations. *Journal of Anthropological Archaeology*, 31, 241–64.

Zeeve, S.R., (1985). Swamp monkeys of the Lomako Forest, Central Zaire. *Primate Conservation*, 5, 32-33.

Zeisel, S. H., Allen, L. H., Coburn, S. P., Erdman, J. W., Failla, M. L., Freake, H. C., et al. (2001). Nutrition: A reservoir for integrative science. *Journal of Nutrition*, 131, 1319–21.

Zhang, J., Hu, J., Lian, J., Fan, Z., Ouyang, X., and Ye, W. (2016). Seeing the forest from drones: Testing the potential of lightweight drones as a tool for long-term forest monitoring. *Biological Conservation*, 198, 60–69.

Zhang, K., Karim, F., Jin, Z., Xiao, H., Yao, Y., Ni, Q., et al. (2022). Diet and feeding behavior of a group of high-altitude rhesus macaques: High adaptation to food shortages and seasonal fluctuations. *Current Zoology*, 9, 304-314.

Zhang, L., Ameca, E. I., Cowlishaw, G., Pettorelli, N., Foden, W., and Mace, G. M. (2019). Global assessment of primate vulnerability to extreme climatic events. *Nature Climate Change*, 9, 554–61.

Zhang, P., Hu, K., Yang, B., and Yang, D. (2016). Snub-nosed monkeys (*Rhinopithecus* spp.): Conservation challenges in the face of environmental uncertainty. *Science Bulletin*, 61, 345–48.

Zhang, S. Y., and Wang, L. X. (1995). Comparison of three fruit census methods in French Guiana. *Journal of Tropical Ecology*, 11, 281–94.

Zhao, H., Yang, J. R., Xu, H., and Zhang, J. (2010). Pseudogenization of the umami taste receptor gene *Tas1r1* in the giant panda coincided with its dietary switch to bamboo. *Molecular Biology and Evolution*, 27, 2669–73.

Zhou, Q. H., Wei, H., Huang, Z. H., and Huang, C. M. (2011). Diet of the Assamese macaque *Macaca assamensis* in limestone habitats of Nonggang, China. *Current Zoology*, 57, 18–25.

Zhou, Q., Wei, H., Tang, H., Huang, Z., Krzton, A., and Huang, C. (2014). Niche separation of sympatric macaques, *Macaca assamensis* and *M. mulatta*, in limestone habitats of Nonggang, China. *Primates*, 55, 125–37.

Zhao H, Li J, Wang X, Pei J, Wang C, Ren Y, et al. (2020) Nutrient strategies of the Sichuan snub-nosed monkey (*Rhinopithecus roxellana*) when confronted with a shortage of food resources in the Qinling Mountains, China. *Global Ecology and Conservation*, 22: e00963.

Zheng, W. H., Li, M., Liu, J. S., and Shao, S. L. (2008). Seasonal acclimatization of metabolism in Eurasian tree sparrows (*Passer montanus*). *Comparative Biochemistry and Physiology Part A: Molecular & Integrative Physiology*, 151(4), 519–25.

Ziegler, T., Hodges, J. K., Winkler, P., and Heistermann, M. (2000). Hormonal correlates of reproductive seasonality in wild female Hanuman langurs (*Presbytis entellus*). *American Journal of Primatology*, 51, 119–34.

Zihlman, A. L., McFarland, R. K., and Underwood, C. E. (2011). Functional anatomy and adaptation of male gorillas (*Gorilla gorilla gorilla*) with comparison to male orangutans (*Pongo pygmaeus*). *Anatomical Record—Advances in Integrative Anatomy and Evolutionary Biology*, 294(11), 1842–55.

Zimmerman, L. M., Bowden, R. M., and Vogel, L. A. (2014). A vertebrate cytokine primer for eco-immunologists. *Functional Ecology*, 28(5), 1061–73.

Zink, K.D., and Lieberman, D.E., (2016). Impact of meat and Lower Palaeolithic food processing techniques on chewing in humans. *Nature*, 531, 500–503.

Zinner, D. (1999). Relationship between feeding time and food intake in hamadryas baboons (*Papio hamadryas*) and the value of feeding time as predictor of food intake. *Zoo Biology*, 18, 495–505.

Ziomkiewicz, A., Babiszewska, M., Apanasewicz, A., Piosek, M., Wychowaniec, P., Cierniak, A., et al. (2021). Psychosocial stress and cortisol stress reactivity predict breast milk composition. *Scientific Reports*, 11(1), 1–14.

Zivkovic, A. M., German, J. B., Lebrilla, C. B., and Mills, D. A. (2011). Human milk glycobiome and its impact on the infant gastrointestinal microbiota. *Proceedings of the National Academy of Sciences of the United States of America*, 108(suppl. 1), 4653–58.

Zoetendal, E. G., Akkermans, A. D. L., Akkermans-va Vliet, W. M., de Visser, J. A. G. M., and De Vos, W. M. (2001). The host genotype affects the bacterial community in the human gastrointestinal tract. *Microbial Ecology in Health and Disease*, 13, 129–34.

Zuberbühler, K. (2000a). Causal cognition in a non-human primate: Field playback experiments with Diana monkeys. *Cognition*, 76, 195–207.

Zuberbühler, K. (2000b). Referential labelling in Diana monkeys. *Animal Behaviour*, 59, 917–27.

Zuberbühler, K. (2001). Predator-specific alarm calls in Campbell's monkeys, *Cercopithecus campbelli*. *Behavioral Ecology and Sociobiology*, 50, 414–22.

Zuberbühler, K., and Janmaat, K. R. L. (2010). Foraging cognition in nonhuman primates. In M. Platt and A. Ghazanfar (Eds.), *Primate Neuroethology* (pp. 64–83). Oxford University Press.

Zuccotti, L. F., Williamson, M. D., Limp, W. F., and Ungar, P. S. (1998). Technical note: Modeling primate occlusal topography using geographic information systems technology. *American Journal of Physical Anthropology*, 107(1), 137–42.

Zucker, E. L., and Clarke, M. R. (2003). Longitudinal assessment of immature-to-adult ratios in two groups of Costa Rican *Alouatta palliata*. *International Journal of Primatology*, 24, 87–101.

Zucker, W. V. (1983). Tannins: Does structure determine function? An ecological perspective. *American Naturalist*, 121, 335–65.

Zvereva, E. L., and Kozlov, M. V. (2006). Consequences of simultaneous elevation of carbon dioxide and temperature for plant-herbivore interactions: A meta-analysis. *Global Change Biology*, 12, 27–41.

List of Contributors

VOLUME EDITORS

Margaret A. H. Bryer
Department of Anthropology
University of Wisconsin–Madison
1180 Observatory Drive
Madison, WI 53706
USA
email: mbryer@wisc.edu

Joanna E. Lambert
Department of Environmental Studies
University of Colorado Boulder
Boulder, CO 80303-0397
USA
email: joanna.lambert@colorado.edu

Jessica M. Rothman
Department of Anthropology
Hunter College of the City University of New York
New York, NY 10065
USA
email: jessica.rothman@hunter.cuny.edu

CHAPTER AUTHORS

Shauhin Alavi
Department for the Ecology of Animal Societies
Max Planck Institute of Animal Behavior
78467 Konstanz
Germany
email: salavi@ab.mpg.de

Katharine R. Amato
Department of Anthropology
1810 Hinman Avenue
Evanston, IL 60208
USA
email: Katherine.amato@northwestern.edu

Landing Badji
Laboratoire de Biologie evolutive, Ecologie et Gestion des Ecosystemes
Departement de Biologie animale
Universite Cheikh Anta Diop de Dakar
BP 48216
10500 Dakar-Medina
Senegal
email: landing9.badji@gmail.com

York University
Glendon Campus
2275 Bayview Avenue
North York, ON M4N 3M6
Canada

Gregory E. Blomquist
Department of Anthropology
236 Swallow Hall
University of Missouri
Columbia, MO 65211
USA
email: blomquistG@missouri.edu

Stephanie L. Bogart
Department of Anthropology
1112 Turlington Hall, Room B137
PO Box 117305
Gainesville, FL 32611-7305
USA
email: sbogart@ufl.edu

Carola Borries
Department of Anthropology and Interdepartmental Doctoral Program in Anthropological Sciences
Stony Brook University, SUNY
Stony Brook, NY, 11794
USA
email: carola.borries@stonybrook.edu

Sarah Bortolamiol
CNRS-UMR 7533
Laboratoire Dynamiques Sociales et Recomposition des Espaces (LADYSS)
Campus Condorcet, 5 Cours des Humanités
93322 Aubervilliers
France
email: bortolamiol.sarah@gmail.com

Mary H. Brown
Re:wild
PO Box 129
Austin, TX 78767
USA
email: mbrown@rewild.org

Margaret A. H. Bryer
Department of Anthropology
University of Wisconsin–Madison
1180 Observatory Drive
Madison, WI 53706
USA
email: mbryer@wisc.edu

Colin A. Chapman
Biology Department
Vancouver Island University
Nanaimo, BC V9R 5S5
Canada
email: colin.chapman.research@gmail.com

Shaanxi Key Laboratory for Animal Conservation
Northwest University
Xi'an
China

David J. Chivers
Selwyn College
Grange Road
University of Cambridge
Cambridge, CB3 9DQ
United Kingdom
email: djc7@cam.ac.uk

Nancy Lou Conklin-Brittain
Department of Human Evolutionary Biology
Harvard University
Boston, MA 02138
USA
email: nancyloubrittain@gmail.com

Amelie Corriveau
Department of Anthropology
McGill University
Montreal, Quebec
Canada

School for Environmental Research
Charles Darwin University
Darwin, Northern Territory
Australia
email: amelie.corriveau@cdu.edu.au

Margaret C. Crofoot
Department for the Ecology of Animal Societies
Max Planck Institute of Animal Behavior
Bücklestraße 5
78467 Konstanz
Germany
email: mcrofoot@ab.mpg.de

Brooke Crowley
Departments of Geosciences and Anthropology
University of Cincinnati
Cincinnati, OH 45221
USA
email: brooke.crowley@uc.edu

Alex R. DeCasien
Section on Developmental Neurogenomics
National Institutes of Mental Health
National Institute of Health
Bethesda, MD 20892
USA
email: alex.decasien@gmail.com

Marie-Lyne Després-Einspenner
Éco corridors laurentiens
517 Rue St Georges
Saint-Jérôme, QC J7Z 5B6
Canada
email: mldespres@gmail.com
email: marie.despres@ecocorridorslaurentiens.org

Andrea DiGiorgio
Princeton Writing Program and Department of Anthropology
307 New South
Princeton University
Princeton, NJ 08544
USA
email: andreald@princeton.edu

Melissa Emery Thompson
Department of Anthropology
University of New Mexico
MSC01-1040 Building 11 Room 240
500 University Boulevard NE
Albuquerque, NM 87131
USA
email: memery@unm.edu

Fabiola Carolina Espinosa-Gómez
Facultad de Medicina Veterinaria y Zootecnia
Universidad Popular Autónoma del Estado de Puebla (UPAEP)
21 Sur 1103, Barrio Santiago
72410 Puebla
Mexico
email: fabiolacarolina.espinosa@upaep.mx

Annika Felton
Southern Swedish Forest Research Centre
Sweden University of Agricultural Sciences
Lomma 234 22
Sweden
email: annika.felton@slu.se

William J. Foley
Research School of Biology
Room C204 Level 2 RN Roberson Building
Australian National University
Canberra
Australia
email: William.foley@anu.edu.au

Joerg U. Ganzhorn
Animal Ecology and Conservation
University of Hamburg
Martin Luther King Platz
Hamburg, 20146
Germany
email: Joerg.ganzhorn@uni-hamburg.de

Paul A. Garber
Department of Anthropology
109 Davenport Hall
607 South Mathews Avenue
University of Illinois
Urbana, IL 61801
USA
email: p-garber@illinois.edu

AJ Hardie
Department of Anthropology
University of Wisconsin–Madison
Madison, WI 53706
USA
email: ajhardie@wisc.edu

Joseph E. Hawes
Institute of Science and Environment
University of Cumbria
Ambleside, Cumbria LA22 9BB
United Kingdom
email: joseph.hawes@cumbria.ac.uk

Eckhard W. Heymann
Verhaltensokologie and Soziobiologie
Deutsches Primatenzentrum—Leibniz Institut fur Primatenforschung
Gottingen
Germany
email: ehyeman@gwdg.de

Russell Hill
Department of Anthropology
Durham University
Dawson Building
South Road
Durham, DH1 3LE
United Kingdom
email: r.a.hill@durham.ac.uk

Primate and Predator Project
Soutpansberg Mountains
South Africa

Department of Biological Sciences, Faculty of Science, Engineering and Agriculture
University of Venda
Thohoyandou 0950
South Africa

Katie Hinde
School of Human Evolution and Social Change
Arizona State University
Tucson, Arizona
USA
email: katiehinde@gmail.com

Center for Evolution and Medicine
Arizona State University
Tucson, Arizona
USA

Brain, Mind and Behavior Unit
California National Primate Research Center

Mitchell T. Irwin
Department of Anthropology
Northern Illinois University
DeKalb, IL 60115
USA
email: mirwin@niu.edu

Charles H. Janson
Division of Biological Sciences
University of Montana
Missoula, MT 59812
USA
email: charles.janson@mso.umt.edu

Cheryl D. Knott
Departments of Anthropology and Biology
232 Bay State Road
Boston University
Boston, MA 02215
USA
email: knott@bu.edu

Andreas Koenig
Department of Anthropology and Interdepartmental Doctoral Program in Anthropological Sciences
Stony Brook University, SUNY
Stony Brook, NY 11794
USA
email: andreas.koenig@stonybrook.edu

Joanna E. Lambert
Department of Environmental Studies
University of Colorado Boulder
Boulder, CO 80303-0397
USA
email: joanna.lambert@colorado.edu

Stacy M. Lindshield
Department of Anthropology, Room STON 311
Purdue University
West Lafayette, IN 47907
USA
email: slindshi@purdue.edu

Peter Lucas
Smithsonian Tropical Research Institute
Balboa
Panama
email: peterwlucas@gmail.com

John Makombo
Department of Conservation
Uganda Wildlife Authority
Kamokya, Kampala
Uganda
email: john.makombo@wildlife.go.ug

Andrew J. Marshall
Department of Anthropology
204B West Hall
University of Michigan
Ann Arbor, MI 48109
USA
email: ajmarsha@umich.edu

Kim R. McConkey
School of Natural Sciences and Engineering
National Institute of Advanced Studies
Indian Institute of Science Campus
Bangalore
India
email: kimmcconkey@gmail.com

School of Geography
University of Nottingham—Malaysia Campus
Semenyih, Selangor
Malaysia

Amanda D. Melin
Department of Anthropology and Archaeology
University of Calgary
2500 University Drive NW, T2N 1N4
email: Amanda.melin@ucalgary.ca

Department of Medical Genetics
University of Calgary
Calgary
Canada

Alberta Children's Hospital Research Institute
University of Calgary
Canada

Lauren A. Milligan
Department of Anthropology
Mira Costa Community College
Oceanside, CA 92056
USA
email: laurennewmark@gmail.com

Lyle L. Moldawer
Sepsis and Critical Illness Research Center
Department of Surgery
University of Florida College of Medicine
Gainesville, FL 32611
USA
email: moldawer@surgery.ufl.edu

Richard Mutegeki
Makerere University Biological Field Station
Kibale National Park
Fort Portal
Uganda
email: richardmutegeki@yahoo.co.uk

Papa Ibnou Ndiaye
Laboratoire de Biologie evolutive, Ecologie et Gestion des Ecosystemes
Departement de Biologie animale
Universite Cheikh Anta Diop de Dakar
Senegal
email: ibnou.ndiaye@ucad.edu.sn

Carlos A. Peres
School of Environmental Sciences
University of East Anglia
Norwich
United Kingdom
email: c.peres@uea.ac.uk

Herman Pontzer
Department of Evolutionary Anthropology
Duke University
Durham, NC 27708
USA
email: herman.pontzer@duke.edu

Jill D. Pruetz
Department of Anthropology, ELA 228
Texas State University
San Marcos, TX 78666
USA
email: pruetz@txstate.edu

David Raubenheimer
Charles Perkins Center and School of Life and Environmental Sciences
University of Sydney
Sydney, NSW
Australia
email: David.raubenheimer@sydney.edu.au

Stephen R. Ross (deceased)
Lester E. Fisher Center for the Study and Conservation of Apes
Lincoln Park Zoo
Chicago, IL 60614
USA

Jessica M. Rothman
Department of Anthropology
Hunter College of the City University of New York
New York, NY, 10065
USA
email: jessica.rothman@hunter.cuny.edu

Andrew C. Smith
Department of Biology
Anglia Ruskin University
Cambridge
United Kingdom
email: Andrew.smith@aru.ac.uk

Matt Sponheimer
Department of Anthropology
UCB 233
University of Colorado, Boulder
Boulder, CO 80309
USA
email: matt.sponheimer@gmail.com

Eleanor M. Stalenberg
Hawkesbury Institute for the Environment
Western Sydney University
Hawkesbury Campus
Bourke Street, Richmond
2753 NSW
Australia
email: e.stalenberg@gmail.com

Tessa Steiniche
Department of Anthropology
Indiana University, Bloomington
Student Building 130
701 Kirkwood Avenue
Bloomington, IN 47405-7100
USA
email: tsteinic@indiana.edu

Karen B. Strier
Department of Anthropology
University of Wisconsin–Madison
Madison, WI 53706
USA
email: kbstrier@wisc.edu

Peter S. Ungar
Department of Anthropology
University of Arkansas
Old Main 330
Fayetteville, AR 72701
USA
email: pungar@uark.edu

Sri Suci Utami Atmoko
Falkutas Biology
Universitas Nasional
Jl. Sawo Manila
Pejaten, Ps. Minggu, Jakarta
Indonesia
email: suciatmoko7@gmail.com

Moreen Uwimbabazi
National Forestry Resource Research Institution
PO Box 1752
Kampala
Uganda
email: muwimbabazi@gmail.com

Kim Valenta
Department of Anthropology
University of Florida
Gainesville, FL 32611-7305
USA
email: valentakim@gmail.com

Adam van Casteren
Department of Human Origins
Max Planck Institute for Evolutionary Anthropology
Leipzig 04103
Germany
email: adam.vancasteren@gmail.com

Carrie C. Veilleux
Department of Anatomy
Midwestern University
19555 N55 Avenue
Glendale, AZ 85308
USA
email: cveill@midwestern.edu

Erin R. Vogel
Department of Anthropology
Rutgers, the State University of New Jersey
New Brunswick, NJ 08901
USA
email: erin.vogel@rutgers.edu

Center for Human Evolutionary Studies
Rutgers, the State University of New Jersey
New Brunswick, NJ 08901
USA

Kristina R. Walkup
Department of Biology
Des Moines Area Community College
Des Moines, IA 50314
USA
email: krwalkup@dmacc.edu

Michael D. Wasserman
Department of Anthropology
Indiana University, Bloomington
Student Building 130
701 Kirkwood Avenue
Bloomington, IN 47405-7100
email: mdwasser@indiana.edu

David P. Watts
Department of Anthropology
Yale University
PO Box 208277
New Haven, CT 06520-8277
USA
email: David.watts@yale.edu

Astri Zulfa
Falkutas Biology
Universitas Nasional
Jl. Sawo Manila
Pejaten, Ps. Minggu, Jakarta
Indonesia
email: zulfa.unas@gmail.com

Index

Page numbers in italics refer to figures.

abundance of food. *See* foods: abundance of
access and equity in science, xviii, 555
acid-detergent: fiber (ADF), 26, 167, 319, 418, 428; insoluble crude protein (ADI-CP), 417, 418, 427, *427*; insoluble nitrogen (ADIN), 418, 427; lignin (ADL), 418, 428–29; residue (ADR), 428
acquisition of food. *See* food acquisition
adaptations: anatomical, 11, 14; behavioral, 7–8, 11, 24, 37, 40–41, 46, 383, 395; co-, 96; coastal, 353; for diets and feeding, 16, 21, 34, 44, 46, 50, 400, 411; and ecosystems, 16; environmental and cultural ecology, 65; fallback, 505–6; local environmental and cultural ecology, 65; morphological, 24, 37; physiological, 24, 27, 46; progeny-specific, 60; and symbiotic fermenting microbes, 24. *See also* dental-dietary adaptations
ADF. *See* acid-detergent fiber (ADF)
ADI-CP. *See under* acid-detergent fiber (ADF)
ADIN. *See under* acid-detergent fiber (ADF)
ADL. *See under* acid-detergent fiber (ADF)
ADR. *See under* acid-detergent fiber (ADF)
Africa: baboons in, 165; coastal adaptation in, 353; diets/foods in, 35, 189, 390; foraging in, 365; forests in, 544–45; hominins in, 436; monkeys in, 168, 376, 540. *See also specific countries*
agriculture: and carrying capacity, 515; and cropland, 544; exposure to, 195; and forests, 544; and industrial food processing, 496; and nutritional ecology, 175; and species loss, 544
alarm calls, 267, 366–70, *367*, 376, 378, 380
Alavi, Shauhin E., xvii, 241, 355, 715
algae, as foods, 294, 304–6, 311, 314, 319, 404
Allenopithecus nigroviridis (Allen's swamp monkey), 90, 328
Allocebus trichoti (hairy-eared dwarf lemur), 39, 45–46, 131
Alouatta (howler monkeys), xi, 11, 499, 551; *caraya*, 90, 541; *palliata*, 1, 4, 7, 14, 23, 31, 90, *104*, 276; *pigra*, 25, 26, 29–30, 481, 536, 541; *sara*, 90; *seniculus*, 23, *133*
Altmann, Stuart, 1, 120, 473, 477, 480–81
Amato, Katherine R., xvii, 135, 137, 715
Amboseli National Park, Kenya, 120, 159, 230, 232, 234, 339–40, 342, 378–79, 549
American Journal of Primatology, 97, 516, 519
Americas, diets/foods in, 35. *See also* Central America; Latin America; North America; South America

amino acids: and body mass and size, *205*; and digestion, 141, 327; extracellular, *205*; and fatty acids, 265; and fruit, lacking in, 5; gluconeogenesis from, 208; and immunity, 203, *205*, 207–8, 211–12; and mother's milk, 53–55; and nutrition, *205*, 207, 320; and proteins, 22, 53, 203, 208, 211–12, 327, 420, 426, 428, 430
anatomy: and adaptations, 11, 14; and behavior, 11, 17, 94, 321; and diets/foods, 35, 50; and ecology, 84; and life history, 462; oral, 8–9; and physiology of mouth, 448; positional, 11
ancestry, xvi–xvii, 15–16, 38, 56–57, 109, 139, 142–44, 257. *See also* extinct species; fossils
angiosperms, 16, 95, 179, 185, 188, 550
animal-source foods (ASFs), 6, 19, 21, 50–51, 97, 400, 439, *475*; and hunting, 327, 339, 342, 345, 348–49, 351, 353–54
anorexia nervosa, female suffering from, 443
Anthropocene, 532
anthropogenic changes: and conservation of primates, xviii, 544–53; and conservation of primates, conclusions, 553; and conservation of primates, future directions, 551–52; and conservation of primates, summary, 552–53;

723

anthropogenic changes (*continued*)
and foraging, xviii, 364, 397, 544–53; and foraging, conclusions, 553; and foraging, future directions, 551–52; and foraging, summary, 552–53; and nutritional intake and balance, 271

Anthropoidea and anthropoids (apes, monkeys, simians), 31, 94, 130–31, 160, 329, 498

anthropology: and biology, 555; and mother's milk research, 66; and nonhuman primates study, 555; primate research in, 555; and stable isotopes, 445. *See also* paleoanthropology

anthropomorphism, 555

ants, as foods, 44, 101, 118, 264, 308, 314, 323–25, 386, 404, 412

Aotus and Aotidae (night or owl monkeys), 42, 48, 50–51, 74, 99–100, 228, 279, 339

AP. *See* available protein (AP)

apes. *See Gorilla* (gorillas); Hominidae and hominids (apes); Hominoidea (apes)

arboreal: animals, 418; guenon species, 267; mammals, 82, 547; monkeys, 348, 379; plants, 117; primates, 2, 11, 82–83, 275, 371, 375; refuges, 366; travel routes, 37, 284, 289, 291, 294; tunnels, ants in, 314; vertebrates, as prey, 38

archaeology, 315, 350–52, 354, 444

Argentina, 66, 384, 389–90, 541; capuchin monkeys in, 259

arthropods, 37–38, 45, 124, 138, 141, 284, 292, 302–3, 327, 400, 404–5, 411–14, 439, 537

ASFs. *See* animal-source foods (ASFs)

Asia, 35, 96, 189, 390, 545. *See also* Borneo; Eurasia; Southeast Asia; *and specific countries*

Ateles (spider monkeys), 11, 26–27, 32; *chamek*, 21, 29, 481, 483, 486; Colombian, xi; *geoffroyi* (Geoffroy's), 7, 103–4, 268, 276, 280, 387, 536; Peruvian, 483, 486

Atelidae and atelids (monkeys), 7, 41, 49, 50, 193, 329, 382, 384, 460

audition. *See* hearing (audition)

autecology, 399

Avahi (woolly lemurs), 40, 107, 125

availability of food. *See* food availability

available protein (AP), 319; and crude protein (CP), 22, 166, 426–28, 429; defined, 166; and foraging, 487; and insoluble protein, 427; intake, 24,

167, 482; and macronutrients, 261; and nonprotein energy, 320; and nutritional chemistry, 417, 418; and unavailable protein, 427

aye-aye lemurs. *See Daubentonia* and Daubentoniidae (aye-aye lemurs)

baboons. *See Papio* (baboons)

Badji, Landing, xvii, 241, 299, 715

Baranga, Deborah, xv

bark, as food, 5, 11, 37, 107, 117, 124, 134, 185, 390, *401*, 505–6

basal metabolic rates (BMRs), 41, 43–47, 71, 75, 82–84, 90, 93–94

bears, 5, 31, 301, 325, 378–79, 483

behavior: and adaptations, 7–8, 14, 24, 37, 40–41, 46, 383; and anatomy, 11, 17, 94, 321; collective, and movement, 363; constraints on, 383; and diets/foods, 35, 114, 555; and ecology, xv, 167, 196, 252, 368, 380, 542; and environment, 17, 134; and evolutionary biology, xv, 92; flexibility, and diets, xvii–xviii, 381–95, 392; flexibility, and diets, summary and conclusions, 394–95; and food choices, 381–83; and innovation, 272–73; and morphology, 17; and nutrition, 262, 463; and physiology, 92, 114, 134, 176–77, 185, 195, 197, 198, 260, 261, 359, 387, 394, 463, 489; positional, 11–12; and social nutrition, xvii, 261–71; and social nutrition, summary and conclusions, 271. *See also* behavioral ecology; feeding behavior; neurobehavior; social behavior

behavioral ecology: and environment, 134; and evolutionary ecology, 106, 134; and foodscapes, 415–16; of motherhood, 66; and primatology, 471; and social nutrition, xvii, 261–71; and social nutrition, summary and conclusions, 271

Berenty Reserve, Madagascar, 153–55, 163, 542

Beza Mahafaly Special Reserve, Madagascar, 154–55, 391, 514, 536

Bingham, Harold, 1

biochemistry, and nutrition, 463

biodiversity: and climate change, 532; and conservation of primates, 112; and ecology, 409, 497; and exotic species, 539–40; and food web theory, xviii;

losses and erosion of, 544, 550, 552; and measuring of foods, 408–9; and metabarcodinog, 404–6; and metabolomics, 172; nutritional, xviii, 463, 497

biogeochemistry, 15, 444

biology: adaptive-functional, xv; and anthropology, 555; behavioral, 57, 60; and causation, xv–xvi; cellular, *111*, 496; and culture, 321; and diet/food choices, 37; and diversity, 463; and ecology, xv, 92, 94, 321, 463–64; and evolution, 94, 464; feeding, xi–xii, xv, *xvi*, xviii; and food environments, 489–90; functional, 463; historical-evolutionary, xv; mechanistic, 463; mechanistic-physiological, xv; molecular, 1, *111*, 496; and natural selection, 463; neuro-, xi; and nonhuman primates study, 555; and nutrition, xviii, 463–64; and paleontology, 222; primate research in, 555; and proximate causation, xv–xvi; socio-, 12, 464; and ultimate causation, xv–xvi. *See also* conservation biology; evolutionary biology; neurobiology

biomechanics, 357, 461

biomedical: primate models, 173; research, 195

Blomquist, Gregory E., xvii, 19, 52, 716

BMRs. *See* basal metabolic rates (BMRs)

body mass and size: and body fat, 232; and brain size, 272; and diet/food choices, 7, 21, 36, 50, 327; of mammals, 227, 499; and nutrients/nutrition, 225–29, 239; and phylogeny, 59, 91, 229

Bogart, Stephanie L., xvii, 241, 299, 716

bone(s) and bone marrow, as foods, 8, 349–51, 354

bonobos. *See Pan paniscus* (bonobo)

Borneo, 9–10, 115–16, *116*, 132, 158, 161, 305, 307, 308, 322, 324, 390, 508, 550; langurs in, 9; macaques in, 324; monkeys in, 521; orangutans in, 115, 127, 224, 324, 460, 518; owls in, 437; plants and mammals from, 435; proboscis monkeys in, 521; seasonality of food availability in, 127; tarsiers in, 437

Borries, Carola, xvii, 68, 135, 222, 716

Bortolamiol, Sarah, xviii, 397, 544, 716

Brachyteles (muriqui monkeys), 41–42, 73, 178, 387, 395, 536

brains: and babies, 19, 82; and behavior, 82; fore-, 272; size of, 61, 65, 83–84, 86–88, 90, 92–94, 272, 297, 323, 351. *See also* cognition and cognitive ecology
breeding. *See* reproduction
Brown, Joel S., 249, 361, 364, 370–74
Brown, Mary H., xvii, 19, 82, 716
Bryer, Margaret A. H., xv, xvii, 241, 261, 715, 716
bushbabies. *See Euoticus* (needle-clawed bushbabies); *Galagoides*, Galagidae, galagos (bushbabies, nagapies, night monkeys); *Otolemur* (greater galagos, thick-tailed bushbabies)
bushmeat, 546–47, 550–53
Bwindi Impenetrable National Park, Uganda, 36; mountain gorillas in, 30, 125, 127, 161, 165, 185–86, 197, 484–85, 488, 505–6, 540; plant food items in, 185–86, 197

Cacajao melanocephalus ouakary (golden-backed uacari), 331, 366
Callicebus torquatus (collared titi monkey), 8, 10, 128
Callimico goeldi (Goeldi's monkey), 43, 47, 330, 334, 403, 413–14
Callithrix (marmoset monkeys), 5, 7, 12, 29–30, 42, 47–48, 81, 104–6, 330, 376, 413; *flaviceps*, 43, 334, 411, 413; *geoffroyi*, 330; hybrids, 414; *jacchus* (common), 29, 70–78, 90, 173, 182, 235, 273–74, 276, 280, 328, 334, 339, 359, 400, 483, 485; *pygmaea* (pygmy), 42, 90, 129. *See also Cebuella* (marmoset monkeys)
Callitrichidae, callitrichids, callitrichines (monkeys), 12, 39, 42–43, 48, 50, 70, 74–75, 78, 81, 117, 125, 235, 330, 334
calories: and animal-source foods (ASFs), 349; intake of, 71–77, 114–15, 118–19, 124–25, 128, 131–34, 208, 210, 213, 216, 222, 340, 349, 479; kilo-, 119, 121, 134; and lipids, 349; from meat, 354; sources of, 28
Cameroon, 9, 156, 161, 164, 324, 551
cancer, 11, 78, 148, 179, 182–83, 206, 213–14, 492
Cant, J. G. H., xv
Cape Floral Region, 352–53
captivity, studies, 68, 80, 112, 161, 165, 167–68, 295

capuchins. *See Cebus* (capuchin monkeys); *Sapajus* (capuchin monkeys)
carbohydrates: and adaptations, 24; cellulose as, 143; in diets and foods, 19, 21; and fats, 23–24, 34, 121, 483, 489–90; and fatty acids, 24; and hemicellulose, 24, 385; intake of, 21, 23, 478, 489, 496; lignin as, 24; and lipids, 33; and macronutrients, xvii, 21–25, 34, 143, 149, 261; nutritional, 19, 24–25; and polysaccharides, 24, 149; and proteins, 23–24, 121, 468, 478, 483, 489–90
Carlito syrichta (Philippine tarsier), 38, 327–29, 334
carnivores and carnivory, 5, 19, 139; and digestion, 7, 142, 145–46; and foraging, 374, 377–78; and hunting, 7, 333–34, 339–40, 347–50, 377–78; and isotopic foods, 433, 438, 443–45; and mother's milk, 59, 61; and senses, 109; and sweetness, taste of, 109; teeth of, 501, 510; tool-assisted, 350; trophic level of, 19; and vertebrates, 39–40. *See also* faunivores and faunivory; insectivores and insectivory; meat eating; omnivores and omnivory
Carpenter, Clarence Ray, xi, 1–2
carrying capacity, for foods, xviii, 1, 515–31; and agriculture, 515; analytical approaches, 526–28; and climate change, 531; and competition, 519; and conservation biology, 515; and critical periods, 530–31; and crucial foods, 530–31; defined, 515–17, 531; as ecological concept, 515; effects of, 518–24; and feeding ecology/studies, 17; and fisheries, 515; and food availability, xviii, 517, 527, 530–31; in future work, 531; and groups, affects, 519–23; and habitats, 17, 516, 531; and human population growth, 515; and individuals, affects, 518–19; and livestock management, 515; mechanisms influencing, 520; and nutritional approaches, 529–30; and population density, 516, 520, 522; and population ecology, 397, 515–16, 523–26, 529, 531; and protein-to-fiber ratios, 528–29; questions about, 531; and sociality, 518; summary and conclusions, 531; term, usage, 516;

understanding as important, 517–18; as varied, 531
Catarrhini and catarrhines (monkeys), 73, 96, 100, 108–9, 113, 130, 143, 293, 307, 498
Cebidae (monkeys), 39, 48, 50, 193, 330
Cebuella (marmoset monkeys), 48, 106, 413; *pygmaea/niveiventris* (pygmy), 42, 274, 330, 376, 414. *See also Callithrix* (marmoset monkeys)
Cebus (capuchin monkeys), 48, 98, 272, 310–13, 317, 323–24, 339, 460, 504; *albifrons unicolor*, 128; *apella*, 76, 87, 90, 193, 224, 483, 485, 512; *capucinus*, 22, 71, 268, 281, 303, 306, 330, 405, 413; *imitator*, 103–4
cellulose: as carbohydrate, 143; and chitin, 8; and hemicellulose, 385, 428; and lignin, 139, 427–28
Cenozoic, adaptive radiation in, 57
Center for Human Evolutionary Studies Rutgers, State University of New Jersey, 721
Central America, 389–90, 545
Cephalopachus bancanus (Horsefield's tarsier), 38–39
Cercocebus (mangabeys): *albigena*, 230; *sanjei*, 73, 76–77, 132, 331; *torquatus*, 387
Cercopithecidae, Cercopithecoidea, Cercopithecinae, cercopithecines (monkeys), 8–9, 11, 27, 29, 38, 49–50, 124, 143, 147, 165, 189, 191, 194, 266, 307, 323, 327, 331, 334, 336, 409, 441, 499, 507
Cercopithecus (monkeys): *albogularis/nictitans mitis*, 168, 368, 368, 372, 376, 540; *ascanius* (redtail), xvi, 8, 14, 25, 27, 30, 32, 185, 216, 267, 269, 336, 386, 533, 537, 550; *diana* (Diana), 332, 378, 384; guenon, 72–73, 250, 266, 267, 336, 371, 384, 409, 506; *mitis stuhlmanni* (blue), 25, 27, 32, 185, 223, 268, 270, 365–66, 445, 537, 540, 550; *neglectus* (De Brazza's), 328; *nititans*, 378
Chapman, Colin A., xviii, 14, 397, 544, 716
Cheirogaleus, Cheirogaleidae, cheirogaleids (dwarf and mouse lemurs), 39, 42–43, 48, 50, 70, 74–79, 98–99, 101, 106, 131, 227–28, 236, 274, 279, 329, 334, 377, 409, 439–40, 538

chemicals and chemistry, and nutritional analyses, 417–18. *See also* biochemistry; biogeochemistry; nutritional chemistry; phytochemicals and phytochemistry; wild plant food chemistry

chimpanzees. See *Pan troglodytes* (chimpanzee)

chitin, xvi, 8, 319, 386, 429

Chivers, David J., xv, xvii, 1, 4–7, 397, 473, 582, 716

Chlorocebus pygerythrus (vervet monkey), 41, 43, 64, *144*, 159, 332, 366, 378, 405, 445, 533

climate change: and anthropogenic changes, 381; and atmospheric CO_2 increases, 537–38; and behavioral flexibility, 381; and biodiversity, 532; causes and consequences of, xii, 542–43; and conservation of primates, 528–42; and dehydration, 542; and dietary flexibility, 392; and ecosystem changes, 541–42; and ecosystems, 541–42; and food availability, 13, 397, 542–43; and foraging, 549–50; as global threat, 546, 552–53; and habitats, 416, 541–42; and hibernation, 538; and management in practice, 538–42; and nutritional ecology, xii, xviii, 397, 532–43; and nutritional ecology, summary and conclusions, 542–43; and primates, 532–38; and PSMs, 537–38; and resilience, 541, 543; and socioecology, 1–2; and thermal stress, 533–35, 542; and water stress, 535–37, 542

Clutton-Brock, T. H. (Tim), xi–xiii, xv, 4, 255

coadaptation, 96

coevolution, 9, 15–16, 96–97, 146–47

cognition and cognitive ecology: and body mass, 272; and decision-making, 273, 297; and diet/food choices, 37, 50; and ecology, 272, 274–75, 297; and encephalization, xvii, 241, 273, 297, 323, 438; and food availability, 241; and foraging, 112, 273, 297–98, 300; and foraging, social, xxvii, 272–98; and hunting, 241, 354; and innovation, 272, 274, 296–97, 298; and instinct, 273; and intelligence, 37, 273–74, 297; and learning, 272–74, 298; and locating foods, 14; and movement/travel, 241, 273, 297, 358; and neurobehavior, 274; and planning, 37, 272–74, 296–97, 298; and primate biodiversity, 112; and problem-solving, 272–73, 296, 298; and social environments, 241; and social foraging, xvii, 272–98; and social foraging, summary and conclusions, 297–98; and social learning, 272–74; and spatial map formation, 14, 273, 275–95, 297–98; and tool use, 241, 272–73; and traveling salesperson model, 278; and visual information to navigate to feeding and sleeping sites, 276. *See also* brains

Colobinae (colobine monkeys), 5, 7, 9, 23–24, 26, 38, 40, 61, 104–5, 109, 138, 143–44, 152, 156, 163–67, 173, 176, 189–92, 194, 225, 231, 234, 327, 333, 345, 390, 448, 499, 530, 547–48, 550

Colobus monkeys, 415, 501; *guareza/guereza*, xi, 25, 30, 32, 90, *144*, 185, 433, 487, 519; *satanas*, 9, 23, 25, 38, 164. See also *Piliocolobus* (red colobuses); *Procolobus* (colobus monkeys)

Colombia, xi, 23, 128, 384, 388

community ecology, 399–402, *416*, 463

community landscape, and ecological community in foodscapes, *416*

competition: and carrying capacity, 519; and community ecology, 399–400; for diets and foods, 383–85, 399–400; and group dynamics, 13, 271; intrapopulation, 13; intraspecific, and social nutrition, 269; and polyspecific associations, 269, 384; sexual, 259–60. *See also* social food competition

Conklin-Brittain, Nancy Lou, xviii, 397, 417, 716

conservation biology, 1, 515, 544–45, 549, 551, 553

conservation of primates: and anthropogenic changes, xviii, 544–53; and anthropogenic changes, conclusions, 553; and anthropogenic changes, future directions, 551–52; and anthropogenic changes, summary, 552–53; and behavioral flexibility, 395; and biodiversity, 112; and climate change, 528–42; and food choices, 33, 34; and foraging, xviii, 544–53; and foraging, conclusions, 553; and foraging, future directions, 546, 551–52; and foraging, summary, 552–53; and nutritional ecology, 497, 538–42; and nutritional food resources, 553; and threats to survival of, 555

Corriveau, Amélie, xviii, 397, 544, 716

CPs. *See* crude proteins (CPs)

crocodiles, 301, 325

Crofoot, Margaret C., xvii, 241, 355, 716

Crook, John, xi, 2, 4

Crowley, Brooke, xviii, 397, 431, 717

crude lipids or fat (ether extract) (EE), 35, 319, 417–18, 423, 425–26, 429

crude proteins (CPs): and available protein (AP), 22, 166, 426–28, 429; fractions, *427*; and wild plant food chemistry, 417–18

crustaceans, as foods, 312

culture and cultural ecology, 65, 321, 496

Daubentonia and Daubentoniidae (aye-aye lemurs), 48, 50; *madagascariensis*, 30–31, 38, 87, 90, 98, 302–3

DeCasien, Alex R., xvii, 19, 68, 82, 717

decision-making. *See* cognition and cognitive ecology

deforestation, 218, 221, 546–48, 551, 553. *See also* reforestation

Denali National Park, Alaska, 441

dental-dietary adaptations: craniodental, 38; and deformability of foods, 454; ecological perspectives, 504–8; and evolution, 513; of extant primates, 498; and fallback adaptations, 505–6, 514; and foodprints, 498–514; and foodprints, final thoughts, 513–14; and foodprints, summary and conclusions, 514; and foraging, 514; and fossil primate diets, xii, xviii, 397, 498–514; and fossil primate diets, final thoughts, 513–14; and fossil primate diets, summary and conclusions, 514; indri, 40; and Liem's paradox, 505–6; morphology of, xviii, 513; pithecid, 38; for seeds, 35; and species-specific dietary patterns, 504–5. *See also* teeth

Després-Einspenner, Marie-Lyne, xvii, 135, 176, 717

DeVore, Irven, 1
diets, 63; animal, 19; Atelidae, *384*; and behavioral flexibility, xvii–xviii, 381–95, *392*; and behavioral flexibility, summary and conclusions, 394–95; categories and classes of, 5–7, 15–16, 35, 37–44, 50–51; chemical constraints of, 135; choices and decisions, 21, 51; constraints on, 135, 383; as diverse, 50–51, 68; early analyses, 1–2; and feeding biology, xv–xvii; finding, building, using, xi, xvii, 19; as flexible, 51, 394; global review and summary of, xvii, 35–51; intraspecific and intrapopulation variation in, 13; measuring in field, 397, 406–7; in multidimensional framework, 32, 34; patterns in, 35; profiles, 35–36, 45–46, 50–51, 191, 249, 404; quality of, 80; similarities and overlaps of, 409; species-specific, 504–5; strategies for, 15, 37, 39, 44–45, 47, 50–51, 68, 80, 382, 399–401, 406, 414, 480; variations in, 35–37, 50–51. *See also* exudates (foods); feeding and feeding ecology; foods; fossil primate diets; nutrition

digestion: and adaptations, 7–8, 40–41; and alimentary canal, 137–38, 140; and alkaloids, 140; alloenzymatic, 135, 137–38, 149; and amino acids, 141, 327; and amylase, 138, 140; anatomy of, 9, 12, 137, 140, 147; autoenzymatic, 135, 137–38, 149; basics of, 140; and behavioral adaptations, 7–8, 40–41; and Bilatera/bilateral symmetry, 137–39; and carbohydrases, 138, 141–42; and carbohydrates, 137, 142–43, 149; and catabolic process (catabolism), 138, 140; and cellulase, 45, 138, 143; cephalic phase of, defined, 138; chemical, 137–38; and chitin, 138, 141–42, 146; as compromise, 7; definitions and key terms related to, *138–39*; and deuterostomes, 137–38, 149; and diet/food choices, 21, 37, 50; in digestive tracts, 135; and disaccharides, 138–39, 142–43; and enzymes, xvii, 137–49; and fatty acids, 141, 144; and fermentation, 7, 9–10, 42–43, 59, 119, 135, 137–49, 343, 537; and foregut fermenter, 7, 138, 140, 143–44, 147, 327, 537; and fungi, 138, 141, 143, 147; gastric phase of, defined, 138; and gastrointestinal tracts, 7–8, 137–38, *144*, 149; and glycosidases, 142, 147; and gut microbiota, 144–49; and gut morphology, 7–8, 46, 147; and health, 148–49; and heterotrophy, 138, 142; and hindgut fermenter, 119, 139–40, 143–44, 147, 343; and humans, 137; inhibitors, xv–xvi, 9, 21, 32, 122; integrated understanding of, xvii, 137–49; integrated understanding of, summary and conclusions, 149; intestinal phase of, defined, 139; key terms and definitions related to, *138–39*; and lipases, 139–41; and lipolytic esterases (lipase), 139, 141; of macronutrients, 417; mechanical, 137, 139; and metabolism, 137; and microbes, xvii, 137–49; and microbiomes, 146–47, 149, 183, 264; and monoglycerides, 141; morphology, 7–8, 46, 50, 137, 147; and neurogenics, 139–40; and nutrition, 148; and oligosaccharides, 139, 142–43, 145; and peristalsis, 8, 137–40; and phenolic inhibitors, 9; and phylogeny, 146–49; and phylosymbioses, 146–47; and physiology, 9; and polysaccharides, 138–39, 141–45, 149; and primary consumers, 139, 142; and proteases, 139, 141; and protostomes, 137, 139; and secondary consumers, 139, 142; of seeds, 35; and toxins, 122, 140; and vagus nerve, 138–40. *See also* fermentation; microbes

DiGiorgio, Andrea L., xvii, 19, 114, 717
disaccharides, 24, 138. *See also* monosaccharides; polysaccharides
DM. *See* dry matter (DM)
dry matter (DM), 26, 418, 421, 423–24
Dunbar, Robin, 4
durometers, 446
Dzanga-Ndoki National Park, Central African Republic, 505

ears. *See* hearing (audition)
eating. *See* diets; feeding; foods; nutrition

ecological cognition. *See* cognition and cognitive ecology
ecological community. *See* community ecology
ecology: and behavior, xv, 167, 196, 252, 368, 380, 542; and biodiversity, 409, 497; and biology, xv, 92, 94, 321, 463–64; and culture, 321; and diets/foods, 35; and ecosystems, *10*; and evolution, xii, xv, 92, 94; and fear, 364; and feeding behavior, 196; and foodscapes, 415–16; and nutrition, 463; paleo-, 494, 498, 508, 511; and paleontology, 513; and primates, grades of, 2, 4; and primates, questions about, xii; and resilience, 395; restoration, 95, 539; and sociality, 261; and species differences, xii. *See also* behavioral ecology; cognition and cognitive ecology; community ecology; culture and cultural ecology; evolutionary ecology; feeding and feeding ecology; movement and movement ecology; nutritional ecology; population ecology; senses and sensory ecology; socioecology; species ecology

Ecology of Arboreal Folivores, The (Montgomery), 175
ecosystems: and adaptations, 16; and complex, 416; conservation of, 95; degradation and disturbance of, 16; and ecology, *10*; and feeding ecology, 16; and nutritional ecology, 541. *See also* habitats
Ecuador, 27, 83, 129, 133, 291, 384
EDCs. *See* endocrine-disrupting chemicals (EDCs)
EE. *See* crude lipids or fat (ether extract) (EE)
Eisenberg, J., 2, 4
elastic modulus, 447–48, 451–53
elephants, 58, 183, 227–28, 350, 437, 441, 499
Emery Thompson, Melissa, xvii, 19, 68, 218, 717
encephalization. *See* cognition and cognitive ecology
endangered species, 134, 395, 497, 532, 539, 551–52
endocrine-disrupting chemicals (EDCs), xi, xvii, 194–96
endocrinology, 132, 135, 177, 197

energetics and energy: and activity budget analyses, 94; and anatomy, 82; and babies, 82–84; and behavior, 82–94; and biology, 94; and BMRs, 82–83, 94; and body mass and size, 68, 70, 84, 86, 94; and brains, 82–94; and calories, 210; and captivity, 92; and chemistry, 430; and diet/food choices, 37, 51; and dry matter (DM), 28, 343; and ecology, 82, 90, 92, 94; and evolution, 93–94; expenditure and allocation, 92–93; and food chemistry, 430; and food choices, 27–29, 33–34; and foraging, 134, 397, 477–79; and humans, 87, 90; intake, xvii, 21, 27, 29, 32, 61, 70–73, 79, 81, 86, 94, 114, 135, 198, 212–13, 216, 236, 265, 268, 270, 278, 317, 320, 343, 345–46, 349, 357, 382–83, 389, 399, 467, 469, 473–74, 477–82, 483, 486, 490–94; and life history, xi, 82, 90, 92, 135; and macronutrients, 27–29, 34, 135, 397; maternal, 62, 225; maximization, 21, 28, 120, 361, 465, 468, 476–79; and metabolics, 94; and micronutrients, 397; and mother's milk, 52; and movement, 359; and natural selection, 68, 92; and nutrients/nutrition, 114, 135, 198, 221, 241, 477–79, 488, 555; and phylogenetics, 82, 84–86, 90; and pregnancy, 69–71; and protein intake, 32; and public health, 83–84; questions about, 29, 343; and ranging behavior, 94; of reproduction, xvii, 68–82; of reproduction, conclusions, 80–81; of reproduction, summary and questions for research, 80; requirements, xi, xvii, 82–94; requirements, summary and conclusions, 84; and seasonality, xvii, 114–34; and senescence, 84, 93, 94; and sex differences, 92–93; and socioecology, 83; and socioeconomics, 83–84; statistical analyses of, 85–86; and stress, 394; term, usage, 479; total expenditure (TEE), 82–94
environmental science, and public health, 195
environment(s): and behavior, 17, 134; and diets/foods, 68, 97, 112, 445; and feeding ecology, 10, 17; fluctuating, 134; and morphology, 17. *See also* social environments

enzymes. *See* digestion
Eocene, 16
epigenetics, xviii, 78, 477. *See also* genomics
epiphytes, 117, 133, 267
equity and access in science, xviii, 555
Erythrocebus patas/baumstarki (patas monkey), 43, 191, 223, 332, 539
Espinosa-Gómez, Fabiola, xviii, 397, 544, 717
Estación Biológica Quebrada Blanco, Peru, 36, 334
ether extract (EE). *See* crude lipids or fat (ether extract) (EE)
Eulemur (brown lemurs), 48, 51, 87, 90, 100, 105, 108, 124, 128, 154–55, 163, 192, 228, 279, 329, 388, 505, 533, 536
Euoticus (needle-clawed bushbabies), 39, 42, 45, 48, 99, 106, 329
Eurasia, 350–51, 353
Europe: forests in, 545; hominins in, 350–51; Neanderthals in, 354; population in, 544; PSFs in, 351; TEE in, 83
evolution: and biology, 94, 464; co-, 9, 15–16, 96–97, 146–47; and ecology, xii, 94; and food choices, 34; grades of, and species ecology, 2, 4; of humans, 1, 33, 64–66, 141, 225, 300, 327–28, 345, 348–54; and nutrition, 463, 496; questions about, xii; and species differences, xii; and species ecology, 2, 4; synthesis theory of, xv. *See also* life history
evolutionary biology: and behavior, xv, 92; and ecology, xv, 92; and physiology, 92
evolutionary ecology, and behavioral ecology, 106, 134
exotic species, 376, 539–41; and biodiversity, 539–40
extant primates, xv; and body mass and size, 48–49; conservation of, xviii; dental-dietary adaptations of, xviii, 498; and exudates (foods), 42–43, 45, 47–51; and food physics, 462; global survey of diets and foods, xvii, 35–51, 48–49; global survey of diets and foods, introduction, 35–37; global survey of diets and foods, summary and conclusions, 50–51; insects in diets of, 142; measuring abundance of, 413–14; and tool use, 300

extinct species, xv; ancient DNA of, 509; foraging strategies of, 513, 514; fossils of, 461, 509; life history of, 223. *See also* ancestry; fossils
exudates (foods), 36; adaptations for, 42–43; availability of, 117, 414; carbohydrates in, 24, 42–43; defined, 402; in diets, 6, 50–51, 112, 292, 399, 400, 402, 413–14; and extant primates, 42–43, 45, 47–51; and foraging, 274, 292; and gouging, 42–43; measuring, 400–403, 401, 410, 413–14; and nutrients, 24, 30; and olfaction, 99, 112; as poor quality food sources, 42–43; and resins, 402–3; seapod, 43; sources of, 43; tree, 30. *See also* exudativores and exudativory; gums, as foods; saps, as foods
exudativores and exudativory, 35–36, 39, 42–43, 47, 117, 399, 413–14. *See also* exudates (foods); gummivores and gummivory; gums, as foods; saps, as foods
eyesight. *See* vision (sight)

fats: and carbohydrates, 23–24, 34, 121, 483, 489–90; in diets and foods, 19; and fatty acids, 213, 426; intake of, 23, 71, 489; and lipids, 27, 417; and macronutrients, 19, 27, 34, 261; nutritional, 19, 27, 426; polyunsaturated, 199, 212; saturated, 213; seasonal storage/metabolism of, 130–31, 134. *See also* crude lipids or fat (ether extract) (EE)
fatty acids: and amino acids, 265; and carbohydrates, 24; and crude lipids, 426; fats, 213, 426; and lipids, 27; long-chain, 65; as macronutrients, 24, 27, 420; and meat eating, 345; as micronutrients, 420; and monoglycerides, 141; polyunsaturated (PUFAs), 212; and saturated fats, 213; short-chain, 24, 139, 144
faunivores and faunivory: and body mass/size, 7, 327; and carnivory, 38–40; and diets/foods, 5–6, 15–16, 35, 37–40, 42, 44, 47, 50, 399; and digestion, 8, 141; and hunting, 327, 334, 342–43; and insectivory, 38–39; and mother's milk, 61; and prey, 409–11; questions about, 444; and sense, 101, 107; and stable isotopes, 435,

439–40, 444; and teeth, 8. *See also* carnivores and carnivory; meat eating

fear ecology. *See* landscape of fear

feeding and feeding ecology: and biology, xi–xii, xv, *xvi*, xviii; and community, *10*; defined, 299, 399; and diets/foods, xv–xvii; and environment(s), *10*, 17; factors determining, *10*; and individual(s), *10*; inter-and intraspecific differences in, xi; neighborhoods, and movement ecology, xvii, 241, 355–63; and nutrition, xv–xvii; and observational field studies, 34; physics of, *447*; and positional behavior, 11–12; subdisciplines of, 17. *See also* diets; feeding behavior; feeding studies; foods; nutrition

feeding behavior, xvii–xviii, 11–13, 21, 555; and availability of food, quantifying, 400; and caloric/nutrient intake, 118; and diets/foods, xi, 1, 9, 46; and ecology, 196; and environment, 17; and food physics, 462; and food quality, 34; and life history, xi; mechanistic explanations for, 46; and morphology, 17, 462; multidisciplinary perspectives, 9; and predation risk, 17; quantifying and measuring, 4, 400–407; and seasonality, 119; understanding of, 46, 118. *See also* feeding and feeding ecology; feeding studies

feeding studies: of ancestral primates, 15–16; and behavior, 17; and chemistry, 151, 168–69; and cognition, 14; and cognitive processes in locating foods, 14; and dietary categories, 5–7, 50; and diets, categories of, 5–7; and diets, early analyses, 1–2; and early dietary analyses, 1–2; and ecological grades of primates, 2, 4; and ecological roles, 15; and ecology, 2, 4, 15; and ecosystem disturbance, 16; and ecosystems, 16; and environment, 17; and environment(s), 17; evolution of, 1–17; evolution of, timeline, *3*; and feeding behavior, 17; and feeding ecology, history of, 17; and food availability, 13, 14; and foraging, optimal, 12–13; future of, 17; and global diet review, 50; and gut morphology, 7–8; and habitat losses, 17; history of, 17; in human-modified regions, 16; and integrative approaches, 17; and morphology, 7–8, 17; and multidisciplinary perspectives, 9; and nutrition, 14–15; and oral anatomy, 8–9; and phylogeny, 15–16; and positional behavior, 11–12; and resource availability, 13; and sociological synthesis, 4–5; and sociology, 4–5; and species interactions, 13–14; subdisciplines of, 17; summary and conclusions, 17; and technology, 17; timeline, *3*. *See also* feeding and feeding ecology; feeding behavior

Felton, Annika, xvii, 19, 21, 717

fermentation: bacterial, 5; and carbohydrates, 24, 149; and digestion, 7, 9–10, 42–43, 59, 119, 135, 137–49, 343, 537; and fatty acids, 24; and microbes, 7, 45, 106, 135, 137, 139, 142–45, 147, 404; microbial, defined, 139. *See also* digestion

ferns, as foods, 179

fiber: hemicellulose of, 428; indigestible, 261; and macronutrients, 261; and proteins, 10, 21, 25–27, 34, 262, 528–29, 547

first diet. *See* mother's milk

flavanones, *171*

flavonoids, 178–79, 181

florivores and florivory, 15, 35, 41–42. *See also* flowers

flowers, as foods, 5–12, 15–16, 22, 24, 35–36, 40–44, 47–51, 99–102, 106, 112, 117–18, 124, 133, 162, 178–79, 183, 186, 274, 291, 394, 400–402, 410–16, 473, 475, 505, 533, 536. *See also* florivores and florivory; nectarivores and nectarivory

Foley, William J., xvii, 135, 150, 717

foliage, as food, xi, 5, 37, 40–41, 45–51, 98, 100, 112, 266, 460. *See also* folivores and folivory; leaves, as foods

folivores and folivory, 5–15, 23–25, 29–32, 35, 37–47, 50, 61–62, 83–84, 92, 99, 103–7, 119, 122, 124, 150–51, 166–69, 171, 175, 185, 188–89, 194, 197, 249–50, 253–54, 382, 385–94, 399, 410, 415, 433, 444, 460, 475–76, 499–501, 505, 510–11, 518, 528–29, 533, 536, 545–53; frugi-, 9. *See also* foliage, as food; leaves, as foods

food acquisition: and food processing, 9; and nutrition, xi, xvii, 241; and oral processing, 448; and social competition, 243–60; and social competition, conclusions, 260; in social environments, xi, xvii, 241

food availability: and behavior, *392*; and behavioral flexibility, 388–90, *392*; and comparative analyses., 239; and feeding behavior, 400; and feeding responses to, 13; and foodscapes, xvii; and measurements, 236–37, 239, 400; and nutrition, 21, 410; over time in two habitats, *527*; and seasonality, xvii, 14, 19, 114–34, *123*, *126*, 198, 400; and seasonality, summary and conclusions, 134; and socioecology, 1–2, 399; and standards for comparative analyses, 239

foodprints: and dental-dietary adaptations, 498–514; and dental-dietary adaptations, final thoughts, 513–14; and dental-dietary adaptations, summary and conclusions, 514; and dental wear, 510–13, *512*; and food debris, 509; and fossil primate diets, xviii, 498–514; and fossil primate diets, final thoughts, 513–14; and fossil primate diets, summary and conclusions, 514. *See also* foodscapes

foods: abundance of, xviii, 72, 74, 118, 130, 237, 243–44, 247–50, 258, 260, 265, 358, 383–91, 395, 400, 410–16, 526, 531; choices of, 13, 21–34, 35–37, 150, 381–83, 399, 498; dirty, 457–60; distribution of, 243, 247–50, 386–87; diversity in, measuring, 408–10; evaluating, 95–113; fallback, 121–22, 505–6, 531; fallback, defined, 506, 529; filler, 14, 122, 531; finding and locating, 14, 19, 95–113; industrial processing, and agriculture, 496; isotopic, 438–40, 444–45; measuring in field, xviii, 397, 399–416, *401*; measuring in field, defined, 400; measuring in field, mechanical properties, 451–57; measuring in field, summary and conclusions, 416; mechanical properties of, xviii, 397, 446–62; mechanical properties of, definitions and terminology, 448–50; mechanical properties of, measuring, 451–57; mechanical properties of, summary and conclusions, 462; and nutrition, xv–xvii, 19, 21, 33–34, 135,

foods (*continued*)
397, *416*, 553; physics of, 450–51, 460, 462; preferred, 121–22, 408–9, 514; processing of, xi, xvii, 8–9, 135, 385, 397, 446, 499, 510; quality of, xii, 12, 16, 25, 28–29, 33–34, 102, 112, 251, 254, 265, 268, 317, 386, 487, 526, 530, 542; quantity of, 293, 406–7, 531; scarcity of, 13, 79, 115, 122, 128, 131, 381, 386, 390, 400, 526–28; selection of, 11, 15, 21–27, 31, 34–35, 95, 102, 121, 151, 166, 382, 461, 483; staple, 14, 122, 165, 185, 250, 454, 548; taxonomic resolution of, 404–6; types of, xvii, *401*, 401–4. *See also* diets; exudates (foods); feeding and feeding ecology; food acquisition; food availability; nutrients; nutrition; *and specific foods*

foodscapes, xviii, 401, 415–16. *See also* foodprints

food science, *111*, 449

food seasonality. *See* seasonality

foraging: and adaptations, anatomical, 11; and adaptive radiation, 95, 109, 112; and anthropogenic changes, xviii, 364, 397, 544–53; and anthropogenic changes, conclusions, 553; and anthropogenic changes, future directions, 551–52; and anthropogenic changes, summary, 552–53; and behavioral ecology, 379; and behavioral flexibility, 386–87; and climate change, 549–50; and cognition, 112, 273, 278, 291–95, 297–98, 300; and community, *10*, 416; and conservation of primates, xviii, 544–53; and conservation of primates, conclusions, 553; and conservation of primates, future directions, 546, 551–52; and conservation of primates, summary, 552–53; defined, 299; and dental-dietary adaptations, 514; and ecology, 95, 114; and embodied capital, 353; and environment, *10*; and episodic memory, *292*; of extinct species, 513, 514; and exudates (foods), 274, 292; facilitation, defined, 245; and fear, defined, 364; and fear, in landscape of, xvii, 241, 364–80, *416*; and fear, landscape of, summary and conclusions, 380; and fear ecology, key terms used in, 365; and food choices, 498; and foodscapes, *416*; and forests, 544, 553; as goal oriented, 273; and habitats, destruction of, 546–48; hypothesis-based, *292*, 296–97; and immunity, 221; and indirect risk, 365; and innovation, 272; and intrinsic risk, 365; juvenile, 62; and linear programming (LP), 473, 479–81; and memory, episodic, *292*; and mixture triangles, 473; multimodal, 97; and natural selection, 105, 107, 120, 471–72; and nutrients/nutrition, 46, 62, 95, 221, 463, 466, 471–79, 483–87, 496, 552–53; optimal, 12–13, 21, 23, 33, 46, 95, 119–20, 278, 291–95, 328, 397, 471–79, 487; phenotypic plasticity in, 241; and physiology, 134; and positional behavior, 11–12; and predation, 365–66, 380, 416; and social landscapes, 415–16; and social organization, 95; and socioecology, 241; success, measuring, 409–10; and tool use, 299–301. *See also* ranging behavior; social foraging; tool-assisted foraging and feeding (TAF)

forests: annual changes in areas by region, 1990–2010, 544–45, *545*; degradation, and foraging, 544, 553; disturbed, *393*; feeding and sleeping sites in, *276*; and food resources, 553; fragmentation of, and habitat destruction, 16, 517, 546–48, 552–53; logging in, and habitat destruction, 16, 517, 546–48, 552–53; montane, 338, 340, 505; and primates, 555; productivity cycles of, 9; and reforestation, 95; regeneration of, *552*; restoration of, 95, 539. *See also* deforestation; habitats; rainforests

Fossey, Dian, 4, 505, 601

fossil primate diets: and dental-dietary adaptations, and foodprints, xii, xviii, 397, 498–514; and dental-dietary adaptations, and foodprints, final thoughts, 513–14; and dental-dietary adaptations, and foodprints, summary and conclusions, 514; reconstructing, 397, 498–514; reconstructing, summary and conclusions, 514. *See also* foodprints

fossils, xviii, 498–99, 504–5, 509–10; of extinct species, 461, 509; of hominins, 15, 461–62; micro-, 509, 514; and tool use, 300. *See also* ancestry; extinct species; fossil primate diets

frugivores and frugivory: and amino acids, 382; arboreal, 117; and Atelidae diet, *384*; and behavior, 382–85, 388–89, 395; and carrying capacity, 526; and climate change, 533, 536, 541; and diets/foods, 5–8, 13, 15, 35–47, 50, 95–96, 399, 410; and estrogenic foods, 185, 188–89, 194; and foraging, 294, 373, 545–50, 553; and movement, 358; and mutualism, 95; and nutrition, 27, 475; and reproduction, 73; and seasonality, 117, 125; and senses, 95–96, 102–6, 109; and social food competition, 249–50, 253, 269–70; and stable isotopes, 44, 435; and teeth, 460–61, 499–501, 510–11; and tool use, 324. *See also* fruits, as foods

fruits, as foods, 2, 5–11, 15–16, 19, 21, 25, 29–30, 32–37, 40–41, 46, 50–51, 72, 95–101, 104–5, 112, 115–18, 122–28, 132–33, 164–65, 178, 188–89, 197, 215, 224, 247, 249, 260, 268, 272, 275, 293–94, 299, 310, 313, 320–26, 343–44, 355, 357, 386, 390, 393, 399–402, 407, 410–11, 416, 429, 433, 435, 442–46, 460–61, 473, 475, 486, 505–8, 535–38, 550. *See also* frugivores and frugivory; seeds, as foods

fungi, as foods, 5, 43, 47, 179–80, 274, 387, 400–401, 403–4, 414

Galagoides, Galagidae, galagos (bushbabies, nagapies, night monkeys), 38–39, 48, 50, 70, 131, 178, 191, 304, 309, 313–15, 327, 329, 334, 338–41, 371, 413. *See also Euoticus* (needle-clawed bushbabies); *Otolemur* (greater galagos, thick-tailed bushbabies)

Ganzhorn, Jörg U., xvii, 4–5, 8, 14, 135, 150, 163, 166, 717

Garber, Paul A., xvii, 5, 12–14, 97–98, 241, 252, 272, 717

Gartlan, Steve, 2, 4

geladas. See *Theropithecus gelada* (gelada)

genetics and genomics: and evolution, 106–7; meta-, 173, 416; and metabolomics, 172–73, 175; olfactory, 109, *111*. *See also* epigenetics; phylogenetics

geometric framework for/of nutrition (GF), 80, 119, 121, 216, 320, 464; defined, 466. *See also* nutritional geometry; nutritional geometry framework (NGF)
geometry, nutritional. *See* nutritional geometry
geophagy, 30, 43, 72, 174
GF. *See* geometric framework for/of nutrition (GF)
gibbons. *See Hylobates*, Hylobatidae, hylobatids (gibbons); *Nomascus* (gibbons); *Symphalangus syndactylus* (siamang gibbon)
Glander, Ken, xv
glycosides: cardiac, 180; cyanogenic, 151, 157, 161, 164, 176. *See also* saponins and saponosides (glycosides)
Go-Between, The (Hartley), 513
Goeldi's monkeys. *See Callimico goeldi* (Goeldi's monkey)
Goodall, Alan, 4
Gorilla (gorillas): cognition of, 273; diets of, 5, 25, 43, 90, 324; as folivores, 188; and stinging nettles, ingestion of, 273; and tool use/TAF, 307; western lowland, 27, 125, 324, 460, 505–6, 509. *See also Gorilla beringei beringei* (mountain gorilla)
Gorilla beringei beringei (mountain gorilla), xi, 23–25, 30, 32, 44, 121–25, 167, 178, 185, 197, 238, 268, 271, 307, 324, 382, 461, 483–88, 505–6, 534, 539, *540*
gouging, 42–43, 400, 414
GPS technology, 297, 359, 374–75, 377, 380, 446
graminivores and graminivory, 40–41, 44, 388. *See also* grasses, as foods
granivores and granivory, 37–38, 40, 193–94, 400. *See also* seeds, as foods
grasses, as foods, 8, 12, 28, 35, 40–41, 46–49, 179, 274, 388, 402, 433–38, 441, 462, 476, 509–14, 551. *See also* graminivores and graminivory
group dynamics: and competition, 13, 271; and food distribution, 13
guenons. *See under Cercopithecus* (monkeys)
gummivores and gummivory, 106, 189, 193–95; and exudativory, 42–43, 117, 399. *See also* exudativores and exudativory; gums, as foods; saps, as foods

gums, as foods, 5–6, 12, 24, 30, 35, 42–43, 47–49, 106, 117, 311, 313, 402–3, 414, 473, 475. *See also* exudates (foods); gummivores and gummivory; saps, as foods
Gunung Palung National Park, West Kalimantan, Borneo, Indonesia, 115–16, 132, 390, 519, 525; female orangutan and seven-month-old infant feeding on fruit in, *116*; gibbons and leaf monkeys in, *522*
gustation. *See* taste (gustation)
gut. *See* digestion
gymnosperms, 179

habitats: anthropogenic changes of, 134, 397; and behavioral flexibility for diets, 386–91; and climate change, 416, 541–42; degradation, destruction, disturbance, fragmentation, losses of, 1, 16–17, 134, 218, 221, 237, 376, 387–88, 393, 395, 416, 517, 521, 546–49, 552–53; and feeding ecology, 17; and food availability, 397; and food choices, 21, 33, 34; heterogeneity of, 386–87; logging in, 546–48; management of, 1; marginal, behavioral flexibility in, 381, 390–91; micro-, 5, 112, 128, 275, 278, 291, 369, 373, 411, 444, 462, 535; and nutritional ecology, 541; restoration of, 33, 34, 541–42; roads in, 195, 291; and socioecology, 1–2; tropical, 21, 188–89, 212, 415–16. *See also* ecosystems; forests; landscape of fear
Hapalemur (lemurs), 41; biomass of, 10; and color vision, 113; feeding of, 327, 441, 555; golden bamboo feeding on giant bamboo shoots, *36*; and phenological cycle, 504; and PSMs, 163, 176; ring-tailed feeding on tamarind leaves, *36*; and stable isotopes, 445; subfossil, 437, *437. See also Avahi* (woolly lemurs); *Cheirogaleus*, Cheirogaleidae, cheirogaleids (dwarf and mouse lemurs); *Daubentonia* and Daubentoniidae (aye-aye lemurs); *Eulemur* (brown lemurs); *Indri*, Indriidae, indriids (lemurs); Lemuridae and lemurids (lemurs); *Lepilemur* and Lepilemuridae (lemurs); *Propithecus* (sifaka lemurs)

Haplorhini and haplorhines (apes, monkeys, tarsiers), 7, 59, 75, 130, 329, 334
haptic sense. *See* touch (haptic sense)
Harcourt, Sandy, 4
Hardie, AJ, xvii–xviii, 241, 381, 717
Hartley, L. P., 513
Hawes, Joseph E., xvii, 19, 35, 68, 717
hearing (audition), 35, 38; for locating/evaluating foods, 95–96, 98, 100, 105–6, 110, 112–13; and mechanosensation, 110, 113
hemicellulose: and carbohydrates, 24, 385; and cellulose, 385, 428; of fibers, 428; of neutral detergent fiber (NDF), 119, 428; and polysaccharides, 24; and saccharides, 142
herbivores and herbivory, xvii, 5, 13, 19–22, 28–29, 31–33, 37, 42, 61, 107, 139, 142, 145–46, 150–52, 165–68, 172, 176, 179–83, 196, 374, 432–33, 438–45, 459, 478, 501–3, 510, 528, 534, 542, 547. *See also* exudativores and exudativory; florivores and florivory; folivores and folivory; graminivores and graminivory; herbs, as foods; omnivores and omnivory; palynivores and palynivory; plants
herbs, as foods, 35, 124, 165, 179, 294, 307, 433
Heymann, Eckhard W., xviii, 397, 399, 718
hibernation, 131, 227, 538
high altitudes, and diets/foods, 390
Hill, Russell A., xvii, 241, 364, 365–66, 718
Hill, W., 7
Hinde, Katie, xvii, 19, 52, 718
Hladik, Annette, 9
Hladik, Marcel, 4, 5–7, 9
Hohmann, Fowler, 165
holistics, 59, 172
Hominidae and hominids (apes), 49–50, 210, 333, 498–99. *See also* apes; Hominoidea (apes)
hominins, 15, 65, 84, 94, 300, 325, 328, 348–51, 353–54, 435–36, 444–45, 461–62, 502, 509. *See also* humans
Hominoidea (apes): and angiosperm radiation, 16; behavior of, xi; and biomedical research, 195; and cognition, 273–74, 279–90, 293, 297; as dexterous, 105; and fallback foods, 530; feeding preferences of, 165–66;

Hominoidea (*continued*)
  habitat use by, 530; and hunting, 333; life history of, 237–38; and low fat diets, 33; and phenological cycle, 504; and phytoestrogens, 189; and PSMs, 152, 163, 165–66, 174; and senses, 11; siamang, xi; small (Hylobatidae), 11; and social foraging, 273–74, 278, 297; and tool use/TAF, 274, 300, 307, 314–15, 326. *See also* Hominidae and hominids (apes)

Homo sapiens. See humans

honey, as food, 43, 263–64, 294, 301–6, 308, 311, 313–14, 317–19, 322–23, 326, 352

hormones: degradation of, 181; and microbiomes, 194–96, 555; and phytochemicals in diets, xvii, 176–97; and phytochemicals in diets, future investigations and research, 194–97; and phytochemicals in diets, summary and conclusions, 197. *See also* phytochemicals

Huffman, Michael, 11

humans: and appetites, 489–90; and biocultural niche construction, 354; and cognition, 272; and digestion, 137; and hormones, 178; and hunting, 328, 345, 348–54; morbidity and mortality in, 209; and mother's milk, 64–66; and nutrition, 225, 489–96; nutritional ecology of, 488–96; as omnivores, 5; and phylogeny, 90; and phytochemicals, 176; population growth, and carrying capacity, 515; processed foods, in diets of, 494, *494*; and reproduction, 68, 81; ultraprocessed foods, in diets of, 494, *494*. *See also* evolution; hominins

hunting: and animal-source foods (ASFs), 327, 339, 342, 345, 348–49, 351, 353–54; and behavioral ecology, 328, 348; and body mass and size, 327; and body mass/size, 327; and cognition, 241, 354; and consumption of vertebrate prey, *329–33*; cooperative, 241, 328, 345–48, 353–54; defined, 328; and evolution, 327–28, 348–53; and humans, 328, 345, 348–53; incidence and frequency, 328–39; and life history, 351–54; and meat eating, 327–28, 339–45, 349, 351–54; in montane forest habitats, 338, 340; and morphology, 327; and mortality, 258; and mutualism, 347, 354; pressure of, 378, 521; and prey, as food, 97, 107, 334–39, *401*; and prey, taxonomic distribution of, 328, 334; and primary consumers, xvii; by primates, xvii, 327–54; by primates, questions and future research, 342–43, 353–54; by primates, summary and conclusions, 354; of primates by primates, 334; in rainforests, 338; and tool use, 327. *See also* bushmeat; meat eating; predation

Huxley, Julian, xv, 621

Hwange National Park, Zimbabwe, 374

Hylobates, Hylobatidae, hylobatids (gibbons), 11, 37, 49–50, 498–99; *albibarbis*, 390, 519, 522; *hoolock*, 37, 49, 51, 333; *Kloss*, 8, 10, 378; *lar*, 1–2, *2*, 508; movement of, 355. *See also Nomascus* (gibbons); *Symphalangus syndactylus* (siamang gibbon)

immunity and immunology: and acute phase response sickness syndrome, *206*; and allostasis, 198, 203; and amino acids, 203, *205*, 207–8, 211–12; and anabolic processes, *207*; and antagonistic effects, 198; and behavior, 198; and ecology, 221; and energy/energetics, 135, 198, 206–10, 221; and fatty acids, 199, 207–8, 212–14; and foraging, 221; and functional redistribution of the body cell mass, *205*; and immunoresponse, 135; and infection, *205, 206*, 211; and inflammation, 198, *206–8*, 210–16, *213*; and injuries, *208*; and land use, 221; and life history, 135; and lipids, 212–14; and macronutrients, 135, 198; methodologies for studying, 217–21; and micronutrients, 198, 214–16; and minerals, 198; and mother's milk, 66–67; and nutraceuticals, defined, 212; and nutrition, xvii, 198–221; and nutrition, summary and conclusions, 221; and pathogen resistance, 221; and physiology, 198, 221; and protein intake, 203–5, 208, 210–12, 221; stress, *219–20*; studying, methodologies for, 198, 217–18, 221; and synergistic effects, 198; and vitamins, 198; and wildlife reserves, 221

inanition. *See* starvation

India, 157, 222, 230, 232, 234, 537

Indonesia, 23, 36, 116, 120, 217, 224, 230, 232, 234, 376, 378, 519

Indri, Indriidae, indriids (lemurs), 4–5, 7, 40–41, 48, 50–51, 100, 153, 163, 192, 253, 521

industrial food processing, and agriculture, 496

infanticide, 79, 244, 249, 251, 254–55, 258–60, 378–79, 517–19

infant mortality, 76–77, 81, 214, 536

infectious diseases. *See under* immunity and immunology

inflammation. *See under* immunity and immunology

Inge, William Ralph, vi

innovation. *See* cognition and cognitive ecology

insectivores and insectivory, 2, 5, 7, 30, 38–40, 43–47, 59, 98, 118, 130, 141–42, 189, 194, 250, 264, 321–22, 399, 409–10, 412, 438, 499, 501, 510, 548. *See also* insects

insects: calories from, 45; and chitin, xvi, 386, 429; in extant primates diets, 142; as foods, 8, 19, 33, 35, 45, 72, 99, 107, 112, 275, 311, 313–15, 322, 324, 326, 411–13, 501; nutrients from, 45. *See also* insectivores and insectivory; *and specific insects*

intelligence. *See* cognition and cognitive ecology

International Journal of Primatology, 516, 519

invertebrates: adaptations of, 38; as dietary class, 38–39; as foods (prey), 27, 35, 51, 97, 107, 112, 327–28, 400–401. *See also* vertebrates

Irwin, Mitchell T., 397, 532, 718

isotopes. *See* stable isotopes

Janson, Charles H., xvii, 241, 243, 718

Jay, Phyllis, 1

Kahuzi-Biega, 335, 338

Kay, Rich, xv, 5

Khao Yai National Park, Thailand, 232, 234; lar gibbons eating in, *2*

Kibale National Park, Uganda, 25–27, 30, 36, 115, 122, 124, 177, 197, 267–70, 293, 373, 389, 391, 429, 506–7, 521, 537, 546, 550–51; agricultural land

adjacent to, 196; chimpanzee in, 188; forest regeneration in, 552; gray-cheeked mangabey forages in same tree as blue monkeys in, 270; plant food items in, 185–87, 187, 197; red colobus monkey in, 187; redtail monkeys in, foraging, xvi
Kinzey, Warren, 4, 446
Kirindy Forest, Madagascar, 131, 154
Kirindy Mitea National Park, western Madagascar, Verreaux's sifaka in dry deciduous forest in, 103
Kleiber's law (Kleiber, Max), 7, 82, 94, 499
Klein, Lewis and Deborah, 4
Knott, Cheryl D., xvii, 19, 114, 718
Koenig, Andreas, xvii, 68, 135, 222, 718
Korup National Park, Cameroon, 551
Kruger National Park, South Africa, 440–41, 445; CAM plants from, 434
Kutai National Park, Borneo, 161, 306

Lagothrix lagotricha (woolly monkeys), 11, 27, 144
Lajuma Research Centre, South Africa, 718
La Lope National Park, Gabon, 161, 444
Lambert, Joanna E., xv, xvii, 14, 19, 21, 37, 117–19, 122, 135, 137, 166, 254, 261, 376, 381, 399, 464, 485, 506, 528, 715, 718
landscape of fear: and anthropogenic changes, 364, 375–76; and behavior, 380; definitions and key terms, 364–65, 416; and ecology, 364, 380; and experiental titrations of risk, 370–74; foraging in, xvii, 241, 364–80; foraging in, summary and conclusions, 380; giving-up density (GUD) in, 366, 370–72, 374, 379–80; and habitats, 366, 380; and human shields, 378–80; and indirect/risk effects, defined, 365; and intra-/interspecific interactions, 241; and intrinsic risk, defined, 365; measuring, 366, 380; and predation, 241, 364–65, 380; prey in, 241; quantifying of, 380; and risk, experimental titrations of, 370–74; and risk disturbance, 376–78; and social nutrition, 271; and socioecology, 525; and telemetry, 374–75, 379; terms, usage, 365; and vigilance behavior, 369–70. *See also* habitats

langurs, xi, 9, 16, 157, 222, 366, 409, 518, 521, 546. *See also Pygathrix* (douc langurs); *Semnopithecus entellus* (gray langur); *Simias concolor* (pig-tailed langur); *Trachypithecus* (lutung monkeys)
latex, as food, 402, 423
Latin America, diets/foods in, 189, 193–94
leaf monkeys, 157, 407, 522, 525. *See also Presbytis* (leaf monkeys); *Trachypithecus* (lutung monkeys)
learning. *See* cognition and cognitive ecology
leaves, as foods, 2, 5–10, 15, 19, 21, 23, 30, 33–35, 40–41, 50, 96–97, 100, 106–7, 112, 122, 125, 128, 133–34, 178, 188–89, 215, 249, 260, 311, 327, 343, 386, 393, 399–402, 407, 410–11, 416, 429, 435, 442–44, 461–62, 473–75, 486, 535, 538. *See also* foliage, as food; folivores and folivory
Lemuridae and lemurids (lemurs), 48, 50, 75, 329. *See also Hapalemur* (lemurs)
lemurs. *See Allocebus trichoti* (hary-eared dwarf lemur); *Avahi* (woolly lemurs); *Daubentonia* and Daubentoniidae (aye-aye lemurs); *Hapalemur* (lemurs); Lemuridae and lemurids (lemurs); *Lepilemur* and Lepilemuridae (lemurs); *Microcebus* (lemurs); *Phaner* (lemurs); *Prolemur simus* (bamboo lemur); *Varecia* (lemurs)
Leontocebus (tamarins): *nigrifrons*, 276, 404, 405; *weddelli*, 275, 283–84, 291, 413
Lepilemur and Lepilemuridae (lemurs), 5, 7, 11, 40, 43, 46, 48, 50–51, 88, 90, 124, 154–55, 163, 167, 192, 327
lichens, as foods, 43, 162, 249, 274, 390, 400–401, 404
Liem's paradox, and dental-dietary fallback adaptations, 505–6
life history: and adaptations, 223–24; and anatomy, 462; and body mass and size, 223, 225–29, 232, 237, 239; and climate change, 538; and ecology, 223–25, 238; and energetics/energy, xi, 82, 90, 92, 135; of extinct species, 223; and feeding behavior, xi; and food availability, 236–39; and gestation length, 239; and hibernation, 227, 538; high plasticity of, 239;

and hunting, 351–54; and immunity, 135; and interbirth intervals, 233–34, 239; and meat eating, 351–53; and mother's milk, xvii, 60–61, 66–67; and nutrients/nutrition, xvii, 222–39, 463; and nutrition, summary and conclusions, 238–39; and paleontology, 462; and parturition, age at first, 231–33, 239; and phenotypic plasticity, 222–24; and reproduction, 68, 80, 222, 233–34, 239; variables and variation, 222–24, 239. *See also* evolution; reproduction
lignin: as carbohydrate, 24; and cellulose, 139, 427–28; and nitrogen, 22
Lindburg, Don, 4
Lindshield, Stacy M., xvii, 241, 299, 718
Linnaeus, Carl, 52
lipids: and ASFs, 349; and calories, 349; and carbohydrates, 33; in diets and foods, 21; and fats, 27, 417; and fatty acids, 27; and immunity, 212–14; and macronutrients, xvii, 21–22, 27; and proteins, 470. *See also* crude lipids or fat (ether extract) (EE)
Loango National Park, Gabon, 293, 335, 336, 338, 344
locomotion. *See* movement
Lophocebus albigena (mangabey), 4, 25–27, 32, 44, 117, 159, 186, 294, 307, 508, 521, 550; behavior of, xi; dental topography of, 500, 512; gray-cheeked, 8, 27, 38, 76, 124, 185, 224, 267, 269–71, 270, 336–37, 357, 391, 506–7, 507, 512, 512–13, 537; red-capped, 387; sooty, 506–7
Lorisidae and lorisids (strepsirrhines), 38–39, 48, 50, 192, 329
Lorisiformes, 329
Lucas, Peter, xviii, 397, 446, 718
lutung monkeys. *See Trachypithecus* (lutung monkeys)

Macaca (macaque monkeys), 132, 256, 317, 324–28, 445; *fascicularis* (long-tailed), 88, 90, 119, 173, 182, 192, 200, 224, 230, 232, 234, 303, 307, 310–13, 319, 328, 332, 440, 508; *fuscata* (Japanese), 15, 25, 30, 76, 118, 159, 165, 192, 200–201, 213, 223, 230, 232, 234, 328, 332, 334, 519, 521, 528, 535; *mulatta* (rhesus), xi, 32, 55, 58, 60, 65, 71, 76–79, 90,

Macaca (*continued*)
124, 147, 165, 182, 209, 214, 218, 274, 377, 476, 483, 487; *radiata*, 88, 90, 332; southern pig-tailed using its cheek pouches, 10; *sylvanus*, 159, 228, 230, 313, 387; *thibetana* (Tibetan), 72, 276

macaques. See *Macaca* (macaque monkeys)

MacDonald, K., 272

macronutrients, 397, 399; and available protein (AP), 261; balancing, 481–83, 483, 487, 493; and carbohydrates, xvii, 21–22, 24–25, 34, 143, 149, 261; comparison of distributions in Paleolithic and modern food environments, 495; constrained variation in, 486; and cultural ecology, 496; and culture, 496; defined, 21, 139; digestion of, 417; and dry matter (DM), 21, 22; and energy, 27–29, 34, 135, 397; and fats, 19, 27, 34, 261; fatty acids as, 24, 27, 420; and fiber, 261; and food choices, xvii, 19, 21–34; and food choices, introduction, 21–22; and food choices, summary and conclusions, 33–34; and humans, 490–91, 493, 495; and immunity, 135, 198; intake, 26, 32, 46, 127–28, 271, 385, 467, 480–81, 483, 490, 495; and lipids, xvii, 21–22, 27; in milk samples, 485; and nutritional ecology, 19, 21, 29–31; and nutritional geometry, 271, 489, 490; and proteins, xvii, 21–26, 261; questions about, xv–xvi, 492; and social nutrition, 261, 271; and supermarket food costs, 495; and wild plant food chemistry, 417

Madagascar, 153–55, 192, 194; aye-aye lemurs in, 30–31, 38–39, 97–98, 110, 302, 454, 456; CAM plants from, 436–37, 437; diets/foods in, 35–40, 36, 42, 46, 189, 390; extinct strepsirrhines in, 435–36; folivores in, 40; frugivores in, lack of, 8, 15, 550; fruit in, 105, 115; high altitudes in, 390; island environment of, 75; lemurs in, 10, 37, 39, 46, 75, 130, 163, 168, 387, 391, 437, 439, 445, 536, 541–42; rainfall in, 537; sifakas in, 30, 42, 103, 537, 540; spiny forests of, 436

Makombo, John B., xviii, 397, 532, 718

mangabeys. See *Cercocebus* (mangabeys); *Lophocebus albigena* (mangabey)

Manombo forest, Madagascar, 130, 541

Manu National Park, Peru, 16, 257

marmosets. See *Callithrix* (marmoset monkeys); *Cebuella* (marmoset monkeys)

Marshall, Andrew J., xviii, 14, 121–22, 397, 506, 515, 719

marsupials, 56–57, 60, 64, 99, 106, 150–51, 166, 168–69, 171–72, 175, 264, 373

Masai-Amboseli Game Reserve, Kenya, 521

Mashatu Game Reserve, Botswana, 36

mating. See reproduction

Mayr, Ernst, xv

McConkey, Kim R., xvii, 1, 397, 719

meat eating, 35; defined, 327; and evolution, 327, 351–53; and fatty acids, 345; and hunting, 327–28, 339–45, 349, 351–54; and micronutrients, 354; and nutrition, 328, 353; proteins from, 354; questions about, 444; and secondary consumers, 342. See also bushmeat; carnivores and carnivory; faunivores and faunivory; hunting

medicines: mother's milk as, 52, 66; and secondary compounds, 174–75. See also self-medication

Melin, Amanda D., xvii, 19, 95, 719

metabarcodinog, 404–6

metabolics, and diet/food choices, 21, 37, 45. See also basal metabolic rates (BMRs)

metabolism, 78, 130–31, 137, 177, 256–57, 345, 399, 432, 463

metabolites, xi, 41, 166–69, 420–21. See also plant secondary metabolites (PSMs); secondary compounds, in foods

metabolomics: and biodiversity, 172; experiments and studies of, 170–72; and genetics and genomics, 172–73, 175; and genomics, 172–73, 175; and molecular networks, 170; and nutritional ecology, 172; and spectral data, 172

Mico (monkeys, marmosets, tamarins), 42, 48, 51, 413

microbes: and diets/foods, 145–46; and digestion, xvii, 137–49; and fermentation, 7, 45, 106, 135, 137, 139, 142–45, 147, 404; and mutualism, 183, 194; and phylogeny, 146–48; and polysaccharides, 143–45; symbiont, 135, 137–40, 143, 149. See also digestion; enzymes

microbiomes: and digestion, 146–47, 149, 183, 264; and grazing by infants, 59; and hormonally active phytochemicals in diets, 177, 182–83, 194–97, 555; and hormones, 194–96, 555; maternal, 72; oral, 509

Microcebus (lemurs), 39, 45–46, 48, 51, 70, 88, 90, 98, 131, 155, 193, 236, 274, 279, 329, 334, 377, 413–14, 439, 538

micronutrients, 399; in ASFs, 327; balancing, 357; and energy, 397; fatty acids as, 420; and food choices, xvii, 19, 21–34; and food choices, summary and conclusions, 33–34; and immunity, 198, 214–16; intake, 214–16; and meat eating, 354; and minerals, xvii, 19, 29, 198, 381; nutritional, 19, 21, 29–31, 271; and nutritional geometry, 271; questions about, xv–xvi; and social nutrition, 261, 271; vitamins as, xvii, 29, 198, 381, 420, 422

milk. See mother's milk

Milligan, Lauren A., xvii, 19, 52, 719

Milton, Katherine, xv, 7

minerals: in diets and foods, 19, 21, 29, 31; and immunity, 198; macro-, 29; meat as source of, 320; and micronutrients, xvii, 19, 29, 198, 381; nutritional, 19, 29, 31; and trace elements, 29. See also vitamins

Miocene, 16

models and modeling, of nutrition, xviii, 397, 463–97

Moldawer, Lyle L., xvii, 135, 198, 719

mollusks, as foods, 312

monkeys: Afro-Eurasian, 307; New World, xi, 4, 8, 11, 36, 152, 163–64, 279, 291, 293, 409, 445, 499; Old World, xi, 4, 9, 11, 152, 156, 163–65, 285, 291, 293, 499. See also *specific monkeys*

monosaccharides, 24, 139. See also disaccharides; polysaccharides

morphology: and adaptations, 24, 37; and behavior, 17; and environment(s), 17; and feeding behavior, 17, 462; and food physics, 462; and hunting, 327; and molar, 8; and nutrition, 463;

and phylogeny, xii; and physiology, 385–86, 394–95, 400
mother's milk: and adaptations, 60; and amino acids, 53–55; and anthropology research, 66; of baboons, 53; and behavior, xvii, 52, 57, 60, 66; and biology, 52, 57, 60, 64, 66; carbohydrates in, 54; and community ecology, 64; composition of, 52, 60–64, 66–67; constituents in, 66; and cultural ecology, 65; and disaccharides, 54–55; and ecology, 60, 66; and energy transmission, 52; evolution of, 52, 56–57, 64–66; fat in, 67; and fatty acids, 53, 55, 64–65; and faunivory, 61; as first diet, xvii, 19, 43, 52–67; as first diet, future research, 52; as first diet, summary and conclusions, 66–67; as food, 52, 66; and grazing, transitions to, 59; and hormones, 66; human, 54, 64–66, 67; and immunofactors, 66–67; and infant care, 66; and lactation, 52, 57–60, 66–67, 69–71; and life history, xvii, 60–61, 66–67; and litter size, 59–60; macronutrients in, 483; and mammals, 69–70; and maternal diet, 66; as medicine, 52, 66; and metabolics, 59; and microbes, 52; and microbiomes, 59, 66; and natural selection, 52; and neurobiology/neurodevelopment, 57–58, 60, 64–65; and nursing behavior, 59, 66; and nutrients/nutrition, 52–54, 61–62, 66–67; oligosaccharides in, 54–55, 64, 66, 67; and phylogenetics/phylogeny, 58, 60–63, 67; and physiology, 66; proteases in, 57; proteins in, 67; and public health (human), 59, 65–66; and radiation, adaptive, 57; and sex ratio of young, 59–60, 66; as signal, 52, 66; and social network analyses, 66; sugar in, 67; synapomorphy of, 52, 66; synthesis of, 52, 54–56, 66–67. *See also* reproduction
mountain gorillas. See *Gorilla beringei beringei* (mountain gorilla)
mouth. *See* oral anatomy; taste (gustation); teeth
movement and movement ecology, 361–62; and biomechanics, 357; causes and consequences of, 358; and cognition, 241, 273, 297, 358; and collective behavior, 363; and consumption/feeding of depleting resources, 361; and decision-making, 355; and diet/food choices, 37, 50; and diversity, 360–61; and ecological cognition, 358; and environment, 358; and environmental changes, 363; and evolution, 355; and feeding neighborhoods, xvii, 241, 355–63; and feeding neighborhoods, conclusions, 363; and feeding neighborhoods, summary, 361, 363; and foraging, xvii, 112, 298, 355, 359–61, *361*; and habitats, 355; internal state/motivation for, 355–57, *356*; mechanistic, 355, 363; and motion capacity, 357; and navigation capacity, 357–58; and nutrition, 355–60, 363; and physiology, 355; questions about, 360, 363; and ranging behavior, 40; and seasonality of food availability, 125–28, *126*; and social foraging, 360–61; technological monitoring of, 363; and telemetry, 357, 359. *See also* ranging behavior
muriqui monkeys. See *Brachyteles* (muriqui monkeys)
Mutegeki, Richard, xvii, 135, 137, 176, 719
mutualism: aphid, 264; and frugivory, 95; and hunting by-products, 347, 354; and lichens, 404; and microbes, 183, 194; and parasitism, 264; and seed dispersal, 97
mycophagy. *See* fungi, as foods

Nasalis larvatus (proboscis monkey), 25, 40, 49, 51, 88, 90, 144, 158, 165, 192, 521
natural selection: adaptive logic of, 463; and biology, 463; and energy/energetics, 68, 92; and foraging, 105, 107, 120, 471–72; and functional biology, 463; and nutrition, 463; and seasonality, 123; and senses, 105, 107–8; and teeth, 502, 504
navigation. *See* movement
NDF. *See* neutral detergent fiber (NDF)
Ndiaye, Papa Ibnou, xvii, 241, 299, 719
Neanderthals, 65, 350–51, 354, 438
nectarivores and nectarivory, 41–42, 442–43. *See also* flowers, as foods; nectars, as foods
nectars, as foods, 5, 21, 30, 35, 47, 49, 106, 124, 274–75, 291–92, 400–402. *See also* nectarivores and nectarivory
neighborhoods. *See under* feeding and feeding ecology
Neotropics, 8, 11, 15, 37–39, 41, 43, 46–47, 50, 73, 115, 128, 369, 411, 518
Nepal, 222, 224, 230, 232, 234
nettles, as foods, 273
neurobehavior, and primate intelligence, 274
neutral detergent fiber (NDF), 26, 119, 319, 418, 428, *484–85*
NGF. *See* nutritional geometry framework (NGF)
Nigeria, 83, 160, 165, 230, 234, 378
Nimba Mountains, Guinea, 327–28
Nissen, Henry, 1
Nomascus (gibbons), 37, 49, 51, 85, 88, 90, 127, 162, 191, 333
nonprotein energy (NPE), 23, 121, 167, 210, 221, 268, 320, 471, 477–79, 481–83, *482*, *483–86*, *486*, 489–92, *491*; defined, 382
nonprotein nitrogen (NPN), 22, 417, 418, 427, 427–29
North America, 459, 545
noses. *See* olfaction and olfactory ecology
NPE. *See* nonprotein energy (NPE)
NPN. *See* nonprotein nitrogen (NPN)
nutrients: balanced and combined, 31–33; balancing, 5, 21, 31–34, 37, 51, 121–22, 124, 133, 164, 203, 216, 221, 264–65, 273, 361, 394, 465, 467, 469, 471, 476–79, 483–88, 526; in diets and foods, 19, 135, 479, 524; and food processing, xi, xvii, 135; intake, 15, 17, 29, 31, 114, 118, 121, 123, 127–28, 130–34, 198, 202, 204–5, 208, 216–17, 264, 291, 354, 385, 465–69, 479, 488, 492–93, 534, 542. *See also* macronutrients; micronutrients; nutrition
nutrition: assay, laboratory methods to, 397; and diets/foods, xv–xvii, 19, 21, 33–34, 135, 397, *416*, 553; integrative capacity of, 463; and interdisciplinary research, 464; methods, practice, application, xviii, 397; mixture hierarchy of, 463; modeling, xviii, 397, 463–97; modeling, research priorities, 487–89; modeling, summary and conclusions, 496–97; pharmaco-, 212; research priorities, 487–89; science, 397, 496. *See also* diets;

nutrition (*continued*)
feeding and feeding ecology; foods; geometric framework for/of nutrition (GF); nutrients; nutritional ecology; nutritional geometry framework (NGF); social nutrition

nutritional biology, and biodiversity and food web theory, xviii

nutritional chemistry, 119, 417

nutritional ecology, xi–xiii; and animal-environment interface, 465; and diet/food choices, 21; models and modeling in, xviii, 397, 463–97; quantitative approaches in, 80; term, usage, xv. *See also* behavioral ecology; cognitive ecology; movement ecology; nutrition

Nutritional Ecology Laboratory, Harvard University, 421

nutritional geometry, 15, 31, 121, 464–71, *467–70*; and agent-based models, 265; amounts-based, 466–71, 483, 489; basic concepts of, 464; and energy eaten, protein and non-protein, *488*; of female chimpanzees, 320; and field research, 487–89; graphical representations of, 479–80; and human appetites, 489–90; and intake targets, *488*; as integrative approach, 464; and macronutrients, 271, 489, 490; and micronutrients, 271; and multidimensional models, 51, 382; and nutrient intake and balancing, 121; and nutrient regulation in relation to dietary imbalance in, *469*; proportions-based, 471, 483, 490–96; and simulation models, 263; and social environments, 268; and social network analyses, 265, 270; state-space models of, 264–65. *See also* geometric framework for/of nutrition (GF); nutritional geometry framework (NGF)

nutritional geometry framework (NGF), 464–96; defined, 466; and dietary constraints, responses to, 483; and dietary profiles, 51; and human appetites, 489; and humans, 488–96; and intake target selection, 481–85; and linear programming (LP), 473–74, 479–81; and mixture triangles, 473; packaging problem of, and dual, 480–81. *See also* geometric framework for/of nutrition (GF); nutritional geometry

nuts, as foods, 8, 10, 180, 272, 295, 299, 302, 305–6, 315–16, 320–21, 326, 343, 426, 506–7, 513–14

Nyungwe National Park, Rwanda, 128, 338

Oates, John, 4
Oceania, forests in, *545*
odors. *See* olfaction and olfactory ecology
olfaction and olfactory ecology: and cellular biology, *111*; chemistry of, *111*; and chemosensation, 109–10, 113; and cognition, 279–80, 282, 284; and evaluating foods, 105; and exudates (foods), 99, 112; and finding foods, 95–96, 98–100, 112–13; and fruits, 104; and genomics, 109, *111*; multidisciplinary contributions toward an integrated understanding of, *111*; of pollinators, 106; and taste, evolution of, 109–10

omnivores and omnivory, 5, 19, 21, 23, 29, 35, 37, 41, 44, 61, 104, 124, 142, 145, 189, 193–94, 244, 269, 433, 443–45; defined, 44, 139. *See also* carnivores and carnivory; herbivores and herbivory

oral anatomy, and diets/foods, 8–9
orangutans, xi, 4, 11–12, 23, 37, 44–46, 58, 70–77, 84–85, 94, 115–16, 119, 122–33, 143, 161, 210, 217–18, 224, 227, 294, 302, 307–8, 316, 322–27, 334, 385, 390, 460, 485–88, 498, 502, 504, 508, 518, 530, 542; female and seven-month-old infant feeding on fruit, *116*

Organization for Tropical Studies (US), xii
Oribi Gorge, South Africa, 438
Otolemur (greater galagos, thick-tailed bushbabies), 39, 42–43, 45, 48, 168, 189, 191, 329, 334, 373
Outspoken Essays (Inge), vi

paleoanthropology, 432, 444, 502, 509. *See also* anthropology
paleontology, 222, 462, 498, 505, 513
paleoprimatology, 432, 435, 508
Paleotropics, 15, 39
palynivores and palynivory, and folivores and folivory, 41–42. *See also* pollen, as food

Panama Canal Zone, howler monkeys in, 1
Pan paniscus (bonobo), 37, 40, 85, 88, 90, 160–61, 165, 189, 190, 209, 238, 287, 294–95, 304, 307, 323–24, 333–34, 339, 341–42, 349, 404–5
Pan troglodytes (chimpanzee): adult female Fongoli uses tool to hunt Galago senegalensis as juvenile daughter looks on, *309*; adult male Fongoli eats baobab fruit after using stone anvil to open fruit, *300*; adult male Fongoli uses tool to termite fish, *315*; behavior of, xi; and climate change, 537; and cognition, 297; diets and foods of, 43, 324, 435; feeding of, 299; feeding on figs, 36; hair, *439*; and hunting/predation, 327–28, *335*, *337*, 348, 354; in Kibale National Park, Uganda, *188*; Nigerian, 165; in Nimba Mountains, Guinea, 327–28; and nutrients, 32; nutritional geometry of female, 320; and PSMs, 176; and seasonality, 128; from Taï National Forest, Cote d'Ivoire, *440*; taxa, 310–13; and tool use/ TAF, 299, 308–9, 314–16, 322–23, 325–26, 327
Papio (baboons): *anubis* (olive), 1, 40–41, 90, 165, 178, 223–24, 230, 234, 328, 332, 334, 340, 375, 534–36; behavioral flexibility of, 388; carrying capacity of, 519; and climate change, 537; cognition of, 273, 297; *cynocephalus*, 69, 88, 90, 190, 202, 215, 223, 230, 232, 234, 307, 313, 332, 334, 339, 440, 519, 528, 537; dental-dietary adaptations of, 510; and diets/foods, 39–41, 44; digestion of, 146; evaluating diets of with stable isotopes, 433, 440, 445; feeding studies of, 7, 14; and foraging, 273, 278, 294, 297, 368, 374–75, 377, 379; habitats of, 365; *hamadryas*, 27, 32, 173, 190, 202, 215, 234, 273, 286, 303, 307, 313, 332, 334, 377, 407; hormonally active phytochemicals in diets of, 181; hunting by, 328, 332, 334, 339–44, 348, 354; IBIs in, 237; immunity of, 202, 215–16; mechanical properties of foods of, 459; mother's milk of, 53; movement of, 355; Nambian, 365; and nutrients in food choices, 27; and nutrition, 202, 215–16, 483, 537; omnivorous,

23; protein intakes of, *482*; and reproduction, 69–73, 76–79, 81; savanna, 253, 433, 537; and seasonality of food availability, 120; secondary components in foods of, 158, 165, 173; social food competition of, 253, 256–57; and stable isotopes, 445; tool use by, 307; *ursinus* (chacma), 32, 36, 58, 72, 90, 165, 224, 262, 273, 307, 328, 342, 365, 367, 440, 459; yellow, 223–24, 226–29, 237, 334, 339–40, 342. See also *Theropithecus gelada* (gelada)

parasites and parasitism, 11, 39, 112, 176, 255–58, 264, 524–25

patas monkeys. See *Erythrocebus patas/baumstarki* (patas monkey)

PE. See petroleum ether (PE)

penetrometers and of penetrometry, 446–47, 451–52

Peres, Carlos A., xvii, 19, 35, 719

petroleum ether (PE), 319, 418, 425

Phaner (lemurs), 42, 47–48, 51, 106, 413

pharmacology, 212

pharmaconutrition, 212

phenolics: defined, 179; and estrogenic compounds, 179; as PSMs, 9, 151–66, 168, 175; in resins, 403

phenological cycle, 504

phenology, 35, 108, 115, 117, 188, 278, 391, 410–11, 549

phenotypes, 66, 92, 107–9, 112, 170, 172–73, 213, 222–24

philopatry, 5, 251, 254–55, 259–60, 392–93

photosynthesis, 142–43, 402–4, 413, 433, 436, 438, 443–44, 509

phylogenetics, 2, 16, 75, 109, 112, 146–47, 171, 189–90, 189–92, 225, 227, *229*, *231*, *233*, *235*, 255, 259, 260, 324, 387, 393

phylogeny, *62–63*; and body mass and size, 59, 91, 229; and diets/foods, 15–16, *51*, 90–91; and morphology, xii

physics: of feeding, mechanical properties, *447*; of foods, 450–51, 460, 462

physiology: and adaptations for carbohydrates, 24; and adaptations for diets, 46; and adaptations for tannins, 27; and anatomy of mouth, 448; and behavior, 92, 114, 134, 176–77, 185, 195, 197, 198, 260, 261, 359, 387, 394, 463, 489; and diet/food choices, 34, 50; and evolutionary biology, 92; and morphology, 385–86, 394–95, 400; and nutrients/nutrition, 463

phytochemicals and phytochemistry: and anthropogenic endocrine-disrupting chemicals, 194–96; and behavior, 176–77, 181–83, 185, 195, 197; and biology, 197; biomonitoring of, 197; categories of, 177–79; and chemical exposure, 194–96; and conservation, 177; and ecology, 177, 197; effects of, 181–83; and endocrinology, 177, 197; and estrogenic activity based on transfection assays in plants, *186*; and estrogenic plant families, *186–87*, *190–94*, 197; and evolution, 177, 197; and feeding ecology, 177; and gastrointestinal microbiota, 182–83; and gut microbiota/microbiomes, 194, 196, 197; and gut morphology, 196; hormonally active in diets, xvii, 176–97; hormonally active in diets, future investigations and research, 194–97; hormonally active in diets, summary and conclusions, 197; and humans, 176; and landscape models of chemical exposure, 194–96; measuring, 184; mechanisms of action of, 180–81; and microbiomes, 177, 194–97, 555; and phylogeny, 185, 188–89, 194, 197; and physiology, 176–77, 181–83, 185, 195, 197, 555; and phytoestrogens, 179–80, 184–85, 188–89, 194, 197; and phytosteroids, *178*, 197; and pollutants, 197; and PSMs, 176–77; and reproduction, 181–83, 197; and self-medication, 181, 189. See also hormones; plant secondary metabolites (PSMs)

Piliocolobus (red colobuses), 25, 40, 42, 49, 51, 127, 157, 164–65, 185, *187*, 190, 334, 336, 548

Pitheciidae, pitheciids, pitheciines (monkeys), 38, 47–51, 156, 193, 282, 331

plants: CAM, 433–34, *437*, 440; estrogenic, *186–87*, *190–94*, 197; Fabaceae, 43, *178*, 180, 185–86, 188–89, 197; as foods, 5–7, 21, 35, 50–51, 116, 400–401, 416–17, 454, 542; Moraceae, 37, 185–86, 188–89, 197, 533; phytosteroid-containing, *178*; plant-source foods (PSFs), 327, 351; and saccharides, 142; as staple foods, 454; toxins in, xv–xvi, 165; woody, 150. See also angiosperms; CAM plants; herbivores and herbivory; wild plant food chemistry; *and specific plants*

plant secondary metabolites (PSMs), 150–66, 173–77; and agriculture, 175; and alkaloids, 176; and behavior, 135; beneficial impacts of, 174–75; and captivity, 161, 165, 167–68, 175; and climate change, 537–38; cyanogenic glycosides, 176; defined, 150–52; and dietary/feeding ecology, 150; digestion inhibitors as, 9; and endocrinology, 135; and feeding behavior/ecology, 150, 175; and genomics, 175; and habitats, 150; and hormonally active phytochemicals in diets, 176–77; and metabolomes, 175; and metabolomics, 175; and neurotransmission, 135; and parasites, 176; phenolics as, 9, 151–66, 168, 175; and physiology, 176, 542; and phytosteroids, 176; previous work with, review of, 152–66; questions about, 175; and saponins, 176; and self-medication, 11, 163, 174–76; and tannins, 151, 168, 175, 176; toxins as, 9; and vertebrates, 150–51, 177. See also phytochemicals and phytochemistry; secondary compounds, in foods

Platyrrhini and platyrrhines (monkeys), 9, 36, 40–41, 44–47, 61, 99–100, 103, 108, 113, 128, 143, 194, 236, 266, 302, 329, 413, 498, 500–501, 503

Plecturocebus (titi monkeys), 72, 402–3, 506

Pleistocene, 354

Plio-Pleistocene: hominins, from Africa, diets evaluated with stable isotopes, 436; monkeys, dental-dietary adaptations of, 514

pollen, as food, 41, 106, 291, 402. See also palynivores and palynivory

Pollock, John, 4

polysaccharides: and carbohydrates, 24, 149; defined, 24, 139; in gums, 403; and hemicellulose, 24; in lichens, 404; and microbes, 143–45. See also disaccharides; monosaccharides

polyspecific associations: cercopithecine, 409; and competition, 269, 384; defined, 266; of redtail monkeys, 14; and social nutrition, 261–62, 265–71

Pontzer, Herman, xvii, 19, 82, 719
population ecology: and carrying capacity, for foods, 397, 515–16, 523–26, 529, 531; and community ecology, 400, 463; and nutrition, 463
possums, common brushtail, *171*
predation, 517; and fear, 241; and feeding behavior, xvii, 17, 400; and food availability, 365–66; and mortality, 258; protection, and foraging, 266; rates, 335, 364–65, 374; and reproduction, 80; risk, xvii, 5, 14, 125, 168, 251, 266–69, 277, 291, 295, 358, 364–73, 376, 378–80, *416*, 480, 547; risk, defined, 365; and sociality, 380; and spatial variation, xvii; on squirrels, 339; in water resource areas, 539. *See also* hunting; prey
pregnancy. *See* reproduction
Presbytis (leaf monkeys), 8–9, 40, 49, 51, 501, 530, 546, 548; *entellus*, 1, 73, 191, 521; *johnii*, 165; *melalophos*, 158; *pileata*, 191; *rubicunda*, 9, 25, 30, 38, 124, 157–58, 165, 167, 390, 519; *thomasi*, 366, 508, 522
prey: camouflage of, 399; exotic species as, 376, 540; as food, 51, 97, 107, 329–39, 400–401, 416; invertebrates as, 35, 51, 97, 107, 112, 400–401; large, 354; mammals as, 313, 334, 339, 350; monkeys as, 348; taxonomic distribution of, 328–34; vertebrates as, 38, 51, 328–33. *See also* predation
primary consumers, xvii, 410; defined, 139. *See also* secondary consumers
Primate and Predator Project, Lajuma Research Centre, South Africa, 718
Primate Behavior (DeVore), 1
Primate Ecology, xi
primates. *See* extant primates; fossil primate diets; *and specific primates*
primatology, 68, 80, 83, 150–51, 174–75, 320, 359, 460, 479, 496, 516, 518–19, 531; and adaptive management, 541; behavioral, and olfaction, 111, *111*; comparative research in, 523; and cross-fertilization of ideas, 175; and field measurements, 451; and food mechanical property data, 462; nutrition work in, 417, 430; paleo-, 432, 435, 508; penetrometers in, 446
problem-solving. *See* cognition and cognitive ecology

proboscis monkeys. *See Nasalis larvatus* (proboscis monkey)
Procolobus (colobus monkeys), 15–16, 40, 49, 51, 157, 164, 178, 190, 334, 377, 519, 521. *See also Colobus* monkeys
Prolemur simus (bamboo lemurs), 41, 163, 536
Propithecus (sifaka lemurs), 32, 100, *103*, 131, 486
prosimians: behavior of, xi; and cognition, 272–73, 278–79, 291, 293, 297; diets/foods of, 189; insectivorous, 141
protein intake: and atrophy, 211–12; and energy, 32; and food choices, 21, 33–34; and immunity, 203–5, 208, 210–12; maximized, 22–24; and nutrition, 470, 478; and social nutrition, 268
proteins: and amino acids, 22, 53, 203, 208, 211–12, 327, 420, 426, 428, 430; balancing, 37; and carbohydrates, 23–24, 121, 468, 478, 483, 489–90; in diets and foods, 21, 33; and fiber, 10, 21, 25–27, 34, 262, 528–29, 547; and lipids, *470*; and macronutrients, xvii, 21–26, 261; maximization, 21, 22–24, 120, 467–68, 477; from meat, 354; nitrogen in, 417; nutritional, 19, 22, 37; and tannins, 25–27. *See also* crude proteins (CPs); protein intake
Pruetz, Jill D., xvii, 241, 299, 719
PSMs. *See* plant secondary metabolites (PSMs)
public health: and endocrine disruption, 195; and energy (TEE), 83–84; and environmental science, 195; and mother's milk, mammalian, 59, 65–66
Pygathrix (douc langurs): *nemaeus*, 89–90; *nigripes*, 40, 42

Quadopedia, 52
quality of foods. *See* foods, quality of

radiation, solar, 537
rainforests, 15, 45, 84, 94, 98, 115–17, 125, 128, 134, 163, 249, 338, 357, 406, 419, 505, 532–33, 535
ranging behavior, xvii–xviii, 4–5, 14, 40, 355–56, 363, 368, 375, 378–79, 491, 536. *See also* foraging; movement and movement ecology
Ranomafana National Park, Madagascar, 36, 387, 537

Raubenheimer, David, xviii, 175, 397, 463, 476, 719
Reader, S. M., 272, 323
reforestation, 95. *See also* deforestation
Rekambo (Loango, Gabon), 336–37, 344
reproduction: and adaptations of mammalian ancestors, 57; and ancestry, 57; and behavior, 68; and body mass/size, 68, 70; and body size, 68; and callitrichids, 81; and captive studies, 68, 80; and conception, 75–76, 81; and cooperative breeding, 74–75, 80–81; costs of, 4, 68–69, 71–75, 78–81, 131, 197; diet and energetics of, xvii, 68–81, 82, 463–64; diet and energetics of, conclusions, 80–81; diet and energetics of, summary and questions for research, 80–81; and diet/food choices, 37; and ecology, 80; and evolution, 68; and females, 68, 75, 80–81; and fetal development, 76, 214; and fetal loss, 70, 76, 80–81; and field studies, 68; and food availability, 80; and foraging, 80–81; and gestation, energetics of, 68, 80; and hormones, 80; and humans, 68, 81; and infant mortality, 76–77, 81, 214, 536; and lactation, energetics of, 68–72, 80–81; and life history, 68, 80, 222, 233–34, 239; and males, 68, 78–81; and maternal health, 78, 80–81; and metabolics, 68; and natural selection, 68; and nutrients/nutrition, 19, 68, 75–78, 80–81, 229–35, 239; and offspring growth/development, 77–78, 81, 239; and ovarian function, 75–76; and physiology, 68, 80; and phytochemistry, 181–83, 197; and predation, 80; and pregnancy, energetics of, 68–71, 80–81; and seasonality, 72–74, 80–81, 130, 131–32; and social organization, 68; and socioecology, 81; variation in, 68; and wild studies, 68. *See also* life history; mother's milk
Reserva Particular do Patrimônio Natural-Feliciano Miguel Abdala, Minas Gerais, Brazil, 384, 386, 389, 395
resilience, 393, 395, 532, 536, 539, 541, 543
resins, 402–3
restoration ecology, 95, 539
Rhinopithecus (snub-nosed monkeys), 30,

43, 49, 51, 130, 158, 165, 189, 191, 273, *276*, 333, 390, 404, 481, 530
Richard, Alison, xv, 4, 555
Rodman, Peter, xv, 4
roots, as foods, 43, 117, 124, *401*, 510. *See also* tubers, as foods
Ross, Stephen R., xvii, 19, 82, 720
Rothman, Jessica M., xv, xviii, 14, 86, 122, 397, 483, 485, 532, 715, 720

saccharides, 142
Saguinus (tamarins), 7, 30, 39, 42–43, 70–71, 90, 106, 249, 275–80, 283–84, 295–97, 328, 412, 414; black-fronted saddleback, *405*; cotton-top, 71, 78; emperor, 282, 295–96; *fuscicollis*, 89–90, 132, 193, 275, 330, 334, 369–70, 413; Geoffroy's saddleback, *276*; lion, golden, 76, 132, 225, 235; lion, smaller-bodied, 42; moustached, *36*, 266–67, 369–70, 519; *mystax*, *36*, 89–90, 193, 266, 283, 330, 334, 404, 519; saddleback, 132, 266–67, 275–77, *276*, 284, 291–93, 295–96, 369–70, *405*; *weddelli*, 275, 283–84
Saimiri (squirrel monkeys), 39–40, 46, 48, 51, 79, 89–90, 99, 105, 127, 132, 193, 228, 249, 330–31, 333–34, 370, 412–13, 445
Sapajus (capuchin monkeys), 39–40, 44, 48, 51, 99, 124, 228, 259, 272, 284–85, 302–3, 306, 310–13, 317, 319–28, 331, 334, 339, 387, 460–61, 504
saponins and saponosides (glycosides), 157, 164, 166, 168, 176, 180
saps, as foods, 6, 24, 35, 42, 47, 49, 117, *129*, 304, 402–3, 413–14, 473, *475*. *See also* exudates (foods); gummivores and gummivory; gums, as foods
science: access and equity in, xviii, 555; environmental, 195; life, 496; materials, 451; metabolic, 82; nutrition, 397, 496; and technology, 555. *See also* chemicals and chemistry; food science; physics
sea cucumbers, as foods, 312
seasonality: and adaptations, 14, 134; and available protein (AP), 119; and behavior, 14, 119–30, *123*, 134, 392; and biology, 7; breeding, 80; and caloric intakes, assessing, 118–19, 134; and captivity, 133; and conservation of primates, 132–34; coping strategies for, 134; defined, 114–16; and dietary changes, flexibility, switching, 123–30, 392; diversity of primate responses to, 123, *123*; and ecological crunches, 14; and energy intake, xvii, 114–34; and energy intake, summary and conclusions, 134; and evolutionary ecology, 134; and exudates (foods), 117; and fallback foods, 121–22; and fat storage/metabolism, 130–31, 134; and feeding behavior, 119; and feeding studies, 14; fission-fusion dynamic as social response to, 130; of flowers, 394; and food availability, xvii, 14, 19, 114–34, *123*, *126*, 198, 400; and food availability, summary and conclusions, 134; and foraging, 119–20, 123–30, 132, 134; and habitats, 123, 128–29, 134; and hibernation, 131; and life history, 224–25; and marginal habitats, 390; measuring in habitats, 114, 116–18, 134; and metabolism, 130–31; and movement, 125–28; and natural selection, 123; and nutrients/nutrition, xv–xvi, 118–19, 121; and phenology, 35; physiological responses to, *123*, 130–32, 134; and preferred foods, 121–22; and social behavior, *123*, 129–30; study of, 114–19
secondary compounds, in foods, xvii, 150–75; alkaloids as, 9, 151–66, 168, 173, 175; and chemical analysis, 151, 399; and chromatography, 169; and cyanogens, 151; and feeding studies, 9–11, 151, 167–68; and hypothesis generation, 167–68; and metabolites, 150, 166–69, 175; and metabolomics, 169–73; and nutrients/nutrition, 166–69; and phenolics, 175; and physiology, 167–68; and proteins, 173–74; and spectrometry, 169; summary and conclusions, 175; and tannins, 151, 168, 173–75. *See also* plant secondary metabolites (PSMs); *and specific compounds*
secondary consumers, 342, 410; defined, 139. *See also* primary consumers
Sector Santa Rosa, northwestern Costa Rica, 98, *104*
seeds, as foods, 1, 5–10, 6, 15–17, 23, 27, 30, 32, 35–38, *36*, 40, 42–44, 47–50, 55, 95–97, 112, 115, 117, 124, 140, 156–57, 162, 164–65, 179–80, 183, 185–86, 213, 249, 291, 302, 304, 308, 317, 320, 322, 390, 400–402, 410, 414, 418–19, 422–23, 429, 433, 451, 460, 473, *475*, 500, 502, 506–8, 513–14, 530, 540, 542, 551; dispersal, 9, 15–17, 42–43, 95–97, 112, 400–402, 542, 551. *See also* fruits, as foods; granivores and granivory
self-medication: and phytochemicals, 181, 189; and PSMs, 11, 163, 174–76; and tannins, 163. *See also* medicines
Semnopithecus entellus (gray langur), xi, 1, 11, *36*, 40, 42, 222, 224, 230–31, 236
senescence, and energy, 84, 93, 94
senses and sensory ecology, 399; and behavioral ecology, 106, 112, 415; and biodiversity, 112; chemo-, 109–10, 113; and cognition, 97; and cognition and cognitive ecology, 97; and communication, 97; and ecology, 97; and environment, 97, 112–13; and environment(s), 97, 112–13; evolution, 107–10; and evolution, 107–10; evolution of, 95, 97, 107–10; and feeding studies, 11; for finding and evaluating foods, xvii, 11, 95–113, 399; for finding and evaluating foods, conclusions, 112–13; for finding and evaluating foods, future studies, 111–12; for finding and evaluating foods, interdisciplinary approaches, 97; for finding and evaluating foods, summary, 112; and foodscapes, 415; and foraging, 95–97, 112–13, 463; and frugivory, 95–96, 102; and genetics/genomics, 97, 106–7, 109–13; and genomics, 97, 106–7, 109–12, 109–13; and habitats, 112; mechano-, 110, 113; multimodal, 112; multiple simultaneously, 97, 112; and natural selection, 105, 107–8; and prey, 101. *See also* hearing (audition); olfaction and olfactory ecology; taste (gustation); touch (haptic sense); vision (sight)
Shaanxi Key Laboratory for Animal Conservation, 716
Shiwalik Forest Division, India, *36*
siamangs. *See* Hylobatidae and hylobatids (gibbons); *Symphalangus syndactylus* (siamang gibbon)

sifakas. See *Propithecus* (sifaka lemurs)
sight. *See* vision (sight)
simians. *See* Anthropoidea and anthropoids; apes; Haplorhini/haplorhines (tarsiers, monkeys, apes); prosimians
Simias concolor (pig-tailed langur), 40, 42
Simpson, George Gaylord, 498
smell, sense of. *See* olfaction and olfactory ecology
Smith, Andrew C., xvii, 19, 35, 720
Smith, Chris, 4
Smuts, B. B., xv
snub-nosed monkeys. See *Rhinopithecus* (snub-nosed monkeys)
social behavior, xvii, 11, 45, 50, 52, 261–71, 386–87
social environments: food acquisition in, xi, xvii, 241; and nutritional geometry, 268; nutrition in, xi, xvii, 241
social food competition, xvii, 241, 243–60, 399–400; assumptions of, 251–57; and behavior, 260, 383–85; categories and types of, 243–51; causes and consequences of, 243; and cognition, 252; conclusions, 260; and contest strategy, 243–44, 246, 250, 258–60, 263–71, 281; and diets/foods, 253–54; and ecology, 258; and fission-fusion societies, 260; and fitness, 243; and food abundance/distribution, 243, 247–50; and food quality, 254; and foraging, 251–52, 260; and genetics, 260; and group size/structure, 243; and habitats, 260; mechanisms of, 251–57; and metabolics, 260; and metabolism, 256–57; and nepotism, 254–55, 259–60; and parasite resistance, 256–57; and philopatry, 251, 254–55, 259–60; and phylogenetics, 260; and physiology, 256–57, 260; questions and research about, 243, 257–58; and reproduction, 260; scramble, 243–46, 250, 253–54, 256, 258, 260, 264–65, 268, 281; and seasonality, 260; and social dominance, 245; and sociality, 245, 257; and social nutrition, xvii; and social structure, 243, 254–56; and socioecology, 243–45, 253–55, 261; and stress, 256–57; summary, 258–59; synthesis of, 251. *See also* competition; social foraging

social foraging: and cognitive ecology, xvii, 272–98; and cognitive ecology, summary and conclusions, 297–98; ecological significance of, 241; evolutionary significance of, 241; in group living, and conflicts of interest, 360–61; and movement, 360–61; questions about, 359–60; and social nutrition, 261–63, 271; and socioecology, 241. *See also* social food competition
social network analyses: and nutritional geometry, 265, 270; and social nutrition, 264–65, 267, 270
social nutrition: and anthropogenic changes, 271; and antifeedant components, 261; and behavior, xvii, 261–71; and behavior, research potential, 271; and behavior, summary and conclusions, 271; and ecology, 261; and empirical nonprimate studies, 263–64; and empirical studies, 261, 271; and energy, 271; and environment, 261; and evolution, 261; and fission-fusion systems, 260; and fitness, 261; and intraspecific competition, 269; and landscapes, 271; and macronutrients, 261, 271; and microbiomes, 264; and micronutrients, 261, 271; models, 264; and multidimensional nutritional niche framework, 271; and nutritional geometry, 268, 271; and physiology, 261; and polyspecific associations, 261–62, 265–71; and protein intake, 268; and social food competition, xvii; and social foraging, 261–63, 271; and sociality, 261; and social network analyses, 264–65, 267, 270; and socioecology, 261, 265, 268, 271
socioecology, 4–5; and behavioral flexibility, 392; and climate change, 1–2; and food availability, 1–2, 399; and foodscapes, 415; and foraging, 241; and landscape of fear, 525; and nutrition, 261, 265; and social food competition, 243–45, 253–55, 261; and social foraging, 241; and social nutrition, 265, 268, 271
soils, as foods, 30, 43, 72, 174
solar radiation, 537
sounds. *See* audition; hearing; hearing (audition)

South Africa, 165, 168, 339, 349, 365–68, 371, 376, 434–41, 445, 540
South America, xviii, 16, 45–47, 115–16, 390, 506, 540, 544–45, 551
Southeast Asia, 9–10, 16, 37, 73, 84, 115, 390, 411, 530, 546, 548
Soutpansberg Mountains, South Africa, 367–68
species ecology, and evolutionary grades, 2, 4
spectrometers and spectrometry, 85, 99, 169–70, 184, 220, 446
spider monkeys. See *Ateles* (spider monkeys)
Sponheimer, Matt, xviii, 15, 341, 397, 402, 431, 510, 720
squirrel monkeys. See *Saimiri* (squirrel monkeys)
stable isotopes: and ancient specimens, 445; and anthropology, 445; carbon, 443; and conservation of primates, 445; and dietary information, 436–38; and diets, brief primer, 431–34; and diets, evaluating, xviii, 15, 397, 431–45; and diets, evaluating, potential high-priority applications, 445; and diets, evaluating, questions about, 434–35, 439, 441, 444–45; and diets, evaluating, summary and conclusions, 445; and disparate isotopic compositions, 445; and ecology, 445; and evolutionary biology, 445; and foodprints, 509–10; and isotopic foods, 438–40, 444–45; limitations of, and practical concerns, 442–43; nitrogen, 434, 439, 443; and other isotopes, 441–42, 443–45; and predation, 354; and studies, 434–36; and variation across space and time, 440–41
Stalenberg, Eleanor M., xvii, 135, 150, 720
starvation, 77, 94, 206–9, 375–77, 434, 533, 536–38
Steiniche, Tessa, xvii, 135, 176, 720
Strepsirrhini and strepsirrhines (galagos, lemurs, lorisids): and biomedical research, 195; BMRs of, 82–83, 94; and digestion, 143; extinct, 435–36; feeding behavior of, 4; as frugivores, 7; as insectivores, 38, 98; and litter mass, 59; and mother's milk, 59, 61, 64; and phylogeny, 194; and phytochemicals, 194–95; as predator, 329;

and reproduction, 70, 75, 81; and seasonality, 119, 128, 130–31; and senses, 98–101; and social nutrition, 266; and TEE, 82–83; teeth of, 435, 499, 501, 503; and tool use, 302. *See also* Lorisidae and lorisids (strepsirrhines)

Strier, Karen B., xii–xviii, 241, 381, 720

Struhsaker, Tom, xv

Sumatra: gibbons in, 10; macaques in, 132, 324; orangutans in, 94, 224, 294, 302, 307–8, 322, 326, 327, 460, 508, 518

surilis. See *Presbytis* (leaf monkeys)

Sussman, Robert, 4

Symphalangus syndactylus (siamang gibbon), 4, 8, 37, 72, 85, 89–90

Systema Naturæ (Linnaeus), 52

TAF. *See* tool-assisted foraging and feeding (TAF)

Taï National Forest, Cote d'Ivoire (Ivory Coast), chimpanzees from, *440*

Taï National Park, Cote d'Ivoire (Ivory Coast), 293, 336, 506–7, *507*, 512, 550

tamarins. See *Leontocebus* (tamarins); *Saguinus* (tamarins)

Tambopata National Reserve, Peru, 36

Tangkoko-Duasudara Nature Reserve, Sulawesi, Indonesia, spectral tarsier eating cockroach at night in, *120*

Tangkoko National Park, Indonesia, 36

tannins: chemical analysis of, 151; and dry matter (DM), 26; physiological adaptations for, 27; and proteins, 25–27; and self-medication, 163; and wild plant food chemistry, 430. *See also* toxicity; toxins

Tarsius, Tarsiidae, Tarsiiformes (tarsiers), 13, 38–39, 44, 48, 61, 98, 107, 118, 127, 130, 141–42, 327–29, 411, 437; *bancanus* (Bornean), 38, 328–29, 334, 339; Gursky's spectral, feeding mon an orthopteran, 36; spectral eating cockroach at night, *120*. See also *Carlito syrichta* (Philippine tarsier); *Cephalopachus bancanus* (Horsefield's tarsier)

taste (gustation): adaptations for, 11; and chemosensation, 109–10, 113; and evaluating foods, 106; and finding foods, 95–96, 112–13; and nutrition, 463; and olfaction, evolution of, 109–10

taxonomy, 7, 328, 334, 400, 404–6, 416

TDF. *See* total dietary fiber (both soluble and insoluble fibers) (TDF)

TDIF. *See* total dietary insoluble fiber (TDIF)

TDSF. *See* total dietary soluble fiber (TDSF)

technology, 17, 105, 109, 118, 273, 323, 352, 375, 555

TEE. *See* energetics and energy, total expenditure (TEE)

teeth: and chitin, 8; crowns, shape, 499–501; and diet, 498; enamel, 457, 459, 502–4, 508; and food processing, 499, 510; function, 498; incisors, size, 498–99; measuring of, 501–4; molars, occlusal wear, 502, 511–12, *512*; molars, size, 499; morphology, 8, 460; and natural selection, 502, 504; shape, 498, 514; size, 498–99, 514; structure of, 498, 501–2, 514; topography, 122, *500*, 500–501; wear, 510–14. *See also* dental-dietary adaptations

telemetry: and landscape of fear, 374–75, 379; and locomotion/movement, 357, 359

Terborgh, John, 14

termites, as foods, 22, 30, 38, 118, 264, 299, 308, 314, 316, 320, 324–25, 386, 404

terpenoids, chemical analysis of, 151

Thailand, 1–2, 232, 234, 317, 324

Theropithecus gelada (gelada), xi, 4, 41, 59, 72, 230–31, 388, 512

Tibet, 72, 276

titi monkeys. See *Callicebus torquatus* (collared titi monkey); *Plecturocebus* (titi monkeys)

TNC. *See* total nonstructural carbohydrates (TNC)

tool-assisted foraging and feeding (TAF): accounts of, *303–6*; and Associative Tool Use, 301–2, 307–9, 314; behaviors, 299–300; and calories, 317; cultural aspects of, 299, 320–21; ecological aspects of, 299, 320–21, 326; and evolution, 300; foods acquired, extracted, processed, or targeted with aid of, *310–13*, *317–19*; and learning abilities, 326; of nonprimates, 301–2, 303–4; and nutrition, 317–20, 326; and nutritional geometry, 326; origins of, 325; of primates, 302, 305–17, 320–25; and problem-solving skills, 299; prolific, 321–23; and proto-tool use (p-TAF), 299–301, 326; and resource density, 326; social aspects of, 320–21; and social tolerance, 326; summary and conclusions, 325–26; term usage, 299; and terrestriality, 326; uses of, various, 316–17

tool use, 362; and cognition, 241, 272–73; defined, 299; and evolution, 300; feeding-related, systematic overview, xvii, 299–326; feeding-related, systematic overview, summary and conclusions, 325–26; and fitness, 300; and foraging, 299–301; and hunting, 327; and predation, 327; and sociality, 316, 323; and social network analyses, 321. *See also* tool-assisted foraging and feeding (TAF)

total dietary: fiber (both soluble and insoluble fibers) (TDF), 418; insoluble fiber (TDIF), 418; soluble fiber (TDSF), 418

total nonstructural carbohydrates (TNC), 418, 429–30

touch (haptic sense): and evaluating foods, 103–5; and finding foods, 95–96, 101, 110, 113; and mechanosensation, 110, 113; and pressure, 113

toughness, and mechanical properties of foods, 449–50, 452–57

toxicity, 4, 112, 122, 150, 166, 174–75. *See also* tannins; toxins

toxins: and alkaloids, 9, 140; in diets/foods, 41, 46, 163; and fear, 373; and food choices, 21; and nutrients/nutrition, 121, 165, 473, 480, 542; in plants, xv–xvi, 165; as PSMs, 9. *See also* tannins; toxicity

trace elements, 29, 345

Trachypithecus (lutung monkeys), 8, 9, 40; *auratus sondaicus*, 23; *cristatus*, 235; *francoisi* (François' langur), 90, 276; *leucocephalus*, 276, 518; *phayrei* (Phayre's leaf monkey), 38, 58, 72, 178, 181, 224, 234–35

transfection assays, 178, 184–87

trapping methods, employed for estimating arthropod abundance, 411–13, *413*

travel. *See* movement and movement ecology

Tropical Biology Association (Europe), xii
Tsimanampetsotsa, Madagascar, 36
Tsimane, indigenous Bolivian group, 84, 352
Tuanan Orangutan Research Program/Project, 127, 217, 234, 508
tubers, as foods, 44, 117, 302, *401*. *See also* roots, as foods

uacaris, golden-backed, 331, 366
Uganda, xvi, 15–16, 27, 29–30, 36, 58, 102, 115, 121, 124–25, 132, 156–61, 164–65, 177, 185–87, 188, 196–97, 230, 237, 267, 269–70, 293, 305, 314, 320–21, 323, 325, 339, 373, 382, 389, 417, 460, *485*, 505–6, 521, 537, 540, 542, 546, 548, 550–52
Ungar, Peter S., xviii, 397, 498, 720
United Kingdom, TEE in, 83–84
urbanization, 194–95, 551–52
Utami Atmoko, Sri Suci, xvii, 135, 198, 720
Uwimbabazi, Moreen, xvii, 241, 261, 720

Valenta, Kim, xviii, 397, 544, 720
van Casteren, Adam, xviii, 397, 446, 720
van Schaik, Carel, 4
Varecia (lemurs), 37, 48, 51, 89–90, 100, 130, 227–28, 235, 387, 541
Veilleux, Carrie C., xvii, 19, 95, 720
Venezuela, 27, 125, 145, 156, 164
vertebrates: arboreal, 38; capture of, 40, 354; as dietary class, 39–40; as foods (prey), 35, 38–40, 51, 112, 315, 327–33, 354, *401*; and PSMs, 150–51, 177. *See also* invertebrates
vervet monkeys. See *Chlorocebus pygerythrus* (vervet monkey)

Virunga National Park, Congo (DR), mountain gorillas in, 484–85, 488, 505–6
vision (sight): color, 96–97, 100, 103, 107–8, 112–13; for finding/evaluating foods, 95–97, 100–103, 107–8, 112–13
vitamins: antioxidant properties of, 214; in diets and foods, 19, 21, 31; and immunity, 198; meat as source of, 320; as micronutrients, xvii, 29, 198, 381, 420, 422; nutritional, 19. *See also* minerals
Vogel, Erin R., xvii, 135, 198, 720

Walkup, Kristina R., xvii, 241, 299, 721
Waser, Peter, 4
Wasserman, Michael D., xvii, 135, 176, 721
Watts, David P., xvii, 241, 327, 721
wet weight (WW), 30, 319, 343, 344, 418, 419
wild plant food chemistry, xviii, 417–30; abbreviations and terms, 417–18; and amino acids, 420, 426, 428, 430; and as-eaten weight (AEW), 418–19; and carbohydrates, 429–30; checklist, 419; and crude lipids or fat (ether extract) (EE), 417–18, 423, 425–26, 429–30; and crude proteins (CPs), 417–18, 426–29; and detergent system of fiber analyses, 428–29; DM coefficient, 424; energy calculations, 429–30; and fatty acids, 420, 426; and fiber, 428–30; fiber analyses, 428–29; and field dry weight (FDW), 418, 419, 421; and free simple sugars (FSS), 418, 430; and fresh weight (FW), 418–19, 421, 424; gravimetric procedures, 424; insoluble protein, and available protein (AP), 427; in laboratory, 421; and macronutrients, 417; and nutritional ecology, 430; permits, 417; and phytoestrogens, 11; resolutions for, 430; samples, checklist, 419; samples, collecting, 417–19; samples, field drying of, 420–21; samples, grinding or milling of, 422–24; samples, preparing for analyses, 422; samples, processing in field, 419–20; samples, shipping of, 421; and sulfuric acid, cellulose (Cs), 418; and sulfuric acid, lignin (Ls), 418; summary and conclusions, 430; and tannins, 430; and total ash determinations, 424–25, *427*
woolly lemurs. See *Avahi* (woolly lemurs)
woolly monkeys. See *Lagothrix lagotricha* (woolly monkey)
World Wildlife Fund, 544
Wrangham, Richard, xv, 4, 14, 121–22, 218, 344, 347, 506
WW. *See* wet weight (WW)

Yasuni National Park, Ecuador: Atelidae diet comparison of, *384*; pygmy marmoset feeding on sap in, *129*; red howler monkey feeding on leaves in, *133*
Yellowstone National Park (US), 366, 369, 378
Yerkes, Robert, 1
yield stress, and mechanical properties of foods, 447–49, 457–58

zoology, 1–2, 210
Zulfa, Astri, xvii, 135, 198, 721